Student Solutions Manual

to accompany

Precalculus
Graphs and Models

Third Edition

Raymond A. Barnett
Merritt College

Michael R. Ziegler
Marquette University

Karl E. Byleen
Marquette University

Dave Sobecki
Miami University Hamilton

Prepared by
Dave Sobecki
Miami University Hamilton

McGraw-Hill
Higher Education

Boston   Burr Ridge, IL   Dubuque, IA   New York   San Francisco   St. Louis
Bangkok   Bogotá   Caracas   Kuala Lumpur   Lisbon   London   Madrid   Mexico City
Milan   Montreal   New Delhi   Santiago   Seoul   Singapore   Sydney   Taipei   Toronto

 **Higher Education**

Student Solutions Manual to accompany
PRECALCULUS: GRAPHS AND MODELS, THIRD EDITION
RAYMOND A. BARNETT, MICHAEL R. ZIEGLER, KARL E. BYLEEN, AND DAVE SOBECKI

Published by McGraw-Hill Higher Education, an imprint of The McGraw-Hill Companies, Inc., 1221 Avenue of the Americas, New York, NY 10020. Copyright © 2009, 2005, & 2000 by The McGraw-Hill Companies, Inc. All rights reserved.

This book is printed on acid-free paper.

1 2 3 4 5 6 7 8 9 0 QPD/QPD 0 9 8

ISBN: 978-0-07-334178-1
MHID: 0-07-334178-9

www.mhhe.com

TABLE OF CONTENTS

CHAPTER 1   Functions, Graphs, and Models

Section 1-1

1. Since the x coordinate, 0, is between Xmin and -7 and Xmax = 9, and the y coordinate, 0, is between Ymin = -4 and Ymax = 11, (0, 0) lies in the viewing window.

3. Since the x coordinate, 10, is not between Xmin = -7 and Xmax = 9, (10, 0) does not lie in the viewing window.

5. Since the x coordinate, -5, is between Xmin = -7 and Xmax = 9, and the y coordinate, -3, is between Ymin = -4 and Ymax = 11, (-5, -3) lies in the viewing window.

7. (A) Since the x coordinates vary between a minimum of -7 and a maximum of 6, Xmin = -7 and Xmax = 6. Since the y coordinates vary between a minimum of -9 and a maximum of 14, Ymin = -9 and Ymax = 14.

   (B)

   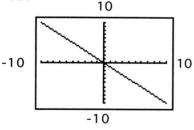

   The viewing window is shown. The cursor can, in general, show the coordinates of the points only approximately. For example, the point (6, -4) is approximated by the cursor location (6, -4.177419) in the screen shown. Pixels are geometrical rectangles whose coordinates are intervals, specified by the coordinates of the point at their center. Since a limited number of pixels are available on the calculator, it is not possible to show points exactly, since a point is a geometric object with no dimensions, and there are an infinite number of points.

9. Answers will vary.

11.

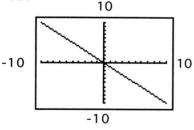

13.

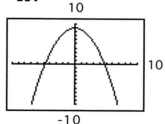

15.

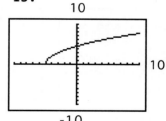

17. Answers can vary. One possible answer is Xmin = -20, Xmax = 10, Ymin = -2, Ymax = 10.

19. Answers can vary. One possible answer is Xmin = -20, Xmax = 10, Ymin = 0, Ymax = 20.

21.

x	$y = 4 + 4x - x^2$
-2	-8
0	4
2	8
4	4
6	-8

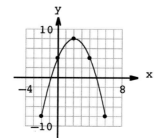

23.

x	$y = 2\sqrt{2x + 10}$
-5	0
-3	4
-1	5.7
1	6.9
3	8
5	8.9

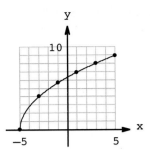

25.

x	$y = 0.5x(4 - x)(x + 2)$
-3	10.5
-1	-2.5
1	4.5
3	7.5
5	-17.5

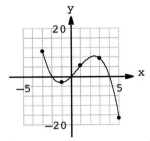

27. The graph of $y = 4 - 3\sqrt[3]{x + 4}$ is shown:

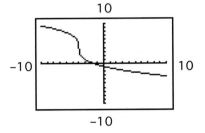

(A) After applying ZOOM and TRACE, the following is obtained. $x = -6.37$.

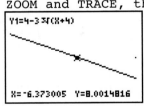
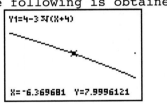

(B) After applying ZOOM and TRACE, the following is obtained. $x = 0.63$.

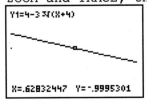
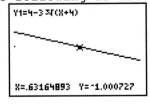

29. The graph of $y = 3 + x + 0.1x^3$ is shown.

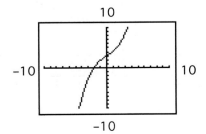

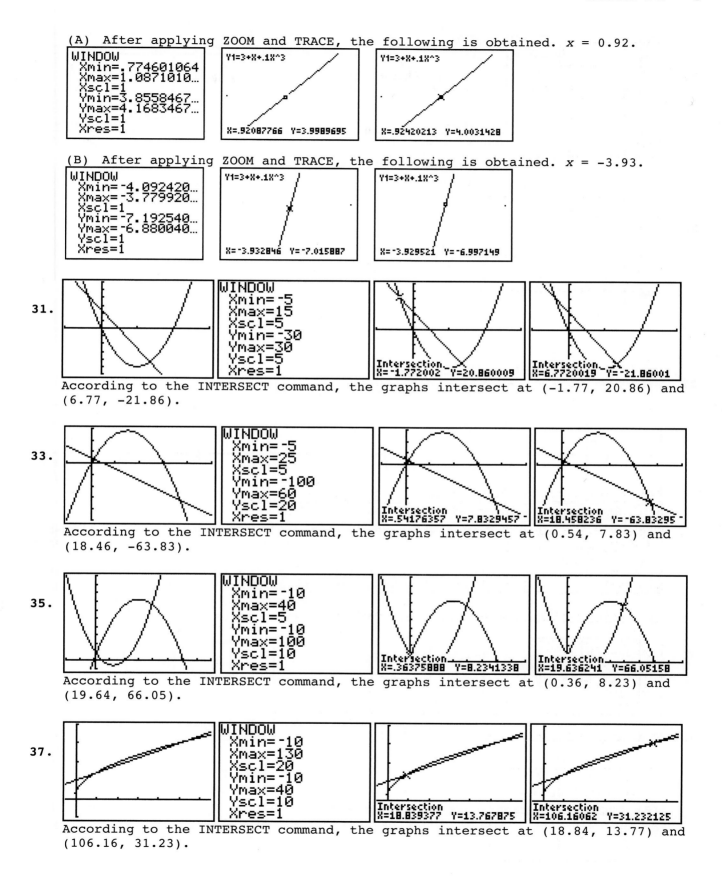

(A) After applying ZOOM and TRACE, the following is obtained. $x = 0.92$.

(B) After applying ZOOM and TRACE, the following is obtained. $x = -3.93$.

31. According to the INTERSECT command, the graphs intersect at $(-1.77, 20.86)$ and $(6.77, -21.86)$.

33. According to the INTERSECT command, the graphs intersect at $(0.54, 7.83)$ and $(18.46, -63.83)$.

35. According to the INTERSECT command, the graphs intersect at $(0.36, 8.23)$ and $(19.64, 66.05)$.

37. According to the INTERSECT command, the graphs intersect at $(18.84, 13.77)$ and $(106.16, 31.23)$.

39.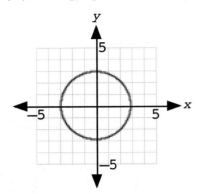

According to the INTERSECT command, the graphs intersect at (-2.43, 2.75) and (53.37, 7.96).

41. (A) The graph is a circle centered at (0, 0) with radius 3.

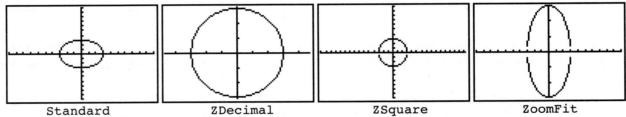

(B), (C) The various zoom options are shown below.

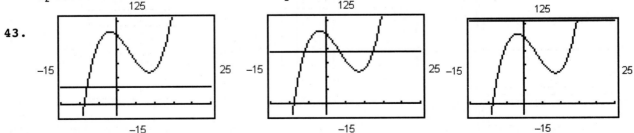

Standard   ZDecimal   ZSquare   ZoomFit

The graph is distorted in a standard viewing window since the scale is the same in each direction, but the window is wider than its height. ZDecimal produces a circle that fills the screen. ZSquare also produces a circle, but a smaller one. ZoomFit produces another distorted graph, making it difficult to tell that the graph is a circle. (Note: In each case, these graphs resulted from choosing the zoom option while in a standard viewing window. Choosing the three zoom options one after another will produce different results.)

43.

(A) The first graph shows $y_1 = 0.1x^3 - x^2 - 5x + 100$, $y_2 = 25$. The only place where the graphs cross is (-7.99, 25), so the only solution is $x = -7.99$.

(B) The second graph shows $y_1 = 0.1x^3 - x^2 - 5x + 100$, $y_2 = 75$. This time, the graphs cross three times: at (-5.85, 75), (3.44, 75) and (12.41, 75). So there are three solutions: $x = -5.85$, $x = 3.44$, and $x = 12.41$.

(C) The third graph shows $y_1 = 0.1x^3 - x^2 - 5x + 100$, $y_2 = 125$. Again, the graphs cross only once, at (14.60, 125), so the only solution is $x = 14.60$.

45.

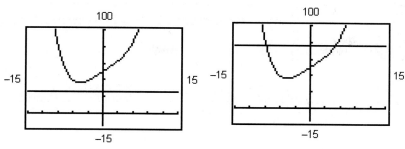

(A) The first graph shows $y_1 = 0.01x^4 + 4x + 50$, $y_2 = 25$. Since the two graphs never cross, the equation has no solution.

(B) The second graph shows $y_1 = 0.01x^4 + 4x + 50$, $y_2 = 75$. This time, the graphs cross twice, at $(-8.81, 75)$ and $(4.86, 75)$, so the two solutions are $x = -8.81$ and $x = 4.86$.

47.

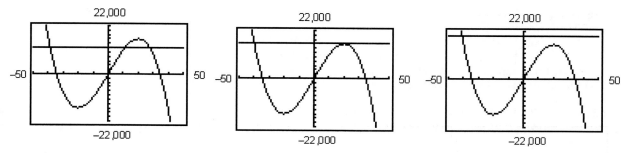

(A) The first graph shows $y_1 = 1,200x - x^3$, $y_2 = 12,000$. The two graphs cross three times, at $(-38.84, 12,000)$, $(11.16, 12,000)$ and $(27.69, 12,000)$, so the three solutions are $x = -38.84$, $x = 11.16$ and $x = 27.69$.

(B) The second graph shows $y_1 = 1,200x - x^3$, $y_2 = 16,000$. The graphs definitely cross once around $x = -40$, but it's inconclusive whether they actually intersect near $x = 20$. That's why the INTERSECT command is helpful--it turns out that they do, in fact, cross twice: at $(-40, 16,000)$ and $(20, 16,000)$. So the two solutions are $x = -40$ and $x = 20$.

(C) The third graph shows $y_1 = 1,200x - x^3$, $y_2 = 20,000$. These graphs cross only at $(-41.07, 20,000)$, so the only solution is $x = -41.07$.

49.

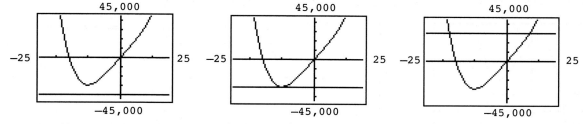

(A) The first graph shows $y_1 = x^4 + 4,000x$, $y_2 = 30,000$. The graphs cross twice at $(-17.84, 30,000)$ and $(6.93, 30,000)$, so there are two solutions: $x = -17.84$ and $x = 6.93$.

(B) The second graph shows $y_1 = x^4 + 4,000x$, $y_2 = -30,000$. It appears that the graphs intersect right at the local minimum; this is confirmed by the INTERSECT command. The point is $(-10, -30,000)$, so the only solution is $x = -10$.

(C) The third graph shows $y_1 = x^4 + 4,000x$, $y_2 = -40,000$. The two graphs never cross, so the equation has no solution.

51. The graphs show
$y_1 = 0.2x^2 + x^3$ and
$y_2 = \dfrac{1}{2}x + 2$. Using the
INTERSECT command, we see
that the graphs intersect at
$x = -4.98$ and $x = 1.28$, so
these are the solutions.

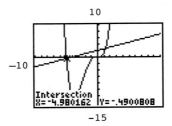 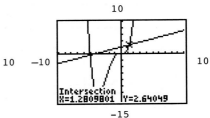

53. The graph shows $y_1 = x^2 - 3x + 1$ and
$y_2 = \sqrt{2x - 7}$. The graphs do not
intersect, so the equation has no
solution.

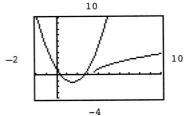

55. When we graph $y_1 = 0.05(x + 4)^3 + 4$ and
$y_2 = 17 - x$ in a standard viewing window, it
looks like the graphs don't intersect, but
they actually do up near height 15. The
INTERSECT command shows that the only
solution is $x = 2.03$.

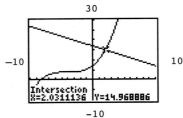

57. After several applications of ZOOM and TRACE, the following portion of the
graph of $y = x^2$ near $y = 2$ is obtained.

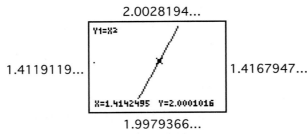

When $y = 2.0001$, $x = 1.4142$. This approximates the calculator value of 1.4142
(to four decimal places).

59. Points are geometric objects without dimensions. Pixels are rectangles. The
coordinates of a point are a pair of numbers. The screen coordinates of a pixel
are intervals on the x and y axes, specified by the coordinates of the point at
the center of the pixel. There are infinitely many geometric points within a
pixel.

61. Graph $y_1 = -16.7x^3 - 400x^2 + 6,367x - 7,000$ and $y_2 = 11,000$, then use the
INTERSECT command.

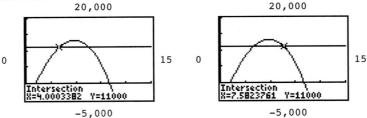

There are two potential solutions, $x = 4$ and $x = 7.58$. But we need x to be a
whole number, so the number of clients that gets profit closest to $11,000 is
4.

63.

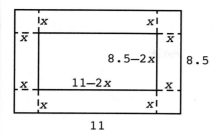

When the corners are cut out and the sides folded up, the dimensions of the box will be 11 - 2x, 8.5 - 2x, and x. So the volume is
$$V = (11 - 2x)(8.5 - 2x)x.$$
The volume is to be 55 sq. in. so we need to solve the equation 55 = (11 - 2x)(8.5 - 2x)x using a graphing calculator. Set y_1 = (11 - 2x)(8.5 - 2x)x, y_2 = 55 and use the INTERSECT command.

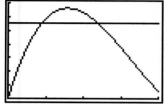 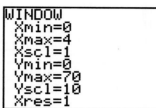 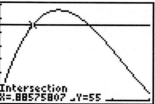 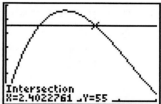

There are 2 solutions: x = 0.89 and x = 2.40.

A 0.89 in. square or a 2.40 in. square can be cut out.

If x = 0.89, 8.5 - 2x = 6.72, 11 - 2x = 9.22.

Dimension for smaller square: 0.89" × 9.22" × 6.72".

If x = 2.40, 8.5 - 2x = 3.70, 11 -2x = 6.20.

Dimension for larger square: 2.40" × 6.20" × 3.70".

65.

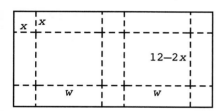

The two dimensions labeled w must be equal and must add up to 24 - 3x; the 24" length of the cardboard, minus 3 cutouts of width x. So each
$$w = \frac{1}{2}(24 - 3x),$$ and the volume of the box is
$$V = x(12 - 2x)\frac{1}{2}(24 - 3x).$$ The volume is to be

100 cu. inches, so our equation is $100 = x(12 - 2x)\frac{1}{2}(24 - 3x)$.

Let $y_1 = x(12 - 2x)\frac{1}{2}(24 - 3x)$ and y_2 = 100 and use the INTERSECT command:

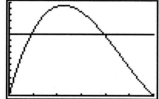 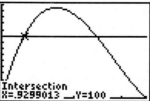 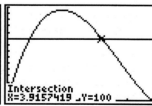

There are 2 solutions: x = 0.93 and x = 3.92.

A 0.93 in. square or a 3.92 in. square can be cut out.

If x = 0.93, 12 - 2x = 10.14, $\frac{1}{2}(24 - 3x)$ = 10.61.

Dimension for smaller square: 0.93" × 10.14" × 10.61".

If x = 3.92, 12 - 2x = 4.16, $\frac{1}{2}(24 - 3x)$ = 6.12.

Dimension for larger square: 3.92" × 4.16" × 6.12".

67. The volume of a right circular cylinder is $V = \pi r^2 h$. The radius plus the height must be 50 feet, so $r + h = 50 \Rightarrow h = 50 - r$. So $V = \pi r^2(50 - r)$. Since the volume is 40,000 cubic feet, we get $40,000 = \pi r^2(50 - r)$. Solve using the INTERSECT command on a graphing calculator: use $y_1 = \pi r^2(50 - r)$ and $y_2 = 40,000$.

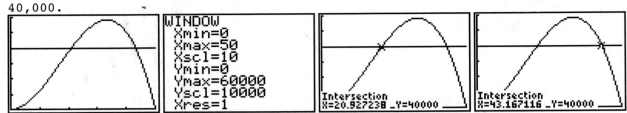

There are 2 solutions: $r = 20.93$ and $r = 43.17$.
If $r = 20.93$, $h = 50 - 20.93 = 29.07$; the cylinder has radius 20.93 feet and height 29.07 feet. If $r = 43.17$, $h = 50 - 43.17 = 6.83$; the cylinder has radius 43.17 feet and height 6.83 feet.

69. The graph of $y = 100 - 0.6\sqrt{x}$ is shown.

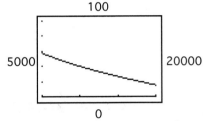

(A) Applying TRACE and ZOOM to the graph we obtain $x = 17,800$ when $y = 20$ as shown.

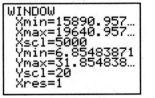 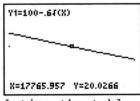 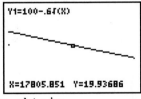

Similarly, completing the table, we obtain

x	17,800	15,600	13,600
y	20	25	30

(B) As the price is increased from \$25 to \$30, demand decreases from 15,600 to 13,600, that is, by 2,000 cases.

(C) As the price is decreased from \$25 to \$20, demand increases from 15,600 to 17,800, that is, by 2,200 cases.

71. $R = xy = x(100 - 0.6\sqrt{x})$
(A) Using the data from problem 69, the table can be completed as follows:

x	17,800	15,600	13,600
y	20	25	30
R	356,000	390,000	408,000

(B) As the price is increased from \$25 to \$30, the revenue increases from \$390,000 to \$408,000, that is, by \$18,000.

(C) As the price is decreased from \$25 to \$20, the revenue decreases from \$390,000 to \$356,000, that is, by \$40,000.

(D) The company should raise the price $5 to increase the revenue.

Section 1-2

1. False. Only relations where each element in the domain corresponds to exactly one element in the range are functions.

3. Answers will vary.

5. This is a function: each element in the domain corresponds to exactly one element in the range. $\{(-1, 3), (0, 2), (1, 1)\}$

7. This is not a function since the domain elements 3 and 5 both correspond to more than one range element. $\{(1, 3), (3, 5), (3, 7), (5, 7), (5, 9)\}$

9. This is a function: each element in the domain corresponds to exactly one element in the range. $\{(-1, 3), (0, 3), (1, 3), (2, 3)\}$

11. Domain: $\{2, 3, 4, 5\}$; Range: $\{4, 6, 8, 10\}$; function

13. Domain: $\{0, 5, 10\}$; Range: $\{-10, -5, 0, 5, 10\}$; not a function

15. Domain: $\{0, 1, 2, 3, 4, 5\}$; Range: $\{1, 2\}$; function

17. A function

19. Not a function (fails vertical line test since the y-axis hits the graph 3 times.)

21. Not a function (fails vertical line test since the graph is vertical itself!)

23. (A) This is a function. Each congressional district has only one representative, so all residents correspond to exactly one representative.

    (B) This is not a function. Every state has two senators that represent the entire state, so every resident corresponds to two senators. This violates the definition of function.

25. This equation does define y as a function of x. If you plug in any number for x, there is only one choice of y that will then make the equation true.

27. This equation does not define y as a function of x. For many choices of x, there are two values of y that will make the equation true. For example, if $x = 0$ the equation is $y^2 = 4$ and y can be either 2 or -2. Since we found a value for x that corresponds to two different values of y, the equation is not a function.

29. $f(-1) = 3(-1) - 5 = -8$

31. $G(-2) = (-2) - (-2)^2 = -6$ 33. $F(-1) + f(3) = 3(-1)^2 + 2(-1) - 4 + 3(3) - 5$
$$= 3 - 2 - 4 + 9 - 5 = 1$$

35. $2F(-2) - G(-1) = 2[3(-2)^2 + 2(-2) - 4] - [(-1) - (-1)^2]$
$$= 2[12 - 4 - 4] - [-1 - 1] = 10$$

37. $\dfrac{f(0) \cdot g(-2)}{F(-3)} = \dfrac{[3(0) - 5] \cdot [4 - (-2)]}{3(-3)^2 + 2(-3) - 4} = \dfrac{(-5)(6)}{27 - 6 - 4} = -\dfrac{30}{17}$

39. $\dfrac{f(4) - f(2)}{2} = \dfrac{[3(4) - 5] - [3(2) - 5]}{2} = \dfrac{7 - 1}{2} = \dfrac{6}{2} = 3$

41. Since $f(x)$ is a polynomial, the domain is all real numbers; $(-\infty, \infty)$.

43. If the denominator of the fraction is zero, the function will be undefined, since division by zero is undefined. For any other values of x, $h(x)$ represents a real number. Solve the equation $4 - x = 0$; $x = 4$. The domain is all real numbers except 4 or $(-\infty, 4) \cup (4, \infty)$.

45. The square root of a number is defined only if the number is nonnegative. The expression $\sqrt{t - 4}$ is defined if $t - 4 \geq 0$ and undefined if $t - 4 < 0$. Note that $t - 4 \geq 0$ when $t \geq 4$, so the domain is all real numbers greater than or equal to 4 or $[4, \infty)$.

47. The formula $\sqrt{7 + 3w}$ is defined only if $7 + 3w \geq 0$, since the square root is only defined if the number inside is nonnegative.
We solve the inequality: $7 + 3w \geq 0$
$$3w \geq -7$$
$$w \geq -\frac{7}{3}$$

The domain is all real numbers greater than or equal to $-\dfrac{7}{3}$ or $\left[\dfrac{7}{3}, \infty\right)$.

49. The fraction $\dfrac{u}{u^2 + 4}$ is defined for any value of u that does <u>not</u> make the denominator zero, so we solve the equation $u^2 + 4 = 0$ to find values that make the function undefined. $u^2 + 4 = 0$
$$u^2 = -4$$
This equation has no solution since the square of a number can't be negative. So there are no values of u that make the fraction undefined and the domain is all real numbers or $(-\infty, \infty)$.

51. The fraction $\dfrac{v + 2}{v^2 - 16}$ is defined for any value of v that does <u>not</u> make the denominator zero, so we solve the equation $v^2 - 16 = 0$ to find values that make the function undefined.
$$v^2 - 16 = 0$$
$$v^2 = 16$$
$$v = 4, -4$$
The domain is all real numbers except 4 and -4, or $(-\infty, -4) \cup (-4, 4) \cup (4, \infty)$.

53. There are two issues to consider: we need to make certain that $x + 4 \geq 0$ so that the number inside the square root is nonnegative, and we need to avoid x-values that make the denominator zero.

First, $x + 4 \geq 0$ whenever $x \geq -4$. So x must be greater than or equal to -4 to avoid a negative under the root. Also, $x - 1 = 0$ when $x = 1$, so x cannot be 1. The domain is all real numbers greater than or equal to -4 except 1, or $[-4, 1) \cup (1, \infty)$.

55. There are two issues to consider: the number inside the square root has to be nonnegative, and the denominator of the fraction has to be nonzero. Since we just have t inside the square root, t has to be greater than or equal to zero. Next, we solve $3 - \sqrt{t} = 0$ to find any t-values that make the denominator zero.
$$3 - \sqrt{t} = 0$$
$$3 = \sqrt{t}$$
$$9 = t$$
So t cannot be 9. The domain is all real numbers greater than or equal to zero except 9, or $[0, 9) \cup (9, \infty)$.

57. $f(x) = 2x - 3$

59. $f(x) = 4x^2 - 2x + 9$

61.
$$F(s) = 3s + 15$$
$$F(2 + h) = 3(2 + h) + 15$$
$$F(2) = 3(2) + 15$$
$$\frac{F(2 + h) - F(2)}{h} = \frac{[3(2 + h) + 15] - [3(2) + 15]}{h} = \frac{[6 + 3h + 15] - [21]}{h}$$
$$= \frac{3h + 21 - 21}{h} = \frac{3h}{h} = 3$$

63.
$$g(x) = 2 - x^2$$
$$g(3 + h) = 2 - (3 + h)^2$$
$$g(3) = 2 - (3)^2$$
$$\frac{g(3 + h) - g(3)}{h} = \frac{[2 - (3 + h)]^2 - [2 - (3)^2]}{h} = \frac{[2 - 9 - 6h - h^2] - [-7]}{h}$$
$$= \frac{-6h - h^2}{h} = \frac{h(-6 - h)}{h} = -6 - h$$

65.
$$L(w) = -2w^2 + 3w - 1$$
$$L(-2 + h) = -2(-2 + h)^2 + 3(-2 + h) - 1$$
$$L(-2) = -2(-2)^2 + 3(-2) - 1$$
$$\frac{L(-2 + h) - L(-2)}{h} = \frac{[-2(-2 + h)^2 + 3(-2 + h) - 1] - [-2(-2)^2 + 3(-2) - 1]}{h}$$
$$= \frac{[-2(4 - 4h + h^2) - 6 + 3h - 1] - [-15]}{h}$$
$$= \frac{-8 + 8h - 2h^2 - 6 + 3h - 1 + 15}{h}$$
$$= \frac{11h - 2h^2}{h} = \frac{h(11 - 2h)}{h} = 11 - 2h$$

67. $g(x) = 3x + 1$

69. $F(x) = \dfrac{x}{8 + \sqrt{x}}$

71. Function f multiplies the domain element by 2 and then subtracts the product of 3 and the square of the element.

73. Function F takes the square root of the sum of the fourth power of the domain element and 9.

75. In this case, the domain element is the expression $(x + h)$, so we identify the operations that are performed on $(x + h)$. The function multiplies the square of $(x + h)$ by 2, then subtracts 4 times $(x + h)$, and finally adds 6. The function equation is $f(x) = 2x^2 - 4x + 6$.

77. In this case, the domain element is the expression $(x + h)$, so we identify the operations that are performed on $(x + h)$. The function multiplies 4 by $(x + h)$, then subtracts 3 times the square root of $(x + h)$, and finally adds 9. The function equation is $f(x) = 4x - 3\sqrt{x} + 9$.

79. (A)
$$f(x) = 3x - 4$$
$$f(x + h) = 3(x + h) - 4$$
$$\frac{f(x + h) - f(x)}{h} = \frac{[3(x + h) - 4] - [3x - 4]}{h}$$
$$= \frac{3x + 3h - 4 - 3x + 4}{h}$$
$$= \frac{3h}{h}$$
$$= 3$$

(B)
$$f(x) = 3x - 4$$
$$f(a) = 3a - 4$$
$$\frac{f(x) - f(a)}{x - a} = \frac{(3x - 4) - (3a - 4)}{x - a}$$
$$= \frac{3x - 4 - 3a + 4}{x - a}$$
$$= \frac{3x - 3a}{x - a}$$
$$= \frac{3(x - a)}{x - a}$$
$$= 3$$

81. (A)
$$f(x) = x^2 - 1$$
$$f(x + h) = (x + h)^2 - 1$$
$$\frac{f(x + h) - f(x)}{h} = \frac{[(x + h)^2 - 1] - [x^2 - 1]}{h}$$
$$= \frac{x^2 + 2xh + h^2 - 1 - x^2 + 1}{h}$$
$$= \frac{2xh + h^2}{h}$$
$$= \frac{h(2x + h)}{h}$$
$$= 2x + h$$

(B)
$$f(x) = x^2 - 1$$
$$f(a) = a^2 - 1$$
$$\frac{f(x) - f(a)}{x - a} = \frac{(x^2 - 1) - (a^2 - 1)}{x - a}$$
$$= \frac{x^2 - 1 - a^2 + 1}{x - a}$$
$$= \frac{x^2 - a^2}{x - a}$$
$$= \frac{(x - a)(x + a)}{x - a}$$
$$= x + a$$

83. (A)
$$f(x) = -3x^2 + 9x - 12$$
$$f(x + h) = -3(x + h)^2 + 9(x + h) - 12$$
$$\frac{f(x + h) - f(x)}{h} = \frac{[-3(x + h)^2 + 9(x + h) - 12] - [-3x^2 + 9x - 12]}{h}$$
$$= \frac{-3(x^2 + 2xh + h^2) + 9x + 9h - 12 + 3x^2 - 9x + 12}{h}$$
$$= \frac{-3x^2 - 6xh - 3h^2 + 9x + 9h - 12 + 3x^2 - 9x + 12}{h}$$
$$= \frac{-6xh - 3h^2 + 9h}{h}$$
$$= \frac{h(-6x - 3h + 9)}{h}$$
$$= -6x - 3h + 9$$

(B)
$$f(x) = -3x^2 + 9x - 12$$
$$f(a) = -3a^2 + 9a - 12$$
$$\frac{f(x) - f(a)}{x - a} = \frac{(-3x^2 + 9x - 12) - (-3a^2 + 9a - 12)}{x - a}$$
$$= \frac{-3x^2 + 9x - 12 + 3a^2 - 9a + 12}{x - a}$$
$$= \frac{-3x^2 + 3a^2 + 9x - 9a}{x - a}$$
$$= \frac{(x - a)(-3x - 3a) + 9(x - a)}{x - a}$$
$$= \frac{(x - a)(-3x - 3a + 9)}{x - a}$$
$$= -3x - 3a + 9$$

85. (A)
$$f(x) = x^3$$
$$f(x + h) = (x + h)^3$$
$$\frac{f(x + h) - f(x)}{h} = \frac{(x + h)^3 - x^3}{h}$$
$$= \frac{x^3 + 3x^2h + 3xh^2 + h^3 - x^3}{h}$$
$$= \frac{3x^2h + 3xh^2 + h^3}{h}$$
$$= \frac{h(3x^2 + 3xh + h^2)}{h}$$
$$= 3x^2 + 3xh + h^2$$

(B)
$$f(x) = x^3$$
$$f(a) = a^3$$
$$\frac{f(x) - f(a)}{x - a} = \frac{x^3 - a^3}{(x - a)}$$
$$= \frac{(x - a)(x^2 + xa + a^2)}{x - a}$$
$$= x^2 + ax + a^2$$

87. First, let's look at the graph of $f(x)$ in a window that shows only x-values close to 1.

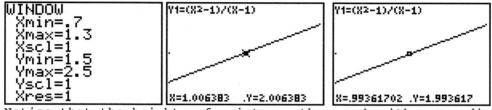

Notice that the heights of points on the graph with x-coordinates very close to 1 are very close to 2.

Next, we'll use the TABLE command to evaluate $f(x)$ for some numbers close to $x = 1$. We start at $x = 0.7$ and increment by 0.1 as shown on the TABLE SETUP screen:

Notice that as x gets closer to 1, the result gets closer to 2. Finally, we'll simplify the fraction by factoring the numerator.

$$\frac{x^2 - 1}{x - 1} = \frac{(x + 1)(x - 1)}{x - 1} = x + 1 \text{ as long as } x \neq 1.$$

It's easy to see that if you plug x-values close to 1 into the expression $x + 1$, the results will be close to 2.

89. (A)

x	0	5,000	10,000	15,000	20,000	25,000	30,000
$B(x)$	212	203	194	185	176	167	158

(B) The boiling point drops 9°F for each 5,000 foot increase in altitude.

91. (A) $S(0) = 16(0)^2 = 0$; $S(1) = 16(1)^2 = 16$; $S(2) = 16(2)^2 = 16(4) = 64$;
$S(3) = 16(3)^2 = 16(9) = 144$ (Note: Remember that the order of operations requires that we apply the exponent first then multiply by 16.)

(B)
$$\frac{S(2 + h) - S(2)}{h} = \frac{16(2 + h)^2 - 16(2)^2}{h} = \frac{16(4 + 4h + h^2) - 16(4)}{h}$$
$$= \frac{64 + 64h + 16h^2 - 64}{h} = \frac{64h + 16h^2}{h} = \frac{h(64 + 16h)}{h} = 64 + 16h$$

(Note: Be careful when evaluating $S(2 + h)$! You need to replace the variable t in the function with $(2 + h)$. $S(2 + h)$ is <u>not</u> the same thing as $S(2) + h$!)

(C) $h = 1$: $64 + 16(1) = 80$; $h = -1$: $64 + 16(-1) = 64 - 16 = 48$
$h = 0.1$: $64 + 16(0.1) = 64 + 1.6 = 65.6$;
$h = -0.1$: $64 + 16(-0.1) = 64 - 1.6 = 62.4$
$h = 0.01$: $64 + 16(0.01) = 64 + 0.16 = 64.16$:
$h = -0.01$: $64 + 16(-0.01) = 64 - 0.16 = 63.84$
$h = 0.001$: $64 + 16(0.001) = 64 + 0.016 = 64.016$
$h = -0.001$: $64 + 16(-0.001) = 64 - 0.016 = 63.984$

(D)   The smaller h gets the closer the result is to 64. The numerator of the fraction, $S(2 + h) - S(2)$, is the difference between how far an object has fallen after $2 + h$ seconds and how far it's fallen after 2 seconds. This difference is how far the object falls in the small period of time from 2 to $2 + h$ seconds. When you divide that distance by the time (h), you get the average velocity of the object between 2 and $2 + h$ seconds. Part (C) shows that this average velocity approaches 64 feet per second as h gets smaller.

93. The rental charges are \$20 per day plus \$0.25 per mile driven.

95. For detailed help on how to use your graphing calculator on this problem, see Example 10 in Section 1-2.
(A) We begin by entering our lists under the STAT EDIT screen. In this model, the independent variable is years after 1997 and the dependent variable is the average admission price. So our first list is 0, 2, 4, 6, and 8, representing the years from 1997 to 2005. The second list is the corresponding prices.

 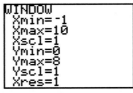 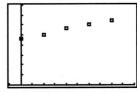

The second screen shows the appropriate plot settings; the third shows a good choice of viewing window; the last is the scatter plot of the data from the table.

To compare the algebraic function, we enter $y_1 = 0.23x + 4.6$ in the equation editor, choose the "ask" option for independent variable in the TABLE SETUP window, then enter x values 0, 2, 4, 6, 8.

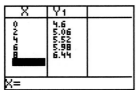

Comparing the two tables, we see that some of the average prices are very close and none differ by more than 14 cents. Finally, we look at the graphs of both the function and the scatter plot:

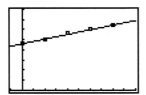

As expected, the graph of the function matches the scatter plot points closely.

(B) To estimate the average price in 2006, we need to evaluate the given function for $x = 9$, since 2006 is 9 years after 1997.
$A(9) = 0.23(9) + 4.6 = 6.67$
The average price in 2006 is \$6.67. Similarly, $A(10) = 0.23(10) + 4.6 = 6.90$, so the average price in 2007 is \$6.90.

97. For detailed help on how to use your graphing calculator on this problem, see Example 10 in Section 1-2.

(A) In this example, the independent variable is years after 1997, so the first list under the STAT EDIT screen is 0, 2, 4, 6, and 8. The dependent variable is sales, so the second list is the sales row: 14.0, 17.3, 21.2, 22.5, and 22.0.

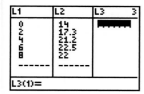  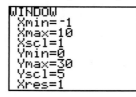 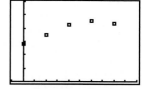

The second screen shows the appropriate plot settings; the third shows a good choice of viewing window; the last is the scatter plot of the data from the table.

To compare to the algebraic function, we enter $y_1 = -0.18x^2 + 2.5x + 14$ in the equation editor, choose the "ask" option for independent variable in the TABLE SETUP window, than enter x values 0, 2, 4, 6, 8.

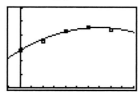

Comparing the two tables, we see that all of the sales figures are similar — the largest difference is less than 1.

Finally, we look at the graphs of both the function and the scatter plot:

As expected, the graph of the function matches the scatter plot points closely.

(B) To estimate sales in 2006, we need to evaluate the given function for $x = 9$, since 2006 is 9 years after 1997.
$$S(9) = -0.18(9)^2 + 2.5(9) + 14 = 21.92.$$
The estimated sales figure for 2006 is \$22 billion (rounded to 2 significant digits). Similarly, $S(11) = -0.18(11)^2 + 2.5(11) + 14 = 19.72$, so the estimated sales figure for 2008 is \$20 billion (rounded to 2 significant digits).

(C) For the period 1997-2001, Merck's sales increased steadily by an average of approximately \$6 billion per year.

99. For detailed help on how to use your graphing calculator on this problem, see Example 10 in Section 1-2.

(A) In this example, the independent variable (r) is R & D expenses, so the first list under the STAT EDIT screen comes from the R & D row of the table: 1.7, 2.1, 2.5, 3.2, 3.8. The dependent variable(s) is sales, so the second list is the sales row: 14.0, 17.3, 21.2, 22.5, and 22.0.

 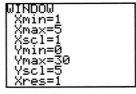 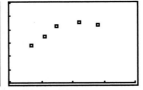

The second screen shows the appropriate plot settings; the third shows a good choice of viewing window; the last is the scatter plot of the data from the table.

To compare to the algebraic function, we enter $y_1 = -3.54x^2 + 23.3x - 15.5$ in the equation editor, choose the "ask" option for independent variable in the TABLE SETUP window, then enter x values 1.7, 2.1, 2.5, 3.2, 3.8.

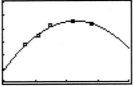

Comparing the two tables, we see that the corresponding sales figures are very close.

Finally, we look at the graphs of both the function and the scatter plot:

As expected, the graph of the function matches the points in the scatter plot very well.

(B) To estimate the sales with $1.2 billion in research and development spending, we evaluate the function for $r = 1.2$:

$$S(1.2) = -3.54(1.2)^2 + 23.3(1.2) - 15.5 = 7.3624$$

To 2 significant digits, the estimated sales figure is $7.4 billion with $1.2 billion spent on research and development.

Similarly, $S(4.2) = -3.54(4.2)^2 + 23.3(4.2) - 15.5 = 19.9144$, so the estimated sales figure is $20 billion with $4.2 billion spent on research and development.

Section 1-3

1. Answers will vary.   **3.** Answers will vary.   **5.** Answers will vary.

7. (A) [-4, 4)   (B) [-3, 3)   (C) 0   (D) 0   (E) [-4, 4)   (F) None
(G) None   (H) None

9. (A) $(-\infty, \infty)$ (B) [-4, ∞)   (C) -3, 1   (D) -3   (E) [-1, ∞)   (F) $(-\infty, -1]$
(G) None   (H) None

11. (A) $(-\infty, 2) \cup (2, \infty)$ (The function is not defined at $x = 2$.)
    (B) $(-\infty, -1) \cup [1, \infty)$ (C) None (D) 1 (E) None (F) $(-\infty, -2] \cup (2, \infty)$
    (G) $[-2, 2)$ (H) $x = 2$

13. $f(-4) = -3$ since the point $(-4,-3)$ is on the graph; $f(0) = 0$ since the point $(0,0)$ is on the graph; $f(4)$ is undefined since there is no point on the graph at $x = 4$.

15. $h(-3) = 0$ since the point $(-3,0)$ is on the graph; $h(0) = -3$ since the point $(0,-3)$ is on the graph; $h(2) = 5$ since the point $(2,5)$ is on the graph.

17. $p(-2) = 1$ since the point $(-2,1)$ is on the graph; $p(2)$ is undefined since there is no point on the graph at $x = 2$; $p(5) = -4$ since the point $(5,-4)$ is on the graph.

19. The graph of $f(x)$ is shown below in a standard window.

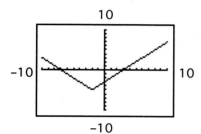

The function is increasing on $[-2, \infty)$ and decreasing on $(-\infty, -2]$. It is not constant on any interval shown.

21. The graph of $j(x)$ is shown below in a standard window.

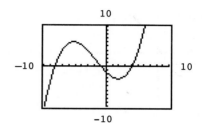

The function is increasing on $(-\infty, -5]$ and $[2, \infty)$ and decreasing on $[-5, 2]$. It is not constant on any interval shown.

23. The graph of $m(x)$ is shown below in a standard window.

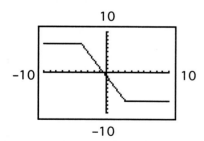

The function is decreasing on $[-4, 3]$ and constant on $[-\infty, -4]$ and $[3, \infty]$. It is not increasing on any interval shown.

25. The graph of $r(x)$ is shown below in a standard window.

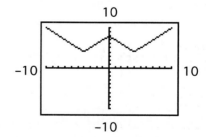

The function is increasing on $[-4, 0]$ and $[4, \infty]$ and decreasing on $[-\infty, -4]$ and $[0, 4]$. It is not constant on any interval shown.

27. To find the x-intercepts we'll use a graphing calculator to draw the graph then use the ZERO command from the calculate menu. (For detailed instructions on using the ZERO command, see Example 1 in Section 1-3.)

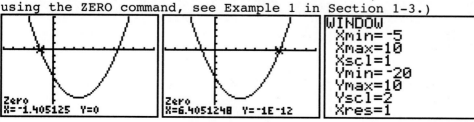

The two x-intercepts are $x = -1.405$ and $x = 6.405$. The y-intercept is easier to find if we just plug zero in for x in the given function: $f(0) = 0^2 - 5(0) - 9 = -9$, so the y-intercept is -9. It appears that the graph of f has a local minimum near $x = 2$ and no local maximum. (To be on the safe side, you may want to look at some bigger viewing windows to make sure there are no other extrema.)

We'll use the feature from the calculate menu. (For detailed instructions on using the MINIMUM command, see Example 4 in Section 1-3.)

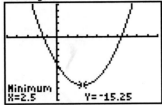

We see from the graph that has a local minimum value of -15.25 at $x = 2.5$.

29. To find the x-intercepts we'll use a graphing calculator to draw the graph then use the ZERO command from the calculate menu. (For detailed instructions on using the ZERO command, see Example 1 in Section 1-3.)

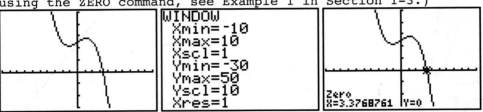

Notice that a fairly large viewing window is necessary to see all of the key features of the graph. We see that the only x-intercept is $x = 3.377$. The y-intercept is easier to find if we just plug zero in for x in the given function: $h(0) = -(0)^3 + 4(0) + 25 = 25$, so the y-intercept is 25. We can see that the graph has one local minimum and one local maximum, which we'll find using the MINIMUM and MAXIMUM commands from the calculate menu. (For detailed instructions on using the MINIMUM and MAXIMUM commands see Example 4 in Section 1-3.)

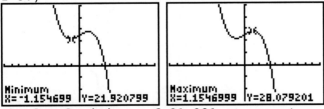

The local minimum of 21.921 occurs at $x = -1.155$; the local maximum of 28.079 occurs at $x = 1.155$.

31. To find the x-intercepts we'll use a graphing calculator to draw the graph then use the ZERO command from the calculate menu. (For detailed instructions on using the ZERO command, see Example 1 in Section 1-3.)

This is a tricky one because the graph is a bit deceiving onscreen. It looks like the graph does not hit the x-axis. But if you look at the formula for the function, you can see that the height of the graph will be zero when $x^2 - 12 = 0$; this happens when $x = \sqrt{12}$ or $-\sqrt{12}$, so in fact there are 2 intercepts which the ZERO command will find.

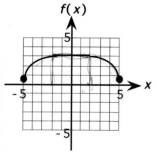

The two x-intercepts are $x = -3.464$ and $x = 3.464$. (These are $-\sqrt{12}$ and $\sqrt{12}$ as we expected.)

The y-intercept is easier to find if we just plug zero in for x in the given function: $m(0) = \sqrt{|0^2 - 12|} = \sqrt{12} \approx 3.464$ so the y-intercept is 3.464.

From the graph we see that there is one local minimum and two local maxima. The good news is that we've found all three already! The local maximum is the y-intercept so it's 3.464 and occurs at $x = 0$. The two local minima are the x-intercepts: both are zero and occur at $x = -3.464$ and $x = 3.464$.

33. One possible answer: **35.** One possible answer: **37.** One possible answer:

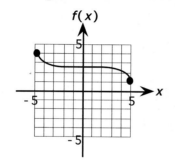

39. Answers will vary.

41. With the aid of a graphing calculator, the graph is seen to be as follows:

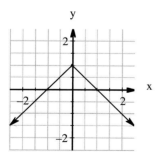

Notice that the top formula, $x + 1$, corresponds to x-values less than or equal to zero, so we need to plug $x = -2$ and $x = -1$ into the top formula.
   $f(-2) = -2 + 1 = -1$;
   $f(-1) = -1 + 1 = 0$
The bottom formula, $-x + 1$, corresponds to x-values greater than zero, so we need to plug $x = 1$ and $x = 2$ into the bottom formula.
   $f(1) = -1 + 1 = 0$; $f(2) = -1$

43. With the aid of a graphing calculator the graph is seen to be as follows:

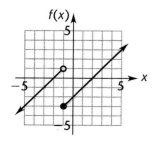

Notice that only x-values less than 1 are supposed to be plugged into the top formula, $x + 2$. So we plug $x = -2$ into the top formula $f(-2) = -2 + 2 = 0$. The remaining three x-values, -1, 1, and 2, are all greater than or equal to -1, so we plug into the bottom formula, $x - 2$.
   $f(-1) = -1 - 2 = -3$;
   $f(1) = 1 - 2 = -1$; $f(2) = 2 - 2 = 0$

45. With the aid of a graphing
calculator the graph is seen
to be as follows:

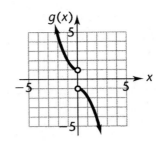

Notice that all x-values less than zero are supposed to be plugged into the top
formula, $x^2 + 1$. So we plug $x = -2$ and $x = -1$ into the top formula:

$f(-2) = (-2)^2 + 1 = 4 + 1 = 5$

$f(-1) = (-1)^2 + 1 = 1 + 1 = 2$

The remaining two x-values, $x = 1$ and $x = 2$, are greater than zero so we plug
them into the bottom formula, $-x^2 - 1$.

$f(1) = -(1)^2 - 1 = -1 - 1 = -2$; $f(2) = -(2)^2 - 1 = -4 - 1 = -5$

Common Error: When evaluating -1^2, order of operations tells us to square 1
first **then** multiply by -1 so the result is -1 not +1.

47. First we look at the graph in an appropriate viewing window:

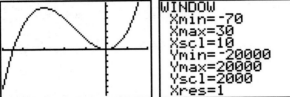

Because such a large viewing window is necessary to see all the features,
ZoomFit is very helpful here.

Domain: All real numbers or $(-\infty, \infty)$. (The function is a polynomial.)

Range: All real numbers or $(-\infty, \infty)$. (Every height gets hit eventually.)

y-intercept: Plug in zero for x. $m(0) = 0^3 + 45(0)^2 - 30 = -30$

x-intercepts: There are three of them. It's not clear from the graph, but since
$(0, -30)$ is on the graph, it must cross the x-axis twice close to $x = 0$.

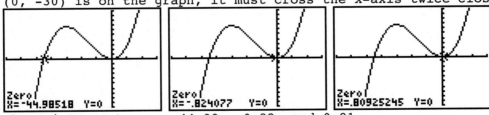

The x-intercepts are -44.99, -0.82, and 0.81.

49. First we look at the graph in an appropriate viewing window:

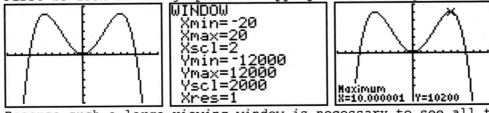

Because such a large viewing window is necessary to see all the features,
ZoomFit is very helpful here.

Domain: All real numbers (The function is a polynomial.)

Range: $(-\infty, 10,200]$ The highest points on the graph are at height 10,200 and
there is no lowest point.

y-intercept: Plug in zero for *x*. $n(0) = 200 + 200(0)^2 - (0)^4 = 200$
x-intercept: The graph crosses the *x*-axis twice. (It looks like it might hit the *x*-axis at zero, but we just found that it's at height 200 there.)

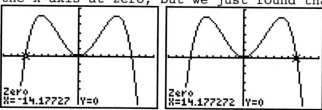

The *x*-intercepts are -14.18 and 14.18.

51. First we look at the graph in an appropriate viewing window:

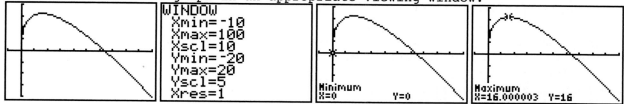

Domain: $[0, \infty)$. This makes sense because any negative numbers will make the root undefined.
Range: $(-\infty, 16]$. The highest point on the graph is at height 16 and there is no lowest point.
y-intercept: Since (0, 0) is on the graph, the *y*-intercept is 0.
x-intercepts: The graph hits the *x*-axis twice:

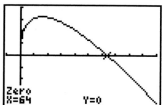

The *x*-intercepts are 0 and 64.

53. First we look at the graph in an appropriate viewing window:

Domain: $[-5, \infty)$. There's nothing to the left of -5.
Range: $[-134.02, \infty)$. The lowest point on the graph is at height -134.02 and there is no highest point.
y-intercept: Using the "value" feature on a graphing calculator we see that $k(0) = -111.80$, so the *y*-intercept is -111.80.
x-intercepts: The graph crosses the *x*-axis twice.

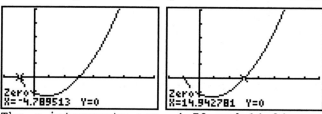

The *x*-intercepts are -4.79 and 14.94.

55.

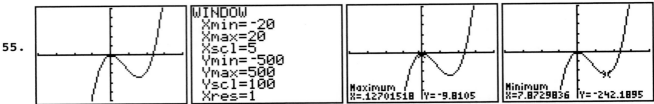

The graph of f is rising and f is increasing on $(-\infty, 0.13]$ and $[7.87, \infty)$. The graph of f is falling and f is decreasing on $[0.13, 7.87]$. Note that these are x-values! $f(0.13) = -9.81$ is a local maximum and $f(7.87) = -242.19$ is a local minimum.

57.

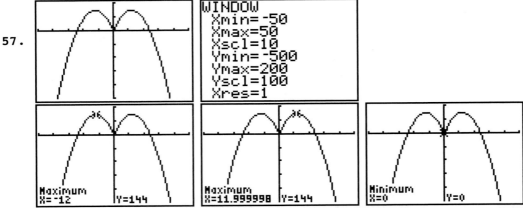

The graph of m is rising and m is increasing on $(-\infty, -12]$ and $[0, 12]$. The graph of m is falling and m is decreasing on $[-12, 0]$ and $[12, \infty)$. Note that these are x-values! $m(-12) = 144$ and $m(12) = 144$ are local maxima and $m(0) = 0$ is a local minimum.

59.

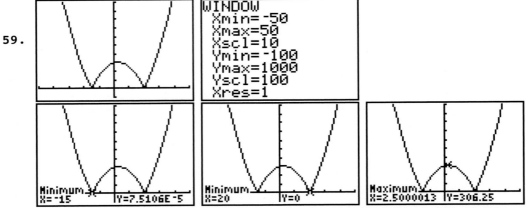

The graph of g is rising and g is increasing on $[-15, 2.5]$ and $[20, \infty)$. The graph of g is falling and g is decreasing on $(-\infty, -15]$ and $[2.5, 20]$. Note that these are x-values! $g(2.5) = 306.25$ is a local maximum and $g(-15) = 0$ and $g(20) = 0$ are local minima.

61. The graph of $f(x)$ is shown as well as the result of applying the MINIMUM command to find the local minimum.

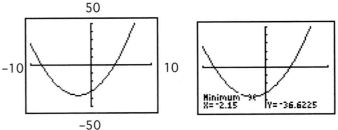

The graph of f decreases on $[-10, -2.15]$ to a local minimum value, $f(-2.15) = -36.62$, and then increases on $[-2.15, 10]$.

63. The results of applying the MAXIMUM and MINIMUM commands to find the local maximum and the local minimum are shown.

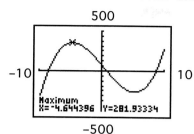 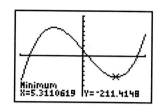

The graph of h increases on [-10, -4.64] to a local maximum value, $f(-4.64) \approx 281.93$, decreases on [-4.64, 5.31] to a local minimum value, $f(5.31) \approx -211.41$, and then increases on [5.31, 10].

65. The results of applying the MAXIMUM and MINIMUM commands to find the local minima and the local maximum are shown.

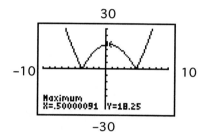 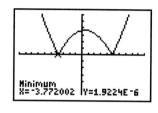 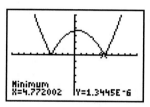

The graph of p decreases on [-10, -3.77] to a local minimum value, $f(-3.77) \approx 0$, increases on [-3.77, 0.50] to a local maximum value, $f(0.50) \approx 18.25$, decreases on [0.50, 4.77] to a local minimum value, $f(4.77) \approx 0$, and then increases on [4.77, 10].

67. One possible answer:

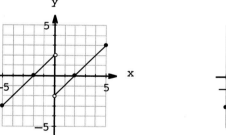

69. One possible answer:

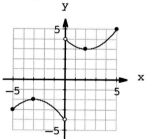

71. One possible answer:

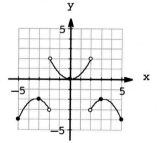

73. The graph of f is shown below.

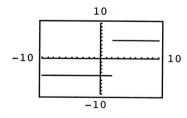

Domain: all real number except $x = 2$.
Range: The only possible values of y are -5 and 5. The range consists of these two values in set notation: {-5, 5}. (This does not mean the interval (-5, 5).)
Discontinuous: at $x = 2$.

75. The graph of f is shown below.

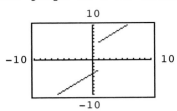

Domain: All real numbers
except $x = 1$;
Range: $(-\infty, -3) \cup (5, \infty)$;
Discontinuous at $x = 1$

77. The graph of f is shown below.

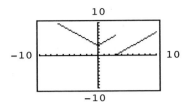

Domain: All real numbers
except $x = 3$;
Range: $(0, \infty)$;
Discontinuous at $x = 3$

79. We note: $f(0) = [\![0/2]\!] = [\![0]\!] = 0$
$\qquad\qquad f(1) = [\![1/2]\!] = 0$
$\qquad\qquad f(2) = [\![2/2]\!] = [\![1]\!] = 1$
$\qquad\qquad f(3) = [\![3/2]\!] = 1 \qquad f(x) = [\![x/2]\!]$ appears to jump at intervals of 2 units.

Generalizing, we can write: $f(x) = \begin{cases} \vdots & & \vdots \\ -2 & \text{if} & -4 \le x < -2 \\ -1 & \text{if} & -2 \le x < 0 \\ 0 & \text{if} & 0 \le x < 2 \\ 1 & \text{if} & 2 \le x < 4 \\ 2 & \text{if} & 4 \le x < 6 \\ \vdots & \text{if} & \vdots \end{cases}$

Domain: All real numbers
Range: All integers
Discontinuous at the even integers

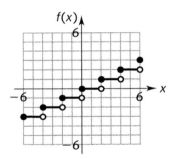

81. Having noted in problem 79 that $[\![\frac{x}{2}]\!]$ jumps at intervals of 2 units, we can
investigate $[\![3x]\!] = [\![x \div \frac{1}{3}]\!]$ to see if it jumps at intervals of $\frac{1}{3}$ unit.
$\qquad f(0) = [\![3 \cdot 0]\!] = [\![0]\!] = 0$
$\qquad f(\frac{1}{6}) = [\![3 \cdot \frac{1}{6}]\!] = [\![\frac{1}{2}]\!] = 0$
$\qquad f(\frac{1}{3}) = [\![3 \cdot \frac{1}{3}]\!] = [\![1]\!] = 1$
$\qquad f(\frac{1}{2}) = [\![3 \cdot \frac{1}{2}]\!] = [\![\frac{3}{2}]\!] = 1$
$\qquad f(\frac{2}{3}) = [\![3 \cdot \frac{2}{3}]\!] = [\![2]\!] = 2$
$\qquad f(\frac{5}{6}) = [\![3 \cdot \frac{5}{6}]\!] = [\![\frac{5}{2}]\!] = 2$

Generalizing, we can write: $f(x) = \begin{cases} \vdots & & \vdots \\ -2 & \text{if} & -\frac{2}{3} \le x < -\frac{1}{3} \\ -1 & \text{if} & -\frac{1}{3} \le x < 0 \\ 0 & \text{if} & 0 \le x < \frac{1}{3} \\ 1 & \text{if} & \frac{1}{3} \le x < \frac{2}{3} \\ 2 & \text{if} & \frac{2}{3} \le x < 1 \\ \vdots & \text{if} & \vdots \end{cases}$

Domain: All real numbers
Range: All integers
Discontinuous at rational numbers of the
form $\frac{k}{3}$ where k is an integer

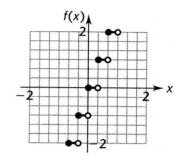

83.   This function is discontinuous at every integer value of x. If x is an integer,
then $x - [\![x]\!] = x - x = 0$. If x is between n and $n + 1$ for some integer n, then
$[\![x]\!] = n$, and $x - [\![x]\!] = x - n$. So the graph between $x = n$ and $x = n + 1$ is a
line segment of the form $y = x - n$, connecting $(n, 0)$ and $(n + 1, 1)$. But
notice that $n + 1$ is always an integer, so f(n + 1) = 0, not 1. We can write:

$f(x) = \begin{cases} \vdots & & \vdots \\ x + 2 & \text{if} & -2 \le x < -1 \\ x + 1 & \text{if} & -1 \le x < 0 \\ x & \text{if} & 0 \le x < 1 \\ x - 1 & \text{if} & 1 \le x < 2 \\ x - 2 & \text{if} & 2 \le x < 3 \\ \vdots & \text{if} & \vdots \end{cases}$

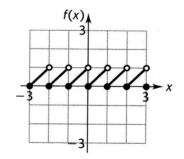

Domain: all real numbers
Range: [0, 1)

85. (A) One possible answer:

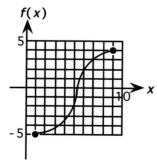

(B) This graph crosses the x axis once.
To meet the conditions specified a graph
must cross the x axis exactly once. If
it crossed more times the function would
have to be decreasing somewhere; if it
did not cross at all the function would
have to be discontinuous somewhere.

87. (A) One possible answer:

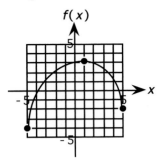

(B) This graph crosses the x axis twice. To meet the conditions specified a graph must cross the x axis at least twice. If it crossed fewer times the function would have to be discontinuous somewhere. However, the graph could cross more times; in fact there is no upper limit on the number of times it can cross the x axis.

89. (A) One possible answer:

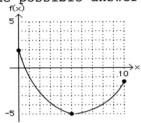

(B) This graph crosses the x axis once. If $f(0)$ were negative, a graph could meet the given conditions and cross the x axis 0 times. If $f(0)$ remained positive, while $f(10)$ were positive, a graph could meet the given conditions and cross the x axis twice. It could not cross more than twice without having another local extremum on the interval.

91. Look at the graph of the revenue function on the given interval and use the MAXIMUM command.

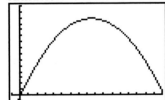
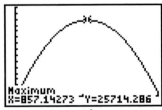

The variable x represents the number of car seats; the x-coordinate of the maximum is 857.14273, so 857 car seats must be sold. (It doesn't make sense to sell .14273 car seats.) The second coordinate represents revenue so the maximum revenue is $25,714.

93.

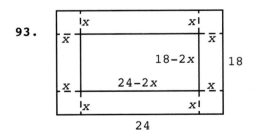

We need to write a function that describes the volume of the box then find the maximum of that function to find the volume of a rectangular box. We multiply the three dimensions. According to the diagram, the dimensions of the base are 24 - 2x and 18 - 2x while the height is x. (The height comes from turning up the flaps.)

So the volume is $V(x) = x(24 - 2x)(18 - 2x)$. Look at the graph of this function and use the MAXIMUM command.

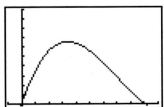
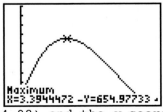

The maximum occurs at the point (3.39, 654.98) and the y-coordinate represents the volume. So the maximum volume is 654.98 cubic inches.

95. Look at the graph of the function $C(x)$ on the given interval and use the MINIMUM command.

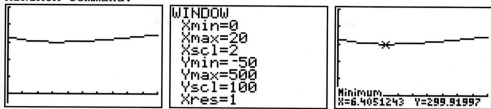

The minimum cost occurs when $x = 6.41$. But notice that the first part of the question asks for the land portion of the pipe, which according to the diagram is $20 - x$. So the land portion should be $20 - 6.41 = 13.59$ miles. The minimum value of $C(x)$ rounds to 300. Since the units for C is thousands of dollars the minimum cost is $300,000.

97.

$$
\begin{aligned}
f(4) &= 10\|0.5 + 0.4\| = 10(0) = 0\\
f(-4) &= 10\|0.5 - 0.4\| = 10(0) = 0\\
f(6) &= 10\|0.5 + 0.6\| = 10(1) = 10\\
f(-6) &= 10\|0.5 - 0.6\| = 10(-1) = -10\\
f(24) &= 10\|0.5 + 2.4\| = 10(2) = 20\\
f(25) &= 10\|0.5 + 2.5\| = 10(3) = 30\\
f(247) &= 10\|0.5 + 24.7\| = 10(25) = 250\\
f(-243) &= 10\|0.5 - 24.3\| = 10(-24) = -240\\
f(-245) &= 10\|0.5 - 24.5\| = 10(-24) = -240\\
f(-246) &= 10\|0.5 - 24.6\| = 10(-25) = -250
\end{aligned}
$$

} rounds numbers to the tens place

99. Since $f(x) = [\![10x + 0.5]\!]/10$ rounds numbers to the nearest tenth, (see text Example 7) we try $[\![100x + 0.5]\!]/100 = f(x)$ to round to the nearest hundredth.

$$
\begin{aligned}
f(3.274) &= [\![327.9]\!]/100 = 3.27\\
f(7.846) &= [\![785.1]\!]/100 = 7.85\\
f(-2.8783) &= [\![-287.33]\!]/100 = -2.88
\end{aligned}
$$

A few examples suffice to convince us that this is probably correct. (A proof would be out of place in this book.)
$f(x) = [\![100x + 0.5]\!]/100.$

101. Since 100 miles are included, only the daily charge of $32 applies for mileage between 0 and 100. So if $R(x)$ is the daily cost of rental where x is miles driven, $R(x) = 32$ if $0 \le x \le 100$. After 100 miles, the charge is an extra $0.16 for each mile: the mileage charge will be 0.16 times the number of miles over 100 which is $x - 100$. So the mileage charge is $0.16(x - 100)$ or $0.16x - 16$. The $32 charge still applies so when $x \ge 100$ the rental charge is $32 + 0.16x - 16$, or $16 + 0.16x$.

$$
R(x) = \begin{cases} 32 & 0 \le x \le 100\\ 16 + 0.16x & x > 100 \end{cases}
$$

103. (A) $C(x) = \begin{cases} 15 & 0 < x \le 1\\ 18 & 1 < x \le 2\\ 21 & 2 < x \le 3\\ 24 & 3 < x \le 4\\ 27 & 4 < x \le 5\\ 30 & 5 < x \le 6 \end{cases}$

(B) The two functions appear to coincide, for example
$$C(3.5) = 24 \qquad f(3.5) = 15 + 3[\![3.5]\!] = 15 + 3 \cdot 3 = 24$$
However,
$$C(1) = 15 \qquad f(1) = 15 + 3[\![1]\!] = 15 \cdot 3 \cdot 1 = 18$$
The functions are not the same, therefore.
In fact, $f(x) \neq C(x)$ at $x = 1, 2, 3, 4, 5, 6$.

105. If $0 \leq x \leq 3{,}000$, $E(x) = 200$

	Base Salary	+	Commission on Sales over \$3,000
If \$3,000 < x < \$8,000, $E(x)$ =	200	+	0.04(x − 3,000)
	= 200	+	0.04x − 120
	= 80	+	0.04x

Common Error:
Commission is not 0.04x
(4% of sales) nor is it
0.04x + 200
(base salary plus 4% of sales).

There is a point of discontinuity at $x = 8{,}000$.

	Salary	+	Bonus
If $x \geq 8{,}000$, $E(x)$ =	80 + 0.04x	+	100
	= 180	+	0.04x

Summarizing, $E(x) = \begin{cases} 200 & \text{if } 0 \leq x \leq 3{,}000 \\ 80 + 0.04x & \text{if } 3{,}000 < x < 8{,}000 \\ 180 + 0.04x & \text{if } 8{,}000 \leq x \end{cases}$

$$E(5{,}750) = 80 + 0.04(5{,}750) = \$310$$
$$E(9{,}200) = 180 + 0.04(9{,}200) = \$548$$

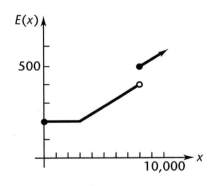

x	$y = 200$	x	$y = 80 + 0.04x$	x	$y = 180 + 0.04x$
0	200	3,000	200	8,000	500
2,000	200	7,000	360	10,000	580

107. A tax of 5.35% on any amount up to \$19,890 is calculated by multiplying .0535 by the income, which is represented by x. So the tax due, $t(x)$, is 0.0535x if x is between 0 and 19,890. For x-values between 19,890 and 65,330, the percentage is computed only on the portion over 19,890. To find that portion we subtract 19,890 from the income (x − 19,890); then multiply by 0.0705 to get the percentage portion of the tax. The total tax is \$1,064 plus the percentage portion, so $t(x) = 1{,}064 + 0.0705(x - 19{,}890)$ if x is between 19,890 and 65,330. The tax for incomes over \$65,330 is computed in a similar manner: \$4,268 plus 7.85% of the portion over 65,330, which is $x - 65{,}330$. We get $t(x) = 4{,}268 + 0.0785(x - 65{,}330)$ if x is over 65,330. Combined we get

$$t(x) = \begin{cases} 0.0535x & 0 \leq x \leq 19{,}890 \\ 1{,}064 + 0.0705(x - 19{,}890) & 19{,}890 < x \leq 65{,}330 \\ 4{,}268 + 0.0785(x - 65{,}330) & x > 65{,}330 \end{cases}$$

or if we multiply out parentheses to simplify,

$$t(x) = \begin{cases} 0.0535x & 0 \le x \le 19{,}890 \\ 0.0705x - 338.25 & 19{,}890 < x \le 65{,}330 \\ 0.0785x - 860.41 & x > 65{,}330 \end{cases}$$

$t(10{,}000) = 0.0535(10{,}000) = 535$; the tax is \$535

$t(30{,}000) = 1{,}064 + 0.0705(30{,}000 - 19{,}890) = 1{,}776.75$; the tax is \$1,776.75

$t(100{,}000) = 4{,}268 + 0.0785(100{,}000 - 65{,}330) = 6{,}989.60$; the tax is \$6,989.60

109. (A) For detailed instructions on inputting the data see Example 10 in Section 1-2. The data input, viewing window, and scatter plot are as follows:

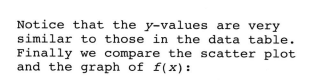

Next, we input the given function and use the table feature to construct a table with the same x values:

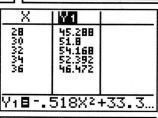

Notice that the y-values are very similar to those in the data table. Finally we compare the scatter plot and the graph of $f(x)$:

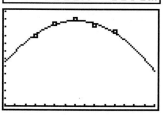

As expected, the graph of $f(x)$ fits the points in the scatter plot well.

(B) $f(31) = -0.518(31)^2 + 33.3(31) - 481 = 53.50$, so the mileage for a tire pressure of 31 lb/m^2 will be 53,500. $f(35) = 49.95$ so the mileage for a tire pressure of 35 lb/in^2 will be 49,950.

(C) As the tire pressure increases from 28 to 32 lb/in^2 the mileage increases from 45,000 mi to 55,000 mi. A local maximum occurs when the tire pressure is 32 lb/in^2 and the mileage is 55,000 mi. As the tire pressure continues to increase from 32 lb/in^2 to 36 lb/in^2 the mileage decreases from 55,000 mi to 47,000 mi.

Section 1-4

1. Answers will vary

3. Answers will vary.

5. The graph of $f(x)$ is shifted up 2 units

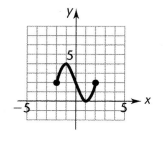

7. The graph of $g(x)$ is shifted up 2 units

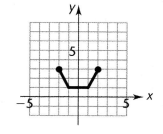

9. The graph of $f(x)$ is shifted right 2 units.

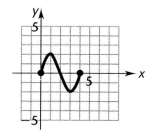

11. The graph of $g(x)$ is shifted left 2 units.

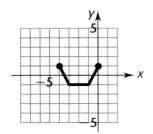

Common Errors:
Confusing the -2 in $f(x - 2)$ with a shift left.
Confusing the positive 2 in $g(x + 2)$ with a shift right.

13. The graph of $f(x)$ is reflected in the x axis.

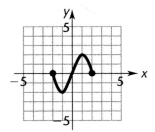

15. The graph of $g(x)$ is vertically stretched by multiplying each y coordinate by 2.

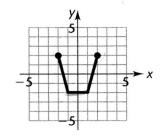

17. $g(2x)$ is a horizontal shrinking of the graph of $g(x)$ by a factor of $\frac{1}{2}$.

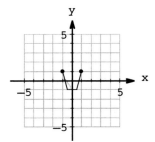

19. The graph of $f(-x)$ is a reflection in the y-axis of the graph of $f(x)$.

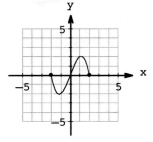

21. $g(-x) = (-x)^3 + (-x) = -x^3 - x = -(x^3 + x) = -g(x)$. Odd

23. $m(-x) = (-x)^4 + 3(-x)^2 = x^4 = 3x^2 = m(x)$. Even

25. $F(-x) = (-x)^5 + 1 = -x^5 + 1$
$-F(x) = -(x^5 + 1) = -x^5 - 1$
Therefore $F(-x) \neq F(x)$. $F(-x) \neq -F(x)$. $F(x)$ is neither even nor odd.

27. $G(-x) = (-x)^4 + 2 = x^4 + 2 = G(x)$. Even

29. $q(-x) = (-x)^2 + (-x) - 3 = x^2 - x - 3$.
$-q(x) = -(x^2 + x - 3) = -x^2 - x + 3$
Therefore $q(-x) \neq q(x)$. $q(-x) \neq -q(x)$. $q(x)$ is neither even nor odd.

31. The graph of $y = x^2$ is shifted 2 units to the right: $y = (x - 2)^2$.
Check: $y = (x - 2)^2$ is graphed on a graphing calculator.

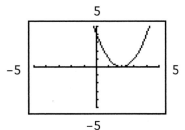

33. The graph of $y = x^3$ is shifted down 2 units: $y = x^3 - 2$.
Check: $y = x^3 - 2$ is graphed on a graphing calculator.

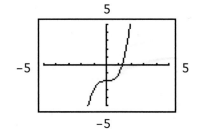

35. The graph of $y = |x|$ is shrunk
vertically by a factor of 0.25:
$y = 0.25|x|$.
Check: $y = 0.25|x|$ is graphed on a
graphing calculator.

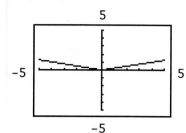

37. The graph of $y = x^3$ is reflected in
the x axis (or the y axis): $y = -x^3$.
Check: $y = -x^3$ is graphed on a
graphing calculator.

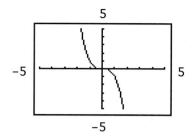

39. $g(x) = \sqrt[3]{x + 4} - 5$.
The graphs of $f(x) = \sqrt[3]{x}$ (thin curve)
and $g(x)$ (thick curve) are shown.

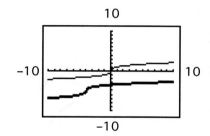

41. $g(x) = -0.5(6 + \sqrt{x})$
The graph of $f(x) = \sqrt{x}$ (thin curve)
and $g(x)$ (thick curve) are shown.

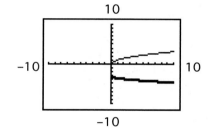

43. $g(x) = -2(x + 4)^2 - 2$
The graph of $f(x) = x^2$ (thin curve)
and $g(x)$ (thick curve) are shown.

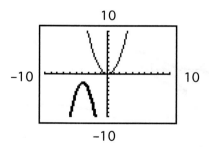

45. $g(x) = \sqrt{-\dfrac{1}{2}(x + 2)}$. The graph of $f(x) = \sqrt{x}$ and $g(x)$ are shown.

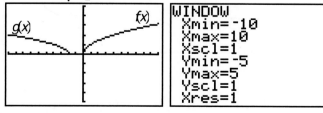

47. The graph of $y = x^2$ is shifted
7 units left and 9 units up.

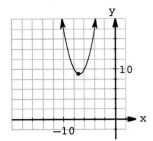

49. The graph of $y = |x|$ is shifted 8
units right and reflected in the
x axis.

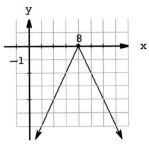

51. The graph of $y = \sqrt{x}$ is reflected in
the x axis and shifted 3 units up.

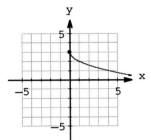

53. The graph of $y = x^2$ is stretched
vertically by a factor of 4 and
reflected in the x axis.

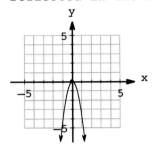

55. The graph of $y = |x|$ is shifted 2
units left and 2 units up:
$y = |x + 2| + 2$.
Check: $y = |x + 2| + 2$ is graphed on
a graphing calculator.

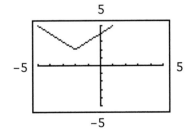

57. The graph of $y = \sqrt{x}$ is reflected in
the x axis and shifted 4 units up:
$y = 4 - \sqrt{x}$
Check: $y = 4 - \sqrt{x}$ is graphed in a
graphing calculator.

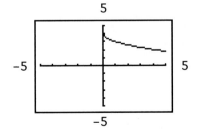

59. The graph of $y = x^2$ is shifted 1 unit
right, reflected in the x axis, and
shifted 4 units up:
$y = 4 - (x - 1)^2$
Check: $y = 4 - (x - 1)^2$ is graphed in
a graphing calculator.

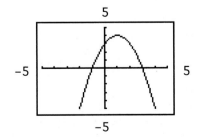

61. The graph of $y = x^3$ is shifted 3 units right, vertically shrunk by a factor of 0.5, and shifted 1 unit up:
$y = 0.5(x - 3)^3 + 1$
Check: $y = 0.5(x - 3)^3 + 1$ is graphed in a graphing calculator.

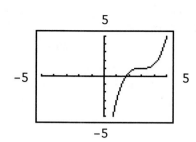

63. (A) The function f is a horizontal shrink of $y = \sqrt[3]{x}$ by a factor of 1/8, while g is a vertical stretch of $y = \sqrt[3]{x}$ by a factor of 2.

(B) The graphs are shown below in a standard window: they are identical.

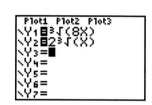 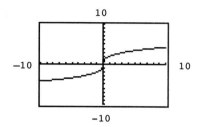

(C) $\sqrt[3]{8x} = \sqrt[3]{8} \cdot \sqrt[3]{x} = 2\sqrt[3]{x}$

65. (A) The graphs are shown below.

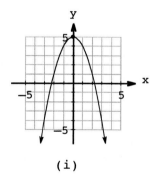 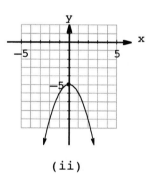

        (i)   (ii)

The graphs are different, so order is significant when performing multiple transformations.

(B) (i): $y = -\left(x^2 - 5\right)$; (ii): $y = -x^2 - 5$

These functions are different. In the second one, order of operations tells us to first multiply by -1, then subtract 5. In the first, the parentheses indicate that this order should be reversed.

67. The graph of $y = f(x - h)$ represents a horizontal shift from the graph of $y = f(x)$. The graph of $y = f(x) + k$ represents a vertical shift from the graph of $y = f(x)$. The graph of $y = f(x - h) + k$ represents both a horizontal and a vertical shift but the order does not matter:
Vertical first then horizontal: $y = f(x) \rightarrow y = f(x) + k \rightarrow y = f(x - h) + k$
Horizontal first then vertical: $y = f(x) \rightarrow y = f(x - h) \rightarrow y = f(x - h) + k$
The same result is achieved; reversing the order does not change the result.

69. Consider the graph of $y = x^2$.
If a vertical shift is performed the equation becomes $y = x^2 + k$.
If a reflection is now performed the equation becomes
$y = -(x^2 + k)$ or $y = -x^2 - k$.
If the reflection is performed first the equation becomes $y = -x^2$.
If the vertical shift is now performed the equation becomes $y = -x^2 + k$.
Since $y = -x^2 - k$ and $y = -x^2 + k$ differ (unless $k = 0$), reversing the order changes the result.

71. The graph of $y = f(x - h)$ represents a horizontal shift from the graph of $y = f(x)$. The graph of $y = -f(x)$ represents a reflection of the graph of $y = f(x)$ in the x axis. The graph of $y = -f(x - h)$ represents both a horizontal shift and a reflection but the order does not matter:
Shift first then reflection: $y = f(x) \;\rightarrow\; y = f(x - h) \;\rightarrow\; y = -f(x - h)$
Reflection first then shift: $y = f(x) \;\rightarrow\; y = -f(x) \;\rightarrow\; y = -f(x - h)$
The same result is achieved; reversing the order does not change the result.

73.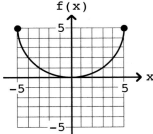

75.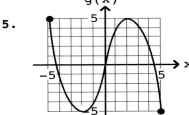

77. (A) $E(x) = \dfrac{1}{2}[f(x) + f(-x)]$

$E(-x) = \dfrac{1}{2}[f(-x) + f\{-(-x)\}] = \dfrac{1}{2}[f(-x) + f(x)] = \dfrac{1}{2}[f(x) + f(-x)] = E(x).$

$E(x)$ is even.

(B) $O(x) = \dfrac{1}{2}[f(x) - f(-x)]$

$O(-x) = \dfrac{1}{2}[f(-x) - f\{-(-x)\}] = \dfrac{1}{2}[f(-x) - f(x)] = -\dfrac{1}{2}[f(x) - f(-x)] = -O(x).$

$O(x)$ is odd.

(C) $E(x) + O(x) = \dfrac{1}{2}[f(x) + f(-x)] + \dfrac{1}{2}[f(x) - f(-x)]$

$\qquad\qquad = \dfrac{1}{2}f(x) + \dfrac{1}{2}f(-x) + \dfrac{1}{2}f(x) - \dfrac{1}{2}f(-x)$

$\qquad\qquad = f(x)$

Conclusion: Any function can be written as the sum of two other functions, one even and the other odd.

79. Graph of $f(x)$

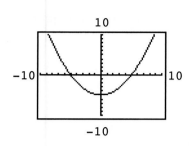

Graph of
$|f(x)| = |0.2x^2 - 5|$

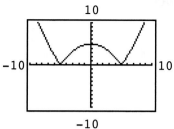

Graph of
$-|f(x)| = -|0.2x^2 - 5|$

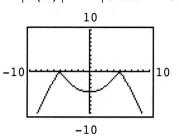

81. Graph of $f(x)$

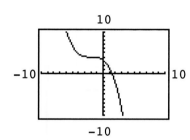

Graph of
$|f(x)| = |4 - 0.1(x + 2)^3|$

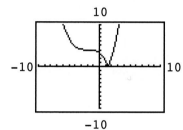

Graph of
$-|f(x)| = -|4 - 0.1(x + 2)^3|$

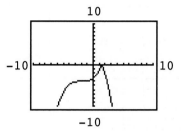

83. The graph of $y = |f(x)|$ is the same as the graph of $y = f(x)$ whenever $f(x) \geq 0$ and is the reflection of the graph of $y = f(x)$ with respect to the x axis whenever $f(x) < 0$.

85. The graph of the function $C(x) = 30,000 + f(x)$ is the same as the given graph of the function $f(x)$ shifted up 30,000 units ($).

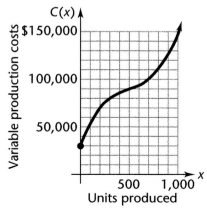

87. $y = 10 + 0.004(x - 10)^3$,
$y = 15 + 0.004(x - 10)^3$,
$y = 20 + 0.004(x - 10)^3$.
Each graph is a vertical translation of the graph of $y = 0.004(x - 10)^3$.

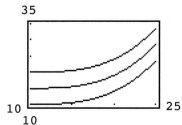

89. $y = \dfrac{1}{2}x^2 - 2$, $y = \dfrac{1}{3}x^2 - 3$, $y = \dfrac{1}{4}x^2 - 4$, $y = \dfrac{1}{5}x^2 - 5$.

Each graph is a vertical shrink followed by a vertical translation of the graph of $y = x^2$.

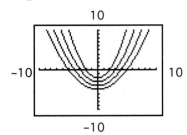

91. $V(t) = \dfrac{64}{c^2}(C - t)^2$ $0 \le t \le C$

t	$C = 1$ $V = 64(1 - t)^2$	$C = 2$ $V = 16(2 - t)^2$	$C = 4$ $V = 4(4 - t)^2$	$C = 8$ $V = (8 - t)^2$
0	64	64	64	64
1	0	16	36	49
2	not defined for $t>1$	0	16	36
4		not defined for $t>0$	0	16
8			not defined for $t > 4$	0

Each graph is a portion of the graph of a horizontal translation followed by a vertical shrink (except for $C = 8$) of the graph of $y = t^2$.

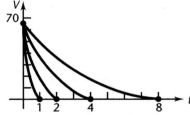

The height of the graph represents the volume of water left in the tank, so we see that for larger values of C, the water stays in the tank longer. We can conclude that larger values of C correspond to a smaller opening.

Section 1-5

1. Answers will vary.

3. No. Elements that make the output of g equal to zero are excluded from the domain of f/g.

5. No. Explanations will vary.

7. Construct a table of values for $f(x)$ and $g(x)$ from the graph, then add to obtain $(f + g)(x)$.

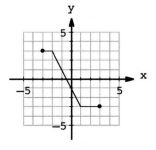

x	-3	-2	-1	0	1	2	3
$f(x)$	1	0	-1	-2	-3	-2	-1
$g(x)$	2	3	2	1	0	-1	-2
$(f + g)(x)$	3	3	1	-1	-3	-3	-3

9. Construct a table of values for $f(x)$ and $g(x)$ from the graph, then multiply to obtain $(fg)(x)$.

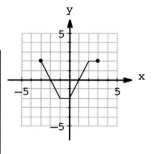

x	-3	-2	-1	0	1	2	3
$f(x)$	1	0	-1	-2	-3	-2	-1
$g(x)$	2	3	2	1	0	-1	-2
$(f + g)(x)$	2	0	-2	-2	0	2	2

11. $(f \circ g)(-1) = f[g(-1)]$. From the graph of g, $g(-1) = 2$. From the graph of f, $f[g(-1)] = f(2) = -2$.

13. $(g \circ f)(-2) = g[f(-2)]$. From the graph of f, $f(-2) = 0$. From the graph of g, $g[f(-2)] = g(0) = 1$.

15. From the graph of g, $g(1) = 0$. From the graph of f, $f[g(1)] = f(0) = -2$.

17. From the graph of f, $f(2) = -2$. From the graph of g, $g[f(2)] = g(-2) = 3$.

19. Using values from the table, $g(-7) = 4$, so $(f \circ g)(-7) = f(g(-7)) = f(4) = 3$. Similarly, $(f \circ g)(0) = f(g(0)) = f(-2) = 9$, and $(f \circ g)(4) = f(g(4)) = f(6) = -10$.

21. $(f + g)(x) = f(x) + g(x) = 4x + x + 1 = 5x + 1$ Domain: $(-\infty, \infty)$
$(f - g)(x) = f(x) - g(x) = 4x - (x + 1) = 3x - 1$ Domain: $(-\infty, \infty)$
$(fg)(x) = f(x)g(x) = 4x(x + 1) = 4x^2 + 4x$ Domain: $(-\infty, \infty)$
$\left(\dfrac{f}{g}\right)(x) = \dfrac{f(x)}{g(x)} = \dfrac{4x}{x + 1}$ Domain: $\{x \mid x \neq -1\}$, or

$(-\infty, -1) \cup (-1, \infty)$

> **Common Error:**
> $f(x) - g(x) \neq 4x - x + 1$. The parentheses are necessary.

23. $(f + g)(x) = f(x) + g(x) = 2x^2 + x^2 + 1 = 3x^2 + 1$ Domain: $(-\infty, \infty)$
$(f - g)(x) = f(x) - g(x) = 2x^2 - (x^2 + 1) = x^2 - 1$ Domain: $(-\infty, \infty)$
$(fg)(x) = f(x)g(x) = 2x^2(x^2 + 1) = 2x^4 + 2x^2$ Domain: $(-\infty, \infty)$
$\left(\dfrac{f}{g}\right)(x) = \dfrac{f(x)}{g(x)} = \dfrac{2x^2}{x^2 + 1}$ Domain: $(-\infty, \infty)$

(since $g(x)$ is never 0.)

25. $(f + g)(x) = f(x) + g(x) = 3x + 5 + x^2 - 1$
$= x^2 + 3x + 4$ Domain: $(-\infty, \infty)$
$(f - g)(x) = f(x) - g(x) = 3x + 5 - (x^2 - 1)$
$= 3x + 5 - x^2 + 1 = -x^2 + 3x + 6$ Domain: $(-\infty, \infty)$
$(fg)(x) = f(x)g(x) = (3x + 5)(x^2 - 1)$
$= 3x^3 - 3x + 5x^2 - 5 = 3x^3 + 5x^2 - 3x - 5$ Domain: $(-\infty, \infty)$
$\left(\dfrac{f}{g}\right)(x) = \dfrac{f(x)}{g(x)} = \dfrac{3x + 5}{x^2 - 1}$ Domain: $\{x \mid x \neq \pm 1\}$, or

$(-\infty, -1) \cup (-1, 1) \cup (1, \infty)$

27. $(f \circ g)(x) = f[g(x)] = f(x^2 - x + 1) = (x^2 - x + 1)^3$ Domain: $(-\infty, \infty)$

    $(g \circ f)(x) = g[f(x)] = g(x^3) = (x^3)^2 - x^3 + 1 = x^6 - x^3 + 1$ Domain: $(-\infty, \infty)$

29. $(f \circ g)(x) = f[g(x)] = f(2x + 3) = |2x + 3 + 1| = |2x + 4|$ Domain: $(-\infty, \infty)$

    $(g \circ f)(x) = g[f(x)] = g(|x + 1|) = 2|x + 1| + 3$ Domain: $(-\infty, \infty)$

31. $(f \circ g)(x) = f[g(x)] = f(2x^3 + 4) = (2x^3 + 4)^{1/3}$ Domain: $(-\infty, \infty)$

    $(g \circ f)(x) = g[f(x)] = g(x^{1/3}) = 2(x^{1/3})^3 + 4 = 2x + 4$ Domain: $(-\infty, \infty)$

33. $(f \circ g)(x) = f[g(x)] = f(2x - 2) = \dfrac{1}{2}(2x - 2) + 1 = x - 1 + 1 = x$

    $(g \circ f)(x) = g[f(x)] = g\left(\dfrac{1}{2}x + 1\right) = 2\left(\dfrac{1}{2}x + 1\right) - 2 = x + 2 - 2 = x$

Graphing f, g, $f \circ g$, and $g \circ f$, we obtain the graph at the right.

The graphs of f and g are reflections of each other in the line $y = x$, which is the graph of $f \circ g$ and $g \circ f$.

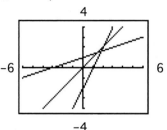

35. $(f \circ g)(x) = f[g(x)] = f\left(-\dfrac{3}{2}x - \dfrac{5}{2}\right) = -\dfrac{2}{3}\left(-\dfrac{3}{2}x - \dfrac{5}{2}\right) - \dfrac{5}{3} = x + \dfrac{5}{3} - \dfrac{5}{3} = x$

    $(g \circ f)(x) = g[f(x)] = g\left(-\dfrac{2}{3}x - \dfrac{5}{3}\right) = -\dfrac{3}{2}\left(-\dfrac{2}{3}x - \dfrac{5}{3}\right) - \dfrac{5}{2} = x + \dfrac{5}{2} - \dfrac{5}{2} = x$

Graphing f, g, $f \circ g$, and $g \circ f$, we obtain the graph at the right.

The graphs of f and g are reflections of each other in the line $y = x$, which is the graph of $f \circ g$ and $g \circ f$.

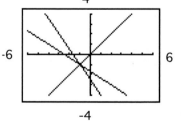

37. $(f + g)(x) = f(x) + g(x) = \sqrt{2 - x} + \sqrt{x + 3}$

    $(f - g)(x) = f(x) - g(x) = \sqrt{2 - x} - \sqrt{x + 3}$

     $(fg)(x) = f(x)g(x) = \sqrt{2 - x}\,\sqrt{x + 3} = \sqrt{(2 - x)(3 + x)} = \sqrt{6 - x - x^2}$

    $\left(\dfrac{f}{g}\right)(x) = \dfrac{f(x)}{g(x)} = \dfrac{\sqrt{2 - x}}{\sqrt{x + 3}} = \sqrt{\dfrac{2 - x}{x + 3}}$

The domains of f and g are:
Domain of $f = \{x \mid 2 - x \geq 0\} = (-\infty, 2]$

Domain of $g = \{x \mid x + 3 \geq 0\} = [-3, \infty)$
The intersection of these domains is $[-3, 2]$.
This is the domain of the functions of $f + g$, $f - g$, and fg.

Since $g(-3) = 0$, $x = -3$ must be excluded from the domain of $\dfrac{f}{g}$, so its domain is $(-3, 2]$.

39. $(f + g)(x) = f(x) + g(x) = \sqrt{x} + 2 + \sqrt{x} - 4 = 2\sqrt{x} - 2$

$(f - g)(x) = f(x) - g(x) = \sqrt{x} + 2 - (\sqrt{x} - 4) = \sqrt{x} + 2 - \sqrt{x} + 4 = 6$

$(fg)(x) = f(x)g(x) = (\sqrt{x} + 2)(\sqrt{x} - 4) = x - 2\sqrt{x} - 8$

$\left(\dfrac{f}{g}\right)(x) = \dfrac{f(x)}{g(x)} = \dfrac{\sqrt{x} + 2}{\sqrt{x} - 4}$

The domains of f and g are both $\{x \mid x \geq 0\} = [0, \infty)$

This is the domain of $f + g$, $f - g$, and fg. We note that in the domain of $\dfrac{f}{g}$,

$g(x)$ cannot be 0, so $\sqrt{x} - 4$ cannot be 0. To find which values to exclude, we solve

$\sqrt{x} - 4 = 0$

$\sqrt{x} = 4$ | **Common Error:** $x \neq \sqrt{4}$ |

$x = 16$

So 16 must be excluded from $\{x \mid x \geq 0\}$ to find the domain of $\dfrac{f}{g}$.

Domain of $\dfrac{f}{g} = \{x \mid x \geq 0, x \neq 16\} = [0, 16) \cup (16, \infty)$.

41. $(f + g)(x) = f(x) + g(x) = \sqrt{x^2 + x - 6} + \sqrt{7 + 6x - x^2}$

$(f - g)(x) = f(x) - g(x) = \sqrt{x^2 + x - 6} - \sqrt{7 + 6x - x^2}$

$(fg)(x) = f(x)g(x) = \sqrt{x^2 + x - 6}\sqrt{7 + 6x - x^2} = \sqrt{-x^4 + 5x^3 + 19x^2 - 29x - 42}$

$\left(\dfrac{f}{g}\right)(x) = \dfrac{f(x)}{g(x)} = \dfrac{\sqrt{x^2 + x - 6}}{\sqrt{7 + 6x - x^2}} = \sqrt{\dfrac{x^2 + x - 6}{7 + 6x - x^2}}$

The domains of f and g are:

Domain of $f = \{x \mid x^2 + x - 6 \geq 0\} = \{x \mid (x + 3)(x - 2) \geq 0\} = (-\infty, -3] \cup [2, \infty)$

Domain of $g = \{x \mid 7 + 6x - x^2 \geq 0\} = \{x \mid (7 - x)(1 + x) \geq 0\} = [-1, 7]$

The intersection of these domains is $[2, 7]$.

This is the domain of the functions $f + g$, $f - g$, and fg.

Since $g(x) = 7 + 6x - x^2 = (7 - x)(1 + x)$, $g(7) = 0$ and $g(-1) = 0$, so 7 must be

excluded from the domain of $\dfrac{f}{g}$. Its domain is $[2, 7)$.

43. $(f \circ g)(x) = f[g(x)] = f(x - 4) = \sqrt{x - 4}$ Domain: $\{x \mid x \geq 4\}$ or $[4, \infty)$

$(g \circ f)(x) = g[f(x)] = g(\sqrt{x}) = \sqrt{x} - 4$ Domain: $\{x \mid x \geq 0\}$ or $[0, \infty)$

45. $(f \circ g)(x) = f[g(x)] = f\left(\dfrac{1}{x}\right) = \dfrac{1}{x} + 2$ Domain: $\{x \mid x \neq 0\}$ or $(-\infty, 0) \cup (0, \infty)$

$(g \circ f)(x) = g[f(x)] = g(x + 2) = \dfrac{1}{x + 2}$ Domain: $\{x \mid x \neq -2\}$ or $(-\infty, -2)$
$\cup (-2, \infty)$

47. $(f \circ g)(x) = f[g(x)] = f\left(\dfrac{1}{x - 1}\right) = \left|\dfrac{1}{x - 1}\right| = \dfrac{1}{|x - 1|}$ Domain:
$\{x \mid x \neq 1\}$ or $(-\infty, 1) \cup (1, \infty)$

$(g \circ f)(x) = g[f(x)] = g(|x|) = \dfrac{1}{|x| - 1}$

To find the domain we must exclude any x values for which $|x| - 1 = 0$ from the domain of f.

$|x| - 1 = 0$

$|x| = 1$

$x = -1$ or 1

Domain of $g \circ f$: $\{x \mid x \neq -1 \text{ or } 1\} = (-\infty, -1) \cup (-1, 1) \cup (1, \infty)$

In Problems #49 through 51, f and g are linear functions. f has slope -2 and y intercept 2, so f(x) = -2x + 2. g has slope 1 and y intercept -2, so g(x) = x - 2.

49. $(f + g)(x) = f(x) + g(x) = (-2x + 2) + (x - 2) = -x$.
The graph of $f + g$ is a straight line with slope -1 passing through the origin. This corresponds to graph (d).

51. $(g - f)(x) = g(x) - f(x) = (x - 2) - (-2x + 2) = x - 2 + 2x - 2 = 3x - 4$.
The graph of $g - f$ is a straight line with slope 3 and y intercept -4. This corresponds to graph (a).

In Problems #53 through 59, there may be other possible solutions.

53. If we let $g(x) = 2x - 7$, then
$h(x) = [g(x)]^4$
Now if we let $f(x) = x^4$, we have
$h(x) = [g(x)]^4 = f[g(x)] = (f \circ g)(x)$

55. If we let $g(x) = 4 + 2x$, then
$h(x) = \sqrt{g(x)}$
Now if we let $f(x) = x^{1/2}$, we have
$h(x) = \sqrt{g(x)} = [g(x)]^{1/2}$
$= f[g(x)] = (f \circ g)(x)$.

57. If we let $f(x) = x^7$, then
$h(x) = 3f(x) - 5$
Now if we let $g(x) = 3x - 5$, we have
$h(x) = 3f(x) - 5 = g[f(x)] = (g \circ f)(x)$

59. If we let $f(x) = x^{-1/2}$, then
$h(x) = 4f(x) + 3$
Now if we let $g(x) = 4x + 3$, we have
$h(x) = 4f(x) + 3 = g[f(x)] = (g \circ f)(x)$

61. Yes, the function $g(x) = x$ satisfies these conditions.
$(f \circ g)(x) = f(g(x)) = f(x)$, so $f \circ g = f$
$(g \circ f)(x) = g(f(x)) = f(x)$, so $g \circ f = f$

63. $(f + g)(x) = f(x) + g(x) = x + \dfrac{1}{x} + x - \dfrac{1}{x} = 2x$

$\boxed{\textbf{Common Error:} \\ \text{Domain is not } (-\infty, \infty). \text{ See below.}}$

$(f - g)(x) = f(x) - g(x) = x + \dfrac{1}{x} - \left(x - \dfrac{1}{x}\right) = \dfrac{2}{x}$

$(fg)(x) = f(x)g(x) = \left(x + \dfrac{1}{x}\right)\left(x - \dfrac{1}{x}\right) = x^2 - \dfrac{1}{x^2}$

$\left(\dfrac{f}{g}\right)(x) = \dfrac{f(x)}{g(x)} = \dfrac{x + \frac{1}{x}}{x - \frac{1}{x}} = \dfrac{x^2 + 1}{x^2 - 1}$

The domains of f and g are both $\{x \mid x \neq 0\} = (-\infty, 0) \cup (0, \infty)$
This is therefore the domain of $f + g$, $f - g$, and fg. To find the domain of $\dfrac{f}{g}$, we must exclude from this domain the set of values of x for which $g(x) = 0$.

$x - \dfrac{1}{x} = 0$
$x^2 - 1 = 0$
$x^2 = 1$
$x = -1, 1$

The domain of $\dfrac{f}{g}$ is $\{x \mid x \neq 0, -1, \text{ or } 1\}$ or

$(-\infty, -1) \cup (-1, 0) \cup (0, 1) \cup (1, \infty)$.

65. $(f + g)(x) = f(x) + g(x) = 1 - \dfrac{x}{|x|} + 1 + \dfrac{x}{|x|} = 2$

$(f - g)(x) = f(x) - g(x) = 1 - \dfrac{x}{|x|} - \left(1 + \dfrac{x}{|x|}\right) = 1 - \dfrac{x}{|x|} - 1 - \dfrac{x}{|x|} = \dfrac{-2x}{|x|}$

$(fg)(x) = f(x)g(x) = \left(1 - \dfrac{x}{|x|}\right)\left(1 + \dfrac{x}{|x|}\right) = (1)^2 - \left(\dfrac{x}{|x|}\right)^2 = 1 - \dfrac{x^2}{|x|^2}$

$$= 1 - \dfrac{x^2}{x^2} = 1 - 1 = 0$$

$\left(\dfrac{f}{g}\right)(x) = \dfrac{f(x)}{g(x)} = \dfrac{1 - \frac{x}{|x|}}{1 + \frac{x}{|x|}} = \dfrac{|x| - x}{|x| + x}$. This can be further simplified

however, when we examine the domain of $\dfrac{f}{g}$ below.

The domains of f and g are both
$\{x \mid x \neq 0\} = (-\infty, 0) \cup (0, \infty)$

This is therefore the domain of $f + g$, $f - g$, and fg. To find the domain of $\dfrac{f}{g}$, we must exclude from this domain the set of values of x for which $g(x) = 0$.

$1 + \dfrac{x}{|x|} = 0$

$|x| + x = 0$

$\quad |x| = -x$

This is true when x is negative.

The domain of $\dfrac{f}{g}$ is the positive numbers, $(0, \infty)$. On this domain,

$|x| = x$, so $\left(\dfrac{f}{g}\right)(x) = \dfrac{|x| - x}{|x| + x} = \dfrac{x - x}{x + x} = \dfrac{0}{2x} = 0$

67. $(f \circ g)(x) = f[g(x)] = f(x^2) = \sqrt{4 - x^2}$
The domain of f is $(-\infty, 4]$. The domain of g is all real numbers. The domain of $f \circ g$ is therefore the set of those real numbers x for which $g(x)$ is in
$(-\infty, 4]$, that is, for which $x^2 \leq 4$, or $-2 \leq x \leq 2$.
Domain of $f \circ g = \{x \mid -2 \leq x \leq 2\} = [-2, 2]$
$(g \circ f)(x) = g[f(x)] = g(\sqrt{4 - x}) = (\sqrt{4 - x})^2 = 4 - x$
The domain of $g \circ f$ is the set of those numbers x in $(-\infty, 4]$ for which $f(x)$ is in $(-\infty, \infty)$, that is, $(-\infty, 4]$.

69. $(f \circ g)(x) = f[g(x)] = f\left(\dfrac{x}{x - 2}\right) = \dfrac{\frac{x}{x-2} + 5}{\frac{x}{x-2}} = \dfrac{x + 5(x - 2)}{x} = \dfrac{x + 5x - 10}{x} = \dfrac{6x - 10}{x}$
The domain of f is $\{x \mid x \neq 0\}$. The domain of g is $\{x \mid x \neq 2\}$. The domain of $f \circ g$ is therefore the set of those numbers in $\{x \mid x \neq 2\}$ for which $g(x)$ is in $\{x \mid x \neq 0\}$. We must exclude from $\{x \mid x \neq 2\}$ those numbers x for which $\dfrac{x}{x - 2} = 0$, or $x = 0$. The domain of $f \circ g$ is $\{x \mid x \neq 0, \ x \neq 2\}$, or $(-\infty, 0) \cup (0, 2) \cup (2, \infty)$.
$(g \circ f)(x) = g[f(x)] = g\left(\dfrac{x + 5}{x}\right) = \dfrac{\frac{x+5}{x}}{\frac{x+5}{x} - 2} = \dfrac{x + 5}{x + 5 - 2x} = \dfrac{x + 5}{5 - x}$
The domain of $g \circ f$ is the set of those numbers in $\{x \mid x \neq 0\}$ for which $f(x)$ is in $\{x \mid x \neq 2\}$. We must exclude from $\{x \mid x \neq 0\}$ those numbers x for which $\dfrac{x + 5}{x} = 2$, or $x + 5 = 2x$, or $x = 5$. The domain of $g \circ f$ is $\{x \mid x \neq 0, \ x \neq 5\}$ or $(-\infty, 0) \cup (0, 5) \cup (5, \infty)$.

71. $(f \circ g)(x) = f[g(x)] = f(\sqrt{9 + x^2}) = \sqrt{25 - (\sqrt{9 + x^2})^2} = \sqrt{25 - (9 + x^2)} = \sqrt{16 - x^2}$
The domain of f is [-5, 5]. The domain of g is $(-\infty, \infty)$. The domain of $f \circ g$ is therefore the set of those real numbers x for which $g(x)$ is in [-5, 5], that is, $\sqrt{9 + x^2} \le 5$, or $9 + x^2 \le 25$, or $x^2 \le 16$, or $-4 \le x \le 4$. The domain of $f \circ g$ is {x | $-4 \le x \le 4$} or [-4, 4].

$(g \circ f)(x) = g[f(x)] = g(\sqrt{25 - x^2}) = \sqrt{9 + (\sqrt{25 - x^2})^2} = \sqrt{9 + 25 - x^2} = \sqrt{34 - x^2}$
The domain of $g \circ f$ is the set of those numbers x in [-5, 5] for which $g(x)$ is real. Since $g(x)$ is real for all x, the domain of $g \circ f$ is [-5, 5].

> **Common Error:** The domain of $g \circ f$ is not evident from the final form $\sqrt{34 - x^2}$. It is not $[-\sqrt{34}, \sqrt{34}]$.

73. $f(x) = \sqrt{5 - x^2}$ Domain of f: $[-\sqrt{5}, \sqrt{5}]$
$g(x) = \sqrt{3 - x}$ Domain of g: $(-\infty, 3]$
$(f \circ g)(x) = f(\sqrt{3 - x}) = \sqrt{5 - (\sqrt{3 - x})^2}$
Graph of $f \circ g$, not simplified:

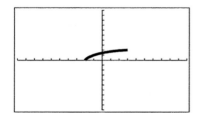

Simplifying $(f \circ g)(x)$, we obtain $\sqrt{5 - (\sqrt{3 - x})^2}$
$= \sqrt{5 - (3 - x)} = \sqrt{5 - 3 + x} = \sqrt{2 + x}$. Apparent graph of $f \circ g$, simplified:

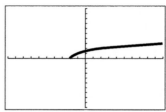

The domain of $f \circ g$ is not apparent from the final form $\sqrt{2 + x}$. It is not $[-2, \infty)$. The domain of $f \circ g$ is the set of those numbers in $(-\infty, 3]$ for which $g(x)$ is in $[-\sqrt{5}, \sqrt{5}]$, that is, $-\sqrt{5} \le \sqrt{3 - x} \le \sqrt{5}$, of $3 - x \le 5$, or $x \ge -2$. The domain of $f \circ g$ is [-2, 3]. This is clearly shown in the first graph so the first graph is correct.

75. $f(x) = \sqrt{x^2 + 5}$ Domain of f: $(-\infty, \infty)$
$g(x) = \sqrt{x^2 - 4}$ Domain of g: $(-\infty, -2] \cup [2, \infty)$
$(f \circ g)(x) = f(\sqrt{x^2 - 4}) = \sqrt{(\sqrt{x^2 - 4})^2 + 5}$
Graph of $f \circ g$, not simplified:

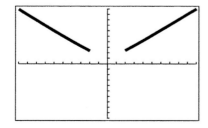

Simplifying $(f \circ g)(x)$, we obtain

$$\sqrt{(\sqrt{x^2 - 4})^2 + 5} = \sqrt{x^2 - 4 + 5} = \sqrt{x^2 + 1}.$$

Apparent graph of $f \circ g$, simplified:

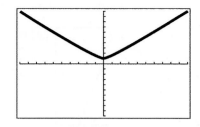

The domain of $f \circ g$ is not apparent from the final form $\sqrt{x^2 + 1}$. It is not $(-\infty, \infty)$. The domain of $f \circ g$ is the set of those numbers in $(-\infty, -2] \cup [2, \infty)$ for which $g(x)$ is real, that is, all of $(-\infty, -2] \cup [2, \infty)$. This is clearly shown in the first graph, so the first graph is correct.

77. $f(x) = \sqrt{x^2 + 7}$ Domain of f: $(-\infty, \infty)$

$g(x) = \sqrt{9 - x^2}$ Domain of g: $[-3, 3]$

$(f \circ g)(x) = f(\sqrt{9 - x^2}) = \sqrt{(\sqrt{9 - x^2})^2 + 7}$

Graph of $f \circ g$, not simplified:

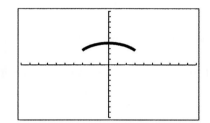

Simplifying $(f \circ g)(x)$, we obtain $\sqrt{(\sqrt{9 - x^2})^2 + 7}$ $= \sqrt{9 - x^2 + 7} = \sqrt{16 - x^2}$. Apparent graph of $f \circ g$, simplified:

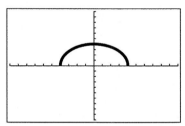

The domain of $f \circ g$ is not apparent from the final form $\sqrt{16 - x^2}$. It is not $[-4, 4]$. The domain of $f \circ g$ is the set of those numbers in $[-3, 3]$ for which $g(x)$ is real, that is, all of $[-3, 3]$. This is clearly shown in the first graph, so the first graph is correct.

79. Profit is the difference of the amount of money taken in (Revenue) and the amount of money spent (Cost), so $P(x) = R(x) - C(x)$

$$= \left(20x - \frac{1}{200} x^2\right) - (2x + 8{,}000)$$

$$= 20x - \frac{1}{200} x^2 - 2x - 8{,}000 \quad \text{(Distribute!)}$$

$$= 18x - \frac{1}{200} x^2 - 8{,}000$$

We have a profit function, but it's a function of the demand (x), not the price (p). We were given $x = 4{,}000 - 20p$, so we can substitute $4{,}000 - 200p$ in for x to get the desired function:

$$P(x) = P(4{,}000 - 200p) = 18(4{,}000 - 200p) - \frac{1}{200}(4{,}000 - 200p)^2 - 8{,}000$$

$$= 72{,}000 - 3{,}600p - \frac{1}{200}(16{,}000{,}000 - 1{,}600{,}000p + 40{,}000p^2) - 8{,}000$$

$$= 72{,}000 - 3{,}600p - 80{,}000 + 8{,}000p - 200p^2 - 8{,}000$$

$$= -16{,}000 + 4{,}400p - 200p^2$$

Now we'll graph the function and use the MAXIMUM command to find the largest profit.

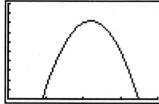

```
WINDOW
 Xmin=0
 Xmax=20
 Xscl=5
 Ymin=0
 Ymax=10000
 Yscl=1000
 Xres=1
```

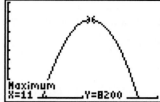

The maximum is (11, 9,200), so the largest profit occurs when the price is $11.

81. We are given $V(r) = 0.1A(r) = 0.1\pi r^2$ and $r(t) = 0.4t^{1/3}$.
We can use composition to express V as a function of the time.

$$(V \circ r)(t) = V[r(t)]$$
$$= 0.1\pi[r(t)]^2$$
$$= 0.1\pi[0.4t^{1/3}]^2$$
$$= 0.1\pi[0.16t^{2/3}]$$
$$= 0.016\pi t^{2/3}$$

So $V(t) = 0.016\pi t^{2/3}$

 83.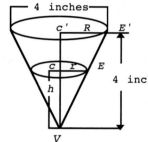

(A) Note: In the figure, triangles VCE and $VC'E'$ are similar. Also

R = radius of cup = $\frac{1}{2}$ diameter of cup = $\frac{1}{2}(4) = 2$ inches.

So $\frac{r}{2} = \frac{h}{4}$ or $r = \frac{1}{2}h$. We can write $r(h) = \frac{1}{2}h$.

(B) Since $V = \frac{1}{3}\pi r^2 h$ and $r = \frac{1}{2}h$, $V = \frac{1}{3}\pi\left(\frac{1}{2}h\right)^2 h = \frac{1}{3}\pi\frac{1}{4}h^2 h = \frac{1}{12}\pi h^3$.

We can write $V(h) = \frac{1}{12}\pi h^3$.

(C) Since $V(h) = \frac{1}{12}\pi h^3$ and $h(t) = 0.5\sqrt{t}$, we can use composition to express V as a function of t.

$$(V \circ h)(t) = V[h(t)]$$
$$= \frac{1}{12}\pi\,[h(t)]^3$$
$$= \frac{1}{12}\pi\,[0.5\sqrt{t}]^3$$
$$= \frac{1}{12}\pi(0.125)(t^{1/2})^3$$
$$= \frac{0.125}{12}\pi t^{3/2}$$

Then $V(t) = \frac{0.125}{12}\pi t^{3/2}$

Section 1-6

1. This is a one-to-one function. All of the first coordinates are distinct, and each first coordinate is paired with a different second coordinate. If all of the ordered pairs are reversed, the situation is the same.

3. This is a function but it is not one-to-one. First coordinates 5 and 2 are both paired with 4, and first coordinates 4 and 3 are both paired with 3. If the ordered pairs are reversed, the result is not a function since first coordinates 3 and 4 will each be paired with two different second coordinates.

5. This is not a function: 1 is a first coordinate that is paired with 2 different second coordinates (as is -3). If the ordered pairs are reversed, the result is also not a function since 4 will be a first coordinate paired with 2 different second coordinates (as will 2).

7. One-to one

9. The range element 7 corresponds to more than one domain element. Not one-to-one.

11. One-to-one

13. Some range elements (0, for example) correspond to more than one domain element. Not one-to-one.

15. One-to-one   17. One-to-one

19. This is a one-to-one function, so the inverse function exists.
$$f^{-1} = \left\{(3, 2), (4, 3), (5, 4), (6, 5)\right\}$$

21. This is not a one-to-one function (the range element -3 has two domain elements corresponding to it), so the inverse function does not exist.

23. This is a one-to-one function, so the inverse function exists.
$$F^{-1} = \left\{(7, a), (11, c), (-9, e), (-13, g)\right\}$$

25. Answers will vary.   27. Answers will vary.

29. Answers will vary.   31. Assume $F(a) = F(b)$
$$\frac{1}{2}a + 1 = \frac{1}{2}b + 1$$
Then $\frac{1}{2}a = \frac{1}{2}b$
$$a = b$$
Therefore F is one-to-one.

33. $H(x) = 4 - x^2$
Since $H(1) = 4 - 1^2 = 3$ and $H(-1) = 4 - (-1)^2 = 3$, both (1, 3) and (-1, 3) belong to H.
H is not one-to-one.

35. Assume $M(a) = M(b)$
$$\sqrt{a + 1} = \sqrt{b + 1}$$
Then $a + 1 = b + 1$
$$a = b$$
M is one-to-one.

37. $f(x) = \dfrac{7}{x}$

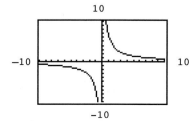

This graph passes the horizontal line test. f is one-to-one.

39. $f(x) = 0.3x^3 - 3.8x^2 + 12x + 11$

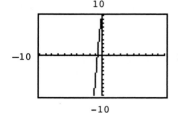 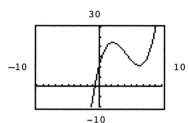

In a standard window (first graph), the graph appears to pass the horizontal line test, but with an expanded view (second graph) we see that it actually does not. f is not one-to-one.

41. $f(x) = \dfrac{x^2 + |x|}{x}$

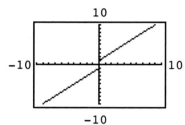

This graph passes the horizontal line test. f is one-to-one.

43. $f(x) = \dfrac{x^2 - 4}{|x - 2|}$

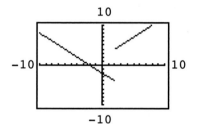

This graph does not pass the horizontal line test. f is not one-to-one.

45. $f(g(x)) = f\left(\dfrac{1}{3}x - \dfrac{5}{3}\right) = 3\left(\dfrac{1}{3}x - \dfrac{5}{3}\right) + 5 = x - 5 + 5 = x$

$g(f(x)) = g(3x + 5) = \dfrac{1}{3}(3x + 5) - \dfrac{5}{3} = x + \dfrac{5}{3} - \dfrac{5}{3} = x$

f and g are inverses

47. $f(g(x)) = f(\sqrt[3]{3 - x} - 1) = 2 - ((\sqrt[3]{3 - x} - 1) + 1)^3 = 2 - (\sqrt[3]{3 - x})^3$
$\qquad\qquad = 2 - (3 - x) = -1 + x$

f and g are not inverses since $(f \circ g)(x)$ is not x.

49. $f(g(x)) = f\left(\dfrac{3 + 4x}{2 - x}\right) = \dfrac{2\left(\frac{3+4x}{2-x}\right) - 3}{\frac{3+4x}{2-x} + 4} = \dfrac{\frac{6+8x}{2-x} - 3}{\frac{3+4x}{2-x} + 4} = \dfrac{\frac{6+8x}{2-x} - \frac{3(2-x)}{2-x}}{\frac{3+4x}{2-x} + \frac{4(2-x)}{2-x}}$

$\qquad = \dfrac{\frac{6+8x-6+3x}{2-x}}{\frac{3+4x+8-4x}{2-x}} = \dfrac{\frac{11x}{2-x}}{\frac{11}{2-x}} = \dfrac{11x}{2-x} \cdot \dfrac{2-x}{11} = x$

$g(f(x)) = g\left(\dfrac{2x - 3}{x + 4}\right) = \dfrac{3 + 4\left(\frac{2x-3}{x+4}\right)}{2 - \left(\frac{2x-3}{x+4}\right)} = \dfrac{3 + \frac{8x-12}{x+4}}{2 - \frac{2x-3}{x+4}} = \dfrac{\frac{3(x+4)}{x+4} + \frac{8x-12}{x+4}}{\frac{2(x+4)}{x+4} - \frac{2x-3}{x+4}}$

$\qquad = \dfrac{\frac{3x+12+8x-12}{x+4}}{\frac{2x+8-2x+3}{x+4}} = \dfrac{\frac{11x}{x+4}}{\frac{11}{x+4}} = \dfrac{11x}{x+4} \cdot \dfrac{x+4}{11} = x$

f and g are inverses.

51. The function h multiplies the input by 3 and then subtracts 7. The opposite of this is adding 7 to the input and then dividing by 3. The equation that corresponds to these steps is $h^{-1}(x) = \dfrac{x + 7}{3}$. (Remember that the order of the steps must be reversed.)

53. The function m adds 11 to the input then takes the cube root. The opposite of this is cubing the input, then subtracting 11. The equation that corresponds to these steps is $m^{-1}(x) = x^3 - 11$. (Remember that the order of the steps must be reversed.)

55. The function s multiplies the input by 3, adds 17, than raises the resulting expression to the fifth power. The opposite of this is taking the fifth root of the input, subtracting 17, then dividing by 3. The equation that corresponds to these steps is $s^{-1}(x) = \dfrac{\sqrt[5]{x} - 17}{3}$. (Remember that the order of the steps must be reversed.)

57. From the given graph of f, we see
Domain of f = [-4, 4]   Range of f = [1, 5]
Then
Range of f^{-1} = [-4, 4]   Domain of f^{-1} = [1, 5]
The graph of f^{-1} is drawn by reflecting the given graph of f in the line $y = x$.

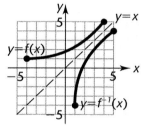

59. From the given graph of f, we see
Domain of f = [-5, 3]   Range of f = [-3, 5]
Then
Range of f^{-1} = [-5, 3]   Domain of f^{-1} = [-3, 5]
The graph of f^{-1} is drawn by reflecting the given graph of f in the line $y = x$.

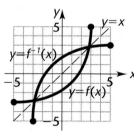

61. $f(x) = 3x + 6$ $g(x) = \dfrac{1}{3}x - 2$

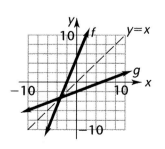

$g[f(x)] = g(3x + 6) = \dfrac{1}{3}(3x + 6) - 2 = x + 2 - 2 = x$

$f[g(x)] = f\left(\dfrac{1}{3}x - 2\right) = 3\left(\dfrac{1}{3}x - 2\right) + 6 = x - 6 + 6 = x$

f and g are inverses

63. $f(x) = 4 + x^2$ $x \geq 0$ $g(x) = \sqrt{x - 4}$

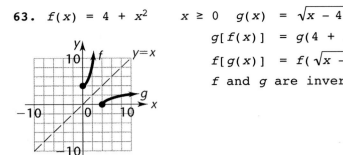

$g[f(x)] = g(4 + x^2) = \sqrt{4 + x^2 - 4} = \sqrt{x^2} = x$ since $x \geq 0$

$f[g(x)] = f(\sqrt{x - 4}) = 4 + (\sqrt{x - 4})^2 = 4 + x - 4 = x$

f and g are inverses

65. $f(x) = -\sqrt{x - 2}$ $g(x) = x^2 + 2,$ $x \leq 0$

$f[g(x)] = f(x^2 + 2) = -\sqrt{x^2 + 2 - 2} = -\sqrt{x^2} = -(-x) = x$

(Note: $\sqrt{x^2} = -x$ since $x \leq 0$)

$g[f(x)] = g(-\sqrt{x - 2}) = (-\sqrt{x - 2})^2 + 2 = x - 2 + 2 = x$

f and g are inverses

> **Common Error:** $(-\sqrt{x - 2})^2 \neq -(x - 2)$ nor $-x - 2$

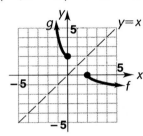

67. $f(x) = 3x$
Solve $y = f(x)$ for x:
$$y = 3x$$
$$x = \frac{1}{3}y = f^{-1}(y)$$
Interchange x and y:
$$y = f^{-1}(x) = \frac{1}{3}x$$
Domain of f^{-1} = Range of $f = (-\infty, \infty)$
Check: $f^{-1}[f(x)] = f^{-1}(3x) = \frac{1}{3}(3x) = x$

$$f[f^{-1}(x)] = f\left(\frac{1}{3}x\right) = 3\left(\frac{1}{3}x\right) = x$$

$f^{-1}(x) = \frac{1}{3}x$

69. $f(x) = 4x - 3$
Solve $y = f(x)$ for x:
$$y = 4x - 3$$
$$y + 3 = 4x$$
$$x = \frac{y + 3}{4} = f^{-1}(y)$$
Interchange x and y:
$$y = f^{-1}(x) = \frac{x + 3}{4}$$
Domain of f^{-1} = Range of $f = (-\infty, \infty)$

Check: $f^{-1}[f(x)] = f^{-1}(4x - 3)$
$$= \frac{4x - 3 + 3}{4}$$
$$= \frac{4x}{4}$$
$$= x$$

$f[f^{-1}(x)] = f\left(\frac{x + 3}{4}\right)$
$$= 4\left(\frac{x + 3}{4}\right) - 3$$
$$= x + 3 - 3$$
$$= x$$

$$f^{-1}(x) = \frac{x + 3}{4}$$

71. $f(x) = \dfrac{1}{10}x + \dfrac{3}{5}$

Solve $y = f(x)$ for x:

$$y = \dfrac{1}{10}x + \dfrac{3}{5}$$

$$10y = x + 6$$
$$x = 10y - 6 = f^{-1}(y)$$

Interchange x and y:
$$y = f^{-1}(x) = 10x - 6$$
Domain of f^{-1} = Range of f = $(-\infty, \infty)$

Check: $f^{-1}[f(x)] = f^{-1}\left(\dfrac{1}{10}x + \dfrac{3}{5}\right)$

$$= 10\left(\dfrac{1}{10}x + \dfrac{3}{5}\right) - 6$$
$$= x + 6 - 6$$
$$= x$$
$$f[f^{-1}(x)] = f(10x - 6)$$
$$= \dfrac{1}{10}(10x - 6) + \dfrac{3}{5}$$
$$= x - \dfrac{6}{10} + \dfrac{3}{5} = x$$

$$f^{-1}(x) = 10x - 6$$

73. $f(x) = \dfrac{2}{x - 1}$

Solve $y = f(x)$ for x:

$$y = \dfrac{2}{x - 1} \qquad x \neq 1$$

$$y(x - 1) = 2$$
$$xy - y = 2$$
$$xy = y + 2$$
$$x = \dfrac{y + 2}{y} = f^{-1}(y)$$

Interchange x and y:
$$y = f^{-1}(x) = \dfrac{x + 2}{x}$$
Domain of f^{-1} = $(-\infty, 0) \cup (0, \infty)$

Check: $f^{-1}[f(x)] = f^{-1}\left(\dfrac{2}{x - 1}\right)$

$$= \dfrac{\frac{2}{x-1} + 2}{\frac{2}{x-1}}$$

$$= \dfrac{2 + 2(x - 1)}{2}$$

$$= \dfrac{2 + 2x - 2}{2} = \dfrac{2x}{2} = x$$

$$f^{-1}(x) = \dfrac{x + 2}{x}$$

Graph of $y = f(x)$:

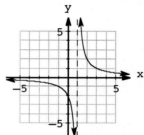

Range of f = $(-\infty, 0) \cup (0, \infty)$

$$f[f^{-1}(x)] = \dfrac{2}{\frac{x+2}{x} - 1}$$

$$= \dfrac{2x}{x + 2 - x} = \dfrac{2x}{2} = x$$

75. $f(x) = \dfrac{x}{x + 2}$

Solve $y = f(x)$ for x:

$$y = \dfrac{x}{x + 2} \qquad x \neq -2$$

$$y(x + 2) = x$$
$$xy + 2y = x$$
$$2y = x - xy$$
$$2y = x(1 - y)$$
$$x = \dfrac{2y}{1 - y} = f^{-1}(y)$$

Interchange x and y:
$$y = f^{-1}(x) = \dfrac{2x}{1 - x}$$
Domain of f^{-1} = $(-\infty, 1) \cup (1, \infty)$

Graph of $y = f(x)$:

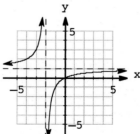

Range of f = $(-\infty, 1) \cup (1, \infty)$

Check: $f^{-1}[f(x)] = \dfrac{2\frac{x}{x+2}}{1 - \frac{x}{x+2}}$ $f[f^{-1}(x)] = \dfrac{\frac{2x}{1-x}}{\frac{2x}{1-x} + 2}$

$= \dfrac{2x}{x + 2 - x}$ $= \dfrac{2x}{2x + 2(1 - x)}$

$= \dfrac{2x}{2}$ $= \dfrac{2x}{2x + 2 - 2x}$

$= x$ $= \dfrac{2x}{2}$

     $= x$

$f^{-1}(x) = \dfrac{2x}{1 - x}$

77. $f(x) = \dfrac{2x + 5}{3x - 4}$

Solve $y = f(x)$ for x:

$y = \dfrac{2x + 5}{3x - 4}$

$(3x - 4)y = 2x + 5$

$3xy - 4y = 2x + 5$

$3xy - 2x = 4y + 5$

$x(3y - 2) = 4y + 5$

$x = \dfrac{4y + 5}{3y - 2} = f^{-1}(y)$

Interchange x and y:

$y = f^{-1}(x) = \dfrac{4x + 5}{3x - 2}$

Domain of $f^{-1} = \left(-\infty, \dfrac{2}{3}\right) \cup \left(\dfrac{2}{3}, \infty\right)$

Graph of $y = f(x)$:

Range of $f = \left(-\infty, \dfrac{2}{3}\right) \cup \left(\dfrac{2}{3}, \infty\right)$

Check: $f^{-1}[f(x)] = \dfrac{4\frac{2x+5}{3x-4} + 5}{3\frac{2x+5}{3x-4} - 2}$ $f[f^{-1}(x)] = \dfrac{2\frac{4x+5}{3x-2} + 5}{3\frac{4x+5}{3x-2} - 4}$

$= \dfrac{4(2x + 5) + 5(3x - 4)}{3(2x + 5) - 2(3x - 4)}$ $= \dfrac{2(4x + 5) + 5(3x - 2)}{3(4x + 5) - 4(3x - 2)}$

$= \dfrac{8x + 20 + 15x - 20}{6x + 15 - 6x + 8}$ $= \dfrac{8x + 10 + 15x - 10}{12x + 15 - 12x + 8}$

$= \dfrac{23x}{23}$ $= \dfrac{23x}{23}$

$= x$ $= x$

$f^{-1}(x) = \dfrac{4x + 5}{3x - 2}$

79. $f(x) = x^3 + 1$

Solve $y = f(x)$ for x:

$y = x^3 + 1$

$y - 1 = x^3$

$x = \sqrt[3]{y - 1} = f^{-1}(y)$

Interchange x and y:

$y = f^{-1}(x) = \sqrt[3]{x - 1}$

Domain of $f^{-1} =$ Range of

$f = (-\infty, \infty)$

Check: $f^{-1}[f(x)] = \sqrt[3]{x^3 + 1 - 1} = \sqrt[3]{x^3} = x$

$f[f^{-1}(x)] = (\sqrt[3]{x - 1})^3 + 1 = x - 1 + 1 = x$

$f^{-1}(x) = \sqrt[3]{x - 1}$

81. $f(x) = 4 - \sqrt[5]{x + 2}$

Solve $y = f(x)$ for x:

$$y = 4 - \sqrt[5]{x + 2}$$

$$y - 4 = -\sqrt[5]{x + 2}$$

$$4 - y = \sqrt[5]{x + 2}$$

$$(4 - y)^5 = x + 2$$

$$x = (4 - y)^5 - 2 = f^{-1}(y)$$

Graph of f:

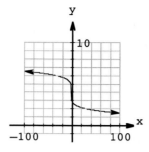

Range of $f = (-\infty, \infty)$

Interchange x and y:

$y = f^{-1}(x) = (4 - x)^5 - 2$ Domain of f^{-1}: $(-\infty, \infty)$

Check:

$$f^{-1}[f(x)] = [4 - (4 - \sqrt[5]{x + 2})]^5 - 2 \qquad f[f^{-1}(x)] = 4 - \sqrt[5]{(4 - x)^5 - 2 + 2}$$

$$= [4 - 4 + \sqrt[5]{x + 2}]^5 - 2 \qquad\qquad = 4 - \sqrt[5]{(4 - x)^5}$$

$$= [\sqrt[5]{x + 2}]^5 - 2 \qquad\qquad\qquad = 4 - (4 - x)$$

$$= x + 2 - 2 \qquad\qquad\qquad\qquad = 4 - 4 + x$$

$$= x \qquad\qquad\qquad\qquad\qquad = x$$

$$f^{-1}(x) = (4 - x)^5 - 2$$

83. $f(x) = \dfrac{1}{2}\sqrt{16 - x}$

Solve $y = f(x)$ for x:

$$y = \frac{1}{2}\sqrt{16 - x}$$

$$\left.\begin{array}{r} 2y = \sqrt{16 - x} \\ 4y^2 = 16 - x \end{array}\right\} \text{these are equivalent only if } y \geq 0$$

$$x = 16 - 4y^2 \quad y \geq 0 \quad f^{-1}(y) = 16 - 4y^2$$

Interchange x and y:

$y = f^{-1}(x) = 16 - 4x^2, \ x \geq 0$ Domain of f^{-1} = Range of $f = [0, \infty)$

Check: $f^{-1}[f(x)] = 16 - 4\left[\dfrac{1}{2}\sqrt{16 - x}\right]^2 \qquad f[f^{-1}(x)] = \dfrac{1}{2}\sqrt{16 - (16 - 4x^2)}$

$$= 16 - 4\left[\frac{1}{4}\sqrt{16 - x}\right] \qquad\qquad = \frac{1}{2}\sqrt{16 - 16 + 4x^2}$$

$$= 16 - (16 - x) \qquad\qquad\qquad = \frac{1}{2}\sqrt{4x^2} \quad \sqrt{4x^2} = 2x \text{ since } x \geq 0$$

$$= 16 - 16 + x \qquad\qquad\qquad = \frac{1}{2}(2x)$$

85. $f(x) = 3 - \sqrt{x - 2}$

Solve $y = f(x)$ for x:

$$y = 3 - \sqrt{x - 2}$$

$$y - 3 = -\sqrt{x - 2}$$

$\left. \begin{array}{l} 3 - y = \sqrt{x - 2} \\ (3 - y)^2 = x - 2 \end{array} \right\}$ these are equivalent only if $y \leq 3$

$$x = (3 - y)^2 + 2 \quad y \leq 3 \quad f^{-1}(y) = (3 - y)^2 + 2$$

Interchange x and y

$y = f^{-1}(x) = (3 - x)^2 + 2 \quad x \leq 3$ Domain of f^{-1} = Range of $f = (-\infty, 3]$

Check:

$f^{-1}[f(x)] = [3 - (3 - \sqrt{x - 2})]^2 + 2 \qquad f[f^{-1}(x)] = 3 - \sqrt{(3 - x)^2 + 2 - 2}$

$\qquad\qquad = (\sqrt{x - 2})^2 + 2 \qquad\qquad\qquad = 3 - \sqrt{(3 - x)^2}$

$$\sqrt{(3 - x)^2} = 3 - x \text{ since } x \leq 3$$

$\qquad\qquad = x - 2 + 2 \qquad\qquad\qquad\qquad = 3 - (3 - x)$

$\qquad\qquad = x \qquad\qquad\qquad\qquad\qquad = 3 - 3 + x$

$\qquad\qquad\qquad\qquad\qquad\qquad\qquad\qquad = x$

$f^{-1}(x) = (3 - x)^2 + 2, \ x \leq 3$

87. Since in passing from a function to its inverse, x and y are interchanged, the x intercept of f is the y intercept of f^{-1} and the y intercept of f is the x intercept of f^{-1}.

89. They are not inverses. The function $f(x) = x^2$ is not one-to-one, so does not have an inverse. If that function is restricted to the domain $[0, \infty)$, then it is one-to-one, and $g(x) = \sqrt{x}$ (which has range $[0, \infty)$) is its inverse.

91. $f(x) = (x - 1)^2 + 2 \quad x \geq 1$

Solve $y = f(x)$ for x:

$$y = (x - 1)^2 + 2 \quad x \geq 1$$

$$y - 2 = (x - 1)^2 \quad x \geq 1$$

$$\sqrt{y - 2} = \sqrt{(x - 1)^2} \quad \text{Since } x \geq 1, \ \sqrt{(x - 1)^2} = x - 1$$

$$\sqrt{y - 2} = x - 1$$

$$x = 1 + \sqrt{y - 2} = f^{-1}(y)$$

Interchange x and y:

$y = f^{-1}(x) = 1 + \sqrt{x - 2}$ Domain: $x \geq 2$

Check:

$f^{-1}[f(x)] = 1 + \sqrt{(x - 1)^2 + 2 - 2} \qquad\qquad f[f^{-1}(x)] = (1 + \sqrt{x - 2} - 1)^2 + 2$

$\qquad\qquad = 1 + \sqrt{(x - 1)^2} \qquad\qquad\qquad\qquad\qquad = (\sqrt{x - 2})^2 + 2$

$$\sqrt{(x - 1)^2} = x - 1 \text{ since } x \geq 1 \qquad = x - 2 + 2$$

$\qquad\qquad = 1 + x - 1 \qquad\qquad\qquad\qquad\qquad\qquad = x$

$\qquad\qquad = x$

$f^{-1}(x) = 1 + \sqrt{x - 2}, \ x \geq 2$

93. $f(x) = x^2 + 2x - 2 \quad x \leq -1$
Solve $y = f(x)$ for x:

$$y = x^2 + 2x - 2 \qquad x \leq -1$$
$$y + 3 = x^2 + 2x + 1 \qquad x \leq -1$$
$$y + 3 = (x + 1)^2 \qquad x \leq -1$$
$$\sqrt{y + 3} = \sqrt{(x + 1)^2} \quad \text{Since } x \leq -1 \quad \sqrt{(x + 1)^2} = -(x + 1)$$
$$\sqrt{y + 3} = -(x + 1)$$
$$-\sqrt{y + 3} = x + 1$$
$$x = -1 - \sqrt{y + 3} = f^{-1}(y)$$

Interchange x and y:
$$y = f^{-1}(x) = -1 - \sqrt{x + 3} \quad x \geq -3 \quad \text{Domain: } [-3, \infty)$$

Check: $f^{-1}[f(x)] = -1 - \sqrt{x^2 + 2x - 2 + 3} \qquad x \leq -1$
$$= -1 - \sqrt{x^2 + 2x + 1} \qquad x \leq -1$$
$$= -1 - \sqrt{(x + 1)^2} \qquad x \leq -1$$
$$\sqrt{(x + 1)^2} = -(x + 1) \text{ since } x \leq -1$$
$$= -1 - [-(x + 1)]$$
$$= -1 + x + 1$$
$$= x$$

$$f[f^{-1}(x)] = (-1 - \sqrt{x + 3})^2 + 2(-1 - \sqrt{x + 3}) - 2$$
$$= 1 + 2\sqrt{x + 3} + x + 3 - 2 - 2\sqrt{x + 3} - 2$$
$$= 4 + x - 4$$
$$= x$$
$$f^{-1}(x) = -1 - \sqrt{x + 3}, \quad x \geq -3$$

95. $f(x) = -\sqrt{9 - x^2} \quad 0 \leq x \leq 3 \qquad f$ is one-to-one. See graph below.

Solve $y = f(x)$ for x:
$$y = -\sqrt{9 - x^2} \quad 0 \leq x \leq 3$$
$$y^2 = 9 - x^2 \quad y \leq 0$$
$$y^2 - 9 = -x^2 \quad y \leq 0$$
$$9 - y^2 = x^2 \quad y \leq 0$$
$$x = \underbrace{\sqrt{9 - y^2}}_{} \quad y \leq 0 \quad f^{-1}(y) = \sqrt{9 - y^2}$$

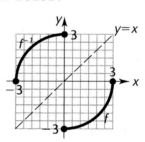

positive square root only because $0 \leq x$

Interchange x and y:
$$y = f^{-1}(x) = \sqrt{9 - x^2} \quad \text{Domain: } -3 \leq x \leq 0$$

Check: $f^{-1}[f(x)] = \sqrt{9 - (-\sqrt{9 - x^2})^2}$

$$= \sqrt{9 - (9 - x^2)}$$

$$= \sqrt{9 - 9 + x^2}$$

$$= \sqrt{x^2}$$

$$\sqrt{x^2} = x \text{ since } x \geq 0 \text{ } in \text{ } the \text{ } domain \text{ } of \text{ } f$$

$$= x$$

$$f[\,f^{-1}(x)\,] = -\sqrt{9 - (-\sqrt{9 - x^2})^2}$$

$$= -\sqrt{9 - (9 - x^2)}$$

$$= -\sqrt{9 - 9 + x^2}$$

$$= -\sqrt{x^2}$$

$$\sqrt{x^2} = -x \text{ since } x \le 0 \text{ in the domain of } f^{-1}$$

$$= -(-x)$$

$$= x$$

$f^{-1}(x) = \sqrt{9 - x^2}$ Domain of f^{-1} = [-3, 0]

Range of f^{-1} = Domain of f = [0, 3]

97. $f(x) = \sqrt{9 - x^2}$ $-3 \le x \le 0$ f is one-to-one. See graph below.

Solve $y = f(x)$ for x

$$y = \sqrt{9 - x^2} \qquad -3 \le x \le 0$$
$$y^2 = 9 - x^2 \qquad\quad y \ge 0$$
$$y^2 - 9 = -x^2 \qquad\quad y \ge 0$$
$$9 - y^2 = x^2 \qquad\quad y \ge 0$$
$$x = -\sqrt{9 - y^2} \quad y \ge 0 \quad f^{-1}(y) = -\sqrt{9 - y^2}$$

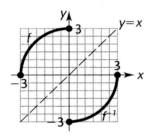

negative square root only because $x \le 0$

Interchange x and y:

$y = f^{-1}(x) = -\sqrt{9 - x^2}$ Domain: $0 \le x \le 3$

Check: $f^{-1}[\,f(x)\,] = -\sqrt{9 - (\sqrt{9 - x^2})^2}$

$$= -\sqrt{9 - (9 - x^2)}$$

$$= -\sqrt{9 - 9 + x^2}$$

$$= -\sqrt{x^2}$$

$$\sqrt{x^2} = -x \text{ since } x \le 0 \text{ in the domain of } f$$

$$= -(-x)$$

$$= x$$

$$f[\,f^{-1}(x)\,] = \sqrt{9 - (\sqrt{9 - x^2})^2}$$

$$= \sqrt{9 - (9 - x^2)}$$

$$= \sqrt{9 - 9 + x^2}$$

$$= \sqrt{x^2}$$

$$\sqrt{x^2} = x \text{ since } x \ge 0 \text{ in the domain of } f^{-1}$$

$$= x$$

$f^{-1}(x) = -\sqrt{9 - x^2}$ Domain: f^{-1} = [0, 3]

Range of f^{-1} = Domain of f = [-3, 0]

99. $f(x) = 1 + \sqrt{1 - x^2}$ $0 \le x \le 1$ f is one-to-one. See graph below.

Solve $y = f(x)$ for x:

$$y = 1 + \sqrt{1 - x^2} \quad 0 \le x \le 1$$
$$y - 1 = \sqrt{1 - x^2}$$
$$(y - 1)^2 = 1 - x^2 \qquad y \ge 1$$
$$y^2 - 2y + 1 = 1 - x^2$$
$$y^2 - 2y = -x^2$$
$$2y - y^2 = x^2$$
$$x = \underbrace{\sqrt{2y - y^2}} \qquad y \ge 1 \quad f^{-1}(y) = \sqrt{2y - y^2}$$

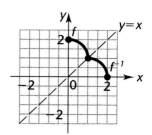

positive square root only because $0 \le x$
Interchange x and y:

$$y = f^{-1}(x) = \sqrt{2x - x^2} \quad \text{Domain: } 1 \le x \le 2$$

Check: $f[f^{-1}(x)] = 1 + \sqrt{1 - (\sqrt{2x - x^2})^2}$

$$= 1 + \sqrt{1 - (2x - x^2)}$$
$$= 1 + \sqrt{1 - 2x + x^2}$$
$$= 1 + \sqrt{(1 - x)^2}$$
$$= 1 + [-(1 - x)] \text{ because } 1 \le x \text{ in the domain of } f^{-1}$$
$$= 1 - 1 + x$$
$$= x$$

$$f^{-1}[f(x)] = \sqrt{2(1 + \sqrt{1 - x^2}) - (1 + \sqrt{1 - x^2})^2}$$

$$= \sqrt{2 + 2\sqrt{1 - x^2} - (1 + 2\sqrt{1 - x^2} + 1 - x^2)}$$

$$= \sqrt{2 + 2\sqrt{1 - x^2} - 1 - 2\sqrt{1 - x^2} - 1 + x^2}$$

$$= \sqrt{x^2}$$

$$= x$$

because $x \ge 0$ in the domain of f

$f^{-1}(x) = \sqrt{2x - x^2}$ Domain of $f^{-1} = [1, 2]$
Range of f^{-1} = Domain of $f = [0, 1]$

101. $f(x) = (2 - x)^2$

(A) $x \le 2$
Solve $y = f(x)$ for x:

$$y = (2 - x)^2 \qquad x \le 2$$
$$\sqrt{y} = 2 - x \quad \text{Positive square root only}$$
$$\qquad\qquad \text{since } 2 - x \ge 0$$
$$\sqrt{y} - 2 = -x$$
$$x = 2 - \sqrt{y} = f^{-1}(y)$$

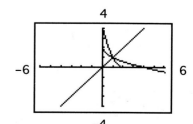

Interchange x and y:

$$y = f^{-1}(x) = 2 - \sqrt{x} \quad \text{Domain: } x \ge 0$$

Check: $f^{-1}[f(x)] = 2 - \sqrt{(2 - x)^2}$
$= 2 - (2 - x)$ since $2 - x \geq 0$ *in the domain of f*
$= 2 - 2 + x$
$= x$

$f[f^{-1}(x)] = [2 - (2 - \sqrt{x})]^2$
$= [2 - 2 + \sqrt{x}]^2$
$= [\sqrt{x}]^2$
$= x$

$f^{-1}(x) = 2 - \sqrt{x}$

(B) $x \geq 2$
Solve $y = f(x)$ for x:
$y = (2 - x)^2$ $x \geq 2$
$-\sqrt{y} = 2 - x$ Negative square root only
since $2 - x \leq 0$

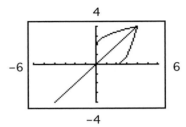

$\sqrt{y} = -2 + x$
$x = 2 + \sqrt{y} = f^{-1}(y)$

Interchange x and y:
$y = f^{-1}(x) = 2 + \sqrt{x}$ Domain: $x \geq 0$

Check: $f^{-1}[f(x)] = 2 + \sqrt{(2 - x)^2}$
$= 2 + [-(2 - x)]$ since $2 - x \leq 0$ *in the domain of f*
$= 2 - 2 + x$
$= x$

$f[f^{-1}(x)] = [2 - (2 + \sqrt{x})]^2$
$= [2 - 2 - \sqrt{x}]^2$
$= [-\sqrt{x}]^2$
$= x$

$f^{-1}(x) = 2 + \sqrt{x}$

103. $f(x) = \sqrt{4x - x^2}$
(A) $0 \leq x \leq 2$
Solve $y = f(x)$ for x:
$y = \sqrt{4x - x^2}$
$y^2 = 4x - x^2$ $y \geq 0$
$-y^2 = x^2 - 4x$
$4 - y^2 = x^2 - 4x + 4$
$4 - y^2 = (x - 2)^2$
$-\underbrace{\sqrt{4 - y^2}} = x - 2$

negative square root only because $x \leq 2$
$x = 2 - \sqrt{4 - y^2}$ $y \geq 0$ $f^{-1}(y) = 2 - \sqrt{4 - y^2}$
Interchange x and y:
$y = f^{-1}(x) = 2 - \sqrt{4 - x^2}$ Domain: $0 \leq x \leq 2$

Check: $f^{-1}[f(x)] = 2 - \sqrt{4 - (\sqrt{4x - x^2})^2}$
$= 2 - \sqrt{4 - (4x - x^2)}$
$= 2 - \sqrt{4 - 4x + x^2}$
$= 2 - \sqrt{(2 - x)^2}$
$= 2 - (2 - x)$ since $2 - x \geq 0$
$= x$

$$f[f^{-1}(x)] = \sqrt{4(2 - \sqrt{4 - x^2}) - (2 - \sqrt{4 - x^2})^2}$$

$$= \sqrt{8 - 4\sqrt{4 - x^2} - (4 - 4\sqrt{4 - x^2} + 4 - x^2)}$$

$$= \sqrt{8 - 4\sqrt{4 - x^2} - 4 + 4\sqrt{4 - x^2} - 4 + x^2}$$

$$= \sqrt{x^2}$$

$$= x \text{ since } 0 \le x$$

$f^{-1}(x) = 2 - \sqrt{4 - x^2}$, $0 \le x \le 2$

(B) $2 \le x \le 4$

Solve $y = f(x)$ for x:

$$y = \sqrt{4x - x^2}$$
$$y^2 = 4x - x^2 \quad y \ge 0$$
$$-y^2 = x^2 - 4x$$
$$4 - y^2 = x^2 - 4x + 4$$
$$4 - y^2 = (x - 2)^2$$
$$\sqrt{4 - y^2} = x - 2$$

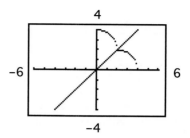

positive square root only because $x \ge 2$

$x = 2 + \sqrt{4 - y^2}$ $\quad y \ge 0$ $\quad f^{-1}(y) = 2 + \sqrt{4 - y^2}$

Interchange x and y:

$y = f^{-1}(x) = 2 + \sqrt{4 - x^2}$ $\quad$ Domain: $0 \le x \le 2$

Check: $f^{-1}[f(x)] = 2 + \sqrt{4 - (\sqrt{4x - x^2})^2}$

$$= 2 + \sqrt{4 - (4x - x^2)}$$
$$= 2 + \sqrt{4 - 4x + x^2}$$
$$= 2 + \sqrt{(2 - x)^2}$$
$$= 2 + [-(2 - x)] \text{ since } 2 \le x \text{ in the domain of } f$$
$$= 2 - 2 + x$$
$$= x$$

$$f[f^{-1}(x)] = \sqrt{4(2 + \sqrt{4 - x^2}) - (2 + \sqrt{4 - x^2})^2}$$

$$= \sqrt{8 + 4\sqrt{4 - x^2} - (4 + 4\sqrt{4 - x^2} + 4 - x^2)}$$

$$= \sqrt{8 + 4\sqrt{4 - x^2} - 4 - 4\sqrt{4 - x^2} - 4 + x^2}$$

$$= \sqrt{x^2}$$

$$= x \text{ since } 0 \le x \text{ in the domain of } f^{-1}$$

$f^{-1}(x) = 2 + \sqrt{4 - x^2}$, $0 \le x \le 2$

105. (A) Graph $q = d(p)$ on the interval $[10, 70]$.
Examining the graph, we see that d is a
decreasing function, and has its maximum
value on $[10, 70]$ at $p = 10$, where

$d(10) = \dfrac{3000}{0.2(10) + 1} = 1,000$, and its minimum

value on $(10, 70)$ at $p = 70$ where

$d(70) = \dfrac{3000}{0.2(70) + 1} = 200$. Therefore the range

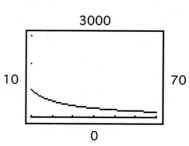

of d is $[200, 1000]$, or $200 \le q \le 1,000$.

(B) Solve for p:

$$q = \frac{3,000}{0.2p + 1}$$

$$q(0.2p + 1) = 3,000$$

$$0.2pq + q = 3,000$$

$$0.2pq = 3,000 - q$$

$$p = \frac{3,000 - q}{0.2q}$$

$$p = \frac{3,000}{0.2q} - \frac{q}{0.2q}$$

$$p = \frac{15,000}{q} - 5$$

$$d^{-1}(q) = \frac{15,000}{q} - 5$$

The domain of d^{-1} is the range of d, or $[200, 1,000]$, or $200 \le q \le 1,000$.
The range of d^{-1} is the domain of d, or $[10, 70]$, or $10 \le p \le 70$.

(C) p and q should not be interchanged. That step is necessary only to maintain the conventional roles of x and y.

107. To find the revenue, we multiply the price (p) and the demand (x). We're asked to find the revenue in terms of x, so we begin by solving the given equation for p:

$$x = 2,000 - 40p$$

$$-40p = x - 2,000$$

$$p = \frac{x - 2,000}{-40} = \frac{-1}{40}x + 50$$

Now we multiply by the demand (x):

$$R(x) = x\left(\frac{-1}{40}x + 50\right) = \frac{-1}{40}x^2 + 50x \quad \text{or} \quad R(x) = -0.025x^2 + 50x$$

CHAPTER 1 REVIEW

1. Since the x coordinates vary between a minimum of -4 and a maximum of 9, Xmin $= -4$ and Xmax $= 9$. Since the y coordinates vary between a minimum of -6 and a maximum of 7, Ymin $= -6$ and Ymax $= 7$. *(1-1)*

2. (A) All of the first coordinates are distinct, so this is a function with domain $\{1, 2, 3\}$. All of the second coordinates are distinct, so the function is one-to-one. The range is $\{1, 4, 9\}$. The inverse function is obtained by reversing the order of the ordered pairs: $\{(1, 1), (4, 2), (9, 3)\}$. It has domain $\{1, 4, 9\}$ and range $\{1, 2, 3\}$.

   (B) This is not a function: both 1 and 2 are first coordinates that get matched with two different second coordinates.

   (C) All of the first coordinates are distinct, so this is a function with domain $\{-2, -1, 0, 1, 2\}$. The range is $\{2\}$. It's most definitely not one-to-one since every first coordinate gets matched with the same second coordinate.

   (D) All of the first coordinates are distinct, so this is a function with domain $\{-2, -1, 0, 1, 2\}$. All of the second coordinates are distinct, so the function is one-to-one. The range is $\{-2, -1, 1, 2, 3\}$.
   The inverse function is obtained by reversing the order of the ordered pairs: $\{(2, -2), (3, -1), (-1, 0), (-2, 1), (1, 2)\}$. It has domain $\{-2, -1, 1, 2, 3\}$ and range $\{-2, -1, 0, 1, 2\}$. *(1-2)*

3. (A) Not a function (fails vertical line test)   (B) A function
   (C) A function   (D) Not a function (fails vertical line test)   *(1-2)*

4. (A) $f(1) = (1)^2 - 2(1) = -1$

   (B) $f(-4) = (-4)^2 - 2(-4) = 24$

   (C) $f(2) \cdot f(-1) = [(2)^2 - 2(2)] \cdot [(-1)^2 - 2(-1)] = 0 \cdot 3 = 0$

   (D) $\dfrac{f(0)}{f(3)} = \dfrac{(0)^2 - 2(0)}{(3)^2 - 2(3)} = \dfrac{0}{3} = 0$ *(1-2)*

5. Construct a table of values of $f(x)$ and $g(x)$ from the graph, then subtract to obtain $(f - g)(x)$.

x	-3	-2	-1	0	1	2	3
$f(x)$	3	2	1	0	1	2	3
$g(x)$	-4	-3	-2	-1	0	1	2
$(f - g)(x)$	7	5	3	1	1	1	1

   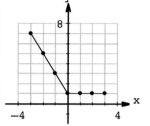

   (1-5)

6. Use the top 3 rows of the table above and multiply to get a table of values for $(fg)(x)$. (Note that from the graphs of f and g, we can see that $f\left(\dfrac{1}{2}\right) = \dfrac{1}{2}$ and $g\left(\dfrac{1}{2}\right) = -\dfrac{1}{2}$,

   so $(fg)\left(\dfrac{1}{2}\right) = -\dfrac{1}{4}$.)

   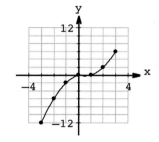

   (1-5)

7. $(f \circ g)(-1) = f[g(-1)]$. From the graph of g, $g(-1) = -2$. From the graph of f, $f[g(-1)] = f(-2) = 2$.   *(1-5)*

8. $(g \circ f)(-2) = g[f(-2)]$. From the graph of f, $f(-2) = 2$. From the graph of g, $g[f(-2)] = g(2) = 1$.   *(1-5)*

9. From the graph of g, $g(1) = 0$. From the graph of f, $f[g(1)] = f(0) = 0$.   *(1-5)*

10. From the graph of f, $f(-3) = 3$. From the graph of g, $g[f(-3)] = g(3) = 2$.   *(1-5)*

11. Some range elements (1 for example) correspond to more than one domain element. Not one-to-one.   *(1-6)*

12. Yes, one-to-one.   *(1-6)*

13. Using values from the table, $g(-11) = -4$, so $(f \circ g)(-11) = f(g(-11)) = f(-4) = 8$. Similarly, $(f \circ g)(1) = f(g(1)) = f(-11) = 12$, and $(f \circ g)(6) = f(g(6)) = f(9) = 1$.   *(1-5)*

14. (A) $f(-x) = (-x)^5 + 6(-x) = -x^5 - 6x = -(x^5 + 6x) = -f(x)$. Odd

    (B) $g(-t) = (-t)^4 + 3(-t)^2 = t^4 + 3t^2 = g(t)$. Even

    (C) $h(-z) = (-z)^5 + 4(-z)^2 = -z^5 + 4z^2$
        $-h(z) = -(z^5 + 4z^2) = -z^5 - 4z^2$
    Therefore, $h(-z) \neq h(z)$. $h(-z) \neq -h(z)$. h is neither even nor odd.   *(1-4)*

15. When $x = -4$, the corresponding y value on the graph is 4. $f(-4) = 4$.
   When $x = 0$, the corresponding y value on the graph is -4. $f(0) = -4$.
   When $x = 3$, the corresponding y value on the graph is 0. $f(3) = 0$.
   When $x = 5$, there is no corresponding y value on the graph.
   $f(5)$ is not defined.   (*1-2, 1-3*)

16. Two values of x correspond to $f(x) = -2$ on the graph. They are $x = -2$ and
   $x = 1$.   (*1-2, 1-3*)

17. Domain: $[-4, 5)$. Range: $[-4, 4]$ (*1-3*)

18. The graph is increasing on $[0, 5)$ and decreasing on $[-4, 0]$.   (*1-3*)

19. The graph is discontinuous at $x = 0$.   (*1-3*)

20. The graph of $f(x)$ is shifted up 1 unit.

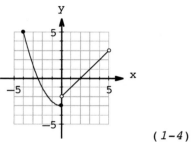

(*1-4*)

21. The graph of $f(x)$ is shifted left 1 unit.

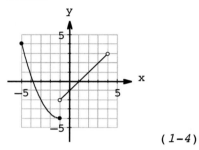

(*1-4*)

22. The graph of $f(x)$ is reflected in the x axis.

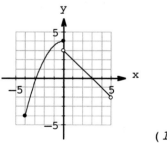

(*1-4*)

23. The graph of $f(x)$ is contracted by a factor of 0.5.

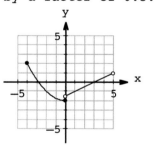

(*1-4*)

24. The graph is horizontally contracted by a factor of $\frac{1}{2}$.

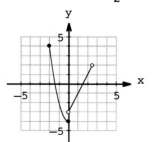

(*1-4*)

25. The graph is reflected about both axes.

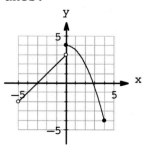

(*1-4*)

26. (A) The graph that is decreasing, then increasing, and has a minimum at $x = 2$ is g.

(B) The graph that is increasing, then decreasing, and has a maximum at $x = -2$ is m.

(C) The graph that is increasing, then decreasing, and has a maximum at $x = 2$ is n.

(D) The graph that is decreasing, then increasing, and has a minimum at $x = -2$ is f. *(1-4)*

27. (A) $\left(\dfrac{f}{g}\right)(x) = \dfrac{f(x)}{g(x)} = \dfrac{x^2 - 4}{x + 3}$ Domain: $\{x \mid x \neq -3\}$, or $(-\infty, -3) \cup (-3, \infty)$

(B) $\left(\dfrac{g}{f}\right)(x) = \dfrac{g(x)}{f(x)} = \dfrac{x + 3}{x^2 - 4}$ Domain: $\{x \mid x^2 - 4 \neq 0\}$, or $\{x \mid x \neq -2, 2\}$ or

$(-\infty, -2) \cup (-2, 2) \cup (2, \infty)$

(C) $(f \circ g)(x) = f[g(x)] = f(x + 3) = (x + 3)^2 - 4 = x^2 + 6x + 5$ Domain: $(-\infty, \infty)$

(D) $(g \circ f)(x) = g[f(x)] = g(x^2 - 4) = x^2 - 4 + 3 = x^2 - 1$ Domain: $(-\infty, \infty)$

(1-5)

28. (A) When $x = 0$, the corresponding y value on the graph is 0, to the nearest integer.

(B) When $x = 1$, the corresponding y value on the graph is 1, to the nearest integer.

(C) When $x = 2$, the corresponding y value on the graph is 2, to the nearest integer.

(D) When $x = -2$, the corresponding y value on the graph is 0, to the nearest integer. *(1-3)*

29. (A) Two values of x correspond to $y = 0$ on the graph. To the nearest integer, they are -2 and 0.

(B) Two values of x correspond to $y = 1$ on the graph. To the nearest integer, they are -1 and 1.

(C) No value of x corresponds to $y = -3$ on the graph.

(D) To the nearest integer, $x = 3$ corresponds to $y = 3$ on the graph. Also, every value of x such that $x < -2$ corresponds to $y = 3$. *(1-3)*

30. Domain: $(-\infty, \infty)$. Range: $(-3, \infty)$.   *(1-3)*

31. $[-2, -1]$, $[1, \infty)$.   *(1-3)*   **32.** $[-1, 1)$ *(1-3)*   **33.** $(-\infty, -2)$ *(1-3)*

34. The graph of q is discontinuous at $x = -2$ and $x = 1$.   *(1-3)*

35. The graph and viewing window are shown below:

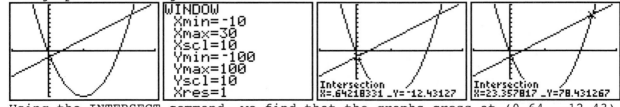

Using the INTERSECT command, we find that the graphs cross at (0.64, -12.43) and (23.36, 78.43). *(1-1)*

36. The graph and viewing window are shown below:

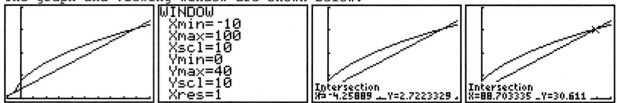

Using the INTERSECT command, we find that the graphs cross at $(-4.26, 2.72)$ and $(88.70, 30.61)$. *(1-1)*

37. (A) We begin by graphing $y_1 = 0.1x^3 - 2x^2 - 6x + 80$. When $b = 0$, the solutions are the x-values where the graph is at height zero, so we use the ZERO command:

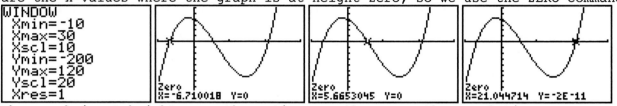

The graph is at height zero three times so there are three solutions: $x = -6.71$, $x = 5.67$, and $x = 21.04$

(B) Graph $y_1 = 0.1x^3 - 2x^2 - 6x + 80$, $y_2 = 100$; then use the INTERSECT command to find where the graphs cross.

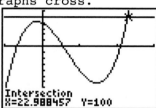

The graphs intersect at $(22.99, 100)$ so the only solution is $x = 22.99$.

(C) Graph $y_1 = 0.1x^3 - 2x^2 - 6x + 80$, $y_2 = -50$, then use the INTERSECT command to find where the graphs cross.

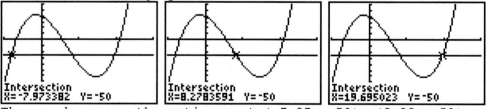

The graphs cross three times, at $(-7.97, -50)$, $(8.28, -50)$, and $(19.70, -50)$. There are three solutions: $x = -7.97$, $x = 8.28$, and $x = 19.70$.

(D) Graph $y_1 = 0.1x^3 - 2x^2 - 6x + 80$, $y_2 = -150$; then use the INTERSECT command to find where the graphs cross.

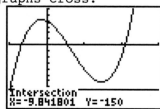

The graphs cross at $(-9.84, -150)$, so the only solution is $x = -9.84$. *(1-1, 1-3)*

38. This equation defines a function. If you solve the equation for y, you get $y = 5 - 0.5x$. This tells us that for any choice of x, we can calculate a unique y that corresponds to it. $(1-2)$

39. This equation does not define a function since most choices of x will result in two corresponding values of y. For example, the pairs $(2, 2)$ and $(2, -2)$ both make the equation a true statement. $(1-2)$

40. (A) The domain of $f(x) = x^2 - 4x + 5$ is all real numbers.

(B) The fraction $\dfrac{t + 2}{t - 5}$ represents a real number for all replacements of t by real numbers except when $t - 5 = 0$ or $t = 5$, since division by 0 is not defined. The domain of g is the set of all real numbers except 5.

(C) $2 + 3\sqrt{w}$ represents a real number only for replacements of w by nonnegative real numbers, since the square root of a negative number is not defined. The domain of h is the set of all real numbers $w \geq 0$, or $[0, \infty)$. $(1-2)$

41.

$$\frac{g(2 + h) - g(2)}{h} = \frac{[2(2 + h)^2 - 3(2 + h) + 6] - [2(2)^2 - 3(2) + 6]}{h}$$

$$= \frac{2(4 + 4h + h^2) - 6 - 3h + 6 - [8 - 6 + 6]}{h}$$

$$= \frac{8 + 8h + 2h^2 - 3h - 8}{h}$$

$$= \frac{5h + 2h^2}{h} = \frac{h(5 + 2h)}{h}$$

$$= 5 + 2h$$

Common Errors:
1. $g(2 + h)$ is not $g(2) + g(h)$.
2. $g(2 + h)$ is not $g(2) + h$.
3. Forgetting to enclose $2(2)^2 - 3(2) + 6$ in parentheses or brackets to indicate the unperformed subtraction.

$(1-2)$

42. $f(x) = 4x^3 - \sqrt{x}$ $(1-2)$

43. The function f multiplies the square of the domain element by 3, adds 4 times the domain element, and then subtracts 6. $(1-2)$

44. The graph is shown below.

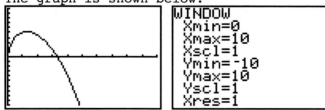

WINDOW
Xmin=0
Xmax=10
Xscl=1
Ymin=-10
Ymax=10
Yscl=1
Xres=1

Use the ZERO command to find the x-intercepts:

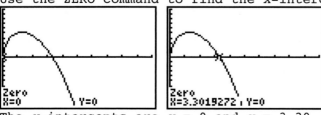

The x-intercepts are $x = 0$ and $x = 3.30$. Since $(0, 0)$ is on the graph, $y = 0$ is the y-intercept.

Te point (0,0) is lower than any nearby points, so is a local minimum. There is a local maximum somewhere near $x = 1$, so we use the MAXIMUM command.

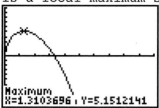

The maximum occurs at $x = 1.31$, $y = 5.15$. The domain is $[0, \infty)$ and the range is $(-\infty, 5.15]$. (The highest point on the graph is at height 5.15 and there is no lowest point.).   (1-3)

45. The graph is shown below.

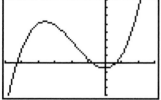

```
WINDOW
 Xmin=-30
 Xmax=15
 Xscl=5
 Ymin=-2000
 Ymax=4000
 Yscl=500
 Xres=1
```

Use the ZERO command to find the x-intercepts and the value feature with $x = 0$ to find the y-intercept.

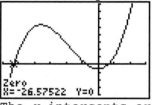

The x-intercepts are -26.58, -3.58, and 3.15. The y-intercept is -300. There is a local maximum and a local minimum, so we use the MAXIMUM and MINIMUM commands to locate the local extrema.

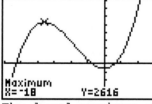
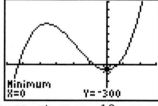

The local maximum occurs at $x = -18$, $y = 2,616$, and the local minimum occurs at $x = 0$, $y = -300$.
The function is a polynomial, so the domain is $(-\infty, \infty)$. There is no highest or lowest point, so the range is also $(-\infty, \infty)$.   (1-3)

46. (A)

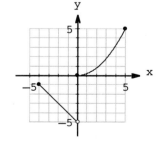

(B) Domain: $[-4, 5]$, range: $(-5, -1] \cup [0, 5]$

(C) The graph is discontinuous at $x = 0$.

(D) Decreasing on $[-4, 0)$, increasing on $[0, 5]$

                                                                    (1-3)

47. The graph of *f* is shown as well as the result of applying the MAXIMUM and MINIMUM commands to find the local maximum and the local minimum.

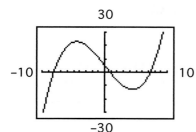 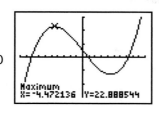 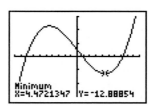

The graph of *f* increases on $[-\infty, -4.47]$ to a local maximum value, $f(-4.47) \approx 22.89$, decreases on $[-4.47, 4.47]$ to a local minimum value, $f(4.47) \approx -12.89$, and then increases on $[4.47, \infty]$.   (*1-3*)

48. (A) The graph of $y = x^2$ is reflected across the *x* axis.

(B) The graph of $y = x^2$ is shifted down 3 units.

(C) The graph of $y = x^2$ is shifted left 3 units.

(D) The graph of $y = x^2$ is shrunk horizontally by $\frac{1}{2}$.   (*1-4*)

49. (A) The graph of $y = x^2$ is shifted 2 units to the right, reflected in the *x* axis, and shifted up 4 units: $y = -(x - 2)^2 + 4$
Check: $y = -(x - 2)^2 + 4$ is graphed on a graphing calculator.

(B) The graph of $y = \sqrt{x}$ is reflected in the *x* axis, stretched by a factor of 4, and shifted up 4 units: $y = 4 - 4\sqrt{x}$
Check: $y = 4 - 4\sqrt{x}$ is graphed on a graphing calculator.

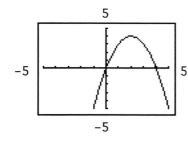 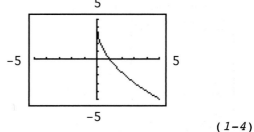

(*1-4*)

50. $g(x) = 8 - 3|x - 4|$
The -4 shifts 4 units right, the 3 stretches vertically by a factor of 3, and the 8 shifts 8 units up.

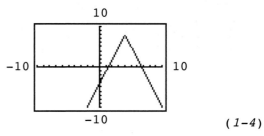

(*1-4*)

51. The equation is $t(x) = \left(\frac{1}{2}(x + 2)\right)^2 - 4$. The $\frac{1}{2}$ stretches horizontally by 2, the +2 shifts 2 units left, and the -4 shifts 4 units down. This function can be simplified: $\left(\frac{1}{2}(x + 2)\right)^2 - 4 =$

$\left(\frac{1}{2}x + 1\right)^2 - 4 = \frac{1}{4}x^2 + x + 1 - 4 = \frac{1}{4}x^2 + x - 3$.

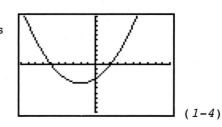

(*1-4*)

52. Find the composition of the 2 functions in both orders.

$(u \circ v)(x) = u(v(x)) = u(0.25x + 2) = 4(0.25x + 2) - 8 = x + 8 - 8 = x$

$(v \circ u)(x) = v(u(x)) = v(4x - 8) = 0.25(4x - 8) + 2 = x - 2 + 2 = x$

The functions are inverses. *(1-6)*

53. The function k cubes the input and then adds 5. The opposite of this is subtracting 5 and taking the cube root. (Remember to reverse the order of the steps!) The inverse function is $k^{-1}(x) = \sqrt[3]{x - 5}$. To verify we find the composition of the two functions in both orders.

$(k \circ k^{-1})(x) = k(k^{-1}(x)) = k(\sqrt[3]{x - 5}) = (\sqrt[3]{x - 5})^3 + 5 = x - 5 + x = x$

$(k^{-1} \circ k)(x) = k^{-1}(k(x)) = k^{-1}(x^3 + 5) = \sqrt[3]{(x^3 + 5) - 5} = \sqrt[3]{x^3} = x.$ *(1-6)*

54. $\dfrac{x}{\sqrt{x} - 3}$ represents a real number only for nonnegative values of x, except when $\sqrt{x} - 3 = 0$, that is, when $\sqrt{x} = 3$ or $x = 9$. The domain of f is the set of all nonnegative real numbers except 9: $x \geq 0$, $x \neq 9$, or $[0, 9) \cup (9, \infty)$. *(1-2)*

55. $f(x) = \sqrt{x} - 8 \qquad g(x) = |x|$

(A) $(f \circ g)(x) = f[g(x)] = f(|x|) = \sqrt{|x|} - 8$

$(g \circ f)(x) = g[f(x)] = g(\sqrt{x} - 8) = |\sqrt{x} - 8|$

(B) The domain of f is $\{x \mid x \geq 0\}$. The domain of g is all real numbers. So the domain of $f \circ g$ is the set of those real numbers x for which $g(x)$ is non-negative, that is, all real numbers.

The domain of $(g \circ f)$ is the set of all those non-negative numbers x for which $f(x)$ is real, that is all $\{x \mid x \geq 0\}$ or $[0, \infty)$. (The output of g is never negative!) *(1-5)*

56. (A) $f(x) = x^3$. The graph passes the horizontal line test, so f is one-to-one.

Alternatively, if we assume $f(a) = f(b)$

$$a^3 = b^3$$
$$a^3 - b^3 = 0$$
$$(a - b)(a^2 + ab + b^2) = 0$$

The only real solutions of this equation are those for which $a - b = 0$, so $a = b$. We conclude that $f(x)$ is one-to-one.

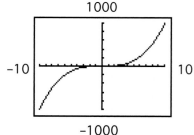

(B) $g(x) = (x - 2)^2$. Since $g(3) = g(1) = 1$, g is not one-to-one.

(C) $h(x) = 2x - 3$
Assume $h(a) = h(b)$

$2a - 3 = 2b - 3$

Then $\qquad 2a = 2b$

$\qquad\qquad a = b$

We conclude that h is one-to-one.

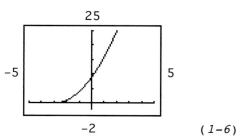

(D) $F(x) = (x + 3)^2 \quad x \geq -3$
The graph passes the horizontal line test, so F is one-to-one. *(1-6)*

(E) When viewed in a standard window (first graph), it looks like the graph passes the horizontal line test. But an expanded window shows that it does not (second graph), so f is not one-to-one.

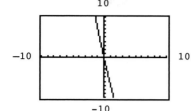

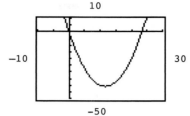

57. $f(x) = 3x - 7$

Assume $f(a) = f(b)$

$$3a - 7 = 3b - 7$$
$$3a = 3b$$
$$a = b$$

f is one-to-one. Solve $y = f(x)$ for x:

$$y = 3x - 7$$
$$y + 7 = 3x$$
$$x = \frac{1}{3}y + \frac{7}{3} = f^{-1}(y)$$

Interchange x and y: $y = f^{-1}(x) = \frac{1}{3}x + \frac{7}{3}$

Domain of f^{-1}: $(-\infty, \infty)$

Range of f^{-1} = Domain of f: $(-\infty, \infty)$

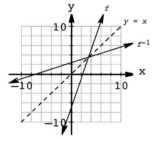

Check: $f^{-1}[f(x)] = \frac{1}{3}(3x - 7) + \frac{7}{3}$

$$= x - \frac{7}{3} + \frac{7}{3} = x$$

$$f[f^{-1}(x)] = 3\left(\frac{1}{3}x + \frac{7}{3}\right) - 7$$

$$= x + 7 - 7 = x$$

$$f^{-1}(x) = \frac{1}{3}x + \frac{7}{3} = \frac{x + 7}{3} \qquad\qquad (1\text{-}6)$$

58. $f(x) = \sqrt{x - 1}$.

Assume $f(a) = f(b)$

$$\sqrt{a - 1} = \sqrt{b - 1}$$
$$a - 1 = b - 1$$
$$a = b$$

f is one-to-one.

Solve $y = f(x)$ for x

$$y = \sqrt{x - 1}$$
$$y^2 = x - 1 \qquad y \geq 0$$
$$x = 1 + y^2 \quad y \geq 0 \quad f^{-1}(y) = 1 + y^2$$

Interchange x and y:
$y = f^{-1}(x) = 1 + x^2$
Domain: $x \geq 0$, $[0, \infty)$

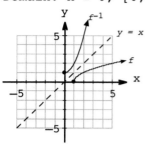

Check:
$f^{-1}[f(x)] = 1 + (\sqrt{x-1})^2 = 1 + x - 1 = x$
$f[f^{-1}(x)] = \sqrt{1 + x^2 - 1} = \sqrt{x^2} = x$
since $x \geq 0$ in the domain of f^{-1}

Range of f^{-1} = domain of f = $[1, \infty)$

(1-6)

59. $f(x) = x^2 - 1 \qquad x \geq 0$

Assume $f(a) = f(b)$
$$a^2 - 1 = b^2 - 1 \qquad a \geq 0,\ b \geq 0$$
$$a^2 - b^2 = 0$$
$$(a - b)(a + b) = 0$$
$a - b = 0 \quad$ or $\quad a + b = 0$
$\quad a = b \qquad\qquad\quad a = -b \qquad$ impossible, since $a \geq 0$ and $b \geq 0$,
$\qquad\qquad\qquad\qquad\qquad\qquad\qquad$ unless $a = b = -b = 0$.

$\quad a = b$
f is one-to-one.

Solve $y = f(x)$ for x:
$y = x^2 - 1 \quad x \geq 0$
$x^2 = y + 1 \quad x \geq 0$

$x = \underbrace{\sqrt{y + 1}} = f^{-1}(y)$
positive square root since $x \geq 0$

Interchange x and y:
$y = f^{-1}(x) = \sqrt{x + 1} \quad$ Domain: $[-1, \infty)$
Range of f^{-1} = domain of f = $[0, \infty)$

Check:
$f^{-1}[f(x)] = \sqrt{x^2 - 1 + 1} \quad x \geq 0$
$\qquad\qquad = \sqrt{x^2} \qquad x \geq 0$
$\qquad\qquad = x$ since $x \geq 0$

$f[f^{-1}(x)] = (\sqrt{x + 1})^2 - 1$
$\qquad\qquad = x + 1 - 1$
$\qquad\qquad = x$

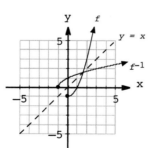

(1-6)

60. (A) One possible answer:

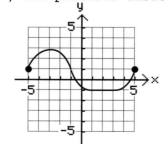

(B) One possible answer:

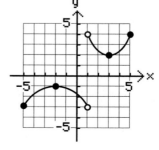

61. The function squares the input, multiplies the result by 2, then subtracts 4 times the input and adds 5. The equation is $g(t) = 2t^2 - 4t + 5$.   (1-2)

62.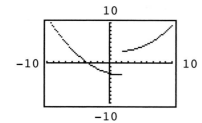

Domain: $x \neq 2$ or $(-\infty, 2) \cup (2, \infty)$.
Range: $y > -3$ or $(-3, \infty)$
Discontinuous at $x = 2$.

(1-3)

63. (A) 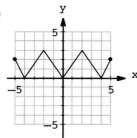   (B)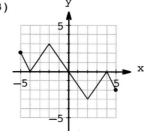

(1-4)

64. (A) $f(x) = 3x^2 - 5x + 7$

$f(x + h) = 3(x + h)^2 - 5(x + h) + 7$

$$\frac{f(x + h) - f(x)}{h} = \frac{[3(x + h)^2 - 5(x + h) + 7] - [3x^2 - 5x + 7]}{h}$$

$$= \frac{3(x^2 + 2xh + h^2) - 5x - 5h + 7 - 3x^2 + 5x - 7}{h}$$

$$= \frac{3x^2 + 6xh + 3h^2 - 5x - 5h + 7 - 3x^2 + 5x - 7}{h}$$

$$= \frac{6xh + 3h^2 - 5h}{h}$$

$$= \frac{h(6x + 3h - 5)}{h}$$

$$= 6x + 3h - 5$$

(B) $f(x) = 3x^2 - 5x + 7$

$f(a) = 3a^2 - 5a + 7$

$$\frac{f(x) - f(a)}{x - a} = \frac{(3x^2 - 5x + 7) - (3a^2 - 5a + 7)}{x - a}$$

$$= \frac{3x^2 - 5x + 7 - 3a^2 + 5a - 7}{x - a}$$

$$= \frac{3x^2 - 3a^2 - 5x + 5a}{x - a}$$

$$= \frac{(x - a)(3x + 3a) - 5(x - a)}{x - a}$$

$$= \frac{(x - a)(3x + 3a - 5)}{x - a}$$

$$= 3x + 3a - 5$$

(1-2)

65. (A) One of many possible graphs is shown. The graph can cross the axis exactly once.

(B) A possible discontinuous graph is shown. The graph can cross the axis once, as in part (A), or not at all, as shown.

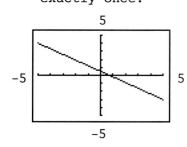

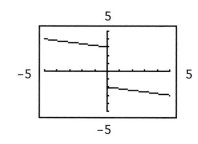

(1-3)

66. (A) We'll begin by computing some values for f.

$$f(-3) = [\![\,|-3|\,]\!] = [\![3]\!] = 3 \qquad f(3) = [\![\,|3|\,]\!] = [\![3]\!] = 3$$
$$f(-2.5) = [\![\,|-2.5|\,]\!] = [\![2.5]\!] = 2 \qquad f(2.5) = [\![\,|2.5|\,]\!] = [\![2.5]\!] = 2$$
$$f(-2) = [\![\,|-2|\,]\!] = [\![2]\!] = 2 \qquad f(2) = [\![\,|2|\,]\!] = [\![2]\!] = 2$$
$$f(-1.5) = [\![\,|-1.5|\,]\!] = [\![1.5]\!] = 1 \qquad f(1.5) = [\![\,|1.5|\,]\!] = [\![1.5]\!] = 1$$
$$f(-1) = [\![\,|-1|\,]\!] = [\![1]\!] = 1 \qquad f(1) = [\![\,|1|\,]\!] = [\![1]\!] = 1$$
$$f(-0.5) = [\![\,|-0.5|\,]\!] = [\![0.5]\!] = 0 \qquad f(0.5) = [\![\,|0.5|\,]\!] = [\![0.5]\!] = 0$$
$$f(0) = [\![\,|0|\,]\!] = [\![0]\!] = 0$$

Generalizing, we can write:

$$f(x) = \begin{cases} 2 & \text{for } -3 < x \le -2 \\ 1 & \text{for } -2 < x \le -1 \\ 0 & \text{for } -1 < x < 1 \\ 1 & \text{for } 1 \le x < 2 \\ 2 & \text{for } 2 \le x < 3 \end{cases}$$

(B)

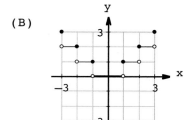

(C) Range: Nonnegative integers

(D) Discontinuous at all integers except 0.

(E) $f(-x) = [\![\,|-x|\,]\!] = [\![x]\!] = [\![\,|x|\,]\!] = f(x)$. Even

(1-3, 1-4)

67. (A) Examining the graph of p, we obtain

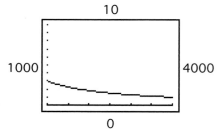

Clearly, d is a decreasing function, passing the horizontal line test, so d is one-to-one.

When $q = 1000$, $p = d(1000) = \dfrac{9}{1 + 0.002(1000)} = 3$

When $q = 4000$, $p = d(4000) = \dfrac{9}{1 + 0.002(4000)} = 1$

The range of d is $1 \le p \le 3$ or $[1, 3]$.

(B) Solve $p = d(q)$ for q.

$$p = \frac{9}{1 + 0.002q}$$

$$p(1 + 0.002q) = 9$$

$$1 + 0.002q = \frac{9}{p}$$

$$0.002q = \frac{9}{p} - 1$$

$$q = \frac{4,500}{p} - 500 = d^{-1}(p)$$

Domain d^{-1} = range d = [1, 3]
Range d^{-1} = domain d = [1000, 4000].

(C) Revenue is price times number sold, so we need to multiply the variable p (price) by the function q (number sold) that we obtained in part B:

$$p \cdot q = p \cdot \left(\frac{4,500}{p} - 500\right) = 4,500 - 500p$$

(D) Revenue is price times number sold, so we need to multiply the function p (price) times the variable q (numbers sold):

$$p \cdot q = \frac{9}{1 + 0.002q} \cdot q = \frac{9q}{1 + 0.002q} \qquad (1\text{-}6)$$

68. Profit is revenue minus cost so we start by subtracting the revenue and cost functions:

$$P(x) = R(x) - C(x) = (50x - 0.1x^2) - (10x + 1,500)$$
$$= 50x - 0.1x^2 - 10x - 1,500 = -0.1x^2 + 40x - 1,500$$

The variable is x but we're asked to find profit in terms of price (p) so we need to substitute $500 - 10p$ in for x.

$$p(500 - 10p) = -0.1(500 - 10p)^2 + 40(500 - 10p) - 1,500$$
$$= -0.1(250,000 - 10,000p + 100p^2) + 20,000 - 400p - 1,500$$
$$= -25,000 + 1,000p - 10p^2 + 20,000 - 400p - 1,500$$
$$= -10p^2 + 600p - 6,500$$

so $P(p) = -10p^2 + 600p - 6,500$

We graph the function and use the MAXIMUM command to find the largest profit.

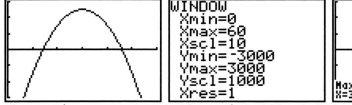

The maximum occurs at the point (30, 2,500), so the maximum profit is \$2,500 when the price is \$30. (1-5)

69. (A) The graph of $y = 55x - 4.88x^2$ is shown.

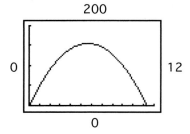

The time the arrow is airborne is the length of the interval between the two x intercepts of the graph, since y which represents height is positive throughout this interval. The first intercept is clearly $x = 0$. Applying the ZERO command to find the second intercept, the result shown is obtained.

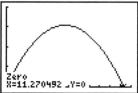

The arrow is airborne 11.3 seconds.

(B) The result of applying the MAXIMUM command to find the maximum value of y is shown.

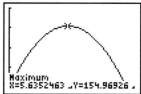

The maximum altitude of the arrow is 155 meters.   (1-3)

70. (A) We begin by graphing $y_1 = 9.5x^2 - 152x + 708.7$ and $y_2 = 300$ on the interval $0 \le x \le 10$. The first coordinate of the point where the two graphs intersect is the time when the jumper is at height 300 feet. Using the INTERSECT command,

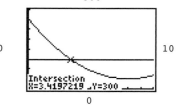

We find that this occurs after 3.4 seconds.

(B) This time we use the MINIMUM command:

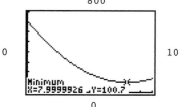

The jumper reaches a minimum height of 100.7 feet after 8 seconds.   (1-3)

71. The dimensions that will make up the length, width, and height of the box are labeled on the diagram below:

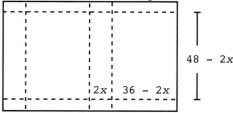

The volume is the product of the three dimensions, so $V(x) = 2x(36 - 2x)(48 - 2x)$.

From the graph, applying the MAXIMUM command we find that the maximum volume is approximately 10,480 in^3 when the flap is 6.8 in wide.

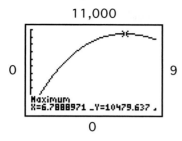

(B) We need to find x when the output of our volume function from part (A) is 9,200, so we add $y_2 = 9,200$ to the graph and use the INTERSECT command.

The volume will be 9,200 cubic inches if the flap is either 4.2″ or 9.7″.

72. (A) This is easy to do using the TBLSET and TABLE commands on a graphing calculator:

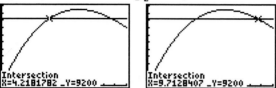

The table of values is:

Elevation	Oxygen Level(%)
0	21.6
2,000	20.1
4,000	18.5
6,000	17.0
8,000	15.5

(B) The oxygen level in the air decreases by about 1.5% for every 2,000 foot increase in altitude.

(C) Evaluating f for 29,035, we get
$$f(29,035) = -0.000767(29,035) + 21.5 = -0.7$$
This answer is not reasonable — there can't be less than 0% oxygen in the air. This happened because the domain of the function f is $0 \le x \le 8,000$, and the altitude 29,035 is well outside of the domain.   (1-2)

73. The 19 in the equation does not depend on the number of minutes x, so that part is the fixed monthly cost for access regardless of minutes used. The $0.012x$ tells us that each minute of calls costs $0.012, or 1.2 cents. So we could say that the provider charges $19 per month for access, plus 1.2 cents per minute of usage.   (1-2)

74.   (A) The graphs are shown below. The lowest graph corresponds to C = 200, and
the highest to C = 1,500.

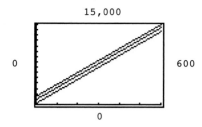

This collection of functions is a series of parallel lines with y intercept
equal to the particular value of C.

(B) To find this facility's fee for 300 guests, we can find $R(300)$:

$$R(300) = C + 22.50(300) = C + 6,750$$

To find the flat fee to keep the price at $7,500, we set $R(300)$ equal to 7,500
and solve:

$C + 6,750 = 7,500$ (Subtract 6,750 from both sides)

$C = 750$

The maximum flat fee is $750.   *(1-3)*

75. (A) The graph of $f(x)$ is shown.   (B) Using the MAXIMUM command to find the
local maximum yields the result shown.

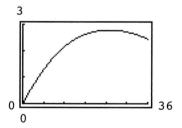

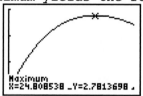

The function increases on [0, 24.8] to a
local maximum of 2.8 cc/sec, and then
decreases on [24.8, 36].   *(1-3)*

76. (A) $f(1) = 1 - (\llbracket \sqrt{1} \rrbracket)^2 = 1 - 1 = 0$
  $f(2) = 2 - (\llbracket \sqrt{2} \rrbracket)^2 = 2 - 1 = 1$
  $f(3) = 3 - (\llbracket \sqrt{3} \rrbracket)^2 = 3 - 1 = 2$
  $f(4) = 4 - (\llbracket \sqrt{4} \rrbracket)^2 = 4 - 4 = 0$
  $f(5) = 5 - (\llbracket \sqrt{5} \rrbracket)^2 = 5 - 4 = 1$

(B) $f(n^2) = n^2 - (\llbracket \sqrt{n^2} \rrbracket)^2$

$= n^2 - (\llbracket n \rrbracket)^2$ since $\sqrt{n^2} = n$ if n is positive

$= n^2 - (n)^2$ since $\llbracket n \rrbracket = n$ if n is a (positive) integer.

$= 0$

$f(6) = 6 - (\llbracket \sqrt{6} \rrbracket)^2 = 6 - 4 = 2$ $f(12) = 12 - (\llbracket 12 \rrbracket)^2 = 12 - 9 = 3$
$f(7) = 7 - (\llbracket 7 \rrbracket)^2 = 7 - 4 = 3$ $f(13) = 13 - (\llbracket 13 \rrbracket)^2 = 13 - 9 = 4$
$f(8) = 8 - (\llbracket 8 \rrbracket)^2 = 8 - 4 = 4$ $f(14) = 14 - (\llbracket 14 \rrbracket)^2 = 14 - 9 = 5$
$f(9) = 9 - (\llbracket 9 \rrbracket)^2 = 9 - 9 = 0$ $f(15) = 15 - (\llbracket 15 \rrbracket)^2 = 15 - 9 = 6$
$f(10) = 10 - (\llbracket 10 \rrbracket)^2 = 10 - 9 = 1$ $f(16) = 16 - (\llbracket 16 \rrbracket)^2 = 16 - 16 = 0$
$f(11) = 11 - (\llbracket 11 \rrbracket)^2 = 11 - 9 = 2$

(C) It determines if an integer is a perfect square integer. If $f(x) = 0$, then x is a perfect square and if $f(x) = 10$, x is not a perfect square integer. (*1-3*)

77. In this example, the independent variable is years since 1900 so the first list under the STAT EDIT screen comes from the top row of Table 1. (Note that x is years after 1900, so the values we input are 12, 32, 52, 72 and 92. The dependent variable is time, so the second list is 5.41, 4.81, 4.51, 4.00, 3.75.)

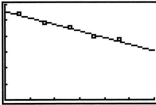

The second screen shows the appropriate plot settings; the third shows a good choice of viewing window; the last is the scatter plot of the data from the table.

To compare the algebraic function, we enter $y_1 = -0.021x + 5.57$ in the equation editor, choose the "ask" option for independent variable in the table setup window, then enter x-values 12, 32, 52, 72, and 92.

Comparing the two tables, we see that the corresponding times match very closely. Finally, we look at the graphs of both the function and the scatter plot:

As expected, the graph of the function matches the points from the scatter plot very well.

(B) Since 2010 is 110 years after 1900, we need to plug in 110 for x:
$$f(110) = -0.021(100) + 5.57 = 3.26$$
The estimated time is 3.26 seconds.   (*1-2*)

78. Amounts less than \$3,000: Tax is 2% of income $(x) = 0.02x$

Amounts between \$3,000 and \$5,000: Tax is \$60 + 3% of income over 3,000 $(x - 3,000)$ so tax = $60 + 0.03(x - 3,000) = 0.03x - 30$

Amounts between \$5,000 and \$17,000: Tax is \$120 + 5% of income over 5,000 $(x - 5,000)$ so tax = $120 + 0.05(x - 5,000) = 0.05x - 130$

Amounts over \$17,000: Tax is \$720 + 5.75% of income over 17,000 $(x - 17,000)$ so tax = $720 + 0.0575(x - 17,000) = 0.0575x - 257.5$

$$\text{Combined, we get } t(x) = \begin{cases} 0.02x & 0 \le x \le 3{,}000 \\ 0.03x - 30 & 3{,}000 < x \le 5{,}000 \\ 0.05x - 130 & 5{,}000 < x \le 17{,}000 \\ 0.0575x - 257.5 & x > 17{,}000 \end{cases}$$

$t(2{,}000) = 0.02(2{,}000) = 40$. The tax on \$2,000 is \$40.

$t(4{,}000) = 0.03(4{,}000) - 30 = 120 - 30 = 90$. The tax on \$4,000 is \$90.

$t(10{,}000) = 0.05(10{,}000) - 130 - 500 - 130 = 370$. The tax on \$10,000 is \$130.

$t(30{,}000) = 0.0575(30{,}000) - 257.5 - 1{,}725 - 257.5 = 1{,}467.5$. The tax on \$30,000 is \$1,467.50.   (*1-3*)

CHAPTER 2   Modeling with Linear and Quadratic Functions
Section 2-1

1. Answers will vary.   3. Answers will vary.

5. Answers will vary.

7. The vertical segment has length 3 so rise = 3. The horizontal segment has length 5 so run = 5. Slope = $\dfrac{\text{rise}}{\text{run}} = \dfrac{3}{5}$. (2, 2) is on the graph.
   Use the point-slope form.

   $$y - 2 = \frac{3}{5}(x - 2)$$ (Multiply both sides by 5)

   $$5y - 10 = 3(x - 2)$$ Common error: Multiplying both $\frac{3}{5}$ and
   $$5y - 10 = 3x - 6$$
   $$-3x + 5y = +4$$ $(x - 2)$ by 5 on the right side.
   $$3x - 5y = -4$$

9. The vertical segment has length 2 so rise = 2. The horizontal segment has length 8 so run = 8.

   $$\text{slope} = \frac{\text{rise}}{\text{run}} = \frac{2}{8} = \frac{1}{4}$$ (-4, 1) is on the graph. Use the point-slope form.

   $$y - 1 = \frac{1}{4}(x - (-4))$$ (Multiply both sides by 4)

   $$4y - 4 = x + 4$$
   $$-x + 4y = 8$$
   $$x - 4y = -8$$

11. The vertical segment has length 3 and goes downward so rise = -3. The horizontal segment has length 5 so run = 5.

   $$\text{slope} = \frac{\text{rise}}{\text{run}} = \frac{-3}{5}$$ (-4, 2) is on the graph. Use the point-slope form.

   $$y - 2 = \frac{-3}{5}(x - (-4))$$ (Multiply both sides by 5)

   $$5y - 10 = -3(x + 4)$$
   $$5y - 10 = -3x - 12$$
   $$3x + 5y = -2$$

13. The x intercept is -2. The y intercept is 2. From the point (-2, 0) to the point (0, 2), the value of y increases by 2 units as the value of x increases by 2 units; slope = $\dfrac{\text{change in } y}{\text{change in } x} = \dfrac{2}{2} = 1$. Using $y = mx + b$, we get $y = x + 2$.

15. The x intercept is -2. The y intercept is -4. From the point (-2, 0) to the point (0, -4) the value of y decreases by 4 units as the value of x increases by 2 units; slope = $\dfrac{\text{change in } y}{\text{change in } x} = \dfrac{-4}{2} = -2$. Using $y = mx + b$, we get $y = -2x - 4$.

17. The x intercept is 3. The y intercept is -1. From the point (0, -1) to the point (3, 0) the value of y increases by 1 unit as the value of x increases by 3 units; slope = $\dfrac{\text{change in } y}{\text{change in } x} = \dfrac{1}{3}$. Using $y = mx + b$, we get $y = \dfrac{1}{3}x - 1$.

19. The equation $y = 2x^2$ is not linear because in linear equations both variables appear only to the first power.

21. The equation $y = \dfrac{x - 5}{3}$ is linear. It can be rewritten as $y = \dfrac{1}{3}x - \dfrac{5}{3}$ which matches the form $y = mx + b$.

23. Simplify the equation:
$$y = \frac{2}{3}(x - 7) - \frac{1}{2}(3 - x) \qquad \text{(Multiply both sides by 6)}$$
$$6y = 4(x - 7) - 3(3 - x)$$
$$6y = 4x - 28 - 9 + 3x$$
$$-7x + 6y = -37$$
The equation can be rewritten in the form $Ax + By = C$, so it is a linear equation.

25. Simplify the equation:
$$y = \frac{1}{4}(2x + 2) + \frac{1}{2}(4 - x) \qquad \text{(Multiply both sides by 4)}$$
$$4y = 2x + 2 + 4 - x$$
$$4y = x + 6$$
$$-x + 4y = 6$$
The equation can be written in the form $Ax + By = C$, so it is a linear equation.

27. This equation is not linear. If we attempt to move x to the numerator by multiplying by $x - 5$, we get
$$y(x - 5) = 3$$
$$xy - 5y = 3$$
In a linear equation, the two variables appear in different terms.

29. $y = -\dfrac{3}{5}x + 4$

    slope: $-\dfrac{3}{5}$

    y intercept: 4

    To find the x intercept, set $y = 0$ and solve for x:
$$0 = -\frac{3}{5}x + 4$$
$$\frac{3}{5}x = 4$$
$$x = \frac{20}{3} \qquad (x \text{ intercept})$$
    To graph, use the intercepts or find a few points by substituting values.

x	y
0	4
5	1
−5	7

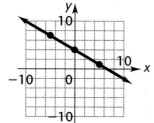

31. $y = -\frac{3}{4}x$ or $y = -\frac{3}{4}x + 0$

slope: $-\frac{3}{4}$

y intercept: 0

To find the x intercept, set $y = 0$ and solve for x:

$0 = -\frac{3}{4}x$

$0 = x$ (x intercept)

To graph, use the intercept and find one or more other points by substituting values.

x	y
0	0
4	-3
-4	3

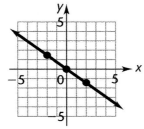

33. $2x - 3y = 15$

$-3y = -2x + 15$

$y = \frac{2}{3}x - 5$

slope: $\frac{2}{3}$

y intercept: -5

To find the x intercept, set $y = 0$ in $2x - 3y = 15$ and solve for x:

$2x - 3(0) = 15$

$2x = 15$

$x = \frac{15}{2}$ (x intercept)

To graph, use the intercepts or find a few points by substituting values.

x	y
0	-5
3	-3
-3	-7

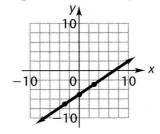

35. $\frac{y}{8} - \frac{x}{4} = 1$

$\frac{y}{8} = \frac{x}{4} + 1$

$y = 8\left(\frac{x}{4} + 1\right)$

$y = 2x + 8$

slope: 2

y intercept: 8

To find the x intercept, set $y = 0$ in $\frac{y}{8} - \frac{x}{4} = 1$ and solve for x:

$\frac{0}{8} - \frac{x}{4} = 1$

$-\frac{x}{4} = 1$

$x = -4$ (x intercept)

To graph, use the intercepts or find other points by substituting values.

x	y
0	8
-4	0
-8	-8

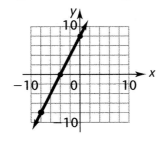

37. $x = -3$

This is a vertical line.
The slope is undefined.
There is no y intercept.
The x intercept is -3.
Graph a vertical line with all x coordinates equal to -3.

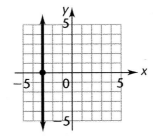

39. $y = 3.5$
This is a horizontal line.
slope = 0
The y intercept is 3.5.
There is no x intercept.
Graph a horizontal line with all
y coordinates equal to 3.5.

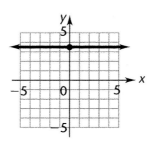

41. The equation of any vertical line looks like $x = k$, where k is the x coordinate of any point on the line. In this case, we get $x = 2$.

43. The equation of any horizontal line looks like $y = k$, where k is the y coordinate of any point on the line. In this case, we get $y = -4$.

45. First, we find the slope:
$$m = \frac{4 - 4}{5 - (-3)} = \frac{0}{8} = 0$$

This tells us that the line is horizontal, and the y coordinate of every point is 4, so the equation is $y = 4$. (You may have noticed that the line is horizontal by observing that both points have the same y coordinate.)

47. First, we find the slope:
$$m = \frac{6 - (-3)}{4 - 4} = \frac{9}{0}, \text{ which is undefined.}$$

This tells us that the line is vertical, and the x coordinate of every point is 4, so the equation is $x = 4$. (You may have noticed that the line is vertical by observing that both points have the same x coordinate.)

49. Slope and y intercept are given; we use slope-intercept form.
$y = 1x + 0$
$y = x$

Subtract x from each side to write in standard form:

$Y = x$

$-x + y = 0$

51. Slope and y intercept are given; we use slope-intercept form.
$y = -\dfrac{2}{3}x + (-4)$
$y = -\dfrac{2}{3}x - 4$

Now we write in standard form.

$y = -\dfrac{2}{3}x - 4$ $\qquad$ Multiply both sides by 3

$3y = -2x - 12$ $\qquad$ Add $2x$ to both sides

$2x + 3y = -12$

53. Use the point-slope form:
$y - 4 = -3(x - 0)$
$y = -3x + 4$

55. Use the point-slope form:

$$y - 4 = -\frac{2}{5}\left(x - (-5)\right)$$

$$y - 4 = -\frac{2}{5}(x + 5)$$

$$y - 4 = -\frac{2}{5}x - 2$$

$$y = -\frac{2}{5}x + 2$$

57. First, find the slope:

$$m = \frac{-2 - 6}{5 - 1} = \frac{-8}{4} = -2$$

Now use the point-slope form: (You can use either point; we chose (1,6))

$$y - 6 = -2(x - 1)$$

$$y - 6 = -2x + 2$$

$$y = -2x + 8$$

59. First, find the slope:

$$m = \frac{0 - 8}{2 - (-4)} = \frac{-8}{6} = -\frac{4}{3}$$

Now use the point-slope form: (You can use either point; we chose (2,0))

$$y - 0 = -\frac{4}{3}(x - 2)$$

$$y = -\frac{4}{3}x + \frac{8}{3}$$

61. The intercepts are the points (-4,0) and (0,3).   First, find the slope:

$$m = \frac{3 - 0}{0 - (-4)} = \frac{3}{4}$$

Now use the point-slope form: (You can use either point; we chose (0,3))

$$y - 3 = \frac{3}{4}(x - 0)$$

$$y = \frac{3}{4}x + 3$$

63. The line $y = 3x - 5$ has slope 3 (coefficient of x), so the line we're looking for does as well (since they are parallel).   Use the point-slope form:

$$y - 4 = 3\left(x - (-3)\right)$$

$$y - 4 = 3(x + 3)$$

$$y - 4 = 3x + 9$$

$$y = 3x + 13$$

65. The line $y = -\frac{1}{3}x$ has slope -1/3 (coefficient of x); the line perpendicular to it will have slope 3 (negative reciprocal of -1/3).   Use the point-slope form:

$$y - (-3) = 3(x - 2)$$

$$y + 3 = 3x - 6$$

$$y = 3x - 9$$

67. To find the slope of the given line, solve for y:

$3x - 2y = 4$

$-2y = -3x + 4$

$y = -\dfrac{1}{2}(-3x + 4)$

$y = \dfrac{3}{2}x - 2$

This line has slope 3/2 (coefficient of x), so the line parallel to it does as well. Use the point-slope form:

$y - 0 = \dfrac{3}{2}(x - 5)$

$y = \dfrac{3}{2}x - \dfrac{15}{2}$

69. To find the slope of the given line, solve for y:

$x + 3y = 9$

$3y = -x + 9$

$y = -\dfrac{1}{3}x + 3$

This line has slope $-1/3$ (coefficient of x), so the line perpendicular to it will have slope 3 (negative reciprocal). Use the point-slope form:

$y - (-4) = 3(x - 0)$

$y + 4 = 3x$

$y = 3x - 4$

71. The graphs are straight lines; all have the same y intercept, 2, but their slopes vary as m varies.

73. (A) First find the slope:

$m = \dfrac{2 - (-3)}{7 - (-1)} = \dfrac{5}{8}$

Now use the point-slope form

$y - 2 = \dfrac{5}{8}(x - 7)$

$y - 2 = \dfrac{5}{8}x - \dfrac{35}{8}$

$y = \dfrac{5}{8}x - \dfrac{35}{8} + \dfrac{16}{8}$

$y = \dfrac{5}{8}x - \dfrac{19}{8}$

$f(x) = \dfrac{5}{8}x - \dfrac{19}{8}$

(B) $m = \dfrac{7 - (-1)}{2 - (-3)} = \dfrac{8}{5}$

$y - 7 = \dfrac{8}{5}(x - 2)$

$y - 7 = \dfrac{8}{5}x - \dfrac{16}{5}$

$y = \dfrac{8}{5}x - \dfrac{16}{5} + \dfrac{35}{5}$

$y = \dfrac{8}{5}x + \dfrac{19}{5}$

$g(x) = \dfrac{8}{5}x + \dfrac{19}{5}$

(C) The graphs are mirror images about the line $y = x$, so the two functions are inverses.

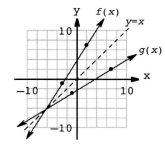

75. From the graph, we see that the two points on this line are (1, 1/2) and (4, 8). Use the slope formula:

$$m = \frac{8 - 1/2}{4 - 1} = \frac{16/2 - 1/2}{3} = \frac{15/2}{3} = \frac{15}{6} = \frac{5}{2}$$

77. First, we need the y coordinate of the point with x coordinate 2:

$$f(2) = \frac{1}{2}(2)^2 = 2$$

The two points on the graph are (1, 1/2) and (2, 2). Use the slope formula:

$$m = \frac{2 - 1/2}{2 - 1} = \frac{4/2 - 1/2}{1} = \frac{3}{2}$$

79. First, we need the y coordinate of the point with x coordinate 5/4:

$$f\left(\frac{5}{4}\right) = \frac{1}{2}\left(\frac{5}{4}\right)^2 = \frac{1}{2} \cdot \frac{25}{16} = \frac{25}{32}$$

The two points on the graph are (1, 1/2) and (5/4, 25/32). Use the slope formula:

$$m = \frac{25/32 - 1/2}{5/4 - 1} = \frac{25/32 - 16/32}{5/4 - 4/4} = \frac{9/32}{1/4} = \frac{9}{32} \cdot \frac{4}{1} = \frac{9}{8}$$

81. (A)

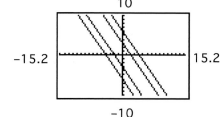

(B) Varying C produces a family of parallel lines.

(C) Solve $Ax + By = C$ for y in terms of x.

$$By = -Ax + C$$
$$y = \frac{-Ax + C}{B}$$
$$y = -\frac{A}{B}x + \frac{C}{B}$$

For fixed A and B, and varying C this is a family of lines with equal slopes, $-\dfrac{A}{B}$. Therefore the lines are parallel.

83. Two points are given; we first find the slope, then use the point-slope form.

$$m = \frac{0 - b}{a - 0} = -\frac{b}{a} \quad \text{(note that } a \neq 0\text{)}$$
$$y - b = -\frac{b}{a}(x - 0)$$
$$y - b = -\frac{bx}{a}$$

Divide both sides by b, then (note that $b \neq 0$)

$$\frac{y - b}{b} = -\frac{x}{a}$$
$$\frac{y}{b} - 1 = -\frac{x}{a}$$
$$\frac{y}{b} = 1 - \frac{x}{a}$$
$$\frac{x}{a} + \frac{y}{b} = 1$$

85.   (A) In a linear function, the slope represents the rate of change.   The slope of $A(x)$ is -2.13, so the rate of change is 2.13 percent per month; the negative indicates that the approval rating was going down.

(B) October of 2001 corresponds to $x = 0$; $A(0) = -2.13(0) + 91 = 91\%$

January of 2003 corresponds to $x = 15$; $A(15) = -2.13(15) + 91 = 59.05\%$

Over the span of 15 months, the approval rating decreased by 31.95; this works out to -31.95 %/15 months = -2.13% per month, which agrees with part (A).

87.   (A) In a linear function, the slope represents the rate of change.   The slope of this function is 1.04, so the rate of change is 1.04 million immigrants per year.   The positive rate indicates that the number of immigrants was going up.

(B) 1996 corresponds to $x = 0$; $N(0) = 1.04(0) + 25.95 = 25.95$.   There were 25.95 million immigrants in 1996.

2004 corresponds to $x = 8$; $N(8) = 1.04(8) + 25.95 = 34.27$.   There were 34.27 million immigrants in 2004.

Over the span of 8 years, the immigrant population increased by 8.32 million; this works out to 8.32 million/8 years = 1.04 million per year, which agrees with part (A).

89.   The rate of -23 miles per hour (negative because the distance away is getting smaller) represents the slope of a linear equation.   The distance of 145 occurred at time $h = 0$, so it is the y intercept (or in this case d intercept) of the function.   This gives us the equation $d = -23h + 145$.

Since 6 AM is 7 hours after 11 PM, we plug in $h = 7$:

$d = -23(7) + 145 = -16$

This negative distance indicates that the storm will have already reached the city.

91.   (A) If F is linearly related to C, then we are looking for an equation whose graph passes through $(C_1, F_1) = (0, 32)$ and $(C_2, F_2) = (100, 212)$. We find the slope, and then use the point-slope form to find the equation.

$$m = \frac{F_2 - F_1}{C_2 - C_1} = \frac{212 - 32}{100 - 0} = \frac{180}{100} = \frac{9}{5}$$

$$F - F_1 = m(C - C_1)$$

$$F - 32 = \frac{9}{5}(C - 0)$$

$$F - 32 = \frac{9}{5}C$$

$$F = \frac{9}{5}C + 32$$

(B) We are asked for F when $C = 20$.

$$F = \frac{9}{5}(20) + 32 = 36 + 32 = 68°$$

We are then asked for C when $F = 86°$

$$86 = \frac{9}{5}C + 32$$

$$430 = 9C + 160$$

$$270 = 9C$$

$$C = 30°$$

(C) The slope is $m = \frac{9}{5}$.

93. (A) If V is linearly related to t, then we are looking for an equation whose graph passes through $(t_1, V_1) = (0, 8{,}000)$ and $(t_2, V_2) = (5, 0)$. We find the slope, and then we use the point-slope form to find the equation.

$$m = \frac{V_2 - V_1}{t_2 - t_1} = \frac{0 - 8{,}000}{5 - 0} = \frac{-8{,}000}{5} = -1{,}600$$

$$
\begin{aligned}
V - V_1 &= m(t - t_1) \\
V - 8{,}000 &= -1{,}600(t - 0) \\
V - 8{,}000 &= -1{,}600t \\
V &= -1{,}600t + 8{,}000 \quad 0 \le t \le 5
\end{aligned}
$$

(B) We are asked for V when $t = 3$

$$
\begin{aligned}
V &= -1{,}600(3) + 8{,}000 \\
V &= -4{,}800 + 8{,}000 \\
V &= 3{,}200
\end{aligned}
$$

(C) The slope is $m = -1{,}600$; the value decreases by \$1,600 per year.

95. (A) The question asks us to use x to represent the number of golf clubs produced, so we can set up ordered pairs as follows: (golf clubs produced, total daily cost). That gives us $(80, 8{,}147)$ and $(100, 9{,}647)$.

$$m = \frac{9{,}647 - 8{,}147}{100 - 80} = \frac{1{,}500}{20} = 75$$

$$
\begin{aligned}
y - 9{,}647 &= 75(x - 100) \\
y - 9{,}647 &= 75x - 7{,}500 \\
y &= 75x + 2{,}147 \\
C(x) &= 75x + 2{,}147
\end{aligned}
$$

(B) The slope represents the variable cost per golf club manufactured; the cost per club is \$75. The y-intercept indicates that the fixed daily costs are \$2,147.

97. (A) If R is linearly related to C, then we are looking for an equation whose graph passes through $(C_1, R_1) = (210, 0.160)$ and $(C_2, R_2) = (231, 0.192)$. We find the slope, and then we use the point-slope form to find the equation.

$$m = \frac{R_2 - R_1}{C_2 - C_1} = \frac{0.192 - 0.160}{231 - 210} = \frac{0.032}{21} = 0.00152$$

$$
\begin{aligned}
R - R_1 &= m(C - C_1) \\
R - 0.160 &= 0.00152(C - 210) \\
R - 0.160 &= 0.00152C - 0.319 \\
R &= 0.00152C - 0.159, \quad C \ge 210
\end{aligned}
$$

(B) We are asked for R when $C = 260$.

$$
\begin{aligned}
R &= 0.00152(260) - 0.159 \\
R &= 0.236
\end{aligned}
$$

(C) The slope is $m = 0.00152$; coronary risk increases 0.00152 per unit increase in cholesterol above the 210 cholesterol level.

99. (A) The temperature is 70 degrees at altitude zero, so the point $(0, 70)$ is on our line. At 1,000 feet, the temperature will have dropped by 5 degrees, to 65, so the point $(1, 65)$ is on the line as well. We can find the slope and use the point-slope form to find the equation:

$$m = \frac{65 - 70}{1 - 0} = -5$$

$$T - 70 = -5\big(A - 0\big)$$

$$T = -5A + 70$$

(B) An altitude of 10,000 feet corresponds to $A = 10$, so we plug in 10 for A:

$T = -5(10) + 70 = 20$

The temperature at 10,000 feet would be 20 degrees.

(C) The slope is -5. This indicates that the temperature decreases at the rate of 5 degrees for each 1,000 feet of altitude.

Section 2-2

1. Answers will vary.

3. Answers will vary.

5. Answers will vary.

7. Answers will vary.

9. $u(x) = 0$ where the graph of u intersects the x axis, at $x = c$ and $x = f$.

11. $u(x) = v(x)$ for values of x that correspond to the intersection of the graphs of u and v, at $x = b$ and $x = e$.

13. Simplify the left side:
$3(x - 2) - 2(x + 1) = 3x - 6 - 2x - 2 = x - 8$
This is equal to the right side, so the equation is an identity.

15. Simplify the left side:
$2(x - 1) - 3(2 - x) = 2x - 2 - 6 + 3x = 5x - 8$
This is not equal to the right side, so the equation is not an identity.
$$5x - 8 = x - 8$$
$$4x = 0$$
$$x = 0$$
This is a conditional equation with solution $x = 0$.

17. Simplify the left side:
$5(x + 2) - 3(x - 1) = 5x + 10 - 3x + 3 = 2x + 13$
This is not equal to the right side, so the equation is not an identity.
$$2x + 13 = 2x + 4$$
$$0 = -9$$
This is never true, so the original equation is a contradiction.

19. $10x - 7 = 4x - 25$
$$6x - 7 = -25$$
$$6x = -18$$
$$x = -3$$

21. $3(x + 2) = 5(x - 6)$
$$3x + 6 = 5x - 30$$
$$-2x = -36$$
$$x = 18$$

23. $5 + 4(t - 2) = 2(t + 7) + 1$
$$5 + 4t - 8 = 2t + 14 + 1$$
$$4t - 3 = 2t + 15$$
$$2t = 18$$
$$t = 9$$

25. $\dfrac{x}{2} - 2 = \dfrac{x}{5} + \dfrac{2}{5}$ LCD = 10

$10 \cdot \dfrac{x}{2} - 10 \cdot 2 = 10 \cdot \dfrac{x}{5} + 10 \cdot \dfrac{2}{5}$
$$5x - 20 = 2x + 4$$
$$3x = 24$$
$$x = 8$$

27. $3 + 5(x - 7) = 3x - 2(16 - x)$
$$3 + 5x - 35 = 3x - 32 + 2x$$
$$5x - 32 = 5x - 32$$
The equation is an identity, so the solution is all real numbers.

29.
$$5 - \frac{2x - 1}{4} = \frac{x + 2}{3} \quad \text{LCD} = 12$$

$$12 \cdot 5 - 12 \frac{(2x - 1)}{4} = 12 \frac{(x + 2)}{3}$$

$$12 \cdot 5 - 3(2x - 1) = 4(x + 2)$$

$$60 - 6x + 3 = 4x + 8$$

$$-6x + 63 = 4x + 8$$

$$-10x = -55$$

$$x = \frac{-55}{-10}$$

Solution: $\frac{11}{2}$ or 5.5

Check:

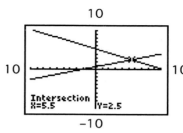

Common Error:
After line 2,
students often write
$$60 - \overset{3}{\cancel{12}} \frac{2x - 1}{\cancel{4}} = \ldots$$
$$60 - 6x - 1 = \ldots$$
Forgetting to
distribute the -3.
Put compound
numerators in
parentheses to avoid
this.

31.
$$\frac{7}{t} + 4 = \frac{2}{t}$$
Excluded value: $t \neq 0 \quad \text{LCD} = t$
$$t \cdot \frac{7}{t} + t \cdot 4 = t \cdot \frac{2}{t}$$
$$7 + 4t = 2$$
$$4t = -5$$
$$t = -\frac{5}{4}$$

33.
$$\frac{1}{m} - \frac{1}{9} = \frac{4}{9} - \frac{2}{3m}$$
Excluded value: $m \neq 0 \quad \text{LCD} = 9m$
$$9m \cdot \frac{1}{m} - 9m \cdot \frac{1}{9} = 9m \cdot \frac{4}{9} - 9m \cdot \frac{2}{3m}$$
$$9 - m = 4m - 6$$
$$-5m = -15$$
$$m = 3$$

35. $(x - 2)(x + 3) = (x - 4)(x + 5)$
$$x^2 + x - 6 = x^2 + x - 20$$
$$-6 = -20$$
There is no solution.

37. $(x + 2)(x - 3) = (x - 4)(x + 5)$
$$x^2 - x - 6 = x^2 + x - 20$$
$$-x - 6 = x - 20$$
$$-2x = -14$$
$$x = 7$$

39. $(x - 2)^2 = (x - 1)(x + 2)$
$$x^2 - 4x + 4 = x^2 + x - 2$$
$$-4x + 4 = x - 2$$
$$-5x = -6$$
$$x = \frac{6}{5}$$

41. $\dfrac{2x}{x - 3} = 2 + \dfrac{6}{x - 3}$
Excluded value: $x \neq 3 \quad \text{LCD} = x - 3$
$$(x - 3)\frac{2x}{x - 3} = (x - 3)2 + (x - 3)\frac{6}{x - 3}$$
$$2x = 2x - 6 + 6$$
$$2x = 2x$$
The solution is all real numbers except 3.

43. $\dfrac{2x}{x - 3} = 7 + \dfrac{4}{x - 3}$
Excluded value: $x \neq 3 \quad \text{LCD} = x - 3$
$$(x - 3)\frac{2x}{x - 3} = 7(x - 3) + (x - 3)\frac{4}{x - 3}$$
$$2x = 7x - 21 + 4$$
$$2x = 7x - 17$$
$$-5x = -17$$
$$x = \frac{17}{5} \text{ or } 3.4$$

45. $\dfrac{2x}{x - 3} = 7 + \dfrac{6}{x - 3} \quad \text{LCD: } x - 3$

$$(x - 3)\frac{2x}{x - 3} = (x - 3)7 + (x - 3)\frac{6}{x - 3}$$
$$2x = 7x - 21 + 6$$
$$-5x = -15$$
$$x = 3 \qquad 3 \text{ is excluded since it makes the original equation}$$
undefined, so there is no solution.

47. $P = 2\ell + 2w$

$P - 2\ell = 2w$

$\dfrac{1}{2}(P - 2\ell) = w$

$w = \dfrac{P - 2\ell}{2} = \dfrac{1}{2}P - \ell$

49. $a_n = a_1 + (n - 1)d$

$a_1 + (n - 1)d = a_n$

$(n - 1)d = a_n - a_1$

$d = \dfrac{a_n - a_1}{n - 1}$

51. $\dfrac{1}{f} = \dfrac{1}{d_1} + \dfrac{1}{d_2}$ LCD $= d_1 d_2 f$

$d_1 d_2 f \dfrac{1}{f} = d_1 d_2 f \dfrac{1}{d_1} + d_1 d_2 f \dfrac{1}{d_2}$

$d_1 d_2 = d_2 f + d_1 f$

$d_2 f + d_1 f = d_1 d_2$

$(d_2 + d_1)f = d_1 d_2$

$f = \dfrac{d_1 d_2}{d_2 + d_1}$

53. $A = 2ab + 2ac + 2bc$

$2ab + 2ac + 2bc = A$

$2ab + 2ac = A - 2bc$

$a(2b + 2c) = A - 2bc$

$a = \dfrac{A - 2bc}{2b + 2c}$

55. $y = \dfrac{2x - 3}{3x + 5}$

$(3x + 5)y = 2x - 3$

$3xy + 5y = 2x - 3$

$5y + 3 = 2x - 3xy$

$5y + 3 = x(2 - 3y)$

$\dfrac{5y + 3}{2 - 3y} = x$

$x = \dfrac{5y + 3}{2 - 3y}$

57. The graphs of $y_1 = x$ and $y_2 = \sqrt{x^2}$ are shown at the right. The graphs are not identical, so the equation $x = \sqrt{x^2}$ is a conditional equation, and not an identity.

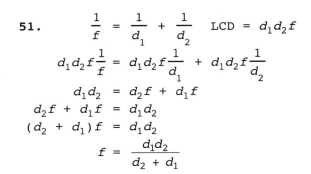

59. The length is twice the width

$\ell \quad = \quad 2 \quad \cdot \quad w$

$\ell = 2w$

61. The width is half the length.

$w \quad = \quad \dfrac{1}{2} \quad \cdot \quad \ell$

$w = \dfrac{1}{2}\ell$

63. The length is 3 more than the width

$\ell \quad = \quad 3 \quad + \quad w$

$\ell = 3 + w$

65. The length is 4 less than the width

$\ell \quad = \quad w - 4$

67. (A) For detailed instructions on using a graphing calculator for linear regression, see Example 8 in Section 2-2 of your text. Enter the x column (1, 5) in list L_1, and the y column (-1, 1) in list L_2.

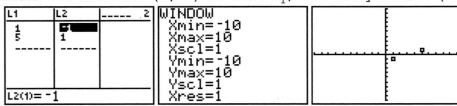

Use the STAT CALC menu to find the linear regression:

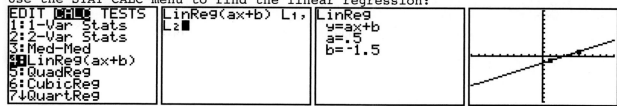

The line is $y = 0.5x - 1.5$

(B) Repeat the steps above using $-1, 1$ as L_1 and $1, 5$ as L_2.

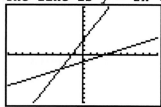

The line is $y = 2x + 3$.

The two lines are symmetrical with respect to the line $y = x$ so the two linear functions are inverses. (This is expected because the difference between tables A and B is that the order of x and y were switched.)

69. $\dfrac{x - \frac{1}{x}}{1 + \frac{1}{x}} = 1$ (Multiply both sides by $1 + \frac{1}{x}$)

$x - \dfrac{1}{x} = 1 + \dfrac{1}{x}$ LCD = x

$\qquad x\left(x - \dfrac{1}{x}\right) = x\left(1 + \dfrac{1}{x}\right)$

$\qquad\qquad x^2 - 1 = x + 1$

$\qquad x^2 - x - 2 = 0$

$(x - 2)(x + 1) = 0$

$x - 2 = 0, \ x + 1 = 0$

$\qquad x = 2, \qquad x = -1$

Check: $x = 2$: $\dfrac{2 - \frac{1}{2}}{1 + \frac{1}{2}} = \dfrac{\frac{3}{2}}{\frac{3}{2}} = 1$

$x = -1$: $\dfrac{-1 - \frac{1}{-1}}{1 + \frac{1}{-1}} = \dfrac{0}{0}$ Not a solution

Solution: $x = 2$

71. $\dfrac{x + 1 - \frac{2}{x}}{1 - \frac{1}{x}} = x + 2$ (Multiply left side by $\frac{x}{x}$)

$\dfrac{x\left(x + 1 - \frac{2}{x}\right)}{x\left(1 - \frac{1}{x}\right)} = x + 2$

$\dfrac{x^2 + x - 2}{x - 1} = x + 2$ (Multiply both sides by $x - 1$)

$x^2 + x - 2 = (x + 2)(x - 1)$

$x^2 + x - 2 = x^2 + x - 2$

This is an identity provided that $x \neq 0$ or 1. (Both 0 and 1 make the left side of the original equation undefined.)

73. Let x = the least of the integers. Then $x + 1$ and $x + 2$ are the others. Their sum looks like $x + (x + 1) + (x + 2)$, so

$$x + (x + 1) + (x + 2) = 84$$
$$3x + 3 = 84$$
$$3x = 81$$
$$x = 27$$

The three consecutive integers are 27, 28, and 29.

75. Let x = the least of the integers. Then the other three are $x + 2$, $x + 4$, and $x + 6$ (they are all separated by two since all are even).
Translate the question into an equation:

The sum of the first three is 2 more than twice the fourth

$$x \quad + (x + 2) + (x + 4) = 2 \quad + \quad 2(x + 6)$$
$$3x + 6 = 2 + 2x + 12$$
$$x = 8$$

The four even integers are 8, 10, 12, and 14.

77. Let ℓ = the length, w = the width. Translate the statements into symbols:

The length is twice the width
$$\ell \quad = \quad 2 \quad \cdot \quad w$$

The perimeter is 60 inches
$$2\ell + 2w = 60 \qquad \text{We know } 2w = \ell, \text{ so}$$
$$2\ell + \ell = 60$$
$$3\ell = 60$$
$$\ell = 20$$
$$w = 10$$

The length is twice the width, so $\ell = 20$.

The dimensions are 10 in × 20 in.

79. Total cost = Fixed cost + variable cost. If x = the number of doughnuts produced, variable cost is $0.12x$ ($\$0.12$ per doughnut times the number of doughnuts produced.) If $C(x)$ represents total cost,
$$C(x) = 124 + 0.12x$$
Plug in 250 for $C(x)$:
$$250 = 124 + 0.12x$$
$$126 = 0.12x$$
$$1{,}050 = x$$
The shop can produce 1,050 doughnuts for $250.

81. Let x = sales of employee
Then $x - 7{,}000$ = sales on which 8% commission is paid
$0.08(x - 7{,}000)$ = (rate of commission) × (sales) = (amount of commission)
$2{,}150 + 0.08(x - 7{,}000)$ = (base salary) + (amount of commission) = earnings
$2{,}150 + 0.08(x - 7{,}000) = 3{,}170$

We're looking for the sales that will earn our employee $3,170, so we set our expression for earnings equal to 3,170 and solve for x.

$$2{,}150 + 0.08x - 560 = 3{,}170$$
$$0.08x + 1{,}590 = 3{,}170$$
$$0.08x = 1{,}580$$
$$x = \frac{1{,}580}{0.08}$$
$$x = \$19{,}750$$

83. Let x = the speed of the current, and arrange the given information into a table.

	distance	rate	time
upstream	1,000	3 − x	
downstream	1,200	3 + x	

We can use the formula $d = r \cdot t$ to find an expression for each time:

$1,000 = (3 - x) \cdot$ (time upstream); time upstream = $\dfrac{1,000}{3 - x}$

$1,200 = (3 + x) \cdot$ (time downstream); time downstream = $\dfrac{1,200}{3 + x}$

The problem tells us that these two times are equal, so we set them equal to each other and solve.

$\dfrac{1,000}{3 - x} = \dfrac{1,200}{3 + x}$ LCD = $(3 + x)(3 - x)$

$(3 + x)(3 - x)\dfrac{1,000}{3 - x} = (3 + x)(3 - x)\dfrac{1,200}{3 + x}$

$(3 + x)1,000 = (3 - x)1,200$

$3,000 + 1,000x = 3,600 - 1,200x$

$2,200x = 600$

$x = \dfrac{600}{2,200} \approx 0.27$

The speed of the current is about 0.27 meters per second.

85. Let d = distance flown north, and arrange the given information in a table. The two distances are the same; the trip north is against a 30 mph wind, so the overall speed is 120 mph; the trip south is with that same wind, so the speed is 180 mph.

	distance	rate	time
north	d	120	
south	d	180	

We can use the formula $d = r \cdot t$ to find an expression for each time:

$d = 120 \cdot$ (time flying north); time flying north = $\dfrac{d}{120}$

$d = 180 \cdot$ (time flying south); time flying south = $\dfrac{d}{180}$

The two times add up to 3 hours, so we set their sum equal to 3 and solve.

$\dfrac{d}{120} + \dfrac{d}{180} = 3$ LCD = 360

$360 \cdot \dfrac{d}{120} + 360 \cdot \dfrac{d}{180} = 360 \cdot 3$

$3d + 2d = 1,080$

$5d = 1,080$

$d = \dfrac{1,080}{5} = 216$

He should fly 216 miles north.

(B) We still use the above ideas, except that rate flying north = rate flying south = 150 miles per hour.

$\dfrac{d}{150} + \dfrac{d}{150} = 3$

$\dfrac{2d}{150} = 3$

$\dfrac{d}{75} = 3$

$d = 225$ miles

87. Let D = distance from earthquake to station

Then $\dfrac{D}{5}$ = time of primary wave.

$\dfrac{D}{3}$ = time of secondary wave.

Time difference = time of *slower* secondary wave - time of *faster* primary wave

$$12 = \frac{D}{3} - \frac{D}{5}$$

$$15(12) = 15\left(\frac{D}{3}\right) - 15\left(\frac{D}{5}\right)$$

$$180 = 5D - 3D$$
$$180 = 2D$$
$$D = 90 \text{ miles}$$

> **Common Error:**
> Time difference is *not* time of fast wave (short) - time of slow wave (long)

89.

Let x = amount of distilled water

$\qquad 50$ = amount of 30% solution

Then $50 + x$ = amount of 25% solution

acid in 30% solution + acid in distilled water = acid in 25% solution

$$0.3(50) + 0 = 0.25(50 + x)$$
$$0.3(50) = 0.25(50 + x)$$
$$15 = 12.5 + 0.25x$$
$$2.5 = 0.25x$$
$$x = 10 \text{ gallons}$$

91.

Let x = amount of 50% solution

$\qquad 5$ = amount of distilled water

Then $x - 5$ = amount of 90% solution

acid in 90% solution + acid in distilled water = acid in 50% solution

$$0.9(x - 5) + 0 = 0.5x$$
$$0.9x - 4.5 = 0.5x$$
$$-4.5 = -0.4x$$
$$x = 11.25 \text{ liters}$$

93. (A) Note: The temperature increased 2.5°C for each additional 100 meters of depth, and one kilometer is ten times as far as 100 meters. So the temperature increased 25 degrees for each additional kilometer of depth.

Let x = the depth (in kilometers); then $x - 3$ = the depth beyond 3 kilometers, and $25(x - 3)$ = the temperature increase for $x - 3$ kilometers of depth.

$\qquad T$ = temperature at 3 kilometers + temperature increase beyond 3 km

$\qquad T = 30 + 25(x - 3)$

(B) We are asked to find T when $x = 15$. Plug $x = 15$ into the formula from part (A):

$\qquad T = 30 + 25(15 - 3)$
$\qquad\ \ = 330°C$

(C) We're asked to find depth (x) when $T = 280$, so we set our expression for T equal to 280 and solve.

$$280 = 30 + 25(x - 3)$$
$$280 = 30 + 25x - 75$$
$$280 = 25x - 45$$
$$325 = 25x$$
$$13 = x$$

The depth is 13 kilometers.

95. For detailed instructions on using a graphing calculator for linear regression, see Example 8 in Section 2-2 of your text.

Enter the years after 1999 (0,1,2,3,4,5, and 6) as L_1, and the Public column as L_2.

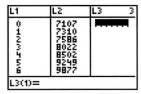  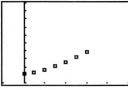

Use the STAT CALC menu to find the linear regression command:

 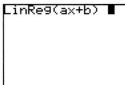 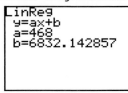 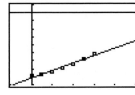

The linear model is $y = 468x + 6,832$.

The year 2008 corresponds to $x = 9$, so plug $x = 9$ into our model:

$y = 468(9) + 6,832 = 11,044$

The estimated average tuition in 2008 is \$11,044. Finally, to find when average tuition will reach \$15,000, set y equal to 15,000 and solve for x:

$15,000 = 468x + 6,832$

$8,168 = 468x$

$x = \dfrac{8,168}{468} \approx 18$

Average tuition is predicted to reach \$15,000 about 18 years after 1999, or in 2017.

97. For detailed instructions on using a graphing calculator for linear regression, see Example 8 in Section 2-2 of your text.
The independent variable is years since 1968, so enter 0, 4, 8, 12, 16, 20, 24, 28, 32, and 36 as L_1 for the years in the table. The first dependent variable is the winning time in the men's 100M freestyle, so enter the times in that column as L_2. Also enter the women's 100M freestyle times as list L_3.

 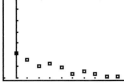 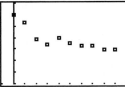

Note that the entire table of values does not fit on one screen, so the first screen shot above is a partial. The first scatter plot is the men's times; the second is the women's.

Use the STAT CALC menu to find the linear regression, first with lists L_1 and L_2, then with lists L_1 and L_3.

The line for men's times is $y = -0.103x + 51.509$. The line for women's times is $y = -0.145x + 58.286$.

To see if the models indicate that the women's times will catch up, we graph both lines and see if they ever meet. We'll need a considerably larger viewing window.

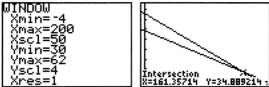

We see that the lines do meet near $x = 161$; 161 years after 1968 is 1968 + 161 = 2129, so in 2129 the times predicted by these models will be equal.
We'll leave it up to you to decide whether or not this will actually occur.
(Note: You can also find where the lines meet by setting the two equations equal to each other and solving for x.)

99. For detailed instructions on using a graphing calculator for linear regression, see Example 8 in Section 2-2 of your text.
The independent variable is price, so enter the values in the first price column as list L_1. The dependent variable is supply, so enter the supply column as list L_2.

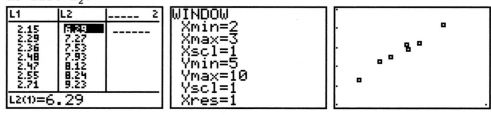

Use the STAT CALC menu to find the linear regression:

The price-supply line is $y = 4.95x - 4.22$.

Repeat with the second price column as L_1 and the demand column as L_2.

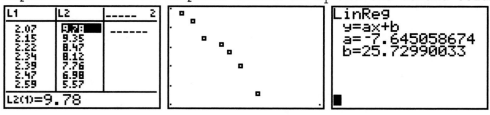

The demand-supply line is $y = -7.65x + 25.7$.

The equilibrium price is the price at which supply and demand are equal, so we graph the two lines and see where they cross.

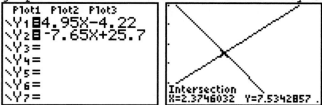

The lines meet at (2.37, 7.53), so the equilibrium price is $2.37 per bushel.

Section 2-3

1. Answers will vary.

3. Answers will vary.

5. Answers will vary.

7. Vertex: $(-3, -4)$; axis: $x = -3$

   Hand graph:

   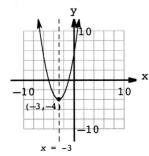

   Graphing calculator graph:

   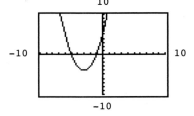

9. Vertex: $\left(\dfrac{3}{2}, -5\right)$; axis: $x = \dfrac{3}{2}$

   Hand graph:

   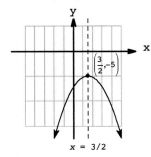

   Graphing calculator graph:

   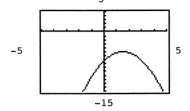

11. Vertex: $(-10, 20)$; axis $x = -10$

    Hand graph:

    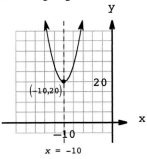

    Graphing calculator graph:

    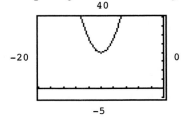

13. The graph of $y = x^2$ is shifted right 2 units and up 1 unit.

15. The graph of $y = x^2$ is shifted left 1 unit and reflected in the x axis.

17. The graph of $y = x^2$ is shifted right 2 units and down 3 units.

19. The graph of $y = x^2$ has been shifted to the right 2 units. This is the graph of $y = (x - 2)^2$, corresponding to the function k.

21. The graph of $y = x^2$ has been shifted to the right 2 units and down 3 units. This is the graph of $y = (x - 2)^2 - 3$, corresponding to the function m.

23. The graph of $y = x^2$ has been reflected in the x axis and shifted to the left 1 unit, corresponding to the function h.

25. Find the number that must be added: $\left(\dfrac{10}{2}\right)^2 = 25$

Complete the square: $x^2 + 10x + 25 = \left(x + 5\right)^2$

27. Find the number that must be added: $\left(\dfrac{-7}{2}\right)^2 = \dfrac{49}{4}$

Complete the square: $x^2 - 7x + \dfrac{49}{4} = \left(x - \dfrac{7}{2}\right)^2$

29. Find the number that must be added: $\left(\dfrac{7/5}{2}\right)^2 = \left(\dfrac{7}{10}\right)^2 = \dfrac{49}{100}$

Complete the square: $x^2 + \dfrac{7}{5}x + \dfrac{49}{100} = \left(x + \dfrac{7}{10}\right)^2$

31. Begin by grouping the first two terms with parentheses:

$$f(x) = \left(x^2 - 4x\right) + 5 \qquad \text{Find the number needed to complete the square}$$
$$= \left(x^2 - 4x + ?\right) + 5 \qquad \left(-4/2\right)^2 = 4; \text{ add and subtract } 4$$
$$= \left(x^2 - 4x + 4\right) + 5 - 4 \qquad \text{Factor parentheses, combine like terms}$$
$$= \left(x - 2\right)^2 + 1$$

The vertex form is $f(x) = \left(x - 2\right)^2 + 1$. The vertex is $(2,1)$ and the axis is $x = 2$.

33. Begin by grouping the first two terms with parentheses, then factoring -1 out of those two terms so that the coefficient of x^2 is 1:

$$h(x) = -1\left(x^2 + 2x\right) - 3 \qquad \text{Find the number needed to complete the square}$$
$$= -1\left(x^2 + 2x + ?\right) - 3 \qquad \left(2/2\right)^2 = 1; \text{ add 1 inside the parentheses}$$
$$= -1\left(x^2 + 2x + 1\right) - 3 + ? \qquad \text{We actually added -1(1), so add 1 as well}$$
$$= -1\left(x^2 + 2x + 1\right) - 3 + 1 \qquad \text{Factor parentheses, combine like terms}$$
$$= -1\left(x + 1\right)^2 - 2$$

The vertex form is $h(x) = -1\left(x + 1\right)^2 - 2$. The vertex is $(-1,-2)$ and the axis is $x = -1$.

35. Begin by grouping the first two terms with parentheses, then factoring 2 out of those two terms so that the coefficient of x^2 is 1:

$$m(x) = 2\left(x^2 - 6x\right) + 22 \qquad \text{Find the number needed to complete the square}$$
$$= 2\left(x^2 - 6x + ?\right) + 22 \qquad \left(-6/2\right)^2 = 9; \text{ add 9 inside the parentheses}$$
$$= 2\left(x^2 - 6x + 9\right) + 22 + ? \qquad \text{We actually added 2(9), so subtract 18 as well}$$
$$= 2\left(x^2 - 6x + 9\right) + 22 - 18 \qquad \text{Factor parentheses, combine like terms}$$
$$= 2\left(x - 3\right)^2 + 4$$

The vertex form is $m(x) = 2\left(x - 3\right)^2 + 4$. The vertex is $(3,4)$ and the axis is $x = 3$.

37. Begin by grouping the first two terms with parentheses, then factoring 1/2 out of those two terms so that the coefficient of x^2 is 1:

$$f(x) = \frac{1}{2}\left(x^2 + 6x\right) - \frac{7}{2} \qquad \text{Find the number needed to complete the square}$$

$$= \frac{1}{2}\left(x^2 + 6x + ?\right) - \frac{7}{2} \qquad (6/2)^2 = 9; \text{ add 9 inside the parentheses}$$

$$= \frac{1}{2}\left(x^2 + 6x + 9\right) - \frac{7}{2} + ? \qquad \text{We actually added } \frac{1}{2}(9); \text{ subtract } \frac{9}{2} \text{ as well}$$

$$= \frac{1}{2}\left(x^2 + 6x + 9\right) - \frac{7}{2} - \frac{9}{2} \qquad \text{Factor the parentheses, combine like terms}$$

$$= \frac{1}{2}\left(x + 3\right)^2 - 8$$

The vertex form is $f(x) = \frac{1}{2}\left(x + 3\right)^2 - 8$. The vertex is $(-3,-8)$ and the axis is $x = -3$.

39. Begin by grouping the first two terms with parentheses, then factoring 2 out of those two terms so that the coefficient of x^2 is 1:

$$f(x) = 2\left(x^2 - 12x\right) + 90 \qquad \text{Find the number needed to complete the square}$$

$$= 2\left(x^2 - 12x + ?\right) + 90 \qquad (-12/2)^2 = 36; \text{ add 36 inside the parentheses}$$

$$= 2\left(x^2 - 12x + 36\right) + 90 + ? \qquad \text{We actually added 2(36); subtract 72 as well}$$

$$= 2\left(x^2 - 12x + 36\right) + 90 - 72 \qquad \text{Factor the parentheses, combine like terms}$$

$$= 2\left(x - 6\right)^2 + 18$$

The vertex form is $f(x) = 2\left(x - 6\right)^2 + 18$. The vertex is $(6,18)$ and the axis is $x = 6$.

41. $x = -\dfrac{b}{2a} = -\dfrac{8}{2(1)} = -4; \quad f(-4) = (-4)^2 + 8(-4) + 8 = 16 - 32 + 8 = -8$

The vertex is $(-4,-8)$. The coefficient of x^2 is positive, so the parabola opens up. The graph is symmetric about its axis, $x = -4$. It decreases until reaching a minimum at $(-4,-8)$, then increases. The range is $[-8,\infty)$.

43. $x = -\dfrac{b}{2a} = -\dfrac{-7}{2(-1)} = -\dfrac{7}{2}$

$$f\left(-\frac{7}{2}\right) = -\left(-\frac{7}{2}\right)^2 - 7\left(-\frac{7}{2}\right) + 4 = -\frac{49}{4} + \frac{49}{2} + 4 = -\frac{49}{4} + \frac{98}{4} + \frac{16}{4} = \frac{65}{4}$$

The vertex is $\left(-\dfrac{7}{2}, \dfrac{65}{4}\right)$. The coefficient of x^2 is negative, so the parabola opens down. The graph is symmetric about its axis, $x = -7/2$. It increases until reaching a maximum at $(-7/2,65/4)$, then decreases. The range is $(-\infty,65/4]$.

45. $x = -\dfrac{b}{2a} = -\dfrac{-18}{2(4)} = \dfrac{18}{8} = \dfrac{9}{4}$

$$f\left(\frac{9}{4}\right) = 4\left(\frac{9}{4}\right)^2 - 18\left(\frac{9}{4}\right) + 25 = 4\left(\frac{81}{16}\right) - \frac{162}{4} + 25 = \frac{81}{4} - \frac{162}{4} + \frac{100}{4} = \frac{19}{4}$$

The vertex is $\left(\dfrac{9}{4}, \dfrac{19}{4}\right)$. The coefficient of x^2 is positive, so the parabola opens up. The graph is symmetric about its axis, $x = 9/4$. It decreases until reaching a minimum at $\left(\dfrac{9}{4}, \dfrac{19}{4}\right)$, then increases. The range is $[19/4,\infty)$.

47. $x = -\dfrac{b}{2a} = -\dfrac{50}{2(-10)} = \dfrac{50}{20} = \dfrac{5}{2}$

$f\left(\dfrac{5}{2}\right) = -10\left(\dfrac{5}{2}\right)^2 + 50\left(\dfrac{5}{2}\right) + 12 = -10\left(\dfrac{25}{4}\right) + 125 + 12 = -\dfrac{125}{2} + \dfrac{250}{2} + \dfrac{24}{2} = \dfrac{149}{2}$

The vertex is $\left(\dfrac{5}{2}, \dfrac{149}{2}\right)$.   The coefficient of x^2 is negative, so the parabola opens down. The graph is symmetric about its axis, $x = 5/2$.   It increases until reaching a maximum at $\left(\dfrac{5}{2}, \dfrac{149}{2}\right)$, then decreases.   The range is $(-\infty, 149/2]$.

49. The vertex of the parabola is at (4, 8). Therefore the equation must have form
$y = a(x - 4)^2 + 8$
Since the x intercept is 6, (6, 0) must satisfy the equation
$0 = a(6 - 4)^2 + 8$
$0 = 4a + 8$
$a = -2$
The equation is
$y = -2(x - 4)^2 + 8$
$y = -2(x^2 - 8x + 16) + 8$
$y = -2x^2 + 16x - 32 + 8$
$y = -2x^2 + 16x - 24$

51. The vertex of the parabola is at (-4, 12). Therefore the equation must have form
$y = a(x + 4)^2 + 12$
Since the y intercept is 4, (0, 4) must satisfy the equation.
$4 = a(0 + 4)^2 + 12$
$4 = 16a + 12$
$-8 = 16a$
$a = -0.5$
The equation is
$y = -0.5(x + 4)^2 + 12$
$y = -0.5(x^2 + 8x + 16) + 12$
$y = -0.5x^2 - 4x - 8 + 12$
$y = -0.5x^2 - 4x + 4$

53. The vertex of the parabola is at (-5, -25). Therefore the equation must have form
$y = a(x + 5)^2 - 25$
Since the parabola passes through (-2, 20), these coordinates must satisfy the equation.
$20 = a(-2 + 5)^2 - 25$
$20 = 9a - 25$
$45 = 9a$
$a = 5$
The equation is $y = 5(x + 5)^2 - 25$
$y = 5(x^2 + 10x + 25) - 25$
$y = 5x^2 + 50x + 125 - 25$
$y = 5x^2 + 50x + 100$

55. The vertex of the parabola is at (1, -4). Therefore the equation must have form
$$y = a(x - 1)^2 - 4$$
Since the parabola passes through (3, 4), these coordinates must satisfy the equation
$$4 = a(3 - 1)^2 - 4$$
$$8 = 4a$$
$$a = 2.$$
The equation is
$$y = 2(x - 1)^2 - 4$$
$$y = 2(x^2 - 2x + 1) - 4$$
$$y = 2x^2 - 4x + 2 - 4$$
$$y = 2x^2 - 4x - 2$$

57. The vertex of the parabola is at (-1, 4). Therefore the equation must have form
$$y = a(x + 1)^2 + 4$$
Since the parabola passes through (1, 2), these coordinates must satisfy the equation.
$$2 = a(1 + 1)^2 + 4$$
$$2 = 4a + 4$$
$$-2 = 4a$$
$$a = -0.5$$
 The equation is
$$y = -0.5(x + 1)^2 + 4$$
$$y = -0.5(x^2 + 2x + 1) + 4$$
$$y = -0.5x^2 - x - 0.5 + 4$$
$$y = -0.5x^2 - x + 3.5$$

59. Notice that the table does not provide the exact coordinates of the vertex, so we can't tell for certain what they are. We know that $f(-1)$ and $f(3)$ are both zero, so the axis of symmetry is halfway between $x = -1$ and $x = 3$. In other words, the x-coordinate of the vertex is 1; the equation looks like
$f(x) = a(x - 1)^2 + k$. Plug in $x = -1$:
$$f(-1) = a(-1 - 1)^2 + k = a(-2)^2 + k = 4a + k = 0 \quad \text{(since (-1, 0) is on the graph)}$$
$$4a + k = 0$$
$$k = -4a$$
Substitute $-4a$ in for k: $f(x) = a(x - 1)^2 - 4a$
Plug in $x = 0$:
$$f(0) = a(0 - 1)^2 - 4a = a - 4a = -3a = -3 \text{ (since (0, -3) is on the graph)}$$
$$-3a = -3$$
$$a = 1$$
$f(x) = (x - 1)^2 - 4$ or $f(x) = x^2 - 2x - 3$

61. Notice that the table does not provide the exact coordinates of the vertex, so we can't tell for certain what they are. We know that $f(-1)$ and $f(5)$ are equal, so the axis of symmetry is halfway between $x = -1$ and $x = 5$. In other words, the x-coordinate of the vertex is 2; the equation looks like
$$f(x) = a(x - 2)^2 + k$$
Plug in $x = -1$: $f(-1) = a(-1 - 2)^2 + k = a(-3)^2 + k = 9a + k = 0$
$$\text{(since (-1, 0) is on the graph)}$$
$$9a + k = 0$$
$$k = -9a$$
Substitute $-9a$ in for k: $f(x) = a(x - 2)^2 - 9a$
Plug in $x = 0$: $f(0) = a(0 - 2)^2 - 9a = a(-2)^2 - 9a = 4a - 9a = -5a = 2.5$
$$\text{(since (0, 2.5) is on the graph)}$$
$$-5a = 2.5$$
$$a = -0.5$$
$f(x) = 0.5(x - 2)^2 + 4.5$ or $f(x) = -0.5x^2 + 2x + 2.5$

63.
$$a(x - h)^2 + k = a\left(x^2 - 2xh + h^2\right) + k$$
$$= ax^2 - 2axh + ah^2 + k$$
$$= ax^2 - (2ah)x + \left(ah^2 + k\right)$$

65. The graphs shown are $f(x) = x^2 + 6x + 1$, $f(x) = x^2 + 1$, and $f(x) = x^2 - 4x + 1$. These correspond to $f(x) = x^2 + kx + 1$ with $k = 6, 0,$ and -4 respectively.

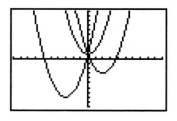

Note that all have the same shape but a different vertex. In fact, all three are translations of the graph $y = x^2$.

67. The graphs of $f(x) = (x - 1)^2$, $g(x) = (x - 1)^2 + 4$, and $h(x) = (x - 1)^2 - 5$ are shown.

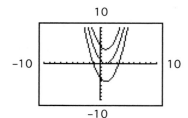

It is clear that $f(x)$ has one x intercept (at $x = 1$), $g(x)$ has no x intercepts and $h(x)$ has two x intercepts. In general, for $a > 0$, the graph of $f(x) = a(x - h)^2 + k$ can be expected to have no intercepts for $k > 0$, one intercept at $x = h$ for $k = 0$, and two intercepts for $k < 0$.

69. $f(x) = a(x - h)^2 + k$
$$f(h + r) = a(h + r - h)^2 + k = ar^2 + k$$
$$f(h - r) = a(h - r - h)^2 + k = a(-r)^2 + k = ar^2 + k$$

For any number r, the inputs $h + r$ and $h - r$ are evenly spaced on either side of h. Since the output is the same for each, we can conclude that the height of the graph is the same for each as well. This tells us that the graph is symmetric about the line $x = h$.

71. The secant line is the line through $(-1, -3)$ and $(3, 5)$. To find its equation we find its slope, then use the point-slope form of the equation of a line.

$$m = \frac{f(x_2) - f(x_1)}{x_2 - x_1} = \frac{5 - (-3)}{3 - (-1)} = \frac{8}{4} = 2$$

Graph:

$$y - f(x_1) = m(x - x_1)$$
$$y - (-3) = 2[x - (-1)]$$
$$y + 3 = 2[x + 1]$$
$$y + 3 = 2x + 2$$
$$y = 2x - 1$$

x	$y = 2x - 1$	$y = x^2 - 4$
-2	-5	0
-1	-3	-3
0	-1	-4
1	1	-3
2	3	0
3	5	5

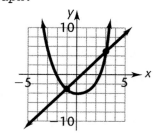

73. (A) The slope of the secant line is given by
$$m = \frac{f(2 + h) - f(2)}{(2 + h) - 2} = \frac{f(2 + h) - f(2)}{h}$$

Since $f(2 + h) = (2 + h)^2 - 3(2 + h) + 5 = 4 + 4h + h^2 - 6 - 3h + 5 = 3 + h + h^2$ and $f(2) = (2)^2 - 3(2) + 5 = 4 - 6 + 5 = 3$ we have

$$m = \frac{f(2 + h) - f(2)}{h} = \frac{(3 + h + h^2) - (3)}{h} = \frac{h + h^2}{h} = \frac{h(1 + h)}{h} = 1 + h$$

(B)

h	$slope = 1 + h$
1	2
0.1	1.1
0.01	1.01
0.001	1.001

The slope seems to be approaching 1.

75. Let one number = x. Then the other number is $x - 30$. The product is a function of x given by $f(x) = x(x - 30) = x^2 - 30x$. This is a quadratic function with $a > 0$, so it has a minimum value at the vertex of its graph (which is a parabola). Find the vertex:

$$x = -\frac{b}{2a} = -\frac{-30}{2(1)} = 15; \quad f(15) = 15^2 - 30(15) = 225 - 450 = -225$$

The minimum product is -225, when $x = 15$ and $x - 30 = -15$. There is no "highest point" on this parabola, so there's no maximum product.

77. Find the vertex, using $a = -1.2$ and $b = 62.5$:

$$x = -\frac{b}{2a} = -\frac{62.5}{2(-1.2)} \approx 26; \quad P(26) = -1.2(26)^2 + 62.5(26) - 491 = 322.8$$

The company should hire 26 employees to make a maximum profit of $322,800.

79. (A) Find the first coordinate of the vertex, using $a = -0.19$ and $b = 1.2$:

$$x = -\frac{b}{2a} = -\frac{1.2}{2(-0.19)} = 3.2$$

The maximum box office revenue was three years after 2000, which is 2003.

(B)   The function specifies yearly totals for revenue, so the domain should be restricted to whole numbers. The exact vertex occurs at $x = 3.2$, so we needed to round down to 3.

81. (A) Since the four sides needing fencing are x, y, $x + 50$, and y, we have $x + y + (x + 50) + y = 250$. Solving for y, we get $y = 100 - x$.

(Since both x and $100 - x$ must be nonnegative, the domain of $A(x)$ is $0 \le x \le 100$.)

Therefore the area
$A(x) = (x + 50)y = (x + 50)(100 - x) = -x^2 + 50x + 5000$.

(B) This is a quadratic function with $a < 0$, so it has a maximum value at the vertex:

$$x = -\frac{b}{2a} = -\frac{50}{2(-1)} = 25; \quad A(25) = -(25)^2 + 50(25) + 5,000 = -625 + 1,250 + 5,000 = 5,625$$

The maximum area is 5,625 square feet when $x = 25$.

(C) When $x = 25$, $y = 100 - 25 = 75$. The dimensions of the corral are then $x + 50$ by y, or 75 ft by 75 ft.

83. According to Example 10, the function describing the height of the sandbag is $h(t) = 10,000 - 16t^2$ (since the initial height is 10,000 feet). We want to know when it reaches ground level, so plug in zero for $h(t)$, then solve for t.

$$0 = 10,000 - 16t^2$$
$$16t^2 = 10,000$$
$$t^2 = 625$$
$$t = 25$$

It hits the ground 25 seconds after it's dropped.

85. According to Example 10, the function describing the height of the diver is $h(t) = h_0 - 16t^2$, where h_0 is the initial height in feet. (This initial height is the height of the cliff.) We know that $h(2.5) = 0$ since it takes 2.5 seconds to reach the water; we plug in 2.5 for t and 0 for $h(t)$, which allows us to solve for h_0.

$$0 = h_0 - 16(2.5)^2$$
$$0 = h_0 - 100$$
$$100 = h_0$$

The cliff is 100 feet high.

87. (A) Since $d(t)$ is a quadratic function with maximum value 484 when $t = 5.5$, an equation for $d(t)$ must be of the form

$$d(t) = a(t - 5.5)^2 + 484$$

Since $d(0) = 0$,

$$0 = d(0) = a(0 - 5.5)^2 + 484$$
$$0 = 30.25a + 484$$
$$a = -16$$

So $d(t) = -16(t - 5.5)^2 + 484$
$$= -16(t^2 - 11t + 30.25) + 484$$
$$= -16t^2 + 176t - 484 + 484$$
$$= -16t^2 + 176t$$

Since the graph of d must be symmetric with respect to $t = 5.5$, and $d(0) = 0$, $d(11)$ must also equal 0. The distance above the ground will be nonnegative only for values of t between 0 and 11, so the domain of the function is $0 \le t \le 11$.

(B) Solve $250 = -16t^2 + 176t$ by graphing $Y_1 = 250$ and $Y_2 = -16x^2 + 176x$ and applying the INTERSECT command.

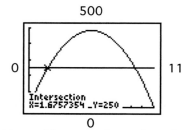 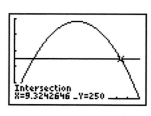

From the graphs, $t = 1.68$ sec and $t = 9.32$ sec, to two decimal places.

89. (A) If coordinates are chosen with origin at the center of the base, the parabola is the graph of a quadratic function $h(x)$ with maximum value 14 when $x = 0$. The equation must be of form

$$h(x) = ax^2 + 14$$

The parabola is 20 feet wide and symmetrical, so the point where it hits the ground must be 10 feet from the center. This tells us that $h(10) = 0$. We can use this information to find a.

$$0 = h(10) = a(10)^2 + 14$$
$$0 = 100a + 14$$
$$a = -0.14$$
$$h(x) = -0.14x^2 + 14 \qquad -10 \le x \le 10$$

(B) Suppose the truck were to drive so as to maximize its clearance, that is, in the center of the roadway. Then half its width, or 4 ft, would extend to each side. But if $x = 4$, $h(x) = -0.14(4)^2 + 14 = 11.76$ ft. The arch is only 11.76 feet high, but the truck is 12 feet high. The truck cannot pass through the arch.

(C) From part (B), if $x = 4$, $h(x) = 11.76$ ft is the height of the tallest truck.

(D) Find x so that $h(x) = 12$. Solve the equation

$$12 = 0.14x^2 + 14$$

$$-2 = -0.14x^2$$

$$x = \sqrt{\frac{2}{0.14}} = 3.78 \qquad \text{(The negative solution doesn't make sense.)}$$

The width of the truck is at most $2x = 2(3.78) = 7.56$ feet

91. (A) The independent variable (x) is number of flashlights, so we enter the numbers from the demand column as L_1. Note that demand is in thousands so 45,800 is entered as 45.8. The dependent variable (p) is price, so we enter the prices as L_2.

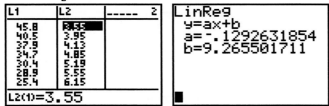

The linear regression function on a graphing calculator tells us that the linear function $a(x) = -0.129x + 9.27$ describes the price in terms of demand.

(B) To find the revenue, we multiply price times number of flashlights sold: this is $r(x) = p \cdot x = (-0.129x + 9.27)x = -0.129x^2 + 9.27x$. We use the maximum feature to find the maximum:

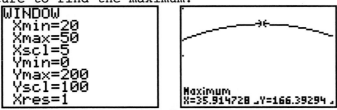

The maximum point is (35.9, 166), so maximum revenue occurs when $x = 35.9$. To find the associated price, we plug back in to our price equation:
$$d(35.9) = -0.129(35.9) + 9.27 = 4.64$$
The price that maximizes revenue is \$4.64.

Section 2-4

1. Answers will vary.

3. Answers will vary.

5. Answers will vary.

7. Imaginary, complex

9. Imaginary, pure imaginary, complex

11. Real, complex

13. $i^2 = \left(\sqrt{-1}\right)^2 = -1$

15. $\left(4i\right)^2 = 4^2 \cdot i^2 = 16(-1) = -16$

17. $\left(-i\sqrt{3}\right)^2 = (-1)^2 \cdot i^2 \cdot \sqrt{3}^2 = 1(-1)(3) = -3$

19. $\left(4i\right)(6i) = 24i^2 = 24(-1) = -24$

21. $(2 + 4i) + (5 + i) = 2 + 4i + 5 + i = 7 + 5i$

23. $(-2 + 6i) + (7 - 3i) = -2 + 6i + 7 - 3i = 5 + 3i$

25. $(6 + 7i) - (4 + 3i) = 6 + 7i - 4 - 3i = 2 + 4i$

27. $(3 + 5i) - (-2 - 4i) = 3 + 5i + 2 + 4i = 5 + 9i$

29. $(4 - 5i) + 2i = 4 - 5i + 2i = 4 - 3i$

31. $-3i(2 - 4i) = -6i + 12i^2 = -6i + 12(-1) = -12 - 6i$

33. $(3 + 3i)(2 - 3i) = 6 - 9i + 6i - 9i^2 = 6 - 3i - 9(-1) = 6 - 3i + 9 = 15 - 3i$

35. $(2 - 3i)(7 - 6i) = 14 - 12i - 21i + 18i^2 = 14 - 33i - 18 = -4 - 33i$

37. $(7 + 4i)(7 - 4i) = 49 - 28i + 28i - 16i^2 = 49 + 16 = 65$

39. $(4 - 3i)^2 = (4 - 3i)(4 - 3i) = 16 - 24i + 9i^2 = 16 - 24i - 9 = 7 - 24i$

41. $\dfrac{1}{2 + i} = \dfrac{1}{(2 + i)} \dfrac{(2 - i)}{(2 - i)} = \dfrac{2 - i}{4 - i^2} = \dfrac{2 - i}{4 + 1} = \dfrac{2 - i}{5} = \dfrac{2}{5} - \dfrac{1}{5} i$

43. $\dfrac{3 + i}{2 - 3i} = \dfrac{(3 + i)}{(2 - 3i)} \dfrac{(2 + 3i)}{(2 + 3i)} = \dfrac{6 + 11i + 3i^2}{4 - 9i^2} = \dfrac{6 + 11i - 3}{4 + 9} = \dfrac{3 + 11i}{13} = \dfrac{3}{13} + \dfrac{11}{13} i$

45. $\dfrac{13 + i}{2 - i} = \dfrac{(13 + i)}{(2 - i)} \dfrac{(2 + i)}{(2 + i)} = \dfrac{26 + 15i + i^2}{4 - i^2} = \dfrac{26 + 15i - 1}{4 + 1} = \dfrac{25 + 15i}{5} = 5 + 3i$

47. $\sqrt{2}\sqrt{8} = \sqrt{16} = 4$ **49.** $\sqrt{2}\sqrt{-8} = \sqrt{2}\cdot i\sqrt{8} = i\sqrt{16} = 4i$

51. $\sqrt{-2}\sqrt{8} = i\sqrt{2}\sqrt{8} = i\sqrt{16} = 4i$ **53.** $\sqrt{-2}\sqrt{-8} = i\sqrt{2}\cdot i\sqrt{8} = i^2\sqrt{16} = (-1)\sqrt{16} = -4$

55. $(2 - \sqrt{-4}) + (5 - \sqrt{-9}) = (2 - i\sqrt{4}) + (5 - i\sqrt{9}) = 2 - 2i + 5 - 3i = 7 - 5i$

57. $(9 - \sqrt{-9}) - (12 - \sqrt{-25}) = (9 - i\sqrt{9}) - (12 - i\sqrt{25})$

$\qquad = (9 - 3i) - (12 - 5i) = 9 - 3i - 12 + 5i = -3 + 2i$

59. $(3 - \sqrt{-4})(-2 + \sqrt{-49}) = (3 - i\sqrt{4})(-2 + i\sqrt{49})$

$\qquad = (3 - 2i)(-2 + 7i) = -6 + 25i - 14i^2 = -6 + 25i + 14$

$\qquad\qquad = 8 + 25i$

61. $\dfrac{5 - \sqrt{-4}}{7} = \dfrac{5 - i\sqrt{4}}{7} = \dfrac{5 - 2i}{7} = \dfrac{5}{7} - \dfrac{2}{7} i$

63. $\dfrac{1}{2 - \sqrt{-9}} = \dfrac{1}{2 - i\sqrt{9}} = \dfrac{1}{2 - 3i} = \dfrac{1}{(2 - 3i)} \dfrac{(2 + 3i)}{(2 + 3i)} = \dfrac{2 + 3i}{4 - 9i^2}$

$\qquad = \dfrac{2 + 3i}{4 + 9} = \dfrac{2 + 3i}{13} = \dfrac{2}{13} + \dfrac{3}{13} i$

65. $\dfrac{2}{5i} = \dfrac{2}{5i}\cdot\dfrac{i}{i} = \dfrac{2i}{5i^2} = \dfrac{2i}{-5} = -\dfrac{2}{5} i$

67. $\dfrac{1 + 3i}{2i} = \dfrac{(1 + 3i)}{(2i)}\cdot\dfrac{i}{i} = \dfrac{i + 3i^2}{2i^2} = \dfrac{i - 3}{-2} = \dfrac{3}{2} - \dfrac{1}{2} i$

69. $(2 - 3i)^2 - 2(2 - 3i) + 9 = 4 - 12i + 9i^2 - 4 + 6i + 9$

$\qquad\qquad = 4 - 12i - 9 - 4 + 6i + 9 = -6i$

71. $f(x) = x^2 - 2x + 2$

(A) $f(1 + i) = (1 + i)^2 - 2(1 + i) + 2$ $\qquad f(1 - i) = (1 - i)^2 - 2(1 - i) + 2$

$\qquad\qquad = 1 + 2i + i^2 - 2 - 2i + 2$ $\qquad\qquad = 1 - 2i + i^2 - 2 + 2i + 2$

$\qquad\qquad = 1 + 2i - 1 - 2 - 2i + 2$ $\qquad\qquad = 1 - 2i - 1 - 2 + 2i + 2$

$\qquad\qquad = 0$ $\qquad\qquad\qquad = 0$

$1 + i$ and $1 - i$ are both zeros of $f(x)$.

(B) There are no real solutions of $x^2 - 2x + 2 = 0$. The only solutions are the complex numbers $1 + i$ and $1 - i$. Therefore there are no real zeros of f. In fact, since $f(x) = (x - 1)^2 + 1$, the graph is a parabola with vertex at $(1, 1)$, and 1 is a minimum value for f. There can be no x intercepts for the graph of f.

73. $i^{18} = \underbrace{i^{16}}\cdot i^2 = (i^4)^4\cdot i^2 = 1^4(-1) = -1$

largest integer in 18 exactly divisible by 4

$i^{32} = (i^4)^8 = 1^8 = 1$

$i^{67} = \underbrace{i^{64}}\cdot i^3 = (i^4)^{16}\cdot i^2\cdot i = 1^{16}(-1)i = -i$

largest integer in 67 exactly divisible by 4

75. According to the definition of equality for complex numbers

$(2x - 1) + (3y + 2)i = 5 - 4i = 5 + (-4)i$

if and only if

$2x - 1 = 5$ and $3y + 2 = -4$

So $2x = 6$ and $3y = -6$

$x = 3 \quad y = -2$

77. $\dfrac{(1 + x) + (y - 2)i}{1 + i} = 2 - i$　　　Multiply both sides by　$1 + i$

$(1 + i)\cdot\dfrac{(1 + x) + (y - 2)i}{1 + i} = (1 + i)(2 - i)$　Simplify

$(1 + x) + (y - 2)i = 2 - i + 2i - i^2$

$(1 + x) + (y - 2)i = 2 + i - (-1)$

$(1 + x) + (y - 2)i = 2 + i + 1$

$(1 + x) + (y - 2)i = 3 + i$

According to the definition of equality for complex numbers

$(1 + x) + (y - 2)i = 3 + i$

if and only if

$1 + x = 3$　and　$y - 2 = 1$

So $x = 2$ and $y = 3$.

79. $(2 + i)z + i = 4i$

$(2 + i)z = 3i$

$z = \dfrac{3i}{2 + i}$

$z = \dfrac{3i}{(2 + i)}\dfrac{(2 - i)}{(2 - i)} = \dfrac{6i - 3i^2}{4 - i^2} = \dfrac{6i - 3(-1)}{4 + 1} = \dfrac{3 + 6i}{5} = 0.6 + 1.2i$

81. $3iz + (2 - 4i) = (1 + 2i)z - 3i$

$3iz - (1 + 2i)z = -3i - (2 - 4i)$

$3iz - z - 2iz = -3i - 2 + 4i$

$iz - z = -2 + i$

$(-1 + i)z = -2 + i$

$z = \dfrac{-2 + i}{-1 + i}$

$z = \dfrac{(-2 + i)}{(-1 + i)}\dfrac{(-1 - i)}{(-1 - i)} = \dfrac{2 + 2i - i - i^2}{1 - i^2} = \dfrac{2 + i + 1}{1 + 1} = \dfrac{3 + i}{2}$

$= 1.5 + 0.5i$

83. b is a square root of x if $b^2 = x$.
Since $(2 - i)^2 = 2^2 - 2(2)i + i^2 = 4 - 4i - 1 = 3 - 4i$, $2 - i$ is a square root of $3 - 4i$.
Similarly, since $(-2 + i)^2 = (-2)^2 + 2(-2)i + i^2 = 4 - 4i - 1 = 3 - 4i$, $-2 + i$ is also a square root of $3 - 4i$.

85. As noted in the text, some of the properties of radicals that are true for real numbers are not true for complex numbers. In particular, for positive real numbers a and b, $\sqrt{-a}\sqrt{-b} \neq \sqrt{(-a)(-b)}$, and so it is incorrect to write $\sqrt{-1}\sqrt{-1} = \sqrt{(-1)(-1)}$.

87. $(a + bi)(a - bi) = a^2 - abi + abi - b^2 i^2 = a^2 - b^2(-1) = a^2 + b^2$. This is a real number.

89. $(a + bi) + (c + di) = a + bi + c + di = a + c + bi + di = (a + c) + (b + d)i$

91. $(a + bi)(a - bi) = a^2 - b^2 i^2 = a^2 + b^2$ or $(a^2 + b^2) + 0i$

93. $(a + bi)(c + di) = ac + adi + bci + bdi^2 = ac + (ad + bc)i - bd$
$\qquad\qquad\qquad = (ac - bd) + (ad + bc)i$

95. $i^{4k} = (i^4)^k = (i^2 \cdot i^2)^k = [(-1)(-1)]^k = 1^k = 1$

97. $S_1 = i$
$S_2 = i + i^2 = -1 + i$
$S_3 = i + i^2 + i^3 = -1 + i + (-i) = -1$
$S_4 = i + i^2 + i^3 + i^4 = -1 + 1 = 0$
Then $S_5 = S_4 + i^5 = S_4 + i = S_4 + S_1 = S_1$
$\qquad S_6 = S_5 + i^6 = S_1 + i^2 = S_2$
$\qquad S_7 = S_6 + i^7 = S_2 + i^3 = S_3$
$\qquad S_8 = S_7 + i^8 = S_3 + i^4 = S_4$
Generalizing, the only possible values for S_n are $S_1 = i$, $S_2 = -1 + i$, $S_3 = -1$, and $S_4 = 0$.

99. 1. Definition of addition
2. Commutative property for addition of real numbers.
3. Definition of addition (read from right to left).

Section 2-5

1. Answers will vary.   **3.** Answers will vary.

5. Answers will vary.

7. $(x - 8)(2x + 3) = 0$
$x - 8 = 0 \qquad$ or $\qquad 2x + 3 = 0$
$\qquad x = 8 \qquad$ or $\qquad\qquad 2x = -3$
$\qquad\qquad\qquad\qquad\qquad\qquad x = -\dfrac{3}{2}$

9. $x^2 - 3x - 10 = 0$
$(x - 5)(x + 2) = 0$
$x - 5 = 0 \qquad$ or $\qquad x + 2 = 0$
$\qquad x = 5 \qquad\qquad\qquad\qquad x = -2$

11.
$$4u^2 = 8u$$
$$4u^2 - 8u = 0$$
$$4u(u - 2) = 0$$
$$4u = 0 \text{ or } u - 2 = 0$$
$$u = 0 \qquad u = 2$$

13.
$$9y^2 = 12y - 4$$
$$9y^2 - 12y + 4 = 0$$
$$(3y - 2)^2 = 0$$
$$3y - 2 = 0$$
$$3y = 2$$
$$y = \frac{2}{3} \text{ (double root)}$$

15.
$$11x = 2x^2 + 12$$
$$0 = 2x^2 - 11x + 12$$
$$2x^2 - 11x + 12 = 0$$
$$(2x - 3)(x - 4) = 0$$
$$2x - 3 = 0 \text{ or } x - 4 = 0$$
$$2x = 3 \text{ or } x = 4$$
$$x = \frac{3}{2}$$

17. $x^2 - 6x - 3 = 0$

$$x^2 - 6x = 3 \qquad \left(-\frac{6}{2}\right)^2 = 9$$
$$x^2 - 6x + 9 = 3 + 9 \qquad \text{(Add 9 to each side)}$$
$$(x - 3)^2 = 12 \qquad \text{(Factor the left side)}$$
$$x - 3 = \pm\sqrt{12}$$
$$x = 3 \pm 2\sqrt{3} \qquad (\sqrt{12} = \sqrt{4}\sqrt{3} = 2\sqrt{3})$$

19. $t^2 - 4t + 8 = 0$

$$t^2 - 4t = -8 \qquad \left(-\frac{4}{2}\right)^2 = 4$$
$$t^2 - 4t + 4 = -8 + 4 \qquad \text{(Add 4 to each side)}$$
$$(t - 2)^2 = -4 \qquad \text{(Factor the left side)}$$
$$t - 2 = \pm\sqrt{-4} \qquad (\sqrt{-4} = i\sqrt{4} = 2i)$$
$$t - 2 = \pm 2i$$
$$t = 2 \pm 2i$$

21. $m^2 + 2m + 9 = 0$

$$m^2 + 2m = -9 \qquad \left(\frac{2}{2}\right)^2 = 1$$
$$m^2 + 2m + 1 = -9 + 1 \qquad \text{(Add 1 to each side)}$$
$$(m + 1)^2 = -8 \qquad \text{(Factor the left side)}$$
$$m + 1 = \pm\sqrt{-8}$$
$$m + 1 = \pm 2\sqrt{2}\,i \qquad \sqrt{-8} = i\sqrt{8} = i\sqrt{4}\sqrt{2} = 2\sqrt{2}\,i$$
$$m = -1 \pm 2\sqrt{2}\,i$$

23. $2d^2 + 5d - 25 = 0$ $\qquad$ (Divide both sides by 2)

$$d^2 + \frac{5}{2}d - \frac{25}{2} = 0 \qquad \left(\text{Add } \frac{25}{2} \text{ to each side}\right)$$

$$d^2 + \frac{5}{2}d = \frac{25}{2} \qquad \left(\frac{1}{2} \cdot \frac{5}{2}\right)^2 = \left(\frac{5}{4}\right)^2 = \frac{25}{16}$$

$$d^2 + \frac{5}{2}d + \frac{25}{16} = \frac{25}{2} + \frac{25}{16} \qquad \left(\text{Add } \frac{25}{16} \text{ to each side}\right)$$

$$\left(d + \frac{5}{4}\right)^2 = \frac{25}{2} + \frac{25}{16} \qquad \text{(Factor the left side)}$$

$$\left(d + \frac{5}{4}\right)^2 = \frac{200}{16} + \frac{25}{16} = \frac{225}{16}$$

$$d + \frac{5}{4} = \pm\sqrt{\frac{225}{16}}$$

$$d + \frac{5}{4} = \pm\frac{15}{4}$$

$$d = -\frac{5}{4} \pm \frac{15}{4}$$

$$d = \frac{-5 + 15}{4} \quad \text{or} \quad d = \frac{-5 - 15}{4}$$

$$d = \frac{10}{4} = \frac{5}{2} \quad \text{or} \quad d = -\frac{20}{4} = -5$$

25. $\quad 2v^2 - 2v + 1 = 0 \qquad$ (Divide both sides by 2)

$\quad\ v^2 - v + \frac{1}{2} = 0 \qquad$ (Subtract ½ from each side)

$\qquad\quad v^2 - v = -\frac{1}{2} \qquad\qquad \left(-\frac{1}{2}\right)^2 = \frac{1}{4}$

$\quad v^2 - v + \frac{1}{4} = -\frac{1}{2} + \frac{1}{4} \qquad$ (Add $\frac{1}{4}$ to each side)

$\qquad \left(v - \frac{1}{2}\right)^2 = -\frac{1}{2} + \frac{1}{4} \qquad$ (Factor the left side)

$\qquad \left(v - \frac{1}{2}\right)^2 = -\frac{2}{4} + \frac{1}{4} = -\frac{1}{4}$

$\qquad\quad v - \frac{1}{2} = \pm\sqrt{-\frac{1}{4}}$

$\qquad\quad v - \frac{1}{2} = \pm\frac{1}{2}i \qquad\qquad \sqrt{-\frac{1}{4}} = i\sqrt{\frac{1}{4}} = \frac{1}{2}i$

$\qquad\qquad\quad v = \frac{1}{2} \pm \frac{1}{2}i$

27. $\quad 4y^2 + 3y + 9 = 0 \qquad$ (Divide both sides by 4)

$\quad\ y^2 + \frac{3}{4}y + \frac{9}{4} = 0 \qquad$ (Subtract $\frac{9}{4}$ from each side)

$\qquad\quad y^2 + \frac{3}{4}y = -\frac{9}{4} \qquad\qquad \left(\frac{1}{2} \cdot \frac{3}{4}\right)^2 = \left(\frac{3}{8}\right)^2 = \frac{9}{64}$

$\quad y^2 + \frac{3}{4}y + \frac{9}{64} = -\frac{9}{4} + \frac{9}{64} \qquad$ (Add $\frac{9}{64}$ to each side)

$\qquad \left(y + \frac{3}{8}\right)^2 = -\frac{9}{4} + \frac{9}{64} \qquad$ (Factor the left side)

$\qquad \left(y + \frac{3}{8}\right)^2 = -\frac{144}{64} + \frac{9}{64} = -\frac{135}{64}$

$\qquad\quad y + \frac{3}{8} = \pm\sqrt{-\frac{135}{64}}$

$\qquad\quad y + \frac{3}{8} = \pm\frac{3i\sqrt{15}}{8} \qquad\qquad \sqrt{-\frac{135}{64}} = i\frac{\sqrt{135}}{\sqrt{64}} = \frac{i\sqrt{9}\sqrt{15}}{8} = \frac{3i\sqrt{15}}{8}$

$\qquad\qquad\quad y = -\frac{3}{8} \pm \frac{3\sqrt{15}}{8}i$

29. $x^2 - 10x - 3 = 0$

$$x = \frac{-b \pm \sqrt{b^2 - 4ac}}{2a} \qquad a = 1, \ b = -10, \ c = -3$$

$$x = \frac{-(-10) \pm \sqrt{(-10)^2 - 4(1)(-3)}}{2(1)}$$

$$x = \frac{10 \pm \sqrt{112}}{2} = \frac{10 \pm 4\sqrt{7}}{2}$$

$$= 5 \pm 2\sqrt{7}$$

Common Error:
It is incorrect to "cancel" this way:
$$\frac{\cancel{10} \pm \sqrt{112}}{\cancel{2}} \neq 5 \pm \sqrt{112}$$

31. $\qquad x^2 + 8 = 4x$
$x^2 - 4x + 8 = 0$

$$x = \frac{-b \pm \sqrt{b^2 - 4ac}}{2a} \qquad a = 1, \ b = -4, \ c = 8$$

$$x = \frac{-(-4) \pm \sqrt{(-4)^2 - 4(1)(8)}}{2(1)} = \frac{4 \pm \sqrt{-16}}{2}$$

$$x = \frac{4 \pm i\sqrt{16}}{2}$$

$$x = \frac{4 \pm 4i}{2}$$

$$x = 2 \pm 2i$$

33. $\qquad 2x^2 + 1 = 4x$
$2x^2 - 4x + 1 = 0$

$$x = \frac{-b \pm \sqrt{b^2 - 4ac}}{2a} \qquad a = 2, \ b = -4, \ c = 1$$

$$= \frac{-(-4) \pm \sqrt{(-4)^2 - 4(2)(1)}}{2(2)}$$

$$x = \frac{4 \pm \sqrt{8}}{4}$$

$$x = \frac{4 \pm 2\sqrt{2}}{4}$$

$$x = \frac{2 \pm \sqrt{2}}{2}$$

Common Errors:
$$\frac{2 \pm \sqrt{2}}{2} \neq \pm\sqrt{2}$$
$$\neq 1 \pm \sqrt{2}$$
These involve incorrect "cancelling".

35. $\qquad 5x^2 + 2 = 2x$
$5x^2 - 2x + 2 = 0$

$$x = \frac{-b \pm \sqrt{b^2 - 4ac}}{2a} \qquad a = 5, \ b = -2, \ c = 2$$

$$x = \frac{-(-2) \pm \sqrt{(-2)^2 - 4(5)(2)}}{2(5)}$$

$$x = \frac{2 \pm \sqrt{-36}}{10}$$

$$x = \frac{2 \pm 6i}{10}$$

$$x = \frac{1}{5} \pm \frac{3}{5}i$$

37. $4\left(x^2 + 2x\right) - 4 = 0$

$$4x^2 + 8x - 4 = 0$$

$$x = \frac{-b \pm \sqrt{b^2 - 4ac}}{2a} \qquad a = 4,\ b = 8,\ c = -4$$

$$= \frac{-8 \pm \sqrt{8^2 - 4(4)(-4)}}{2(4)}$$

$$x = \frac{-8 \pm \sqrt{128}}{8}$$

$$= \frac{-8 \pm 8\sqrt{2}}{8}$$

$$= -1 \pm \sqrt{2}$$

(Note: If you begin by dividing both sides by 4, the calculations are simpler)

39. $2.4x^2 + 6.4x - 4.3 = 0 \qquad a = 2.4,\ b = 6.4,\ c = -4.3$
$b^2 - 4ac = 6.4^2 - 4(2.4)(-4.3) = 82.24$
Since the discriminant is positive, there are two real zeros.

41. $6.5x^2 - 7.4x + 3.4 = 0 \qquad a = 6.5,\ b = -7.4,\ c = 3.4$
$b^2 - 4ac = (-7.4)^2 - 4(6.5)(3.4) = -33.64$
Since the discriminant is negative, there are two imaginary zeros.

43. $0.3x^2 + 3.6x + 10.8 = 0 \qquad a = 0.3,\ b = 3.6,\ c = 10.8$
$b^2 - 4ac = 3.6^2 - 4(0.3)(10.8) = 0$
Since the discriminant is zero, there is one real zero.

45. The graph is below:

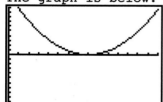

It looks like there's only one zero, but to be sure we find the minimum value:

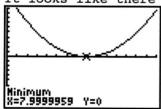

The minimum is at height zero so there is one real zero.

47. The graph clearly intersects the *x*-axis twice, so there are two real zeros.

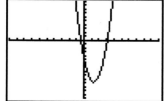

49. It appears that the graph intersects the *x*-axis once:

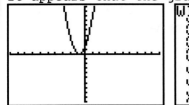

To be sure, we find the minimum value:

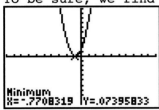

And it's a good thing we did! The minimum height is positive so the graph does not intersect the *x*-axis and there are two imaginary zeros.

51. $x^2 - 6x - 3 = 0$ The left side doesn't factor, so use the quadratic formula.

$$x = \frac{-b \pm \sqrt{b^2 - 4ac}}{2a} \qquad a = 1, \; b = -6, \; c = -3$$

$$= \frac{6 \pm \sqrt{(-6)^2 - 4(1)(-3)}}{2(1)}$$

$$= \frac{6 \pm \sqrt{48}}{2}$$

$$= \frac{6 \pm 4\sqrt{3}}{2}$$

$$= 3 \pm 2\sqrt{3}$$

Check: $3 + 2\sqrt{3} \approx 6.46$
$\qquad\quad\; 3 - 2\sqrt{3} \approx -0.46$

Graph $y = x^2 - 6x - 3$ and find the intercepts using the ZERO command.

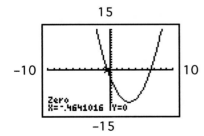
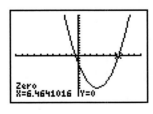

53. $2y^2 - 6y + 3 = 0$ The left side doesn't factor, so use the quadratic formula.

$$y = \frac{-b \pm \sqrt{b^2 - 4ac}}{2a} \qquad a = 2, \; b = -6, \; c = 3$$

$$= \frac{6 \pm \sqrt{(-6)^2 - 4(2)(3)}}{2(2)}$$

$$= \frac{6 \pm \sqrt{12}}{4}$$

$$= \frac{6 \pm 2\sqrt{3}}{4}$$

$$= \frac{3}{2} \pm \frac{1}{2}\sqrt{3}$$

Check: $\dfrac{3}{2} + \dfrac{1}{2}\sqrt{3} \approx 2.366$

$\dfrac{3}{2} - \dfrac{1}{2}\sqrt{3} \approx 0.634$

Graph $y = 2x^2 - 6x + 3$ and find the intercepts using the ZERO command.

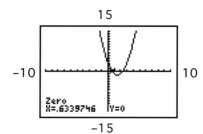 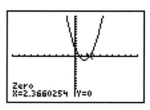

55. $3x^2 - 2x - 2 = 0$ The left side doesn't factor, so use the quadratic formula.

$x = \dfrac{-b \pm \sqrt{b^2 - 4ac}}{2a}$ $a = 3,\ b = -2,\ c = -2$

$= \dfrac{2 \pm \sqrt{(-2)^2 - 4(3)(-2)}}{2(3)}$

$= \dfrac{2 \pm \sqrt{28}}{6}$

$= \dfrac{2 \pm 2\sqrt{7}}{6}$

$= \dfrac{1 \pm \sqrt{7}}{3}$

Check: $\dfrac{1 + \sqrt{7}}{3} \approx 1.215$

$\dfrac{1 - \sqrt{7}}{3} \approx -0.549$

Graph $y = 3x^2 - 2x - 2$ and find the intercepts using the ZERO command.

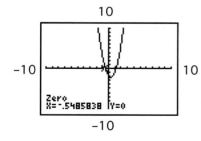 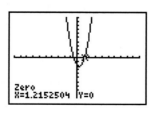

57. $12x^2 + 7x = 10$

$12x^2 + 7x - 10 = 0$

$(4x + 5)(3x - 2) = 0$ Polynomial is factorable.

$4x + 5 = 0$ or $3x - 2 = 0$

$4x = -5$ $3x = 2$

$x = -\dfrac{5}{4}$ $x = \dfrac{2}{3}$

$-\dfrac{5}{4} = -1.25$ $\dfrac{2}{3} \approx 0.67$

Graph $y = 12x^2 + 7x - 10$ and find the intercepts using the ZERO command.

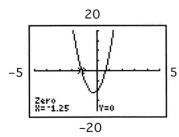

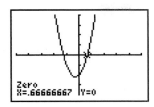

59.
$$x^2 = 3x + 1$$
$$x^2 - 3x - 1 = 0 \quad \text{Polynomial is not factorable, use quadratic formula.}$$

$$x = \frac{-b \pm \sqrt{b^2 - 4ac}}{2a} \quad \begin{array}{l} a = 1, \\ b = -3, \\ c = -1 \end{array}$$

$$x = \frac{-(-3) \pm \sqrt{(-3)^2 - 4(1)(-1)}}{2(1)}$$

$$x = \frac{3 \pm \sqrt{13}}{2}$$

Check: $\dfrac{3 + \sqrt{13}}{2} \approx 3.30$

$\dfrac{3 - \sqrt{13}}{2} \approx -0.30$

Graph $y = x^2 - 3x - 1$ and find the intercepts using the ZERO command.

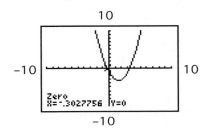

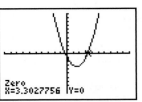

61. $s = \dfrac{1}{2} gt^2$

$\dfrac{1}{2} gt^2 = s$

$gt^2 = 2s$

$t^2 = \dfrac{2s}{g}$

$t = \sqrt{\dfrac{2s}{g}}$

63.
$$P = EI - RI^2$$
$$RI^2 - EI + P = 0$$

$$I = \frac{-b \pm \sqrt{b^2 - 4ac}}{2a} \quad a = R, \ b = -E, \ c = P$$

$$I = \frac{-(-E) \pm \sqrt{(-E)^2 - 4(R)(P)}}{2(R)}$$

$$I = \frac{E + \sqrt{E^2 - 4RP}}{2R} \quad \text{(positive square root)}$$

65. The quadratic formula is a good choice:

$x^2 - \sqrt{7}\,x + 2 = 0$

$$x = \frac{-b \pm \sqrt{b^2 - 4ac}}{2a} \quad a = 1, \ b = -\sqrt{7}, \ c = 2$$

$$x = \frac{-(-\sqrt{7}) \pm \sqrt{(-\sqrt{7})^2 - 4(1)(2)}}{2(1)}$$

$$x = \frac{\sqrt{7} \pm \sqrt{7 - 8}}{2} = \frac{\sqrt{7} \pm \sqrt{-1}}{2} = \frac{\sqrt{7} \pm i}{2}$$

The graph can be used to see that there are two imaginary zeros but not to confirm the exact answer we got.

67. The quadratic formula is a good choice:

$$x^2 - 2\sqrt{3}\,x + 3 = 0$$

$$x = \frac{-b \pm \sqrt{b^2 - 4ac}}{2a} \qquad a = 1,\ b = -2\sqrt{3},\ c = 3$$

$$x = \frac{-(-2\sqrt{3}) \pm \sqrt{(-2\sqrt{3})^2 - 4(1)(3)}}{2(1)}$$

$$x = \frac{2\sqrt{3} \pm \sqrt{12 - 12}}{2} = \frac{2\sqrt{3} \pm 0}{2} = \sqrt{3}$$

We check graphically by graphing the function $y = x^2 - 2\sqrt{3}\,x + 3$ and using the ZERO command:

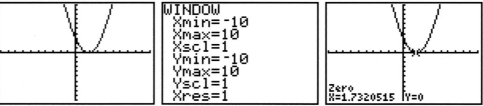

The only zero is at $x = 1.7320515$ which is easily confirmed to be $\sqrt{3}$.

69. The quadratic formula is a good choice:

$$x^2 + \sqrt{3}\,x - 4 = 0$$

$$x = \frac{-b \pm \sqrt{b^2 - 4ac}}{2a} \qquad a = 1,\ b = \sqrt{3},\ c = -4$$

$$x = \frac{-\sqrt{3} \pm \sqrt{(\sqrt{3})^2 - 4(1)(-4)}}{2(1)}$$

$$x = \frac{-\sqrt{3} \pm \sqrt{3 + 16}}{2}$$

$$x = \frac{-\sqrt{3} \pm \sqrt{19}}{2}$$

We check graphically by graphing the function $y = x^2 + \sqrt{3}\,x - 4$ and using the ZERO command.

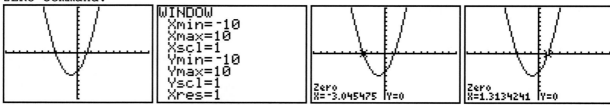

The two zeros are at $x = -3.045$ and $x = 1.313$.

$$\frac{-\sqrt{3} + \sqrt{19}}{2} \approx 1.313; \quad \frac{-\sqrt{3} - \sqrt{19}}{2} \approx -3.045$$

71. Begin by multiplying both sides by the least common denominator, x^2.

$$x^2 \cdot 1 + x^2 \cdot \frac{9}{x^2} = x^2 \cdot \frac{5}{x}$$

$$x^2 + 9 = 5x$$

$$x^2 - 5x + 9 = 0$$

$$x = \frac{-b \pm \sqrt{b^2 - 4ac}}{2a} \qquad a = 1,\ b = -5,\ c = 9$$

$$x = \frac{-(-5) \pm \sqrt{(-5)^2 - 4(1)(9)}}{2(1)} = \frac{5 \pm \sqrt{25 - 36}}{2} = \frac{5 \pm \sqrt{-11}}{2} = \frac{5 \pm i\sqrt{11}}{2}$$

The graph can be used to see that there are two imaginary zeros but not to confirm the exact answer we got.

73. Begin by multiplying both sides by the least common denominator, x^2.

$$x^2 \cdot 1 + x^2 \cdot \frac{9}{x^2} = x^2 \cdot \frac{6}{x}$$

$$x^2 + 9 = 6x$$

$$x^2 - 6x + 9 = 0 \quad \text{(The left side factors)}$$

$$(x - 3)^2 = 0$$

$$x - 3 = 0$$

$$x = 3$$

We check graphically by graphing the function $y = 1 + \dfrac{9}{x^2} - \dfrac{6}{x}$ and using the ZERO command:

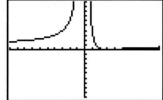 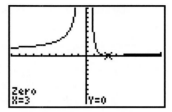

The only solution is $x = 3$.

75. Begin by multiplying both sides by the least common denominator, x^2.

$$x^2 \cdot 1 + x^2 \cdot \frac{9}{x^2} = x^2 \cdot \frac{7}{x}$$

$$x^2 + 9 = 7x$$

$$x^2 - 7x + 9 = 0 \quad \text{(Doesn't factor)}$$

$$x = \frac{-b \pm \sqrt{b^2 - 4ac}}{2a} \qquad a = 1, \ b = -7, \ c = 9$$

$$x = \frac{-(-7) \pm \sqrt{(-7)^2 - 4(1)(9)}}{2(1)} = \frac{7 \pm \sqrt{49 - 36}}{2} = \frac{7 \pm \sqrt{13}}{2}$$

We check graphically by graphing the functions $y_1 = 1 + \dfrac{9}{x^2}$ and $y_2 = \dfrac{7}{x}$ then using the INTERSECT command.

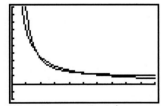  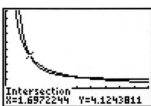 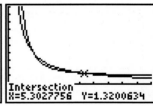

The two solutions are $x = 1.697$ and $x = 5.303$.

$$\frac{7 + \sqrt{13}}{2} = 5.303, \quad \frac{7 - \sqrt{13}}{2} = 1.697$$

77. We begin by multiplying both sides by the least common denominator which is $(x + 4)(x - 4)$.

$$(x + 4)(x - 4) \cdot 3 + (x + 4)(x - 4)\frac{5}{x - 4} = (x + 4)(x - 4)\frac{7}{x + 4}$$

$$3(x^2 - 16) + 5(x + 4) = 7(x - 4)$$

$$3x^2 - 48 + 5x + 20 = 7x - 28$$

$$3x^2 + 5x - 28 = 7x - 28$$

$$3x^2 - 2x = 0 \quad \text{(The left side factors.)}$$

$$x(3x - 2) = 0$$

$$x = 0 \quad \text{or} \quad 3x - 2 = 0$$

$$3x = 2$$

$$x = \frac{2}{3}$$

We check graphically by graphing the functions $y_1 = 3 + \dfrac{5}{x - 4}$ and $y_2 = \dfrac{7}{x + 4}$ and using the INTERSECT command:

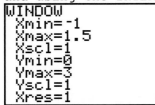 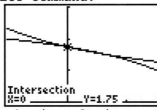 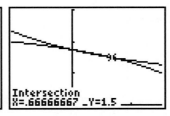

This confirms our algebraic solution.

79. We begin by multiplying both sides by the least common denominator which is $(x - 5)(x + 5)$.

$$(x - 5)(x + 5) \cdot \frac{8}{x - 5} = (x - 5)(x + 5)\frac{3}{x + 5} - 2(x - 5)(x + 5)$$

$$8(x + 5) = 3(x - 5) - 2(x^2 - 25)$$

$$8x + 40 = 3x - 15 - 2x^2 + 50$$

$$8x + 40 = -2x^2 + 3x + 35$$

$$2x^2 + 5x + 5 = 0 \qquad \text{(Doesn't factor)}$$

$$x = \frac{-b \pm \sqrt{b^2 - 4ac}}{2a} \qquad a = 2,\ b = 5,\ c = 5$$

$$x = \frac{-5 \pm \sqrt{5^2 - 4(2)(5)}}{4}$$

$$x = \frac{-5 \pm \sqrt{25 - 40}}{4}$$

$$x = \frac{-5 \pm \sqrt{-15}}{4}$$

$$x = \frac{-5 \pm i\sqrt{15}}{4}$$

The graph can be used to see that there are two imaginary zeros but not to confirm the exact answer we got.

81. In this problem, $a = 1$, $b = 4$, $c = c$. The discriminant is
$b^2 - 4ac = (4)^2 - 4(1)(c) = 16 - 4c$.
If $16 - 4c > 0$, $16 > 4c$ or $c < 4$, then there are two distinct real roots.
If $16 - 4c = 0$, $c = 4$, then there is one real double root,
and if $16 - 4c < 0$, $16 < 4c$ or $c > 4$, then there are two distinct imaginary roots.

83. $x^2 + 3ix - 2 = 0$

$$x = \frac{-b \pm \sqrt{b^2 - 4ac}}{2a} \qquad a = 1,\ b = 3i,\ c = -2$$

$$x = \frac{-3i \pm \sqrt{(3i)^2 - 4(1)(-2)}}{2(1)}$$

$$x = \frac{-3i \pm \sqrt{-9 + 8}}{2}$$

$$x = \frac{-3i \pm \sqrt{-1}}{2}$$

$$x = \frac{-3i \pm i}{2}$$

$$x = \frac{-3i + i}{2} \quad \text{or} \quad x = \frac{-3i - i}{2}$$

$$x = -i \qquad \text{or} \quad x = -2i$$

85. $x^2 + 2ix = 3$
$x^2 + 2ix - 3 = 0$

$x = \dfrac{-b \pm \sqrt{b^2 - 4ac}}{2a}$ $a = 1,\ b = 2i,\ c = -3$

$x = \dfrac{-2i \pm \sqrt{(2i)^2 - 4(1)(-3)}}{2(1)}$

$x = \dfrac{-2i \pm \sqrt{-4 + 12}}{2}$

$x = \dfrac{-2i \pm \sqrt{8}}{2}$

$x = \dfrac{-2i \pm 2\sqrt{2}}{2}$

$x = \dfrac{2(-i \pm \sqrt{2})}{2}$

$x = -i \pm \sqrt{2}$
$x = \sqrt{2} - i,\ -\sqrt{2} - i$

87. $x^3 - 1 = 0$
$(x - 1)(x^2 + x + 1) = 0$
$x - 1 = 0$ or $x^2 + x + 1 = 0$

$x = 1$ $\qquad x = \dfrac{-b \pm \sqrt{b^2 - 4ac}}{2a}$ $a = 1,\ b = 1,\ c = 1$

$\qquad\qquad x = \dfrac{-1 \pm \sqrt{(1)^2 - 4(1)(1)}}{2(1)}$

$\qquad\qquad x = \dfrac{-1 \pm \sqrt{1 - 4}}{2}$

$\qquad\qquad x = \dfrac{-1 \pm \sqrt{-3}}{2}$

$\qquad\qquad x = \dfrac{-1 \pm i\sqrt{3}}{2}$ or $-\dfrac{1}{2} \pm \dfrac{1}{2}i\sqrt{3}$

89. If a quadratic equation has two roots, they are $\dfrac{-b + \sqrt{b^2 - 4ac}}{2a}$ and $\dfrac{-b - \sqrt{b^2 - 4ac}}{2a}$. If $a,\ b,\ c$ are rational, then so are $-b$, $2a$, and $b^2 - 4ac$. There are then three cases: if $\sqrt{b^2 - 4ac}$ is rational, then $\dfrac{-b + \sqrt{b^2 - 4ac}}{2a}$ and $\dfrac{-b - \sqrt{b^2 - 4ac}}{2a}$ are both rational. If $\sqrt{b^2 - 4ac}$ is irrational, then $\dfrac{-b + \sqrt{b^2 - 4ac}}{2a}$ and $\dfrac{-b - \sqrt{b^2 - 4ac}}{2a}$ are both irrational. If $\sqrt{b^2 - 4ac}$ is imaginary, then $\dfrac{-b + \sqrt{b^2 - 4ac}}{2a}$ and $\dfrac{-b + \sqrt{b^2 - 4ac}}{2a}$ are both imaginary. There is no other possibility. We conclude that one root cannot be rational while the other is irrational.

91. $\quad r_1 \; = \; \dfrac{-b + \sqrt{b^2 - 4ac}}{2a} \qquad r_2 \; = \; \dfrac{-b - \sqrt{b^2 - 4ac}}{2a}$

$r_1 r_2 \; = \; \dfrac{(-b + \sqrt{b^2 - 4ac})}{2a} \; \dfrac{(-b - \sqrt{b^2 - 4ac})}{2a}$

$\qquad = \; \dfrac{(-b)^2 - (\sqrt{b^2 - 4ac})^2}{4a^2} \; = \; \dfrac{b^2 - (b^2 - 4ac)}{4a^2} \; = \; \dfrac{b^2 - b^2 + 4ac}{4a^2} \; = \; \dfrac{4ac}{4a^2} \; = \; \dfrac{c}{a}$

93. The ± in front still yields the same two numbers even if a is negative.

95. Let x = one number.
Since their sum is 21, x + the
other number = 21, and the other
number = 21 − x.
Then, since their product is 104,
$$x(21 - x) = 104$$
$$21x - x^2 = 104$$
$$0 = x^2 - 21x + 104$$
$$x^2 - 21x + 104 = 0$$
$$(x - 13)(x - 8) = 0$$
$$x - 13 = 0 \quad \text{or} \quad x - 8 = 0$$
$$x = 13 \qquad\qquad x = 8$$
The numbers are 8 and 13.

97. Let x = first of the two consecutive
even integers.
Then $x + 2$ = second of these integers
Since their product is 168,
$$x(x + 2) = 168$$
$$x^2 + 2x = 168$$
$$x^2 + 2x - 168 = 0$$
$$(x - 12)(x + 14) = 0$$
$$x - 12 = 0 \quad \text{or} \quad x + 14 = 0$$
$$x = 12 \qquad\qquad x = -14$$
If $x = 12$, the two consecutive
positive even integers must be 12 and
14. We discard the other solution,
since the numbers must be positive.

99. Sketch a figure.

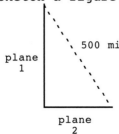

Let t = time since 6:00 AM (in hours)
Then (distance of plane 1) = (rate of plane 1) × (time flown by plane 1)
$$= 200\,t$$
(distance of plane 2) = (rate of plane 2) × (time flown by plane 2)
$$= 170(t - 0.5)$$

By the Pythagorean Theorem, the planes will be 500
miles apart, and no longer able to communicate with
one another, when
$$(200t)^2 + [170(t - 0.5)]^2 = 500^2$$
$$40,000t^2 + 28,900(t - 0.5)^2 = 250,000$$
$$40,000t^2 + 28,900(t^2 - t + 0.25) = 250,000$$
$$40,000t^2 + 28,900t^2 - 28,900t + 7,225 = 250,000$$
$$68,900t^2 - 28,900t - 242,775 = 0$$

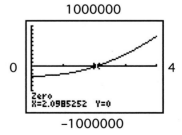

Graphing $y = 68,900x^2 - 28,900x - 242,775$ and applying the ZERO command, we get
solution $t = 2.098$ hrs = 126 min, or 2 hrs, 6 min.
The planes will no longer be able to communicate at 8:06 AM.

101. From the diagram, the dimensions of the planting area are 30 - 2x and 20 - 2x.
Then   Area = 400 = (30 - 2x)(20 - 2x)

$$400 = 600 - 100x + 4x^2$$
$$0 = 4x^2 - 100x + 200$$
$$0 = x^2 - 25x + 50$$

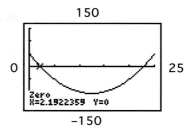

Graphing $y = x^2 - 25x + 50$ and
applying the ZERO command, we get
$x = 2.19$.

The walkway should be 2.19 ft wide.

(There is another intercept for the graph, but here $x > 20$, so this solution
does not make sense.)

103. Begin with a sketch of the garden.

The area is given by $\ell \cdot w$, and
we're told that the area is 1,200,
so we get the equation
$\ell \cdot w = 1,200$.

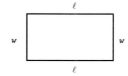

We need to eliminate one of the variables. The total amount of fence is 150,
so adding up the lengths of the sides, we get
$w + \ell + w + \ell = 150$
$2w + 2\ell = 150$
Solve for w and substitute the result in for w in the above equation.
$2w = 150 - 2\ell$
$w = 75 - \ell$
The equation becomes
$\ell(75 - \ell) = 1,200$
$75\ell - \ell^2 = 1,200$
$-\ell^2 + 75\ell - 1,200 = 0$
Graphing $y = -\ell^2 + 75\ell - 1,200$ and applying the ZERO command, we get solutions
$\ell = 23.1$ and 51.9.

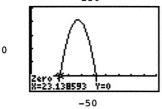
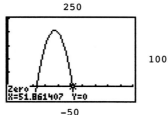

If ℓ is 23.1, then $w = 75 - 23.1 = 51.9$, so there is actually only one solution:
23.1 feet by 51.9 feet.

105. (A) Label the text figure as follows:

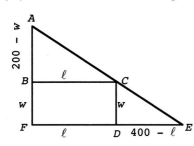

Then, using the hint, in the similar triangles
ABC and *AFE*, we have
$$\frac{\ell}{200 - w} = \frac{400}{200}$$
$$\frac{\ell}{200 - w} = 2$$
$$\ell = 2(200 - w)$$
$$\ell = 400 - 2w$$
Then $A(w) = \ell w = (400 - 2w)w = 400w - 2w^2$.

Since $w \geq 0$ and $\ell = 400 - 2w \geq 0$, the domain of
the function is $0 \leq w \leq 200$.

(B) Solve $150,000 \le 400w - 2w^2$
$$2w^2 - 400w + 15,000 \le 0$$
$$w^2 - 200w + 7,500 \le 0$$
$$(w - 50)(w - 150) \le 0$$

The solution of this inequality consists of all values of w for which the graph of $y = w^2 - 200w + 7,500$ is on or below the w axis. The w intercepts of the graph are 50 and 150.

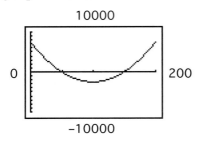

Examining the graph, we see that the solution of the inequality is $50 \le w \le 150$.

(C) The cross-sectional area was found in part (A) to be
$$A(w) = 400w - 2w^2$$
Graphing this and applying the MAXIMUM command, we obtain the graph at the right.

The maximum cross-sectional area of such a building is 20,000 square feet when $w = 100$ ft. A building with area 25,000 square feet is impossible.

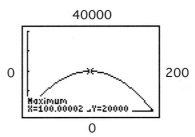

107. Let x = distance from the warehouse to Factory A. Since the distance from the warehouse to Factory B via Factory A is known (it is the difference in odometer readings: 52937—52846) to be 91 miles, then
    $91 - x$ = distance from Factory A to Factory B.
The distance from Factory B to the warehouse is known (it is the difference in odometer readings: 53002—52937) to be 65 miles. Applying the Pythagorean theorem, we have

$$x^2 + (91 - x)^2 = 65^2$$
$$x^2 + 8281 - 182x + x^2 = 4225$$
$$2x^2 - 182x + 4056 = 0$$
$$x^2 - 91x + 2028 = 0$$
$$(x - 52)(x - 39) = 0$$
$$x - 52 = 0 \quad \text{or} \quad x - 39 = 0$$
$$x = 52 \text{ mi} \qquad x = 39 \text{ mi}$$

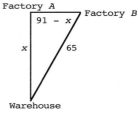

Since we are told that the distance from the warehouse to Factory A was greater than the distance from Factory A to Factory B, we discard the solution $x = 39$, which would lead to $91 - x = 52$ miles, a contradiction.
    52 miles.

109. (A) The independent variable is years since 1960, so we enter 0, 5, 10, 15, 20, 25, 30, 35, and 40 as L_1. The dependent variable is beer consumption, so we enter the values in the beer column as L_2.

(Note that the whole list is not visible on this figure.)
Next we choose the quadratic regression command from the STAT CALC menu.

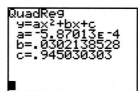

The quadratic regression model is $y = -0.000\,587x^2 + 0.0302x + 0.945$.

(B) We need to use our quadratic regression model and find x when $y = 0.99$. We graph $y_1 = -0.000\,587x^2 + 0.0302x + 0.945$ and $y_2 = 0.99$, then use the INTERSECT command:

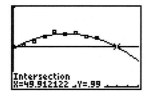

The intersection occurs at $x = 49.91$: we round to 50 and find that beer consumption will return to 1960 levels 50 years after 1960, or the year 2010.

(C) Use the VALUE command to find that in 2005, when $x = 45$, predicted beer consumption is 1.16 gallons.

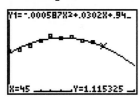

111. (A) The independent variable is years since 1950, so we enter list L_1 with zero for 1950, 5 for 1955, and increment by 5 until reaching 50 for 2000. The dependent variable is production, so we enter the production column as L_2.

(Note that the whole list is not visible on this figure.)

Next we choose the quadratic regression command from the STAT CALC menu

 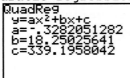

The quadratic regression model is $y = -0.328x^2 + 18.3x + 339$.

(B) We need to use our quadratic regression model and find x when $y = 370$. We graph $y_1 = -0.328x^2 + 18.3x + 339$ and $y_2 = 370$ then use the INTERSECT command:

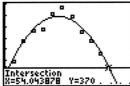 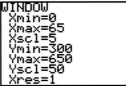

The intersection occurs at $x = 54.04$; we round to 54 and conclude that production will return to 1950 levels 54 years after 1950, or 2004.

(C) Use the VALUE command to find that in 2005, when $x = 55$, production is predicted to be 353 billion.

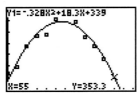

113. The independent variable is speed, so we enter the data in the speed column as L_1. The dependent variable is the skid mark length for Auto A, so we enter the length in the second column L_2.

Next we choose the quadratic regression command from the STAT CALC menu:

The quadratic regression model is $y = 0.0476x^2 + 0.0452x - 0.357$.

(B) We need to use our quadratic regression model and find x when $y = 200$. We graph $y_1 = 0.0476x^2 + 0.0452x - 0.357$ and $y_2 = 200$ then use the INTERSECT command:

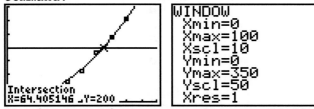

The intersection occurs when $x = 64.405$. We round to 64 and conclude that Auto A was traveling 64 mph.

115. (A) The independent variable is speed of Boat A, so we enter the first column as L_1. The dependent variable is the associated mileage, so we enter the second column as L_2.

Next we use the quadratic regression command from the STAT CALC menu:

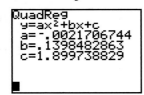

The quadratic regression model is $y = -0.00217x^2 + 0.140x + 1.90$.

(B) The rental cost is \$10 per hour plus \$2.30 per gallon for gas. This can be expressed as $C = 10t + 2.3g$ where t is the number of hours and g is the number of gallons. We can write each of t and g in terms of the speed (x) to get a model for cost in terms of speed. The time of the trip will be the distance divided by the speed since distance = speed · time. The trip is 100 miles, so $t = \dfrac{100}{x}$. The gasoline usage can be calculated by dividing the length (100 miles) by the miles per gallon (y), so $g = \dfrac{100}{y}$. But our quadratic model tells us that $y = -0.00217x^2 + 0.140x + 1.90$, so

$$g = \frac{100}{-0.00217x^2 + 0.140x + 1.90}.$$

Our rental cost model is now

$$C = 10\left(\frac{100}{x}\right) + 2.3\left(\frac{100}{-0.00217x^2 + 0.140x + 1.90}\right)$$

$$C = \frac{1000}{x} + \frac{230}{-0.00217x^2 + 0.140x + 1.90}$$

We can use the MINIMUM command to find when this is minimized:

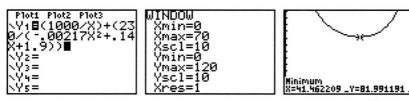

The minimum cost occurs when the speed is 41.5 mph. Plugging $x = 41.5$ back into our quadratic regression model y, we find that the mileage at 41.5 mph is 3.97 mpg. We can use our expressions for t, g, and C to answer the remaining questions.

$$t = \frac{100}{x} = \frac{100}{41.5} = 2.41 \text{ hours}$$

$$g = \frac{100}{y} = \frac{100}{3.97} = 25.2 \text{ gallons}$$

$$C = 10t + 2.3g = 10(2.41) + 2.3(25.2) = 24.1 + 57.96 = \$82.06$$

Exercise 2-6

1. This statement is true since $\sqrt{5}$ and $-\sqrt{5}$ are the only two solutions of the equation $x^2 = 5$.

3. This statement is false. It is not correct to "distribute" exponents. The correct way to simplify the left side is $(2x - 1)^2 = (2x - 1)(2x - 1) = 4x^2 - 4x + 1$.

5. This statement is false.
   $(\sqrt{x - 1} + 1)^2$ is not equal to $x - 1 + 1$ or x, in general.
   In fact
   $(\sqrt{x - 1} + 1)^2 = x - 1 + 2\sqrt{x - 1} + 1 = x + 2\sqrt{x - 1}$.
   This is only equal to x when $x = 1$.

7. This statement is false. If $x^3 = 2$ then $x = \sqrt[3]{2}$ or x is equal to one of two non-real complex numbers whose cube is 2. If $x = 8$, x^3 must equal 512.

9. This is of quadratic type since the power of the first term is twice the power of the second term. With $u = x^{-3}$, the equation becomes
   $2u^2 - 4u = 0$

11. This is not of quadratic type: the power of the first term is 3 times the power of the second term.

13. This is of quadratic type since the power of the middle term is half the power of the third term. With the substitution $u = \frac{1}{x^2}$, the equation becomes
    $\frac{10}{9} + 4u - 7u^2 = 0$

15. Answers will vary.   17. Answers will vary.

19. $\sqrt{4x - 7} = 5$
    $4x - 7 = 25$
    $4x = 32$
    $x = 8$

    Check:

    $\sqrt{4(8) - 7} = \sqrt{32 - 7} = \sqrt{25} = 5$ √

    Solution: $x = 8$

Graphical check:

Graph $y_1 = \sqrt{4x - 7}$ and $y_2 = 5$ and use the INTERSECT command:

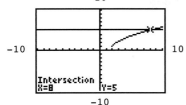

21. $\sqrt{5x + 6} + 6 = 0$

$\qquad \sqrt{5x + 6} = -6$

The left side represents the positive square root of some number and can't be negative, so there is no solution. If we hadn't noticed this, we would have continued:

$5x + 6 = 36$
$\quad\ 5x = 30$
$\quad\ \ x = 6$

Check: $\sqrt{5(6) + 6} + 6 = \sqrt{36} + 6 = 12 \neq 0$

There is no solution.

Graphical check:

Graph $y_1 = \sqrt{5x + 6} + 6$:

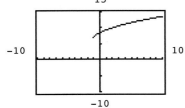

The graph is never at height zero, so there is no solution.

23. $\sqrt[3]{x + 5} = 3$
$\quad\ x + 5 = 27$
$\qquad\ \ x = 22$
Check:
$\sqrt[3]{22 + 5} = \sqrt[3]{27} = 3 \quad \sqrt{}$

Solution: $x = 22$

Graphical check: Graph $y_1 = \sqrt[3]{x + 5}$ and $y_2 = 3$ and use the INTERSECT command to find their intersection.

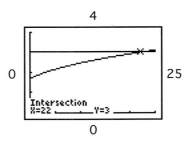

25. $\sqrt{x + 5} + 7 = 0$

$\sqrt{x + 5} = -7$

The left side represents the positive square root of some number and can't be negative, so there is no solution. If we hadn't noticed this, we would have continued:

$x + 5 = 49$

$x = 44$

Check: $\sqrt{44 + 5} + 7 = \sqrt{49} + 7 = 14 \neq 0$

There is no solution.

Graphical check: Graph $y_1 = \sqrt{x + 5} + 7$:

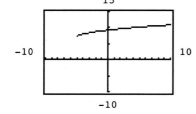

The graph is never at height zero, so there is no solution.

27. $y^4 - 2y^2 - 8 = 0$

Let $u = y^2$, then

$u^2 - 2u - 8 = 0$

$(u - 4)(u + 2) = 0$

$u = 4, -2$

$y^2 = 4 \qquad y^2 = -2$

$y = \pm 2 \qquad y = \pm i\sqrt{2}$

Graphical check: Graph $y = x^4 - 2x^2 - 8$ and use the ZERO command to find the x intercepts.

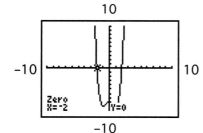 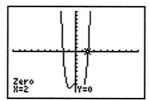

29. $3x = \sqrt{x^2 - 2}$

$9x^2 = x^2 - 2$

$8x^2 = -2$

$x^2 = -\dfrac{1}{4}$

$x = \pm\sqrt{-\dfrac{1}{4}} = \pm\dfrac{1}{2}i$

Since the solutions are imaginary, we can't check graphically.

31. $\sqrt{x^2 - 5x} = \sqrt{x - 8}$

$x^2 - 5x = x - 8$

$x^2 - 6x + 8 = 0$

$(x - 4)(x - 2) = 0$

$x = 4, 2$

Check:

$\sqrt{4^2 - 5(4)} = \sqrt{-4}; \quad \sqrt{4 - 8} = \sqrt{-4} \quad \checkmark$

$\sqrt{2^2 - 5(2)} = \sqrt{-6}; \quad \sqrt{2 - 8} = \sqrt{-6} \quad \checkmark$

Solution: $x = 4, 2$

Both solutions result in imaginary values for the equation, so we can't check graphically.

33. $\sqrt{5n + 9} = n - 1$

$5n + 9 = n^2 - 2n + 1$

$0 = n^2 - 7n - 8$

$n^2 - 7n - 8 = 0$

$(n - 8)(n + 1) = 0$

$n = 8, -1$

Check:

$\sqrt{5(8) + 9} = \sqrt{49} = 7; \quad 8 - 1 = 7 \quad \checkmark$

$\sqrt{5(-1) + 9} = \sqrt{4} = 2; \quad -1 - 1 = -2 \quad$ Not a solution

Solution: $n = 8$

Graphical check:
Graph $y_1 = \sqrt{5x + 9}$ and $y_2 = x - 1$ and use the INTERSECT command to find the intersection.

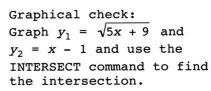

35. $\sqrt{3x + 4} = 2 + \sqrt{x}$

$3x + 4 = 4 + 4\sqrt{x} + x$

$2x = 4\sqrt{x}$

$x = 2\sqrt{x}$

$x^2 = 4x$

$x^2 - 4x = 0$

$x(x - 4) = 0$

$x = 0, 4$

Check:

$\sqrt{3(0) + 4} = \sqrt{44} = 2; \quad 2 + \sqrt{0} = 4 \quad \checkmark$

$\sqrt{3(4) + 4} = \sqrt{16} = 4; \quad 2 + \sqrt{4} = 4 \quad \checkmark$

Solution: $x = 0, 4$

Graphical check: Graph $y_1 = \sqrt{3x + 4}$ and $y_2 = 2 + \sqrt{x}$ and use the INTERSECT command:

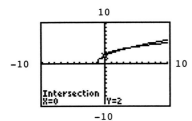 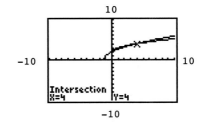

37. $2x^{2/3} + 3x^{1/3} - 2 = 0$
Let $u = x^{1/3}$, then
$$2u^2 + 3u - 2 = 0$$
$$(2u - 1)(u + 2) = 0$$
$$u = \frac{1}{2}, \ -2$$

$x^{1/3} = \dfrac{1}{2}$ ⠀⠀ $x^{1/3} = -2$

$x = \dfrac{1}{8}$ ⠀⠀⠀ $x = -8$

Graphical check: Graph $y = 2x^{2/3} + 3x^{1/3} - 2$ and use the INTERSECT command to find the x intercepts.

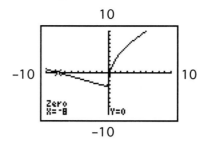
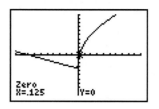

39. $(m^2 - m)^2 - 4(m^2 - m) = 12$
Let $u = m^2 - m$, then
$$u^2 - 4u = 12$$
$$u^2 - 4u - 12 = 0$$
$$(u - 6)(u + 2) = 0$$
$$u = 6, \ -2$$

$m^2 - m = 6$ ⠀⠀⠀⠀ $m^2 - m = -2$
$m^2 - m - 6 = 0$ ⠀ $m^2 - m + 2 = 0$

$(m - 3)(m + 2) = 0$ ⠀⠀⠀ $m = \dfrac{-(-1) \pm \sqrt{(-1)^2 - 4(1)(2)}}{2(1)}$

$m = 3, \ -2$ ⠀⠀⠀⠀⠀ $m = \dfrac{1 \pm \sqrt{-7}}{2}$ or $\dfrac{1 \pm i\sqrt{7}}{2}$

⠀⠀⠀⠀⠀⠀⠀⠀⠀ $m = \dfrac{1}{2} \pm \dfrac{\sqrt{7}}{2}i$

Graphical check: Graph $y1 = (x^2 - x)^2 - 4(x^2 - x)$ and $y2 = 12$ and use the INTERSECT command to find the intersections.

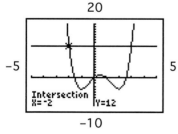
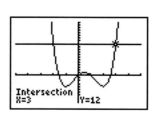

41.
$$\sqrt{u - 2} = 2 + \sqrt{2u + 3}$$
$$u - 2 = 4 + 4\sqrt{2u + 3} + 2u + 3$$
$$u - 2 = 2u + 7 + 4\sqrt{2u + 3}$$
$$-u - 9 = 4\sqrt{2u + 3}$$
$$u^2 + 18u + 81 = 16(2u + 3)$$
$$u^2 + 18u + 81 = 32u + 48$$
$$u^2 - 14u + 33 = 0$$
$$(u - 3)(u - 11) = 0$$
$$u = 3, \ 11$$
Check: ⠀ $\sqrt{3 - 2} \overset{?}{=} 2 + \sqrt{2(3) + 3}$
⠀⠀⠀⠀⠀⠀ $1 \neq 5$
⠀⠀⠀ $\sqrt{11 - 2} \overset{?}{=} 2 + \sqrt{2(11) + 3}$
⠀⠀⠀⠀⠀⠀ $3 \neq 7$
No solution.

> **Common Error:**
> $u - 2 = 4 + 2u + 3$ is not an equivalent equation to the given equation.
> $(2 + \sqrt{2u + 3})^2 \neq 4 + 2u + 3$

Graphical check: Graph $y_1 = \sqrt{x - 2}$ and $y_2 = 2 + \sqrt{2x + 3}$:

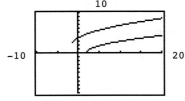

The graphs do not intersect.

43. $\sqrt{3y - 2} = 3 - \sqrt{3y + 1}$

$3y - 2 = 9 - 6\sqrt{3y + 1} + 3y + 1$

$3y - 2 = 3y + 10 - 6\sqrt{3y + 1}$

$-12 = -6\sqrt{3y + 1}$

$2 = \sqrt{3y + 1}$

$4 = 3y + 1$

$3 = 3y$

$y = 1$

Check: $\sqrt{3(1) - 2} \overset{?}{=} 3 - \sqrt{3(1) + 1}$

$1 \overset{\checkmark}{=} 1$

Solution: $y = 1$

Graphical check: Graph $y1 = \sqrt{3x - 2}$ and $y2 = 3 - \sqrt{3x + 1}$ and use the INTERSECT command to find the intersection.

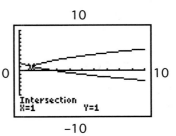

45. $\sqrt{7x - 2} - \sqrt{x + 1} = \sqrt{3}$

$\sqrt{7x - 2} = \sqrt{x + 1} + \sqrt{3}$

$7x - 2 = x + 1 + 2\sqrt{3}\sqrt{x + 1} + 3$

$7x - 2 = x + 4 + 2\sqrt{3(x + 1)}$

$6x - 6 = 2\sqrt{3(x + 1)}$

$3x - 3 = \sqrt{3x + 3}$

$9x^2 - 18x + 9 = 3x + 3$

$9x^2 - 21x + 6 = 0$

$3x^2 - 7x + 2 = 0$

$(3x - 1)(x - 2) = 0$

$x = \frac{1}{3}, \ 2$

Check: $\sqrt{7\left(\frac{1}{3}\right) - 2} - \sqrt{\frac{1}{3} + 1} \overset{?}{=} \sqrt{3}$

$\sqrt{\frac{1}{3}} - \sqrt{\frac{4}{3}} \overset{?}{=} \sqrt{3}$

$-\frac{1}{3}\sqrt{3} \neq \sqrt{3}$

$\sqrt{7(2) - 2} - \sqrt{2 + 1} \overset{?}{=} \sqrt{3}$

$\sqrt{12} - \sqrt{3} \overset{?}{=} \sqrt{3}$

$\sqrt{3} \overset{\checkmark}{=} \sqrt{3}$

Solution: $x = 2$

Graphical check:
Graph $y1 = \sqrt{7x - 2} - \sqrt{x + 1}$ and $y2 = \sqrt{3}$ and use the INTERSECT command to find the intersection.

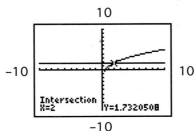

47. $\sqrt{4x^2 + 12x + 1} - 6x = 9$

$\sqrt{4x^2 + 12x + 1} = 6x + 9$

$4x^2 + 12x + 1 = 36x^2 + 108x + 81$

$0 = 32x^2 + 96x + 80$

$0 = 2x^2 + 6x + 5$

$x = \dfrac{-b \pm \sqrt{b^2 - 4ac}}{2a}$ $\quad a = 2 \quad b = 6 \quad c = 5$

$x = \dfrac{-6 \pm \sqrt{6^2 - 4(2)(5)}}{2(2)}$

$x = \dfrac{-6 \pm \sqrt{36 - 40}}{4}$

$x = \dfrac{-6 \pm \sqrt{-4}}{4}$

$x = \dfrac{-6 \pm 2i}{4}$

$x = -\dfrac{3}{2} \pm \dfrac{1}{2}i$

Check: $-\dfrac{3}{2} + \dfrac{1}{2}i$: $\sqrt{4\left(-\dfrac{3}{2} + \dfrac{1}{2}i\right)^2 + 12\left(-\dfrac{3}{2} + \dfrac{1}{2}i\right) + 1} - 6\left(-\dfrac{3}{2} + \dfrac{1}{2}i\right) \overset{?}{=} 9$

$\sqrt{(-3 + i)^2 - 18 + 6i + 1} + 9 - 3i \overset{?}{=} 9$

$\sqrt{9 - 6i + i^2 - 18 + 6i + 1} + 9 - 3i \overset{?}{=} 9$

$\sqrt{9 - 6i - 1 - 18 + 6i + 1} + 9 - 3i \overset{?}{=} 9$

$\sqrt{-9} + 9 - 3i \overset{?}{=} 9$

$3i + 9 - 3i \overset{\checkmark}{=} 9$

A solution

$-\dfrac{3}{2} - \dfrac{1}{2}i$: $\sqrt{4\left(-\dfrac{3}{2} - \dfrac{1}{2}i\right)^2 + 12\left(-\dfrac{3}{2} - \dfrac{1}{2}i\right) + 1} - 6\left(-\dfrac{3}{2} - \dfrac{1}{2}i\right) \overset{?}{=} 9$

$\sqrt{(-3 - i)^2 - 18 - 6i + 1} + 9 + 3i \overset{?}{=} 9$

$\sqrt{9 + 6i + i^2 - 18 - 6i + 1} + 9 + 3i \overset{?}{=} 9$

$\sqrt{9 + 6i - 1 - 18 - 6i + 1} + 9 + 3i \overset{?}{=} 9$

$\sqrt{-9} + 9 + 3i \overset{?}{=} 9$

$3i + 9 + 3i \overset{?}{=} 9$

$9 + 6i \neq 9$ Not a solution

Solution: $x = -\dfrac{3}{2} + \dfrac{1}{2}i$

Since the solutions are complex, we can't check graphically.

49. $3n^{-2} - 11n^{-1} - 20 = 0$
Let $u = n^{-1}$, then
$3u^2 - 11u - 20 = 0$
$(3u + 4)(u - 5) = 0$

$u = -\dfrac{4}{3},\ 5$

$n^{-1} = -\dfrac{4}{3} \qquad n^{-1} = 5$

$n = -\dfrac{3}{4} \qquad n = \dfrac{1}{5}$

Graphical check: Graph $y = 3x^{-2} - 11x^{-1} - 20$ and use the ZERO command to find the zeros.

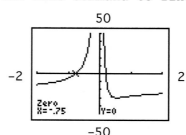

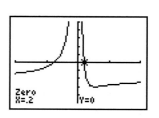

51. $9y^{-4} - 10y^{-2} + 1 = 0$

Let $u = y^{-2}$, then

$9u^2 - 10u + 1 = 0$

$(9u - 1)(u - 1) = 0$

$u = \dfrac{1}{9}, \ 1$

$y^{-2} = \dfrac{1}{9} \qquad y^{-2} = 1$

$y^2 = 9 \qquad\quad y^2 = 1$

$y = \pm 3 \qquad\ \ y = \pm 1$

Graphical check: Graph $y = 9x^{-4} - 10x^{-2} + 1$ and use the ZERO command to find the zeros. (Only the negative zeros are shown here.)

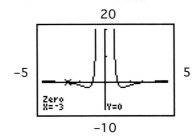

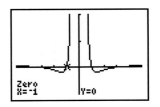

53. $y^{1/2} - 3y^{1/4} + 2 = 0$

Let $u = y^{1/4}$, then

$u^2 - 3u + 2 = 0$

$(u - 1)(u - 2) = 0$

$u = 1, \ 2$

$y^{1/4} = 1 \qquad y^{1/4} = 2$

$y = 1 \qquad\quad y = 16$

Common Error:
$y \neq 2^{1/4}$
Both sides must be raised to the fourth power to eliminate the $y^{1/4}$.

Graphical check: Graph $y = x^{1/2} - 3x^{1/4} + 2$ and use the ZERO command to find the zeros.

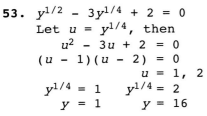

55. $(m - 5)^4 + 36 = 13(m - 5)^2$

Let $u = (m - 5)^2$, then

$u^2 + 36 = 13u$

$u^2 - 13u + 36 = 0$

$(u - 4)(u - 9) = 0$

$u = 4, \ 9$

$(m - 5)^2 = 4 \qquad\quad (m - 5)^2 = 9$

$m - 5 = \pm 2 \qquad\ \ m - 5 = \pm 3$

$m = 5 \pm 2 \qquad\quad m = 5 \pm 3$

$m = 3, \ 7 \qquad\qquad m = 2, \ 8$

Graphical check: Graph $y1 = (x - 5)^4 + 36$ and $y2 = 13(x - 5)^2$ and use the INTERSECT command to find the intersections.
(Only $x = 2$ and $x = 3$ are shown here.)

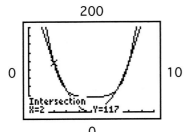

57. The first step was an attempt to square both sides, but it was done incorrectly. The square of the expression $\sqrt{x + 3} + 5$ is NOT $x + 3 + 25$. To solve the equation correctly, you should begin by isolating the root, THEN squaring both sides.

59.
$$\sqrt{5 - 2x} - \sqrt{x + 6} = \sqrt{x + 3}$$
$$\sqrt{5 - 2x} = \sqrt{x + 6} + \sqrt{x + 3}$$
$$5 - 2x = x + 6 + 2\sqrt{x + 6}\sqrt{x + 3} + x + 3$$
$$5 - 2x = 2x + 9 + 2\sqrt{x + 6}\sqrt{x + 3}$$
$$-4x - 4 = 2\sqrt{x + 6}\sqrt{x + 3}$$
$$-2x - 2 = \sqrt{(x + 6)(x + 3)}$$
$$4x^2 + 8x + 4 = (x + 6)(x + 3)$$
$$4x^2 + 8x + 4 = x^2 + 9x + 18$$
$$3x^2 - x - 14 = 0$$
$$(3x - 7)(x + 2) = 0$$
$$x = \frac{7}{3}, \; -2$$

Check:
$$\sqrt{5 - 2\left(\frac{7}{3}\right)} - \sqrt{\frac{7}{3} + 6} \overset{?}{=} \sqrt{\frac{7}{3} + 3}$$
$$\sqrt{\frac{1}{3}} - \sqrt{\frac{25}{3}} \overset{?}{=} \sqrt{\frac{16}{3}}$$
$$-\frac{4\sqrt{3}}{3} \neq \frac{4\sqrt{3}}{3}$$

$$\sqrt{5 - 2(-2)} - \sqrt{-2 + 6} \overset{?}{=} \sqrt{-2 + 3}$$
$$1 \overset{\checkmark}{=} 1$$

Solution: $x = -2$

Graphical check:
Graph $y1 = \sqrt{5 - 2x} - \sqrt{x + 6}$ and $y2 = \sqrt{x + 3}$ and use the INTERSECT command to find the intersection.

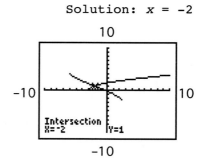

61.
$$2 + 3y^{-4} = 6y^{-2} \qquad \text{Multiply both sides by } y^4 \quad (y \neq 0)$$
$$2y^4 + 3 = 6y^2$$
$$2y^4 - 6y^2 + 3 = 0$$

Let $u = y^2$, then
$$2u^2 - 6u + 3 = 0$$

$$u = \frac{-b \pm \sqrt{b^2 - 4ac}}{2a} \qquad a = 2, \; b = -6, \; c = 3$$

$$u = \frac{-(-6) \pm \sqrt{(-6)^2 - 4(2)(3)}}{2(2)}$$

$$u = \frac{6 \pm \sqrt{36 - 24}}{4} = \frac{6 \pm \sqrt{12}}{4} = \frac{2(3 \pm \sqrt{3})}{4} = \frac{3 \pm \sqrt{3}}{2}$$

$$y^2 = \frac{3 \pm \sqrt{3}}{2}$$

$$y = \pm\sqrt{\frac{3 \pm \sqrt{3}}{2}} \quad \text{(four roots)}$$

Note:
$$\sqrt{\frac{3 + \sqrt{3}}{2}} \approx 1.538$$

$$\sqrt{\frac{3 - \sqrt{3}}{2}} \approx 0.7962$$

Graphical check:
Graph: $y1 = 2 + 3x^{-4}$ and
$y2 = 6x^{-2}$ and use the
INTERSECT command to find
the intersections. (Only
the positive roots are
shown here.)

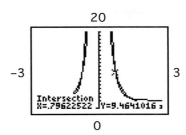
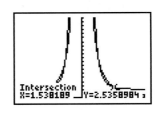

63. By squaring: $m - 7\sqrt{m} + 12 = 0$

$$m + 12 = 7\sqrt{m}$$
$$m^2 + 24m + 144 = 49m$$
$$m^2 - 25m + 144 = 0$$
$$(m - 9)(m - 16) = 0$$
$$m = 9, 16$$

Check: $9 - 7\sqrt{9} + 12 \overset{?}{=} 0$ $\qquad$ $16 - 7\sqrt{16} + 12 \overset{?}{=} 0$
$\qquad\qquad\qquad\quad 0 \overset{\checkmark}{=} 0$ $\qquad\qquad\qquad\qquad 0 \overset{\checkmark}{=} 0$

Solution: $m = 9, 16$

By substitution: $m - 7\sqrt{m} + 12 = 0$
Let $u = \sqrt{m}$, then
$$u^2 - 7u + 12 = 0$$
$$(u - 4)(u - 3) = 0$$
$$u = 3, 4$$
$$\sqrt{m} = 3 \qquad\qquad \sqrt{m} = 4$$
$$m = 9 \qquad\qquad\quad m = 16$$

These answers have already been checked.

Graphical check:
Graph: $y_1 = x - 7\sqrt{x} + 12$ and use the ZERO command.

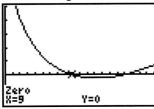
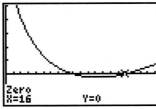
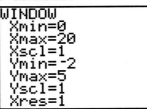

65. By squaring: $t - 11\sqrt{t} + 18 = 0$
$$t + 18 = 11\sqrt{t}$$
$$t^2 + 36t + 324 = 121t$$
$$t^2 - 85t + 324 = 0$$
$$(t - 4)(t - 81) = 0$$
$$t = 4, 81$$

Check: $4 - 11\sqrt{4} + 18 \overset{?}{=} 0$
$\qquad\qquad\qquad 0 \overset{\checkmark}{=} 0$

$\qquad 81 - 11\sqrt{81} + 18 \overset{?}{=} 0$
$\qquad\qquad\qquad 0 \overset{\checkmark}{=} 0$

Solution: $t = 4, 81$

By substitution: $t - 11\sqrt{t} + 18 = 0$
Let $u = \sqrt{t}$, then
$$u^2 - 11u + 18 = 0$$
$$(u - 9)(u - 2) = 0$$
$$u = 2, \ 9$$

$u = 2$	$u = 9$
$\sqrt{t} = 2$	$\sqrt{t} = 9$
$t = 4$	$t = 81$

These answers have already been checked.

Graphical check:

Graph: $y_1 = x - 11\sqrt{x} + 18$ and use the ZERO command.

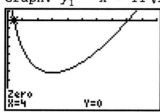 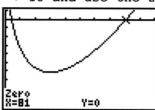 

67. Algebraically
$$2\sqrt{x + 5} = 0.01x + 2.04$$
$$200\sqrt{x + 5} = x + 204$$
$$40,000(x + 5) = (x + 204)^2$$
$$40,000x + 200,000 = x^2 + 408x + 41,616$$
$$0 = x^2 - 39,592x - 158,384$$

Although this is factorable it's really hard because of the size of the coefficients, so the quadratic formula is used

$$x = \frac{-b \pm \sqrt{b^2 - 4ac}}{2a} \qquad a = 1$$
$$b = -39,592$$
$$c = -158,384$$

$$x = \frac{-(-39,592) \pm \sqrt{(-39,592)^2 - 4(1)(-158,384)}}{2(1)}$$

$$x = \frac{39,592 \pm 39,600}{2}$$

$$x = -4, \ 39,596$$

Check: $2\sqrt{-4 + 5} \overset{?}{=} 0.01(-4) + 2.04$
$$2 \overset{\checkmark}{=} 2$$
$$2\sqrt{39,596 + 5} \overset{?}{=} 0.01(39,596) + 2.04$$
$$398 \overset{\checkmark}{=} 398$$

Solution: $x = -4, \ 39,596$

Graphically:
Graph $y1 = 2\sqrt{x + 5}$ and $y2 = 0.01x + 2.04$ and use the INTERSECT command to find the intersections.

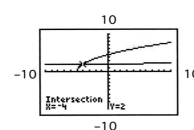

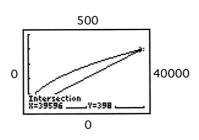

Algebraic method:
   Advantages: straightforward calculation, both solutions arise naturally.
   Disadvantage: messy arithmetic.
Graphical method:
   Advantage: no arithmetic.
   Disadvantages: it is extremely unlikely that one would spot the second solution at $x = 39{,}596$; one must essentially know the answer beforehand; it is not at all obvious what window to choose.

69. Algebraically
$$2x^{-2/5} - 5x^{-1/5} + 1 = 0$$
Let $u = x^{-1/5}$, then
$$2u^2 - 5u + 1 = 0$$

$$u = \frac{-b \pm \sqrt{b^2 - 4ac}}{2a} \qquad \begin{array}{l} a = 2 \\ b = -5 \\ c = 1 \end{array}$$

$$u = \frac{-(-5) \pm \sqrt{(-5)^2 - 4(2)(1)}}{2(2)}$$

$$u = \frac{5 \pm \sqrt{17}}{4}$$

$$x^{-1/5} = \frac{5 + \sqrt{17}}{4} \qquad x^{-1/5} = \frac{5 - \sqrt{17}}{4}$$

$$x^{1/5} = \frac{4}{5 + \sqrt{17}} \qquad x^{1/5} = \frac{4}{5 - \sqrt{17}}$$

$$x = \left(\frac{4}{5 + \sqrt{17}}\right)^5 \qquad x = \left(\frac{4}{5 - \sqrt{17}}\right)^5$$

$$x \approx 0.016203 \qquad x \approx 1974.98$$

Graphically:
Graph $y1 = 2x^{-2/5} - 5x^{-1/5} + 1$ and use the ZERO command to find the zeros.

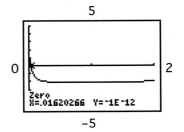

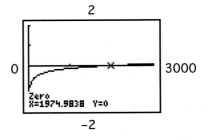

Advantages and disadvantages: See discussion under problem 67.

71.

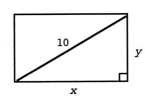

Area = length $\cdot$ width, so $45 = x \cdot y$.
According to the Pythagorean Theorem, $x^2 + y^2 = 10^2$; solve for y:
$$x^2 + y^2 = 100$$
$$y^2 = 100 - x^2$$
$$y = \sqrt{100 - x^2}$$
$$45 = x \cdot y = x\sqrt{100 - x^2}$$

Graph $y_1 = 45$, $y_2 = x\sqrt{100 - x^2}$ and use the INTERSECT command.

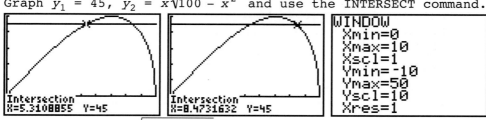

If $x = 5.3$, $y = \sqrt{100 - 5.3^2} = 8.5$.

If $x = 8.5$, $y = \sqrt{100 - 8.5^2} = 5.3$

The only solution is 5.3 in × 8.5 in.

73. As in Example 6 in Section 2-6, we know that the time until the splash is heard can be broken down into two segments.

t_1 = time for stone to reach water

t_2 = time for sound of splash to travel back to surface

We've seen before that the distance covered by a falling object is given by $16t^2$, so if x is the depth of the well, then $x = 16t^2$. From Example 6 we know that the distance traveled by the sound of the splash is $x = 1,100t_2$.

Solve each equation for the time.

$$x = 16t_1^2 \qquad\qquad x = 1,100t_2$$

$$\frac{x}{16} = t_1^2 \qquad\qquad t_2 = \frac{x}{1,100}$$

$$t_1 = \frac{\sqrt{x}}{4}$$

The total time, which is 14 seconds, is the sum of the two time segments.

$$\frac{\sqrt{x}}{4} + \frac{x}{1,100} = 14$$

Graph $y_1 = \dfrac{\sqrt{x}}{4} + \dfrac{x}{1,100}$, $y_2 = 14$ and use the INTERSECT command.

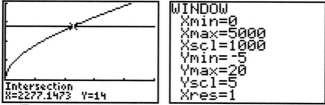

The solution is 2,277, so the well is 2,277 feet deep.

75. (A)

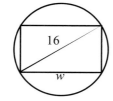

The diameter of the log cross-section is the diagonal of the beam cut out. We can use the Pythagorean Theorem to find the height of the beam.

$$w^2 + \text{height}^2 = 16^2$$

$$\text{height}^2 = 256 - w^2$$

$$\text{height} = \sqrt{256 - w^2}$$

The area is width·height, so $A = w\sqrt{256 - w^2}$. (Note that w must be positive and cannot be 16" or more, so $0 < w < 16$.)

(B) Plug 120 in for A and solve for w:

$$120 = w$$

Graph $y_1 = 120$, $y_2 = x\sqrt{256 - x^2}$ and use the INTERSECT command.

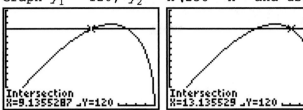

The two solutions are $x = 9.1$ and $x = 13.1$.

If $w = 9.1$, height $= \sqrt{256 - 9.1^2} = 13.1$

If $w = 13.1$, height $= \sqrt{256 - 13.1^2} = 9.1$

The only solution is 13.1 in by 9.1 in.

(C) Graph $y_1 = x\sqrt{256 - x^2}$ and use the MAXIMUM command.

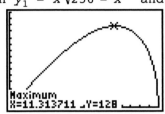

The maximum occurs when the width is 11.3 in; the maximum area is 128 in^2.

77.

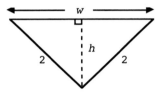

Let w = width of trough
h = altitude of triangular end

Examining the triangular end of the trough sketched above, we see that

$h^2 + \left(\frac{1}{2}w\right)^2 = 2^2$. The area is given by $A = \frac{1}{2}wh$. Since the volume of the trough V is given by $V = A \cdot 6$, we have

$$9 = 6A$$
$$9 = 6\left(\frac{1}{2}wh\right)$$
$$9 = 3wh$$
$$3 = wh$$

Since $h^2 = 2^2 - \left(\frac{1}{2}w\right)^2$

$$h^2 = 2^2 - \frac{1}{4}w^2$$
$$h = \sqrt{4 - \frac{1}{4}w^2}$$

We solve

$$3 = w\sqrt{4 - \frac{1}{4}w^2}$$
$$9 = w^2\left(4 - \frac{1}{4}w^2\right)$$
$$9 = 4w^2 - \frac{1}{4}(w^2)^2$$
$$36 = 16w^2 - (w^2)^2$$
$$(w^2)^2 - 16w^2 + 36 = 0$$

$$w^2 = \frac{-b \pm \sqrt{b^2 - 4ac}}{2a} \qquad a = 1, \ b = -16, \ c = 36$$

$$w^2 = \frac{-(-16) \pm \sqrt{(-16)^2 - 4(1)(36)}}{2(1)}$$

$$w^2 = \frac{16 \pm \sqrt{256 - 144}}{2}$$

$$w^2 = \frac{16 \pm \sqrt{112}}{2}$$

$$w^2 = 8 \pm 2\sqrt{7}$$

$$w = \sqrt{8 \pm 2\sqrt{7}}$$

$$w = 1.65 \text{ ft or } 3.65 \text{ ft}$$

Section 2-7

1. Answers will vary.

3. Answers will vary

5. $u(x) > 0$ where the graph of u is above the x axis; this occurs when $c < x < f$ or when or (c, f).

7. $v(x) \geq u(x)$ where the graph of v is above or intersects the graph of u; this occurs when $x \leq b$ or $x \geq e$, or $(-\infty, b] \cup [e, \infty)$.

9. $u(x) \leq 0$ where the graph of u is above or intersects the x-axis; this occurs when $x \leq c$ or $x \geq f$, or $(-\infty, c] \cup [f, \infty)$.

11. $v(x) - u(x) > 0$ is equivalent to $v(x) > u(x)$. This occurs where the graph of v is above the graph of u; $x < b$ or $x > e$, or $(-\infty, b) \cup (e, \infty)$.

13. The distance from x to 3 is less than 5.
    $|x - 3| < 5$

15. The distance from y to -1 is greater than 6.
    $|y - (-1)| > 6$
    $|y + 1| > 6$

17. The distance from a to 3 is less than or equal to 5.
    $|a - 3| \leq 5$

19. The distance from d to -2 is greater than or equal to 4.
    $|d - (-2)| \geq 4$
    $|d + 2| \geq 4$

21. The distance from y to the origin is less than or equal to 7.

23. The distance from w to the origin is greater than 7.

25. $7x - 8 < 4x + 7$
    $3x < 15$
    $x < 5$ or $(-\infty, 5)$

27. $-5t < -10$
    $t > 2$ or $(2, \infty)$

    (Note that the direction of the inequality changed.)

29. $3 - m < 4(m - 3)$
    $3 - m < 4m - 12$
    $-5m < -15$
    $m > 3$ or $(3, \infty)$

    (Note that the direction of the inequality changed.)

31. Find zeros of the left side: $(x - 3)(x + 4) = 0$; $x = 3$ or $x = -4$
Choose test numbers and make a table:

Interval	Test Number	Result
$x < -4$	-5	$(-5 - 3)(-5 + 4) = 8$; positive
$-4 < x < 3$	0	$(0 - 3)(0 + 4) = -12$; negative
$x > 3$	4	$(4 - 3)(4 + 4) = 8$; positive

The expression $(x - 3)(x + 4)$ is negative on $-4 < x < 3$, or $(-4, 3)$.

33. Find zeros of the left side: $x(2x - 7) = 0$; $x = 0$ or $2x - 7 = 0$
$$2x = 7$$
$$x = 7/2$$

Choose test numbers and make a table:

Interval	Test Number	Result
$x < 0$	-1	$(-1)(2(-1) - 7) = 9$; positive
$0 < x < 7/2$	1	$(1)(2(1) - 7) = -5$; negative
$x > 7/2$	4	$(4)(2(4) - 7) = 4$; positive

The expression $x(2x - 7)$ is zero or positive on $x \leq 0$ or $x \geq 7/2$, or $(-\infty, 0] \cup [7/2, \infty)$.

35. The distance from s to 5 is less than 3, so $s - 5$ is between -3 and 3.
$$|s - 5| < 3$$
$$-3 < s - 5 < 3$$
$$2 < s < 8$$
$$(2, 8)$$

37. The distance from s to 5 is more than 3, so $s - 5$ is either less than -3 or greater than 3.
$$|s - 5| > 3$$
$$s - 5 < -3 \text{ or } s - 5 > 3$$
$$s < 2 \text{ or } s > 8$$
$$(-\infty, 2) \cup (8, \infty)$$

39. $-4 < 5t + 6 \leq 21$
$-10 < 5t \leq 15$
$-2 < t \leq 3$ or $(-2, 3]$

41. $-12 < \dfrac{3}{4}(2 - x) \leq 24$
$\dfrac{4}{3}(-12) < 2 - x \leq \dfrac{4}{3}(24)$
$-16 < 2 - x \leq 32$
$-18 < -x \leq 30$
$18 > x \geq -30$
$-30 \leq x < 18$ or $[-30, 18)$

43. $\dfrac{q}{7} - 3 > \dfrac{q - 4}{3} + 1$
$21\left(\dfrac{q}{7} - 3\right) > 21\left(\dfrac{(q - 4)}{3} + 1\right)$
$3q - 63 > 7(q - 4) + 21$
$3q - 63 > 7q - 28 + 21$
$3q - 63 > 7q - 7$
$-4q > 56$
$q < -14$ or $(-\infty, -14)$

45. Rearrange so that zero is on the right: $x^2 + 3x - 10 < 0$

Find the zeros of the left side: $x^2 + 3x - 10 = 0$

$$(x + 5)(x - 2) = 0$$

$$x = -5 \quad \text{or} \quad x = 2$$

Choose test numbers and make a table:

Interval	Test Number	Result
$x < -5$	-6	$(-6)^2 + 3(-6) - 10 = 8$; positive
$-5 < x < 2$	0	$(0)^2 + 3(0) - 10 = -10$; negative
$x > 2$	3	$(3)^2 + 3(3) - 10 = 8$; positive

The expression $x^2 + 3x - 10$ is negative on $-5 < x < 2$, or $(-5, 2)$.

47. Rearrange so that zero is on the right: $x^2 - 10x + 21 > 0$

Find the zeros of the left side: $x^2 - 10x + 21 = 0$

$$(x - 7)(x - 3) = 0$$

$$x = 7 \quad \text{or} \quad x = 3$$

Choose test numbers and make a table:

Interval	Test Number	Result
$x < 3$	0	$(0)^2 - 10(0) + 21 = 21$; positive
$3 < x < 7$	5	$(5)^2 - 10(5) + 21 = -4$; negative
$x > 7$	8	$(8)^2 - 10(8) + 21 = 5$; positive

The expression $x^2 - 10x + 21$ is positive on $x < 3$ or $x > 7$, or $(-\infty, 3) \cup (7, \infty)$.

49. Rearrange so that zero is on the right: $x^2 - 8x \leq 0$

Find the zeros of the left side: $x^2 - 8x = 0$

$$x(x - 8) = 0$$

$$x = 0 \quad \text{or} \quad x = 8$$

Choose test numbers and make a table:

Interval	Test Number	Result
$x < 0$	-1	$(-1)^2 - 8(-1) = 9$; positive
$0 < x < 8$	1	$(1)^2 - 8(1) = -7$; negative
$x > 8$	10	$(10)^2 - 8(10) = 20$; positive

The expression $x^2 - 8x$ is zero or negative on $0 \leq x \leq 8$, or $[0, 8]$.

51. Rearrange so that zero is on the right: $x^2 - 2x + 1 < 0$

Find the zeros of the left side: $x^2 - 2x + 1 = 0$

$$(x - 1)(x - 1) = 0$$

$$x = 1$$

Choose test numbers and make a table:

Interval	Test Number	Result
$x < 1$	0	$(0)^2 - 2(0) + 1 = 1$; positive
$x > 1$	2	$(2)^2 - 2(2) + 1 = 1$; positive

The expression $x^2 - 2x + 1$ is never negative, so the inequality has no solution.

53. Rearrange so that zero is on the right: $x^2 - 4x - 21 \geq 0$

Find the zeros of the left side: $x^2 - 4x - 21 = 0$
$$(x - 7)(x + 3) = 0$$
$$x = 7 \quad \text{or} \quad x = -3$$

Choose test numbers and make a table:

Interval	Test Number	Result
$x < -3$	-4	$(-4)^2 - 4(-4) - 21 = 11;$ positive
$-3 < x < 7$	0	$(0)^2 - 4(0) - 21 = -21;$ negative
$x > 7$	8	$(8)^2 - 4(8) - 21 = 11;$ positive

The expression $x^2 - 4x - 21$ is zero or positive on $x \leq -3$ or $x \geq 7$, or $(-\infty, -3] \cup [7, \infty)$.

55. Find the zeros of the left side (this requires the quadratic formula):
$$x^2 - 5x + 3 = 0 \qquad a = 1, \; b = -5, \; c = 3$$
$$x = \frac{-(-5) \pm \sqrt{(-5)^2 - 4(1)(3)}}{2(1)}$$
$$= \frac{5 \pm \sqrt{13}}{2} \approx 4.3, 0.7$$

Choose test numbers and make a table:

Interval	Test Number	Result
$x < 0.7$	0	$(0)^2 - 5(0) + 3 = 3;$ positive
$0.7 < x < 4.3$	1	$(1)^2 - 5(1) + 3 = -1;$ negative
$x > 4.3$	5	$(5)^2 - 5(5) + 3 = 3;$ positive

The expression $x^2 - 5x + 3$ is positive on $x < 0.7$ or $x > 4.3$, or $(-\infty, 0.7) \cup (4.3, \infty)$.

57. Rearrange so that the right side is zero: $-2x^2 + x + 7 \geq 0$
Find the zeros of the left side (this requires the quadratic formula):
$$-2x^2 + x + 7 = 0 \qquad a = -2, \; b = 1, \; c = 7$$
$$x = \frac{-1 \pm \sqrt{1^2 - 4(-2)(7)}}{2(-2)}$$
$$= \frac{-1 \pm \sqrt{57}}{-4} \approx -1.6, 2.1$$

Choose test numbers and make a table:

Interval	Test Number	Result
$x < -1.6$	-2	$-2(-2)^2 + (-2) + 7 = -3;$ negative
$-1.6 < x < 2.1$	0	$-2(0)^2 + (0) + 7 = 7;$ positive
$x > 2.1$	3	$-2(3)^2 + (3) + 7 = -8;$ negative

The expression $-2x^2 + x + 7$ is zero or positive on $-1.6 \leq x \leq 2.1$, or $[-1.6, 2.1]$.

59. $|3x - 7| \leq 4$
$$-4 \leq 3x - 7 \leq 4$$
$$3 \leq 3x \leq 11$$
$$1 \leq x \leq \frac{11}{3}$$

$\left[1, \dfrac{11}{3}\right]$

61. $|4 - 2t| > 6$
$$4 - 2t > 6 \quad \text{or} \quad 4 - 2t < -6$$
$$-2t > 2 \quad \text{or} \quad -2t < -10$$
$$t < -1 \quad \text{or} \quad t > 5$$
$$(-\infty, -1) \cup (5, \infty)$$

63. $|0.2u + 1.7| \geq 0.5$
$$0.2u + 1.7 \geq 0.5 \quad \text{or} \quad 0.2u + 1.7 \leq -0.5$$
$$0.2u \geq -1.2 \quad \text{or} \quad 0.2u \leq -2.2$$
$$u \geq -6 \quad \text{or} \quad u \leq -11$$
$$\left(-\infty, -11\right] \cup \left[-6, \infty\right)$$

65.   If $a - b = 1$, $a = b + 1$ and a is greater than b. $a > b$.

67.   If $a < 0$, $b < 0$, then if
$$\frac{b}{a} > 1$$
$a \cdot \dfrac{b}{a} < a \cdot 1$ (since a is negative, multiplying both sides by a reverses the inequality)
$$b < a$$
$$a > b$$

69.   The distance from x to 3 is between 0 and 0.1. The number x will be within 0.1 of 3 if it is between 2.9 and 3.1. But we also need the distance between x and 3 to be non-zero, so x cannot be 3. So the solution is all numbers between 2.9 and 3.1 except 3, or $(2.9, 3) \cup (3, 3.1)$.

71.   The distance from x to c is between 0 and $2c$. The number x will be within $2c$ units of c if it is between $c - 2c$ and $c + 2c$. In other words, x must be between $-c$ and $3c$. But we also need the distance between x and c to be nonzero, so x and c cannot be equal. So the solution is all numbers between $-c$ and $3c$ except c, or $\left(-c, c\right) \cup \left(c, 3c\right)$.

73.   Find the zeros of f: $2x^2 - 5x - 12 = 0$
$$\left(2x + 3\right)\left(x - 4\right) = 0$$
$$2x + 3 = 0 \quad or \quad x - 4 = 0$$
$$2x = -3 \qquad\qquad x = 4$$
$$x = -3/2$$

Choose test numbers and make a table:

Interval	Test Number	Result
$x < -3/2$	-2	$2(-2)^2 - 5(-2) - 12 = 6$; positive
$-3/2 < x < 4$	1	$2(1)^2 - 5(1) - 12 = -15$; negative
$x > 4$	5	$2(5)^2 - 5(5) - 12 = 13$; positive

The function $f(x) = 2x^2 - 5x - 12$ is positive on $x < -3/2$ or $x > 4$, or $\left(-\infty, -3/2\right) \cup \left(4, \infty\right)$. It is negative on the interval $(-3/2, 4)$.

75. Find the zeros of h: $x^2 + 9x + 2 = 0$ $a = 1$, $b = 9$, $c = 2$

$$x = \frac{-9 \pm \sqrt{9^2 - 4(1)(2)}}{2(1)}$$

$$= \frac{-9 \pm \sqrt{73}}{2} \approx -0.2, -8.8$$

Choose test numbers and make a table:

Interval	Test Number	Result
$x < -8.8$	-10	$(-10)^2 + 9(-10) + 2 = 12$; positive
$-8.8 < x < -0.2$	-1	$(-1)^2 + 9(-1) + 2 = -6$; negative
$x > -0.2$	0	$(0)^2 + 9(0) + 2 = 2$; positive

The function $h(x) = x^2 + 9x + 2$ is positive on $x < -8.8$ or $x > -0.2$, or $(-\infty, -8.8) \cup (-0.2, \infty)$. It is negative on the interval $(-8.8, -0.2)$.

77. One example is $x^2 \geq 0$. Any inequality of the form $(\ \cdot\ )^2 \geq 0$ will do it, as will any inequality of the form $(\ \cdot\ )^2 > a$, where a is any negative number.

79. The difference between A and 12.436 has an absolute value less than the error of 0.001.
$|A - 12.436| < 0.001$
$-0.001 < A - 12.436 < 0.001$
$12.435 < A < 12.437$ or, in interval notation, $(12.435, 12.437)$

81. Let x = number of calculators sold
Then Revenue = (price per calculator) × (number of calculators sold) = $63x$
       Cost = fixed cost + variable cost
            = 650,000 + (Cost per calculator) × (number sold)
            = $650,000 + 47x$

(A) We want Revenue > Cost
$$63x > 650,000 + 47x$$
$$16x > 650,000$$
$$x > 40,625$$
More than 40,625 calculators must be sold for the company to make a profit.

(B) We want Revenue = Cost
$$63x = 650,000 + 47x$$
$$16x = 650,000$$
$$x = 40,625$$

(C) 40,625 calculators sold represents the break-even point, the boundary between profit and loss.

83. (A) The company might try to increase sales and keep the price the same (see part B). It might try to increase the price and keep the sales the same (see part C). Either of these strategies would need further analysis.

(B) Here the cost has been changed to $650,000 + 50.5x$, but the revenue is still $63x$.
       Revenue > Cost
$$63x > 650,000 + 50.5x$$
$$12.5x > 650,000$$
$$x > 52,000 \text{ calculators}$$

(C) Let p = the new price. Here the cost is still 650,000 + 50.5x as in part (B) where x is now known to be 40,625; now cost = 650,000 + 50.5(40,625). The revenue is (price per calculator) × (number of calculators) = p(40,625).

Revenue > Cost

$$p(40,625) > 650,000 + 50.5(40,625)$$

$$p > \frac{650,000 + 50.5(40,625)}{40,625}$$

$$p > 66.50$$

The price could be raised by \$3.50 to \$66.50.

85. The printer will show a profit if revenue exceeds cost; that is, if $R(x) > C(x)$, or

$$10x - 0.05x^2 > 200 + 2.25x$$
$$-200 + 7.75x - 0.05x^2 > 0$$

The solution of this inequality consists of all values of x for which the graph of $y = -200 + 7.75x - 0.05x^2$ is above the x axis. Graphing and using the ZERO command, we get

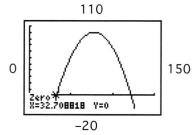 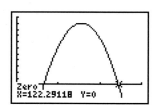

To the nearest integer, the printer will show a profit for 33 < x < 122.

87. The Fahrenheit range is described by 60 < F < 80

Since $F = \frac{9}{5}C + 32$, we write

$$60 < \frac{9}{5}C + 32 < 80$$

$$28 < \frac{9}{5}C < 48$$

$$\frac{5}{9} \cdot 28 < \frac{5}{9} \cdot \frac{9}{5}C < \frac{5}{9} \cdot 48$$

$$16° < C < 27° \text{ (to the nearest degree)}$$

The Celsius range is 16°C to 27°C.

89. (A) Since the quadratic distance function d has a maximum value of 256 feet after 4 seconds, the vertex of the graph is (4, 256) and the vertex form of the function $d(t)$ is

$$d(t) = a(t - 4)^2 + 256$$

To find a we use the fact that $d(0) = 0$ (the object was propelled from ground level).

$$d(0) = a(0 - 4)^2 + 256 = 0$$
$$16a + 256 = 0$$
$$16a = -256$$
$$a = -16$$

The quadratic function is

$$d(t) = -16(t - 4)^2 + 256$$
$$d(t) = -16(t^2 - 8t - 16) + 256$$
$$d(t) = -16t^2 + 128t - 256 + 256$$
$$d(t) = -16t^2 + 128t$$

(B) To find when the object is more than 240 feet above the ground, we solve the inequality $-16t^2 + 128t > 240$, or $-16t^2 + 128t - 240 > 0$.

$-16t^2 + 128t - 240 > 0$ .... (Divide both sides by -16)

$t^2 - 8t + 15 < 0$ .... (Direction of inequality was reversed!)

Find the zeros of the left side: $(t - 5)(t - 3) = 0$

$$t = 5 \text{ or } t = 3$$

Choose test numbers and make a table:

Interval	Test Number	Result
$t < 3$	1	$(1)^2 - 8(1) + 15 = 8$; positive
$3 < t < 5$	4	$(4)^2 - 8(4) + 15 = -1$; negative
$t > 5$	6	$(6)^2 - 8(1) + 15 = 8$; positive

The solution of the inequality is $3 < t < 5$, so the object is above height 240 for times between 3 and 5 seconds after being propelled.

91. The temperature increases at a constant rate so the model will be a linear function. We're given enough information to find two points on the line. The temperature at 1 km is 30°C so (1, 30) is one point. The temperature increases 2.8°C for each 100 meters, so at a depth of 2 km it will have increased 28°C; at 2 km depth, the temperature is 58°C, so (2, 58) is a second point.

$$m = \frac{58 - 30}{2 - 1} = 28$$
$$y - 30 = 28(x - 1)$$
$$y - 30 = 28x - 28$$
$$y = 28x + 2$$

The model is $T(x) = 28x + 2$

We're asked to solve the inequality $150 < 28x + 2 < 200$.

$$150 < 28x + 2 < 200$$
$$148 < 28x < 198$$
$$5.286 < x < 7.071$$

Depths between 5.286 km and 7.071 km will have temperatures between 150°C and 200°C.

93. The independent variable is the altitude so we enter the numbers in the first height column as L_1. The dependent variable is the temperature so we enter the first temperature column as L_2.

L1	L2	----- 2
2574	8	------
3087	5	
3962	-1	
4877	-8	
4992	-8	
6096	-16	
7315	-24	

$L_2 = \{8, 5, -1, -8, -\ldots$

(Note that the entire list is not visible on this diagram.)
We use the linear regression command on the STAT CALC menu:

```
EDIT CALC TESTS
1:1-Var Stats
2:2-Var Stats
3:Med-Med
4:LinReg(ax+b)
5:QuadReg
6:CubicReg
7↓QuartReg
```

```
LinReg(ax+b) L1,
L2■
```

```
LinReg
y=ax+b
a=-.0073828955
b=28.21590646
```

The linear model is $y = -0.00738x + 28.2$.

We're asked to solve the inequality below:
$$-30 < -0.00738x + 28.2 < -10$$
$$-58.2 < -0.00738x < -38.2$$
$$7,890 > \quad x \quad > 5,180 \quad \text{(Note that the direction reversed)}$$

The temperature will be between $-10°C$ and $-30°C$ at altitudes between 5,180 meters and 7,890 meters.

95. The independent variable is altitude so we enter the first height column as L_1.
The dependent variable is pressure so we enter the first pressure column as L_2.

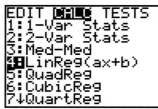

(Note that the entire list is not visible on this diagram.)

We use the linear regression command on the STAT CALC menu:

 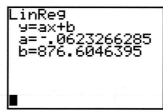

The linear model is $y = -0.0623x + 877$.

We are asked to solve the inequality below:
$$350 < -0.0623x + 877 < 650$$
$$-527 < -0.0623x < -227$$
$$8,460 > \quad x \quad > 3,640$$

The pressure will be between 350 hpa and 650 hpa between 3,640 and 8,460 meters.

97. (A) The independent variable is demand so we enter the orange juice demand column as L_1. The dependent variable is price so we enter the orange juice price column as L_2.

We use the linear regression command on the STAT CALC menu.

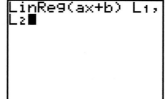 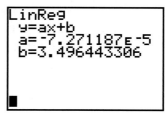

The linear model is $p = d(x) = -0.000\ 0727x + 3.50$.

To find the domain we first note that x must be at least zero. Also, the price cannot be negative. Using the graph and the ZERO command, we find that price is negative when demand is greater than 48,100 (rounded to 3 significant digits), so the domain is $0 \leq x \leq 48,100$.

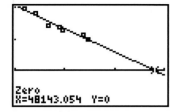

The revenue function is

$R(x) = xp$

$= x(-0.000\ 0727x + 3.50)$

$= -0.000\ 0727x^2 + 3.50x$

The domain is the same as the domain of $d(x)$.

The cost function is

$C(x) = 20,000 + 0.50x$

Fixed costs + \$0.50 per gallon produced

The domain is $x \geq 0$. (There's no stated limitation on how much they can sell.)

(C) The company will break even when revenue = cost so we graph $y_1 = R(x) = -0.000\ 0727x^2 + 3.50x$ and $y_2 = C(x) = 20,000 + 0.50x$ and find the intersection.

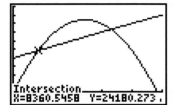 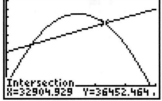 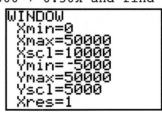

The revenue and cost graphs intersect at $x = 8,360$ and $x = 32,900$ so the company will break even if they sell 8,360 gallons or 32,900 gallons. They'll make a profit if they sell between these two levels, and will lose money if they sell less than 8,360 gallons or more than 32,900 gallons.

(D) The profit is calculated by subtracting cost from revenue so the profit function is

$P(x) = R(x) - C(x) =$

$P(x) = -0.000\ 0727x^2 + 3.50x - (20,000 + 0.50x)$

$P(x) = -0.000\ 0727x^2 + 3.00x - 20,000$

Graph the profit function and use the MAXIMUM command

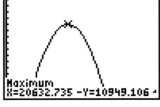

The maximum profit of \$10,900 occurs when 20,600 gallons are sold.

CHAPTER 2 REVIEW

1. The points at two of the vertices of the triangle are (-4, 3) and (1, 1). The line moves 2 units down between these points so the rise is -2. The line moves 5 units to the right between these points so the run is 5. The slope is therefore $m = -\dfrac{2}{5}$.

$$y - 1 = -\frac{2}{5}(x - 1)$$

$$y - 1 = -\frac{2}{5}x + \frac{2}{5} \qquad \text{(Multiply both sides by 5)}$$

$$5y - 5 = -2x + 2$$

$$2x + 5y = 7 \qquad\qquad\qquad\qquad\qquad\qquad (2\text{-}1)$$

2. $3x + 2y = 9$
$$2y = -3x + 9$$
$$y = -\frac{3}{2}x + \frac{9}{2}$$

Slope: $-\dfrac{3}{2}$

To graph, find a few points by substituting values.

x	y
0	$\frac{9}{2}$
3	0
1	3

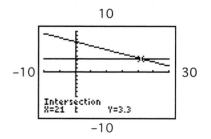

$(2\text{-}1)$

3. The line passes through the two given points, (6, 0) and (0, 4), so its slope is given by
$$m = \frac{0 - 4}{6 - 0} = \frac{-4}{6} = -\frac{2}{3}$$
The equation of the line is, therefore, using the point-slope form,
$$y - 0 = -\frac{2}{3}(x - 6)$$
or $\quad 3y = -2(x - 6)$
or $\quad 3y = -2x + 12.$
$$2x + 3y = 12 \qquad\qquad (2\text{-}1)$$

4. $y = mx + b \qquad m = -\dfrac{2}{3} \qquad b = 2$

$$y = -\frac{2}{3}x + 2 \qquad\qquad\qquad (2\text{-}1)$$

5. *vertical:* $x = -3$, slope not defined; *horizontal:* $y = 4$, slope = 0 $\qquad (2\text{-}1)$

6. (A) $0.05x + 0.25(30 - x) = 3.3$
$$0.05x + 7.5 - 0.25x = 33$$
$$-0.2x + 7.5 = 3.3$$
$$-0.2x = -4.2$$
$$x = 21$$

(B) $\dfrac{5x}{3} - \dfrac{4 + x}{2} = \dfrac{x - 2}{4} + 1 \quad$ LCD = 12
$$4(5x) - 6(4 + x) = 3(x - 2) + 12$$
$$20x - 24 - 6x = 3x - 6 + 12$$
$$14x - 24 = 3x + 6$$
$$11x = 30$$
$$x = \frac{30}{11} \approx 2.727$$

Check:

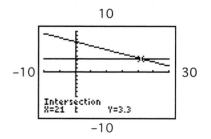

Check:

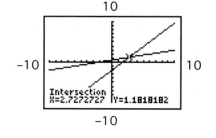

$(2\text{-}2)$

7. (A) $f(x) = -x^2 - 2x + 3$

$$= -(x^2 + 2x + ?) + 3 \qquad \frac{1}{2}(2) = 1 \qquad 1^2 = 1$$

$$= -(x^2 + 2x + 1) + 3 + 1$$

$$= -(x + 1)^2 + 4$$

(B) The graph of the function is the same as the graph of $y = x^2$ reflected in the x axis, shifted left 1 unit, and up 4 units.

(C) Solve $0 = -x^2 - 2x + 3$

$$0 = x^2 + 2x - 3$$

$$0 = (x + 3)(x - 1)$$

$$x + 3 = 0 \quad \text{or} \quad x - 1 = 0$$

$$x = -3 \qquad\qquad x = 1$$

x intercepts: $-3, 1$

Check:

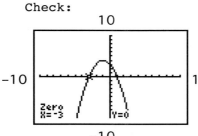 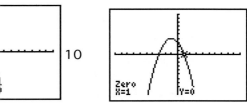

$(2\text{-}3,\ 2\text{-}5)$

8. (A) $f(x) = x^2 - 3x - 2$

$$= (x^2 - 3x + ?) - 2 \qquad \frac{1}{2}(-3) = -\frac{3}{2} \qquad \left(-\frac{3}{2}\right)^2 = \frac{9}{4}$$

$$= \left(x^2 - 3x + \frac{9}{4}\right) - 2 - \frac{9}{4}$$

$$= \left(x - \frac{3}{2}\right)^2 - \frac{17}{4}$$

(B) The graph of the function is the same as the graph of $y = x^2$ shifted right $\frac{3}{2}$ units and down $\frac{17}{4}$ units.

(C) $x^2 - 3x - 2 = 0$

$$x = \frac{-b \pm \sqrt{b^2 - 4ac}}{2a} \qquad a = 1,\ b = -3,\ c = -2$$

$$x = \frac{-(-3) \pm \sqrt{(-3)^2 - 4(1)(-2)}}{2(1)}$$

$$x = \frac{3 \pm \sqrt{9 + 8}}{2}$$

$$x = \frac{3 \pm \sqrt{17}}{2}$$

x intercepts: $\dfrac{3 + \sqrt{17}}{2} \approx 3.5616$

$\dfrac{3 - \sqrt{17}}{2} \approx -0.5616$

Check:

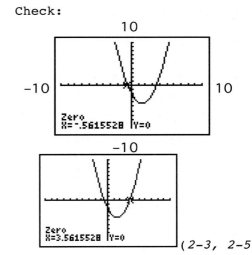

$(2\text{-}3,\ 2\text{-}5)$

9. Use the vertex formula:

$$x = -\frac{b}{2a} = -\frac{9}{2(-3)} = -\frac{9}{-6} = \frac{3}{2}$$

To find the second coordinate, plug back into f:

$$f\left(\frac{3}{2}\right) = -3\left(\frac{3}{2}\right)^2 + 9\left(\frac{3}{2}\right) - 4 = -3\left(\frac{9}{4}\right) + \frac{27}{2} - 4 = -\frac{27}{4} + \frac{54}{4} - \frac{16}{4} = \frac{11}{4}$$

The vertex is $\left(\dfrac{3}{2}, \dfrac{11}{4}\right)$; since $a = -3$ is negative, the parabola opens down and the vertex is a maximum.

$(2\text{-}3)$

10. (A) $(-3 + 2i) + (6 - 8i) = -3 + 2i + 6 - 8i = 3 - 6i$

(B) $(3 - 3i)(2 + 3i) = 6 + 3i - 9i^2 = 6 + 3i + 9 = 15 + 3i$

(C) $\dfrac{13 - i}{5 - 3i} = \dfrac{(13 - i)}{(5 - 3i)}\dfrac{(5 + 3i)}{(5 + 3i)}$

$\qquad = \dfrac{65 + 34i - 3i^2}{25 - 9i^2} = \dfrac{65 + 34i + 3}{25 + 9} = \dfrac{68 + 34i}{34} = 2 + i$

(D) $(7 + 6i) - (8 + 9i) = 7 + 6i - 8 - 9i = -1 - 3i$ *(2-4)*

11. $4 - 3(2x + 7) = 5(4 - x) + 11x - 3$

$\quad 4 - 6x - 21 = 20 - 5x + 11x - 3$

$\qquad -6x - 17 = 17 + 6x$

$\qquad\quad -12x = 34$

$\qquad\qquad x = -\dfrac{34}{12} = -\dfrac{17}{6} \approx -2.8$

Check: Graph $y_1 = 4 - 3(2x + 7)$ and $y_2 = 5(4 - x) + 11x - 3$, use the INTERSECT command:

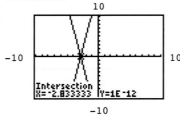

(2-2)

12. $(2x + 5)^2 = 23$

$\quad 2x + 5 = \pm\sqrt{23}$

$\qquad 2x = -5 \pm \sqrt{23}$

$\qquad x = \dfrac{-5 \pm \sqrt{23}}{2} \approx -0.1, -4.9$

Check: Graph $y_1 = (2x + 5)^2$ and $y_2 = 23$, use the INTERSECT command:

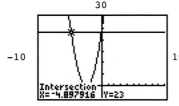 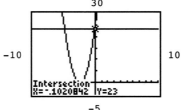

(2-5)

13. $x^2 - 2x + 3 = x^2 + 3x - 7$ Check:

$\quad -2x + 3 = 3x - 7$

$\quad -5x + 3 = -7$

$\qquad -5x = -10$

$\qquad\quad x = 2$

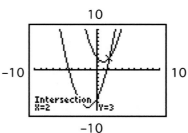

(2-1, 2-5)

14. $2x^2 - 7 = 0$ Check:

$$2x^2 = 7$$

$$x^2 = \frac{7}{2}$$

$$x = \pm\sqrt{\frac{7}{2}}$$

$$x = \pm\frac{\sqrt{14}}{2} \approx \pm1.8708$$

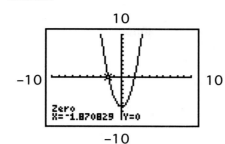
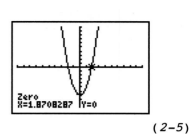

(*2-5*)

15. $2x^2 = 4x$ Check:

$$2x^2 - 4x = 0$$

$$2x(x - 2)$$

$$2x = 0 \quad x - 2 = 0$$

$$x = 0 \qquad x = 2$$

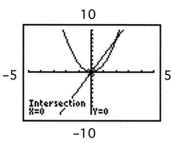
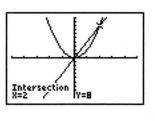

(*2-5*)

16. $2x^2 = 7x - 3$

$$2x^2 - 7x + 3 = 0$$

$$(2x - 1)(x - 3) = 0$$

$$2x - 1 = 0 \quad x - 3 = 0$$

$$x = \frac{1}{2} \qquad x = 3$$

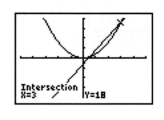

(*2-5*)

17. $m^2 + m + 1 = 0$

$$m = \frac{-b \pm \sqrt{b^2 - 4ac}}{2a} \qquad a = 1$$

$$b = 1$$

$$c = 1$$

$$m = \frac{-1 \pm \sqrt{(1)^2 - 4(1)(1)}}{2(1)}$$

$$m = \frac{-1 \pm \sqrt{-3}}{2}$$

$$m = \frac{-1 \pm i\sqrt{3}}{2}$$

$$m = -\frac{1}{2} \pm \frac{\sqrt{3}}{2}i \qquad\qquad (2\text{-}5)$$

18.

$$y^2 = \frac{3}{2}(y + 1)$$

$$2y^2 = 3(y + 1)$$

$$2y^2 = 3y + 3$$

$$2y^2 - 3y - 3 = 0$$

$$y = \frac{-b \pm \sqrt{b^2 - 4ac}}{2a}$$

$$a = 2$$
$$b = -3$$
$$c = -3$$

$$y = \frac{-(-3) \pm \sqrt{(-3)^2 - 4(2)(-3)}}{2(2)}$$

$$y = \frac{3 \pm \sqrt{33}}{4}$$

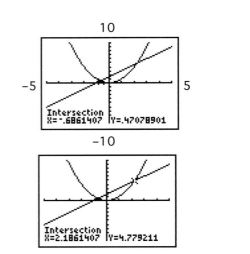

$(2-5)$

19. $\sqrt{4 + 7x} = 5$

$4 + 7x = 25$

$7x = 21$

$x = 3$

Algebraic check: $\sqrt{4 + 7(3)} = \sqrt{25} = 5$ ✓

Graphical check:

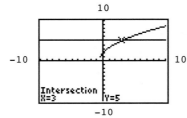

$(2-6)$

20. $\sqrt{5x - 6} - x = 0$

$\sqrt{5x - 6} = x$

$5x - 6 = x^2$

$0 = x^2 - 5x + 6$

$x^2 - 5x + 6 = 0$

$(x - 3)(x - 2) = 0$

$x = 2, \ 3$

Check: $\sqrt{5(2) - 6} - 2 \overset{?}{=} 0$

$0 \overset{✓}{=} 0$

$\sqrt{5(3) - 6} - 3 \overset{?}{=} 0$

$0 \overset{✓}{=} 0$

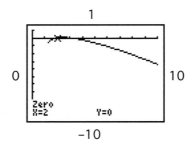   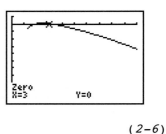

Solution: $x = 2, \ 3$

$(2-6)$

21. $\sqrt{x^2 + 2} = \sqrt{4x + 14}$

$x^2 + 2 = 4x + 14$

$x^2 - 4x - 12 = 0$

$(x - 6)(x + 2) = 0$

$x = 6$ or $x = -2$

Algebraic check: $\sqrt{6^2 + 2} = \sqrt{38};$ $\sqrt{4(6) + 14} = \sqrt{38}$ ✓

$\sqrt{(-2)^2 + 2} = \sqrt{6};$ $\sqrt{4(-2) + 14} = \sqrt{6}$ ✓

Graphical check:

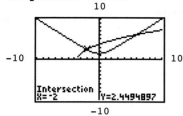

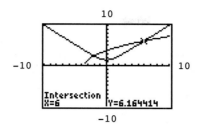

(*2-6*)

22. $3(2 - x) - 2 \leq 2x - 1$

$6 - 3x - 2 \leq 2x - 1$

$-3x + 4 \leq 2x - 1$

$-5x \leq -5$

$x \geq 1$

$[1, \infty)$

(*2-7*)

23. Rearrange so that zero is on the right: $x^2 + x - 20 < 0$

Find the zeros of the left side: $x^2 + x - 20 = 0$

$(x + 5)(x - 4) = 0$

$x = -5 \quad \text{or} \quad x = 4$

Choose test numbers and make a table:

Interval	Test Number	Result
$x < -5$	-6	$(-6)^2 + (-6) - 20 = 10;$ positive
$-5 < x < 4$	0	$(0)^2 + (0) - 20 = -20;$ negative
$x > 4$	5	$(5)^2 + (5) - 20 = 10;$ positive

The expression $x^2 + x - 20$ is negative on $-5 < x < 4$, or $(-5, 4)$.

(*2-7*)

24. Rearrange so that zero is on the right: $x^2 - 4x - 12 > 0$

Find zeros of the left side: $x^2 - 4x - 12 = 0$

$(x - 6)(x + 2) = 0$

$x = 6, -2$

Choose test numbers and make a table:

Interval	Test Number	Result
$x < -2$	-3	$(-3)^2 - 4(-3) - 12 = 9;$ positive
$-2 < x < 6$	0	$(0)^2 - 4(0) - 12 = -12;$ negative
$x > 6$	10	$(10)^2 - 4(10) - 12 = 48;$ positive

The expression $x^2 - 4x - 12$ is positive on $x < -2$ or on $x > 6$, or $(-\infty, -2) \cup (6, \infty)$.

(*2-7*)

25. (A) A line with positive slope is rising as x moves from left to right. It is the graph of an increasing function.

(B) A line with negative slope is falling as x moves from left to right. It is the graph of a decreasing function.

(C) A parabola that opens upward and has vertex at (h, k) is falling to the left of the vertex and rising to the right. It is the graph of a function that is decreasing on $(-\infty, h]$ and increasing on $[h, \infty)$.

(D) A parabola that opens downward and has vertex at (h, k) is rising to the left of the vertex and falling to the right. It is the graph of a function that is increasing on $(-\infty, h]$ and decreasing on $[h, \infty)$.   (*2-1, 2-3*)

26. Since two points are given, we find the slope, then apply the point-slope form.

$$m = \frac{-3 - 3}{0 - (-4)} = \frac{-6}{4} = -\frac{3}{2}$$

$$y - 3 = -\frac{3}{2}[x - (-4)]$$

$$2(y - 3) = -3(x + 4)$$

$$2y - 6 = -3x - 12$$

$$3x + 2y = -6$$ (*2-1*)

27. The line $6x + 3y = 5$, or
$3y = -6x + 5$,

or $y = -2x + \frac{5}{3}$, has slope -2.

(A) We require a line through (-2, 1), with slope -2. Applying the point-slope form, we have

$$y - 1 = -2[x - (-2)]$$
$$y - 1 = -2x - 4$$
$$y = -2x - 3$$

(B) A line perpendicular to one with slope -2 has slope $\frac{1}{2}$, so we need a line

with slope $m = \frac{1}{2}$. Again applying the point-slope form, we have

$$y - 1 = \frac{1}{2}[x - (-2)]$$

$$y - 1 = \frac{1}{2}x + 1$$

$$y = \frac{1}{2}x + 2$$ (*2-1*)

28. $|y + 9| < 5$ **29.** $|2x - 8| \geq 3$

$-5 < y + 9 < 5$ $2x - 8 \geq 3$ or $2x - 8 \leq -3$

$-14 < y < -4$ (*2-7*) $2x \geq 11$ $2x \leq 5$

                        $x \geq 5.5$ or $x \leq 2.5$ (*2-2*)

30. Rearrange so that zero is on the right: $2x^2 - 7x + 1 \geq 0$

Find zeros of the left side: $2x^2 - 7x + 1 = 0$ $a = 2, b = -7, c = 1$

$$x = \frac{-(-7) \pm \sqrt{(-7)^2 - 4(2)(1)}}{2(2)}$$

$$= \frac{7 \pm \sqrt{41}}{4} \approx 3.4, 0.1$$

Choose test numbers and make a table:

Interval	Test Number	Result
$x < 0.1$	0	$2(0)^2 - 7(0) + 1 = 1$; positive
$0.1 < x < 3.4$	1	$2(1)^2 - 7(1) + 1 = -4$; negative
$x > 3.4$	5	$2(5)^2 - 7(5) + 1 = 16$; positive

The expression $2x^2 - 7x + 1$ is zero or positive on $x \leq 0.1$ or on $x \geq 3.4$. (*2-7*)

31. The distance between y and 5 is no more than 2.

$$|y - 5| \leq 2$$
$$-2 \leq y - 5 \leq 2$$
$$3 \leq y \leq 7 \quad \text{or} \quad [3, 7] \qquad\qquad (2\text{-}7)$$

32. The distance between t and -6 is greater than 9.

$$|t - (-6)| > 9, \quad \text{or} \quad |t + 6| > 9$$
$$t + 6 > 9 \quad \text{or} \quad t + 6 < -9$$
$$t > 3 \quad \text{or} \qquad t < -15$$
$$(-\infty, -15) \cup (3, \infty) \qquad\qquad (2\text{-}7)$$

33. $0.1x^2 + x + 1.5 = 0 \quad a = 0.1, \; b = 1, \; c = 1.5$
$b^2 - 4ac = 1^2 - 4(0.1)(1.5) = 1 - 0.6 = 0.4$
The discriminant is positive so there are 2 real zeros. See check below. $(2\text{-}5)$

34. $0.1x^2 + x + 2.5 = 0 \quad a = 0.1, \; b = 1, \; c = 2.5$
$b^2 - 4ac = 1^2 - 4(0.1)(2.5) = 1 - 1 = 0$
The discriminant is zero so there is one real zero. See check below. $(2\text{-}5)$

35. $0.1x^2 + x + 3.5 = 0 \quad a = 0.1, \; b = 1, \; c = 3.5$
$b^2 - 4ac = 1^2 - 4(0.1)(3.5) = 1 - 1.4 = -0.4$
The discriminant is negative so there are two imaginary roots.

Check for questions 30 — 32:

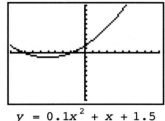 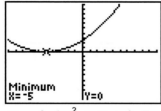 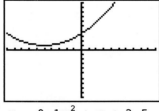

$\quad y = 0.1x^2 + x + 1.5 \quad\quad y = 0.1x^2 + x + 2.5 \quad\quad y = 0.1x^2 + x + 3.5 \qquad (2\text{-}5)$

36. (A) $f(x) = 0.5x^2 - 4x + 5$
$\qquad\qquad = 0.5(x^2 - 8x) + 5$
$\qquad\qquad = 0.5(x^2 - 8x + 16) + 5 - 0.5(16)$
$\qquad\qquad = 0.5(x - 4)^2 - 3$

Therefore the line $x = 4$ is the axis of the parabola and $(4, -3)$ is its vertex.

(B) The parabola opens upward, since $0.5 = a > 0$. So, the parabola is decreasing on $(-\infty, 4]$ and increasing on $[4, \infty)$. The range is $[-3, \infty)$.

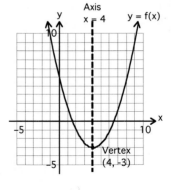

(C) The minimum value of -3 occurs at $x = 4$. There is no maximum. $\qquad\qquad (2\text{-}3)$

37. g is a linear function. Its graph passes through $(x_1, y_1) = (-1, 0)$ and $(x_2, y_2) = (1, 4)$. Therefore the slope of the line is

$$m = \frac{y_2 - y_1}{x_2 - x_1} = \frac{4 - 0}{1 - (-1)} = 2. \text{ We use the point-slope form:}$$

$$y - y_1 = m(x - x_1)$$
$$y - 0 = 2[x - (-1)]$$
$$y = 2x + 2$$

The function is given by $g(x) = 2x + 2$.
f is a quadratic function. Its equation must be of the form
$f(x) = y = a(x - h)^2 + k$. The vertex of the parabola is at $(1, 2)$.

Therefore the equation must have form
$$y = a(x - 1)^2 + 2$$
Since the parabola passes through (3, 0), these coordinates must satisfy the equation.
$$0 = a(3 - 1)^2 + 2$$
$$0 = 4a + 2$$
$$-2 = 4a$$
$$a = -0.5$$
The equation is
$$y = f(x) = -0.5(x - 1)^2 + 2$$
$$f(x) = -0.5(x^2 - 2x + 1) + 2$$
$$f(x) = -0.5x^2 + x - 0.5 + 2$$
$$f(x) = -0.5x^2 + x + 1.5$$
(2-1, 2-3)

38. (A) $(3 + i)^2 - 2(3 + i) + 3 = 9 + 6i + i^2 - 6 - 2i + 3$
$$= 9 + 6i - 1 - 6 - 2i + 3 = 5 + 4i$$
(B) $i^{27} = i^{26}i = (i^2)^{13}i = (-1)^{13}i = (-1)i = -i$ *(2-4)*

39. (A) $(2 - \sqrt{-4}) - (3 - \sqrt{-9}) = (2 - i\sqrt{4}) - (3 - i\sqrt{9}) = (2 - 2i) - (3 - 3i)$
$$= 2 - 2i - 3 + 3i = -1 + i$$

(B) $\dfrac{2 - \sqrt{-1}}{3 + \sqrt{-4}} = \dfrac{2 - i\sqrt{1}}{3 + i\sqrt{4}} = \dfrac{2 - i}{3 + 2i} = \dfrac{(2 - i)}{(3 + 2i)}\dfrac{(3 - 2i)}{(3 - 2i)} = \dfrac{6 - 7i + 2i^2}{9 - 4i^2} = \dfrac{6 - 7i - 2}{9 + 4}$

$$= \dfrac{4 - 7i}{13} = \dfrac{4}{13} - \dfrac{7}{13}i$$

(C) $\dfrac{4 + \sqrt{-25}}{\sqrt{-4}} = \dfrac{4 + i\sqrt{25}}{i\sqrt{4}} = \dfrac{4 + 5i}{2i} = \dfrac{4 + 5i}{2i}\dfrac{i}{i} = \dfrac{4i + 5i^2}{2i^2} = \dfrac{4i - 5}{-2} = \dfrac{5}{2} - 2i$

(D) $-\sqrt{-16} \cdot \sqrt{-25} = 4i \cdot 5i = 20i^2 = 20(-1) = -20$ *(2-4)*

40. $\left(x + \dfrac{5}{2}\right)^2 = \dfrac{5}{4}$

$x + \dfrac{5}{2} = \pm\sqrt{\dfrac{5}{4}}$

$x + \dfrac{5}{2} = \pm\dfrac{\sqrt{5}}{2}$

$x = -\dfrac{5}{2} \pm \dfrac{\sqrt{5}}{2}$

$x = \dfrac{-5 \pm \sqrt{5}}{2}$

Graphical check: $\dfrac{-5 + \sqrt{5}}{2} \approx -1.382$

$\dfrac{-5 - \sqrt{5}}{2} \approx -3.618$

Graph $y_1 = \left(x + \dfrac{5}{2}\right)^2$ and $y_2 = \dfrac{5}{4}$ and find the intersections using the INTERSECT command.

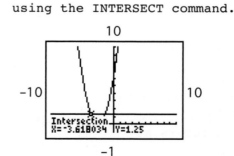
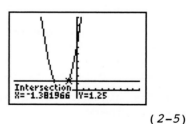

(2-5)

41. $1 + \dfrac{3}{u^2} = \dfrac{2}{u}$ Excluded value: $u \neq 0$

$u^2 + 3 = 2u$

$u^2 - 2u = -3$

$u^2 - 2u + 1 = -2$

$(u - 1)^2 = -2$

$u - 1 = \pm\sqrt{-2}$

$u = 1 \pm \sqrt{-2}$

$u = 1 \pm i\sqrt{2}$ The solutions are imaginary and cannot be confirmed graphically.

(2-6)

42. $2x + 3\sqrt{4x^2 - 4x + 9} = 1$

$$3\sqrt{4x^2 - 4x + 9} = 1 - 2x$$
$$9(4x^2 - 4x + 9) = 4x^2 - 4x + 1$$
$$36x^2 - 36x + 81 = 4x^2 - 4x + 1$$
$$32x^2 - 32x + 80 = 0$$
$$2x^2 - 2x + 5 = 0$$

$$x = \frac{-b \pm \sqrt{b^2 - 4ac}}{2a} \qquad a = 2$$
$$b = -2$$
$$c = 5$$

$$x = \frac{-(-2) \pm \sqrt{(-2)^2 - 4(2)(5)}}{2(2)} = \frac{2 \pm \sqrt{-36}}{4} = \frac{2 \pm 6i}{4} = \frac{1 \pm 3i}{2} \quad \text{or} \quad \frac{1}{2} \pm \frac{3}{2}i$$

Check: $\dfrac{1}{2} + \dfrac{3}{2}i$:

$$2\left(\frac{1}{2} + \frac{3}{2}i\right) + 3\sqrt{4\left(\frac{1}{2} + \frac{3}{2}i\right)^2 - 4\left(\frac{1}{2} + \frac{3}{2}i\right) + 9} \overset{?}{=} 1$$

$$1 + 3i + 3\sqrt{1 + 6i + 9i^2 - 2 - 6i + 9} \overset{?}{=} 1$$

$$1 + 3i + 3\sqrt{1 + 6i - 9 - 2 - 6i + 9} \overset{?}{=} 1$$

$$1 + 3i + 3\sqrt{-1} \overset{?}{=} 1$$
$$1 + 3i + 3i \overset{?}{=} 1$$
$$1 + 6i \neq 1$$

Not a solution

Check: $\dfrac{1}{2} - \dfrac{3}{2}i$: $\quad 2\left(\frac{1}{2} - \frac{3}{2}i\right) + 3\sqrt{4\left(\frac{1}{2} - \frac{3}{2}i\right)^2 - 4\left(\frac{1}{2} - \frac{3}{2}i\right) + 9} \overset{?}{=} 1$

$$1 - 3i + 3\sqrt{1 - 6i + 9i^2 - 2 + 6i + 9} \overset{?}{=} 1$$

$$1 - 3i + 3\sqrt{1 - 6i - 9 - 2 + 6i + 9} \overset{?}{=} 1$$

$$1 - 3i + 3\sqrt{-1} \overset{?}{=} 1$$

$$1 - 3i + 3i \overset{\checkmark}{=} 1 \quad \text{A solution}$$

Solution: $\dfrac{1}{2} - \dfrac{3}{2}i$ The solutions cannot be confirmed graphically.   *(2-6)*

43. $2x^{2/3} - 5x^{1/3} - 12 = 0$ Graphical check: Graph $y = 2x^{2/3} - 5x^{1/3} - 12$ and
Let $u = x^{1/3}$, then   find the zeros using the ZERO command.
$$2u^2 - 5u - 12 = 0$$
$$(2u + 3)(u - 4) = 0$$
$$u = -\frac{3}{2}, \ 4$$

$x^{1/3} = -\dfrac{3}{2} \qquad x^{1/3} = 4$

$x = -\dfrac{27}{8} \qquad x = 64$

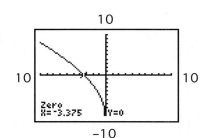

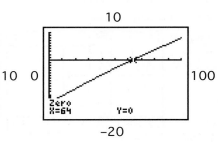

(2-6)

44. $m^4 + 5m^2 - 36 = 0$
Let $u = m^2$, then
$\quad u^2 + 5u - 36 = 0$
$\quad (u + 9)(u - 4) = 0$
$\qquad\qquad\qquad u = -9,\ 4$
$m^2 = -9 \qquad m^2 = 4$
$\ m = \pm 3i \qquad\ m = \pm 2$

Graphical check: Graph $y = x^4 + 5x^2 - 36$ and find the zeros using a built-in routine.

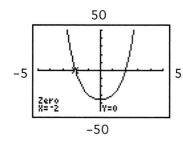
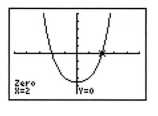

(2-6)

45. $\sqrt{y - 2} - \sqrt{5y + 1} = -3$

$\quad -\sqrt{5y + 1} = -3 - \sqrt{y - 2}$

$\quad \left(-\sqrt{5y + 1}\right)^2 = \left(-3 - \sqrt{y - 2}\right)^2$

$\qquad 5y + 1 = 9 + 6\sqrt{y - 2} + y - 2$

$\qquad 5y + 1 = y + 7 + 6\sqrt{y - 2}$

$\qquad 4y - 6 = 6\sqrt{y - 2}$

$\qquad 2y - 3 = 3\sqrt{y - 2}$

$\quad 4y^2 - 12y + 9 = 9(y - 2)$

$\quad 4y^2 - 12y + 9 = 9y - 18$

$\quad 4y^2 - 21y + 27 = 0$

$\quad (4y - 9)(y - 3) = 0$

$\qquad\qquad y = \frac{9}{4},\ 3$

> **Common Error:**
> $y - 2 - 5y + 1 = 9$
> is not equivalent to the equation formed by squaring both members of the given equation.

Check: $\sqrt{\frac{9}{4} - 2} - \sqrt{5\left(\frac{9}{4}\right) + 1} \overset{?}{=} -3$

$\qquad\qquad \sqrt{\frac{1}{4}} - \sqrt{\frac{49}{4}} \overset{?}{=} -3$

$\qquad\qquad\qquad\qquad -3 \overset{\checkmark}{=} -3$

$\qquad \sqrt{3 - 2} - \sqrt{5(3) + 1} \overset{?}{=} -3$

$\qquad\qquad\qquad\qquad -3 \overset{\checkmark}{=} -3$

Solution: $\frac{9}{4},\ 3$

Graphical check: Graph $y1 = \sqrt{x - 2} - \sqrt{5x + 1}$ and $y2 = -3$ and find the zeros using a built-in routine solutions using the INTERSECT command.

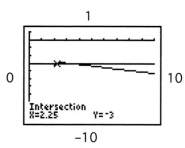
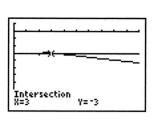

(2-6)

46. Enter the x column as L_1 and the y column as L_2 then choose the linear regression command on the STAT CALC menu.

(A)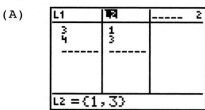

The regression line is $y = 2x - 5$.

(B)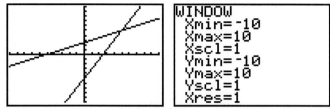

The regression line is $y = .5x + 2.5$

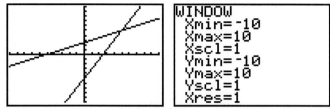

The graphs are symmetric with respect to the line $y = x$; the two functions are inverses.   (2-2)

47. It's possible for a quadratic function to have only imaginary zeros. For example, $f(x) = x^2 + 4$ has zeros $\pm 2i$, $g(x) = x^2 + 4x + 5$ has zeros $-2 \pm i$. If a quadratic function $f(x) = ax^2 + bx + c$ has only imaginary zeros, they will be

$x = \dfrac{-b \pm i\sqrt{4ac - b^2}}{2a}$ and therefore will be a pair of complex conjugate numbers. (2-5)

48. The graph of any quadratic function is a parabola. If the function has only imaginary zeros, it still can be graphed. The parabola will simply not intersect the x axis.   (2-5)

49. In this problem, $a = 1$, $b = -6$, $c = c$. The discriminant is $b^2 - 4ac = (-6)^2 - 4(1)(c) = 36 - 4c$:
If $36 - 4c > 0$, (in which case $36 > 4c$ or $c < 9$), there are two distinct real roots.
If $36 - 4c = 0$, (in which case $c = 9$), there is one real double root.
If $36 - 4c < 0$, (in which case $36 < 4c$ or $c > 9$), there are two distinct imaginary roots.   (2-5)

50.
$$P = M - Mdt$$
$$M - Mdt = P$$
$$M(1 - dt) = P$$
$$M = \dfrac{P}{1 - dt} \qquad (2\text{-}2)$$

51.
$$P = EI - RI^2$$
$$RI^2 - EI + P = 0$$
$$I = \dfrac{-b \pm \sqrt{b^2 - 4ac}}{2a} \qquad \begin{array}{l} a = R \\ b = -E \\ c = P \end{array}$$
$$I = \dfrac{-(-E) \pm \sqrt{(-E)^2 - 4(R)(P)}}{2(R)}$$
$$I = \dfrac{E \pm \sqrt{E^2 - 4PR}}{2R} \qquad (2\text{-}5)$$

52. The given inequality can be simplified as follows:
$$a + b < b - a$$
$$2a + b < b$$
$$2a < 0$$
$$a < 0$$

This tells us that the truth of the inequality has nothing to do with b; it is true as long as a is negative for any value of b.   (2-7)

53. If $a > b$ and b is negative, then $\dfrac{a}{b} < \dfrac{b}{b}$, that is, $\dfrac{a}{b} < 1$, since dividing both sides by b reverses the order of the inequality. $\dfrac{a}{b}$ is less than 1.   (2-7)

54. The distance from x to 6 is between zero and d; this distance is less than d if x is between $6 - d$ and $6 + d$. But in order for that distance to be nonzero, x cannot be equal to 6. So the solution is all numbers between $6 - d$ and $6 + d$ except 6, or $(6 - d, 6) \cup (6, 6 + d)$.   (2-7)

55. $(a + bi)\left(\dfrac{a}{a^2 + b^2} - \dfrac{b}{a^2 + b^2}\,i\right)$

$= \dfrac{(a + bi)}{1}\left(\dfrac{a}{a^2 + b^2} - \dfrac{bi}{a^2 + b^2}\right)$

$= \dfrac{a(a + bi)}{a^2 + b^2} - \dfrac{bi(a + bi)}{a^2 + b^2}$

$= \dfrac{a^2 + abi - abi - b^2 i^2}{a^2 + b^2}$

$= \dfrac{a^2 + b^2}{a^2 + b^2} = 1$ (2-4)

56. If $m = 0$, the equations are reduced to $-y = b$ (a horizontal line) and $x = b$ (a vertical line). In this case, the graphs are perpendicular. Otherwise, $m \neq 0$. Solving for y yields

$$mx - y = b \qquad\qquad x + my = b$$
$$mx = y + b \qquad\qquad my = -x + b$$
$$y = mx - b \qquad\qquad y = -\dfrac{1}{m}x + \dfrac{b}{m}$$

The first line has slope m and the second has slope $-\dfrac{1}{m}$. The graphs are perpendicular in this case also.   (2-1)

57. $2(2 + 3i)^2 - 8(2 + 3i) + 26$

$= 2(4 + 12i + 9i^2) - 16 - 24i + 26$

$= 8 + 24i - 18 - 16 - 24i + 26$

$= 0$

The imaginary zeros of quadratic equations come in conjugate pairs, so the other one must be the conjugate of $2 + 3i$, or $2 - 3i$.   (2-5)

58. The other root of the equation must be $5 - 2i$ (conjugate pairs), so the left side of the equation must factor as $(x - (5 + 2i))(x - (5 - 2i))$. Multiply out the parentheses:

$x^2 - (5 + 2i)x - (5 - 2i)x + (5 - 2i)(5 + 2i)$

$= x^2 - 5x - 2ix - 5x + 2ix + 25 - 10i + 10i - 4i^2$

$= x^2 - 10x + 25 - 4(-1)$

$= x^2 - 10x + 29$

So $b = -10$ and $c = 29$.   (2-5)

59. Algebraically:

$3x^{-2/5} - 4x^{-1/5} + 1 = 0$

Let $u = x^{-1/5}$, then

$3u^2 - 4u + 1 = 0$

$(3u - 1)(u - 1) = 0$

$u = \dfrac{1}{3},\ 1$

Graphically: Graph $y = 3x^{-2/5} - 4x^{-1/5} + 1$ and use the ZERO command to find the zeros.

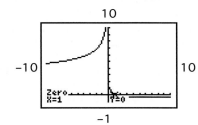

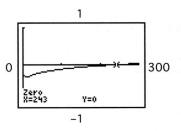

$x^{-1/5} = \dfrac{1}{3}$	$x^{-1/5} = 1$
$x^{1/5} = 3$	$x^{1/5} = 1$
$x = 243$	$x = 1$

$(2\text{-}6)$

60. $x^3 + 1 = 0$

Factor using the sum of cubes formula:

$(x + 1)(x^2 - x + 1) = 0$

$x + 1 = 0 \qquad x^2 - x + 1 = 0$

$x = -1 \qquad x = \dfrac{-b \pm \sqrt{b^2 - 4ac}}{2a} \qquad \begin{aligned} a &= 1 \\ b &= -1 \\ c &= 1 \end{aligned}$

$x = \dfrac{-(-1) \pm \sqrt{(-1)^2 - 4(1)(1)}}{2(1)}$

$= \dfrac{1 \pm \sqrt{-3}}{2}$

$= \dfrac{1 \pm i\sqrt{3}}{2}$

Solutions: $x = -1,\ \dfrac{1 \pm i\sqrt{3}}{2}$

$(2\text{-}5)$

61. Let the three integers be represented by x, $x + 1$, and $x + 2$. Then

$x + (x + 1) + (x + 2) = 144$

(The sum of the integers is 144.)

$3x + 3 = 144$

$3x = 141$

$x = 47$

The integers are 47, 48, and 49.

$(2\text{-}2)$

62. Let the three consecutive integers be represented by x, $x + 2$, and $x + 4$. (Since all are even, they must be separated by 2.) Then

$x + 2(x + 2) = 2(x + 4)$

(First plus two times second is twice the third.)

$x + 2x + 4 = 2x + 8$

$3x + 4 = 2x + 8$

$x = 4$

The integers are 4, 6, and 8.

$(2\text{-}2)$

63. $b = 5h$ $\qquad (2\text{-}2)$

64. $h = \dfrac{1}{4}b$ $\qquad (2\text{-}2)$

65. (A) The slope of this linear function is 3.4 (coefficient of x), so the snow is falling at 3.4 inches per hour.

(B) To find the amount of snow at the time it started to snow, we plug in zero for x:

$S(0) = 3.4(0) + 11.1 = 11.1$ inches

$(2\text{-}1)$

66. (A) First, we need to find the rate of growth in terms of months. We were given an increase of 75,000 per day, which corresponds to $75,000 \times 365 = 27,375,000$ per year, or $27,375,000/12 = 2,281,250$ per month. This rate of growth represents the slope of our linear function, and the given initial value (60 million) is the y-intercept. Using the slope-intercept form, we get

$$B(x) = 2,281,250x + 60,000,000$$

(B) We set our function equal to 100,000,000 and solve for x:
$$2,281,250x + 60,000,000 = 100,000,000$$
$$2,281,250x = 40,000,000$$

$$x = \frac{40,000,000}{2,281,250} \approx 17.5$$

The number of blogs is predicted to hit 100,000,000 17.5 months after August 1, 2006, which is mid-January 2008. *(2-1, 2-2)*

67. Find the vertex of the function:
$$x = -\frac{b}{2a} = -\frac{8.54}{2(-0.23)} = 18.6$$

A student should study 18.6 hours to maximize their score. *(2-3)*

68. 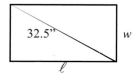 The area is given by $A = \ell \cdot w$. We can use the Pythagorean Theorem to find a relationship between ℓ and w:
$$\ell^2 + w^2 = 32.5^2$$
$$\ell^2 = 1056.25 - w^2$$
$$\ell = \sqrt{1056.25 - w^2}$$

The area is 375 square inches so
$$375 = \ell \cdot w \ \sqrt{1056.25 - w^2}$$

Graph $y_1 = 375$, $y_2 = x\sqrt{1056.25 - x^2}$ and use the INTERSECT command.

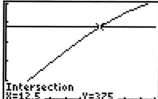

The intersection occurs when $x = 12.5$ so the width is 12.5".
$\ell = \sqrt{1056.25 - 12.5^2} = 30$ so the length is 30.0". *(2-5)*

69. The distance in feet covered by a falling object is given by $d(t) = 16t^2$. If the hammer reaches the ground after 3.5 seconds, $d(3.5)$ is the height of the building.
$$d(3.5) = 16(3.5)^2 = 196$$
The building is 196 feet tall. *(2-5)*

70. $C(x) = 0.001x^2 - 9.5x + 30,000$ is a quadratic function.

(A) Graph C and apply the MINIMUM command to find the minimum.

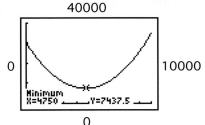

The minimum weekly cost is $7,437.50 when $x = 4,750$ calculators.

(B) Graph $y1 = C(x)$ and $y2 = 12,000$ and apply the INTERSECT command to find the intersections.

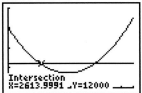 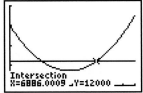

The weekly cost of $12,000 is achieved at $x = 2,614$ or $x = 6,886$ calculators.

(C) Since the minimum weekly cost is $7,437.50, a cost of $6000 is impossible. There would be no intersection of $y = C(x)$ and $y = 6,000$ (not shown).   *(2-3)*

71. $C(x) = 0.001x^2 - 9.5x + 30,000$. The revenue function is $R(x) = 3x$. The break-even points occur when $C(x) = R(x)$. Graph $y1 = C(x)$ and $y2 = R(x)$ and apply the INTERSECT command to find the intersections.

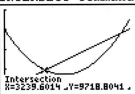 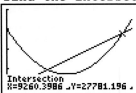

The break-even points are $x = 3,240$ or $x = 9,260$ calculators.   *(2-7)*

72. Referring to the graphs in problem 71, a profit will occur when $R(x) > C(x)$, that is, when the line is above the parabola, or $3,240 < x < 9,260$. A loss will occur when $R(x) < C(x)$, that is, when the parabola is above the line, for nonnegative values of x, $0 \le x < 3,240$ or $x > 9,260$ calculators.   *(2-7)*

73. (A) If V is linearly related to t, then we are looking for an equation whose graph passes through $(t_1, V_1) = (0, 12,000)$ and $(t_2, V_2) = (8, 2,000)$. We find the slope, and then we use the point-slope form to find the equation.

$$m = \frac{V_2 - V_1}{t_2 - t_1} = \frac{2,000 - 12,000}{8 - 0} = \frac{-10,000}{8} = -1,250$$

$$V - V_1 = m(t - t_1)$$
$$V - 12,000 = -1,250(t - 0)$$
$$V - 12,000 = -1,250t$$
$$V = -1,250t + 12,000$$

(B) We are asked for V when $t = 5$
$$v = -1,250(5) + 12,000$$
$$V = -6,250 + 12,000$$
$$V = \$5,750$$ *(2-1)*

74. (A) If R is linearly related to C, then we are looking for an equation whose graph passes through $(C_1, R_1) = (30, 48)$ and $(C_2, R_2) = (20, 32)$. We find the slope, and then we use the point-slope form to find the equation.

$$m = \frac{R_2 - R_1}{C_2 - C_1} = \frac{32 - 48}{20 - 30} = \frac{-16}{-10} = 1.6$$

$$R - R_1 = m(C - C_1)$$
$$R - 48 = 1.6(C - 30)$$
$$R - 48 = 1.6C - 48$$
$$R = 1.6C$$

(B) We are asked for R when $C = 105$.

$$R = 1.6(105)$$
$$= \$168$$

(2-1)

75. If $0 \le x \le 3,000$, $E(x) = 400$

$$\underbrace{\begin{pmatrix} \text{Base} \\ \text{Salary} \end{pmatrix}}_{} + \underbrace{\begin{pmatrix} \text{Commission on} \\ \text{Sales over } \$3,000 \end{pmatrix}}_{}$$

If $x > 3,000$,
$$\begin{aligned} E(x) &= 400 + 0.1(x - 3,000) \\ &= 400 + 0.1x - 300 \\ &= 0.1x + 100 \end{aligned}$$

Summarizing,

$$E(x) = \begin{cases} 400 & \text{if } 0 \le x \le 3,000 \\ 0.1x + 100 & \text{if } x > 3,000 \end{cases}$$

$$E(2,000) = \$400$$
$$E(5,000) = 0.1(5,000) + 100 = 500 + 100 = \$600$$

(2-1)

76. (A) From the figure, we can see that $A = x(y + y) = 2xy$. To get this written only in terms of x, we'll need another way to express variable y. Notice that the fence is made up of four pieces of length y and three pieces of length x, and a total of 120 feet of fence is used, so

$$4y + 3x = 120.$$

We can solve this equation for y:

$$4y = 120 - 3x$$

$$y = 30 - \frac{3}{4}x$$

Now we can substitute this in for y in our original area expression:

$$A = 2x\left(30 - \frac{3}{4}x\right)$$

$$A(x) = 60x - \frac{3}{2}x^2$$

(B) The domain will be based on the fact that both x and y have to be positive, since they are lengths. One obvious restriction is $x > 0$. Since we also need $y > 0$, we get

$$30 - \frac{3}{4}x > 0$$

$$30 > \frac{3}{4}x$$

$$\frac{4}{3} \cdot 30 > x$$

$$40 > x$$

So the domain is $0 < x < 40$.

(C) To find the maximum, we find the first coordinate of the vertex of the quadratic function $A(x) = -\dfrac{3}{2}x^2 + 60x$:

$x = -\dfrac{b}{2a} = -\dfrac{60}{2(-3/2)} = 20;$ when $x = 20$, $y = 30 - \dfrac{3}{4}(20) = 30 - 15 = 15$, so the area will be largest when $x = 20$ and $y = 15$. *(2-3)*

77. (A) $H = 0.7(220 - A)$

(B) We are asked to find H when $A = 20$.
$H = 0.7(220 - 20)$
$H = 140$ beats per minute.

(C) We are asked to find A when $H = 126$.
$126 = 0.7(220 - A)$
$126 = 154 - 0.7A$
$-28 = -0.7A$
$A = 40$ years old. *(2-1)*

78. Let x = width of page
y = height of page
Then $xy = 480$, and $y = \dfrac{480}{x}$.

Since the printed portion is surrounded by margins of 2 cm on each side, we have
$x - 4$ = width of printed portion
$y - 4$ = height of printed portion

The printed area is given by $(x - 4)(y - 4)$, and we're told that this area should be 320 sq. cm., so

$(x - 4)(y - 4) = 320$, that is

$(x - 4)\left(\dfrac{480}{x} - 4\right) = 320$

Solving this, we obtain:

$x\left(\dfrac{480}{x}\right) - 4x - 4\left(\dfrac{480}{x}\right) + 16 = 320$

$480 - 4x - \dfrac{1,920}{x} + 16 = 320$

$-4x - \dfrac{1,920}{x} = -176$ LCD: x $(x \neq 0)$

$-4x^2 - 1,920 = -176x$

$0 = 4x^2 - 176x + 1,920$

$0 = x^2 - 44x + 480$

$0 = (x - 20)(x - 24)$

$x - 20 = 0$ or $x - 24 = 0$
$x = 20$ $x = 24$
$\dfrac{480}{x} = 24$ $\dfrac{480}{x} = 20$

The dimensions of the page are 20 cm by 24 cm. *(2-5)*

79.

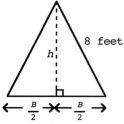

In the isosceles triangle we note:

$$\frac{1}{2}Bh = A = 24$$

So

$$h = \frac{48}{B}$$

Applying the Pythagorean theorem, we have

$$h^2 + \left(\frac{B}{2}\right)^2 = 8^2$$

$$\left(\frac{48}{B}\right)^2 + \left(\frac{B}{2}\right)^2 = 8^2$$

$$\frac{2,304}{B^2} + \frac{B^2}{4} = 64$$

$$4B^2\left(\frac{2,304}{B^2}\right) + 4B^2\left(\frac{B^2}{4}\right) = 4B^2(64)$$

$$9,216 + B^4 = 256B^2$$

$$(B^2)^2 - 256B^2 + 9,216 = 0$$

$$B^2 = \frac{-b \pm \sqrt{b^2 - 4ac}}{2a} \qquad a = 1, \ b = -256, \ c = 9,216$$

$$B^2 = \frac{-(-256) \pm \sqrt{(-256)^2 - 4(1)(9,216)}}{2(1)}$$

$$B^2 = \frac{256 \pm \sqrt{65,536 - 36,864}}{2}$$

$$B^2 = \frac{256 \pm \sqrt{28,672}}{2}$$

$$B^2 = 128 \pm 32\sqrt{7}$$

$$B = \sqrt{128 \pm 32\sqrt{7}}$$

$$B = 14.58 \text{ ft or } 6.58 \text{ ft} \qquad\qquad (2\text{-}6)$$

80. Sketch a figure:

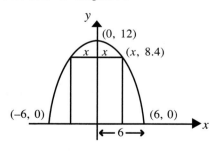

The equation of the parabolic entrance way in the coordinate system with vertex at $(0, 12)$ shown must be of the form $y = ax^2 + 12$. Since $(6, 0)$ is on the parabola, the coordinates of the point must satisfy the equation, so

$$0 = a(6)^2 + 12$$

$$a = -\frac{1}{3}$$

The equation of the parabola is $y = -\frac{1}{3}x^2 + 12$. To find the width of the door that has height 8.4, let $y = 8.4$ and solve for x.

$$8.4 = -\frac{1}{3}x^2 + 12$$

$$-3.6 = -\frac{1}{3}x^2$$

$$x^2 = 10.8$$

$$x = \sqrt{10.8} \approx 3.3 \text{ ft.}$$

The width of the door is $2x = 6.6$ ft. $\qquad\qquad (2\text{-}3)$

81. (A)

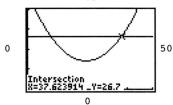

Enter data.   Compute regression equation.

The quadratic regression model is
$$y(x) = 0.0615x^2 - 2.88x + 48.0.$$

(B)

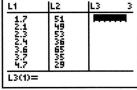

Transfer the regression   Graph the regression equation and
equation to the equation   determine when the level of 26.7
editor.   will be reached a second time.

The model predicts that the percentage of marijuana users will return to the
1979 level when $t \approx 38$, or 2008.   (2-3)

82. (A)

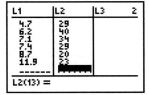

Enter data.   Compute regression equation.

The linear regression model is $y = -3.10x + 54.6$.

(B)

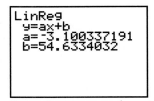

Transfer the regression   Graph the regression equation and
equation to the equation   determine the percentage that corresponds
editor.   to the assessed value of $300.

The model predicts a percentage of 45.33% votes for democrats.   (2-1)

83. For the first model, the independent variable is supply so enter the supply
column as L_1. The dependent variable is price so enter the price column as L_2.
Then choose linear regression from the STAT CALC menu.

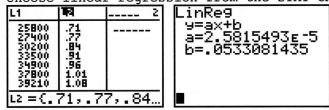

The price-supply model is $y = 0.000\ 0258x + 0.0533$.

For the second model, the independent variable is demand so we enter the demand column as L_1 leaving price as L_2.

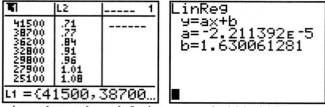

The price-demand model is $y = -0.000\ 0221x + 1.63$.

The equilibrium price occurs when supply and demand are equal. We graph the supply and demand curves and find where they intersect.

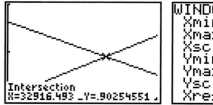

The equilibrium price is the y-coordinate of the intersection, which is $0.90.(2-2)$

84. (A) The revenue function is obtained from multiplying the demand function (price) by pounds sold (x).
$$R(x) = (-0.000\ 0221x + 1.63)x = -0.000\ 0221x^2 + 1.63x$$

The number of pounds must be positive and revenue must be positive also. The graph of $R(x)$ shows that the domain of $R(x)$ in this setting is $0 \le x \le 73{,}800$.

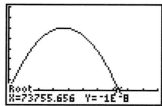

The domain of the cost function is $x \ge 0$: pounds produced cannot be negative and we're not given a limit on production.

The cost is fixed costs ($15,000) plus $0.20 times pounds produced (x).
$$C(x) = 15{,}000 + 0.20x$$

(B) The company will break-even if revenue and cost are equal so we find where the graphs of $R(x)$ and $C(x)$ intersect.

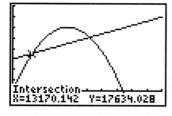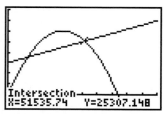

The x-values at the points of intersection are 13,200 and 51,500 (rounded to 3 significant digits), so the company will break-even with production of 13,200 lbs. or 51,500 lbs. They will make a profit if production is between those levels and will lose money if production is less than 13,200 lbs. or greater then 51,500 lbs.

(C) Profit = Revenue - Cost

$P(x) = R(x) - C(x) = -0.0000221x^2 + 1.63x - (15{,}000 + 0.20x)$

$P(x) = -0.0000221x^2 + 1.43x - 15{,}000$ (Distribute the negative!)

Graph $y_1 = P(x)$ as above and use the MAXIMUM command.

The maximum profit of $8,130 occurs when 32,400 pounds are sold.

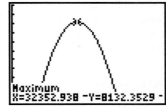

(2-5,2-7)

85. (A) The independent variable is speed and the dependent variable is mileage, so we enter the speed column as L_1 and the mileage column as L_2. Then we choose quadratic regression from the STAT CALC menu.

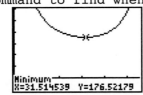

The quadratic model is $y = -0.00174x^2 + 0.0865x + 0.980$.

(B) The rental cost is $15 per hour plus $2.55 per gallon for gas. This can be expressed as $C = 15t + 2.55g$ where t is the number of hours and g is gallons of gas used. We can write each of t and g in terms of speed (x) to get a model for cost in terms of speed.

The time of the trip will be the distance divided by the speed since distance = speed times time. The trip is 100 miles so $t = \dfrac{100}{x}$. The gasoline usage can be calculated by dividing the length (100 miles) by the miles per gallon (y). So $g = \dfrac{100}{y}$. Our quadratic model tells us that $y = -0.00174x^2 + 0.0865x + 0.980$.

So, $g = \dfrac{100}{-0.00174x^2 + 0.0865x + 0.980}$

Our rental cost is now $C = 15\left(\dfrac{100}{x}\right) + 2.55\left(\dfrac{100}{-0.00174x^2 + 0.0865x + 0.980}\right)$

$$C = \frac{1500}{x} + \frac{255}{-0.00174x^2 + 0.0865x + 0.980}$$

We use the MINIMUM command to find when this is minimized.

The minimum cost occurs when the speed is 31.5 mph. Plugging $x = 31.5$ into our quadratic regression model we find that the mileage at 31.5 mph is 1.98 mpg. We use our expressions for t, g, and C to answer the remaining questions.

$t = \dfrac{100}{x} = \dfrac{100}{31.5} = 3.17$ hours

$g = \dfrac{100}{y} = \dfrac{100}{1.98} = 50.5$ gallons

$C = 15t + 2.55g = 15(2.98) + 2.55(52.1) = \$178$

(2-5)

Chapters 1 & 2 Cumulative Review

1. (A)

(B) Since the x coordinates vary between a minimum of −3 and a maximum of 3, Xmin = −3 and Xmax = 3. Since the y coordinates vary between a minimum of −4 and a maximum of 4, Ymin = −4 and Ymax = 4.

(C) Not a function (two range values correspond to domain value 1)

(1-1, 1-2)

2. (A) Use slope-intercept form with $m = \dfrac{6 - 2}{5 - 3} = \dfrac{4}{2} = 2.$

$$y = 2x + b$$

Since (3, 2) is on the line, its coordinates satisfy the equation:

$$2 = 2 \cdot 3 + b$$
$$b = -4$$

Equation: $y = 2x - 4$

(B) A line perpendicular to $y = 2x - 4$, which has slope 2, will have slope satisfying $2m = -1$, or $m = -\dfrac{1}{2}$. We use the point-slope form

$$y - 6 = -\frac{1}{2}(x - 5)$$
$$y - 6 = -\frac{1}{2}x + \frac{5}{2}$$
$$y = -\frac{1}{2}x + \frac{5}{2} + 6$$
$$y = -\frac{1}{2}x + \frac{17}{2}$$

(C)

x	$y = 2x - 4$	$y = -\dfrac{1}{2}x + \dfrac{17}{2}$
0	−4	17 / 2
1	−2	8
5	6	6

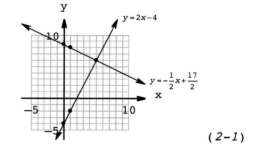

(2-1)

3. $2x - 3y = 6$

$$-3y = -2x + 6$$
$$y = \frac{2}{3}x - 2$$

slope: $\dfrac{2}{3}$ y intercept: −2 x intercept: 3 (if $y = 0$, $2x = 6$, and $x = 3$)

x	y
−3	−4
0	−2
0	0

(2-1)

4. (A) $f(-2) + g(-3) = (-2)^2 - 2(-2) + 5 + 3(-3) - 2 = 4 + 4 + 5 - 9 - 2 = 2$

(B) $f(1) \cdot g(1) = (1^2 - 2 \cdot 1 + 5) \cdot (3 \cdot 1 - 2) = 4 \cdot 1 = 4$

(C) $\dfrac{g(0)}{f(0)} = \dfrac{3 \cdot 0 - 2}{0^2 - 2 \cdot 0 + 5} = \dfrac{-2}{5}$ *(1-2)*

5. (A) Stretched vertically by a factor of 2
(B) Shifted right 2 units
(C) Shifted down 2 units   *(1-4)*

6. Domain: $[-2, 3]$ Range: $[-1, 2]$ *(1-2)*

7. Notice that $f(2) = 2$, while $f(-2) = -1$. We have found an x value for which $f(x)$ and $f(-x)$ are not equal, so f is not even. We have also found an x value for which $f(x) \neq -f(-x)$, so f is not odd, either.   *(1-4)*

8. (A) The graph of $f(x)$ is shifted left one unit and reflected across the x axis.

(B) The graph of $f(x)$ is stretched vertically by a factor of two and shifted down two units.

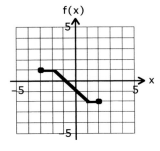

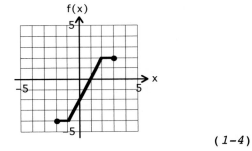

(1-4)

9. $5 - 3(x + 4) = x + 2(11 - 2x)$
$5 - 3x - 4 = x + 22 - 4x$
$-3x + 1 = -3x + 22$
$1 = 22$

This is a contradiction, so there is no solution.

Check: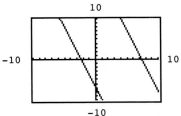

The lines are parallel and never intersect, so there is no solution.   *(2-2)*

Check:

10. $(4x + 15)^2 = 9$
$4x + 15 = \pm 3$
$4x = -15 \pm 3 = -18, -12$
$x = -\dfrac{18}{4}, -\dfrac{12}{4}$
$x = -\dfrac{9}{2}, -3$

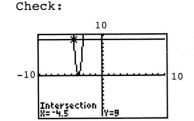

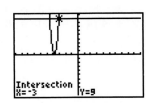

11.
$$\frac{7x}{5} - \frac{3 + 2x}{2} = \frac{x - 10}{3} + 2$$
$$30\frac{7x}{5} - 30\frac{(3 + 2x)}{2} = 30\frac{(x - 10)}{3} + 2(30)$$
$$42x - 15(3 + 2x) = 10(x - 10) + 60$$
$$42x - 45 - 30x = 10x - 100 + 60$$
$$12x - 45 = 10x - 40$$
$$2x = 5$$
$$x = \frac{5}{2}$$

Check:

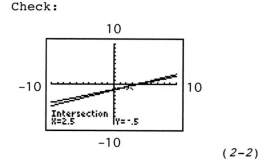

(2-2)

12.
$$3x^2 = -12x$$
$$3x^2 + 12x = 0$$
$$3x(x + 4) = 0$$
$$3x = 0 \quad \text{or} \quad x + 4 = 0$$
$$x = 0 \qquad\qquad x = -4$$

Check:

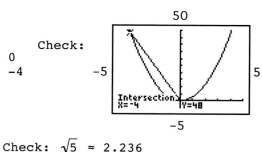
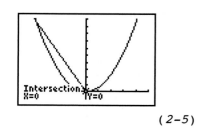

(2-5)

13.
$$4x^2 - 20 = 0$$
$$4x^2 = 20$$
$$x^2 = 5$$
$$x = \pm\sqrt{5}$$

Check: $\sqrt{5} \approx 2.236$

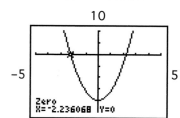
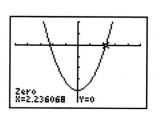

(2-5)

14.
$$x^2 - 6x + 2 = 0$$
$$x^2 - 6x = -2$$
$$x^2 - 6x + 9 = 7$$
$$(x - 3)^2 = 7$$
$$x - 3 = \pm\sqrt{7}$$
$$x = 3 \pm \sqrt{7}$$

Check: $3 + \sqrt{7} \approx 5.646$
$3 - \sqrt{7} \approx 0.354$

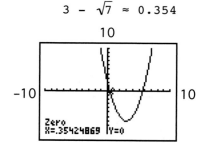
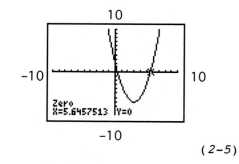

(2-5)

15.
$$x - \sqrt{12 - x} = 0$$
$$x = \sqrt{12 - x}$$
$$x^2 = 12 - x$$
$$x^2 + x - 12 = 0$$
$$(x + 4)(x - 3) = 0$$
$$x = -4, 3$$

Check: $-4 - \sqrt{12 - (-4)} = -4 - \sqrt{16} = -8$ Not a solution
$3 - \sqrt{12 - 3} = 3 - \sqrt{9} = 0$ √

Solution: $x = 3$

Check:

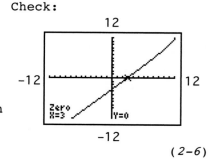

(2-6)

16. $2(3 - y) + 4 \leq 5 - y$
$6 - 2y + 4 \leq 5 - y$
$-2y + 10 \leq 5 - y$
$-y \leq -5$
$y \geq 5$ *(2-7)*

17. $-3 \leq 2 - 3x < 11$
$-5 \leq -3x < 9$
$\dfrac{5}{3} \geq x > -3$ (Note that the inequalities reversed)

Solution: $-3 < x \leq \dfrac{5}{3}$, or $\left(-3, \dfrac{5}{3}\right]$ *(2-7)*

18. $|x - 2| < 7$
$-7 < x - 2 < 7$
$-5 < x < 9$ *(2-7)*

19. Rearrange so that the right side is zero:
$x^2 + 3x - 10 \geq 0$

Find the zeros of the left side:
$(x + 5)(x - 2) = 0$;   zeros are -5 and 2

Choose test numbers and make a table:

Interval	Test Number	Result
$x < -5$	-6	$(-6)^2 + 3(-6) - 10 = 8$; positive
$-5 < x < 2$	0	$(0)^2 + 3(0) - 10 = -10$; negative
$x > 2$	4	$(4)^2 + 3(4) - 10 = 18$; positive

Solution: $x \leq -5$ or $x \geq 2$, or $(-\infty, -5] \cup [2, \infty)$ *(2-7)*

20. (A) $f(x) = x^2 - 4x - 1$
$= (x^2 - 4x\ \ \ \ \ ) - 1$
$= (x^2 - 4x + 4) - 1 - 4$
$= (x - 2)^2 - 5$

(B) The graph of f is the same as the graph of $y = x^2$ shifted to the right 2 units and down 5 units.

(C) Solve $0 = x^2 - 4x - 1$
$0 = (x - 2)^2 - 5$
$(x - 2)^2 = 5$
$x - 2 = \pm\sqrt{5}$
$x = 2 \pm \sqrt{5}$

Check: $2 + \sqrt{5} \approx 4.236$
$2 - \sqrt{5} \approx -0.236$

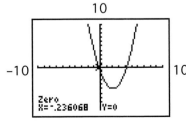
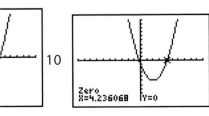

(2-3, 2-5)

21. (A) $(2 - 3i) - (-5 + 7i) = 2 - 3i + 5 - 7i = 7 - 10i$

(B) $(1 + 4i)(3 - 5i) = 3 + 7i - 20i^2 = 3 + 7i + 20 = 23 + 7i$

(C) $\dfrac{5 + i}{2 + 3i} = \dfrac{(5 + i)}{(2 + 3i)}\dfrac{(2 - 3i)}{(2 - 3i)} = \dfrac{10 - 13i - 3i^2}{4 - 9i^2}$

$\qquad = \dfrac{10 - 13i + 3}{4 + 9} = \dfrac{13 - 13i}{13} = 1 - i$ *(2-4)*

22. (A) All real numbers $(-\infty, \infty)$

(B) From the graph, the possible function values include -2 (only) and all numbers greater than or equal to 1. In set and interval notation: $\{-2\} \cup [1, \infty)$.

(C) $f(-3) = 1$ $f(-2) = 2$ (not -2) $f(2) = -2$ (not 2).
Then $f(-3) + f(-2) + f(2) = 1 + 2 + (-2) = 1$

(D) $[-3, -2]$ and $[2, \infty)$

(E) f is discontinuous at $x = -2$ and at $x = 2$.   *(1-4)*

23. $(f \circ g)(x) = f[g(x)] = f\left(\dfrac{x + 3}{x}\right) = \dfrac{1}{\frac{x+3}{x} - 2} = \dfrac{x(1)}{x\left(\frac{x+3}{x} - 2\right)} = \dfrac{x}{x + 3 - 2x} = \dfrac{x}{3 - x}$

The domain of f is all real numbers except 2.
The domain of g is all non-zero real numbers.
The domain of $f \circ g$ is the set of all non-zero real numbers for which $g(x) \neq 2$.

$g(x) = 2$ only if $\dfrac{x + 3}{x} = 2$, that is $x + 3 = 2x$, or $x = 3$.

The domain of $f \circ g$ is the set of all non-zero real numbers except 3, that is, $(-\infty, 0) \cup (0, 3) \cup (3, \infty)$.

> Common Error:
> The domain of $f \circ g$ cannot be found by looking at the final form $\dfrac{x}{3 - x}$. It is not $\{x \mid x \neq 3\}$.

(1-5)

24. $f(x) = 2x + 5$
Assume $f(a) = f(b)$
$\qquad 2a + 5 = 2b + 5$
$\qquad\quad 2a = 2b$
$\qquad\quad\ a = b$
f is one-to-one.

Solve $y = f(x)$ for x:
$\quad y = 2x + 5$
$y - 5 = 2x$
$\quad x = \dfrac{y - 5}{2} = f^{-1}(y)$

Interchange x and y
$y = f^{-1}(x) = \dfrac{x - 5}{2}$ or $\dfrac{1}{2}x - \dfrac{5}{2}$ Domain: $(-\infty, \infty)$ *(1-6)*

25. $f(x) = \sqrt{x + 4}$
Assume $f(a) = f(b)$
$\qquad \sqrt{a + 4} = \sqrt{b + 4}$
$\qquad\ a + 4 = b + 4$
$\qquad\qquad a = b$
f is one-to-one.

(A) Solve $y = f(x)$ for x
$y = \sqrt{x + 4}$
$y^2 = x + 4 \quad y \geq 0$
$\ x = y^2 - 4 \quad y \geq 0 \quad f^{-1}(y) = y^2 - 4$

Interchange x and y:

$y = f^{-1}(x) = x^2 - 4$ Domain: $x \geq 0$

Check: $f^{-1}[f(x)] = (\sqrt{x+4})^2 - 4 = x + 4 - 4 = x$

$\qquad f[f^{-1}(x)] = \sqrt{x^2 - 4 + 4} = \sqrt{x^2} = x$ since $x \geq 0$ in the domain of f^{-1}.

(B) Domain of $f = [-4, \infty)$ = Range of f^{-1}
    Range of $f = [0, \infty)$ = Domain of f^{-1}

(C)

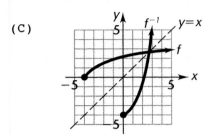

$(1\text{-}6)$

26. (A) The graph of f passes the horizontal line test.
f is one-to-one.

(B) Since $g(0) = g(-1) = 0$, g is not one-to-one.

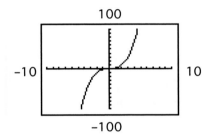

$(1\text{-}6)$

27. The line $3x + 2y = 12$, or
$\qquad 2y = -3x + 12$,

or $y = -\dfrac{3}{2}x + 6$, has slope $-\dfrac{3}{2}$.

(A) We require a line through $(-6, 1)$ with slope $-\dfrac{3}{2}$. Applying the point-slope

form, we have $y - 1 = -\dfrac{3}{2}(x + 6)$

$\qquad\qquad\qquad y - 1 = -\dfrac{3}{2}x - 9$

$\qquad\qquad\qquad\quad y = -\dfrac{3}{2}x - 8$

(B) The slope of a line parallel to one with slope $-\dfrac{3}{2}$ will have slope $\dfrac{2}{3}$

(negative reciprocals). Again applying the point-slope form, we have

$$y - 1 = \frac{2}{3}(x + 6)$$

$$y - 1 = \frac{2}{3}x + 4$$

$$y = \frac{2}{3}x + 5 \qquad\qquad (2\text{-}1)$$

28. $f(x) = x^2 - 2x - 8$. Use the vertex formula:

$$x = -\frac{b}{2a} = -\frac{-2}{2(1)} = 1; \quad f(1) = 1^2 - 2(1) - 8 = -9$$

The vertex is $(1, -9)$ and the axis is $x = 1$. the parabola opens up since $x > 0$, so the vertex is a minimum; the minimum value is -9 and occurs at $x = 1$. The range is then $[-9, \infty)$.

y intercept: Set $x = 0$, then $f(0) = -8$ is the y intercept.

x intercepts: Set $f(x) = 0$, then

$$0 = x^2 - 2x - 8$$
$$0 = (x - 4)(x + 2)$$

$x = 4$ or $x = -2$ are the x intercepts.

Graph: Locate axis and vertex, then plot several points on either side of the axis.

x	$f(x)$
-3	7
-2	0
-1	-5
0	-8
1	-9
2	-8
3	-5
4	0
5	7

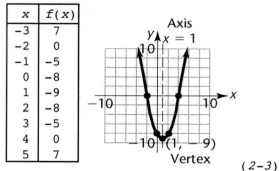

(2-3)

29. $|4x - 9| > 3$

$$4x - 9 < -3 \quad \text{or} \quad 4x - 9 > 3$$
$$4x < 6 \qquad\qquad 4x > 12$$
$$x < \frac{3}{2} \quad \text{or} \qquad x > 3$$

$$\left(-\infty, \frac{3}{2}\right) \cup (3, \infty) \qquad\qquad (2\text{-}7)$$

30. Multiply both sides by 2 to clear fractions:

$$x^2 - 2x - 16 > 0$$

Find the zeros of the left side (quadratic formula with $a = 1$, $b = -2$, $c = -16$)

$$x = \frac{-(-2) \pm \sqrt{(-2)^2 - 4(1)(-16)}}{2(1)}$$

$$x = \frac{2 \pm \sqrt{68}}{2} = \frac{2 \pm 2\sqrt{17}}{2} = 1 \pm \sqrt{17} \approx 5.1, -3.1$$

Choose test numbers and make a table:

Interval	Test Number	Result
$x < -3.1$	-4	$(-4)^2 - 2(-4) - 16 = 8;$ positive
$-3.1 < x < 5.1$	0	$(0)^2 - 2(0) - 16 = -16;$ negative
$x > 5.1$	6	$(6)^2 - 2(6) - 16 = 8;$ positive

Solution: $x < -3.1$ or $x > 5.1$, or $(-\infty, -3.1) \cup (5.1, \infty)$ $\qquad\qquad (2\text{-}7)$

31. (A) $(2 - 3i)^2 - (4 - 5i)(2 - 3i) - (2 + 10i)$

$$= (2)^2 - 2(2)(3i) + (3i)^2 - (8 - 12i - 10i + 15i^2) - 2 - 10i$$

$$= 4 - 12i + 9i^2 - (8 - 22i + 15i^2) - 2 - 10i$$

$$= 4 - 12i + 9i^2 - 8 + 22i - 15i^2 - 2 - 10i$$

$$= 4 - 12i - 9 - 8 + 22i + 15 - 2 - 10i$$

$$= 0 + 0i \quad \text{or} \quad 0$$

(B) $\dfrac{3}{5} + \dfrac{4}{5}i + \dfrac{1}{\frac{3}{5} + \frac{4}{5}i} = \dfrac{3}{5} + \dfrac{4}{5}i + \dfrac{1}{\left(\frac{3}{5} + \frac{4}{5}i\right)} \dfrac{\left(\frac{3}{5} - \frac{4}{5}i\right)}{\left(\frac{3}{5} - \frac{4}{5}i\right)}$

$\qquad\qquad\qquad = \dfrac{3}{5} + \dfrac{4}{5}i + \dfrac{\frac{3}{5} - \frac{4}{5}i}{\frac{9}{25} - \frac{16}{25}i^2}$

$\qquad\qquad\qquad = \dfrac{3}{5} + \dfrac{4}{5}i + \dfrac{\frac{3}{5} - \frac{4}{5}i}{\frac{9}{25} + \frac{16}{25}}$

$\qquad\qquad\qquad = \dfrac{3}{5} + \dfrac{4}{5}i + \dfrac{3}{5} - \dfrac{4}{5}i$

$\qquad\qquad\qquad = \dfrac{6}{5}$

(C) $i^{35} = i^{32}i^3 = (i^4)^8(-i) = 1^8(-i) = -i$ $\qquad\qquad\qquad$ (*2-4*)

32. (A) $(5 + 2\sqrt{-9}) - (2 - 3\sqrt{-16}) = (5 + 2i\sqrt{9}) - (2 - 3i\sqrt{16})$

$\qquad\qquad\qquad\qquad = (5 + 6i) - (2 - 12i)$

$\qquad\qquad\qquad\qquad = 5 + 6i - 2 + 12i$

$\qquad\qquad\qquad\qquad = 3 + 18i$

(B) $\dfrac{2 + 7\sqrt{-25}}{3 - \sqrt{-1}} = \dfrac{2 + 7i\sqrt{25}}{3 - i}$

$\qquad = \dfrac{2 + 35i}{3 - i}$

$\qquad = \dfrac{(2 + 35i)(3 + i)}{(3 - i)(3 + i)}$

$\qquad = \dfrac{6 + 107i + 35i^2}{9 - i^2}$

$\qquad = \dfrac{6 + 107i - 35}{9 + 1}$

$\qquad = \dfrac{-29 + 107i}{10}$

$\qquad = -2.9 + 10.7i$

(C) $\dfrac{12 - \sqrt{-64}}{\sqrt{-4}} = \dfrac{12 - i\sqrt{64}}{i\sqrt{4}}$

$\qquad = \dfrac{12 - 8i}{2i}$

$\qquad = \dfrac{12 - 8i}{2i} \dfrac{i}{i}$

$\qquad = \dfrac{12i - 8i^2}{2i^2}$

$\qquad = \dfrac{12i + 8}{-2}$

$\qquad = -4 - 6i$ $\qquad$ (*2-4*)

33.

x	$y = x - 1$	x	$y = x^2 + 1$
-1	-2	0	1
-2	-3	1	2
		2	5

Domain: all real numbers
Range: $(-\infty, -1) \cup [1, \infty)$
Discontinuous at: $x = 0$

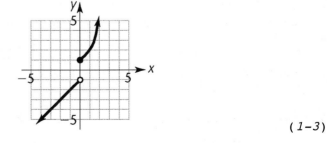

(*1-3*)

34. The graph of $y = |x|$ is vertically shrunk by $\dfrac{1}{2}$, reflected in the x axis, shifted two units to the right and three units up; $y = -\dfrac{1}{2}|x - 2| + 3$.　　(*1-4*)

35. The vertex is (-2, -3), so the equation is of the form

$$y = a(x + 2)^2 - 3$$

What remains is to find a. To do so, we can use the other point we know, (-1, -2). This point has to satisfy the equation above, so

$$-2 = a(-1 + 2)^2 - 3 \qquad \text{(Solve for } a)$$
$$-2 = a - 3$$
$$1 = a$$

The equation is $y = (x + 2)^2 - 3$. (2-3)

36.
$$1 + \frac{14}{y^2} = \frac{6}{y} \qquad \text{Excluded value: } y \neq 0$$

$$y^2(1) + y^2\frac{14}{y^2} = y^2\frac{6}{y}$$

$$y^2 + 14 = 6y$$
$$y^2 - 6y + 14 = 0$$
$$y^2 - 6y = -14$$
$$y^2 - 6y + 9 = -5$$
$$(y - 3)^2 = -5$$
$$y - 3 = \pm\sqrt{-5}$$
$$y = 3 \pm \sqrt{-5}$$
$$y = 3 \pm i\sqrt{5}$$

(2-6)

37. $4x^{2/3} - 4x^{1/3} - 3 = 0$
Let $u = x^{1/3}$, then
$$4u^2 - 4u - 3 = 0$$
$$(2u - 3)(2u + 1) = 0$$
$$u = \frac{3}{2}, \ -\frac{1}{2}$$

$$x^{1/3} = \frac{3}{2} \qquad x^{1/3} = -\frac{1}{2}$$

$$x = \frac{27}{8} \qquad x = -\frac{1}{8}$$

Check: Graph $y = 4x^{2/3} - 4x^{1/3} - 3$ and use the ZERO command to find the zeros.

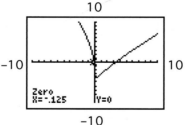 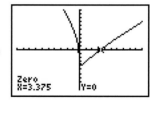

(2-6)

38. $u^4 + u^2 - 12 = 0$
Let $w = u^2$, then
$$w^2 + w - 12 = 0$$
$$(w + 4)(w - 3) = 0$$
$$w = -4, \ 3$$
$$u^2 = -4 \qquad u^2 = 3$$
$$u = \pm 2i \qquad u = \pm\sqrt{3}$$

Check: Graph $y = x^4 + x^2 - 12$ and use the ZERO command to find the zeros.

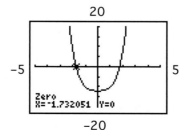 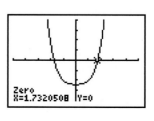

(2-6)

39.
$$\sqrt{8t - 2} - 2\sqrt{t} = 1$$
$$\sqrt{8t - 2} = 2\sqrt{t} + 1$$
$$8t - 2 = 4t + 4\sqrt{t} + 1$$
$$4t - 3 = 4\sqrt{t}$$
$$16t^2 - 24t + 9 = 16t$$
$$16t^2 - 40t + 9 = 0$$
$$(4t - 1)(4t - 9) = 0$$
$$t = \frac{1}{4}, \frac{9}{4}$$

Check: $\sqrt{8\left(\frac{1}{4}\right) - 2} - 2\sqrt{\frac{1}{4}} = \sqrt{0} - 2\left(\frac{1}{2}\right) = -1 \neq 1$
Not a solution

$\sqrt{8\left(\frac{9}{4}\right) - 2} - 2\sqrt{\frac{9}{4}} = \sqrt{16} - 2\left(\frac{3}{2}\right) = 4 - 3 = 1$ √

Solution: $t = \frac{9}{4}$

Graphical check: Graph $y1 = \sqrt{8x - 2} - 2\sqrt{x}$ and $y2 = 1$ and use the INTERSECT command to find the intersection.

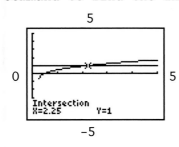

$(2-6)$

40.
$$6x = \sqrt{9x^2 - 48}$$
$$36x^2 = 9x^2 - 48$$
$$27x^2 = -48$$
$$x^2 = -\frac{48}{27}$$
$$x^2 = -\frac{16}{9}$$
$$x = \pm\sqrt{-\frac{16}{9}}$$
$$x = \pm\frac{4}{3}i$$

Check: $\frac{4}{3}i$:

$$6\left(\frac{4}{3}i\right) \overset{?}{=} \sqrt{9\left(\frac{4}{3}i\right)^2 - 48}$$
$$8i \overset{?}{=} \sqrt{-16 - 48}$$
$$8i \overset{\checkmark}{=} \sqrt{-64}$$
A solution

Check: $-\frac{4}{3}i$:

$$6\left(-\frac{4}{3}i\right) \overset{?}{=} \sqrt{9\left(-\frac{4}{3}i\right)^2 - 48}$$
$$-8i \overset{?}{=} \sqrt{-16 - 48}$$
$$-8i \neq \sqrt{-64}$$
Not a solution

Solution: $\frac{4}{3}i$ $(2-6)$

41. In this problem, $a = 1$, $b = b$, $c = 1$, and the discriminant $b^2 - 4ac = b^2 - 4 \cdot 1 \cdot 1 = b^2 - 4$. The number and types of roots depend on the sign of $b^2 - 4 = (b - 2)(b + 2)$. A glance at the graph of $y = x^2 - 4$ (not shown) shows that $b^2 - 4$ is positive if $b > 2$ or $b < -2$, and that there are two distinct real roots of $x^2 + bx + 1 = 0$ in this case. $b^2 - 4$ is equal to zero if $b = -2$ or $b = 2$, and there is one real double root of $x^2 + bx + 1 = 0$ in this case. $b^2 - 4$ is negative if $-2 < b < 2$, and there are two distinct imaginary roots of $x^2 + bx + 1 = 0$ in this case. $(2-5)$

42. Odd: $f(x) = x$ Even: $f(x) = x^2$
If a function is both even and odd, then $f(x) = f(-x) = -f(x)$; adding $f(x)$ to both sides, we get $2f(x) = 0$ and $f(x) = 0$ for all x in its domain. $(1-4)$

43. A quadratic equation $ax^2 + bx + c = 0$ with real coefficients will have imaginary roots only if $b^2 - 4ac =$ is negative. In this case it will have two distinct imaginary roots that are conjugates, so it is not possible for it to have one real and one imaginary root or a double imaginary root. $(2-5)$

44. $\dfrac{g(2+h)-g(2)}{h} = \dfrac{[-2(2+h)^2+3(2+h)-1]-[-2(-2)^2+3(-2)-1]}{h}$

$\qquad = \dfrac{-2(4+4h+h^2)+6+3h-1-[-8-6-1]}{h}$

$\qquad = \dfrac{-8-8h-2h^2+6+3h-1+8+6+1}{h}$

$\qquad = \dfrac{-5h-2h^2}{h} = -5-2h$ $\hspace{3cm}$ $(1\text{-}3)$

45. The graph of $y = x^{1/3}$ has been vertically stretched by a factor of 2 and shifted 1 unit to the left and 1 unit down. $y = 2\sqrt[3]{x+1}-1$. $\hspace{1cm}$ $(1\text{-}4)$

46. (A) $A = 2\pi r(r+h)$
$\qquad A = 2\pi r^2 + 2\pi rh$
$\qquad A - 2\pi r^2 = 2\pi rh$
$\qquad h = \dfrac{A-2\pi r^2}{2\pi r}$

(B) $A = 2\pi r(r+h)$
$\qquad A = 2\pi r^2 + 2\pi rh$
$\qquad r^2 + hr = \dfrac{A}{2\pi}$
$\qquad r^2 + hr + \dfrac{h^2}{4} = \dfrac{h^2}{4} + \dfrac{A}{2\pi}$
$\qquad \left(r+\dfrac{h}{2}\right)^2 = \dfrac{h^2}{4} + \dfrac{A}{2\pi}$
$\qquad r+\dfrac{h}{2} = \pm\sqrt{\dfrac{h^2}{4} + \dfrac{A}{2\pi}}$
$\qquad r = -\dfrac{h}{2} + \sqrt{\dfrac{h^2}{4} + \dfrac{A}{2\pi}}$

The negative root $r = -\dfrac{h}{2} - \sqrt{\dfrac{h^2}{4} + \dfrac{A}{2\pi}}$ is discarded because r, the radius of the cylinder, must be positive. $\hspace{2cm}$ $(2\text{-}2, 2\text{-}5)$

47. (A) $g(x) = \sqrt{4-x^2}$

The domain of g is those values of x for which $4-x^2 \geq 0$.
Graphing $y = 4-x^2$, we obtain the graph at the right.

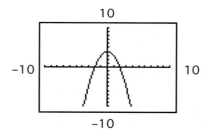

The domain of g is $-2 \leq x \leq 2$, or $[-2, 2]$.

(B) $\left(\dfrac{f}{g}\right)(x) = \dfrac{f(x)}{g(x)} = \dfrac{x^2}{\sqrt{4-x^2}}$. The domain of $\dfrac{f}{g}$ is the intersection of the domains of f (all real numbers) and g ($[-2, 2]$) with the exclusion of those points (-2 and 2) where $g(x) = 0$. So the domain of $f/g = (-2, 2)$.

(C) $(f \circ g)(x) = f[g(x)] = f(\sqrt{4-x^2}) = (\sqrt{4-x^2})^2 = 4-x^2$.
The domain of $f \circ g$ is the set of all real numbers in the domain of g ($[-2, 2]$) for which $f(x)$ is real, that is, all numbers in $[-2, 2]$. $\hspace{1cm}$ $(1\text{-}5)$

48. (A) $f(x) = x^2 - 2x - 3$ $x \geq 1$

f passes the horizontal line test (see graph below, in part C), so f is one-to-one.

Solve $y = f(x)$ for x:
$$y = x^2 - 2x - 3 \quad x \geq 1$$
$$x^2 - 2x = y + 3 \quad\quad x \geq 1$$
$$x^2 - 2x + 1 = y + 4 \quad\quad x \geq 1$$
$$(x - 1)^2 = y + 4 \quad\quad x \geq 1$$
$$x - 1 = \underbrace{\sqrt{y + 4}} \quad\quad x \geq 1$$

positive square root only because $x \geq 1$
$$x = 1 + \sqrt{y + 4} = f^{-1}(y)$$

Interchange x and y:
$$y = f^{-1}(x) = 1 + \sqrt{x + 4} \quad \text{Domain: } x \geq -4 \quad \text{(Check omitted)}$$

(B) Domain of $f^{-1} = [-4, \infty)$
Range of f^{-1} = Domain of $f = [1, \infty)$.

(C)

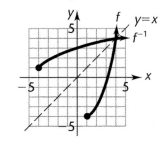

$(1-6)$

49. $x^2 - x + 2 = \left(\dfrac{1}{2} - \dfrac{i}{2}\sqrt{7}\right)^2 - \left(\dfrac{1}{2} - \dfrac{i}{2}\sqrt{7}\right) + 2$

$$= \left(\dfrac{1}{2}\right)^2 - 2\left(\dfrac{1}{2}\right)\left(\dfrac{i}{2}\sqrt{7}\right) + \left(\dfrac{i}{2}\sqrt{7}\right)^2 - \dfrac{1}{2} + \dfrac{i}{2}\sqrt{7} + 2$$

$$= \dfrac{1}{4} - \dfrac{i}{2}\sqrt{7} + \dfrac{7i^2}{4} - \dfrac{1}{2} + \dfrac{i}{2}\sqrt{7} + 2$$

$$= \dfrac{1}{4} - \dfrac{7}{4} - \dfrac{1}{2} + 2$$

$$= 0 \quad\quad\quad (2-4)$$

50. Simplify the given inequality:

$a - b < b - a$	Add a to both sides
$2a - b < b$	Add b to both sides
$2a < 2b$	Divide both sides by 2
$a < b$	

The original inequality is true for any real numbers a and b with $a < b$.

$(2-2)$

51. $\dfrac{a + bi}{a - bi} = \dfrac{(a + bi)(a + bi)}{(a - bi)(a + bi)} = \dfrac{(a + bi)^2}{a^2 - (bi)^2} = \dfrac{a^2 + 2abi + b^2i^2}{a^2 - b^2i^2}$

$$= \dfrac{a^2 + 2abi - b^2}{a^2 + b^2}$$

$$= \dfrac{a^2 - b^2 + 2abi}{a^2 + b^2}$$

$$= \dfrac{a^2 - b^2}{a^2 + b^2} + \dfrac{2ab}{a^2 + b^2}i \quad\quad (2-4)$$

52.

$$3x^2 = 2\sqrt{2}\,x - 1$$

$$3x^2 - 2\sqrt{2}\,x + 1 = 0$$

$$x = \frac{-b \pm \sqrt{b^2 - 4ac}}{2a} \qquad a = 3$$

$$b = -2\sqrt{2}$$

$$c = 1$$

$$x = \frac{-(-2\sqrt{2}) \pm \sqrt{(-2\sqrt{2})^2 - 4(3)\,(1)}}{2(3)}$$

$$x = \frac{2\sqrt{2} \pm \sqrt{8 - 12}}{6}$$

$$x = \frac{2\sqrt{2} \pm \sqrt{-4}}{6}$$

$$x = \frac{2\sqrt{2} \pm 2i}{6}$$

$$x = \frac{2(\sqrt{2} \pm i)}{6}$$

$$x = \frac{\sqrt{2} \pm i}{3} \qquad\qquad (2\text{-}5)$$

53.

$$1 + 13x^{-2} + 36x^{-4} = 0$$

$$1 + \frac{13}{x^2} + \frac{36}{x^4} = 0 \qquad \text{LCD} = x^4 \qquad x \neq 0$$

$$x^4 + 13x^2 + 36 = 0$$

$$(x^2 + 4)(x^2 + 9) = 0$$

$$x^2 + 4 = 0 \quad \text{or} \quad x^2 + 9 = 0$$

$$x^2 = -4 \qquad\qquad x^2 = -9$$

$$x = \pm 2i \qquad\qquad x = \pm 3i \qquad\qquad (2\text{-}6)$$

54.

$$\sqrt{16x^2 + 48x + 39} - 2x = 3$$

$$\sqrt{16x^2 + 48x + 39} = 3 + 2x$$

$$16x^2 + 48x + 39 = 9 + 12x + 4x^2$$

$$12x^2 + 36x + 30 = 0$$

$$2x^2 + 6x + 5 = 0$$

$$x = \frac{-b \pm \sqrt{b^2 - 4ac}}{2a} \qquad a = 2$$

$$b = 6$$

$$c = 5$$

$$x = \frac{-6 \pm \sqrt{6^2 - 4(2)\,(5)}}{2(2)}$$

$$x = \frac{-6 \pm \sqrt{36 - 40}}{4}$$

$$x = \frac{-6 \pm \sqrt{-4}}{4}$$

$$x = \frac{-6 \pm 2i}{4}$$

$$x = \frac{-3 \pm i}{2}$$

Check $\dfrac{-3 + i}{2}$:

$$\sqrt{16\left(\tfrac{-3+i}{2}\right)^2 + 48\left(\tfrac{3+i}{2}\right) + 39} \;-\; 2\left(\tfrac{-3+i}{2}\right) \overset{?}{=} 3$$

$$\sqrt{4(9 - 6i + i^2) + 24(-3 + i) + 39} \;+\; 3 \;-\; i \overset{?}{=} 3$$

$$\sqrt{36 - 24i + 4i^2 - 72 + 24i + 39} \;+\; 3 \;-\; i \overset{?}{=} 3$$

$$\sqrt{36 - 4 - 72 + 39} \;+\; 3 \;-\; i \overset{?}{=} 3$$

$$\sqrt{-1} \;+\; 3 \;-\; i \overset{?}{=} 3$$

$$i \;+\; 3 \;-\; i \overset{?}{=} 3$$

$$i \;+\; 3 \;-\; i \overset{\checkmark}{=} 3 \quad \text{A solution}$$

Check $\dfrac{-3 - i}{2}$:

$$\sqrt{16\left(\tfrac{-3-i}{2}\right)^2 + 48\left(\tfrac{3-i}{2}\right) + 39} \;-\; 2\left(\tfrac{-3-i}{2}\right) \overset{?}{=} 3$$

$$\sqrt{4(9 + 6i + i^2) + 24(-3 - i) + 39} \;+\; 3 \;+\; i \overset{?}{=} 3$$

$$\sqrt{36 + 24i + 4i^2 - 72 - 24i + 39} \;+\; 3 \;+\; i \overset{?}{=} 3$$

$$\sqrt{36 - 4 - 72 + 39} \;+\; 3 \;+\; i \overset{?}{=} 3$$

$$\sqrt{-1} \;+\; 3 \;+\; i \overset{?}{=} 3$$

$$i \;+\; 3 \;+\; i \overset{?}{=} 3$$

$$2i \;+\; 3 \neq 3 \quad \text{Not a solution}$$

Solution: $x = \dfrac{-3 + i}{2}$ (or $-1.5 + 0.5i$)   *(2-6)*

55. $3x^{-2/5} - x^{-1/5} - 1 = 0$

Let $u = x^{-1/5}$, then

$3u^2 - u - 1 = 0$

$$u = \frac{-b \pm \sqrt{b^2 - 4ac}}{2a} \qquad \begin{aligned} a &= 3 \\ b &= -1 \\ c &= -1 \end{aligned}$$

$$u = \frac{-(-1) \pm \sqrt{(-1)^2 - 4(3)(-1)}}{2(3)}$$

$$u = \frac{1 \pm \sqrt{13}}{6}$$

$$x^{-1/5} = \frac{1 + \sqrt{13}}{6} \qquad\qquad x^{-1/5} = \frac{1 - \sqrt{13}}{6}$$

$$x^{1/5} = \frac{6}{1 + \sqrt{13}} \qquad\qquad x^{1/5} = \frac{6}{1 - \sqrt{13}}$$

$$x = \left(\frac{6}{1 + \sqrt{13}}\right)^5 \qquad\qquad x = \left(\frac{6}{1 - \sqrt{13}}\right)^5$$

$$x \approx 3.75 \qquad\qquad\qquad x \approx -64.75$$

Check: Graph $y = 3x^{-2/5} - x^{-1/5} - 1$ and use the ZERO command to find the zeros.

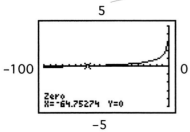

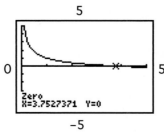

(2-6)

56. a is a square root of x if $a^2 = x$. So $5 + i$ is a square root of $24 + 10i$ because $(5 + i)^2 = 25 + 10i + i^2 = 25 + 10i - 1 = 24 + 10i$. $-5 - i$ is also a square root of $24 + 10i$ because $(-5 - i)^2 = 25 + 10i + i^2 = 25 + 10i - 1 = 24 + 10i$. To find the square roots of $24 + 10i$, set $(a + bi)^2 = 24 + 10i$, simplify the left side, and apply the definition of equality for complex numbers. This will yield the system of equations $a^2 - b^2 = 24$, $2ab = 10$, which can be solved for real a and b. 　　　　　　　　　　　　　　　　　　　　　　　　　　　　　　　　　　　　$(2\text{-}4)$

57. (A)
$$\frac{f(x + h) - f(x)}{h} = \frac{[0.5(x + h)^2 - 3(x + h) - 7] - [0.5x^2 - 3x - 7]}{h}$$
$$= \frac{0.5(x^2 + 2xh + h^2) - 3x - 3h - 7 - 0.5x^2 + 3x + 7}{h}$$
$$= \frac{0.5x^2 + xh + 0.5h^2 - 3x - 3h - 7 - 0.5x^2 + 3x + 7}{h}$$
$$= \frac{xh + 0.5h^2 - 3h}{h}$$
$$= x + 0.5h - 3$$

(B)
$$\frac{f(x) - f(a)}{x - a} = \frac{(0.5x^2 - 3x - 7) - (0.5a^2 - 3a - 7)}{x - a}$$
$$= \frac{0.5x^2 - 3x - 7 - 0.5a^2 + 3a + 7}{x - a}$$
$$= \frac{0.5(x^2 - a^2) - 3(x - a)}{x - a}$$
$$= \frac{(x - a)[0.5(x + a) - 3]}{x - a}$$
$$= 0.5(x + a) - 3$$
$$= 0.5x + 0.5a - 3$$
　　　　　　　　　　　　　　　　　　　　　　　　　　　　　　　　　　　　$(1\text{-}2)$

58. A simple example of a function that satisfies all these conditions, $y = 0.24x^2 - 2.6x + 4$, is graphed as shown.

It would not be possible for a function to have fewer than the two x intercepts shown and still be continuous. Minimum: 2
However, a graph could cross the x axis arbitrarily often (an even number of times) and still satisfy all the stated conditions. There is no maximum number of intercepts.

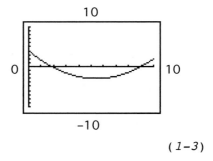

　　　　　　　　　　　　　　　　　　　　　　　　　　　　　　　　　　　　$(1\text{-}3)$

59. $f(x) = |x + 2| + |x - 2|$
If $x < -2$, then $|x + 2| = -(x + 2)$ and $|x - 2| = -(x - 2)$, so
$f(x) = -(x + 2) + -(x - 2)$
　　　$= -x - 2 - x + 2$
　　　$= -2x$
If $-2 \le x \le 2$ then $|x + 2| = x + 2$ but $|x - 2| = -(x - 2)$, so
$f(x) = x + 2 + -(x - 2)$
　　　$= x + 2 - x + 2$
　　　$= 4$

If $x > 2$, then $|x + 2| = x + 2$ and $|x - 2| = x - 2$, so

$f(x) = x + 2 + x - 2$
$\qquad = 2x$

Domain: $(-\infty, \infty)$

x	$f(x) = -2x$	x	$f(x) = 4$	x	$f(x) = 2x$
-5	10	-2	4	3	6
-3	6	0	4	5	10
		2	4		

Piecewise definition for f:

$$f(x) = \begin{cases} -2x & \text{if} & x < -2 \\ 4 & \text{if} & -2 \le x \le 2 \\ 2x & \text{if} & x > 2 \end{cases}$$

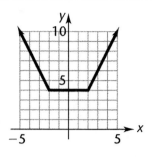

Range: $[4, \infty)$ (*1-3*)

60. In problems 79 and 81, Section 1-3, we noted that $[\![x/2]\!]$ jumps at intervals of 2 units, and $[\![3x]\!] = [\![x \div \frac{1}{3}]\!]$ jumps at intervals of $\frac{1}{3}$ unit. We expect $[\![2x]\!]$ to jump at intervals of $\frac{1}{2}$ unit.

$f(x) = 2x - [\![2x]\!]$

$f(0) = 2 \cdot 0 - [\![2 \cdot 0]\!] = 0 - [\![0]\!] = 0 - 0 = 0$

$f(\frac{1}{4}) = 2 \cdot \frac{1}{4} - [\![2 \cdot \frac{1}{4}]\!] = \frac{1}{2} - [\![\frac{1}{2}]\!] = \frac{1}{2} - 0 = \frac{1}{2}$

$f(\frac{1}{2}) = 2 \cdot \frac{1}{2} - [\![2 \cdot \frac{1}{2}]\!] = 1 - [\![1]\!] = 1 - 1 = 0$

$f(\frac{3}{4}) = 2 \cdot \frac{3}{4} - [\![2 \cdot \frac{3}{4}]\!] = \frac{3}{2} - [\![\frac{3}{2}]\!] = \frac{3}{2} - 1 = \frac{1}{2}$

$f(1) = 2 \cdot 1 - [\![2 \cdot 1]\!] = 2 - [\![2]\!] = 2 - 2 = 0$

$f(x) = 2x$ if $0 \le x \le \frac{1}{2}$, $f(x) = 2x - 1$ if $\frac{1}{2} \le x < 1$

Generalizing, we can write: $f(x) = \begin{cases} \vdots & & \vdots \\ 2x + 2 & \text{if} & -1 \le x < -\frac{1}{2} \\ 2x + 1 & \text{if} & -\frac{1}{2} \le x < 0 \\ 2x & \text{if} & 0 \le x < \frac{1}{2} \\ 2x - 1 & \text{if} & \frac{1}{2} \le x < 1 \\ 2x - 2 & \text{if} & 1 \le x < \frac{3}{2} \\ 2x - 3 & \text{if} & \frac{3}{2} \le x < 2 \\ \vdots & & \vdots \end{cases}$

Domain: All real numbers
Range: $[0, 1)$
Discontinuous at $x = k/2$, k an integer

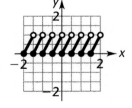

61. $x^3 - 8 = 0$

$(x - 2)(x^2 + 2x + 4) = 0$

$\begin{array}{ll} x - 2 = 0 & x^2 + 2x + 4 = 0 \\ \quad x = 2 & x^2 + 2x \qquad = -4 \\ & x^2 + 2x + 1 = -4 + 1 \\ & \qquad (x + 1)^2 = -3 \\ & \qquad x + 1 = \pm\sqrt{-3} \\ & \qquad \quad x = -1 \pm i\sqrt{3} \end{array}$

Solution: $x = 2, -1 \pm i\sqrt{3}$

(*1-3*)

(*2-5*)

62. (A) The rate at which the amount of water is changing (-13,200 gallons per hour, which is the rate for three pumps each working at 4,400 gallons per hour) is the slope of a linear function; the original amount (400,000) is the y intercept. The equation is $y = mx + b = -13,200x + 400,000$

(B) Set the amount of water, represented here by y, equal to zero and solve:
$$0 = -13,200x + 400,000$$
$$13,200x = 400,000$$
$$x = \frac{400,000}{13,200} \approx 30.3$$

It will take 30.3 hours.

(C) $100,000 < -13,2000x + 400,000 < 200,000$
$-300,000 < -13,200x < -200,000$
$22.7 > x > 15.2$ (Note change in direction)

The amount of water will be between 100,000 and 200,000 gallons from 15.2 to 22.7 hours after pumping begins. $(2-1, 2-7)$

63. "Break even" means Cost = Revenue
   Let x = number of books sold
Revenue = number of books sold × price per book
$$= x(9.65)$$
   Cost = Fixed Cost + Variable Cost
$$= 41,800 + \text{number of books} \times \text{cost per book}$$
$$= 41,800 + x(4.90)$$
$$9.65x = 41,800 + 4.90x$$
$$4.75x = 41,800$$
$$x = \frac{41,800}{4.75}$$
$$x = 8,800 \text{ books} \qquad (2-2)$$

64. The distance of p from 200 must be no greater than 10.
$|p - 200| \leq 10$ $(2-2)$

65. (A) A profit will result if revenue is greater than cost; that is, if
$$R > C$$
$$15p - 2p^2 > 88 - 12p$$
$$88 + 27p - 2p^2 > 0$$
$$2p^2 - 27p + 88 < 0$$
$$(2p - 11)(p - 8) < 0$$

The solution of this inequality consists of all values of x for which the graph of $y = 2x^2 - 27x + 88$ is below the x axis. The x intercepts of the graph are 5.5 and 8.

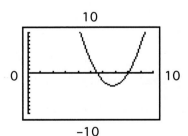

Examining the graph, we see that $2p^2 - 27p + 88 < 0$ and a profit will occur ($R > C$) for $\$5.5 < p < \$8$ or ($\$5.5, \$8$).

(B) A loss will result if cost is greater than revenue, that is, if
$$C > R$$
$$88 - 12p > 15p - 2p^2$$
$$2p^2 - 27p + 88 > 0$$

Referring to the graph in part (A), we see that $2p^2 - 27p + 88 > 0$, and a loss will occur ($C > R$), for $p < \$5.5$ or $p > \$8$. Since a negative price doesn't make sense, we delete any number to the left of 0. A loss will occur for $\$0 \leq p < \$5.5$ or $p > \$8$. $[\$0, \$5.5) \cup (\$8, \infty)$ $(2-3, 2-7)$

66. (A) If v is linearly related to t, then we are looking for an equation whose graph passes through $(t_1, v_1) = (0, 20000)$ and $(t_2, v_2) = (8, 4000)$. We find the slope, m, and then we use the point-slope form to find the equation.

$$m = \frac{v_2 - v_1}{t_2 - t_1} = \frac{4,000 - 20,000}{8 - 0} = -2,000$$

$$v - v_1 = m(t_2 - t_1)$$
$$v - 20,000 = -2,000(t - 0)$$
$$v - 20,000 = -2,000t$$
$$v = -2,000t + 20,000$$

(B) Solve for t
$$v = -2,000t + 20,000$$
$$v - 20,000 = -2,000t$$
$$t = \frac{v - 20,000}{-2,000}$$
$$t = -0.0005v + 10$$

The inverse function tells us how long it would take to reach a specified value. *(1-6)*

67.   Let x = the distance from port A to port B

Then $115 - x$ = the distance from port B to port C

Applying the Pythagorean theorem, we have

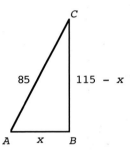

$$x^2 + (115 - x)^2 = 85^2$$
$$x^2 + 13,225 - 230x + x^2 = 7,225$$
$$2x^2 - 230x + 6,000 = 0$$
$$x^2 - 115x + 3,000 = 0$$
$$(x - 40)(x - 75) = 0$$
$$x = 40,\ 75$$
$$115 - x = 75,\ 40$$

Based on the given information, there are two possible solutions:
     40 miles from A to B and 75 miles from B to C *or*
     75 miles from A to B and 40 miles from B to C *(2-5)*

68. If x is linearly related to p, then we are looking for an equation whose graph passes through $(p_1, x_1) = (3.79, 1,160)$ and $(p_2, x_2) = (3.59, 1,340)$. We find the slope, and then we use the point-slope form to find the equation.

$$m = \frac{x_2 - x_1}{p_2 - p_1} = \frac{1,340 - 1,160}{3.59 - 3.79} = \frac{180}{-0.2} = -900$$

$$x - x_1 = m(p - p_1)$$
$$x - 1,160 = -900(p - 3.79)$$
$$x - 1,160 = -900p + 3,411$$
$$x = -900p + 4,571$$

If the price is lowered to \$3.29, we are asked for x when $p = 3.29$
$$x = -900(3.29) + 4,571$$
$$= 1,610 \text{ bottles}$$ *(2-2)*

69. For the first 60 calls: If $0 \le x \le 60$, $C(x) = 0.06x$
For the next 90 calls: If $60 < x \le 150$:

$$\begin{pmatrix} \text{Cost of first} \\ 60 \text{ calls} \end{pmatrix} + \begin{pmatrix} \text{Cost of next } x - 60 \\ \text{calls at } 0.05 \text{ per call} \end{pmatrix}$$

$$\begin{aligned} C(x) &= 0.06(60) + 0.05(x - 60) \\ &= 3.60 + 0.05x - 3 \\ &= 0.05x + 0.6 \end{aligned}$$

For the next 150 calls, if $150 < x \le 300$,

$$\begin{pmatrix} \text{Cost of first} \\ 150 \text{ calls} \end{pmatrix} + \begin{pmatrix} \text{Cost of next } x - 150 \\ \text{calls at } 0.04 \text{ per call} \end{pmatrix}$$

$$\begin{aligned} C(x) &= 0.05(150) + 0.6 + 0.04(x - 150) \\ &= 7.5 + 0.6 + 0.04x - 6 \\ &= 0.04x + 2.1 \end{aligned}$$

Finally, if $x > 300$,

$$\begin{pmatrix} \text{Cost of first} \\ 300 \text{ calls} \end{pmatrix} + \begin{pmatrix} \text{Cost of next } x - 300 \\ \text{calls at } 0.03 \text{ per call} \end{pmatrix}$$

$$\begin{aligned} C(x) &= 0.04(300) + 2.1 + 0.03(x - 300) \\ &= 12 + 2.1 + 0.03x - 9 \\ &= 0.03x + 5.1 \end{aligned}$$

Summarizing, $C(x) = \begin{cases} 0.06x & \text{if } 0 \le x \le 60 \\ 0.05x + 0.6 & \text{if } 60 < x \le 150 \\ 0.04x + 2.1 & \text{if } 150 < x \le 300 \\ 0.03x + 5.1 & \text{if } 300 < x \end{cases}$

x	$C(x) = 0.06x$	x	$C(x) = 0.05x + 0.6$	x	$C(x) = 0.04x + 2.1$	x	$C(x) = 0.03x + 5.1$
0	0	90	5.1	160	8.5	310	14.4
30	1.8	140	7.6	200	10.1	400	17.1
60	3.6						

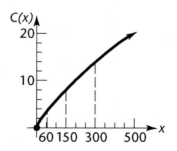

$(1-3)$

70. (A) Let x = width of pen
Then $2x + \ell = 80$

So $\ell = 80 - 2x$ = length of pen

$$A(x) = x\ell = x(80 - 2x) = 80x - 2x^2$$

(B) Since all distances must be positive, $x > 0$ and $80 - 2x > 0$, so $80 > 2x$ or $x < 40$. Therefore $0 < x < 40$ or $(0, 40)$ is the domain of $A(x)$.

(C) Note: 0, 40 were excluded from the domain for geometrical reasons; but can be used to help draw the graph since the function $80x - 2x^2$ has domain including these values.

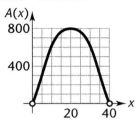

x	$A(x)$
0	0
10	600
20	800
30	600
40	0

Use the vertex formula: $x = -\dfrac{b}{2a} = -\dfrac{80}{2(-2)} = 20$. The maximum value of A occurs when x is 20. The other dimension is $80 - 2(20) = 40$, so the dimensions are 20 feet by 40 feet. *(1-3, 2-3)*

71. (A) $f(1) = 1 - 2[\![1/2]\!] = 1 - 2(0) = 1$
$f(2) = 2 - 2[\![2/2]\!] = 2 - 2(1) = 0$
$f(3) = 3 - 2[\![3/2]\!] = 3 - 2(1) = 1$
$f(4) = 4 - 2[\![4/2]\!] = 4 - 2(2) = 0$

(B) If n is an integer, n is either odd or even.
If n is even, it can be written as $2k$, where k is an integer.
Then $f(n) = f(2k) = 2k - 2[\![2k/2]\!] = 2k - 2[\![k]\!] = 2k - 2k = 0$.
Otherwise, n is odd, and n can be written as $2k + 1$, where k is an integer.
Then $f(n) = f(2k + 1) = 2k + 1 - 2[\![(2k + 1)/2]\!] = 2k + 1 - 2[\![k + \tfrac{1}{2}]\!]$
$= 2k + 1 - 2k$
$= 1$

Summarizing, if n is an integer, $f(n) = \begin{cases} 1 \text{ if } n \text{ is an odd integer} \\ 0 \text{ if } n \text{ is an even integer} \end{cases}$ *(1-3)*

72. (A) 30,000 bushels

(B) The demand decreases by 10,000 bushels to 20,000 bushels.

(C) The demand increases by 10,000 bushels to 40,000 bushels.

(D) As the price varies from $3.00 to $3.50 per bushel, the demand varies from 15,000 to 50,000 bushels, increasing with decreasing price and decreasing with increasing price.

(E) From the graph, the following table is constructed:

q	20	25	30	35	40
P	340	332	325	320	315

```
L1      L2     L3    2
20      340    ------
25      332
30      325
35      320
40      315
------  ------
L2(6) =
```

```
QuadReg
 y=ax²+bx+c
 a=.0228571429
 b=-2.611428571
 c=383.0285714
```

Enter the data   Compute the regression equation.

The quadratic regression model is $y = -0.0229x^2 - 2.61x + 383$. *(1-1, 2-1)*

73. (A) The independent variable is number of years since 1970, so we enter 0, 5, 10, 15, 20, 25, 30, and 35 as L_1. The dependent variable is egg consumption, so we enter 309, 276, 271, 255, 233, 234, 252, and 255 as L_2. Then we choose quadratic regression from the STAT CALC menu.

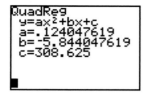

The quadratic regression model is $y = 0.124x^2 - 5.84x + 309$.

(B) We're asked to find x when y is 309 and again when y is 400. Graph $y_1 = 0.124x^2 - 5.84x + 309$, $y_2 = 309$, and $y_3 = 400$, and use the INTERSECT command.

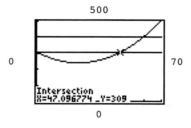 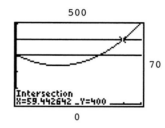

The quadratic model is at height 309 at $x = 47$ so consumption will return to the 1970 level 47 years after 1970 or 2017. The quadratic model is at height 400 when $x = 59$, so consumption will return to the 1945 level 59 years after 1970 or 2029.

(C) Egg consumption declined from 309 eggs per capita in 1970 to a low of 233 in 1990 then began to increase, reaching 255 in 2005. *(2-5)*

74. (A) The independent variable is speed and the dependent variable is length of skid marks, so we enter the speed column as L_1 and the length column as L_2. Then we choose quadratic regression from the STAT CALC menu

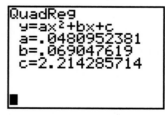

The quadratic model is $y = 0.0481x^2 + 0.0690x + 2.21$.

(B) We're asked to find x (speed) when y (length) is 220. Graph $y_1 = 0.0481x^2 + 0.0690x + 2.21$, $y_2 = 220$, and find the intersection.

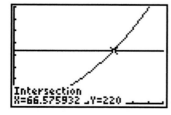 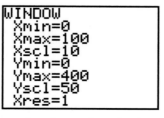

The speed is 66.6 mph when the length of the skid mark is 220 feet. *(2-5)*

75. (A) The independent variable is speed and the dependent variable is mileage, so we enter the speed column as L1 and the mileage column as L2 then use the quadratic regression command from the STAT CALC menu.

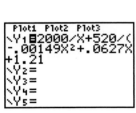 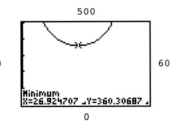

The quadratic model is $y = -0.00149x^2 + 0.0627x + 1.21$.

(B) The rental cost is $10 per hour plus $2.60 per gallon of gas. This can be written as $C = 10t + 2.60g$ where t is the number of hours and g is gallons of gas used. We can write each of t and g in terms of speed (x) to get a model for cost in terms of speed.

The time of the trip is the distance divided by the speed, since distance = speed times time. The trip is 200 miles so $t = \dfrac{200}{x}$.

The gasoline usage can be calculated by dividing the length (200 miles) by the miles per gallon (y); $g = \dfrac{200}{y}$. Substituting in our quadratic model for y:

$$g = \frac{200}{-0.00149x^2 + 0.0627x + 1.21}$$

Our rental cost is

$$C = 10t + 2.60g = 10\left(\frac{200}{x}\right) + 2.60\left(\frac{200}{-0.00149x^2 + 0.0627x + 1.21}\right)$$

$$C = \frac{2,000}{x} + \frac{520}{-0.00149x^2 + 0.0627x + 1.21}$$

We use the MINIMUM command to find the minimum cost.

The minimum cost occurs when the speed is 26.9 mph. Plugging $x = 26.9$ back into our quadratic regression model, we find that the mileage at 26.9 mph is 1.82 mpg. We use our expressions for t, g, and C to answer the remaining questions:

$$t = \frac{200}{x} = \frac{200}{26.9} = 7.43 \text{ hours}$$

$$g = \frac{200}{y} = \frac{200}{1.82} = 110 \text{ gallons}$$

$$C = 10t + 2.60g = 10(7.43) + 2.60(110) = \$360$$

(2-5)

CHAPTER 3   Polynomial and Rational Functions

Section 3-1

1. Answers will vary.

3. Answers will vary.

5. True. The general quadratic function $f(x) = ax^2 + bx + c$ fits the definition of polynomial.

7. False. The coefficients can be imaginary as well.

9. False. The zeros of a polynomial are x intercepts of its graph only if they are real numbers.

11. The degree is odd and the coefficient is positive, so $f(x)$ increases without bound as $x \to \infty$ and decreases without bound as $x \to -\infty$. This matches graph c.

13. The degree is even and the coefficient is positive, so $h(x)$ increases without bound both as $x \to \infty$ and $x \to -\infty$. This matches graph d.

15. The real zeros are the x-intercepts: -1 and 3. The turning point is $(1, 4)$. $P(x) \to -\infty$ as $x \to \infty$ and $P(x) \to -\infty$ as $x \to -\infty$.

17. The real zeros are the x-intercepts: -2 and 1. The turning points are $(-1, 4)$ and $(1, 0)$. $P(x) \to -\infty$ as $x \to -\infty$ and $P(x) \to \infty$ as $x \to \infty$.

19. The graph of a polynomial always increases or decreases without bound as $x \to \infty$ and $x \to -\infty$. This graph does not.

21. The graph of a polynomial always has a finite number of turning points — at most one less than the degree. This graph has infinitely many turning points.

23. This is a polynomial with degree 3.

25. This is not a polynomial because the variable is in the denominator of the last term.

27. This is not a polynomial because the variable is under a radical.

29. This is a polynomial. If the parentheses are multiplied out, it will fit the definition of polynomial, and the degree is 3.

31. Set each factor equal to zero and solve:
$$x = 0 \qquad x^2 - 9 = 0 \qquad x^2 + 4 = 0$$
$$x^2 = 9 \qquad\qquad x^2 = -4$$
$$x = 3, -3 \qquad\quad x = 2i, -2i$$
The zeros are 0, 3, -3, $2i$, and $-2i$; of these 0, 3, and -3 are x-intercepts.

33. Set each factor equal to zero and solve:
$$x + 5 = 0 \qquad x^2 + 9 = 0 \qquad x^2 + 16 = 0$$
$$x = -5 \qquad\quad x^2 = -9 \qquad\qquad x^2 = -16$$
$$x = 3i, -3i \qquad\quad x = 4i, -4i$$
The zeros are -5, $3i$, $-3i$, $4i$, and $4i$. Only -5 is an x-intercept.

35. $P(x) = x^3 - 5x^2 + 2x + 6 \qquad n = 3 \qquad a_n = 1$

(A) Since $a_n > 0$ and n is odd $P(x) \to \infty$ as $x \to \infty$ and $P(x) \to -\infty$ as $x \to -\infty$. Since the degree of $P(x)$ is 3, $P(x)$ can have a maximum of three x intercepts and two local extrema.

(B) Graphing and applying the ZERO command yields:

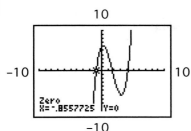
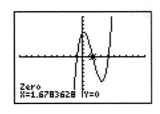
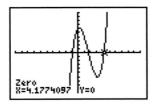

The three x intercepts are $x = -0.86$, 1.68, and 4.18. Next, we use the MAXIMUM and MINIMUM commands.

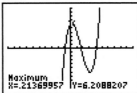

The local maximum is $P(0.21) \approx 6.21$.   The local minimum is $P(3.12) \approx -6.06$.

37. $P(x) = -x^3 + 4x^2 + x + 5$ $n = 3$ $a_n = -1$

(A) Since $a_n < 0$ and n is odd $P(x) \to -\infty$ as $x \to \infty$ and $P(x) \to \infty$ as $x \to -\infty$. Since the degree of $P(x)$ is 3, $P(x)$ can have a maximum of three x intercepts and two local extrema.

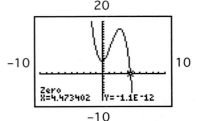

(B) Graphing and applying the ZERO command yields the graph at the right.

There is actually only one x intercept at $x = 4.47$. Next, we use the MAXIMUM and MINIMUM commands.

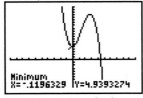
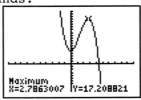

The local minimum is $P(-0.12) \approx 4.94$ The local maximum is $P(2.79) \approx 17.21$

39. $P(x) = x^4 + x^3 - 5x^2 - 3x + 12$ $n = 4$ $a_n = 1$

(A) Since $a_n > 0$ and n is even $P(x) \to \infty$ as $x \to \infty$ and $P(x) \to \infty$ as $x \to -\infty$. Since the degree of $P(x)$ is 4, $P(x)$ can have a maximum of four x intercepts and three local extrema.

(B) As seen in the graphs below, there are no x intercepts. Graphing and applying the MAXIMUM and MINIMUM commands yields:

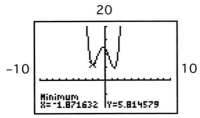
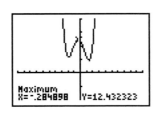
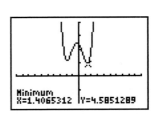

Local minimum at
$P(-1.87) \approx 5.81$

Local maximum at
$P(-0.28) \approx 12.43$

Local minimum at
$P(1.41) \approx 4.59$

41.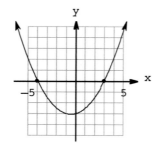

43.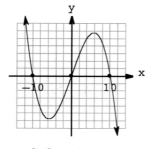

45.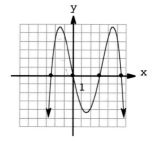

47. $p(x) = x(x - 2)(x + 3)$ (factored form)

$\qquad = x(x^2 + x - 6)$

$\qquad = x^3 + x^2 - 6x$ (standard form)

49. $p(x) = x(x + \frac{1}{2})(x - 4)$ (to get the specified zeros)

$\quad\; p(x) = x(2x + 1)(x - 4)$ (multiply $(x + 1/2)$ by 2 to get integer coefficients)

$\qquad = x(2x^2 - 7x - 4)$

$\qquad = 2x^3 - 7x^2 - 4x$ (standard form)

51. $p(x) = (x - 2)(x - 3)(x - 4)$ (factored form)

$\qquad = (x - 2)(x^2 - 7x + 12)$

$\qquad = x^3 - 2x^2 - 7x^2 + 14x + 12x - 24$

$\qquad = x^3 - 9x^2 + 26x - 24$ (standard form)

53. $p(x) = x(x - i)(x + i)$ (factored form)

$\qquad = x(x^2 - i^2)$

$\qquad = x(x^2 + 1)$

$\qquad = x^3 + x$ (standard form)

55. $P(x) = x^3$ is an example of a third-degree polynomial with one x intercept $(x = 0)$.

57. No such polynomial exists; the graph of a third-degree polynomial must cross the x axis at least once.

59. This is the same as the right hand behavior of the function $p(x) = x^2$; even degree, positive leading coefficient, so $\lim\limits_{x \to \infty} x^2 = \infty$.

61. This is the same as the left hand behavior of the function $p(x) = x^3 - x^2 + x$; odd degree, positive leading coefficient, so $\lim\limits_{x \to -\infty}\left(x^3 - x^2 + x\right) = -\infty$.

63. This is the same as the right hand behavior of the function $p(x) = 4 - x - 3x^4$; even degree, negative leading coefficient, so $\lim\limits_{x \to \infty}\left(4 - x - 3x^4\right) = -\infty$.

65. This is the same as the left hand behavior of the function $p(x) = -x^4 + x^3 + 5x^2$; even degree, negative leading coefficient, so $\lim\limits_{x \to -\infty}\left(-x^4 + x^3 + 5x^2\right) = -\infty$

67. Graphing and applying the ZERO command yields:

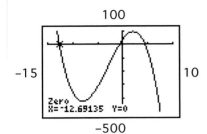
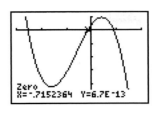
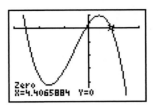

x intercepts at $x = -12.69, -0.72, 4.41$. Next, we use the MAXIMUM and MINIMUM commands.

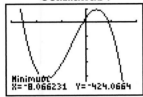
Local minimum at
$P(-8.07) \approx -424.07$

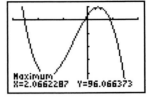
Local maximum at
$P(2.07) \approx 96.07$

69. Graphing and applying the ZERO command yields:

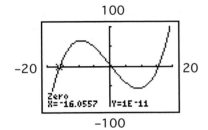
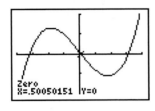
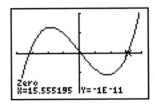

x intercepts at $-16.06, 0.50, 15.56$. Next, we use the MAXIMUM and MINIMUM commands.

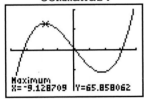
Local maximum at
$P(-9.13) \approx 65.86$

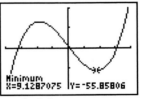
Local minimum at
$P(9.13) \approx -55.86$

71. Graphing and applying the ZERO commands yields:

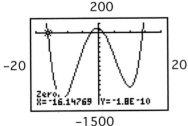
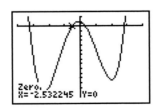
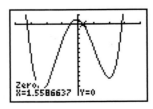
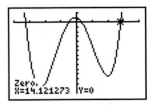

x intercepts at $-16.15, -2.53, 1.56, 14.12$. Next, we use the MAXIMUM and MINIMUM commands.

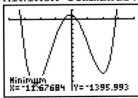
Local minimum at
$P(-11.68) \approx -1,395.99$

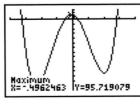
Local maximum at
$P(-0.50) \approx 95.72$

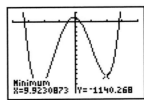
Local minimum at
$P(9.92) \approx -1,140.27$

73. Graphing and applying the ZERO command yields:

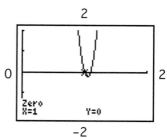 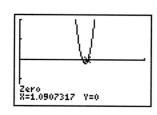

x intercepts at 1, 1.09. Next, we use the MAXIMUM and MINIMUM commands.

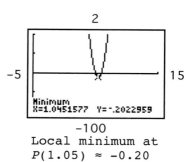 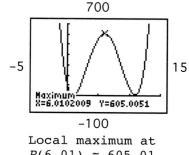 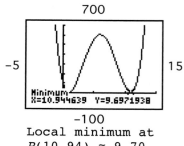

Local minimum at   Local maximum at   Local minimum at
$P(1.05) \approx -0.20$ $P(6.01) \approx 605.01$ $P(10.94) \approx 9.70$

75. (A) If the degree is even the graph has to approach either ∞ or $-\infty$ both to the right and left, so there has to be at least one turning point. $f(x) = x^4$, graphed below, has only 1 turning point.

A degree 4 polynomial can't have any more than 3 turning points. $f(x) = x^4 - 5x^2$, graphed below, has 3 turning points.

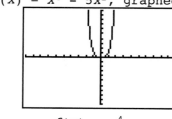 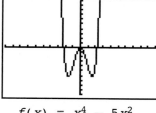

$f(x) = x^4$ $f(x) = x^4 - 5x^2$

(B) An even degree polynomial can have no x-intercepts as illustrated by $f(x) = x^4 + 5$ (graphed below). A polynomial of degree 4 can have at most 4 x-intercepts. For example, $f(x) = x^4 - 5x^2 + 5$, graphed below, has 4 x-intercepts.

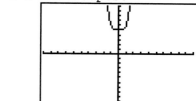 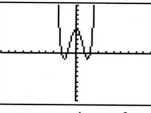

$f(x) = x^4 + 5$ $f(x) = x^4 - 5x^2 + 5$

77. No. If $f(x)$ is an even function, then $f(x) = f(-x)$ for any x. But $f(x) = x^2 + x$ is an even degree polynomial and $f(1) = 2$ while $f(-1) = 0$.

79. (A)

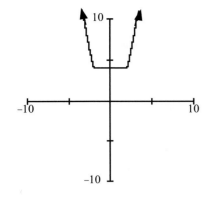

(B) All the points on the horizontal portion of the graph satisfy the definition of local minimum, but none are turning points.
The set $\{c \mid -2 \le x \le 2\}$ represents all such numbers c.

81. It is possible. A third degree polynomial can have at most two turning points, and it has to approach ∞ from either the left or right, and $-\infty$ from the other side; the graph of such a function with real zeros 0 and 2 is shown at the right.

(It is the graph of $p(x) = x^3 - 4x^2 + 4x$.)

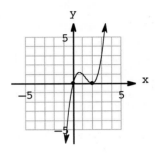

83. (A) The revenue $R(x)$ is given by revenue = (number sold)(price per object)

$$R(x) = xp(x)$$
$$= x(0.0004x^2 - x + 569)$$
$$= 0.0004x^3 - x^2 + 569x$$

(B) Graphing and applying the MAXIMUM command yields the graph at the right.

The number of air conditioners that maximizes the revenue is 364. The maximum revenue is \$93,911.

The price is $p(x) = \dfrac{R(x)}{x} = \dfrac{93,911}{364} = \$258.$

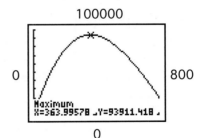

85.

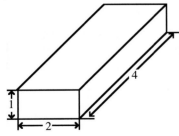

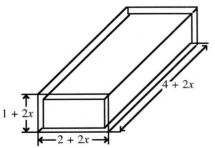

(A) The volume of the shielding is equal to the volume of the shielded box (right hand picture) minus the volume of the unshielded box (left hand picture). So

$$\begin{pmatrix}\text{Volume of}\\ \text{shielding}\end{pmatrix} = \begin{pmatrix}\text{Volume of}\\ \text{shielded box}\end{pmatrix} - \begin{pmatrix}\text{Volume of}\\ \text{unshielded box}\end{pmatrix}$$

$$V = (1 + 2x)(2 + 2x)(4 + 2x) - 1 \cdot 2 \cdot 4$$

$$= (2 + 6x + 4x^2)(4 + 2x) - 8$$

$$= 8 + 24x + 16x^2 + 4x + 12x^2 + 8x^3 - 8$$

$$= 28x + 28x^2 + 8x^3$$

(B) Set our expression for volume equal to 3 and solve for *x*:

$(1 + 2x)(2 + 2x)(4 + 2x) - 8 = 3$

Graphing and applying the INTERSECT command yields the graph at the right.

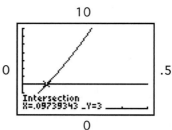

The thickness of the shielding is 0.097 ft.

87. (A) The independent variable is years since 1960, so we enter 0, 10, 20, 30, 35, 40, and 45 as L_1. The dependent variable is total expenditures, so we enter the total expenditures column as L_2. Then we choose the cubic regression command from the STAT CALC menu.

L1	L2	L3 3
0	27.5	
10	74.9	
20	253.9	
30	714	
35	1016.5	
40	1353.3	
45	1987.7	

L3(1)=

```
EDIT CALC TESTS
1:1-Var Stats
2:2-Var Stats
3:Med-Med
4:LinReg(ax+b)
5:QuadReg
6:CubicReg
7↓QuartReg
```

```
CubicReg
y=ax³+bx²+cx+d
a=.0177980059
b=.0774524168
c=3.424120513
d=23.23533045
```

The cubic regression polynomial is $y = 0.0178x^3 + 0.0775x^2 + 3.42x + 23.2$.

(B) 2010 is 50 years after 1960, so we plug 50 into our regression model.

$$y = 0.0178(50)^3 + 0.0775(50)^2 + 3.42(50) + 23.2$$
$$= 2{,}612.95$$

Total expenditures in 2010 will be $2,613.0 billion.

(C) This time, we choose the quadratic regression command from the STAT CALC menu:

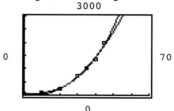

```
QuadReg
y=ax²+bx+c
a=1.28906087
b=-17.04592302
c=62.7765295
```

The quadratic regression model is $y = 1.29x^2 - 17.0x + 62.8$. The graph shows both the cubic and quadratic models; the cubic model is the one to the left near the top of the screen, which shows that it predicts more rapid long-term increase than the quadratic.

89. (A) The independent variable is years since 1950, so we enter 0, 10, 20, 30, 40, 50, and 55 as L_1. The dependent variable is marriage rate, so we enter the marriage column as L_2. Then we use the cubic regression and quartic regression commands from the STAT CALC menu.

L1	L2	L3 3
0	11.1	
10	8.5	
20	10.6	
30	10.6	
40	9.8	
50	8.5	
55	7.5	

L3(1)=

```
EDIT CALC TESTS
1:1-Var Stats
2:2-Var Stats
3:Med-Med
4:LinReg(ax+b)
5:QuadReg
6:CubicReg
7↓QuartReg
```

```
CubicReg
y=ax³+bx²+cx+d
a=-1.513892E-4
b=.0109178757
c=-.2065964991
d=10.73247864
```

```
QuarticReg
y=ax⁴+bx³+…+e
a=9.2016053E-6
b=-.001158449
c=.0449230795
d=-.5579819787
e=11.01726938
```

The cubic regression polynomial is $y = -0.000\ 151x^3 + 0.0109x^2 - 0.207x + 10.7$, and the quartic is $y = 0.000\ 00920x^4 - 0.001\ 16x^3 + 0.0449x^2 - 0.558x + 11.0$.

(B) 2008 is 58 years after 1950, so we plug 58 into our regression models.

$$y = -0.000\ 151(58)^3 + 0.0109(58)^2 - 0.207(58) + 10.7$$
$$= 5.9 \text{ marriages per 1,000 population}$$

$$y = 0.000\ 00920(58)^4 - 0.00116(58)^3 + 0.0449(58)^2 - 0.558(58) + 11.0$$
$$= 7.9 \text{ marriages per 1,000 population}$$

(C) The first graph below shows the cubic model and the scatter plot; the second shows the quartic model:

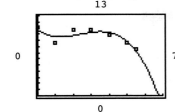
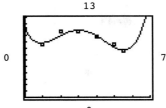

The quartic model fits the data points better, but the cubic probably provides a more realistic prediction for 2008. After the last data point, the quartic model predicts a sudden increase in the marriage rate that is not at all supported by the data; the trend appears likely to continue downward.

(D) A quadratic model would do a poor job with this data because it can only change direction once, but the data changes direction twice.

Section 3-2

1. $x^2 - 5x + 12 = (x - 2)(x - 3) + 6$. The dividend is $x^2 - 5x + 12$, the divisor is $x - 3$, the quotient is $x - 2$, and the remainder is 6.

3. Answers will vary.

5. Answers will vary.

7.
$$\begin{array}{r} 2m + 1 \\ 2m - 1 \overline{\smash{\big)}\ 4m^2 + 0m - 1} \\ \underline{-\left(4m^2 - 2m\right)} \\ 2m - 1 \\ \underline{-\left(2m - 1\right)} \\ 0 \end{array}$$
$2m + 1$

9.
$$\begin{array}{r} 4x - 5 \\ 2x + 1 \overline{\smash{\big)}\ 8x^2 - 6x + 6} \\ \underline{-\left(8x^2 + 4x\right)} \\ -10x + 6 \\ \underline{-\left(-10x - 5\right)} \\ 11 \end{array}$$
$4x - 5 + \dfrac{11}{2x + 1}$

11.
$$\begin{array}{r} x^2 + x + 1 \\ x - 1 \overline{\smash{\big)}\ x^3 - 0x^2 + 0x - 1} \\ \underline{-\left(x^3 - x^2\right)} \\ x^2 + 0x \\ \underline{-\left(x^2 - x\right)} \\ x - 1 \\ \underline{-\left(x - 1\right)} \\ 0 \end{array}$$
$x^2 + x + 1$

13.
$$\begin{array}{r} 2y^2 - 5y + 13 \\ y + 2 \overline{\smash{\big)}\ 2y^3 - y^2 + 3y - 1} \\ \underline{-\left(2y^3 + 4y^2\right)} \\ -5y^2 + 3y \\ \underline{-\left(-5y^2 - 10y\right)} \\ 13y - 1 \\ \underline{-\left(13y + 26\right)} \\ -27 \end{array}$$
$2y^2 - 5y + 13 - \dfrac{27}{y + 2}$

15.
$$\begin{array}{r} \ 1 \quad 3 \quad -7 \\ \ \quad 2 \quad 10 \\ \hline 2\,\overline{\big|\ 1 \quad 5 \quad 3} \end{array}$$
$\dfrac{x^2 + 3x - 7}{x - 2} = x + 5 + \dfrac{3}{x - 2}$

17.
$$\begin{array}{r} \ 4 \quad 10 \quad -9 \\ \ \quad -12 \quad 6 \\ \hline -3\,\overline{\big|\ 4 \quad -2 \quad -3} \end{array}$$
$\dfrac{4x^2 + 10x - 9}{x + 3} = 4x - 2 - \dfrac{3}{x + 3}$

19.

$$
\begin{array}{r}
\ 2\quad\ 0\ \ -3\quad\ 1 \\
\ 4\quad\ 8\quad 10 \\
\hline
2\ \underline{|\ 2\quad\ 4\quad\ 5\quad 11}
\end{array}
$$

$$
\frac{2x^3 - 3x + 1}{x - 2} = 2x^2 + 4x + 5 + \frac{11}{x - 2}
$$

> **Common Error:**
> The first row is *not*
> 2 -3 1
> The 0 must be inserted for the missing power.

21. $P(-2) = 3(-2)^2 - (-2) - 10 = 12 + 2 - 10 = 4$

$$
\begin{array}{r}
\ 3\ \ -1\ \ -10 \\
\ -6\quad\ 14 \\
\hline
-2\ \underline{|\ 3\ \ -7\quad\ 4}
\end{array}
$$

$P(-2) = 4$

23. $P(2) = 2(2)^3 - 5(2)^2 + 7(2) - 7 = 16 - 20 + 14 - 7 = 3$

$$
\begin{array}{r}
\ 2\ \ -5\quad\ 7\ \ -7 \\
\ 4\ \ -2\quad 10 \\
\hline
2\ \underline{|\ 2\ \ -1\quad\ 5\quad\ 3}
\end{array}
$$

$P(2) = 3$

25. $P(-4) = (-4)^4 - 10(-4)^2 + 25(-4) - 2 = 256 - 160 - 100 - 2 = -6$

$$
\begin{array}{r}
\ 1\quad\ 0\ \ -10\quad 25\ \ -2 \\
\ -4\quad 16\ \ -24\ \ -4 \\
\hline
-4\ \underline{|\ 1\ \ -4\quad\ 6\quad\ 1\ \ -6}
\end{array}
$$

$P(-4) = -6$

27. $x - 1$ will be a factor of $P(x)$ if $P(1) = 0$. Since $P(x) = x^{18} - 1$, $P(1) = 1^{18} - 1 = 0$. Therefore $x - 1$ is a factor of $x^{18} - 1$.

29. $x + 1$ will be a factor of $P(x)$ if $P(-1) = 0$. Since $P(x) = 3x^3 - 7x^2 - 8x + 2$, $P(-1) = 3(-1)^3 - 7(-1)^2 - 8(-1) + 2 = -3 - 7 + 8 + 2 = 0$. Therefore $x + 1$ is a factor of $3x^3 - 7x^2 - 8x + 2$.

31.

$$
\begin{array}{r}
2x^2 + x + 3 \\
-3x + 4\ \overline{)\ -6x^3 + 5x^2 - 5x + 12} \\
-\underline{\left(-6x^3 + 8x^2\right)} \\
-3x^2 - 5x \\
-\underline{\left(-3x^2 + 4x\right)} \\
-9x + 12 \\
-\underline{\left(-9x + 12\right)} \\
0
\end{array}
$$

$2x^2 + x + 3$

33.

$$
\begin{array}{r}
x^2 - 2x - 5 \\
2x - 1\ \overline{)\ 2x^3 - 5x^2 - 8x + 1} \\
-\underline{\left(2x^3 - x^2\right)} \\
-4x^2 - 8x \\
-\underline{\left(-4x^2 + 2x\right)} \\
-10x + 1 \\
-\underline{\left(-10x + 5\right)} \\
-4
\end{array}
$$

$x^2 - 2x - 5 - \dfrac{4}{2x - 1}$

35.

$$
\begin{array}{r}
x - 4 \\
x^2 + 0x - 2 \overline{\smash{)}\, x^3 - 4x^2 - 7x + 1} \\
- \left(x^3 + 0x^2 - 2x\right) \\
\hline
- 4x^2 - 5x + 1 \\
- \left(-4x^2 + 0x + 8\right) \\
\hline
- 5x - 7
\end{array}
$$

$$x - 4 + \dfrac{-5x - 7}{x^2 - 2}$$

37.

$$
\begin{array}{r}
3x^2 - 3x - 5 \\
x^2 + x + 2 \overline{\smash{)}\, 3x^4 + 0x^3 - 2x^2 - 3x + 5} \\
- \left(3x^4 + 3x^3 + 6x^2\right) \\
\hline
- 3x^3 - 8x^2 - 3x \\
- \left(-3x^3 - 3x^2 - 6x\right) \\
\hline
- 5x^2 + 3x + 5 \\
- \left(-5x^2 - 5x - 10\right) \\
\hline
8x + 15
\end{array}
$$

$$3x^2 - 3x - 5 + \dfrac{8x + 15}{x^2 + x + 2}$$

39.

$$
\begin{array}{r|rrrrr}
 & 3 & 0 & 0 & -1 & -4 \\
 & & -3 & 3 & -3 & 4 \\
\hline
-1 & 3 & -3 & 3 & -4 & 0
\end{array}
$$

$$3x^3 - 3x^2 + 3x - 4$$

41.

$$
\begin{array}{r|rrrrrr}
 & 1 & 0 & 0 & 0 & 0 & 1 \\
 & & -1 & 1 & -1 & 1 & -1 \\
\hline
-1 & 1 & -1 & 1 & -1 & 1 & 0
\end{array}
$$

$$x^4 - x^3 + x^2 - x + 1$$

43.

$$
\begin{array}{r|rrrrr}
 & 3 & 2 & 0 & -4 & -1 \\
 & & -9 & 21 & -63 & 201 \\
\hline
-3 & 3 & -7 & 21 & -67 & 200
\end{array}
$$

$$3x^3 - 7x^2 + 21x - 67 + \dfrac{200}{x + 3}$$

45.

$$
\begin{array}{r|rrrrrrr}
 & 2 & -13 & 0 & 75 & 2 & 0 & -50 \\
 & & 10 & -15 & -75 & 0 & 10 & 50 \\
\hline
5 & 2 & -3 & -15 & 0 & 2 & 10 & 0
\end{array}
$$

$$2x^5 - 3x^4 - 15x^3 + 2x + 10$$

47.

$$
\begin{array}{r|rrrrr}
 & 4 & 2 & -6 & -5 & 1 \\
 & & -2 & 0 & 3 & 1 \\
\hline
-\frac{1}{2} & 4 & 0 & -6 & -2 & 2
\end{array}
$$

$$4x^3 - 6x - 2 + \dfrac{2}{x + 1/2}$$

49.

$$
\begin{array}{r|rrrr}
 & 4 & 4 & -7 & -6 \\
 & & -6 & 3 & 6 \\
\hline
-\frac{3}{2} & 4 & -2 & -4 & 0
\end{array}
$$

$$4x^2 - 2x - 4$$

51.

$$
\begin{array}{r|rrrrr}
 & 3 & -2 & 2 & -3 & 1 \\
 & & 1.2 & -0.32 & 0.672 & -0.9312 \\
\hline
0.4 & 3 & -0.8 & 1.68 & -2.328 & 0.0688
\end{array}
$$

$$3x^3 - 0.8x^2 + 1.68x - 2.328 + \dfrac{0.0688}{x - 0.4}$$

53.

$$
\begin{array}{r|rrrrrr}
 & 3 & 2 & 5 & 0 & -7 & -3 \\
 & & -2.4 & 0.32 & -4.256 & 3.4048 & 2.87616 \\
\hline
-0.8 & 3 & -0.4 & 5.32 & -4.256 & -3.5952 & -0.12384
\end{array}
$$

$$3x^4 - 0.4x^3 + 5.32x^2 - 4.256x - 3.5952 - \dfrac{0.12384}{x + 0.8}$$

55. $P\left(\dfrac{5}{2}\right) = 4\left(\dfrac{5}{2}\right)^3 - 12\left(\dfrac{5}{2}\right)^2 - 7\left(\dfrac{5}{2}\right) + 10 = \dfrac{125}{2} - 75 - \dfrac{35}{2} + 10$

$= \dfrac{125}{2} - \dfrac{150}{2} - \dfrac{35}{2} + \dfrac{20}{2} = -\dfrac{40}{2} = -20$

$$
\begin{array}{r|rrrr}
 & 4 & -12 & -7 & 10 \\
 & & 10 & -5 & -30 \\
\hline
5/2 & 4 & -2 & -12 & -20
\end{array}
\qquad P(5/2) = -20
$$

57. $P\left(\dfrac{1}{2}\right) = 3\left(\dfrac{1}{2}\right)^3 + 5\left(\dfrac{1}{2}\right)^2 - \dfrac{1}{2} + 2 = \dfrac{3}{8} + \dfrac{5}{4} - \dfrac{1}{2} + 2$

$= \dfrac{3}{8} + \dfrac{10}{8} - \dfrac{4}{8} + \dfrac{16}{8} = \dfrac{25}{8}$

$$
\begin{array}{r|rrrr}
 & 3 & 5 & -1 & 2 \\
 & & 3/2 & 13/4 & 9/8 \\
\hline
1/2 & 3 & 13/2 & 9/4 & 25/8
\end{array}
\qquad P(1/2) = 25/8
$$

59. $P(2i) = (2i)^3 + 1 = 2^3 \cdot i^3 + 1 = 8(-i) + 1 = 1 - 8i$

$$
\begin{array}{r|rrrr}
 & 1 & 0 & 0 & 1 \\
 & & 2i & -4 & -8i \\
\hline
2i & 1 & 2i & -4 & 1-8i
\end{array}
\qquad P(2i) = 1 - 8i
$$

61.
$$
\begin{array}{r|rrrr}
 & 1 & 0 & -7 & 6 \\
 & & 1 & 1 & -6 \\
\hline
1 & 1 & 1 & -6 & 0
\end{array}
$$

The remainder is zero, so $x = 1$ is a zero. The quotient is $x^2 + x - 6$, so we need to set this equal to zero and solve.

$x^2 + x - 6 = 0$
$(x + 3)(x - 2) = 0$

$x + 3 = 0 \quad or \quad x - 2 = 0$
$x = -3 \qquad\qquad x = 2$

The three zeros are $x = 1, 2,$ and -3.

63.
$$
\begin{array}{r|rrrr}
 & 3 & -8 & -5 & 6 \\
 & & 9 & 3 & -6 \\
\hline
3 & 3 & 1 & -2 & 0
\end{array}
$$

The remainder is zero, so $x = 3$ is a zero. The quotient is $3x^2 + x - 2$, so we need to set this equal to zero and solve.

$3x^2 + x - 2 = 0$
$(3x - 2)(x + 1) = 0$

$3x - 2 = 0 \quad or \quad x + 1 = 0$
$3x = 2 \qquad\qquad x = -1$
$x = 2/3$

The three zeros are $x = 3, 2/3,$ and -1.

65.
$$
\begin{array}{r|rrrr}
 & 1 & -5 & 4 & -20 \\
 & & 5 & 0 & 20 \\
\hline
5 & 1 & 0 & 4 & 0
\end{array}
$$

The remainder is zero, so $x = 5$ is a zero. The quotient is $x^2 + 4$, so we need to set this equal to zero and solve.

$$x^2 + 4 = 0$$
$$x^2 = -4$$
$$x = \pm 2i$$

The three zeros are $x = 5$, $2i$, and $-2i$.

67.

```
        1   -2   -3    10
            -2    8   -10
   -2 | 1   -4    5     0
```

The remainder is zero, so $x = -2$ is a zero. The quotient is $x^2 - 4x + 5$, so we need to set this equal to zero and solve.

$$x^2 - 4x + 5 = 0$$

$$x = \frac{-(-4) \pm \sqrt{(-4)^2 - 4(1)(5)}}{2(1)}$$

$$x = \frac{4 \pm \sqrt{-4}}{2} = \frac{4 \pm 2i}{2} = 2 \pm i$$

The three zeros are $x = -2$, $2 + i$, and $2 - i$.

69.

```
        1        -3          1   -3
                  i    -3i - 1    3
   i | 1   -3 + i       -3i       0
```

$$x^2 + (-3 + i)x - 3i$$

71. We apply the remainder theorem to evaluate using synthetic division.

(A)
```
        1       2i     -10
              2 - i      5
 2 - i | 1   2 + i      -5
```
$P(2 - i) = -5$

(B)
```
        1          2i          -10
                5 - 5i      10 - 40i
 5 - 5i | 1     5 - 3i         -40i
```
$P(5 - 5i) = -40i$

(C)
```
        1       2i     -10
              3 - i     10
 3 - i | 1   3 + i       0
```
$P(3 - i) = 0$

(D)
```
        1        2i     -10
              -3 - i     10
 -3 - i | 1  -3 + i       0
```
$P(-3 - i) = 0$

73. We will divide by $x - 2$ using synthetic division, and try to figure out what value of k will make the remainder zero (which guarantees that $x = 2$ is a zero).

```
        1   -3    4      k
             2   -2      4
    2 | 1   -1    2    k + 4
```

If $k = -4$, the remainder is zero, and $x = 2$ is a zero of P.

75. We will use the same approach as in Problem 73.

```
       -4   -6       k          -6
            12     -18     -3k + 54
   -3 | -4    6   k - 18   -3k + 48
```

Now we need the value of k for which $-3k + 48 = 0$.

$$-3k + 48 = 0 \quad \Rightarrow \quad -3k = -48 \quad \Rightarrow \quad k = 16$$

77. (A)
$$a_2(x) + (a_1 + a_2r)$$

$$x - r \overline{)a_2x^2 + a_1x \qquad\qquad\qquad + a_0}$$

$$\underline{a_2x^2 - a_2rx}$$
$$(a_1 + a_2r)x \qquad\qquad + a_0$$
$$\underline{(a_1 + a_2r)x - r(a_1 + a_2r)}$$
$$a_0 + r(a_1 + a_2r)$$

$$
\begin{array}{ccc}
a_2 & a_1 & a_0 \\
 & a_2r & (a_1 + a_2r)r \\
\hline
r \;\; a_2 & a_1 + a_2r & a_0 + (a_1 + a_2r)r
\end{array}
$$

In both cases the coefficient of x is a_2, the constant term $a_2r + a_1$, and the remainder is $(a_2r + a_1)r + a_0$.

(B) The remainder expanded is $a_2r^2 + a_1r + a_0 = P(r)$.

79.
$$P(x) = \{[(2x - 3)x + 2]x - 5\}x + 7$$
$$P(-2) = \{[\{2(-2) - 3\}(-2) + 2](-2) - 5\}(-2) + 7$$
$$= \{[\{-7\}(-2) + 2](-2) - 5\}(-2) + 7$$
$$= \{[16](-2) - 5\}(-2) + 7$$
$$= \{-37\}(-2) + 7$$
$$= 81$$
$$P(1.7) = \{[\{2(1.7) - 3\}(1.7) + 2](1.7) - 5\}(1.7) + 7$$
$$= \{[\{0.4\}(1.7) + 2](1.7) - 5\}(1.7) + 7$$
$$= \{[2.68](1.7) - 5\}(1.7) + 7$$
$$= \{-0.444\}(1.7) + 7$$
$$= 6.2452 \text{ or } 6.2 \text{ to two significant digits.}$$

81. (A) $P(x) = (x - r)Q(x) + R$ where $Q(x)$ is a polynomial with degree one less than the degree of P.

(B) Plug in $x = r$: $P(r) = (r - r)Q(r) + R = 0 + R = R$
Conclusion: $P(r)$ is equal to the remainder when dividing by $x - r$.

83. No. The two remaining zeros could be imaginary, as in $P(x) = x^3 + x = x(x^2 + 1)$. One of the zeros is $x = 0$, and the two remaining zeros are i and $-i$.

Section 3-3

1.
$$
\begin{array}{r|rrrr}
 & 2 & -4 & -18 & 1 \\
 & & 8 & 16 & -8 \\
\hline
4 & 2 & 4 & -2 & -7
\end{array}
$$

The coefficients in the quotient row are not all positive, so $x = 4$ is not an upper bound for the zeros.

$$
\begin{array}{r|rrrr}
 & 2 & -4 & -18 & 1 \\
 & & -6 & 30 & -36 \\
\hline
-3 & 2 & -10 & 12 & -35
\end{array}
$$

The coefficients in the quotient row alternate in sign, so $x = -3$ is a lower bound for the zeros.

3.

```
        1    4   -3  -10
             2   12   18
    2 | 1    6    9    8
```

The coefficients in the quotient row are all positive, so $x = 2$ is an upper bound for the zeros.

```
         1    4   -3  -10
             -4    0   12
   -4 |  1    0   -3    2
```

The coefficients in the quotient row do not alternate in sign, so $x = -4$ is not a lower bound for the zeros.

5.

```
        1    2   -3    4   -5
             1    3    0    4
    1 | 1    3    0    4   -1
```

The coefficients in the quotient row are not all positive, so $x = 1$ is not an upper bound for the zeros.

```
        1     2   -3    4    -5
             -3    3    0   -12
   -3 | 1    -1    0    4   -17
```

The coefficients in the quotient row do not alternate in sign, so $x = -3$ is not a lower bound for the zeros.

7. The degree is 2 so there are at most two zeros. Graph $y = x^2 + 5x - 2$ and use the ZERO command.

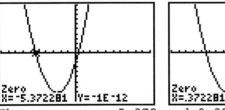

The zeros are -5.372 and 0.372.

9. The degree is 3 so there are at most three zeros. Graph $y = 2x^3 - 5x + 2$ and use the ZERO command.

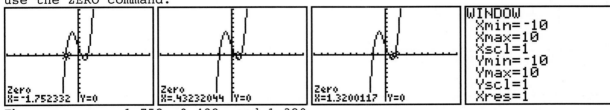

The zeros are -1.752, 0.432, and 1.320.

11. Yes. There are at most three zeros for this third degree polynomial, and the graph has three x intercepts.

13. Yes. The upper and lower bound theorem tells us that $x = 4$ is an upper bound for the zero, which means any number bigger than 4 (10 in particular) is also an upper bound for the zeros.

15. Answers will vary.

17. This inequality is true wherever the graph of $P(x)$ is above the x-axis or intersects the x-axis. This is $[-2, -1] \cup \{1\} \cup [3, \infty)$.

19. This inequality is true only when $P(x)$ is above the x-axis but not where it intersects the x-axis. This is $(-2, -1) \cup (3, \infty)$.

21. See the graph of $P(x) = x^2 + 5x - 2$ in Problem 7. It is positive (above the x-axis) on $(-\infty, -5.372) \cup (0.372, \infty)$.

23. See the graph of $P(x) = 2x^3 - 5x + 2$ in Problem 9. It is negative (below the x-axis) or zero (on the x-axis) on $(-\infty, -1.752] \cup [0.432, 1.320]$.

25. We form a synthetic division table.

	1	0	-3	1	
0	1	0	-3	1	
1	1	1	-2	-1	
2	1	2	1	3	an upper bound
-1	1	-1	-2	3	
-2	1	-2	1	-1	a lower bound

2 is an upper bound; -2 is a lower bound.

27. We form a synthetic division table.

	1	-3	4	2	-9	
0	1	-3	4	2	-9	
1	1	-2	2	4	-5	
2	1	-1	2	6	3	
3	1	0	4	14	33	an upper bound
-1	1	-4	8	-6	-3	
-2	1	-5	14	-26	43	a lower bound

3 is an upper bound; -2 is a lower bound.

29. We form a synthetic division table.

	1	0	-3	3	2	-2	
0	1	0	-3	3	2	-2	
1	1	1	-2	1	3	1	
2	1	2	1	5	12	22	an upper bound
-1	1	-1	-2	5	-3	1	
-2	1	-2	1	1	0	-2	
-3	1	-3	6	-15	47	-143	a lower bound

2 is an upper bound; -3 is a lower bound.

31. (A) We form a synthetic division table.

	1	-2	3	-8
0	1	-2	3	-8
1	1	-1	2	-6
2	1	0	3	-2
3	1	1	6	10
-1	1	-3	6	-14

From the table, 3 is an upper bound and -1 is a lower bound.

(B) The only interval in which a real zero is indicated is (2, 3). The remainders have different signs for $x = 2$ and $x = 3$. We search for this real zero. We organize our calculations in a table.

Sign Change Interval (a, b)	Midpoint m	Sign of P		
		P(a)	P(m)	P(b)
(2, 3)	2.5	-	+	+
(2, 2.5)	2.25	-	+	+
(2, 2.25)	2.125	-	-	+
(2.125, 2.25)	2.1875	-	-	+
(2.1875, 2.25)	2.21875	-	-	+
(2.21875, 2.25)	2.234375	-	-	+
(2.234375, 2.25)	2.2421875	-	-	+
(2.2421875, 2.25)	2.24609375	-	-	+
(2.24609375, 2.25)	We stop here	-		+

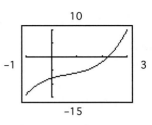

Since each endpoint rounds to 2.25, a real zero lies on this last interval and is given by 2.25 to two decimal place accuracy. A glance at the graph of $P(x)$ on [-1, 3] confirms that this is the only real zero.

33. (A) We form a synthetic division table.

	1	1	-5	7	-22
0	1	1	-5	7	-22
1	1	2	-3	4	-18
2	1	3	1	9	-4
3	1	4	7	28	62
-1	1	0	-5	12	-34
-2	1	-1	-3	13	-48
-3	1	-2	1	4	-34
-4	1	-3	7	-21	62

From the table, 3 is an upper bound and -4 is a lower bound.

(B) There are real zeros in the intervals (2, 3) and (-4, -3) indicated in the table. The remainders have different signs for each pair of x values. We search for the zero in (2, 3). We organize our calculations in a table.

Sign Change Interval (a, b)	Midpoint m	Sign of P		
		P(a)	P(m)	P(b)
(2, 3)	2.5	-	+	+
(2, 2.5)	2.25	-	+	+
(2, 2.25)	2.125	-	+	+
(2, 2.125)	2.0625	-	-	+
(2.0625, 2.125)	2.09375	-	-	+
(2.09375, 2.125)	2.109375	-	-	+
(2.109375, 2.125)	2.1171875	-	-	+
(2.1171875, 2.125)	2.12109375	-	+	+
(2.1171875, 2.12109375)	We stop here	-		+

Since each endpoint rounds to 2.12, a real zero lies on this last interval and is given by 2.12 to two decimal place accuracy. A similar search (details omitted) leads to -3.51 as the other indicated real zero. A glance at the graph of $P(x)$ on [-4, 3] confirms that these are the only zeros.

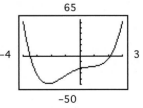

35. (A) We form a synthetic division table.

	1	0	-3	0	-4	4
0	1	0	-3	0	-4	4
1	1	1	-2	-2	-6	-2
2	1	2	1	2	0	4
-1	1	-1	-2	2	-6	10
-2	1	-2	1	-2	0	4
-3	1	-3	6	-18	50	-146

From the table, 2 is an upper bound and -3 is a lower bound.

(B) There are real zeros in the intervals (0, 1), (1, 2), and (-3, -2) indicated in the table. The remainders have different signs for each pair of x values. We search for the real zero in (0, 1). We organize our calculations in a table.

| Sign Change Interval | Midpoint | Sign of P | | |
(a, b)	m	P(a)	P(m)	P(b)
(0, 1)	0.5	+	+	-
(0.5, 1)	0.75	+	-	-
(0.5, 0.75)	0.625	+	+	-
(0.625, 0.75)	0.6875	+	+	-
(0.6875, 0.75)	0.71875	+	+	-
(0.71875, 0.75)	0.734375	+	+	-
(0.734375, 0.75)	0.7421875	+	+	-
(0.7421875, 0.75)	0.74609875	+	+	-
(0.74609875, 0.75)	We stop here	+		-

Since each endpoint rounds to 0.75, a real zero lies on this last interval and is given by 0.75 to two decimal place accuracy. Similar searches (details omitted) lead to -2.09 and 1.88 as the other indicated real zeros. A glance at the graph of $P(x)$ on [-3, 3] confirms that these are the only zeros.

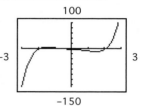

37. (A) We form a synthetic division table.

	1	1	3	1	2	-5
0	1	1	3	1	2	-5
1	1	2	5	6	8	3
-1	1	0	3	-2	4	-9

From the table, 1 is an upper bound and -1 is a lower bound.

(B) There is a real zero in the interval (0, 1) indicated in the table. The remainders have different signs for $x = 0$ and $x = 1$. We search for this real zero. We organize our calculations in a table.

| Sign Change Interval | Midpoint | Sign of P | | |
(a, b)	m	P(a)	P(m)	P(b)
(0, 1)	0.5	-	-	+
(0.5, 1)	0.75	-	-	+
(0.75, 1)	0.875	-	+	+
(0.75, 0.875)	0.8125	-	-	+
(0.8125, 0.875)	0.84375	-	+	+
(0.8125, 0.84375)	0.828125	-	-	+
(0.828125, 0.84375)	0.8359375	-	+	+
(0.828125, 0.8359375)	0.83203125	-	-	+
(0.83203125, 0.8359375)	0.833984375	-	-	+
(0.833984375, 0.8359375)	0.8349609375	-	+	+
(0.833984375, 0.8349609375)	We stop here	-		+

Since each endpoint rounds to 0.83, a real zero lies on this last interval and is given by 0.83 to two decimal place accuracy. A glance at the graph of $P(x)$ on [-2, 2] confirms that this is the only real zero.

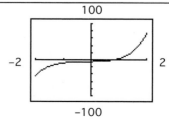

39. (A) $P(3) = 3^3 - 2(3)^2 - 5(3) + 4 = -2$ Different signs, so there is
        $P(4) = 4^3 - 2(4)^2 - 5(4) + 4 = 16$ a zero between 3 and 4.

   (B) Midpoint of [3, 4]: 3.5; $P(3.5) = 4.875$; New interval: [3, 3.5]
       Midpoint of [3, 3.5]: 3.25; $P(3.25) = 0.953$; New interval: [3, 3.25]
       Midpoint of [3, 3.25]: 3.125; $P(3.125) = -0.639$;
                                           New interval: [3.125, 3.25]
       Midpoint of [3.125, 3.25]: 3.1875; $P(3.1875) = 0.128$;
                                           New interval: [3.125, 3.1875]
       Midpoint of [3.125, 3.1875]: 3.15625; $P(3.15625) = -0.263$

The sign is different for $P(3.15625)$ and $P(3.1875)$; both of these x-values
round to 3.2, so 3.2 is the zero to one decimal place accuracy. It took 5
intervals.

Calculator Note: Here's a convenient way to repeatedly plug different numbers
into the same function, as we did in Problem 39. Enter the function as y_1 in
the equation editor and use the "value" command from the calculate menu. All
you have to do is key in the x-value you want to plug in.

---

41. (A) $P(-2) = -9$ Different signs, so there's a zero between -2 and -1.
        $P(-1) = 3$

   (B) Midpoint of [-2, -1] = -1.5; $P(-1.5) = -1.375$; New interval: [-1.5, -1]
       Midpoint of [-1.5, -1] = -1.25; $P(-1.25) = 1.17$;
                                           New interval: [-1.5, -1.25]
       Midpoint of [-1.5, -1.25] = -1.375; $P(-1.375) = -0.006$;
                                           New interval: [-1.375, -1.25]
       Midpoint of [-1.375, -1.25] = -1.3125; $P(-1.3125) = 0.606$;
                                           New interval: [-1.375, -1.3125]
       Midpoint of [-1.375, -1.3125] = -1.34375; $P(-1.34375) = 0.306$;
                                           New interval: [-1.375, -1.34375]
       Midpoint of [-1.375, -1.34375] = -1.359375; $P(-1.359375) = 0.15$

$P(-1.375)$ and $P(-1.359375)$ have opposite signs and both x-values round to -1.4,
so -1.4 is the zero to one decimal place accuracy. It took 6 intervals.

43. (A) $P(3) = -2$
        $P(4) = 59$ Different signs, so there's a zero between 3 and 4.

   (B) Midpoint of [3, 4] = 3.5; $P(3.5) = 17.1$; New interval: [3, 3.5]
       Midpoint of [3, 3.5] = 3.25; $P(3.25) = 5.2$; New interval: [3, 3.25]
       Midpoint of [3, 3.25] = 3.125; $P(3.125) = 1.1$; New interval: [3, 3.125]
       Midpoint of [3, 3.125] = 3.0625; $P(3.0625) = -0.6$

$P(3.0625)$ and $P(3.125)$ have opposite signs and both x-values round to 3.1, so
3.1 is the zero to one decimal place accuracy. It took 4 intervals.

45. (A) $P(-1) = -3$
        $P(0) = 3$ Different signs, so there's a zero between -1 and 0.

   (B) Midpoint of [-1, 0] = -0.5; $P(-0.5) = .19$; New interval: [-1, -0.5]
       Midpoint of [-1, -0.5] = -0.75; $P(-0.75) = -1.5$;
                                           New interval: [-0.75, -0.5]
       Midpoint of [-0.75, -0.5] = -0.625; $P(-0.625) = -0.7$;
                                           New interval: [-0.625, -0.5]
       Midpoint of [-0.625, -0.5] = -0.5625; $P(-0.5625) = -0.3$;
                                           New interval: [-0.5625, -0.5]
       Midpoint of [-0.5625, -0.5] = -0.53125; $P(-0.53125) = -0.02$

$P(-0.5)$ and $P(-0.53125)$ have opposite signs and both x-values round to -0.5,
so -0.5 is the zero to one decimal place accuracy. It took 5 intervals.

47. Since 1 is a zero of multiplicity 2, $P(x)$ does not change sign at $x = 1$. The bisection method fails. (The TI-84, using a more complicated routine, can find the zero.)

49. Since 3 is a zero of multiplicity 4, $P(x)$ does not change sign at $x = 3$. The bisection method fails. (The TI-84 cannot find this zero.)

51. The graph in a standard window shows two zeros that are turning points. The degree is 4 so there could be more zeros. We'll make a synthetic division table to see if $x = 10$ and $x = -10$ are bounds for the real zeros.

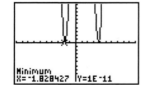

		1	-4	-10	28	49
-10		1	-14	130	-1272	12769
10		1	6	50	528	5329

The signs alternate for -10 so it's a lower bound for the real zeros, and all signs are positive for 10 so it's an upper bound. In other words, the only real zeros are the two we see on the standard screen.

To find the zeros, we use the MINIMUM command:

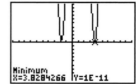

The zeros are —1.83 and 3.83

53. The graph in a standard window shows three zeros, two of which are turning points. The degree is 5 so there could be more zeros. We'll make a synthetic division table to see if $x = 10$ and $x = -10$ are bounds for the real zeros.

		1	-6	4	24	-16	-32
-10		1	-16	164	-1616	16144	-161472
10		1	4	44	464	4624	46208

The signs alternate for -10 so it's a lower bound for the real zeros, and all signs are positive for 10 so it's an upper bound. In other words, the only real zeros are the two we see on the standard screen.

To find the zero that's not a turning point, we use the ZERO command; and to find the two that are, we use the MAXIMUM and MINIMUM commands:

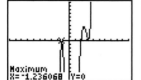 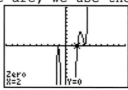 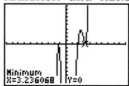

The zeros are -1.24, 2, and 3.24.

55. On a standard viewing window it's difficult to see what's happening close to $x = 2$, so we examine that area more closely.

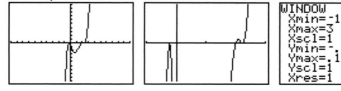

Now we can see that there are three zeros, two of which are turning points. The degree is five so there could be other zeros. A synthetic division table will help us check.

	1	-6	11	-4	-3.75	-0.5
-2	1	-8	27	-58	112.25	-225
10	1	4	51	506	5056.25	50562

The signs alternate for -2 so it's a lower bound for the zeros. The signs are all positive for 10 so it's an upper bound for the zeros and we know that we have seen all of them. To find the zero that's not a turning point we use the ZERO command; and to find the two that are, we use the MAXIMUM and MINIMUM commands.

 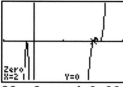 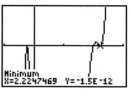

The zeros are -0.22, 2, and 2.22.

57. <u>Algebraic Solution</u>
Find zeros of the left side:
$$(2x - 5)(x + 3)(x + 7) = 0$$
$$x = 5/2, -3, -7$$

Choose test numbers and make a table:

Interval	Test Number	Result
$x < -7$	-10	$(2(-10) - 5)(-10 + 3)(-10 + 7) = -525;$ negative
$-7 < x < -3$	-5	$(2(-5) - 5)(-5 + 3)(-5 + 7) = 60;$ positive
$-3 < x < 5/2$	0	$(2(0) - 5)(0 + 3)(0 + 7) = -105;$ negative
$x > 5/2$	5	$(2(5) - 5)(5 + 3)(5 + 7) = 480;$ positive

Solution: $(-7, -3) \cup (5/2, \infty)$
<u>Graphical Solution</u>

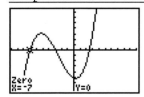 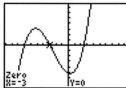 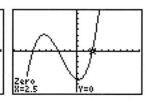

The graph is positive (above the x axis) on $(-7, -3) \cup (5/2, \infty)$.

59. <u>Algebraic Solution</u>
Rearrange so that zero is on the right side:
$$x^2 - 2 > 0$$
Find the zeros of the left side:
$$x^2 - 2 = 0 \quad \Rightarrow x^2 = 2 \quad \Rightarrow x = \pm\sqrt{2} \approx \pm 1.414$$

Interval	Test Number	Result
$x < -1.414$	-2	$(-2)^2 - (-2) = 8;$ positive
$-1.414 < x < 1.414$	0	$0^2 - 2 = -2;$ negative
$x > 1.414$	2	$2^2 - 2 = 2;$ positive

Solution: $(-\infty, -1.414) \cup (1.414, \infty)$

Graphical Solution

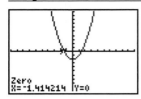 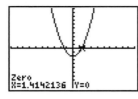

```
WINDOW
Xmin=-10
Xmax=10
Xscl=1
Ymin=-10
Ymax=10
Yscl=1
Xres=1
```

The graph is positive (above the x axis) on $(-\infty, -1.414) \cup (1.414, \infty)$.

61. Algebraic Solution

Rearrange so that zero is on the right side:

$x^3 - 9x \leq 0$

Find zeros of the left side:

$x^3 - 9x = 0$

$x(x^2 - 9) = x(x + 3)(x - 3) = 0$

Zeros: $x = 0, -3, 3$

Choose test numbers and make a table:

Interval	Test Number	Result
$x < -3$	-5	$(-5)^3 - 9(-5) = -80$; negative
$-3 < x < 0$	-1	$(-1)^3 - 9(-1) = 8$; positive
$0 < x < 3$	1	$(1)^3 - 9(1) = -8$; negative
$x > 3$	5	$(5)^3 - 9(5) = 80$; positive

Solution: $(-\infty, -3] \cup [0, 3]$

Graphical Solution

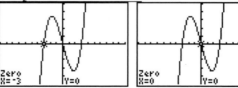 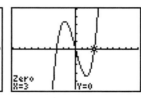

```
WINDOW
Xmin=-10
Xmax=10
Xscl=1
Ymin=-15
Ymax=15
Yscl=3
Xres=1
```

The graph is negative or zero (below or on the x axis) on $(-\infty, -3] \cup [0, 3]$.

63. Algebraic Solution

Rearrange so that zero is on the right side:

$x^3 + 5x^2 + 6x \leq 0$

Find zeros of the left side:

$x^3 + 5x^2 + 6x = 0$

$x(x^2 + 5x + 6) = x(x + 3)(x + 2) = 0$

Zeros: $x = 0, -3, -2$

Choose test numbers and make a table:

Interval	Test Number	Result
$x < -3$	-5	$(-5)^3 + 5(-5)^2 + 6(-5) = -30$; negative
$-3 < x < -2$	-2.5	$(-2.5)^3 + 5(-2.5)^2 + 6(-2.5) = 0.625$; positive
$-2 < x < 0$	-1	$(-1)^3 + 5(-1)^2 + 6(-1) = -2$; negative
$x > 0$	1	$(1)^3 + 5(1)^2 + 6(1) = 12$; positive

Solution: $(-\infty, -3] \cup [-2, 0]$

Graphical Solution

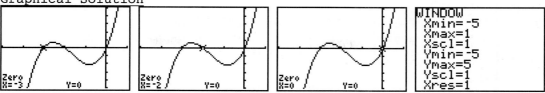

The graph is negative or zero (below or on the x axis) on $(-\infty, -3] \cup [-2, 0]$.

Problems 65–69 are solved graphically because finding the zeros algebraically is difficult, or maybe impossible.

65. $x^2 + 7x - 3 \le x^3 + x + 4$

$-x^3 + x^2 + 6x - 7 \le 0$

Find the zeros of $P(x) = -x^3 + x^2 + 6x - 7$ graphically:

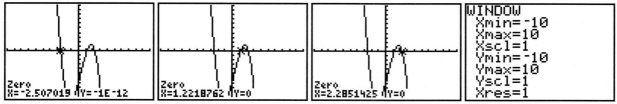

The zeros are -2.507, 1.222, and 2.285. $P(x)$ is negative or zero (below, or on, the x-axis) on $[-2.507, 1.222] \cup [2.285, \infty)$.

67. $x^4 < 8x^3 - 17x^2 + 9x - 2$

$x^4 - 8x^3 + 17x^2 - 9x + 2 < 0$

Find the zeros of $P(x) = x^4 - 8x^3 + 17x^2 - 9x + 2$ graphically:

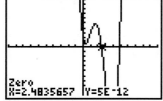 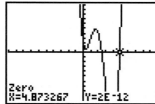 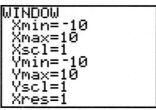

The zeros are 2.484 and 4.873. $P(x)$ is negative (below the x-axis) on $(2.484, 4.873)$.

69. $(x^2 + 2x - 2)^2 \ge 2$

$(x^2 + 2x - 2)^2 - 2 \ge 0$

Find the zeros of $P(x) = (x^2 + 2x - 2)^2 - 2$ graphically (standard window):

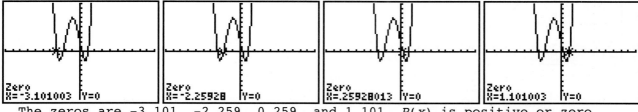

The zeros are -3.101, -2.259, 0.259, and 1.101. $P(x)$ is positive or zero (above, or on, the x-axis) on $(-\infty, -3.101] \cup [-2.259, 0.259] \cup [1.101, \infty)$.

71. (A) 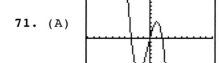 There are three real zeros. Because this is a third degree polynomial, there can be at most three zeros, so we can see all of them in this window.

(B) The largest zero looks to be around $x = 2$, so $x = 3$ is an upper bound for the real zeros.

(C)
$$
\begin{array}{r|rrrr}
 & -1 & -1 & 6 & 0 \\
 & & -3 & -12 & -18 \\
\hline
3 & -1 & -4 & -6 & -18
\end{array}
$$

All coefficients in the quotient row are negative; when the leading coefficient of a polynomial is negative, $x = r$ is an upper bound for the real zeros if dividing by $x - r$ using synthetic division makes all coefficients in the quotient row negative.

73. (A) We form a synthetic division table.

	1	-24	-25	10
0	1	-24	-25	10
10	1	-14	-165	-1640
20	1	-4	-105	-2090
30	1	6	155	4660
-10	1	-34	315	-3140

From the table, 30 is an upper bound and -10 is a lower bound.

(B) Graphing $P(x)$ using the window suggested by part (A), we obtain

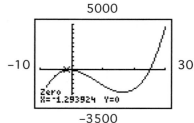 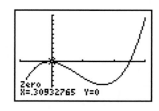 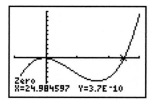

The zeros are -1.29, 0.31, and 24.98.

75. (A) We form a synthetic division table.

	1	12	-900	0	5,000
0	1	12	-900	0	5,000
10	1	22	-680	-680	-63,000
20	1	32	-260	-5200	-99,000
30	1	42	360	10,800	329,000
-10	1	2	-920	9,200	-87,000
-20	1	-8	-740	14,800	-291,000
-30	1	-18	-360	10,800	-319,000
-40	1	-28	220	-8,800	357,000

From the table, 30 is an upper bound and -40 is a lower bound.

(B) Graphing $P(x)$ using the window suggested in part (A), we obtain

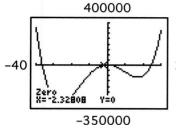
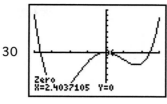
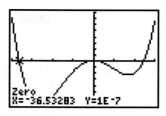
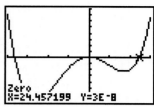

The zeros are -36.53, -2.33, 2.40, and 24.46.

77. (A) We form a synthetic division table.

	1	0	-100	-1,000	-5,000
0	1	0	-100	-1,000	-5,000
10	1	10	0	-1,000	-15,000
20	1	20	300	5,000	95,000
-10	1	-10	0	-1,000	5,000

From the table, 20 is an upper bound and -10 is a lower bound.

(B) Graphing $P(x)$ using the window suggested in part (A), we obtain

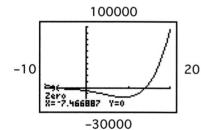
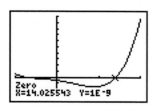

The real zeros are -7.47 and 14.03.

79. (A) We form a synthetic division table.

	4	-40	-1,475	7,875	-10,000
0	4	-40	-1,475	7,875	-10,000
10	4	0	-1,475	-6,875	-78,750
20	4	40	-675	-5,625	-122,500
30	4	80	925	35,625	1,058,750
-10	4	-80	-675	14,625	-156,250
-20	4	-120	925	-10,625	202,500

From the table, 30 is an upper bound and -20 is a lower bound.

(B) Graphing $P(x)$ using the window suggested in part (A), we obtain the graph shown at the right.

It appears that there are zeros between -20 and -10, and between 20 and 30, and there may be a zero of even multiplicity between 0 and 10.

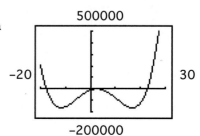

Applying ZERO and MAXIMUM commands, we obtain

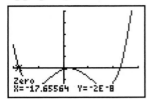 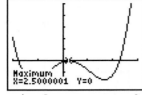 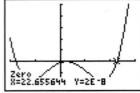

-17.66 and 20.66 are simple zeros and 2.50 is probably a double zero. To check that it is really a zero, we note

```
        4    -40    -1475     7875    -10000
             10      -75     -3875     10000
    ────────────────────────────────────────
2.50│ 4    -30    -1550     4000         0
```

2.50 is definitely a zero. You can check that it is a double zero using a second synthetic division.

81. (A) We form a synthetic division table.

	0.01	-0.1	-12	0	0	9,000
0	0.01	-0.1	-12	0	0	9,000
10	0.01	0	-12	-120	-1,200	-3,000
20	0.01	0.1	-10	-200	-4,000	-71,000
30	0.01	0.2	-6	-180	-5,400	-153,000
40	0.01	0.3	0	0	0	9,000
-10	0.01	-0.2	-10	100	-1,000	19,000
-20	0.01	-0.3	-6	120	-2,400	57,000
-30	0.01	-0.4	0	0	0	9,000
-40	0.01	-0.5	8	-320	12,800	-503,000

From the table, 40 is an upper bound and -40 is a lower bound.

(B) Graphing $P(x)$ using a window suggested by part (A), we obtain

100000

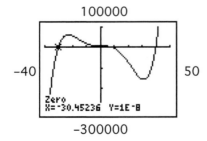 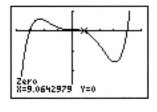 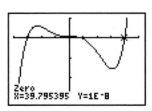

-40   50

-300000

The real zeros are -30.45, 9.06, and 39.80.

A graph in a smaller window confirms that there are no other real zeros between -10 and 10, and therefore most likely no other real zeros.

20000

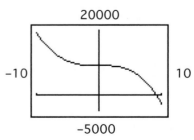

-10   10

-5000

83. Yes. According to the Remainder Theorem, if the remainder is 10 when $P(x)$ is divided by $x + 4$, then $P(-4) = 10$. Similarly, if the remainder is -8 when dividing by $x + 5$, then $P(-5) = -8$. Since $P(-5)$ and $P(-4)$ have opposite signs, there must be a zero between -5 and -4.

85. (A) $P(x) = (x - r)Q(x) + R$ where $Q(x)$ is a polynomial with degree one less than the degree of P, and R is a number. If all numbers in the quotient row are positive after performing the division, it means that all of the coefficients of Q are positive, and so is the number R.

(B) If x is greater than r, then of course $x - r$ is positive. We know that the sign of $Q(x)$ has to be positive because: (1) r is positive, so r raised to any power is positive as well; (2) all of the coefficients of Q are positive; and (3) Q is a polynomial, so it's the sum of some finite number of terms that look like a coefficient times x to a power. All of the powers are positive, and all of the coefficients are positive, so their sum is positive as well.

(C) From (B), we know that both $x - r$ and $Q(x)$ are positive, so $(x - r)Q(x)$ is positive. We also know that R is positive, so from part (A), $P(x)$ is the sum of two positive numbers, and must be positive.

(D) We started by assuming that x is ANY number greater than r, and concluded in part (C) that $P(x)$ cannot be zero. In other words, no number greater than r is a zero for P, so r is an upper bound for the real zeros of P.

87. (A) We need to find the zeros of $P(x)$, which we will do graphically:

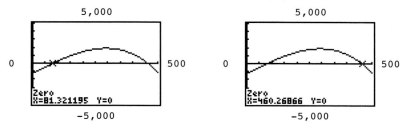

The variable x represents the number of toasters produced, so the answer must be a whole number. To break even, the company must make 81 or 460 toasters.

(B) We will add the horizontal line $y = 1,200$ to the graph, and find where the profit function is above this height:

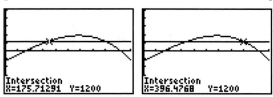

The profit is at or above \$1,200 per week for production levels between 176 and 396 toasters.

89. Let (x, x^2) be a point on the graph of $y = x^2$. Then the distance from $(1, 2)$ to (x, x^2) must equal 1 unit. Applying the distance formula, we have,

$$\sqrt{(x - 1)^2 + (x^2 - 2)^2} = 1$$
$$(x - 1)^2 + (x^2 - 2)^2 = 1$$
$$x^2 - 2x + 1 + x^4 - 4x^2 + 4 = 1$$
$$x^4 - 3x^2 - 2x + 4 = 0$$

Let $P(x) = x^4 - 3x^2 - 2x + 4$

The only rational zero is 1.

$$
\begin{array}{r|rrrr}
 & 1 & 0 & -3 & -2 & 4 \\
 &   & 1 & 1 & -2 & -4 \\
\hline
1 & 1 & 1 & -2 & -4 & 0 \\
\end{array}
$$

$P(x) = (x - 1)(x^3 + x^2 - 2x - 4)$

We examine $Q(x) = x^3 + x^2 - 2x - 4$.
Graphing $y = Q(x)$ we obtain the graph
shown at the right.

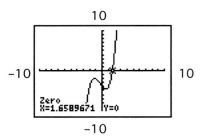

$Q(x)$ (and $P(x)$) has a second zero at
$x = 1.659$. Therefore, the two real zeros
of $P(x)$ are 1 and 1.659. The two required
points are $(1, 1)$ and $(1.659, 1.659^2) =$
$(1.659, 2.752)$.

91.

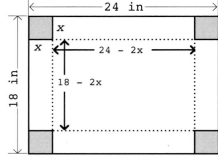

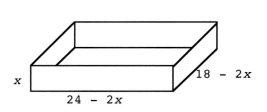

From the above figures it should be clear that
$V = $ length $\times$ width $\times$ height $= (24 - 2x)(18 - 2x)x$ $0 < x < 9$ (So that all the
dimensions are positive)

We solve $(24 - 2x)(18 - 2x)x = 600$

$$432x - 84x^2 + 4x^3 = 600$$
$$4x^3 - 84x^2 + 432x - 600 = 0$$
$$x^3 - 21x^2 + 108x - 150 = 0$$

Let $P(x) = x^3 - 21x^2 + 108x - 150$. Graphing $y = P(x)$ on the interval $0 < x < 9$,
we obtain

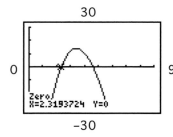

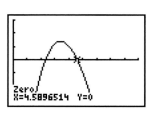

The zeros of $P(x)$ are 2.319 and 4.590, $0 < x < 9$.
$x = 2.319$ inches or 4.590 inches.

93.   We note:

$$\begin{pmatrix} \text{Volume} \\ \text{of} \\ \text{tank} \end{pmatrix} = \begin{pmatrix} \text{Volume of} \\ \text{two hemispheres} \\ \text{of radius } x \end{pmatrix} + \begin{pmatrix} \text{Volume of cylinder} \\ \text{with radius } x, \\ \text{height } 10 - 2x \end{pmatrix}$$

$$20\pi = \frac{4}{3}\pi x^3 + \pi x^2 (10 - 2x)$$

$$20 = \frac{4}{3}x^3 + 10x^2 - 2x^3$$

$$60 = 4x^3 + 30x^2 - 6x^3$$

$$2x^3 - 30x^2 + 60 = 0$$

$$x^3 - 15x^2 + 30 = 0 \quad \text{From physical considerations, } x > 0 \text{ and } 10 - 2x > 0,$$
$$\text{so we are interested only in solutions in } (0, 5).$$

Let $P(x) = x^3 - 15x^2 + 30$

Graphing $y = P(x)$ on the interval $0 < x < 5$, we obtain

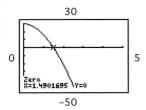

The only zero of $P(x)$, $0 < x < 5$, is 1.490.

$x = 1.490$ feet.

95. The independent variable is years after 1996, so we enter 0, 2, 4, 6, 7, 8, and 9 as L_1, and the Hispanic row from the table as L_2. Then we use the cubic regression command from the STAT CALC menu:

The cubic model is $y = 0.0142x^3 - 0.219x^2 + 1.55x + 42.7$. The inequality we need to solve is $0.0142x^3 - 0.219x^2 + 1.55x + 42.7 > 60$. To solve, we enter the cubic regression model as y_1 and the horizontal line $y_2 = 60$, and see where the cubic model is above height 60.

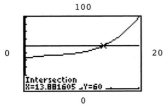

The model predicts that the Hispanic homeownership rate will pass 60% about 14 years after 1996, or 2010.

97. The independent variable is years after 1996, so we enter 0, 2, 4, 6, 7, 8, and 9 as L_1, and the White row from the table as L_2. Then we use the linear regression command from the STAT CALC menu:

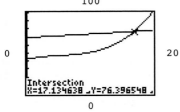

The linear model is $y = 0.420x + 69.2$. The inequality we need to solve is $0.0142x^3 - 0.219x^2 + 1.55x + 42.7 > 0.420x + 69.2$ (using the model from Problem 95). To solve, we enter the cubic regression model from Problem 95 as as y_1 and the linear model from this problem as y_2, and see where the cubic model is above the linear.

The model predicts that the Hispanic homeownership rate will pass the White rate about 17 years after 1996, or 2013.

99. The independent variable is years after 1970, so we enter 0, 10, 20, 25, 30, and 35 as L_1, and the Males column as L_2. Then we use the cubic regression command from the STAT CALC menu:

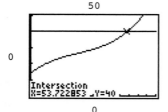

The cubic model is $y = 0.000281x^3 - 0.0220x^2 + 0.853x + 14.1$. The inequality we need to solve is $0.000281x^3 - 0.0220x^2 + 0.853x + 14.1 > 40$. To solve, we enter the cubic model as y_1 and $y_2 = 40$, and see when the cubic model gets above height 40.

The model predicts that the percentage of males in the U.S. that have completed at least four years of college will pass 40% after about 54 years, or in 2024.

Section 3-4

1. -8 (multiplicity 3), 6 (multiplicity 2); degree of $P(x)$ is 5

3. -4 (multiplicity 3), 3 (multiplicity 2); -1; degree of $P(x)$ is 6.

5. $2i$ (multiplicity 3), $-2i$ (multiplicity 4), 2 (multiplicity 5), -2 (multiplicity 5); degree of $P(x) = 17$.

7. The first factor can be factored further:
$P(x) = x(x + 3)(x - 3)(x^2 + 9)(x + 9)^2$
0 (multiplicity 1), -3 (multiplicity 1), 3 (multiplicity 1), $3i$ (multiplicity 1), $-3i$ (multiplicity 1), -9 (multiplicity 2); degree of $P(x)$ is 7.

9. Answers will vary.   **11.** Answers will vary.

13. False. Explanations will vary.

15. $P(x) = (x - 3)^2(x + 4)$; degree 3

17. $P(x) = (x + 7)^3[x - (-3 + \sqrt{2})][x - (-3 - \sqrt{2})]$; degree 5

19. $P(x) = [x - (2 - 3i)][x - (2 + 3i)](x + 4)^2$; degree 4

21. Since -2, 1, and 3 are zeros, $P(x) = (x + 2)(x - 1)(x - 3)$ is the lowest degree polynomial that has this graph. The degree of $P(x)$ is 3.

23. Since -2 and 1 are zeros, each with multiplicity 2, $P(x) = (x + 2)^2(x - 1)^2$ is the lowest degree polynomial that has this graph. The degree of $P(x)$ is 4.

25. Since -3, -2, 0, 1, and 2 are zeros, $P(x) = (x + 3)(x + 2)x(x - 1)(x - 2)$ is the lowest degree polynomial that has this graph. The degree of $P(x)$ is 5.

27. Possible factors of 6 (the leading coefficient) are ±1, ±2, ±3, ±6. Possible factors of 1 (the constant term) are ±1. To find all possible rational zeros, we form all possible fractions with numerators from the first list and denominators from the second:
$$\pm\frac{1}{1}, \pm\frac{2}{1}, \pm\frac{3}{1}, \pm\frac{6}{1}.$$

The possible rational zeros are ±1, ±2, ±3, ±6.

29. Possible factors of 4 (the leading coefficient) are ±1, ±2, ±4. Possible factors of 3 (the constant term) are ±1, ±3. To find all possible rational zeros, we form all possible fractions with numerators from the first list and denominators from the second:

$$\pm\frac{1}{1}, \pm\frac{2}{1}, \pm\frac{4}{1}, \pm\frac{1}{3}, \pm\frac{2}{3}, \pm\frac{4}{3}$$

The possible rational zeros are ±1, ±2, ±4, $\pm\frac{1}{3}$, $\pm\frac{2}{3}$, $\pm\frac{4}{3}$.

31. Possible factors of 3 (the leading coefficient) are ±1, ±3. Possible factors of 12 (the constant term) are ±1, ±2, ±3, ±4, ±6, ±12. To find all possible rational zeros, we form all possible fractions with numerators from the first list and denominators from the second:

$$\pm\frac{1}{1}, \pm\frac{3}{1}, \pm\frac{1}{2}, \pm\frac{3}{2}, \pm\frac{1}{3}, \pm\frac{3}{3}, \pm\frac{1}{4}, \pm\frac{3}{4}, \pm\frac{1}{6}, \pm\frac{3}{6}, \pm\frac{1}{12}, \pm\frac{3}{12}$$

Reducing fractions and removing duplicates, we find that the possible rational zeros are ±1, ±3, $\pm\frac{1}{2}$, $\pm\frac{3}{2}$, $\pm\frac{1}{3}$, $\pm\frac{1}{4}$, $\pm\frac{3}{4}$, $\pm\frac{1}{6}$, $\pm\frac{1}{12}$.

33. (A) Let $u = x^2$; then $x^4 + 5x^2 + 4 = u^2 + 5u + 4 = (u + 4)(u + 1)$. Replacing u with x^2, we get:

$$P(x) = (x^2 + 4)(x^2 + 1)$$

(B) The zeros of $x^2 + 4$ are $2i$ and $-2i$, so it factors as $(x - 2i)(x + 2i)$. The zeros of $x^2 + 1$ are i and $-i$, so it factors as $(x - i)(x + i)$.

$$P(x) = (x - 2i)(x + 2i)(x - i)(x + i)$$

35. (A) Factor by grouping:

$$P(x) = (x^3 - x^2) + (25x - 25) = x^2(x - 1) + 25(x - 1) = (x^2 + 25)(x - 1)$$

(B) The zeros of $x^2 + 25$ are $5i$ and $-5i$ so it factors as $(x - 5i)(x + 5i)$.

$$P(x) = (x - 5i)(x + 5i)(x - 1)$$

37. $P(x) = x^3 + 9x^2 + 24x + 16$. Since -1 is a zero of $P(x)$, we can write

```
       1    9    24    16
           -1    -8   -16
  -1⌋ 1    8    16     0
```

$$P(x) = (x + 1)(x^2 + 8x + 16)$$

Now we can factor $x^2 + 8x + 16$ as $(x + 4)(x + 4)$, so
$P(x) = (x + 1)(x + 4)(x + 4)$.

39. $P(x) = x^4 - 1$. Since 1 is a zero of $P(x)$, we can write

```
      1    0    0    0   -1
           1    1    1    1
  1⌋ 1    1    1    1    0
```

$P(x) = (x - 1)(x^3 + x^2 + x + 1)$. Since -1 is also a zero of $P(x)$, we can write

```
      1    1    1    1
          -1    0   -1
  -1⌋ 1    0    1    0
```

$P(x) = (x - 1)(x + 1)(x^2 + 1)$. Since $x^2 + 1 = x^2 - (-1) = x^2 - i^2$, we can write
$P(x) = (x - 1)(x + 1)(x - i)(x + i)$

41. $P(x) = 2x^3 - 17x^2 + 90x - 41$. Since $\frac{1}{2}$ is a zero of $P(x)$, we can write

$$
\begin{array}{r}
\phantom{\frac{1}{2}|}\;\; 2 \quad -17 \quad\;\; 90 \quad\;\; -41 \\
\phantom{\frac{1}{2}|}\;\;\qquad\quad 1 \quad -8 \quad\;\; 41 \\
\hline
\tfrac{1}{2}\,|\;\; 2 \quad -16 \quad\;\; 82 \qquad\;\; 0
\end{array}
$$

$$P(x) = \left(x - \frac{1}{2}\right)(2x^2 - 16x + 82) = \left(x - \frac{1}{2}\right)2(x^2 - 8x + 41) = (2x - 1)(x^2 - 8x + 41)$$

Since $x^2 - 8x + 41$ is a quadratic, we can find its zeros by the quadratic formula
$x^2 - 8x + 41 = 0$ $a = 1$
                                 $b = -8$
                                 $c = 41$

$$
\begin{aligned}
x &= \frac{-b \pm \sqrt{b^2 - 4ac}}{2a} \\[2mm]
&= \frac{-(-8) \pm \sqrt{(-8)^2 - 4(1)(41)}}{2(1)} \\[2mm]
&= \frac{8 \pm \sqrt{-100}}{2} \\[2mm]
&= \frac{8 \pm 10i}{2} \\[2mm]
&= 4 \pm 5i
\end{aligned}
$$

Since the zeros of $x^2 - 8x + 41$ are $4 + 5i$ and $4 - 5i$, by the factor theorem
$$x^2 - 8x + 41 = [x - (4 + 5i)][x - (4 - 5i)]$$
and $P(x) = (2x - 1)[x - (4 + 5i)][x - (4 - 5i)]$

43. The possible rational zeros are ± 1, ± 2, ± 3, ± 5, ± 6, ± 10, ± 15, ± 30.

Examining a portion of the graph of $y = P(x)$ we see that there are possible integer zeros at 2, 3, and -5. We test the likely candidates and find

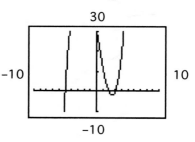

$$
\begin{array}{r}
\;\; 1 \quad\;\; 0 \quad -19 \quad\;\; 30 \\
\qquad\quad 2 \quad\;\;\; 4 \quad -30 \\
\hline
2\,|\;\; 1 \quad\;\; 2 \quad -15 \qquad\;\; 0
\end{array}
$$

2 is a zero.

So $P(x) = (x - 2)(x^2 + 2x - 15)$
            $= (x - 2)(x - 3)(x + 5)$
The zeros of $P(x)$ are 2, 3, -5.

45. $P(x) = x^4 - \dfrac{21}{10}x^3 + \dfrac{2}{5}x = \dfrac{1}{10}(10x^4 - 21x^3 + 4x) = \dfrac{1}{10}x(10x^3 - 21x^2 + 4)$.

Because x is a factor, 0 is a zero.
We examine $Q(x) = 10x^3 - 21x^2 + 4$.
Possible factors of 4 are ± 1, ± 2, ± 4.
Possible factors of 10 are ± 1, ± 2, ± 5, ± 10.
The possible rational zeros of

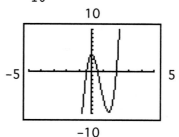

$Q(x)$ are ± 1, ± 2, ± 4, $\pm \dfrac{1}{2}$, $\pm \dfrac{1}{5}$, $\pm \dfrac{2}{5}$, $\pm \dfrac{4}{5}$, $\pm \dfrac{1}{10}$.

Examining the graph of $y = Q(x)$ we see that there are zeros between -1 and 0, and between 0 and 1, and a possible integer zero at 2. Testing 2, we obtain

$$\begin{array}{r} 10 \quad -21 \quad 0 \quad 4 \\ 20 \quad -2 \quad -4 \\ \hline 2\overline{\vert 10} \quad\quad -1 \quad -2 \quad 0 \end{array}$$

2 is a zero.

So $P(x) = \dfrac{1}{10} x(x - 2)(10x^2 - x - 2)$

$ = \dfrac{1}{10} x(x - 2)(5x + 2)(2x - 1)$

The zeros of $P(x)$ are 0, 2, $-\dfrac{2}{5}$, $\dfrac{1}{2}$.

47. $P(x) = x^4 - 5x^3 + \dfrac{15}{2} x^2 - 2x - 2 = \dfrac{1}{2}(2x^4 - 10x^3 + 15x^2 - 4x - 4).$

Possible factors of -4 are ±1, ±2, ±4.
Possible factors of 2 are ±1, ±2.
The possible rational zeros are
±1, ±2, ±4, $\pm\dfrac{1}{2}$.

Examining the graph of $y = P(x)$ we see that
there are zeros between -1 and 0, and between
1 and 2, and a possible double zero at 2.
Testing 2, we obtain

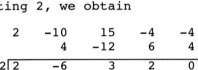

$$\begin{array}{r} 2 \quad -10 \quad 15 \quad -4 \quad -4 \\ 4 \quad -12 \quad 6 \quad 4 \\ \hline 2\overline{\vert 2} \quad\quad -6 \quad 3 \quad 2 \quad 0 \end{array}$$

2 is a zero. Testing whether it is a double zero, we examine the reduced
polynomial $2x^3 - 6x^2 + 3x + 2$.

$$\begin{array}{r} 2 \quad -6 \quad 3 \quad 2 \\ 4 \quad -4 \quad -2 \\ \hline 2\overline{\vert 2} \quad -2 \quad -1 \quad 0 \end{array}$$

So 2 is a double zero of $P(x)$, and we have
$P(x) = \dfrac{1}{2}(x - 2)^2(2x^2 - 2x - 1)$

To find the remaining zeros, we solve $2x^2 - 2x - 1 = 0$.
Applying the quadratic formula with $a = 2$, $b = -2$, $c = -1$, we obtain

$x = \dfrac{-(-2) \pm \sqrt{(-2)^2 - 4(2)(-1)}}{2(2)} = \dfrac{2 \pm \sqrt{12}}{4} = \dfrac{1 \pm \sqrt{3}}{2}$ or $\dfrac{1}{2} \pm \dfrac{1}{2}\sqrt{3}$.

So the zeros of $P(x)$ are 2 (double), $\dfrac{1}{2} \pm \dfrac{1}{2}\sqrt{3}$.

49. Let $u = x^2$. Then
$P(x) = x^4 + 11x^2 + 30 = u^2 + 11u + 30 = (u + 6)(u + 5) = (x^2 + 6)(x^2 + 5).$
The zeros of $x^2 + 6$ are $i\sqrt{6}$ and $-i\sqrt{6}$; the zeros of $x^2 + 5$ are $i\sqrt{5}$ and $-i\sqrt{5}$.
So the zeros of $P(x)$ are $\pm i\sqrt{6}$, $\pm i\sqrt{5}$.

51. Possible factors of 5 are ±1, ±5. Possible factors of 3 are ±1, ±3. The possible rational zeros are ±1, ±5, $\pm\dfrac{1}{3}$, $\pm\dfrac{5}{3}$.

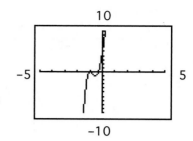

Examining the graph of $y = P(x)$, we see that there is a zero between -1 and 0, and a possible double zero at -1. Testing -1, we obtain

$$
\begin{array}{r|rrrrrr}
 & 3 & -5 & -8 & 16 & 21 & 5 \\
 & & -3 & 8 & 0 & -16 & -5 \\
\hline
-1 & 3 & -8 & 0 & 16 & 5 & 0
\end{array}
$$

-1 is a zero.

Testing whether it is a double zero, we examine the reduced polynomial $3x^4 - 8x^3 + 16x + 5$.

$$
\begin{array}{r|rrrrr}
 & 3 & -8 & 0 & 16 & 5 \\
 & & -3 & 11 & -11 & -5 \\
\hline
-1 & 3 & -11 & 11 & 5 & 0
\end{array}
$$

So -1 is a double zero. The only remaining rational candidate between -1 and 0 is $-\dfrac{1}{3}$. Testing $-\dfrac{1}{3}$ in the reduced polynomial $3x^3 - 11x^2 + 11x + 5$, we obtain

$$
\begin{array}{r|rrrr}
 & 3 & -11 & 11 & 5 \\
 & & -1 & 4 & -5 \\
\hline
-\frac{1}{3} & 3 & -12 & 15 & 0
\end{array}
$$

$-\dfrac{1}{3}$ is a zero.

So $P(x) = (x + 1)^2\left(x + \dfrac{1}{3}\right)(3x^2 - 12x + 15) = (x + 1)^2\left(x + \dfrac{1}{3}\right)3(x^2 - 4x + 5)$

To find the remaining zeros, we solve $x^2 - 4x + 5 = 0$ using the quadratic formula ($a = 1$, $b = -4$, $c = 5$)

$$x = \frac{-(-4) \pm \sqrt{(-4)^2 - 4(1)(5)}}{2(1)}$$

$$x = \frac{4 \pm \sqrt{-4}}{2} = \frac{4 \pm 2i}{2} = 2 \pm i$$

So the zeros of $P(x)$ are -1 (double), $-\dfrac{1}{3}$, $2 \pm i$.

53. Let $P(x) = 2x^3 - 5x^2 + 1$.

Possible rational zeros: ±1, $\pm\dfrac{1}{2}$.

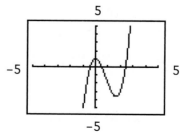

Examining the graph of $y = P(x)$, we see that there is a zero between -1 and 0, between 0 and 1, and between 1 and 2.

We test the likely candidates, $\dfrac{1}{2}$ and $-\dfrac{1}{2}$, and find

$$
\begin{array}{r|rrrr}
 & 2 & -5 & 0 & 1 \\
 & & 1 & -2 & -1 \\
\hline
\frac{1}{2} & 2 & -4 & -2 & 0
\end{array}
$$

$\dfrac{1}{2}$ is a zero.

So $P(x) = \left(x - \dfrac{1}{2}\right)(2x^2 - 4x - 2) = (2x - 1)(x^2 - 2x - 1)$

To find the remaining zeros, we solve $x^2 - 2x - 1 = 0$, using the quadratic formula:

$x^2 - 2x - 1 \qquad a = 1, \ b = -2, \ c = -1$

$x = \dfrac{-(-2) \pm \sqrt{(-2)^2 - 4(1)(-1)}}{2(1)}$

$= \dfrac{2 \pm \sqrt{8}}{2} = \dfrac{2 \pm 2\sqrt{2}}{2} = 1 \pm \sqrt{2}$

The zeros are $\dfrac{1}{2}$, $1 \pm \sqrt{2}$. These are the roots of the equation.

55. Let $P(x) = x^4 + 4x^3 - x^2 - 20x - 20$. Possible rational zeros: $\pm 1, \ \pm 2, \ \pm 4, \ \pm 5, \ \pm 10, \ \pm 20$.

Examining the graph of $y = P(x)$, we see that there is a zero between 2 and 3, and possibly a double zero near -2. We test -2.

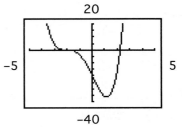

$$
\begin{array}{r|rrrrr}
 & 1 & 4 & -1 & -20 & -20 \\
 & & -2 & -4 & 10 & 20 \\
\hline
-2 & 1 & 2 & -5 & -10 & 0 \\
\end{array}
$$

-2 is a zero. Testing whether it is a double zero, we examine the reduced polynomial $x^3 + 2x^2 - 5x - 10$.

$$
\begin{array}{r|rrrr}
 & 1 & 2 & -5 & -10 \\
 & & -2 & 0 & 10 \\
\hline
-2 & 1 & 0 & -5 & 0 \\
\end{array}
$$

-2 is a double zero.

So $P(x) = (x + 2)^2(x^2 - 5) = (x + 2)^2(x - \sqrt{5})(x + \sqrt{5})$. So the zeros of the polynomial are -2 (double), $\pm\sqrt{5}$. These are the roots of the equation.

57. Let $P(x) = x^4 - 2x^3 - 5x^2 + 8x + 4$.
Possible rational zeros: $\pm 1, \ \pm 2, \ \pm 4$.

Examining the graph of $y = P(x)$ we see that there is a zero between -1 and 0 and between 2 and 3. The possible integer zeros are -2 and 2, which we test.

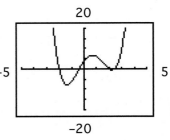

$$
\begin{array}{r|rrrrr}
 & 1 & -2 & -5 & 8 & 4 \\
 & & 2 & 0 & -10 & -4 \\
\hline
2 & 1 & 0 & -5 & -2 & 0 \\
\end{array}
$$

2 is a zero. Next, we examine the reduced polynomial $x^3 - 5x - 2$.

$$
\begin{array}{r|rrrr}
 & 1 & 0 & -5 & -2 \\
 & & -2 & 4 & 2 \\
\hline
-2 & 1 & -2 & -1 & 0 \\
\end{array}
$$

-2 is a zero.

So $P(x) = (x - 2)(x + 2)(x^2 - 2x - 1)$. The zeros of $x^2 - 2x - 1$ are $1 \pm \sqrt{2}$ (see problem 53). So the zeros of the polynomial are ± 2, $1 \pm \sqrt{2}$. These are the roots of the equation.

59. Factoring is the simplest method of solving. Let $u = x^2$.

$$x^4 + 10x^2 + 9 = 0$$
$$u^2 + 10u + 9 = 0$$
$$(u + 9)(u + 1) = 0$$
$$u = -9, \qquad u = -1$$
$$x^2 = -9, \qquad x^2 = -1$$
$$x = 3i, \ -3i, \ i, \ -i$$

61. Let $P(x) = 2x^5 - 3x^4 - 2x + 3$. Possible rational zeros: ± 1, ± 3, $\pm\dfrac{1}{2}$, $\pm\dfrac{3}{2}$.

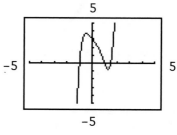

Examining the graph of $y = P(x)$ we see that there is a zero between 1 and 2, and possible zeros at -1 and 1. We test the likely candidates and find

```
      2    -3    0     0    -2     3
            2   -1    -1    -1    -3
   1 | 2   -1   -1    -1    -3     0
```

1 is a zero. We examine the reduced polynomial and find

```
      2    -1    -1    -1    -3
           -2     3    -2     3
  -1 | 2   -3     2    -3     0
```

-1 is a zero. We examine the new reduced polynomial and find

```
      2    -3     2    -3
             3     0     3
 3/2 | 2     0     2     0
```

$\dfrac{3}{2}$ is a zero.

So $P(x) = (x - 1)(x + 1)\left(x - \dfrac{3}{2}\right)(2x^2 + 2)$

$$= (x - 1)(x + 1)\left(x - \dfrac{3}{2}\right)2(x^2 + 1)$$

$$= (x - 1)(x + 1)(2x - 3)(x^2 + 1)$$

$$= (x - 1)(x + 1)(2x - 3)(x - i)(x + i)$$

The zeros are ± 1, $\dfrac{3}{2}$, $\pm i$. These are the roots of the equation.

63. The possible rational zeros are ± 1, ± 2, ± 4, $\pm\dfrac{1}{2}$, $\pm\dfrac{1}{3}$, $\pm\dfrac{2}{3}$, $\pm\dfrac{4}{3}$, $\pm\dfrac{1}{6}$.

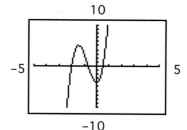

Examining the graph of $y = P(x)$, we see that there is a zero between -1 and 0, between 0 and 1, and a possible integer zero at -2. Testing -2, we obtain

```
      6    13     0    -4
          -12    -2     4
  -2 | 6    1    -2     0
```

-2 is a zero.

$$P(x) = (x + 2)(6x^2 + x - 2) = (x + 2)(3x + 2)(2x - 1)$$

65. The possible rational zeros are ±1, ±2, ±4.
Examining the graph of $y = P(x)$, we see that there
is a zero between -1 and 0, between 2 and 3, and a
possible integer zero at -4. Testing -4, we obtain

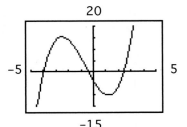

$$
\begin{array}{r}
1 2 -9 -4 \\
 -4 8 4 \\
\hline
-4|1 -2 -1 0
\end{array}
$$

   -4 is a zero.

So $P(x) = (x + 4)(x^2 - 2x - 1)$. The zeros of $x^2 - 2x - 1$ are $1 \pm \sqrt{2}$ (see
problem 53). So $P(x) = (x + 4)[x - (1 + \sqrt{2})][x - (1 - \sqrt{2})]$

67. The possible rational zeros are ±1, ±2, $\pm\dfrac{1}{2}$, $\pm\dfrac{1}{4}$.

Examining the graph of $y = P(x)$, we see that there
is a zero between -1 and 0, between 0 and 1, and
possible integer zeros at -1 and 2. Testing 2, we
obtain

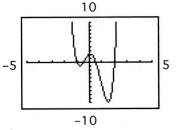

$$
\begin{array}{r}
4 -4 -9 1 2 \\
 8 8 -2 -2 \\
\hline
2|4 4 -1 -1 0
\end{array}
$$

2 is a zero. Testing -1, we examine the reduced polynomial $4x^3 + 4x^2 - x - 1$.

$$
\begin{array}{r}
4 4 -1 -1 \\
 -4 0 1 \\
\hline
-1|4 0 -1 0
\end{array}
$$

-1 is a zero.

So $P(x) = (x - 2)(x + 1)(4x^2 - 1) = (x - 2)(x + 1)(2x - 1)(2x + 1)$.

69.
$$
\begin{aligned}
[x - (4 - 5i)][x - (4 + 5i)] &= [x - 4 + 5i][x - 4 - 5i] \\
&= [(x - 4) + 5i][(x - 4) - 5i] \\
&= (x - 4)^2 - 25i^2 \\
&= x^2 - 8x + 16 + 25 \\
&= x^2 - 8x + 41
\end{aligned}
$$

71.
$$
\begin{aligned}
[x - (3 + 4i)][x - (3 - 4i)] &= [x - 3 - 4i][x - 3 + 4i] \\
&= [(x - 3) - 4i][(x - 3) + 4i] \\
&= (x - 3)^2 - 16i^2 \\
&= x^2 - 6x + 9 + 16 \\
&= x^2 - 6x + 25
\end{aligned}
$$

73.
$$
\begin{aligned}
[x - (a + bi)][x - (a - bi)] &= [x - a - bi][x - a + bi] \\
&= [(x - a) - bi][(x - a) + bi] \\
&= (x - a)^2 - b^2i^2 \\
&= (x - a)^2 + b^2 \\
&= x^2 - 2ax + a^2 + b^2
\end{aligned}
$$

75. (A) $2 - i$

(B) $\left(x - (2 + i)\right.$ and $\left(x - (2 - i)\right)$

(C)
$$
\begin{aligned}
(x - (2 + i)(x - (2 - i)) &= (x - 2 - i)(x - 2 + i) \\
&= \big((x - 2) - i\big)\big((x - 2) + i\big) \\
&= \left(x - 2\right)^2 - i^2 \\
&= x^2 - 4x + 4 - (-1) \\
&= x^2 - 4x + 5
\end{aligned}
$$

(D)

$$
\begin{array}{r}
x - 3 \\
x^2 - 4x + 5\overline{\smash{\big)}\, x^3 - 7x^2 + 17x - 15} \\
-\left(x^3 - 4x^2 + 5x\right) \\
\hline
-3x^2 + 12x - 15 \\
-\left(-3x^2 + 12x - 15\right) \\
\hline
0
\end{array}
$$

The remaining factor is $x - 3$, so the remaining zero is $x = 3$.

77. If $3 - i$ is a zero, then $3 + i$ is a zero. So $Q(x) = [x - (3 - i)][x - (3 + i)]$ divides $P(x)$ evenly. Applying the result of Problem 73, $Q(x) = x^2 - 6x + 9 + 1 = x^2 - 6x + 10$. Dividing, we see

$$
\begin{array}{r}
x + 1 \\
x^2 - 6x + 10\overline{\smash{\big)}\, x^3 - 5x^2 + 4x + 10} \\
x^3 - 6x^2 + 10x \\
\hline
x^2 - 6x + 10 \\
x^2 - 6x + 10 \\
\hline
0
\end{array}
$$

So $P(x) = (x + 1)Q(x)$ and the two other zeros are -1 and $3 + i$.

79. If $-5i$ is a zero, then so is $5i$. So $Q(x) = (x - 5i)(x + 5i)$ divides $P(x)$ evenly. $Q(x) = x^2 - 25i^2 = x^2 + 25$. Dividing, we see

$$
\begin{array}{r}
x - 3 \\
x^2 + 25\overline{\smash{\big)}\, x^3 - 3x^2 + 25x - 75} \\
x^3 + 25x \\
\hline
- 3x^2 - 75 \\
- 3x^2 - 75 \\
\hline
0
\end{array}
$$

So $P(x) = (x - 3)Q(x)$ and the two other zeros are $5i$ and 3.

81. If $2 + i$ is a zero, then $2 - i$ is a zero. So $Q(x) = [x - (2 + i)][x - (2 - i)]$ divides $P(x)$ evenly. Applying the result of Problem 73, $Q(x) = x^2 - 4x + 4 + 1 = x^2 - 4x + 5$. Dividing, we see

$$
\begin{array}{r}
x^2 - 2 \\
x^2 - 4x + 5\overline{\smash{\big)}\, x^4 - 4x^3 + 3x^2 + 8x - 10} \\
x^4 - 4x^3 + 5x^2 \\
\hline
- 2x^2 + 8x - 10 \\
- 2x^2 + 8x - 10 \\
\hline
0
\end{array}
$$

So $P(x) = Q(x)(x^2 - 2)$. $x^2 - 2$ has two zeros: $\sqrt{2}$ and $-\sqrt{2}$.
Summarizing, $P(x)$ has 4 zeros: $2 + i$, $2 - i$, $\sqrt{2}$, $-\sqrt{2}$.

83. $\sqrt{6}$ is a root of $x^2 = 6$ or $x^2 - 6 = 0$. The possible rational roots of this equation are ±1, ±2, ±3, ±6. Since none of them satisfies $x^2 = 6$, there are no rational roots, and $\sqrt{6}$ is not rational.

85. $\sqrt[3]{5}$ is a root of $x^3 = 5$ or $x^3 - 5 = 0$. The possible rational roots of this equation are ±1, ±5. Since none of them satisfies $x^3 = 5$, there are no rational roots, and $\sqrt[3]{5}$ is not rational.

87. The possible rational zeros are ±1, ±2, ±3, ±6. Examining the graph, we can see that there are two real zeros, one between -3 and -2, and the other between 2 and 3. Since none of the possible rational zeros actually are zeros, there must not be any rational zeros.

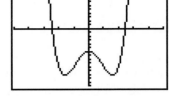

89. The only possible rational zeros are ±1, but $P(1) = -3$ and $P(-1) = 1$, so there are no rational zeros. A look at the graph in a standard window shows 3 real zeros which is the maximum number possible for a third degree polynomial.

91. Here is a graph of
$P(x) = 3x^3 - 37x^2 + 84x - 24$.

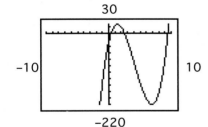

The rational zero theorem gives ±1, ±2, ±3, ±4, ±6, ±8, ±12, ±24, $\pm\dfrac{1}{3}$, $\pm\dfrac{2}{3}$, $\pm\dfrac{4}{3}$, $\pm\dfrac{8}{3}$ as possible rational zeros. However, the graph suggests that $P(x)$ has no integer zeros; there are no negative zeros, and the positive zeros would appear to lie between 0 and 1, 2 and 3, and 9 and 10. This suggests that we consider only the possible zeros $\dfrac{1}{3}$, $\dfrac{2}{3}$, and $\dfrac{8}{3}$. Testing $\dfrac{1}{3}$, we obtain

$$
\begin{array}{r|rrrr}
 & 3 & -37 & 84 & -24 \\
 &   & 1 & -12 & 24 \\
\hline
\tfrac{1}{3} & 3 & -36 & 72 & 0
\end{array}
$$

$\dfrac{1}{3}$ is a zero.

So $P(x) = \left(x - \dfrac{1}{3}\right)(3x^2 - 36x + 72) = \left(x - \dfrac{1}{3}\right)3(x^2 - 12x + 24)$

To find the remaining zeros, we solve $x^2 - 12x + 24 = 0$ using the quadratic formula:

$x^2 - 12x + 24 = 0 \qquad a = 1,\ b = -12,\ c = 24$

$x = \dfrac{-(-12) \pm \sqrt{(-12)^2 - 4(1)(24)}}{2(1)}$

$= \dfrac{12 \pm \sqrt{48}}{2} = \dfrac{12 \pm 4\sqrt{3}}{2} = 6 \pm 2\sqrt{3}$

So the zeros of $P(x)$ are $\dfrac{1}{3}$, $6 \pm 2\sqrt{3}$.

93. Here is a graph of
$P(x) = 4x^4 + 4x^3 + 49x^2 + 64x - 240$.

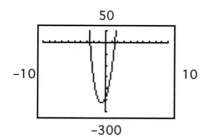

The rational zero theorem gives ±1, ±2, ±3, ±4, ±5, ±6, ±8, ±10, ±12, ±15, ±16, ±20, ±24, ±30, ±40, ±48, ±60, ±80, ±120, ±240, $\pm\frac{1}{2}$, $\pm\frac{3}{2}$, $\pm\frac{5}{2}$, $\pm\frac{15}{2}$, $\pm\frac{1}{4}$, $\pm\frac{3}{4}$, $\pm\frac{5}{4}$, $\pm\frac{15}{4}$ as possible rational zeros. However, the graph suggests that $P(x)$ has no integer zeros, and the zeros appear to lie between -3 and -2, and 1 and 2.

This suggests that we consider only the possible zeros $\frac{3}{2}$, $\frac{5}{4}$, and $-\frac{5}{2}$.

Testing $\frac{3}{2}$, we obtain

$$
\begin{array}{r|rrrrr}
 & 4 & 4 & 49 & 64 & -240 \\
 & & 6 & 15 & 96 & 240 \\
\hline
\frac{3}{2} & 4 & 10 & 64 & 160 & 0 \\
\end{array}
$$

$\frac{3}{2}$ is a zero.

So $P(x) = \left(x - \frac{3}{2}\right)(4x^3 + 10x^2 + 64x + 160) = \left(x - \frac{3}{2}\right)2(2x^3 + 5x^2 + 32x + 80)$

We consider the reduced polynomial $Q(x) = 2x^3 + 5x^2 + 32x + 80$. The graph suggests that the other real zero is negative, so we try $-\frac{5}{2}$ next.

$$
\begin{array}{r|rrrr}
 & 2 & 5 & 32 & 80 \\
 & & -5 & 0 & -80 \\
\hline
-\frac{5}{2} & 2 & 0 & 32 & 0 \\
\end{array}
$$

$-\frac{5}{2}$ is a zero.

So $P(x) = \left(x - \frac{3}{2}\right)2\left(x + \frac{5}{2}\right)(2x^2 + 32) = 4\left(x - \frac{3}{2}\right)\left(x + \frac{5}{2}\right)(x^2 + 16)$.

The remaining zeros of $P(x)$ are the zeros of $x^2 + 16$, that is, $\pm 4i$. So the zeros of $P(x)$ are $\frac{3}{2}$, $-\frac{5}{2}$, $\pm 4i$.

95. Here is a graph of
$P(x) = 4x^4 - 44x^3 + 145x^2 - 192x + 90$.

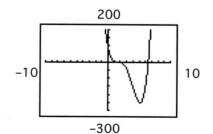

The rational zero theorem gives ±1, ±2, ±3, ±5, ±6, ±9, ±10, ±15, ±18, ±30, ±45, ±90, $\pm\frac{1}{2}$, $\pm\frac{3}{2}$, $\pm\frac{5}{2}$, $\pm\frac{9}{2}$, $\pm\frac{15}{2}$, $\pm\frac{45}{2}$, $\pm\frac{1}{4}$, $\pm\frac{3}{4}$, $\pm\frac{5}{4}$, $\pm\frac{9}{4}$, $\pm\frac{15}{4}$, $\pm\frac{45}{4}$ as possible rational zeros. However, the graph suggests that the only possible integer zeros are 1 and 2; there are no negative zeros, and the zeros appear to lie between 0 and 3, and between 6 and 7. This suggests that we consider only the possible zeros 1, 2, $\frac{1}{2}$, $\frac{3}{2}$, $\frac{5}{2}$, $\frac{1}{4}$, $\frac{3}{4}$, $\frac{5}{4}$, $\frac{9}{4}$. Testing the possible candidates, we obtain $P(1) = 3$, $P(2) = -2$, and

$$\begin{array}{r} 4 \quad -44 \quad\;\; 145 \quad\;\; -192 \quad\;\; 90 \\ 6 \quad\;\; -57 \quad\;\;\; 132 \quad\; -90 \end{array}$$

$$\frac{3}{2}\,\overline{)\,4 \quad -38 \quad\;\; 88 \quad\; -60 \quad\quad\; 0\;}$$

$\dfrac{3}{2}$ is a zero.

So $P(x) = \left(x - \dfrac{3}{2}\right)(4x^3 - 38x^2 + 88x - 60) = \left(x - \dfrac{3}{2}\right)2(2x^3 - 19x^2 + 44x - 30)$

We consider the reduced polynomial $Q(x) = 2x^3 - 19x^2 + 44x - 30$. The remaining possibilities from the reduced list are $\dfrac{1}{2}, \dfrac{3}{2}, \dfrac{5}{2}$. Testing for a possible double zero at $\dfrac{3}{2}$, we obtain

$$\begin{array}{r} 2 \quad -19 \quad\;\; 44 \quad\; -30 \\ 3 \quad\; -24 \quad\;\; 30 \end{array}$$

$$\frac{3}{2}\,\overline{)\,2 \quad -16 \quad\;\; 20 \quad\quad\; 0\;}$$

$\dfrac{3}{2}$ is a double zero.

So $P(x) = \left(x - \dfrac{3}{2}\right)^2 2(2x^2 - 16x + 20) = \left(x - \dfrac{3}{2}\right)^2 4(x^2 - 8x + 10)$

To find the remaining zeros, we solve $x^2 - 8x + 10 = 0$, using the quadratic formula.

$x^2 - 8x + 10 = 0 \qquad a = 1, \ b = -8, \ c = 10$

$$x = \frac{-(-8) \pm \sqrt{(-8)^2 - 4(1)(10)}}{2(1)}$$

$$= \frac{8 \pm \sqrt{24}}{2} = \frac{8 \pm 2\sqrt{6}}{2} = 4 \pm \sqrt{6}$$

So the zeros of $P(x)$ are $\dfrac{3}{2}$ (double), $4 \pm \sqrt{6}$.

97. (A) Since there are 3 zeros of $x^3 - 1$, there are 3 cube roots of 1.

(B) $x^3 - 1 = (x - 1)(x^2 + x + 1)$. The other cube roots of 1 will be solutions to $x^2 + x + 1 = 0$. Applying the quadratic formula with $a = b = c = 1$, we have

$x = \dfrac{-1 \pm \sqrt{(1)^2 - 4(1)(1)}}{2(1)} = \dfrac{-1 \pm \sqrt{-3}}{2} = \dfrac{-1 \pm i\sqrt{3}}{2}$. The other cube roots of 1 are

$-\dfrac{1}{2} + \dfrac{\sqrt{3}}{2}i$ and $-\dfrac{1}{2} - \dfrac{\sqrt{3}}{2}i$.

99. $P(x)$ can have at most n and must have at least one real zero. Each zero of $P(x)$ represents a point where $P(x) = y = 0$ so the graph of $P(x)$ will cross the x axis at and only at zeros of $P(x)$. Thus there can be a maximum of n axis crossings and there is a minimum of 1 axis crossing.

101. $P(2 + i) = (2 + i)^2 + 2i(2 + i) - 5$

$\qquad\qquad = 4 + 4i + i^2 + 4i + 2i^2 - 5$

$\qquad\qquad = 4 + 4i - 1 + 4i - 2 - 5$

$\qquad\qquad = -4 + 8i$

So $P(2 + i) \neq 0$ and $2 + i$ is not a zero of $P(x)$. This does not contradict the theorem, since $P(x)$ is not a polynomial with real coefficients (the coefficient of x is the imaginary number $2i$).

103. The right and left behaviors of an odd polynomial are opposite: if the leading coefficient is positive, the graph approaches $-\infty$ to the left, and ∞ to the right, and if the leading coefficient is negative, the graph approaches ∞ to the left and $-\infty$ to the right. In either case, some of the graph is above the x axis and some is below. The graphs of polynomials are always continuous, so the graph has to cross the x axis at least once.

105. If the constant term is zero, then every term has at least one power of the variable. In that case you can factor out the largest power of the variable that appears in every term; this will leave a reduced polynomial with a nonzero constant term, and then the rational zero theorem can be applied as usual. One of the zeros of the polynomial will be zero, and the rest will be the zeros of the reduced polynomial.

107. Let x = the amount of increase.

Then old volume = $1 \times 2 \times 3 = 6$

new volume = $(x + 1)(x + 2)(x + 3) = x^3 + 6x^2 + 11x + 6$

Since (new volume) = $10 \cdot$ (old volume), we must solve

$$x^3 + 6x^2 + 11x + 6 = 10(6)$$
$$x^3 + 6x^2 + 11x + 6 = 60$$
$$P(x) = x^3 + 6x^2 + 11x - 54 = 0$$

The possible rational zeros are ± 1, ± 2, ± 3, ± 6, ± 9, ± 18, ± 27, ± 54. Testing in order, we obtain $P(1) = -36$ $P(2) = 0$

$$
\begin{array}{r|rrrr}
 & 1 & 6 & 11 & -54 \\
 &   & 2 & 16 & 54 \\
\hline
2 & 1 & 8 & 27 & 0 \\
\end{array}
$$

Now we know that $x^3 + 6x^2 + 11x - 54 = (x - 2)(x^2 + 8x + 27)$. Since $x^2 + 8x + 27 = 0$ has no positive zeros, 2 is the only positive solution of the equation. The increase must equal 2 feet.

109.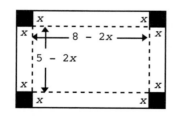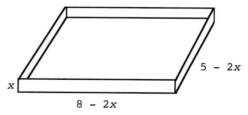

From the figure, we can see that
Volume = $x(5 - 2x)(8 - 2x) = 14$

Since x, $5 - 2x$, and $8 - 2x$ must all be positive, the domain of x is $0 < x < \dfrac{5}{2}$ or $(0, 2.5)$. We solve $x(5 - 2x)(8 - 2x) = 14$, or $4x^3 - 26x^2 + 40x = 14$, for x in this domain.

$$4x^3 - 26x^2 + 40x = 14$$
$$4x^3 - 26x^2 + 40x - 14 = 0$$
$$P(x) = 2x^3 - 13x^2 + 20x - 7 = 0$$

Possible rational zeros: ± 1, ± 7, $\pm\dfrac{1}{2}$, $\pm\dfrac{7}{2}$. Testing the possibilities, we obtain

$$P(1) = 2 \qquad P(7) = 182 \qquad P\left(\frac{1}{2}\right) = 0$$

$$\begin{array}{r} \phantom{\tfrac{1}{2}|}\ \ 2 \quad -13 \quad\ \ 20 \quad\ -7 \\ \phantom{\tfrac{1}{2}|\ 2}\quad\ \ 1 \quad\ -6 \quad\ \ \ 7 \\ \hline \tfrac{1}{2}\,|\ \ 2 \quad -12 \quad\ \ 14 \quad\ \ \ 0 \end{array}$$

$\dfrac{1}{2}$ is a zero.

So $2x^3 - 13x^2 + 20x - 7 = \left(x - \dfrac{1}{2}\right)(2x^2 - 12x + 14) = (2x - 1)(x^2 - 6x + 7)$

To find the remaining zeros, we solve $x^2 - 6x + 7 = 0$, using the quadratic formula:

$x^2 - 6x + 7 = 0 \qquad a = 1,\ b = -6,\ c = 7$

$x = \dfrac{-(-6) \pm \sqrt{(-6)^2 - 4(1)(7)}}{2(1)}$

$ = \dfrac{6 \pm \sqrt{8}}{2} = \dfrac{6 \pm 2\sqrt{2}}{2} = 3 \pm \sqrt{2}$

The zeros are $\dfrac{1}{2}$, $3 - \sqrt{2}$, $3 + \sqrt{2}$, or 0.5, 1.59, 4.41 to two significant digits. We discard $3 + \sqrt{2}$ or 4.41, since it is not in the interval (0, 2.5). The square should be 0.5 × 0.5 inches or 1.59 × 1.59 inches.

Section 3-5

1. Answers will vary.   **3.** Answers will vary.

5. It is not accurate. The graph of a function can never intersect a vertical asymptote, but it can intersect a horizontal asymptote.

7. This graph has a vertical asymptote $x = 2$, and a horizontal asymptote $y = -2$. This corresponds to $g(x)$.

9. This graph has a vertical asymptote $x = 2$, and a horizontal asymptote $y = 2$. This corresponds to $h(x)$.

11. (A) ∞ (B) $-\infty$ (C) 2   (D) 2

13. (A) $-\infty$ (B) ∞ (C) 2   (D) 2

In Problems 15 − 33, we will use the symbol n(x) to represent the numerator of a rational function, and d(x) to represent the denominator.

15. $\dfrac{2x - 4}{x + 1}$ *Domain:* $d(x) = x + 1$ *zero:* $x = -1$ *domain:* $(-\infty,\ -1) \cup (-1,\ \infty)$
x intercepts: $n(x) = 2x - 4$ *zero:* $x = 2$ *x intercept:* 2

17. $\dfrac{x^2 - 1}{x^2 - 16}$ *Domain:* $d(x) = x^2 - 16$ *zeros:* $x^2 - 16 = 0$

$$x^2 = 16$$
$$x = \pm 4$$

    domain: $(-\infty,\ -4) \cup (-4,\ 4) \cup (4,\ \infty)$

x intercepts: $n(x) = x^2 - 1$ *zeros:* $x^2 - 1 = 0$
$$x^2 = 1$$
$$x = \pm 1$$

        x intercepts: -1, 1

19. $\dfrac{x^2 - x - 6}{x^2 - x - 12}$ *Domain:* $d(x) = x^2 - x - 12$ zeros: $x^2 - x - 12 = 0$

$$(x + 3)(x - 4) = 0$$
$$x = -3, \ 4$$

domain: $(-\infty, \ -3) \ \cup \ (-3, \ 4) \ \cup \ (4, \ \infty)$

x intercepts: $n(x) = x^2 - x - 6$ zeros: $x^2 - x - 6 = 0$

$$(x + 2)(x - 3) = 0$$
$$x = -2, \ 3$$

x intercepts: $-2, \ 3$

21. $\dfrac{x}{x^2 + 4}$ *Domain:* $d(x) = x^2 + 4$ no real zeros

Domain: all real numbers

23. $\dfrac{4 - x^2}{x}$ *Domain:* $d(x) = x$ zero: 0

Domain: $(-\infty, \ 0) \ \cup \ (0, \ \infty)$

x intercepts: $n(x) = 4 - x^2$ zeros: $4 - x^2 = 0$

$$(2 + x)(2 - x) = 0$$
$$x = -2, \ 2$$

x intercepts: $-2, \ 2$

25. $\dfrac{2x}{x - 4}$ *vertical asymptotes:* $d(x) = x - 4$ zero: $x = 4$ | **Common Error:** $x = 0$ is not a vertical asymptote. |

vertical asymptote $x = 4$

horizontal asymptotes: Since $n(x)$ and $d(x)$ have the same degree, the line $y = 2$ is a horizontal asymptote (The ratio of the leading coefficients is 2/1, or 2).

27. $\dfrac{2x^2 + 3x}{3x^2 - 48}$ *vertical asymptotes:* $d(x) = 3x^2 - 48$ zeros: $3x^2 - 48 = 0$

$$3(x + 4)(x - 4) = 0$$
$$x = -4, \ 4$$

vertical asymptotes: $x = -4, \ x = 4$

horizontal asymptotes: Since $n(x)$ and $d(x)$ have the same degree, the line $y = \dfrac{2}{3}$ is a horizontal asymptote (The ratio of the leading coefficients is 2/3).

29. $\dfrac{2x}{x^4 + 1}$ *vertical asymptote:* $d(x) = x^4 + 1$ No real zeros:

No vertical asymptotes

horizontal asymptotes: Since the degree of $n(x)$ is less than the degree of $d(x)$, the x axis is a horizontal asymptote
horizontal asymptote: $y = 0$

31. $\dfrac{6x^4}{3x^2 - 2x - 5}$ *vertical asymptotes:* $d(x) = 3x^2 - 2x - 5$ zeros: $3x^2 - 2x - 5 = 0$

$$(3x - 5)(x + 1) = 0$$
$$x = \dfrac{5}{3}, \ -1$$

vertical asymptotes: $x = -1, \ x = \dfrac{5}{3}$

horizontal asymptotes: Since the degree of $n(x)$ is greater than the degree of $d(x)$, there are no horizontal asymptotes.

33. $\dfrac{4x^2 + 4x - 24}{x^2 - 2x} = \dfrac{4(x^2 + x - 6)}{x(x - 2)} = \dfrac{4(x + 3)(x - 2)}{x(x - 2)} = \dfrac{4(x + 3)}{x}$, $x \neq 2$

vertical asymptote: $x = 0$
Horizontal asymptotes: Since $n(x)$ and $d(x)$ have the same degree, the horizontal asymptote is $y = 4$ (The ratio of the leading coefficients is 4/1 or 4).

Common error: $x = 2$ is not an asymptote. The graph has a hole at (2, 10) and the function is not defined at $x = 2$, however, the function does not exhibit asymptotic behavior near $x = 2$.

35. The graph has more than one horizontal asymptote.

37. The graph has a sharp corner at (0, 0).

39. $\dfrac{x^2 + 2x}{x} = \dfrac{x(x + 2)}{x} = x + 2$ if $x \neq 0$. The graph of f is the same as the graph of g except that f has a hole at (0, 2).

41. $\dfrac{x + 2}{x^2 + 10x + 16} = \dfrac{x + 2}{(x + 2)(x + 8)} = \dfrac{1}{x + 8}$ if $x \neq -2$. The graph of f is the same as the graph of g except that f has a hole at $\left(-2, \dfrac{1}{6}\right)$.

43. $f(x) = \dfrac{1}{x - 4} = \dfrac{n(x)}{d(x)}$

Intercepts. There are no real zeros of $n(x) = 1$. No x intercept

$f(0) = -\dfrac{1}{4}$ $y = -\dfrac{1}{4}$ y intercept

Vertical asymptotes. $d(x) = x - 4$ zeros: 4 $x = 4$

Horizontal asymptotes. Since the degree of $n(x)$ is less that the degree of $d(x)$, the x axis is a horizontal asymptote.
Complete the sketch. Plot a few points.

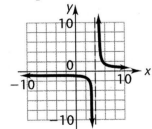

x	$f(x)$
4.5	2
5	1
6	.5
3.5	-2
3	-1
2	-.5

Common Error:
Entering $1/x - 4$ in the calculator.
Parentheses are needed: $1/(x - 4)$.

45. $f(x) = \dfrac{x}{x + 1} = \dfrac{n(x)}{d(x)}$

Intercepts. Real zeros of $n(x) = x$ $x = 0$ x intercept
$\qquad\qquad\qquad\qquad\qquad f(0) = 0$ $y = 0$ y intercept
The graph crosses the coordinate axes only at the origin.

Vertical asymptotes. $d(x) = x + 1$ zeros: -1 $x = -1$
Horizontal asymptotes. Since $n(x)$ and $d(x)$ have the same degree, the line $y = 1$ is a horizontal asymptote (The ratio of the leading coefficients is 1/1, or 1).
Complete the sketch. Plot a few additional points.

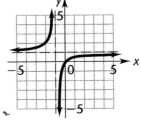

47. $h(x) = \dfrac{x}{2x - 2} = \dfrac{n(x)}{d(x)}$

Intercepts. Real zeros of $n(x) = x$ $x = 0$ x intercept
$\qquad\qquad\qquad\qquad\qquad h(0) = 0$ $y = 0$ y intercept
The graph crosses the coordinate axes only at the origin.
Vertical asymptotes. $d(x) = 2x - 2$ zeros: 1 $x = 1$

Horizontal asymptotes. Since $n(x)$ and $d(x)$ have the same degree, the line
$y = \dfrac{1}{2}$ is a horizontal asymptote (The ratio of the
leading coefficients is 1/2).
Complete the sketch. Plot a few additional points.

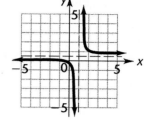

49. $f(x) = \dfrac{2x - 4}{x + 3} = \dfrac{n(x)}{d(x)}$

Intercepts. Real zeros of $n(x) = 2x - 4$ $x = 2$ x intercept
$\qquad\qquad\qquad\qquad f(0) = -\dfrac{4}{3}$ y intercept
Vertical asymptotes. $d(x) = x + 3$ zeros: -3 $x = -3$

Horizontal asymptotes. Since $n(x)$ and $d(x)$ have the same degree, the line $y = 2$
is a horizontal asymptote (The ratio of the
leading coefficients is 2/1, or 2).
Complete the sketch. Plot a few additional points.

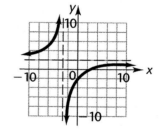

51. $g(x) = \dfrac{1 - x^2}{x^2} = \dfrac{n(x)}{d(x)}$

Intercepts. Real zeros of $n(x) = 1 - x^2$ $1 - x^2 = 0$
$\qquad\qquad\qquad\qquad\qquad\qquad\qquad x^2 = 1$
$\qquad\qquad\qquad\qquad\qquad\qquad\qquad x = \pm 1$ x intercepts
$\qquad\qquad g(0)$ is not defined $\qquad\qquad$ no y intercepts
Vertical asymptotes. $d(x) = x^2$ zeros: 0 $x = 0$
Horizontal asymptotes. Since $n(x)$ and $d(x)$ have the same degree, the line
$y = -1$ is a horizontal asymptote (The ratio of the
leading coefficients is -1/1, or -1).
Complete the sketch. Plot a few additional points.

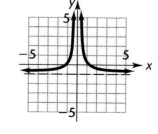

53. $f(x) = \dfrac{9}{x^2 - 9} = \dfrac{n(x)}{d(x)}$

Intercepts. There are no real zeros of $n(x) = 9$. No x intercept
$\qquad\qquad\qquad\qquad\qquad\qquad f(0) = -1$ y intercept

Vertical asymptotes. $d(x) = x^2 - 9$ zeros: $x^2 - 9 = 0$
$\qquad\qquad\qquad\qquad\qquad\qquad\qquad\qquad x^2 = 9$
$\qquad\qquad\qquad\qquad\qquad\qquad\qquad\qquad x = \pm 3$

Horizontal asymptotes. Since the degree of $n(x)$ is less than the degree of $d(x)$, the x axis is a horizontal asymptote.

Complete the sketch. Plot a few additional points.

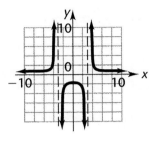

55. $f(x) = \dfrac{x}{x^2 - 1} = \dfrac{n(x)}{d(x)}$

Intercepts. Real zeros of $n(x) = x$ $x = 0$ x intercept
$\qquad\qquad\qquad\qquad\qquad f(0) = 0$ $y = 0$ y intercept
The graph crosses the coordinate axes only at the origin.

Vertical asymptotes. $d(x) = x^2 - 1$ zeros: $x^2 - 1 = 0$
$\qquad\qquad\qquad\qquad\qquad\qquad\qquad\qquad x^2 = 1$
$\qquad\qquad\qquad\qquad\qquad\qquad\qquad\qquad x = \pm 1$

Horizontal asymptotes. Since the degree of $n(x)$ is less than the degree of $d(x)$, the x axis is a horizontal asymptote.

Complete the sketch. Plot a few additional points.

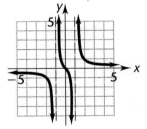

57. $g(x) = \dfrac{2}{x^2 + 1} = \dfrac{n(x)}{d(x)}$

Intercepts. There are no real zeros of $n(x) = 2$.   No x intercept
$\qquad\qquad\qquad\qquad\qquad\qquad g(0) = 2$ y intercept

Vertical asymptotes. There are no real zeros of $d(x) = x^2 + 1$
$\qquad$ No vertical asymptotes

Horizontal asymptotes. Since the degree of $n(x)$ is less than the degree of $d(x)$, the x axis is a horizontal asymptote.

Complete the sketch. Plot a few additional points.

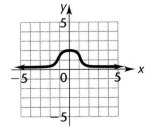

59. $f(x) = \dfrac{12x^2}{(3x + 5)^2} = \dfrac{n(x)}{d(x)}$

Intercepts. Real zeros of $n(x) = 12x^2$ $x = 0$ x intercept
$f(0) = 0$ $y = 0$ y intercept
The graph crosses the coordinate axes only at the origin.
Vertical asymptotes. $d(x) = (3x + 5)^2$ zeros: $(3x + 5)^2 = 0$
$3x + 5 = 0$
$x = -\dfrac{5}{3}$

Horizontal asymptotes. Since $n(x)$ and $d(x)$ have the same degree, the line

$y = \dfrac{12}{3^2} = \dfrac{4}{3}$ is a horizontal asymptote (The

denominator in expanded form is $9x^2 + 30x + 25$, so
the ratio of the leading coefficients is 12/9 or
4/3).

Complete the sketch. Plot a few additional points.

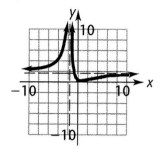

61. $f(x) = \dfrac{x^2 - 1}{x^2 + 7x + 10} = \dfrac{n(x)}{d(x)}$

Intercepts. Real zeros of $n(x) = x^2 - 1$ $x^2 - 1 = 0$
$x^2 = 1$
$x = \pm 1$ x intercepts
$f(0) = -\dfrac{1}{10}$ y intercept

Vertical asymptotes. Real zeros of $d(x) = x^2 + 7x + 10$ $x^2 + 7x + 10 = 0$
$(x + 2)(x + 5) = 0$
$x = -2, -5$

Horizontal asymptotes. Since $n(x)$ and $d(x)$ have the same degree, the line $y = 1$
is a horizontal asymptote (The ratio of
the leading coefficients is 1/1, or 1).

Complete the sketch. Plot a few
additional points.

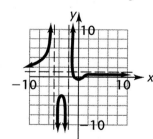

63. $f(x) = \dfrac{x^2 - 3x - 10}{x - 5} = \dfrac{(x - 5)(x + 2)}{x - 5} = x + 2$ $x \neq 5$.

The graph is the same as the graph of $y = x + 2$
except that f has a hole at (5, 7). There are no
asymptotes. The line has x intercept $x = -2$ and
y intercept $y = 2$.

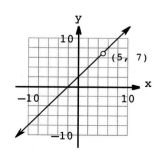

65. To have zeros −2, −1, 1, and 2, the numerator must have factors $(x + 2)$, $(x + 1)$, $(x - 1)$, and $(x - 2)$. To have a horizontal asymptote at $y = 3$, the degree of the denominator should be the same as the numerator, and the leading coefficient of the numerator should be 3 times as large. To have no vertical asymptote, the denominator should have no real zeros.

$$f(x) = \frac{3(x + 2)(x - 2)(x + 1)(x - 1)}{x^4 + 1} \quad \text{or} \quad f(x) = \frac{3(x^2 - 4)(x^2 - 1)}{x^4 + 1} \quad \text{will work.}$$

67. To get $y = 2x + 5$ as an oblique asymptote, our function should look like $f(x) = 2x + 5 + \dfrac{r(x)}{q(x)}$ where the degree of $q(x)$ is greater than the degree of $r(x)$. If $q(x) = x - 10$, we'll have $x = 10$ as vertical asymptote. In that case $r(x)$ will have to be constant so that it's power is less than $q(x)$; $r(x) = 100$ will do.

$$f(x) = 2x + 5 + \frac{100}{x - 10}$$
$$= \frac{(2x + 5)(x - 10)}{x - 10} + \frac{100}{x - 10}$$
$$= \frac{(2x + 5)(x - 10) + 100}{x - 10}$$

69. $\dfrac{x}{x - 2} \leq 0$

Algebraic Solution

Common Error:
It is not correct to "multiply both sides" by $x - 2$. The expression $x - 2$ can be either positive or negative, so the direction of the inequality would be uncertain.

Let $f(x) = \dfrac{p(x)}{q(x)} = \dfrac{x}{x - 2}$

The zero of $p(x) = x$ is 0. The zero of $q(x) = x - 2$ is 2. These two zeros partition the x axis into the three intervals shown in the table. A test number is chosen from each interval to determine the sign of $f(x)$.

Interval	Test number x	$f(x)$	Sign of f
$(-\infty, 0)$	−1	$\dfrac{1}{3}$	+
$(0, 2)$	1	−1	−
$(2, \infty)$	3	3	+

The left side is equal to zero at $x = 0$, but not at $x = 2$ (zero denominator). We conclude that the solution set is $[0, 2)$.

Graphical Solution

Graph $y_1 = \dfrac{x}{x - 2}$ and find where the graph is at or below height zero:

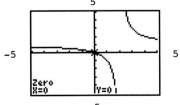

The graph is at or below height zero on $[0, 2)$.

71. $\dfrac{x^2 - 16}{5x - 2} > 0$

Algebraic Solution

Let $f(x) = \dfrac{p(x)}{q(x)} = \dfrac{x^2 - 16}{5x - 2}$

The zeros of $p(x) = x^2 - 16 = (x + 4)(x - 4)$ are -4 and 4.

The zero of $q(x) = 5x - 2$ is $\dfrac{2}{5}$.

These three zeros partition the x axis into the four intervals shown in the table. A test number is chosen from each interval to determine the sign of $f(x)$.

Interval	Test number x	$f(x)$	Sign of f
$(-\infty, -4)$	-5	$-\dfrac{1}{3}$	−
$\left(-4, \dfrac{2}{5}\right)$	0	8	+
$\left(\dfrac{2}{5}, 4\right)$	1	−5	−
$(4, \infty)$	5	$\dfrac{9}{23}$	+

We conclude that the solution set is $\left(-4, \dfrac{2}{5}\right) \cup (4, \infty)$.

Graphical Solution

Graph $y_1 = \dfrac{x^2 - 16}{5x - 2}$ and find where the graph above height zero.

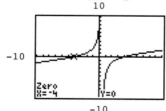 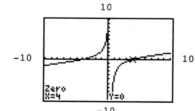

The graph is above height zero on $\left(-4, 2/5\right) \cup \left(4, \infty\right)$. (You have to note that the vertical asymptote is at $x = 2/5$; this isn't obvious at all from the graph!)

73. $\dfrac{x^2 + 4x - 20}{3x} \geq 4$

Algebraic Solution

$\dfrac{x^2 + 4x - 20}{3x} - 4 \geq 0$

$\dfrac{x^2 + 4x - 20}{3x} - \dfrac{12x}{3x} \geq 0$

$\dfrac{x^2 - 8x - 20}{3x} \geq 0$

Let $f(x) = \dfrac{p(x)}{q(x)} = \dfrac{x^2 - 8x - 20}{3x}$.

The zeros of $p(x) = x^2 - 8x - 20 = (x + 2)(x - 10)$ are -2 and 10. The zero of $q(x)$ is 0. These three zeros partition the x axis into the four intervals shown in the table. A test number is chosen from each interval to determine the sign of $f(x)$.

Interval	Test number x	$f(x)$	Sign of f
$(-\infty, -2)$	-3	$-\dfrac{13}{9}$	$-$
$(-2, 0)$	-1	$\dfrac{11}{3}$	$+$
$(0, 10)$	1	-9	$-$
$(10, \infty)$	11	$\dfrac{13}{33}$	$+$

The left side is equal to zero at -2 and 10, but not at 0 (zero denominator). We conclude that the solution set is $[-2, 0) \cup [10, \infty)$.

__Graphical Solution__

Graph $y_1 = \dfrac{x^2 + 4x - 20}{3x}$, $y_2 = 4$ and find where the height of the first graph is at or above the height of the second.

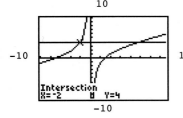 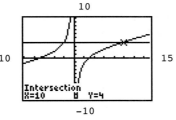

The first graph is at or above the second on $[-2, 0) \cup [10, \infty)$.

75. $\dfrac{5x}{x^2 - 1} < \dfrac{9}{x}$

__Algebraic Solution__

$$\dfrac{5x}{x^2 - 1} - \dfrac{9}{x} < 0$$

$$\dfrac{5x^2 - 9(x^2 - 1)}{x(x^2 - 1)} < 0$$

$$\dfrac{-4x^2 + 9}{x(x^2 - 1)} < 0$$

Let $f(x) = \dfrac{p(x)}{q(x)} = \dfrac{-4x^2 + 9}{x(x^2 - 1)}$.

The zeros of $p(x) = -4x^2 + 9 = 9 - 4x^2 = (3 + 2x)(3 - 2x)$ are $-\dfrac{3}{2}$ and $\dfrac{3}{2}$. The zeros of $q(x) = x(x^2 - 1) = x(x + 1)(x - 1)$ are 0, 1, and -1. These five zeros partition the x axis into the six intervals shown in the table. A test number is chosen from each interval to determine the sign of $f(x)$.

Interval	Test number x	$f(x)$	Sign of f
$\left(-\infty, -\dfrac{3}{2}\right)$	-2	$\dfrac{7}{6}$	$+$
$\left(-\dfrac{3}{2}, -1\right)$	$-\dfrac{5}{4}$	$-\dfrac{176}{45}$	$-$
$(-1, 0)$	$-\dfrac{1}{2}$	$\dfrac{64}{3}$	$+$
$(0, 1)$	$\dfrac{1}{2}$	$-\dfrac{64}{3}$	$-$
$\left(1, \dfrac{3}{2}\right)$	$\dfrac{5}{4}$	$\dfrac{176}{45}$	$+$
$\left(\dfrac{3}{2}, \infty\right)$	2	$-\dfrac{7}{6}$	$-$

We conclude that the solution set is $\left(-\dfrac{3}{2}, -1\right) \cup (0, 1) \cup \left(\dfrac{3}{2}, \infty\right)$.

Graphical Solution

Graph $y_1 = \dfrac{5x}{x^2 - 1}$, $y_2 = \dfrac{9}{x}$ and find where the height of the first graph is below the height of the second. (The second function is the thicker graph.)

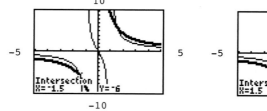

The first graph is below the second on $(-1.5, -1)$ (note the vertical asymptote is at $x = -1$), on $(0, 1)$ (note the 2nd vertical asymptote is at $x = 1$), and on $(1.5, \infty)$, so the solution is $(-1.5, -1) \cup (0, 1) \cup (1.5, \infty)$.

77. $\dfrac{x^2 + 7x + 3}{x + 2} > 0$

Algebraic Solution

zeros of the numerator $x^2 + 7x + 3 = 0$ $a = 1$, $b = 7$, $c = 3$

$$x = \frac{-7 \pm \sqrt{7^2 - 4(1)(3)}}{2(1)} = \frac{-7 \pm \sqrt{37}}{2} = \frac{-7 \pm 6.083}{2} = -6.541, \ -0.459$$

zeros of the denominator: $x = -2$

The three zeros partition the x axis into the four intervals shown in the table below. A test number is chosen from each interval to determine the sign of the rational expression.

Interval	Test number x	$f(x)$	Sign of f
$(-\infty, -6.541)$	-10	$-\dfrac{33}{8}$	$-$
$(-6.541, -2)$	-5	$\dfrac{7}{3}$	$+$
$(-2, -0.459)$	-1	-3	$-$
$(-0.459, \infty)$	0	$\dfrac{3}{2}$	$+$

The left side is positive on $(-6.541, -2) \cup (-0.459, \infty)$. We need to exclude all endpoints, so this is the solution.

Graphical Solution

Graph $y_1 = \dfrac{x^2 + 7x + 3}{x + 2}$ and find where it is above height zero

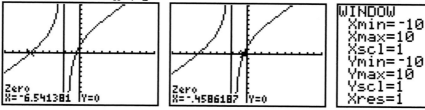

The graph is above the x axis on $(-6.541, -2) \cup (-0.459, \infty)$.

79. $\dfrac{5}{x^2} - \dfrac{1}{x+3} < 0$ (Simplify the left side)

<u>Algebraic Solution</u>

$$\dfrac{5(x+3)}{x^2(x+3)} - \dfrac{x^2}{x^2(x+3)} < 0$$

$$\dfrac{5x + 15 - x^2}{x^2(x+3)} < 0$$

$$\dfrac{-x^2 + 5x + 15}{x^2(x+3)} < 0$$

zeros of the numerator: $-x^2 + 5x + 15 = 0$ $a = -1,\ b = 5,\ c = 15$

$$x = \dfrac{-5 \pm \sqrt{5^2 - 4(-1)(15)}}{2(-1)} = \dfrac{-5 \pm \sqrt{85}}{-2} = \dfrac{-5 \pm 9.220}{-2} = 7.110,\ -2.110$$

zeros of the denominator: $x^2(x+3) = 0$

$$x = 0,\ -3$$

The four zeros partition the x axis into the five intervals shown in the table below. A test number is chosen from each interval to determine the sign of the rational expression.

Interval	Test number x	$f(x)$	Sign of f
$(-\infty, -3)$	-5	$\dfrac{7}{10}$	$+$
$(-3, -2.110)$	-2.5	-1.2	$-$
$(-2.110,\ 0)$	-1	$\dfrac{9}{2}$	$+$
$(0,\ 7.110)$	1	$\dfrac{19}{4}$	$+$
$(7.110, \infty)$	8	$-\dfrac{9}{704}$	$-$

The left side is negative on $(-3, -2.110) \cup (7.110, \infty)$. We need to exclude all endpoints so this is the solution.

<u>Graphical Solution</u>

Graph $y_1 = \dfrac{5}{x^2} - \dfrac{1}{x+3}$ and find where it is below height zero

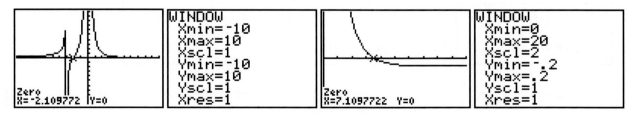

The graph is below the x axis on $(-3, -2.110) \cup (7.110, \infty)$.

81. $\dfrac{9}{x} - \dfrac{5}{x^2} \le 1$ (Simplify)

<u>Algebraic Solution</u>

$$\frac{9}{x} - \frac{5}{x^2} - 1 \le 0$$

$$\frac{9x}{x^2} - \frac{5}{x^2} - \frac{x^2}{x^2} \le 0$$

$$\frac{-x^2 + 9x - 5}{x^2} \le 0$$

zeros of the numerator: $-x^2 + 9x - 5 = 0$ $a = -1,\ b = 9,\ c = -5$

$$x = \frac{-9 \pm \sqrt{9^2 - 4(-1)(-5)}}{2(-1)} = \frac{-9 \pm \sqrt{61}}{-2} = \frac{-9 \pm 7.810}{-2} = 0.595,\ 8.405$$

zeros of the denominator: $x = 0$

These three zeros partition the x axis into the four intervals shown in the table below. A test number is chosen from each interval to determine the sign of the rational expression.

Interval	Test number x	$f(x)$	Sign of f
$(-\infty, 0)$	-1	-15	$-$
$(0, 0.595)$	0.5	-3	$-$
$(0.595, 8.405)$	1	3	$+$
$(0.8405, \infty)$	10	$-\dfrac{3}{20}$	$-$

The left side is negative on $(-\infty, 0)$, $(0, 0.595)$, and $(8.405, \infty)$. It's equal to zero when $x = 0.595$ and 8.405, so we include those endpoints, but we exclude $x = 0$ since the expression is undefined there. The solution is $(-\infty, 0) \cup (0, 0.595] \cup [8.405, \infty)$.

<u>Graphical Solution</u>

Graph $y_1 = \dfrac{5}{x^2} - \dfrac{1}{x + 3}$, $y_2 = 1$ and find where the first graph is below the second.

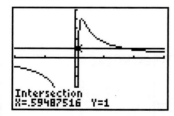 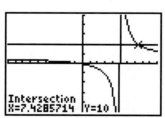

```
WINDOW
 Xmin=-10
 Xmax=15
 Xscl=5
 Ymin=-5
 Ymax=5
 Yscl=1
 Xres=1
```

The graph of $y_1 = \dfrac{9}{x} - \dfrac{5}{x^2}$ is below the graph of $y_2 = 1$ on $(-\infty, 0)$, $(0, 0.595)$ and $(8.405, \infty)$ and they intersect at $x = 0.595$ and $x = 8.405$.

83. $\dfrac{3x + 2}{x - 5} > 10$ (Simplify)

<u>Algebraic Solution</u>

$$\dfrac{3x + 2}{x - 5} - 10 > 0$$

$$\dfrac{3x + 2}{x - 5} - \dfrac{10(x - 5)}{x - 5} > 0$$

$$\dfrac{3x + 2 - 10x + 50}{x - 5} > 0$$

$$\dfrac{-7x + 52}{x - 5} > 0$$

zeros of the numerator: $-7x + 52 = 0$
$$-7x = -52$$
$$x = 7.429$$

zeros of the denominator: $x = 5$

These two zeros partition the x axis into the 3 intervals shown in the table below. A test number is chosen from each interval to determine the sign of the rational expression.

Interval	Test number x	$f(x)$	Sign of f
$(-\infty, 5)$	0	$-\dfrac{52}{5}$	$-$
$(5, 7.429)$	6	10	$+$
$(7.429, \infty)$	8	$-\dfrac{4}{3}$	$-$

The left side is positive on $(5, 7.429)$. We need to exclude the endpoints, so this is the solution.

<u>Graphical Solution</u>

Graph $y_1 = \dfrac{3x + 2}{x - 5}$, $y_2 = 10$ and find where the first graph is above the second.

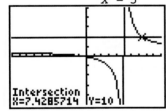

```
WINDOW
 Xmin=-10
 Xmax=10
 Xscl=1
 Ymin=-25
 Ymax=25
 Yscl=5
 Xres=1
```

```
Intersection
X=7.4285714  Y=10
```

The graph of $y_1 = \dfrac{3x + 2}{x - 5}$ is above the graph of $y_2 = 10$ on $(5, 7.429)$.

85. $\dfrac{4}{x + 1} \geq \dfrac{7}{x}$ (Simplify)

<u>Algebraic Solution</u>

$$\dfrac{4}{x + 1} - \dfrac{7}{x} \geq 0$$

$$\dfrac{4x}{x(x + 1)} - \dfrac{7(x + 1)}{x(x + 1)} \geq 0$$

$$\dfrac{4x - 7x - 7}{x(x + 1)} \geq 0$$

$$\dfrac{-3x - 7}{x(x + 1)} \geq 0$$

zeros of the numerator: $-3x - 7 = 0$
$$-3x = 7$$
$$x = -2.333$$

zeros of the denominator: $x(x + 1) = 0$
$$x = 0, -1$$

These three zeros partition the x axis into the four intervals shown in the table below. A test number is chosen from each interval to determine the sign of the rational expression.

Interval	Test number x	$f(x)$	Sign of f
$(-\infty, -2.333)$	-3	$\dfrac{1}{3}$	$+$
$(-2.333, -1)$	-2	$-\dfrac{1}{2}$	$-$
$(-1, 0)$	$-\dfrac{1}{2}$	22	$+$
$(0, \infty)$	1	-5	$-$

The left side is positive on $(-\infty, -2.333) \cup (-1, 0)$. We include $x = -2.333$ because it makes the left side zero, but exclude 0 and -1 as they make the expression undefined. The solution is $(-\infty, -2.333] \cup (-1, 0)$.

Graphical Solution

Graph $y_1 = \dfrac{4}{x + 1}$, $y_2 = \dfrac{7}{x}$ and find where the first graph is at or above the second. (The thicker graph is $7/x$.)

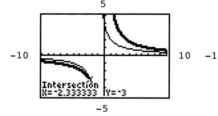 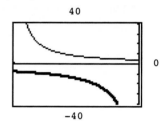

The thinner graph, $y_1 = \dfrac{4}{x + 1}$, is at or above the thicker on $(-\infty, -2.333]$, and again from $x = -1$ to $x = 0$. (That's hard to tell from the first graph; the second shows both graphs on just that interval) So the solution is $(-\infty, -2.333] \cup (-1, 0)$.

87. $f(x) = \dfrac{2x^2}{x - 1} = \dfrac{n(x)}{d(x)}$

Vertical asymptotes. Real zeros of $d(x) = x - 1$ $x = 1$
Horizontal asymptote. Since the degree of $n(x)$ is greater than the degree of $d(x)$, there is no horizontal asymptote.
Oblique asymptote.

$$
\begin{array}{r}
2x + 2 \\
x - 1 \overline{) 2x^2 } \\
\underline{2x^2 - 2x} \\
2x \\
\underline{2x - 2} \\
2
\end{array}
$$

We can write f as $f(x) = 2x + 2 + \dfrac{2}{x - 1}$, so the line $y = 2x + 2$ is an oblique asymptote.

89. $p(x) = \dfrac{x^3}{x^2 + 1} = \dfrac{n(x)}{d(x)}$

Vertical asymptotes. There are no real zeros of $d(x) = x^2 + 1$.
No vertical asymptotes.
Horizontal asymptotes. Since the degree of $n(x)$ is greater than the degree of $d(x)$, there is no horizontal asymptote.
Oblique asymptote:

$$
\begin{array}{r}
x \\
x^2 + 1 \overline{)\, x^3 } \\
\underline{x^3 + x} \\
-x
\end{array}
$$

We can write p as $p(x) = x + \dfrac{-x}{x^2 + 1}$, so the line $y = x$ is an oblique asymptote.

91. $r(x) = \dfrac{2x^2 - 3x + 5}{x} = \dfrac{n(x)}{d(x)}$

Vertical asymptotes. Real zeros of $d(x) = x$ $x = 0$
Horizontal asymptote. Since the degree of $n(x)$ is greater than the degree of $d(x)$, there is no horizontal asymptote.

Oblique asymptote. $\dfrac{2x^2 - 3x + 5}{x} = \dfrac{2x^2}{x} - \dfrac{3x}{x} + \dfrac{5}{x}$

$$= 2x - 3 + \dfrac{5}{x}$$

We can write r as $r(x) = 2x - 3 + \dfrac{5}{x}$, so the line $y = 2x - 3$ is an oblique asymptote.

93. Here is a computer-generated graph of $f(x)$.

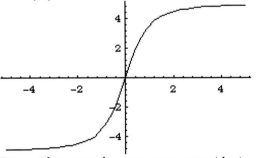

From the graph, we can see that $f(x) \to 5$ as $x \to \infty$ and $f(x) \to -5$ as $x \to -\infty$; the lines $y = 5$ and $y = -5$ are horizontal asymptotes.

95. Here is a computer-generated graph of $f(x)$.

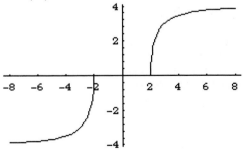

From the graph, we can see that $f(x) \to 4$ as $x \to \infty$ and $f(x) \to 4$ as $x \to -\infty$; the lines $y = 4$ and $y = -4$ are horizontal asymptotes.

97. $f(x) = \dfrac{x^2 + 1}{x} = \dfrac{n(x)}{d(x)}$

Intercepts. There are no real zeros of $n(x) = x^2 + 1$.   No x intercept
 $f(0)$ is not defined No y intercept
Vertical asymptotes. Real zeros of $d(x) = x$. $x = 0$
Horizontal asymptote. Since the degree of $n(x)$ is greater than the degree of $d(x)$, there is no horizontal asymptote.

Oblique asymptote. $f(x) = \dfrac{x^2 + 1}{x} = \dfrac{x^2}{x} + \dfrac{1}{x} = x + \dfrac{1}{x}$

The line $y = x$ is an oblique asymptote.

Complete the sketch. Plot a few points.

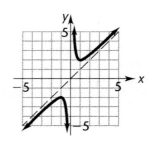

x	$f(x)$
$\frac{1}{2}$	$\frac{5}{2}$
1	2
2	$\frac{5}{2}$
$-\frac{1}{2}$	$-\frac{5}{2}$
-1	-2
-2	$-\frac{5}{2}$

99. $k(x) = \dfrac{x^2 - 4x + 3}{2x - 4} = \dfrac{n(x)}{d(x)}$

Intercepts. Real zeros of $n(x) = x^2 - 4x + 3$

$$x^2 - 4x + 3 = 0$$
$$(x - 1)(x - 3) = 0$$
$$x = 1, 3 \quad x \text{ intercepts}$$

$$k(0) = -\frac{3}{4} \quad y \text{ intercept}$$

Vertical asymptotes. Real zeros of $d(x) = 2x - 4$

$$2x - 4 = 0$$
$$2x = 4$$
$$x = 2$$

Horizontal asymptote. Since the degree of $n(x)$ is greater than the degree of $d(x)$, there is no horizontal asymptote.

Oblique asymptote:

$$
\begin{array}{r}
\frac{1}{2}x - 1 \\
2x - 4 \overline{\smash{)}\, x^2 - 4x + 3} \\
\underline{x^2 - 2x} \\
-2x + 3 \\
\underline{-2x + 4} \\
- 1
\end{array}
$$

We can write k as $k(x) = \dfrac{1}{2}x - 1 + \dfrac{-1}{2x - 4}$, so the line $y = \dfrac{1}{2}x - 1$ is an oblique asymptote.

Complete the sketch. Plot a few additional points.

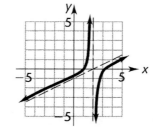

101. $F(x) = \dfrac{8 - x^3}{4x^2} = \dfrac{n(x)}{d(x)}$

Intercepts. Real zeros of $n(x) = 8 - x^3$

$$8 - x^3 = 0$$
$$8 = x^3$$
$$2 = x$$
$$x = 2 \quad x \text{ intercept}$$

$F(0)$ is not defined. No y intercept.

Vertical asymptotes. Real zeros of $d(x) = 4x^2$. $x = 0$

Horizontal asymptote. Since the degree of $n(x)$ is greater than the degree of $d(x)$, there is no horizontal asymptote.

Oblique asymptote. $F(x) = \dfrac{8 - x^3}{4x^2} = \dfrac{8}{4x^2} - \dfrac{x^3}{4x^2} = -\dfrac{1}{4}x + \dfrac{2}{x^2}$.

The line $y = -\dfrac{1}{4}x$ is an oblique asymptote.

Complete the sketch. Plot a few additional points.

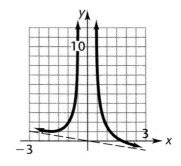

103. $f(x) = \dfrac{x^2 - 4}{x - 2}$. $f(x)$ is not defined if $x - 2 = 0$,
that is, $x = 2$
$$\text{Domain: } (-\infty, \ 2) \ \cup \ (2, \ \infty)$$
$f(x) = \dfrac{(x - 2)(x + 2)}{(x - 2)}$
$f(x) = x + 2;\ x \neq 2$
The graph is a straight line with slope 1 and
y intercept 2, except that the point (2, 4) is
not on the graph.
There are no asymptotes.

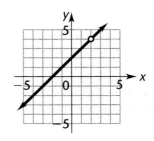

105. $r(x) = \dfrac{x + 2}{x^2 - 4}$. $f(x)$ is not defined if $x^2 - 4 = 0$, that is,
$$x^2 = 4$$
$$x = \pm 2$$
$$\text{Domain: } (-\infty, \ -2) \ \cup \ (-2, \ 2) \ \cup \ (2, \ \infty)$$

$r(x) = \dfrac{x + 2}{(x + 2)(x - 2)}$

$r(x) = \dfrac{1}{x - 2}$

The graph is the same as the graph of the function $\dfrac{1}{x - 2}$, except that the point
$\left(-2, -\dfrac{1}{4}\right)$ is not on the graph.

Intercepts: $y = -\dfrac{1}{2}$. No x intercept.

Vertical asymptote: $x = 2$
Horizontal asymptote: $y = 0$
Complete the sketch. Plot a few
additional points.

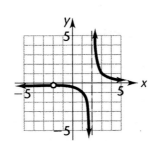

x intercepts: $n(x) = x$ zero: $x = 0$
x intercept: 0

107. $N(t) = \dfrac{50t}{t+4}$ $t \geq 0$

Intercepts: Real zeros of $50t$: $t = 0$ $N(0) = 0$
Vertical asymptotes: None, since -4, the only zero of $t + 4$, is not in the domain of N.

Horizontal asymptote: $N = 50$. As $t \to \infty$, $N \to 50$
This number tells us that in the long run, an average employee will be able to assemble 50 components per day.

Complete the sketch. Plot a few additional points.

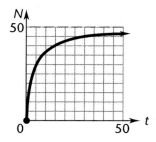

109. $N(t) = \dfrac{5t + 30}{t}$ $t \geq 1$

Intercepts: Real zeros of $5t + 30$, $t \geq 1$. None, since -6, the only zero of $5t + 30$, is not in the domain of N.
Vertical asymptotes: None, since 0, the only zero of t, is not in the domain of N.

Horizontal asymptote:
$N = 5$. As $t \to \infty$,
$N \to 5$

This tells us that in the long run, the average student will remember 5 of the symbols.
Complete the sketch. Plot a few points.

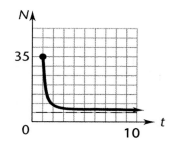

t	$N(t)$
1	35
3	15
5	11
10	8

111. (A) $\overline{C}(n) = \dfrac{C(n)}{n} = \dfrac{2,500 + 175n + 25n^2}{n} = 25n + 175 + \dfrac{2,500}{n}$

(B) To find the minimum value, we will graph $\overline{C}(n)$ and use the MINIMUM command:

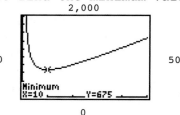

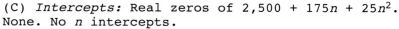

The minimum occurs after 10 years.

(C) *Intercepts:* Real zeros of $2,500 + 175n + 25n^2$.
None. No n intercepts.
0 is not in the domain of n, so there are no $\overline{C}$ intercepts.
Vertical asymptotes: Real zeros of n. The line $n = 0$ is a vertical asymptote.
Sign behavior: $\overline{C}$ is always positive since $n \geq 0$.
Horizontal asymptote: None, since the degree of $C(n)$ is greater than the degree of n.

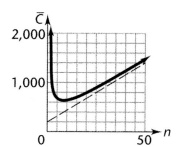

Oblique asymptote: The line $\overline{C} = 25n + 175$ is an oblique asymptote.

113. (A) Begin with a diagram:

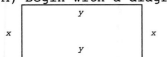

The total amount of fence needed is $L = 2x + 2y$, adding up the lengths of the four sides. But we were asked for an expression in terms of x, so we need to eliminate the variable y. We know that the area is 225 sq. ft., and we can see from the diagram that the area is also length × width, or xy. So

$$xy = 225 \quad \Rightarrow \quad y = \frac{225}{x}$$

and we can write the total amount of fence as

$$L(x) = 2x + 2\left(\frac{225}{x}\right) = 2x + \frac{450}{x} \text{ or } \frac{2x^2 + 450}{x}.$$

(B) At the very least, x has to be positive. Also, if the two sides of length x use up all of the fence, there will be no enclosed area, so we need $2x < 225$, or $x < 112.5$. The domain is $0 < x < 112.5$.

(C) To find the minimum, we will graph $L(x)$ on the appropriate domain, and use the MINIMUM command:

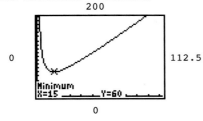

The minimum occurs when the width x is 15, in which case the length y is 225/15 = 15. So the dimensions are 15 ft × 15 ft.

(D) *Intercepts:* Real zeros of $2x^2 + 450$. None; no x intercepts. 0 is not in the domain of L, so there are no L intercepts.
Vertical asymptotes: Real zeros of x. The line $x = 0$ is a vertical asymptote.
Horizontal asymptote: None, since the degree of $2x^2 + 450$ is greater than the degree of x.
Oblique asymptote: The line $L = 2x$ is an oblique asymptote.

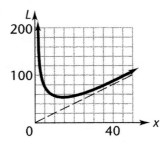

Section 3-6

1. Answers will vary.   **3.** Answers will vary.

5. Answers will vary.

7. $F = \dfrac{k}{x}$ **9.** $R = kST$ **11.** $L = km^3$ **13.** $A = kc^2d$ **15.** $P = kx$ **17.** $h = \dfrac{k}{\sqrt{s}}$

19. $R = k\dfrac{m}{d^2}$ **21.** $D = k\dfrac{xy^2}{z}$

23. The quantities are inversely proportional, so if F increases, x will decrease.

25. The quantities are directly proportional, so if m increases, L will increase.

27. Write $u = k\sqrt{v}$. Substitute $u = 3$ and $v = 4$ and solve for k.

$$3 = k\sqrt{4}$$
$$3 = 2k$$
$$k = \frac{3}{2}$$

The equation of variation is $u = \frac{3}{2}\sqrt{v}$. When $v = 10$, $u = \frac{3}{2}\sqrt{10}$.

29. Write $L = \frac{k}{M^2}$. Substitute $L = 9$ and $M = 9$ and solve for k.

$$9 = \frac{k}{9^2}$$
$$k = 9^3 = 729$$

The equation of variation is $L = \frac{729}{M^2}$. When $M = 6$, $L = \frac{729}{6^2} = \frac{81}{4}$.

31. Write $Q = k\frac{mn^2}{P}$. Substitute $Q = 2$, $m = 3$, $n = 6$, and $P = 12$ and solve for k.

$$2 = k\frac{3(6)^2}{12}$$
$$2 = 9k$$
$$k = \frac{2}{9}$$

The equation of variation is $Q = \frac{2}{9}\frac{mn^2}{P}$.
When $m = 4$, $n = 18$, and $P = 2$,

$$Q = \frac{2}{9}\frac{4(18)^2}{2} = 144$$

33. The variation is direct. As one quantity increases, the other does as well.

35. The variation is inverse. As one quantity increases, the other decreases.

37. $t = \frac{k}{T}$ **39.** $L = k\frac{wh^2}{x}$ **41.** $N = K\frac{F}{d}$

43. Write $f = kx$. Then $f_1 = kx_1$ and $f_2 = kx_2$. Therefore

$$\frac{kx_1}{kx_2} = \frac{f_1}{f_2}$$
$$\frac{x_1}{x_2} = \frac{f_1}{f_2}$$

45. Write $w = \frac{k}{d^2}$. Substitute $w = 100$ and $d = 4,000$ and solve for k.

$$100 = \frac{k}{(4,000)^2}$$
$$k = 100(4,000)^2 = 1.6 \times 10^9$$

The equation of variation is $w = \frac{1.6 \times 10^9}{d^2}$.

When she is 400 miles above the earth's surface, $d = 4,000 + 400 = 4,400$. Substitute to find

$$w = \frac{1.6 \times 10^9}{(4,400)^2} = 83 \text{ lbs to the nearest pound}$$

47. Write $I = k\dfrac{E}{R}$. Substitute $I = 22$, $E = 110$, and $R = 5$ and solve for k.

$$22 = k\dfrac{110}{5}$$

$$k = 1$$

The equation of variation is $I = \dfrac{E}{R}$. When $E = 220$ and $R = 11$,

$$I = \dfrac{220}{11} = 20 \text{ amperes}$$

49. Write $P = kv^3$. Let $P_1 = kv_1^{\,3}$ be the original horsepower. To represent doubling the speed, we let $v_2 = 2v_1$; then $P_2 = kv_2^{\,3} = k(2v_1)^3 = 8kv_1^{\,3} = 8P_1$. The new horsepower ($P_2$) is 8 times the original (P_1), so horsepower must be multiplied by 8 to double the speed of the boat.

51. Write $f = k\dfrac{\sqrt{T}}{L}$. Let $f_1 = k\dfrac{\sqrt{T_1}}{L_1}$ be the original frequency. To represent increasing tension by a factor of 4 and doubling the length, we let $T_2 = 4T_1$, and $L_2 = 2L_1$; then

$$f_2 = k\dfrac{\sqrt{T_2}}{L_2} = k\dfrac{\sqrt{4T_1}}{2L_1} = k\dfrac{\sqrt{T_1}}{L_1} = f_1. \text{ The new frequency } (f_2) \text{ is the same as the}$$

original (f_1), so there would be no net effect.

53. Write $t = k\dfrac{r}{v}$. Substitute $t = 1.42$, $r = 4{,}050$, and $v = 18{,}000$ and solve for k.

$$1.42 = k\dfrac{4{,}050}{18{,}000}$$

$$k = 1.42\dfrac{18{,}000}{4{,}050}$$

$$k = 6.311$$

The equation of variation is $t = 6.311\dfrac{r}{v}$. When $r = 4{,}300$ and $v = 18{,}500$,

$$t = 6.311\dfrac{4{,}300}{18{,}500} = 1.47 \text{ hours}$$

55. Write $d = kh$. Substitute $d = 4$ and $h = 500$ and solve for k.

$$4 = k(500)$$

$$k = \dfrac{1}{125}$$

The equation of variation is $d = \dfrac{1}{125}h$. When $h = 2{,}500$,

$$d = \dfrac{1}{125}2{,}500 = 20 \text{ days.}$$

57. Write $L = kv^2$. Let $L_1 = kv_1^{\,2}$ be the original length. To represent doubling the speed, we let $v_2 = 2v_1$. Then

$$L_2 = kv_2^{\,2} = k(2v_1)^2 = 4kv_1^{\,2} = 4L_1.$$

The new length (L_2) is 4 times the original length (L_1), so the length would be quadrupled.

59. Write $P = kAv^2$. Substitute $P = 120$, $A = 100$, and $v = 20$ and solve for k.
$$120 = k(100)20^2$$
$$120 = 40,000k$$
$$k = \frac{3}{1000}$$

The equation of variation is $P = \frac{3}{1000}Av^2$. When $A = 200$ and $v = 30$,
$$P = \frac{3}{1000}(200)(30)^2 = 540 \text{ lb}$$

61. (A) $\Delta S = kS$

(B) Substitute $\Delta S = 1$ and $S = 50$ and solve for k.
$$1 = k(50)$$
$$k = \frac{1}{50}$$

The equation of variation is $\Delta S = \frac{1}{50}S$. When $S = 500$, $\Delta S = \frac{1}{50}(500) = 10$ oz.

(C) Substitute $\Delta S = 1$ and $S = 60$ and solve for k.
$$1 = k(60)$$
$$k = \frac{1}{60}$$

The equation of variation is $\Delta S = \frac{1}{60}S$. When $S = 480$, $\Delta S = \frac{1}{60}(480) = 8$ candlepower.

63. Write $V = kr^3$. Let $V_1 = kr_1^3$ be the original volume. To represent doubling the radius, we let $r_2 = 2r_1$; then
$$V_2 = kr_2^3 = k(2r_1)^3 = 8kr_1^3 = 8V_1$$
The new volume (V_1) is 8 times the original volume (V_1).

65. Write frequency $= f$, length $= x$, $f = \frac{k}{x}$. Substitute $f = 16$ and $x = 32$ and solve for k.
$$16 = \frac{k}{32}$$
$$k = 512$$

The equation of variation is $f = \frac{512}{x}$. When $x = 16$, $f = \frac{512}{16} = 32$ times per second.

CHAPTER 3 REVIEW

1. (A) This is not a polynomial since the variable appears in the denominator of the last term.

(B) This is a polynomial with degree 3. *(3-1)*

2. The zeros are -1 and 3; the turning points are (-1, 0), (1, 2), and (3, 0); $P(x) \to \infty$ as $x \to -\infty$ and $P(x) \to \infty$ as $x \to \infty$. *(3-1)*

3.
$$\begin{array}{r} 2 \quad 3 \quad 0 \quad -1 \\ -4 \quad 2 \quad -4 \\ \hline -2\overline{\smash{\big)}2 \quad -1 \quad 2 \quad -5} \end{array}$$

$2x^3 + 3x^2 - 1$
$= (x + 2)(2x^2 - x + 2) - 5$ *(3-2)*

4.
$$\begin{array}{r} 1 \quad -4 \quad 0 \quad 9 \quad 0 \quad -8 \\ 3 \quad -3 \quad -9 \quad 0 \quad 0 \\ \hline 3\overline{\smash{\big)}1 \quad -1 \quad -3 \quad 0 \quad 0 \quad -8} \end{array}$$

$P(3) = -8$ *(3-2)*

5. 2, -4, -1 *(3-1)*

6. The odd degree and positive leading coefficient tell us that the graph should approach ∞ as $x \to \infty$ (out to the right) and approach $-\infty$ as $x \to -\infty$ (out to the left). The zeros provide the x intercepts of the graph.

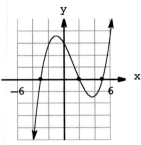

(3-1)

7. $P(x) = (x - 5)(x + 3)(x - 2)$ is one possible answer. In standard form,

$P(x) = (x - 5)(x^2 + x - 6)$

$= x^3 - 5x^2 + x^2 - 5x - 6x + 30$

$= x^3 - 4x^2 - 11x + 30$ (3-1)

8. Since complex zeros come in conjugate pairs, $1 - i$ is a zero. (3-4)

9. (A) Since the graph has x intercepts -2, 0, and 2, these are zeros of $P(x)$. Therefore, $P(x) = (x + 2)x(x - 2) = x^3 - 4x$.
(B) $P(x) \to \infty$ as $x \to \infty$ and $P(x) \to -\infty$ as $x \to -\infty$. (3-1)

10. We form a synthetic division table:

	1	-4	0	2	
-2	1	-6	12	-22	both are lower bounds, since
-1	1	-5	5	-3	both rows alternate in sign
3	1	-1	-3	-7	
4	1	0	0	2	upper bound

(3-3)

11. We investigate $P(1)$ and $P(2)$ by forming a synthetic division table.

	2	-3	1	-5
1	2	-1	0	-5
2	2	1	3	1

Since $P(1)$ and $P(2)$ have opposite signs, there is at least one real zero between 1 and 2. (3-3)

12. The factors of 6 are ±1, ±2, ±3, ±6. The leading coefficient is 1, so these are all possible rational zeros. (3-4)

13. Using the possibilities found in problem 12, we form a synthetic division table:

	1	-4	1	6	
1	1	-3	-2	4	
2	1	-2	-3	0	2 is a zero

Then $x^3 - 4x^2 + x + 6 = (x - 2)(x^2 - 2x - 3) = (x - 2)(x - 3)(x + 1)$. The rational zeros are 2, 3, -1. (3-4)

14. (A) $f(x) = \dfrac{2x - 3}{x + 4} = \dfrac{n(x)}{d(x)}$

The domain of f is the set of all real numbers x such that $d(x) = x + 4 \neq 0$, that is, $(-\infty, -4) \cup (-4, \infty)$. f has an x intercept where $n(x) = 2x - 3 = 0$, that is, $x = \dfrac{3}{2}$.

(B) $g(x) = \dfrac{3x}{x^2 - x - 6} = \dfrac{n(x)}{d(x)}$

The domain of g is the set of all real numbers x such that
$d(x) = x^2 - x - 6 \neq 0$, that is, $(x + 2)(x - 3) \neq 0$, that is, $x \neq -2$, 3, or
$(-\infty, -2) \cup (-2, 3) \cup (3, \infty)$. g has an x intercept where $n(x) = 3x = 0$,
that is, $x = 0$. *(3-5)*

15. (A) Horizontal asymptote: since $n(x)$ and $d(x)$ have the same degree, the line
$y = 2$ is a horizontal asymptote (The ratio of the leading coefficients is 2/1,
or 2). Vertical asymptotes: zeros of $d(x)$
$x + 4 = 0$
$x = -4$

(B) Horizontal asymptote: since the degree of $n(x)$ is less than the degree of
$d(x)$, the line $y = 0$ is a horizontal asymptote.
Vertical asymptotes: zeros of $d(x)$
$x^2 - x - 6 = 0$
$(x + 2)(x - 3) = 0$
$x = -2, \ x = 3$ *(3-5)*

16. $F = k\sqrt{x}$ *(3-6)* 17. $G = kxy^2$ *(3-6)* 18. $H = \dfrac{k}{z^3}$ *(3-6)*

19. $R = kx^2y^2$ *(3-6)* 20. $S = \dfrac{k}{u^2}$ *(3-6)* 21. $T = k\dfrac{v}{w}$ *(3-6)*

22. When two quantities are directly proportional, an increase in one corresponds
to an increase in the other as well. In this case, the larger surface area
will increase the amount of light captured. *(3-6)*

23. When two quantities are inversely proportional, an increase in one corresponds
to a decrease in the other. In this case, an increase in the number of
children will correspond with a decrease in cleanliness. *(3-6)*

24. The graph of a polynomial has to approach ∞ or $-\infty$ as $x \to \infty$ and $x \to -\infty$.
This graph approaches zero. *(3-1)*

25. $P(x) = x^3 - 3x^2 - 3x + 4$

(A)

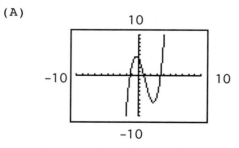

The graph of $P(x)$ has three x
intercepts and two turning points.
$P(x) \to \infty$ as $x \to \infty$ and $P(x) \to -\infty$
as $x \to -\infty$.

(B) Applying the ZERO command yields:

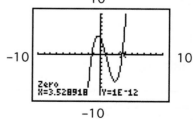

The largest x intercept is 3.53. *(3-1)*

26. We use synthetic division:

$$\frac{1}{4}\begin{array}{r|rrrrr} & 8 & -14 & -13 & -4 & 7 \\ & & 2 & -3 & -4 & -2 \\ \hline & 8 & -12 & -16 & -8 & 5 \end{array}$$

We can write P as

$P(x) = \left(x - \dfrac{1}{4}\right)(8x^3 - 12x^2 - 16x - 8) + 5$

$P\left(\dfrac{1}{4}\right) = 5$ *(3-2)*

27.

$$\begin{array}{r} x^2 - 9x + 18 \\ x^2 + 2x \overline{\big)\ x^4 - 7x^3 + 0x^2 - 5x + 1} \end{array}$$

$$-\left(x^4 + 2x^3\right)$$

$$-9x^3 + 0x^2$$
$$-\left(-9x^3 - 18x^2\right)$$

$$18x^2 - 5x$$
$$-\left(18x^2 + 36x\right)$$

$$-41x + 1$$

$$x^2 - 9x + 18 + \frac{-41x + 1}{x^2 + 2x} \qquad\qquad (3\text{-}2)$$

28.

$$\begin{array}{rrrr} 4 & -8 & -3 & -3 \\ & -2 & 5 & -1 \end{array}$$

$$-\tfrac{1}{2}\overline{\big)\ 4 \quad -10 \quad\ 2 \quad -4}$$

$$P\!\left(-\frac{1}{2}\right) = -4 \qquad\qquad (3\text{-}2)$$

29. The quadratic formula tells us that $x^2 - 2x - 1 = 0$ if

$$x = \frac{-b \pm \sqrt{b^2 - 4ac}}{2a} \qquad a = 1,\ b = -2,\ c = -1$$

$$x = \frac{-(-2) \pm \sqrt{(-2)^2 - 4(1)\,(-1)}}{2(1)}$$

$$= \frac{2 \pm \sqrt{8}}{2}$$

$$= 1 \pm \sqrt{2}$$

Since $1 \pm \sqrt{2}$ are zeros of $x^2 - 2x - 1$, its factors are $x - (1 + \sqrt{2})$ and
$x - (1 - \sqrt{2})$, that is, $x^2 - 2x - 1 = [x - (1 + \sqrt{2})][x - (1 - \sqrt{2})]$ $\qquad (3\text{-}2)$

30. $x + 1$ will be a factor of $P(x)$ if $P(-1) = 0$.
$P(-1) = 9(-1)^{26} - 11(-1)^{17} + 8(-1)^{11} - 5(-1)^4 - 7 = 9 + 11 - 8 - 5 - 7 = 0$, so
the answer is yes, $x + 1$ is a factor. $\qquad\qquad (3\text{-}2)$

31. First, divide out $x = 2$ using synthetic division:

$$\begin{array}{r} \begin{array}{rrrr} 2 & -9 & 3 & 14 \\ & 4 & -10 & -14 \end{array} \\ 2\ \overline{\big)\ 2 \quad -5 \quad -7 \quad\ 0} \end{array}$$

Now find the zeros of the reduced polynomial, $2x^2 - 5x - 7$ by factoring:

$$2x^2 - 5x - 7 = (2x - 7)(x + 1) = 0$$

$$2x - 7 = 0 \quad\text{or}\quad x + 1 = 0$$
$$2x = 7 \qquad\qquad\quad x = -1$$
$$x = 7/2 \qquad\qquad\qquad\qquad\qquad\qquad\qquad (3\text{-}2)$$

32. The possible rational zeros are ±1, ±2, ±4, ±8, ±$\frac{1}{2}$. We form a synthetic division table:

	2	-3	-18	-8
1	2	-1	-19	-27
2	2	1	-16	-40
4	2	5	2	0

So $2x^3 - 3x^2 - 18x - 8 = (x - 4)(2x^2 + 5x + 2)$
$$= (x - 4)(2x + 1)(x + 2)$$

Zeros: 4, $-\frac{1}{2}$, -2   (3-4)

33. From the given zeros, we get factors of $(x - 4)$, $(x + 1/2)$, and $(x + 2)$. But this can't be the correct factorization because the leading coefficient of the polynomial is 2, and there are no fractional coefficients. We remedy this by multiplying the second factor by 2, making it $(2x + 1)$.

$P(x) = (x - 4)(2x + 1)(x + 2)$ (3-4)

34. The possible rational zeros are ±1, ±5. We form a synthetic division table:

	1	-3	0	5
1	1	-2	-2	3
5	1	2	10	55
-1	1	-4	4	1
-5	1	-8	40	-195

There are no rational zeros, since all possibilities fail.   (3-4)

35. $P(x) = 2x^4 - x^3 + 2x - 1$.

Possible rational zeros: ±1, ±$\frac{1}{2}$

Examining the graph of $y = P(x)$, we see that there is a zero between 0 and 1, and possibly a zero at -1.

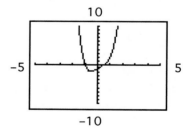

We test the likely candidates, and find

$$\begin{array}{r|rrrrr} & 2 & -1 & 0 & 2 & -1 \\ & & -2 & 3 & -3 & 1 \\ \hline -1 & 2 & -3 & 3 & -1 & 0 \end{array}$$

-1 is a zero. So $P(x) = (x + 1)(2x^3 - 3x^2 + 3x - 1) = (x + 1)Q(x)$

Testing $\frac{1}{2}$ in the reduced polynomial $Q(x)$, we obtain

$$\begin{array}{r|rrrr} & 2 & -3 & 3 & -1 \\ & & 1 & -1 & 1 \\ \hline \frac{1}{2} & 2 & -2 & 2 & 0 \end{array}$$

$\frac{1}{2}$ is a zero.

So $P(x) = (x + 1)\left(x - \frac{1}{2}\right)(2x^2 - 2x + 2) = (x + 1)(2x - 1)(x^2 - x + 1)$

To find the remaining zeros, we solve $x^2 - x + 1 = 0$, by the quadratic formula.
$x^2 - x + 1 = 0$

$$x = \frac{-b \pm \sqrt{b^2 - 4ac}}{2a} \qquad a = 1,\ b = -1,\ c = 1$$

$$x = \frac{-(-1) \pm \sqrt{(-1)^2 - 4(1)(1)}}{2(1)}$$

$$x = \frac{1 \pm \sqrt{-3}}{2}$$

$$x = \frac{1 \pm i\sqrt{3}}{2}$$

The four zeros are -1, $\dfrac{1}{2}$, and $\dfrac{1 \pm i\sqrt{3}}{2}$ (3-4)

36. $(x + 1)\left(x - \dfrac{1}{2}\right) 2\left(x - \dfrac{1 + i\sqrt{3}}{2}\right)\left(x - \dfrac{1 - i\sqrt{3}}{2}\right) = (x + 1)(2x - 1)\left(x - \dfrac{1 + i\sqrt{3}}{2}\right)\left(x - \dfrac{1 - i\sqrt{3}}{2}\right)$

(3-4)

37. The degree is 9. The zeros are 1 (multiplicity 3), -1 (multiplicity 4), i and $-i$. (3-4)

38. (A) Let $u = x^2$. Then
$$P(x) = x^4 + 5x^2 - 36$$
$$= u^2 + 5u - 36 = (u + 9)(u - 4) = (x^2 + 9)(x^2 - 4) = (x^2 + 9)(x + 2)(x - 2)$$

(B) $x^2 + 9$ has zeros $-3i$ and $3i$, so it factors as $(x + 3i)(x - 3i)$.
$$P(x) = (x + 3i)(x - 3i)(x + 2)(x - 2)$$ (3-4)

39. (A) Examining the graph of $P(x)$, we see that there may be zeros of even multiplicity between -1 and 0, and between 4 and 5, and a possible integer zero at 2.

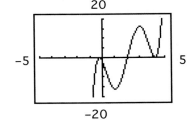

Testing 2, we obtain

```
   1   -10    30   -20   -15    -2
          2   -16    28    16     2
  ─────────────────────────────────
2| 1    -8    14     8     1     0
```

2 is a zero. Applying the MAXIMUM and MINIMUM commands, we obtain

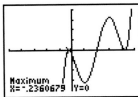 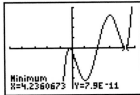

There are zeros of even multiplicity at -0.24 and 4.24. Since $P(x)$ is a fifth-degree polynomial, they must be double zeros and 2 must be a simple zero.

(B) -0.24 can be approximated with the MAXIMUM command; 2 can be approximated by the bisection method; 4.24 can be approximated with the MINIMUM command. (3-3)

40. (A) We form a synthetic division table.

	1	-2	-30	0	-25	
0	1	-2	-30	0	-25	
1	1	-1	-31	-31	-56	
2	1	0	-30	-60	-145	
3	1	1	-27	-81	-268	
4	1	2	-22	-88	-377	
5	1	3	-15	-75	-400	
6	1	4	-6	-36	-241	
7	1	5	5	35	220	7 is an upper bound
-1	1	-3	-27	27	-52	
-2	1	-4	-22	44	-113	
-3	1	-5	-15	45	-160	
-4	1	-6	-6	24	-121	
-5	1	-7	5	-25	100	-5 is a lower bound

(B) We search for the real zero in (6, 7) indicated in the table. We organize our calculations in a table.

Sign Change Interval (a, b)	Midpoint m	$P(a)$	$P(m)$	$P(b)$
(6, 7)	6.5	-	-	+
(6.5, 7)	6.75	-	+	+
(6.5, 6.75)	6.625	-	+	+
(6.5, 6.625)	6.5625	-	-	+
(6.5625, 6.625)	We stop here	-		+

Since each endpoint rounds to 6.6, a real zero lies on this interval and is given by 6.6 to one decimal place accuracy. 4 intervals were required.

(C) Graphing and applying the ZERO command, we obtain:

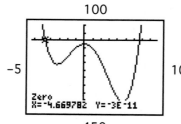
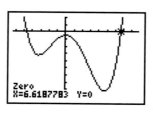

The zeros are -4.67 and 6.62.   (3-3)

41. $f(x) = \dfrac{x - 1}{2x + 2} = \dfrac{n(x)}{d(x)}$

(A) The domain of f is the set of all real numbers x such that $d(x) = 2x + 2 \neq 0$, that is $(-\infty, -1) \cup (-1, \infty)$.

f has an x intercept where $n(x) = x - 1 = 0$, that is, $x = 1$. $f(0) = -\dfrac{1}{2}$, so f has a y intercept at $y = -\dfrac{1}{2}$

(B) Vertical asymptote: $x = -1$. Horizontal asymptote: since $n(x)$ and $d(x)$ have the same degree, the line $y = \dfrac{1}{2}$ is a horizontal asymptote.

(C) Complete the sketch:

Graphing calculator: Hand sketch:

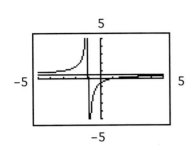   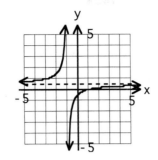

(*3-5*)

42. Rearrange so that zero is on the left:
$$3x^3 - 4x^2 - 15x \geq 0$$

Find the zeros of the left side:
$$3x^3 - 4x^2 - 15x = 0$$
$$x\left(3x^2 - 4x - 15\right) = 0$$
$$x(3x + 5)(x - 3) = 0$$
$$x = 0, \ -5/3, \ 3$$

Choose test numbers and make a table:

Interval	Test Number	Result
$x < -5/3$	-2	$3(-2)^3 - 4(-2)^2 - 15(-2) = -10$; negative
$-5/3 < x < 0$	-1	$3(-1)^3 - 4(-1)^2 - 15(-1) = 8$; positive
$0 < x < 3$	1	$3(1)^3 - 4(1)^2 - 15(1) = -16$; negative
$x > 3$	4	$3(4)^3 - 4(4)^2 - 15(4) = 68$; positive

Solution: $\left[-5/3, 0\right] \cup \left[3, \infty\right)$

(*3-3*)

43. (A) Find the zeros of $P(x) = x^3 - 5x + 4$ graphically:

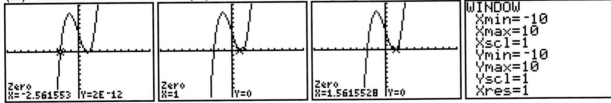

The zeros are -2.562, 1, and 1.562. (Two are very close together, so you may need to look closely at that portion of the graph to see that there are two.) $P(x)$ is negative (below the x-axis) on $(-\infty, -2.562) \cup (1, 1.562)$.

(B) Keeping $y_1 = x^3 - 5x + 4$, graph $y_2 = 2$.

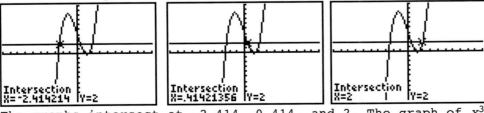

The graphs intersect at -2.414, 0.414, and 2. The graph of $x^3 - 5x + 4$ is below the graph of 2 on $(-\infty, -2.414) \cup (0.414, 2)$.

(*3-3*)

44. This is not the graph of a rational function because of what happens at $x = 0$; a rational function can only be discontinuous at an x value because of a vertical asymptote or a hole in the graph. *(3-5)*

45. Write $B = \dfrac{k}{\sqrt{C}}$. Substitute $B = 5$ and $C = 4$ and solve for k.

$$5 = \frac{k}{\sqrt{4}}$$
$$k = 10$$

The equation of variation is $B = \dfrac{10}{\sqrt{c}}$. When $c = 25$,

$$B = \frac{10}{\sqrt{25}} = \frac{10}{5} = 2$$ *(3-6)*

46. Write $D = kxy$. Substitute $D = 10$, $x = 3$, and $y = 2$ and solve for k.

$$10 = k(3)(2)$$
$$k = \frac{5}{3}$$

The equation of variation is $D = \dfrac{5}{3}xy$. When $x = 9$ and $y = 8$,

$$D = \frac{5}{3}(9)(8) = 120$$ *(3-6)*

47.

$$
\begin{array}{ccccc}
 & 1 & 0 & 3 & 2 \\
 & & 1 + i & 2i & 1 + 5i \\
\hline
1 + i\,] & 1 & 1 + i & 3 + 2i & 3 + 5i
\end{array}
$$
(See calculations below)

$$(1 + i)^2 = (1 + i)(1 + i) = 1 + 2i + i^2 = 1 + 2i - 1 = 2i$$
$$(1 + i)(3 + 2i) = 3 + 5i + 2i^2 = 3 + 5i - 2 = 1 + 5i$$

$$P(x) = [x^2 + (1 + i)x + (3 + 2i)][x - (1 + i)] + 3 + 5i$$ *(3-2)*

48. $P(x) = \left(x + \dfrac{1}{2}\right)^2 (x + 3)(x - 1)^3$. The degree is 6. *(3-4)*

49. $P(x) = (x + 5)[x - (2 - 3i)][x - (2 + 3i)]$. The degree is 3. *(3-4)*

50. The possible rational zeros are ± 1, ± 2, ± 4, $\pm \dfrac{1}{2}$.

Examining the graph of $y = P(x)$, we see that there are zeros between -1 and 0, between 0 and 1, between 2 and 3, and possible integer zeros at -2 and 2. Testing 2, we obtain

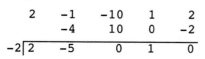

$$
\begin{array}{r|rrrrr}
2 & 2 & -5 & -8 & 21 & 0 & -4 \\
 & & 4 & -2 & -20 & 2 & 4 \\
\hline
 & 2 & -1 & -10 & 1 & 2 & 0
\end{array}
$$

2 is a zero.
So $P(x) = (x - 2)(2x^4 - x^3 - 10x^2 + x + 2)$.
Testing -2 in the reduced polynomial, we obtain

$$
\begin{array}{r|rrrrr}
-2 & 2 & -1 & -10 & 1 & 2 \\
 & & -4 & 10 & 0 & -2 \\
\hline
 & 2 & -5 & 0 & 1 & 0
\end{array}
$$

-2 is a zero.
$P(x) = (x - 2)(x + 2)(2x^3 - 5x^2 + 1)$. $2x^3 - 5x^2 + 1$ has been shown (see Exercise 3-3, problem 49 for details) to have zeros $\dfrac{1}{2}$, $1 \pm \sqrt{2}$.

Then $P(x)$ has zeros $\dfrac{1}{2}$, ± 2, $1 \pm \sqrt{2}$. *(3-4)*

51. $(x - 2)(x + 2)\left(x - \dfrac{1}{2}\right)2[x - (1 - \sqrt{2})][x - (1 + \sqrt{2})]$

$= (x - 2)(x + 2)(2x - 1)[x - (1 - \sqrt{2})][x - (1 + \sqrt{2})]$ (3-4)

52. Graphing $y = P(x)$, we obtain:

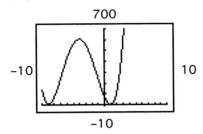

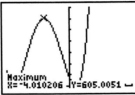

Clearly there is a local maximum near -4. Applying the MAXIMUM command, we obtain

To find the other required points, we need different windows. Redrawing and applying the MINIMUM and ZERO commands, we obtain

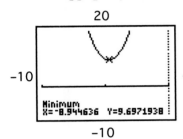

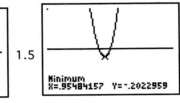

Summarizing, $P(x)$ has zeros at 0.91 and 1. It has a local minimum $P(-8.94) \approx 9.7$. It has a local maximum $P(-4.01) \approx 605.01$. It has a local minimum $P(0.95) \approx -0.20$.   (3-3)

53. $P(x)$ changes sign three times. Therefore, it has three zeros and its minimal degree is 3.   (3-1)

54. Since $1 + 2i$ is a zero, its conjugate $1 - 2i$ is also a zero. Then
$[x - (1 - 2i)][x - (1 + 2i)] = [(x - 1) + 2i][(x - 1) - 2i] = (x - 1)^2 - 4i^2$
$= x^2 - 2x + 5$ is a factor. Since $P(x)$ is a cubic polynomial, it must be of the form $a(x - r)(x^2 - 2x + 5)$. Since the constant term of this polynomial, $-5ar$, must be an integer, r must be a rational number. So there can be no irrational zeros.   (3-4)

55. (A) The cube roots of 27 are solutions to the equation $x^3 = 27$, or $x^3 - 27 = 0$. The polynomial $x^3 - 27$ is a third degree polynomial and has three zeros, so there must be three cube roots of 27.

(B) Divide out x = 3 using synthetic division:

$$
\begin{array}{r|rrrr}
 & 1 & 0 & 0 & -27 \\
 & & 3 & 9 & 27 \\
\hline
3 & 1 & 3 & 9 & 0
\end{array}
$$

We can find the zeros of the reduced polynomial $x^2 + 3x + 9$ using the quadratic formula with $a = 1$, $b = 3$, $c = 9$:

$$x = \frac{-3 \pm \sqrt{3^2 - 4(1)(9)}}{2(1)} = \frac{-3 \pm \sqrt{-27}}{2} = \frac{-3 \pm 3\sqrt{3}i}{2} = -\frac{3}{2} \pm \frac{3\sqrt{3}}{2}i \qquad (3\text{-}4)$$

56. (A) We form a synthetic division table.

	1	2	-500	0	-4,000
0	1	2	-500	0	-4,000
10	1	12	-380	-3,800	-42,000
20	1	22	-60	-1,200	-28,000
30	1	32	460	13,800	410,000
-10	1	-8	-420	4,200	-46,000
-20	1	-18	-140	2,800	-60,000
-30	1	-28	340	-10,200	302,000

From the table, 30 is an upper bound and -30 is a lower bound.

(B) Graphing $P(x)$ in the window suggested by part (A), we obtain

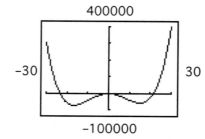
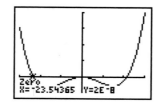
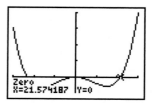

The zeros are -23.54 and 21.57. Examining the graph more closely near $x = 0$, we obtain

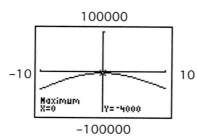

There is no other real zero.   (3-3)

57. $f(x) = \dfrac{x^2 + 2x + 3}{x + 1} = \dfrac{n(x)}{d(x)}$

Intercepts. There are no real zeros of $n(x) = x^2 + 2x + 3$. No x intercept
$f(0) = 3$ y intercept
Vertical asymptotes. Real zeros of $d(x) = x + 1$ $x = -1$ is a vertical asymptote
Horizontal asymptotes. Since the degree of $n(x)$ is greater than the degree of $d(x)$, there is no horizontal asymptote.
Oblique asymptote:

$$
\begin{array}{r}
x + 1 \\
x + 1 \overline{)\, x^2 + 2x + 3} \\
\underline{x^2 + x} \\
x + 3 \\
\underline{x + 1} \\
2
\end{array}
$$

$f(x) = x + 1 + \dfrac{2}{x + 1}$.

The line $y = x + 1$ is an oblique asymptote.

Complete the sketch:

Graphing calculator: Hand sketch:

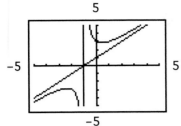 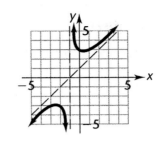

(*3-5*)

58. Graphing $y = f(x)$, we obtain the graph at the right.

From the graph, we can see that $f(x) \to 2$ as $x \to \infty$ and $f(x) \to -2$ as $x \to -\infty$; the lines $y = 2$ and $y = -2$ are horizontal asymptotes.

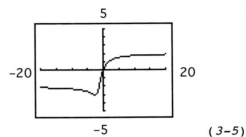

(*3-5*)

59. (A) $\dfrac{x^2 - 3}{x^3 - 3x + 1} \le 0$

zeros of the numerator: $x^2 - 3 = 0$
$$x^2 = 3$$
$$x = \sqrt{3}, \ -\sqrt{3} \text{ or } 1.732, \ -1.732$$

zeros of the denominator: Find graphically:

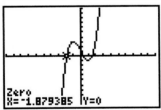 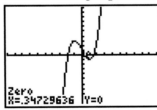 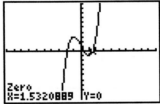

The zeros are -1.879, 0.347, 1.532.

Interval	Test Number	Result
$(-\infty, -1.879)$	-2	-1; negative
$(-1.879, -1.732)$	-1.8	0.423; positive
$(-1.732, 0.347)$	0	-3; negative
$(0.347, 1.532)$	1	2; positive
$(1.532, 1.732)$	1.6	-1.49; negative
$(1.732, \infty)$	2	0.333; positive

The expression is negative on $(-\infty, -1.879)$, $(-1.732, 0.347)$, and $(1.532, 1.732)$. We should include -1.732 and 1.732 (since the expression is zero there) and exclude -1.879, 0.347, and 1.532 (since the expression is undefined there.)

The solution is $(-\infty, -1.879) \cup [-1.732, 0.347) \cup (1.532, 1.732]$.

Graphical check

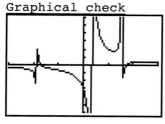

The graph is below the x-axis on $(-\infty, -1.879) \cup (-1.732, 0.347) \cup (1.532, 1.732)$, and intersects the x-axis at -1.732 and 1.732.

(B) $\dfrac{x^2 - 3}{x^3 - 3x + 1} > \dfrac{5}{x^2}$

$$\dfrac{x^2 - 3}{x^3 - 3x + 1} - \dfrac{5}{x^2} > 0$$

$$\dfrac{x^2(x^2 - 3)}{x^2(x^3 - 3x + 1)} - \dfrac{5(x^3 - 3x + 1)}{x^2(x^3 - 3x + 1)} > 0$$

$$\dfrac{x^4 - 3x^2 - 5x^3 + 15x - 5}{x^2(x^3 - 3x + 1)} > 0$$

zeros of the numerator: Find graphically:

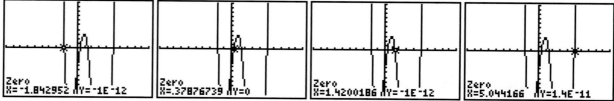

The zeros are -1.843, 0.379, 1.420, and 5.044.

zeros of the denominator:

$x^2(x^3 - 3x + 1 = 0$
$x^2 = 0 \quad x^3 - 3x + 1 = 0$
$x = 0 \qquad\qquad x = -1.879, 0.347, 1.532 \quad$ (From part A)

Interval	Test Number	Result
$(-\infty, -1.879)$	-2	-2.25; negative
$(-1.879, -1.843)$	-1.85	0.47; positive
$(-1.843, 0)$	-1	-5.7; negative
$(0, 0.347)$	0.1	-504.3; negative
$(0.347, 0.379)$	0.35	363; positive
$(0.379, 1.420)$	1	-3; negative
$(1.420, 1.532)$	1.5	3.8; positive
$(1.532, 5.044)$	2	-0.92; negative
$(5.044, \infty)$	6	0.03; positive

The expression is positive on $(-1.879, -1.843) \cup (0.347, 0.379) \cup (1.420, 1.532) \cup (5.044, \infty)$

(3-5)

60. The potential rational zeros are ± 1, ± 2, and ± 4. A look at the graph shows 3 zeros, none of which occur at integer values, so none of the candidates for rational zeros are actually zeros.

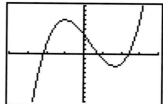

(3-4)

61. To have real zeros -3, 0, and 2, we should have factors $(x + 3)$, x, and $(x - 2)$ in the numerator. To have vertical asymptotes $x = -1$ and $x = 4$, we should have factors $(x + 1)$ and $(x - 4)$ in the numerator. At the moment, our function is

$$f(x) = \frac{x(x + 3)(x - 2)}{(x + 1)(x - 4)}$$

But the degree of the numerator is greater than the degree of the denominator, so there will be no horizontal asymptote. We can fix that by squaring one of the factors in the denominator. Multiplying the numerator by 5 will then make $y = 5$ a horizontal asymptote.

$$f(x) = \frac{5x(x + 3)(x - 2)}{(x + 1)^2(x - 4)}$$

(There are other possible answers.) *(3-5)*

62. (A) Graph $P(x)$ on the domain $0 \leq x \leq 20$ and use the ZERO command:

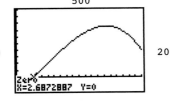

She should be open 2.7 hours to break even.

(B) Add $y_2 = 325$ to the graph and use the INTERSECT command:

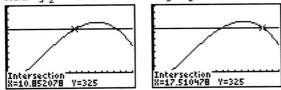

To make a weekly profit of at least \$325, she should be open between 10.9 and 17.5 hours. *(3-3)*

63. Write $F = \dfrac{k}{r}$. Let $F_1 = \dfrac{k}{r_1}$ be the original force. To represent doubling the radius, we let $r_2 = 2r_1$. Then

$$F_2 = \frac{k}{r_2} = \frac{k}{2r_1} = \frac{1}{2}\frac{k}{r_1} = \frac{1}{2}F_1.$$

The force would be $\dfrac{1}{2}$ the original. *(3-6)*

64. $v = k\dfrac{\sqrt{T}}{\sqrt{w}}$ *(3-6)*

65. Write $A = kWt$. Substitute $A = A_0$, $W = 10$ and $t = 8$ and solve for k.

$$A_0 = k(10)(8)$$

$$k = \frac{A_0}{80}$$

The equation of variation for this job is $A = \dfrac{A_0}{80}Wt$.

Substitute $A = A_0$ (again) and $W = 4$ and solve for t.

$$A_0 = \frac{A_0}{80}(4)t$$

$$A_0 = \frac{A_0}{20}t$$

$$t = 20 \text{ days}$$

(3-6)

66. Write $I = kpr$. Substitute $I = 8$, $p = 100$, and $r = 0.04$ and solve for k.
$$8 = k(100)(0.04)$$
$$k = 2$$
The equation of variation is $I = 2pr$. When $p = 150$ and $r = 0.03$,
$$I = 2(150)(0.03)$$
$$I = \$9.00$$

(3-6)

67. In the given figure, let y = height of door
$$2x = \text{width of door}$$
Then Area of door = $48 = 2xy$

Since (x, y) is a point on the parabola $y = 16 - x^2$, its coordinates satisfy the equation of the parabola.
$$48 = 2x(16 - x^2)$$
$$48 = 32x - 2x^3$$
$$2x^3 - 32x + 48 = 0$$
$$x^3 - 16x + 24 = 0$$
The possible rational solutions of this equation are ±1, ±2, ±3, ±4, ±6, ±8, ±12, ±24. Testing the likely candidates, we obtain

```
      1    0   -16    24
           2    4    -24
  2│  1    2   -12     0
```

2 is a solution.
The equation can be factored
$(x - 2)(x^2 + 2x - 12) = 0$.
To find the remaining zeros, we solve $x^2 + 2x - 12 = 0$, by completing the square.
$$x^2 + 2x = 12$$
$$x^2 + 2x + 1 = 13$$
$$(x + 1)^2 = 13$$
$$x + 1 = \pm\sqrt{13}$$
$$x = -1 + \sqrt{13} \text{ (discarding the negative solution)} \approx 2.61$$
The positive zeros are $x = 2$, 2.61. The dimensions of the door are either $2x = 4$ feet by $16 - x^2 = 12$ feet, or $2x = 5.2$ feet by $16 - x^2 = 9.2$ feet. (3-3)

68. We note:

$$\begin{pmatrix} \text{Volume} \\ \text{of} \\ \text{silo} \end{pmatrix} = \begin{pmatrix} \text{Volume of} \\ \text{hemisphere} \\ \text{of radius } x \end{pmatrix} + \begin{pmatrix} \text{Volume of} \\ \text{cylinder with} \\ \text{radius } x, \text{height 18} \end{pmatrix}$$

$$486\pi = \frac{2}{3}\pi x^3 \qquad + \pi x^2 \cdot 18$$

$$486 = \frac{2}{3}x^3 + 18x^2$$

$$0 = x^3 + 27x^2 - 729$$

Examining the graph of
$y = P(x)$, we obtain, for positive x:

The radius is 4.789 feet.

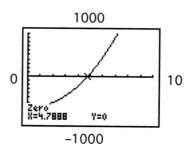

(3-3)

69.

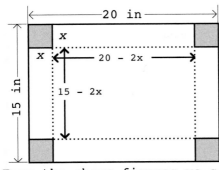
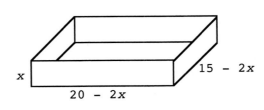

From the above figures we see that
    V = length × width × height = $(20 - 2x)(15 - 2x)x$
To ensure that all dimensions are positive $0 < x < 7.5$
We solve $(20 - 2x)(15 - 2x)x = 300$
$$300x - 70x^2 + 4x^3 = 300$$
$$4x^3 - 70x^2 + 300x - 300 = 0$$
Examining the graph of $y = P(x)$ for $0 < x < 7.5$, we obtain

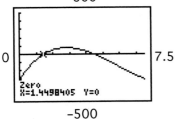
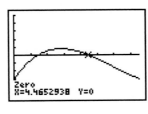

$x = 1.450$ inches or $x = 4.465$ inches.   (*3–3*)

70. Let (x, x^2) be a point on the graph of $y = x^2$. Then the distance from $(1, 4)$ to (x, x^2) must equal 3 units. Applying the distance formula, we have,

$$\sqrt{(x - 1)^2 + (x^2 - 4)^2} = 3$$
$$(x - 1)^2 + (x^2 - 4)^2 = 9$$
$$x^2 - 2x + 1 + x^4 - 8x^2 + 16 = 9$$
$$x^4 - 7x^2 - 2x + 8 = 0$$

Let $P(x) = x^4 - 7x^2 - 2x + 8$. The possible rational zeros are ±1, ±2, ±4, ±8.

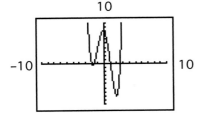

Examining the graph of $y = P(x)$, we see that there is a zero between -2 and -1, a zero between 2 and 3, and possible integer zeros at -2 and 1. Testing the likely candidates, we obtain

$$
\begin{array}{r|rrrrr}
 & 1 & 0 & -7 & -2 & 8 \\
 &   & 1 & 1 & -6 & -8 \\
\hline
1 & 1 & 1 & -6 & -8 & 0 \\
\end{array}
$$

1 is a zero. So $P(x) = (x - 1)(x^3 + x^2 - 6x - 8)$. Testing -2 in the reduced polynomial, we obtain

$$
\begin{array}{r}
1 \quad1 \quad -6 \quad -8 \\
-2 \quad2 \quad8 \\
\hline
-2)1 \quad -1 \quad -4 \quad0
\end{array}
$$

-2 is a zero. The remaining zeros are found by solving $x^2 - x - 4 = 0$. Applying the quadratic formula with $a = 1$, $b = -1$, $c = -4$, we obtain

$$x = \frac{-(-1) \pm \sqrt{(-1)^2 - 4(1)(-4)}}{2(1)} = \frac{1 \pm \sqrt{17}}{2}$$

To three decimal place accuracy, $x = -1.562$ or 2.562. There are four real zeros of $P(x)$, 1, -2, -1.562, and 2.562. There are four points on the graph of $y = x^2$ that are 3 units from (1, 4) and their coordinates are (1, 1), (-2, 4), (-1.562, 2.440), and (2.562, 6.564)

(3-3)

71. (A)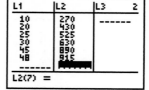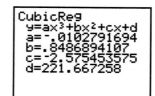

Enter the data.

Compute the regression equation.

The cubic model is $y = -0.0103x^3 + 0.849x^2 - 2.56x + 222$.

(B)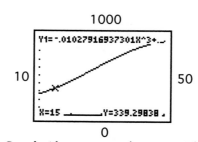

1000

Transfer the regression equation to the equation editor.

Graph the regression equation and determine the value of y corresponding to $x = 15$.

The model predicts that 339 refrigerators would be sold if 15 ads were placed.

(C)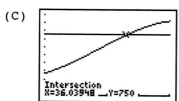

Use the INTERSECT command to determine the value of x that corresponds to $y = 750$.

The model predicts that 750 refrigerators would be sold if 36 ads were placed.

(3-1)

72. (A) Since the percentage figures in the second column decrease, then increase, then decrease again with increasing age, this might correspond to a function with both a local minimum and a local maximum. A cubic model can have both.

(B) The independent variable is age and the dependent variable is percentage, so we enter the age column as L_1 and the percentage column as L_2. Then choose cubic regression from the STAT CALC menu.

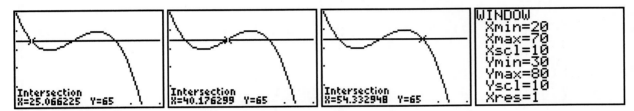

(Note that the entire list is not visible on the graphic.)

The cubic regression model is $y = -0.00424x^3 + 0.507x^2 - 19.3x + 297$.

(C) Graph $y_1 = -0.00424x^3 + 0.507x^2 - 19.3x + 297$, $y_2 = 65$, and use the INTERSECT command.

The percentage is 65 for ages 25, 40, and 54.   *(3-1)*

CHAPTER 4   Exponential and Logarithmic Functions

Section 4-1

1. Answers will vary.   **3.** Answers will vary.   **5.** Answers will vary.

7. (A) The graph of $y = (0.2)^x$ is decreasing and passes through the point $(-1, 0.2^{-1}) = (-1, 5)$. This corresponds to graph g.

(B) The graph of $y = 2^x$ is increasing and passes through the point $(1, 2)$. This corresponds to graph n.

(C) The graph of $y = \left(\dfrac{1}{3}\right)^x$ is decreasing and passes through the point $(-1, 3)$. This corresponds to graph f.

(D) The graph of $y = 4^x$ is increasing and passes through the point $(1, 4)$. This corresponds to graph m.

9. 16.24   **11.** 7.524   **13.** 1.649   **15.** 4.469

17. Make a table of values:

x	-3	-2	-1	0	1	2
y	$1/27$	$1/9$	$1/3$	0	3	9

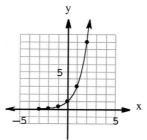

19. Make a table of values:

x	-3	-2	-1	0	1	2
y	27	9	3	1	$1/3$	$1/9$

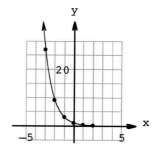

21. $10^{3x-1}10^{4-x} = 10^{(3x-1)+(4-x)} = 10^{2x+3}$ **23.** $\dfrac{3x}{3^{1-x}} = 3^{x-(1-x)} = 3^{x-1+x} = 3^{2x-1}$

25. $\left(\dfrac{4^x}{5^y}\right)^{3z} = \dfrac{\left(4^x\right)^{3z}}{\left(5^y\right)^{3z}} = \dfrac{4^{3xz}}{5^{3yz}}$ **27.** $\dfrac{e^{5x}}{e^{2x+1}} = e^{5x-(2x+1)} = e^{5x-2x-1} = e^{3x-1}$

29. (A) Although $1 + \dfrac{1}{x}$ approaches 1, $1 + \dfrac{1}{x}$ is not equal to 1 for any x, so reasoning as if it were 1 is incorrect.

(B) As $x \to \infty$, $\left(1 + \dfrac{1}{x}\right)^x \to e$.

31. The graph of $y = 2^x$ is shifted 3 units right and 1 unit down.

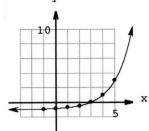

33. The graph of $y = \left(\dfrac{1}{3}\right)^x$ is shifted 5 units left and 10 units down.

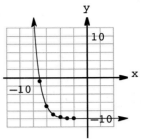

35. The graph of $y = e^x$ is shifted 2 units up.

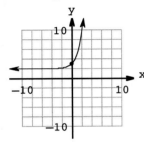

37. The graph of $y = e^x$ is reflected in the x axis, shifted 2 units left, and stretched vertically by a factor of 2.

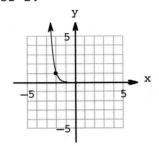

39. $5^{3x} = 5^{4x-2}$ if and only if
$$3x = 4x - 2$$
$$-x = -2$$
$$x = 2$$

41. $\qquad 7^{x^2} = 7^{2x+3}$ if and only if
$$x^2 = 2x + 3$$
$$x^2 - 2x - 3 = 0$$
$$(x - 3)(x + 1) = 0$$
$$x = -1,\ 3$$

43. $(1 - x)^5 = (2x - 1)^5$ if and only if
$$1 - x = 2x - 1$$
$$-3x = -2$$
$$x = \frac{2}{3}$$

45. $\qquad 9^{x^2} = 3^{3x-1}$
$$(3^2)^{x^2} = 3^{3x-1}$$
$$3^{2x^2} = 3^{3x-1}$$
if and only if
$$2x^2 = 3x - 1$$
$$2x^2 - 3x + 1 = 0$$
$$(2x - 1)(x - 1) = 0$$
$$x = \frac{1}{2},\ 1$$

47. $4^{x^2} = 8^x$
$$\left(2^2\right)^{x^2} = \left(2^3\right)^x$$
$$2^{2x^2} = 2^{3x} \quad \text{if and only if}$$
$$2x^2 = 3x$$
$$2x^2 - 3x = 0$$
$$x(2x - 3) = 0$$
$$x = 0 \quad or \quad 2x - 3 = 0$$
$$2x = 3$$
$$x = 3/2$$

49. $2xe^{-x} = 0$ if $2x = 0$ or $e^{-x} = 0$.
Since e^{-x} is never 0, the only
solution is $x = 0$.

51. $x^2e^x - 5xe^x = 0$
$xe^x(x - 5) = 0$
$x = 0$ or $e^x = 0$ or $x - 5 = 0$
$\phantom{x = 0 \text{ or }}$ never $\phantom{\text{or}}$ $x = 5$
$x = 0, 5$

53. $\quad a^2 = a^{-2}$

$\quad a^2 = \dfrac{1}{a^2}$

$\quad a^4 = 1 \quad (a \neq 0)$
$a^4 - 1 = 0$
$(a - 1)(a + 1)(a^2 + 1) = 0$
$a = 1$ or $a = -1$
This does not violate the exponential property mentioned because $a = 1$ and a
negative are excluded from consideration in the statement of the property.

55. The graph of $y = 1^x$ is a horizontal straight
line since $1^x = 1$ for all real x. This graph
is neither increasing nor decreasing and has
no asymptotes. It is not the graph of an
exponential function.

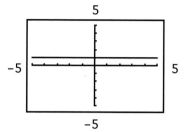

57.

x	−3	−2	−1	0	1	2	3
y	1	1	1	1	1	1	1

The resulting function is simply the constant function $y = 1$.

59. The graph of a polynomial (other than the constant polynomial $P(x) = C$ for any
real number C), doesn't have a horizontal asymptote.

61. $\dfrac{-2x^3e^{-2x} - 3x^2e^{-2x}}{x^6} = \dfrac{x^2e^{-2x}(-2x - 3)}{x^6} = \dfrac{e^{-2x}(-2x - 3)}{x^4}$

63. $(e^x + e^{-x})^2 + (e^x - e^{-x})^2 = (e^x)^2 + 2(e^x)(e^{-x}) + (e^{-x})^2 + (e^x)^2 - 2(e^x)(e^{-x}) + (e^{-x})^2$

$\phantom{(e^x + e^{-x})^2 + (e^x - e^{-x})^2} = e^{2x} + 2 + e^{-2x} + e^{2x} - 2 + e^{-2x}$

$\phantom{(e^x + e^{-x})^2 + (e^x - e^{-x})^2} = 2e^{2x} + 2e^{-2x}$

> **Common Errors:**
> $(e^x)^2 \neq e^{x^2}$
> $e^{2x} + e^{2x} \neq e^{4x}$

65. Examining the graph of $y = f(x)$, we obtain

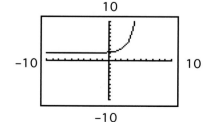
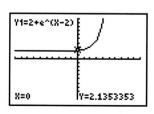

There are no local extrema and no x intercepts. The y intercept is 2.14.
As $x \to -\infty$, $y \to 2$, so the line $y = 2$ is a horizontal asymptote.

67. Examining the graph of $y = m(x)$, we obtain

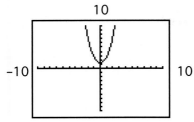
10

-10

10

-10

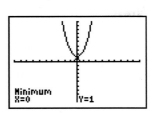
Minimum
X=0   Y=1

There is a local minimum at $m(0) = 1$, and 1 is the y intercept. There is no x intercept and no horizontal asymptote.

69. Examining the graph of $y = S(x)$, we obtain

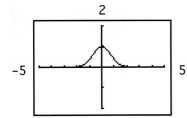
2

-5

5

-2

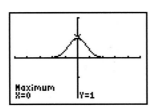
Maximum
X=0   Y=1

There is a local maximum at $S(0) = 1$, and 1 is the y intercept. There is no x intercept. As $x \to \infty$ or $x \to -\infty$, $y \to 0$, so the line $y = 0$ (the x axis) is a horizontal asymptote.

71. Examining the graph of $y = F(x)$, we obtain

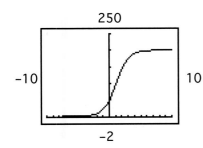
250

-10

10

-2

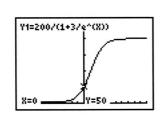
Y1=200/(1+3/e^(X))

X=0   Y=50

There are no local extrema and no x intercepts.
When $x = 0$, $F(0) = \dfrac{200}{1 + 3e^{-0}} = 50$ is the y intercept. As $x \to -\infty$, $y \to 0$, so the line $y = 0$ (the x axis) is a horizontal asymptote.
As $x \to \infty$, $y \to 200$, so the line $y = 200$ is also a horizontal asymptote.

73.

10

-10

10

-10

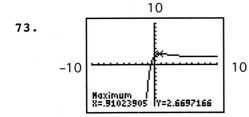
Maximum
X=.91023905   Y=2.6697166

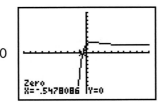
Zero
X=-.5478086   Y=0

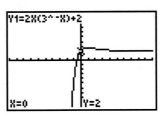
Y1=2X(3^-X)+2

X=0   Y=2

The local maximum is $m(0.91) \approx 2.67$. The x intercept is -0.55, and the y intercept is 2. The line $y = 2$ appears to be a horizontal asymptote. As $x \to -\infty$, $m(x) \to -\infty$.

75.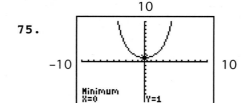

The local minimum is $f(0) = 1$, so zero is the y intercept. There are no x intercepts or horizontal asymptotes; $f(x) \to \infty$ as $x \to \infty$ and $x \to -\infty$.

77. Examining the graph of $y = f(x)$, we obtain

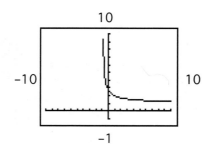

As $x \to 0$, $f(x) = (1 + x)^{1/x}$ seems to approach a value near 3. A table of values near $x = 0$ yields

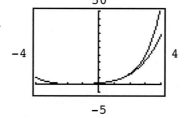

Although $f(0)$ is not defined, as $x \to 0$, $f(x)$ seems to approach a number near 2.718. In fact, it approaches e, since as $x \to 0$, $u = \dfrac{1}{x} \to \infty$, and $f(x) = \left(1 + \dfrac{1}{u}\right)^u$ must approach e as $u \to \infty$.

79.
81.

83. Here are graphs of $f_1(x) = \dfrac{x}{e^x}$, $f_2(x) = \dfrac{x^2}{e^x}$, and $f_3(x) = \dfrac{x^3}{e^x}$. In each case as $x \to \infty$, $f_n(x) \to 0$. The line $y = 0$ is a horizontal asymptote.
As $x \to -\infty$, $f_1(x) \to -\infty$ and $f_3(x) \to -\infty$, while $f_2(x) \to \infty$. It appears that as $x \to -\infty$, $f_n(x) \to \infty$ if n is even and $f_n(x) \to -\infty$ if n is odd.
f_1: f_2: f_3:

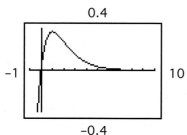
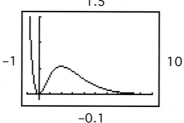
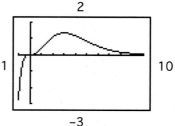

As confirmation of these observations, we show the graph of $f_4 = \dfrac{x^4}{e^x}$ (not required).

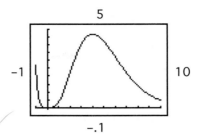

85. Make a table of values, substituting in each requested x value:

x	1.4	1.41	1.414	1.4142	1.41421	1.414214
2^x	2.639016	2.657372	2.664750	2.665119	2.665138	2.665145

The approximate value of $2^{\sqrt{2}}$ is 2.665145 to six decimal places. Using a calculator to compute directly, we get 2.665144.

87. (A) We can use the compound interest formula. Since interest is compounded weekly, m (the number of times per year interest is compounded) is 52. We're given a principal of \$4,000 and an interest rate of 11%, which corresponds to $r = 0.11$. After ½ year, interest will have been compounded 26 times.

$$A = P\left(1 + \frac{r}{m}\right)^n ; \quad m = 52; \quad P = 4,000; \quad r = 0.11; \quad n = 26$$

$$A = 4,000\left(1 + \frac{0.11}{52}\right)^{26} = \$4,225.92$$

(B) Everything is the same except n, which in 10 years is $52 \cdot 10 = 520$.

$$A = 4,000\left(1 + \frac{0.11}{365/7}\right)^{520} = \$12,002.71$$

89. We use the Continuous Compound Interest Formula
$A = Pe^{rt}$
$P = 5,250 \quad r = 0.1138 \quad A = 5,250e^{0.1138t}$
(A) $t = 6.25 \quad A = 5,250e^{0.1138(6.25)} = \$10,691.81$
(B) $t = 17 \quad A = 5,250e^{0.1138(17)} = \$36,336.69$

91. We use the compound interest formula
$$A = P\left(1 + \frac{r}{m}\right)^n$$
For the first account, $P = 3,000$, $r = 0.08$, $m = 365$. Let y_1 be the amount in the first account; then
$y_1 = 3,000(1 + 0.08/365)^x$ where x is the number of compounding periods (days).
For the second account, $P = 5,000$, $r = 0.05$, $m = 365$. Let y_2 be the amount in the second account; then
$y_2 = 5,000(1 + 0.05/365)^x$

Examining the graphs of y_1 and y_2, we obtain:

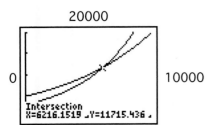

X	Y₁	Y₂
6215	11712	11714
6216	11715	11715
6217	11718	11717
6218	11720	11718
6219	11723	11720
6220	11725	11722
6221	11728	11723

X=6218

The graphs intersect at $x = 6216.15$ days.
Comparing the amounts in the accounts, we see that the first account is worth more than the second for $x \geq 6,217$ days.

93. We use the compound interest formula

$$A = P\left(1 + \frac{r}{m}\right)^n$$

For the first account, $P = 10,000$, $r = 0.089$, $m = 365$. Let y_1 be the amount in the first account, then $y_1 = 10,000(1 + 0.089/365)^x$

where x is the number of compounding periods (days). For the second account, $P = 10,000$, $r = 0.09$, $m = 4$. Let y_2 be the amount in the second account,

then $y_2 = 10,000(1 + 0.09/4)^{4x/365}$ where

x is the number of days. Examining the graphs of y_1 and y_2, we obtain the graph at the right.

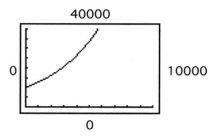

The two graphs are indistinguishable from one another. Examining a table of values, we obtain:

X	Y₁	Y₂
0	10000	10000
91.25	10225	10225
182.5	10455	10455
273.75	10690	10690
365	10931	10931
456.25	11177	11177
547.5	11428	11428

X=0

X	Y₁	Y₂
1186.3	13354	13354
1277.5	13655	13655
1368.8	13961	13962
1460	14275	14276
1551.3	14597	14597
1642.5	14925	14926
1733.75	15261	15262

X=1733.75

The two accounts are extremely close in value, but after 1,277.5 days (14 quarters) the second account begins to be noticeably larger than the first. The first account will never be larger than the second.

95. We use the Continuous Compound Interest Formula and solve for P to obtain a formula for finding the principal.

$A = Pe^{rt}$

$P = \dfrac{A}{e^{rt}}$ or $P = Ae^{-rt}$

$A = 30,000 \quad r = 0.09 \quad t = 10$

$P = 30,000e^{(-0.09)(10)}$

$P = \$12,197.09$

97. Gill Savings: Use the Continuous Compound Interest Formula

$A = Pe^{rt} \quad P = 1,000 \quad r = 0.083 \quad t = 2.5$

$A = 1,000e^{(0.083)(2.5)}$

$A = \$1,230.60$

Richardson S & L: Use the Compound Interest Formula

$A = P\left(1 + \dfrac{r}{m}\right)^n \quad P = 1,000 \quad r = 0.084 \quad n = (4)(2.5)$

$A = 1,000\left(1 + \dfrac{0.084}{4}\right)^{(4)(2.5)}$

$A = \$1,231.00$

U.S.A. Savings: Use the Compound Interest Formula

$A = P\left(1 + \dfrac{r}{m}\right)^n \quad P = 1,000 \quad r = 0.0825 \quad n = (365)(2.5)$

$A = 1,000\left(1 + \dfrac{0.0825}{365}\right)^{(365)(2.5)}$

$A = \$1,229.03$

99. Use the compound interest formula with $r = 0.0825$, $m = 365$ (daily means interest is compounded 365 times per year), $A = 100{,}000$ (we're given \$100,000 as the amount in the account after a certain amount of time), and $n = 365(17) = 6{,}205$ (interest compounded 365 times per year for 17 years).

$$A = P\left(1 + \frac{r}{m}\right)^n \quad \Rightarrow \quad 100{,}000 = P\left(1 + \frac{0.0825}{365}\right)^{6{,}205}$$

Now solve for P:

$$P = \frac{100{,}000}{\left(1 + \dfrac{0.0825}{365}\right)^{6{,}205}} = \$24{,}602$$

Section 4-2

1. Answers will vary.   **3.** Answers will vary.

5. Use the doubling time model $A = A_0(2)^{t/d}$ with $A_0 = 200$, $d = 5$.

   $A = 200(2)^{t/5}$

7. Use the continuous growth model $A = A_0 e^{rt}$ with $A_0 = 2{,}000$, $r = 0.02$.

   $A = 2{,}000 e^{0.02t}$

9. Use the half-life model $A = A_0\left(\dfrac{1}{2}\right)^{t/h}$ with $A_0 = 100$, $h = 6$.

   $A = 100\left(\dfrac{1}{2}\right)^{t/6}$

11. Use the exponential decay model $A = A_0 e^{-kt}$ with $A_0 = 4$, $k = 0.124$.

   $A = 4e^{-0.124t}$

13.

n	L
1	2
2	4
3	8
4	16
5	32
6	64
7	128
8	256
9	512
10	1,024

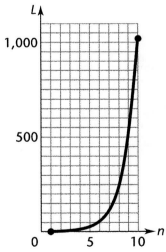

15. Use the doubling time model $A = A_0\left(\dfrac{1}{2}\right)^{t/h}$ with $A_0 = 10$, $d = 2.4$

   $A = 10(2)^{t/2.4}$ (Notice that the doubling time is given in days, so we have to convert any times given to days.)

   (A) Plug in $t = 7$ (1 week is 7 days): $A = 10(2)^{7/2.4} = 75.5$
   In one week there will be about 76 flies.

   (B) Plug in $t = 14$ (2 weeks is 14 days): $A = 10(2)^{14/2.4} = 570.2$
   In two weeks there will be about 570 flies.

17. Use the doubling time model $A = A_0\left(\dfrac{1}{2}\right)^{t/h}$ with $A_0 = 2,200,\quad d = 2$.

$A = 2,200(2)^{t/2}$ where t is years after 1970.

(A) For $t = 20$: $A = 2,200(2)^{20/2} = 2,252,800$

(B) For $t = 35$: $A = 2,200(2)^{35/2} = 407,800,360$

19. Use the half-life model $A = A_0\left(\dfrac{1}{2}\right)^{t/h}$ with $A_0 = 25,\quad h = 12$

$A = 25\left(\dfrac{1}{2}\right)^{t/12}$

(A) For $t = 5$, $A = 25\left(\dfrac{1}{2}\right)^{5/12} = 19$ pounds

(B) For $t = 20$, $A = 25\left(\dfrac{1}{2}\right)^{20/12} = 7.9$ pounds

21. We use the Unlimited Growth model
$A = A_0 e^{kt}$
$A_0 = 6.5 \times 10^9 \quad r = 0.0114 \quad t = 10$
$A = 6.5 \times 10^9 e^{(0.0114)(10)}$
$A = 7.3 \times 10^9,\ 7.3$ billion

23. We use the Unlimited Growth model.
$A = A_0 e^{kt}$
For Russia, $A_0 = 1.43 \times 10^8$, $r = -0.0037$, $A_1 = 1.43 \times 10^8 e^{-0.0037t}$
For Nigeria, $A_0 = 1.29 \times 10^8$, $r = 0.0256$, $A_2 = 1.29 \times 10^8 e^{0.0256t}$
Below is a graph of A_1 and A_2.

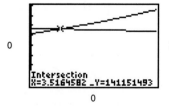

From the graph, assuming $t = 0$ in 2005, it appears that the two populations will be equal when t is approximately 3.5, in 2008. After that the population of Nigeria will be greater than that of Russia.

25. A table of values can be generated by a graphing calculator and yields

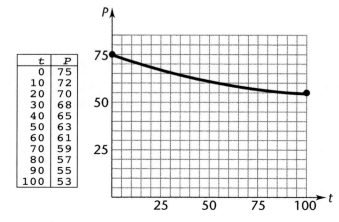

t	P
0	75
10	72
20	70
30	68
40	65
50	63
60	61
70	59
80	57
90	55
100	53

27. $I = I_0 e^{-0.00942d}$

(A) $d = 50$ $I = I_0 e^{-0.00942(50)} = 0.62 I_0$ 62%

(B) $d = 100$ $I = I_0 e^{-0.00942(100)} = 0.39 I_0$ 39%

29. (A) We use the Unlimited Growth model $A = A_0 e^{kt}$ with $A_0 = 39.4$ million, $k = 0.032$,
   and $t = 6$ since 2010 is 6 years after 2004.
$$A = 39.4 \times 10^6 e^{0.032(6)}$$
$$A = 47.7 \text{ million people living with HIV in 2010.}$$

(B) Here $A_0 = 39.4$ million, $k = 0.032$, and $t = 11$ since 2015 is 11 years
   after 2004.
$$A = 39.4 \times 10^6 e^{0.032(11)}$$
$$A = 56.0 \text{ million people living with HIV in 2015.}$$

31. $T = T_m + (T_0 - T_m) e^{-kt}$
$T_m = 40°$ $T_0 = 72°$ $k = 0.4$ $t = 3$
$T = 40 + (72 - 40) e^{-0.4(3)}$
$T = 50°$

33. As t increases without bound, $e^{-0.2t}$ approaches 0, so $q = 0.0009(1 - e^{-0.2t})$
approaches 0.0009. This tells us that 0.0009 coulomb is the maximum charge on
the capacitor.

35. (A) Examining the graph of $N(t)$, we obtain the graphs below.

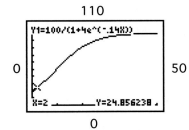

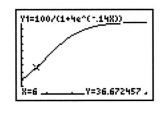

After 2 years, 25 deer will be
present. After 6 years, 37 deer will
be present.

(B) Applying the INTERSECT command, we obtain the graph
at the right.

It will take 10 years for the herd to grow to 50 deer.

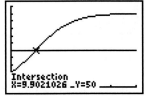

(C) As t increases without bound, $e^{-0.14t}$ approaches 0,
and $N = \dfrac{100}{1 + 4e^{-0.14t}}$ approaches 100. This tells us that
100 is the number of deer the island can support.

37. The independent variable is years after purchase, so enter 1, 2, 3, 4, 5, and 6
as L_1. The dependent variable is value, so enter the Value column as L_2. Then
use the exponential regression command from the STAT CALC menu.

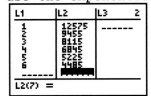

The model gives $y = 14,910(0.8163)^x$.

When $x = 0$, $y = \$14,910$ is the estimated purchase price. Applying the VALUE command, we obtain the graph at the right.

When $x = 10$, the estimated value of the van is $\$1,959$.

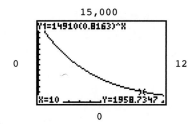

39. (A) The independent variable is years since 1980, so enter 0, 5, 10, 15, 18, and 19 as L_1. The dependent variable is power generation in North America, so enter the North America column as L_2. Then use the logistic regression command from the STAT CALC menu.

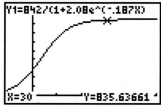

The model is

$$y = \frac{842}{1 + 2.08e^{-0.187x}}$$

(B) 2010 is 30 years after 1980, so we evaluate our model for $x = 30$ using the VALUE command.

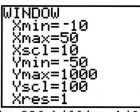

In 2010, approximately 836 billion kilowatt hours will be generated.

Section 4-3

1. Answers will vary.   **3.** Answers will vary.   **5.** Answers will vary.

7. $81 = 3^4$ **9.** $0.001 = 10^{-3}$ **11.** $\dfrac{1}{36} = 6^{-2}$ **13.** $\log_4 8 = \dfrac{3}{2}$

15. $\log_{32} \dfrac{1}{2} = -\dfrac{1}{5}$ **17.** $\log_{2/3} \dfrac{8}{27} = 3$

19. Make a table of values for each function:

x	$f(x) = 3^x$	x	$f^{-1}(x) = \log_3 x$
-3	$1/27$	$1/27$	-3
-2	$1/9$	$1/9$	-2
-1	$1/3$	$1/3$	-1
0	1	1	0
1	3	3	1
2	9	9	2
3	27	27	3

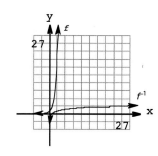

21. Make a table of values for each function:

x	$f(x) = (2/3)^x$	x	$f^{-1}(x) = \log_{2/3} x$
-3	27 / 8	27 / 8	-3
-2	9 / 4	9 / 4	-2
-1	3 / 2	2 / 3	-1
0	1	1	0
1	2 / 3	2 / 3	1
2	4 / 9	4 / 9	2
3	8 / 27	8 / 27	3

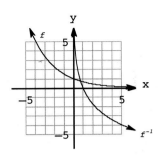

23. 0   **25.** 1   **27.** 4

29. $\log_{10} 0.01 = \log_{10} 10^{-2} = -2$ **31.** $\log_3 27 = \log_3 3^3 = 3$

33. $\log_{1/2} 2 = \log_{1/2} \left(\dfrac{1}{2}\right)^{-1} = -1$ **35.** 5   **37.** $\log_5 \sqrt[3]{5} = \log_5 5^{1/3} = \dfrac{1}{3}$

39. 4.6923   **41.** 3.9905

43. $\log_7 13 = \dfrac{\ln 13}{\ln 7} = 1.3181$ using the change of base formula

45. $\log_5 120.24 = \dfrac{\ln 120.24}{\ln 5} = 2.9759$ using the change of base formula

47. $x = 10^{5.3027} = 200{,}800$ **49.** $x = 10^{-3.1773} = 6.648 \times 10^{-4} = 0.000\ 6648$

51. $x = e^{3.8655} = 47.73$ **53.** $x = e^{-0.3916} = 0.6760$

55. Write $\log_2 x = 2$ in equivalent exponential form. $x = 2^2 = 4$

57. $\log_4 16 = \log_4 4^2 = 2$

$y = 2$

59. Write $\log_b 16 = 2$ in equivalent exponential form.

$16 = b^2$

$b^2 = 16$

$b = 4$ since bases are required to be positive

61. Write $\log_b 1 = 0$ in equivalent exponential form: $1 = b^0$

This statement is true if b is any real number except 0. However, bases are required to be positive and 1 is not allowed, so the original statement is true if b is any positive real number except 1.

63. Write $\log_4 x = \dfrac{1}{2}$ in equivalent exponential form: $x = 4^{1/2} = 2$

65. $\log_{1/3} 9 = \log_{1/3} 3^2 = \log_{1/3} \dfrac{1}{\left(\frac{1}{3}\right)^2} = \log_{1/3} \left(\dfrac{1}{3}\right)^{-2} = -2$

67. Write $\log_b 1{,}000 = \dfrac{3}{2}$ in equivalent exponential form:

$1{,}000 = b^{3/2}$

$10^3 = b^{3/2}$

$(10^3)^{2/3} = (b^{3/2})^{2/3}$ (If two numbers are equal the results are equal if they are raised to the same exponent.)

$10^{3(2/3)} = b^{3/2(2/3)}$

$10^2 = b$

$b = 100$

69. Write $\log_8 x = -\dfrac{4}{3}$ in equivalent exponential form.

$$8^{-4/3} = x$$
$$x = (8^{1/3})^{-4} = 2^{-4} = \dfrac{1}{16}$$

71. Write $\log_{16} 8 = y$ in equivalent exponential form.

$$16^y = 8$$
$$(2^4)^y = 2^3$$
$$2^{4y} = 2^3 \text{ if and only if}$$
$$4y = 3$$
$$y = \dfrac{3}{4}$$

73. 4.959   **75.** 7.861   **77.** 2.280

79. $\log\left(\dfrac{x}{y}\right) = \log x - \log y$

81. $\log\left(x^4 y^3\right) = \log x^4 + \log y^3 = 4 \log x + 3 \log y$

83. $\ln x - \ln y = \ln\left(\dfrac{x}{y}\right)$

85. $2 \ln x + 5 \ln y - \ln z = \ln x^2 + \ln y^5 - \ln z = \ln\left(\dfrac{x^2 y^5}{z}\right)$

87. $\log(xy) = \log x + \log y = -2 + 3 = 1$

89. $\log\left(\dfrac{\sqrt{x}}{y^3}\right) = \log x^{1/2} - \log y^3 = \dfrac{1}{2} \log x - 3 \log y = \dfrac{1}{2}(-2) - 3(3) = -1 - 9 = -10$

91. The graph of g is the same as the graph of f shifted upward 3 units; g is increasing.
Domain: $(0, \infty)$
Vertical asymptote: $x = 0$

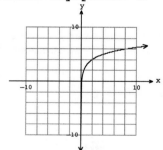

93. The graph of g is the same as the graph of f shifted 2 units to the right; g is decreasing.
Domain: $(2, \infty)$
Vertical asymptote: $x = 2$

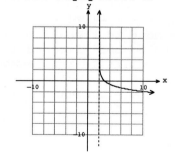

95. The graph of g is the same as the graph of f reflected through the x axis and shifted downward 1 unit; g is decreasing.
Domain: $(0, \infty)$
Vertical asymptote: $x = 0$

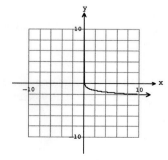

97. The graph of g is the same as the graph of f reflected through the x axis, stretched vertically by a factor of 3, and shifted upward 5 units. g is decreasing.
Domain: $(0, \infty)$
Vertical asymptote: $x = 0$

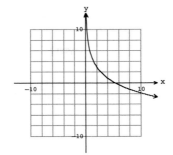

99. Write $y = \log_5 x$
In exponential form:
$\quad 5^y = x$
Interchange x and y:
$\quad 5^x = y$
Therefore $f^{-1}(x) = 5^x$.

101. Write $y = 4 \log_3(x + 3)$
$$\frac{y}{4} = \log_3(x + 3)$$
In exponential form:
$\quad 3^{y/4} = x + 3$
$\quad x = 3^{y/4} - 3$
Interchange x and y:
$\quad y = 3^{x/4} - 3$
Therefore $f^{-1}(x) = 3^{x/4} - 3$

103. (A) Write $y = \log_3(2 - x)$
In exponential form:
$\quad 3^y = 2 - x$
$\quad x = 2 - 3^y$
Interchange x and y:
$\quad y = 2 - 3^x$
Therefore, $f^{-1}(x) = 2 - 3^x$

(B) The graph is the same as the graph of $y = 3^x$ reflected through the x axis and shifted 2 units upward.

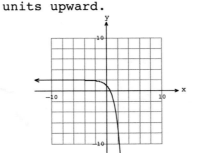

(C)

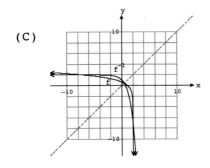

105. The inequality sign in the last step reverses because $\log \dfrac{1}{3}$ is negative.

107.

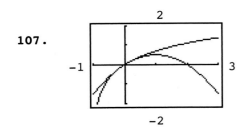

109.

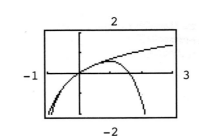

111. Let $u = \log_b M$ and $v = \log_b N$. Changing each equation to exponential form, $b^u = M$ and $b^v = N$. Then we can write M/N as

$$\frac{M}{N} = \frac{b^u}{b^v} = b^{u-v}$$ using a familiar property of exponents. Now change this equation to logarithmic form:

$$\log_b\left(\frac{M}{N}\right) = u - v$$

Finally, recall the way we defined u and v in the first line of our proof:

$$\log_b\left(\frac{M}{N}\right) = \log_b M - \log_b N$$

Section 4-4

1. Answers will vary.   **3.** Answers will vary.

5. We use the decibel formula
$$D = 10 \log \frac{I}{I_0}$$

(A) $I = I_0$

$$D = 10 \log \frac{I_0}{I_0}$$
$$D = 10 \log 1$$
$$D = 0 \text{ decibels}$$

(B) $I_0 = 1.0 \times 10^{-12}$ $I = 1.0$

$$D = 10 \log \frac{1.0}{1.0 \times 10^{-12}}$$
$$D = 120 \text{ decibels}$$

7. Use the decibel formula $D = 10 \log \dfrac{I}{I_0}$:

Let I_1 be the intensity of the first sound, and let I_2 be the intensity of the second sound. Then $I_2 = 1{,}000 I_1$.

$$D_1 = 10 \log \frac{I}{I_0}; \quad D_2 = 10 \log \frac{I_2}{I_0}$$

$$D_2 - D_1 = 10 \log \frac{I_2}{I_0} - 10 \log \frac{I_1}{I_0}$$

$$= 10 \log\left(\frac{I_2/I_0}{I_1/I_0}\right)$$

$$= 10 \log \frac{I_2}{I_1}$$

$$= 10 \log \frac{1{,}000 I_1}{I_1}$$

$$= 10 \log 1{,}000$$

$$= 30 \text{ decibels}$$

9. We use the magnitude formula
$$M = \frac{2}{3} \log \frac{E}{E_0}$$

$E_0 = 10^{4.40}$ $E = 1.99 \times 10^{17}$

$$M = \frac{2}{3} \log \frac{1.99 \times 10^{17}}{10^{4.40}}$$

$$= 8.6$$

11. We use the magnitude formula

$$M = \frac{2}{3} \log \frac{E}{E_0}$$

For the Long Beach earthquake,   For the Anchorage earthquake,

$$6.3 = \frac{2}{3} \log \frac{E_1}{E_0} \qquad\qquad 8.3 = \frac{2}{3} \log \frac{E_2}{E_0}$$

$$9.45 = \log \frac{E_1}{E_0} \qquad\qquad 12.45 = \log \frac{E_2}{E_0}$$

(Change to exponential form)   (Change to exponential form)

$$\frac{E_1}{E_0} = 10^{9.45} \qquad\qquad \frac{E_2}{E_0} = 10^{12.45}$$

$$E_1 = E_0 \cdot 10^{9.45} \qquad\qquad E_2 = E_0 \cdot 10^{12.45}$$

Now we can compare the energy levels by dividing the more powerful (Anchorage) by the less (Long Beach):

$$\frac{E_2}{E_1} = \frac{E_0 \cdot 10^{12.45}}{E_0 \cdot 10^{9.45}} = 10^3$$

$E_2 = 10^3 E_1$, or 1,000 times as powerful

13. Use the magnitude formula $M = \frac{2}{3} \log \frac{E}{E_0}$ with $E = 1.34 \times 10^{14}$, $E_0 = 10^{4.40}$:

$$M = \frac{2}{3} \log \frac{1.34 \times 10^{14}}{10^{4.40}} = 6.5$$

15. Use the magnitude formula $M = \frac{2}{3} \log \frac{E}{E_0}$ with $E = 2.38 \times 10^{21}$, $E_0 = 10^{4.40}$:

$$M = \frac{2}{3} \log \frac{2.38 \times 10^{21}}{10^{4.40}} = 11.3$$

17. We use the rocket equation.

$$v = c \ln \frac{W_t}{W_b}$$

$$v = 2.57 \ln (19.8)$$

$$v = 7.67 \text{ km/s}$$

19. (A) $pH = -\log[H^+] = -\log(4.63 \times 10^{-9}) = 8.3$. Since this is greater than 7, the substance is basic.

(B) $pH = -\log[H^+] = -\log(9.32 \times 10^{-4}) = 3.0$ Since this is less than 7, the substance is acidic.

21. Since $pH = -\log[H^+]$, we have

$$5.2 = -\log[H^+], \text{ or}$$

$$[H^+] = 10^{-5.2} = 6.3 \times 10^{-6} \text{ moles per liter}$$

23. $m = 6 - 2.5 \log \dfrac{L}{L_0}$

(A) We find m when $L = L_0$

$m = 6 - 2.5 \log \dfrac{L_0}{L_0}$

$m = 6 - 2.5 \log 1$

$m = 6$

(B) We compare L_1 for $m = 1$ with L_2 for $m = 6$

$1 = 6 - 2.5 \log \dfrac{L_1}{L_0}$

$-5 = -2.5 \log \dfrac{L_1}{L_0}$

$2 = \log \dfrac{L_1}{L_0}$

$\dfrac{L_1}{L_0} = 10^2$

$L_1 = 100 L_0$

$6 = 6 - 2.5 \log \dfrac{L_2}{L_0}$

$0 = -2.5 \log \dfrac{L_2}{L_0}$

$0 = \log \dfrac{L_2}{L_0}$

$\dfrac{L_2}{L_0} = 1$

$L_2 = L_0$

Comparing the two, $\dfrac{L_1}{L_2} = \dfrac{100 L_0}{L_0} = 100$. The star of magnitude 1 is 100 times brighter.

25. (A) Enter the x values shown in the table as L_1. Enter the yield values as L_2. Use the logarithmic regression model from the STAT CALC menu.

The model is $y = -525.1353183 + 142.7911807 \ln x$. Evaluating this for $x = 103$ (year 2003) yields 136.7 bushels per acre. Evaluating for $x = 110$ (year 2010) yields 146.0 bushels per acre.

(B) The actual yield was higher than predicted by the model. Entering this as further data would tend to raise the curve slightly and increase the predicted estimate for 2010.

Section 4-5

1. Answers will vary.

3. Answers will vary.

5. Answers will vary.

7. Algebraic Solution:

$10^{-x} = 0.0347$

$-x = \log_{10} 0.0347$

$x = -\log_{10} 0.0347$

$x = 1.46$

Graphical Solution:

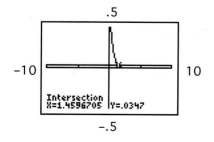

9. Algebraic Solution:

$10^{3x+1} = 92$

$3x + 1 = \log_{10} 92$

$3x = \log_{10} 92 - 1$

$x = \dfrac{\log_{10} 92 - 1}{3}$

$x = 0.321$

Graphical Solution:

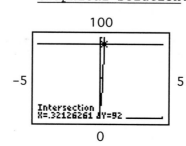

11. Algebraic Solution:

$e^x = 3.65$

$x = \ln 3.65$

$x = 1.29$

Graphical Solution:

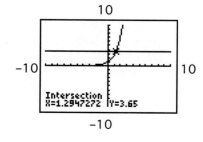

13. Algebraic Solution:
$$e^{2x-1} = 405$$
$$\ln 5^x = \ln 18$$
$$2x - 1 = \ln 405$$
$$2x = 1 + \ln 405$$
$$x = \frac{1 + \ln 405}{2}$$
$$x = 3.50$$
Graphical Solution:

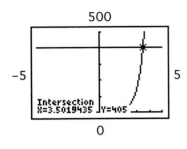

15. Algebraic Solution:
$$5^x = 18$$
$$x \ln 5 = \ln 18$$
$$x = \frac{\ln 18}{\ln 5}$$
$$x = 1.80$$
Graphical Solution:

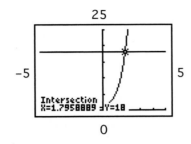

17. Algebraic Solution:
$$2^{-x} = 0.238$$
$$\ln 2^{-x} = \ln 0.238$$
$$-x \ln 2 = \ln 0.238$$
$$-x = \frac{\ln 0.238}{\ln 2}$$
$$x = -\frac{\ln 0.238}{\ln 2}$$
$$x = 2.07$$
Graphical Solution:

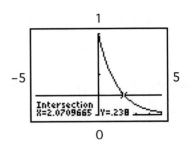

19. Algebraic Solution:
$$\log_5(2x - 7) = 2$$

Change to exponential form:
$$5^2 = 2x - 7$$
$$25 = 2x - 7$$
$$32 = 2x$$
$$x = 16$$

Graphical Solution: (Notice that we key in the left side using natural log and the change of base formula)

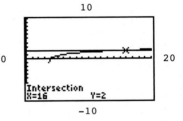

21. Algebraic Solution:
$$\log_3\left(x^2 - 8x\right) = 2$$

Change to exponential form:
$$3^2 = x^2 - 8x$$
$$0 = x^2 - 8x - 9$$
$$0 = (x - 9)(x + 1)$$
$$x = 9, -1$$

Graphical Solution: (Notice that we key in the left side using natural log and the change of base formula)

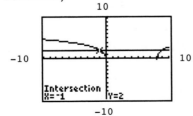
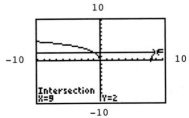

23. $\log 5 + \log x = 2$
$$\log(5x) = 2$$

Change to
exponential form
$$5x = 10^2$$
$$5x = 100$$
$$x = 20$$

25. $\log x + \log(x - 3) = 1$
$$\log[x(x - 3)] = 1$$

Change to exponential form
$$x(x - 3) = 10^1$$
$$x^2 - 3x = 10$$
$$x^2 - 3x - 10 = 0$$
$$(x - 5)(x + 2) = 0$$
$$x = 5 \text{ or } -2$$

> **Common Error:**
> $\log(x - 3) \neq \log x - \log 3$

Check:

$\log 5 + \log(5 - 3) \overset{\checkmark}{=} 1$
$\log(-2) + \log(-2 - 3)$
is not defined.
Solution: $x = 5$

27. $\log(x + 1) - \log(x - 1) = 1$
$$\log \frac{x + 1}{x - 1} = 1$$

Change to exponential form
$$\frac{x + 1}{x - 1} = 10^1$$
$$\frac{x + 1}{x - 1} = 10$$
$$x + 1 = 10(x - 1)$$
$$x + 1 = 10x - 10$$
$$11 = 9x$$
$$x = \frac{11}{9}$$

> **Common Error:**
> $\frac{x + 1}{x - 1} \neq \log 1$

Check:

$$\log\left(\frac{11}{9} + 1\right) - \log\left(\frac{11}{9} - 1\right) \overset{?}{=} 1$$
$$\log \frac{20}{9} - \log \frac{2}{9} \overset{?}{=} 1$$
$$\log 10 \overset{\checkmark}{=} 1$$

29. $\ln(4x - 3) = \ln(x + 1)$
$$4x - 3 = x + 1$$
$$3x = 4$$
$$x = \frac{4}{3}$$

Check: $\ln\left(4\left(\frac{4}{3}\right) - 3\right) = \ln\left(\frac{16}{3} - \frac{9}{3}\right) = \ln\left(\frac{7}{3}\right)$

$\ln\left(\frac{4}{3} + 1\right) = \ln\left(\frac{4}{3} + \frac{3}{3}\right) = \ln\left(\frac{7}{3}\right)$ $\checkmark$

31. $\log_2\left(x^2 - 2x\right) = \log_2(3x - 6)$

$$x^2 - 2x = 3x - 6$$
$$x^2 - 5x + 6 = 0$$
$$(x - 3)(x - 2) = 0$$
$$x = 3, 2$$

Check: $x = 3$: $\log_2\left(3^2 - 2(3)\right) = \log_2 3$

$\log_2\left(3(3) - 6\right) = \log_2 3$ $\checkmark$

$x = 2$: $\log_2\left(2^2 - 2(2)\right) = \log_2 0$ (undefined)

The only solution is $x = 3$.

33. <u>Algebraic Solution:</u>
$$2 = 1.05^x$$
$$\ln 2 = \ln 1.05^x$$
$$\ln 2 = x \ln 1.05$$
$$\frac{\ln 2}{\ln 1.05} = x$$
$$x = 14.2$$

<u>Graphical Solution:</u>

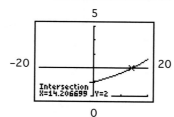

35. Algebraic Solution:

$$e^{-1.4x} = 13$$
$$\ln e^{-1.4x} = \ln 13$$
$$-1.4x = \ln 13$$
$$x = \frac{\ln 13}{-1.4}$$
$$x = -1.83$$

Graphical Solution:

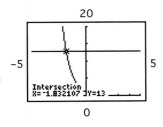

37. Algebraic Solution:

$$5 + 3^x = 10$$
$$3^x = 5$$
$$\ln 3^x = \ln 5$$
$$x \cdot \ln 3 = \ln 5$$
$$x = \frac{\ln 5}{\ln 3} \approx 1.46$$

Graphical Solution:

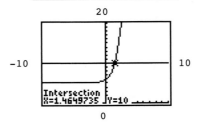

39. Algebraic Solution:

$$10^{2x+5} - 7 = 13$$
$$10^{2x+5} = 20$$
$$2x + 5 = \log 20$$
$$2x = \log 20 - 5$$
$$x = \frac{\log 20 - 5}{2} \approx -1.85$$

Graphical Solution:

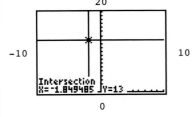

41. Algebraic Solution:

$$123 = 500e^{-0.12x}$$

$$\frac{123}{500} = e^{-0.12x}$$

$$\ln\left(\frac{123}{500}\right) = \ln e^{-0.12x}$$

$$\ln\left(\frac{123}{500}\right) = -0.12x$$

$$\frac{\ln\left(\frac{123}{500}\right)}{-0.12} = x$$

$$x = 11.7$$

Graphical Solution:

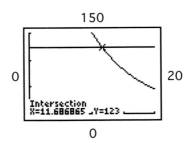

43.

$$e^{-x^2} = 0.23$$
$$\ln e^{-x^2} = \ln 0.23$$
$$-x^2 = \ln 0.23$$
$$x^2 = -\ln 0.23$$
$$x = \pm\sqrt{-\ln 0.23}$$
$$x = \pm 1.21$$

Check:

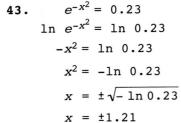

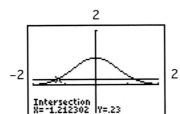

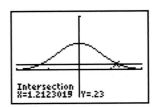

45. $\log x - \log 5 = \log 2 - \log(x - 3)$

$$\log \frac{x}{5} = \log \frac{2}{x - 3}$$

$$\frac{x}{5} = \frac{2}{x - 3}$$

Excluded value: $x \neq 3$

$$5(x - 3)\frac{x}{5} = 5(x - 3)\frac{2}{x - 3}$$

$$(x - 3)x = 10$$

$$x^2 - 3x = 10$$

$$x^2 - 3x - 10 = 0$$
$$(x - 5)(x + 2) = 0$$
$$x = 5, -2$$

Check:

$\log 5 - \log 5 \overset{\checkmark}{=} \log 2 - \log 2$
$\log(-2)$ is not defined

Solution: $x = 5$

47. $\ln x = \ln(2x - 1) - \ln(x - 2)$

$$\ln x = \ln \frac{2x - 1}{x - 2}$$

$$x = \frac{2x - 1}{x - 2}$$

Excluded value: $x \neq 2$

$$x(x - 2) = (x - 2)\frac{2x - 1}{x - 2}$$

$$x(x - 2) = 2x - 1$$
$$x^2 - 2x = 2x - 1$$
$$x^2 - 4x + 1 = 0$$

$$x = \frac{-b \pm \sqrt{b^2 - 4ac}}{2a}$$

$$a = 1, \ b = -4, \ c = 1$$

$$x = \frac{-(-4) \pm \sqrt{(-4)^2 - 4(1)(1)}}{2(1)}$$

$$x = \frac{4 \pm \sqrt{12}}{2} = 2 \pm \sqrt{3}$$

Check:

$\ln(2 + \sqrt{3}) \overset{?}{=} \ln[2(2 + \sqrt{3}) - 1]$
$\qquad\qquad - \ln[(2 + \sqrt{3}) - 2]$

$\ln(2 + \sqrt{3}) \overset{?}{=} \ln(3 + 2\sqrt{3}) - \ln\sqrt{3}$

$\ln(2 + \sqrt{3}) \overset{?}{=} \ln\left(\dfrac{3 + 2\sqrt{3}}{\sqrt{3}}\right)$

$\ln(2 + \sqrt{3}) \overset{\checkmark}{=} \ln(\sqrt{3} + 2)$

$\ln(x - 2)$ is not defined if
$x = 2 - \sqrt{3}$
Solution: $x = 2 + \sqrt{3}$

49.

$$\log(2x + 1) = 1 - \log(x - 1)$$
$$\log(2x + 1) + \log(x - 1) = 1$$
$$\log[(2x + 1)(x - 1)] = 1$$
$$(2x + 1)(x - 1) = 10$$
$$2x^2 - x - 1 = 10$$
$$2x^2 - x - 11 = 0$$

$$x = \frac{-b \pm \sqrt{b^2 - 4ac}}{2a}$$

$$a = 2, \ b = -1, \ c = -11$$

$$x = \frac{-(-1) \pm \sqrt{(-1)^2 - 4(2)(-11)}}{2(2)}$$

$$x = \frac{1 \pm \sqrt{89}}{4}$$

Check: $\log\left(2\,\dfrac{1+\sqrt{89}}{4}+1\right) \overset{?}{=} 1 - \log\left(\dfrac{1+\sqrt{89}}{4}-1\right)$

$\log\left(\dfrac{1+\sqrt{89}+2}{2}\right) \overset{?}{=} 1 - \log\left(\dfrac{1+\sqrt{89}-4}{4}\right)$

$\log\left(\dfrac{3+\sqrt{89}}{2}\right) \overset{?}{=} 1 - \log\left(\dfrac{\sqrt{89}-3}{4}\right)$

$\log\left(\dfrac{3+\sqrt{89}}{2}\right) \overset{?}{=} \log 10 - \log\left(\dfrac{\sqrt{89}-3}{4}\right)$

$\overset{?}{=} \log\left(\dfrac{40}{\sqrt{89}-3}\right)$

$\overset{?}{=} \log\left[\dfrac{40\left(\sqrt{89}+3\right)}{89-9}\right]$

$\overset{\checkmark}{=} \log\left(\dfrac{\sqrt{89}+3}{2}\right)$

$\log(x-1)$ is not defined if $x = \dfrac{1-\sqrt{89}}{4}$. Solution: $x = \dfrac{1+\sqrt{89}}{4}$.

51.
$$(\ln x)^3 = \ln x^4$$
$$(\ln x)^3 = 4 \ln x$$
$$(\ln x)^3 - 4 \ln x = 0$$
$$\ln x[(\ln x)^2 - 4] = 0$$
$$\ln x(\ln x - 2)(\ln x + 2) = 0$$

$\ln x = 0 \qquad \ln x - 2 = 0 \qquad \ln x + 2 = 0$
$\quad x = 1 \qquad\quad \ln x = 2 \qquad\quad \ln x = -2$
$\qquad\qquad\qquad\quad x = e^2 \qquad\qquad\quad x = e^{-2}$

Check:

$(\ln 1)^3 \overset{?}{=} \ln 1^4 \qquad (\ln e^2)^3 \overset{?}{=} \ln(e^2)^4 \qquad (\ln e^{-2})^3 \overset{?}{=} \ln(e^{-2})^4$
$\quad 0 \overset{\checkmark}{=} 0 \qquad\qquad 8 \overset{\checkmark}{=} 8 \qquad\qquad\quad -8 \overset{\checkmark}{=} -8$

Solution: $x = 1,\ e^2,\ e^{-2}$

53. $\ln(\ln x) = 1$
Change to
exponential form
$\quad \ln x = e^1$
$\quad \ln x = e$
$\qquad x = e^e$

55. $x^{\log x} = 100x$
We start by taking logarithms of both sides.
$$\log(x^{\log x}) = \log 100x$$
$$\log x \log x = \log 100 + \log x$$
$$(\log x)^2 = 2 + \log x$$
$$(\log x)^2 - \log x - 2 = 0$$
$$(\log x - 2)(\log x + 1) = 0$$

$\log x - 2 = 0 \qquad \log x + 1 = 0$
$\quad \log x = 2 \qquad\quad \log x = -1$
$\qquad x = 10^2 \qquad\qquad x = 10^{-1}$
$\qquad x = 100 \qquad\qquad x = 0.1$

Check: $100^{\log 100} \overset{?}{=} 100 \cdot 100 \qquad (0.1)^{\log(0.1)} \overset{?}{=} 100(0.1)$
$\qquad\qquad\quad 10^4 \overset{\checkmark}{=} 10^4 \qquad\qquad\qquad\quad 0.1^{-1} \overset{?}{=} 10$
$\qquad\qquad\qquad\qquad\qquad\qquad\qquad\qquad\qquad 10 \overset{\checkmark}{=} 10$

57. (A) In solving an exponential equation, we often can get the variable out of the exponent by using logarithms. If we do that in this case, however, we have made no progress:

$$e^{x/2} = 5 \ln x$$
$$\ln e^{x/2} = \ln(5 \ln x)$$
$$\frac{x}{2} = \ln 5 + \ln(\ln x)$$

This equation is no easier than the original to solve. Other algebraic methods seem to be equally useless.

(B) Examining the graphs of $y = e^{x/2}$ and $y = 5 \ln x$, it appears that there are two solutions of the equation.

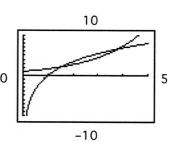

59. (A) In solving an exponential equation, we often can get the variable out of the exponent by using logarithms. If we do that in this case, however, we have made no progress:

$$3^x + 2 = 7 + x - e^{-x}$$
$$3^x = -e^{-x} + x + 5$$
$$\log_3 3^x = \log_3(-e^{-x} + x + 5)$$
$$x = \log_3(-e^{-x} + x + 5)$$

This equation is no easier than the original to solve.

(B) Graphing and applying a built-in routine, we obtain

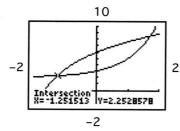 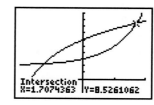

There are two solutions.

61.
$$A = Pe^{rt}$$
$$\frac{A}{P} = e^{rt}$$
$$\ln \frac{A}{P} = rt$$
$$\frac{1}{t} \ln \frac{A}{P} = r$$
$$r = \frac{1}{t} \ln \frac{A}{P}$$

63.
$$D = 10 \log \frac{I}{I_0}$$
$$\frac{D}{10} = \log \frac{I}{I_0}$$
$$\frac{I}{I_0} = 10^{D/10}$$
$$I = I_0(10^{D/10})$$

65.
$$M = 6 - 2.5 \log \frac{I}{I_0}$$
$$6 - M = 2.5 \log \frac{I}{I_0}$$
$$\frac{6 - M}{2.5} = \log \frac{I}{I_0}$$
$$\frac{I}{I_0} = 10^{(6-M)/2.5}$$
$$I = I_0[10^{(6-M)/2.5}]$$

67.

$$I = \frac{E}{R}(1 - e^{-Rt/L})$$

$$RI = E(1 - e^{-Rt/L})$$

$$\frac{RI}{E} = 1 - e^{-Rt/L}$$

$$\frac{RI}{E} - 1 = -e^{-Rt/L}$$

$$-\left(\frac{RI}{E} - 1\right) = e^{-Rt/L}$$

$$-\frac{RI}{E} + 1 = e^{-Rt/L}$$

$$1 - \frac{RI}{E} = e^{-Rt/L}$$

$$\ln\left(1 - \frac{RI}{E}\right) = -\frac{Rt}{L}$$

$$-\frac{L}{R}\ln\left(1 - \frac{RI}{E}\right) = t$$

$$t = -\frac{L}{R}\ln\left(1 - \frac{RI}{E}\right)$$

69.

$$y = \frac{e^x + e^{-x}}{2}$$

$$2y = e^x + e^{-x}$$

$$2y = e^x + \frac{1}{e^x}$$

$$2ye^x = (e^x)^2 + 1$$

$$0 = (e^x)^2 - 2ye^x + 1$$

This equation is quadratic in e^x

$$e^x = \frac{-b \pm \sqrt{b^2 - 4ac}}{2a}$$

$$a = 1, \quad b = -2y, \quad c = 1$$

$$e^x = \frac{-(-2y) \pm \sqrt{(-2y)^2 - 4(1)(1)}}{2(1)}$$

$$e^x = \frac{2y \pm \sqrt{4y^2 - 4}}{2}$$

$$e^x = \frac{2(y \pm \sqrt{y^2 - 1})}{2}$$

$$e^x = y \pm \sqrt{y^2 - 1}$$

$$x = \ln(y \pm \sqrt{y^2 - 1})$$

71.

$$y = \frac{e^x - e^{-x}}{e^x + e^{-x}}$$

$$y = \frac{e^x - \frac{1}{e^x}}{e^x + \frac{1}{e^x}}$$

$$y = \frac{e^x e^x - \frac{1}{e^x}e^x}{e^x e^x + \frac{1}{e^x}e^x}$$

$$y = \frac{e^{2x} - 1}{e^{2x} + 1}$$

$$y(e^{2x} + 1) = e^{2x} - 1$$

$$ye^{2x} + y = e^{2x} - 1$$

$$1 + y = e^{2x} - ye^{2x}$$

$$1 + y = (1 - y)e^{2x}$$

$$e^{2x} = \frac{1 + y}{1 - y}$$

$$2x = \ln\frac{1 + y}{1 - y}$$

$$x = \frac{1}{2}\ln\frac{1 + y}{1 - y}$$

73. Graphing $y = 2^{-x} - 2x$ and applying the ZERO command, we obtain

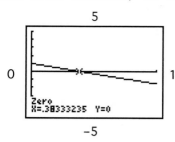

The required solution of $2^{-x} - 2x = 0$, $0 \le x \le 1$, is $x = 0.38$.

75. Graphing $y = x3^x - 1$ and applying the ZERO command, we obtain

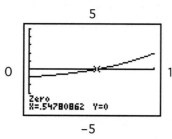

The required solution of $x3^x - 1 = 0$, $0 \le x \le 1$, is $x = 0.55$.

77. Graphing $y = e^{-x} - x$ and applying the ZERO command, we obtain

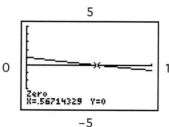

The required solution of $e^{-x} - x = 0$, $0 \le x \le 1$, is $x = 0.57$.

79. Graphing $y = xe^x - 2$ and applying the ZERO command, we obtain

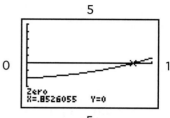

The required solution of $xe^x - 2 = 0$, $0 \le x \le 1$, is $x = 0.85$.

81. Graphing $y = \ln x + 2x$ and applying the ZERO command, we obtain

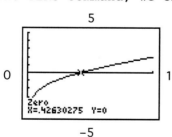

The required solution of $\ln x + 2x = 0$, $0 \le x \le 1$, is 0.43.

83. Graphing $y = \ln x + e^x$ and applying the ZERO command, we obtain

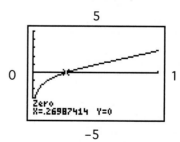

The required solution of $\ln x + e^x = 0$, $0 \le x \le 1$, is 0.27.

85. To find the doubling time we replace A in $A = P(1 + 0.15)^n$ with $2P$ and solve for n.

$$2P = P(1.15)^n$$
$$2 = (1.15)^n$$
$$\ln 2 = n \ln 1.15$$
$$n = \frac{\ln 2}{\ln 1.15}$$

$n = 5$ years to the nearest year

87. We solve $A = Pe^{rt}$ for r, with $A = 2,500$, $P = 1,000$, $t = 10$

$$2,500 = 1,000e^{r(10)}$$
$$2.5 = e^{10r}$$
$$10r = \ln (2.5)$$
$$r = \frac{1}{10} \ln 2.5$$
$$r = 0.0916 \text{ or } 9.16\%$$

89. (A) We're given $P_0 = 10.5$ (we could use 10.5 million, but if you look carefully at the calculations below, you'll see that the millions will cancel out anyhow), 11.3 for P, and 2 for t (since May 2007 is two years after May 2005).

$$11.3 = 10.5e^{r \cdot 2}$$
$$\frac{11.3}{10.5} = e^{2r}$$
$$\ln\left(\frac{11.3}{10.5}\right) = 2r$$
$$r = \frac{\ln\left(\frac{11.3}{10.5}\right)}{2} \approx 0.0367$$

The annual growth rate is 3.67%.

(B) $P = 10.5e^{0.0367t}$; plug in 20 for P and solve for t.

$$20 = 10.5e^{0.0367t}$$
$$\frac{20}{10.5} = e^{0.0367t}$$
$$\ln\left(\frac{20}{10.5}\right) = 0.0367t$$
$$t = \frac{\ln\left(\frac{20}{10.5}\right)}{0.0367} \approx 17.6$$

The illegal immigrant population is predicted to reach 20 million near the end of 2022, which is 17.6 years after May 2005.

91. We solve $P = P_0e^{rt}$ for t with
$P = 2P_0$, $r = 0.0114$.

$$2P_0 = P_0e^{0.0114t}$$
$$2 = e^{0.0114t}$$
$$\ln 2 = 0.0114t$$
$$t = \frac{\ln 2}{0.0114}$$
$$t = 61 \text{ years to the nearest year}$$

93. We're given $A_0 = 5$, $A = 1$, $t = 6$:

$$A = A_0\left(\frac{1}{2}\right)^{t/h}$$
$$1 = 5\left(\frac{1}{2}\right)^{6/h}$$
$$\frac{1}{5} = \left(\frac{1}{2}\right)^{6/h}$$
$$\ln\frac{1}{5} = \ln\left(\frac{1}{2}\right)^{6/h}$$
$$\ln\frac{1}{5} = \frac{6}{h}\ln\left(\frac{1}{2}\right)$$
$$h \cdot \ln\frac{1}{5} = 6\ln\left(\frac{1}{2}\right)$$
$$h = \frac{6\ln(1/2)}{\ln(1/5)} \approx 2.58$$

The half-life is about 2.58 hours.

95. Let A_0 represent the amount of Carbon-14 originally present. Then the amount left in 2003 was $0.289A_0$. Plug this in for A, and solve for t:

$$0.289A_0 = A_0e^{-0.000\,124t}$$

$$0.289 = e^{-.000\,124t}$$

$$\ln 0.289 = \ln e^{-.000\,124t}$$

$$\ln 0.289 = -.000\,124t$$

$$t = \frac{\ln 0.289}{-.000\,124} \approx 10,010$$

The sample was about 10,010 years old.

97. Let A_0 represent the amount of Carbon-14 originally present. Then the amount left in 2004 was $0.883A_0$. Plug this in for A, and solve for t:

$$0.883A_0 = A_0e^{-0.000\,124t}$$

$$0.883 = e^{-.000\,124t}$$

$$\ln 0.883 = \ln e^{-.000\,124t}$$

$$\ln 0.883 = -.000\,124t$$

$$t = \frac{\ln 0.883}{-.000\,124} \approx 1,003$$

It was 1,003 years old in 2004, so it was made in 1001.

99. We solve $q = 0.0009(1 - e^{-0.2t})$ for t with $q = 0.0007$

$$0.0007 = 0.0009(1 - e^{-0.2t})$$

$$\frac{0.0007}{0.0009} = 1 - e^{-0.2t}$$

$$\frac{7}{9} = 1 - e^{-0.2t}$$

$$-\frac{2}{9} = -e^{-0.2t}$$

$$\frac{2}{9} = e^{-0.2t}$$

$$\ln \frac{2}{9} = -0.2t$$

$$t = \frac{\ln \frac{2}{9}}{-0.2}$$

$$t = 7.52 \text{ seconds}$$

101. First, we solve $T = T_m + (T_0 - T_m)e^{-kt}$ for k, with $T = 61.5°$, $T_m = 40°$, $T_0 = 72°$, $t = 1$

$$61.5 = 40 + (72 - 40)e^{-k(1)}$$

$$21.5 = 32e^{-k}$$

$$\frac{21.5}{32} = e^{-k}$$

$$\ln \frac{21.5}{32} = -k$$

$$k = -\ln \frac{21.5}{32}$$

$$k = 0.40$$

Now we solve $T = T_m + (T_0 - T_m)e^{-0.40t}$ for t, with $T = 50°$, $T_m = 40°$, $T_0 = 72°$

$$50 = 40 + (72 - 40)e^{-0.40t}$$

$$10 = 32e^{-0.40t}$$

$$\frac{10}{32} = e^{-0.40t}$$

$$\ln \frac{10}{32} = -0.40t$$

$$t = \frac{\ln^{10/32}}{-0.40}$$

$$t = 2.9 \text{ hours}$$

103. (A) Plug in $M = 7.0$ and solve for E:

$$7 = \frac{2}{3} \log \frac{E}{10^{4.40}}$$

$$\frac{21}{2} = \log \frac{E}{10^{4.4}}$$

$$10^{21/2} = \frac{E}{10^{4.4}}$$

$$E = 10^{21/2} \cdot 10^{4.4} \approx 7.94 \times 10^{14} \text{ joules}$$

(B) $\dfrac{7.94 \times 10^{14} \text{ joules}}{2.88 \times 10^{14} \text{ joules/day}} = 2.76 \text{ days}$

105. First, find the energy released by one magnitude 7.5 earthquake:

$$7.5 = \frac{2}{3} \log \frac{E}{10^{4.40}}$$

$$11.25 = \log \frac{E}{10^{4.4}}$$

$$10^{11.25} = \frac{E}{10^{4.4}}$$

$$E = 10^{11.25} \cdot 10^{4.4} \approx 4.47 \times 10^{15} \text{ joules}$$

Now multiply by ten to get the energy released by ten such earthquakes:

$$10 \cdot 4.47 \times 10^{15} = 4.47 \times 10^{16}$$

Finally, divide by the energy consumption per year:

$$\frac{4.47 \times 10^{16} \quad \text{joules}}{1.05 \times 10^{17} \quad \text{joules/year}} = 0.426$$

So this energy could power the U.S. for 0.426 years, or about 155 days.

CHAPTER 4 REVIEW

1. (A) The graph of $y = \log_2 x$ passes through $(1, 0)$ and $(2, 1)$. This corresponds to graph m.

(B) The graph of $y = 0.5^x$ passes through $(0, 1)$ and $(1, 0.5)$. This corresponds to graph f.

(C) The graph of $y = \log_{0.5} x$ passes through $(1, 0)$ and $(0.5, 1)$. This corresponds to graph n.

(D) The graph of $y = 2^x$ passes through $(0, 1)$ and $(1, 2)$. This corresponds to graph g. *(4-1, 4-3)*

2. $\log m = n$ *(4-3)* **3.** $\ln x = y$ *(4-3)* **4.** $x = 10^y$ *(4-3)* **5.** $y = e^x$ *(4-3)*

6. (A) Make a table of values:

x	-2	-1	0	1	2	3
$\left(\dfrac{4}{3}\right)^x$	$\dfrac{9}{16}$	$\dfrac{3}{4}$	1	$\dfrac{4}{3}$	$\dfrac{16}{9}$	$\dfrac{64}{27}$

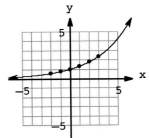

(4-1)

(B) The function in part (B) is the inverse of the one graphed in part (A), so its graph is a reflection about the line $y = x$ of the graph to the left. To plot points, just switch the x and y coordinates of the points from the table in part (A).

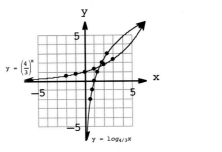

(4-3)

7. $\dfrac{7^{x+2}}{7^{2-x}} = 7^{(x+2)-(2-x)}$

$= 7^{x+2-2+x}$

$= 7^{2x}$ *(4-1)*

8. $\left(\dfrac{e^x}{e^{-x}}\right)^x = [e^{x-(-x)}]^x$

$= (e^{2x})^x = e^{2x \cdot x}$

$= e^{2x^2}$ *(4-1)*

9. $\log_2 x = 3$

$x = 2^3$

$x = 8$ *(4-3)*

10. $\log_x 25 = 2$
$25 = x^2$
$x = 5$
since bases are
restricted positive
$(4-3)$

11. $\log_3 27 = x$
$\log_3 3^3 = x$
$x = 3$ $(4-3)$

12. $10^x = 17.5$
$x = \log_{10} 17.5$
$x = 1.24$ $(4-5)$

13. $e^x = 143,000$
$x = \ln 143,000$
$x = 11.9$ $(4-5)$

14. $\ln x = -0.01573$
$x = e^{-0.01573}$
$x = 0.984$ $(4-3)$

15. $\log x = 2.013$
$x = 10^{2.013}$
$x = 103$ $(4-3)$

16. 1.145 $(4-3)$

17. Not defined. ($-e$ is not in the domain of the logarithm function.) $(4-3)$

18. 2.211 $(4-3)$

19. 11.59 $(4-1)$

20. $\log x + 3 \log y - \dfrac{1}{2} \log z$
$= \log x + \log y^3 - \log z^{1/2}$
$= \log\left(\dfrac{xy^3}{\sqrt{z}}\right)$ $(4-3)$

21. $\ln \dfrac{x^3}{y} = \ln x^3 - \ln y = 3 \ln x - \ln y$ $(4-3)$

22. $\ln(2x - 1) = \ln(x + 3)$
$2x - 1 = x + 3$
$x = 4$

Check:
$\ln(2 \cdot 4 - 1) \overset{?}{=} \ln(4 + 3)$
$\ln 7 \overset{\checkmark}{=} \ln 7$
$(4-5)$

23. $\log(x^2 - 3) = 2 \log(x - 1)$
$\log(x^2 - 3) = \log(x - 1)^2$
$x^2 - 3 = (x - 1)^2$
$x^2 - 3 = x^2 - 2x + 1$
$-3 = -2x + 1$
$-4 = -2x$
$x = 2$

Check:
$\log(2^2 - 3) \overset{?}{=} 2 \log(2 - 1)$
$\log 1 \overset{?}{=} 2 \log 1$
$0 \overset{\checkmark}{=} 0$ $(4-5)$

24. $e^{x^2-3} = e^{2x}$
$x^2 - 3 = 2x$
$x^2 - 2x - 3 = 0$
$(x - 3)(x + 1) = 0$
$x = 3, \quad -1$
$(4-5)$

25. $4^{x-1} = 2^{1-x}$
$(2^2)^{x-1} = 2^{1-x}$
$2^{2(x-1)} = 2^{1-x}$
$2(x - 1) = 1 - x$
$2x - 2 = 1 - x$
$3x = 3$
$x = 1$ $(4-5)$

26. $4 - 3^x = 2$
$-3^x = -2$
$3^x = 2$
$\ln 3^x = \ln 2$
$x \ln 3 = \ln 2$
$x = \dfrac{\ln 2}{\ln 3}$ $(4-5)$

27. $5 + \dfrac{1}{2} e^x = \dfrac{17}{2}$
$\dfrac{1}{2} e^x = \dfrac{17}{2} - 5 = \dfrac{17}{2} - \dfrac{10}{2} = \dfrac{7}{2}$
$e^x = 2 \cdot \dfrac{7}{2} = 7$
$x = \ln 7$ $(4-5)$

28. $2x^2 e^{-x} = 18e^{-x}$
$2x^2 e^{-x} - 18e^{-x} = 0$
$2e^{-x}(x^2 - 9) = 0$
$2e^{-x}(x - 3)(x + 3) = 0$
$2e^{-x} = 0 \qquad x - 3 = 0 \qquad x + 3 = 0$
never $\qquad\quad x = 3 \qquad\quad x = -3$
Solution: $x = 3, -3$ $(4-5)$

29. $\log_{1/4} 16 = x$
$\log_{1/4} 4^2 = x$
$\log_{1/4} \left(\dfrac{1}{4}\right)^{-2} = x$
$x = -2$ $(4-5)$

30. $\log_x 9 = -2$

$x^{-2} = 9$

$\dfrac{1}{x^2} = 9$

$1 = 9x^2$

$\dfrac{1}{9} = x^2$

$x = \pm\sqrt{\dfrac{1}{9}}$

$x = \dfrac{1}{3}$

since bases are restricted positive

$(4-5)$

31. $\log_{16} x = \dfrac{3}{2}$

$16^{3/2} = x$

$64 = x$

$x = 64$

$(4-5)$

32. $\log_x e^5 = 5$

$e^5 = x^5$

$x = e \qquad (4-5)$

33. $10^{\log_{10} x} = 33$

$\log_{10} x = \log_{10} 33$

$x = 33 \qquad (4-5)$

34. $\ln x = 0$

$e^0 = x$

$x = 1 \qquad (4-5)$

35. $x = 2(10^{1.32})$

$x = 41.8 \qquad (4-1)$

36. $x = \log_5 23$

$x = \dfrac{\log 23}{\log 5} \text{ or } \dfrac{\ln 23}{\ln 5}$

$x = 1.95 \qquad (4-3)$

37. $\ln x = -3.218$

$x = e^{-3.218}$

$x = 0.0400 \qquad (4-3)$

38. $x = \log(2.156 \times 10^{-7})$

$x = -6.67 \qquad (4-3)$

39. $x = \dfrac{\ln 4}{\ln 2.31}$

$x = 1.66 \qquad (4-3)$

40. $25 = 5(2)^x$

$\dfrac{25}{5} = 2^x$

$5 = 2^x$

$\ln 5 = \ln 2^x$

$\ln 5 = x \ln 2$

$\dfrac{\ln 5}{\ln 2} = x$

$x = 2.32 \qquad (4-5)$

41. $4{,}000 = 2{,}500 e^{0.12x}$

$\dfrac{4{,}000}{2{,}500} = e^{0.12x}$

$\ln\left(\dfrac{4{,}000}{2{,}500}\right) = \ln e^{0.12x}$

$0.12x = \ln \dfrac{4{,}000}{2{,}500}$

$x = \dfrac{1}{0.12} \ln \dfrac{4{,}000}{2{,}500}$

$x = 3.92 \qquad (4-5)$

42. $0.01 = e^{-0.05x}$

$\ln 0.01 = \ln e^{-0.05x}$

$-0.05x = \ln 0.01$

$x = \dfrac{\ln 0.01}{-0.05}$

$x = 92.1 \qquad (4-5)$

43. $5^{2x-3} = 7.08$

$\log 5^{2x-3} = \log 7.08$

$(2x - 3)\log 5 = \log 7.08$

$2x - 3 = \dfrac{\log 7.08}{\log 5}$

$x = \dfrac{1}{2}\left[3 + \dfrac{\log 7.08}{\log 5}\right]$

$x = 2.11 \qquad (4-5)$

44. $\dfrac{e^x - e^{-x}}{2} = 1$

$e^x - e^{-x} = 2$

$e^x - \dfrac{1}{e^x} = 2$

$e^x e^x - e^x\left(\dfrac{1}{e^x}\right) = 2e^x$

$(e^x)^2 - 1 = 2e^x$

$(e^x)^2 - 2e^x - 1 = 0$

This equation is quadratic in e^x: $\quad e^x = \dfrac{-b \pm \sqrt{b^2 - 4ac}}{2a}$

$$a = 1, \ b = -2, \ c = -1$$

$$e^x = \dfrac{-(-2) \pm \sqrt{(-2)^2 - 4(1)(-1)}}{2}$$

$$e^x = \dfrac{2 \pm \sqrt{8}}{2}$$

$$e^x = 1 \pm \sqrt{2}$$

$$x = \ln(1 \pm \sqrt{2}) \qquad 1 - \sqrt{2} \text{ is negative, so not in the domain of the logarithm function.}$$

$$x = \ln(1 + \sqrt{2})$$

$$x = 0.881 \tag{4-5}$$

45. $\log 3x^2 - \log 9x = 2$

$$\log \dfrac{3x^2}{9x} = 2$$

Check:

$$\dfrac{3x^2}{9x} = 10^2 \qquad\qquad \log(3 \cdot 300^2) - \log(9 \cdot 300) \stackrel{?}{=} 2$$

$$\dfrac{x}{3} = 100 \qquad\qquad \log(270{,}000) - \log(2{,}700) \stackrel{?}{=} 2$$

$$x = 300 \qquad\qquad\qquad\qquad \log \dfrac{270{,}000}{2{,}700} \stackrel{?}{=} 2$$

$$\log 100 \stackrel{\checkmark}{=} 2 \tag{4-5}$$

46. $\log x - \log 3 = \log 4 - \log(x + 4)$

$$\log \dfrac{x}{3} = \log \dfrac{4}{x + 4}$$

$$\dfrac{x}{3} = \dfrac{4}{x + 4} \qquad \text{excluded value:}$$

$$\qquad\qquad\qquad\qquad\qquad x \ne -4$$

$$3(x + 4)\dfrac{x}{3} = 3(x + 4)\dfrac{4}{x + 4}$$

$$(x + 4)x = 12$$

$$x^2 + 4x = 12$$

$$x^2 + 4x - 12 = 0$$

$$(x + 6)(x - 2) = 0$$

$$x = -6 \quad x = 2$$

Check: $\log(-6)$ is not defined

$$\log 2 - \log 3 \stackrel{?}{=} \log 4 - \log(2 + 4)$$

$$\log \dfrac{2}{3} \stackrel{?}{=} \log \dfrac{4}{6}$$

$$\log \dfrac{2}{3} \stackrel{\checkmark}{=} \log \dfrac{2}{3}$$

Solution: $x = 2$ $\qquad\qquad$ (4-5)

47. $\ln(x + 3) - \ln x = 2 \ln 2$

$$\ln \dfrac{x + 3}{x} = \ln 2^2$$

$$\dfrac{x + 3}{x} = 2^2$$

$$\dfrac{x + 3}{x} = 4$$

$$x + 3 = 4x$$

$$3 = 3x$$

$$x = 1$$

Check:

$$\ln(1 + 3) - \ln 1 \stackrel{?}{=} 2 \ln 2$$

$$\ln 4 - 0 \stackrel{?}{=} 2 \ln 2$$

$$\ln 4 \stackrel{\checkmark}{=} \ln 4 \tag{4-5}$$

48. $\ln(2x + 1) - \ln(x - 1) = \ln x$

$$\ln \dfrac{2x + 1}{x - 1} = \ln x$$

$$\dfrac{2x + 1}{x - 1} = x \quad \text{Excluded value: } x \ne 1$$

$$(x - 1)\dfrac{2x + 1}{x - 1} = x(x - 1)$$

$$2x + 1 = x^2 - x$$

$$0 = x^2 - 3x - 1$$

$$x = \frac{-b \pm \sqrt{b^2 - 4ac}}{2a} \quad a = 1, \ b = -3, \ c = -1$$

$$x = \frac{-(-3) \pm \sqrt{(-3)^2 - 4(1)(-1)}}{2(1)}$$

$$x = \frac{3 \pm \sqrt{13}}{2}$$

Check: $\ln\left(\dfrac{3 - \sqrt{13}}{2}\right)$ is not defined

$$\ln\left(2 \cdot \frac{3 + \sqrt{13}}{2} + 1\right) - \ln\left(\frac{3 + \sqrt{13}}{2} - 1\right) \overset{?}{=} \ln\left(\frac{3 + \sqrt{13}}{2}\right)$$

$$\ln(3 + \sqrt{13} + 1) - \ln\left(\frac{3 + \sqrt{13} - 2}{2}\right) \overset{?}{=} \ln\left(\frac{3 + \sqrt{13}}{2}\right)$$

$$\ln(4 + \sqrt{13}) - \ln\left(\frac{1 + \sqrt{13}}{2}\right) \overset{?}{=} \ln\left(\frac{3 + \sqrt{13}}{2}\right)$$

$$\ln\left(\frac{4 + \sqrt{13}}{1} \cdot \frac{2}{1 + \sqrt{13}}\right) \overset{?}{=} \ln\left(\frac{3 + \sqrt{13}}{2}\right)$$

$$\ln\left(\frac{\left(4 + \sqrt{13}\right)2}{1 + \sqrt{13}}\right) \overset{?}{=} \ln\left(\frac{3 + \sqrt{13}}{2}\right)$$

$$\ln\left(\frac{\left(4 + \sqrt{13}\right)2\left(1 - \sqrt{13}\right)}{\left(1 + \sqrt{13}\right)\left(1 - \sqrt{13}\right)}\right) \overset{?}{=} \ln\left(\frac{3 + \sqrt{13}}{2}\right)$$

$$\ln\left(\frac{2\left(4 - 3\sqrt{13} - 13\right)}{1 - 13}\right) \overset{?}{=} \ln\left(\frac{3 + \sqrt{13}}{2}\right)$$

$$\ln\left(\frac{-18 - 6\sqrt{13}}{-12}\right) \overset{?}{=} \ln\left(\frac{3 + \sqrt{13}}{2}\right)$$

$$\ln\left(\frac{3 + \sqrt{13}}{2}\right) \overset{\sqrt{}}{=} \ln\left(\frac{3 + \sqrt{13}}{2}\right)$$

Solution: $x = \dfrac{3 + \sqrt{13}}{2}$ 　　　　　　　　　　　　　　　　(4-5)

49.

$$(\log x)^3 = \log x^9$$
$$(\log x)^3 = 9 \log x$$
$$(\log x)^3 - 9 \log x = 0$$
$$\log x[(\log x)^2 - 9] = 0$$
$$\log x(\log x - 3)(\log x + 3) = 0$$
$$\log x = 0 \quad \log x - 3 = 0 \quad \log x + 3 = 0$$
$$x = 1 \quad \log x = 3 \quad \log x = -3$$
$$x = 10^3 \quad\quad x = 10^{-3}$$

Check:

$$(\log 1)^3 \overset{?}{=} \log 1^9$$
$$0 \overset{\sqrt{}}{=} 0$$
$$(\log 10^3)^3 \overset{?}{=} \log(10^3)^9$$
$$27 \overset{\sqrt{}}{=} 27$$
$$(\log 10^{-3})^3 \overset{?}{=} \log(10^{-3})^9$$
$$-27 \overset{\sqrt{}}{=} -27$$

Solution: $x = 1, 10^3, 10^{-3}$ (4-5)

50. $\ln(\log x) = 1$
$\log x = e$
$x = 10^e$ 　　(4-5)

51. $(e^x + 1)(e^{-x} - 1) - e^x(e^{-x} - 1) = e^x e^{-x} - e^x + e^{-x} - 1 - e^x e^{-x} + e^x$
$$= 1 - e^x + e^{-x} - 1 - 1 + e^x = e^{-x} - 1 \qquad (4-1)$$

52. $(e^x + e^{-x})(e^x - e^{-x}) - (e^x - e^{-x})^2 = (e^x)^2 - (e^{-x})^2 - [(e^x)^2 - 2e^x e^{-x} + (e^{-x})^2]$
$$= e^{2x} - e^{-2x} - [e^{2x} - 2 + e^{-2x}]$$
$$= e^{2x} - e^{-2x} - e^{2x} + 2 - e^{-2x}$$
$$= 2 - 2e^{-2x} \qquad (4-1)$$

53. The graph is shown below, along with a second screen showing the y intercept.

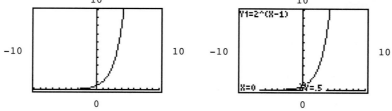

The domain is all real numbers, and the range is $(0, \infty)$. There is no x intercept, and the y intercept is 1/2. There is no vertical asymptote, and the horizontal asymptote is $y = 0$. $\qquad (4-1)$

54. The graph is shown below, along with a second screen showing the y intercept.

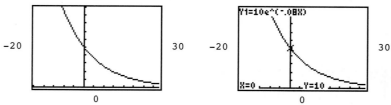

The domain is all real numbers, and the range is $(0, \infty)$. There is no x intercept, and the y intercept is 10. There is no vertical asymptote, and the horizontal asymptote is $y = 0$. $\qquad (4-1)$

55. The graph is shown below, along with a second screen showing the x intercept.

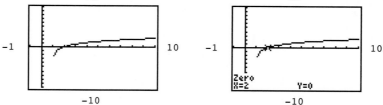

The domain is $[1, \infty)$, and the range is all real numbers. There is no y intercept, and the x intercept is 2. There is no horizontal asymptote, and the vertical asymptote is $x = 1$. $\qquad (4-3)$

56. The graph is shown below, along with a second screen showing the y intercept.

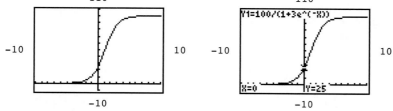

The domain is all real numbers, and the range is $(0, 100)$. There is no x intercept, and the y intercept is 25. There is no vertical asymptote, and the horizontal asymptotes are $y = 0$ and $y = 100$. $\qquad (4-1)$

57. Reflecting a graph in the line $y = x$ results in the graph of the inverse of the original function. In this case, the result is the inverse of $y = e^x$, which is $y = \ln x$. If the graph of $y = e^x$ is reflected in the x axis, y is replaced by $-y$ and the graph becomes the graph of $-y = e^x$ or $y = -e^x$.

If the graph of $y = e^x$ is reflected in the y axis, x is replaced by $-x$ and the graph becomes the graph of $y = e^{-x}$ or $y = \dfrac{1}{e^x}$ or $y = \left(\dfrac{1}{e}\right)^x$. *(4-1)*

58. Examining the graph of $f(x) = 4 - x^2 + \ln x$, we obtain

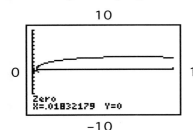 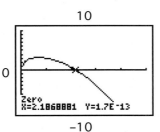

The zeros are at 0.018 and 2.187. *(4-5)*

59. Graphing $y_1 = 10^{x-3}$ and $y_2 = 8 \log x$, we obtain

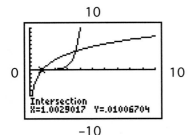 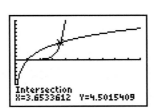

The graphs intersect at (1.003, 0.010) and (3.653, 4.502). *(4-5)*

60.
$$D = 10 \log \frac{I}{I_0}$$
$$\frac{D}{10} = \log \frac{I}{I_0}$$
$$10^{D/10} = \frac{I}{I_0}$$
$$I_0 10^{D/10} = I$$
$$I = I_0(10^{D/10})$$

(4-5)

61.
$$y = \frac{1}{\sqrt{2\pi}} e^{-x^2/2}$$
$$\sqrt{2\pi}\, y = e^{-x^2/2}$$
$$-\frac{x^2}{2} = \ln(\sqrt{2\pi}\, y)$$
$$x^2 = -2 \ln(\sqrt{2\pi}\, y)$$
$$x = \pm\sqrt{-2 \ln(\sqrt{2\pi}\, y)} \qquad (4-5)$$

62.
$$x = -\frac{1}{k} \ln \frac{I}{I_0}$$
$$-kx = \ln \frac{I}{I_0}$$
$$\frac{I}{I_0} = e^{-kx}$$
$$I = I_0(e^{-kx})$$

(4-5)

63.

$$r = P \frac{i}{1 - (1 + i)^{-n}}$$

$$\frac{r}{P} = \frac{i}{1 - (1 + i)^{-n}}$$

$$\frac{P}{r} = \frac{1 - (1 + i)^{-n}}{i}$$

$$\frac{Pi}{r} = 1 - (1 + i)^{-n}$$

$$\frac{Pi}{r} - 1 = -(1 + i)^{-n}$$

$$1 - \frac{Pi}{r} = (1 + i)^{-n}$$

$$\ln\left(1 - \frac{Pi}{r}\right) = \ln(1 + i)^{-n}$$

$$\ln\left(1 - \frac{Pi}{r}\right) = -n \ln (1 + i)$$

$$\frac{\ln\left(1 - \frac{Pi}{r}\right)}{- \ln(1 + i)} = n$$

$$n = - \frac{\ln\left(1 - \frac{Pi}{r}\right)}{\ln(1 + i)} \qquad (4\text{-}5)$$

64. (A) For $x > -1$, $y = e^{-x/3}$ decreases from $e^{1/3}$ to 0 while $\ln(x + 1)$ increases from $-\infty$ to ∞. Consequently, the graphs can intersect at exactly one point.

(B) Graphing $y_1 = e^{-x/3}$ and $y_2 = 4 \ln(x + 1)$ we obtain

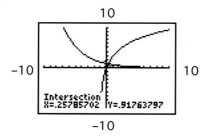

The solution of $e^{-x/3} = 4 \ln(x + 1)$ is $x = 0.258$. $\qquad (4\text{-}5)$

65.

$$\ln y = -5t + \ln c$$

$$\ln y - \ln c = -5t$$

$$\ln \left(\frac{y}{c}\right) = -5t$$

$$\frac{y}{c} = e^{-5t}$$

$$y = ce^{-5t} \qquad (4\text{-}5)$$

66.

x	$y = \log_2 x$	$x = \log_2 y$	y
1	0	0	1
2	1	1	2
4	2	2	4
8	3	3	8

Domain $f = (0, \infty) = $ Range f^{-1}
Range $f = (-\infty, \infty) = $ Domain f^{-1}

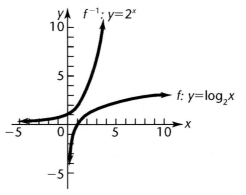

$(4\text{-}3)$

67. If $\log_1 x = y$, then we would have to have $1^y = x$; that is, $1 = x$ for arbitrary positive x, which is impossible. $(4\text{-}3)$

68. Let $u = \log_b M$ and $v = \log_b N$ and then change each equation to exponential form:
$$b^u = M; \quad b^v = N$$

Now multiply the two left sides and use a familiar property of exponents:
$$b^u b^v = b^{u+v} = MN$$

Next, apply the logarithm base b to both sides:
$$\log_b b^{u+v} = \log_b(MN)$$

The left side is now just $u + v$, which is $\log_b M + \log_b N$. Switching the order of the equation, we get
$$\log_b(MN) = \log_b M + \log_b N \qquad (4\text{-}3)$$

69. We solve $P = P_0(1.03)^t$ for t, using $P = 2P_0$.
$$2P_0 = P_0(1.03)^t$$
$$2 = (1.03)^t$$
$$\ln 2 = t \ln 1.03$$
$$\frac{\ln 2}{\ln 1.03} = t$$
$$t = 23.4 \text{ years} \qquad (4\text{-}2)$$

70. We solve $P = P_0 e^{0.03t}$ for t using $P = 2P_0$.
$$2P_0 = P_0 e^{0.03t}$$
$$2 = e^{0.03t}$$
$$\ln 2 = 0.03t$$
$$\frac{\ln 2}{0.03} = t$$
$$t = 23.1 \text{ years} \qquad (4\text{-}2)$$

71. $A_0 = $ original amount
$0.01A_0 = 1$ percent of original amount

We solve $A = A_0 e^{-0.000124t}$ for t, using $A = 0.01A_0$.
$$0.01A_0 = A_0 e^{-0.000124t}$$
$$0.01 = e^{-0.000124t}$$
$$\ln 0.01 = -0.000124t$$
$$\frac{\ln 0.01}{-0.000124} = t$$
$$t = 37,100 \text{ years} \qquad (4\text{-}2)$$

72. (A) When $t = 0$, $N = 1$. As t increases by $1/2$, N doubles.
$$N = 1 \cdot (2)^{t \div 1/2}$$
$$N = 2^{2t} \text{ (or } N = 4^t)$$

(B) We solve $N = 4^t$ for t, using $N = 10^9$:
$$10^9 = 4^t$$
$$9 = t \log 4$$
$$t = \frac{9}{\log 4}$$
$$t = 15 \text{ days} \qquad (4\text{-}2)$$

73. We use $A = Pe^{rt}$ with $P = 1$, $r = 0.03$, and $t = 2010$.
$$A = 1e^{0.03(2010)}$$
$$A = 1.5 \times 10^{26} \text{ dollars}$$
$$(4\text{-}1)$$

74. (A)

t	P
0	1,000
5	670
10	449
15	301
20	202
25	135
30	91

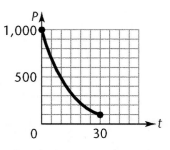

(B) As t tends to infinity, P appears to tend to 0. $(4-1)$

75. $M = \dfrac{2}{3} \log \dfrac{E}{E_0}$ $E_0 = 10^{4.40}$

We use $E = 1.99 \times 10^{14}$

$M = \dfrac{2}{3} \log \dfrac{1.99 \times 10^{14}}{10^{4.40}}$

$M = \dfrac{2}{3} \log(1.99 \times 10^{9.6})$

$M = \dfrac{2}{3}(\log 1.99 + 9.6)$

$M = \dfrac{2}{3}(0.299 + 9.6)$

$M = 6.6$ $(4-4)$

76. We solve $M = \dfrac{2}{3} \log \dfrac{E}{E_0}$ for E, using

$E_0 = 10^{4.40}$, $M = 8.3$

$8.3 = \dfrac{2}{3} \log \dfrac{E}{10^{4.40}}$

$\dfrac{3}{2}(8.3) = \log \dfrac{E}{10^{4.40}}$

$12.45 = \log \dfrac{E}{10^{4.40}}$

$\dfrac{E}{10^{4.40}} = 10^{12.45}$

$E = 10^{4.40} \cdot 10^{12.45}$

$E = 10^{16.85}$ or 7.08×10^{16} joules
$(4-4)$

77. We use the given formula twice, with
$I_2 = 100,000 I_1$

$D_1 = 10 \log \dfrac{I_1}{I_0}$ $D_2 = 10 \log \dfrac{I_2}{I_0}$

$D_2 - D_1 = 10 \log \dfrac{I_2}{I_0} - 10 \log \dfrac{I_1}{I_0} = 10 \log \left(\dfrac{I_2}{I_0} \div \dfrac{I_1}{I_0} \right) = 10 \log \dfrac{I_2}{I_1}$

$= 10 \log \dfrac{100,000 I_1}{I_1} = 10 \log 100,000 = 50$ decibels

The level of the louder sound is 50 decibels more. $(4-4)$

78. $I = I_0 e^{-kd}$
To find k, we solve for k using
$I = \dfrac{1}{2} I_0$ and $d = 73.6$

$\dfrac{1}{2} I_0 = I_0 e^{-k(73.6)}$

$\dfrac{1}{2} = e^{-73.6k}$

$-73.6k = \ln \dfrac{1}{2}$

$k = \dfrac{\ln \frac{1}{2}}{-73.6}$

$k = 0.00942$

We now find the depth at which 1% of the surface light remains. We solve
$I = I_0 e^{-0.00942d}$ for d with $I = 0.01 I_0$

$0.01 I_0 = I_0 e^{-0.00942d}$

$0.01 = e^{-0.00942d}$

$-0.00942d = \ln 0.01$

$d = \dfrac{\ln 0.01}{-0.00942}$

$d = 489$ feet $(4-4)$

79. We solve $N = \dfrac{30}{1 + 29e^{-1.35t}}$ for t with $N = 20$.

$$20 = \dfrac{30}{1 + 29e^{-1.35t}}$$

$$\dfrac{1}{20} = \dfrac{1 + 29e^{-1.35t}}{30}$$

$$1.5 = 1 + 29e^{-1.35t}$$

$$0.5 = 29e^{-1.35t}$$

$$\dfrac{0.5}{29} = e^{-1.35t}$$

$$-1.35t = \ln \dfrac{0.5}{29}$$

$$t = \dfrac{\ln \frac{0.5}{29}}{-1.35}$$

$$t = 3 \text{ years} \qquad\qquad (4\text{-}2)$$

80. (A) The independent variable is years since 1980, so enter 0, 5, 10, 15, 20, and 25 as L_1. The dependent variable is Medicare expenditures, so enter that column as L_2. Then use the exponential regression command on the STAT CALC menu.

 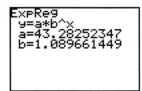

The exponential model is $y = 43.3(1.09)^x$.

To find total expenditures in 2010 and 2020, we plug in 30 and 40 for x:

$$y(30) = 43.3(1.09)^{30} = 575; \qquad y(40) = 43.3(1.09)^{40} = 1,360$$

Expenditures are predicted to be \$575 billion in 2010 and \$1,360 billion in 2020.

(B) Graph $y_1 = 43.3(1.09)^x$ and $y_2 = 900$ and use the INTERSECT command:

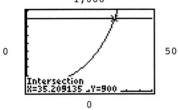

Expenditures are predicted to reach \$900 billion in 2015. $\qquad (4\text{-}2)$

81. (A)

L1	L2	L3	2
75	522	------	
80	659		
85	1152		
90	1373		
95	1690		

L2(6) =

Enter the data.

LnReg
y=a+blnx
a= -21796.9294
b=5153.244133

Compute the regression equation.

Plot1 Plot2 Plot3
\Y1■-21796.92940
2933+5153.244132
94321n(X)
\Y2=
\Y3=
\Y4=
\Y5=

Transfer the data to the equation editor.

3000

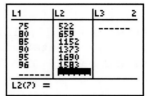

0 | | 115

0

The model predicts a yield of 1,724 million bushels in 1996 and a yield of 2,426 bushels in 2010.

(B) Since this data indicates a lower *y* value for *x* = 96, it would tend to move the curve lower, and should lead to a lower estimate for 2010. This is indeed what happens:

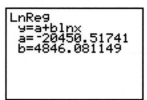

Enter the new data.

Compute the new regression equation.

Transfer the new data to the equation editor.

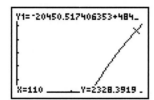

The model now predicts a yield of 2,328 million bushels for 2010.   (*4-4*)

Chapters 3 & 4 Cumulative Review

1. (A) Since the graph has x intercepts -1, 1, and 2, and -1 is at least a double zero (the graph is tangent to the x axis at $x = -1$), the lowest degree equation would be $P(x) = (x + 1)^2 (x - 1)(x - 2)$.

    (B) $P(x) \to \infty$ as $x \to \infty$ and as $x \to -\infty$.   (3-1)

2. The given zeros are the x intercepts of the graph. The negative leading coefficient tells us that the graph should approach $-\infty$ as x approaches ∞.

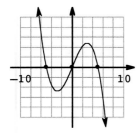

(3-1)

3. (A) The graph of $y = \left(\dfrac{3}{4}\right)^x$ passes through $(0, 1)$ and $\left(1, \dfrac{3}{4}\right)$.

    This corresponds to graph m.

    (B) The graph of $y = \left(\dfrac{4}{3}\right)^x$ passes through $(0, 1)$ and $\left(1, \dfrac{4}{3}\right)$.

    This corresponds to graph g.

    (C) The graph of $y = \left(\dfrac{3}{4}\right)^x + \left(\dfrac{4}{3}\right)^x$ passes through $(0, 2)$.

    This corresponds to graph n.

    (D) The graph of $y = \left(\dfrac{4}{3}\right)^x - \left(\dfrac{3}{4}\right)^x$ passes through $(0, 0)$.

    This corresponds to graph f.   (4-1)

4.
$$
\begin{array}{r}
\quad 3 \quad\; 5 \qu\; -18 \quad -3 \\
\; -9 \quad\; 12 \quad\; 18 \\
\hline
-3\overline)\;3 \quad -4 \quad\; -6 \quad\; 15
\end{array}
$$

5. -2, 3, 5 (3-1)

$3x^3 + 5x^2 - 18x - 3 = (x + 3)(3x^2 - 4x - 6) + 15$ (3-2)

6. We investigate $P(1)$ and $P(2)$ by forming a synthetic division table.

	4	-5	-3	-1
1	4	-1	-4	-5
2	4	3	3	5

Since $P(1)$ and $P(2)$ have opposite signs, there is at least one real zero between 1 and 2.   (3-3)

7. The possible rational zeros are ± 1, ± 2, ± 4, ± 8. We form a synthetic division table.

	1	1	-10	8	
1	1	2	-8	0	1 is a zero

$x^3 + x^2 - 10x + 8 = (x - 1)(x^2 + 2x - 8) = (x - 1)(x - 2)(x + 4)$.
The rational zeros are 1, 2, -4.   (3-4)

8. (A) $x = \log_{10} y$ or $x = \log y$ (B) $x = e^y$ (4-3)

9. (A) $(2e^x)^3 = 2^3(e^x)^3 = 8e^{3x}$ (B) $\dfrac{e^{3x}}{e^{-2x}} = e^{3x-(-2x)} = e^{5x}$ (*4-1*)

10. (A) $\log_3 x = 2$ (B) $\log_3 81 = x$ (C) $\log_x 4 = -2$

$\qquad\qquad x = 3^2 \qquad\qquad \log_3 3^4 = x \qquad\qquad x^{-2} = 4$

$\qquad\qquad x = 9 \qquad\qquad\quad\ x = 4 \qquad\qquad\quad \dfrac{1}{x^2} = 4$

$$1 = 4x^2$$

$$x^2 = \dfrac{1}{4}$$

$$x = \dfrac{1}{2}$$

since bases are restricted positive   (*4-3*)

11. (A) $10^x = 2.35$ (B) $e^x = 87{,}500$

$\qquad\quad x = \log 2.35 \qquad\qquad x = \ln 87{,}500$

$\qquad\quad x = 0.371 \qquad\qquad\ x = 11.4$

(C) $\log x = -1.25$ (D) $\ln x = 2.75$

$\qquad\quad x = 10^{-1.25} \qquad\qquad x = e^{2.75}$

$\qquad\quad x = 0.0562 \qquad\qquad x = 15.6$ (*4-3*)

12. $E = k\dfrac{P}{x^3}$ (*3-6*)   **13.** $F = k\dfrac{q_1 q_2}{r^2}$ (*3-6*)

14. The graph of a nonconstant polynomial cannot have a horizontal asymptote. (*3-1*)

15. The graph of a rational function that has a horizontal asymptote has to approach the horizontal asymptote both as $x \to -\infty$ and $x \to \infty$.   (*3-5*)

16. $f(x) = 3 \ln x - \sqrt{x}$ (*4-3*)

17. The function f multiplies the base e raised to the power of one-half of the domain element by 100 and then subtracts 50.   (*4-1*)

18. $f(x) = \dfrac{2x + 8}{x + 2}$

(A) The domain of f is the set of all real numbers x such that the denominator, $x + 2 \neq 0$; that is $(-\infty, -2) \cup (-2, \infty)$ or $x \neq -2$. f has an x intercept where the numerator, $2x + 8 = 0$; that is, $x = -4$. $f(0) = 4$, so f has a y intercept at $y = 4$.

(B) *Vertical asymptote:* $x = -2$
   Horizontal asymptote: Since the numerator and denominator have the same degree, the line $y = 2$ is a horizontal asymptote.

Complete the sketch:
Graphing calculator: Hand sketch:

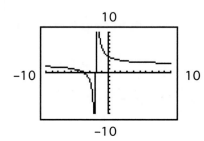   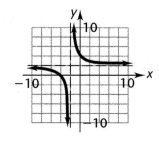

(*3-5*)

19. $P(x)$ can be factored further:

$$P(x) = x(x^2 + 4)(x + 4)$$

$$x^2 + 4 = 0 \quad \Rightarrow \quad x^2 = -4 \quad \Rightarrow \quad x = \pm\sqrt{-4} = \pm 2i$$

The zeros are 0, $2i$, $-2i$, and -4. Only 0 and -4 are x intercepts.   (*3-1*)

20. <u>Algebraic solution</u>:

The real zeros of $(x^3 + 4x)(x + 4)$ are 0 and -4.

Interval	Test Number	Result
$x < -4$	-5	$\left((-5)^3 + 4(-5)\right)(-5 + 4) = 145$; positive
$-4 < x < 0$	-1	$\left((-1)^3 + 4(-1)\right)(-1 + 4) = -15$; negative
$x > 0$	1	$\left((1)^3 + 4(1)\right)(1 + 4) = 25$; positive

$(x^3 + 4x)(x + 4)$ is negative on $(-4, 0)$ and is equal to zero when $x = -4$ or 0, so the solution is $[-4, 0]$.

<u>Graphical solution</u>:

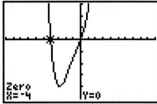 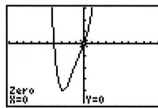 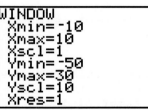

The graph of $y = (x^3 + 4x)(x + 4)$ is below the x axis or intersects the x axis on $[0, 4]$.   (*3-3*)

21.

$$
\begin{array}{r}
\begin{array}{rrrr}
2 & -5 & 3 & 2 \\
 & 1 & -2 & \frac{1}{2} \\
\end{array} \\
\hline
\tfrac{1}{2}\,\big|\,
\begin{array}{rrrr}
2 & -4 & 1 & \frac{5}{2}
\end{array}
\end{array}
$$

$$P\left(\frac{1}{2}\right) = \frac{5}{2}$$ (*3-2*)

22. Use synthetic division to divide out $x + 1$:

$$
\begin{array}{r|rrrr}
 & 3 & -7 & -18 & -8 \\
 & & -3 & 10 & 8 \\
\hline
-1 & 3 & -10 & -8 & 0 \\
\end{array}
$$

The reduced polynomial is $3x^2 - 10x - 8$, which factors as $(3x + 2)(x - 4)$; the remaining zeros are 4 and $-2/3$.   (*3-2*)

23. $x - 1$ will be a factor of $P(x)$ if $P(1) = 0$

$$P(1) = 1^{25} - 1^{20} + 1^{15} + 1^{10} - 1^5 + 1$$

$$= 1 - 1 + 1 + 1 - 1 + 1 = 2 \neq 0,$$

so $x - 1$ is not a factor. $x + 1$ will be a factor of $P(x)$ if $P(-1) = 0$

$$P(-1) = (-1)^{25} - (-1)^{20} + (-1)^{15} + (-1)^{10} - (-1)^5 + 1$$

$$= -1 - 1 - 1 + 1 + 1 + 1 = 0,$$

so $x + 1$ is a factor of $P(x)$.   (*3-2*)

24. (A) We form a synthetic division table:

	1	0	-8	0	
-3	1	-3	1	-3	12 = $P(-3)$
-2	1	-2	-4	8	-13 = $P(-2)$
-1	1	-1	-7	7	-4 = $P(-1)$
0	1	0	-8	0	3 = $P(0)$
1	1	1	-7	-7	-4 = $P(1)$
2	1	2	-4	-8	-13 = $P(2)$
3	1	3	1	3	12 = $P(3)$

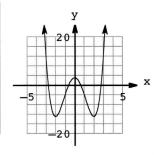

The graph of $P(x)$ has four x intercepts and three turning points; $P(x) \to \infty$ as $x \to \infty$ and as $x \to -\infty$

(B) There are real zeros in the intervals $(-3, -2)$, $(-1, 0)$, $(0, 1)$, and $(2, 3)$ indicated in the table. We search for the real zero in $(2, 3)$. Examining the graph of P in a graphing calculator we obtain the graph at the right.

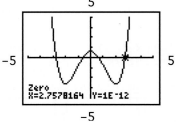

The largest x intercept is 2.76.

$(3-1)$

25. Examining the graph of $P(x)$, we see that there may be a zero of even multiplicity between -1 and 0, a zero of odd multiplicity near 2, and a zero of even multiplicity between 2 and 3.

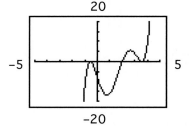

Applying the MAXIMUM command between -1 and 0, the ZERO command near 2, and the MINIMUM command between 2 and 3, we obtain

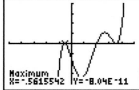 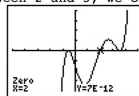 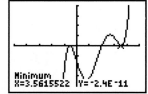

-0.56 is a zero of even multiplicity, 2 is a zero of odd multiplicity, and 3.56 is a zero of even multiplicity. Since $P(x)$ is a fifth-degree polynomial, -0.56 and 3.56 must be double zeros and 2 must be a simple zero. $(3-3)$

26. (A) We form a synthetic division table:

	1	2	-20	0	-30
0	1	2	-20	0	-30
1	1	3	-17	-17	-47
2	1	4	-12	-24	-78
3	1	5	-5	-15	-75
4	1	6	4	16	34
-1	1	1	-21	21	-51
-2	1	0	-20	40	-110
-3	1	-1	-17	51	-183
-4	1	-2	-12	48	-222
-5	1	-3	-5	25	-155
-6	1	-4	4	-24	114

From the table, 4 is an upper bound and -6 is a lower bound.

(B) We search for the largest real zero in (3, 4). We organize our calculations in a table.

Sign Change Interval (a, b)	Midpoint m	Sign of P P(a)	P(m)	P(b)
(3, 4)	3.5	−	−	+
(3.5, 4)	3.75	−	−	+
(3.75, 4)	3.875	−	+	+
(3.75, 3.875)	3.8125	−	+	+
(3.75, 3.8125)	We stop here	−	−	+

Since each endpoint rounds to 3.8, a real zero lies on this last interval and is given by 3.8 to one decimal place accuracy. Four further intervals were required.

(C) Examining the graph of $P(x)$, we obtain the graphs at the right.

The real zeros are −5.68 and 3.80.

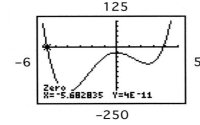 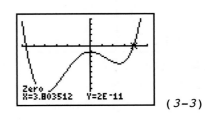

(3-3)

27. The possible rational zeros are ± 1, ± 3, ± 5, ± 15, $\pm\frac{1}{2}$, $\pm\frac{3}{2}$, $\pm\frac{5}{2}$, $\pm\frac{15}{2}$, $\pm\frac{1}{4}$, $\pm\frac{3}{4}$, $\pm\frac{5}{4}$, $\pm\frac{15}{4}$. We form a synthetic division table.

	4	−20	29	−15	
1	4	−16	13	−2	
3	4	−8	5	0	3 is a zero

So $P(x) = (x - 3)(4x^2 - 8x + 5)$

To find the remaining zeros, we solve $4x^2 - 8x + 5 = 0$ by the quadratic formula.

$4x^2 - 8x + 5 = 0$

$$x = \frac{-b \pm \sqrt{b^2 - 4ac}}{2a} \qquad a = 4,\ b = -8,\ c = 5$$

$$x = \frac{-(-8) \pm \sqrt{(-8)^2 - 4(4)(5)}}{2(4)}$$

$$x = \frac{8 \pm \sqrt{64 - 80}}{8}$$

$$x = \frac{8 \pm \sqrt{-16}}{8}$$

$$x = \frac{8 \pm 4i}{8}$$

$$x = 1 \pm \frac{1}{2}i$$

The zeros are 3, $1 \pm \frac{1}{2}i$

(3-4)

28. The possible rational zeros are ±1, ±2, ±3, ±4, ±6, ±12. We form a synthetic division table.

	1	5	1	-15	-12	
1	1	6	7	-8	-20	
2	1	7	15	15	18	a zero between 1 and 2; 2 is an upperbound
-1	1	4	-3	-12	0	-1 is a zero

We now examine $x^3 + 4x^2 - 3x - 12$. This factors by grouping into $(x + 4)(x^2 - 3)$, however, if we don't notice this, we find the remaining possible rational zeros to be -1, -2, -3, -4, -6, -12. We form a synthetic division table.

	1	4	-3	-12	
-1	1	3	-6	-6	not a double zero
-2	1	2	-7	2	a zero between -1 and -2
-3	1	1	-6	6	
-4	1	0	-3	0	-4 is a zero

So $P(x) = (x + 1)(x + 4)(x^2 - 3) = (x + 1)(x + 4)(x - \sqrt{3})(x + \sqrt{3})$. The four zeros are -1, -4, $\pm\sqrt{3}$.   *(3-4)*

29.
$$2^{x^2} = 4^{x+4}$$
$$2^{x^2} = (2^2)^{x+4}$$
$$2^{x^2} = 2^{2(x+4)}$$
$$x^2 = 2(x + 4)$$
$$x^2 = 2x + 8$$
$$x^2 - 2x - 8 = 0$$
$$(x - 4)(x + 2) = 0$$
$$x - 4 = 0 \qquad x + 2 = 0$$
$$x = 4 \qquad\quad x = -2 \qquad (4\text{-}5)$$

30.
$$\frac{13}{2} - 3^x = \frac{1}{2}$$
$$-3^x = \frac{1}{2} - \frac{13}{2} = -\frac{12}{2} = -6$$
$$3^x = 6$$
$$\ln 3^x = \ln 6$$
$$x \ln 3 = \ln 6$$
$$x = \frac{\ln 6}{\ln 3} \qquad (4\text{-}5)$$

31.
$$2x^2 e^{-x} + xe^{-x} = e^{-x}$$
$$2x^2 e^{-x} + xe^{-x} - e^{-x} = 0$$
$$e^{-x}(2x^2 + x - 1) = 0$$
$$e^{-x}(2x - 1)(x + 1) = 0$$
$$e^{-x} = 0 \quad 2x - 1 = 0 \quad x + 1 = 0$$
$$\text{never} \qquad x = \frac{1}{2} \qquad x = -1$$
Solutions: $\frac{1}{2}$, -1   *(4-5)*

32.
$$e^{\ln x} = 2.5$$
$$x = 2.5 \qquad (4\text{-}5)$$

33.
$$\log_x 10^4 = 4$$
$$x^4 = 10^4$$
$$x = 10 \qquad (4\text{-}5)$$

34.
$$\log_9 x = -\frac{3}{2}$$
$$9^{-3/2} = x$$
$$x = \frac{1}{27} \qquad (4\text{-}5)$$

35. $\ln(x + 4) - \ln(x - 4) = 2 \ln 3$

$\ln \dfrac{x + 4}{x - 4} = \ln 3^2$

$\dfrac{x + 4}{x - 4} = 3^2$

$\dfrac{x + 4}{x - 4} = 9 \qquad x \neq 4$

$x + 4 = 9(x - 4)$

$x + 4 = 9x - 36$

$-8x = -40$

$x = 5$

Check: $\ln(5 + 4) - \ln(5 - 4) \overset{?}{=} 2 \ln 3$

$\ln 9 - \ln 1 \overset{?}{=} 2 \ln 3$

$\ln 9 - 0 \overset{\surd}{=} \ln 9$

Solution: $x = 5$ $\hspace{2cm}$ (*4-5*)

36. $\ln(2x^2 + 2) = 2 \ln(2x - 4)$

$\ln(2x^2 + 2) = \ln(2x - 4)^2$

$2x^2 + 2 = (2x - 4)^2$

$2x^2 + 2 = 4x^2 - 16x + 16$

$0 = 2x^2 - 16x + 14$

$0 = 2(x - 7)(x - 1)$

$x = 7, 1$

Check: $x = 7$

$\ln(2 \cdot 7^2 + 2) \overset{?}{=} 2 \ln(2 \cdot 7 - 4)$

$\ln 100 \overset{?}{=} 2 \ln 10$

$\ln 100 \overset{\surd}{=} \ln 100$

$x = 1$

$\ln(2 \cdot 1^2 + 2) \overset{?}{=} 2 \ln(2 \cdot 1 - 4)$

$\ln(4) \neq 2 \ln(-2)$

Solution: $x = 7$ $\hspace{2cm}$ (*4-5*)

37. $\log x + \log(x + 15) = 2$

$\log[x(x + 15)] = 2$

$x(x + 15) = 10^2$

$x^2 + 15x = 100$

$x^2 + 15x - 100 = 0$

$(x - 5)(x + 20) = 0$

$x = 5, -20$

Check: $x = 5$

$\log 5 + \log(5 + 15) \overset{?}{=} 2$

$\log 5 + \log 20 \overset{?}{=} 2$

$\log 100 \overset{\surd}{=} 2$

$x = -20$

$\log(-20) + \log(-20 + 15) \neq 2$

Solution: $x = 5$ $\hspace{2cm}$ (*4-5*)

38. $\log(\ln x) = -1$

$\ln x = 10^{-1}$

$\ln x = 1/10$

$x = e^{1/10}$ $\hspace{2cm}$ (*4-5*)

39. $4(\ln x)^2 = \ln x^2$

$4(\ln x)^2 = 2 \ln x$

$4(\ln x)^2 - 2 \ln x = 0$

$2 \ln x(2 \ln x - 1) = 0$

$2 \ln x = 0 \qquad 2 \ln x - 1 = 0$

$\ln x = 0 \qquad\qquad \ln x = \dfrac{1}{2}$

$x = 1 \qquad\qquad\quad x = e^{1/2}$

Check: $x = 1$

$4(\ln 1)^2 \overset{?}{=} \ln 1^2$

$0 \overset{\surd}{=} 0$

$x = e^{1/2}$

$4(\ln e^{1/2})^2 \overset{?}{=} \ln(e^{1/2})^2$

$4(1/2)^2 \overset{?}{=} \ln(e^{2(1/2)})$

$4(1/4) \overset{?}{=} \ln e$

$1 \overset{\surd}{=} 1$

Solution: $1, e^{1/2}$ $\hspace{2cm}$ (*4-5*)

40. $x = \log_3 41$

We use the change of base formula

$x = \dfrac{\log 41}{\log 3}$

$x = 3.38$ $(4\text{-}5)$

41. $\ln x = 1.45$

$x = e^{1.45}$

$x = 4.26$ $(4\text{-}5)$

42. $4(2^x) = 20$

$2^x = 5$

$x \log 2 = \log 5$

$x = \dfrac{\log 5}{\log 2}$

$x = 2.32$ $(4\text{-}5)$

43. $10e^{-0.5x} = 1.6$

$e^{-0.5x} = 0.16$

$-0.5x = \ln 0.16$

$x = \dfrac{\ln 0.16}{-0.5}$

$x = 3.67$ $(4\text{-}5)$

44.

$\dfrac{e^x - e^{-x}}{e^x + e^{-x}} = \dfrac{1}{2}$

$\dfrac{e^x - \frac{1}{e^x}}{e^x + \frac{1}{e^x}} = \dfrac{1}{2}$

$\dfrac{e^x\left(e^x - \frac{1}{e^x}\right)}{e^x\left(e^x + \frac{1}{e^x}\right)} = \dfrac{1}{2}$

$\dfrac{(e^x)^2 - 1}{(e^x)^2 + 1} = \dfrac{1}{2}$

$\dfrac{e^{2x} - 1}{e^{2x} + 1} = \dfrac{1}{2}$

$2(e^{2x} + 1)\,\dfrac{e^{2x} - 1}{e^{2x} + 1} = 2(e^{2x} + 1)\dfrac{1}{2}$

$2(e^{2x} - 1) = e^{2x} + 1$

$2e^{2x} - 2 = e^{2x} + 1$

$e^{2x} = 3$

$2x = \ln 3$

$x = \dfrac{1}{2} \ln 3$

$x = 0.549$ $(4\text{-}5)$

45. The graph is shown at the right.

$f(x) = 3^{1-x}$

Domain: $(-\infty, \infty)$

Range: $(0, \infty)$

x intercept: none

y intercept: $f(0) = 3^{1-0} = 3$

Horizontal asymptote:

as $x \to \infty$, $f(x) \to 0$

$y = 0$

           $(4\text{-}1)$

46. The graph is shown below, along with commands for finding the intercepts.

$g(x) = \ln(2 - x)$

Domain: $2 - x > 0$ $(-\infty, 2)$

          $x < 2$

Range: $(-\infty, \infty)$

y intercept: 0.693 (or $\ln 2$); x intercept: 1

Vertical asymptote: as $x \to 2^-$, $g(x) \to -\infty$; $x = 2$ is a vertical asymptote

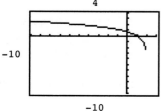 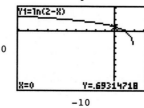 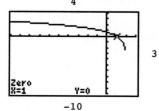

                                                $(4\text{-}3)$

47. The graph is shown to the right.

$A(t) = 100e^{-0.3t}$

Domain: $(-\infty, \infty)$

Range: $(0, \infty)$

x intercept: none

y intercept: $A(0) = 100e^{-0.3(0)} = 100$

Horizontal asymptote:

as $t \to \infty$, $A(t) \to 0$

$y = 0$

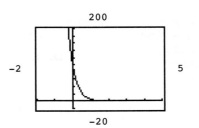

$(4-1)$

48. The graph is shown below, along with commands for finding the intercepts.

$h(x) = -2e^{-x} + 3$

Domain: $(-\infty, \infty)$

Range: Since $e^{-x} > 0$

$\qquad -2e^{-x} < 0$

$\qquad -2e^{-x} + 3 < 3$

$\qquad (-\infty, 3)$

y intercept: 1; x intercept: -0.41

y intercept: $h(0) = -2e^{-0} + 3 = 1$

Horizontal asymptote:

as $x \to \infty$, $h(x) \to 3$; $y = 3$ is a horizontal asymptote

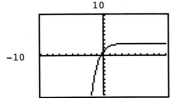 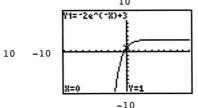 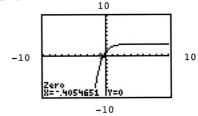

$(4-1)$

49. The graph is shown below, along with a window that shows the y intercept.

$N(t) = \dfrac{6}{2 + e^{-0.1t}}$

Domain: $(-\infty, \infty)$

Range: $(0, 3)$

x intercept: none

y intercept: 2

Horizontal asymptotes: As $t \to \infty$, $N(t) \to 3$, so $y = 3$ is a horizontal asymptote.

$\qquad\qquad\qquad\qquad$ As $t \to -\infty$, $N(t) \to 0$, so $y = 0$ is a horizontal asymptote.

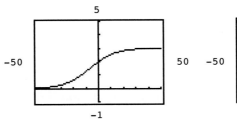 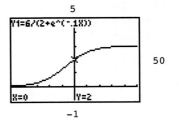

$(4-1)$

50. Reflecting a graph in the line $y = x$ results in the graph of the inverse function. In this case, the inverse function of $y = \ln x$ is $y = e^x$. If the graph of $y = \ln x$ is reflected in the x axis, y is replaced by $-y$ and the graph becomes the graph of $-y = \ln x$ or $y = -\ln x$.

If the graph of $y = \ln x$ is reflected in the y axis, x is replaced by $-x$ and the graph becomes the graph of $y = \ln(-x)$.

$(4-3)$

51. (A) For $x > 0$, $y = e^{-x}$ decreases from 1 to 0 while $\ln x$ increases from $-\infty$ to ∞. Consequently, the graphs can intersect at exactly one point.

(B) Graphing $y1 = e^{-x}$ and $y2 = \ln x$, we obtain

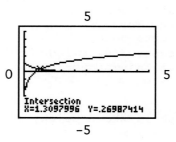

The solution is 1.31.   (4-

52. (A) Let $u = x^2$. Then
$$P(x) = x^4 + 9x^2 + 18$$
$$P(x) = u^2 + 9u + 18$$
$$P(x) = (u + 6)(u + 3)$$
$$P(x) = (x^2 + 6)(x^2 + 3)$$

(B) $x^2 + 6$ has $i\sqrt{6}$ and $-i\sqrt{6}$ as zeros, so it factors as
$$x^2 + 6 = (x + i\sqrt{6})(x - i\sqrt{6}).$$ $x^2 + 3$ has $i\sqrt{3}$ and $-i\sqrt{3}$ as zeros, so it factors as
$$x^2 + 3 = (x + i\sqrt{3})(x - i\sqrt{3}).$$
$$P(x) = (x + i\sqrt{6})(x - i\sqrt{6})(x + i\sqrt{3})(x - i\sqrt{3})$$ (3-4)

53. (A) Let $u = x^2$. Then
$$P(x) = x^4 - 23x^2 - 50$$
$$P(x) = u^2 - 23u - 50$$
$$P(x) = (u - 25)(u + 2)$$
$$P(x) = (x^2 - 25)(x^2 + 2)$$
$$P(x) = (x + 5)(x - 5)(x^2 + 2)$$

(B) $x^2 + 2$ has zeros $i\sqrt{2}$ and $-i\sqrt{2}$ so it factors as
$$x^2 + 2 = (x + i\sqrt{2})(x - i\sqrt{2}).$$
$$P(x) = (x + 5)(x - 5)(x + i\sqrt{2})(x - i\sqrt{2})$$ (3-4)

54. Write $G = kx^2$. Substitute $G = 10$ and $x = 5$ and solve for k.
$$10 = k(5)^2$$
$$10 = 25k$$
$$k = 0.4$$
The equation of variation is $G = 0.4x^2$.
When $x = 7$
$$G = 0.4(7)^2 = 19.6$$ (3-6)

55. Write $H = \dfrac{k}{r^3}$. Substitute $H = 162$ and $r = 2$ and solve for k.
$$162 = \frac{k}{2^3}$$
$$k = 8(162)$$
$$k = 1,296$$

The equation of variation is $H = \dfrac{1,296}{r^3}$.
When $r = 3$
$$H = \frac{1,296}{3^3} = 48$$ (3-6)

56. $f(x) = \dfrac{x^2 + 4x + 8}{x + 2}$

Intercepts. There are no real zeros of the numerator. No x intercept

$f(0) = 4$ y intercept

Vertical asymptotes. The only zero of the denominator is $x = -2$.

Horizontal asymptote. Since the degree of the numerator is greater than the degree of the denominator, there is no horizontal asymptote.

Oblique asymptote.

$$
\begin{array}{r}
x + 2 \\
x + 2 \overline{)\ x^2 + 4x + 8} \\
\underline{x^2 + 2x } \\
2x + 8 \\
\underline{2x + 4} \\
4
\end{array}
$$

So, $f(x) = x + 2 + \dfrac{4}{x + 2}$

and the line $y = x + 2$ is an oblique asymptote.

Complete the sketch.
Graphing calculator:

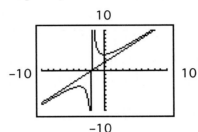

Hand sketch:

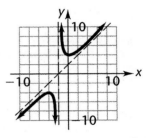

$(3\text{-}5)$

57. <u>Algebraic solution:</u>

zeros of numerator: $x^3 - x = 0$

$$x(x^2 - 1) = 0$$
$$x(x + 1)(x - 1) = 0$$
$$x = 0,\ -1,\ 1$$

zeros of the denominator: $x^3 - 8 = 0$

$$x^3 = 8$$
$$x = 2$$

These four zeros divide the real line into the five intervals shown below.

Interval	Test Number	Result	Sign of f
$(-\infty, -1)$	-2	0.375	$+$
$(-1, 0)$	$-1/2$	-0.05	$-$
$(0, 1)$	$1/2$	0.05	$+$
$(1, 2)$	$3/2$	-0.4	$-$
$(2, \infty)$	3	1.3	$+$

The expression is positive on $(-\infty, -1) \cup (0, 1) \cup (2, \infty)$. It's zero at -1, 0, and 1, so we include them but exclude 2 as the expression is undefined.

Solution: $(-\infty, -1] \cup [0, 1] \cup (2, \infty)$

Graphical solution:

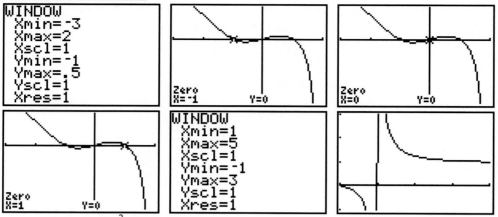

The graph of $\dfrac{x^3 - x}{x^3 - 8}$ is above, or intersects, the x axis on

$(-\infty, -1] \cup [0, 1] \cup (2, \infty)$. *(3-5)*

58. A preliminary graph of $y = P(x)$ is shown at the right.

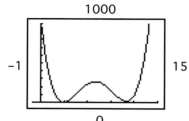

Examining the behavior near the maxima and minima, we obtain

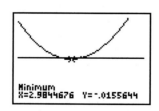
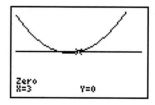

There are zeros at $x = 2.97$ and $x = 3$ and a local minimum at $P(2.98) \approx -0.02$.

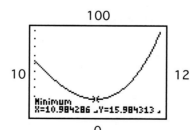

There is also a local maximum at $P(7.03) \approx 264.03$ and a local minimum at $P(10.98) \approx 15.98$. *(3-3)*

59. $[x - (-1)]^2(x - 0)^3[x - (3 + 5i)][x - (3 - 5i)]$

$$= (x + 1)^2 x^3 (x - 3 - 5i)(x - 3 + 5i)$$

degree 7

(3-4)

60. Yes, for example:

$P(x) = (x + i)(x - i)(x + \sqrt{2})(x - \sqrt{2}) = x^4 - x^2 - 2$ has irrational zeros $\sqrt{2}$ and $-\sqrt{2}$.   (3-4)

61. (A) We form a synthetic division table:

	1	9	-500	0	20,000
0	1	9	-500	0	20,000
10	1	19	-310	-3,100	-11,000
20	1	29	80	1,600	52,000
-10	1	-1	-490	4,900	-29,000
-20	1	-11	-280	5,600	-92,000
-30	1	-21	130	-3,900	137,000

From the table, 20 is an upper bound and -30 is a lower bound.

(B) Examining the graph of $y = P(x)$, we obtain

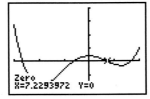
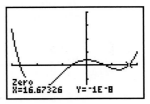

The zeros of $P(x)$ are -26.68, -6.22, 7.23, and 16.67.   (3-3)

62. The possible rational zeros of $P(x)$ are ±1, ±2, ±3, ±4, ±6, ±12. Examining the graph of $y = P(x)$, it appears that there may be a zero of even multiplicity near -1 and a zero of odd multiplicity near 2. We test the likely candidates and find

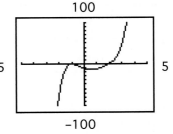

```
     1   -4    3   10   -10   -12
              2   -4   -2    16    12
    2| 1   -2   -1    8     6     0
```

2 is a zero. $P(x) = (x - 2)(x^4 - 2x^3 - x^2 + 8x + 6)$. Testing -1 in $x^4 - 2x^3 - x^2 + 8x + 6$ we obtain

```
     1   -2   -1    8    6
             -1    3   -2   -6
   -1| 1   -3    2    6    0
```

-1 is a zero. $P(x) = (x - 2)(x + 1)(x^3 - 3x^2 + 2x + 6)$. Testing -1 again in $x^3 - 3x^2 + 2x + 6$ we obtain

```
     1   -3    2    6
             -1    4   -6
   -1| 1   -4    6    0
```

-1 is a double zero of $P(x)$. $P(x) = (x - 2)(x + 1)^2(x^2 - 4x + 6)$.

We complete the solution by solving $x^2 - 4x + 6 = 0$ using the quadratic formula with $a = 1$, $b = -4$, and $c = 6$.

$$x = \frac{-(-4) \pm \sqrt{(-4)^2 - 4(1)(6)}}{2(1)} = \frac{4 \pm \sqrt{-8}}{2} = \frac{4 \pm 2i\sqrt{2}\,i}{2} = 2 \pm i\sqrt{2}$$

The zeros of $P(x)$ are 2, -1 (double), and $2 \pm i\sqrt{2}$.

$$P(x) = (x - 2)(x + 1)^2[x - (2 + i\sqrt{2})][x - (2 - i\sqrt{2})]$$
$$= (x - 2)(x + 1)^2(x - 2 - i\sqrt{2})(x - 2 + i\sqrt{2})$$

(3-4)

63. The possible rational zeros of $P(x)$ are ±1, ±2, ±4. Examining the graph of $y = P(x)$, we obtain the graph at the right.

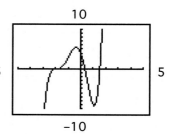

It appears that there may be zeros of unclear multiplicity near -2, as well as zeros between 1 and 2. The only possible rational zero allowed by the graph from the above list is -2. Testing it, we obtain

```
     1    4    1   -11   -8    4
         -2   -4    6    10   -4
   -2│1    2   -3   -5    2    0
```

-2 is a zero. Testing whether it is a double zero, we obtain

```
     1    2   -3   -5    2
         -2    0    6   -2
   -2│1    0   -3    1    0
```

So $P(x) = (x + 2)^2(x^3 - 3x + 1)$. Examining the graph of $x^3 - 3x + 1$, we obtain

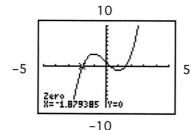 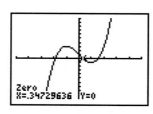 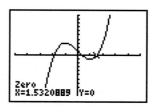

The zeros of $P(x)$ are -2 (double), -1.88, 0.35, and 1.53.

(3-4)

64. Since the real zeros are 5 and 8, we should have factors of $x - 5$ and $x - 8$ in the numerator. A factor of $x - 1$ in the denominator provides vertical asymptote 1, but we need the degree of the numerator and denominator to be equal in order to have a non-zero horizontal asymptote. So we'll make the denominator $(x - 1)^2$. Finally, a factor of 3 in the numerator makes $y = 3$ the horizontal asymptote: $f(x) = \dfrac{3(x - 8)(x - 5)}{(x - 1)^2}$.

(3-5)

65.
$$A = P\frac{(1 + i)^n - 1}{i}$$
$$Ai = P[(1 + i)^n - 1]$$
$$\frac{Ai}{P} = (1 + i)^n - 1$$
$$1 + \frac{Ai}{P} = (1 + i)^n$$
$$\ln\left(1 + \frac{Ai}{P}\right) = n\,\ln(1 + i)$$
$$n = \frac{\ln\left(1 + \frac{Ai}{P}\right)}{\ln(1 + i)} \qquad (4\text{-}5)$$

66.
$$\ln y = 5x + \ln A$$
$$\ln y - \ln A = 5x$$
$$\ln\left(\frac{y}{A}\right) = 5x$$
$$\frac{y}{A} = e^{5x}$$
$$y = Ae^{5x} \qquad (4\text{-}5)$$

67.
$$y = \frac{e^x - 2e^{-x}}{2}$$
$$2y = e^x - 2e^{-x}$$
$$2y = e^x - \frac{2}{e^x}$$
$$2ye^x = e^xe^x - e^x\left(\frac{2}{e^x}\right)$$
$$2ye^x = (e^x)^2 - 2$$
$$0 = (e^x)^2 - 2ye^x - 2$$

This equation is quadratic in e^x

$$e^x = \frac{-b \pm \sqrt{b^2 - 4ac}}{2a} \qquad a = 1,\ b = -2y,\ c = -2$$

$$e^x = \frac{-(-2y) \pm \sqrt{(-2y)^2 - 4(1)(-2)}}{2(1)}$$

$$e^x = \frac{2y \pm \sqrt{4y^2 + 8}}{2}$$

$$e^x = y \pm \sqrt{y^2 + 2}$$

Note: Since $0 < 2$, $y^2 < y^2 + 2$, $\sqrt{y^2} < \sqrt{y^2 + 2}$ and $y < \sqrt{y^2 + 2}$ for all real y.

So $y - \sqrt{y^2 + 2}$ is always negative, which means it can't provide a solution.

Also, $y + \sqrt{y^2 + 2}$ is always positive.

$$x = \ln(y + \sqrt{y^2 + 2}) \qquad (4\text{-}5)$$

68. Algebraic solution:

$$\frac{4x}{x^2 - 1} - 3 < 0$$

$$\frac{4x}{x^2 - 1} - \frac{3(x^2 - 1)}{x^2 - 1} < 0$$

$$\frac{4x - 3x^2 + 3}{x^2 - 1} < 0$$

$$\frac{-3x^2 + 4x + 3}{x^2 - 1} < 0$$

zeros of the numerator: $a = -3$, $b = 4$, $c = 3$

$$x = \frac{-4 \pm \sqrt{16 - 4(-3)(3)}}{2(-3)}$$

$$= \frac{-4 \pm \sqrt{52}}{-6}$$
$$= -0.535, \ 1.869$$

zeros of the denominator:
$$x^2 - 1 = 0$$
$$x^2 = 1$$
$$x = \pm 1$$

These four zeros divide the real line into the five intervals shown below.

Interval	Test Number	Result	Sign of f
$(-\infty, -1)$	-2	-5.9	$-$
$(-1, -0.535)$	-0.8	5.9	$+$
$(-0.535, 1)$	0.8	-11.9	$-$
$(1, 1.869)$	$3/2$	1.8	$+$
$(1.869, \infty)$	3	-1.5	$-$

The expression is negative on $(-\infty, -1) \cup (-0.535, 1) \cup (1.869, \infty)$.

Graphical solution:
Graph $y_1 = \frac{4x}{x^2 - 1}$, $y_2 = 3$ and find where the graph of y_1 is below the graph of y_2.

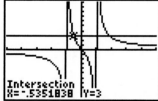 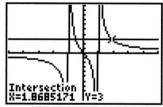 

This occurs on $(-\infty, -1) \cup (-0.535, 1) \cup (1.869, \infty)$. *(3-5)*

69. (A) Graph $y_1 = -4.8x^3 + 47x^2 - 35x - 40$ on the interval $[0,12]$ and use the MAXIMUM command.

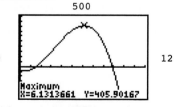

The snack bar should be open for 6.1 hours to maximize profit.

(B) Add $y_2 = 300$ to the graph and use the INTERSECT command.

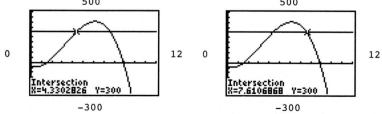

It should be open 4.3 or 7.6 hours to make a profit of $300.

(C) Use the ZERO command.

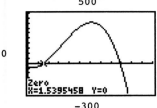
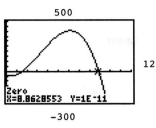

They will make a profit if open between 1.5 and 8.9 hours.   *(3-1, 3-3)*

70. (A) Evaluate T for $w = 1$: $T(1) = 540 - \dfrac{450(1)}{1 + 2} = 390$ seconds

(B) The degrees of the numerator and denominator of the fraction are the same, so that term of the function approaches 450/1 (ratio of leading coefficients), or 450. The entire function approaches $540 - 450 = 90$. This tells you that in the long run, her best time will approach 90 seconds.

(C) The vertical asymptote occurs at $w = -2$. Since w represents weeks after learning how to solve the puzzle, negative values are irrelevant.   *(3-5)*

71. We are given y = length of container and x = width of one end; girth is the sum of the four ends, so $4x$ = girth.

Length + girth = $y + 4x = 10$
So $y = 10 - 4x$
Since Volume = $8 = x^2 y$, we have
$8 = x^2(10 - 4x)$
$8 = 10x^2 - 4x^3$
$4x^3 - 10x^2 + 8 = 0$
$2x^3 - 5x^2 + 4 = 0$
The possible rational solutions of this

equation are ± 1, ± 2, ± 4, $\pm\dfrac{1}{2}$.

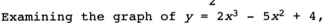

Examining the graph of $y = 2x^3 - 5x^2 + 4$,
it appears that there may be an integer zero at 2, a zero between -1 and 0, and a zero between 1 and 2. Testing 2, we obtain

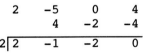

2 is a zero.
So, the equation can be factored
$(x - 2)(2x^2 - x - 2) = 0$
To find the remaining zeros, we solve $2x^2 - x - 2 = 0$ by the quadratic formula.
$2x^2 - x - 2 = 0$

$$x = \frac{-b \pm \sqrt{b^2 - 4ac}}{2a} \qquad a = 2,\ b = -1,\ c = -2$$

$$x = \frac{-(-1) \pm \sqrt{(-1)^2 - 4(2)(-2)}}{2(2)}$$

$$x = \frac{1 + \sqrt{17}}{4} \text{ (discarding the negative solution)}$$

The positive zeros are $x = 2$, 1.28.
The dimensions of the package are $x = 2$ feet and $y = 2$ feet, or $x = 1.28$ feet and $y = 4.88$ feet.   *(3-4)*

72.

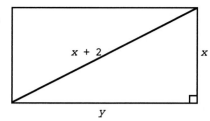

Labeling the rectangle as in the diagram, we have

$$\text{Area} = xy = 6$$

$$\text{So } y = \frac{6}{x}$$

Applying the Pythagorean theorem, we have

$$(x + 2)^2 = x^2 + y^2$$

$$(x + 2)^2 = x^2 + \left(\frac{6}{x}\right)^2$$

$$x^2 + 4x + 4 = x^2 + \frac{36}{x^2}$$

$$4x + 4 = \frac{36}{x^2}$$

$$4x^3 + 4x^2 = 36$$

$$4x^3 + 4x^2 - 36 = 0$$

$$x^3 + x^2 - 9 = 0$$

There are no rational zeros of $P(x) = x^3 + x^2 - 9$. Examining the graph of $y = P(x)$, we see that the only real zero is $x = 1.79$.
Then $y = \dfrac{6}{1.79} = 3.35$. The dimensions of the rectangle are 1.79 feet by 3.35 feet.

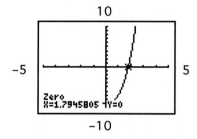

(*3-4*)

73. $t^2 = kd^3$ (*3-6*)

74. Write $v = \dfrac{k}{\sqrt{w}}$. Substitute $v = 0.3$ and $w = w_1$ and solve for k.

$$0.3 = \frac{k}{\sqrt{w_1}}$$

$$k = 0.3\sqrt{w_1} \text{ where } w_1 = \text{weight of oxygen molecule.}$$

The equation of variation is

$$v = \frac{0.3\sqrt{w_1}}{\sqrt{w_2}}$$

If $w_2 = \dfrac{w_1}{16}$ (weight of hydrogen molecule) then

$$v = \frac{0.3\sqrt{w_1}}{\sqrt{w_1/16}} = \frac{0.3\sqrt{w_1}}{\sqrt{w_1}/4} = 4(0.3) = 1.2 \text{ mile/second}$$

(*3-6*)

75. We use the Doubling Time Growth Model:

$$P = P_0 2^{t/d}$$

Substituting $P_0 = 60$ million and $d = 23$, we have

$$P = 60(2^{t/23}) \text{ million}$$

(A) For $t = 5$, $P = 60(2^{5/23})$ million
$$= 69.8 \text{ million}$$

(B) For $t = 30$, $P = 60(2^{30/23})$ million
$$= 148 \text{ million}$$

(*3-6*)

76. We solve $P = P_0(1.07)^t$ for t, using $P = 2P_0$

$$2P_0 = P_0(1.07)^t$$
$$2 = (1.07)^t$$
$$\ln 2 = t \ln 1.07$$
$$\frac{\ln 2}{\ln 1.07} = t$$
$$t = 10.2 \text{ years} \qquad (4\text{-}1)$$

77. We solve $P = P_0 e^{0.07t}$ for t, using $P = 2P_0$

$$2P_0 = P_0 e^{0.07t}$$
$$2 = e^{0.07t}$$
$$\ln 2 = 0.07t$$
$$t = \frac{\ln 2}{0.07}$$
$$t = 9.90 \text{ years} \qquad (4\text{-}1)$$

78. First, we solve $M = \frac{2}{3} \log\left(\dfrac{E}{E_0}\right)$ for E.

$$M = \frac{2}{3} \log\left(\frac{E}{E_0}\right)$$
$$\frac{3M}{2} = \log \frac{E}{E_0}$$
$$\frac{E}{E_0} = 10^{3m/2}$$
$$E = E_0(10^{3m/2})$$

We now compare E_1 for $M = 8.3$ with E_2 for $M = 7.1$.

$$E_1 = E_0(10^{3 \cdot 8.3/2}) \qquad\qquad E_2 = E_0(10^{3 \cdot 7.1/2})$$
$$E_1 = E_0(10^{12.45}) \qquad\qquad E_2 = E_0(10^{10.65})$$

$$\frac{E_1}{E_2} = \frac{E_0(10^{12.45})}{E_0(10^{10.65})} = 10^{12.45-10.65} = 10^{1.8}$$

$$E_1 = 10^{1.8} E_2 \text{ or } 63.1 E_2.$$

The 1906 earthquake was 63.1 times as powerful.   ($4\text{-}4$)

79. We solve $D = 10 \log \dfrac{I}{I_0}$ for I, with $D = 88$, $I_0 = 10^{-12}$

$$88 = 10 \log \frac{I}{10^{-12}}$$
$$8.8 = \log \frac{I}{10^{-12}}$$
$$10^{8.8} = \frac{I}{10^{-12}}$$
$$I = 10^{8.8} \cdot 10^{-12}$$
$$I = 10^{-3.2}$$
$$I = 6.31 \times 10^{-4} \text{ w/m}^2 \qquad (4\text{-}4)$$

80. Enter the data (note that the whole list is not visible in this window):

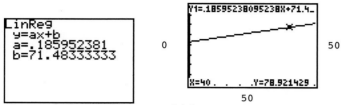

(A) Compute the linear regression equation, graph, and compute the value of the life expectancy that corresponds to $x = 40$ (year 2010).

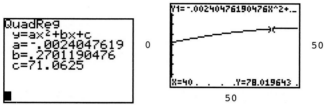

The model predicts a life expectancy of 78.9 years in 2010.

(B) Compute the quadratic regression equation, graph, and compute the value of the life expectancy that corresponds to $x = 40$.

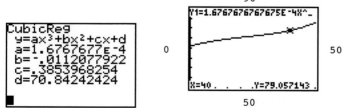

The model predicts a life expectancy of 78.0 years in 2010.

(C) Compute the cubic regression equation, graph, and compute the value of the life expectancy that corresponds to $x = 40$.

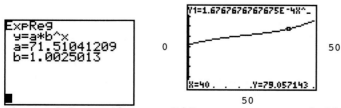

The model predicts a life expectancy of 79.1 years in 2010.

(D) Compute the exponential regression equation, graph, and compute the value of the life expectancy that corresponds to $x = 40$

The model predicts a life expectancy of 79.1 years in 2010. (*3-1, 4-2*)

81. The quadratic model (78.0 years) is closest to the Census Bureau projection. (*3-1, 4-2*)

CHAPTER 5   Trigonometric Functions

Section 5-1

1. Answers will vary.   **3.** Answers will vary.   **5.** Answers will vary.

7. Since 1 rotation corresponds to 360°, $\frac{1}{9}$ rotation corresponds to $\frac{1}{9}(360°) = 40°$

9. Since 1 rotation corresponds to 360°, $\frac{3}{4}$ rotation corresponds to $\frac{3}{4}(360°) = 270°$

11. Since 1 rotation corresponds to 360°, $\frac{9}{8}$ rotation corresponds to $\frac{9}{8}(360°) = 405°$

13. $\theta = \dfrac{s}{r} = \dfrac{24 \text{ centimeters}}{4 \text{ centimeters}} = 6$ radians   **15.** $\theta = \dfrac{s}{r} = \dfrac{30 \text{ feet}}{12 \text{ feet}} = 2.5$ radians

17. Careful about units! We need to either write both in terms of meters, or both in terms of centimeters. $\theta = \dfrac{s}{r} = \dfrac{5 \text{ cm}}{100 \text{ cm}} = 0.05$ radians

19. Since 1 rotation corresponds to 2π radians, $\frac{1}{8}$ rotation corresponds to $\frac{1}{8}(2\pi) = \frac{\pi}{4}$ radians

21. Since 1 rotation corresponds to 2π radians, $\frac{3}{4}$ rotation corresponds to $\frac{3}{4}(2\pi) = \frac{3\pi}{2}$ radians

23. Since 1 rotation corresponds to 2π radians, $\frac{13}{12}$ rotation corresponds to $\frac{13}{12}(2\pi) = \frac{13\pi}{6}$ radians

25. To convert from degrees to radians, we multiply by one in the form $\dfrac{\pi \text{ rad}}{180°}$:

$\dfrac{\pi \text{ rad}}{180°} 30° = \dfrac{\pi}{6}$ rad

$\dfrac{\pi \text{ rad}}{180°} 60° = \dfrac{\pi}{3}$ rad

$\dfrac{\pi \text{ rad}}{180°} 90° = \dfrac{\pi}{2}$ rad

$\dfrac{\pi \text{ rad}}{180°} 120° = \dfrac{2\pi}{3}$ rad

$\dfrac{\pi \text{ rad}}{180°} 150° = \dfrac{5\pi}{6}$ rad

$\dfrac{\pi \text{ rad}}{180°} 180° = \pi$ rad

27. To convert from degrees to radians, we multiply by one in the form $\dfrac{\pi \text{ rad}}{180°}$:

$\dfrac{\pi \text{ rad}}{180°}(-45°) = -\dfrac{\pi}{4}$ rad

$\dfrac{\pi \text{ rad}}{180°}(-90°) = -\dfrac{\pi}{2}$ rad

$\dfrac{\pi \text{ rad}}{180°}(-135°) = -\dfrac{3\pi}{4}$ rad

$\dfrac{\pi \text{ rad}}{180°}(-180°) = -\pi$ rad

29. To convert from degrees to radians, we multiply by one in the form $\dfrac{\pi \ \text{rad}}{180°}$:

$$\frac{\pi \ \text{rad}}{180°} 72° = \frac{2\pi}{5} \ \text{rad}$$

$$\frac{\pi \ \text{rad}}{180°} 144° = \frac{4\pi}{5} \ \text{rad}$$

$$\frac{\pi \ \text{rad}}{180°} 216° = \frac{6\pi}{5} \ \text{rad}$$

$$\frac{\pi \ \text{rad}}{180°} 288° = \frac{8\pi}{5} \ \text{rad}$$

$$\frac{\pi \ \text{rad}}{180°} 360° = 2\pi \ \text{rad}$$

31. To convert from radians to degrees, we multiply by one in the form $\dfrac{180°}{\pi \ \text{rad}}$:

$$\frac{180°}{\pi} \frac{\pi}{3} = 60°$$

$$\frac{180°}{\pi} \frac{2\pi}{3} = 120°$$

$$\frac{180°}{\pi} \pi = 180°$$

$$\frac{180°}{\pi} \frac{4\pi}{3} = 240°$$

$$\frac{180°}{\pi} \frac{5\pi}{3} = 300°$$

$$\frac{180°}{\pi} 2\pi = 360°$$

33. To convert from radians to degrees, we multiply by one in the form $\dfrac{180°}{\pi \ \text{rad}}$:

$$\frac{180°}{\pi} \left(-\frac{\pi}{2}\right) = -90°$$

$$\frac{180°}{\pi} (-\pi) = -180°$$

$$\frac{180°}{\pi} \left(-\frac{3\pi}{2}\right) = -270°$$

$$\frac{180°}{\pi} (-2\pi) = -360°$$

35. To convert from radians to degrees, we multiply by one in the form $\dfrac{180°}{\pi \ \text{rad}}$:

$$\frac{180°}{\pi \ \text{rad}} \frac{\pi}{5} = 36°$$

$$\frac{180°}{\pi \ \text{rad}} \frac{2\pi}{5} = 72°$$

$$\frac{180°}{\pi \ \text{rad}} \frac{3\pi}{5} = 108°$$

$$\frac{180°}{\pi \ \text{rad}} \frac{4\pi}{5} = 144°$$

$$\frac{180°}{\pi \ \text{rad}} \pi = 180°$$

37. True. "Standard position" means that both angles have the positive x-axis as their initial sides. If they have the same initial side and the same measure, they must have the same terminal side as well.

39. True. Angles that are complementary have measures that add up to 90°, so if both are positive and add to 90° the measure of each has to be between 0° and 90°. That's the definition of acute.

41. False. The terminal side of the angle −315° (or $-\frac{7\pi}{4}$) is in quadrant I.

43. $38°41' = \left(38 + \dfrac{41}{60}\right)° = 38.683°$

45. $5°51'33'' = \left(5 + \dfrac{51}{60} + \dfrac{33}{3600}\right)°$
$= 5.859°$

47. $354°8'29'' = \left(5 + \dfrac{8}{60} + \dfrac{29}{3600}\right)°$
$= 354.141°$

49. $27.6° = 27°(0.6 \cdot 60)'$
$= 27°36'$

51. $3.042° = 3°(0.042\cdot 60)'$
$= 3°2.52'$
$= 3°2'(0.52\cdot 60)''$
$= 3°2'31''$

53. $403.223° = 403°(0.223\cdot 60)'$
$= 403°13.38'$
$= 403°13'(0.38\cdot 60)''$
$= 403°13'23''$

55. To convert from degrees to radians, we multiply by one in the form $\dfrac{\pi \text{ rad}}{180°}$:

we have $\dfrac{\pi \text{ rad}}{180°} 64° = 1.117 \text{ rad}$

57. To convert from degrees to radians, we multiply by one in the form $\dfrac{\pi \text{ rad}}{180°}$:

we have $\dfrac{\pi \text{ rad}}{180°} 108.413° = 1.892 \text{ rad}$

59. First we convert 13°25'14" to decimal degrees:

$$13°25'14" = \left(13 + \frac{25}{60} + \frac{14}{3600}\right)°$$
$$= 13.421°$$

To convert from degrees to radians, we multiply by one in the form $\dfrac{\pi \text{ rad}}{180°}$:

$$\frac{\pi \text{ rad}}{180°}(13.421°) = 0.234 \text{ rad}$$

61. To convert from radians to degrees, we multiply by one in the form $\dfrac{180°}{\pi \text{ rad}}$:

we have $\dfrac{180°}{\pi}(0.93) = 53.29°$

63. To convert from radians to degrees, we multiply by one in the form $\dfrac{180°}{\pi \text{ rad}}$:

we have $\dfrac{180°}{\pi}(1.13) = 64.74°$

65. To convert from radians to degrees, we multiply by one in the form $\dfrac{180°}{\pi \text{ rad}}$:

we have $\dfrac{180°}{\pi}(-2.35) = -134.65°$

For Problems 67— 86 the following sketch is useful:

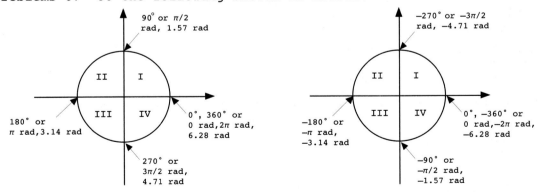

67. From the sketch, we find, since 180° < 187° < 270°, 187° is a quadrant III angle.

69. Since –270° < –200° < –180°, –200° is a quadrant II angle.

71. Since 3.14 < 4 < 4.71, 4 is a quadrant III angle.

73. 270° is a quadrantal angle.

75. Since –1.57 < –1 < 0, –1 is a quadrant IV angle.

77. Since $\dfrac{3\pi}{2} < \dfrac{5\pi}{3} < 2\pi$, $\dfrac{5\pi}{3}$ is a quadrant IV angle.

79. Since $-\dfrac{3\pi}{2} < \dfrac{-7\pi}{6} < -\pi$, $\dfrac{-7\pi}{6}$ is a quadrant II angle.

81. $-\pi$ is a quadrantal angle.

83. 820° is coterminal with 820° − 2(360°) = 100°. Since 90° < 100° < 180°, 100° (and therefore 820°) is a quadrant II angle.

85. $\dfrac{13\pi}{4}$ is coterminal with $\dfrac{13\pi}{4} - 2\pi = \dfrac{13\pi - 8\pi}{4} = \dfrac{5\pi}{4}$. Since $\pi < \dfrac{5\pi}{4} < \dfrac{3\pi}{2}$, $\dfrac{5\pi}{4}$ (and therefore $\dfrac{13\pi}{4}$) is a quadrant III angle.

87. A central angle of radian measure 1 is an angle subtended by an arc of the same length as the radius of the circle.

89. Of all the coterminal angles ($150° + n360°$), $150° + 360° = 510°$ is in the correct interval.

91. Of all the coterminal angles ($-80° + n360°$), $-80° + 360° = 280°$ is in the correct interval.

93. Of all the coterminal angles ($210° + n360°$), $210° − 720° = -510°$ and $210° − 1080° = -870°$ are in the correct interval.

95. Of all the coterminal angles $\left(\dfrac{\pi}{4} + n2\pi\right)$; $\dfrac{\pi}{4} + 2\pi = \dfrac{9\pi}{4}$ and $\dfrac{\pi}{4} + 4\pi = \dfrac{17\pi}{4}$ are in the correct interval.

97. Of all the coterminal angles $\left(-\dfrac{7\pi}{6} + n2\pi\right)$; $-\dfrac{7\pi}{6} + 2\pi = \dfrac{5\pi}{6}$, $-\dfrac{7\pi}{6} + 4\pi = \dfrac{17\pi}{6}$, and $-\dfrac{7\pi}{6} + 6\pi = \dfrac{29\pi}{6}$ are in the correct interval.

99. Of all the coterminal angles $\left(\dfrac{\pi}{2} + n2\pi\right)$; $\dfrac{\pi}{2} - 2\pi = -\dfrac{3\pi}{2}$ and $\dfrac{\pi}{2}$ itself are in the correct interval.

101. Since both arcs come from the same circle, they of course have the same radius, which we'll call r. Using the formula $\theta = s/r$, we get

$$\theta_1 = \dfrac{s_1}{r} \quad \text{and} \quad \theta_2 = \dfrac{s_2}{r}$$

Solve each equation for r:

$$r = \dfrac{\theta_1}{s_1} \quad \text{and} \quad r = \dfrac{\theta_2}{s_2}.$$

Both fractions are equal to r, so $\dfrac{\theta_1}{s_1} = \dfrac{\theta_2}{s_2}$.

103. If this rule of thumb is correct, then the ratio of statute mile measure to nautical mile measure is $\dfrac{8 \text{ statute miles}}{7 \text{ nautical miles}}$. This reduces (by dividing 8/7) to $\dfrac{1.143 \text{ statute miles}}{1 \text{ nautical mile}}$. In Example 6, we are told that one nautical mile corresponds to about 1.151 statute miles, so the 8/7 rule is close to accurate.

105. To calculate angular speed, we need radians and seconds.

200 revolutions × 2π radians = 400π radians

1 minute = 60 sec.

angular speed = $\dfrac{400\pi \text{ radians}}{60 \text{ sec}}$ = 20.94 $\dfrac{\text{rad}}{\text{sec}}$

To calculate linear speed, we need the circumference of the wheel.

$c = \pi d = 6\pi$

In 200 revolutions, the linear distance is 200 × 6π = 1200π feet.

linear speed = $\dfrac{1200\pi \text{ feet}}{60 \text{ sec}}$ = 62.83 ft/sec

107.

From the figure, we can see that the minute (larger) hand is displaced π radians from noon, while the hour (smaller) hand is displaced $\theta = \dfrac{4\frac{1}{2}}{12}$ of a full revolution, or $\dfrac{4\frac{1}{2}}{12} \cdot 2\pi$ radians from noon. Therefore, the larger angle between the hands has measure $\theta + \pi$, or

$\dfrac{4\frac{1}{2}}{12} \cdot 2\pi + \pi = \dfrac{3}{4}\pi + \pi = \dfrac{7}{4}\pi$ radians.

109. We use $\theta = \dfrac{s}{r}$ with $r = \dfrac{1}{2}(10)$ = 5 centimeters and s = 10 meters × 100 $\dfrac{\text{centimeters}}{\text{meter}}$

= 1000 centimeters. Then $\theta = \dfrac{1,000}{5}$ = 200 radians.

111. In one year the line sweeps out one full revolution, or 2π radians. In one week the line sweeps out $\dfrac{1}{52}$ of one full revolution, or $\dfrac{1}{52} \cdot 2\pi = \dfrac{\pi}{26}$ radians.

$\dfrac{\pi}{26}$ = 0.12 radian to two decimal places.

113. Following example 4, we reason that points on the circumference of each wheel travel the same distance. Then $s_1 = r_1\theta_1 = r_2\theta_2 = s_2$. The radius of the front wheel is $\dfrac{1}{2}(40)$ or 20 centimeters. The radius of the back wheel is $\dfrac{1}{2}(60)$ or 30 centimeters. So

$20\theta_1 = 8(30)$

$\theta_1 = \dfrac{240}{20}$

θ_1 = 12 radians

115. Front wheel: d = 40 cm; $c = \pi d = 40\pi$ cm. One complete revolution is 2π radians. So 2π rad = 40π cm or 1 cm = $\dfrac{1}{20}$ radian. In one hour, the linear distance is 10 km which is 1,000,000 cm. 1,000,000 cm = $\dfrac{1}{20} \cdot$ 1,000,000 radian = 50,000 radians

$\dfrac{50,000 \text{ radians}}{\text{hr}} \times \dfrac{1 \text{ hr}}{3,600 \text{ sec}}$ = 13.9 rad/sec

Back wheel: $d = 60$ cm; $c = 60\pi$ cm

$$2\pi \text{ rad} = 60\pi \text{ cm} \quad \text{or} \quad 1 \text{ cm} = \frac{1}{30} \text{ radian}$$

$$1,000,000 \text{ cm} = \frac{1}{30} \cdot 1,000,000 \text{ radian} = 33,333.3 \text{ radians}$$

$$\frac{33,333.3 \text{ radians}}{\text{hr}} \times \frac{1 \text{ hr}}{3,600 \text{ sec}} = 9.26 \text{ rad/sec}$$

117. The central angle between Havana and Cleveland is

$$\theta = 41°30' - 23°08' = 18°22' \text{ or } 18.37°$$

Let s be the length of the arc between these two cities and recall that 21,600 nautical miles is the circumference of the earth.

$$\frac{s}{21,600} = \frac{18.37°}{360°}$$
$$360s = 396,792$$
$$s = \frac{396,792}{360} \approx 1,102 \text{ nautical miles}$$

119. First, we find the distance between the cities. The central angle between Washington and Lima is

$$\theta = 38°53' + 12°06' = 50°59' \text{ or } 50.98° \quad \text{(We added because one is in the northern hemisphere, the other in the southern)}$$

Let s be the length of the arc between these two cities and recall that 21,600 nautical miles is the circumference of the earth.

$$\frac{s}{21,600} = \frac{50.98°}{360°}$$
$$360s = 1,101,168$$
$$s = \frac{1,101,168}{360} \approx 3,059 \text{ nautical miles}$$

The linear speed of the plane is

$$v = \frac{s}{t} = \frac{3,059}{6} \approx 510 \text{ knots.}$$

The angular speed is

$$\omega = \frac{\theta}{t} = \frac{50.98°}{6} = 8.5° \text{ per hour.}$$

121. We apply $\dfrac{s}{c} = \dfrac{\theta°}{360°}$ with $s = 500$ mi, and $\theta° = 7.5°$

$$\text{Then} \quad \frac{500}{c} = \frac{7.5°}{360°} \quad \text{or} \quad \frac{500}{c} = \frac{7.5}{360}$$
$$360c \cdot \frac{500}{c} = 360c \cdot \frac{7.5}{360}$$
$$180,000 = 7.5c$$
$$c = 24,000 \text{ mi}$$

123. The 7.5° angle and θ have a common side. (An extended vertical pole in Alexandria will pass through the center of the earth.) The sun's rays are essentially parallel when they arrive at the earth. So, the other two sides of the angles are parallel, since a sun ray to the bottom of the well, when extended, will pass through the center of the earth. From geometry we know that the alternate interior angles made by a line intersecting two parallel lines

125. We use $c \approx s = r\theta$ with $r = 9.3 \times 10^7$ mi and $\theta = 9.3 \times 10^{-3}$ rad.
Then $c \approx (9.3 \times 10^7)(9.3 \times 10^{-3}) = 865,000$ mi

127. We use $c \approx s = r\theta$. $r = 750$ ft. θ must be converted to radians to use the formula, so, using $\theta_R = \dfrac{\pi \text{ rad}}{180°} 2.5°$ we have $\theta_R = \dfrac{\pi}{72}$ rad.

Then $c \approx 750 \cdot \dfrac{\pi}{72} = 33$ ft.

Section 5-2

1. Answers will vary.   **3.** Answers will vary.   **5.** Answers will vary.

For Problems 7-21, the following sketch is useful:

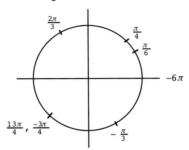

7. $w\left(\dfrac{3\pi}{2}\right) = (0, -1)$ **9.** $w(-6\pi) = (1, 0)$ **11.** $w\left(\dfrac{\pi}{4}\right) = \left(\dfrac{1}{\sqrt{2}}, \dfrac{1}{\sqrt{2}}\right)$

13. $w\left(\dfrac{\pi}{6}\right) = \left(\dfrac{\sqrt{3}}{2}, \dfrac{1}{2}\right)$ **15.** $w\left(-\dfrac{\pi}{3}\right) = \left(\dfrac{1}{2}, -\dfrac{\sqrt{3}}{2}\right)$ **17.** $w\left(\dfrac{2\pi}{3}\right) = \left(-\dfrac{1}{2}, \dfrac{\sqrt{3}}{2}\right)$

19. $w\left(-\dfrac{3\pi}{4}\right) = \left(\dfrac{-1}{\sqrt{2}}, \dfrac{-1}{\sqrt{2}}\right)$ **21.** $w\left(\dfrac{13\pi}{4}\right) = \left(\dfrac{-1}{\sqrt{2}}, \dfrac{-1}{\sqrt{2}}\right)$

It's helpful to remember these rules for finding the coordinates of the circular points when x is a multiple of π:

1) If x has a denominator of 6, like $\dfrac{5\pi}{6}$, the coordinates are $\left(\pm\dfrac{\sqrt{3}}{2}, \pm\dfrac{1}{2}\right)$. Locate the quadrant to decide on the appropriate signs.

2) If x has a denominator of 4, like $-\dfrac{3\pi}{4}$, the coordinates are $\left(\pm\dfrac{1}{\sqrt{2}}, \pm\dfrac{1}{\sqrt{2}}\right)$. Locate the quadrant to decide on the appropriate signs.

3) If x has a denominator of 3, like $\dfrac{7\pi}{3}$, the coordinates are $\left(\pm\dfrac{1}{2}, \pm\dfrac{\sqrt{3}}{2}\right)$. Locate the quadrant to decide on the appropriate signs.

4) If the denominator is 2, or there is no denominator, like $\dfrac{3\pi}{2}$ or -6π, x is on one of the coordinate axes and you should be able to find the coordinates by just locating x on the unit circle.

IMPORTANT: These rules are only valid if x is a <u>reduced</u> <u>fraction</u>! For example, $\dfrac{6\pi}{4}$ has denominator 4 but can be reduced to $\dfrac{3\pi}{2}$.

23. $w\left(\dfrac{3\pi}{2}\right) = (0, -1)$ so $\sin\left(\dfrac{3\pi}{2}\right) = -1$ **25.** $w(-6\pi) = (1, 0)$ so $\cos(-6\pi) = 1$

27. $w\left(\dfrac{\pi}{4}\right) = \left(\dfrac{1}{\sqrt{2}}, \dfrac{1}{\sqrt{2}}\right)$ so $\sec\left(\dfrac{\pi}{4}\right) = \dfrac{1}{\frac{1}{\sqrt{2}}} = \sqrt{2}$ **29.** $w\left(\dfrac{\pi}{6}\right) = \left(\dfrac{\sqrt{3}}{2}, \dfrac{1}{2}\right)$ so $\tan\left(\dfrac{\pi}{6}\right) = \dfrac{\frac{1}{2}}{\frac{\sqrt{3}}{2}} = \dfrac{1}{\sqrt{3}}$

31. $w\left(-\dfrac{\pi}{3}\right) = \left(\dfrac{1}{2}, -\dfrac{\sqrt{3}}{2}\right)$ so $\sin\left(-\dfrac{\pi}{3}\right) = -\dfrac{\sqrt{3}}{2}$ **33.** $w\left(\dfrac{2\pi}{3}\right) = \left(-\dfrac{1}{2}, \dfrac{\sqrt{3}}{2}\right)$ so $\csc\left(\dfrac{2\pi}{3}\right) = \dfrac{1}{\frac{\sqrt{3}}{2}} = \dfrac{2}{\sqrt{3}}$

35. $w\left(-\dfrac{3\pi}{4}\right) = \left(\dfrac{-1}{\sqrt{2}}, \dfrac{-1}{\sqrt{2}}\right)$ so $\cos\left(-\dfrac{3\pi}{4}\right) = \dfrac{-1}{\sqrt{2}}$

37. $w\left(\dfrac{13\pi}{4}\right) = \left(\dfrac{-1}{\sqrt{2}}, \dfrac{-1}{\sqrt{2}}\right)$ so $\cot\left(\dfrac{13\pi}{4}\right) = \dfrac{\frac{-1}{\sqrt{2}}}{\frac{-1}{\sqrt{2}}} = 1$

39. $\sin x = b = \dfrac{12}{13}$ **41.** $\sec x = \dfrac{1}{a} = \dfrac{1}{5/13} = \dfrac{13}{5}$

43. $\cot x = \dfrac{a}{b} = \dfrac{-15/17}{8/17} = -\dfrac{15}{8}$ **45.** $\csc x = \dfrac{1}{b} = \dfrac{1}{8/17} = \dfrac{17}{8}$

47. $\cos x = a = 0$ **49.** $\tan x = \dfrac{b}{a} = \dfrac{-1}{0}$; undefined

For Problems 51-61, we will use definitions for the trigonometric functions provided in the remarks before Example 2 in section 5-2 of your textbook.

51. The point (0,5) lies on a circle of radius 5, so $\cos x = \dfrac{a}{r} = \dfrac{0}{5} = 0$.

53. The definition of tangent does not include the radius, so we don't need to find it here.
$$\tan x = \dfrac{b}{a} = \dfrac{3}{4}$$

55. First, we find the radius of the circle that the point (0.4, 0.09) lies on. The distance from that point to (0,0) is
$$d = \sqrt{(0.4 - 0)^2 + (0.09 - 0)^2} = \sqrt{0.16 + 0.0081} = 0.41$$
$$\csc x = \dfrac{0.41}{0.09} = \dfrac{41}{9}$$

57. The definition of cotangent does not include the radius, so we don't need to find it here.
$$\cot x = \dfrac{a}{b} = -\dfrac{8}{15}$$

59. First, we find the radius of the circle that the point (-9, -40) lies on. The distance from that point to (0,0) is
$$d = \sqrt{(-9 - 0)^2 + (-40 - 0)^2} = \sqrt{81 + 1600} = 41$$
$$\sec x = \dfrac{r}{a} = -\dfrac{41}{9}$$

61. First, we find the radius of the circle that the point (2,-1) lies on. The distance from that point to (0,0) is
$$d = \sqrt{(2 - 0)^2 + (-1 - 0)^2} = \sqrt{4 + 1} = \sqrt{5}$$
$$\sin x = \dfrac{b}{r} = -\dfrac{1}{\sqrt{5}}$$

63. Since $\cos x = a$, it is negative in quadrants II and III.

65. Since sin $x = b$, it is positive in quadrants I and II.

67. Since cot $x = \dfrac{a}{b}$, it is negative if a and b have opposite signs. This occurs in quadrants II and IV.

69. -0.6573 | **Common Error:** cos 2.288 $\neq$ 0.9992; calculator must be in *radian* mode. |

71. -14.60 **73.** 1.000   **75.** 0.8138   **77.** 0.5290

79. 0.4226 (calculator in degree mode)   **81.** -1.573 (calculator in radian mode)

83. 0.8439   **85.** -0.3363

87. $\sin(113°27'13") = \sin\left(113 + \dfrac{27}{60} + \dfrac{13}{3600}\right)° = 0.9174$

89. False. The domain of the wrapping function is the set of all real numbers.

91. False. The wrapping function is not one-to-one. $W(0) = W(2\pi)$ but $0 \neq 2\pi$.

93. True. $a^2 + b^2 = 1$ describes all points (a, b) on the unit circle.

95. False. The reciprocal of sin x is csc x, unless sin $x = 0$.

97. False. The secant function is not one-to-one. sec 0 = sec 2π, but $0 \neq 2\pi$.

99. False. The domain of sin x is all real numbers; the domain of csc x excludes all integer multiples of π.

101. True. All integer multiples of π are turning points.

103. True. $\dfrac{\pi}{2} + k\pi$, k any integer, represents all zeros.

105.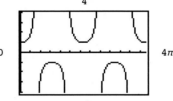

Since there is no secant button, we enter $y = \sec x$ as $y = 1/\cos x$. There are no zeros and 3 turning points in the interval $[0, 4\pi]$. They occur when the graph is at height 1 or -1. Since sec $x = \dfrac{1}{\cos x}$, sec $x = \pm 1$ when cos $x = \pm 1$. This occurs for $x = \pi$, 2π, and 3π, so the turning points are $(\pi, -1)$, $(2\pi, 1)$, and $(3\pi, -1)$.

107.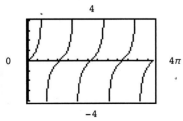

There are 5 zeros. Since tan $x = \dfrac{\sin x}{\cos x}$, the zeros occur when sin $x = 0$ in the interval $[0, 4\pi]$. The x values are 0, π, 2π, 3π, and 4π. There are no turning points.

For Problems 109–118, the following sketches are useful.

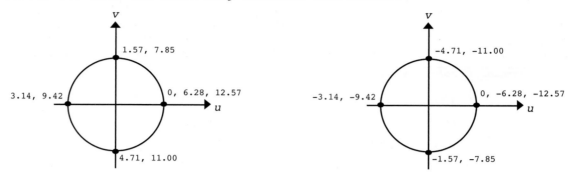

Sketch 1: counterclockwise wrapping   Sketch 2: clockwise wrapping

109. Using sketch 1 and 1.57 < 2 < 3.14, we see that $W(2)$ is in the second quadrant. So, if $W(2) = (a, b)$, then $a < 0$ and $b > 0$, that is, a is negative and b is positive.

111. Using sketch 1 and 1.57 < 3 < 3.14, we see that $W(3)$ is in the second quadrant. So, if $W(3) = (a, b)$, then $a < 0$ and $b > 0$, that is, a is negative and b is positive.

113. Using sketch 1 and 4.71 < 5 < 6.28, we see that $W(5)$ is in the fourth quadrant. So, if $W(5) = (a, b)$, then $a > 0$ and $b < 0$, that is, a is positive and b is negative.

115. Using sketch 2 and –3.14 < –2.5 < –1.57, we see that $W(-2.5)$ is in the third quadrant. So, if $W(-2.5) = (a, b)$, then $a < 0$ and $b < 0$, that is, a and b are negative.

117. Using sketch 2 and –6.28 < –6.1 < –4.71, we see that $W(-6.1)$ is in the first quadrant. So, if $W(-6.1) = (a, b)$, then $a > 0$ and $b > 0$, that is, a and b are positive.

119.

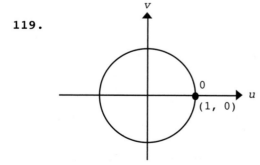

From the sketch, if $W(x) = (1, 0)$, $0 \le x < 2\pi$, then $x = 0$. Since $W(x) = W(x + 2k\pi)$, k any integer, if there are no restrictions on x, then $x = 0 + 2k\pi$ or $x = 2k\pi$, k any integer.

121.

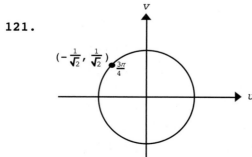

From the sketch, if $W(x) = \left(-\dfrac{1}{\sqrt{2}}, \dfrac{1}{\sqrt{2}}\right)$, $0 \le x < 2\pi$, then $x = \dfrac{3\pi}{4}$. Since $W(x) = W(x + 2k\pi)$, k any integer, if there are no restrictions on x, then $x = \dfrac{3\pi}{4} + 2k\pi$, k any integer.

123. $W(x)$ is the coordinates of a point on a unit circle that is $|x|$ units from $(1, 0)$, in a counterclockwise direction if x is positive and in a clockwise direction if x is negative. $W(x + 4\pi)$ has the same coordinates as $W(x)$, since we return to the same point every time we go around the unit circle any integer multiple of 2π units (the circumference of the circle) in either direction.

125.

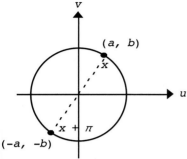

Since (see sketch) $W(x)$ and $W(x + \pi)$ are symmetrically placed with respect to the origin, $W(x) = (a, b)$ requires $W(x + \pi) = (-a, -b)$. T

127.

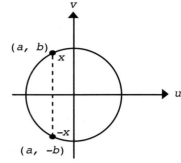

Since (see sketch) $W(x)$ and $W(-x)$ are symmetrically placed with respect to the horizontal axis, $W(x) = (a, b)$ requires $W(-x) = (a, -b)$. F

129. Since for all real numbers x, $W(x) = W(x + 2\pi)$, $W(x) = (a, b)$ requires $W(x + 2\pi) = (a, b)$. T

131. Locate $W(0.4) = (a, b) = (0.9, 0.4)$ on the figure.

(A) $\sin 0.4 = b = 0.4$ (B) $\cos 0.4 = a = 0.9$ (C) $\tan 0.4 = \dfrac{b}{a} = \dfrac{0.4}{0.9} = 0.4$

133. (A) Locate $W(2.2) = (a, b) = (-0.6, 0.8)$ on the figure.

$\sec 2.2 = \dfrac{1}{a} = \dfrac{1}{-0.6} = -2$ (to one significant digit)

(B) Locate $W(5.9) = (a, b) = (0.9, -0.4)$ on the figure.

$\tan 5.9 = \dfrac{b}{a} = \dfrac{-0.4}{0.9} = -0.4$

(C) Locate $W(3.8) = (a, b) = (-0.8, -0.6)$ on the figure.

$\cot 3.8 = \dfrac{a}{b} = \dfrac{-0.8}{-0.6} = 1$(to one significant digit)

135. $\sin x < 0$ in quadrants III and IV; $\cot x < 0$ in quadrants II and IV; therefore, both are true in quadrant IV.

137. $\cos x < 0$ in quadrants II and III; $\sec x > 0$ in quadrants I and IV; therefore, it is not possible to have both true for the same value of x.

139. $\cos x = a$ is always defined. There are no values for which it is undefined.

141. $\tan x = \dfrac{b}{a}$ is undefined if and only if $a = 0$. This occurs at points with x coordinate zero; that is, points on the vertical axis. The only values of x between 0 and 2π for which $W(x)$ is on the vertical axis are $\dfrac{\pi}{2}$ and $\dfrac{3\pi}{2}$.

143. $\sec x = \dfrac{1}{a}$ is undefined if and only if $a = 0$. This occurs at points with x coordinate zero; that is, points on the vertical axis. The only values of x between 0 and 2π for which $W(x)$ is on the vertical axis are $\dfrac{\pi}{2}$ and $\dfrac{3\pi}{2}$.

145. (A) Find the slope of the line through (a, b) and (b, a):

$$m = \frac{a - b}{b - a} = -1$$

This is the negative reciprocal of the slope of the line $y = x$, which is 1, so those two lines are perpendicular.

(B) Find the midpoint of (a, b) and (b, a):

$$\left(\frac{a + b}{2}, \frac{b + a}{2} \right)$$

The coordinates are identical, so that point is on the line $y = x$.

147. Find the slope:

$$m = \frac{b - 0}{a - 0} = \frac{b}{a}$$

Use the point-slope form:

$$y - b = \frac{b}{a}(x - a) \qquad \text{(Multiply both sides by } a\text{)}$$

$$ay - ab = b(x - a) \qquad \text{(Distribute on the right side)}$$

$$ay - ab = bx - ab \qquad \text{(Write in standard form)}$$

$$bx - ay = 0$$

149. Given $n = 12$, $r = 5$, we use the given formula $A = \frac{1}{2}nr^2 \sin \frac{2\pi}{n}$ to obtain:

$$A = \frac{1}{2}(12)(5)^2 \sin \frac{2\pi}{12}$$

$$= 150 \sin \frac{\pi}{6}$$

$$= 150\left(\frac{1}{2}\right) = 75 \text{ square meters}$$

151. Given $n = 3$, $r = 4$, we use the given formula $A = \frac{1}{2}nr^2 \sin \frac{2\pi}{n}$ to obtain:

$$A = \frac{1}{2}(3)(4)^2 \sin \frac{2\pi}{3}$$

$$= 24 \sin \frac{2\pi}{3}$$

$$= 24\left(\frac{\sqrt{3}}{2}\right) = 12\sqrt{3} \approx 20.78 \text{ square inches}$$

153. $a_1 = 0.5$

$a_2 = a_1 + \cos a_1 = 0.5 + \cos 0.5 = 1.377583$

$a_3 = a_2 + \cos a_2 = 1.377583 + \cos 1.377583 = 1.569596$

$a_4 = a_3 + \cos a_3 = 1.569596 + \cos 1.569596 = 1.570796$

$a_5 = a_4 + \cos a_4 = 1.570796 + \cos 1.570796 = 1.570796$

$\dfrac{\pi}{2} = 1.570796$ to six decimal places.

Section 5-3

1. Yes. Explanations will vary.

3. Answers will vary.

5. Answers will vary.

7. $\sin \theta = \dfrac{\text{opposite}}{\text{hypotenuse}} = \dfrac{8}{17}$

9. $\csc \theta = \dfrac{\text{hypotenuse}}{\text{opposite}} = \dfrac{17}{8}$

11. $\tan \theta = \dfrac{\text{opposite}}{\text{adjacent}} = \dfrac{8}{15}$

13. $\dfrac{15}{17} = \dfrac{\text{adjacent}}{\text{hypotenuse}} = \cos \theta$

15. $\dfrac{17}{15} = \dfrac{\text{hypotenuse}}{\text{adjacent}} = \sec \theta$

17. $\dfrac{15}{8} = \dfrac{\text{adjacent}}{\text{opposite}} = \cot \theta$

19. $60.55°$ **21.** $82.90°$ **23.** $37.09°$

25.

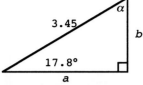

Solve for α:
$\alpha = 90° - 17.8° = 72.2°$

Solve for b: $\sin \beta = \dfrac{b}{c}$

$\sin 17.8° = \dfrac{b}{3.45}$

$\quad b = 3.45 \sin 17.8°$
$\quad\quad = 1.05$

Solve for a: $\cos \beta = \dfrac{a}{c}$

$\cos 17.8° = \dfrac{a}{3.45}$

$\quad a = 3.45 \cos 17.8°$
$\quad\quad = 3.28$

27.

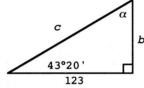

Solve for α:
$\alpha = 90° - 43°20' = 46°40'$

Solve for b: $\tan \beta = \dfrac{b}{a}$

$\tan 43°20' = \dfrac{b}{123}$

$\quad b = 123 \tan 43°20'$
$\quad\quad = 116$

Solve for c: $\sec \beta = \dfrac{c}{a}$

$\sec 43°20' = \dfrac{c}{123}$

$\quad c = 123 \sec 43°20'$
$\quad\quad = 169$

29.

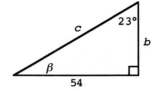

Solve for β:
$\beta = 90° - 23°0' = 67°0'$

Solve for b: $\tan \beta = \dfrac{b}{a}$

$\tan 67°0' = \dfrac{b}{54}$

$\quad b = 54 \tan 67°0'$
$\quad\quad = 127$

Solve for c: $\csc \alpha = \dfrac{c}{a}$

$\csc 23°0' = \dfrac{c}{54}$

$\quad c = 54 \csc 23°0'$
$\quad\quad = 138$

31.

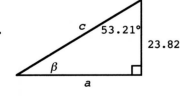

Solve for β:
$\beta = 90° - 53.21° = 36.79°$

Solve for a: $\tan \alpha = \dfrac{a}{b}$

$\tan 53.21° = \dfrac{a}{23.82}$

$\quad a = 23.82 \tan 53.21°$
$\quad\quad = 31.85$

Solve for c: $\sec \alpha = \dfrac{c}{b}$

$\sec 53.21° = \dfrac{c}{23.82}$

$\quad c = 23.82 \sec 53.21°$
$\quad\quad = 39.77$

33.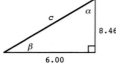

Solve for β: $\tan \beta = \dfrac{b}{a}$

$\tan \beta = \dfrac{8.46}{6.00}$

$\beta = \tan^{-1} \dfrac{8.46}{6.00}$

$\quad = 54.7°$ or $54°40'$

Solve for α:
$\alpha = 90° - 54°40' = 35°20'$

Solve for c: $\csc \beta = \dfrac{c}{b}$

$\csc 54°40' = \dfrac{c}{8.46}$

$\qquad c = 8.46 \csc 54°40'$

$\qquad\quad = 10.4$

35.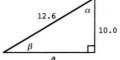

Solve for β: $\sin \beta = \dfrac{b}{c}$

$\sin \beta = \dfrac{10.0}{12.6}$

$\beta = \sin^{-1} \dfrac{10.0}{12.6}$

$\quad = 52.5°$ or $52°30'$

Solve for α:
$\alpha = 90° - 52°30' = 37°30'$

Solve for a: $\cot \beta = \dfrac{a}{b}$

$\cot 52°30' = \dfrac{a}{10.0}$

$\qquad a = 10.0 \cot 52°30'$

$\qquad\quad = 7.67$

37. False. Knowing only the three angles can't tell us anything about the lengths of the sides. The two triangles at the right have the exact same angles but totally different side lengths.

39. True.

According to the diagram,

$\sin \alpha = \dfrac{\text{Opp}}{\text{Hyp}} = \dfrac{a}{c}$ and $\cos \beta = \dfrac{\text{Adj}}{\text{Hyp}} = \dfrac{a}{c}$.

41. False.

According to the diagram,

$\sec \alpha = \dfrac{\text{Hyp}}{\text{Adj}} = \dfrac{5}{3}$ and $\cos \beta = \dfrac{\text{Adj}}{\text{Hyp}} = \dfrac{4}{5}$.

The following figures are used in Problems 44-54.

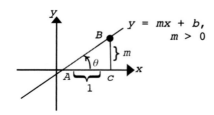 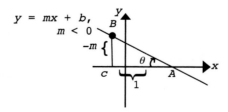

43. See the figure. A line of positive slope m will form a right triangle ABC as shown. $\tan \theta = \dfrac{m}{1}$.

Here $\tan \theta = \dfrac{1}{2}$ since the line $y = \dfrac{1}{2} x + 3$ has slope $\dfrac{1}{2}$

$\qquad \theta = \tan^{-1} \dfrac{1}{2} = 26.6°$

45. See the figure. A line of positive slope m will form a right triangle ABC as shown. $\tan \theta = \dfrac{m}{1}$.

Here $\tan \theta = 5$ since the line $y = 5x + 21$ has slope 5

$\theta = \tan^{-1} 5 = 78.7°$

47. See the figure. A line of negative slope m will form a right triangle ABC as shown. $\tan \theta = \dfrac{-m}{1}$.

Here $\tan \theta = -\dfrac{-2}{1} = 2$ since the line $y = -2x + 7$ has slope -2

$\theta = \tan^{-1} 2 = 63.4°$

49. See the figure. An angle θ can be formed by either of two lines, one with positive slope m and the other with negative slope m.
Therefore, $m = \pm\tan \theta$
Here $m = \pm\tan 20° = \pm0.36$

51. See the figure. An angle θ can be formed by either of two lines, one with positive slope m and the other with negative slope m.
Therefore, $m = \pm\tan \theta$
Here $m = \pm\tan 80° = \pm5.67$

53. See the figure. An angle θ can be formed by either of two lines, one with positive slope m and the other with negative slope m.
Therefore, $m = \pm\tan \theta$

Here $m = \pm\tan\left(\dfrac{\pi}{30}\right) = \pm0.11$

55. (A) In triangle OAD, $\cos \theta = \dfrac{\text{Adj}}{\text{Hyp}} = \dfrac{OA}{1} = OA$

(B) In triangle OED, angle $EOD = 90° - \theta$, angle $OED = 90° - (90° - \theta) = \theta$.
So $\cot OED = \dfrac{\text{Adj}}{\text{Opp}} = \dfrac{DE}{1} = DE = \cot \theta$

(C) In triangle ODC, $\sec \theta = \dfrac{\text{Hyp}}{\text{Adj}} = \dfrac{OC}{1} = OC$

57. (A) As θ approaches 90°, $OA = \cos \theta$ approaches 0.
(B) As θ approaches 90°, $DE = \cot \theta$ approaches 0.
(C) As θ approaches 90°, $OC = \sec \theta$ increases without bound.

59. (A) As θ approaches 0°, $AD = \sin \theta$ approaches 0.
(B) As θ approaches 0°, $CD = \tan \theta$ approaches 0.
(C) As θ approaches 0°, $OE = \csc \theta$ increases without bound.

61.

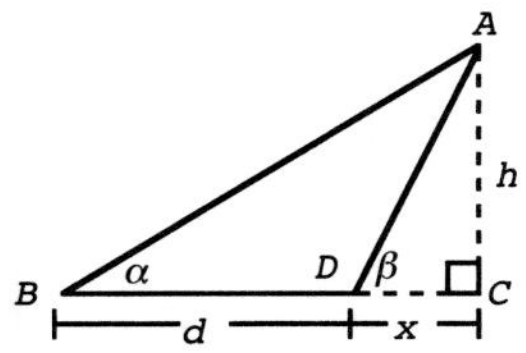

Label as shown at the left.
In right triangle ADC, $\cot \beta = \dfrac{x}{h}$
In right triangle ABC, $\cot \alpha = \dfrac{d + x}{h}$

Then $x = h \cot \beta$ and $d + x = h \cot \alpha$

$d = h \cot \alpha - x$

$d = h \cot \alpha - h \cot \beta$

$d = h(\cot \alpha - \cot \beta)$

$h = \dfrac{d}{\cot \alpha - \cot \beta}$

63. Sketch a figure:

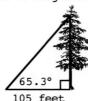

105 feet

Let h = height of tree
From the figure, we can see that

$$\tan 65.3° = \frac{h}{105}$$
$$h = 105 \tan 65.3°$$
$$= 228 \text{ feet}$$

65. Sketch a figure:

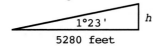

5280 feet

Let h = how far train climbs.

$$\tan 1°23' = \frac{h}{5,280}$$
$$h = 5,280 \tan 1°23'$$
$$= 127.5 \text{ feet}$$

67. Sketch a figure:

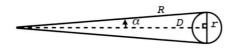

Note $\alpha = \frac{1}{2}(32') = 16'$

diameter $= d = 2r$

$$\tan \alpha = \frac{r}{D}$$

$$\tan 16' = \frac{r}{239,000}$$
$$r = 239,000 \tan 16'$$
$$d = 2(239,000)\tan 16'$$
$$= 2,225 \text{ miles}$$

Alternatively, we can write

$$\sin \alpha = \frac{r}{R}$$

$$\sin 16' = \frac{r}{239,000}$$
$$r = 239,000 \sin 16'$$
$$d = 2(239,000)\sin 16'$$
$$= 2,225 \text{ miles}$$

Although it is not clear whether 239,000 miles is to be interpreted as D, R, or $D - r$, at this accuracy it does not matter.

69. Sketch a figure:

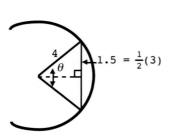

We will find $\frac{1}{2}\theta$ and double it.

$$\sin \frac{1}{2}\theta = \frac{1.5}{4}$$
$$\frac{1}{2}\theta = \sin^{-1}\frac{1.5}{4}$$
$$\theta = 2\sin^{-1}\frac{1.5}{4}$$
$$= 44°$$

71. We use $g = \frac{v}{t \sin \theta}$ with $v = 4.1$,

$t = 3.0$, $\theta = 8.0°$

$$g = \frac{4.1}{3.0 \sin 8.0°}$$

$$g = 9.8 \text{ meters/second}^2$$

73.

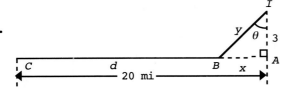

(A) We note that the cable consists of water section *IB*, and shore section *CB* = 20 mi - *AB*.

Let *y* = *IB* and *d* = *CB* = 20 - *x*
In right triangle *ABI*

$$\sec \theta = \frac{y}{3 \text{ mi}} \quad \text{and} \quad \tan \theta = \frac{x}{3 \text{ mi}}$$
$$y = 3 \sec \theta \text{ mi} \qquad x = 3 \tan \theta \text{ mi}$$

So the cost of the cable is

$$\left(\begin{array}{c}\text{Cost of Water} \\ \text{Section Per Mile}\end{array}\right)\left(\begin{array}{c}\text{Number of} \\ \text{Water miles} = y\end{array}\right) + \left(\begin{array}{c}\text{Cost of Shore} \\ \text{Section Per Mile}\end{array}\right)\left(\begin{array}{c}\text{Number of} \\ \text{Shore miles} = 20 - x\end{array}\right)$$

$$C(\theta) = (25{,}000 \, \frac{\text{dollars}}{\text{mi}})(3 \sec \theta \text{ mi}) + (15{,}000 \, \frac{\text{dollars}}{\text{mi}})(20 - 3 \tan \theta \text{ mi})$$
$$C(\theta) = 75{,}000 \sec \theta + 300{,}000 - 45{,}000 \tan \theta$$

(B)

θ	$C(\theta)$
10°	$368,222
20°	$363,435
30°	$360,622
40°	$360,146
50°	$363,050

75. Sketch a figure:

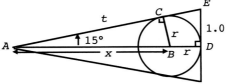

In triangle *ABC*, we have $\dfrac{r}{x} = \sin 15°$

In triangle *ADE*, we have $\dfrac{1.0}{r + x} = \tan 15°$

Eliminating *x*, we see that

$$\frac{r + x}{1.0} = \frac{1}{\tan 15°} = \cot 15° \quad \text{(reciprocal identity)}$$
$$x = \cot 15° - r$$

Substituting, we have

$$\frac{r}{\cot 15° - r} = \sin 15°$$

$$r = \sin 15°(\cot 15° - r)$$

$$r = \sin 15° \cot 15° - \sin 15° r$$

$$r(1 + \sin 15°) = \sin 15° \cot 15°$$

$$r = \frac{\sin 15° \cot 15°}{1 + \sin 15°}$$

$$r = 0.77 \text{ meters}$$

Section 5-4

1. Answers will vary.   3. Answers will vary.   5. Answers will vary.

7. sine: 2π; cotangent: π; cosecant: 2π

9. (A) Since the range of the cosine function is $[-1, 1]$, the largest and smallest y values on its graph are, respectively, 1 and -1. The largest deviation of the function from the x axis is therefore 1 unit.

   (B) Since the range of the tangent function is all real numbers, the graph deviates indefinitely far from the x axis.

   (C) Since the range of the cosecant function is all real numbers $y \geq 1$ or $y \leq -1$, the graph deviates indefinitely far from the x axis.

11. (A) -2π, $-\pi$, 0, π, 2π (B) $-\dfrac{3\pi}{2}$, $-\dfrac{\pi}{2}$, $\dfrac{\pi}{2}$, $\dfrac{3\pi}{2}$ (C) No x intercepts

13. (A) Defined for all real x (B) $-\dfrac{3\pi}{2}$, $-\dfrac{\pi}{2}$, $\dfrac{\pi}{2}$, $\dfrac{3\pi}{2}$ (C) -2π, $-\pi$, 0, π, 2π

15. (A) There are no vertical asymptotes   (B) $-\dfrac{3\pi}{2}$, $-\dfrac{\pi}{2}$, $\dfrac{\pi}{2}$, $\dfrac{3\pi}{2}$

   (C) -2π, $-\pi$, 0, π, 2π

17. (A) A shift of $\pi/2$ to the left will transform the cosecant graph into the secant graph. [The answer is not unique--see part (B).]

   (B) The graph of $y = -\csc(x - \pi/2)$ is a $\pi/2$ shift to the right and a reflection in the x axis of the graph of $y = \csc x$. The result is the graph of $y = \sec x$. The graph of $y = -\csc\left(x + \dfrac{\pi}{2}\right)$ is a $\pi/2$ shift to the left and a reflection in the x axis of the graph of $y = \csc x$. The result is not the graph of $y = \sec x$.

19.

   $\theta = 300°$

   $\alpha = 360° - 300° = 60°$

21.

   $\theta = \dfrac{7\pi}{6}$

   $\alpha = \dfrac{7\pi}{6} - \pi = \dfrac{\pi}{6}$

23.

   $\theta = -\dfrac{5\pi}{3}$

   $\alpha = -\dfrac{5\pi}{3} + 2\pi = \dfrac{\pi}{3}$

25.

   $\theta = 170°$

   $\alpha = 180° - 170° = 10°$

27. Since $\sin \theta = \dfrac{3}{5} > 0$ and $\cos \theta < 0$, θ is a quadrant II angle. We sketch a

reference triangle. Since $\sin \theta = \dfrac{b}{r} = \dfrac{3}{5}$, we know that $b = 3$ and $r = 5$.

Use the Pythagorean theorem to find a.

$a^2 + 3^2 = 5^2$

$\qquad a^2 = 16$

$\qquad a = -4$

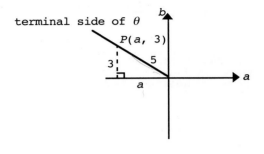

(a must be negative because θ is a quadrant II
angle)

Using $(a, b) = (-4, 3)$ and $r = 5$, we have

$$\cos \theta = \frac{a}{r} = \frac{-4}{5} = -\frac{4}{5} \qquad \tan \theta = \frac{b}{a} = \frac{3}{-4} = -\frac{3}{4}$$

$$\sec \theta = \frac{r}{a} = \frac{5}{-4} = -\frac{5}{4} \qquad \cot \theta = \frac{a}{b} = \frac{-4}{3} = -\frac{4}{3}$$

$$\csc \theta = \frac{r}{b} = \frac{5}{3}$$

29. Since $\cos \theta = -\dfrac{\sqrt{5}}{3} < 0$ and $\cot \theta > 0$, θ is a quadrant III angle. We sketch a

reference triangle. Since $\cos \theta = \dfrac{a}{r} = -\dfrac{\sqrt{5}}{3}$ we know that $a = -\sqrt{5}$ and $r = 3$.

Use the Pythagorean theorem to find b.

$(-\sqrt{5})^2 + b^2 = 3^2$

$\qquad b^2 = 4$

$\qquad b = -2$

(b must be negative because θ is a quadrant
III angle)

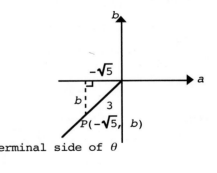

Using $(a, b) = (-\sqrt{5}, -2)$ and $r = 3$, we have

$$\sin \theta = \frac{b}{r} = -\frac{2}{3} \qquad\qquad \sec \theta = \frac{r}{a} = -\frac{3}{\sqrt{5}}$$

$$\tan \theta = \frac{b}{a} = \frac{-2}{-\sqrt{5}} = \frac{2}{\sqrt{5}} \qquad \cot \theta = \frac{a}{b} = \frac{-\sqrt{5}}{-2} = \frac{\sqrt{5}}{2}$$

$$\csc \theta = \frac{r}{b} = \frac{3}{-2} = -\frac{3}{2}$$

31. Since $\tan \theta = \dfrac{1}{2} > 0$, $\sin \theta$ and $\cos \theta$ have the same sign. Also, $\sec \theta < 0$, so

$\cos \theta < 0$ as well. This tells us that θ is a quadrant III angle. We sketch a

reference triangle. Since $\tan \theta = \dfrac{b}{a} = \dfrac{1}{2}$, we know that $a = -2$ and $b = -1$.

Use the Pythagorean theorem to find r.

$(-1)^2 + (-2)^2 = r^2$

$\qquad r^2 = 5$

$\qquad r = \sqrt{5}$

Using $(a, b) = (-2, -1)$ and $r = \sqrt{5}$, we have

$$\cos \theta = \frac{a}{r} = \frac{-2}{\sqrt{5}} = -\frac{2}{\sqrt{5}} \qquad \sin \theta = \frac{b}{r} = \frac{-1}{\sqrt{5}} = -\frac{1}{\sqrt{5}}$$

$$\sec \theta = \frac{r}{a} = \frac{\sqrt{5}}{-2} = -\frac{\sqrt{5}}{2} \qquad \cot \theta = \frac{a}{b} = \frac{-2}{-1} = 2$$

$$\csc \theta = \frac{r}{b} = \frac{\sqrt{5}}{-1} = -\sqrt{5}$$

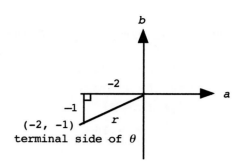

33. Since $\sec \theta = 10$, we know that $\cos \theta = \frac{1}{10} > 0$; since $\tan \theta < 0$, since and cosine have different signs, so θ is a quadrant IV angle. We sketch a reference triangle. Since $\cos \theta = \frac{a}{r} = \frac{1}{10}$, we know that $a = 1$ and $r = 10$. Use the Pythagorean Theorem to find b.

$$1^2 + b^2 = 10^2$$

$$b^2 = 99$$

$$b = -\sqrt{99}$$

(b has to be negative because θ is a quadrant IV angle)

Using $(a, b) = \left(1, -\sqrt{99}\right)$ and $r = 10$, we have

$$\cos \theta = \frac{1}{10} \qquad\qquad \sin \theta = \frac{b}{r} = \frac{-\sqrt{99}}{10} = -\frac{\sqrt{99}}{10}$$

$$\tan \theta = \frac{b}{a} = \frac{-\sqrt{99}}{1} = -\sqrt{99} \qquad \cot \theta = \frac{a}{b} = \frac{1}{-\sqrt{99}} = -\frac{1}{\sqrt{99}}$$

$$\csc \theta = \frac{r}{b} = \frac{10}{-\sqrt{99}} = -\frac{10}{\sqrt{99}}$$

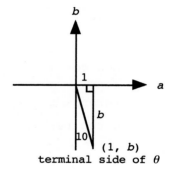

35. $(a, b) = (6, 8)$

$$r = \sqrt{a^2 + b^2} = \sqrt{6^2 + 8^2} = \sqrt{100} = 10$$

$$\sin \theta = \frac{b}{r} = \frac{8}{10} = \frac{4}{5} \qquad \csc \theta = \frac{r}{b} = \frac{10}{8} = \frac{5}{4}$$

$$\cos \theta = \frac{a}{r} = \frac{6}{10} = \frac{3}{5} \qquad \sec \theta = \frac{r}{a} = \frac{10}{6} = \frac{5}{3}$$

$$\tan \theta = \frac{b}{a} = \frac{8}{6} = \frac{4}{3} \qquad \cot \theta = \frac{a}{b} = \frac{6}{8} = \frac{3}{4}$$

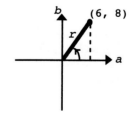

37. $(a, b) = (-1, \sqrt{3})$

$$r = \sqrt{a^2 + b^2} = \sqrt{(-1)^2 + (\sqrt{3})^2} = \sqrt{4} = 2$$

$$\sin \theta = \frac{b}{r} = \frac{\sqrt{3}}{2} \qquad\qquad \csc \theta = \frac{r}{b} = \frac{2}{\sqrt{3}}$$

$$\cos \theta = \frac{a}{r} = \frac{-1}{2} = -\frac{1}{2} \qquad \sec \theta = \frac{r}{a} = \frac{2}{-1} = -2$$

$$\tan \theta = \frac{b}{a} = \frac{\sqrt{3}}{-1} = -\sqrt{3} \qquad \cot \theta = \frac{a}{b} = \frac{-1}{\sqrt{3}} = -\frac{1}{\sqrt{3}}$$

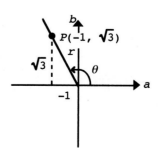

39. $(a, b) = (112, 15)$

$r = \sqrt{a^2 + b^2} = \sqrt{112^2 + 15^2} = \sqrt{12,769} = 113$

$\sin\theta = \dfrac{b}{r} = \dfrac{15}{113}$ $\qquad$ $\csc\theta = \dfrac{r}{b} = \dfrac{113}{15}$

$\cos\theta = \dfrac{a}{r} = \dfrac{112}{113}$ $\qquad$ $\sec\theta = \dfrac{r}{a} = \dfrac{113}{112}$

$\tan\theta = \dfrac{b}{a} = \dfrac{15}{112}$ $\qquad$ $\cot\theta = \dfrac{a}{b} = \dfrac{112}{15}$

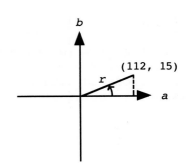

41. $(a, b) = (3, -2)$

$r = \sqrt{a^2 + b^2} = \sqrt{3^2 + (-2)^2} = \sqrt{13}$

$\sin\theta = \dfrac{b}{r} = -\dfrac{2}{\sqrt{13}}$ $\qquad$ $\csc\theta = \dfrac{r}{b} = -\dfrac{\sqrt{13}}{2}$

$\cos\theta = \dfrac{a}{r} = \dfrac{3}{\sqrt{13}}$ $\qquad$ $\sec\theta = \dfrac{r}{a} = \dfrac{\sqrt{13}}{3}$

$\tan\theta = \dfrac{b}{a} = -\dfrac{2}{3}$ $\qquad$ $\cot\theta = \dfrac{a}{b} = -\dfrac{3}{2}$

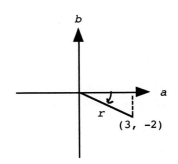

43. Let $f(x) = \dfrac{\tan x}{x}$

$\begin{aligned} f(-x) &= \dfrac{\tan(-x)}{(-x)} \\ &= \dfrac{-\tan x}{-x} \\ &= \dfrac{\tan x}{x} \\ &= f(x) \end{aligned}$

$y = \dfrac{\tan x}{x}$ is even.

45. Let $f(x) = \dfrac{\csc x}{x}$

$\begin{aligned} f(-x) &= \dfrac{\csc(-x)}{-x} \\ &= \dfrac{\frac{1}{\sin(-x)}}{-x} \\ &= \dfrac{\frac{1}{-\sin x}}{-x} \\ &= \dfrac{-\csc x}{-x} \\ &= \dfrac{\csc x}{x} \\ &= f(x) \end{aligned}$

$y = \dfrac{\csc x}{x}$ is even.

47. Let $f(x) = \sin x \cos x$

$\begin{aligned} f(-x) &= \sin(-x)\cos(-x) \\ &= -\sin x \cos x \\ &= -f(x) \end{aligned}$

$y = \sin x \cos x$ is odd.

49. Let $f(x) = x^2 \sin x$

$\begin{aligned} f(-x) &= (-x)^2 \sin(-x) \\ &= x^2(-\sin x) \\ &= -x^2 \sin x \\ &= -f(x) \end{aligned}$

$y = x^2 \sin x$ is odd.

51. $\cos\theta = \dfrac{a}{r} = \dfrac{-1}{2}$ because $r > 0$.

a is negative in the II and III quadrants. The smallest positive θ is associated with a 60° reference triangle in the II quadrant as shown in the figure at the right.

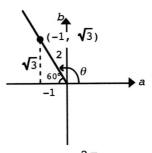

$\theta = 120°$ or $\dfrac{2\pi}{3}$ radians

53. $\sin \theta = \dfrac{b}{r} = \dfrac{-1}{2}$ because $r > 0$.

b is negative in the III and IV quadrants. The smallest positive θ is associated with a 30° reference triangle in the III quadrant as drawn.

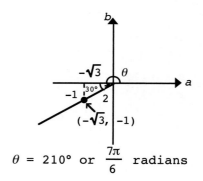

$\theta = 210°$ or $\dfrac{7\pi}{6}$ radians

55. $\csc \theta = \dfrac{r}{b} = \dfrac{2}{-\sqrt{3}}$ because $r > 0$.

b is negative in the III and IV quadrants. The smallest positive θ is associated with a 60° reference triangle in the III quadrant as drawn.

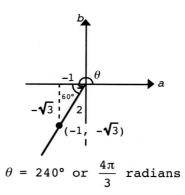

$\theta = 240°$ or $\dfrac{4\pi}{3}$ radians

57. Points (a, b) on the vertical axis have $a = 0$. In this case, functions for which a is in the denominator are not defined. These functions are tangent and secant.

59. $\cos \theta = \dfrac{a}{r} = -\dfrac{\sqrt{3}}{2} = \dfrac{-\sqrt{3}}{2}$

$(a, b) = (-\sqrt{3}, 1)$ or $(-\sqrt{3}, -1)$

θ is associated with a 30° reference triangle in the II quadrant or the III quadrant as drawn:

$\theta = 150°$ or $210°$

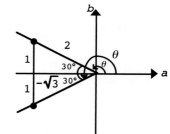

61. $\tan \theta = \dfrac{b}{a} = \dfrac{1}{1} = \dfrac{-1}{-1}$

$(a, b) = (1, 1)$ or $(-1, -1)$

θ is associated with a $\dfrac{\pi}{4}$ reference triangle in the I quadrant or the III quadrant as drawn:

$\theta = \dfrac{\pi}{4}$ or $\dfrac{5\pi}{4}$

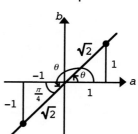

63. False. $\sec x$ and $\csc x$ are never 0.

65. True. The definition of periodicity implies
$f(x + 2p) = f(x + p + p) = f(x + p) = f(x)$
for all x.

67. False. $f(x) = x + 1$ is neither even nor odd.

69. False. $\sin x$ and $\cos x$ have period 2π; however, $\dfrac{\sin x}{\cos x} = \tan x$ has period π.

71. True. If $f(-x) = -f(x)$ and $g(-x) = -g(x)$, then
$(fg)(-x) = f(-x)g(-x) = [-f(x)][-g(x)] = f(x)g(x) = (fg)(x)$ for all x in the
domain of $fg(x)$.

73. A function $f(x)$ is periodic if there is some non-zero number p for which
$f(x + p) = f(x)$ for all x. Applying this to a function of the form
$f(x) = ax + b$, we get:
$$f(x + p) = f(x)$$
$$a(x + p) + b = ax + b$$
$$ax + ap + b = ax + b$$
$$ap = 0$$

We know p is not zero, so a must be zero. But then $f(x) = ax + b$ is just
$f(x) = b$ for any real number b. In other words, f is a constant function.

75. A function $f(x)$ is even if $f(-x) = f(x)$ for all x. Applying this to a function
of the form $f(x) = ax + b$, we get:
$$f(-x) = f(x)$$
$$a(-x) + b = ax + b$$
$$-ax = ax$$
$$2ax = 0$$

Since this has to hold for EVERY x, a has to be zero. In this case,
$f(x) = ax + b$ is just $f(x) = b$ for any real number b. In other words, f is a
constant function.

77. (A)

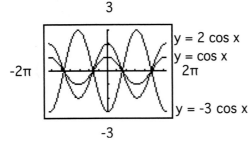

(B) The x intercepts do not change.

(C) The deviation of $y = \cos x$ from the
x axis is 1 unit; the deviation of
$y = 2 \cos x$ from the x axis is 2 units;
the deviation of $y = -3 \cos x$ from the
x axis is 3 units.

(D) The deviation of the graph from the
x axis is changed by changing A. The
deviation appears to be $|A|$.

79. (A)

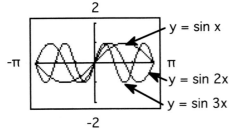

(B) 1 period of $y = \sin x$ appears.
2 periods of $y = \sin 2x$ appear.
3 periods of $y = \sin 3x$ appear.

(C) n periods of $y = \sin nx$ would
appear.

81. (A)

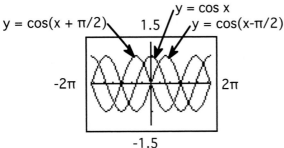

(B) The graph of $y = \cos x$ is
shifted $|C|$ units to the right
if $C < 0$ and $|C|$ units to the left
if $C > 0$.

83. For each case, the number is not in the domain of the function and an error message of some type will appear.

85. Here are graphs of $f(x) = \sin x$ and $g(x) = x$, $-1 \le x \le 1$.

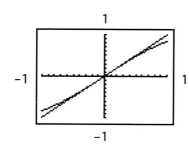

(A) The graphs become more indistinguishable the closer x is to the origin.

(B)

x	-0.3	-0.2	-0.1	0.0	0.1	0.2	0.3
$\sin x$	-0.296	-0.199	-0.100	0.000	0.100	0.199	0.296

87. (A) Since $\theta_R = \dfrac{s}{r}$ and r = radius of circle = 4, we have $\theta_R = \dfrac{7}{4}$ or 1.75 radians

(B) Since $\sin \theta = \dfrac{b}{r}$ and $\cos \theta = \dfrac{a}{r}$, we can write

$$a = r \cos \theta = 4 \cos \frac{7}{4} = -0.713$$

$$b = r \sin \theta = 4 \sin \frac{7}{4} = 3.936$$

$(a, b) = (-0.713, 3.936)$

89. We know that $s = r\theta$. $(a, b) = (6\sqrt{3}, 6)$.
From the Pythagorean theorem,
$$r = \sqrt{a^2 + b^2} = \sqrt{(6\sqrt{3})^2 + 6^2} = 12$$
Since $\tan \theta = \dfrac{b}{a} = \dfrac{6}{6\sqrt{3}} = \dfrac{1}{\sqrt{3}}$, $\theta = \dfrac{\pi}{6}$.
Then $s = r\theta = 12\left(\dfrac{\pi}{6}\right) = 2\pi$ units.

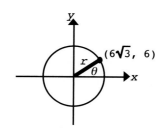

91. $I = k \cos \theta$
For $\theta = 0°$, $I = k \cos 0° = k$
For $\theta = 30°$, $I = k \cos 30° = 0.866k$
For $\theta = 60°$, $I = k \cos 60° = 0.5k$

93. From the figure, we can note the following relationships:
$$a^2 + b^2 = r^2 \qquad r = 1 \qquad \frac{a}{1} = \cos \theta \qquad \frac{b}{1} = \sin \theta$$
Using the Pythagorean theorem in the right triangle whose hypotenuse is the rod connecting the piston to the wheel, we have
$$(y - b)^2 + a^2 = 4^2$$
$$(y - b)^2 = 4^2 - a^2$$
$$y - b = \sqrt{4^2 - a^2}$$
$$y = b + \sqrt{4^2 - a^2}$$
Since $a = \cos \theta$ and $b = \sin \theta$ and $\theta = 6\pi t$, we have
$$y = \sin 6\pi t + \sqrt{16 - (\cos 6\pi t)^2}$$

95. $A = n \tan\left(\dfrac{180°}{n}\right)$

(A) For $n = 8$, $A = 8 \tan \dfrac{180°}{8} = 3.31371$

For $n = 100$, $A = 100 \tan \dfrac{180°}{100} = 3.14263$

For $n = 1000$, $A = 1000 \tan \dfrac{180°}{1000} = 3.14160$

For $n = 10{,}000$, $A = 10{,}000 \tan \dfrac{180°}{10{,}000} = 3.14159$

(B) as $n \to \infty$, A seems to approach $\pi\,(= 3.1415926\ldots)$, the area of the circle.

97. (A) Using the formula given:

For $\theta = 88.7°$, $m = \tan\theta = \tan 88.7° = 44.07$

For $\theta = 162.3°$, $m = \tan\theta = \tan 162.3° = -0.32$

(B) Using the formula for inclination, the slope m is given by $m = \tan 137°$

$= -0.93$. We now use the point-slope form of the equation of a line.

$y - y_0 = m(x - x_0)$

$y - 5 = -0.93[x - (-4)]$

$y - 5 = -0.93x - 3.72$

$y = -0.93x + 1.28$

Section 5-5

1. Answers will vary.   **3.** Answers will vary.   **5.** Answers will vary.

7. Amplitude = $|A| = |3| = 3$ Period = $\dfrac{2\pi}{B} = \dfrac{2\pi}{1} = 2\pi$

9. Amplitude = $|A| = \left|-\dfrac{1}{2}\right| = \dfrac{1}{2}$ Period = $\dfrac{2\pi}{B} = \dfrac{2\pi}{1} = 2\pi$

11. Amplitude = $|A| = |1| = 1$ Period = $\dfrac{2\pi}{B} = \dfrac{2\pi}{3}$

13. Amplitude is not defined for the cotangent function. Period = $\dfrac{\pi}{B} = \dfrac{\pi}{4}$

15. Amplitude is not defined for the tangent function. Period = $\dfrac{\pi}{B} = \dfrac{\pi}{8\pi} = \dfrac{1}{8}$

17. Amplitude is not defined for the cosecant function. Period = $\dfrac{2\pi}{B} = \dfrac{2\pi}{1/2} = 4\pi$

19. Amplitude = $|A| = |1| = 1$. Period = $\dfrac{2\pi}{B} = \dfrac{2\pi}{\pi} = 2$

The basic sine function $\sin t$ has zeros when $t = k\pi$, k an integer.

So the zeros occur when $\pi x = k\pi$, or $x = k$. Then x falls in the interval

$-2 \le x \le 2$ when $x = -2, -1, 0, 1,$ or 2.

21. Amplitude is not defined for the cotangent function. Period $= \dfrac{\pi}{1/2} = 2\pi$.

The basic cotangent function $\cot t$ has zeros when $t = \dfrac{\pi}{2} + k\pi$, k an integer.

So the zeros occur when $\dfrac{x}{2} = \dfrac{\pi}{2} + k\pi$, or $x = \pi + 2k\pi$. Then x falls in the interval $0 < x < 4\pi$ when $k = 1$ or 2, in which case $x = \pi$ or 3π.

23. Amplitude $= |A| = |3| = 3$. Period $= \dfrac{2\pi}{B} = \dfrac{2\pi}{2} = 1$

The basic cosine function $\cos t$ has turning points when $t = k\pi$, k an integer.

So the turning points occur when $2x = k\pi$, or $x = \dfrac{k\pi}{2}$. Then x falls in the interval $-\pi \le x \le \pi$ when $k = -2, -1, 0, 1,$ or 2, in which case $x = -\pi, -\dfrac{\pi}{2}, 0, \dfrac{\pi}{2}$,

or π. The turning points are therefore $(-\pi,\ 3)$, $\left(-\dfrac{\pi}{2}, -3\right)$, $(0,\ 3)$, $\left(\dfrac{\pi}{2}, -3\right)$,

$(\pi,\ 3)$.

25. Amplitude is not defined for the secant function. Period $= \dfrac{2\pi}{B} = \dfrac{2\pi}{\pi} = 2$.

The basic secant function $\sec t$ has turning points when $t = k\pi$, k an integer.
So the turning points occur when $\pi x = k\pi$, or $x = k$. Then x falls in the interval $-1 \le x \le 3$ when $x = -1, 0, 1, 2,$ or 3. The turning points are therefore $(-1, -2)$, $(0, 2)$, $(1, -2)$, $(2, 2)$, $(3, -2)$.

27. $A = 3$ $\quad P = \dfrac{\pi}{2} = \dfrac{2\pi}{4}$. So $B = 4$, and

$y = 3 \sin 4x$, $-\dfrac{\pi}{4} \le x \le \dfrac{\pi}{2}$.

29. $|A| = 10$ $\quad P = 2 = \dfrac{2\pi}{\pi}$. So $B = \pi$, and $A = -10$, since the graph has the form of the standard sine curve turned upside down. $y = -10 \sin \pi x$ $-1 \le x \le 2$.

31. $A = 5$ $\quad P = 8\pi = 2\pi \cdot 4 = 2\pi \div \dfrac{1}{4}$.

So $B = \dfrac{1}{4}$, and

$y = 5 \cos \dfrac{1}{4}x$ $\quad -4\pi \le x \le 8\pi$.

33. $|A| = 0.5$, $P = 8 = 2\pi \cdot \dfrac{4}{\pi} = 2\pi \div \dfrac{\pi}{4}$.

So $B = \dfrac{\pi}{4}$, and $A = -0.5$, since the graph has the form of the standard cosine curve turned upside down.

$y = -0.5 \cos \dfrac{\pi x}{4}$ $\quad -4 \le x \le 8$

35. $y = 4 \cos x$

Amplitude $= |A| = 4$ $\quad$ Period $= 2\pi$

Phase Shift $= 0$

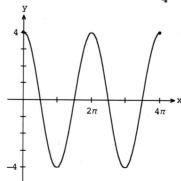

37. $y = \frac{1}{2} \sin\left(x + \frac{\pi}{4}\right)$ completes one cycle as $x + \frac{\pi}{4}$ varies from $x + \frac{\pi}{4} = 0$ to

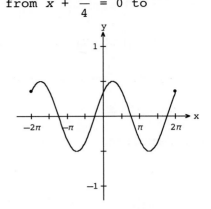

$x + \frac{\pi}{4} = 2\pi$, that is as x varies from $-\frac{\pi}{4}$ to $-\frac{\pi}{4} + 2\pi$.

Amplitude: $|A| = \frac{1}{2}$ Period: 2π Phase shift: $-\frac{\pi}{4}$

Divide the interval $\left[-\frac{\pi}{4}, -\frac{\pi}{4} + 2\pi\right]$ into four equal

parts and sketch one cycle of $y = \frac{1}{2} \sin\left(x + \frac{\pi}{4}\right)$. Then

extend the graph to cover $[-2\pi, 2\pi]$, deleting the
small portion beyond $x = 2\pi$.

39. $y = \cot\left(x - \frac{\pi}{6}\right)$ completes one period as $x - \frac{\pi}{6}$ varies from $x - \frac{\pi}{6} = 0$ to

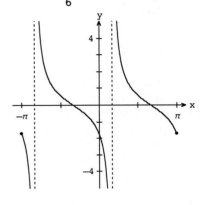

$x - \frac{\pi}{6} = \pi$, that is as x varies from $\frac{\pi}{6}$ to $\frac{\pi}{6} + \pi$.

Period: π Phase shift: $\frac{\pi}{6}$

Sketch one cycle of $y = \cot\left(x - \frac{\pi}{6}\right)$, the graph of

$y = \cot x$ shifted $\frac{\pi}{6}$ units to the right, on the

interval $\left(\frac{\pi}{6}, \frac{\pi}{6} + \pi\right)$. Then extend the graph to cover

$[-\pi, \pi]$, deleting the portion beyond π.

41. $y = 3 \tan 2x$ completes one period as $2x$ varies from $2x = -\frac{\pi}{2}$ to $2x = \frac{\pi}{2}$

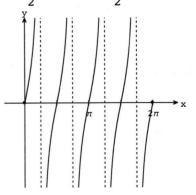

that is, as x varies from $-\frac{\pi}{4}$ to $\frac{\pi}{4}$.

Period: $\frac{\pi}{2}$ Phase shift: 0

Sketch one cycle of $y = 3 \tan 2x$, the graph of
$y = \tan x$ stretched vertically by a factor of 3 and
shrunk horizontally by a factor of 2, on the interval
$\left(-\frac{\pi}{4}, \frac{\pi}{4}\right)$. Then extend the graph to cover $[0, 2\pi]$.

43. $y = 2\pi \sin \frac{\pi x}{2}$ completes one cycle as $\frac{\pi x}{2}$ varies from

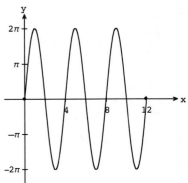

$\frac{\pi x}{2} = 0$ to $\frac{\pi x}{2} = 2\pi$

that is, as x varies from 0 to 4.
Amplitude $= |A| = 2\pi$ Period: 4   Phase shift: 0

Divide the interval $[0, 4]$ into four equal parts and

sketch one cycle of $y = 2\pi \sin \frac{\pi x}{2}$, then extend the

graph to cover $[0, 12]$.

45. $y = -3 \sin\left[2\pi\left(x + \dfrac{1}{2}\right)\right]$ completes one cycle as $2\pi\left(x + \dfrac{1}{2}\right)$ varies from

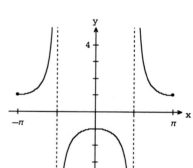

$$2\pi\left(x + \dfrac{1}{2}\right) = 0 \quad \text{to} \quad 2\pi\left(x + \dfrac{1}{2}\right) = 2\pi$$

$$x + \dfrac{1}{2} = 0 \qquad\qquad x + \dfrac{1}{2} = 1$$

$$x = -\dfrac{1}{2} \qquad\qquad x = \dfrac{1}{2}$$

Amplitude $= |A| = |-3| = 3$ Period: $\dfrac{1}{2} - \left(-\dfrac{1}{2}\right) = 1$

Phase shift: $-\dfrac{1}{2}$

Divide the interval $\left[-\dfrac{1}{2}, \dfrac{1}{2}\right]$ into four equal parts and

sketch one cycle of $y = -3 \sin\left[2\pi\left(x + \dfrac{1}{2}\right)\right]$—an upside-down sine curve. Then extend

the graph to $[-1, 2]$.

47. $y = \sec(x + \pi)$ has period 2π and completes one period
as $x + \pi$ varies from $x + \pi = 0$ to $x + \pi = 2\pi$, that is,
as x varies from $-\pi$ to π.
Phase shift: $-\pi$
The required graph is this one period, the graph of
$y = \sec x$ shifted π units to the left.

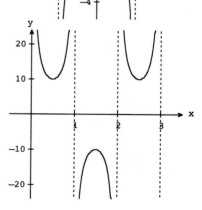

49. $y = 10 \csc \pi x$ completes one period as πx varies from
$\pi x = 0$ to $\pi x = 2\pi$, that is, as x varies from 0 to 2.
Period: 2 Phase shift: 0
Sketch one period of $y = 10 \csc \pi x$, the graph of
$y = \csc x$ stretched vertically by a factor of 10 and

shrunk horizontally by a factor of $\dfrac{1}{\pi}$, on this

interval. Then extend the graph to $[0, 3]$.

51. True. If $x = 0$, $y = A \sin B(0) = 0$. So the point $(0, 0)$ is on the graph.

53. False. The function $y = \sin\left(x - \dfrac{\pi}{4}\right)$ is neither even nor odd.

55. True. Every function of form $A \sin(Bx + C)$ or $A \cos(Bx + C)$ has period $\dfrac{2\pi}{B}$.

57. True. $|A \sin(Bx + C)| \le |A|$ and $|A \cos(Bx + C)| \le |A|$ for all x.

59. True. $A \sin(3Bx + C)$ and $A \cos(3Bx + C)$ define simple harmonics.

61. False. $y = 3 - \sin x$ cannot be represented as $y = A \sin(Bx + C)$.

63. Here is a graph of
$y = \cos^2 x - \sin^2 x,$
$-\pi \le x \le \pi.$

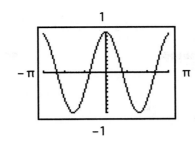

The graph has amplitude 1
and period π. It appears
to be the graph of $y =$
$A \cos Bx$ with $A = 1$, and
$B = 2\pi \div P = 2\pi \div \pi = 2,$
that is, $y = \cos 2x.$

65. Here is a graph of
$y = 2 \sin^2 x,$
$-\pi \le x \le \pi.$

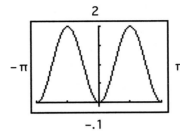

The graph has amplitude
$\dfrac{2 - 0}{2} = 1$ and period $\pi.$
It appears to be the graph
of $y = \cos 2x$ turned upside
down and shifted up one
unit, that is,
$y = -\cos 2x + 1$ or
$y = 1 - \cos 2x.$

67. Here is a graph of
$y = \cot x - \tan x,$
$-\dfrac{\pi}{2} \le x \le \dfrac{\pi}{2}$

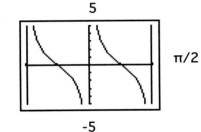

The graph appears to have the form
$y = A \cot Bx$. Since the period is $\dfrac{\pi}{2}$,
set $\dfrac{\pi}{2} = \dfrac{\pi}{B}$ to obtain $B = 2$. The graph of
$y = A \cot 2x$ shown appears to pass through
$\left(\dfrac{\pi}{8}, 2\right)$, so $2 = A \cot 2\left(\dfrac{\pi}{8}\right)$
$$2 = A \cot \dfrac{\pi}{4}$$
$$2 = A$$
The equation of the graph can be written
$y = 2 \cot 2x.$

69. Here is a graph of
$y = \csc x + \cot x,$
$-2\pi \le x \le 2\pi.$

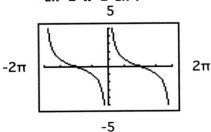

The graph appears to have the form
$y = A \cot Bx$. Since the period is 2π, set $2\pi = \dfrac{\pi}{B}$
to obtain $B = \dfrac{1}{2}$. The graph of $y = A \cot \dfrac{1}{2}x$ shown
appears to pass through $\left(\dfrac{\pi}{2}, 1\right)$, so
$$1 = A \cot \dfrac{1}{2}\left(\dfrac{\pi}{2}\right)$$
$$1 = A \cot \dfrac{\pi}{4}$$
$$1 = A$$
The equation of the graph can be written
$y = \cot \dfrac{1}{2}x.$

71. Here is a graph of
$y = \sin 3x + \cos 3x \cot 3x$,
$-\dfrac{2\pi}{3} \le x \le \dfrac{2\pi}{3}$.

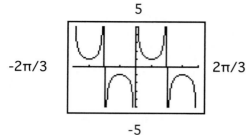

The graph appears to have the form
$y = A \csc Bx$. Since the period is $\dfrac{2\pi}{3}$,
set $\dfrac{2\pi}{3} = \dfrac{2\pi}{B}$ to obtain $B = 3$. The graph of
$y = A \csc 3x$ shown appears to pass through
$\left(\dfrac{\pi}{6}, 1\right)$, so

$$1 = A \csc 3\left(\dfrac{\pi}{6}\right)$$
$$1 = A \csc \dfrac{\pi}{2}$$
$$1 = A$$

The equation of the graph can be written
$y = \csc 3x$

73. Here is a graph of
$y = \dfrac{\sin 4x}{1 + \cos 4x}$, $-\dfrac{\pi}{2} \le x \le \dfrac{\pi}{2}$.

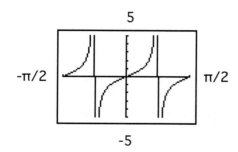

The graph appears to have the form
$y = A \tan Bx$. Since the period is $\dfrac{\pi}{2}$, set $\dfrac{\pi}{B} = \dfrac{\pi}{2}$ to obtain $B = 2$. The graph of $y = A \tan 2x$
shown appears to pass through $\left(\dfrac{\pi}{8}, 1\right)$, so

$$1 = A \tan 2\left(\dfrac{\pi}{8}\right)$$
$$1 = A \tan \dfrac{\pi}{4}$$
$$1 = A$$

The equation of the graph can be written
$y = \tan 2x$.

75. $|A| = 4$, so $A = 4$ or -4. The graph completes one full cycle as x varies over
the interval $[-1, 3]$. Since $-\dfrac{C}{B}$ is required to be between 0 and 2, we cannot
simply set $-\dfrac{C}{B} = -1$. We must (mentally) extend the curve so that the phase
shift is positive. Then the (extended) graph is that of an upside down sine
curve that completes one full cycle as x varies over the interval $[1, 5]$.

$$A = -4$$
$$-\dfrac{C}{B} = 1 \qquad -\dfrac{C}{B} + \dfrac{2\pi}{B} = 5$$
$$C = -B \qquad \dfrac{2\pi}{B} = 4 \qquad B = \dfrac{2\pi}{4} = \dfrac{\pi}{2} \qquad C = -\dfrac{\pi}{2}$$

The equation is then
$y = A \sin(Bx + C)$
$$y = -4 \sin\left(\dfrac{\pi}{2} x - \dfrac{\pi}{2}\right)$$

77. $|A| = $, so $A = \dfrac{1}{2}$ or $-\dfrac{1}{2}$. The graph completes one full cycle of the cosine function as x varies over the (mentally extended) intervals $[-\pi,\ 7\pi]$ or $[3\pi,\ 11\pi]$. Since the phase shift is required to be between 0 and 4π, we must set $-\dfrac{C}{B} = 3\pi$. Then the (extended) graph has the form of a standard cosine curve.

$$A = \frac{1}{2} \qquad -\frac{C}{B} + \frac{2\pi}{B} = 11\pi$$

$$-\frac{C}{B} = 3\pi \qquad \frac{2\pi}{B} = 8\pi \qquad B = \frac{2\pi}{8\pi} = \frac{1}{4} \qquad C = -3\pi B = -\frac{3\pi}{4}$$

$$C = -3\pi B$$

The equation is then

$$y = A\cos(Bx + C)$$

$$y = \frac{1}{2}\cos\left(\frac{1}{4}x - \frac{3\pi}{4}\right)$$

79. First, we will choose a cosine function since this will give us the smallest possible phase shift: there is a turning point closer to the y axis than any zero. This looks like a cosine graph shifted one unit to the left, so the phase shift is −1. The amplitude is $A = 3$ and the period is $P = 12$ (the two peaks are at $x = -1$ and $x = 11$). We can calculate B using the period formula:

$$P = \frac{2\pi}{B} \quad\Rightarrow\quad 12 = \frac{2\pi}{B} \quad\Rightarrow\quad 12B = 2\pi \quad\Rightarrow\quad B = \frac{2\pi}{12} = \frac{\pi}{6}$$

We know the phase shift is −1, so we can calculate C using the phase shift formula, phase shift $= -\dfrac{C}{B}$:

$$-1 = -\frac{C}{B} = -\frac{C}{\pi/6} \quad\Rightarrow\quad -\frac{\pi}{6} = -C \quad\Rightarrow\quad C = \frac{\pi}{6}$$

We now have $A = 3$, $B = \pi/6$, and $C = \pi/6$, so the equation is

$$y = 3\cos\left(\frac{\pi}{6}x + \frac{\pi}{6}\right)$$

81. First, we will choose a sine function since this will give us the smallest possible phase shift: there is a zero closer to the y axis than any turning point. This looks like an upside down sine graph shifted one unit to the right, so the phase shift is 1. The amplitude is $|A| = 2$, and the period is $P = 12$ (the first two zeros are at $x = 1$ and $x = 7$, and the distance between them is half the period). We can calculate B using the period formula:

$$P = \frac{2\pi}{B} \quad\Rightarrow\quad 12 = \frac{2\pi}{B} \quad\Rightarrow\quad 12B = 2\pi \quad\Rightarrow\quad B = \frac{2\pi}{12} = \frac{\pi}{6}$$

We know the phase shift is 1, so we can calculate C using the phase shift formula, phase shift $= -\dfrac{C}{B}$:

$$1 = -\frac{C}{B} = -\frac{C}{\pi/6} \quad\Rightarrow\quad \frac{\pi}{6} = -C \quad\Rightarrow\quad C = -\frac{\pi}{6}$$

We now have $A = -2$ (upside down sine curve), $B = \pi/6$, and $C = -\pi/6$, so the equation is

$$y = -2\sin\left(\frac{\pi}{6}x - \frac{\pi}{6}\right)$$

83. $y = 3.5 \sin\left[\dfrac{\pi}{2}(t + 0.5)\right]$

$A = 3.5$

Solve $\dfrac{\pi}{2}(t + 0.5) = 0 \qquad \dfrac{\pi}{2}(t + 0.5) = 2\pi$

$\qquad\qquad t + 0.5 = 0 \qquad\qquad t + 0.5 = 4$

$\qquad\qquad\qquad t = -0.5 \qquad\qquad\qquad t = -0.5 + 4 = 3.5$

$\qquad\qquad\qquad\qquad$ Phase shift $\qquad$ Period $P = 4$

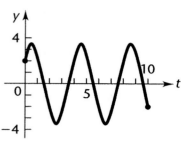

The graph completes one full cycle as t varies over the interval $[-0.5, 3.5]$.

85. $y = 50 \cos[2\pi(t - 0.25)]$

$A = 50$

Solve $2\pi(t - 0.25) = 0 \qquad 2\pi(t - 0.25) = 2\pi$

$\qquad\qquad t - 0.25 = 0 \qquad\qquad t - 0.25 = 1$

$\qquad\qquad\qquad t = 0.25 \qquad\qquad\qquad t = 0.25 + 1 = 1.25$

$\qquad\qquad\qquad\qquad$ Phase shift $\qquad$ Period $P = 1$

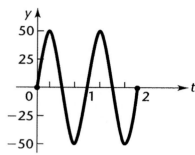

The graph completes one full cycle as t varies over the interval $[0.25, 1.25]$.

87. Here is a graph of $y = \sqrt{2}\, \sin x + \sqrt{2}\, \cos x$, $-2\pi \le x \le 2\pi$. It appears that this is a sine curve shifted to the left, with $A = 2$ and, since $P = \dfrac{2\pi}{B}$ and P appears to be 2π, $B = \dfrac{2\pi}{P} = \dfrac{2\pi}{2\pi} = 1$.

The x intercept closest to the origin, to three decimal places, is -0.785. To find C, substitute $B = 1$ and $x = -0.785$ into the phase-shift formula $x = -\dfrac{C}{B}$ and solve for C:

$$x = -\frac{C}{B}$$
$$-0.785 = -\frac{C}{1}$$
$$C = 0.785$$

The equation is $y = 2 \sin(x + 0.785)$.

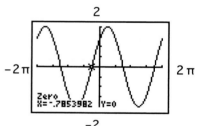

89. Here is a graph of $y = \sqrt{3}\, \sin x - \cos x$, $-2\pi \le x \le 2\pi$. It appears that this is a sine curve shifted to the right, with $A = 2$ and, since $P = \dfrac{2\pi}{B}$ and P appears to be 2π, $B = \dfrac{2\pi}{P} = \dfrac{2\pi}{2\pi} = 1$.

The x intercept closest to the origin, to three decimal places, is 0.524. To find C, substitute $B = 1$ and $x = 0.524$ into the phase-shift formula $x = -\dfrac{C}{B}$ and solve for C:

$$x = -\frac{C}{B}$$
$$0.524 = -\frac{C}{1}$$
$$C = -0.524$$

The equation is $y = 2 \sin(x - 0.524)$.

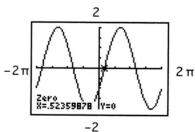

91. Here is a graph of $y = 4.8 \sin 2x - 1.4 \cos 2x$, $-\pi \le x \le \pi$.

It appears that this is a sine curve shifted to the right, with $A = 5$ and, since $P = \dfrac{2\pi}{B}$ and P appears to be π, $B = \dfrac{2\pi}{P} = \dfrac{2\pi}{\pi} = 2$.

The x intercept closest to the origin, to three decimal places, is 0.142. To find C, substitute $B = 2$ and $x = 0.142$ into the phase-shift formula $x = -\dfrac{C}{B}$ and solve for C:

$$x = -\frac{C}{B}$$
$$0.142 = -\frac{C}{2}$$
$$C = -0.284$$

The equation is $y = 5 \sin(2x - 0.284)$.

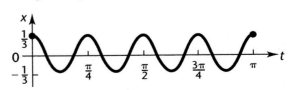

93.

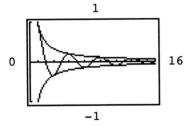

The amplitude is decreasing with time. This is often referred to as a *damped sine wave*. Examples are the vertical motion of a car after going over a bump (which is damped by the suspension system) and the slowing down of a pendulum that is released away from the vertical line of suspension (air resistance and friction).

95.

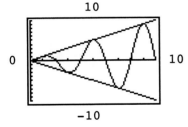

The amplitude is increasing with time. In physical and electrical systems this is referred to as *resonance*. Some examples are the swinging of a bridge during high winds and the movement of tall buildings during an earthquake. Some bridges and buildings are destroyed when the resonance reaches the elastic limits of the structure.

97.

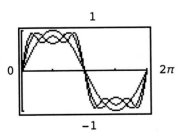

99. $A = \dfrac{1}{3}$

$P = \dfrac{2\pi}{8} = \dfrac{\pi}{4}$

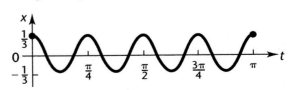

101. When $t = 0$ $y = -8$. So $-8 = A \cos B(0)$, that is, $A = -8$. Since the period is 0.5 seconds, $\dfrac{2\pi}{B} = 0.5$, $B = \dfrac{2\pi}{0.5} = 4\pi$. The equation is $y = -8 \cos 4\pi t$.

103. The graph is the same as the graph of $y = \cos \dfrac{n\pi}{26}$, shifted 1.5 units up.

$A = 1$

$P = 2\pi \div \dfrac{\pi}{26} = 52$

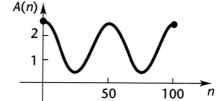

The graph shows the seasonal changes of sulfur dioxide pollutant in the atmosphere; more is produced during winter months because of increased heating.

105. (A) The amplitude is 15; the period is

$P = \dfrac{2\pi}{B} = \dfrac{2\pi}{120\pi} = \dfrac{1}{60}$; the phase shift is

$-\dfrac{C}{B} = -\dfrac{\pi/2}{120\pi} = -\dfrac{1}{240}$.

(B)

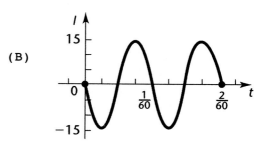

(C) This first minimum occurs 1/4 of the way through the first cycle on the graph that begins at $x = 0$. So it occurs at 1/4 of the period, which is $t = 1/240$.

107. If the disk rotates through an angle θ in t seconds, we see

$$\theta = 3\frac{\text{revolutions}}{\text{second}} \cdot 2\pi\frac{\text{radians}}{\text{revolution}} \cdot t \text{ seconds}$$

$$= 6\pi t \text{ radians}$$

Then $\dfrac{y}{R} = \sin \theta = \sin 6\pi t$

$y = R \sin 6\pi t$

Since the disk has radius 3, $y = 3 \sin 6\pi t$.

This function has $A = 3$, $P = \dfrac{2\pi}{6\pi} = \dfrac{1}{3}$.

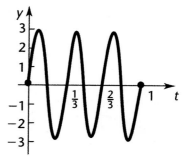

109. (A) In triangle *MNP*, we can write

$$\cos \theta = \frac{20}{c}$$

$$c \cos \theta = 20$$

$$c = \frac{20}{\cos \theta}$$

$$c = 20 \sec \theta$$

Since $\theta = \dfrac{\pi t}{2}$, the equation for c in terms of t is

$c = 20 \sec \dfrac{\pi t}{2}$. The equation is valid for $0 \le t < 1$, since $\sec \dfrac{\pi t}{2}$ is undefined when $t = 1$.

(B) The graph has an asymptote at $t = 1$. Sketch the portion of $y = \sec \dfrac{\pi t}{2}$ on $[0, 1)$.

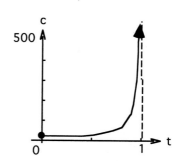

(C) The length of the light beam starts at 20 ft and increases slowly at first, then increases rapidly without end.

111. (A)

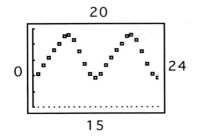

(B) From the table, Max $y = 19{:}35$ and Min $y = 16{:}51$.

$$A = \frac{\text{Max } y - \text{Min } y}{2} = \frac{19\frac{35}{60} - 16\frac{51}{60}}{2} = 1.37$$

$$B = \frac{2\pi}{\text{Period}} = \frac{2\pi}{12} = \frac{\pi}{6}$$

$C = -1.75$ from the graph

$k = \text{Min } y + A = 16\frac{51}{60} + 1.37 = 18.22$

$y = 18.22 + 1.37 \sin(\frac{\pi x}{6} - 1.75)$

(C)

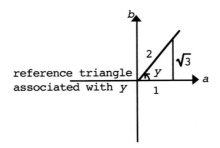

(D)

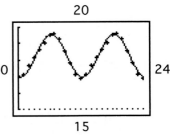

Compute the regression equation.

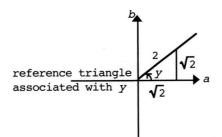

Graph the data and the regression equation.

Section 5-6

1. Answers will vary.   **3.** Answers will vary.   **5.** Answers will vary.

7. $y = \cos^{-1} 0$ is equivalent to
$\cos y = 0 \quad 0 \le y \le \pi$
$y = \dfrac{\pi}{2}$

9. $y = \arcsin \dfrac{\sqrt{3}}{2}$ is equivalent to

$\sin y = \dfrac{\sqrt{3}}{2} \quad -\dfrac{\pi}{2} \le y \le \dfrac{\pi}{2}$

$y = \dfrac{\pi}{3}$

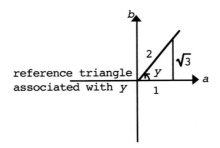

reference triangle associated with y

11. $y = \arctan \sqrt{3}$ is equivalent to
$\tan y = \sqrt{3} \quad -\dfrac{\pi}{2} < y < \dfrac{\pi}{2}$
$y = \dfrac{\pi}{3}$
(see sketch for problem 9)

13. $y = \sin^{-1} \dfrac{\sqrt{2}}{2}$ is equivalent to

$\sin y = \dfrac{\sqrt{2}}{2} \quad -\dfrac{\pi}{2} \le y \le \dfrac{\pi}{2}$

$y = \dfrac{\pi}{4}$

reference triangle associated with y

15. $y = \arccos 1$ is equivalent to
$\cos y = 1 \quad 0 \le y \le \pi$
$y = 0$

17. $y = \sin^{-1} \dfrac{1}{2}$ is equivalent to

$\sin y = \dfrac{1}{2}$ $-\dfrac{\pi}{2} \le y \le \dfrac{\pi}{2}$

$y = \dfrac{\pi}{6}$

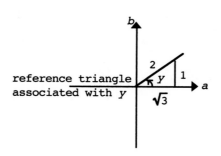

reference triangle associated with y

19. 1.144 **21.** 1.561 **23.** arccos x is not defined if $x > 1$.

25. $y = \arcsin\left(-\dfrac{\sqrt{2}}{2}\right)$ is equivalent to

$\sin y = -\dfrac{\sqrt{2}}{2}$ $-\dfrac{\pi}{2} \le y \le \dfrac{\pi}{2}$

$y = -\dfrac{\pi}{4}$

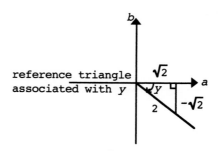

reference triangle associated with y

27. $y = \tan^{-1}(-\sqrt{3})$ is equivalent to

$\tan y = -\sqrt{3}$ $-\dfrac{\pi}{2} < y < \dfrac{\pi}{2}$

$y = -\dfrac{\pi}{3}$

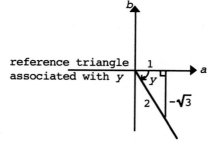

reference triangle associated with y

29. $\tan(\tan^{-1} 25) = 25$ by the tangent-inverse tangent identity.

31. $\cos^{-1}(\cos 2.3) = 2.3$ by the cosine-inverse cosine identity.

33. Let $y = \cos^{-1} \dfrac{\sqrt{3}}{2}$; then $\cos y = \dfrac{\sqrt{3}}{2}$, $0 \le y \le \pi$. Draw the reference triangle

associated with y; then $\sin y = \sin\left[\cos^{-1} \dfrac{\sqrt{3}}{2}\right]$ can be determined directly from

the triangle.

$\cos y = \dfrac{a}{r} = \dfrac{\sqrt{3}}{2}$ $a = \sqrt{3}$ $r = 2$

$(\sqrt{3})^2 + b^2 = 2^2$

$b^2 = 1$

$b = 1$

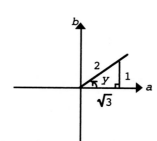

(positive since y is a quadrant I or II angle)

Therefore

$\sin y = \sin\left[\cos^{-1} \dfrac{\sqrt{3}}{2}\right] = \dfrac{1}{2}$.

35. Let $y = \tan(-1)$; then $\tan y = -1$, $-\dfrac{\pi}{2} < y < \dfrac{\pi}{2}$. Draw the reference triangle associated with y; then $\csc y = \csc[\tan^{-1}(-1)]$ can be determined directly from the triangle.

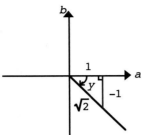

$$\tan y = \dfrac{b}{a} = \dfrac{-1}{1} \qquad b = -1 \qquad a = 1$$
$$(-1)^2 + 1^2 = r^2$$
$$r^2 = 2$$
$$r = \sqrt{2}$$
$$\csc y = \dfrac{r}{b} = \dfrac{\sqrt{2}}{-1} = -\sqrt{2}$$

37. $\sin^{-1}(\sin \pi) = \sin^{-1}(0) = 0$

39. $\cos^{-1}\left(\cos\left(\dfrac{4\pi}{3}\right)\right) = \cos^{-1}\left(-\dfrac{1}{2}\right) = \dfrac{2\pi}{3}$

41. -1.472 **43.** $\cot[\cos^{-1}(-0.7003)] = \dfrac{1}{\tan[\cos^{-1}(-0.7003)]} = -0.9810$ **45.** 2.645

47. In problem 25 we found $\sin^{-1}\left(-\dfrac{\sqrt{2}}{2}\right) = \arcsin\left(-\dfrac{\sqrt{2}}{2}\right) = -\dfrac{\pi}{4}$. Using the relation

$\theta_D = \dfrac{180°}{\pi \text{ rad}} \theta_R$, we have $\sin^{-1}\left(-\dfrac{\sqrt{2}}{2}\right) = \dfrac{180°}{\pi}\left(-\dfrac{\pi}{4}\right) = -45°$.

49. In problem 27 we found $\arctan(-\sqrt{3}) = \tan^{-1}(-\sqrt{3}) = -\dfrac{\pi}{3}$. Using the relation

$\theta_D = \dfrac{180°}{\pi \text{ rad}} \theta_R$, we have $\arctan(-\sqrt{3}) = \dfrac{180°}{\pi}\left(-\dfrac{\pi}{3}\right) = -60°$.

51. In problem 29 we found $\cos^{-1}(-1) = \pi$. Since π rad $= 180°$, we have $\cos^{-1}(-1) = 180°$.

Calculator in degree mode for problems 53 — 58.

53. $43.51°$ **55.** $-21.48°$ **57.** $-89.93°$

59. $\sin^{-1}(\sin 2) = 1.1416 \neq 2$. For the identity $\sin^{-1}(\sin x) = x$ to hold, x must be in the restricted domain of the sine function; that is, $-\dfrac{\pi}{2} \leq x \leq \dfrac{\pi}{2}$. The number 2 is not in the restricted domain.

61. True. A periodic function cannot be one-to-one, since $f(x + p) = f(x)$ but $x + p \neq x$.

63. False. None of them are periodic.

65. True. $\sin^{-1}(-x) = -\sin^{-1}(x)$ for all x, $-1 \leq x \leq 1$

67. True. $\sin^{-1} x$, $\tan^{-1} x$, and $\sec^{-1} x$ are increasing functions on any interval on which they are defined, $\cos^{-1} x$, $\cot^{-1} x$, and $\csc^{-1} x$ similarly decreasing.

69.

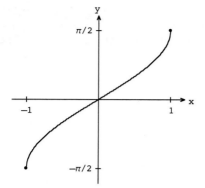

71.

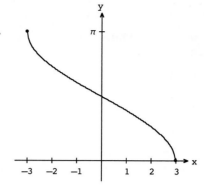

73.

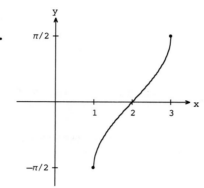

75.

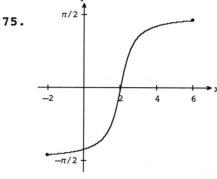

77. (A)

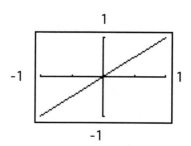

(B) The domain of inverse cosine is restricted to $-1 \leq x \leq 1$; no graph will appear for other x.

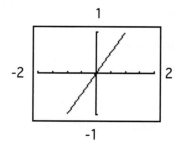

79. Let $y = \sin^{-1} x$. Then $x = \sin y$, $-\dfrac{\pi}{2} \leq y \leq \dfrac{\pi}{2}$. If $\sin y = \dfrac{b}{r} = x = \dfrac{x}{1}$, then let $b = x$, $r = 1$.

$$a^2 + b^2 = r^2$$
$$a^2 + x^2 = 1^2$$
$$a^2 = 1 - x^2$$
$$a = \sqrt{1 - x^2} \quad (a \text{ must be positive since } y \text{ is a I or IV quadrant angle})$$
$$\cos y = \cos(\sin^{-1} x) = \dfrac{a}{r} = \dfrac{\sqrt{1 - x^2}}{1} = \sqrt{1 - x^2}$$

81. Let $y = \arctan x$. Then $x = \tan y$, $-\dfrac{\pi}{2} < y < \dfrac{\pi}{2}$. If $\tan y = \dfrac{b}{a} = x = \dfrac{x}{1}$, then let $b = x$, $a = 1$.

$$a^2 + b^2 = r^2$$
$$1^2 + x^2 = r^2$$
$$r = \sqrt{1 + x^2} \quad (r \text{ is always taken positive})$$
$$\cos y = \cos(\tan^{-1} x) = \dfrac{a}{r} = \dfrac{1}{\sqrt{1 + x^2}}$$

83. $f(x) = \sin(x + 2)$, $-2 - \pi/2 \le x \le -2 + \pi/2$. With this restriction, f is one-to-one and has an inverse (this can be checked with the graph). Solve $y = f(x)$ for x:

$y = \sin(x + 2)$

$\sin^{-1} y = x + 2$

$x = \sin^{-1} y - 2$

Interchange x and y:

$y = \sin^{-1} x - 2$; $f^{-1}(x) = \sin^{-1} x - 2$

Find the range of f:

$-2 - \pi/2 \le x \le -2 + \pi/2$

$-\pi/2 \le x + 2 \le \pi/2$

$\sin(-\pi/2) \le \sin(x + 2) \le \sin(\pi/2)$

$-1 \le \sin(x + 2) \le 1$

The range of f is $[-1, 1]$, so that is the domain of f^{-1}. So the inverse is $f^{-1}(x) = \sin^{-1} x - 2$, $-1 \le x \le 1$.

85. $f(x) = 5 + 2\cos x$, $0 \le x \le \pi$. With this restriction, f is one-to-one and has an inverse (this can be checked with the graph). Solve $y = f(x)$ for x:

$y = 5 + 2\cos x$

$y - 5 = 2\cos x$

$\dfrac{y - 5}{2} = \cos x$

$\cos^{-1}\left(\dfrac{y - 5}{2}\right) = x$

Interchange x and y:

$y = \cos^{-1}\left(\dfrac{x - 5}{2}\right)$; $f^{-1}(x) = \cos^{-1}\left(\dfrac{x - 5}{2}\right)$

Find the range of f:

$0 \le x \le \pi$

$-1 \le \cos x \le 1$

$-2 \le 2\cos x \le 2$

$3 \le 5 + 2\cos x \le 7$

The range of f is $[3, 7]$, so that is the domain of f^{-1}. So the inverse is $f^{-1}(x) = \cos^{-1}\left(\dfrac{x - 5}{2}\right)$, $3 \le x \le 7$.

87. $f(x) = 4 + 2\cos(x - 3)$. $3 \le x \le 3 + \pi$. With this restriction, f is one-to-one and has an inverse (this can be checked with the graph). Solve $y = f(x)$ for x:

$y = 4 + 2\cos(x - 3)$

$y - 4 = 2\cos(x - 3)$

$\dfrac{y - 4}{2} = \cos(x - 3)$

$\cos^{-1}\left(\dfrac{y - 4}{2}\right) = x - 3$

$x = 3 + \cos^{-1}\left(\dfrac{y - 4}{2}\right)$

Interchange x and y:

$$y = 3 + \cos^{-1}\left(\frac{x - 4}{2}\right); \quad f^{-1}(x) = 3 + \cos^{-1}\left(\frac{x - 4}{2}\right)$$

Find the range of f:

$3 \leq x \leq 3 + \pi$
$0 \leq x - 3 \leq \pi$
$-1 \leq \cos(x - 3) \leq 1$
$-2 \leq 2\cos(x - 3) \leq 2$
$2 \leq 4 + 2\cos(x - 3) \leq 6$

The range of f is [2, 6], so that is the domain of f^{-1}. The inverse is

$$f^{-1}(x) = 3 + \cos^{-1}\left(\frac{x - 4}{2}\right), \quad 2 \leq x \leq 6.$$

89. (A)

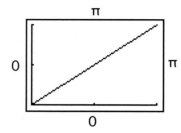

(B) The domain for $\cos x$ is $(-\infty, \infty)$ and the range is $[-1, 1]$, which is the domain for $\cos^{-1} x$. So $y = \cos^{-1}(\cos x)$ has a graph over the interval $(-\infty, \infty)$, but $\cos^{-1}(\cos x) = x$ only on the restricted domain of $\cos x$, $[0, \pi]$.

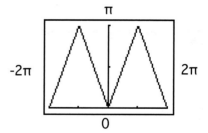

91. For a 28mm lens, $x = 28$, so $\theta = 2\tan^{-1}\dfrac{21.634}{28} = 1.31567$ radians. In decimal degrees, $\theta_D = \dfrac{180°}{\pi}(1.31567 \text{ rad}) = 75.38°$.

For a 100 mm lens, $x = 100$, and $\theta = 2\tan^{-1}\dfrac{21.634}{100} = 0.42611$ radians. In decimal degrees, $\theta_D = \dfrac{180°}{\pi}(0.42611 \text{ rad}) = 24.41°$.

93. (A)

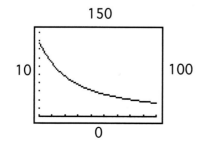

(B) 59.44 mm

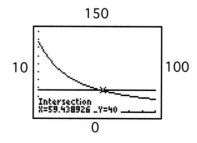

95.

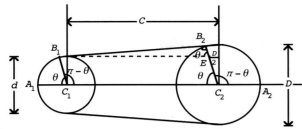

From the above figure, we can observe the following:

Length of belt = $2[arc\ \overparen{A_1 B_1} + B_1 B_2 + arc\ \overparen{B_2 A_2}]$

$arc\ A_1 B_1 = \dfrac{d}{2}(\theta)$

$arc\ B_2\ A_2 = \dfrac{D}{2}(\pi - \theta)$

To find $B_1 B_2$ we note:

$C_1 C_2$ has length C

$B_1 E$ is constructed parallel to $C_1 C_2$. EC_2 is parallel to $B_1 C_1$. Then $EB_1 C_1 C_2$ is a parallelogram. EB_1 has length C. $EB_1 B_2$ is a right triangle. So

(1) $\cos\ \theta = \dfrac{B_2 E}{EB_1} = \dfrac{\frac{D}{2} - \frac{d}{2}}{C} = \dfrac{D - d}{2C}$

(2) $\sin\ \theta = \dfrac{B_1 B_2}{EB_1}$, so $B_1 B_2 = EB_1\ \sin\ \theta = C\ \sin\ \theta$

Finally,
Length of belt

$L = 2[arc\ \overparen{A_1 B_1} + B_1 B_2 + arc\ \overparen{B_2 A_2}]$

$\quad = 2\left[\dfrac{d}{2}\theta + C\ \sin\ \theta + \dfrac{D}{2}(\pi - \theta)\right]$

$\quad = d\theta + 2C\ \sin\ \theta + D(\pi - \theta)$

$L = \pi D + (d - D)\theta + 2C\ \sin\ \theta$

and (from (1) above)

$\theta = \cos^{-1}\dfrac{D - d}{2C}$

Substituting the given values, we have
$D = 4$, $d = 2$, $C = 6$ (calculator in radian mode)

$\theta = \cos^{-1}\dfrac{4 - 2}{2 \cdot 6} = \cos^{-1}\dfrac{1}{6}$

$L = 4\pi + (2 - 4)\cos^{-1}\dfrac{1}{6} + 2 \cdot 6\ \sin\left(\cos^{-1}\dfrac{1}{6}\right)$

$\quad \approx 21.59$ inches

97. (A)

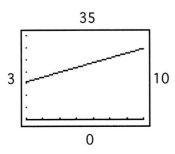

(B) 7.22 inches

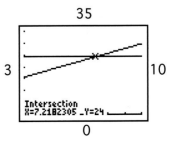

99. (A) Following the hint, we draw AC. Then, since the central angle in a circle subtended by an arc is twice any inscribed angle subtended by the same arc, angle ACB has measure 2θ, and $d = r \cdot 2\theta = 2r\theta$

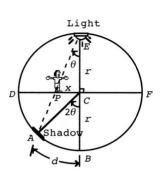

In triangle ECP, $\tan \theta = \dfrac{x}{r}$, so $\theta = \tan^{-1} \dfrac{x}{r}$

$$d = 2r \tan^{-1} \frac{x}{r}$$

(B) Substituting the given values, we have
$$r = 100 \quad x = 40$$

$$d = 2 \cdot 100 \ \tan^{-1} \frac{40}{100}$$

$$= 200 \ \tan^{-1} \frac{40}{100}$$

$$= 200 \ \tan^{-1}(0.4) \quad \text{(calculator in radian mode)}$$

$$= 76.10 \ \text{feet}$$

CHAPTER 5 REVIEW

1. $\theta = \dfrac{s}{r} = \dfrac{15 \text{ centimeters}}{6 \text{ centimeters}} = \dfrac{15}{6} = 2.5 \text{ radians}$ *(5-3)*

2. $s = r\theta = (3 \text{ centimeters})(2.5 \text{ radians}) = 7.5 \text{ centimeters}$ *(5-3)*

3. Solve for α:

$$\alpha = 90° - 35.2° = 54.8°$$

Solve for a:

$$\cos \beta = \frac{a}{c}$$

$$\cos 35.2° = \frac{a}{20.2}$$

$$a = 20.2 \cos 35.2°$$

$$= 16.5 \text{ ft}$$

Solve for b:

$$\sin \beta = \frac{b}{c}$$

$$\sin 35.2° = \frac{b}{20.2}$$

$$b = 20.2 \sin 35.2°$$

$$= 11.6 \text{ ft} \qquad \textit{(5-3)}$$

4. (A)

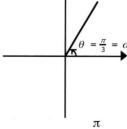

$$\alpha = \frac{\pi}{3}$$

(B)

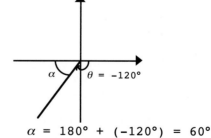

$$\alpha = 180° + (-120°) = 60°$$

(C)

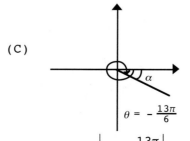

$$\alpha = \left| 2\pi - \frac{13\pi}{6} \right| = \frac{\pi}{6}$$

(D)

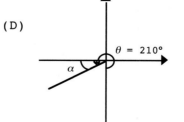

$$a = 210° - 180° = 30° \qquad \textit{(5-4)}$$

5. (A) $\dfrac{b}{r} = \sin\theta < 0$ if $b < 0$. This occurs in quadrants III, IV.

(B) $\dfrac{a}{r} = \cos\theta < 0$ if $a < 0$. This occurs in quadrants II, III.

(C) $\dfrac{b}{a} = \tan\theta < 0$ if a and b have opposite signs. This occurs in quadrants II, IV. *(5-2)*

6. Think of $\left(\dfrac{12}{13}, -\dfrac{5}{13}\right)$ as (a, b). Then:

(A) $\cos x = a = 12/13$.

(B) $\csc x = \dfrac{1}{b} = \dfrac{1}{-5/13} = -\dfrac{13}{5}$

(C) $\cot x = \dfrac{a}{b} = \dfrac{12/13}{-5/13} = -\dfrac{12}{5}$ *(5-2)*

7.

$$a^2 + b^2 = r^2$$
$$4^2 + (-3)^2 = r^2$$
$$25 = r^2$$
$$r = 5$$
$$\sin\theta = \dfrac{b}{r} = \dfrac{-3}{5} = -\dfrac{3}{5}$$
$$\sec\theta = \dfrac{r}{a} = \dfrac{5}{4}$$
$$\cot\theta = \dfrac{a}{b} = \dfrac{4}{-3} = -\dfrac{4}{3}$$

$P(a, b) = (4, -3)$ *(5-4)*

8.

$\theta°$	θ rad	$\sin\theta$	$\cos\theta$	$\tan\theta$	$\csc\theta$	$\sec\theta$	$\cot\theta$
$0°$	0	0	1	0	ND *	1	ND
$30°$	$\pi/6$	$1/2$	$\sqrt{3}/2$	$1/\sqrt{3}$	2	$2/\sqrt{3}$	$\sqrt{3}$
$45°$	$\pi/4$	$1/\sqrt{2}$	$1/\sqrt{2}$	1	$\sqrt{2}$	$\sqrt{2}$	1
$60°$	$\pi/3$	$\sqrt{3}/2$	$1/2$	$\sqrt{3}$	$2/\sqrt{3}$	2	$1/\sqrt{3}$
$90°$	$\pi/2$	1	0	ND	1	ND	0
$180°$	π	0	-1	0	ND	-1	ND
$270°$	$3\pi/2$	-1	0	ND	-1	ND	0
$360°$	2π	0	1	0	ND	1	ND

*ND = not defined   *(5-1, 5-2)*

9. (A) 2π (B) 2π (C) π *(5-4)*

10. (A) Domain = all real numbers, Range = $[-1, 1]$

(B) Domain is set of all real numbers except $x = \dfrac{2k+1}{2}\pi$, k an integer, Range = all real numbers *(5-4)*

11.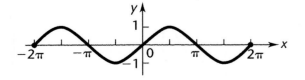

(5-4)

12.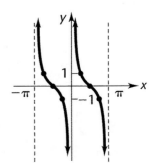

(5-4)

13. The central angle in a circle subtended by an arc of half the length of the radius. *(5-1)*

14. If the graph of $y = \sin x$ is shifted $\frac{\pi}{2}$ units to the left, the result will be the graph of $y = \cos x$. *(5-4)*

15. $132°52'41'' = \left(132 + \dfrac{52}{60} + \dfrac{41}{3,600}\right)° = 132.878°$ *(5-1)*

16. $13.762 \text{ rad} \cdot \dfrac{180°}{\pi \text{ rad}} = 788.5045177°$

    $= 788°\left(0.5045177 \cdot 60\right)'$

    $= 788°30.271062'$

    $= 788°30'\left(0.271062 \cdot 60\right)''$

    $= 788°30'16''$ *(5-1)*

17. First convert to decimal degrees:

    $64°28'14'' = \left(64 + \dfrac{28}{60} + \dfrac{14}{3,600}\right)° = 64.47°$

    $64.47° \cdot \dfrac{\pi \text{ rad}}{180°} = 1.125 \text{ rad}$ *(5-1)*

18. Multiply by one in the form $180°/\pi$ rad:

    $1.37 \text{ rad} \cdot \dfrac{180°}{\pi \text{ rad}} = 78.50°$ *(5-1)*

19. Solve for β:

    $\tan \beta = \dfrac{b}{a}$

    $\tan \beta = \dfrac{13.3}{15.7}$

    $\beta = \tan^{-1} \dfrac{13.3}{15.7}$

    $\quad = 40.3°$

    Solve for α:

    $\alpha = 90° - 40.3° = 49.7°$

    Solve for c:

    $\sec \beta = \dfrac{c}{a}$

    $\sec 40.3° = \dfrac{c}{15.7}$

    $c = 15.7 \sec 40.3°$

    $\quad = 20.6 \text{ cm}$ *(5-3)*

20. (A) Since $-270° < -210° < -180°$, this is a quadrant II angle.

    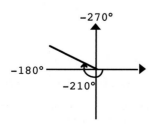

    (C) Since $3.14 < 4.2 < 4.71$, this is a quadrant III angle.

    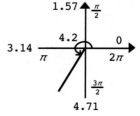

    (B) Since $\dfrac{5\pi}{2}$ is coterminal with $\dfrac{\pi}{2}$, this is a quadrantal angle. *(5-1)*

21. (A) Since -240° + 360° = 120°, this is coterminal with 120°.

   (B) Since $-\dfrac{7\pi}{6} + 2\pi = \dfrac{5\pi}{6}$, which is equivalent to 150°, this is not coterminal with 120°.

   (C) Since 840° - 2(360°) = 120°, this is coterminal with 120°. $\hspace{2em}$ (5-1)

22. (B) and (C), since 3 radians is equivalent to the real number 3, and cosine is periodic with period 2π. (A) is not the same as cos 3, since 3° is equivalent to $\dfrac{\pi}{180°}3°$, not 3. $\hspace{2em}$ (5-2)

23. (A) $\dfrac{b}{a}$ = tan x is not defined if $a = 0$. This occurs if $\theta = \dfrac{\pi}{2}, \dfrac{3\pi}{2}$.

   (B) $\dfrac{a}{b}$ = cot x is not defined if $b = 0$. This occurs if $\theta = 0, \pi$.

   (C) $\dfrac{r}{b}$ = csc x is not defined if $b = 0$. This occurs if $\theta = 0, \pi$. $\hspace{2em}$ (5-4)

24. Since the coordinates of a point on a unit circle are given by (a, b) = (cos x, sin x), we evaluate (cos(-8.305), sin(-8.305))--using a calculator set in radian mode--to obtain P = (-0.436, -0.900). Note that $x = -8.305$, since P is moving clockwise. The quadrant in which $P = (a, b)$ lies can be determined by the signs of a and b. In this case P is in the third quadrant, since a is negative and b is negative. $\hspace{2em}$ (5-1, 5-2)

25. Since tan θ = 4 > 0, sin θ and cos θ have the same sign. Also, cos θ < 0, so sin θ < 0 as well. This tells us that θ is a quadrant III angle. We sketch a reference triangle. Since tan $\theta = \dfrac{b}{a} = 4$, we know that $a = -1$ and $b = -4$. Use the Pythagorean theorem to find r.

$$r^2 = (-1)^2 + (-4)^2 = 17$$
$$r = \sqrt{17}$$

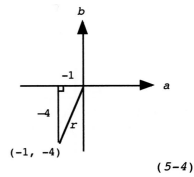

Using $(a, b) = (-1, -4)$ and $r = \sqrt{17}$, we have

$$\cos \theta = \dfrac{a}{r} = -\dfrac{1}{\sqrt{17}} \hspace{3em} \sin \theta = \dfrac{b}{r} = -\dfrac{4}{\sqrt{17}}$$

$$\sec \theta = \dfrac{r}{a} = -\dfrac{\sqrt{17}}{1} = -\sqrt{17} \hspace{2em} \cot \theta = \dfrac{a}{b} = \dfrac{-1}{-4} = \dfrac{1}{4}$$

$$\csc \theta = \dfrac{r}{b} = -\dfrac{\sqrt{17}}{4}$$

$\hspace{25em}$ (5-4)

26. The zeros of cosine occur at the $\pi/2$ values: on the given interval, the zeros are $-\dfrac{3\pi}{2}, -\dfrac{\pi}{2}, \dfrac{\pi}{2}, \dfrac{3\pi}{2}$. The turning points occur at multiples of π (Note that the amplitude is 3, and A is positive, so the point at $x = 0$ is a maximum and is at height 3.). On the given interval the turning points are $(-\pi, -3), (0, 3), (\pi, 3)$. $\hspace{2em}$ (5-2)

27. $(a, b) = (1, 0)$ $r = 1$

$\tan 0 = \dfrac{b}{a} = \dfrac{0}{1} = 0$

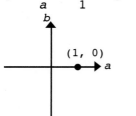

$(5-2)$

28. $(a, b) = (0, 1)$ $r = 1$

$\sec 90° = \dfrac{r}{a} = \dfrac{1}{0}$ Not defined

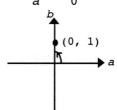

$(5-2)$

29. $y = \cos^{-1} 1$ is equivalent to
$\cos y = 1$ $0 \le y \le \pi$
$y = 0$ $(5-6)$

30.

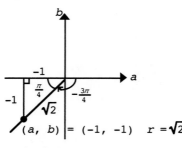

$(a, b) = (-1, -1)$ $r = \sqrt{2}$

$\cos\left(-\dfrac{3\pi}{4}\right) = \dfrac{a}{r} = \dfrac{-1}{\sqrt{2}} = -\dfrac{1}{\sqrt{2}} = -\dfrac{\sqrt{2}}{2}$

$(5-2)$

31. $y = \sin^{-1} \dfrac{\sqrt{2}}{2}$ is equivalent to

$\sin y = \dfrac{\sqrt{2}}{2}$ $-\dfrac{\pi}{2} \le y \le \dfrac{\pi}{2}$

$\sin y = \dfrac{b}{r} = \dfrac{\sqrt{2}}{2}$ $b = \sqrt{2}$ $r = 2$

$a^2 + (\sqrt{2})^2 = 2^2$
$a^2 = 2$
$a = \sqrt{2}$

(positive since y is a quadrant I or IV angle)

$y = \dfrac{\pi}{4}$

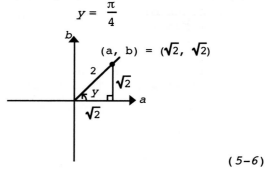

$(5-6)$

32.

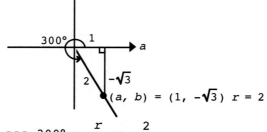

$(a, b) = (1, -\sqrt{3})$ $r = 2$

$\csc 300° = \dfrac{r}{b} = \dfrac{2}{-\sqrt{3}}$

$= -\dfrac{2}{\sqrt{3}}$ or $\dfrac{-2\sqrt{3}}{3}$ $(5-2)$

33. $y = \arctan \sqrt{3}$ is equivalent to
$\tan y = \sqrt{3}$ $-\dfrac{\pi}{2} < y < \dfrac{\pi}{2}$

$\tan y = \dfrac{b}{a} = \dfrac{\sqrt{3}}{1}$ $b = \sqrt{3}$ $a = 1$

$r^2 = 1^2 + (\sqrt{3})^2$
$r^2 = 4$
$r = 2$
$y = \dfrac{\pi}{3}$

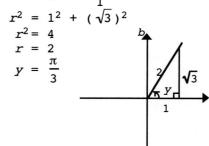

$(5-6)$

34.

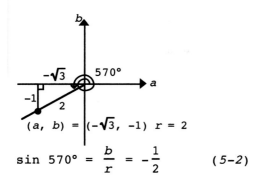

$$\sin 570° = \frac{b}{r} = -\frac{1}{2} \qquad (5\text{-}2)$$

35. $y = \tan^{-1}(-1)$ is equivalent to

$$\tan y = -1 \qquad -\frac{\pi}{2} < y < \frac{\pi}{2}$$

$$\tan y = \frac{b}{a} = \frac{-1}{1} \qquad b = -1 \quad a = 1$$

$$r^2 = (-1)^2 + 1^2$$

$$r^2 = 2$$

$$r = \sqrt{2}$$

$$y = -\frac{\pi}{4}$$

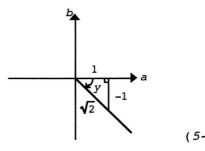

$(5\text{-}6)$

36.

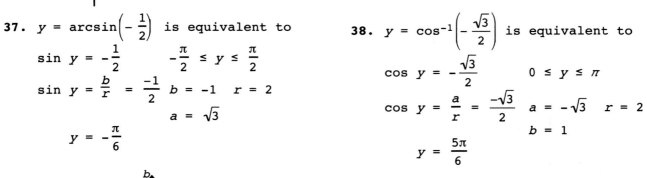

$$\cot\left(-\frac{4\pi}{3}\right) = \frac{a}{b} = \frac{-1}{\sqrt{3}} = -\frac{1}{\sqrt{3}} \text{ or } -\frac{\sqrt{3}}{3}$$

$(5\text{-}2)$

37. $y = \arcsin\left(-\frac{1}{2}\right)$ is equivalent to

$$\sin y = -\frac{1}{2} \qquad -\frac{\pi}{2} \le y \le \frac{\pi}{2}$$

$$\sin y = \frac{b}{r} = \frac{-1}{2} \quad b = -1 \quad r = 2$$

$$a = \sqrt{3}$$

$$y = -\frac{\pi}{6}$$

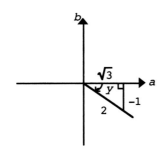

$(5\text{-}6)$

38. $y = \cos^{-1}\left(-\frac{\sqrt{3}}{2}\right)$ is equivalent to

$$\cos y = -\frac{\sqrt{3}}{2} \qquad 0 \le y \le \pi$$

$$\cos y = \frac{a}{r} = \frac{-\sqrt{3}}{2} \quad a = -\sqrt{3} \quad r = 2$$

$$b = 1$$

$$y = \frac{5\pi}{6}$$

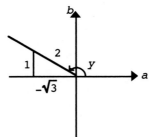

$(5\text{-}6)$

39. $\cos(\cos^{-1} 0.33) = 0.33$ by the cosine-inverse cosine identity. $\hspace{1cm}(5\text{-}6)$

40. Let $y = \tan^{-1}(-1)$, then $\tan y = -1$, $-\frac{\pi}{2} < y < \frac{\pi}{2}$. Using the drawing in problem 35, we have $(a, b) = (1, -1)$, $r = \sqrt{2}$.

$\csc y = \csc[\tan^{-1}(-1)] = \frac{r}{b} = \frac{\sqrt{2}}{-1} = -\sqrt{2}$ (5-6)

41. Let $y = \arccos\left(-\frac{1}{2}\right)$, then

$\cos y = -\frac{1}{2}$, $0 \le y \le \pi$. Draw the reference triangle associated with y, then $\sin y = \sin\left[\arccos\left(-\frac{1}{2}\right)\right]$ can be determined directly from the triangle.

$\cos y = \frac{a}{r} = -\frac{1}{2} = \frac{-1}{2}$ $a = -1$ $r = 2$

$b = \sqrt{3}$

$\sin y = \frac{b}{r} = \frac{\sqrt{3}}{2}$

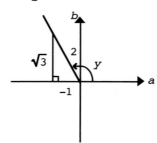

(5-6)

42. Let $y = \sin^{-1}\frac{-4}{5}$, then $\sin y = \frac{-4}{5}$,

$-\frac{\pi}{2} \le y \le \frac{\pi}{2}$. Draw the reference triangle associated with y, then $\tan y = \tan\left[\sin^{-1}\left(-\frac{4}{5}\right)\right]$ can be determined directly from the triangle.

$\sin y = \frac{b}{r} = \frac{-4}{5}$ $b = -4$ $r = 5$ $a = 3$

$\tan y = \frac{b}{a} = \frac{-4}{3}$

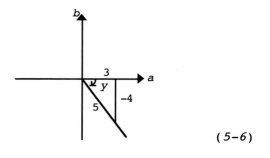

(5-6)

43. 0.4431   (5-2)

44. $\tan\left(93 + \frac{46}{60} + \frac{17}{3600}\right)^\circ = -15.17$ (5-2)

45. $\sec(-2.073) = \dfrac{1}{\cos(-2.073)} = -2.077$ (5-2)

46. -0.9750 (5-6)

47. arccos x is not defined if $x < -1$ (5-6)

48. 1.557   (5-6)

49. 1.095   (5-6)

50. Since $\tan 1.345 = 4.353 > 1$, $\sin^{-1}(\tan 1.345)$ is not defined.   (5-6)

51. (A) $\theta = \arcsin^{-1}\left(-\frac{1}{2}\right)$ is equivalent to

$\sin\theta = -\frac{1}{2}$ $-\frac{\pi}{2} \le \theta \le \frac{\pi}{2}$

$\theta = -\frac{\pi}{6}$

To convert to degrees, multiply by one in the form $\frac{180°}{\pi}$ radians.

$\sin^{-1}\left(-\frac{1}{2}\right) = \frac{180°}{\pi} \cdot \left(-\frac{\pi}{6}\right) = -30°$

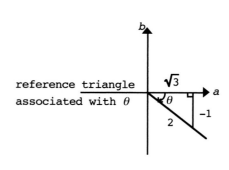

reference triangle associated with θ

(B) $\theta = \arccos\left(-\dfrac{1}{2}\right)$ is equivalent to

$$\cos\theta = -\dfrac{1}{2} \qquad 0 \le \theta \le \pi$$

$$\theta = \dfrac{2\pi}{3}$$

We again multiply by $\dfrac{180°}{\pi}$ radians.

$$\arccos\left(-\dfrac{1}{2}\right) = \dfrac{180°}{\pi} \cdot \dfrac{2\pi}{3} = 120°. \hfill (5\text{-}6)$$

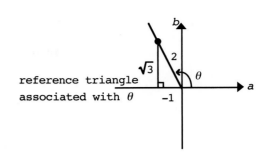

reference triangle
associated with θ

52. Calculator in degree mode: (A) $\theta = 151.20°$ (B) $\theta = 82.28°$ (5-6)

53. $\cos^{-1}[\cos(-2)] = 2$ For the identity $\cos^{-1}(\cos x) = x$ to hold, x must be in the restricted domain of the cosine function; that is, $0 \le x \le \pi$. The number -2 is not in the restricted domain.   (5-6)

54. Amplitude $= |-2| = 2$

    Period $= \dfrac{2\pi}{\pi} = 2$

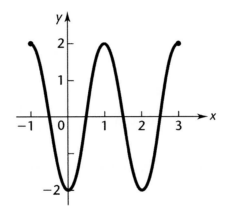

(5-5)

55. $y = -2 + 3\sin\left(\dfrac{x}{2}\right)$

For the graph of $y = 3\sin\dfrac{x}{2}$, we note: $A = 3$, $P = 2\pi \div \dfrac{1}{2} = 4\pi$, phase shift $= 0$. We graph $y = 3\sin\dfrac{x}{2}$, then vertically translate the graph down 2 units.

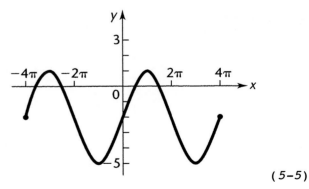

(5-5)

56. $A = 6$ $P = \pi = 2\pi \div 2$, so $B = 2$

    $y = 6\cos 2x; \quad -\dfrac{\pi}{2} \le x \le \pi$ (5-5)

57. $|A| = 0.5$ $P = 2 = 2\pi \div \pi$, so $B = \pi$ $A = -0.5$, since the graph has the form of the standard sine curve turned upside down.

    $y = -0.5\sin\pi x; \quad -1 \le x \le 2$ (5-5)

58. If the graph of $y = \tan x$ is shifted $\dfrac{\pi}{2}$ units to the right and reflected in the x axis, the result will be the graph of $y = \cot x$.   (5-4)

59. (A) $\sin(-x) \cot(-x) = \sin(-x) \dfrac{\cos(-x)}{\sin(-x)}$ Quotient Identity

                 $= \cos(-x)$ Algebra

                 $= \cos x$ Identities for negatives

  (B) $\dfrac{\sin^2 x}{1 - \sin^2 x} = \dfrac{\sin^2 x}{\cos^2 x}$ Pythagorean Identity

               $= \left(\dfrac{\sin x}{\cos x}\right)^2$ Algebra

               $= \tan^2 x$ Quotient Identity   *(5-4)*

60. $y = 3 \sin\left(\dfrac{x}{2} + \dfrac{\pi}{2}\right)$

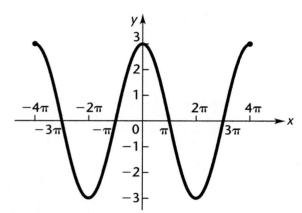

$A = 3$

Solve $\dfrac{x}{2} + \dfrac{\pi}{2} = 0$ $\dfrac{x}{2} + \dfrac{\pi}{2} = 2\pi$

        $\dfrac{x}{2} = -\dfrac{\pi}{2}$ $\dfrac{x}{2} = -\dfrac{\pi}{2} + 2\pi$

        $x = -\pi$ $x = -\pi + 4\pi$

                 Phase shift   Period

The graph completes one full cycle as x varies over the interval $[-\pi, 3\pi]$.

                                          (5-5)

61. $y = -2 \cos\left(\dfrac{\pi}{2} x - \dfrac{\pi}{4}\right)$

amplitude $= |A| = |-2| = 2$

Solve $\dfrac{\pi}{2} x - \dfrac{\pi}{4} = 0$ $\dfrac{\pi}{2} x - \dfrac{\pi}{4} = 2\pi$

        $\dfrac{\pi}{2} x = \dfrac{\pi}{4}$ $\dfrac{\pi}{2} x = \dfrac{\pi}{4} + 2\pi$

         $x = \dfrac{\pi}{4} \div \dfrac{\pi}{2}$ $x = \left(\dfrac{\pi}{4} + 2\pi\right) \div \dfrac{\pi}{2}$

         $x = \dfrac{1}{2}$ $x = \dfrac{1}{2} + 4$

              Phase shift $= \frac{1}{2}$ Period $= 4$ *(5-5)*

62. Domain $= [-1, 1]$

Range $= [0, \pi]$

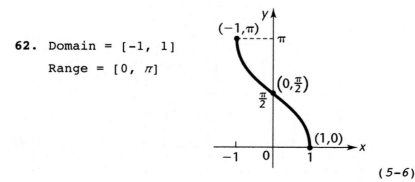

                                       (5-6)

63. Here is the graph of $y = \dfrac{1}{1 + \tan^2 x}$ on the interval $[-2\pi,\ 2\pi]$.

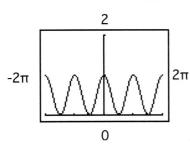

The graph has amplitude $\dfrac{1 - 0}{2} = \dfrac{1}{2}$ and period π. It appears to be the graph of $y = \dfrac{1}{2}\cos 2x$ shifted up $\dfrac{1}{2}$ unit, that is, $y = \dfrac{1}{2}\cos 2x + \dfrac{1}{2}$.

(5-5)

64. (A) Here is the graph of $y = \dfrac{2\sin^2 x}{\sin 2x}$ on the interval $[-2\pi,\ 2\pi]$.

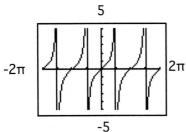

The graph appears to have the form $y = A \tan Bx$. Since the period is π, $B = 1$. The graph of $y = A \tan x$ shown appears to pass through $\left(\dfrac{\pi}{4},\ 1\right)$, so

$$1 = A \tan \dfrac{\pi}{4}$$
$$1 = A$$

The equation of the graph can be written $y = \tan x$.

(B) Here is the graph of $y = \dfrac{2\cos^2 x}{\sin 2x}$ on the interval $[-2\pi,\ 2\pi]$.

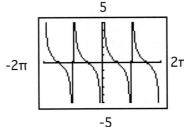

The graph appears to have the form $y = A \cot Bx$. Since the period is π, $B = 1$. The graph of $y = A \cot x$ shown appears to pass through $\left(\dfrac{\pi}{4},\ 1\right)$, so

$$1 = A \cot \dfrac{\pi}{4}$$
$$1 = A$$

The equation of the graph can be written $y = \cot x$.   *(5-5)*

65. (A) $f(-x) = \dfrac{1}{1 + \tan^2(-x)}$

$= \dfrac{1}{1 + (-\tan x)^2}$

$= \dfrac{1}{1 + \tan^2 x}$

$= f(x)$

$f(x)$ is even.

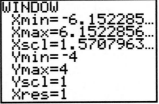

The graph is symmetric about the y-axis so $f(x)$ is even.

(B) $g(-x) = \dfrac{1}{1 + \tan(-x)}$

$= \dfrac{1}{1 - \tan x}$

This is not equal to $g(x)$ nor is it equal to $-g(x)$ which would be $\dfrac{1}{-1 - \tan x}$. So $g(x)$ is neither even nor odd.

The graph is not symmetric about the y-axis nor is is symmetric about the origin. *(5-4)*

66. False.

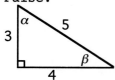

According to the diagram,

$$\sin \alpha = \frac{\text{Opp}}{\text{Hyp}} = \frac{4}{5} \text{ while csc } \beta = \frac{\text{Hyp}}{\text{Opp}} = \frac{5}{4}.$$

(5-3)

67. True. The two acute angles in a right triangle add up to 90° so if they're equal, both are 45°.

$$\sin 45° = \frac{1}{\sqrt{2}} \approx 0.71 \qquad \cos 45° = \frac{1}{\sqrt{2}} \approx 0.71$$

$$\tan 45° = 1 \qquad \cot 45° = 1$$

$$\sec 45° = \sqrt{2} \approx 1.4 \qquad \csc 45° = \sqrt{2} \approx 1.4$$

(5-3)

68. (A) Since $\theta_R = \frac{s}{r}$ and r = radius of circle = distance of A from center = 8,

we have $\theta_R = \frac{20}{8}$ = 2.5 radians.

(B) Since $\sin \theta = \frac{b}{r}$ and $\cos \theta = \frac{a}{r}$, we can write

$$a = r \cos \theta = 8 \cos 2.5 = -6.41 \text{ (calculator in radian mode)}$$
$$b = r \sin \theta = 8 \sin 2.5 = 4.79$$
$$(a, b) = (-6.41, 4.79)$$

(5-1, 5-2)

69. (A) $\cos x = \frac{a}{r} = -\frac{1}{2} = \frac{-1}{2}$. a is negative in the II and III quadrants. The least positive x is associated with a $\frac{\pi}{3}$ reference triangle in the II quadrant as drawn: $x = \frac{2\pi}{3}$

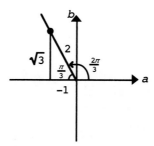

(B) $\csc x = -\sqrt{2} = \frac{\sqrt{2}}{-1} = \frac{r}{b}$, b is negative in the III and IV quadrants. The least positive x is associated with a $\frac{\pi}{4}$ reference triangle in the III quadrant as drawn: $x = \frac{5\pi}{4}$

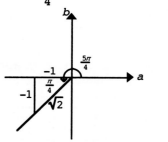

(5-2)

70. The dashed curve is the graph of $y = \cos x$. The solid curve is the required graph of $y = \sec x$, $-\frac{\pi}{2} < x < \frac{3\pi}{2}$

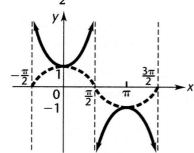

(5-4)

71.

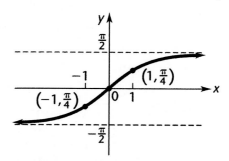

Domain = all real numbers

Range = $\left(-\frac{\pi}{2}, \frac{\pi}{2}\right)$

(5-6)

72. In this function, $B = \pi$ and $C = \pi/2$. Using the formula for period of tangent,
$$P = \frac{\pi}{B} = \frac{\pi}{\pi} = 1.$$

Using the formula for phase shift,
$$\text{Phase shift} = -\frac{C}{B} = -\frac{\pi/2}{\pi} = -\frac{1}{2} \tag{5-5}$$

73. In this function, $B = 1/2$ and $C = -\pi/4$. Using formula for period of cosecant,
$$P = \frac{2\pi}{B} = \frac{2\pi}{1/2} = 4\pi.$$

Using the formula for phase shift,
$$\text{Phase shift} = -\frac{C}{B} = -\frac{-\pi/4}{1/2} = \frac{\pi}{2} \tag{5-5}$$

74. From the figure, we can see that
$\cos(-x) = \cos x$ (P and P_1 have the same x-coordinate)
$\sin(-x) = -\sin x$ (P and P_1 have opposite y-coordinates)
Therefore $\tan(-x) = \dfrac{\sin(-x)}{\cos(-x)} = \dfrac{-\sin x}{\cos x}$
$$= -\tan x$$
It follows that the graph of
(A) sine has origin symmetry
(B) cosine has y axis symmetry
(C) tangent has origin symmetry

$(5-4)$

75. Let $y = \sin^{-1} x$. Then $x = \sin y$, $-\dfrac{\pi}{2} \leq y \leq \dfrac{\pi}{2}$. If $\sin y = \dfrac{b}{r} = \dfrac{x}{1}$, then let
$b = x$, $r = 1$
$a^2 + x^2 = 1^2$
$\quad a^2 = 1 - x^2$
$\quad\quad a = \sqrt{1 - x^2}$

We choose the positive sign because y is a I or IV quadrant angle.
$$\sec y = \sec(\sin^{-1} x) = \frac{r}{a} = \frac{1}{\sqrt{1 - x^2}} \tag{5-6}$$

76. For each case, the number is not in the domain of the function and an error message of some type will appear. $(5-2, 5-6)$

77. $|A| = 2$, so $A = 2$ or -2. The graph completes one full cycle as x varies over the interval $\left[-\dfrac{5}{4}, \dfrac{3}{4}\right]$. Since $-\dfrac{C}{B}$ is required to be between -1 and 0, we cannot simply set $-\dfrac{C}{B} = -\dfrac{5}{4}$. We must (mentally) extend the curve so that the (extended) graph is that of a standard sine curve that completes one full cycle as x varies over the interval $\left[-\dfrac{1}{4}, \dfrac{7}{4}\right]$. Then
$A = 2$
$-\dfrac{C}{B} = -\dfrac{1}{4} \qquad -\dfrac{C}{B} + \dfrac{2\pi}{B} = \dfrac{7}{4}$

$$C = \frac{B}{4} \qquad \frac{2\pi}{B} = 2 \quad B = \frac{2\pi}{2} = \pi \quad C = \frac{\pi}{4}$$

The equation is then

$$y = A \sin(Bx + C)$$

$$y = 2 \sin\left(\pi x + \frac{\pi}{4}\right)$$

(5-5)

78. Here is the graph of
$y = 1.2 \sin 2x + 1.6 \cos 2x$
on the interval $[-2\pi, 2\pi]$.

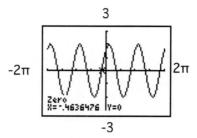

It appears that this is a sine curve shifted to the left, with $A = 2$ and, since $P = \frac{2\pi}{B}$ and P appears to be π, $B = \frac{2\pi}{P} = \frac{2\pi}{\pi} = 2$. From the graphing calculator, we find that the x intercept closest to the origin, to three decimal places, is -0.464. To find C, substitute $B = 2$ and $x = -0.464$ into the phase-shift formula

$x = -\frac{C}{B}$ and solve for C.

$$x = -\frac{C}{B}$$

$$-0.464 = -\frac{C}{2}$$

$$C = 0.928$$

The equation is $y = 2 \sin(2x + 0.928)$.

(5-5)

79. The graph of $y = 2 \tan t$ has asymptotes at multiples of $\pi/2$:
$\pm\frac{\pi}{2}, \pm\frac{3\pi}{2}, \pm\frac{5\pi}{2}, \cdots$. We need to find which of these fall in the given interval for $t = 3x$.

$$3x = \pm\frac{\pi}{2} \qquad\qquad 3x = \pm\frac{3\pi}{2} \qquad\qquad 3x = \pm\frac{5\pi}{2}$$

$$x = \pm\frac{\pi}{6} \qquad\qquad x = \pm\frac{3\pi}{6} = \pm\frac{\pi}{2} \qquad\qquad x = \pm\frac{5\pi}{6}$$

Any values beyond these are outside the given interval. *(5-5)*

80. In one year the line sweeps out one full revolution, or 2π radians in 365 days. In 73 days the line sweeps out $\frac{73}{365}$ of one full revolution, or $\frac{73}{365} \cdot 2\pi = \frac{2\pi}{5}$ radians. *(5-1)*

81. Sketch a figure.
From the figure, we can see that
θ is $\frac{1}{8}$ of a full circle, so

$$\theta = \frac{1}{8}(360°) = 45°$$

$$\frac{s}{2} = r \sin 45° = r\frac{1}{\sqrt{2}}$$

Then $P = 4s = 8\left(\frac{s}{2}\right) = 8r\left(\frac{1}{\sqrt{2}}\right) = 8(5.00)\left(\frac{1}{\sqrt{2}}\right) = 28.3$ cm *(5-2)*

82. $d = 40$ feet, so $c = \pi d = 40\pi$ feet

In 80 revolutions, the wheel turns a total of $80 \times 2\pi$ radians $= 160\pi$ radians.

$$\frac{160\pi \text{ rad}}{1 \text{ min}} \times \frac{1 \text{ min}}{60 \text{ sec}} = 8.38 \frac{\text{rad}}{\text{sec}}$$

In 80 revolutions, the linear distance traveled is $80 \times 40\pi$ feet $= 3200\pi$ feet.

$$\frac{3200\pi \text{ feet}}{1 \text{ min}} \times \frac{1 \text{ min}}{60 \text{ sec}} = 167.55 \frac{\text{ft}}{\text{sec}} \qquad (5\text{-}1)$$

83. When $t = 0$, $I = 30$, so $30 = A \cos B(0)$, that is, $A = 30$. Since the period is $\frac{1}{60}$ second, $\frac{2\pi}{B} = \frac{1}{60}$, $B = 2\pi(60) = 120\pi$. The equation is $I = 30 \cos 120\pi t$. $(5\text{-}5)$

84.

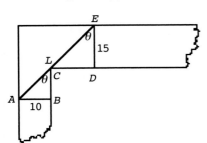

(A) In the figure as labeled, $L = AC + CE$.

In triangle ABC, since $\sin \theta = \frac{10}{AC}$, $AC \sin \theta = 10$,

$$AC = \frac{10}{\sin \theta} = 10 \csc \theta$$

In triangle CDE, since $\cos \theta = \frac{15}{CE}$, $CE \cos \theta = 15$,

$$CE = \frac{15}{\cos \theta} = 15 \sec \theta$$

Then $L = AC + CE = 10 \csc \theta + 15 \sec \theta \qquad 0 < \theta < \frac{\pi}{2}$

(B)

θ radians	0.4	0.5	0.6	0.7	0.8	0.9	1.0
L feet	42.0	38.0	35.9	35.1	35.5	36.9	39.6

From the table, the shortest distance L to the nearest foot, is 35 feet. This is the length of the longest log that can make the corner.

(C) Here is a graph of $L = 10 \csc \theta + 15 \sec \theta$ from a graphing calculator.

The minimum value of L is shown as 35.1 feet, to one decimal place.

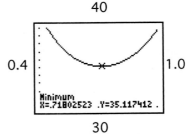

(D) As $\theta \to 0$, $L = 10 \csc \theta + 15 \sec \theta$ approaches an asymptote of $\csc \theta$; L increases without bound.

As $\theta \to \frac{\pi}{2}$, $L = 10 \csc \theta + 15 \sec \theta$ approaches an asymptote of $\sec \theta$; L increases without bound. $(5\text{-}2, 5\text{-}3)$

85. (A) $|A| = \dfrac{R_{max} - R_{min}}{2} = \dfrac{7 - 1}{2} = 3$, so $A = 3$ or -3. The graph appears to be shifted up from a graph of an upside down cosine curve that completes one full cycle as t varies over the interval $[0, 12]$, so $A = -3$.

Then $P = 12$ and we can find B using $P = \dfrac{2\pi}{B}$:

$$12 = \frac{2\pi}{B} \quad \Rightarrow \quad 12B = 2\pi \quad \Rightarrow \quad B = \frac{2\pi}{12} = \frac{\pi}{6}$$

$$k = \frac{R_{max} + R_{min}}{2} = \frac{7 + 1}{2} = 4.$$

$$R(t) = 4 - 3 \cos \frac{\pi}{6} t.$$

(B) The graph shows the seasonal changes in soft drink consumption. Most is consumed in August and the least in February. $(5\text{-}5)$

86. (A)

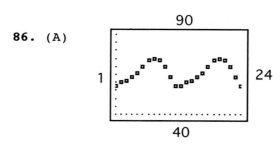

(B) $|A| = \dfrac{y_{max} - y_{min}}{2} = \dfrac{75 - 58}{2} = 8.5$. So $A = 8.5$ or -8.5.

$k = \dfrac{y_{max} + y_{min}}{2} = \dfrac{75 + 58}{2} = 66.5$. Then $P = 12$, and we can find B using

$P = \dfrac{2\pi}{B}$.

$12 = \dfrac{2\pi}{B} \quad \Rightarrow \quad 12B = 2\pi \quad \Rightarrow \quad B = \dfrac{2\pi}{12} = \dfrac{\pi}{6}$

The x intercept closest to the origin is estimated from the graph as 4.5.

To find C, substitute $B = \dfrac{\pi}{6}$ and $x = 4.5$ into the phase shift formula $x = -\dfrac{C}{B}$

and solve for C.

$x = -\dfrac{C}{B}$

$4.5 = -C \div \dfrac{\pi}{6}$

$C = -4.5\left(\dfrac{\pi}{6}\right)$

$C = -2.4$

With this value of C, the graph is seen to be shifted up from the graph of a standard sine curve, so $A = 8.5$. The equation is

$y = 66.5 + 8.5 \sin\left(\dfrac{\pi}{6} x - 2.4\right)$.

(C)

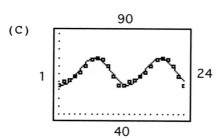

(D)

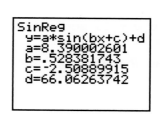

Compute the regression equation.

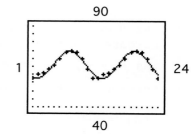

Graph the data and the regression equation.

(5-5)

CHAPTER 6   Trigonometric Identities and Conditional Equations

Section 6-1

1. Answers will vary.

3. Yes, but it would have to be a value that is not in the domain of at least one of the two sides.

5. The graphs are identical on the interval $[-10, 10]$, but the equation is not an identity. The two graphs are different for $x < -15$.

7. $\sin \theta \sec \theta = \sin \theta \dfrac{1}{\cos \theta}$ Reciprocal Identity

   $= \dfrac{\sin \theta}{\cos \theta}$ Algebra

   $= \tan \theta$ Quotient Identity

9. $\cot u \sec u \sin u = \dfrac{\cos u}{\sin u} \dfrac{1}{\cos u} \sin u$ Reciprocal and Quotient Identities

   $= 1$ Algebra

11. $\dfrac{\sin(-x)}{\cos(-x)} = \dfrac{-\sin x}{\cos x}$ Identities for Negatives

   $= -\tan x$ Quotient Identity

13. $\dfrac{\tan \alpha \cot \alpha}{\csc \alpha} = \dfrac{\tan \alpha \frac{1}{\tan \alpha}}{\csc \alpha}$ Reciprocal Identity

   $= \dfrac{1}{\csc \alpha}$ Algebra

   $= \dfrac{1}{\frac{1}{\sin \alpha}}$ Reciprocal Identity

   $= \sin \alpha$ Algebra

15. $\csc u(\cos u + \sin u) = \dfrac{1}{\sin u}(\cos u + \sin u)$ Reciprocal Identity

   $= \dfrac{\cos u}{\sin u} + \dfrac{\sin u}{\sin u}$ Algebra

   $= \dfrac{\cos u}{\sin u} + 1$ Algebra

   $= \cot u + 1$ Quotient Identity

   Key Algebraic Steps:
   $$\frac{1}{b}(a + b) = \frac{a}{b} + \frac{b}{b} = \frac{a}{b} + 1$$

17. $\dfrac{\cos x - \sin x}{\sin x \cos x} = \dfrac{\cos x}{\sin x \cos x} - \dfrac{\sin x}{\sin x \cos x}$ Algebra

   $= \dfrac{1}{\sin x} - \dfrac{1}{\cos x}$ Algebra

   $= \csc x - \sec x$ Reciprocal Identities

19.
$$\frac{\sin^2 t}{\cos t} + \cos t = \frac{\sin^2 t}{\cos t} + \frac{\cos t}{1} \qquad \text{Algebra}$$

$$= \frac{\sin^2 t}{\cos t} + \frac{\cos^2 t}{\cos t} \qquad \text{Algebra}$$

$$= \frac{\sin^2 t + \cos^2 t}{\cos t} \qquad \text{Algebra}$$

$$= \frac{1}{\cos t} \qquad \text{Pythagorean Identity}$$

$$= \sec t \qquad \text{Reciprocal Identity}$$

Key Algebraic Steps:
$$\frac{a^2}{b} + b = \frac{a^2}{b} + \frac{b}{1}$$
$$= \frac{a^2}{b} + \frac{b^2}{b}$$
$$= \frac{a^2 + b^2}{b}$$

21.
$$\frac{\cos x}{1 - \sin^2 x} = \frac{\cos x}{\sin^2 x + \cos^2 x - \sin^2 x} \qquad \text{Pythagorean Identity}$$

$$= \frac{\cos x}{\cos^2 x} \qquad \text{Algebra}$$

$$= \frac{1}{\cos x} \qquad \text{Algebra}$$

$$= \sec x \qquad \text{Reciprocal Identity}$$

23. $(1 - \cos u)(1 + \cos u) = 1 - \cos^2 u \qquad \text{Algebra}$
$$= \sin^2 u + \cos^2 u - \cos^2 u \qquad \text{Pythagorean Identity}$$
$$= \sin^2 u \qquad \text{Algebra}$$

25. $1 - 2 \sin^2 x = \cos^2 x + \sin^2 x - 2 \sin^2 x \qquad \text{Pythagorean Identity}$
$$= \cos^2 x - \sin^2 x \qquad \text{Algebra}$$

27. $(\sec t + 1)(\sec t - 1) = \sec^2 t - 1 \qquad \text{Algebra}$
$$= \tan^2 t + 1 - 1 \qquad \text{Pythagorean Identity}$$
$$= \tan^2 t \qquad \text{Algebra}$$

29. $\csc^2 x - \cot^2 x = 1 + \cot^2 x - \cot^2 x \qquad \text{Pythagorean Identity}$
$$= 1 \qquad \text{Algebra}$$

31.
$$\frac{\cos x + \tan x}{\sin x} = \frac{\cos x}{\sin x} + \frac{\tan x}{\sin x} \qquad \text{Algebra}$$

$$= \cot x + \frac{\frac{\sin x}{\cos x}}{\sin x} \qquad \text{Quotient Identities}$$

$$= \cot x + \frac{\sin x}{\cos x} \div \sin x \qquad \text{Algebra}$$

$$= \cot x + \frac{\sin x}{\cos x} \cdot \frac{1}{\sin x} \qquad \text{Algebra}$$

$$= \cot x + \frac{1}{\cos x} \qquad \text{Algebra}$$

$$= \cot x + \sec x \qquad \text{Reciprocal Identities}$$

Key Algebraic Steps:

$$a + \frac{\frac{b}{c}}{b} = a + \frac{b}{c} \div b$$

$$= a + \frac{b}{c} \cdot \frac{1}{b}$$

$$= a + \frac{1}{c}$$

33. Plug in $x = -1$.
Left side: $\sqrt{(-1)^2} = \sqrt{1} = 1$
Right side: -1
The equation is not an identity.

35. Plug in $x = 2$.
Left side: $2^4 + 2^3 - 2^2 - 2 = 18$
Right side: $2^5 + 2^4 - 2^3 - 2^2 = 36$
The equation is not an identity.

37. Plug in $x = \frac{\pi}{4}$.

Left side: $\sin \frac{\pi}{4} \cos \frac{\pi}{4} = \frac{1}{\sqrt{2}} \cdot \frac{1}{\sqrt{2}} = \frac{1}{2}$

Right side: $\tan \frac{\pi}{4} = 1$
The equation is not an identity.

39. Plug in $x = \frac{\pi}{6}$.

Left side: $\cos \frac{\pi}{6} = \frac{\sqrt{3}}{2}$

Right side: $1 - \sin \frac{\pi}{6} = 1 - \frac{1}{2} = \frac{1}{2}$
The equation is not an identity.

41.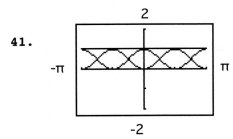

The straight line is the graph of
$y = \sin^2 x + \cos^2 x$, illustrating
the identity $\sin^2 x + \cos^2 x = 1$.

43.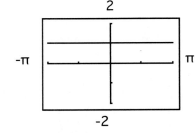

Both graphs are the same horizontal
line, illustrating the identity
$$\frac{\cos x}{\cot x \sin x} = 1$$

45. Simplify the left side:
$$\frac{x^2 - 9}{x + 3} = \frac{\overline{(x + 3)(x - 3)}}{\overline{x + 3}} = x - 3$$
The equation is an identity.

47. Simplify the left side:
$$\sqrt{x^2 + 4x + 4} \;=\; \sqrt{(x + 2)(x + 2)} \;=\; \sqrt{(x + 2)^2} \;=\; |x + 2|$$
The equation is not an identity: the left side must be positive, but the right can be negative. For example, if we plug in $x = -4$, we get
$$\sqrt{(-4)^2 + 4(-4) + 4} \;\overset{?}{=}\; -4 + 2$$
$$\sqrt{4} \;\overset{?}{=}\; -2$$
$$2 \;\overset{?}{=}\; -2$$

49. The domain of $\sqrt{3 - x}$ is $x \le 3$, while the domain of $\sqrt{x - 3}$ is $x \ge 3$, so the only x value in the domain of the entire left side is $x = 3$. When $x = 3$,
$$\sqrt{3 - 3} + \sqrt{3 - 3} = 0$$
is a true statement, so the equation is true for every x in its domain, and is an identity.

51. Evaluate both sides for $x = 4$:
Left side: $\sqrt{4 - 4} = 0$
Right side: $\sqrt{4 - 3} = 1$

The equation is not an identity.

53. Plug in $x = \dfrac{\pi}{4}$: $\quad \sin \dfrac{\pi}{4} - \cos \dfrac{\pi}{4} \;\overset{?}{=}\; 1$
$$\dfrac{1}{\sqrt{2}} - \dfrac{1}{\sqrt{2}} \;\overset{?}{=}\; 1$$
obviously this is false so the equation is not an identity.

55. Since $\sin^2 x + \cos^2 x = 1$, we suspect that this equation is not an identity.
Plug in $x = 0$: $\sin^2(0) - \cos^2(0) \;\overset{?}{=}\; 1$
$$0 - 1 \;\overset{?}{=}\; 1$$
This is false so the equation is not an identity.

> Remember: If you can find even one x-value that 1) makes the equation false, and 2) is in the domain of each side, then the equation is NOT an identity.

57. Yes. The equation is true for all values of x that are in the domain of both sides.

59. Yes. The equation is true for all values of x that are in the domain of both sides.

61.
$$\dfrac{1 - (\sin x - \cos x)^2}{\sin x} = \dfrac{1 - (\sin^2 x - 2\sin x \cos x + \cos^2 x)}{\sin x} \qquad \text{Algebra}$$
$$= \dfrac{1 - (-2\sin x \cos x + \sin^2 x + \cos^2 x)}{\sin x} \qquad \text{Algebra}$$
$$= \dfrac{1 - (-2\sin x \cos x + 1)}{\sin x} \qquad \text{Pythagorean Identity}$$
$$= \dfrac{2\sin x \cos x}{\sin x} \qquad \text{Algebra}$$
$$= 2\cos x \qquad \text{Algebra}$$

Key Algebraic Steps:
$$(a - b)^2 = a^2 - 2ab + b^2 = -2ab + a^2 + b^2$$

63. $\dfrac{\cot\theta+1}{\csc\theta} = \dfrac{\frac{\cos\theta}{\sin\theta}+1}{\frac{1}{\sin\theta}}$　　　　Reciprocal and Quotient Identities

$$= \dfrac{\sin\theta\,\frac{\cos\theta}{\sin\theta}+1\cdot\sin\theta}{\sin\theta\,\frac{1}{\sin\theta}}$$　　　　Algebra

$$= \cos\theta + \sin\theta$$　　　　Algebra

Key Algebraic Steps:

$$\dfrac{\frac{b}{a}+1}{\frac{1}{a}} = \dfrac{a\,\frac{b}{a}+1a}{a\,\frac{1}{a}} = b+a$$

65. $\dfrac{1+\cos y}{1-\cos y} = \dfrac{(1+\cos y)(1-\cos y)}{(1-\cos y)(1-\cos y)}$　　　　Algebra

$$= \dfrac{1-\cos^2 y}{(1-\cos y)^2}$$　　　　Algebra

$$= \dfrac{\sin^2 y + \cos^2 y - \cos^2 y}{(1-\cos y)^2}$$　　　　Pythagorean Identity

$$= \dfrac{\sin^2 y}{(1-\cos y)^2}$$　　　　Algebra

67. $\tan^2 x - \sin^2 x = \dfrac{\sin^2 x}{\cos^2 x} - \sin^2 x$　　　　Quotient Identity

$$= \sin^2 x\dfrac{1}{\cos^2 x} - \sin^2 x\cdot 1$$　　　　Algebra

$$= \sin^2 x\left(\dfrac{1}{\cos^2 x} - 1\right)$$　　　　Algebra

$$= \sin^2 x(\sec^2 x - 1)$$　　　　Reciprocal Identity

$$= \sin^2 x\,\tan^2 x$$　　　　Pythagorean Identity

$$= \tan^2 x\,\sin^2 x$$　　　　Algebra

69. $\dfrac{\csc\theta}{\cot\theta+\tan\theta} = \dfrac{\frac{1}{\sin\theta}}{\frac{\cos\theta}{\sin\theta}+\frac{\sin\theta}{\cos\theta}}$　　　　Reciprocal and Quotient Identities

$$= \dfrac{\cos\theta\sin\theta\,\frac{1}{\sin\theta}}{\cos\theta\sin\theta\,\frac{\cos\theta}{\sin\theta}+\cos\theta\sin\theta\,\frac{\sin\theta}{\cos\theta}}$$　　　　Algebra

$$= \dfrac{\cos\theta}{\cos^2\theta+\sin^2\theta}$$　　　　Algebra

$$= \dfrac{\cos\theta}{1}$$　　　　Pythagorean Identity

$$= \cos\theta$$

Key Algebraic Steps:

$$\dfrac{\frac{1}{a}}{\frac{b}{a}+\frac{a}{b}} = \dfrac{ab\cdot\frac{1}{a}}{ab\cdot\frac{b}{a}+ab\cdot\frac{a}{b}} = \dfrac{b}{b^2+a^2}$$

71. $\ln \tan x = \ln\left(\dfrac{\sin x}{\cos x}\right)$ Quotient Identity

$\qquad = \ln \sin x - \ln \cos x$ Algebra

73. $\ln \cot x = \ln\left(\dfrac{1}{\tan x}\right)$ Reciprocal Identity

$\qquad = \ln 1 - \ln \tan x$ Algebra

$\qquad = 0 - \ln \tan x$ Algebra

$\qquad = -\ln \tan x$ Algebra

75. $\dfrac{\sec A - 1}{\sec A + 1} = \dfrac{\frac{1}{\cos A} - 1}{\frac{1}{\cos A} + 1}$ Reciprocal Identity

$\qquad = \dfrac{\cos A \frac{1}{\cos A} - 1 \cdot \cos A}{\cos A \frac{1}{\cos A} + 1 \cdot \cos A}$ Algebra

$\qquad = \dfrac{1 - \cos A}{1 + \cos A}$ Algebra

Key Algebraic Steps:

$$\frac{\frac{1}{a} - 1}{\frac{1}{a} + 1} = \frac{a \cdot \frac{1}{a} - 1 \cdot a}{a \cdot \frac{1}{a} + 1 \cdot a} = \frac{1 - a}{1 + a}$$

77. $\sin^4 w - \cos^4 w = (\sin^2 w)^2 - (\cos^2 w)^2$ Algebra

$\qquad = (\sin^2 w + \cos^2 w)(\sin^2 w - \cos^2 w)$ Algebra

$\qquad = 1(\sin^2 w - \cos^2 w)$ Pythagorean Identity

$\qquad = \sin^2 w - \cos^2 w$ Algebra

$\qquad = 1 - \cos^2 w - \cos^2 w$ Pythagorean Identity

$\qquad = 1 - 2\cos^2 w$ Algebra

Key Algebraic Steps:

$$a^4 - b^4 = (a^2)^2 - (b^2)^2 = (a^2 + b^2)(a^2 - b^2)$$

79. $\sec x - \dfrac{\cos x}{1 + \sin x} = \dfrac{1}{\cos x} - \dfrac{\cos x}{1 + \sin x}$ Reciprocal Identity

$\qquad = \dfrac{1 + \sin x - \cos^2 x}{(1 + \sin x)\cos x}$ Algebra

$\qquad = \dfrac{\sin^2 x + \cos^2 x + \sin x - \cos^2 x}{(1 + \sin x)\cos x}$ Pythagorean Identity

$\qquad = \dfrac{\sin^2 x + \sin x}{(1 + \sin x)\cos x}$ Algebra

$\qquad = \dfrac{(1 + \sin x)\sin x}{(1 + \sin x)\cos x}$ Algebra

$\qquad = \dfrac{\sin x}{\cos x}$ Algebra

$\qquad = \tan x$ Quotient Identity

Key Algebraic Steps:

$$\frac{1}{a} - \frac{a}{1+b} = \frac{1+b-a^2}{(1+b)a}$$

$$\frac{b^2+b}{(1+b)a} = \frac{(1+b)b}{(1+b)a} = \frac{b}{a}$$

81. $\dfrac{\cos^2 z - 3\cos z + 2}{\sin^2 z}$ $= \dfrac{\cos^2 z - 3\cos z + 2}{\sin^2 z + \cos^2 z - \cos^2 z}$ Algebra

$$= \frac{\cos^2 z - 3\cos z + 2}{1 - \cos^2 z}$$ Pythagorean Identity

$$= \frac{(\cos z - 1)(\cos z - 2)}{(1 - \cos z)(1 + \cos z)}$$ Algebra

$$= \frac{-(\cos z - 2)}{1 + \cos z}$$ Algebra

$$= \frac{2 - \cos z}{1 + \cos z}$$ Algebra

Key Algebraic Steps:

$$\frac{a^2 - 3a + 2}{1 - a^2} = \frac{(a-1)(a-2)}{(1-a)(1+a)} = \frac{-(a-2)}{1+a} = \frac{2-a}{1+a}$$

83. $\dfrac{\cos^3 \theta - \sin^3 \theta}{\cos \theta - \sin \theta} = \dfrac{(\cos \theta - \sin \theta)(\cos^2 \theta + \cos \theta \sin \theta + \sin^2 \theta)}{\cos \theta - \sin \theta}$ Algebra

$$= \cos^2 \theta + \cos \theta \sin \theta + \sin^2 \theta$$ Algebra

$$= 1 + \cos \theta \sin \theta$$ Pythagorean Identity

$$= 1 + \sin \theta \cos \theta$$ Algebra

Key Algebraic Steps:

$$\frac{a^3 - b^3}{a - b} = \frac{(a-b)(a^2 + ab + b^2)}{a-b}$$
$$= a^2 + ab + b^2$$

Common Error:
$$\frac{a^3 - b^3}{a - b} \neq a^2 + b^2$$

85. $(\sec x - \tan x)^2 = \left(\dfrac{1}{\cos x} - \dfrac{\sin x}{\cos x}\right)^2$ Quotient and Reciprocal Identities

$$= \left(\frac{1 - \sin x}{\cos x}\right)^2$$ Algebra

$$= \frac{(1 - \sin x)^2}{\cos^2 x}$$ Algebra

$$= \frac{(1 - \sin x)^2}{1 - \sin^2 x}$$ Pythagorean Identity

$$= \frac{(1 - \sin x)(1 - \sin x)}{(1 - \sin x)(1 + \sin x)}$$ Algebra

$$= \frac{1 - \sin x}{1 + \sin x}$$ Algebra

87. $\dfrac{\csc^4 x - 1}{\cot^2 x} = \dfrac{(\csc^2 x)^2 - (1)^2}{\cot^2 x}$ Algebra

$= \dfrac{(\csc^2 x - 1)(\csc^2 x + 1)}{\cot^2 x}$ Algebra

$= \dfrac{(1 + \cot^2 x - 1)(\csc^2 x + 1)}{\cot^2 x}$ Pythagorean Identity

$= \dfrac{\cot^2 x(\csc^2 x + 1)}{\cot^2 x}$ Algebra

$= \csc^2 x + 1$ Algebra

$= 1 + \cot^2 x + 1$ Pythagorean Identity

$= 2 + \cot^2 x$ Algebra

89. $\dfrac{1 + \sin v}{\cos v} = \dfrac{(1 + \sin v)(1 - \sin v)}{\cos v(1 - \sin v)}$ Algebra

$= \dfrac{1 - \sin^2 v}{\cos v(1 - \sin v)}$ Algebra

$= \dfrac{\cos^2 v}{\cos v(1 - \sin v)}$ Pythagorean Identity

$= \dfrac{\cos v}{1 - \sin v}$ Algebra

91. Graph both sides of the equation in the same viewing window.

$\dfrac{\sin(-x)}{\cos(-x)\tan(-x)} = -1$ is not an identity, since the graphs do not match.

Try $x = -\dfrac{\pi}{4}$

Left side: $\dfrac{\sin\left[-\left(-\frac{\pi}{4}\right)\right]}{\cos\left[-\left(-\frac{\pi}{4}\right)\right]\tan\left[-\left(-\frac{\pi}{4}\right)\right]} = \dfrac{\frac{1}{\sqrt{2}}}{\frac{1}{\sqrt{2}} \cdot 1} = 1$

Right side: $= -1$
This verifies that the equation is not an identity.

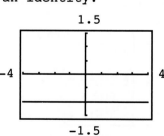

93. Graph both sides of the equation in the same viewing window.

$\dfrac{\sin x}{\cos x \tan(-x)} = -1$ appears to be an identity, which we now verify:

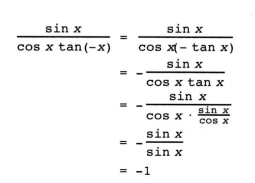

$\dfrac{\sin x}{\cos x \tan(-x)} = \dfrac{\sin x}{\cos x(-\tan x)}$ Identities for Negatives

$= -\dfrac{\sin x}{\cos x \tan x}$ Algebra

$= -\dfrac{\sin x}{\cos x \cdot \frac{\sin x}{\cos x}}$ Quotient Identity

$= -\dfrac{\sin x}{\sin x}$ Algebra

$= -1$ Algebra

95. Graph both sides of the equation in the same viewing window.

$\sin x + \dfrac{\cos^2 x}{\sin x} = \sec x$ is not an identity, since the graphs do not match.

Try $x = -\dfrac{\pi}{4}$

Left side: $\sin\left(-\dfrac{\pi}{4}\right) + \dfrac{\cos^2\left(-\dfrac{\pi}{4}\right)}{\sin\left(-\dfrac{\pi}{4}\right)} = -\dfrac{1}{\sqrt{2}} + \dfrac{\left(\dfrac{1}{\sqrt{2}}\right)^2}{\left(-\dfrac{1}{\sqrt{2}}\right)} = -\dfrac{1}{\sqrt{2}} - \dfrac{1}{\sqrt{2}} = -\sqrt{2}$

Right side: $\sec\left(-\dfrac{\pi}{4}\right) = \sqrt{2}$

This verifies that the equation is not an identity.

97. Graph both sides of the equation in the same viewing window.

$\sin x + \dfrac{\cos^2 x}{\sin x} = \csc x$ appears to be an identity, which we now verify:

$$
\begin{aligned}
\sin x + \dfrac{\cos^2 x}{\sin x} &= \dfrac{\sin x}{1} + \dfrac{\cos^2 x}{\sin x} && \text{Algebra}\\[2mm]
&= \dfrac{\sin^2 x}{\sin x} + \dfrac{\cos^2 x}{\sin x} && \text{Algebra}\\[2mm]
&= \dfrac{\sin^2 x + \cos^2 x}{\sin x} && \text{Algebra}\\[2mm]
&= \dfrac{1}{\sin x} && \text{Pythagorean Identity}\\[2mm]
&= \csc x && \text{Reciprocal Identity}
\end{aligned}
$$

99. Graph both sides of the equation in the same viewing window.

$\dfrac{\tan x}{\sin x - 2\tan x} = \dfrac{1}{\cos x - 2}$ appears to be an identity, which we now verify:

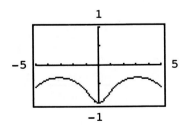

$$
\begin{aligned}
\dfrac{\tan x}{\sin x - 2\tan x} &= \dfrac{\dfrac{\sin x}{\cos x}}{\sin x - 2\dfrac{\sin x}{\cos x}} && \text{Quotient Identity}\\[4mm]
&= \dfrac{\cos x \cdot \dfrac{\sin x}{\cos x}}{\sin x \cos x - 2\cos x \cdot \dfrac{\sin x}{\cos x}} && \text{Algebra}
\end{aligned}
$$

$$= \frac{\sin x}{\sin x \cos x - 2 \sin x} \qquad \text{Algebra}$$

$$= \frac{\sin x}{\sin x(\cos x - 2)} \qquad \text{Algebra}$$

$$= \frac{1}{\cos x - 2} \qquad \text{Algebra}$$

101. Graph both sides of the equation in the same viewing window.

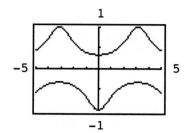

$\dfrac{\tan x}{\sin x + 2 \tan x} = \dfrac{1}{\cos x - 2}$ is not an identity, since the graphs do not match.

Try $x = \dfrac{\pi}{3}$

Left side: $\dfrac{\tan \frac{\pi}{3}}{\sin \frac{\pi}{3} + 2 \tan \frac{\pi}{3}} = \dfrac{\sqrt{3}}{\frac{\sqrt{3}}{2} + 2\sqrt{3}} = \dfrac{2\sqrt{3}}{\sqrt{3} + 4\sqrt{3}} = \dfrac{2\sqrt{3}}{5\sqrt{3}} = \dfrac{2}{5}$

Right side: $\dfrac{1}{\cos \frac{\pi}{3} - 2} = \dfrac{1}{\frac{1}{2} - 2} = \dfrac{1}{-\frac{3}{2}} = -\dfrac{2}{3}$

This verifies that the equation is not an identity.

103. $\dfrac{2 \sin^2 x + 3 \cos x - 3}{\sin^2 x} = \dfrac{2(1 - \cos^2 x) + 3 \cos x - 3}{\sin^2 x}$ Pythagorean Identity

$$= \frac{2 - 2 \cos^2 x + 3 \cos x - 3}{\sin^2 x} \qquad \text{Algebra}$$

$$= \frac{-2 \cos^2 x + 3 \cos x - 1}{\sin^2 x} \qquad \text{Algebra}$$

$$= \frac{-2 \cos^2 x + 3 \cos x - 1}{1 - \cos^2 x} \qquad \text{Pythagorean Identity}$$

$$= \frac{-(2 \cos^2 x - 3 \cos x + 1)}{-(\cos^2 x - 1)} \qquad \text{Algebra}$$

$$= \frac{-(2 \cos x - 1)(\cos x - 1)}{-(\cos x + 1)(\cos x - 1)} \qquad \text{Algebra}$$

$$= \frac{2 \cos x - 1}{\cos x + 1} \qquad \text{Algebra}$$

$$= \frac{2 \cos x - 1}{1 + \cos x} \qquad \text{Algebra}$$

Key Algebraic Steps:

$$\frac{-2a^2 + 3a - 1}{1 - a^2} = \frac{-(2a^2 - 3a + 1)}{-(a^2 - 1)}$$

$$= \frac{-(2a - 1)(a - 1)}{-(a + 1)(a - 1)}$$

$$= \frac{2a - 1}{a + 1}$$

$$= \frac{2a - 1}{1 + a}$$

105. $\dfrac{\tan u + \sin u}{\tan u - \sin u} - \dfrac{\sec u + 1}{\sec u - 1} = \dfrac{\frac{\sin u}{\cos u} + \sin u}{\frac{\sin u}{\cos u} - \sin u} - \dfrac{\sec u + 1}{\sec u - 1}$ Quotient Identity

$$= \frac{\frac{1}{\cos u}\sin u + 1 \cdot \sin u}{\frac{1}{\cos u}\sin u - 1 \cdot \sin u} - \frac{\sec u + 1}{\sec u - 1}$$ Algebra

$$= \frac{\sin u\left(\frac{1}{\cos u} + 1\right)}{\sin u\left(\frac{1}{\cos u} - 1\right)} - \frac{\sec u + 1}{\sec u - 1}$$ Algebra

$$= \frac{\frac{1}{\cos u} + 1}{\frac{1}{\cos u} - 1} - \frac{\sec u + 1}{\sec u - 1}$$ Algebra

$$= \frac{\sec u + 1}{\sec u - 1} - \frac{\sec u + 1}{\sec u - 1}$$ Reciprocal Identity

$$= 0$$ Algebra

Key Algebraic Steps:

$$\frac{\frac{b}{a} + b}{\frac{b}{a} - b} = \frac{b \cdot \frac{1}{a} + b}{b \cdot \frac{1}{a} - b}$$

$$= \frac{b\left(\frac{1}{a} + 1\right)}{b\left(\frac{1}{a} - 1\right)}$$

$$= \frac{\frac{1}{a} + 1}{\frac{1}{a} - 1}$$

107. $\dfrac{\tan \beta + \cot \alpha}{\tan \beta \cot \alpha} = \dfrac{\tan \beta}{\tan \beta \cot \alpha} + \dfrac{\cot \alpha}{\tan \beta \cot \alpha}$ Algebra

$$= \frac{1}{\cot \alpha} + \frac{1}{\tan \beta}$$ Algebra

$$= \tan \alpha + \cot \beta$$ Reciprocal Identities

109.

Statement	Reason
$\cot^2 x + 1 = \left(\dfrac{\cos x}{\sin x}\right)^2 + 1$	$\cot x = \dfrac{\cos x}{\sin x}$
$= \dfrac{\cos^2 x}{\sin^2 x} + 1$	Algebra
$= \dfrac{\cos^2 x + \sin^2 x}{\sin^2 x}$	Algebra
$= \dfrac{1}{\sin^2 x}$	$\cos^2 x + \sin^2 x = 1$
$= \left(\dfrac{1}{\sin x}\right)^2$	Algebra
$= \csc^2 x$	$\dfrac{1}{\cos x} = \sec x$

111. Here is a graph of $f(x)$ on the interval $[-2\pi,\ 2\pi]$.

The graph appears identical with the graph of $g(x) = \cot x$. We attempt to verify that $\dfrac{1 - \sin^2 x}{\tan x} + \sin x \cos x$ $= \cot x$ is an identity.

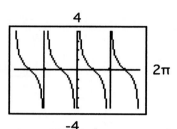

$$\dfrac{1 - \sin^2 x}{\tan x} + \sin x \cos x = \dfrac{1}{\tan x} - \dfrac{\sin^2 x}{\tan x} + \sin x \cos x \qquad \text{Algebra}$$

$$= \cot x - \dfrac{\sin^2 x}{\frac{\sin x}{\cos x}} + \sin x \cos x \qquad \begin{array}{l}\text{Quotient}\\\text{Identity}\end{array}$$

$$= \cot x - \dfrac{\sin^2 x \cos x}{\sin x} + \sin x \cos x \qquad \text{Algebra}$$

$$= \cot x - \sin x \cos x + \sin x \cos x \qquad \text{Algebra}$$

$$= \cot x \qquad \text{Algebra}$$

$f(x) = g(x)$ is an identity for $g(x) = \cot x$.

113. Here is a graph of $f(x)$ on the interval $[-2\pi,\ 2\pi]$.

The graph appears similar to the graph of the cosecant function, but shifted down 1 unit, and with peculiar behavior at $-\dfrac{\pi}{2}$ and $\dfrac{3\pi}{2}$ where $f(x)$ is not defined, although $\csc x$ is defined. We set $g(x) = \csc x - 1$ and attempt to verify that

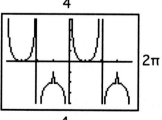

$$\dfrac{\cos^2 x}{1 + \sin x - \cos^2 x} = \csc x - 1 \text{ is an identity.}$$

$$\dfrac{\cos^2 x}{1 + \sin x - \cos^2 x} = \dfrac{1 - \sin^2 x}{1 + \sin x - (1 - \sin^2 x)} \qquad \text{Pythagorean Identity}$$

$$= \dfrac{1 - \sin^2 x}{\sin x + \sin^2 x} \qquad \text{Algebra}$$

$$= \frac{(1 - \sin x)(1 + \sin x)}{\sin x(1 + \sin x)} \qquad \text{Algebra}$$

$$= \frac{1 - \sin x}{\sin x} \qquad \text{Algebra}$$

$$= \frac{1}{\sin x} - \frac{\sin x}{\sin x} \qquad \text{Algebra}$$

$$= \csc x - 1 \qquad \text{Reciprocal Identity; Algebra}$$

$f(x) = g(x)$ is an identity for $g(x) = \csc x - 1$.

115. Here is a graph of $f(x)$ on the interval $[-2\pi, 2\pi]$.

The graph appears identical with the graph of $g(x) = 3 \cos x$. We attempt to verify that
$$\frac{1 + \cos x - 2\cos^2 x}{1 - \cos x} + \frac{\sin^2 x}{1 + \cos x}$$
$= 3 \cos x$ is an identity.

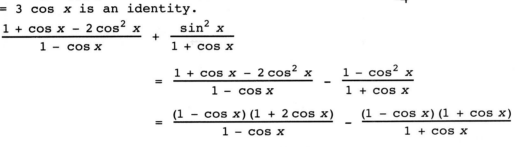

$$\frac{1 + \cos x - 2\cos^2 x}{1 - \cos x} + \frac{\sin^2 x}{1 + \cos x}$$

$$= \frac{1 + \cos x - 2\cos^2 x}{1 - \cos x} - \frac{1 - \cos^2 x}{1 + \cos x}$$

$$= \frac{(1 - \cos x)(1 + 2\cos x)}{1 - \cos x} - \frac{(1 - \cos x)(1 + \cos x)}{1 + \cos x}$$

$$= 1 + 2 \cos x - (1 - \cos x)$$

$$= 1 + 2 \cos x - 1 + \cos x$$

$$= 3 \cos x$$

$f(x) = g(x)$ is an identity for $g(x) = 3 \cos x$.

117. Since $\sqrt{1 - \cos^2 x} = \sqrt{\sin^2 x}$, the equation will be true when $\sqrt{\sin^2 x} = -\sin x$, that is, when $\sin x$ is negative. This occurs in Quadrants III, IV.

119. Since $\sqrt{1 - \cos^2 x} = \sqrt{\sin^2 x}$, the equation will be true when $\sqrt{\sin^2 x} = \sin x$, that is, when $\sin x$ is positive. This occurs in Quadrants I, II.

121. Since $\sqrt{1 - \sin^2 x} = \sqrt{\cos^2 x} = |\cos x|$ is an identity, this will hold in all quadrants.

123. Since $\sqrt{1 - \sin^2 x} = \sqrt{\cos^2 x} = \cos x$ if $\cos x$ is positive, this will hold when $\dfrac{\sin x}{\sqrt{1 - \sin^2 x}} = \dfrac{\sin x}{\sqrt{\cos^2 x}} = \dfrac{\sin x}{\cos x} = \tan x$, that is, when $\cos x$ is positive. This occurs in Quadrants I, IV.

125. $\sqrt{a^2 - u^2} = \sqrt{a^2 - (a \sin x)^2}$

$$= \sqrt{a^2 - a^2 \sin^2 x}$$

$$= \sqrt{a^2(1 - \sin^2 x)}$$

$$= \sqrt{a^2}\sqrt{1 - \sin^2 x}$$

$$= \sqrt{a^2}\sqrt{\cos^2 x}$$

> **Common Error:**
> $\sqrt{1 - \sin^2 x} \neq 1 - \sin x$

Since $a > 0$, $\sqrt{a^2} = a$.

$\cos x$ will be ≥ 0 if x is a quadrant I or IV angle. But the restriction $-\dfrac{\pi}{2} < x < \dfrac{\pi}{2}$ requires that x is such an angle. Therefore $\sqrt{\cos^2 x} = \cos x$

$\sqrt{a^2 - u^2} = a \cos x$

127. $\sqrt{a^2 + u^2} = \sqrt{a^2 + (a \tan x)^2}$

$$= \sqrt{a^2 + a^2 \tan^2 x}$$

$$= \sqrt{a^2(1 + \tan^2 x)}$$

$$= \sqrt{a^2}\sqrt{1 + \tan^2 x}$$

$$= \sqrt{a^2}\sqrt{\sec^2 x}$$

Since $a > 0$, $\sqrt{a^2} = a$.

$\sec x$ will be ≥ 0 if x is a quadrant I or IV angle. But the restriction $0 < x < \dfrac{\pi}{2}$ requires that x is a quadrant I angle. Therefore $\sqrt{\sec^2 x} = \sec x$

$\sqrt{a^2 + u^2} = a \sec x$

Section 6-2

1. Answers will vary.

3. Answers will vary.

5. Yes. Explanations will vary.

7. Plug in $x = 1$, $y = 1$.
Left side: $(1 + 1)^2 = 2^2 = 4$
Right side: $1^2 + 1^2 = 2$
The equation is not an identity.

9. Plug in $x = 2$, $y = \dfrac{\pi}{6}$.

Left side: $2 \sin \dfrac{\pi}{6} = 1$

Right side: $\sin\left(2 \cdot \dfrac{\pi}{6}\right) = \dfrac{\sqrt{3}}{2}$

The equation is not an identity.

11. Plug in $x = \dfrac{\pi}{6}$, $y = \dfrac{\pi}{6}$.

Left side: $\cos\left(\dfrac{\pi}{6} + \dfrac{\pi}{6}\right) = \cos \dfrac{\pi}{3} = \dfrac{1}{2}$

Right side: $\cos \dfrac{\pi}{6} + \cos \dfrac{\pi}{6} = \dfrac{\sqrt{3}}{2} + \dfrac{\sqrt{3}}{2} = \sqrt{3}$

The equation is not an identity.

13. Plug in $x = \dfrac{\pi}{3}$, $y = \dfrac{\pi}{6}$.

Left side: $\tan\left(\dfrac{\pi}{3} - \dfrac{\pi}{6}\right) = \tan\dfrac{\pi}{6} = \dfrac{1}{\sqrt{3}}$

Right side: $\tan\dfrac{\pi}{3} - \tan\dfrac{\pi}{6} = \sqrt{3} - \dfrac{1}{\sqrt{3}} = \dfrac{2}{\sqrt{3}}$

The equation is not an identity.

15. Expand and simplify the left side:

$$\tan(x - \pi) = \dfrac{\tan x - \tan \pi}{1 + \tan x \tan \pi} \qquad \text{Difference Identity}$$

$$= \dfrac{\tan x - 0}{1 + \tan x \cdot 0} \qquad \tan \pi = \dfrac{\sin \pi}{\cos \pi} = \dfrac{0}{-1} = 0$$

$$= \dfrac{\tan x}{1} \qquad \text{Algebra}$$

$$= \tan x \qquad \text{Algebra}$$

The equation is an identity.

17. Expand and simplify the left side:

$$\sin(x - \pi) = \sin x \cos \pi - \cos x \sin \pi \qquad \text{Difference Identity}$$

$$= \sin x(-1) - \cos x(0) \qquad \cos \pi = -1; \; \sin \pi = 0$$

$$= -\sin x \qquad \text{Algebra}$$

The equation is not an identity.

19. Expand and simplify the left side:

$$\csc(2\pi - x) = \dfrac{1}{\sin(2\pi - x)} \qquad\qquad \csc x = \dfrac{1}{\sin x}$$

$$= \dfrac{1}{\sin 2\pi \cos x - \cos 2\pi \sin x} \qquad \text{Difference Identity}$$

$$= \dfrac{1}{0 \cdot \cos x - 1 \cdot \sin x} \qquad \sin 2\pi = 0; \; \cos 2\pi = 1$$

$$= \dfrac{1}{-\sin x} \qquad\qquad \text{Algebra}$$

$$= -\csc x \qquad\qquad \csc x = \dfrac{1}{\sin x}$$

The equation is not an identity.

21. Expand and simplify the left side:

$$\sin\left(x - \dfrac{\pi}{2}\right) = \sin x \cos \dfrac{\pi}{2} - \cos x \sin \dfrac{\pi}{2} \qquad \text{Difference Identity}$$

$$= \sin x \cdot 0 - \cos x \cdot 1 \qquad \cos \dfrac{\pi}{2} = 0; \; \sin \dfrac{\pi}{2} = 1$$

$$= -\cos x \qquad \text{Algebra}$$

The equation is an identity.

23. $\cot\left(\dfrac{\pi}{2} - x\right) = \dfrac{\cos\left(\frac{\pi}{2} - x\right)}{\sin\left(\frac{\pi}{2} - x\right)}$ Quotient Identity

$\qquad = \dfrac{\cos\frac{\pi}{2}\cos x + \sin\frac{\pi}{2}\sin x}{\sin\frac{\pi}{2}\cos x - \cos\frac{\pi}{2}\sin x}$ Difference Identities

$\qquad = \dfrac{0\cos x + 1\sin x}{1\cos x - 0\sin x}$ Known Values

$\qquad = \dfrac{\sin x}{\cos x}$ Algebra

$\qquad = \tan x$ Quotient Identity

25. $\csc\left(\dfrac{\pi}{2} - x\right) = \dfrac{1}{\sin\left(\frac{\pi}{2} - x\right)}$ Reciprocal Identity

$\qquad = \dfrac{1}{\sin\frac{\pi}{2}\cos x - \cos\frac{\pi}{2}\sin x}$ Difference Identity

$\qquad = \dfrac{1}{1\cos x - 0\sin x}$ Known Values

$\qquad = \dfrac{1}{\cos x}$ Algebra

$\qquad = \sec x$ Reciprocal Identity

27. $\sin(30° - x) = \sin 30°\cos x - \cos 30°\sin x$ Difference Identity

$\qquad = \dfrac{1}{2}\cos x - \dfrac{\sqrt{3}}{2}\sin x$ Known Values

$\qquad = \dfrac{1}{2}(\cos x - \sqrt{3}\sin x)$ Algebra

29. $\sin(180° - x) = \sin 180°\cos x - \cos 180°\sin x$ Difference Identity
$\qquad\qquad\quad = 0\cos x - (-1)\sin x$ Known Values
$\qquad\qquad\quad = \sin x$ Algebra

31. $\tan\left(x + \dfrac{\pi}{3}\right) = \dfrac{\tan x + \tan\frac{\pi}{3}}{1 - \tan x\tan\frac{\pi}{3}}$ Sum Identity

$\qquad = \dfrac{\tan x + \sqrt{3}}{1 - \sqrt{3}\tan x}$ Known Values

33. $\sin 15° = \sin\left(45° - 30°\right)$

$\qquad = \sin 45°\cos 30° - \cos 45°\sin 30°$

$\qquad = \dfrac{1}{\sqrt{2}}\cdot\dfrac{\sqrt{3}}{2} - \dfrac{1}{\sqrt{2}}\cdot\dfrac{1}{2}$

$\qquad = \dfrac{\sqrt{3} - 1}{2\sqrt{2}}$

35. $\cos 75° = \cos\left(45° + 30°\right)$

$\qquad = \cos 45°\cos 30° - \sin 45°\sin 30°$

$\qquad = \dfrac{1}{\sqrt{2}}\cdot\dfrac{\sqrt{3}}{2} - \dfrac{1}{\sqrt{2}}\cdot\dfrac{1}{2}$

$\qquad = \dfrac{\sqrt{3} - 1}{2\sqrt{2}}$

37. $\tan\left(\dfrac{7\pi}{12}\right) = \tan\left(\dfrac{\pi}{3} + \dfrac{\pi}{4}\right)$

$= \dfrac{\tan\dfrac{\pi}{3} + \tan\dfrac{\pi}{4}}{1 - \tan\dfrac{\pi}{3}\tan\dfrac{\pi}{4}}$

$= \dfrac{\sqrt{3} + 1}{1 - \sqrt{3}\cdot 1} = \dfrac{1 + \sqrt{3}}{1 - \sqrt{3}}$

39. $\cos\left(\dfrac{11\pi}{12}\right) = \cos\left(\dfrac{2\pi}{3} + \dfrac{\pi}{4}\right)$

$= \cos\dfrac{2\pi}{3}\cos\dfrac{\pi}{4} - \sin\dfrac{2\pi}{3}\sin\dfrac{\pi}{4}$

$= -\dfrac{1}{2}\cdot\dfrac{1}{\sqrt{2}} - \dfrac{\sqrt{3}}{2}\cdot\dfrac{1}{\sqrt{2}} = \dfrac{-1 - \sqrt{3}}{2\sqrt{2}}$

41. $\cos 74° \cos 44° + \sin 74° \sin 44° = \cos(74° - 44°) = \cos 30° = \dfrac{\sqrt{3}}{2}$

43. $\dfrac{\tan 27° + \tan 18°}{1 - \tan 27° \tan 18°} = \tan(27° + 18°) = \tan 45° = 1$

45.

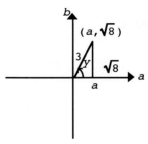

$a = \sqrt{5^2 - (-3)^2} = 4$

$\cos x = \dfrac{4}{5}$

$\tan x = -\dfrac{3}{4}$

$\sin(x - y) = \sin x \cos y - \cos x \sin y$

$= \left(-\dfrac{3}{5}\right)\dfrac{1}{3} - \dfrac{4}{5}\left(\dfrac{\sqrt{8}}{3}\right)$

$= \dfrac{-3 - 4\sqrt{8}}{15}$

$a = \sqrt{3^2 - (\sqrt{8})^2} = 1$

$\cos y = \dfrac{1}{3}$

$\tan y = \dfrac{\sqrt{8}}{1} = \sqrt{8}$

$\tan(x + y) = \dfrac{\tan x + \tan y}{1 - \tan x \tan y}$

$= \dfrac{-\dfrac{3}{4} + \sqrt{8}}{1 - \left(-\dfrac{3}{4}\right)\sqrt{8}}$

$= \dfrac{-3 + 4\sqrt{8}}{4 + 3\sqrt{8}} = \dfrac{4\sqrt{8} - 3}{4 + 3\sqrt{8}}$

47.

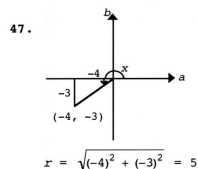

$r = \sqrt{(-4)^2 + (-3)^2} = 5$

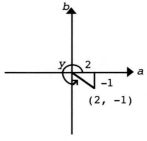

$r = \sqrt{2^2 + (-1)^2} = \sqrt{5}$

$\sin x = -\dfrac{3}{5}$

$\cos x = -\dfrac{4}{5}$

$\sin(x - y) = \sin x \cos y - \cos x \sin y$

$\qquad = \left(-\dfrac{3}{5}\right)\left(\dfrac{2}{\sqrt{5}}\right) - \left(-\dfrac{4}{5}\right)\left(-\dfrac{1}{\sqrt{5}}\right)$

$\qquad = -\dfrac{6}{5\sqrt{5}} - \dfrac{4}{5\sqrt{5}}$

$\qquad = \dfrac{-10}{5\sqrt{5}}$

$\qquad = \dfrac{-2}{\sqrt{5}}$

$\sin y = \dfrac{-1}{\sqrt{5}}$

$\cos y = \dfrac{2}{\sqrt{5}}$

$\tan(x + y) = \dfrac{\tan x + \tan y}{1 - \tan x \tan y}$

$\qquad = \dfrac{\frac{3}{4} + \left(-\frac{1}{2}\right)}{1 - \left(\frac{3}{4}\right)\left(-\frac{1}{2}\right)}$

$\qquad = \dfrac{6 - 4}{8 + 3}$

$\qquad = \dfrac{2}{11}$

49. $\cos 2x = \cos(x + x)$ Algebra

$\qquad\quad = \cos x \cos x - \sin x \sin x$ Sum Identity

$\qquad\quad = \cos^2 x - \sin^2 x$ Algebra

51. $\cot(x + y) = \dfrac{\cos(x + y)}{\sin(x + y)}$ Quotient Identity

$\qquad\qquad\quad = \dfrac{\cos x \cos y - \sin x \sin y}{\sin x \cos y + \cos x \sin y}$ Sum Identities

$\qquad\qquad\quad = \dfrac{\dfrac{\cos x \cos y}{\sin x \sin y} - \dfrac{\sin x \sin y}{\sin x \sin y}}{\dfrac{\sin x \cos y}{\sin x \sin y} + \dfrac{\cos x \sin y}{\sin x \sin y}}$ Algebra

$\qquad\qquad\quad = \dfrac{\dfrac{\cos x}{\sin x} \cdot \dfrac{\cos y}{\sin y} - 1}{\dfrac{\cos y}{\sin y} + \dfrac{\cos x}{\sin x}}$ Algebra

$\qquad\qquad\quad = \dfrac{\cot x \cot y - 1}{\cot y + \cot x}$ Quotient Identity

53. $\tan 2x = \tan(x + x)$ Algebra

$\qquad\quad = \dfrac{\tan x + \tan x}{1 - \tan x \tan x}$ Sum Identity

$\qquad\quad = \dfrac{2 \tan x}{1 - \tan^2 x}$ Algebra

55. $\dfrac{\sin(v + u)}{\sin(v - u)} = \dfrac{\sin v \cos u + \cos v \sin u}{\sin v \cos u - \cos v \sin u}$ Sum and Difference Identities

$\qquad\qquad\quad = \dfrac{\dfrac{\sin v \cos u}{\sin v \sin u} + \dfrac{\cos v \sin u}{\sin v \sin u}}{\dfrac{\sin v \cos u}{\sin v \sin u} - \dfrac{\cos v \sin u}{\sin v \sin u}}$ Algebra

$\qquad\qquad\quad = \dfrac{\dfrac{\cos u}{\sin u} + \dfrac{\cos v}{\sin v}}{\dfrac{\cos u}{\sin u} - \dfrac{\cos v}{\sin v}}$ Algebra

$\qquad\qquad\quad = \dfrac{\cot u + \cot v}{\cot u - \cot v}$ Quotient Identity

57. $\cot x - \tan y = \dfrac{\cos x}{\sin x} - \dfrac{\sin y}{\cos y}$ Quotient Identities

$= \dfrac{\cos x \cos y}{\sin x \cos y} - \dfrac{\sin x \sin y}{\sin x \cos y}$ Algebra

$= \dfrac{\cos x \cos y - \sin x \sin y}{\sin x \cos y}$ Algebra

$= \dfrac{\cos(x + y)}{\sin x \cos y}$ Sum Identity

59. $\tan(x - y) = \dfrac{\tan x - \tan y}{1 + \tan x \tan y}$ Difference Identity

$= \dfrac{\dfrac{1}{\cot x} - \dfrac{1}{\cot y}}{1 + \dfrac{1}{\cot x}\dfrac{1}{\cot y}}$ Reciprocal Identity

$= \dfrac{\cot x \cot y\left(\dfrac{1}{\cot x} - \dfrac{1}{\cot y}\right)}{\cot x \cot y\left(1 + \dfrac{1}{\cot x}\dfrac{1}{\cot y}\right)}$ Algebra

$= \dfrac{\cot y - \cot x}{\cot x \cot y + 1}$ Algebra

61. $\dfrac{\cos(x + h) - \cos x}{h} = \dfrac{\cos x \cos h - \sin x \sin h - \cos x}{h}$ Sum Identity

$= \dfrac{\cos x \cos h - \cos x - \sin x \sin h}{h}$ Algebra

$= \dfrac{\cos x(\cos h - 1) - \sin x \sin h}{h}$ Algebra

$= \cos x\left(\dfrac{\cos h - 1}{h}\right) - \sin x \dfrac{\sin h}{h}$ Algebra

63. $\sin(x - y) = \sin(5.288 - 1.769) = -0.3685$
$\sin x \cos y - \cos x \sin y = \sin 5.288 \cos 1.769 - \cos 5.288 \sin 1.769 = -0.3685$
$\tan(x + y) = \tan(5.288 + 1.769) = 0.9771$
$\dfrac{\tan x + \tan y}{1 - \tan x \tan y} = \dfrac{\tan 5.288 + \tan 1.769}{1 - \tan 5.288 \tan 1.769} = 0.9771$

65. $\sin(x - y) = \sin(42.08° - 68.37°) = -0.4429$
$\sin x \cos y - \cos x \sin y = \sin 42.08° \cos 68.37° - \cos 42.08° \sin 68.37°$
$= -0.4429$
$\tan(x + y) = \tan(42.08° + 68.37°) = -2.682$
$\dfrac{\tan x + \tan y}{1 - \tan x \tan y} = \dfrac{\tan 42.08° + \tan 68.37°}{1 - \tan 42.08° \tan 68.37°} = -2.682$

67. Evaluate each side for a particular set of values of x and y for which each side is defined. If the left side is not equal to the right side, then the equation is not an identity. For example, for $x = 2$ and $y = 1$, both sides are defined, but are not equal.

69. Let $y_1 = \sin\left(x + \dfrac{\pi}{6}\right)$

Then $y_2 = \sin x \cos \dfrac{\pi}{6} + \cos x \sin \dfrac{\pi}{6}$

$= \sin x \cdot \dfrac{\sqrt{3}}{2} + \cos x \cdot \dfrac{1}{2} = \dfrac{\sqrt{3}}{2}\sin x + \dfrac{1}{2}\cos x.$

The graphs coincide as shown at the right.

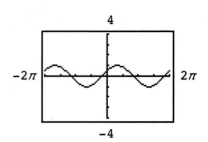

71. Let $y_1 = \cos\left(x - \dfrac{3\pi}{4}\right)$

Then $y_2 = \cos x \cos \dfrac{3\pi}{4} + \sin x \sin \dfrac{3\pi}{4}$

$= \cos x\left(-\dfrac{\sqrt{2}}{2}\right) + \sin x\left(\dfrac{\sqrt{2}}{2}\right)$

$= -\dfrac{\sqrt{2}}{2}\cos x + \dfrac{\sqrt{2}}{2}\sin x$

The graphs coincide as shown at the right.

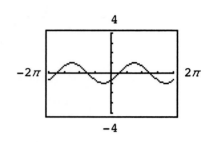

73. Let $y_1 = \tan\left(x + \dfrac{2\pi}{3}\right)$

Then $y_2 = \dfrac{\tan x + \tan \dfrac{2\pi}{3}}{1 - \tan x \tan \dfrac{2\pi}{3}}$

$= \dfrac{\tan x + (-\sqrt{3})}{1 - \tan x(-\sqrt{3})} = \dfrac{\tan x - \sqrt{3}}{1 + \sqrt{3}\tan x}$

The graphs coincide as shown at the right.

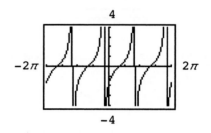

75. Let $u = \cos^{-1}\left(-\dfrac{4}{5}\right)$, $v = \sin^{-1}\left(-\dfrac{3}{5}\right)$

Then we are asked to evaluate $\sin(u + v)$ which is $\sin u \cos v + \cos u \sin v$ from the sum identity. We know $\sin v = \sin\left[\sin^{-1}\left(-\dfrac{3}{5}\right)\right] = -\dfrac{3}{5}$ and

$\cos u = \cos\left[\cos^{-1}\left(-\dfrac{4}{5}\right)\right] = -\dfrac{4}{5}$ from the function-inverse function identities.

It remains to find $\cos v$ and $\sin u$. Note: $0 \le u \le \pi$ and $-\dfrac{\pi}{2} \le v \le \dfrac{\pi}{2}$

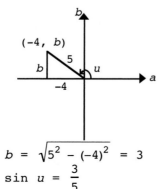

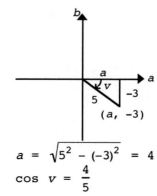

$b = \sqrt{5^2 - (-4)^2} = 3$

$\sin u = \dfrac{3}{5}$

$a = \sqrt{5^2 - (-3)^2} = 4$

$\cos v = \dfrac{4}{5}$

Then $\sin\left[\cos^{-1}\left(-\dfrac{4}{5}\right) + \sin^{-1}\left(-\dfrac{3}{5}\right)\right]$ = $\sin(u + v)$

$= \sin u \cos v + \cos u \sin v$

$= \left(\dfrac{3}{5}\right)\left(\dfrac{4}{5}\right) + \left(-\dfrac{4}{5}\right)\left(-\dfrac{3}{5}\right)$

$= \dfrac{12}{25} + \dfrac{12}{25}$

$= \dfrac{24}{25}$

77. We could proceed as in problem 75. Alternatively, we can shorten the process by recognizing $\arccos \dfrac{1}{2} = \dfrac{\pi}{3}$ and $\arcsin(-1) = -\dfrac{\pi}{2}$. Then $\sin[\arccos \dfrac{1}{2} + \arcsin(-1)]$

$= \sin\left(\dfrac{\pi}{3} + -\dfrac{\pi}{2}\right) = \sin \dfrac{\pi}{3} \cos\left(-\dfrac{\pi}{2}\right) + \cos \dfrac{\pi}{3} \sin\left(-\dfrac{\pi}{2}\right) = \dfrac{\sqrt{3}}{2}(0) + \left(\dfrac{1}{2}\right)(-1) = -\dfrac{1}{2}.$

79. Let $u = \sin^{-1} x$, $v = \cos^{-1} y$. Then $x = \sin u$, $-\dfrac{\pi}{2} \le u \le \dfrac{\pi}{2}$, $y = \cos v$, $0 \le y \le \pi$.

Then $\cos u = \sqrt{1 - \sin^2 u}$ (in Quadrants I, IV) $= \sqrt{1 - x^2}$

$\sin v = \sqrt{1 - \cos^2 v}$ (in Quadrants I, II) $= \sqrt{1 - y^2}$

$\sin(\sin^{-1} x + \cos^{-1} y) = \sin(u + v) = \sin u \cos v + \cos u \sin v$

$= xy + \sqrt{1 - x^2} \sqrt{1 - y^2}$

81. $\cos(x + y + z) = \cos[(x + y) + z]$

$= \cos(x + y)\cos z - \sin(x + y)\sin z$

$= (\cos x \cos y - \sin x \sin y)\cos z$

$\qquad - (\sin x \cos y + \cos x \sin y)\sin z$

$= \cos x \cos y \cos z - \sin x \sin y \cos z$

$\qquad - \sin x \cos y \sin z - \cos x \sin y \sin z$

83. Let $y_1 = \cos 1.2x \cos 0.8x - \sin 1.2x \sin 0.8x$

Then $y_2 = \cos(1.2x + 0.8x) = \cos 2x$

The graphs coincide as shown at the right.

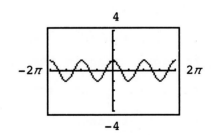

85. $\tan(\theta_2 - \theta_1) = \dfrac{\tan \theta_2 - \tan \theta_1}{1 + \tan \theta_2 \tan \theta_1}$ Difference Identity

$= \dfrac{m_2 - m_1}{1 + m_2 m_1}$ Given

$= \dfrac{m_2 - m_1}{1 + m_1 m_2}$ Algebra

87.

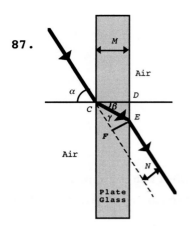

Note: In the text figure we have drawn EF perpendicular to CF, the track of the incident ray. $\triangle CDE$ and $\triangle CEF$ are right triangles. Angle $DCF = \beta + \gamma = \alpha$, so $\gamma = \alpha - \beta$. Denote EC by x. Then,

in $\triangle CDE$ $\quad \dfrac{M}{x} = \cos \beta = \sin(90° - \beta)$

in $\triangle CEF$ $\quad \dfrac{N}{x} = \sin \gamma = \sin(\alpha - \beta)$

Therefore $x = \dfrac{M}{\sin(90° - \beta)} = \dfrac{N}{\sin(\alpha - \beta)}$

$M \sin(\alpha - \beta) = N \sin(90° - \beta)$	Algebra
$M(\sin \alpha \cos \beta - \cos \alpha \sin \beta) = N(\sin 90° \cos \beta - \cos 90° \sin \beta)$	Difference Identities
$M(\sin \alpha \cos \beta - \cos \alpha \sin \beta) = N(1 \cos \beta - 0 \sin \beta)$	Known Values
$M(\sin \alpha \cos \beta - \cos \alpha \sin \beta) = N \cos \beta$	Algebra
$M \sin \alpha \cos \beta - M \cos \alpha \sin \beta = N \cos \beta$	Algebra
$M \sin \alpha \cos \beta - N \cos \beta = M \cos \alpha \sin \beta$	Algebra
$\cos \beta (M \sin \alpha - N) = M \cos \alpha \sin \beta$	Algebra
$\dfrac{M \sin \alpha - N}{M \cos \alpha} = \dfrac{\sin \beta}{\cos \beta}$	Algebra
$\tan \beta = \dfrac{M \sin \alpha - N}{M \cos \alpha}$	Quotient Identity
$\tan \beta = \dfrac{M \sin \alpha}{M \cos \alpha} - \dfrac{N}{M \cos \alpha}$	Algebra
$\tan \beta = \dfrac{\sin \alpha}{\cos \alpha} - \dfrac{N}{M} \dfrac{1}{\cos \alpha}$	Algebra
$\tan \beta = \tan \alpha - \dfrac{N}{M} \sec \alpha$	Quotient and Reciprocal Identities

89. (A) From the text figure:

In right triangle ABE, we have (1) $\cot \alpha = \dfrac{AB}{AE} = \dfrac{AB}{h}$

In right triangle BCD, we have (2) $\cot \alpha = \dfrac{BC}{CD} = \dfrac{BC}{H}$

In right triangle $EE'D$, we have (3) $\tan \beta = \dfrac{E'D}{EE'} = \dfrac{H - h}{AC} = \dfrac{H - h}{AB + BC}$

From (3), $H - h = (AB + BC)\tan \beta$

From (1) and (2), $AB = h \cot \alpha$ and $BC = H \cot \alpha$

Substituting, we have (4) $H - h = (h \cot \alpha + H \cot \alpha)\tan \beta$, or

$\qquad\qquad\qquad = (h + H)\cot \alpha \tan \beta$

Solving for H in terms of h yields:

$$H - h = h \cot \alpha \tan \beta + H \cot \alpha \tan \beta$$

$$H - H \cot \alpha \tan \beta = h \cot \alpha \tan \beta + h$$

$$H(1 - \cot \alpha \tan \beta) = h(\cot \alpha \tan \beta + 1)$$

$$H = h \, \frac{1 + \cot \alpha \tan \beta}{1 - \cot \alpha \tan \beta}$$

(B) $H = h \dfrac{1 + \frac{\cos\alpha}{\sin\alpha}\frac{\sin\beta}{\cos\beta}}{1 - \frac{\cos\alpha}{\sin\alpha}\frac{\sin\beta}{\cos\beta}}$ Quotient Identities

$\quad H = h \dfrac{\left(1 + \frac{\cos\alpha}{\sin\alpha}\frac{\sin\beta}{\cos\beta}\right)\cdot\sin\alpha\cos\beta}{\left(1 - \frac{\cos\alpha}{\sin\alpha}\frac{\sin\beta}{\cos\beta}\right)\cdot\sin\alpha\cos\beta}$ Algebra

$\quad H = h \dfrac{\sin\alpha\cos\beta + \cos\alpha\sin\beta}{\sin\alpha\cos\beta - \cos\alpha\sin\beta}$ Algebra

$\quad H = h \dfrac{\sin(\alpha + \beta)}{\sin(\alpha - \beta)}$ Sum and Difference Identities

(C) Substitute the given values to obtain

$\quad H = 4.90\ \dfrac{\sin(46.23° + 46.15°)}{\sin(46.23° - 46.15°)}$

$\qquad = 4.90\ \dfrac{\sin(92.38°)}{\sin(0.08°)}$

$\quad H = 3{,}510$ ft (to three significant digits)

Section 6-3

1. Answers will vary.   **3.** Answers will vary.   **5.** Answers will vary.

7. $\cos 2(30°) = \cos 60° = \dfrac{1}{2}$

$\cos^2 30° - \sin^2 30° = \left(\dfrac{\sqrt{3}}{2}\right)^2 - \left(\dfrac{1}{2}\right)^2 = \dfrac{3}{4} - \dfrac{1}{4} = \dfrac{2}{4} = \dfrac{1}{2}$

9. $\tan 2\left(\dfrac{\pi}{3}\right) = \tan\dfrac{2\pi}{3} = -\sqrt{3}$

$\dfrac{2}{\cot\frac{\pi}{3} - \tan\frac{\pi}{3}} = \dfrac{2}{\frac{1}{\sqrt{3}} - \sqrt{3}} = \dfrac{2\sqrt{3}}{\sqrt{3}\left(\frac{1}{\sqrt{3}} - \sqrt{3}\right)} = \dfrac{2\sqrt{3}}{1 - 3} = \dfrac{2\sqrt{3}}{-2} = -\sqrt{3}$

11. $\sin\dfrac{\pi}{2} = 1;$ $\quad \sqrt{\dfrac{1 - \cos\pi}{2}} = \sqrt{\dfrac{1 - (-1)}{2}} = \sqrt{\dfrac{2}{2}} = \sqrt{1} = 1$

13. $\sin 22.5° = \sin\left(\dfrac{45°}{2}\right)$ (We use + since 22.5° is in quadrant I)

$\qquad = +\sqrt{\dfrac{1 - \cos 45°}{2}}$ $\qquad\qquad = \sqrt{\dfrac{1 - \frac{1}{\sqrt{2}}}{2}}$

$\qquad\qquad\qquad\qquad\qquad\qquad\qquad = \sqrt{\dfrac{\sqrt{2} - 1}{2\sqrt{2}}}$

$\qquad\qquad\qquad\qquad\qquad\qquad\qquad = \sqrt{\dfrac{\sqrt{2}(\sqrt{2} - 1)}{2\cdot 2}}$

$\qquad\qquad\qquad\qquad\qquad\qquad\qquad = \sqrt{\dfrac{2 - \sqrt{2}}{4}}$

$\qquad\qquad\qquad\qquad\qquad\qquad\qquad = \dfrac{\sqrt{2 - \sqrt{2}}}{2}$

15. $\cos 67.5° = \cos\left(\dfrac{135°}{2}\right)$

$= +\sqrt{\dfrac{1 + \cos 135°}{2}}$

(We use + since 67.5° is in quadrant I)

$= \sqrt{\dfrac{1 + \left(-\dfrac{\sqrt{2}}{2}\right)}{2}}$

$= \sqrt{\dfrac{1 - \dfrac{\sqrt{2}}{2}}{2}}$

$= \sqrt{\dfrac{2 - \sqrt{2}}{4}}$

$= \dfrac{\sqrt{2 - \sqrt{2}}}{2}$

17. $\tan 105° = \tan\dfrac{210°}{2}$

$= \dfrac{\sin 210°}{1 + \cos 210°}$

$= \dfrac{-\dfrac{1}{2}}{1 - \dfrac{\sqrt{3}}{2}}$

$= \dfrac{-\dfrac{1}{2}}{\dfrac{2-\sqrt{3}}{2}}$

$= \dfrac{-1}{2 - \sqrt{3}}$

$= \dfrac{1}{\sqrt{3} - 2}$

19. $\sin\dfrac{5\pi}{8} = \sin\left(\dfrac{5\pi/4}{2}\right)$

$= +\sqrt{\dfrac{1 - \cos\dfrac{5\pi}{4}}{2}}$

(We choose + since 5π/8 is in Quadrant II)

$= \sqrt{\dfrac{1 - \left(-\dfrac{\sqrt{2}}{2}\right)}{2}}$

$= \sqrt{\dfrac{\dfrac{2+\sqrt{2}}{2}}{2}}$

$= \sqrt{\dfrac{2 + \sqrt{2}}{4}} = \dfrac{\sqrt{2 + \sqrt{2}}}{2}$

21. $\cos\dfrac{5\pi}{12} = \cos\left(\dfrac{5\pi/6}{2}\right)$

$= +\sqrt{\dfrac{1 + \cos\dfrac{5\pi}{6}}{2}}$

(We choose + since 5π/12 is in Quadrant I)

$= \sqrt{\dfrac{1 + \left(-\dfrac{\sqrt{3}}{2}\right)}{2}}$

$= \sqrt{\dfrac{\dfrac{2-\sqrt{3}}{2}}{2}}$

$= \sqrt{\dfrac{2 - \sqrt{3}}{4}} = \dfrac{\sqrt{2 - \sqrt{3}}}{2}$

23. $\tan\dfrac{7\pi}{12} = \tan\left(\dfrac{7\pi/6}{2}\right)$

$= \dfrac{\sin\dfrac{7\pi}{6}}{1 + \cos\dfrac{7\pi}{6}}$

$= \dfrac{-\dfrac{1}{2}}{1 + \left(-\dfrac{\sqrt{3}}{2}\right)}$

$= \dfrac{-\dfrac{1}{2}}{\dfrac{2-\sqrt{3}}{2}}$

$= \dfrac{-1}{2 - \sqrt{3}}$

$= \dfrac{1}{\sqrt{3} - 2}$

25.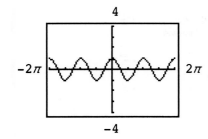

27.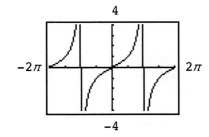

29. $(\sin x + \cos x)^2 = \sin^2 x + 2 \sin x \cos x + \cos^2 x$ Algebra

$= 1 + 2 \sin x \cos x$ Pythagorean Identity

$= 1 + \sin 2x$ Double-angle Identity

31. $\dfrac{1}{2}(1 - \cos 2x) = \dfrac{1}{2}[1 - (1 - 2 \sin^2 x)]$ Double-angle Identity

$= \dfrac{1}{2}[1 - 1 + 2 \sin^2 x)]$ Algebra

$= \dfrac{1}{2}(2 \sin^2 x)$ Algebra

$= \sin^2 x$ Algebra

33. $\tan x \sin 2x = \dfrac{\sin x}{\cos x} 2 \sin x \cos x$ Quotient and Double-angle Identities

$= 2 \sin^2 x$ Algebra

$= 2 \sin^2 x - 1 + 1$ Algebra

$= 1 - (1 - 2 \sin^2 x)$ Algebra

$= 1 - \cos 2x$ Double-angle Identity

35. $\sin^2 \dfrac{x}{2} = \left(\pm\sqrt{\dfrac{1 - \cos x}{2}}\right)^2$ Half-angle Identity

$= \left(\sqrt{\dfrac{1 - \cos x}{2}}\right)^2$ Algebra

$= \dfrac{1 - \cos x}{2}$ Algebra

37. $\cot 2x = \dfrac{1}{\tan 2x}$ Reciprocal Identity

$= \dfrac{1}{\dfrac{2 \tan x}{1 - \tan^2 x}}$ Double-angle Identity

$= \dfrac{1 - \tan^2 x}{2 \tan x}$ Algebra

39.

$$\cot \frac{\theta}{2} = \frac{\cos \frac{\theta}{2}}{\sin \frac{\theta}{2}}$$

Quotient Identity

$$= \frac{\pm\sqrt{\frac{1+\cos \theta}{2}}}{\pm\sqrt{\frac{1-\cos \theta}{2}}}$$

Half-angle Identities

$$= \pm\sqrt{\frac{1 + \cos \theta}{1 - \cos \theta}}$$

Algebra

$$\left|\cot \frac{\theta}{2}\right| = \sqrt{\frac{1 + \cos \theta}{1 - \cos \theta}}$$

Algebra

$$= \sqrt{\frac{1 + \cos \theta}{1 - \cos \theta} \cdot \frac{1 - \cos \theta}{1 - \cos \theta}}$$

Algebra

$$= \sqrt{\frac{1 - \cos^2 \theta}{(1 - \cos \theta)^2}}$$

Algebra

$$= \sqrt{\frac{\sin^2 \theta}{(1 - \cos \theta)^2}}$$

Pythagorean Identity

$$= \left|\frac{\sin \theta}{1 - \cos \theta}\right|$$

Algebra

Since $1 - \cos \theta \geq 0$ and $\sin \theta$ has the same sign as $\cot \frac{\theta}{2}$, we may drop the absolute value signs to obtain

$$\cot \frac{\theta}{2} = \frac{\sin \theta}{1 - \cos \theta}$$

$\left[\right.$To show that $\sin \theta$ has the same sign as $\cot \frac{\theta}{2}$, we note the following cases:

If $0 < \theta < \pi$, $\sin \theta > 0$ then $0 < \frac{\theta}{2} < \frac{\pi}{2}$, $\cot \frac{\theta}{2} > 0$.

If $\pi = \theta$, $\cot \frac{\theta}{2} = \cot \frac{\pi}{2} = 0$, $\frac{\sin \theta}{1 - \cos \theta} = \frac{\sin \pi}{1 - \cos \pi} = \frac{0}{1 - (-1)} = 0$

If $\pi < \theta < 2\pi$, $\sin \theta < 0$, then $\frac{\pi}{2} < \frac{\theta}{2} < \pi$, $\cot \frac{\theta}{2} < 0$.

The truth of the statement for other values of θ follows since $\sin(\theta + 2k\pi) = \sin \theta$ and $\cot \frac{\theta + 2k\pi}{2} = \cot \frac{\theta}{2}.\left.\right]$

Key Algebraic Steps:

If $y = \pm\sqrt{\frac{1 + a}{1 - a}}$, then

$$|y| = \sqrt{\frac{1 + a}{1 - a}} = \sqrt{\frac{(1 + a)(1 - a)}{(1 - a)(1 - a)}} = \sqrt{\frac{1 - a^2}{(1 - a)^2}}$$

41. $\cos 2u = \cos^2 u - \sin^2 u$ Double-angle Identity

$$= \frac{\cos^2 u - \sin^2 u}{1}$$ Algebra

$$= \frac{\cos^2 u - \sin^2 u}{\cos^2 u + \sin^2 u}$$ Pythagorean Identity

$$= \frac{\dfrac{\cos^2 u}{\cos^2 u} - \dfrac{\sin^2 u}{\cos^2 u}}{\dfrac{\cos^2 u}{\cos^2 u} + \dfrac{\sin^2 u}{\cos^2 u}}$$ Algebra

$$= \frac{1 - \dfrac{\sin^2 u}{\cos^2 u}}{1 + \dfrac{\sin^2 u}{\cos^2 u}}$$ Algebra

$$= \frac{1 - \tan^2 u}{1 + \tan^2 u}$$ Quotient Identity

43. $2 \csc 2x = 2 \cdot \dfrac{1}{\sin 2x}$ Reciprocal Identity

$$= \frac{2}{\sin 2x}$$ Algebra

$$= \frac{2}{2 \sin x \cos x}$$ Double-angle Identity

$$= \frac{1}{\sin x \cos x}$$ Algebra

$$= \frac{\cos^2 x + \sin^2 x}{\sin x \cos x}$$ Pythagorean Identity

$$= \frac{\dfrac{\cos^2 x}{\cos^2 x} + \dfrac{\sin^2 x}{\cos^2 x}}{\dfrac{\sin x \cos x}{\cos^2 x}}$$ Algebra

$$= \frac{1 + \dfrac{\sin^2 x}{\cos^2 x}}{\dfrac{\sin x}{\cos x}}$$ Algebra

$$= \frac{1 + \tan^2 x}{\tan x}$$ Quotient Identity

45. $\dfrac{1 - \tan^2 \frac{\alpha}{2}}{1 + \tan^2 \frac{\alpha}{2}} = \dfrac{1 - \left(\pm\sqrt{\dfrac{1-\cos\alpha}{1+\cos\alpha}}\right)^2}{1 + \left(\pm\sqrt{\dfrac{1-\cos\alpha}{1+\cos\alpha}}\right)^2}$ Half-angle Identity

$$= \frac{1 - \dfrac{1-\cos\alpha}{1+\cos\alpha}}{1 + \dfrac{1-\cos\alpha}{1+\cos\alpha}}$$ Algebra

$$= \frac{(1 + \cos\alpha)\left(1 - \dfrac{1-\cos\alpha}{1+\cos\alpha}\right)}{(1 + \cos\alpha)\left(1 + \dfrac{1-\cos\alpha}{1+\cos\alpha}\right)}$$ Algebra

$$= \frac{1 + \cos\alpha - (1 - \cos\alpha)}{1 + \cos\alpha + 1 - \cos\alpha} \qquad \text{Algebra}$$

$$= \frac{1 + \cos\alpha - 1 + \cos\alpha}{1 + \cos\alpha + 1 - \cos\alpha} \qquad \text{Algebra}$$

$$= \frac{2\cos\alpha}{2} \qquad \text{Algebra}$$

$$= \cos\alpha$$

Key Algebraic Steps:

$$\frac{1 - \left(\pm\sqrt{\frac{1-x}{1+x}}\right)^2}{1 + \left(\pm\sqrt{\frac{1-x}{1+x}}\right)^2} = \frac{1 - \frac{1-x}{1+x}}{1 + \frac{1-x}{1+x}} = \frac{(1+x)\left(1 - \frac{1-x}{1+x}\right)}{(1+x)\left(1 + \frac{1-x}{1+x}\right)} = \frac{1 + x - (1-x)}{1 + x + 1 - x}$$

$$= \frac{1 + x - 1 + x}{1 + x + 1 - x} = \frac{2x}{2} = x$$

47. Plug in $x = \dfrac{\pi}{6}$.

Left side: $\tan\left(2 \cdot \dfrac{\pi}{6}\right) = \tan\dfrac{\pi}{3} = \sqrt{3}$

Right side: $2\tan\dfrac{\pi}{6} = \dfrac{2}{\sqrt{3}}$

The equation is not an identity.

49. Plug in $x = \dfrac{\pi}{3}$.

Left side: $\sin\dfrac{\pi/3}{2} = \sin\dfrac{\pi}{6} = \dfrac{1}{2}$

Right side: $\dfrac{1}{2}\sin\dfrac{\pi}{3} = \dfrac{\sqrt{3}}{4}$

The equation is not an identity.

51. Plug in $x = \dfrac{4\pi}{3}$.

Left side: $\cos\dfrac{4\pi/3}{2} = \cos\dfrac{2\pi}{3} = -\dfrac{1}{2}$

Right side: $\sqrt{\dfrac{1 + \cos 4\pi/3}{2}} = \sqrt{\dfrac{1 - \frac{1}{2}}{2}} = \dfrac{1}{2}$

The equation is not an identity.

53. Plug in $x = \dfrac{4\pi}{3}$.

Left side: $\tan\dfrac{4\pi/3}{2} = \tan\dfrac{2\pi}{3} = -\sqrt{3}$

Right side: $\sqrt{\dfrac{1 - \cos\frac{4\pi}{3}}{1 + \cos\frac{4\pi}{3}}} = \sqrt{\dfrac{1 - \left(-\frac{1}{2}\right)}{1 + \left(-\frac{1}{2}\right)}} = \sqrt{3}$

The equation is not an identity.

55. The equation is not an identity. Plug in $x = \dfrac{\pi}{4}$:

$$\sin\left(4 \cdot \dfrac{\pi}{4}\right) \overset{?}{=} 4 \sin \dfrac{\pi}{4} \cos \dfrac{\pi}{4}$$

$$\sin \pi \overset{?}{=} 4 \cdot \dfrac{1}{\sqrt{2}} \cdot \dfrac{1}{\sqrt{2}}$$

$$0 \overset{?}{=} 4 \cdot \dfrac{1}{2}$$

This is false.

57. Expand the left side using a double-angle formula:

$$\cot 2x = \dfrac{1}{\tan(2x)} \qquad\qquad \cot x = \dfrac{1}{\tan x}$$

$$= \dfrac{1}{\dfrac{2\tan x}{1\,\tan^2 x}} \qquad\qquad \text{Double-angle formula}$$

$$= \dfrac{1\,\tan^2 x}{2\tan x} \qquad\qquad \text{Algebra}$$

Now simplify the left side:

$$\dfrac{\tan x(\cot^2 x - 1)}{2} = \dfrac{\tan x\left(\dfrac{1}{\tan^2 x} - 1\right)}{2} \qquad \cot^2 x = \dfrac{1}{\tan^2 x}$$

$$= \dfrac{\dfrac{1}{\tan x} - \tan x}{2} \qquad\qquad \text{Algebra}$$

$$= \dfrac{\tan x}{\tan x} \cdot \dfrac{\left(\dfrac{1}{\tan x} - \tan x\right)}{2} \qquad \text{Algebra}$$

$$= \dfrac{1 - \tan^2 x}{2\tan x} \qquad\qquad \text{Algebra}$$

Both sides can be simplified to the same expression, so the equation is an identity.

59. The equation is not an identity. Plug in $x = \pi$.

$$\cos(2\pi) \overset{?}{=} 1 - 2\cos^2 \pi$$

$$1 \overset{?}{=} 1 - 2(-1)^2$$

$$1 \overset{?}{=} 1 - 2$$

This is false.

61.

$$a = -\sqrt{5^2 - 3^2} = -4$$

$$\cos x = -\dfrac{4}{5} \qquad \tan x = -\dfrac{3}{4}$$

$$\sin 2x = 2\sin x \cos x = 2\left(\dfrac{3}{5}\right)\left(-\dfrac{4}{5}\right) = -\dfrac{24}{25}$$

$$\cos 2x = 1 - 2\sin^2 x = 1 - 2\left(\dfrac{3}{5}\right)^2 = 1 - \dfrac{18}{25} = \dfrac{7}{25}$$

$$\tan 2x = \dfrac{\sin 2x}{\cos 2x} = \left(-\dfrac{24}{25}\right) \div \left(\dfrac{7}{25}\right) = -\dfrac{24}{25} \cdot \dfrac{25}{7} = -\dfrac{24}{7}$$

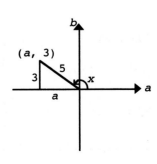

63.

$$r = \sqrt{12^2 + (-5)^2} = 13$$

$$\sin x = -\frac{5}{13} \qquad \cos x = \frac{12}{13}$$

$$\sin 2x = 2 \sin x \cos x = 2\left(-\frac{5}{13}\right)\left(\frac{12}{13}\right) = -\frac{120}{169}$$

$$\cos 2x = 2 \cos^2 x - 1 = 2\left(\frac{12}{13}\right)^2 - 1 = \frac{288}{169} - 1 = \frac{119}{169}$$

$$\tan 2x = \frac{\sin 2x}{\cos 2x} = \left(-\frac{120}{169}\right) \div \left(\frac{119}{169}\right) = -\frac{120}{169} \cdot \frac{169}{119} = -\frac{120}{119}$$

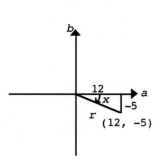

65. Since $\cos x > 0$ and $\tan x < 0$, x is in quadrant IV.

$$b = -\sqrt{3^2 - 2^2} = -\sqrt{5}$$

$$\sin x = -\frac{\sqrt{5}}{3} \qquad \tan x = -\frac{\sqrt{5}}{2}$$

$$\sin 2x = 2 \sin x \cos x = 2\left(-\frac{\sqrt{5}}{3}\right)\left(\frac{2}{3}\right) = -\frac{4\sqrt{5}}{9}$$

$$\cos 2x = 1 - 2 \sin^2 x = 1 - 2\left(\frac{5}{9}\right) = \frac{9}{9} - \frac{10}{9} = -\frac{1}{9}$$

$$\tan 2x = \frac{2 \tan x}{1 - \tan^2 x} = \frac{2\left(-\frac{\sqrt{5}}{2}\right)}{1 - \frac{5}{4}} = \frac{-\sqrt{5}}{\frac{4}{4} - \frac{5}{4}} = \frac{-\sqrt{5}}{-1/4} = 4\sqrt{5}$$

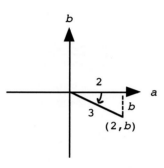

67. Since $\tan x > 0$ and $\sin x < 0$, x is in quadrant III.

$$r = \sqrt{(-2)^2 + (-1)^2} = \sqrt{5}$$

$$\sin x = -\frac{2}{\sqrt{5}} \qquad \cos x = -\frac{1}{\sqrt{5}}$$

$$\sin 2x = 2 \sin x \cos x = 2\left(-\frac{2}{\sqrt{5}}\right)\left(-\frac{1}{\sqrt{5}}\right) = \frac{4}{5}$$

$$\cos 2x = 1 - 2 \sin^2 x = 1 - 2\left(\frac{4}{5}\right) = \frac{5}{5} - \frac{8}{5} = -\frac{3}{5}$$

$$\tan 2x = \frac{2 \tan x}{1 - \tan^2 x} = \frac{2 \cdot 2}{1 - 4} = -\frac{4}{3}$$

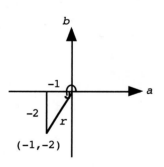

69. Since $\pi < x < \frac{3\pi}{2}$, $\frac{\pi}{2} < \frac{x}{2} < \frac{3\pi}{4}$,

$\sin \frac{x}{2}$ will be positive, $\cos \frac{x}{2}$,

$\tan \frac{x}{2}$ will be negative.

$$a = -\sqrt{3^2 - (-1)^2} = -\sqrt{8}$$

$$\cos x = \frac{-\sqrt{8}}{3}$$

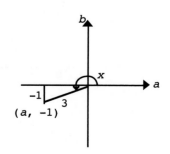

$$\sin\frac{x}{2} = \sqrt{\frac{1-\cos x}{2}} = \sqrt{\frac{1-\left(-\frac{\sqrt{8}}{3}\right)}{2}} = \sqrt{\frac{1+\frac{\sqrt{8}}{3}}{2}} = \sqrt{\frac{3+\sqrt{8}}{6}} = \sqrt{\frac{3+2\sqrt{2}}{6}}$$

$$\cos\frac{x}{2} = -\sqrt{\frac{1+\cos x}{2}} = -\sqrt{\frac{1+\left(-\frac{\sqrt{8}}{3}\right)}{2}} = -\sqrt{\frac{1-\frac{\sqrt{8}}{3}}{2}} = -\sqrt{\frac{3-\sqrt{8}}{6}} = -\sqrt{\frac{3-2\sqrt{2}}{6}}$$

$$\tan\frac{x}{2} = -\sqrt{\frac{1-\cos x}{1+\cos x}} = -\sqrt{\frac{1-\left(-\frac{\sqrt{8}}{3}\right)}{1+\left(-\frac{\sqrt{8}}{3}\right)}} = -\sqrt{\frac{3+\sqrt{8}}{3-\sqrt{8}}} = -\sqrt{\frac{(3+\sqrt{8})(3+\sqrt{8})}{(3-\sqrt{8})(3+\sqrt{8})}}$$

$$= -\sqrt{(3+\sqrt{8})^2} = -(3+\sqrt{8}) = -3-\sqrt{8} = -3-2\sqrt{2}$$

71. Since $-\pi < x < -\frac{\pi}{2}$, $-\frac{\pi}{2} < \frac{x}{2} < -\frac{\pi}{4}$, $\cos\frac{x}{2}$ will be positive, $\sin\frac{x}{2}$, $\tan\frac{x}{2}$ will be negative.

$$r = \sqrt{(-3)^2 + (-4)^2} = 5$$

$$\sin x = -\frac{4}{5} \qquad \cos x = -\frac{3}{5}$$

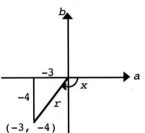

$$\sin\frac{x}{2} = -\sqrt{\frac{1-\cos x}{2}} = -\sqrt{\frac{1-\left(-\frac{3}{5}\right)}{2}}$$

$$= -\sqrt{\frac{1+\frac{3}{5}}{2}} = -\sqrt{\frac{\frac{8}{5}}{2}} = -\sqrt{\frac{4}{5}} = -\frac{2\sqrt{5}}{5}$$

$$\cos\frac{x}{2} = \sqrt{\frac{1+\cos x}{2}} = \sqrt{\frac{1+\left(-\frac{3}{5}\right)}{2}} = \sqrt{\frac{\frac{2}{5}}{2}} = \sqrt{\frac{1}{5}} = \frac{\sqrt{5}}{5}$$

$$\tan\frac{x}{2} = \frac{\sin\frac{1}{2}x}{\cos\frac{1}{2}x} = \frac{-2\sqrt{5}}{5} \div \frac{\sqrt{5}}{5} = -\frac{2\sqrt{5}}{5} \cdot \frac{5}{\sqrt{5}} = -2$$

73. Since $\cos x < 0$ and $\sin x < 0$, x is in quadrant III.

$\pi < x < \frac{3\pi}{2}$, $\frac{\pi}{2} < \frac{x}{2} < \frac{3\pi}{4}$, which is part of QII.

$\sin\frac{x}{2}$ will be positive, $\cos\frac{x}{2}$ will be negative,

and $\tan\frac{x}{2}$ will be negative.

$$\sin\frac{x}{2} = \sqrt{\frac{1-\cos x}{2}} = \sqrt{\frac{1-\left(-\frac{8}{17}\right)}{2}} = \sqrt{\frac{\frac{17}{17}+\frac{8}{17}}{2}} = \sqrt{\frac{25}{34}} = \frac{5}{\sqrt{34}}$$

$$\cos\frac{x}{2} = -\sqrt{\frac{1+\cos x}{2}} = -\sqrt{\frac{1+\left(-\frac{8}{17}\right)}{2}} = -\sqrt{\frac{\frac{17}{17}-\frac{8}{17}}{2}} = -\sqrt{\frac{9}{34}} = -\frac{3}{\sqrt{34}}$$

$$\tan\frac{x}{2} = -\sqrt{\frac{1-\cos x}{1+\cos x}} = -\sqrt{\frac{1-\left(-\frac{8}{17}\right)}{1+\left(-\frac{8}{17}\right)}} = -\sqrt{\frac{\frac{17}{17}+\frac{8}{17}}{\frac{17}{17}-\frac{8}{17}}} = -\sqrt{\frac{25/17}{9/17}} = -\sqrt{\frac{25}{9}} = -\frac{5}{3}$$

75. Since $\tan x > 0$ and $\cos x > 0$, x is in quadrant I.

$0 < x < \dfrac{\pi}{2}$, $0 < \dfrac{x}{2} < \dfrac{\pi}{4}$, which is part of QI.

All of $\sin \dfrac{x}{2}$, $\cos \dfrac{x}{2}$, and $\tan \dfrac{x}{2}$ will be positive.

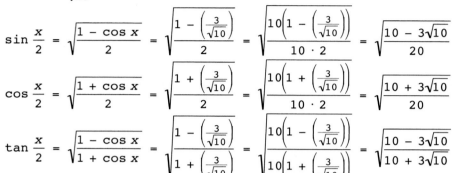

$r = \sqrt{1^2 + 3^2} = \sqrt{10}$

$\cos x = \dfrac{3}{\sqrt{10}}$

$\sin \dfrac{x}{2} = \sqrt{\dfrac{1 - \cos x}{2}} = \sqrt{\dfrac{1 - \left(\dfrac{3}{\sqrt{10}}\right)}{2}} = \sqrt{\dfrac{10\left(1 - \left(\dfrac{3}{\sqrt{10}}\right)\right)}{10 \cdot 2}} = \sqrt{\dfrac{10 - 3\sqrt{10}}{20}}$

$\cos \dfrac{x}{2} = \sqrt{\dfrac{1 + \cos x}{2}} = \sqrt{\dfrac{1 + \left(\dfrac{3}{\sqrt{10}}\right)}{2}} = \sqrt{\dfrac{10\left(1 + \left(\dfrac{3}{\sqrt{10}}\right)\right)}{10 \cdot 2}} = \sqrt{\dfrac{10 + 3\sqrt{10}}{20}}$

$\tan \dfrac{x}{2} = \sqrt{\dfrac{1 - \cos x}{1 + \cos x}} = \sqrt{\dfrac{1 - \left(\dfrac{3}{\sqrt{10}}\right)}{1 + \left(\dfrac{3}{\sqrt{10}}\right)}} = \sqrt{\dfrac{10\left(1 - \left(\dfrac{3}{\sqrt{10}}\right)\right)}{10\left(1 + \left(\dfrac{3}{\sqrt{10}}\right)\right)}} = \sqrt{\dfrac{10 - 3\sqrt{10}}{10 + 3\sqrt{10}}}$

77. (A) 2θ is a second quadrant angle, since θ is a first quadrant angle and $\tan 2\theta$ is negative for 2θ in the second quadrant and not for 2θ in the first.

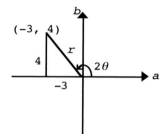

(B) Construct a reference triangle for 2θ in the second quadrant with $(a, b) = (-3, 4)$. Use the Pythagorean theorem to find $r = 5$.

$\sin 2\theta = \dfrac{4}{5}$ and $\cos 2\theta = -\dfrac{3}{5}$.

(C) The double angle identities $\cos 2\theta = 1 - 2 \sin^2 \theta$ and $\cos 2\theta = 2 \cos^2 \theta - 1$.

(D) Use the identities in part (C) in the form

$\sin \theta = \sqrt{\dfrac{1 - \cos 2\theta}{2}}$ and $\cos \theta = \sqrt{\dfrac{1 + \cos 2\theta}{2}}$

The positive radicals are used because θ is in quadrant one.

(E) $\sin \theta = \sqrt{\dfrac{1 - \left(-\dfrac{3}{5}\right)}{2}} = \sqrt{\dfrac{5 + 3}{10}} = \sqrt{\dfrac{8}{10}} = \sqrt{\dfrac{4}{5}} = \dfrac{2}{\sqrt{5}} = \dfrac{2\sqrt{5}}{5}$

$\cos \theta = \sqrt{\dfrac{1 + \left(-\dfrac{3}{5}\right)}{2}} = \sqrt{\dfrac{5 - 3}{10}} = \sqrt{\dfrac{2}{10}} = \sqrt{\dfrac{1}{5}} = \dfrac{1}{\sqrt{5}} = \dfrac{\sqrt{5}}{5}$

79. (A) $\tan[2(252.06°)] = -0.72335$

$\dfrac{2 \tan x}{1 - \tan^2 x} = \dfrac{2 \tan(252.06°)}{1 - \tan^2(252.06°)}$

(B) $\cos \dfrac{252.06°}{2} = -0.58821$

$-\sqrt{\dfrac{1 + \cos 252.06°}{2}} = -0.58821$

$= -0.72335$

81. (A) $\tan[2(0.93457)] = -3.2518$

$\dfrac{2 \tan(0.93457)}{1 - \tan^2(0.93457)} = -3.2518$

(B) $\cos \dfrac{0.93457}{2} = 0.89279$

$\sqrt{\dfrac{1 + \cos 0.93457}{2}} = 0.89279$

83.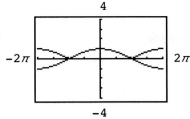

The graphs appear to coincide on the interval $[-\pi, \pi]$.

85.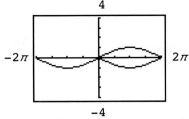

The graphs appear to coincide on the interval $[-2\pi, 0]$.

87.
$$\begin{aligned}
\cos 3x &= \cos(2x + x) && \text{Algebra}\\
&= \cos 2x \cos x - \sin 2x \sin x && \text{Sum Identity}\\
&= (2 \cos^2 x - 1)\cos x - 2 \sin x \cos x \sin x && \text{Double-angle Identities}\\
&= 2 \cos^3 x - \cos x - 2 \sin^2 x \cos x && \text{Algebra}\\
&= 2 \cos^3 x - \cos x - 2(1 - \cos^2 x)\cos x && \text{Pythagorean Identity}\\
&= 2 \cos^3 x - \cos x - 2 \cos x + 2 \cos^3 x && \text{Algebra}\\
&= 4 \cos^3 x - 3 \cos x && \text{Algebra}
\end{aligned}$$

89.
$$\begin{aligned}
\cos 4x &= \cos 2(2x) && \text{Algebra}\\
&= 2 \cos^2 2x - 1 && \text{Double-angle Identity}\\
&= 2(2 \cos^2 x - 1)^2 - 1 && \text{Double-angle Identity}\\
&= 2(4 \cos^4 x - 4 \cos^2 x + 1) - 1 && \text{Algebra}\\
&= 8 \cos^4 x - 8 \cos^2 x + 2 - 1 && \text{Algebra}\\
&= 8 \cos^4 x - 8 \cos^2 x + 1 && \text{Algebra}
\end{aligned}$$

91. Let $u = \cos^{-1} \dfrac{3}{5}$. Then $\cos u = \dfrac{3}{5}$, $0 < u < \pi$. $\cos\left(2 \cos^{-1} \dfrac{3}{5}\right) = \cos 2u$

$$= 2 \cos^2 u - 1 = 2\left(\dfrac{3}{5}\right)^2 - 1 = -\dfrac{7}{25}$$

93. Let $u = \cos^{-1}\left(-\dfrac{4}{5}\right)$. Then $\cos u = -\dfrac{4}{5}$, $0 < u < \pi$.

$b = \sqrt{5^2 - (-4)^2} = 3$

$\tan u = -\dfrac{3}{4}$

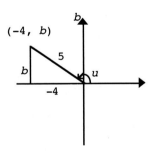

$$\begin{aligned}
\tan\left[2 \cos^{-1}\left(-\dfrac{4}{5}\right)\right] &= \tan 2u = \dfrac{2 \tan u}{1 - \tan^2 u} = \dfrac{2\left(-\dfrac{3}{4}\right)}{1 - \left(-\dfrac{3}{4}\right)^2}\\[2mm]
&= \dfrac{-\dfrac{3}{2}}{1 - \dfrac{9}{16}} = \dfrac{-\dfrac{3}{2}}{\dfrac{7}{16}} = \left(-\dfrac{3}{2}\right) \div \dfrac{7}{16}\\[2mm]
&= -\dfrac{3}{2} \cdot \dfrac{16}{7} = -\dfrac{24}{7}
\end{aligned}$$

95. Let $u = \cos^{-1}\left(-\dfrac{3}{5}\right)$. Then $\cos u = -\dfrac{3}{5}$, $0 < u < \pi$. Since $0 < \dfrac{1}{2}u < \dfrac{\pi}{2}$,

$\cos \dfrac{1}{2}u$ is positive. So

$$\cos\left[\dfrac{1}{2}\cos^{-1}\left(-\dfrac{3}{5}\right)\right] = \cos \dfrac{1}{2}u = \sqrt{\dfrac{1 + \cos u}{2}} = \sqrt{\dfrac{1 + -\dfrac{3}{5}}{2}} = \sqrt{\dfrac{\dfrac{2}{5}}{2}} = \sqrt{\dfrac{1}{5}} = \dfrac{\sqrt{5}}{5}$$

97. Here is the graph of $f(x) = \csc x - \cot x$ on the interval $[-2\pi, 2\pi]$.

The graph appears to have the form $y = A \tan Bx$. Since the period is 2π, $\frac{\pi}{B} = 2\pi$, and $B = \frac{1}{2}$. The graph of $A \tan \frac{1}{2}x$ shown appears to pass through $\left(\frac{\pi}{2}, 1\right)$, so $1 = A \tan \frac{1}{2} \cdot \frac{\pi}{2}$

$$1 = A \tan \frac{\pi}{4}$$

$$1 = A$$

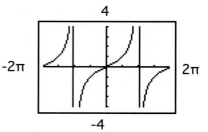

The equation of the graph appears to be $y = \tan \frac{1}{2}x$. We verify that $\csc x - \cot x = \tan \frac{1}{2}x$ is an identity:

$\csc x - \cot x = \dfrac{1}{\sin x} - \dfrac{\cos x}{\sin x}$ Quotient and Reciprocal Identities

$\qquad\qquad = \dfrac{1 - \cos x}{\sin x}$ Algebra

$\qquad\qquad = \tan \dfrac{1}{2}x$ Half-angle Identity

99. Here is the graph of $f(x) = \dfrac{1 - 2\cos 2x}{2\sin x - 1}$ on the interval $[-2\pi, 2\pi]$.

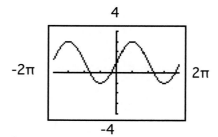

The graph appears to be a standard sine curve shifted up 1 unit. It has amplitude $\dfrac{3 - (-1)}{2} = 2$ and period 2π, so it appears to be the graph of $y = 1 + 2 \sin x$. We verify that $\dfrac{1 - 2\cos 2x}{2\sin x - 1} = 1 + 2 \sin x$ is an identity:

$\dfrac{1 - 2\cos 2x}{2\sin x - 1} = \dfrac{1 - 2(1 - 2\sin^2 x)}{2\sin x - 1}$ Double-angle Identity

$\qquad\qquad = \dfrac{1 - 2 + 4\sin^2 x}{2\sin x - 1}$ Algebra

$\qquad\qquad = \dfrac{4\sin^2 x - 1}{2\sin x - 1}$ Algebra

$\qquad\qquad = \dfrac{(2\sin x - 1)(2\sin x + 1)}{2\sin x - 1}$ Algebra

$\qquad\qquad = 2 \sin x + 1$ Algebra

101. Here is the graph of $f(x) = \dfrac{1}{\cot x \sin 2x - 1}$ on the interval $[-2\pi, 2\pi]$.

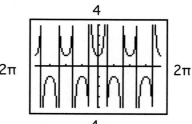

The graph appears to have the form $y = A \sec Bx$. Since the period is π, $\dfrac{2\pi}{B} = \pi$, and $B = 2$.

The graph of $y = A \sec 2x$ shown appears to pass through $(0, 1)$, so

$$1 = A \sec 2 \cdot 0$$
$$1 = A \sec 0$$
$$1 = A$$

The equation of the graph appears to be $y = \sec 2x$. We verify that

$$\frac{1}{\cot x \sin 2x - 1} = \sec 2x \text{ is an identity.}$$

$$\frac{1}{\cot x \sin 2x - 1} = \frac{1}{\frac{\cos x}{\sin x} \sin 2x - 1} \qquad \text{Quotient Identity}$$

$$= \frac{1}{\frac{\cos x}{\sin x} 2 \sin x \cos x - 1} \qquad \text{Double-angle Identity}$$

$$= \frac{1}{2 \cos^2 x - 1} \qquad \text{Algebra}$$

$$= \frac{1}{\cos 2x} \qquad \text{Double-angle Identity}$$

$$= \sec 2x \qquad \text{Reciprocal Identity}$$

103. From the figure we see:

ACD is a right triangle, so $\cos \theta = \dfrac{7}{8}$, $\theta = \cos^{-1} \dfrac{7}{8}$

ABC is a right triangle, with angle $BAC = 2\theta$. Then $\cos 2\theta = \dfrac{7}{x}$. Using the hint,

$\cos 2\theta = 2 \cos^2 \theta - 1$

$\qquad = 2\left(\dfrac{7}{8}\right)^2 - 1$

$\qquad = 2 \cdot \dfrac{49}{64} - 1$

$\qquad = \dfrac{34}{64}$

$\qquad = \dfrac{17}{32}$

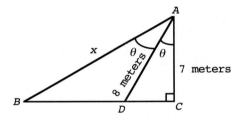

Since $\cos 2\theta = \dfrac{7}{x}$, we get

$\dfrac{17}{32} = \dfrac{7}{x}$

$17x = 32(7)$

$\quad x = \dfrac{32(7)}{17} = \dfrac{224}{17}$

To three decimal places

$\quad x = \dfrac{224}{17} = 13.176$ meters

$\quad \theta = \cos^{-1} \dfrac{7}{8} = 28.955°$ (calculator in degree mode)

105. (A) Since $2 \sin \theta \cos \theta = \sin 2\theta$ by the double-angle identity, we can write

$d = \dfrac{2v_0^2 \sin \theta \cos \theta}{32 \text{ ft/sec}^2}$

$d = \dfrac{2v_0^2 (2 \sin \theta \cos \theta)}{32 \text{ ft/sec}^2}$

$d = \dfrac{v_0^2 \sin 2\theta}{32 \text{ ft/sec}^2}$

(B) For fixed v_0, the quantity $\dfrac{v_0^2 \sin 2\theta}{32}$ will be maximum when $\sin 2\theta$ achieves its maximum value, namely 1. Then

$$\sin 2\theta = 1$$
$$2\theta = 90°$$
$$\theta = 45°$$

its maximum value, namely 1. Then

$$\sin 2\theta = 1$$
$$2\theta = 90°$$
$$\theta = 45°$$

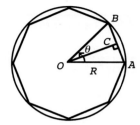

107. (A) Here is a drawing of a typical situation. (The case $n = 8$ is shown, but the reasoning is independent of n):

We note: Area of polygon = n(Area of triangle OAB)

$$= 2n(\text{Area of triangle } OAC)$$

$$= 2n\left(\frac{1}{2} \cdot OC \cdot AC\right)$$

$$= n \cdot OC \cdot AC$$

In triangle OAC, angle $AOC = \dfrac{1}{2}\theta = \dfrac{1}{2} \cdot \dfrac{2\pi}{n} = \dfrac{\pi}{n}$

$$\frac{OC}{R} = \cos AOC = \cos \frac{\pi}{n} \qquad \frac{AC}{R} = \sin AOC = \sin \frac{\pi}{n}$$

Then $OC = R \cos \dfrac{\pi}{n}$, $AC = R \sin \dfrac{\pi}{n}$

Area of polygon = $n\, R \cos \dfrac{\pi}{n} \; R \sin \dfrac{\pi}{n}$

$$= nR^2 \sin \frac{\pi}{n} \cos \frac{\pi}{n}$$

$$= \frac{1}{2} nR^2 \cdot 2 \sin \frac{\pi}{n} \cos \frac{\pi}{n}$$

$$= \frac{1}{2} nR^2 \sin \frac{2\pi}{n} \quad \text{by the double-angle identity for sine}$$

(B) TABLE 1

n	10	100	1,000	10,000
A_n	2.93893	3.13953	3.14157	3.14159

(C) A_n appears to approach π, the area of the circle with radius 1.

(D) A_n will not exactly equal the area of the circumscribing circle for any n no matter how large n is chosen; however, A_n can be made as close to the area of the circumscribing circle as we like by making n sufficiently large.

Section 6-4

1. Answers will vary.   3. The identity for sine of a sum.

5. The double-angle identities for sine and cosine.

7. $\sin x \cos y = \dfrac{1}{2}[\sin(x + y) + \sin(x - y)]$

$\sin 3m \cos m = \dfrac{1}{2}[\sin(3m + m) + \sin(3m - m)]$

$$= \frac{1}{2}(\sin 4m + \sin 2m)$$

$$= \frac{1}{2}\sin 4m + \frac{1}{2}\sin 2m$$

9. $\sin x \sin y = \dfrac{1}{2}[\cos(x - y) - \cos(x + y)]$

$\sin u \sin 3u = \dfrac{1}{2}[\cos(u - 3u) - \cos(u + 3u)]$

$\qquad\qquad = \dfrac{1}{2}(\cos(-2u) - \cos 4u)$

$\qquad\qquad = \dfrac{1}{2}(\cos 2u - \cos 4u)$

$\qquad\qquad = \dfrac{1}{2}\cos 2u - \dfrac{1}{2}\cos 4u$

11. $\sin x + \sin y = 2 \sin \dfrac{x + y}{2} \cos \dfrac{x - y}{2}$

$\sin 3t + \sin t = 2 \sin \dfrac{3t + t}{2} \cos \dfrac{3t - t}{2}$

$\qquad\qquad\quad = 2 \sin 2t \cos t$

13. $\cos x - \cos y = -2 \sin \dfrac{x + y}{2} \sin \dfrac{x - y}{2}$

$\cos 5w - \cos 9w = -2 \sin \dfrac{5w + 9w}{2} \sin \dfrac{5w - 9w}{2}$

$\qquad\qquad\qquad = -2 \sin 7w \sin(-2w)$

$\qquad\qquad\qquad = 2 \sin 7w \sin 2w$

15. $\sin x \cos y = \dfrac{1}{2}[\sin(x + y) + \sin(x - y)]$

$\sin 195° \cos 75° = \dfrac{1}{2}[\sin(195° + 75°) + \sin(195° - 75°)]$

$\qquad\qquad\quad\; = \dfrac{1}{2}[\sin 270° + \sin 120°]$

$\qquad\qquad\quad\; = \dfrac{1}{2}\left(-1 + \dfrac{\sqrt{3}}{2}\right)$

$\qquad\qquad\quad\; = \dfrac{1}{2}\left(\dfrac{\sqrt{3} - 2}{2}\right)$

$\qquad\qquad\quad\; = \dfrac{\sqrt{3} - 2}{4}$

17. $\cos x \cos y = \dfrac{1}{2}\left[\cos(x + y) + \cos(x - y)\right]$

$\cos 157.5° \cos 67.5° = \dfrac{1}{2}\left[\cos\left(157.5° + 67.5°\right) + \cos\left(157.5° - 67.5°\right)\right]$

$\qquad\qquad\qquad\; = \dfrac{1}{2}\left[\cos 225° + \cos 90°\right]$

$\qquad\qquad\qquad\; = \dfrac{1}{2}\left[-\dfrac{1}{\sqrt{2}} + 0\right] = -\dfrac{1}{2\sqrt{2}} \cdot \dfrac{\sqrt{2}}{\sqrt{2}} = -\dfrac{\sqrt{2}}{4}$

19. $\cos x \sin y = \dfrac{1}{2}\left[\sin(x + y) - \sin(x - y)\right]$

$\cos 37.5° \sin 7.5° = \dfrac{1}{2}\left[\sin\left(37.5° + 7.5°\right) - \sin\left(37.5° - 7.5°\right)\right]$

$= \dfrac{1}{2}\left[\sin 45° - \sin 30°\right]$

$= \dfrac{1}{2}\left[\dfrac{\sqrt{2}}{2} - \dfrac{1}{2}\right] = \dfrac{\sqrt{2} - 1}{4}$

21. $\sin x \sin y = \dfrac{1}{2}\left[\cos(x - y) - \cos(x + y)\right]$

$\sin \dfrac{5\pi}{8} \sin \dfrac{\pi}{8} = \dfrac{1}{2}\left[\cos\left(\dfrac{5\pi}{8} - \dfrac{\pi}{8}\right) - \cos\left(\dfrac{5\pi}{8} + \dfrac{\pi}{8}\right)\right]$

$= \dfrac{1}{2}\left[\cos \dfrac{\pi}{2} - \cos \dfrac{3\pi}{4}\right]$

$= \dfrac{1}{2}\left[0 - \left(-\dfrac{\sqrt{2}}{2}\right)\right] = \dfrac{\sqrt{2}}{4}$

23. $\cos x \cos y = \dfrac{1}{2}\left[\cos(x - y) + \cos(x + y)\right]$

$\cos \dfrac{11\pi}{12} \cos \dfrac{\pi}{12} = \dfrac{1}{2}\left[\cos\left(\dfrac{11\pi}{12} - \dfrac{\pi}{12}\right) + \cos\left(\dfrac{11\pi}{12} + \dfrac{\pi}{12}\right)\right]$

$= \dfrac{1}{2}\left[\cos \dfrac{5\pi}{6} + \cos \pi\right]$

$= \dfrac{1}{2}\left[-\dfrac{\sqrt{3}}{2} - 1\right]$

$= \dfrac{1}{2}\left[-\dfrac{\left(\sqrt{3} + 2\right)}{2}\right] = -\dfrac{\left(\sqrt{3} + 2\right)}{4}$

25. $\sin x \cos y = \dfrac{1}{2}\left[\sin(x + y) + \sin(x - y)\right]$

$\sin \dfrac{13\pi}{24} \cos \dfrac{5\pi}{24} = \dfrac{1}{2}\left[\sin\left(\dfrac{13\pi}{24} + \dfrac{5\pi}{24}\right) + \sin\left(\dfrac{13\pi}{24} - \dfrac{5\pi}{24}\right)\right]$

$= \dfrac{1}{2}\left[\sin \dfrac{3\pi}{4} + \sin \dfrac{\pi}{3}\right]$

$= \dfrac{1}{2}\left[\dfrac{\sqrt{2}}{2} + \dfrac{\sqrt{3}}{2}\right] = \dfrac{\sqrt{2} + \sqrt{3}}{4}$

27. $\sin x - \sin y = 2 \cos \dfrac{x + y}{2} \sin \dfrac{x - y}{2}$

$\sin 195° - \sin 105° = 2 \cos \dfrac{195° + 105°}{2} \sin \dfrac{195° - 105°}{2}$

$= 2 \cos 150° \sin 45°$

$= 2 \cdot -\dfrac{\sqrt{3}}{2} \cdot \dfrac{\sqrt{2}}{2} = -\dfrac{\sqrt{6}}{2}$

29. $\cos x - \cos y = -2 \sin \dfrac{x+y}{2} \sin \dfrac{x-y}{2}$

$\cos 75° - \cos 15° = -2 \sin \dfrac{75° + 15°}{2} \sin \dfrac{75° - 15°}{2}$

$= -2 \sin 45° \sin 30°$

$= -2 \cdot \dfrac{\sqrt{2}}{2} \cdot \dfrac{1}{2} = -\dfrac{\sqrt{2}}{2}$

31. $\cos x + \cos y = 2 \cos \dfrac{x+y}{2} \cos \dfrac{x-y}{2}$

$\cos \dfrac{17\pi}{12} + \cos \dfrac{\pi}{12} = 2 \cos \dfrac{\dfrac{17\pi}{12} + \dfrac{\pi}{12}}{2} \cos \dfrac{\dfrac{17\pi}{12} - \dfrac{\pi}{12}}{2}$

$= 2 \cos \dfrac{3\pi}{4} \cos \dfrac{2\pi}{3}$

$= 2 \cdot -\dfrac{\sqrt{2}}{2} \cdot -\dfrac{1}{2} = \dfrac{\sqrt{2}}{2}$

33. $\sin x + \sin y = 2 \sin \dfrac{x+y}{2} \cos \dfrac{x-y}{2}$

$\sin \dfrac{\pi}{12} + \sin \dfrac{5\pi}{12} = 2 \sin \dfrac{\dfrac{\pi}{12} + \dfrac{5\pi}{12}}{2} \cos \dfrac{\dfrac{\pi}{12} - \dfrac{5\pi}{12}}{2}$

$= 2 \sin \dfrac{\pi}{4} \cos\left(-\dfrac{\pi}{6}\right)$

$= 2 \dfrac{\sqrt{2}}{2} \dfrac{\sqrt{3}}{2} = \dfrac{\sqrt{6}}{2}$

35.

$\cos(x + y) = \cos x \cos y - \sin x \sin y$

$\cos(x - y) = \cos x \cos y + \sin x \sin y$

$\overline{\cos(x + y) + \cos(x - y) = 2 \cos x \cos y}$ (adding the above)

$\cos x \cos y = \dfrac{1}{2}[\cos(x + y) + \cos(x - y)]$

37. Let $x = u + v$ and $y = u - v$ and solve for u and v in terms of x and y:

$x = u + v$

$\underline{y = u - v}$

$x + y = 2u \qquad x - y = 2v$

$u = \dfrac{x+y}{2} \qquad v = \dfrac{x-y}{2}$

Substitute these results into $\sin u \sin v = \dfrac{1}{2}[\cos(u - v) - \cos(u + v)]$ to obtain:

$\sin \dfrac{x+y}{2} \sin\left(\dfrac{x-y}{2}\right) = \dfrac{1}{2}[\cos y - \cos x]$

or

$-\sin \dfrac{x+y}{2} \sin \dfrac{x-y}{2} = \dfrac{1}{2}[\cos x - \cos y]$

or

$\cos x - \cos y = -2 \sin \dfrac{x+y}{2} \sin \dfrac{x-y}{2}$

39. $\dfrac{\sin 2t + \sin 4t}{\cos 2t - \cos 4t} = \dfrac{2 \sin \frac{2t+4t}{2} \cos \frac{2t-4t}{2}}{-2 \sin \frac{2t+4t}{2} \sin \frac{2t-4t}{2}}$ Sum-product Identities

$\qquad\qquad\qquad = \dfrac{2 \sin 3t \cos(-t)}{-2 \sin 3t \sin(-t)}$ Algebra

$\qquad\qquad\qquad = \dfrac{2 \sin 3t \cos t}{2 \sin 3t \sin t}$ Identities for Negatives

$\qquad\qquad\qquad = \dfrac{\cos t}{\sin t}$ Algebra

$\qquad\qquad\qquad = \cot t$ Quotient Identity

41. $\dfrac{\sin x - \sin y}{\cos x - \cos y} = \dfrac{2 \cos \frac{x+y}{2} \sin \frac{x-y}{2}}{-2 \sin \frac{x+y}{2} \sin \frac{x-y}{2}}$ Sum-product Identities

$\qquad\qquad\qquad = -\dfrac{\cos \frac{x+y}{2}}{\sin \frac{x+y}{2}}$ Algebra

$\qquad\qquad\qquad = -\cot \dfrac{x+y}{2}$ Quotient Identity

43. $\dfrac{\cos x + \cos y}{\sin x - \sin y} = \dfrac{2 \cos \frac{x+y}{2} \cos \frac{x-y}{2}}{2 \cos \frac{x+y}{2} \sin \frac{x-y}{2}}$ Sum-product Identities

$\qquad\qquad\qquad = \dfrac{\cos \frac{x-y}{2}}{\sin \frac{x-y}{2}}$ Algebra

$\qquad\qquad\qquad = \cot \dfrac{x-y}{2}$ Quotient Identity

45. $\dfrac{\cos x + \cos y}{\cos x - \cos y} = \dfrac{2 \cos \frac{x+y}{2} \cos \frac{x-y}{2}}{-2 \sin \frac{x+y}{2} \sin \frac{x-y}{2}}$ Sum-product Identities

$\qquad\qquad\qquad = -\dfrac{\cos \frac{x+y}{2} \cos \frac{x-y}{2}}{\sin \frac{x+y}{2} \sin \frac{x-y}{2}}$ Algebra

$\qquad\qquad\qquad = -\cot \dfrac{x+y}{2} \cot \dfrac{x-y}{2}$ Quotient Identity

47. Let $x = y = 0$

Left side: $\sin 0 \cos 0 = 0$

Right side: $\sin 0 + \cos 0 = 1$

The equation is not an identity.

49. Let $x = 0$ and $y = \dfrac{\pi}{2}$.

Left side: $\sin 0 \sin \dfrac{\pi}{2} = 0$

Right side: $\sin\left(0 + \dfrac{\pi}{2}\right) = 1$

The equation is not an identity.

51. Let $x = \dfrac{\pi}{3}$ and $y = \dfrac{\pi}{3}$.

Left side: $\cos \dfrac{\pi}{3} + \cos \dfrac{\pi}{3} = 1$

Right side: $\cos \dfrac{\pi}{3} \cos \dfrac{\pi}{3} = \dfrac{1}{4}$

The equation is not an identity.

53. Let $x = \dfrac{\pi}{2}$ and $y = 0$.

Left side: $\sin \dfrac{\pi}{2} - \sin 0 = 1$

Right side: $\cos \dfrac{\frac{\pi}{2} + 0}{2} \sin \dfrac{\frac{\pi}{2} - 0}{2} = \cos \dfrac{\pi}{4} \sin \dfrac{\pi}{4} = \dfrac{1}{\sqrt{2}} \cdot \dfrac{1}{\sqrt{2}} = \dfrac{1}{2}$

The equation is not an identity.

55. $\sin x - \sin y = 2 \cos \dfrac{x + y}{2} \sin \dfrac{x - y}{2}$ Sum-product Identity

$\sin 3x - \sin x = 2 \cos \dfrac{3x + x}{2} \sin \dfrac{3x - x}{2}$ Sum-product Identity

$\qquad = 2 \cos \dfrac{4x}{2} \sin \dfrac{2x}{2}$ Algebra

$\qquad = 2 \cos 2x \sin x$ Algebra

The equation is an identity.

57. Try using a sum-product identity:

$\cos x - \cos y = -2 \sin \dfrac{x + y}{2} \sin \dfrac{x - y}{2}$

$\cos 3x - \cos x = -2 \sin \dfrac{3x + x}{2} \sin \dfrac{3x - x}{2}$

$\qquad = -2 \sin 2x \sin x$

The equation is not an identity — the right side lacks the negative sign.

59. $\cos x + \cos y = 2 \cos \dfrac{x + y}{2} \cos \dfrac{x - y}{2}$ Sum-product Identity

$\cos x + \cos 5x = 2 \cos \dfrac{x + 5x}{2} \cos \dfrac{x - 5x}{2}$ Sum-product Identity

$\qquad = 2 \cos \dfrac{6x}{2} \cos -\dfrac{4x}{2}$ Algebra

$\qquad = 2 \cos 3x \cos(-2x)$ Algebra

$\qquad = 2 \cos 3x \cos 2x$ $\cos(-2x) = \cos(2x)$

The equation is an identity.

61. (A) $\cos 172.63° \sin 20.177° = -0.34207$

$\qquad \dfrac{1}{2}[\sin(172.63° + 20.177°) - \sin(172.63° - 20.177°)] = -0.34207$

(B) $\cos 172.63° + \cos 20.177° = -0.05311$

$2 \cos \dfrac{172.63° + 20.177°}{2} \cos \dfrac{172.63° - 20.177°}{2} = -0.05311$

(calculator in degree mode)

63. (A) $\cos 1.1255 \sin 3.6014 = -0.19115$

$\dfrac{1}{2}[\sin(1.1255 + 3.6014) - \sin(1.1255 - 3.6014)] = -0.19115$

(B) $\cos 1.1255 + \cos 3.6014 = -0.46541$

$2 \cos \dfrac{1.1255 + 3.6014}{2} \cos \dfrac{1.1255 - 3.6014}{2} = -0.46541$

(calculator in radian mode)

65. $\sin 2x + \sin x = 2 \sin \dfrac{2x + x}{2} \cos \dfrac{2x - x}{2}$

$= 2 \sin \dfrac{3x}{2} \cos \dfrac{x}{2}$

Graph $y_1 = \sin 2x + \sin x$ and

$y_2 = 2 \sin \dfrac{3x}{2} \cos \dfrac{x}{2}$

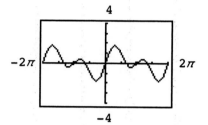

67. $\cos 1.7x - \cos 0.3x$

$= -2 \sin \dfrac{1.7x + 0.3x}{2} \sin \dfrac{1.7x - 0.3x}{2}$

$= -2 \sin x \sin 0.7x$

Graph $y_1 = \cos 1.7x - \cos 0.3x$

$y_2 = -2 \sin x \sin 0.7x$

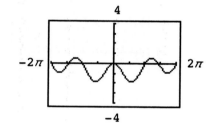

69. $\sin 3x \cos x = \dfrac{1}{2}[\sin(3x + x) + \sin(3x - x)]$

$= \dfrac{1}{2}(\sin 4x + \sin 2x)$

Graph $y_1 = \sin 3x \cos x$ and

$y_2 = \dfrac{1}{2}(\sin 4x + \sin 2x)$

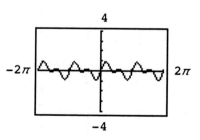

71. $\sin 2.3x \sin 0.7x$

$= \dfrac{1}{2}[\cos(2.3x - 0.7x) - \cos(2.3x + 0.7x)]$

$= \dfrac{1}{2}(\cos 1.6x - \cos 3x)$

Graph $y_1 = \sin 2.3x \sin 0.7x$

$y_2 = \dfrac{1}{2}(\cos 1.6x - \cos 3x)$

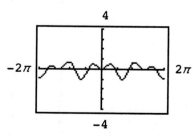

73. $\cos x \cos y \cos z = \cos x \dfrac{1}{2}[\cos (y + z) + \cos(y - z)]$ Product-sum Identity

$\qquad = \dfrac{1}{2} \cos x \cos(y + z) + \dfrac{1}{2} \cos x \cos(y - z)$ Algebra

$\qquad = \dfrac{1}{2}\left\{\dfrac{1}{2}[\cos(x + y + z) + \cos(x - \{y + z\})]\right\}$

$\qquad\quad + \dfrac{1}{2}\left\{\dfrac{1}{2}[\cos(x + y - z) + \cos(x - \{y - z\})]\right\}$ Product-sum Identity

$\qquad = \dfrac{1}{4} \cos(x + y + z) + \dfrac{1}{4} \cos(x - y - z)$

$\qquad\quad + \dfrac{1}{4} \cos(x + y - z) + \dfrac{1}{4} \cos(x - y + z)$ Algebra

$\qquad = \dfrac{1}{4}[\cos(x + y - z) + \cos(x - y - z)$

$\qquad\quad + \cos(z + x - y) + \cos(x + y + z)]$ Algebra

$\qquad = \dfrac{1}{4}[\cos(x + y - z) + \cos\{-(y + z - x)\}$

$\qquad\quad + \cos (z + x - y) + \cos(x + y + z)]$ Algebra

$\qquad = \dfrac{1}{4}[\cos(x + y - z) + \cos(y + z - x) + \cos(z + x - y)$

$\qquad\quad + \cos(x + y + z)]$ Identity for Negatives

75. (A)

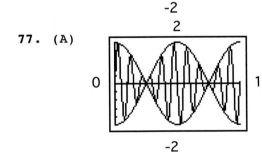

(B) $2 \cos(28\pi x) \cos(2\pi x)$

$\qquad = 2 \cdot \dfrac{1}{2}[\cos(28\pi x + 2\pi x) + \cos(28\pi x - 2\pi x)]$

$\qquad = \cos 30\pi x + \cos 26\pi x$

Graphing $y_1 = \cos 30\pi x + \cos 26\pi x$ will yield the same result as before.

77. (A)

(B) $2 \sin(20\pi x) \cos(2\pi x)$

$\qquad = 2 \cdot \dfrac{1}{2}[\sin(20\pi x + 2\pi x) + \sin(20\pi x - 2\pi x)]$

$\qquad = \sin 22\pi x + \sin 18\pi x$

Graphing $y_1 = \sin 22\pi x + \sin 18\pi x$ will yield the same result as before.

79. (A) $\cos x - \cos y = -2 \sin \dfrac{x + y}{2} \sin \dfrac{x - y}{2}$

In this case

$\cos 128\pi t - \cos 144\pi t = -2 \sin \dfrac{128\pi t + 144\pi t}{2} \sin \dfrac{128\pi t - 144\pi t}{2}$

$\cos 128\pi t - \cos 144\pi t = -2 \sin 136\pi t \sin(-8\pi t)$

$\cos 128\pi t - \cos 144\pi t = 2 \sin 136\pi t \sin 8\pi t$

Multiplying both sides by 0.5, we have

$0.5 \cos 128\pi t - 0.5 \cos 144\pi t = \sin 136\pi t \sin 8\pi t$

(B) $y = 0.5 \cos 128\pi t$ $y = -0.5 \cos 144\pi t$

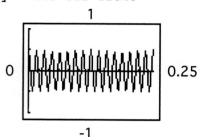

$y = 0.5 \cos 128\pi t - 0.5 \cos 144\pi t$ $y = \sin 8\pi t \sin 136\pi t$

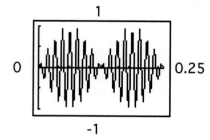 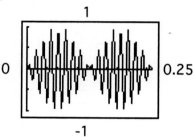

The latter two graphs are the same, illustrating the identity proved in part (A).

Section 6-5

1. Answers will vary.

3. $\sin x + 1 = 0$
   $\qquad \sin x = -1$
   Sketch a graph of $y = \sin x$ and $y = -1$ on the interval $[0, 2\pi)$.

   $x = \dfrac{3\pi}{2}$

   The checking steps in this and subsequent problems are left to the student.

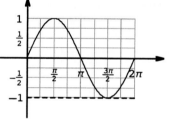

5. We have found all solutions of $\sin x + 1 = 0$ over one period in problem 3. Since the sine function is periodic with period 2π, all solutions are given by

   $x = \dfrac{3\pi}{2} + 2k\pi,$ k any integer

7. $\tan \theta - 1 = 0$
   $\qquad \tan \theta = 1$
   Sketch a graph of $y = \tan \theta$ and $y = 1$ on the interval $[0, 360°)$
   $\qquad \theta = 45°,\ 225°$

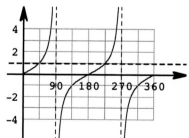

9. $2 \sin x - 1 = 0$
   $\qquad \sin x = \dfrac{1}{2}$

   Sketch a graph of $y = \sin x$ and $y = \dfrac{1}{2}$ on the interval $[0, 2\pi)$
   $x = \dfrac{\pi}{6},\ \dfrac{5\pi}{6}$

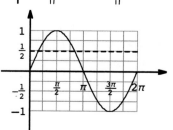

11. We have found all solutions of $2 \sin x - 1 = 0$ over one period in problem 9. Since the sine function is periodic with period 2π, all solutions are given by

$$\left.\begin{array}{l} x = \frac{\pi}{6} + 2k\pi \\ x = \frac{5\pi}{6} + 2k\pi \end{array}\right\} k \text{ any integer}$$

13. $2 \sin \theta + \sqrt{3} = 0$

$$\sin \theta = -\frac{\sqrt{3}}{2}$$

Sketch a graph of $y = \sin \theta$ and $y = -\frac{\sqrt{3}}{2}$

on the interval $[0, 360°)$.

$\theta = 240°, 300°$

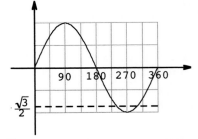

15. We have found all solutions of $2 \sin \theta + \sqrt{3} = 0$ over one period in problem 13. Since the sine function is periodic with period $360°$, all solutions are given by

$$\left.\begin{array}{l} \theta = 240° + k360° \\ \theta = 300° + k360° \end{array}\right\} k \text{ any integer}$$

17. $\tan x - \sqrt{3} = 0$

$\tan x = \sqrt{3}$

Sketch a graph of $y = \tan x$ and $y = \sqrt{3}$ on the interval $[0, 2\pi)$.

$$x = \frac{\pi}{3}, \frac{4\pi}{3}$$

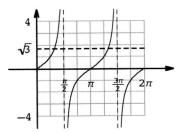

19. We have found all solutions of $\tan x - \sqrt{3} = 0$ over two periods in problem 17. Since the tangent function is periodic with period π, all solutions are given by $x = \frac{\pi}{3} + k\pi$, k any integer.

21. $7 \cos x - 3 = 0 \qquad 0 \le x < 2\pi$

$$\cos x = \frac{3}{7}$$

Sketch a graph of $y = \cos x$ and $y = \frac{3}{7}$,

x on the interval $[0, 2\pi)$.

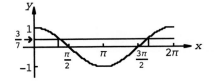

$$x = \begin{cases} \cos^{-1} \frac{3}{7} = 1.1279 \text{ First quadrant solution} \\ 2\pi \cos^{-1} \frac{3}{7} = 5.1553 \text{ Fourth quadrant solution} \end{cases}$$

23. $2 \tan \theta - 7 = 0 \qquad 0° \le \theta < 180°$

$$\tan \theta = \frac{7}{2}$$

Sketch a graph of $y = \tan \theta$ and $y = \frac{7}{2}$

on the interval $[0°, 180°)$

$\theta = \tan^{-1} \frac{7}{2} = 74.0546°$

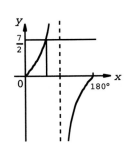

25. 1.3224 sin x + 0.4732 = 0

$\qquad \sin x = -\dfrac{0.4732}{1.3224}$

Solve over one period [0, 2π):

$\qquad \sin x = -0.3578$

Sketch a graph of $y = \sin x$ and
$y = -0.3578$, on the interval [0, 2π)

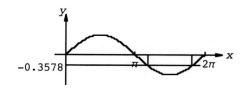

$x = \begin{cases} \pi - \sin^{-1}(-0.3578) = 3.5075 & \text{Third quadrant solution} \\ 2\pi + \sin^{-1}(-0.3578) = 5.9172 & \text{Fourth quadrant solution} \end{cases}$

Since the same function is periodic with period 2π, all solutions are given by

$\qquad x = \begin{cases} 3.5075 + 2k\pi \\ 5.9172 + 2k\pi \end{cases}$ k any integer

27. Here is a graph of $y_1 = 1 - x$ and
$y_2 = 2 \sin x$, on the interval
$0 \le x \le 2\pi$.

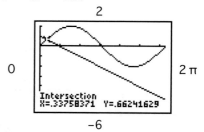

To four decimal places, the only
solution is 0.3376.

Check: $1 - x = 1 - 0.3376 = 0.6624$
$\qquad\quad 2 \sin x = 2 \sin(0.3376) = 0.6624$

29. Here is a graph of $y_1 = \tan \dfrac{x}{2}$ and
$y_2 = 8 - x$, on the interval
$0 \le x \le \pi$.

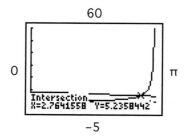

To four decimal places, the only
solution is 2.7642.

Check: $\tan \dfrac{x}{2} = \tan \dfrac{2.7642}{2} = 5.2358$
$\qquad\quad 8 - x = 8 - 2.7642 = 5.2358$

31. $\qquad\qquad 2 \sin^2 \theta + \sin 2\theta = 0$

$\quad 2 \sin^2 \theta + 2 \sin \theta \cos \theta = 0$

$\qquad 2 \sin \theta(\sin \theta + \cos \theta) = 0$

Either $\qquad\qquad\quad 2 \sin \theta = 0$

$\qquad\qquad\qquad\qquad \sin \theta = 0$

Solutions over $0° \le \theta < 360°$ are
$\theta = 0°,\ 180°$

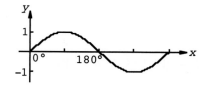

If θ is allowed to range over all possible values, the solutions are
$\theta = 0° + k\ 180°$ or $\theta = k\ 180°$ k any integer.

or $\sin \theta + \cos \theta = 0$

$\quad\quad\quad \sin \theta = -\cos \theta$

$\quad\quad\quad \dfrac{\sin \theta}{\cos \theta} = -1$

$\quad\quad\quad \tan \theta = -1$

Solutions over $0° < \theta < 180°$ are

$\theta = 135°$

If θ is allowed to range over all
possible values, solutions are

$\theta = 135° + k\ 180°$ k any integer

Solutions: $k\ 180°$, $135° + k\ 180°$ k any integer

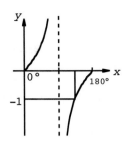

33.

$\quad\quad\quad \tan x = -2 \sin x \quad\quad\quad 0 \leq x < 2\pi$

$\quad\quad\quad \dfrac{\sin x}{\cos x} = -2 \sin x$

$\quad\quad\quad \sin x = -2 \sin x \cos x \quad \cos x \neq 0$

$2 \sin x \cos x + \sin x = 0$

$\quad \sin x(2 \cos x + 1) = 0$

$\sin x = 0 \quad\quad\quad 2 \cos x + 1 = 0$

$\quad x = 0,\ \pi \quad\quad\quad 2 \cos x = -1$

$\quad\quad\quad\quad\quad\quad \cos x = -\dfrac{1}{2}$

$\quad\quad\quad\quad\quad\quad x = \dfrac{2\pi}{3},\ \dfrac{4\pi}{3}$

Solutions: $0,\ \dfrac{2\pi}{3},\ \pi,\ \dfrac{4\pi}{3}$

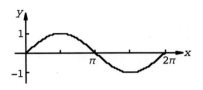

35. $\sec \dfrac{x}{2} + 2 = 0 \quad 0 \leq x \leq 2\pi$ is equivalent to

$\sec \dfrac{x}{2} + 2 = 0 \quad 0 \leq \dfrac{x}{2} \leq \pi$

$\quad \sec \dfrac{x}{2} = -2$

$\quad \cos \dfrac{x}{2} = -\dfrac{1}{2}$

$\quad \dfrac{x}{2} = \dfrac{2\pi}{3}$

$\quad x = \dfrac{4\pi}{3}$

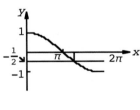

37. $\quad 2 \cos^2 \theta + 3 \sin \theta = 0 \quad\quad 0° \leq \theta < 360°$

$2(1 - \sin^2 \theta) + 3 \sin \theta = 0$

$\quad 2 - 2 \sin^2 \theta + 3 \sin \theta = 0$

$-2 \sin^2 \theta + 3 \sin \theta + 2 = 0$

$\quad 2 \sin^2 \theta - 3 \sin \theta - 2 = 0$

$(2 \sin \theta + 1)(\sin \theta - 2) = 0$

$\quad\quad\quad\quad 2 \sin \theta + 1 = 0 \quad\quad \sin \theta - 2 = 0$

$\quad\quad\quad\quad\quad 2 \sin \theta = -1 \quad\quad \sin \theta = 2$

$\quad\quad\quad\quad\quad\quad \sin \theta = -\dfrac{1}{2} \quad\quad \text{No solution}$

$\quad\quad\quad\quad\quad\quad\quad \theta = 210°,\ 330°$

Solutions: $\theta = 210°,\ 330°$

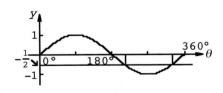

39.

$$\cos 2\theta + \cos \theta = 0 \qquad 0° \le \theta < 360°$$
$$2 \cos^2 \theta - 1 + \cos \theta = 0$$
$$2 \cos^2 \theta + \cos \theta - 1 = 0$$
$$(2 \cos \theta - 1)(\cos \theta + 1) = 0$$

$$2 \cos \theta - 1 = 0 \qquad \cos \theta + 1 = 0$$
$$2 \cos \theta = 1 \qquad \cos \theta = -1$$
$$\theta = 180°$$

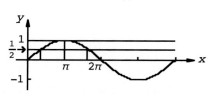

$$\cos \theta = \frac{1}{2}$$
$$\theta = 60°, \ 300°$$

Solutions: $\theta = 60°, \ 180°, \ 300°$

41.

$$2 \sin^2 \frac{x}{2} - 3 \sin \frac{x}{2} + 1 = 0 \qquad 0 \le x \le 2\pi \quad \text{is equivalent to}$$

$$2 \sin^2 \frac{x}{2} - 3 \sin \frac{x}{2} + 1 = 0 \qquad 0 \le \frac{x}{2} \le \pi$$

$$(2 \sin \frac{x}{2} - 1)(\sin \frac{x}{2} - 1) = 0$$

$$2 \sin \frac{x}{2} - 1 = 0 \qquad \boxed{\begin{array}{l} \textbf{Common Error:} \\[4pt] 2 \sin \frac{x}{2} \ne \sin x \end{array}} \qquad \sin \frac{x}{2} - 1 = 0$$

$$2 \sin \frac{x}{2} = 1 \qquad\qquad\qquad\qquad\qquad \sin \frac{x}{2} = 1$$

$$\sin \frac{x}{2} = \frac{1}{2} \qquad\qquad\qquad\qquad\qquad \frac{x}{2} = \frac{\pi}{2}$$

$$\frac{x}{2} = \frac{\pi}{6}, \ \frac{5\pi}{6} \qquad\qquad\qquad\qquad x = \pi$$

$$x = \frac{\pi}{3}, \ \frac{5\pi}{3}$$

Solutions: $x = \dfrac{\pi}{3}, \ \dfrac{5\pi}{3}, \ \pi$

43. Since $\cos^2 x + \sin^2 x = 1$ is an identity, the solutions of this equation are all x, $0 \le x < 2\pi$.

45. Since $|2 \sin \theta| \le 2$ and $\cos \theta \ge -1$ for all θ, the left side of this equation cannot be greater than 2 and the right side cannot be less than 4. The equation has no solutions.

47.

$$6 \sin^2 \theta + 5 \sin \theta = 6 \qquad 0° \le \theta \le 90°$$
$$6 \sin^2 \theta + 5 \sin \theta - 6 = 0$$
$$(3 \sin \theta - 2)(2 \sin \theta + 3) = 0$$

$$3 \sin \theta - 2 = 0 \quad 0° \le \theta \le 90° \qquad 2 \sin \theta + 3 = 0 \quad 0° \le \theta \le 90°$$

$$3 \sin \theta = 2 \qquad\qquad\qquad 2 \sin \theta = -3$$

$$\sin \theta = \frac{2}{3} \qquad\qquad\qquad \sin \theta = -\frac{3}{2}$$

$$\theta = \sin^{-1} \frac{2}{3} \ \begin{array}{l}\text{(calculator in}\\ \text{degree mode)}\end{array} \qquad \text{No solution}$$

$$\theta = 41.81°$$

Solution: $\theta = 41.81°$

49.
$$3 \cos^2 x - 8 \cos x = 3 \qquad 0 \le x \le \pi$$
$$3 \cos^2 x - 8 \cos x - 3 = 0$$
$$(3 \cos x + 1)(\cos x - 3) = 0$$
$$3 \cos x + 1 = 0 \quad 0 \le x \le \pi \qquad \cos x - 3 = 0 \quad 0 \le x \le \pi$$
$$3 \cos x = -1 \qquad\qquad\qquad \cos x = 3$$
$$\cos x = -\frac{1}{3} \qquad\qquad\qquad \text{No solution}$$
$$x = \cos^{-1}\left(-\frac{1}{3}\right) \quad \text{(calculator in radian mode)}$$
$$x = 1.911$$
Solution: $x = 1.911$

51.
$$2 \sin x = \cos 2x \qquad\qquad 0 \le x < 2\pi$$
$$2 \sin x = 1 - 2 \sin^2 x$$
$$2 \sin^2 x + 2 \sin x - 1 = 0 \qquad\qquad \text{Quadratic in } \sin x$$
$$\sin x = \frac{-b \pm \sqrt{b^2 - 4ac}}{2a} \qquad a = 2,\ b = 2,\ c = -1$$
$$\sin x = \frac{-2 \pm \sqrt{2^2 - 4(2)(-1)}}{2(2)} = \frac{-2 \pm \sqrt{12}}{4}$$
$$\sin x = 0.3660 \qquad\qquad \sin x = -1.366$$
$$\qquad\qquad\qquad\qquad\qquad \text{No solution}$$
$$x = \begin{cases} \sin^{-1} 0.3660 \\ \pi - \sin^{-1} 0.3660 \end{cases} = \begin{cases} 0.3747 \\ 2.767 \end{cases}$$
Solutions: $x = 0.3747,\ 2.767$

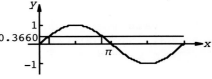

53. $2 \sin^2 x = 1 - 2 \sin x$
$2 \sin^2 x + 2 \sin x - 1 = 0 \quad x$ any real number
See Problem 51.
$$x = \begin{cases} \sin^{-1} 0.3660 \\ \pi - \sin^{-1} 0.3660 \end{cases} = \begin{cases} 0.3747 \\ 2.767 \end{cases} \quad \text{are the solutions over one period } 0 \le x < 2\pi$$

If x can range over all real numbers,
$$x = \begin{cases} 0.3747 + 2k\pi \\ 2.767 + 2k\pi \end{cases} \quad k \text{ any integer}$$

55. Examining the graphs of $y_1 = 2 \sin x$ and $y_2 = \cos 2x$, $0 \le x < 2\pi$, we obtain

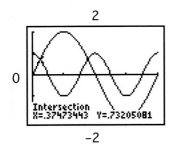

To four decimal places, the solutions are 0.3747 and 2.7669.

Check: 0.3747: 2 sin x = 0.7321　　2.7669: 2 sin x = 2 sin 2.7669 = 0.7321
　　　　　　　cos 2x = 0.7321　　　　　　　　cos 2x = cos 2.7669 = 0.7321

In later problems, the checking is left to the student.

57. Examining the graphs of the two sides, $y_1 = 2 \sin^2 x$ and $y_2 = 1 - 2 \sin x$, $0 \le x \le 2\pi$, we obtain the graphs at the right.

To four decimal places, the solutions are 0.3747 and 2.7669. The solutions over all real x are given by

$$x = \begin{cases} 0.3747 + 2k\pi \\ 2.7669 + 2k\pi \end{cases} k \text{ any integer}$$

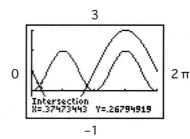 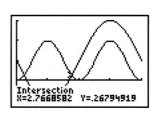

59. Examining the graphs of the two sides, $y_1 = \cos 2x$ and $y_2 = x^2 - 2$, we obtain the graphs at the right.

The graphs intersect at $x = -1.1530$ and $x = 1.1530$. The cosine graph is above the parabola, and $\cos 2x > x^2 - 2$, on the interval $(-1.1530, 1.1530)$.

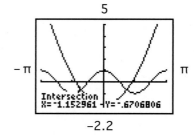 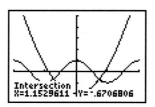

61. Examining the graphs of the two sides, $y_1 = \cos(2x + 1)$ and $y_2 = 0.5x - 2$, we obtain

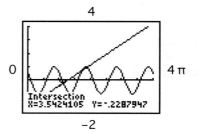

There is clearly one point of intersection at $x = 3.5424$. Examining the graphs again on a smaller interval, we obtain

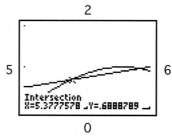 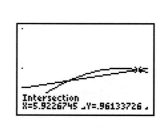

Combining information from all three graphs, the graph of the cosine function is below or intersects the straight line, $\cos(2x + 1) \le 0.5x - 2$, on the two intervals [3.5424, 5.3778] and [5.9227, ∞).

63. Examining the graphs of $y_1 = e^{\sin x}$ and $y_2 = 2x - 1$, we obtain the graph at the right.

The solution is 1.8183.

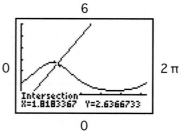

65. $\tan^{-1}(-5.377)$ has exactly one value, -1.387; the equation $\tan x = -5.377$ has infinitely many solutions, which are found by adding $k\pi$, k any integer, to each solution in one period of $\tan x$.

67.
$$\cos x - \sin x = 1 \qquad\qquad 0 \le x < 2\pi$$
$$\pm\sqrt{1 - \sin^2 x} - \sin x = 1$$
$$\pm\sqrt{1 - \sin^2 x} = 1 + \sin x$$
$$1 - \sin^2 x = (1 + \sin x)^2 \qquad \text{Squaring both sides}$$
$$1 - \sin^2 x = 1 + 2 \sin x + \sin^2 x$$

> **Common Error:** $(1 + \sin x)^2 \ne 1 + \sin^2 x$

$$0 = 2 \sin x + 2 \sin^2 x$$
$$0 = 2 \sin x(1 + \sin x)$$

$2 \sin x = 0$	$1 + \sin x = 0$
$\sin x = 0$	$\sin x = -1$
$x = 0,\ \pi$	$x = \dfrac{3\pi}{2}$

In squaring both sides we may have introduced extraneous solutions; so it is necessary to check solutions of these equations in the original equation.

$$x = 0 \qquad\qquad x = \pi \qquad\qquad x = \frac{3\pi}{2}$$

$\cos x - \sin x = 1$	$\cos x - \sin x = 1$	$\cos x - \sin x = 1$
$\cos 0 - \sin 0 \overset{?}{=} 1$	$\cos \pi - \sin \pi \overset{?}{=} 1$	$\cos \dfrac{3\pi}{2} - \sin \dfrac{3\pi}{2} \overset{?}{=} 1$
$1 - 0 \overset{\checkmark}{=} 1$	$-1 - 0 \ne 1$	$0 - (-1) \overset{\checkmark}{=} 1$
A solution	Not a solution	A solution

Solutions: $x = 0,\ \dfrac{3\pi}{2}$

69.
$$\tan x - \sec x = 1 \qquad\qquad 0 \le x < 2\pi$$
$$\pm\sqrt{\sec^2 x - 1} - \sec x = 1$$
$$\pm\sqrt{\sec^2 x - 1} = 1 + \sec x$$
$$\sec^2 x - 1 = (1 + \sec x)^2 \qquad \text{Squaring both sides}$$
$$\sec^2 x - 1 = 1 + 2 \sec x + \sec^2 x$$
$$0 = 2 + 2 \sec x$$
$$-2 \sec x = 2$$
$$\sec x = -1$$
$$x = \pi$$

In squaring both sides we may have introduced extraneous solutions; so it is necessary to check this apparent solution in the original equation.

$$\tan x - \sec x = 1$$
$$\tan \pi - \sec \pi \overset{?}{=} 1$$
$$0 - (-1) \overset{\checkmark}{=} 1$$

Solution: $x = \pi$

71. Examining the graphs of
$y_1 = \sin \dfrac{1}{x}$ and
$y_2 = 1.5 - 5x$,
$0.04 \le x \le 0.2$,
we obtain the graphs
at the right.

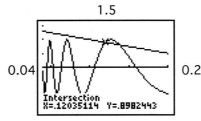
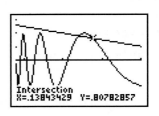

The graphs intersect, and $\sin \dfrac{1}{x} = 1.5 - 5x$, at $x = 0.1204$ and $x = 0.1384$.

438

73. (A) Here are graphs of $f(x) = \sin \dfrac{1}{x}$ on $[0.1, 0.5]$ and $[0.2, 5]$.

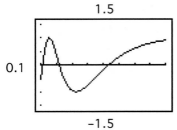
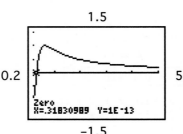

The largest zero for f is 0.3183. As x increases without bound, $\dfrac{1}{x}$ tends to 0 through positive numbers, and $\sin \dfrac{1}{x}$ tends to 0 through positive numbers. $y = 0$ is a horizontal asymptote for the graph of f.

(B) Here is a computer-generated graph of $f(x) = \sin \dfrac{1}{x}$ on $[0.015, 0.1]$.

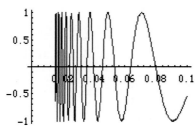

The student may provide other exploratory graphs.
Infinitely many zeros exist between 0 and b, for any b, however small. The exploration graphs suggest this conclusion, which is reinforced by the following reasoning: Note that for each interval $(0, b]$, however small, as x tends to zero through positive numbers, $\dfrac{1}{x}$ increases without bound, and as $\dfrac{1}{x}$ increases without bound, $\sin \dfrac{1}{x}$ will cross the x axis an unlimited number of times. The function f does not have a smallest zero, because, between 0 and b, no matter how small b is, there is always an unlimited number of zeros.

75. $I = 30 \sin 120\pi t \quad I = -10$
$-10 = 30 \sin 120\pi t$
$\sin 120\pi t = -\dfrac{10}{30}$
$120\pi t = \pi + \sin^{-1} \dfrac{10}{30}$ (third quadrant) will yield the least positive solution of the equation.
$t = \dfrac{1}{120\pi}\left(\pi + \sin^{-1} \dfrac{10}{30}\right)$
$= 0.009235$ sec (calculator in radian mode)

77. We want the least positive θ so that $I \cos^2 \theta$ is 40% of I, that is
$I \cos^2 \theta = 0.40 I \qquad 0° \le \theta \le 180°$
$\cos^2 \theta = 0.40$
$\cos \theta = \pm\sqrt{0.40}$
$\theta = \cos^{-1}\sqrt{0.40}$ will yield the least positive solution of the equation
$\theta = 50.77°$ (calculator in degree mode)

79. We are to solve $3.09 \times 10^7 = \dfrac{3.44 \times 10^7}{1 - 0.206 \cos \theta}$

For convenience, we can divide both sides of this equation by 10^7

$$3.09 = \frac{3.44}{1 - 0.206 \cos \theta}$$
$$3.09(1 - 0.206 \cos \theta) = 3.44$$
$$3.09 - (3.09)(0.206)\cos \theta = 3.44$$
$$-(3.09)(0.206)\cos \theta = 3.44 - 3.09$$
$$\cos \theta = \frac{3.44 - 3.09}{-(3.09)(0.206)}$$
$$\cos \theta = -0.5498$$

$\theta = 180° - \cos^{-1}(0.5498)$ (second quadrant) will yield the least positive solution of this equation.
or
$\theta = \cos^{-1}(-0.5498)$
$\theta = 123°$

81. Use $A = \dfrac{1}{2}R^2 (\theta - \sin \theta)$ with $R = 8$ and $A = 48$.

$$48 = \frac{1}{2} \cdot 8^2 (\theta - \sin \theta)$$
$$48 = 32(\theta - \sin \theta)$$
$$1.5 = \theta - \sin \theta$$

Examining the graph of $y_1 = 1.5$ and $y_2 = \theta - \sin \theta$, $0 \le \theta \le \pi$, we obtain

To three decimal places, $\theta = 2.267$ radians.

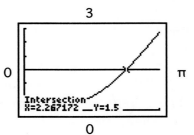

83. Using the figure in the text, we can write:

$$\sin \theta = \frac{a}{R} \qquad (1)$$
$$a^2 + (R - b)^2 = R^2 \qquad (2)$$

applying the Pythagorean theorem to the right triangle with sides R, a, and $R - b$.

$$L = R \cdot 2\theta \qquad (3)$$

Given a and b, we can solve equation (2) for R:

$$a^2 + R^2 - 2Rb + b^2 = R^2$$
$$-2Rb = -a^2 - b^2$$
$$R = \frac{a^2 + b^2}{2b}$$

We can solve equation (1) for θ to obtain $\theta = \sin^{-1} \dfrac{a}{R}$ (since θ is an acute angle) From equation (3),

$$L = R \cdot 2\theta = \frac{a^2 + b^2}{2b} \cdot 2 \sin^{-1} \frac{a}{\frac{a^2 + b^2}{2b}}$$

$$L = \frac{a^2 + b^2}{b} \sin^{-1} \frac{2ab}{a^2 + b^2} \qquad (4)$$

(A) Substituting the given values $a = 5.5$ and $b = 2.5$,

$$L = \frac{(5.5)^2 + (2.5)^2}{2.5} \sin^{-1} \frac{2(5.5)(2.5)}{(5.5)^2 + (2.5)^2}$$
$$L = 12.4575 \text{ mm.}$$

(B) Substituting $L = 12.4575$ and $a = 5.4$ mm yields

$$12.4575 = \frac{(5.4)^2 + b^2}{b} \sin^{-1} \frac{2(5.4)b}{(5.4)^2 + b^2}$$

$$12.4575 = \frac{29.16 + b^2}{b} \sin^{-1} \frac{10.8b}{29.16 + b^2} \quad (5)$$

Examining the graph of $y_1 = 12.4575$ and

$$y_2 = \frac{29.16 + x^2}{x} \sin^{-1} \frac{10.8x}{29.16 + x^2}, \quad 2 \le x \le 3,$$

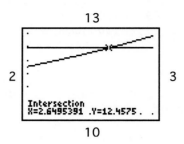

we obtain the graph at the right.

The solution is $b = 2.6496$ mm.

85. $r = 2 \sin \theta \quad 0° \le \theta \le 360°$

$r = \sin 2\theta$

We solve this system of equations by equating the right sides:

$2 \sin \theta = \sin 2\theta$

$2 \sin \theta = 2 \sin \theta \cos \theta$

$\quad\quad 0 = 2 \sin \theta \cos \theta - 2 \sin \theta$

$\quad\quad 0 = 2 \sin \theta (\cos \theta - 1)$

$2 \sin \theta = 0 \quad\quad\quad\quad\quad \cos \theta - 1 = 0$

$\quad \sin \theta = 0 \quad\quad\quad\quad\quad\quad \cos \theta = 1$

$\quad\quad \theta = 0°, 180°, 360° \quad\quad\quad \theta = 0°, 360°$

If we substitute these values of θ in either of the original equations, we obtain

$r = 2 \sin 0° \quad\quad r = 2 \sin 180° \quad\quad r = 2 \sin 360°$

$r = 0 \quad\quad\quad\quad\quad r = 0 \quad\quad\quad\quad\quad r = 0$

The solutions of the system of equations are

$(r, \theta) = (0, 0°), (r, \theta) = (0, 180°),$ and $(r, \theta) = (0, 360°)$

87. $2xy = 1$

$2(u \cos \theta - v \sin \theta)(u \sin \theta + v \cos \theta) = 1$ (substitution)

$2(u \cos \theta u \sin \theta + u \cos \theta v \cos \theta - v \sin \theta u \sin \theta - v \sin \theta v \cos \theta) = 1$

$\quad\quad\quad\quad\quad\quad\quad\quad\quad\quad\quad\quad\quad\quad\quad\quad\quad$ (multiplication)

$2u^2 \cos \theta \sin \theta + uv(2 \cos^2 \theta - 2 \sin^2 \theta) - 2v^2 \sin \theta \cos \theta = 1$

We are to find the least positive θ so that the coefficient of the uv term will be zero.

$2 \cos^2 \theta - 2 \sin^2 \theta = 0$

$\quad 2(\cos^2 \theta - \sin^2 \theta) = 0$

$\quad\quad\quad\quad 2 \cos 2\theta = 0$

$\quad\quad\quad\quad\quad \cos 2\theta = 0$

$\quad\quad\quad\quad\quad\quad\quad 2\theta = \cos^{-1} 0$ yields the least positive θ

$\quad\quad\quad\quad\quad\quad\quad 2\theta = 90°$

$\quad\quad\quad\quad\quad\quad\quad\quad \theta = 45°$

CHAPTER 6 REVIEW

1. $\tan x + \cot x = \dfrac{\sin x}{\cos x} + \dfrac{\cos x}{\sin x}$ Quotient Identities

    $= \dfrac{\sin^2 x + \cos^2 x}{\cos x \sin x}$ Algebra

    $= \dfrac{1}{\cos x \sin x}$ Pythagorean Identity

    $= \dfrac{1}{\cos x}\dfrac{1}{\sin x}$ Algebra

    $= \sec x \csc x$ Reciprocal Identity   (6-1)

2. $\sec^4 x - 2 \sec^2 x \tan^2 x + \tan^4 x = (\sec^2 x - \tan^2 x)^2$ Algebra

    $= (1 + \tan^2 x - \tan^2 x)^2$ Pythagorean Identity

    $= 1^2$ Algebra

    $= 1$ (6-1)

3. $\dfrac{1}{1 - \sin x} + \dfrac{1}{1 + \sin x} = \dfrac{1 + \sin x + 1 - \sin x}{(1 - \sin x)(1 + \sin x)}$ Algebra

    $= \dfrac{2}{1 - \sin^2 x}$ Algebra

    $= \dfrac{2}{\sin^2 x + \cos^2 x - \sin^2 x}$ Pythagorean Identity

    $= \dfrac{2}{\cos^2 x}$ Algebra

    $= 2 \sec^2 x$ Reciprocal Identity   (6-1)

4. $\cos\left(x - \dfrac{3\pi}{2}\right) = \cos x \cos \dfrac{3\pi}{2} + \sin x \sin \dfrac{3\pi}{2}$ Difference Identity

    $= \cos x(0) + \sin x(-1)$ Known Values

    $= -\sin x$ Algebra   (6-2)

5. $\sin x \cos y = \dfrac{1}{2}[\sin(x + y) + \sin(x - y)]$

    $\sin 5\alpha \cos 3\alpha = \dfrac{1}{2}[\sin(5\alpha + 3\alpha) + \sin(5\alpha - 3\alpha)]$

    $= \dfrac{1}{2}[\sin 8\alpha + \sin 2\alpha]$

    $= \dfrac{1}{2}\sin 8\alpha + \dfrac{1}{2}\sin 2\alpha$ (6-4)

6. $\cos x - \cos y = -2 \sin \dfrac{x + y}{2} \sin \dfrac{x - y}{2}$

    $\cos 7x - \cos 5x = -2 \sin \dfrac{7x + 5x}{2} \sin \dfrac{7x - 5x}{2}$

    $= -2 \sin 6x \sin x$ (6-4)

7. $\sin\left(x + \dfrac{9\pi}{2}\right) = \sin x \cos \dfrac{9\pi}{2} + \cos x \sin \dfrac{9\pi}{2}$ Sum Identity

       $= \sin x \cos \dfrac{\pi}{2} + \cos x \sin \dfrac{\pi}{2}$ Periodicity of sine and cosine

       $= \sin x(0) + \cos x(1)$ Known Values

       $= \cos x$ (6-2)

8. $\sqrt{2} \cos \theta + 1 = 0$
Solve over one period $[0°, 360°)$:

     $\cos \theta = -\dfrac{1}{\sqrt{2}}$

Sketch a graph of $y = \cos \theta$ and

$y = -\dfrac{1}{\sqrt{2}}$, θ in $[0°, 360°)$.

$\theta = 135°, 225°$

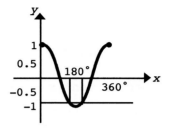

Since the cosine function is periodic with period 360°, all solutions are given by $\theta = 135° + k360°$, $\theta = 225° + k360°$, k any integer.   (6-5)

9. $\sin x \tan x - \sin x = 0$
Solve over one period, $[0, 2\pi)$ for $\sin x$, $[0, \pi)$ for $\tan x$:

$\sin x(\tan x - 1) = 0$
$\sin x = 0$ $\tan x - 1 = 0$
                  $\tan x = 1$
                  $x = \dfrac{\pi}{4}$
$x = 0, \pi$

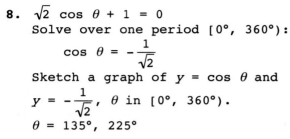

If x is allowed to range over all values, the solutions are $x = 0 + 2k\pi$, $\pi + 2k\pi$, k any integer, and $x = \dfrac{\pi}{4} + k\pi$, k any integer. These solutions can also be written as $x = k\pi$ or $x = \dfrac{\pi}{4} + k\pi$, k any integer.

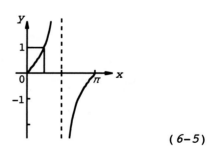

                                       (6-5)

10. $\sin x = 0.7088$ all real x
Solve over one period $[0, 2\pi)$:
Sketch a graph of $y = \sin x$ and $y = 0.7088$, x in $[0, 2\pi)$.

$x = \begin{cases} \sin^{-1} 0.7088 = 0.7878 & \text{First quadrant solution} \\ \pi - \sin^{-1} 0.7088 = 2.3538 & \text{Second quadrant solution} \end{cases}$ (calculator in radian mode)

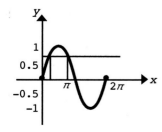

Since the sine function is periodic with period 2π, all solutions are given by

$x = \begin{cases} 0.7878 + 2k\pi \\ 2.3538 + 2k\pi \end{cases}$ k any integer

                                         (6-5)

11. $\cos \theta = 0.2557$
    Solve over one period, $[0°, 360°)$:

    Sketch a graph of $y = \cos \theta$ and $y = 0.2557$, θ in $[0°, 360°)$.

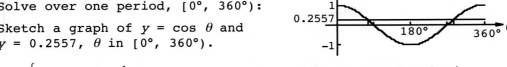

$$x = \begin{cases} \cos^{-1} 0.2557 = 75.1849° & \text{First quadrant solution} \\ 360° - \cos^{-1} 0.2557 = 284.8151° & \text{Fourth quadrant solution} \end{cases}$$
(calculator in degree mode)

Since the cosine function is periodic with period 360°, all solutions are given by

$$x = \begin{cases} 75.1849° + k360° \\ 284.8151° + k360° \end{cases} \quad k \text{ any integer}$$
(6-5)

12. $\cot x = -0.1692 \quad -\dfrac{\pi}{2} < x < \dfrac{\pi}{2}$

    Sketch a graph of $y = \cot x$ and

    $y = -0.1692$, x in $\left(-\dfrac{\pi}{2}, \dfrac{\pi}{2}\right)$

    $x = -\cot^{-1}(0.1692)$ Fourth quadrant solution,
    (calculator in radian mode)

    $\quad = -1.4032$

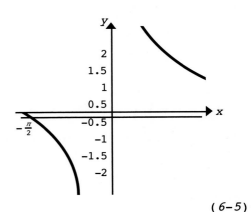

(6-5)

13. $3 \tan(11 - 3x) = 23.46 \quad -\dfrac{\pi}{2} < x < \dfrac{\pi}{2}$

    $\tan(11 - 3x) = \dfrac{23.46}{3}$

    $11 - 3x = \tan^{-1}\left(\dfrac{23.46}{3}\right)$

    $-3x = \tan^{-1}\left(\dfrac{23.46}{3}\right) - 11$

    $x = \dfrac{\tan^{-1}\left(\frac{23.46}{3}\right) - 11}{-3}$

    $x = 3.1855$ (6-5)

14. (A) Graph both sides of the equation in the same viewing window.
    $(\sin x + \cos x)^2 = 1 - 2 \sin x \cos x$ is not an identity, since the graphs do not match.
    Try $x = \dfrac{\pi}{4}$.

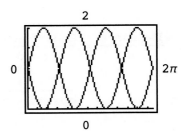

Left side: $\left(\sin \dfrac{\pi}{4} + \cos \dfrac{\pi}{4}\right)^2 = \left(\dfrac{1}{\sqrt{2}} + \dfrac{1}{\sqrt{2}}\right)^2 = \left(\dfrac{2}{\sqrt{2}}\right)^2 = 2$

Right side: $1 - 2 \sin \dfrac{\pi}{4} \cos \dfrac{\pi}{4} = 1 - 2\left(\dfrac{1}{\sqrt{2}}\right)\left(\dfrac{1}{\sqrt{2}}\right) = 0$

This verifies that the equation is not an identity.

(B) Graph both sides of the equation in the same viewing window.

$\cos^2 x - \sin^2 x = 1 - 2 \sin^2 x$ appears to be an identity, which we now verify:

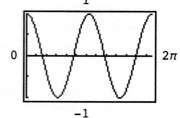

$$\cos^2 x - \sin^2 x = 1 - \sin^2 x - \sin^2 x \qquad \text{Pythagorean Identity}$$
$$= 1 - 2 \sin^2 x \qquad \text{Algebra} \qquad (6\text{-}1)$$

15. $\dfrac{1 - 2 \cos x - 3 \cos^2 x}{\sin^2 x} = \dfrac{1 - 2 \cos x - 3 \cos^2 x}{1 - \cos^2 x}$ Pythagorean Identity

$$= \frac{(1 - 3 \cos x)(1 + \cos x)}{(1 + \cos x)(1 - \cos x)} \qquad \text{Algebra}$$

$$= \frac{1 - 3 \cos x}{1 - \cos x} \qquad \text{Algebra} \qquad (6\text{-}1)$$

16. $(1 - \cos x)(\csc x + \cot x) = (1 - \cos x)\left(\dfrac{1}{\sin x} + \dfrac{\cos x}{\sin x}\right)$ Reciprocal and Quotient Identities

$$= (1 - \cos x)\frac{(1 + \cos x)}{\sin x} \qquad \text{Algebra}$$

$$= \frac{(1 - \cos x)(1 + \cos x)}{\sin x} \qquad \text{Algebra}$$

$$= \frac{1 - \cos^2 x}{\sin x} \qquad \text{Algebra}$$

$$= \frac{\sin^2 x}{\sin x} \qquad \text{Pythagorean Identity}$$

$$= \sin x \qquad \text{Algebra}$$

Key Algebraic Steps: $(1 - a)\left(\dfrac{1}{b} + \dfrac{a}{b}\right) = (1 - a)\left(\dfrac{1 + a}{b}\right)$

$$= \frac{(1 - a)(1 + a)}{b}$$

$$= \frac{1 - a^2}{b} \qquad (6\text{-}1)$$

17. $\dfrac{1 + \sin x}{\cos x} = \dfrac{(1 + \sin x)(1 - \sin x)}{\cos x(1 - \sin x)}$ Algebra

$$= \frac{1 - \sin^2 x}{\cos x(1 - \sin x)} \qquad \text{Algebra}$$

$$= \frac{\cos^2 x}{\cos x(1 - \sin x)} \qquad \text{Pythagorean Identity}$$

$$= \frac{\cos x}{1 - \sin x} \qquad \text{Algebra} \qquad (6\text{-}1)$$

18. $\dfrac{1 - \tan^2 x}{1 + \tan^2 x} = \dfrac{1 - \tan^2 x}{\sec^2 x}$ Pythagorean Identity

$$= \dfrac{1 - \dfrac{\sin^2 x}{\cos^2 x}}{\dfrac{1}{\cos^2 x}}$$ Reciprocal and Quotient Identities

$$= \dfrac{\cos^2 x \left[1 - \dfrac{\sin^2 x}{\cos^2 x}\right]}{\cos^2 x \dfrac{1}{\cos^2 x}}$$ Algebra

$$= \dfrac{\cos^2 x - \sin^2 x}{1}$$ Algebra

$$= \cos^2 x - \sin^2 x$$ Algebra

$$= \cos 2x$$ Double-angle Identity   *(6-3)*

19. $\cot \dfrac{x}{2} = \dfrac{1}{\tan \dfrac{x}{2}}$ Reciprocal Identity

$$= \dfrac{1}{\dfrac{1 - \cos x}{\sin x}}$$ Half-angle Identity

$$= \dfrac{\sin x}{1 - \cos x}$$ Algebra   *(6-3)*

20. $\cot x - \tan x = \dfrac{\cos x}{\sin x} - \dfrac{\sin x}{\cos x}$ Quotient Identities

$$= \dfrac{\cos^2 x - \sin^2 x}{\sin x \cos x}$$ Algebra

$$= \dfrac{2(\cos^2 x - \sin^2 x)}{2 \sin x \cos x}$$ Algebra

$$= \dfrac{2[\cos^2 x - (1 - \cos^2 x)]}{2 \sin x \cos x}$$ Pythagorean Identity

$$= \dfrac{2[2 \cos^2 x - 1]}{2 \sin x \cos x}$$ Algebra

$$= \dfrac{4 \cos^2 x - 2}{2 \sin x \cos x}$$ Algebra

$$= \dfrac{4 \cos^2 x - 2}{\sin 2x}$$ Double-angle Identity   *(6-3)*

21. $\left(\dfrac{1 - \cot x}{\csc x}\right)^2 = \dfrac{(1 - \cot x)^2}{\csc^2 x}$ Algebra

$$= \dfrac{1 - 2 \cot x + \cot^2 x}{\csc^2 x}$$ Algebra

$$= \dfrac{1 + \cot^2 x - 2 \cot x}{\csc^2 x}$$ Algebra

$$= \dfrac{\csc^2 x - 2 \cot x}{\csc^2 x}$$ Pythagorean Identity

$$= \dfrac{\csc^2 x}{\csc^2 x} - \dfrac{2 \cot x}{\csc^2 x}$$ Algebra

$$= 1 - \dfrac{2 \cot x}{\csc^2 x}$$ Algebra

$$= 1 - \dfrac{2 \dfrac{\cos x}{\sin x}}{\dfrac{1}{\sin^2 x}}$$ Quotient and Reciprocal Identities

$$= 1 - 2 \sin x \cos x$$ Algebra

$$= 1 - \sin 2x$$ Double-angle Identity   *(6-3)*

22. $\tan m + \tan n = \dfrac{\sin m}{\cos m} + \dfrac{\sin n}{\cos n}$ Quotient Identity

$\quad\quad\quad\quad\quad\quad = \dfrac{\sin m \cos n + \cos m \sin n}{\cos m \cos n}$ Algebra

$\quad\quad\quad\quad\quad\quad = \dfrac{\sin(m + n)}{\cos m \cos n}$ Sum Identity   (*6-2*)

23. $\tan(x + y) = \dfrac{\tan x + \tan y}{1 - \tan x \tan y}$ Sum Identity

$\quad\quad\quad\quad\quad = \dfrac{\dfrac{1}{\cot x} + \dfrac{1}{\cot y}}{1 - \dfrac{1}{\cot x}\dfrac{1}{\cot y}}$ Reciprocal Identity

$\quad\quad\quad\quad\quad = \dfrac{\cot y + \cot x}{\cot x \cot y - 1}$ Algebra

$\quad\quad\quad\quad\quad = \dfrac{\cot x + \cot y}{\cot x \cot y - 1}$ Algebra   (*6-2*)

24. $\tan 75° = \tan\left(45° + 30°\right)$

$\quad\quad\quad = \dfrac{\tan 45° + \tan 30°}{1 - \tan 45° \tan 30°}$

$\quad\quad\quad = \dfrac{1 + \dfrac{1}{\sqrt{3}}}{1 - 1 \cdot \dfrac{1}{\sqrt{3}}}$

$\quad\quad\quad = \dfrac{\dfrac{\sqrt{3} + 1}{\sqrt{3}}}{\dfrac{\sqrt{3} - 1}{\sqrt{3}}} = \dfrac{\sqrt{3} + 1}{\sqrt{3} - 1}$ (*6-2*)

25. $\cos \dfrac{\pi}{12} = \cos\left(\dfrac{\pi}{3} - \dfrac{\pi}{4}\right)$

$\quad\quad\quad = \cos \dfrac{\pi}{3} \cos \dfrac{\pi}{4} + \sin \dfrac{\pi}{3} \sin \dfrac{\pi}{4}$

$\quad\quad\quad = \dfrac{1}{2} \cdot \dfrac{\sqrt{2}}{2} + \dfrac{\sqrt{3}}{2} \cdot \dfrac{\sqrt{2}}{2}$

$\quad\quad\quad = \dfrac{\sqrt{2}}{4} + \dfrac{\sqrt{6}}{4} = \dfrac{\sqrt{2} + \sqrt{6}}{4}$ (*6-2*)

26. $\sin 105° = \sin\left(\dfrac{210°}{2}\right)$

$\quad\quad\quad = +\sqrt{\dfrac{1 - \cos 210°}{2}}$

(We choose + because 105° is in Quadrant II)

$\quad\quad\quad = \sqrt{\dfrac{1 - \left(-\dfrac{\sqrt{3}}{2}\right)}{2}}$

$\quad\quad\quad = \sqrt{\dfrac{\dfrac{2 + \sqrt{3}}{2}}{2}} = \sqrt{\dfrac{2 + \sqrt{3}}{4}} = \dfrac{\sqrt{2 + \sqrt{3}}}{2}$ (*6-3*)

27. $\cos\dfrac{7\pi}{8} = \cos\left(\dfrac{\frac{7\pi}{4}}{2}\right)$

$\qquad\qquad = -\sqrt{\dfrac{1 + \cos\frac{7\pi}{4}}{2}}$

(We choose $-$ because $\dfrac{7\pi}{8}$ is in Quadrant II)

$\qquad\qquad = -\sqrt{\dfrac{1 + \dfrac{\sqrt{2}}{2}}{2}}$

$\qquad\qquad = -\sqrt{\dfrac{\frac{2+\sqrt{2}}{2}}{2}} = -\sqrt{\dfrac{2 + \sqrt{2}}{4}} = -\dfrac{\sqrt{2 + \sqrt{2}}}{2} \qquad (6\text{--}3)$

28. $\qquad\cos x \sin y = \dfrac{1}{2}[\sin(x + y) - \sin(x - y)]$

$\cos 195° \sin 75° = \dfrac{1}{2}[\sin(195° + 75°) - \sin(195° - 75°)]$

$\qquad\qquad = \dfrac{1}{2}[\sin 270° - \sin 120°]$

$\qquad\qquad = \dfrac{1}{2}\left(-1 - \dfrac{\sqrt{3}}{2}\right)$

$\qquad\qquad = \dfrac{1}{2}\dfrac{-2 - \sqrt{3}}{2}$

$\qquad\qquad = \dfrac{-2 - \sqrt{3}}{4} \qquad\qquad (6\text{--}4)$

29. $\qquad\cos x + \cos y = 2\cos\dfrac{x + y}{2}\cos\dfrac{x - y}{2}$

$\cos 195° + \cos 105° = 2\cos\dfrac{195° + 105°}{2}\cos\dfrac{195° - 105°}{2}$

$\qquad\qquad = 2\cos 150° \cos 45°$

$\qquad\qquad = 2\left(-\dfrac{\sqrt{3}}{2}\right)\left(\dfrac{\sqrt{2}}{2}\right)$

$\qquad\qquad = -\dfrac{\sqrt{6}}{2} \qquad\qquad (6\text{--}4)$

30. $\qquad\sin x \sin y = \dfrac{1}{2}\left[\cos(x - y) - \cos(x + y)\right]$

$\sin\dfrac{11\pi}{24}\sin\dfrac{5\pi}{24} = \dfrac{1}{2}\left[\cos\left(\dfrac{11\pi}{24} - \dfrac{5\pi}{24}\right) - \cos\left(\dfrac{11\pi}{24} + \dfrac{5\pi}{24}\right)\right]$

$\qquad\qquad = \dfrac{1}{2}\left[\cos\dfrac{\pi}{4} - \cos\dfrac{2\pi}{3}\right]$

$\qquad\qquad = \dfrac{1}{2}\left[\dfrac{\sqrt{2}}{2} - \left(-\dfrac{1}{2}\right)\right] = \dfrac{1}{2}\left(\dfrac{\sqrt{2} + 1}{2}\right) = \dfrac{\sqrt{2} + 1}{4} \qquad\qquad (6\text{--}4)$

31. $\sin x - \sin y = 2 \cos \dfrac{x + y}{2} \sin \dfrac{x - y}{2}$

$$\sin \frac{5\pi}{12} - \sin \frac{\pi}{12} = 2 \cos \frac{\dfrac{5\pi}{12} + \dfrac{\pi}{12}}{2} \sin \frac{\dfrac{5\pi}{12} - \dfrac{\pi}{12}}{2}$$

$$= 2 \cos \frac{\pi}{4} \sin \frac{\pi}{6} = 2 \cdot \frac{\sqrt{2}}{2} \cdot \frac{1}{2} = \frac{\sqrt{2}}{2}$$

(6-4)

32. We suspect that this is not an identity since $\cot^2 x = \csc^2 x - 1$ is a Pythagorean Identity. (Note the difference in sign.) To verify, plug in $x = \dfrac{\pi}{2}$.

$$\cot^2 \left(\frac{\pi}{2}\right) \overset{?}{=} \csc^2 \left(\frac{\pi}{2}\right) + 1$$

$$\frac{\cos^2 \left(\frac{\pi}{2}\right)}{\sin^2 \left(\frac{\pi}{2}\right)} \overset{?}{=} \frac{1}{\sin^2 \left(\frac{\pi}{2}\right)} + 1$$

$$\frac{0}{1} \overset{?}{=} 1 + 1$$

Not true, so the equation is not an identity. (6-1)

33. Graphing $y_1 = \cos 3x$, $y_2 = \cos x(\cos 2x - 2 \sin^2 x)$, we see that the graphs match so we suspect that this is an identity.

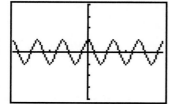

```
WINDOW
 Xmin=-6.152285…
 Xmax=6.1522856…
 Xscl=1.5707963…
 Ymin=-4
 Ymax=4
 Yscl=1
 Xres=1
```

$\cos 3x = \cos(2x + x)$	Algebra
$= \cos 2x \cos x - \sin 2x \sin x$	Sum Identity
$= \cos 2x \cos x - 2 \sin x \cos x \sin x$	Double-angle Identity for sine
$= \cos 2x \cos x - 2 \sin^2 x \cos x$	Algebra
$= \cos x(\cos 2x - 2 \sin^2 x)$	Factoring

The identity is verified. (6-2)

34.

$\sin \left(x + \dfrac{3\pi}{2}\right) = \sin x \cos \dfrac{3\pi}{2} + \cos x \sin \dfrac{3\pi}{2}$	Sum Identity
$= \sin x(0) + \cos x(-1)$	Evaluation
$= -\cos x$	Algebra

The equation is not an identity due to an incorrect sign. (6-2)

35.

$\cos \left(x - \dfrac{3\pi}{2}\right) = \cos x \cos \dfrac{3\pi}{2} + \sin x \sin \dfrac{3\pi}{2}$	Difference Identity
$= \cos x(0) + \sin x(-1)$	Evaluation
$= -\sin x$	Algebra

The equation is not an identity due to an incorrect sign. (6-2)

36. $4 \sin^2 x - 3 = 0 \qquad 0 \le x < 2\pi$

$4 \sin^2 x = 3$

$\sin^2 x = \dfrac{3}{4}$

$\sin x = \pm \dfrac{\sqrt{3}}{2}$

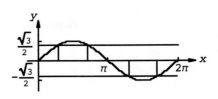

$\sin x = \dfrac{\sqrt{3}}{2} \qquad\qquad \sin x = -\dfrac{\sqrt{3}}{2}$

$x = \dfrac{\pi}{3}, \dfrac{2\pi}{3} \qquad\qquad x = \dfrac{4\pi}{3}, \dfrac{5\pi}{3}$

Solutions: $x = \dfrac{\pi}{3}, \dfrac{2\pi}{3}, \dfrac{4\pi}{3}, \dfrac{5\pi}{3}$

(6-5)

37. $2 \sin^2 \theta + \cos \theta = 1 \qquad 0° \le \theta \le 180°$

$2(1 - \cos^2 \theta) + \cos \theta = 1$

$2 - 2 \cos^2 \theta + \cos \theta = 1$

$1 - 2 \cos^2 \theta + \cos \theta = 0$

$-2 \cos^2 \theta + \cos \theta + 1 = 0$

$2 \cos^2 \theta - \cos \theta - 1 = 0$

$(2 \cos \theta + 1)(\cos \theta - 1) = 0$

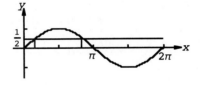

$2 \cos \theta + 1 = 0 \qquad\qquad \cos \theta - 1 = 0$

$\cos \theta = -\dfrac{1}{2} \qquad\qquad \cos \theta = 1$

$\theta = 120° \qquad\qquad\qquad \theta = 0°$ *(6-5)*

Solutions: $\theta = 0°, 120°$

38. $2 \sin^2 x - \sin x = 0$

Solve over one period $[0, 2\pi)$:

$\sin x(2 \sin x - 1) = 0$

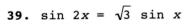

$\sin x = 0 \qquad 2 \sin x - 1 = 0$

$x = 0, \pi \qquad\quad \sin x = \dfrac{1}{2}$

$x = \dfrac{\pi}{6}, \dfrac{5\pi}{6}$

Since the sine function is periodic with period 2π, all solutions are given by

$x = 0 + 2k\pi, \; x = \pi + 2k\pi, \; x = \dfrac{\pi}{6} + 2k\pi, \; x = \dfrac{5\pi}{6} + 2k\pi, \; k$ any integer.

The first two can also be written together as $x = k\pi, \; k$ any integer. *(6-5)*

39. $\sin 2x = \sqrt{3} \sin x$

Solve over one period $[0, 2\pi)$:

$\sin 2x - \sqrt{3} \sin x = 0$

$2 \sin x \cos x - \sqrt{3} \sin x = 0$

$\sin x(2 \cos x - \sqrt{3}) = 0$

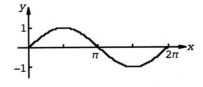

$\sin x = 0 \qquad\qquad 2 \cos x - \sqrt{3} = 0$

$x = 0, \pi \qquad\qquad \cos x = \dfrac{\sqrt{3}}{2}$

$x = \dfrac{\pi}{6}, \dfrac{11\pi}{6}$

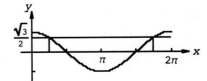

Since the sine and cosine functions are periodic with period 2π, all solutions are given by

$x = 0 + 2k\pi, \; x = \pi + 2k\pi, \; x = \dfrac{\pi}{6} + 2k\pi, \; x = \dfrac{11\pi}{6} + 2k\pi, \; k$ any integer.

The first two can also be written together as $x = k\pi, \; k$ any integer. *(6-5)*

40. $2 \sin^2 \theta + 5 \cos \theta + 1 = 0$
Solve over one period $[0°, 360°)$

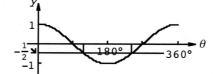

$2(1 - \cos^2 \theta) + 5 \cos \theta + 1 = 0$
$2 - 2 \cos^2 \theta + 5 \cos \theta + 1 = 0$
$-2 \cos^2 \theta + 5 \cos \theta + 3 = 0$
$2 \cos^2 \theta - 5 \cos \theta - 3 = 0$
$(2 \cos \theta + 1)(\cos \theta - 3) = 0$

$2 \cos \theta + 1 = 0$	$\cos \theta - 3 = 0$
$\cos \theta = -\dfrac{1}{2}$	$\cos \theta = 3$
$\theta = 120°, 240°$	No solution

Since the cosine function is periodic with period 360°, all solutions are given by $\theta = 120° + k360°$, $\theta = 240° + k360°$, k any integer.   *(6-5)*

41. $\tan \theta = 0.2557$ all θ
Solve over one period $[0°, 180°)$
Sketch a graph of $y = \tan \theta$ and
$y = 0.2557$,
on the interval $[0°, 180°)$
$\theta = \tan^{-1} 0.2557 = 14.34°$ Since
the tangent function is periodic
with period 180°, all solutions are
given by

$\theta = 14.34° + k180°$

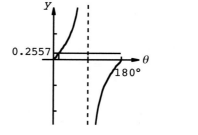

(6-5)

42. $\sin^2 x + 2 = 4 \sin x$
Solve over one period, $[0, 2\pi)$:
$\sin^2 x - 4 \sin x + 2 = 0$ Quadratic in $\sin x$

$\sin x = \dfrac{-b \pm \sqrt{b^2 - 4ac}}{2a}$ $a = 1, b = -4, c = 2$

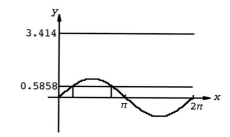

$\sin x = \dfrac{-(-4) \pm \sqrt{(-4)^2 - 4(1)(2)}}{2(1)}$

$\sin x = \dfrac{4 \pm \sqrt{8}}{2}$

$\sin x = \dfrac{4 + \sqrt{8}}{2}$	$\sin x = \dfrac{4 - \sqrt{8}}{2}$
$\sin x = 3.414$	$\sin x = 0.5858$

No solution

$x = \begin{cases} \sin^{-1} 0.5858 \\ \pi - \sin^{-1} 0.5858 \end{cases} = \begin{cases} 0.6259 \text{ First quadrant solution} \\ 2.516 \text{ Second quadrant solution} \end{cases}$

Since the sine function is periodic with period 2π, all solutions are given by

$x = \begin{cases} 0.6259 + 2k\pi \\ 2.516 + 2k\pi \end{cases}$ k any integer   *(6-5)*

43. $\tan^2 x = 2 \tan x + 1$ $0 \leq x < \pi$
$\tan^2 x - 2 \tan x - 1 = 0$ Quadratic in $\tan x$

$\tan x = \dfrac{-b \pm \sqrt{b^2 - 4ac}}{2a}$ $a = 1, b = -2, c = -1$

$\tan x = \dfrac{-(-2) \pm \sqrt{(-2)^2 - 4(1)(-1)}}{2(1)}$ $\tan x = \dfrac{2 + \sqrt{8}}{2}$ $\tan x = \dfrac{2 - \sqrt{8}}{2}$

$$\tan\ x\ =\ \frac{(2)\ \pm\ \sqrt{8}}{2}$$

$$\tan\ x =\ 2.414 \qquad \tan\ x =\ -0.4142$$

$$x\ =\ \tan^{-1}\ 2.414 \qquad x\ =\ \pi\ +\ \tan^{-1}(-0.4142)$$
$$x\ =\ 1.178 \qquad\qquad x\ =\ 2.749$$

Solutions: 1.178, 2.749

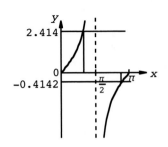

(6-5)

44. Examining the graph of $y_1 = 3 \sin 2x$ and $y_2 = 2x - 2.5$, we obtain the graph at the right.

To four decimal places, the solution in $[0,\ 2\pi]$ is given by $x = 1.4903$, and there is no other solution.

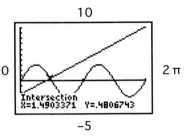

(6-5)

45. From the graph in the previous problem, the graph of $y = 3 \sin 2x$ is above the graph of $y = 2x - 2.5$, that is, $3 \sin 2x > 2x - 2.5$, for x in the interval $(-\infty,\ 1.4903)$.

(6-5)

46. Examining the graph of $y_1 = 2 \sin^2 x - \cos 2x$ and $y_2 = 1 - x^2$, $-\pi \le x \le \pi$, we obtain

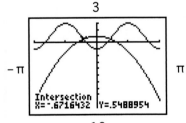 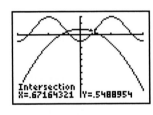

To four decimal places, the solutions in $[-\pi,\ \pi]$ are given by -0.6716 and 0.6716, and there are no other solutions.

(6-5)

47. From the graph in the previous problem, the graph of $y = 2 \sin^2 x - \cos 2x$ is below or intersects the graph of $y = 1 - x^2$, that is, $2 \sin^2 x - \cos 2x \le 1 - x^2$, for x in the interval $[-0.6716,\ 0.6716]$.

(6-5)

48. (A) Testing $x = 0$ and $y = \frac{\pi}{4}$ yields

$$\tan\left(0 + \frac{\pi}{4}\right) \overset{?}{=}\ \tan 0 + \tan \frac{\pi}{4}$$

$$\tan \frac{\pi}{4} \overset{?}{=}\ 0 + \tan \frac{\pi}{4}$$

$$1 \overset{\checkmark}{=}\ 0 + 1$$

Yes, $x = 0$ and $y = \frac{\pi}{4}$ is a solution.

(B) Consider $x = \frac{\pi}{3}$ and $y = \frac{\pi}{3}$. Then the left side $= \tan\left(\frac{\pi}{3} + \frac{\pi}{3}\right) = \tan\frac{2\pi}{3} =$ $-\sqrt{3}$. The right side $= \tan\frac{\pi}{3} + \tan\frac{\pi}{3} = \sqrt{3} + \sqrt{3} = 2\sqrt{3}$. Since the left side is not equal to the right side for at least one set of values for which both are defined, the equation is a conditional equation. (*6-1*)

49. $\sin^{-1} 0.3351$ has exactly one value, while the equation $\sin x = 0.3351$ has infinitely many solutions. (*6-5*)

50. (A) Graph both sides of the equation in the same viewing window.
$\dfrac{\tan x}{\sin x + 2\tan x} = \dfrac{1}{\cos x - 2}$ is not an identity, since the graphs do not match.

Try $x = \dfrac{\pi}{4}$.

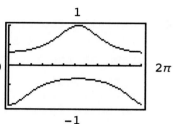

Left side: $\dfrac{\tan\frac{\pi}{4}}{\sin\frac{\pi}{4} + 2\tan\frac{\pi}{4}} = \dfrac{1}{\frac{1}{\sqrt{2}} + 2\cdot 1} = \dfrac{\sqrt{2}}{1 + 2\sqrt{2}}$

Right side: $\dfrac{1}{\cos\frac{\pi}{4} - 2} = \dfrac{1}{\frac{1}{\sqrt{2}} - 2} = \dfrac{\sqrt{2}}{1 - 2\sqrt{2}}$

This verifies that the equation is not an identity.

(B) Graph both sides of the equation in the same viewing window.

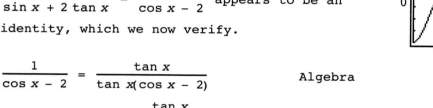

$\dfrac{\tan x}{\sin x + 2\tan x} = \dfrac{1}{\cos x - 2}$ appears to be an identity, which we now verify.

$\dfrac{1}{\cos x - 2} = \dfrac{\tan x}{\tan x(\cos x - 2)}$ Algebra

$\qquad = \dfrac{\tan x}{\tan x \cos x - 2\tan x}$ Algebra

$\qquad = \dfrac{\tan x}{\frac{\sin x}{\cos x}\cos x - 2\tan x}$ Quotient Identity

$\qquad = \dfrac{\tan x}{\sin x - 2\tan x}$ Algebra (*6-1*)

51. Let $y_1 = \cos\left(x - \dfrac{\pi}{3}\right)$

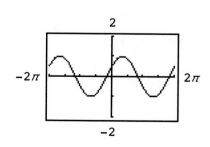

Then $y_2 = \cos x \cos\dfrac{\pi}{3} + \sin x \sin\dfrac{\pi}{3}$

$\qquad = \cos x\left(\dfrac{1}{2}\right) + \sin x\left(\dfrac{\sqrt{3}}{2}\right)$

$\qquad = \dfrac{1}{2}\cos x + \dfrac{\sqrt{3}}{2}\sin x.$

The graphs coincide as shown at the right. (*6-2*)

52. (A) $\tan \dfrac{x}{2} = 2 \sin x \qquad 0 \le x < 2\pi$

$\dfrac{1 - \cos x}{\sin x} = 2 \sin x$

$1 - \cos x = 2 \sin^2 x \quad \sin x \ne 0$

$1 - \cos x = 2(1 - \cos^2 x)$

$1 - \cos x = 2 - 2 \cos^2 x$

$2 \cos^2 x - \cos x - 1 = 0$

$(2 \cos x + 1)(\cos x - 1) = 0$

$2 \cos x + 1 = 0 \qquad\qquad \cos x - 1 = 0$

$\cos x = -\dfrac{1}{2} \qquad\qquad \cos x = 1$

$x = \dfrac{2\pi}{3}, \ \dfrac{4\pi}{3} \qquad\qquad x = 0$

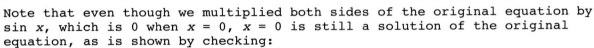

Note that even though we multiplied both sides of the original equation by sin x, which is 0 when x = 0, x = 0 is still a solution of the original equation, as is shown by checking:

Left side: $\tan \dfrac{0}{2} = \tan 0 = 0$

Right side: $2 \sin 0 = 2 \cdot 0 = 0$

(B) Examining the graphs of $y_1 = \tan \dfrac{x}{2}$ and $y_2 = 2 \sin x$, $0 \le x \le 2\pi$ (drawn in dot mode), we obtain

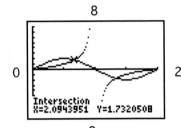

The graph confirms the solution 0 (checked in the previous part of the problem). The other intersections are found at $x = 2.0944$ and $x = 4.1888$. $\qquad$ (6-5)

53. Examining the graph of $y_1 = 3 \cos(x - 1)$ and $y_2 = 2 - x^2$, $-\pi \le x \le \pi$, we obtain

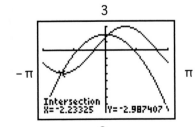
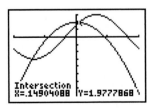

To three decimal places, the solutions on $[-\pi, \pi]$ are 0.149 and -2.233, and there are no other solutions. $\qquad$ (6-5)

54. Since $\dfrac{\pi}{2} \leq x \leq \pi$, $\dfrac{\pi}{4} \leq \dfrac{x}{2} \leq \dfrac{\pi}{2}$, so $\sin \dfrac{x}{2}$ is positive.

$\sin x = \dfrac{3}{5}$ $\cos x = -\dfrac{4}{5}$

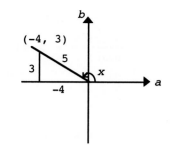

(A) $\sin \dfrac{x}{2} = \sqrt{\dfrac{1 - \cos x}{2}} = \sqrt{\dfrac{1 - \left(-\dfrac{4}{5}\right)}{2}} = \sqrt{\dfrac{\dfrac{9}{5}}{2}} = \sqrt{\dfrac{9}{10}}$

$\qquad\qquad = \dfrac{3}{\sqrt{10}}$ or $\dfrac{3\sqrt{10}}{10}$

(B) $\cos 2x = \cos^2 x - \sin^2 x = \left(-\dfrac{4}{5}\right)^2 - \left(\dfrac{3}{5}\right)^2 = \dfrac{16}{25} - \dfrac{9}{25} = \dfrac{7}{25}$ (*6–3*)

55. Let $u = \tan^{-1}\left(-\dfrac{3}{4}\right)$.

Then $-\dfrac{3}{4} = \tan u$ $-\dfrac{\pi}{2} < u < \dfrac{\pi}{2}$.

$\sin u = -\dfrac{3}{5}$ $\cos u = \dfrac{4}{5}$

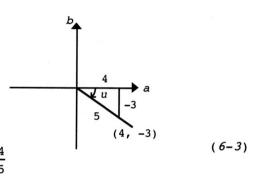

$\sin\left[2\tan^{-1}\left(-\dfrac{3}{4}\right)\right] = \sin 2u = 2 \sin u \cos u$

$\qquad\qquad\qquad = 2\left(-\dfrac{3}{5}\right)\left(\dfrac{4}{5}\right) = -\dfrac{24}{25}$ (*6–3*)

56. Let $u = \sin^{-1}\dfrac{3}{5}$

Then $\dfrac{3}{5} = \sin u$, $-\dfrac{\pi}{2} \leq u \leq \dfrac{\pi}{2}$.

Ordinarily we would also set $v = \cos^{-1}\dfrac{4}{5}$, but a glance at the reference triangle indicates that $\cos^{-1}\dfrac{4}{5} = \sin^{-1}\dfrac{3}{5} = u$

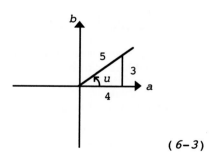

$\sin\left(\sin^{-1}\dfrac{3}{5} + \cos^{-1}\dfrac{4}{5}\right) = \sin(u + u) = \sin 2u$

$\qquad\qquad\qquad\qquad = 2 \sin u \cos u$

$\qquad\qquad\qquad\qquad = 2\left(\dfrac{3}{5}\right)\left(\dfrac{4}{5}\right) = \dfrac{24}{25}$ (*6–3*)

57. (A) $\cos^2 2x = \cos 2x + \sin^2 2x$ $0 \leq x < \pi$ is equivalent to
$\cos^2 2x = \cos 2x + \sin^2 2x$ $0 \leq 2x < 2\pi$

$\cos^2 2x = \cos 2x + 1 - \cos^2 2x$

$2\cos^2 2x - \cos 2x - 1 = 0$

$(2\cos 2x + 1)(\cos 2x - 1) = 0$

$2\cos 2x + 1 = 0$ or $\cos 2x - 1 = 0$

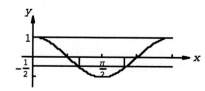

$\cos 2x = -\dfrac{1}{2}$ $\qquad\qquad$ $\cos 2x = 1$

$2x = \dfrac{2\pi}{3}, \dfrac{4\pi}{3}$ $\qquad\qquad$ $2x = 0$

$x = \dfrac{\pi}{3}, \dfrac{2\pi}{3}$ $\qquad\qquad$ $x = 0$

Solutions: $x = 0, \dfrac{\pi}{3}, \dfrac{2\pi}{3}$

(B) Examining the graph of $y_1 = \cos^2 2x$ and $y_2 = \cos 2x + \sin^2 2x$, $0 \le x < \pi$, we obtain

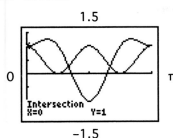

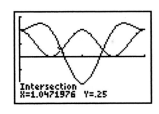

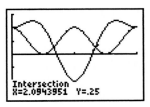

The solutions on $[0, \pi]$ are 0, and, to four decimal places, 1.0472 and 2.0944.

(6-5)

58. (A) Here is a computer-generated graph of $f(x) = \sin \dfrac{1}{x-1}$, $0 \le x \le 2$.

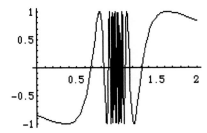

The function appears to have a smallest zero in the interval [0.6, 0.7] and a largest zero in the interval [1.3, 1.4]. Zooming in on these zeros, we obtain the following graphs.

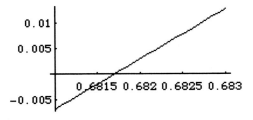

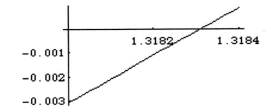

To four decimal places, the smallest zero is 0.6817 and the largest is 1.3183.

(B) As x increases without bound, $\dfrac{1}{x-1}$ tends to 0 through positive numbers and $\sin \dfrac{1}{x-1}$ tends to 0 through positive numbers. $y = 0$ is a horizontal asymptote for the graph of f.

(C) The exploratory graphs are left to the student. There are infinitely many zeros in any interval containing $x = 1$. The number $x = 1$ is not a zero because $\sin \dfrac{1}{x-1}$ is not defined at $x = 1$.

(6-5)

59. We note that $\tan \theta = \dfrac{3}{x}$ and $\tan 2\theta = \dfrac{3+6}{x} = \dfrac{9}{x}$ (see figure).

Then, $\tan 2\theta = \dfrac{2\tan\theta}{1 - \tan^2\theta}$

$$\frac{9}{x} = \frac{2\left(\frac{3}{x}\right)}{1 - \left(\frac{3}{x}\right)^2} = \frac{\frac{6}{x}}{1 - \frac{9}{x^2}} = \frac{x^2 \cdot \frac{6}{x}}{x^2 \cdot 1 - x^2 \cdot \frac{9}{x^2}} = \frac{6x}{x^2 - 9}$$

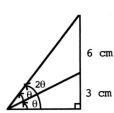

6 cm

3 cm

$$x(x^2 - 9)\frac{9}{x} = x(x^2 - 9)\frac{6x}{x^2 - 9} \qquad x \neq 0, 3, -3$$

$$9(x^2 - 9) = 6x \cdot x$$

$$9x^2 - 81 = 6x^2$$

$$3x^2 - 81 = 0$$

$$x^2 = 27$$

$$x = \sqrt{27} \quad \text{(we discard the negative solution)}$$

To 3 decimal places, $x = 5.196$ cm.

Since $\tan \theta = \dfrac{3}{x} = \dfrac{3}{\sqrt{27}} = \dfrac{1}{\sqrt{3}}$, $\theta = 30.000°$. (6-3)

60. $I = 50 \sin 120\pi(t - 0.001) \qquad I = 40$

$$40 = 50 \sin 120\pi(t - 0.001)$$

$$\sin 120\pi(t - 0.001) = \frac{40}{50}$$

$$120\pi(t - 0.001) = \sin^{-1}\frac{40}{50} \quad \text{will yield the least positive solution of the equation}$$

$$t - 0.001 = \frac{1}{120\pi} \sin^{-1}\frac{40}{50}$$

$$t = 0.001 + \frac{1}{120\pi} \sin^{-1}\frac{40}{50}$$

$$= 0.00346 \text{ sec} \quad \text{(calculator in radian mode)} \qquad (6-5)$$

61. (A) $0.6 \cos 184\pi t - 0.6 \cos 208\pi t = 0.6(\cos 184\pi t - \cos 208\pi t)$ Algebra

$$= 0.6\left[(-2)\sin\frac{184\pi t + 208\pi t}{2} \sin\frac{184\pi t + 208\pi t}{2}\right]$$ Sum-Product Identity

$$= -1.2 \sin 196\pi t \sin(-12\pi t)$$ Algebra

$$= -1.2 \sin 196\pi t(-\sin 12\pi t)$$ Identities for Negatives

$$= 1.2 \sin 12\pi t \sin 196\pi t$$ Algebra

(B) The required graphs are as shown.

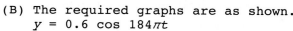

$y = 0.6 \cos 184\pi t$ $y = -0.6 \cos 208\pi t$

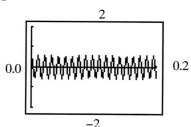
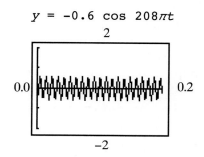

$$y = 0.6 \cos 184\pi t - 0.6 \cos 208\pi t \qquad\qquad y = 1.2 \sin 12\pi t \sin 196\pi t$$

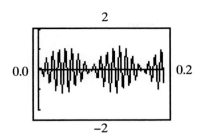

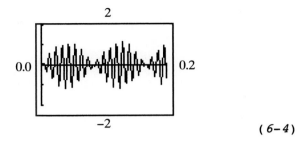

(6-4)

62.

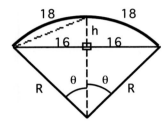

From the figure, $R\theta = 18$ and $\sin \theta = \dfrac{16}{R}$. The definition of radian measure is

$\theta = \dfrac{S}{R}$ where S is the arc length and R the radius. From these two equations,

solving each for R in terms of θ and setting the results equal to each other, we obtain

$$R = \frac{18}{\theta} \qquad R = \frac{16}{\sin \theta}$$

$$\frac{18}{\theta} = \frac{16}{\sin \theta}$$

$$18 \sin \theta = 16\theta$$

$$\sin \theta = \frac{8}{9}\theta \quad \text{as required.}$$

Examining the graph of $y_1 = \sin \theta$

and $y_2 = \dfrac{8}{9}\theta$, $0 \le \theta \le \dfrac{\pi}{2}$, we

obtain

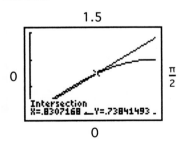

To four decimal places, $\theta = 0.8307$.

Then $R = \dfrac{18}{\theta} = \dfrac{18}{0.8307} = 21.668$ ft.

Since $\dfrac{R - h}{16} = \cot \theta$, $h = R - 16 \cot \theta$

$$= 7.057 \text{ ft.}$$

(6-5)

CHAPTER 7   Additional Topics in Trigonometry

Section 7-1

1. Answers will vary.   **3.** Answers will vary.   **5.** Answers will vary.

7. Answers will vary.

9.

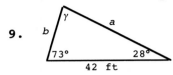

Solve for γ:

$$\alpha + \beta + \gamma = 180°$$
$$\gamma = 180° - (\alpha + \beta)$$
$$= 180° - (73° + 28°)$$
$$= 79°$$

Solve for b:

$$\frac{\sin \beta}{b} = \frac{\sin \gamma}{c}$$
$$b = \frac{c \sin \beta}{\sin \gamma}$$
$$= \frac{42 \sin 28°}{\sin 79°}$$
$$= 20 \text{ ft}$$

Solve for a:

$$\frac{\sin \alpha}{a} = \frac{\sin \gamma}{c}$$
$$a = \frac{c \sin \alpha}{\sin \gamma}$$
$$= \frac{42 \sin 73°}{\sin 79°}$$
$$= 41 \text{ ft}$$

11.

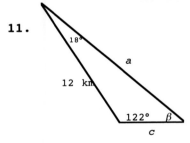

Solve for β:

$$\alpha + \beta + \gamma = 180°$$
$$\beta = 180° - (\alpha + \gamma)$$
$$= 180° - (122° + 18°)$$
$$= 40°$$

Solve for a:

$$\frac{\sin \alpha}{a} = \frac{\sin \beta}{b}$$
$$a = \frac{b \sin \alpha}{\sin \beta}$$
$$= \frac{12 \sin 122°}{\sin 40°}$$
$$= 16 \text{ km}$$

Solve for c:

$$\frac{\sin \beta}{b} = \frac{\sin \gamma}{c}$$
$$c = \frac{b \sin \gamma}{\sin \beta}$$
$$= \frac{12 \sin 18°}{\sin 40°}$$
$$= 5.8 \text{ km}$$

13.

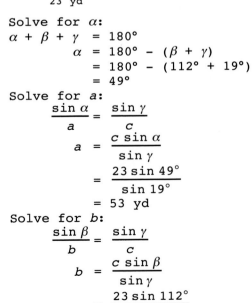

Solve for α:

$$\alpha + \beta + \gamma = 180°$$
$$\alpha = 180° - (\beta + \gamma)$$
$$= 180° - (112° + 19°)$$
$$= 49°$$

Solve for a:

$$\frac{\sin \alpha}{a} = \frac{\sin \gamma}{c}$$
$$a = \frac{c \sin \alpha}{\sin \gamma}$$
$$= \frac{23 \sin 49°}{\sin 19°}$$
$$= 53 \text{ yd}$$

Solve for b:

$$\frac{\sin \beta}{b} = \frac{\sin \gamma}{c}$$
$$b = \frac{c \sin \beta}{\sin \gamma}$$
$$= \frac{23 \sin 112°}{\sin 19°}$$
$$= 66 \text{ yd}$$

15.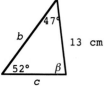

Solve for β:

$\alpha + \beta + \gamma = 180°$

$\beta = 180° - (\alpha + \gamma)$

$= 180° - (52° + 47°)$

$= 81°$

Solve for c:

$\dfrac{\sin \alpha}{a} = \dfrac{\sin \gamma}{c}$

$c = \dfrac{a \sin \gamma}{\sin \alpha}$

$= \dfrac{13 \sin 47°}{\sin 52°}$

$= 12$ cm

Solve for b:

$\dfrac{\sin \alpha}{a} = \dfrac{\sin \beta}{b}$

$b = \dfrac{a \sin \beta}{\sin \alpha}$

$= \dfrac{13 \sin 81°}{\sin 52°}$

$= 16$ cm

17.

We are given two sides and a non-included angle (SSA). α is acute.

$h = b \sin \alpha = 4 \sin 30° = 2$

$a = 2 = h$

One triangle can be constructed.

19.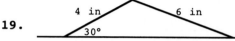

We are given two sides and a non-included angle (SSA). α is acute.

$6 = a \geq b = 4$

One triangle can be constructed.

21.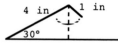

We are given two sides and a non-included angle (SSA). α is acute.

$h = b \sin \alpha = 4 \sin 30° = 2$

$0 < a = 1 < 2 = h$

No triangle can be constructed.

23.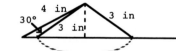

We are given two sides and a non-included angle (SSA). α is acute.

$h = b \sin \alpha = 4 \sin 30° = 2$

$h < a < b$

Two triangles can be constructed.

25.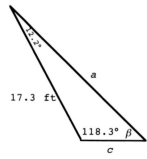

Solve for β:

$\alpha + \beta + \gamma = 180°$

$\beta = 180° - (\alpha + \gamma)$

$= 180° - (118.3° + 12.2°)$

$= 49.5°$

Solve for a:

$\dfrac{\sin \alpha}{a} = \dfrac{\sin \beta}{b}$

$a = \dfrac{b \sin \alpha}{\sin \beta}$

$= \dfrac{17.3 \sin 118.3°}{\sin 49.5°}$

$= 20.0$ ft

Solve for c:

$\dfrac{\sin \beta}{b} = \dfrac{\sin \gamma}{c}$

$c = \dfrac{b \sin \gamma}{\sin \beta}$

$= \dfrac{17.3 \sin 12.2°}{\sin 49.5°}$

$= 4.81$ ft

27.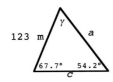

Solve for γ:
$$\alpha + \beta + \gamma = 180°$$
$$\gamma = 180° - (\alpha + \beta)$$
$$= 180° - (67.7° + 54.2°)$$
$$= 58.1°$$

Solve for c:
$$\frac{\sin \beta}{b} = \frac{\sin \gamma}{c}$$
$$c = \frac{b \sin \gamma}{\sin \beta}$$
$$= \frac{123 \sin 58.1°}{\sin 54.2°}$$
$$= 129 \text{ m}$$

Solve for a:
$$\frac{\sin \alpha}{a} = \frac{\sin \beta}{b}$$
$$a = \frac{b \sin \alpha}{\sin \beta}$$
$$= \frac{123 \sin 67.7°}{\sin 54.2°}$$
$$= 140 \text{ m}$$

29.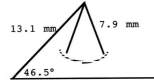

α is acute
$h = b \sin \alpha = 13.1 \sin 46.5° = 9.50$
$a = 7.9 < 9.50 = h$
No triangle is possible.

If we try to draw a triangle with these values, we will be unsuccessful.
No solution.

31.

We're given two sides and a non-included angle (SSA).
α is acute
$h = b \sin \alpha = 29.6 \sin 15.9° = 8.11$
$a = 22.4$; we have $h < a < b$, so two triangles are possible.

Solve for β:
$$\frac{\sin \alpha}{a} = \frac{\sin \beta}{b}$$
$$\sin \beta = \frac{b \sin \alpha}{a}$$
$$= \frac{29.6 \sin 15.9°}{22.4}$$
$$= 0.362$$
$\beta = \sin^{-1}(0.362) = 21.2°$
$\beta' = 180° - 21.2° = 158.8°$

$\boxed{\text{Triangle I}}$ $\beta = 21.2°$

Solve for γ:
$$\alpha + \beta + \gamma = 180°$$
$$= 180° - (\alpha + \beta)$$
$$= 180° - (15.9° + 21.2°)$$
$$= 142.9°$$

$\boxed{\text{Triangle II}}$ $\beta' = 158.8°$

Solve for γ':
$$\alpha + \beta' + \gamma' = 180°$$
$$\gamma' = 180° - (\alpha + \beta')$$
$$= 180° - (15.9° + 158.8°)$$
$$= 5.3°$$

Solve for c:
$$\frac{\sin \alpha}{a} = \frac{\sin \gamma}{c}$$
$$c \sin \alpha = a \sin \gamma$$
$$c = \frac{a \sin \gamma}{\sin \alpha}$$
$$= \frac{22.4 \sin 142.9°}{\sin 15.9°}$$
$$= 49.3 \text{ in}$$

Solve for c':
$$\frac{\sin \alpha}{a} = \frac{\sin \gamma'}{c'}$$
$$c' \sin \alpha = a \sin \gamma'$$
$$c' = \frac{a \sin \gamma'}{\sin \alpha}$$
$$= \frac{22.4 \sin 5.3°}{\sin 15.9°}$$
$$= 7.55 \text{ in}$$

33.

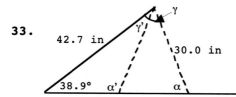

We're given two sides and a non-included angle (SSA).
β is acute.
$$h = a \sin \beta = 42.7 \sin 38.9° = 26.8$$
$h < b < a$, so there are 2 triangles.

Solve for α:
$$\frac{\sin \alpha}{a} = \frac{\sin \beta}{b}$$
$$\sin \alpha = \frac{a \sin \beta}{b}$$
$$= \frac{42.7 \sin 38.9°}{30.0}$$
$$= 0.894$$
$$\alpha = \sin^{-1}(0.894) = 63.4°$$
$$\alpha' = 180 - 63.4° = 116.6°$$

Triangle I	$\alpha = 63.4°$

Solve for γ:
$$\alpha + \beta + \gamma = 180°$$
$$\gamma = 180° - (\alpha + \beta)$$
$$= 180° - (63.4° + 38.9°)$$
$$= 77.7°$$

Solve for c:
$$\frac{\sin \beta}{b} = \frac{\sin \gamma}{c}$$
$$c \sin \beta = b \sin \gamma$$
$$c = \frac{b \sin \gamma}{\sin \beta}$$
$$= \frac{30.0 \sin 77.7°}{\sin 38.9°}$$
$$= 46.7 \text{ in}$$

Triangle II	$\alpha' = 116.6°$

Solve for γ':
$$\alpha' + \beta + \gamma' = 180°$$
$$\gamma' = 180° - (\alpha + \beta)$$
$$= 180° - (116.6° + 38.9°)$$
$$= 24.5°$$

Solve for c':
$$\frac{\sin \beta}{b} = \frac{\sin \gamma'}{c'}$$
$$c' \sin \beta = b \sin \gamma'$$
$$c' = \frac{b \sin \gamma'}{\sin \beta}$$
$$= \frac{30.0 \sin 24.5°}{\sin 38.9°}$$
$$= 19.8 \text{ in}$$

35.

α is obtuse
$$101 = a < b = 152$$
No triangle is possible.

If we try to draw a triangle with these values, we will be unsuccessful. No solution.

37. From a sketch, we see that there is only one triangle possible:

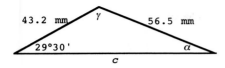

Solve for α:
$$\frac{\sin \alpha}{a} = \frac{\sin \beta}{b}$$
$$\sin \alpha = \frac{a \sin \beta}{b} = \frac{43.2 \sin 29°30'}{56.5}$$
$$\alpha = \sin^{-1}\left(\frac{43.2 \sin 29°30'}{56.5}\right)$$
$$= 22°10'$$

Solve for γ:
$$\alpha + \beta + \gamma = 180°$$
$$\gamma = 180° - (\alpha + \beta)$$
$$= 180° - (22°10' + 29°30')$$
$$= 128°20'$$

Solve for c:
$$\frac{\sin \beta}{b} = \frac{\sin \gamma}{c}$$
$$c = \frac{b \sin \gamma}{\sin \beta}$$
$$= \frac{56.5 \sin 128°20'}{\sin 29°30'}$$
$$= 89.9 \text{ mm}$$

39.

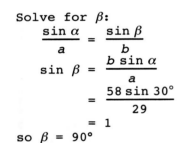

We're given two sides and a non-included angle (SSA). α is acute.
$$h = b \sin \alpha = 58 \sin 30° = 29$$
$$h = a, \text{ so there is 1 triangle.}$$

Solve for β:
$$\frac{\sin \alpha}{a} = \frac{\sin \beta}{b}$$
$$\sin \beta = \frac{b \sin \alpha}{a}$$
$$= \frac{58 \sin 30°}{29}$$
$$= 1$$
so $\beta = 90°$

Solve for c:
$$\frac{\sin \alpha}{a} = \frac{\sin \gamma}{c}$$
$$c \sin \alpha = a \sin \gamma$$
$$c = \frac{a \sin \gamma}{\sin \alpha}$$
$$= \frac{29 \sin 60°}{\sin 30°}$$
$$= 50 \text{ ft}$$

Solve for γ:
$$\alpha + \beta + \gamma = 180°$$
$$\gamma = 180° - (\alpha + \beta)$$
$$= 180° - (30° + 90°)$$
$$= 60°$$

41.

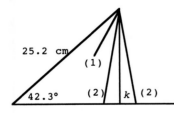

Based on the outcomes for the SSA case in section 7-1 of your textbook, we can see that k in this problem corresponds to the altitude of any possible triangle. (See the diagram.)
$$\sin 42.3° = \frac{k}{25.2 \text{ cm}}$$

$$k = (25.2 \text{ cm})\sin 42.3° = 17.0 \text{ cm}$$

If $0 < a < k$, there is no solution: (1) in diagram.
If $a = k$, there is one solution.
If $k < a < b$, there are two solutions: (2) in diagram.

43. Using the given and the calculated data, we have:
$$(a - b)\cos \frac{\gamma}{2} = c \sin \frac{\alpha - \beta}{2}$$
$$(41 - 20)\cos \frac{79°}{2} = 42 \sin \frac{73° - 28°}{2}$$
$$16.204 \approx 16.073$$

45. Sketch a figure:

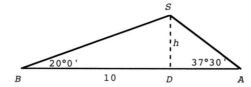

Using the law of sines, in triangle *ABS*:

$$\frac{AS}{\sin ABS} = \frac{AB}{\sin ASB}$$

We are given angle *ABS* = 20°0' and *AB* = 10. To find angle *ASB*, we use
ABS + *ASB* + *BAS* = 180°:

ASB = 180° - (*ABS* + *BAS*)
     = 180° - (20°0' + 37°30')
     = 180° - 57°30'
     = 122°30'

Next, we use the law of sines to find side *AS*:

$$\frac{AS}{\sin 20°0'} = \frac{10}{\sin 122°30'}$$

$$AS = \frac{10 \sin 20°0'}{\sin 122°30'}$$

     = 4.06 miles = distance of ship from point *A*.

To find
h = *SD* = distance of ship from shore, we note in right triangle *ADS*

$$\frac{h}{AS} = \sin DAS$$

$$h = AS \sin DAS$$

  = 4.06 sin 37°30'
  = 2.47 miles

47. First, we redraw the figure to label the other two angles in the small
triangle, as well as the height of the tree.

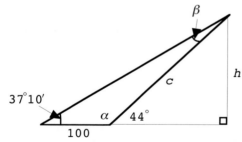

Angle α is supplementary to a 44° angle, so
$\alpha = 180° - 44° = 136°$. We can then find angle β
by subtracting from 180°:

$$\beta = 180° - \left(37°10' + 136°\right) = 6°50'$$

Finally, we can use the law of sines to find the side labeled *c*:

$$\frac{\sin 37°10'}{c} = \frac{\sin 6°50'}{100}$$

$$c = \frac{100 \sin 37°10'}{\sin 6°50'}$$

Now we can use the right triangle to find *h*:

$$\sin 44° = \frac{h}{c} = \frac{h}{\dfrac{100 \sin 37°10'}{\sin 6°50'}}$$

$$h = \frac{100 \sin 37°10'}{\sin 6°50'} \sin 44° \approx 353 \text{ feet}$$

49.

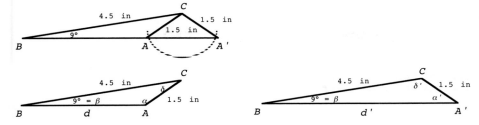

There are two possible triangles, so two possible values of the required distance. Call them d and d'.

The law of sines gives two possible values for the angle that the crankshaft makes with the center line; we denote them α and α' in the figures.

We start by calculating angle BAC, or α, and angle $BA'C$, or α', from the law of sines.

$$\frac{BC}{\sin \alpha} = \frac{AC}{\sin CBA}$$

$$\sin \alpha = \frac{BC \sin CBA}{AC}$$

$$= \frac{4.5 \sin 9°}{1.5}$$

$$= 0.4693$$

The two possibilities are

α obtuse	α acute
$\alpha = 180° - \sin^{-1} 0.4693$	$\alpha' = \sin^{-1} 0.4693$
$= 152°$	$= 28°$

The two possibilities for angle BCA, or δ, become

$\delta = 180° - (\alpha + \beta)$	$\delta' = 180° - (\alpha' + \beta)$
$= 180° - (152° + 9°)$	$= 180° - (28° + 9°)$
$= 19°$	$= 143°$

Applying the law of sines again to calculate d and d' from these two values of δ, we have

$\dfrac{d}{\sin \delta} = \dfrac{AC}{\sin \beta}$	$\dfrac{d'}{\sin \delta'} = \dfrac{AC}{\sin \beta}$
$d = \dfrac{AC \sin \delta}{\sin \beta}$	$d' = \dfrac{AC \sin \delta}{\sin \beta}$
$= \dfrac{1.5 \sin 19°}{\sin 9°}$	$= \dfrac{1.5 \sin 143°}{\sin 9°}$
$= 3.1$ in	$= 5.8$ in

51. Sketch a figure.

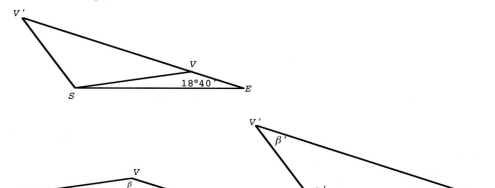

We are given $SE = 1.495 \times 10^8$, $SV = SV' = 1.085 \times 10^8$, angle $SEV = 18°40'$. There are two possible triangles, so two possible values of the required distance. Call them EV and EV'.

The law of sines gives two possible values for angle VSE; we denote them α and α' in the figures.

We start by calculating angle EVS, or β, and angle $EV'S$, or β', from the law of sines.

$$\frac{SE}{\sin \beta} = \frac{SV}{\sin SEV}$$

$$\sin \beta = \frac{SE \sin SEV}{SV}$$

$$= \frac{1.495 \times 10^8 \sin 18°40'}{1.085 \times 10^8}$$

$$= 0.4410$$

The two possibilities are

β obtuse

$\qquad \beta = 180° - \sin^{-1} 0.4410$

$\qquad\quad = 153°50'$

β acute

$\qquad \beta' = \sin^{-1} 0.4410$

$\qquad\quad = 26.2° = 26°10'$

The two possibilities for angle VSE, or α, become

$\qquad \alpha = 180° - (\beta + \gamma)$

$\qquad\quad = 180° - (153°50' + 18°40')$

$\qquad\quad = 7°30'$

$\qquad \alpha' = 180° - (\beta' + \gamma')$

$\qquad\quad = 180° - (26°10' + 18°40')$

$\qquad\quad = 135°10'$

Applying the law of sines again to calculate EV and EV' from these two values of α, we get

$$\frac{EV}{\sin \alpha} = \frac{SV}{\sin SEV}$$

$$\qquad EV = \frac{SV \sin \alpha}{\sin SEV}$$

$$\qquad\quad = \frac{1.085 \times 10^8 \sin(7°30')}{\sin(18°40')}$$

$$\qquad\quad = 4.42 \times 10^7 \text{ kilometers}$$

$$\frac{EV'}{\sin \alpha'} = \frac{SV'}{\sin SEV}$$

$$\qquad EV' = \frac{SV' \sin \alpha'}{\sin SEV}$$

$$\qquad\quad = \frac{1.085 \times 10^8 \sin(135°10')}{\sin(18°40')}$$

$$\qquad\quad = 2.39 \times 10^8 \text{ kilometers}$$

53. Redrawing the figure and labeling, we have:

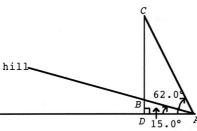

In the figure, note: Triangle ADC is a right triangle, so $\angle ACB = 90° - 62.0° = 28.0°$. Triangle ABC is not a right triangle; but, from the law of sines,

$$\frac{\sin CAB}{BC} = \frac{\sin ACB}{AB}$$

$\angle CAB = \angle CAD - \angle BAD = 62.0° - 15.0° = 47.0°$

$$BC = \frac{AB \sin CAB}{\sin ACB} = \frac{(102 \text{ ft}) \sin 47.0°}{\sin 28.0°} = 159 \text{ ft.}$$

55. First, we draw a figure similar to the one in problem 47.

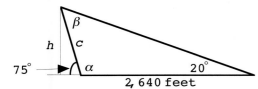

Angle α is supplementary to a 75° angle, so $\alpha = 180° - 75° = 105°$. We can then find angle β by subtracting from 180°:

$$\beta = 180° - \left(20° + 105°\right) = 55°$$

Now we can use the law of sines to find the side labeled c:

$$\frac{\sin 20°}{c} = \frac{\sin 55°}{2,640}$$

$$c = \frac{2,640 \sin 20°}{\sin 55°}$$

Now we can use the right triangle to find h:

$$\sin 75° = \frac{h}{c} = \frac{h}{\dfrac{2,640 \sin 20°}{\sin 55°}}$$

$$h = \frac{2,640 \sin 20°}{\sin 55°} \sin 75° \approx 1,100 \text{ feet to two significant digits.}$$

57. Labeling the diagram as shown, we note: triangle OAB is isosceles, so $\alpha = \beta$. Then $\alpha + \beta + \theta = \alpha + \alpha + 98.9° = 180°$

$$2\alpha = 81.1°$$
$$\alpha = 40.55°$$

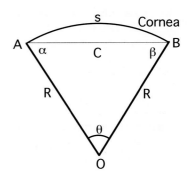

Now apply the law of sines to find R.

$$\frac{\sin \alpha}{R} = \frac{\sin \theta}{AB}$$

$$R = \frac{AB \sin \alpha}{\sin \theta} = \frac{(11.8 \text{ mm}) \sin 40.55°}{\sin 98.9°} = 7.76 \text{ mm}$$

To find s, we use the formula $s = R\theta_{\text{rad}} = R\dfrac{\pi}{180°}\theta_{\text{deg}}$:

$$s = (7.76 \text{ mm}) \frac{\pi}{180°} (98.9°) = 13.4 \text{ mm}$$

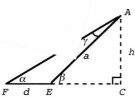

ₗng the figure and labeling, we have:

Triangle ACE is a right triangle, so $\dfrac{h}{a} = \sin\beta$ and $h = a\sin\beta$. Triangle AEF is not a right triangle, but from the law of sines,

$$\frac{\sin\gamma}{d} = \frac{\sin\alpha}{a}$$

So $\qquad a = \dfrac{d\sin\alpha}{\sin\gamma}$

We can find γ since the exterior angle of a triangle has measure equal to the sum of the two nonadjacent interior angles.

$$\alpha + \gamma = \beta$$
$$\gamma = \beta - \alpha$$

$$h = a\sin\beta$$
$$= \frac{d\sin\alpha}{\sin\gamma}\sin\beta$$
$$= d\frac{\sin\alpha\sin\beta}{\sin\gamma}$$
$$h = d\frac{\sin\alpha\sin\beta}{\sin(\beta - \alpha)}$$

Section 7-2

1. Answers will vary.   **3.** Answers will vary.   **5.** Answers will vary.

7. Angle γ is acute. A triangle can have at most one obtuse angle. Since α is acute, then, if the triangle has an obtuse angle it must be the angle opposite the longer of the two sides, b and c. Since γ is the angle opposite the shorter of the two sides (c), it must be acute.

9.

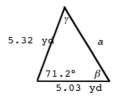

5.32 yd a

71.2° β

5.03 yd

We are given two sides and the included angle (SAS).
Solve for a:
We use the law of cosines

$$a^2 = b^2 + c^2 - 2bc\cos\alpha$$
$$= (5.32)^2 + (5.03)^2 - 2(5.32)(5.03)\cos 71.2°$$
$$= 36.355898\ldots$$
$$a = 6.03 \text{ yd.}$$

> **Common Error:**
> "$a^2 + b^2 = c^2$"
> The Pythagorean relationship
> $a^2 + b^2 = c^2$
> applies only in right triangles. It is not applicable to oblique triangles.

Solve for β:
Since a is the longest side, the angle opposite a, that is, α, must be the largest angle. So both β and γ must be less than 71.2°, and are therefore both acute.

We solve for β using the law of sines.

$$\frac{\sin\alpha}{a} = \frac{\sin\beta}{b}$$
$$\sin\beta = \frac{b\sin\alpha}{a} = \frac{5.32\sin 71.2°}{6.03}$$
$$\beta = \sin^{-1}\left(\frac{5.32\sin 71.2°}{6.03}\right)$$
$$= 56.6°$$

Solve for γ:

$$\alpha + \beta + \gamma = 180°$$
$$\gamma = 180° - (\alpha + \beta)$$
$$= 180° - (71.2° + 56.6°)$$
$$= 52.2°$$

11.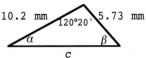

We are given two sides and the included angle (SAS).

Solve for c:
We use the law of cosines
$$c^2 = a^2 + b^2 - 2ab \cos \gamma$$
$$= (5.73)^2 + (10.2)^2 - 2(5.73)(10.2)\cos 120°20'$$
$$= 195.90685…$$
$$c = 14.0 \text{ mm}$$

Solve for β:
Note that since the sum of β and α is $180° - 120°20' = 59°40'$, each must be less than $90°$. That is, both β and α are acute. We use the law of sines.

$$\frac{\sin \beta}{b} = \frac{\sin \gamma}{c}$$
$$\sin \beta = \frac{b \sin \gamma}{c} = \frac{10.2 \sin 120°20'}{14.0}$$
$$\beta = \sin^{-1}\left(\frac{10.2 \sin 120°20'}{14.0}\right)$$
$$= 39°0'$$

Solve for α:

$$\alpha + \beta + \gamma = 180°$$
$$\alpha = 180° - (\beta + \gamma)$$
$$= 180° - (39°0' + 120°20')$$
$$= 20°40'$$

13. If the triangle has an obtuse angle, then it must be the angle opposite the longest side; in this case, β.

15.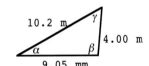

We are given three sides of the triangle (SSS). We solve for the largest angle first (the one across from the longest side), using the law of cosines.

Solve for β:
$$b^2 = a^2 + c^2 - 2ac \cos \beta$$
$$\cos \beta = \frac{a^2 + c^2 - b^2}{2ac}$$
$$\beta = \cos^{-1}\left(\frac{a^2 + c^2 - b^2}{2ac}\right)$$
$$= \cos^{-1}\left(\frac{(4.00)^2 + (9.05)^2 - (10.02)^2}{2(4.00)(9.05)}\right)$$
$$= 94.9°$$

Solve for α:
We use the law of sines. Since the sum of α and γ is $180° - 94.9° = 85.1°$, both α and γ must be acute.

$$\frac{\sin \alpha}{a} = \frac{\sin \beta}{b}$$

$$\sin \alpha = \frac{a \sin \beta}{b}$$

$$= \frac{4.00 \sin 94.9°}{10.2}$$

$$\alpha = \sin^{-1}\left(\frac{4.00 \sin 94.9°}{10.2}\right)$$

$$= 23.0°$$

Solve for γ:

$$\alpha + \beta + \gamma = 180°$$
$$\gamma = 180° - (\alpha + \beta)$$
$$= 180° - (23.0° + 94.9°)$$
$$= 62.1°$$

17. 5.30 km   6.00 km   5.52 km

We are given three sides of the triangle (SSS). We solve for the largest angle first (the one across from the longest side), using the law of cosines.

Solve for α:
$$a^2 = b^2 + c^2 - 2bc \cos \alpha$$
$$\cos \alpha = \frac{b^2 + c^2 - a^2}{2bc}$$
$$\alpha = \cos^{-1}\left(\frac{b^2 + c^2 - a^2}{2bc}\right)$$
$$= \cos^{-1}\left(\frac{(5.30)^2 + (5.52)^2 - (6.00)^2}{2(5.30)(5.52)}\right)$$
$$= 67.3°$$

Solve for β:
We use the law of sines. Both β and γ must be less than $67.3°$, so they are acute.

$$\frac{\sin \alpha}{a} = \frac{\sin \beta}{b}$$
$$\sin \beta = \frac{b \sin \alpha}{a}$$
$$= \frac{5.30 \sin 67.3°}{6.00}$$
$$\beta = \sin^{-1}\left(\frac{5.30 \sin 67.3°}{6.00}\right)$$
$$= 54.6°$$

Solve for γ:
$$\alpha + \beta + \gamma = 180°$$
$$\gamma = 180° - (\alpha + \beta)$$
$$= 180° - (67.3° + 54.6°)$$
$$= 58.1°$$

19.

88.3° 94.5°
23.7 cm

It is impossible to form a triangle with this data, since angles α and γ together add up to more than $180°$. No solution.

21.

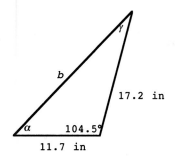

We are given two sides and the included angle (SAS). We use the law of cosines to find the third side, then the law of sines to find a second angle.

Solve for b:
$$b^2 = a^2 + c^2 - 2ac \cos \beta$$
$$= (17.2)^2 + (11.7)^2 - 2(17.2)(11.7)\cos 104.5°$$
$$= 533.50294…$$
$$b = 23.1 \text{ in}$$

Solve for α:
Note that since the sum of α and γ is $180° - 104.5° = 75.5°$, both must be acute.

$$\frac{\sin \alpha}{a} = \frac{\sin \beta}{b}$$
$$\sin \alpha = \frac{a \sin \beta}{b}$$
$$= \frac{17.2 \sin 104.5°}{23.1}$$
$$\alpha = \sin^{-1}\left(\frac{17.2 \sin 104.5°}{23.1}\right)$$
$$= 46.1°$$

Solve for γ:
$$\alpha + \beta + \gamma = 180°$$
$$\gamma = 180° - (\alpha + \beta)$$
$$= 180° - (46.1° + 104.5°)$$
$$= 29.4°$$

23.

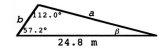

We are given two angles and a non-included side (AAS). We use the law of sines.

Solve for β:
$$\alpha + \beta + \gamma = 180°$$
$$\beta = 180° - (\alpha + \gamma)$$
$$= 180° - (57.2° + 112.0°)$$
$$= 10.8°$$

Solve for b:
$$\frac{\sin \beta}{b} = \frac{\sin \gamma}{c}$$
$$b = \frac{c \sin \beta}{\sin \gamma}$$
$$= \frac{24.8 \sin 10.8°}{\sin 112.0°}$$
$$= 5.01 \text{ m}$$

Solve for α:
$$\frac{\sin \alpha}{a} = \frac{\sin \gamma}{c}$$
$$a = \frac{c \sin \alpha}{\sin \gamma}$$
$$= \frac{24.8 \sin 57.2°}{\sin 112.0°}$$
$$= 22.5 \text{ m}$$

25. We are given two sides and a non-included angle (SSA). From a rough sketch, we see that there is only one triangle possible.

We use the law of sines.

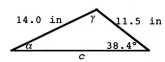

Solve for α:
$$\frac{\sin \alpha}{a} = \frac{\sin \beta}{b}$$
$$\sin \alpha = \frac{a \sin \beta}{b}$$
$$= \frac{11.5 \sin 38.4°}{14.0}$$
$$\alpha = \sin^{-1}\left(\frac{11.5 \sin 38.4°}{14.0}\right) \quad \alpha \text{ is acute}$$
$$= 30.7°$$

Solve for γ:
$$\alpha + \beta + \gamma = 180°$$
$$\gamma = 180° - (\alpha + \beta)$$
$$= 180° - (30.7° + 38.4°)$$
$$= 110.9°$$

Solve for c:
$$\frac{\sin \alpha}{a} = \frac{\sin \gamma}{c}$$
$$c = \frac{a \sin \gamma}{\sin \alpha}$$
$$= \frac{11.5 \sin 110.9°}{\sin 30.7°} = 21.0 \text{ in}$$

27.

We are given three sides (SSS). We solve for the largest angle, β (largest because it is opposite the largest side, b) using the law of cosines. We then solve for a second angle using the law of sines.

Solve for β:
$$b^2 = a^2 + c^2 - 2ac \cos \beta$$
$$\cos \beta = \frac{a^2 + c^2 - b^2}{2ac}$$
$$\beta = \cos^{-1}\left(\frac{a^2 + c^2 - b^2}{2ac}\right)$$
$$= \cos^{-1}\left(\frac{(32.9)^2 + (20.4)^2 - (42.4)^2}{2(32.9)(20.4)}\right)$$
$$= 102.9°$$

Solve for α:
We use the law of sines. Since the sum of α and γ is $180° - 102.9° = 77.1°$, both α and γ must be acute.
$$\frac{\sin \alpha}{a} = \frac{\sin \beta}{b}$$
$$\sin \alpha = \frac{a \sin \beta}{b}$$
$$= \frac{32.9 \sin 102.9°}{42.4}$$
$$\alpha = \sin^{-1}\left(\frac{32.9 \sin 102.9°}{42.4}\right)$$
$$= 49.1°$$

Solve for γ:
$$\alpha + \beta + \gamma = 180°$$
$$\gamma = 180° - (\alpha + \beta)$$
$$= 180° - (49.1° + 102.9°)$$
$$= 28.0°$$

29.

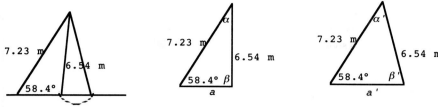

We are given two sides and a non-included angle (SSA). If we try to draw a triangle with these values, we find that two triangles are possible. We use the law of sines to find the two possible values of a second angle.

Solve for β and β':

$$\frac{\sin \beta}{b} = \frac{\sin \gamma}{c}$$

$$\sin \beta = \frac{b \sin \gamma}{c} = \frac{7.23 \sin 58.4°}{6.54} = 0.9416$$

Angle β can be either obtuse or acute:

$$\begin{array}{ll} \beta = 180° - \sin^{-1} 0.9416 & \beta' = \sin^{-1} 0.9416 \\ \quad = 180° - 70.3° & \quad = 70.3° \\ \quad = 109.7° & \end{array}$$

Solve for α and α':

$$\begin{array}{ll} \alpha = 180° - (\beta + \gamma) & \alpha' = 180° - (\beta' + \gamma) \\ \quad = 180° - (109.7° + 58.4) & \quad = 180° - (70.3° + 58.4°) \\ \quad = 11.9° & \quad = 51.3° \end{array}$$

Solve for a and a':

$$\frac{\sin \alpha}{a} = \frac{\sin \gamma}{c} \qquad\qquad \frac{\sin \alpha'}{a'} = \frac{\sin \gamma}{c}$$

$$a = \frac{c \sin \alpha}{\sin \gamma} \qquad\qquad a' = \frac{c \sin \alpha'}{\sin \gamma}$$

$$\quad = \frac{6.54 \sin 11.9°}{\sin 58.4°} \qquad\qquad \quad = \frac{6.54 \sin 51.3°}{\sin 58.4°}$$

$$\quad = 1.58 \text{ m} \qquad\qquad\qquad \quad = 5.99 \text{ m}$$

In summary:
Triangle I: $\beta = 109.7°$, $\alpha = 11.9°$, $a = 1.58$ m
Triangle II: $\beta' = 70.3°$, $\alpha' = 51.3°$, $a' = 5.99$ m

31.

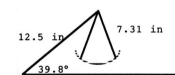

β is acute

$h = a \sin \beta = 12.5 \sin 39.8° = 8.00$

$b = 7.31 < 8.00 = h$

We are given two sides and a non-included angle (SSA). If we try to draw a triangle with these values, we will be unsuccessful. No solution.

33.

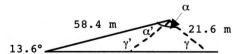

We're given two sides and a non-included angle (SSA).
β is acute

$\quad h = c \sin \beta = 58.4 \sin 13.6° = 13.7$

$\quad h < b < c$; This in combination with the diagram indicates that there
$\qquad$ are 2 triangles.

Find γ:
$$\frac{\sin \beta}{b} = \frac{\sin \gamma}{c}$$
$$\sin \gamma = \frac{c \sin \beta}{b}$$
$$= \frac{58.4 \sin 13.6°}{21.6}$$
$$= 0.63575$$
$$\gamma = \sin^{-1}(0.63575) = 39.5°$$
$$\gamma' = 180° - 39.5° = 140.5°$$

| Triangle I | $\gamma = 39.5°$ |

Find α:
$$\alpha + \beta + \gamma = 180°$$
$$\alpha = 180° - (\beta + \gamma)$$
$$= 180° - (13.6° + 39.5°)$$
$$= 126.9°$$

Find a:
$$\frac{\sin \beta}{b} = \frac{\sin \alpha}{a}$$
$$a \sin \beta = b \sin \alpha$$
$$a = \frac{b \sin \alpha}{\sin \beta}$$
$$= \frac{21.6 \sin 126.9°}{\sin 13.6°}$$
$$= 73.5 \text{ m}$$

| Triangle II | $\gamma' = 140.5°$ |

Find α':
$$\alpha' + \beta + \gamma' = 180°$$
$$\alpha' = 180° - (\beta + \gamma)$$
$$= 180° - (13.6° + 140.5°)$$
$$= 25.9°$$

Find a':
$$\frac{\sin \beta}{b} = \frac{\sin \alpha'}{a'}$$
$$a' \sin \beta = b \sin \alpha'$$
$$a' = \frac{b \sin \alpha'}{\sin \beta}$$
$$= \frac{21.6 \sin 25.9°}{\sin 13.6°} = 40.1 \text{ m}$$

35.

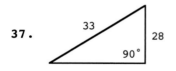

This is a right triangle, so we can simply use the formula $A = \frac{1}{2} bh$:

$$A = \frac{1}{2}(25.1)(13.4) \approx 168 \quad \text{sq. ft. to two significant digits}$$

37.

This is a right triangle, so as soon as we know the base, we can simply use the formula $A = \frac{1}{2} bh$. We can use the Pythagorean Theorem to find the base (which is side c of the triangle).

$$c^2 + 28^2 = 33^2 \quad \Rightarrow \quad c = \sqrt{33^2 - 28^2} \approx 17$$

$$A = \frac{1}{2} bh = \frac{1}{2}(17)(28) \approx 240 \text{ sq. yds. to two significant digits}$$

39.

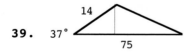

First, we find the height using right triangle ratios:

$$\sin 37° = \frac{h}{14} \quad \Rightarrow \quad h = 14 \sin 37°$$

$$A = \frac{1}{2} bh = \frac{1}{2}(75)(14 \sin 37°) = 320 \text{ sq. m. to two significant digits}$$

41. See Explore-Discuss 3 in Section 7-2 of your text for a discussion of Heron's formula. The semiperimeter s is given by
$$s = \frac{12 + 5.0 + 13}{2} = 15''$$

Using Heron's formula, we get
$$\sqrt{15(15 - 12)(15 - 5.0)(15 - 13)} = 30 \text{ sq. in.}$$

43.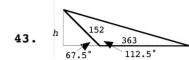

First, we note that the angle marked 67.5° is supplementary to the given 112.5° angle, which is how we calculated its measure. We can find the height using right triangle ratios:

$$\sin 67.5° = \frac{h}{152} \quad \Rightarrow \quad h = 152 \sin 67.5°$$

$$A = \frac{1}{2} bh = \frac{1}{2}(363)\left(152 \sin 67.5°\right) = 25,500 \text{ sq. ft. to two}$$

significant digits

45. See Explore-Discuss 3 in Section 7-2 of your text for a discussion of Heron's formula. The semiperimeter s is given by

$$s = \frac{6.5 + 3.9 + 4.8}{2} = 7.6 \text{ km}$$

Using Heron's formula, we get

$$\sqrt{7.6(7.6 - 6.5)(7.6 - 3.9)(7.6 - 4.8)} = 9.3 \text{ sq. km.}$$

47.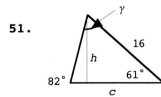

We first find angle γ by subtracting from 180°:

$$\gamma = 180° - \left(72° + 48°\right) = 60°$$

Next, we use the law of sines to find side a:

$$\frac{\sin 60°}{2.6} = \frac{\sin 72°}{a} \quad \Rightarrow \quad a = \frac{2.6 \sin 72°}{\sin 60°}$$

Now we can use right triangle ratios to find h:

$$\sin 48° = \frac{h}{\dfrac{2.6 \sin 72°}{\sin 60°}} \quad \Rightarrow \quad h = \frac{2.6 \sin 72°}{\sin 60°} \sin 48°$$

Finally, we find the area:

$$A = \frac{1}{2} bh = \frac{1}{2}(2.6)\left(\frac{2.6 \sin 72°}{\sin 60°} \sin 48°\right) = 2.8 \text{ sq. m. to two significant digits}$$

49. See Explore-Discuss 3 in Section 7-2 of your text for a discussion of Heron's formula. The semiperimeter s is given by

$$s = \frac{237 + 513 + 455}{2} = 602.5 \text{ yds.}$$

Using Heron's formula, we get

$$\sqrt{602.5(602.5 - 237)(602.5 - 513)(602.5 - 455)} = 53,900 \text{ sq. yds. to three}$$

significant digits

51.

We first find angle γ by subtracting from 180°:

$$\gamma = 180° - \left(82° + 61°\right) = 37°$$

Next, we use the law of sines to find side c:

$$\frac{\sin 82°}{16} = \frac{\sin 37°}{c} \quad \Rightarrow \quad c = \frac{16 \sin 37°}{\sin 82°}$$

Now we can use right triangle ratios to find h:

$$\sin 61° = \frac{h}{16} \quad \Rightarrow \quad h = 16 \sin 61°$$

Finally, we find the area:

$$A = \frac{1}{2}\,bh = \frac{1}{2}\left(\frac{16\sin 37^\circ}{\sin 82^\circ}\right)\!\left(16\sin 61^\circ\right) = 68 \text{ sq. in. to two significant digits}$$

53. False. The two triangles below both have perimeter 12. The first has area $A = \frac{1}{2}\,(4)(3) = 6$, while the second has area $A = \frac{1}{2}\,(4)\!\left(\sqrt{12}\right) = 2\sqrt{12} \approx 6.93$.

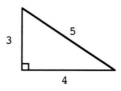 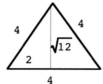

55. True. The area of a square depends on only one variable, the length of the sides. So two squares with the same area have the same side lengths, which means they'll have the same perimeter as well.

57. The law of cosines states that $c^2 = a^2 + b^2 - 2ab\cos\gamma$ for any triangle. If $\gamma = 90^\circ$, $\cos\gamma = 0$; then
$$c^2 = a^2 + b^2 - 2ac\cos 90^\circ$$
$$c^2 = a^2 + b^2 - 0$$
$$c^2 = a^2 + b^2$$

59. We write the law of cosines in its three equivalent forms:

$$
\begin{array}{rcl}
c^2 &=& a^2 + b^2 \qquad\qquad - 2ab\cos\gamma\\
a^2 &=& \qquad b^2 + c^2 \qquad\qquad - 2bc\cos\alpha\\
b^2 &=& a^2 \qquad + c^2 \qquad\qquad\qquad - 2ac\cos\beta\\
\hline
a^2 + b^2 + c^2 &=& 2a^2 + 2b^2 + 2c^2 - 2ab\cos\gamma - 2bc\cos\alpha - 2ac\cos\beta
\end{array}
$$
(Adding)

Subtracting $2a^2 + 2b^2 + 2c^2$ from both sides, we obtain:
$$-a^2 - b^2 - c^2 = -2ab\cos\gamma - 2bc\cos\alpha - 2ac\cos\beta$$

Dividing both sides by $-2abc$, we obtain
$$\frac{-a^2 - b^2 - c^2}{-2abc} = \frac{-2ab\cos\gamma}{-2abc} + \frac{-2bc\cos\alpha}{-2abc} + \frac{-2ac\cos\beta}{-2abc}$$

Removing common factors and rearranging the terms on the right, we obtain
$$\frac{a^2 + b^2 + c^2}{2abc} = \frac{\cos\alpha}{a} + \frac{\cos\beta}{b} + \frac{\cos\gamma}{c}$$
as required.

61. From the law of cosines, we have
$$
\begin{aligned}
(AB)^2 &= (AC)^2 + (BC)^2 - 2(AC)(BC)\cos ACB\\
&= 91^2 + 71^2 - 2(91)(71)\cos 96^\circ\\
&= 14{,}672.7168\ldots\\
AB &= 120 \text{ yds}
\end{aligned}
$$

63. Sketch a figure:

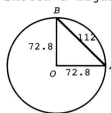

To find $\angle AOB$, we apply the law of cosines:

$$AB^2 = OA^2 + OB^2 - 2(OA)(OB)\ \cos\ AOB$$

$$\cos\ AOB = \frac{OA^2 + OB^2 - AB^2}{2(OA)(OB)} = \frac{72.8^2 + 72.8^2 - 112^2}{2(72.8)(72.8)}$$

$$\angle AOB = \cos^{-1}\left(\frac{72.8^2 + 72.8^2 - 112^2}{2(72.8)(72.8)}\right)$$

$$= 100.6°$$

65. Sketch a figure:

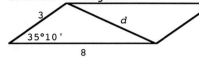

From the law of cosines, we have

$d^2 = 3^2 + 8^2 - 2(3)(8)\cos 35°10'$

$= 33.760960…$

$d = 5.81$ feet

67. Sketch a figure:

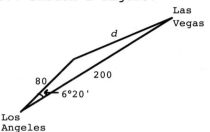

From the law of cosines, we have

$d^2 = 80^2 + 200^2 - 2(80)(200)\cos(6°20')$

$= 14595.296…$

$d = 121$ miles

69. Sketch a figure:

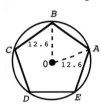

The perimeter $ABCDE$ is 5 times the length of side AB.

Angle $AOB = \dfrac{1}{5}(360°) = 72°$

We use the law of cosines in triangle AOB to determine side AB.

$(AB)^2 = (12.6)^2 + (12.6)^2 - 2(12.6)(12.6)\cos 72°$

$= 219.40092…$

$AB = 14.812$

Perimeter $= 5(AB) = 74.1$ meters

71.

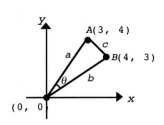

From the given information and the distance formula (Section A-9), we can determine the lengths of the three sides of the triangle.

$$d(P_1,\ P_2) = \sqrt{(x_2 - x_1)^2 + (y_2 - y_1)^2}$$

$$c = d(A,\ B) = \sqrt{(4 - 3)^2 + (3 - 4)^2} = \sqrt{1 + 1} = \sqrt{2}$$

$$b = d(O,\ B) = \sqrt{(4 - 0)^2 + (3 - 0)^2} = \sqrt{16 + 9} = 5$$

$$a = d(O,\ A) = \sqrt{(3 - 0)^2 + (4 - 0)^2} = \sqrt{9 + 16} = 5$$

From the law of cosines we can now determine θ.

$$c^2 = a^2 + b^2 - 2ab\ \cos\ \theta$$

$$\cos\ \theta = \frac{a^2 + b^2 - c^2}{2ab}$$

$$\theta = \cos^{-1}\frac{a^2 + b^2 - c^2}{2ab}$$

$$= \cos^{-1}\left(\frac{5^2 + 5^2 - (\sqrt{2})^2}{2(5)(5)}\right)$$

$$= \cos^{-1} 0.96 = 0.284\ \text{rad}$$

73. From the figure we can see that the sides of the triangle are:

$a = 2.03 + 5.00 = 7.03$
$b = 2.03 + 8.20 = 10.23$
$c = 8.20 + 5.00 = 13.20$

Solve for α:
We use the law of cosines.

$a^2 = b^2 + c^2 - 2bc \cos \alpha$

$\cos \alpha = \dfrac{b^2 + c^2 - a^2}{2bc}$

$\alpha = \cos^{-1}\left(\dfrac{b^2 + c^2 - a^2}{2bc}\right)$

$\alpha = \cos^{-1}\left(\dfrac{(10.23)^2 + (13.20)^2 - (7.03)^2}{2(10.23)(13.20)}\right)$

$= \cos^{-1} 0.8497$

$= 31°50'$

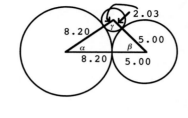

Solve for β:
We use the law of sines.

$\dfrac{\sin \alpha}{a} = \dfrac{\sin \beta}{b}$

$\sin \beta = \dfrac{b \sin \alpha}{a}$

$\beta = \sin^{-1}\left(\dfrac{b \sin \alpha}{a}\right) \qquad \beta$ is acute (see figure.)

$= \sin^{-1}\left(\dfrac{(10.23)(\sin 31°50©)}{7.03}\right)$

$= \sin^{-1} 0.7673$

$= 50°10'$

Solve for γ:

$\gamma = 180° - (\alpha + \beta)$
$= 180° - (31°50' + 50°10')$
$= 98°0'$

75. The three sides of triangle ABC are each in turn the hypotenuse of a right triangle formed with two edges of the solid. By the Pythagorean theorem,

$AB^2 = 4.3^2 + 8.1^2 = 84.10 \qquad AB = 9.17$ cm
$AC^2 = 2.8^2 + 8.1^2 = 73.45 \qquad AC = 8.57$ cm
$BC^2 = 2.8^2 + 4.3^2 = 26.33 \qquad BC = 5.13$ cm

To find $\angle CAB$, we apply the law of cosines.

$BC^2 = AC^2 + AB^2 - 2(AC)(AB)\cos CAB$

$\cos CAB = \dfrac{AC^2 + AB^2 - BC^2}{2(AC)(AB)} = \dfrac{73.45 + 84.10 - 26.33}{2(8.57)(9.17)}$

$\angle CAB = \cos^{-1}\left(\dfrac{73.45 + 84.10 - 26.33}{2(8.57)(9.17)}\right)$

$= 33°$

77. Redrawing the figure and re-labeling somewhat:

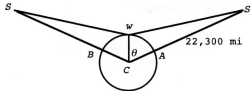

Angle $\theta = \dfrac{1}{2}$ angle $ACB = \dfrac{1}{2}(130°) = 65°$

CW = radius of earth = 3,964 mi

$CS = CA + AS = 3,964 + 22,300 = 26,264$ mi

From the law of cosines applied to triangle SWC, we have

$$SW^2 = CS^2 + CW^2 - 2(CS)(CW)\cos \theta$$
$$= (26,264)^2 + (3,964)^2 - 2(26,264)(3,964)\cos 65°$$
$$\approx 617,513,000$$
$$SW = 24,800 \text{ mi}$$

Section 7-3

1. Answers will vary.   **3.** Answers will vary.   **5.** Answers will vary.

7. The coordinates of P are given by
$$(x_p,\ y_p) = (x_b - x_a,\ y_b - y_a)$$
$$= (10 - 4,\ 11 - 6)$$
$$= (6,\ 5)$$
$$\mathbf{OP} = \langle 6, 5 \rangle$$

9. The coordinates of P are given by
$$(x_p,\ y_p) = (x_b - x_a,\ y_b - y_a)$$
$$= ((-4) - 3,\ 5 - (-9))$$
$$= (-7,\ 14)$$
$$\mathbf{OP} = \langle -7, 14 \rangle$$

11. $\mathbf{OP} = \mathbf{AB} = \langle -6, 7 \rangle$

13. The coordinates of P are given by
$$(x_p,\ y_p) = (x_b - x_a,\ y_b - y_a)$$
$$= (0 - 5,\ 0 - 8)$$
$$= (-5,\ -8)$$
$$\mathbf{OP} = \langle -5, -8 \rangle$$

15. $|\mathbf{v}| = \sqrt{(-10)^2 + 0^2} = \sqrt{100} = 10$

17. $|\mathbf{v}| = \sqrt{5^2 + (-12)^2} = \sqrt{169} = 13$

19. $|\mathbf{v}| = \sqrt{(-24)^2 + 7^2} = \sqrt{625} = 25$

21. $|\mathbf{v}| = \sqrt{1^2 + 1^2} = \sqrt{2}$

Figures for problems
23, 25, 27, 29:

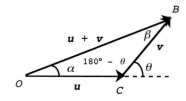 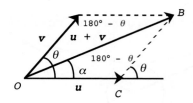

23. $\theta = 68°$, so $\angle OCB = 180° - \theta = 180° - 68° = 112°$
We can find $|\boldsymbol{u} + \boldsymbol{v}|$ using the law of cosines:

$$|\boldsymbol{u} + \boldsymbol{v}|^2 = |\boldsymbol{u}|^2 + |\boldsymbol{v}|^2 - 2|\boldsymbol{u}||\boldsymbol{v}|\cos(OCB)$$
$$= 66^2 + 22^2 - 2(66)(22)\cos 112°$$
$$= 5927.8575…$$

$$|\boldsymbol{u} + \boldsymbol{v}| = \sqrt{5927.8575…}$$
$$= 77g$$

To find α, we use the law of sines:
$$\frac{\sin\alpha}{|\boldsymbol{v}|} = \frac{\sin(OCB)}{|\boldsymbol{u} + \boldsymbol{v}|} \qquad \frac{\sin\alpha}{22} = \frac{\sin 112°}{77}$$
$$\sin\alpha = \frac{22}{77}\sin 112°$$
$$\alpha = \sin^{-1}\left(\frac{22}{77}\sin 112°\right) = 15°$$

25. $\theta = 53°$, so $\angle OCB = 180° - \theta = 180° - 53° = 127°$
We can find $|\boldsymbol{u} + \boldsymbol{v}|$ using the law of cosines:

$$|\boldsymbol{u} + \boldsymbol{v}|^2 = |\boldsymbol{u}|^2 + |\boldsymbol{v}|^2 - 2|\boldsymbol{u}||\boldsymbol{v}|\cos(OCB)$$
$$= 21^2 + 3.2^2 - 2(21)(3.2)\cos 127°$$
$$= 532.12394…$$
$$|\boldsymbol{u} + \boldsymbol{v}| = \sqrt{532.12394…}$$
$$= 23 \text{ knots}$$

To find α, we use the law of sines:

$$\frac{\sin\alpha}{|\boldsymbol{v}|} = \frac{\sin(OCB)}{|\boldsymbol{u} + \boldsymbol{v}|}$$
$$\frac{\sin\alpha}{3.2} = \frac{\sin 127°}{23}$$
$$\sin\alpha = \frac{3.2}{23}\sin 127°$$

$$\alpha = \sin^{-1}\left(\frac{3.2}{23}\sin 127°\right)$$
$$= 6°$$

27. $\theta = 79°$, so $\angle OCB = 180° - \theta = 180° - 79° = 101°$. Also,
$\angle OBC = \beta = \theta - \alpha = 79° - 25° = 54°$.

We can now find $|\boldsymbol{u}|$ and $|\boldsymbol{v}|$ from the law of sines.
$$\frac{\sin\beta}{|\boldsymbol{u}|} = \frac{\sin(OCB)}{|\boldsymbol{u} + \boldsymbol{v}|} \qquad\qquad \frac{\sin\alpha}{|\boldsymbol{v}|} = \frac{\sin(OCB)}{|\boldsymbol{u} + \boldsymbol{v}|}$$
$$|\boldsymbol{u}| = \frac{|\boldsymbol{u} + \boldsymbol{v}|\sin\beta}{\sin(OCB)} \qquad\qquad |\boldsymbol{v}| = \frac{|\boldsymbol{u} + \boldsymbol{v}|\sin\alpha}{\sin(OCB)}$$
$$= \frac{14\sin 54°}{\sin 101°} \qquad\qquad\qquad = \frac{14\sin 25°}{\sin 101°}$$
$$= 12 \text{ kg} \qquad\qquad\qquad\qquad = 6.0 \text{ kg}$$

29. $\theta = 69.4°$, so $\angle OCB = 180° - \theta = 180° - 69.4° = 110.6°$. Also, $\angle OBC = \beta = \theta - \alpha = 69.4° - 42.3° = 27.1°$.

We can now find $|\boldsymbol{u}|$ and $|\boldsymbol{v}|$ from the law of sines:

$$\frac{\sin \beta}{|\boldsymbol{u}|} = \frac{\sin(OCB)}{|\boldsymbol{u} + \boldsymbol{v}|} \qquad\qquad \frac{\sin \alpha}{|\boldsymbol{v}|} = \frac{\sin(OCB)}{|\boldsymbol{u} + \boldsymbol{v}|}$$

$$|\boldsymbol{u}| = \frac{|\boldsymbol{u} + \boldsymbol{v}|\sin \beta}{\sin(OCB)} \qquad\qquad |\boldsymbol{v}| = \frac{|\boldsymbol{u} + \boldsymbol{v}|\sin \alpha}{\sin(OCB)}$$

$$\quad = \frac{223 \sin 27.1°}{\sin 110.6°} \qquad\qquad\qquad = \frac{223 \sin 42.3°}{\sin 110.6°}$$

$$\quad = 109 \text{ mi/hr} \qquad\qquad\qquad\qquad = 160 \text{ mi/hr}$$

31. (A) $\boldsymbol{u} + \boldsymbol{v} = \langle 2, 1 \rangle + \langle -1, 3 \rangle = \langle 1, 4 \rangle$

    (B) $\boldsymbol{u} - \boldsymbol{v} = \langle 2, 1 \rangle - \langle -1, 3 \rangle = \langle 2, 1 \rangle + \langle 1, -3 \rangle = \langle 3, -2 \rangle$

    (C) $2\boldsymbol{u} - \boldsymbol{v} + 3\boldsymbol{w} = 2\langle 2, 1 \rangle - \langle -1, 3 \rangle + 3\langle 3, 0 \rangle$
$$= \langle 4, 2 \rangle + \langle 1, -3 \rangle + \langle 9, 0 \rangle$$
$$= \langle 14, -1 \rangle$$

33. (A) $\boldsymbol{u} + \boldsymbol{v} = \langle -4, -1 \rangle + \langle 2, 2 \rangle = \langle -2, 1 \rangle$

    (B) $\boldsymbol{u} - \boldsymbol{v} = \langle -4, -1 \rangle - \langle 2, 2 \rangle = \langle -4, -1 \rangle + \langle -2, -2 \rangle = \langle -6, -3 \rangle$

    (C) $2\boldsymbol{u} - \boldsymbol{v} + 3\boldsymbol{w} = 2\langle -4, -1 \rangle - \langle 2, 2 \rangle + 3\langle 0, 1 \rangle$
$$= \langle -8, -2 \rangle + \langle -2, -2 \rangle + \langle 0, 3 \rangle$$
$$= \langle -10, -1 \rangle$$

35. $\boldsymbol{v} = \langle -3, 4 \rangle = \langle -3, 0 \rangle + \langle 0, 4 \rangle$
$$= -3\langle 1, 0 \rangle + 4\langle 0, 1 \rangle$$
$$= -3\boldsymbol{i} + 4\boldsymbol{j}$$

37. $\boldsymbol{v} = \langle 3, 0 \rangle = 3\langle 1, 0 \rangle = 3\boldsymbol{i}$

39. $\boldsymbol{v} = \overrightarrow{AB} = \langle -3 - 2, 1 - 3 \rangle$
$$= \langle -5, -2 \rangle$$
$$= \langle -5, 0 \rangle + \langle 0, -2 \rangle$$
$$= -5\langle 1, 0 \rangle - 2\langle 0, 1 \rangle$$
$$= -5\boldsymbol{i} - 2\boldsymbol{j}$$

41. $\boldsymbol{u} + \boldsymbol{v} = 3\boldsymbol{i} - 2\boldsymbol{j} + 2\boldsymbol{i} + 4\boldsymbol{j} = 5\boldsymbol{i} + 2\boldsymbol{j}$

43. $2\boldsymbol{u} - 3\boldsymbol{v} = 2(3\boldsymbol{i} - 2\boldsymbol{j}) - 3(2\boldsymbol{i} + 4\boldsymbol{j})$
$$= 6\boldsymbol{i} - 4\boldsymbol{j} - 6\boldsymbol{i} - 12\boldsymbol{j}$$
$$= -16\boldsymbol{j}$$

45. $2\boldsymbol{u} - \boldsymbol{v} - 2\boldsymbol{w} = 2(3\boldsymbol{i} - 2\boldsymbol{j}) - (2\boldsymbol{i} + 4\boldsymbol{j}) - 2(2\boldsymbol{i})$
$$= 6\boldsymbol{i} - 4\boldsymbol{j} - 2\boldsymbol{i} - 4\boldsymbol{j} - 4\boldsymbol{i}$$
$$= -8\boldsymbol{j}$$

47. $|\boldsymbol{v}| = \sqrt{4^2 + 3^2} = \sqrt{25} = 5$

$$\boldsymbol{u} = \frac{1}{|\boldsymbol{v}|}\boldsymbol{v} = \frac{1}{5}\langle 4, 3 \rangle = \left\langle \frac{4}{5}, \frac{3}{5} \right\rangle$$

49. $|\boldsymbol{v}| = \sqrt{(-1)^2 + 1^2} = \sqrt{2}$

$$\boldsymbol{u} = \frac{1}{|\boldsymbol{v}|}\boldsymbol{v} = \frac{1}{\sqrt{2}}\langle -1, 1 \rangle = \left\langle -\frac{1}{\sqrt{2}}, \frac{1}{\sqrt{2}} \right\rangle$$

51. $|\boldsymbol{v}| = \sqrt{(-8)^2 + 0^2} = \sqrt{64} = 8$

$\boldsymbol{u} = \dfrac{1}{|\boldsymbol{v}|}\boldsymbol{v} = \dfrac{1}{8}\langle -8, 0\rangle = \langle -1, 0\rangle$

53. $|\boldsymbol{v}| = \sqrt{5^2 + (\sqrt{11})^2} = \sqrt{36} = 6$

$\boldsymbol{u} = \dfrac{1}{|\boldsymbol{v}|}\boldsymbol{v} = \dfrac{1}{6}(5\boldsymbol{i} + \sqrt{11}\,\boldsymbol{j}) = \dfrac{5}{6}\boldsymbol{i} + \dfrac{\sqrt{11}}{6}\boldsymbol{j} = \left\langle \dfrac{5}{6}, \dfrac{\sqrt{11}}{6}\right\rangle$

55. False. $\langle 0, 1\rangle$ and $\langle 1, 0\rangle$ both have magnitude 1, but are not equal.

57. True. If $A(x_a,\ y_a) = B(x_a,\ y_a)$, then $\langle x_a - x_a, y_a - y_a\rangle = \langle 0, 0\rangle$.

59. False. $|\boldsymbol{i}| = 1$; $|\boldsymbol{i} + \boldsymbol{i}| = |2\boldsymbol{i}| = 2$

61. False. $|\boldsymbol{0}| = 0$

63. This is almost true, but is false. The statement would be true if k is required to be positive, but if k is negative, $k\mathbf{v}$ will have the opposite direction as $\mathbf{v}$. For example, $-2\langle 1, 1\rangle = \langle -2, -2\rangle$ has the opposite direction as $\langle 1, 1\rangle$.

65. $\begin{aligned}[t]
\boldsymbol{u} + (\boldsymbol{v} + \boldsymbol{w}) &= \langle a,\ b\rangle + (\langle c,\ d\rangle + \langle e,\ f\rangle)\\
&= \langle a,\ b\rangle + \langle c + e,\ d + f\rangle && \text{Definition of vector addition}\\
&= \langle a + (c + e),\ b + (d + f)\rangle && \text{Definition of vector addition}\\
&= \langle (a + c) + e,\ (b + d) + f\rangle && \text{Associative property for addition of}\\
&&& \text{real numbers}\\
&= \langle a + c,\ b + d\rangle + \langle e,\ f\rangle && \text{Definition of vector addition}\\
&= (\langle a,\ b\rangle + \langle c,\ d\rangle) + \langle e,\ f\rangle && \text{Definition of vector addition}\\
&= (\boldsymbol{u} + \boldsymbol{v}) + \boldsymbol{w}
\end{aligned}$

67. $\begin{aligned}[t]
\boldsymbol{u} + \boldsymbol{O} &= \langle a,\ b\rangle + \langle 0,\ 0\rangle\\
&= \langle a + 0,\ b + 0\rangle && \text{Definition of vector addition}\\
&= \langle a,\ b\rangle && \text{Additive identity property for}\\
&&& \text{real numbers}\\
&= \boldsymbol{u}
\end{aligned}$

69. $\begin{aligned}[t]
(m + n)\boldsymbol{u} &= (m + n)\langle a,\ b\rangle\\
&= \langle (m + n)a,\ (m + n)b\rangle && \text{Definition of scalar multiplication}\\
&= \langle ma + na,\ mb + nb\rangle && \text{Distributive property for real numbers}\\
&= \langle ma,\ mb\rangle + \langle na,\ nb\rangle && \text{Definition of vector addition}\\
&= m\langle a,\ b\rangle + n\langle a,\ b\rangle && \text{Definition of scalar multiplication}\\
&= m\boldsymbol{u} + n\boldsymbol{u}
\end{aligned}$

71. $\begin{aligned}[t]
m(n\boldsymbol{u}) &= m(n\langle a,\ b\rangle)\\
&= m\langle na,\ nb\rangle && \text{Definition of scalar multiplication}\\
&= \langle m(na),\ m(nb)\rangle && \text{Definition of scalar multiplication}\\
&= \langle (mn)a,\ (mn)b\rangle && \text{Associative property for multiplication of real numbers}\\
&= (mn)\langle a,\ b\rangle && \text{Definition of scalar multiplication}\\
&= (mn)\boldsymbol{u}
\end{aligned}$

73. We will add the two velocity vectors by placing them tip to tail. We will leave out the resultant vector (which represents the actual velocity) in the first diagram to find an important piece of information:

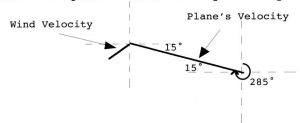

The coordinate systems have been put in at the tail of each vector to help visualize the compass headings. The angle at the tail of the plane's velocity that is marked 15° is so because the plane's compass heading is 285°, and this angle is the difference between that compass heading and a 270° angle. We can then see that the other marked angle is 15° using alternate interior angles since the two horizontal dotted lines are parallel. Now we put in the resultant vector:

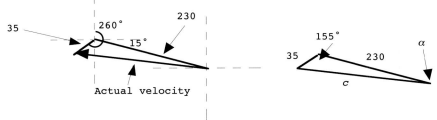

In the right hand diagram, we included only the information needed to solve the triangle and find the resultant vector. The angle at the top is 155° because it is 260° (wind's compass direction) minus 90° + 15° (as seen in the diagram). First, we solve for c, giving us the magnitude of the actual velocity vector, using the law of cosines:

$$c^2 = 35^2 + 230^2 - 2(35)(230) \cos 155°$$
$$= 68,716.5$$
$$c \approx 260 \text{ mph}$$

Next, we solve for α using the law of sines:

$$\frac{\sin \alpha}{35} = \frac{\sin 155°}{260}$$
$$\sin \alpha = \frac{35 \sin 155°}{260}$$
$$\alpha = \sin^{-1}\left(\frac{35 \sin 155°}{260}\right) \approx 3° \text{ to the nearest degree}$$

Finally, we note that the actual heading is 3° short of the plane's compass heading, so we get 285° − 3° = 282°. The plane, relative to the ground, is traveling at 260 mph at a heading of 282°.

75.

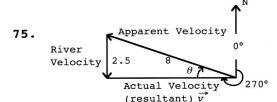

Using the Pythagorean theorem, we find the magnitude of the resultant vector to be

$$|\boldsymbol{v}| = \sqrt{8^2 - 2.5^2} = 7.6 \text{ knots}$$

To find θ, we see that

$$\sin\,\theta = \frac{2.5}{8}$$

$$\theta = \sin^{-1}\left(\frac{2.5}{8}\right) = 18°$$

Compass heading $= 270° + \theta = 270° + 18° = 288°$

Actual speed: 7.6 knots.

77. 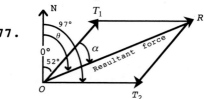 In the figure, we are given $OT_1 = 2,300$, $T_1R = OT_2 = 1,900$. $\angle T_1OT_2 = 97° - 52° = 45°$, so $\angle OT_1R = 180° - 45° = 135°$. Since we know two sides and the included angle in triangle OT_1R, we can use the law of cosines to find OR, the magnitude of the resultant force.

$$\begin{aligned}
OR^2 &= OT_1{}^2 + T_1R^2 - 2(OT_1)(T_1R)\cos(\angle OT_1R)\\
&= (2,300)^2 + (1,900)^2 - 2(2,300)(1,900)\cos 135°\\
&= 15,080,113.\ldots\\
OR &= \sqrt{15,080,113.\ldots}\\
&= 3,900 \text{ pounds}
\end{aligned}$$

We use the law of sines to find α

$$\frac{\sin\,\alpha}{T_1R} = \frac{\sin(OT_1R)}{OR}$$

$$\frac{\sin\,\alpha}{1,900} = \frac{\sin 135°}{3,900}$$

$$\sin\,\alpha = \frac{1,900}{3,900}\sin 135°$$

$$\alpha = \sin^{-1}\left(\frac{1,900}{3,900}\sin 135°\right) = 20°$$

Then the compass direction $\theta = 52° + \alpha = 72°$.

79.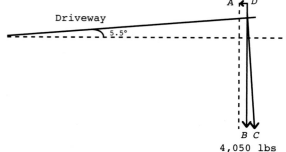

4,050 lbs

(A) To keep the car at D from rolling down the hill, we need a force with the magnitude of $\boldsymbol{DA}$ but oppositely directed. To find $|\boldsymbol{DA}|$, we observe that DAB is a right triangle with $\angle ABD = 5.5°$.

$$\sin 5.5° = \frac{|\boldsymbol{DA}|}{4,050}$$

$$\begin{aligned}
|\boldsymbol{DA}| &= 4,050 \sin 5.5°\\
&= 388 \text{ lb}
\end{aligned}$$

(B) To find $|\boldsymbol{DC}|$, the magnitude of the force perpendicular to the driveway, we note that DCB is a right triangle with $\angle BDC = 5.5°$.

$$\cos 5.5° = \frac{|\boldsymbol{DC}|}{4,050}$$

$$\begin{aligned}
|\boldsymbol{DC}| &= 4,050 \cos 5.5°\\
&= 4,030 \text{ lb}
\end{aligned}$$

81.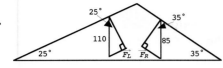

The weight of an object acts perpendicularly to the ground; the forces acting down the incline are the two marked $\overrightarrow{F_L}$ and $\overrightarrow{F_R}$. The bigger of these will decide the direction.

$$F_L = 110 \sin 25° \qquad\qquad F_R = 85 \sin 35°$$
$$= 46.5° \qquad\qquad\qquad \approx 48.8 \text{ lb}$$

Since F_R is greater than F_L, they will slide to the right.

83. First, form a force diagram with all force vectors in standard position at the origin.

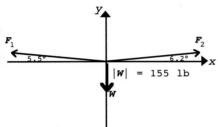

Let $\mathbf{F}_1$ = the tension in left rope
$\mathbf{F}_2$ = the tension in right rope

Write each force vector in terms of $\mathbf{i}$ and $\mathbf{j}$ unit vectors.

$\mathbf{F}_1 = |\mathbf{F}_1|(-\cos 5.5°)\mathbf{i} + |\mathbf{F}_1|(\sin 5.5°)\mathbf{j}$
$\mathbf{F}_2 = |\mathbf{F}_2|(\cos 6.2°)\mathbf{i} + |\mathbf{F}_2|(\sin 6.2°)\mathbf{j}$

$\mathbf{W} = -155\mathbf{j}$

For the system to be in static equilibrium, we must have
$\mathbf{F}_1 + \mathbf{F}_2 + \mathbf{W} = \mathbf{O}$

which becomes, on addition,
$[-|\mathbf{F}_1|(\cos 5.5°) + |\mathbf{F}_2|(\cos 6.2°)]\mathbf{i} + [|\mathbf{F}_1|(\sin 5.5°) + |\mathbf{F}_2|(\sin 6.2°) - 155]\mathbf{j}$
$$= 0\mathbf{i} + 0\mathbf{j}$$

Since two vectors are equal if and only if their corresponding components are equal, we are led to the following system of equations in $|\mathbf{F}_1|$ and $|\mathbf{F}_2|$:

$$-|\mathbf{F}_1|\cos 5.5° + |\mathbf{F}_2|\cos 6.2° = 0$$
$$|\mathbf{F}_1|\sin 5.5° + |\mathbf{F}_2|\sin 6.2° - 155 = 0$$

To solve this system of equations using the elimination method, begin by multiplying the first equation by $\dfrac{\sin 5.5°}{\cos 5.5°}$, which is also $\tan 5.5°$:

$$-|\mathbf{F}_1|\cos 5.5° \frac{\sin 5.5°}{\cos 5.5°} + |\mathbf{F}_2|\cos 6.2° \tan 5.5° = 0$$
$$|\mathbf{F}_1|\sin 5.5° + |\mathbf{F}_2|\sin 6.2° - 155 = 0$$

Now add the two equations:
$$|\mathbf{F}_2|\cos 6.2° \tan 5.5 + |\mathbf{F}_2|\sin 6.2° - 155 = 0$$

Solve this equation for $|\mathbf{F}_2|$:

$$|\mathbf{F}_2|\left(\cos 6.2° \tan 5.5° + \sin 6.2°\right) = 155$$

$$|\mathbf{F}_2| = \frac{155}{\cos 6.2° \tan 5.5° + \sin 6.2°} = 761 \text{ lb. to the right}$$

Now plug this back into the first equation and solve for $|F_1|$:

$$-|F_1| \cos 5.5° + (761) \cos 6.2° = 0$$

$$-|F_1| = \frac{-761 \cos 6.2°}{\cos 5.5°} = -760; \quad |F_1| = 760 \text{ lb. to the left}$$

85. First, form a force diagram with all force vectors in standard position at the origin.

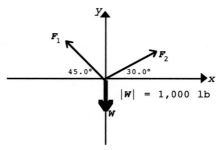

Let F_1 = the tension in left cable
F_2 = the tension in right cable

Write each force vector in terms of i and j unit vectors.

$F_1 = |F_1|(-\cos 45.0°)i + |F_1|(\sin 45.0°)j$
$F_2 = |F_2|(\cos 30.0°)i + |F_2|(\sin 30.0°)j$
$W = -1,000j$

For the system to be in static equilibrium, we must have
$F_1 + F_2 + W = O$

which becomes, on addition,
$[-|F_1|(\cos 45.0°) + |F_2|(\cos 30.0°)]i + [|F_1|(\sin 45.0°)$
$+ |F_2|(\sin 30.0°) - 1,000]j = 0i + 0j$

Since two vectors are equal if and only if their corresponding components are equal, we are led to the following system of equations in $|F_1|$ and $|F_2|$:

$$-|F_1|\cos 45.0° + |F_2|\cos 30.0° = 0$$
$$|F_1|\sin 45.0° + |F_2|\sin 30.0° - 1,000 = 0$$

To solve, use the same technique as the system in Problem 83.

$$|F_1| = \frac{1,000}{\cos 30° \tan 45.0° + \sin 30.0°} = 732 \text{ lb to the right}$$

$$|F_2| = 732 \frac{\cos 30°}{\cos 45°} = 897 \text{ lb to the left.}$$

87. In the force diagram, figure (b), we are given $|c|$ = 400 lb. In triangle *ABC*, figure (a), we note: $\cos ABC = \frac{1}{2}$, $ABC = 60°$. Then write each force in terms of i and j unit vectors.

$1a = |a|i$

$b = |b|(-\cos 60°)i + |b|\sin 60° \, j = -\frac{1}{2}|b|i + \frac{\sqrt{3}}{2}|b|j$

$c = -400j$

For the system to be in static equilibrium, we must have
$$a + b + c = O$$

which becomes, on addition

$$\left[|a| - \frac{1}{2}|b|\right]i + \left[\frac{\sqrt{3}}{2}|b| - 400\right]j = 0i + 0j$$

Since two vectors are equal if and only if their corresponding components are equal, we are led to the following system of equations in $|a|$ and $|b|$:

$$|a| - \frac{1}{2}|b| = 0$$

$$\frac{\sqrt{3}}{2}|b| - 400 = 0$$

To solve this system, solve the second equation for $|b|$ then plug that value into the first equation and solve for $|a|$.

$|b| = \frac{2}{\sqrt{3}}(400) = 462$ lb.   This corresponds to a tension force of 462 lb in member CB.

$|a| = \frac{1}{2}|b| = 231$ lb   This corresponds to a compression force of 231 lb in member AB.

89. First, form a force diagram with all force vectors in standard position at the origin.

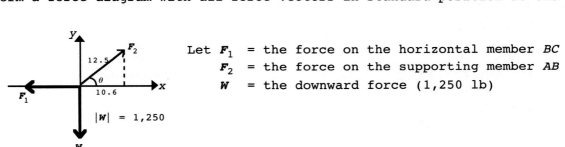

Let F_1 = the force on the horizontal member BC
F_2 = the force on the supporting member AB
W = the downward force (1,250 lb)

We note: $\cos \theta = \frac{10.6}{12.5}$ $\theta = \cos^{-1}\left(\frac{10.6}{12.5}\right) = 32.0°$

Then write each force vector in terms of i and j unit vectors.

$F_1 = -|F_1|i$
$F_2 = |F_2|(\cos 32.0°)i + |F_2|(\sin 32.0°)j$
$W = -1,250j$

For the system to be in static equilibrium, we must have
$F_1 + F_2 + W = O$

which becomes, on addition,
$[-|F_1| + |F_2|(\cos 32.0°)]i + [|F_2|(\sin 32.0°) - 1,250]j = 0i + 0j$

Since two vectors are equal if and only if their corresponding components are equal, we are led to the following system of equations in $|F_1|$ and $|F_2|$:

$-|F_1| + |F_2|(\cos 32.0°) = 0$
$|F_2|(\sin 32.0°) - 1,250 = 0$

To solve this system, solve the second equation for $|F_2|$ then plug that value into the first equation and solve for $|F_1|$.

$|F_2| = \frac{1,250}{\sin 32.0°} = 2,360$ lb

$|F_1| = |F_2|\cos 32.0° = 2,000$ lb

The force in the member *AB* is directed oppositely to the diagram— a compression of 2,360 lb. The force in the member *BC* is also directed oppositely to the diagram— a tension of 2,000 lb.

Section 7-4

1. Answers will vary.   **3.** Answers will vary.   **5.** Answers will vary.

7.

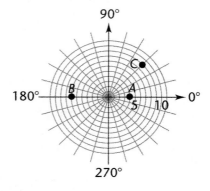

9.

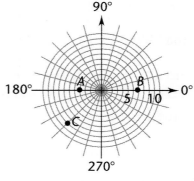

11.

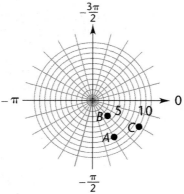

13.

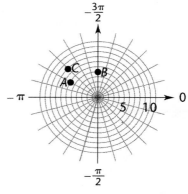

15.   See figure.

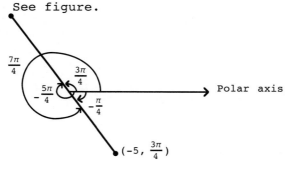

The point with coordinates $\left(-5, \dfrac{3\pi}{4}\right)$ can equally well be described as $\left(5, -\dfrac{\pi}{4}\right)$ or $\left(5, \dfrac{7\pi}{4}\right)$ or $\left(-5, -\dfrac{5\pi}{4}\right)$.

$\left(5, -\dfrac{\pi}{4}\right)$: The polar axis is rotated $\dfrac{\pi}{4}$ radians clockwise (negative direction) and the point is located 5 units from the pole along the positive polar axis;

$\left(5, \dfrac{7\pi}{4}\right)$: The polar axis is rotated $\dfrac{7\pi}{4}$ radians counterclockwise (positive direction) and the point is located 5 units from the pole along the positive polar axis;

$\left(-5, -\dfrac{5\pi}{4}\right)$: The polar axis is rotated $\dfrac{5\pi}{4}$ radians clockwise (negative direction) and the point is located 5 units from the pole along the negative polar axis.

17.

θ	0	$\frac{\pi}{6}$	$\frac{\pi}{4}$	$\frac{\pi}{3}$	$\frac{\pi}{2}$	$\frac{2\pi}{3}$	$\frac{3\pi}{4}$	$\frac{5\pi}{6}$	π
r	0	5	$5\sqrt{2}$ ≈ 7.1	$5\sqrt{3}$ ≈ 8.7	10	$5\sqrt{3}$ ≈ 8.7	$5\sqrt{2}$ ≈ 7.1	5	0

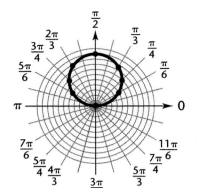

Check:

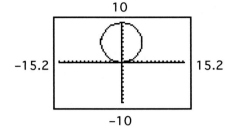

19. The graph consists of all points whose distance from the pole is 8: a circle with center at the pole, and radius 8.

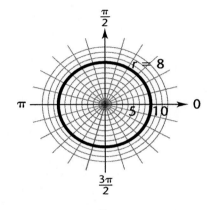

21. The graph consists of all points on a line which forms an angle of $\frac{\pi}{3}$ with the polar axis, and passes through the pole.

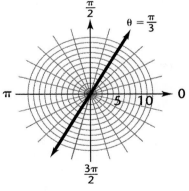

In problems 23—27, use a calculator set in radian mode.

23. $(r, \theta) = \left(6, \frac{\pi}{6}\right)$

$x = r \cos \theta = 6 \cos \frac{\pi}{6} = 5.196$

$y = r \sin \theta = 6 \sin \frac{\pi}{6} = 3.000$

$(x, y) = (5.196, 3.000)$

25. $(r, \theta) = \left(-2, \frac{7\pi}{8}\right)$

$x = r \cos \theta = -2 \cos \frac{7\pi}{8} = 1.848$

$y = r \sin \theta = -2 \sin \frac{7\pi}{8} = -0.765$

$(x, y) = (1.848, -0.765)$

27. $(r, \theta) = (-4.233, -2.084)$

$x = r \cos \theta = -4.233 \cos(-2.084) = 2.078$

$y = r \sin \theta = -4.233 \sin(-2.084) = 3.688$

$(x, y) = (2.078, 3.688)$

In problems 29—33, use a calculator set in degree mode.

29. $(x, y) = (3.5, 7.1)$

$r = \sqrt{x^2 + y^2} = \sqrt{3.5^2 + 7.1^2} = 7.9$

$\tan \theta = \dfrac{y}{x} = \dfrac{7.1}{3.5}$

θ is a first quadrant angle and is to be chosen as that $-180° < \theta \le 180°$.

$\theta = \tan^{-1}\left(\dfrac{7.1}{3.5}\right) = 64°$

$(r, \theta) = (7.9, 64°)$

31. $(x, y) = (22, -14)$

$r = \sqrt{x^2 + y^2} = \sqrt{22^2 + (-14)^2} = 26$

$\tan \theta = \dfrac{y}{x} = \dfrac{-14}{22}$

θ is a fourth quadrant angle and is to be chosen so that $-180° < \theta \le 180°$

$\theta = \tan^{-1}\left(\dfrac{-14}{22}\right) = -32°$

$(r, \theta) = (26, -32°)$

33. $(x, y) = (-7.33, -2.04)$

$r = \sqrt{x^2 + y^2} = \sqrt{(-7.33)^2 + (-2.04)^2} = 7.61$

$\tan \theta = \dfrac{y}{x} = \dfrac{-2.04}{-7.33}$

θ is a third quadrant angle and must be chosen so that $-180° < \theta \le 180°$.

$\theta = -180° + \tan^{-1}\dfrac{-2.04}{-7.33} = -164.4°$.

$(r, \theta) = (7.61, -164.4°)$

35.

θ	$\sin \theta$	$4 \sin \theta$	
0 to $\frac{\pi}{2}$	0 to 1	0 to 4	
$\frac{\pi}{2}$ to 0	1 to 0	4 to 0	
π to $\frac{3\pi}{2}$	0 to -1	0 to -4	Curve is traced out a second time in this region although coordinates seem different.
$\frac{3\pi}{2}$ to 2π	-1 to 0	-4 to 0	

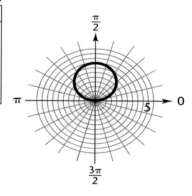

Check:

```
       5
-7.6 |-------|  7.6
       -5
```

37.

θ	2θ	$\sin 2\theta$	$10 \sin 2\theta$
0 to $\frac{\pi}{4}$	0 to $\frac{\pi}{2}$	0 to 1	0 to 10
$\frac{\pi}{4}$ to $\frac{\pi}{2}$	$\frac{\pi}{2}$ to π	1 to 0	10 to 0
$\frac{\pi}{2}$ to $\frac{3\pi}{4}$	π to $\frac{3\pi}{2}$	0 to -1	0 to -10
$\frac{3\pi}{4}$ to π	$\frac{3\pi}{2}$ to 2π	-1 to 0	-10 to 0
π to $\frac{5\pi}{4}$	2π to $\frac{5\pi}{2}$	0 to 1	0 to 10
$\frac{5\pi}{4}$ to $\frac{3\pi}{2}$	$\frac{5\pi}{2}$ to 3π	1 to 0	10 to 0
$\frac{3\pi}{2}$ to $\frac{7\pi}{4}$	3π to $\frac{7\pi}{2}$	0 to -1	0 to -10
$\frac{7\pi}{4}$ to 2π	$\frac{7\pi}{2}$ to 4π	-1 to 0	-10 to 0

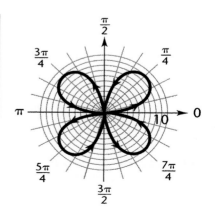

Check:

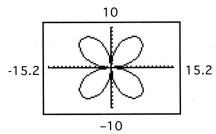

39.

θ	3θ	$\cos 3\theta$	$5 \cos 3\theta$
0 to $\frac{\pi}{6}$	0 to $\frac{\pi}{2}$	1 to 0	5 to 0
$\frac{\pi}{6}$ to $\frac{\pi}{3}$	$\frac{\pi}{2}$ to π	0 to -1	0 to -5
$\frac{\pi}{3}$ to $\frac{\pi}{2}$	π to $\frac{3\pi}{2}$	-1 to 0	-5 to 0
$\frac{\pi}{2}$ to $\frac{2\pi}{3}$	$\frac{3\pi}{2}$ to 2π	0 to 1	0 to 5
$\frac{2\pi}{3}$ to $\frac{5\pi}{6}$	2π to $\frac{5\pi}{2}$	1 to 0	5 to 0
$\frac{5\pi}{6}$ to π	$\frac{5\pi}{2}$ to 3π	0 to -1	0 to -5

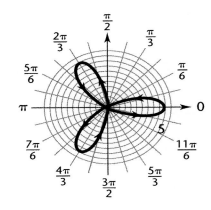

Check:

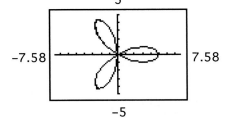

41.

θ	$\sin \theta$	$2 \sin \theta$	$2 + 2 \sin \theta$
0 to $\frac{\pi}{2}$	0 to 1	0 to 2	2 to 4
$\frac{\pi}{2}$ to π	1 to 0	2 to 0	4 to 2
π to $\frac{3\pi}{2}$	0 to -1	0 to -2	2 to 0
$\frac{3\pi}{2}$ to π	-1 to 0	-2 to 0	0 to 2

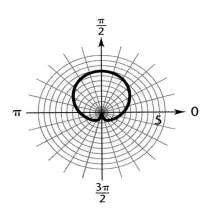

Check:

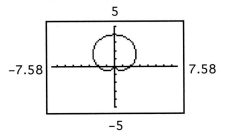

43.

θ	$\sin \theta$	$4 \sin \theta$	$2 + 4 \sin \theta$
0 to $\frac{\pi}{2}$	0 to 1	0 to 4	2 to 6
$\frac{\pi}{2}$ to π	1 to 0	4 to 0	6 to 2
π to $\frac{3\pi}{2}$	0 to -1	0 to -4	2 to -2
$\frac{3\pi}{2}$ to π	-1 to 0	-4 to 0	-2 to 2

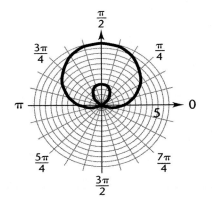

Check:

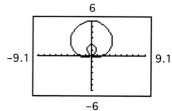

45. $r = 2 + 2 \sin \theta$ $r = 4 + 2 \sin \theta$ $r = 2 + 4 \sin \theta$

47. (A) $r = 4 \sin \theta$ $r = 4 \sin 3\theta$ $r = 4 \sin 5\theta$

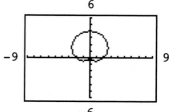

(B) and (C) Since there is one "leaf" for $r = 4 \sin \theta$, three leaves for $r = 4 \sin 3\theta$, and 5 leaves for $r = 4 \sin 5\theta$, reasonable guesses would be seven leaves for $r = 4 \sin 7\theta$ and n leaves for $r = a \sin n\theta$, n odd, $a > 0$.

49. (A) $r = 4 \sin 2\theta$ $r = 4 \sin 4\theta$ $r = 4 \sin 6\theta$

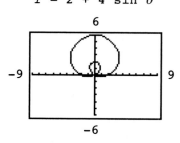

(B) and (C) Since there are four leaves for $r = 4 \sin 2\theta$, eight leaves for $r = 4 \sin 4\theta$, and twelve leaves for $r = 4 \sin 6\theta$, reasonable guesses would be 16 leaves for $r = 4 \sin 8\theta$ and $2n$ leaves for $r = a \sin n\theta$, n even, $a > 0$.

51.
$$y^2 = 5y - x^2$$
$$x^2 + y^2 = 5y$$
Use $r^2 = x^2 + y^2$
and $y = r \sin \theta$
$$r^2 = 5r \sin \theta$$
$$r^2 - 5r \sin \theta = 0$$
$$r(r - 5 \sin \theta) = 0$$
$$r = 0 \text{ or } r - 5 \sin \theta = 0$$
The graph of $r = 0$ is the pole. Since the pole is included in the graph of $r - 5 \sin \theta = 0$, we can discard $r = 0$ and keep only
$r - 5 \sin \theta = 0$ or $r = 5 \sin \theta$

53.
$$y = x$$
$$y = r \sin \theta \text{ and } x = r \cos \theta, \text{ so we get}$$
$$r \sin \theta = r \cos \theta$$
$$r \sin \theta - r \cos \theta = 0$$
$$r(\sin \theta - \cos \theta) = 0$$
$$r = 0 \text{ or } \sin \theta - \cos \theta = 0$$

The graph of $r = 0$ is the pole. Since the pole is included in the graph of $\sin \theta - \cos \theta = 0$, we can discard $r = 0$ and keep only
$$\sin \theta - \cos \theta = 0$$
$$\sin \theta = \cos \theta$$
$$\text{or } \tan \theta = 1 \quad \text{or} \quad \theta = \frac{\pi}{4}$$

55.
$$y^2 = 4x$$
$$y = r \sin \theta \text{ and } x = r \cos \theta, \text{ so we get}$$
$$(r \sin \theta)^2 = 4r \cos \theta$$
$$r^2 \sin^2 \theta = 4r \cos \theta$$
$$r^2 \sin^2 \theta - 4r \cos \theta = 0$$
$$r(r \sin^2 \theta - 4 \cos \theta) = 0$$
$$r = 0 \text{ or } r \sin^2 \theta - 4 \cos \theta = 0$$

The graph of $r = 0$ is the pole. Since the pole is included in the graph of $r \sin^2 \theta - 4 \cos \theta = 0$, we can discard $r = 0$ and keep only
$$r \sin^2 \theta - 4 \cos \theta = 0$$
$$r \sin^2 \theta = 4 \cos \theta$$
$$r = \frac{4 \cos \theta}{\sin^2 \theta}$$
$$\text{or } r = 4 \frac{\cos \theta}{\sin \theta} \frac{1}{\sin \theta}$$
$$\text{or } r = 4 \cot \theta \csc \theta$$

57. $r(3 \cos \theta - 4 \sin \theta) = -1$
$3r \cos \theta - 4r \sin \theta = -1$
$x = r \cos \theta$ and $y = r \sin \theta$, so we get $3x - 4y = -1$

59. $r = -2 \sin \theta$
We multiply both sides by r, which adds the pole to the graph. But the pole is already part of the graph, so we have changed nothing.
$r^2 = -2r \sin \theta$
But $r^2 = x^2 + y^2$ and $r \sin \theta = y$, so we get $x^2 + y^2 = -2y$.

61.
$$\theta = \frac{\pi}{4}$$
$$\tan \theta = \tan \frac{\pi}{4}$$
$$\tan \theta = 1$$
$$\text{Use } \tan \theta = \frac{y}{x}, \text{ then}$$
$$\frac{y}{x} = 1$$
$$y = x$$
(Multiplying by x adds the origin to the graph, but the origin is already part of the graph.)

63. Here are computer-generated graphs of $r = 1 + 2 \sin (n\theta)$ for $n = 1$, $n = 2$, $n = 3$, and $n = 4$.

$n = 1$

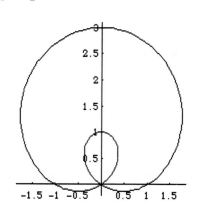

$n = 2$

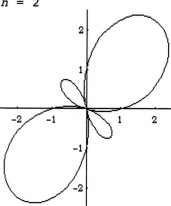

$n = 3$

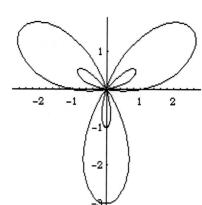

$n = 4$

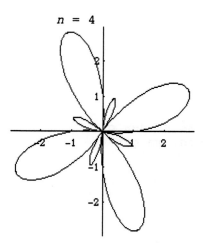

Generalizing from these, we can say that for each value of n, there are n large petals and n small petals. For n odd, the small petals are within the large petals; for n even, the small petals are between the large petals.

65. We solve the system

$r = 4 \cos \theta$

$r = -4 \sin \theta$

by equating the right sides:

$4 \cos \theta = -4 \sin \theta$

$\cos \theta = -\sin \theta$

$-1 = \tan \theta$

The only solution of this equation,

$0 \leq \theta \leq \pi$, is $\theta = \dfrac{3\pi}{4}$. If we substitute

this in either of the original equations, we get

$r = 4 \cos \dfrac{3\pi}{4} = -4 \sin \dfrac{3\pi}{4} = 4\left(\dfrac{-\sqrt{2}}{2}\right) = -2\sqrt{2}$

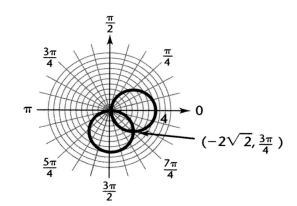

A sketch of the two graphs shows that the pole is on both graphs; however, the pole has no ordered pairs of coordinates that simultaneously satisfy both equations. As $\left(0, \dfrac{\pi}{2}\right)$ it satisfies the first; as $(0,0)$ it satisfies the second; it is not a solution of the system.

Solution: $\left(-2\sqrt{2}, \dfrac{3\pi}{4}\right)$

67. We solve the system
$r = 6 \cos \theta \qquad 0° \le \theta \le 360°$
$r = 6 \sin 2\theta$
by equating the right sides:
$6 \cos \theta = 6 \sin 2\theta$
$\quad \cos \theta = \sin 2\theta$
$\quad \cos \theta = 2 \sin \theta \cos \theta$
$\qquad 0 = 2 \sin \theta \cos \theta - \cos \theta$
$\qquad 0 = \cos \theta (2 \sin \theta - 1)$

$\cos \theta = 0 \qquad\qquad 2 \sin \theta - 1 = 0$
$\quad \theta = 90°, 270° \qquad\qquad \sin \theta = \dfrac{1}{2}$
$\qquad\qquad\qquad\qquad\qquad\qquad \theta = 30°, 150°$

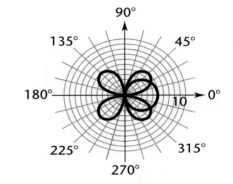

If we substitute these values of θ in either of the original equations, we get the corresponding values of r:

$\theta = 90° \quad r = 6 \cos 90° = 0$
$\theta = 270° \quad r = 6 \cos 270° = 0$
$\theta = 30° \quad r = 6 \cos 30° = 3\sqrt{3}$
$\theta = 150° \quad r = 6 \cos 150° = -3\sqrt{3}$

The four solutions of the system are $(0, 90°)$, $(0, 270°)$, $(3\sqrt{3}, 30°)$, and $(-3\sqrt{3}, 150°)$. Note that two of these $(0, 90°)$ and $(0, 270°)$ name the same point (the pole). A sketch of the two graphs shows three points of intersection.

69. $\qquad d = \sqrt{r_1^2 + r_2^2 - 2r_1 r_2 \cos(\theta_2 - \theta_1)}$

$(r_1, \theta_1) = \left(4, \dfrac{\pi}{4}\right) \quad (r_2, \theta_2) = \left(1, \dfrac{\pi}{2}\right)$

$\qquad d = \sqrt{4^2 + 1^2 - 2(4)(1) \cos\left(\dfrac{\pi}{2} - \dfrac{\pi}{4}\right)}$

$\qquad d = \sqrt{17 - 8 \cos \dfrac{\pi}{4}}$
$\qquad d = 3.368 \text{ units}$

71. 6 knots at 30°, 13 knots at 75°, 12 knots at 135°, 9 knots at 180°

73. (A) $e = 0.4$

$$r = \frac{8}{1 - 0.4\cos\theta}$$

(B) $e = 1$

$$r = \frac{8}{1 - \cos\theta}$$

(C) $e = 1.6$

$$r = \frac{8}{1 - 1.6\cos\theta}$$

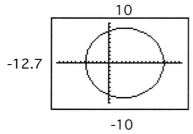

An ellipse

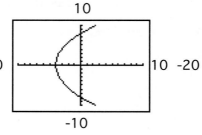

A parabola

A hyperbola

75. (A)

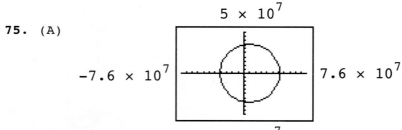

At aphelion, the distance is 4.34×10^7 mi; at perihelion it is 2.85×10^7 mi.

(B) Faster at perihelion. Since the distance from the sun to Mercury is less at perihelion than at aphelion, the planet must move faster near perihelion in order for the line joining Mercury to the sun to sweep out equal areas in equal intervals of time.

Section 7-5

1. Answers will vary.   **3.** Answers will vary.   **5.** Answers will vary.

7.
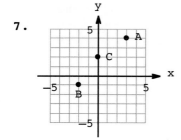

9.
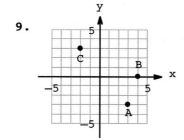

11.
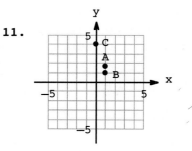

13.
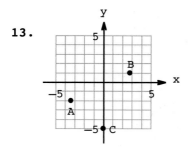

15. **(A)** A sketch shows that $\sqrt{3} + i$ is associated with a special 30°-60° triangle. By inspection, $r = 2$, $\theta = 30°$, and

$$\sqrt{3} + i = 2(\cos 30° + i \sin 30°)$$

$$= 2e^{30°i}$$

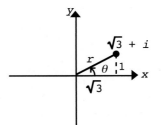

(B) A sketch shows that $-1 - i$ is associated with a special 45° triangle. By inspection, $r = \sqrt{2}$, $\theta = -135°$, and

$$-1 - i = \sqrt{2}[\cos(-135°) + i \sin(-135°)]$$

$$= \sqrt{2}\, e^{-135°i}$$

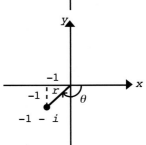

(C) A sketch shows that $5 - 6i$ is not associated with a special triangle.

$$r = \sqrt{5^2 + (-6)^2} = 7.81$$

$$\theta = \tan^{-1} \frac{-6}{5} = -50.19°$$

$$5 - 6i = 7.81[\cos(-50.19°) + i \sin(-50.19°)]$$

$$= 7.81e^{-50.19°i}$$

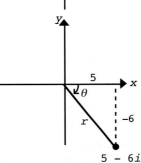

17. **(A)** A sketch shows that $-i\sqrt{3}$ is associated with the special quadrantal angle $-\frac{\pi}{2}$. By inspection,

$r = \sqrt{3}$, $\theta = -\frac{\pi}{2}$, and

$$-i\sqrt{3} = \sqrt{3}\left[\cos\left(-\frac{\pi}{2}\right) + i \sin\left(-\frac{\pi}{2}\right)\right]$$

$$= \sqrt{3}\, e^{(-\pi/2)i}$$

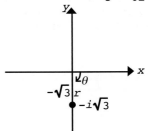

(B) A sketch shows that $-\sqrt{3} - i$ is associated with a special 30°-60° triangle. By inspection, $r = 2$, $\theta = -\frac{5\pi}{6}$, and

$$-\sqrt{3} - i = 2\left[\cos\left(-\frac{5\pi}{6}\right) + i \sin\left(-\frac{5\pi}{6}\right)\right]$$

$$= 2e^{(-5\pi/6)i}$$

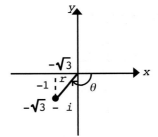

(C) A sketch shows that $-8 + 5i$ is not associated with a special triangle.

$$r = \sqrt{(-8)^2 + 5^2} = 9.43$$

$$\theta = \pi + \tan^{-1} \frac{5}{(-8)} = 2.58$$

$$-8 + 5i = 9.43(\cos 2.58 + i \sin 2.58)$$

$$= 9.43e^{2.58i}$$

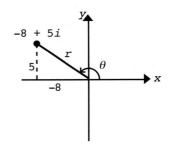

19. (A) $x + iy = 2e^{(\pi/3)i}$

$$= 2\left(\cos\frac{\pi}{3} + i\sin\frac{\pi}{3}\right)$$

$$= 2\left(\frac{1}{2} + i\frac{\sqrt{3}}{2}\right)$$

$$= 1 + i\sqrt{3}$$

(B) $x + iy = \sqrt{2}e^{(-45°)i}$

$$= \sqrt{2}[\cos(-45°) + i\sin(-45°)]$$

$$= \sqrt{2}\left[\frac{1}{\sqrt{2}} + i\left(-\frac{1}{\sqrt{2}}\right)\right]$$

$$= 1 - i$$

(C) $x + iy = 3.08e^{2.44i}$

$$= 3.08(\cos 2.44 + i\sin 2.44)$$

$$= -2.35 + 1.99i$$

21. (A) $x + iy = 6e^{(\pi/6)i}$

$$= 6\left(\cos\frac{\pi}{6} + i\sin\frac{\pi}{6}\right)$$

$$= 6\left(\frac{\sqrt{3}}{2} + i\frac{1}{2}\right)$$

$$= 3\sqrt{3} + 3i$$

(B) $x + iy = \sqrt{7}e^{(-90°)i}$

$$= \sqrt{7}[\cos(-90°) + i\sin(-90°)]$$

$$= \sqrt{7}(0 - 1i)$$

$$= -i\sqrt{7}$$

(C) $4.09e^{(-122.88°)i} = 4.09[\cos(-122.88°) + i\sin(-122.88°)]$

$$= -2.22 - 3.43i$$

23. $z_1z_2 = e^{195°i} \cdot e^{55°i} = e^{195°i+55°i} = e^{250°i}$

$$\frac{z_1}{z_2} = \frac{e^{195°i}}{e^{55°i}} = e^{195°i-55°i} = e^{140°i}$$

25. $z_1z_2 = 7e^{82°i} \cdot 2e^{31°i}$

$$= 7 \cdot 2e^{i(82°+31°)} = 14e^{113°i}$$

$$\frac{z_1}{z_2} = \frac{7e^{82°i}}{2e^{31°i}}$$

$$= \frac{7}{2}e^{i(82°-31°)} = 3.5e^{51°i}$$

27. $z_1z_2 = 5e^{52°i} \cdot 2e^{83°i}$

$$= 5 \cdot 2e^{i(52°+83°)} = 10e^{135°i}$$

$$\frac{z_1}{z_2} = \frac{5e^{52°i}}{2e^{83°i}}$$

$$= \frac{5}{2}e^{i(52°-83°)} = 2.5e^{(-31°)i}$$

29. $z_1z_2 = 3.05e^{1.76i} \cdot 11.94e^{2.59i}$

$$= 3.05 \cdot 11.94e^{(1.76+2.59)i} = 36.42e^{4.35i}$$

$$\frac{z_1}{z_2} = \frac{3.05e^{1.76i}}{11.94e^{2.59i}}$$

$$= \frac{3.05}{11.94}e^{(1.76-2.59)i} = 0.26e^{-0.83i}$$

31. $\left(e^{20°i}\right)^5 = e^{5\cdot 20°i} = e^{100°i}$

33. $(2e^{30°i})^3 = 2^3e^{3\cdot 30°i} = 8e^{90°i}$

35. $(\sqrt{2}e^{10°i})^6 = (\sqrt{2})^6e^{6\cdot 10°i} = (2^{1/2})^6e^{60°i} = 2^3e^{60°i} = 8e^{60°i}$

37. First, find the polar form of $1 + i\sqrt{3}$:

$$r = \sqrt{1^2 + \sqrt{3}^2} = \sqrt{4} = 2; \quad \theta = \tan^{-1}\frac{\sqrt{3}}{1} = 60°; \quad 1 + i\sqrt{3} = 2e^{60°i}$$

$$(1 + i\sqrt{3})^3 = (2e^{60°i})^3 = 2^3e^{3\cdot 60°i} = 8e^{180°i}$$

39. First, find the polar form of $-\sqrt{3} - i$:

$$r = \sqrt{\left(-\sqrt{3}\right)^2 + \left(-1\right)^2} = \sqrt{4} = 2; \quad \tan^{-1}\frac{1}{\sqrt{3}} = 30°; \theta = -150° \quad \text{(Quadrant III)};$$

$$-\sqrt{3} - i = 2e^{-150°i}$$

$$(-\sqrt{3} - i)^4 = (2e^{-150°i})^4 = 2^4 e^{4(-150°i)} = 16e^{-600°i}$$

$$= 16[\cos(-600°) + i\sin(-600°)] = 16\left(-\frac{1}{2} + i\frac{\sqrt{3}}{2}\right) = -8 + 8i\sqrt{3}$$

41. First, find the polar form of $1 - i$:

$$r = \sqrt{1^2 + \left(-1\right)^2} = \sqrt{2}; \quad \theta = \tan^{-1}\frac{-1}{1} = -45°; \quad 1 - i = \sqrt{2}e^{-45°i}$$

$$(1 - i)^8 = (\sqrt{2}e^{-45°i})^8 = (\sqrt{2})^8 e^{8(-45°i)} = 16e^{-360°i}$$

$$= 16[\cos(-360°) + i\sin(-360°)] = 16(1 + 0i) = 16$$

43. First, find the polar form of $-\frac{1}{2} + \frac{\sqrt{3}}{2}i$:

$$r = \sqrt{\left(-\frac{1}{2}\right)^2 + \left(\frac{\sqrt{3}}{2}\right)^2} = \sqrt{1} = 1; \quad \tan^{-1}\frac{\sqrt{3}/2}{-1/2} = -60°; \quad \theta = 120° \quad \text{(Quadrant II)};$$

$$-\frac{1}{2} + \frac{\sqrt{3}}{2}i = e^{120°i}$$

$$\left(-\frac{1}{2} + \frac{\sqrt{3}}{2}i\right)^3 = (e^{120°i})^3 = e^{360°i} = \cos 360° + i\sin 360° = 1 + 0i = 1$$

45. Using the nth-root theorem, all three cube roots of $8e^{30°i}$ are given by

$$8^{1/3}e^{(30°/3 + k360°/3)i} = 8^{1/3}e^{(10° + k120°)i} \quad k = 0, 1, 2$$

$$w_1 = 8^{1/3}e^{(10° + 0 \cdot 120°)i} = 2e^{10°i}$$

$$w_2 = 8^{1/3}e^{(10° + 1 \cdot 120°)i} = 2e^{130°i}$$

$$w_3 = 8^{1/3}e^{(10° + 2 \cdot 120°)i} = 2e^{250°i}$$

47. Using the nth-root theorem, all four fourth roots of $81e^{60°i}$ are given by

$$81^{1/4}e^{(60°/4 + k360°/4)i} = 81^{1/4}e^{(15° + k90°)i} \quad k = 0, 1, 2, 3$$

$$w_1 = 81^{1/4}e^{(15° + 0 \cdot 90°)i} = 3e^{15°i}$$

$$w_2 = 81^{1/4}e^{(15° + 1 \cdot 90°)i} = 3e^{105°i}$$

$$w_3 = 81^{1/4}e^{(15° + 2 \cdot 90°)i} = 3e^{195°i}$$

$$w_4 = 81^{1/4}e^{(15° + 3 \cdot 90°)i} = 3e^{285°i}$$

49. First write $1 - i$ in polar form.

$$1 - i = \sqrt{2}e^{(-45°)i}$$

Using the nth-root theorem, all five fifth roots of $\sqrt{2}e^{(-45°)i}$ are given by

$$(\sqrt{2})^{1/5}e^{(-45°/5 + k360°/5)i} = (2^{1/2})^{1/5}e^{(-9° + k72°)i} \quad k = 0, 1, 2, 3, 4$$

$$w_1 = 2^{1/10}e^{(-9° + 0 \cdot 72°)i} = 2^{1/10}e^{(-9°)i}$$

$$w_2 = 2^{1/10}e^{(-9° + 1 \cdot 72°)i} = 2^{1/10}e^{63°i}$$

$$w_3 = 2^{1/10}e^{(-9° + 2 \cdot 72°)i} = 2^{1/10}e^{135°i}$$

$$w_4 = 2^{1/10}e^{(-9° + 3 \cdot 72°)i} = 2^{1/10}e^{207°i}$$

$$w_5 = 2^{1/10}e^{(-9° + 4 \cdot 72°)i} = 2^{1/10}e^{279°i}$$

51. First write 8 in polar form.

$$8 = 8 + 0i = 8e^{0°i}$$

Using the nth-root theorem, all three cube roots of $8e^{0°i}$ are given by

$$8^{1/3}e^{(0°/3 + k360°/3)i} = 8^{1/3}e^{k120°i} \qquad k = 0, 1, 2$$

$$w_1 = 2e^{0 \cdot 120°i} = 2e^{0°i}$$

$$w_2 = 2e^{1 \cdot 120°i} = 2e^{120°i}$$

$$w_3 = 2e^{2 \cdot 120°i} = 2e^{240°i}$$

w_1, w_2, w_3 lie on a circle of radius 2, equally spaced so that the angle between successive roots is 120°.

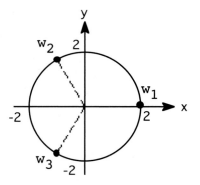

53. First write -16 in polar form.

$$-16 = -16 + 0i = 16e^{180°i}$$

Using the nth-root theorem, all four fourth roots of $16e^{180°i}$ are given by

$$16^{1/4}e^{(180°/4 + k \cdot 360°/4)i} = 16^{1/4}e^{(45° + k \cdot 90°)i}$$

$$k = 0, 1, 2, 3$$

$$w_1 = 2e^{(45° + 0 \cdot 90°)i} = 2e^{45°i}$$

$$w_2 = 2e^{(45° + 1 \cdot 90°)i} = 2e^{135°i}$$

$$w_3 = 2e^{(45° + 2 \cdot 90°)i} = 2e^{225°i}$$

$$w_4 = 2e^{(45° + 3 \cdot 90°)i} = 2e^{315°i}$$

w_1, w_2, w_3, w_4 lie on a circle of radius 2, equally spaced so that the angle between successive roots is 90°.

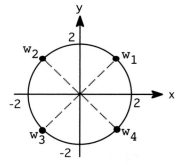

55. First write i in polar form.

$$i = 0 + 1i = 1e^{90°i}$$

Using the nth-root theorem, all sixth roots of $1e^{90°i}$ are given by

$$1^{1/6}e^{(90°/6 + k \cdot 360°/6)i} = 1e^{(15° + k \cdot 60°)i}$$

$$k = 0, 1, 2, 3, 4, 5$$

$$w_1 = 1e^{(15° + 0 \cdot 60°)i} = 1e^{15°i}$$

$$w_2 = 1e^{(15° + 1 \cdot 60°)i} = 1e^{75°i}$$

$$w_3 = 1e^{(15° + 2 \cdot 60°)i} = 1e^{135°i}$$

$$w_4 = 1e^{(15° + 3 \cdot 60°)i} = 1e^{195°i}$$

$$w_5 = 1e^{(15° + 4 \cdot 60°)i} = 1e^{255°i}$$

$$w_6 = 1e^{(15° + 5 \cdot 60°)i} = 1e^{315°i}$$

These roots lie on a circle of radius 1, equally spaced so that the angle between successive roots is 60°.

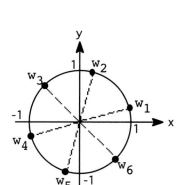

57. (A) Substituting $1 + i$ for x in $x^4 + 4 = 0$ yields

$$(1 + i)^4 + 4 \overset{?}{=} 0$$

$$((1 + i)^2)^2 + 4 \overset{?}{=} 0$$

$$(1 + 2i + i^2)^2 + 4 \overset{?}{=} 0$$

$$(2i)^2 + 4 \overset{?}{=} 0$$

$$-4 + 4 \overset{\checkmark}{=} 0$$

$1 + i$ is a root of $x^4 + 4 = 0$.

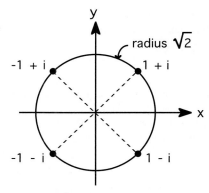

(B) The four roots are equally spaced around the circle. Since there are four roots, the angle between successive roots on the circle is $\frac{360°}{4} = 90°$.

(C) Substituting each root for x in $x^4 + 4 = 0$ yields

$-1 + i$: $(-1 + i)^4 + 4 = ((-1 + i)^2)^2 + 4 = (1 - 2i + i^2)^2 + 4 = (-2i)^2 + 4$
$$= -4 + 4 = 0$$

$-1 - i$: $(-1 - i)^4 + 4 = ((-1 - i)^2)^2 + 4 = (1 + 2i + i^2)^2 + 4 = (2i)^2 + 4$
$$= -4 + 4 = 0$$

$1 - i$: $(1 - i)^4 + 4 = ((1 - i)^2)^2 + 4 = (1 - 2i + i^2)^2 + 4 = (-2i)^2 + 4$
$$= -4 + 4 = 0$$

59. $x^3 + 64 = 0$
$$x^3 = -64$$
So we are looking for the three cube roots of -64.

First write -64 in polar form:
$$-64 = 64e^{180°i}$$
Using the nth-root theorem, all three cube roots of -64 are given by
$$64^{1/3}e^{(180°/3 + k \cdot 360°/3)i} = 4e^{(60° + k \cdot 120°)i} \quad k = 0,\ 1,\ 2$$
$$w_1 = 4e^{60°i} = 4(\cos 60° + i \sin 60°) = 2 + 2i\sqrt{3}$$
$$w_2 = 4e^{180°i} = 4(\cos 180° + i \sin 180°) = -4$$
$$w_3 = 4e^{300°i} = 4(\cos 300° + i \sin 300°) = 2 - 2i\sqrt{3}$$

61. $x^3 - 27 = 0$
$$x^3 = 27$$
So we are looking for the three cube roots of 27.

First write 27 in polar form:
$$27 = 27e^{0°i}$$
Using the nth-root theorem, all three cube roots of 27 are given by
$$27^{1/3}e^{(0°/3 + k \cdot 360°/3)i} = 3e^{k \cdot 120°i} \quad k = 0,\ 1,\ 2$$
$$w_1 = 3e^{0°i} = 3(\cos 0° + i \sin 0°) = 3$$
$$w_2 = 3e^{120°i} = 3(\cos 120° + i \sin 120°) = -\frac{3}{2} + \frac{3\sqrt{3}}{2}i$$
$$w_3 = 3e^{240°i} = 3(\cos 240° + i \sin 240°) = -\frac{3}{2} - \frac{3\sqrt{3}}{2}i$$

63. True. If $z_1 = r_1e^{i0°}$ and $z_2 = r_2e^{i0°}$, then $z_1z_2 = r_1r_2e^{i0°}$.

65. False. 16 has fourth roots 2, -2, $2i$, $-2i$.

67. True. If $z^2 = 1$, then $(z^2)^3 = 1^3$ and $z^6 = 1$.

69. True. If $w^3 = w^4$, then $w^4 - w^3 = 0$, so either $w^3 = 0$ (not allowed here) or $w = 1$.

71. False. If $w^3 = w^6$, then $w^6 - w^3 = 0$, so either $w^3 = 0$ (not allowed here) or $w^3 - 1 = 0$, in which case $w^3 = 1$ and w is a cube root of 1. However, w need not equal 1. $\dfrac{-1 + \sqrt{3}i}{2}$ is also a cube root of 1.

73. If w is real and $w^n = z$ then $w = re^{i0°}$ and $w^n = r^n e^{i0°}$; w^n must be real.

75. To show $\left(r^{1/n} e^{(\theta/n + k \cdot 360°/n)i}\right)^n = re^{i\theta}$ we apply DeMoivre's Theorem:

$$\left(r^{1/n} e^{(\theta/n + k \cdot 360°/n)i}\right)^n = (r^{1/n})^n e^{n(\theta/n + k \cdot 360°/n)i}$$
$$= re^{(\theta + k360°)i}$$
$$= r[\cos(\theta + k360°) + i\sin(\theta + k \cdot 360°)]$$
$$= r(\cos\theta + i\sin\theta) \quad \text{since } k \text{ is an integer and sine and}$$
$$\qquad\qquad\qquad\qquad\qquad\qquad \text{cosine are periodic with period } 360°$$
$$= re^{i\theta}$$

77. The complex zeros of $P(x) = x^5 - 32$ are the values of x for which $x^5 - 32 = 0$, or $x^5 = 32$. So we are looking for the five fifth roots of 32. First write 32 in polar form:
$$32 = 32e^{0°i}$$
Using the nth-root theorem, all five fifth roots of 32 are given by
$$32^{1/5} e^{(0°/5 + k \cdot 360°/5)i} = 2e^{k \cdot 72°i} \qquad k = 0, 1, 2, 3, 4$$

$$w_1 = 2e^{0°i}$$
$$w_2 = 2e^{72°i}$$
$$w_3 = 2e^{144°i}$$
$$w_4 = 2e^{216°i}$$
$$w_5 = 2e^{288°i}$$

79. $x^5 + 1 = 0$
$$x^5 = -1$$
So we are looking for the five fifth roots of -1. First write -1 in polar form:
$$-1 = 1e^{180°i}$$
Using the nth-root theorem, all five fifth roots of -1 are given by
$$1^{1/5} e^{(180°/5 + k \cdot 360°/5)i} = 1e^{(36° + k \cdot 72°)i} \qquad k = 0, 1, 2, 3, 4$$
$$w_1 = 1e^{(36° + 0 \cdot 72°)i} = e^{36°i}$$
$$w_2 = 1e^{(36° + 1 \cdot 72°)i} = e^{108°i}$$
$$w_3 = 1e^{(36° + 2 \cdot 72°)i} = e^{180°i}$$
$$w_4 = 1e^{(36° + 3 \cdot 72°)i} = e^{252°i}$$
$$w_5 = 1e^{(36° + 4 \cdot 72°)i} = e^{324°i}$$

81. The linear factors of $x^6 + 64$ will be $x - x_i$, where x_i are zeros of $x^6 + 64$, that is $x^6 + 64 = 0$ or $x^6 = -64$. So the x_i's are the six sixth roots of -64, or $64e^{180°i}$.
Using the nth-root theorem, all six sixth roots of -64 are given by
$$64^{1/6} e^{(180°/6 + k \cdot 360°/6)i} = 2e^{(30° + k \cdot 60°)i} \qquad k = 0, 1, 2, 3, 4, 5$$

$$w_1 = 2e^{30°i} = 2(\cos 30° + i \sin 30°) = 2\left(\frac{\sqrt{3}}{2} + i\frac{1}{2}\right) = \sqrt{3} + i$$

$$w_2 = 2e^{90°i} = 2(\cos 90° + i \sin 90°) = 2(0 + i1) = 2i$$

$$w_3 = 2e^{150°i} = 2(\cos 150° + i \sin 150°) = 2\left(-\frac{\sqrt{3}}{2} + i\frac{1}{2}\right) = -\sqrt{3} + i$$

$$w_4 = 2e^{210°i} = 2(\cos 210° + i \sin 210°) = 2\left(-\frac{\sqrt{3}}{2} - i\frac{1}{2}\right) = -\sqrt{3} - i$$

$$w_5 = 2e^{270°i} = 2(\cos 270° + i \sin 270°) = 2(0 - i1) = -2i$$

$$w_6 = 2e^{330°i} = 2(\cos 330° + i \sin 330°) = 2\left(\frac{\sqrt{3}}{2} - i\frac{1}{2}\right) = \sqrt{3} - i$$

$$x^6 + 64 = (x - 2i)(x + 2i)[x - (-\sqrt{3} + i)][x - (-\sqrt{3} - i)]$$
$$[x - (\sqrt{3} + i)][x - (\sqrt{3} - i)]$$

CHAPTER 7 REVIEW

1.

We are given two sides and a non-included angle (SSA). α is acute
$11 = a \geq b = 3.7$
One triangle can be constructed.

(7-1)

2. It is impossible to draw or form a triangle with this data, since angles α and β together add up to more than 180°. No triangle can be constructed.

(7-1)

3.

We are given two sides and a non-included angle (SSA). α is acute.
$h = b \sin \alpha = 22 \sin 54° = 17.8$
$h < a < b$
Two triangles can be constructed. (7-1)

4. Angle β is acute. A triangle can have at most one obtuse angle. Since α is acute, then, if the triangle has an obtuse angle it must be the angle opposite the longer of the two sides, b and c. So β, the angle opposite the shorter of the two sides, b, must be acute. (7-2)

5. We are given two angles and the included side (ASA). We use the law of sines.

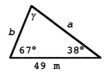

Solve for γ:
$$\alpha + \beta + \gamma = 180°$$
$$\gamma = 180° - (\alpha + \beta)$$
$$= 180° - (67° + 38°)$$
$$= 75°$$

Solve for a:
$$\frac{\sin \alpha}{a} = \frac{\sin \gamma}{c}$$
$$a = \frac{c \sin \alpha}{\sin \gamma}$$
$$= \frac{49 \sin 67°}{\sin 75°}$$
$$= 47 \text{ m}$$

Solve for b:

$$\frac{\sin \beta}{b} = \frac{\sin \gamma}{c}$$

$$b = \frac{c \sin \beta}{\sin \gamma}$$

$$= \frac{49 \sin 38°}{\sin 75°}$$

$$= 31 \text{ m}$$

(7-1)

6.

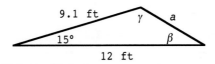

We are given two sides and the included angle (SAS). We use the law of cosines to find the third side, then the law of sines to find a second angle.

Solve for a:

$$a^2 = b^2 + c^2 - 2bc \cos \alpha$$

$$= (9.1)^2 + (12)^2 - 2(9.1)(12)\cos 15°$$

$$= 15.8518$$

$$a = 4.00 \text{ ft}$$

Solve for β:

$$\frac{\sin \alpha}{a} = \frac{\sin \beta}{b}$$

$$\sin \beta = \frac{b \sin \alpha}{a}$$

$$= \frac{9.1 \sin 15°}{4.00}$$

$$= 0.5888$$

$$\beta = \sin^{-1}(0.5888) = 36°$$

There is another solution of $\sin \beta = 0.5915$ which deserves brief consideration, that is, $\beta = 180° - \sin^{-1}(0.5915) = 144°$. This would lead to a contradiction, however, since the largest side of the triangle, c, must be opposite the largest angle, γ, and this would lead to two obtuse angles in the triangle.

Solve for γ:

$$\alpha + \beta + \gamma = 180°$$

$$\gamma = 180° - (\alpha + \beta)$$

$$= 180° - (15° + 36°)$$

$$= 129°$$

(7-1, 7-2)

7. We are given two sides and a non-included angle (SSA). From a rough sketch we see that there is only one triangle possible.

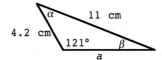

We use the law of sines.
Solve for β:

$$\frac{\sin \beta}{b} = \frac{\sin \gamma}{c}$$

$$\sin \beta = \frac{b \sin \gamma}{c}$$

$$= \frac{4.2 \sin 121°}{11}$$

$$\beta = \sin^{-1} \frac{4.2 \sin 121°}{11} \quad \beta \text{ is acute}$$

$$= 19°$$

Solve for α:

$$\alpha + \beta + \gamma = 180°$$

$$\alpha = 180° - (\beta + \gamma)$$

$$= 180° - (19° + 121°)$$

$$= 40°$$

Solve for a:

$$\frac{\sin \alpha}{a} = \frac{\sin \gamma}{c}$$

$$a = \frac{c \sin \alpha}{\sin \gamma}$$

$$= \frac{11 \sin 40°}{\sin 121°}$$

$$= 8.2 \text{ cm}$$

(7-1)

8.

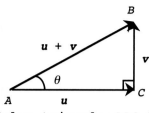

To find $|\boldsymbol{u} + \boldsymbol{v}|$: Apply the Pythagorean theorem to triangle ABC.

$$
\begin{aligned}
|\boldsymbol{u} + \boldsymbol{v}|^2 = AB^2 &= AC^2 + BC^2 \\
&= |\boldsymbol{u}|^2 + |\boldsymbol{v}|^2 \\
&= 160^2 + 55^2 \\
&= 28,625 \\
|\boldsymbol{u} + \boldsymbol{v}| &= \sqrt{28,625} = 170 \text{ mi/hr}
\end{aligned}
$$

Solve triangle ABC for θ

$$
\tan \theta = \frac{BC}{AC} = \frac{|\boldsymbol{v}|}{|\boldsymbol{u}|}
$$

$$
\theta = \tan^{-1} \frac{|\boldsymbol{v}|}{|\boldsymbol{u}|} \quad \theta \text{ is acute}
$$

$$
= \tan^{-1} \frac{55}{160} = 19°
$$

(7-3)

9. The algebraic vector $\langle a, b \rangle$ corresponds to the standard geometric vector with terminal point $P = (a, b)$ and initial point $O = (0, 0)$. The coordinates of the point $P = (a, b)$ are given by

$$
a = x_B - x_A = 5 - 2 = 3
$$
$$
b = y_B - y_A = -1 - 6 = -7
$$

So $\langle a, b \rangle = \langle 3, -7 \rangle$

(7-3)

10. $|\boldsymbol{v}| = \sqrt{(-3)^2 + (-5)^2} = \sqrt{34}$ (7-3)

11. The graph is the set of all points on a line forming an angle of $\dfrac{\pi}{6}$ with the polar axis.

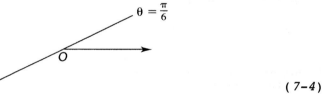

(7-4)

12. The graph is the set of all points at distance 6 from the pole—a circle of radius 6 with center at the pole.

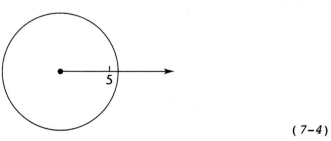

(7-4)

13.

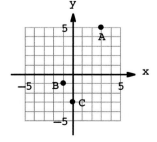

(7-5)

14. See figure.

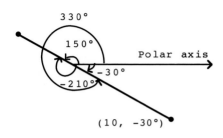

The point with coordinates (10, -30°) can equally well be described as (-10°, -210°) or (-10, 150°) or (10, 330°). We can describe each as follows. (-10, -210°): The polar axis is rotated 210° clockwise (negative direction) and the point is located 10 units from the pole along the negative polar axis. (-10, 150°): The polar axis is rotated 150° counterclockwise (positive direction) and the point is located 10 units from the pole along the negative polar axis. (10, 330°): The polar axis is rotated 330° counterclockwise and the point is located 10 units from the pole along the positive polar axis. (*7-4*)

15.

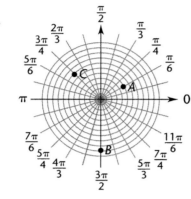

(*7-5*)

16. (A) A sketch shows that $1 - i\sqrt{3}$ is associated with a special 30°-60° triangle. By inspection, $r = 2$, $\theta = -60°$, and

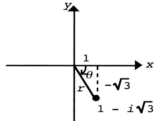

$$1 - i\sqrt{3} = 2[\cos(-60°) + i\sin(-60°)]$$
$$= 2e^{(-60°)i}$$

(B) $4e^{(-30°)i} = 4[\cos(-30°) + i\sin(-30°)]$
$$= 4\left[\frac{\sqrt{3}}{2} + i\left(-\frac{1}{2}\right)\right]$$
$$= 2\sqrt{3} - 2i$$

(*7-5*)

17. (A) $\left[-\frac{1}{2} - \frac{\sqrt{3}}{2}i\right]^3 = (e^{-120°i})^3 = e^{-360°i} = \cos(-360°) + i\sin(-360°) = 1$

(B) The calculation is shown using $a + bi$ mode on a TI-84:

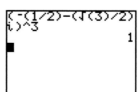

(*7-5*)

18. $(2e^{15°i})^4 = 2^4 e^{4(15°i)} = 16e^{60°i} = 16(\cos 60° + i \sin 60°) = 16\left(\dfrac{1}{2} + i \dfrac{\sqrt{3}}{2}\right)$

$\qquad = 8 + 8i\sqrt{3}$ $\hspace{8cm}$ (7-5)

19. If the triangle has an obtuse angle, then it must be the angle opposite the longest side; in this case, α. $\hspace{5cm}$ (7-2)

20.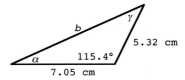
We are given two sides and the included angle (SAS). We use the law of cosines to find the third side, then the law of sines to find a second angle.

Solve for b:

$\quad b^2 = a^2 + c^2 - 2ac \cos \beta$

$\qquad = (5.32)^2 + (7.05)^2 - 2(5.32)(7.05)\cos 115.4°$

$\qquad = 110.18018…$

$\quad b = 10.5$ cm

Solve for α:

Note that since the sum of α and γ is $180° - 115.4° = 64.6°$, both must be acute.

$\qquad \dfrac{\sin \alpha}{a} = \dfrac{\sin \beta}{b}$

$\qquad \sin \alpha = \dfrac{a \sin \beta}{b}$

$\qquad\qquad = \dfrac{5.32 \sin 115.4°}{10.5}$

$\qquad \alpha = \sin^{-1}\left(\dfrac{5.32 \sin 115.4°}{10.5}\right)$

$\qquad\quad = 27.2°$

Solve for γ:

$\quad \alpha + \beta + \gamma = 180°$

$\qquad\qquad \gamma = 180° - (\alpha + \beta)$

$\qquad\qquad\quad = 180° - (27.2° + 115.4°)$

$\qquad\qquad\quad = 37.4°$ $\hspace{8cm}$ (7-2)

21.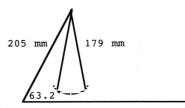

α is acute

$h = b \sin \alpha = 205 \sin 63.2° = 183$

$a = 179 < 183 = h$

We are given two sides and a non-included angle (SSA). If we try to draw a triangle with these values, we will be unsuccessful.
No solution. $\hspace{8cm}$ (7-2)

22.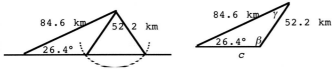

We are given two sides and a non-included angle (SSA). If we try to draw a triangle with these values, we find that two triangles are possible. Following the instructions, we choose the solution for which β is obtuse.

Solve for β:
$$\frac{\sin \beta}{b} = \frac{\sin \alpha}{a}$$
$$\sin \beta = \frac{b \sin \alpha}{a} = \frac{84.6 \sin 26.4°}{52.2} = 0.7206$$
β is chosen obtuse
$$\beta = 180° - \sin^{-1} 0.7206$$
$$= 180° - 46.1°$$
$$= 133.9°$$

Solve for γ:
$$\gamma = 180° - (\alpha + \beta)$$
$$= 180° - (26.4° + 133.9°)$$
$$= 19.7°$$

Solve for c:
$$\frac{\sin \alpha}{a} = \frac{\sin \gamma}{c}$$
$$c = \frac{a \sin \gamma}{\sin \alpha}$$
$$= \frac{52.2 \sin 19.7°}{\sin 26.4°}$$
$$= 39.6 \text{ km} \qquad (7-1)$$

23.

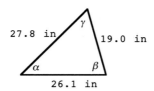

27.8 in γ 19.0 in

α β

26.1 in

We are given three sides (SSS). We solve for the largest angle, β (largest because it is opposite the largest side, b) using the law of cosines. We then solve for a second angle using the law of sines.

Solve for β:
$$b^2 = a^2 + c^2 - 2ac \cos \beta$$
$$\cos \beta = \frac{a^2 + c^2 - b^2}{2ac}$$
$$\beta = \cos^{-1}\left(\frac{a^2 + c^2 - b^2}{2ac}\right)$$
$$= \cos^{-1}\left(\frac{(19.0)^2 + (26.1)^2 - (27.8)^2}{2(19.0)(26.1)}\right) = 74.2°$$

Solve for α:
We use the law of sines. Since α must be smaller than β, α is acute.
$$\frac{\sin \alpha}{a} = \frac{\sin \beta}{b}$$
$$\sin \alpha = \frac{a \sin \beta}{b}$$
$$= \frac{19.0 \sin 74.2°}{27.8}$$
$$\alpha = \sin^{-1}\left(\frac{19.0 \sin 74.2°}{27.8}\right)$$
$$= 41.1°$$

Solve for γ:
$$\alpha + \beta + \gamma = 180°$$
$$\gamma = 180° - (\alpha + \beta)$$
$$= 180° - (41.1° + 74.2°)$$
$$= 64.7° \qquad\qquad (7-1, 7-2)$$

24. The sum of all of the force vectors must be the zero vector for the object to remain at rest. *(7-3)*

25.

$\theta = 57.2°$, and $\angle OCB = 180° - \theta = 180° - 57.2° = 122.8°$

We can find $|\boldsymbol{u} + \boldsymbol{v}|$ using the law of cosines:

$$|\boldsymbol{u} + \boldsymbol{v}|^2 = |\boldsymbol{u}|^2 + |\boldsymbol{v}|^2 - 2|\boldsymbol{u}||\boldsymbol{v}|\cos(OCB)$$
$$= (75.2)^2 + (34.2)^2 - 2(75.2)(34.2)\cos 122.8°$$
$$= 9611.0537\ldots$$
$$|\boldsymbol{u} + \boldsymbol{v}| = \sqrt{9611.0537\ldots}$$
$$= 98.0 \text{ kg}$$

To find α, we use the law of sines:

$$\frac{\sin \alpha}{|\boldsymbol{v}|} = \frac{\sin(OCB)}{|\boldsymbol{u} + \boldsymbol{v}|}$$

$$\frac{\sin \alpha}{34.2} = \frac{\sin 122.8°}{98.0}$$

$$\sin \alpha = \frac{34.2}{98.0} \sin 122.8°$$

$$\alpha = \sin^{-1}\left(\frac{34.2}{98.0} \sin 122.8°\right) = 17.1°$$ *(7-3)*

26. (A) $\boldsymbol{u} = \langle -3, 9 \rangle = \langle -3, 0 \rangle + \langle 0, 9 \rangle$ (B) $\boldsymbol{v} = \langle 0, -2 \rangle = -2\langle 0, 1 \rangle = -2\boldsymbol{j}$
$$= -3\langle 1, 0 \rangle + 9\langle 0, 1 \rangle$$
$$= -3\boldsymbol{i} + 9\boldsymbol{j}$$ *(7-3)*

27. (A) $\boldsymbol{u} - \boldsymbol{v} = \langle -2, 3 \rangle - \langle 2, -4 \rangle = \langle -2, 3 \rangle + \langle -2, 4 \rangle = \langle -4, 7 \rangle$

(B) $3\boldsymbol{u} - \boldsymbol{v} + 2\boldsymbol{w} = 3\langle -2, 3 \rangle - \langle 2, -4 \rangle + 2\langle -3, 0 \rangle$
$$= \langle -6, 9 \rangle + \langle -2, 4 \rangle + \langle -6, 0 \rangle$$
$$= \langle -14, 13 \rangle$$ *(7-3)*

28. (A) $\boldsymbol{u} - \boldsymbol{v} = (\boldsymbol{i} - 2\boldsymbol{j}) - (3\boldsymbol{i} + 2\boldsymbol{j})$
$$= \boldsymbol{i} - 2\boldsymbol{j} - 3\boldsymbol{i} - 2\boldsymbol{j}$$
$$= -2\boldsymbol{i} - 4\boldsymbol{j}$$

(B) $3\boldsymbol{u} - \boldsymbol{v} + 2\boldsymbol{w} = 3(\boldsymbol{i} - 2\boldsymbol{j}) - (3\boldsymbol{i} + 2\boldsymbol{j}) + 2(-\boldsymbol{j})$
$$= 3\boldsymbol{i} - 6\boldsymbol{j} - 3\boldsymbol{i} - 2\boldsymbol{j} - 2\boldsymbol{j}$$
$$= -10\boldsymbol{j}$$ *(7-3)*

29. $|\boldsymbol{v}| = \sqrt{(-1)^2 + (-3)^2} = \sqrt{10}$

Then $\boldsymbol{u} = \dfrac{1}{|\boldsymbol{v}|}\boldsymbol{v} = \dfrac{1}{\sqrt{10}}\langle -1, -3 \rangle = \left\langle \dfrac{-1}{\sqrt{10}}, \dfrac{-3}{\sqrt{10}} \right\rangle$ *(7-3)*

30.

Since the triangle is right, we can calculate the area directly.

$$A = \frac{1}{2} bh = \frac{1}{2}(66)(24) = 790 \text{ sq. ft. to two significant digits}$$ *(7-2)*

31.

We can find h using right triangle trig ratios:

$$\sin 15° = \frac{h}{34.5}; \quad h = 34.5 \sin 15°$$

$$A = \frac{1}{2} bh = \frac{1}{2}(29.8)\left(34.5 \sin 15°\right) = 133 \text{ sq. m. to}$$
three significant digits *(7-2)*

32. See Explore-Discuss 3 in Section 7-2 of your text for a discussion of Heron's formula. The semiperimeter s is given by

$$s = \frac{84 + 113 + 38}{2} = 117.5 \text{ yds}$$

Using Heron's formula, we get

$$\sqrt{117.5(117.5 - 84)(117.5 - 113)(117.5 - 38)} = 1,200 \text{ sq. yds. to two significant digits}$$

(7-2)

33.

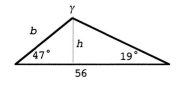

We first find angle γ by subtracting from $180°$:

$$\gamma = 180° - \left(47° + 19°\right) = 114°$$

Next we use the law of sines to find side b:

$$\frac{\sin 114°}{56} = \frac{\sin 19°}{b} \quad \Rightarrow \quad b = \frac{56 \sin 19°}{\sin 114°} \text{ in.}$$

Now we can use right triangle ratios to find h:

$$\sin 19° = \frac{h}{\dfrac{56 \sin 19°}{\sin 114°}} \quad \Rightarrow \quad h = \frac{56 \sin 19°}{\sin 114°} \sin 47° \text{ in.}$$

Finally, we find the area:

$$A = \frac{1}{2} bh = \frac{1}{2}(56)\left(\frac{56 \sin 19°}{\sin 114°} \sin 47°\right) = 410 \text{ sq. in. to two significant digits} \quad (7-2)$$

34. We set up a table that indicates how r varies as we let θ vary through each set of quadrant values.

θ	$\cos \theta$	$4 \cos \theta$	$6 + 4 \cos \theta$
0 to $\frac{\pi}{2}$	1 to 0	4 to 0	10 to 6
$\frac{\pi}{2}$ to π	0 to -1	0 to -4	6 to 2
π to $\frac{3\pi}{2}$	-1 to 0	-4 to 0	2 to 6
$\frac{3\pi}{2}$ to 2π	0 to 1	0 to 4	6 to 10

Check:

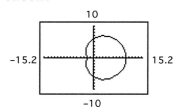

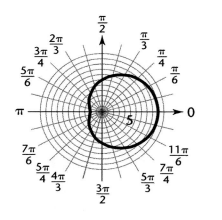

(7-4)

35. We set up a table that indicates how r varies as we let θ vary through each set of quadrant values.

θ	$\sin \theta$	$8 \sin \theta$	$8 + 8 \sin \theta$
0 to $\frac{\pi}{2}$	0 to 1	0 to 8	8 to 16
$\frac{\pi}{2}$ to π	1 to 0	8 to 0	16 to 8
π to $\frac{3\pi}{2}$	0 to -1	0 to -8	8 to 0
$\frac{3\pi}{2}$ to 2π	-1 to 0	-8 to 0	0 to 8

Check:

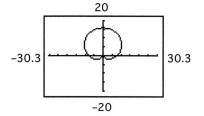

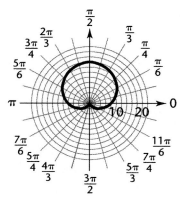

(7-4)

36. We set up a table that indicates how *r* varies as we let 2θ vary through each set of quadrant values.

θ	2θ	$\cos 2\theta$	$10 \cos 2\theta$
0 to $\frac{\pi}{4}$	0 to $\frac{\pi}{2}$	1 to 0	10 to 0
$\frac{\pi}{4}$ to $\frac{\pi}{2}$	$\frac{\pi}{2}$ to π	0 to -1	0 to -10
$\frac{\pi}{2}$ to $\frac{3\pi}{4}$	π to $\frac{3\pi}{2}$	-1 to 0	-10 to 0
$\frac{3\pi}{4}$ to π	$\frac{3\pi}{2}$ to 2π	0 to 1	0 to 10
π to $\frac{5\pi}{4}$	2π to $\frac{5\pi}{2}$	1 to 0	10 to 0
$\frac{5\pi}{4}$ to $\frac{3\pi}{2}$	$\frac{5\pi}{2}$ to 3π	0 to -1	0 to -10
$\frac{3\pi}{2}$ to $\frac{7\pi}{4}$	3π to $\frac{7\pi}{2}$	-1 to 0	-10 to 0
$\frac{7\pi}{4}$ to 2π	$\frac{7\pi}{2}$ to 4π	0 to 1	0 to 10

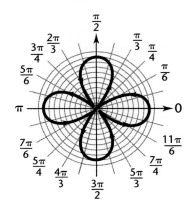

(*7-4*)

Check:

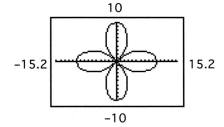

37. We set up a table that indicates how *r* varies as we let 3θ vary through each set of quadrant values.

θ	3θ	$\sin 3\theta$	$8 \sin 3\theta$
0 to $\frac{\pi}{6}$	0 to $\frac{\pi}{2}$	0 to 1	0 to 8
$\frac{\pi}{6}$ to $\frac{\pi}{3}$	$\frac{\pi}{2}$ to π	1 to 0	8 to 0
$\frac{\pi}{3}$ to $\frac{\pi}{2}$	π to $\frac{3\pi}{2}$	0 to -1	0 to -8
$\frac{\pi}{2}$ to $\frac{2\pi}{3}$	$\frac{3\pi}{2}$ to 2π	-1 to 0	-8 to 0
$\frac{2\pi}{3}$ to $\frac{5\pi}{6}$	2π to $\frac{5\pi}{2}$	0 to 1	0 to 8
$\frac{5\pi}{6}$ to π	$\frac{5\pi}{2}$ to 3π	1 to 0	8 to 0
π to 2π	repeats curve already drawn		

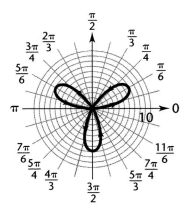

Check:

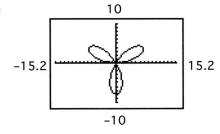

(*7-4*)

38. 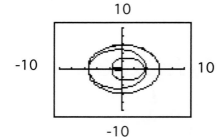 *(7-4)*

39. 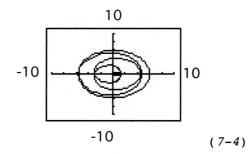 *(7-4)*

40. $n = 1$ $n = 2$ $n = 3$

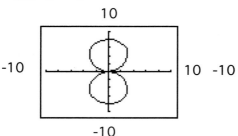   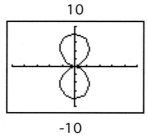   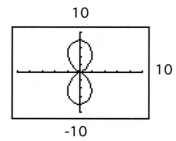

Generalizing from these graphs, we expect 2 leaves for all n.   *(7-4)*

41. (A) $e = 0.55$
$$r = \frac{3}{1 - 0.55 \cos \theta}$$

(B) $e = 1$
$$r = \frac{3}{1 - \cos \theta}$$

(C) $e = 1.7$
$$r = \frac{3}{1 - 1.7 \cos \theta}$$

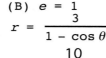   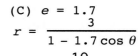   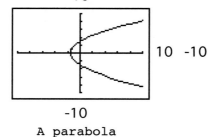

An ellipse   A parabola   A hyperbola   *(7-4)*

42. $x^2 + y^2 = 6x$
Use $r^2 = x^2 + y^2$ and $x = r \cos \theta$.
$r^2 = 6r \cos \theta$
$r^2 - 6r \cos \theta = 0$
$r(r - 6 \cos \theta) = 0$
$r = 0$ or $r - 6 \cos \theta = 0$
The graph of $r = 0$ is the pole. Since the pole is included in the graph of
$r - 6 \cos \theta = 0$, we can discard $r = 0$ and keep only
$r - 6 \cos \theta = 0$ or $r = 6 \cos \theta$ *(7-4)*

43. $r = 5 \cos \theta$
We multiply both sides by r, which adds the pole to the graph. But the pole is
already part of the graph, so we have changed nothing.
$r^2 = 5r \cos \theta$
But $r^2 = x^2 + y^2$ and $r \cos \theta = x$, so $x^2 + y^2 = 5x$.   *(7-5)*

44. A sketch shows that $-1 + i$ is associated with a special 45° triangle. By inspection, $r = \sqrt{2}$, $\theta = 135°$, and

$$-1 + i = \sqrt{2}(\cos 135° + i \sin 135°)$$
$$= \sqrt{2}\,e^{135°i}$$

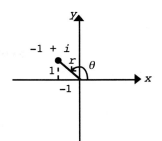

A sketch shows that $-1 - i\sqrt{3}$ is associated with a special 30°–60° triangle. By inspection, $r = 2$, $\theta = -120°$, and

$$-1 - i\sqrt{3} = 2[\cos(-120°) + i \sin(-120°)]$$
$$= 2e^{(-120°)i}$$

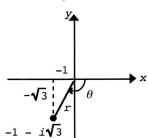

A sketch shows that 5 is associated with the special quadrantal angle 0°. By inspection, $r = 5$, $\theta = 0°$, and

$$5 = 5 + 0i = 5(\cos 0° + i \sin 0°)$$
$$= 5e^{0°i}$$

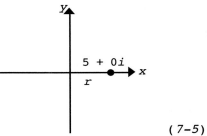

$(\textit{7-5})$

45. $z_1 = x + iy = \sqrt{2}\,e^{(\pi/4)i}$

$\qquad = \sqrt{2}\left(\cos\dfrac{\pi}{4} + i\sin\dfrac{\pi}{4}\right)$

$\qquad = \sqrt{2}\left(\dfrac{1}{\sqrt{2}} + i\,\dfrac{1}{\sqrt{2}}\right)$

$\qquad = 1 + i$

$z_2 = x + iy = 3e^{210°i}$

$\qquad = 3(\cos 210° + i \sin 210°)$

$\qquad = 3\left[-\dfrac{\sqrt{3}}{2} + i\left(-\dfrac{1}{2}\right)\right]$

$\qquad = \dfrac{-3\sqrt{3}}{2} - \dfrac{3}{2}i$

$z_3 = x + iy = 2e^{(-2\pi/3)i} = 2\left[\cos\left(-\dfrac{2\pi}{3}\right) + i\sin\left(-\dfrac{2\pi}{3}\right)\right]$

$\qquad\qquad\qquad\qquad\quad = 2\left[-\dfrac{1}{2} + i\left(-\dfrac{\sqrt{3}}{2}\right)\right]$

$\qquad\qquad\qquad\qquad\quad = -1 - i\sqrt{3}$

$(\textit{7-5})$

46. (A) $z_1 z_2 = 8e^{25°i} \cdot 4e^{19°i}$

$\qquad\qquad = 8 \cdot 4e^{i(25° + 19°)} = 32e^{44°i}$

$\quad$ (B) $\dfrac{z_1}{z_2} = \dfrac{8e^{25°i}}{4e^{19°i}}$

$\qquad\qquad = \dfrac{8}{4}e^{(25° - 19°)i} = 2e^{6°i}$

$(\textit{7-5})$

47. (A) $(1 + i\sqrt{3})^4 = (2e^{60°i})^4 = 2^4 e^{4 \cdot 60°i} = 16e^{240°i}$

$\qquad\qquad = 16[\cos 240° + i \sin 240°] = 16\left[-\dfrac{1}{2} + i\left(-\dfrac{\sqrt{3}}{2}\right)\right] = -8 - 8i\sqrt{3}$

(B) The calculation is shown using $a + bi$ mode on a TI-84 (note that $8\sqrt{3} \approx 13.86$):

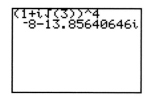

(7-5)

48. First write i in polar form.
$$i = 0 + 1i = 1e^{90°i}$$
Using the nth-root theorem, all three cube roots of $1e^{90°i}$ are given by
$$1^{1/3}e^{(90°/3 + k360°/3)i} = 1^{1/3}e^{(30° + k120°)i} \qquad k = 0, 1, 2$$
$$w_1 = 1^{1/3}e^{(30° + 0 \cdot 120°)i} = e^{30°i}$$
$$w_2 = 1^{1/3}e^{(30° + 1 \cdot 120°)i} = e^{150°i}$$
$$w_3 = 1^{1/3}e^{(30° + 2 \cdot 120°)i} = e^{270°i}$$

Now convert to rectangular form:
$$e^{30°i} = 1(\cos 30° + i \sin 30°)$$
$$= \frac{\sqrt{3}}{2} + \frac{1}{2}i$$
$$e^{150°i} = 1(\cos 150° + i \sin 150°)$$
$$= -\frac{\sqrt{3}}{2} + \frac{1}{2}i$$
$$e^{270°i} = 1(\cos 270° + i \sin 270°)$$
$$= -i$$

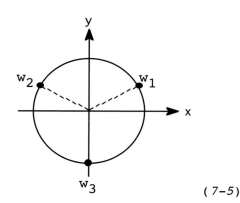

These roots lie on a circle of radius 1, equally spaced so that the angle between successive roots is 120°.

(7-5)

49. First write $-4\sqrt{3} + 4i$ in polar form.
$$r = \sqrt{\left(-4\sqrt{3}\right)^2 + \sqrt{4}^2} = \sqrt{64} = 8; \quad \tan^{-1}\frac{4}{-4\sqrt{3}} = -30°; \quad \theta = 150° \text{ (Quadrant II)};$$
$$-4\sqrt{3} + 4i = 8e^{150°i}$$
Using the nth-root theorem, all three cube roots of $8e^{150°i}$ are given by
$$8^{1/3}e^{(150°/3 + k360°/3)i} = 8^{1/3}e^{(50° + k \cdot 120°)i} \qquad k = 0, 1, 2$$
$$w_1 = 8^{1/3}e^{(50° + 0 \cdot 120°)i} = 2e^{50°i}$$
$$w_2 = 8^{1/3}e^{(50° + 1 \cdot 120°)i} = 2e^{170°i}$$
$$w_3 = 8^{1/3}e^{(50° + 2 \cdot 120°)i} = 2e^{290°i}$$

(7-5)

50. $(4e^{15°i})^2 = 4^2 e^{2 \cdot 15°i} = 16e^{30°i} = 16(\cos 30° + i \sin 30°) = 16\left(\frac{\sqrt{3}}{2} + i \cdot \frac{1}{2}\right)$

$$= 8\sqrt{3} + 8i$$

We conclude that $4e^{15°i}$ is a square root of $8\sqrt{3} + 8i$.

(7-5)

51. $(x, y) = (5.17, -2.53)$
$$r = \sqrt{x^2 + y^2} = \sqrt{5.17^2 + (-2.53)^2} = 5.76$$
$$\tan \theta = \frac{y}{x} = \frac{-2.53}{5.17}$$
θ is a fourth quadrant angle and is to be chosen so that $-180° < \theta \leq 180°$.
$$\theta = \tan^{-1}\frac{-2.53}{5.17} = -26.08°$$
$$(r, \theta) = (5.76, -26.08°)$$

(7-4)

52. $(r, \theta) = (5.81, -2.72)$

$\quad\quad x = r \cos \theta = 5.81 \cos(-2.72) = -5.30$

$\quad\quad y = r \sin \theta = 5.81 \sin(-2.72) = -2.38$

$\quad (x, y) = (-5.30, -2.38)$ (7-4)

53. $r = \sqrt{(-3.18)^2 + (4.19)^2} = 5.26$

$\quad \theta = 180° + \tan^{-1}\dfrac{4.19}{-3.18} = 127.20°$ since $(-3.18, 4.19)$ is in the second quadrant.

$\quad -3.18 + 4.19i = 5.26(\cos 127.20° + i \sin 127.20°) = 5.26e^{127.20°i}$ (7-5)

54. $x + iy = 7.63e^{(-162.27°)i}$

$\quad\quad\quad\quad = 7.63[\cos(-162.27°) + i \sin(-162.27°)]$

$\quad\quad\quad\quad = -7.27 - 2.32i$ (7-5)

55. (A) There are a total of three cube roots and they are spaced equally around a circle of radius 2, so that the angle between successive roots is 120°.

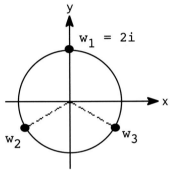

(B) Since $w_1 = 2e^{90°i}$, $w_2 = 2e^{(90° + 120°)i} = 2e^{210°i}$, and $w_3 = 2e^{(210° + 120°)i} = 2e^{330°i}$

$\quad\quad w_2 = 2(\cos 210° + i \sin 210°) = 2\left[-\dfrac{\sqrt{3}}{2} + i\left(-\dfrac{1}{2}\right)\right] = -\sqrt{3} - i$

$\quad\quad w_3 = 2(\cos 330° + i \sin 330°) = 2\left[\dfrac{\sqrt{3}}{2} + i\left(-\dfrac{1}{2}\right)\right] = \sqrt{3} - i$

(C) $(2i)^3 = 8i^3 = 8(-i) = -8i$

$\quad (-\sqrt{3} - i)^3 = (-\sqrt{3} - i)(-\sqrt{3} - i)(-\sqrt{3} - i)$

$\quad\quad\quad\quad\quad\quad = (3 + 2i\sqrt{3} + i^2)(-\sqrt{3} - i)$

$\quad\quad\quad\quad\quad\quad = (2 + 2i\sqrt{3})(-\sqrt{3} - i)$

$\quad\quad\quad\quad\quad\quad = -2\sqrt{3} - 2i - 6i + 2\sqrt{3}$

$\quad\quad\quad\quad\quad\quad = -8i$

$\quad (\sqrt{3} - i)^3 = (\sqrt{3} - i)(\sqrt{3} - i)(\sqrt{3} - i)$

$\quad\quad\quad\quad\quad\quad = (3 - 2i\sqrt{3} + i^2)(\sqrt{3} - i)$

$\quad\quad\quad\quad\quad\quad = (2 - 2i\sqrt{3})(\sqrt{3} - i)$

$\quad\quad\quad\quad\quad\quad = 2\sqrt{3} - 2i - 6i + 2\sqrt{3}i^2$

$\quad\quad\quad\quad\quad\quad = -8i$ (7-5)

56.

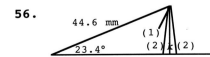

From the diagram, we can see that k = altitude of any possible triangle. Then

$$\sin \alpha = \frac{k}{b} \quad\quad k = b \sin \alpha$$

$$k = (44.6 \text{ mm})\sin 23.4° = 17.7 \text{ mm}$$

If $0 < a < k$, there is no solution: (1) in above diagram.

If $a = k$, there is one solution.

If $k < a < b$, there are two solutions: (2) in above diagram.   (7-1)

57. See Exercise 7-2, Problem 57, for a solution. (*7-2*)

58. (A) $u + v = \langle a,\ b \rangle + \langle c,\ d \rangle$

$\qquad = \langle a + c,\ b + d \rangle$ Definition of vector addition

$\qquad = \langle c + a,\ d + b \rangle$ Commutative property for addition of real numbers

$\qquad = \langle c,\ d \rangle + \langle a,\ b \rangle$ Definition of vector addition

$\qquad = v + u$

(B) $m(u + v) = m(\langle a,\ b \rangle + \langle c,\ d \rangle)$

$\qquad = m \langle a + c,\ b + d \rangle$ Definition of vector addition

$\qquad = \langle m(a + c)\ ,\ m(b + d) \rangle$ Definition of scalar multiplication

$\qquad = \langle ma + mc,\ mb + md \rangle$ Distributive property for real numbers

$\qquad = \langle ma,\ mb \rangle + \langle mc,\ md \rangle$ Definition of vector addition

$\qquad = m \langle a,\ b \rangle + m \langle c,\ d \rangle$ Definition of scalar multiplication

$\qquad = mu + mv$ (*7-3*)

59. We set up a table that indicates how r varies as we let $\dfrac{\theta}{2}$ vary through each set of quadrant values.

θ	$\dfrac{\theta}{2}$	$\cos \dfrac{\theta}{2}$	$4 \cos \dfrac{\theta}{2}$	$4 + 4 \cos \dfrac{\theta}{2}$
0 to $\dfrac{\pi}{2}$	0 to $\dfrac{\pi}{4}$	1 to $\dfrac{\sqrt{2}}{2}$	4 to $2\sqrt{2}$	8 to $4 + 2\sqrt{2}$
$\dfrac{\pi}{2}$ to π	$\dfrac{\pi}{4}$ to $\dfrac{\pi}{2}$	$\dfrac{\sqrt{2}}{2}$ to 0	$2\sqrt{2}$ to 0	$4 + 2\sqrt{2}$ to 4
π to $\dfrac{3\pi}{2}$	$\dfrac{\pi}{2}$ to $\dfrac{3\pi}{4}$	0 to $-\dfrac{\sqrt{2}}{2}$	0 to $-2\sqrt{2}$	4 to $4 - 2\sqrt{2}$
$\dfrac{3\pi}{2}$ to 2π	$\dfrac{3\pi}{4}$ to π	$-\dfrac{\sqrt{2}}{2}$ to -1	$-2\sqrt{2}$ to -4	$4 - 2\sqrt{2}$ to 0
2π to $\dfrac{5\pi}{2}$	π to $\dfrac{5\pi}{4}$	-1 to $-\dfrac{\sqrt{2}}{2}$	-4 to $-2\sqrt{2}$	0 to $4 - 2\sqrt{2}$
$\dfrac{5\pi}{2}$ to 3π	$\dfrac{5\pi}{4}$ to $\dfrac{3\pi}{2}$	$-\dfrac{\sqrt{2}}{2}$ to 0	$-2\sqrt{2}$ to 0	$4 - 2\sqrt{2}$ to 4
3π to $\dfrac{7\pi}{2}$	$\dfrac{3\pi}{2}$ to $\dfrac{7\pi}{4}$	0 to $\dfrac{\sqrt{2}}{2}$	0 to $2\sqrt{2}$	4 to $4 + 2\sqrt{2}$
$\dfrac{7\pi}{2}$ to 4π	$\dfrac{7\pi}{4}$ to 2π	$\dfrac{\sqrt{2}}{2}$ to 1	$2\sqrt{2}$ to 4	$4 + 2\sqrt{2}$ to 8

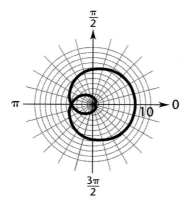

(B)

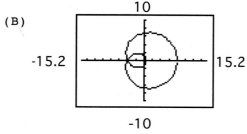

(*7-4*)

60. (A)

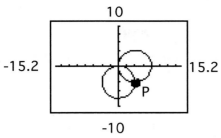

The coordinates of P represent a simultaneous solution.

(B) We solve the system
$$r = -8 \sin \theta \qquad 0 \le \theta \le \pi$$
$$r = 8 \cos \theta$$
by equating the right sides:
$$-8 \sin \theta = 8 \cos \theta$$
$$-\sin \theta = \cos \theta$$
$$-1 = \tan \theta$$
The only solution of this equation,
$0 \le \theta \le \pi$, is $\theta = \dfrac{3\pi}{4}$. If we substitute this
in either of the original equations, we get
$$r = -8 \sin \frac{3\pi}{4} = 8 \cos \frac{3\pi}{4} = 8\left(-\frac{\sqrt{2}}{2}\right) = -4\sqrt{2}$$

The coordinates of P are $\left(-4\sqrt{2}, \dfrac{3\pi}{4}\right)$.

(C) The pole is not a simultaneous solution, because the two graphs go through the pole at different values of θ. As $(0, 0)$ the pole satisfies the first equation; as $\left(0, \dfrac{\pi}{2}\right)$ it satisfies the second; it is not a solution of the system.

(7-4)

61. The solution to $x^8 - 1 = 0$ or $x^8 = 1$ will be the eight eighth roots of 1.
First write 1 in polar form:
$$1 = 1e^{0°i}$$
Using the *n*th-root theorem, all eight eighth roots of 1 are given by
$$1^{1/8}e^{(0°/8 + k \cdot 360°/8)i} = e^{k \cdot 45°i} \qquad k = 0, 1, 2, 3, 4, 5, 6, 7$$
$$w_1 = e^{0 \cdot 45°i} = e^{0°i} = \cos 0° + i \sin 0° = 1$$
$$w_2 = e^{1 \cdot 45°i} = e^{45°i} = \cos 45° + i \sin 45° = \frac{\sqrt{2}}{2} + i\frac{\sqrt{2}}{2}$$
$$w_3 = e^{2 \cdot 45°i} = e^{90°i} = \cos 90° + i \sin 90° = i$$
$$w_4 = e^{3 \cdot 45°i} = e^{135°i} = \cos 135° + i \sin 135° = -\frac{\sqrt{2}}{2} + i\frac{\sqrt{2}}{2}$$
$$w_5 = e^{4 \cdot 45°i} = e^{180°i} = \cos 180° + i \sin 180° = -1$$
$$w_6 = e^{5 \cdot 45°i} = e^{225°i} = \cos 225° + i \sin 225° = -\frac{\sqrt{2}}{2} - i\frac{\sqrt{2}}{2}$$
$$w_7 = e^{6 \cdot 45°i} = e^{270°i} = \cos 270° + i \sin 270° = -i$$
$$w_8 = e^{7 \cdot 45°i} = e^{315°i} = \cos 315° + i \sin 315° = \frac{\sqrt{2}}{2} - i\frac{\sqrt{2}}{2}$$

(7-5)

62. The linear factors of $x^3 - 8i$ will be $x - x_i$, where x_i's are zeros of $x^3 - 8i$; that is, $x^3 - 8i = 0$ or $x^3 = 8i$. So the x_i's are the three cube roots of $8i$, or $8e^{90°i}$.

Using the *n*th-root theorem, all three cube roots of $8e^{90°i}$ are given by
$$8^{1/3}e^{(90°/3 + k \cdot 360°/3)i} = 2e^{(30° + k \cdot 120°)i} \qquad k = 0, 1, 2$$
$$w_1 = 2e^{(30° + 0 \cdot 120°)i} = 2e^{30°i} = 2(\cos 30° + i \sin 30°) = 2\left(\frac{\sqrt{3}}{2} + i\frac{1}{2}\right) = \sqrt{3} + i$$
$$w_2 = 2e^{(30° + 1 \cdot 120°)i} = 2e^{150°i} = 2(\cos 150° + i \sin 150°) = 2\left(-\frac{\sqrt{3}}{2} + i\frac{1}{2}\right) = -\sqrt{3} + i$$
$$w_3 = 2e^{(30° + 2 \cdot 120°)i} = 2e^{270°i} = 2(\cos 270° + i \sin 270°) = 2(0 - 1i) = -2i$$
So $x^3 - 8i = [x - (\sqrt{3} + i)][x - (\sqrt{3} + i)](x + 2i)$

(7-5)

63. Sketch a figure.

Using $D = rt$, we have

Plane 1: $r = 256$ miles per hour, $t = 2$ hours,
$D = 256(2) = 512$ miles $= SA$

Plane 2: $r = 304$ miles per hour, $t = 2$ hours,
$D = 304(2) = 608$ miles $= SB$

The angle α between east and south-east is 45°

Given two sides and the included angle, we find d from the law of cosines:
$$d^2 = (SA)^2 + (SB)^2 - 2(SA)(SB)\cos \alpha$$
$$= (512)^2 + (608)^2 - 2(512)(608)\cos 45°.$$
$$= 191,568.97\ldots$$
$$d = 438 \text{ miles}$$

(7-3)

64. Sketch a figure.

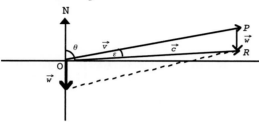

The actual course $\vec{c}$ will be the resultant of $\vec{v}$, the plane's heading, and $\vec{w}$, the wind velocity. In triangle OPR, we know $OP = |\vec{v}| = 450$, $PR = |\vec{w}| = 65$, and angle OPR, which is congruent to θ, the heading of 75°.

Given two sides and the included angle, we use the law of cosines to find
$|\vec{c}| = OR$
$$|\vec{c}|^2 = OP^2 + PR^2 - 2(OP)(PR)\cos OPR$$
$$= (450)^2 + (65)^2 - 2(450)(65)\cos 75°$$
$$= 191,584.09\ldots$$
$$|\vec{c}| = 438 \text{ miles per hour.}$$

To find the actual direction of the plane, we must find angle POR, or ε. Then the heading will be $\theta + \varepsilon$. We use the law of sines.

$$\frac{\sin \varepsilon}{PR} = \frac{\sin OPR}{OR}$$
$$\frac{\sin \varepsilon}{65} = \frac{\sin 75°}{438}$$
$$\sin \varepsilon = \frac{65 \sin 75°}{438}$$
$$\sin \varepsilon = 0.1434$$
$$\varepsilon = \sin^{-1} 0.1434 \text{ since } \varepsilon \text{ is acute}$$
$$\varepsilon = 8°$$

The heading of the plane's actual direction $= \theta + \varepsilon = 75° + 8° = 83°$.   (7-3)

65.

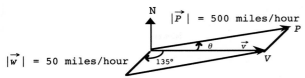

We wish to find $|\vec{v}|$ and θ such that $\vec{v}$, the resultant of $\vec{P}$, the plane's velocity, and $\vec{w}$, the wind's velocity, is in the direction due east. Applying the law of sines to triangle OPV in which we have given two sides, OP and PV, and the angle OVP opposite one of them, we obtain

$$\frac{\sin\theta}{PV} = \frac{\sin OVP}{OP}$$

$$\sin\theta = \frac{PV\sin OVP}{OP}$$

$$= \frac{50\sin 135°}{500} = 0.0707$$

Since θ must be acute, there is only one possibility:

$\theta = \sin^{-1} 0.0707$

$\quad = 4°$

The heading of the plane will therefore be 90° – 4°, or 86°. We apply the law of sines again to find $|\vec{v}|$, noting that

$\angle OPV = 180° - (\angle OVP + \theta)$

$\quad\quad\;\; = 180° - (135° + 4°)$

$\quad\quad\;\; = 41°$

$$\frac{|\vec{v}|}{\sin OPV} = \frac{OP}{\sin OVP}$$

$$|\vec{v}| = \frac{OP\sin OPV}{\sin OVP}$$

$$|\vec{v}| = \frac{500\sin 41°}{\sin 135°}$$

$|\vec{v}| = 464$ miles per hour

$(7\text{-}3)$

66.

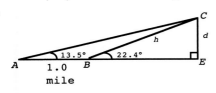

We note: In triangle ABC, angle ABC is the supplement of given angle CBE. Angle ABC has measure 180° – 22.4° = 157.6°. In triangle ABC we know two angles and the included side, so we can find angle ACB and side BC. Side BC is then the hypotenuse of right triangle BCE, so that we can find d.

Solve for angle ACB: $\quad$ 13.5° + 157.6° + $\angle ACB$ = 180°

$\angle ACB = 180° - 13.5° - 157.6°$

$\quad\quad\quad\;\; = 8.9°$

Solve for $CB = h$: $\quad \dfrac{\sin A}{h} = \dfrac{\sin ACB}{AB}$

$\quad\quad h = \dfrac{AB\sin A}{\sin ACB}$

$\quad\quad\quad = \dfrac{1.0\sin 13.5°}{\sin 8.9°}$

$\quad\quad\quad = 1.51$ miles

Solve for d: $\sin CBE = \dfrac{d}{h}$

$d = h\sin CBE$

$d = 1.51\sin 22.4°$

$d = 0.6$ miles

$(7\text{-}1)$

67.

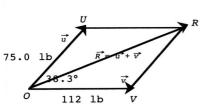

In triangle OVR, angle OVR has measure 180° – 38.3° = 141.7°. We have $OV = 112$, $VR = 75.0$, so we can use the law of cosines to find $OR = |\vec{R}|$

$|\vec{R}|^2 = (OV)^2 + (VR)^2 - 2(OV)(VR)\cos\alpha$

$\quad\quad = (112)^2 + (75.0)^2 - 2(112)(75.0)\cos 141.7°$

$\quad\quad = 31,353.243…$

$|\vec{R}| = 177$ pounds

To find the direction of $\vec{R}$ relative to v, we find angle VOR from the law of sines.

$$\frac{\sin \alpha}{\left|\vec{R}\right|} = \frac{\sin VOR}{VR}$$

$$\sin VOR = \frac{VR \sin \alpha}{\left|\vec{R}\right|}$$

$$\sin VOR = \frac{75.0 \sin 141.7°}{177}$$

$$\sin VOR = 0.2626$$

$$\angle VOR = \sin^{-1} 0.2626 \text{ since this angle must be acute}$$

$$\angle VOR = 15.2°$$

(7–3)

68.

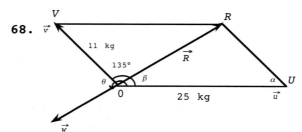

From the figure we can see that:

The two forces u and v act together as their resultant R. Needed to balance them, then, is a force w in precisely the opposite direction to R with the same magnitude as R. We calculate $\left|R\right|$, the magnitude of R, then its direction β relative to u from triangle OUR. Then the direction of w relative to u is
$\theta = 180° + \beta$.

Solve for $\left|R\right|$: We use the law of cosines, with $\alpha = 180° - 135° = 45°$.

$$\left|R\right|^2 = (OU)^2 + (UR)^2 - 2(OU)(UR)\cos \alpha$$
$$= 25^2 + 11^2 - 2(25)(11)\cos 45°$$
$$= 357.09127…$$
$$\left|R\right| = 19 \text{ kg}$$

Solve for β: We use the law of sines.

$$\frac{\sin \beta}{UR} = \frac{\sin \alpha}{\left|R\right|}$$

$$\sin \beta = \frac{UR \sin \alpha}{\left|R\right|}$$

$$= \frac{11 \sin 45°}{19} = 0.4094$$

$$\beta = \sin^{-1} 0.4094 \text{ since } \beta \text{ must be acute}$$

$$\beta = 24°$$

The direction of w is $\theta = 180° + 24° = 204°$ relative to u.

(7–3)

69. First, form a force diagram with all forces in standard position at the origin.

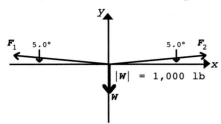

Let F_1 = the tension in left cable

F_2 = the tension in right cable

Write each force vector in terms of i and j unit vectors.

$F_1 = |F_1|(-\cos 5.0°)i + |F_1|(\sin 5.0°)j$

$F_2 = |F_2|(\cos 5.0°)i + |F_2|(\sin 5.0°)j$

$W = -1,000j$

For the system to be in static equilibrium, we must have

$F_1 + F_2 + W = O$

which becomes, on addition,

$[-|F_1|(\cos 5.0°) + |F_2|(\cos 5.0°)]i + [|F_1|(\sin 5.0°)$
$\quad\quad\quad + |F_2|(\sin 5.0°) - 1,000]j = 0i + 0j$

Since two vectors are equal if and only if their corresponding components are equal, we are led to the following system of equations in $|F_1|$ and $|F_2|$:

$-|F_1|\cos 5.0° + |F_2|\cos 5.0° = 0$

$|F_1|\sin 5.0° + |F_2|\sin 5.0° - 1,000 = 0$

To solve this system, first solve the first equation for $|F_1|$, then plug the result in for $|F_1|$ in the second equation. The result is:

$|F_1| = |F_2|$ (as we expect from the symmetry of the situation)

$|F_1| = \dfrac{1,000}{2\sin 5.0°} = 5,740$ lb = tension in each half of the cable.   (7-3)

70. (A)

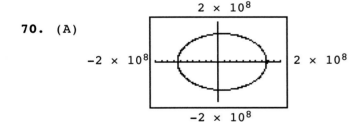

From the trace function of the utility as applied to the graph, the maximum distance (aphelion) is 1.56×10^8 miles, and the minimum distance (perihelion) is 1.29×10^8 miles.

(B) r is maximum when the denominator $1 - 0.0934 \cos \theta$ is minimum. This occurs when $\cos \theta$ is as large as possible, that is, $\cos \theta = 1$.

Then $r = \dfrac{1.41 \times 10^8}{1 - 0.0934} = 1.56 \times 10^8$ miles (aphelion)

r is minimum when the denominator $1 - 0.0934 \cos \theta$ is maximum. This occurs when $\cos \theta = -1$.

Then $r = \dfrac{1.41 \times 10^8}{1 + 0.0934} = 1.29 \times 10^8$ miles (perihelion)   (7-4)

Chapters 5, 6 & 7 Cumulative Review

1. $s = r\theta = (6 \text{ meters})(0.31 \text{ radians}) = 1.86 \text{ meters}$ *(5-1)*

2. Solve for θ:
$\quad \theta = 90° - 32.7° = 57.3°$

Solve for a:
$$\tan \theta = \frac{b}{a}$$
$$\tan 57.3° = \frac{12.2}{a}$$
$$a = \frac{12.2}{\tan 57.3°}$$
$$= 7.83 \text{ cm}$$

Solve for c:
$$\sin \theta = \frac{b}{c}$$
$$\sin 57.3° = \frac{12.2}{c}$$
$$c = \frac{12.2}{\sin 57.3°}$$
$$c = 14.5 \text{ cm} \qquad (5-3)$$

3. (A) $\dfrac{b}{r} = \sin \theta > 0$ if $b > 0$. This occurs in quadrants I, II.

(B) $\dfrac{a}{r} = \cos \theta > 0$ if $a > 0$. This occurs in quadrants I, IV.

(C) $\dfrac{b}{a} = \tan \theta > 0$ if a and b have the same sign. This occurs in quadrants I, III.   *(5-2)*

4.
$$a^2 + b^2 = r^2$$
$$(-3)^2 + 4^2 = r^2$$
$$25 = r^2$$
$$r = 5$$

(A) $\cos \theta = \dfrac{a}{r} = \dfrac{-3}{5} = -\dfrac{3}{5}$

(B) $\csc \theta = \dfrac{r}{b} = \dfrac{5}{4}$

(C) $\tan \theta = \dfrac{b}{a} = \dfrac{4}{-3} = -\dfrac{4}{3}$

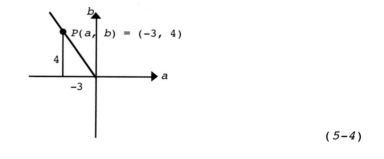

     (5-4)

5. (A)

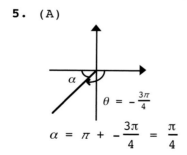

$\alpha = \pi + -\dfrac{3\pi}{4} = \dfrac{\pi}{4}$

(B)

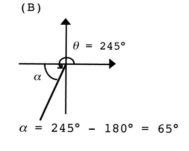

$\alpha = 245° - 180° = 65°$

(C)

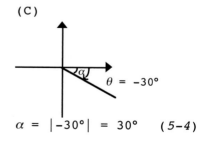

$\alpha = |-30°| = 30°$ *(5-4)*

6. (A) Domain: all real numbers; Range: $-1 \le y \le 1$; Period: 2π

(B) Domain: all real numbers; Range: $-1 \le y \le 1$; Period: 2π

(C) Domain: all real numbers except $x = \dfrac{\pi}{2} + k\pi$, k an integer; Range: all real numbers; Period: π

     (5-4)

7.

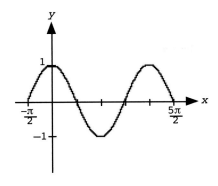

(*5-4*)

8.

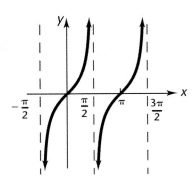

(*5-4*)

9. See figure.

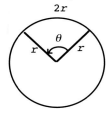

A central angle in a circle with radian measure 2 is the angle subtended by an arc of length twice the measure of the radius.

(*5-1*)

10. See figure for problem 7 above. If the graph of $y = \cos x$ is shifted $\frac{\pi}{2}$ units to the right, the result will be the graph of $y = \sin x$. (*5-4*)

11. $\cot \theta \sec \theta = \dfrac{\cos \theta}{\sin \theta} \dfrac{1}{\cos \theta}$ Quotient and Reciprocal Identities

$\qquad\qquad = \dfrac{1}{\sin \theta}$ Algebra

$\qquad\qquad = \csc \theta$ Reciprocal Identity   (*6-1*)

12. $\sec x - \cos x = \dfrac{1}{\cos x} - \cos x$ Reciprocal Identity

$\qquad\qquad = \dfrac{1}{\cos x} - \dfrac{\cos^2 x}{\cos x}$ Algebra

$\qquad\qquad = \dfrac{1 - \cos^2 x}{\cos x}$ Algebra

$\qquad\qquad = \dfrac{\sin^2 x + \cos^2 x - \cos^2 x}{\cos x}$ Pythagorean Identity

$\qquad\qquad = \dfrac{\sin^2 x}{\cos x}$ Algebra

$\qquad\qquad = \dfrac{\sin x}{\cos x} \sin x$ Algebra

$\qquad\qquad = \tan x \sin x$ Quotient Identity   (*6-1*)

13. $\sin\left(x - \dfrac{\pi}{2}\right) = \sin x \cos \dfrac{\pi}{2} - \cos x \sin \dfrac{\pi}{2}$ Sum Identity

$\qquad\qquad = \sin x(0) - \cos x(1)$ Known Values

$\qquad\qquad = -\cos x$ Algebra   (*6-2*)

14. $\csc 2x = \dfrac{1}{\sin 2x}$ Reciprocal Identity

       $= \dfrac{1}{2\sin x \cos x}$ Double-angle Identity

       $= \dfrac{1}{2} \dfrac{1}{\sin x} \dfrac{1}{\cos x}$ Algebra

       $= \dfrac{1}{2}\csc x \sec x$ Reciprocal Identities   (*6-3*)

15. (A) Graph both sides of the equation in the same viewing window.

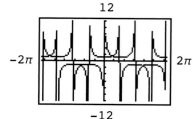

    $\dfrac{\sin^2 x}{\cos x} + \cos x = \csc x$ is not an identity, since the graphs do not match.

Try $x = \dfrac{3\pi}{4}$.

Left side: $\dfrac{\sin^2\left(\frac{3\pi}{4}\right)}{\cos\left(\frac{3\pi}{4}\right)} + \cos\left(\dfrac{3\pi}{4}\right) = \dfrac{\frac{1}{2}}{-\frac{1}{\sqrt{2}}} + \left(-\dfrac{1}{\sqrt{2}}\right) = -\sqrt{2}$

Right side: $\csc\left(\dfrac{3\pi}{4}\right) = \sqrt{2}$

This verifies that the equation is not an identity.

(B) Graph both sides of the equation in the same viewing window.

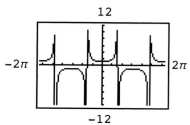

    $\dfrac{\sin^2 x}{\cos x} + \cos x = \sec x$ appears to be an identity, which we now verify:

$\dfrac{\sin^2 x}{\cos x} + \cos x = \dfrac{\sin^2 x}{\cos x} + \dfrac{\cos^2 x}{\cos x}$ Algebra

              $= \dfrac{\sin^2 x + \cos^2 x}{\cos x}$ Algebra

              $= \dfrac{1}{\cos x}$ Pythagorean Theorem

              $= \sec x$ Reciprocal Identity

                                    (*6-1*)

16. Angle α is acute. A triangle can have at most one obtuse angle. Since β is acute, then, if the triangle has an obtuse angle it must be the angle opposite the longer of the two sides, a and c. So α, the angle opposite the shorter of the two sides, a, must be acute.   (*7-2*)

17. $\sin x = 0.3188$ $0 \le x < 2\pi$
Sketch a graph of $y = \sin x$ and $y = 0.3188$
$x = [0, 2\pi)$.
$x =$
$\begin{cases} \sin^{-1} 0.3188 = 0.3245 \text{ First quadrant solution} \\ \pi - \sin^{-1} 0.3188 = 2.8171 \text{ Second quadrant solution} \end{cases}$

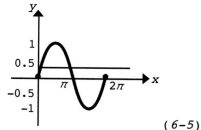

                                    (*6-5*)

18. $\tan \theta = -4.076$ $-90° < \theta < 90°$
Sketch a graph of $y = \tan \theta$ and $y = -4.076$
$\theta = (-90°, 90°)$
$\theta = \tan^{-1}(-4.076) = -76.2154°$

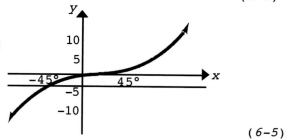

                                    (*6-5*)

19. We are given two sides and the included angle (SAS). We use the law of cosines to find the third side, then the law of sines to find a second angle.

Solve for b:
$$b^2 = a^2 + c^2 - 2ac \cos \beta$$
$$= 12^2 + 13^2 - 2(12)(13)\cos 121°$$
$$= 473.69188…$$
$$b = 22 \text{ ft}$$

Solve for α:
Note that since the sum of α and γ is $180° - 121° = 59°$, both must be acute.

$$\frac{\sin \alpha}{a} = \frac{\sin \beta}{b}$$
$$\sin \alpha = \frac{a \sin \beta}{b}$$
$$= \frac{12 \sin 121°}{22}$$
$$\alpha = \sin^{-1}\left(\frac{12 \sin 121°}{22}\right)$$
$$= 28°$$

Solve for γ:
$$\alpha + \beta + \gamma = 180°$$
$$\gamma = 180° - (\alpha + \beta)$$
$$= 180° - (28° + 121°)$$
$$= 31°$$

(7-1, 7-2)

20. The algebraic vector $\langle a, b \rangle$ corresponds to the standard geometric vector with terminal point $P = (a, b)$ and initial point $O = (0, 0)$. The coordinates of the point $P = (a, b)$ are given by
$$a = x_B - x_A = 3 - (-3) = 6$$
$$b = y_B - y_A = -1 - 2 = -3$$
$$\langle a, b \rangle = \langle 6, -3 \rangle$$

(7-3)

21. See figure.

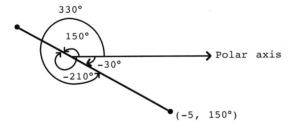

(-5, 150°)

The point with coordinates (-5, 150°) can equally well be described as (5, -30°) or (-5, -210°) or (5, 330°). We can describe each as follows: (5, -30°): The polar axis is rotated 30° clockwise (negative direction) and the point is located 5 units from the pole along the positive polar axis. (-5, -210°): The polar axis is rotated 210° clockwise (negative direction) and the point is located 5 units from the pole along the negative polar axis. (5, 330°): The polar axis is rotated 330° counterclockwise (positive direction) and the point is located 5 units from the pole along the positive polar axis.

(7-4)

22. We set up a table that indicates how r varies as we let θ vary through each set of quadrant values.

θ	$\cos \theta$	$6 \cos \theta$	
0 to $\frac{\pi}{2}$	1 to 0	6 to 0	
$\frac{\pi}{2}$ to π	0 to -1	0 to -6	
π to $\frac{3\pi}{2}$	-1 to 0	-6 to 0	Curve is traced out a second
$\frac{3\pi}{2}$ to 2π	0 to 1	0 to 6	time in this region though coordinates seem different.

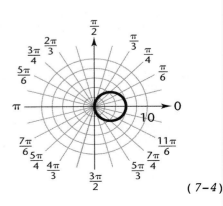

(7-4)

23.

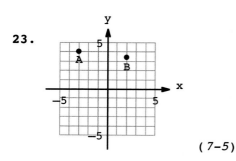

(7-5)

24. $(2e^{10°i})^3 = 2^3 e^{3 \cdot 10°i} = 8e^{30°i}$
$= 8(\cos 30° + i \sin 30°)$
$= 8\left(\dfrac{\sqrt{3}}{2} + i\dfrac{1}{2}\right)$
$= 4\sqrt{3} + 4i$ (7-5)

25. 30° is not coterminal with 150°; 30° is a I quadrant angle, but 150° is a II quadrant angle.

Since $-\dfrac{7\pi}{6} + 2\pi = \dfrac{5\pi}{6}$, which is equivalent to 150°, $-\dfrac{7\pi}{6}$ is coterminal with 150°.

Since 870° - 2(360°) = 150°, 870° is coterminal with 150°.   (5-2)

26. $\theta_D = \dfrac{180°}{\pi}\theta_R = \dfrac{180°}{\pi}(1.31) = 75.06°$ (5-1)

27. (A) and (C), since 8 radians is equivalent to the real number 8, and cosine is periodic with period 2π. (B) is not the same as cos 8, since 8° is equivalent to $\dfrac{\pi}{180°}8°$, not 8.   (5-2)

28.

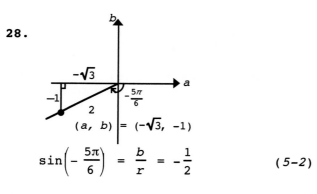

$\sin\left(-\dfrac{5\pi}{6}\right) = \dfrac{b}{r} = -\dfrac{1}{2}$ (5-2)

29.

$(a, b) = (0, 1)$ $r = 1$
$\tan \dfrac{\pi}{2} = \dfrac{b}{a} = \dfrac{1}{0}$ Not defined

30.

$\cot \dfrac{7\pi}{4} = \dfrac{a}{b} = \dfrac{1}{-1} = -1$ (5-2)

31.

$(a, b) = (\sqrt{3}, -1)$ $r = 2$

$\sec 330° = \dfrac{r}{a} = \dfrac{2}{\sqrt{3}}$ or $\dfrac{2\sqrt{3}}{3}$ (5-2)

32. $y = \cos^{-1}(-1)$ is equivalent to
$\cos y = -1$ $0 \le y \le \pi$
$y = \pi$ (5-6)

33. $\sin^{-1} x$ is not defined if $x > 1$.   (5-6)

34. $y = \arccos\left(-\frac{1}{2}\right)$ is equivalent to

$\cos y = -\frac{1}{2}$ $0 \le y \le \pi$

$\cos y = \frac{a}{r} = -\frac{1}{2}$ $a = -1$ $r = 2$

$(-1)^2 + b^2 = 2^2$

         $b^2 = 3$

         $b = \sqrt{3}$

(positive since y is a quadrant I or II angle.)

       $y = \frac{2\pi}{3}$

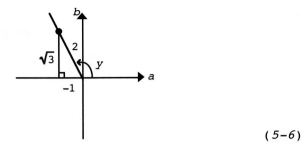

(5-6)

35. $\sin(\sin^{-1} 0.55) = 0.55$ by the sine-inverse sine identity.   (5-6)

36. Let $y = \sin^{-1}\left(-\frac{4}{5}\right)$

then $\sin y = -\frac{4}{5}$ $-\frac{\pi}{2} \le y \le \frac{\pi}{2}$

Draw the reference triangle associated with y, then $\cos y = \cos\left[\sin^{-1}\left(-\frac{4}{5}\right)\right]$ can be determined directly from the triangle.

$\sin y = \frac{b}{r} = \frac{-4}{5}$ $b = -4$ $r = 5$ $a = 3$

$\cos y = \frac{a}{r} = \frac{3}{5}$

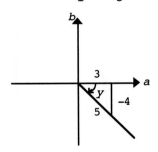

(5-6)

37. Let $y = \tan^{-1}(-2)$

then $\tan y = -2$ $-\frac{\pi}{2} < y < \frac{\pi}{2}$

Draw the reference triangle associated with y, then $\cos y = \cos[\tan^{-1}(-2)]$ can be determined directly from the triangle.

$\tan y = \frac{b}{a} = -2 = \frac{-2}{1}$

$b = -2$ $a = 1$ $r = \sqrt{5}$

$\cos y = \frac{a}{r} = \frac{1}{\sqrt{5}}$ or $\frac{\sqrt{5}}{5}$

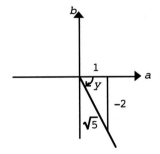

(5-6)

38. (A) 9.871   (B) -3.748   (C) -1.559

(D) Since $\tan 2.314 = -1.088 < -1$, $\cos^{-1}(\tan 2.314)$ is not defined.

                                     (5-2, 5-6)

39. $y = 2 - 2\cos\frac{\pi x}{2}$

For the graph of $y = 2\cos\frac{\pi x}{2}$, we note: $A = 2$

$P = 2\pi \div \frac{\pi}{2} = 4$, Phase shift = 0. We graph

$y = 2\cos\frac{\pi x}{2}$ turned upside down, then vertically translate the graph up 2 units.

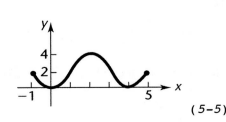

(5-5)

40. (A) $y = \cos^{-1}\left(-\dfrac{\sqrt{3}}{2}\right)$ is equivalent to

(B) $-19.755°$

$$\cos y = -\dfrac{\sqrt{3}}{2} \quad 0 \le y \le \pi$$

$$y = \dfrac{5\pi}{6}$$

Using the relation $\theta_D = \dfrac{180°}{\pi \text{ rad}}\,\theta_R$,

we have

$$\cos^{-1}\left(-\dfrac{\sqrt{3}}{2}\right) = \dfrac{180°}{\pi}\left(\dfrac{5\pi}{6}\right) = 150°$$

reference triangle associated with y

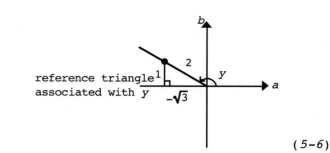

(5-6)

41. $\sin^{-1}(\sin 3) = 0.142$. For the identity $\sin^{-1}(\sin x) = x$ to hold, x must be in the restricted domain of the sine function; that is, $-\dfrac{\pi}{2} \le x \le \dfrac{\pi}{2}$. The number 3 is not in the restricted domain. *(5-6)*

42. Since the coordinates of a point on a unit circle are given by $(a, b) = (\cos x, \sin x)$, we evaluate $P = (\cos(11.205), \sin(11.205))$—using a calculator set in radian mode — to obtain $P = (0.208, -0.978)$. The quadrant in which $P = (a, b)$ lies can be determined by the signs of a and b. In this case P is in the fourth quadrant, since a is positive and b is negative. *(5-1, 5-2)*

43. The equation has infinitely many solutions [$x = \tan^{-1}(-24.5) + k\pi$, k any integer]; $\tan^{-1}(-24.5)$ has a unique value (-1.530 to three decimal places). *(5-6)*

44. $k = 3$. $A = \dfrac{1}{2}(5 - 1) = 2$ $P = 2 = \dfrac{2\pi}{\pi}$. So $B = \pi$, $y = 3 + 2 \sin \pi x$ *(5-5)*

45. $y = 3 \sin(2x - \pi)$
$A = 3$
Solve $2x - \pi = 0$ $\quad 2x - \pi = 2\pi$
$\qquad\qquad 2x = \pi$ $\qquad\quad 2x = \pi + 2\pi$
$\qquad\qquad\; x = \dfrac{\pi}{2}$ $\qquad\qquad x = \dfrac{\pi}{2} + \pi$

$\uparrow$————————$\uparrow$ $\quad \uparrow$
Phase shift $\qquad$ Period $P = \pi$

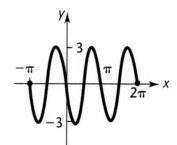

The graph completes one full cycle as x varies over the interval $\left[\dfrac{\pi}{2}, \dfrac{3\pi}{2}\right]$ *(5-5)*

46. One cycle of $y = 2 \tan\left(\dfrac{\pi x}{2} - \dfrac{\pi}{2}\right)$ is completed as $\dfrac{\pi x}{2} - \dfrac{\pi}{2}$ varies from 0 to π.

Solve each equation for x:
$\dfrac{\pi x}{2} - \dfrac{\pi}{2} = 0$ $\qquad\qquad \dfrac{\pi x}{2} - \dfrac{\pi}{2} = \pi$
$\qquad \dfrac{\pi x}{2} = \dfrac{\pi}{2}$ $\qquad\qquad\quad \dfrac{\pi x}{2} = \dfrac{\pi}{2} + \pi$
$\qquad\quad x = 1$ $\qquad\qquad\qquad\; x = 1 + 2$
Phase shift $= 1$ $\qquad\qquad$ Period $= 2$

Sketch the graph for one period [1, 3], then extend over the interval (0, 4).

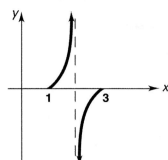 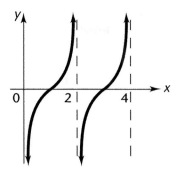

(5-5)

47. The dashed curve is $y = \sin x$. The solid curve is $y = \csc x$.

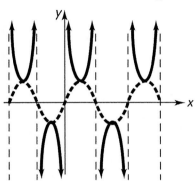

(5-4)

48. See figure.

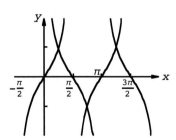

If the graph of $y = \cot x$ is shifted to the left $\dfrac{\pi}{2}$ units and reflected in the x axis, the result will be the graph of $y = \tan x$.

(5-4)

49. Here is the graph of $y = \dfrac{1}{\cot^2 x + 1}$ on the interval $[0, 2\pi]$.

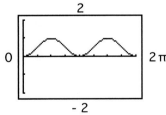

The graph has amplitude $\dfrac{1 - 0}{2} = \dfrac{1}{2}$ and period π. It appears to be the graph of $y = -\dfrac{1}{2} \cos 2x$ shifted up $\dfrac{1}{2}$ unit, that is, $y = \dfrac{1}{2} - \dfrac{1}{2} \cos 2x$. (5-5)

50. Here is the graph of $y = \dfrac{2 - 2\sin^2 x}{\sin 2x}$ on the interval $[0, 2\pi]$.

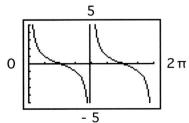

The graph appears to have the form $y = A \cot Bx$. Since the period is π, $B = 1$. The graph of $y = A \cot x$ shown appears to pass through the point $\left(\dfrac{\pi}{4}, 1\right)$, so

$$1 = A \cot \dfrac{\pi}{4}$$
$$1 = A$$

The equation of the graph can be written $y = \cot x$. (5-5)

51. (A) Check $x = 0$ Check $x = \pi$

Left side: Left side:

   $\sin 2x = \sin 2 \cdot 0 = \sin 0 = 0$ $\sin 2x = \sin 2\pi = 0$

Right side: Right side:

   $2 \sin x = 2 \sin 0 = 2 \cdot 0 = 0$ $2 \sin x = 2 \sin \pi = 2 \cdot 0 = 0$

0 is a solution π is a solution

(B) The equation is not an identity because both sides are defined at $x = \dfrac{\pi}{2}$;

for example, but $\dfrac{\pi}{2}$ is not a solution.

Left side: $\sin 2x = \sin 2\left(\dfrac{\pi}{2}\right) = \sin \pi = 0$

Right side: $2 \sin x = 2 \sin \dfrac{\pi}{2} = 2 \cdot 1 = 2$

The equation is conditional.   *(6-1)*

52. $\dfrac{\sin u}{1 + \cos u} + \cot u = \dfrac{\sin u}{1 + \cos u} + \dfrac{\cos u}{\sin u}$ Quotient Identity

$\qquad = \dfrac{\sin^2 u}{\sin u(1 + \cos u)} + \dfrac{\cos u(1 + \cos u)}{\sin u(1 + \cos u)}$ Algebra

$\qquad = \dfrac{\sin^2 u + \cos u(1 + \cos u)}{\sin u(1 + \cos u)}$ Algebra

$\qquad = \dfrac{\sin^2 u + \cos u + \cos^2 u}{\sin u(1 + \cos u)}$ Algebra

$\qquad = \dfrac{\sin^2 u + \cos^2 u + \cos u}{\sin u(1 + \cos u)}$ Algebra

$\qquad = \dfrac{1 + \cos u}{\sin u(1 + \cos u)}$ Pythagorean Identity

$\qquad = \dfrac{1}{\sin u}$ Algebra

$\qquad = \csc u$ Reciprocal Identity

Key Algebraic Steps:

$\qquad \dfrac{a}{1 + b} + \dfrac{b}{a} = \dfrac{a^2}{a(1 + b)} + \dfrac{b(1 + b)}{a(1 + b)}$

$\qquad\qquad = \dfrac{a^2 + b(1 + b)}{a(1 + b)}$

$\qquad\qquad = \dfrac{a^2 + b + b^2}{a(1 + b)}$

$\qquad\qquad = \dfrac{a^2 + b^2 + b}{a(1 + b)}$ *(6-1)*

53. $\sec x + \tan x = \dfrac{1}{\cos x} + \dfrac{\sin x}{\cos x}$ Reciprocal and Quotient Identities

$\qquad = \dfrac{1 + \sin x}{\cos x}$ Algebra

$\qquad = \dfrac{(1 + \sin x)(1 - \sin x)}{\cos x(1 - \sin x)}$ Algebra

$$= \frac{1 - \sin^2 x}{\cos x(1 - \sin x)} \qquad \text{Algebra}$$

$$= \frac{\cos^2 x + \sin^2 x - \sin^2 x}{\cos x(1 - \sin x)} \qquad \text{Pythagorean Identity}$$

$$= \frac{\cos^2 x}{\cos x(1 - \sin x)} \qquad \text{Algebra}$$

$$= \frac{\cos x}{1 - \sin x} \qquad \text{Algebra} \qquad (6\text{-}1)$$

54. $\tan \dfrac{x}{2} = \dfrac{1 - \cos x}{\sin x} \qquad$ Half-angle Identity

$$= \frac{1}{\sin x} - \frac{\cos x}{\sin x} \qquad \text{Algebra}$$

$$= \csc x - \cot x \qquad \text{Reciprocal and Quotient Identities} \qquad (6\text{-}3)$$

55. $\csc^2 \dfrac{x}{2} = \dfrac{1}{\sin^2 \frac{x}{2}} \qquad$ Reciprocal Identity

$$= \frac{1}{\left(\pm\sqrt{\frac{1-\cos x}{2}}\right)^2} \qquad \text{Half-angle Identity}$$

$$= \frac{1}{\frac{1-\cos x}{2}} \qquad \text{Algebra}$$

$$= 1 \div \frac{1 - \cos x}{2} \qquad \text{Algebra}$$

$$= 1 \cdot \frac{2}{1 - \cos x} \qquad \text{Algebra}$$

$$= \frac{2}{1 - \cos x} \qquad \text{Algebra}$$

$$= \frac{2}{1 - \cos x}\frac{1 + \cos x}{1 + \cos x} \qquad \text{Algebra}$$

$$= \frac{2(1 + \cos x)}{1 - \cos^2 x} \qquad \text{Algebra}$$

$$= \frac{2(1 + \cos x)}{\sin^2 x + \cos^2 x - \cos^2 x} \qquad \text{Pythagorean Identity}$$

$$= \frac{2(1 + \cos x)}{\sin^2 x} \qquad \text{Algebra}$$

$$= 2 \cdot \frac{1}{\sin x}\frac{1 + \cos x}{\sin x} \qquad \text{Algebra}$$

$$= 2 \cdot \frac{1}{\sin x}\left(\frac{1}{\sin x} + \frac{\cos x}{\sin x}\right) \qquad \text{Algebra}$$

$$= 2 \csc x(\csc x + \cot x) \qquad \text{Reciprocal and Quotient Identities}$$

Key Algebraic Steps:

$$\frac{1}{\frac{1-a}{2}} = 1 \div \frac{1 - a}{2} = 1 \cdot \frac{2}{1 - a} = \frac{2}{1 - a} = \frac{2}{1 - a}\frac{1 + a}{1 + a} = \frac{2(1 + a)}{1 - a^2} \qquad (6\text{-}3)$$

56. $\dfrac{2}{1 + \cos 2x} = \dfrac{2}{1 + 2\cos^2 x - 1}$ Double-angle Identity

$\qquad\qquad\qquad\quad = \dfrac{2}{2\cos^2 x}$ Algebra

$\qquad\qquad\qquad\quad = \dfrac{1}{\cos^2 x}$ Algebra

$\qquad\qquad\qquad\quad = \sec^2 x$ Reciprocal Identity $(6\text{-}3)$

57. $\dfrac{\cos x + \cos y}{\sin x - \sin y} = \dfrac{2\cos \frac{x+y}{2}\cos \frac{x-y}{2}}{2\cos \frac{x+y}{2}\sin \frac{x-y}{2}}$ Sum-product Identity

$\qquad\qquad\qquad\quad = \dfrac{\cos \frac{x-y}{2}}{\sin \frac{x-y}{2}}$ Algebra

$\qquad\qquad\qquad\quad = \cot \dfrac{x - y}{2}$ Quotient Identity $(6\text{-}4)$

58. (A) Graph both sides of the equation in the same viewing window.

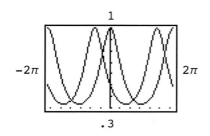

$\dfrac{\tan x}{2\tan x - \sin x} = \dfrac{1}{2 + \sin x}$ is not an identity, since the graphs do not match.

Try $x = \dfrac{\pi}{4}$

Left side: $\dfrac{\tan \frac{\pi}{4}}{2\tan \frac{\pi}{4} - \sin \frac{\pi}{4}} = \dfrac{1}{2 \cdot 1 - \frac{1}{\sqrt{2}}} = \dfrac{\sqrt{2}}{2\sqrt{2} - 1}$

Right side: $\dfrac{1}{2 + \sin \frac{\pi}{4}} = \dfrac{1}{2 + \frac{1}{\sqrt{2}}} = \dfrac{\sqrt{2}}{2\sqrt{2} + 1}$

This verifies that the equation is not an identity.

(B) Graph both sides of the equation in the same viewing window.

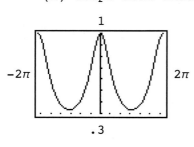

$\dfrac{\tan x}{2\tan x - \sin x} = \dfrac{1}{2 - \cos x}$ appears to be an identity, which we now verify.

$\dfrac{1}{2 - \cos x} = \dfrac{\tan x}{\tan x(2 - \cos x)}$ Algebra

$\qquad\qquad = \dfrac{\tan x}{2\tan x - \cos x \tan x}$ Algebra

$\qquad\qquad = \dfrac{\tan x}{2\tan x - \cos x \cdot \frac{\sin x}{\cos x}}$ Quotient Identity

$\qquad\qquad = \dfrac{\tan x}{2\tan x - \sin x}$ Algebra $(6\text{-}1)$

59.

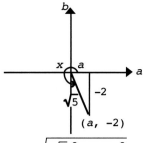

$a = \sqrt{(\sqrt{5})^2 - (-2)^2} = 1$

$\cos x = \dfrac{1}{\sqrt{5}}$

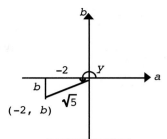

$b = -\sqrt{(\sqrt{5})^2 - (-2)^2} = -1$

$\sin y = \dfrac{-1}{\sqrt{5}}$

$\cos(x - y) = \cos x \cos y + \sin x \sin y$

$\qquad = \dfrac{1}{\sqrt{5}}\left(-\dfrac{2}{\sqrt{5}}\right) + \left(\dfrac{-2}{\sqrt{5}}\right)\left(-\dfrac{1}{\sqrt{5}}\right)$

$\qquad = \dfrac{-2}{5} + \dfrac{2}{5}$

$\qquad = 0$

$(6\text{-}2)$

60.

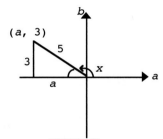

$a = -\sqrt{5^2 - 3^2} = -4$

$\cos x = -\dfrac{4}{5}$

$\sin 2x = 2 \sin x \cos x$

$\qquad = 2\left(\dfrac{3}{5}\right)\left(-\dfrac{4}{5}\right)$

$\qquad = \dfrac{-24}{25}$

Since $\dfrac{\pi}{2} \le x \le \pi$, $\dfrac{\pi}{4} \le \dfrac{x}{2} \le \dfrac{\pi}{2}$,

$\cos \dfrac{x}{2}$ will be positive

$\cos \dfrac{x}{2} = \sqrt{\dfrac{1 + \cos x}{2}}$

$\qquad = \sqrt{\dfrac{1 + \left(-\dfrac{4}{5}\right)}{2}} = \sqrt{\dfrac{\dfrac{1}{5}}{2}} = \sqrt{\dfrac{1}{10}}$ or $\dfrac{\sqrt{10}}{10}$

$(6\text{-}3)$

61.

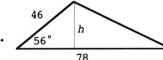

We can find h using right triangle trig ratios:

$\sin 56° = \dfrac{h}{46} \quad \Rightarrow \quad h = 46 \sin 56°$

$A = \dfrac{1}{2} bh = \dfrac{1}{2}(78)\left(46 \sin 56°\right) = 1{,}500$ sq. ft. to two

significant digits $(7\text{-}2)$

62. See Explore-Discuss 3 in Section 7-2 of your text for a discussion of Heron's formula. The semiperimeter s is given by

$s = \dfrac{4.92 + 7.33 + 5.76}{2} = 9.005$ km

Using Heron's formula, we get

$\sqrt{9.005(9.005 - 4.92)(9.005 - 7.33)(9.005 - 5.76)} = 14.1$ sq. m. to three

significant digits $\qquad (7\text{-}2)$

63.
$$2 \sin^2 \theta + \sin \theta = 1 \qquad 0° \leq \theta < 360°$$
$$2 \sin^2 \theta + \sin \theta - 1 = 0$$
$$(2 \sin \theta - 1)(\sin \theta + 1) = 0$$
$$2 \sin \theta - 1 = 0 \qquad \sin \theta + 1 = 0$$
$$\sin \theta = \frac{1}{2} \qquad \sin \theta = -1$$
$$\theta = 30°, 150° \qquad \theta = 270°$$

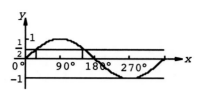

Solutions: $\theta = 30°, 150°, 270°$

$(6\text{-}5)$

64.
$$\sin 2x = \sin x$$
$$\sin 2x - \sin x = 0$$
$$2 \sin x \cos x - \sin x = 0$$
$$\sin x(2 \cos x - 1) = 0$$
$$\sin x = 0 \qquad 2 \cos x - 1 = 0$$
$$x = k\pi \qquad \cos x = \frac{1}{2}$$
k any integer

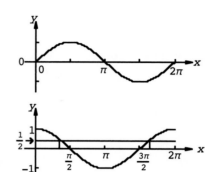

Solutions over one period
$$x = \frac{\pi}{3}, \frac{5\pi}{3}$$

Solutions over all real x:
$$x = \frac{\pi}{3} + 2k\pi, \quad \frac{5\pi}{3} + 2k\pi = -\frac{\pi}{3} + 2\pi + 2k\pi$$
$$\left. \begin{aligned} &= -\frac{\pi}{3} + (2k + 2)\pi \\ \text{or } &-\frac{\pi}{3} + 2k\pi \end{aligned} \right\} k \text{ any integer}$$

Solutions: $x = k\pi, \ \frac{\pi}{3} + 2k\pi, \ -\frac{\pi}{3} + 2k\pi, \ k$ any integer. $(6\text{-}5)$

65. (A) $\cot x = -2 \cos x$
$$\frac{\cos x}{\sin x} = -2 \cos x$$
$$\frac{\cos x}{\sin x} + 2 \cos x = 0$$
$$\frac{\cos x}{\sin x} + \frac{2 \sin x \cos x}{\sin x} = 0$$
$$\frac{\cos x + 2 \sin x \cos x}{\sin x} = 0$$

We multiply both sides by $\sin x$, keeping in mind that this could introduce extraneous solutions, so it is necessary to check the solutions in the original equation.

$$\cos x + 2 \sin x \cos x = 0$$
$$\cos x(1 + 2 \sin x) = 0$$
$$\cos x = 0 \qquad 1 + 2 \sin x = 0$$
$$x = \frac{\pi}{2}, \frac{3\pi}{2} \qquad \sin x = -\frac{1}{2}$$
$$x = \frac{7\pi}{6}, \frac{11\pi}{6}$$

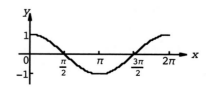

Check:

$$x = \frac{\pi}{2}$$

$$x = \frac{3\pi}{2}$$

$$x = \frac{7\pi}{6}$$

$$x = \frac{11\pi}{6}$$

$$\cot \frac{\pi}{2} \overset{?}{=} -2 \cos \frac{\pi}{2}$$

$$\cot \frac{3\pi}{2} \overset{?}{=} -2 \cos \frac{3\pi}{2}$$

$$\cot \frac{7\pi}{6} \overset{?}{=} -2 \cos \frac{7\pi}{6}$$

$$\cot \frac{11\pi}{6} \overset{?}{=} -2 \cos \frac{11\pi}{6}$$

$$0 \overset{\surd}{=} 0$$

$$0 \overset{\surd}{=} 0$$

$$\sqrt{3} \overset{\surd}{=} -2\left(-\frac{\sqrt{3}}{2}\right)$$

$$-\sqrt{3} \overset{\surd}{=} -2\left(\frac{\sqrt{3}}{2}\right)$$

A solution   A solution   A solution   A solution

Solutions: $\dfrac{\pi}{2}$, $\dfrac{3\pi}{2}$, $\dfrac{7\pi}{6}$, $\dfrac{11\pi}{6}$

(B) Examining the graphs of $y_1 = \cot x$ (drawn as $\cos x/\sin x$ in dot mode) and $y_2 = -2 \cos x$, we obtain

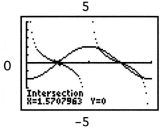
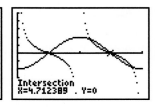
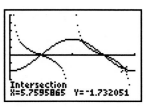

To three decimal places, the graphs intersect and $\cot x = -2 \cos x$, for $x = 1.571$, 3.665, 4.712, and 5.760.   (6-5)

66. Examining the graph of $y_1 = 2 \cos x$ and $y_2 = x - \cos 2x$, $-2\pi \le x \le 2\pi$, we obtain the graph at the right. To three decimal places, the graphs intersect and $2 \cos x = x - \cos 2x$, for $x = 0.926$ and there are no other solutions.

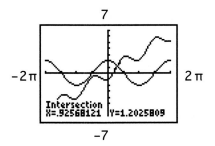

(6-5)

67.

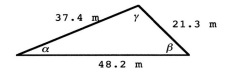

We are given three sides (SSS). We solve for the largest angle, γ (largest because it is opposite the longest side, c) using the law of cosines. We then solve for a second angle using the law of sines.

Solve for γ:

$$c^2 = a^2 + b^2 - 2ab \cos \gamma$$

$$\cos \gamma = \frac{a^2 + b^2 - c^2}{2ab}$$

$$\gamma = \cos^{-1}\left(\frac{a^2 + b^2 - c^2}{2ab}\right)$$

$$= \cos^{-1}\left(\frac{(21.3)^2 + (37.4)^2 - (48.2)^2}{2(21.3)(37.4)}\right)$$

$$= 107.2°$$

Solve for α:
We use the law of sines. Since the sum of α and β is 180° - 107.2° = 72.8°, both must be acute.

$$\frac{\sin \alpha}{a} = \frac{\sin \gamma}{c}$$

$$\sin \alpha = \frac{a \sin \gamma}{c}$$

$$= \frac{21.3 \sin 107.2°}{48.2}$$

$$\alpha = \sin^{-1}\left(\frac{21.3 \sin 107.2°}{48.2}\right)$$

$$= 25.0°$$

Solve for β:

$$\alpha + \beta + \gamma = 180°$$

$$\beta = 180° - (\alpha + \gamma)$$

$$= 180° - (25.0° + 107.2°)$$

$$= 47.8° \qquad (7\text{-}1,\ 7\text{-}2)$$

68.

α is obtuse

$$20.3 = a < b = 25.4$$

We are given two sides and a non-included angle (SSA). If we try to draw a triangle with these values, we will be unsuccessful.
No solution.

$(7\text{-}1)$

69.

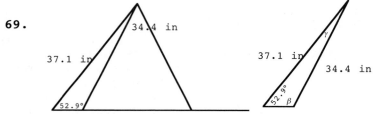

We are given two sides and a non-included angle (SSA). If we try to draw a triangle with these values, we find that two triangles are possible. Following the instructions, we choose the solution for which β is obtuse.

Solve for β:

$$\frac{\sin \beta}{b} = \frac{\sin \alpha}{a}$$

$$\sin \beta = \frac{b \sin \alpha}{a}$$

$$= \frac{37.1 \sin 52.9°}{34.4}$$

$$= 0.8602 \qquad \beta \text{ is chosen obtuse}$$

$$\beta = 180° - \sin^{-1} 0.8602$$

$$= 180° - 59.3°$$

$$= 120.7°$$

Solve for γ:

$$\alpha + \beta + \gamma = 180°$$

$$\gamma = 180° - (\alpha + \beta)$$

$$= 180° - (52.9° + 120.7°)$$

$$= 6.4°$$

Solve for c:

$$\frac{\sin \alpha}{a} = \frac{\sin \gamma}{c}$$

$$c = \frac{a \sin \gamma}{\sin \alpha}$$

$$= \frac{34.4 \sin 6.4°}{\sin 52.9°}$$

$$= 4.81 \text{ in} \qquad (7\text{-}1)$$

70. β must be acute. A triangle can have at most one obtuse angle, and since γ is acute, the obtuse angle, if present, must be opposite the longer of the two sides a and b.   *(7-2)*

71.

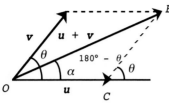

$\theta = 48.3°$, so $\angle OCB = 180° - \theta = 180° - 48.3° = 131.7°$. We can find $|u + v|$ using the law of cosines:

$$|u + v|^2 = |u|^2 + |v|^2 - 2|u||v|\cos(OCB)$$
$$= (25.3)^2 + (13.4)^2 - 2(25.3)(13.4)\cos 131.7°$$
$$= 1270.7028\ldots$$
$$|u + v| = 35.6 \text{ lb}$$

To find α, we use the law of sines:

$$\frac{\sin \alpha}{|v|} = \frac{\sin(OCB)}{|u + v|}$$

$$\frac{\sin \alpha}{13.4} = \frac{\sin 131.7°}{35.6}$$

$$\sin \alpha = \frac{13.4}{35.6} \sin 131.7°$$

$$\alpha = \sin^{-1}\left(\frac{13.4}{35.6} \sin 131.7°\right) = 16.3° \qquad \textit{(7-1, 7-2, 7-3)}$$

72. (A) $2u - v + 3w = 2\langle -1, \ 2\rangle - \langle 0, \ -2\rangle + 3\langle 1, \ -1\rangle$
$$= \langle -2, \ 4\rangle + \langle 0, \ 2\rangle + \langle 3, \ -3\rangle$$
$$= \langle 1, \ 3\rangle$$

(B) $2u - v + 3w = 2(2i - j) - (i + 3j) + 3(2j)$
$$= 4i - 2j - i - 3j + 6j$$
$$= 3i + j \qquad \textit{(7-3)}$$

73. $x^2 + y^2 = 8y$
Use $r^2 = x^2 + y^2$ and $y = r \sin \theta$
$$r^2 = 8r \sin \theta$$
$$r^2 - 8r \sin \theta = 0$$
$$r(r - 8 \sin \theta) = 0$$
$$r = 0 \quad \text{or} \quad r - 8 \sin \theta = 0$$

The graph of $r = 0$ is the pole. Since the pole is included in the graph of $r - 8 \sin \theta = 0$, we can discard $r = 0$ and keep only
$$r - 8 \sin \theta = 0 \text{ or}$$
$$r = 8 \sin \theta \qquad \textit{(7-4)}$$

74. $r = -4 \cos \theta$
We multiply both sides by r, which adds the pole to the graph. But the pole is already part of the graph, so we have changed nothing.
$$r^2 = -4r \cos \theta$$
But $r^2 = x^2 + y^2$ and $r \cos \theta = x$. So we get $x^2 + y^2 = -4x$.
   (7-4)

75. We set up a table that indicates how r varies as we let θ vary through each set of quadrant values.

θ	$\cos \theta$	$4 \cos \theta$	$4 + 4 \cos \theta$
0 to $\frac{\pi}{2}$	1 to 0	4 to 0	8 to 4
$\frac{\pi}{2}$ to π	0 to -1	0 to -4	4 to 0
π to $\frac{3\pi}{2}$	-1 to 0	-4 to 0	0 to 4
$\frac{3\pi}{2}$ to 2π	0 to 1	0 to 4	4 to 8

Check:

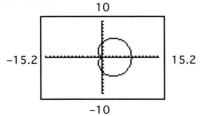

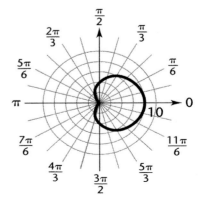

(7-4)

76. We set up a table that indicates how r varies as we let 3θ vary through each set of quadrant values.

θ	3θ	$\sin 3\theta$	$6 \sin 3\theta$
0 to $\frac{\pi}{6}$	0 to $\frac{\pi}{2}$	0 to 1	0 to 6
$\frac{\pi}{6}$ to $\frac{\pi}{3}$	$\frac{\pi}{2}$ to π	1 to 0	6 to 0
$\frac{\pi}{3}$ to $\frac{\pi}{2}$	π to $\frac{3\pi}{2}$	0 to -1	0 to -6
$\frac{\pi}{2}$ to $\frac{2\pi}{3}$	$\frac{3\pi}{2}$ to 2π	-1 to 0	-6 to 0
$\frac{2\pi}{3}$ to $\frac{5\pi}{6}$	2π to $\frac{5\pi}{2}$	0 to 1	0 to 6
$\frac{5\pi}{6}$ to π	$\frac{5\pi}{2}$ to 3π	1 to 0	6 to 0
π to 2π	repeats curve already drawn		

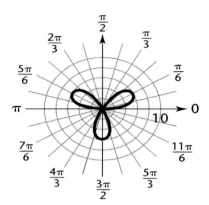

Check:

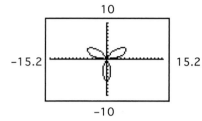

(7-4)

77. $n = 1$

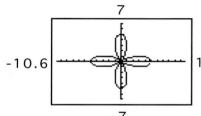

$n = 2$

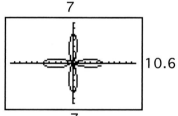

$n = 3$

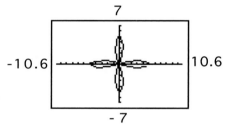

Generalizing from these graphs, we expect 4 leaves for all n. (7-4)

78.

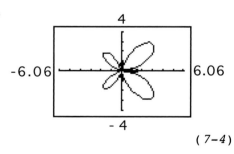

(7-4)

79. $(x, y) = (-2.78, -3.19)$

$r = \sqrt{x^2 + y^2} = \sqrt{(-2.78)^2 + (-3.19)^2} = 4.23$

$\tan \theta = \dfrac{y}{x} = \dfrac{-3.19}{-2.78}$

θ is a third quadrant angle and is to be chosen so that $-180° < \theta \le 180°$.

$\theta = \tan^{-1} \dfrac{-3.19}{-2.78} - 180° = -131.07°$

$(r, \theta) = (4.23, -131.07°)$ (7-4)

80. $(r, \theta) = (6.22, -4.08)$

$x = r \cos \theta = 6.22 \cos(-4.08) = -3.68$

$y = r \sin \theta = 6.22 \sin(-4.08) = 5.02$

$(x, y) = (-3.68, 5.02)$ (7-4)

81. $2e^{(-\pi/6)i} = 2\left[\cos\left(-\dfrac{\pi}{6}\right) + i \sin\left(-\dfrac{\pi}{6}\right)\right]$

$= 2\left[\dfrac{\sqrt{3}}{2} + i\left(-\dfrac{1}{2}\right)\right]$

$= \sqrt{3} - i$ (7-5)

82. A sketch shows that $-1 + i\sqrt{3}$ is associated with a special 30°-60° triangle. By inspection, $r = 2$, $\theta = 120°$, and

$-1 + i\sqrt{3} = 2(\cos 120° + i \sin 120°)$

$= 2e^{120°i}$

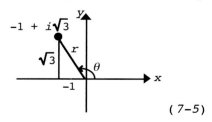

(7-5)

83. First, write $1 - i\sqrt{3}$ in polar form:

$$r = \sqrt{1^2 + \left(-\sqrt{3}\right)^2} = \sqrt{4} = 2; \quad \theta = \tan^{-1}\frac{-\sqrt{3}}{1} = -60°; \quad 1 - i\sqrt{3} = 2e^{-60°i}$$

$(1 - i\sqrt{3})^6 = (2e^{-60°i})^6 = 2^6 e^{6(-60°i)} = 64e^{-360°i} = 64[\cos(-360°) + i\sin(-360°)]$

$\qquad\qquad = 64(1) = 64 \text{ or } 64 + 0i$ (7-5)

84. First write $-i$ in polar form

$\quad -i = 0 - i = 1e^{-90°i}$

Using the nth-root theorem, all three cube roots of $1e^{-90°i}$ are given by

$\quad 1^{1/3}e^{(-90°/3 + k \cdot 360°/3)i} = 1^{1/3}e^{(-30° + k120°)i} \qquad k = 0, 1, 2$

$w_1 = 1^{1/3}e^{(-30° + 0 \cdot 120°)i} = e^{-30°i} = \cos(-30°) + i\sin(-30°) = \dfrac{\sqrt{3}}{2} - i\dfrac{1}{2}$

$w_2 = 1^{1/3}e^{(-30° + 1 \cdot 120°)i} = e^{90°i} = \cos 90° + i\sin 90° = i$

$w_3 = 1^{1/3}e^{(-30° + 2 \cdot 120°)i} = e^{210°i} = \cos 210° + i\sin 210° = -\dfrac{\sqrt{3}}{2} - i\dfrac{1}{2}$

These roots lie on a circle of radius 1, equally spaced so that the angle between successive roots is 120°.

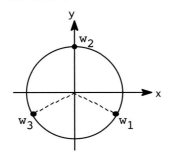

(7-5)

85. $-4.88 - 3.17i$ is not associated with a special triangle.

$r = \sqrt{(-4.88)^2 + (-3.17)^2} = 5.82$

$\theta = \tan^{-1}\dfrac{-3.17}{-4.88} - 180° = -146.99°$

$-4.88 - 3.17i = 5.82[\cos(-146.99°) + i\sin(-146.99°)]$

$\qquad\qquad = 5.82e^{(-146.99°)i}$

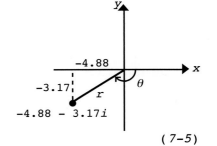

(7-5)

86. $x + iy = 6.97e^{163.87°i}$

$\qquad = 6.97(\cos 163.87° + i\sin 163.87°)$

$\qquad = -6.70 + 1.94i$ (7-5)

87. (A) There are a total of four fourth roots and they are spaced equally around a circle of radius $\sqrt{2}$, so that the angle between successive roots is 90°.

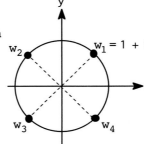

(B) Since $w_1 = \sqrt{2}\,e^{45°i}$, $w_2 = \sqrt{2}\,e^{(45° + 90°)i} = \sqrt{2}\,e^{135°i}$, $w_3 = \sqrt{2}\,e^{(135° + 90°)i} = \sqrt{2}\,e^{225°i}$, and $w_4 = \sqrt{2}\,e^{(225° + 90°)i} = \sqrt{2}\,e^{315°i}$

Then $w_2 = \sqrt{2}(\cos 135° + i \sin 135°) = -1 + i$

$w_3 = \sqrt{2}(\cos 225° + i \sin 225°) = -1 - i$

$w_4 = \sqrt{2}(\cos 315° + i \sin 315°) = 1 - i$

(C) $(1 + i)^4 = [(1 + i)^2]^2 = (1 + 2i + i^2)^2 = (2i)^2 = -4$

$(-1 + i)^4 = [(-1 + i)^2]^2 = (1 - 2i + i^2)^2 = (-2i)^2 = -4$

$(-1 - i)^4 = [(-1 - i)^2]^2 = (1 + 2i + i^2)^2 = (2i)^2 = -4$

$(1 - i)^4 = [(1 - i)^2]^2 = (1 - 2i + i^2)^2 = (-2i)^2 = -4$ *(7-5)*

88. On this unit circle,

$P = (a, b) = (\cos \theta, \sin \theta)$

$= (\cos(s \text{ rad}), \sin(s \text{ rad}))$

$= (\cos 1.2, \sin 1.2)$

$= (0.362, 0.932)$

(5-2)

89. We graph $y = \sec x$, then translate the graph up one unit.

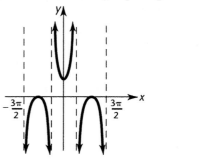

(5-4)

90. $|A| = 3$. The graph completes one full cycle as x varies over the (mentally extended) intervals $\left[-\dfrac{7}{8}, \dfrac{1}{8}\right]$ or $\left[\dfrac{1}{8}, \dfrac{9}{8}\right]$. Since the phase shift is required to be between 0 and 1, we must set $-\dfrac{C}{B} = \dfrac{1}{8}$. Then the graph has the form of a standard cosine curve.

$A = 3$

$-\dfrac{C}{B} = \dfrac{1}{8}$ $\qquad -\dfrac{C}{B} + \dfrac{2\pi}{B} = \dfrac{9}{8}$

$\dfrac{2\pi}{B} = 1 \qquad B = 2\pi \qquad C = -\dfrac{1}{8}B = -\dfrac{\pi}{4}$

The equation is then $y = A \cos(Bx + C)$

$y = 3 \cos\left(2\pi x - \dfrac{\pi}{4}\right)$

The amplitude is 3, the period is 1, and the phase shift is $\dfrac{1}{8}$. *(5-5)*

91. Here is a graph of $y = 1.6 \sin 2x - 1.2 \cos 2x$ on the interval $[-\pi, \pi]$.

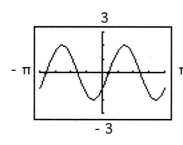

It appears that this is a sine curve shifted to the right, with $A = 2$ and, since $P = \dfrac{2\pi}{B}$ and P appears to be π, $B = \dfrac{2\pi}{P} = \dfrac{2\pi}{\pi} = 2$.

From the graphing utility, we find that the x intercept closest to the origin, to three decimal places, is 0.322. To find C, substitute $B = 2$ and $x = 0.322$ into the phase shift formula $x = -\dfrac{C}{B}$ and solve for C:

$x = -\dfrac{C}{B}$

$0.322 = -\dfrac{C}{2}$

$C = -0.644$

The equation is then $y = 2 \sin(2x - 0.644)$. *(5-5)*

92. Let $y = \cos^{-1} x$, then $x = \cos y$ $0 \le y \le \pi$

Then $\csc y = \dfrac{1}{\sin y} = \dfrac{1}{\sqrt{1 - \cos^2 y}}$ from the reciprocal and Pythagorean

identities. (The positive sign is chosen for the square root since $\sin y$ is positive in Quadrants I and II.)

$\csc y = \csc(\cos^{-1} x) = \dfrac{1}{\sqrt{1 - \cos^2 y}} = \dfrac{1}{\sqrt{1 - x^2}}.$ *(5-6)*

93. Let $u = \cot^{-1} \dfrac{3}{4}$. Then $\cot u = \dfrac{3}{4}$, $0 < u < \pi$.

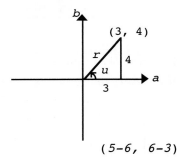

$r = \sqrt{3^2 + 4^2} = 5$

$\sin u = \dfrac{4}{5} \qquad \cos u = \dfrac{3}{5}$

$\sin\left[2 \cot^{-1}\left(\dfrac{3}{4}\right)\right] = \sin 2u = 2 \sin u \cos u = 2\left(\dfrac{4}{5}\right)\left(\dfrac{3}{5}\right) = \dfrac{24}{25}$

(5-6, 6-3)

94. $b = \sqrt{5^2 - (-3)^2} = 4$

$\cos x = -\dfrac{3}{5}$

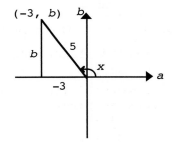

(A) Since $\dfrac{\pi}{2} \le x \le \pi$, $\dfrac{\pi}{4} \le \dfrac{x}{2} \le \dfrac{\pi}{2}$, $\sin \dfrac{x}{2}$ will be positive.

$$\sin \dfrac{x}{2} = \sqrt{\dfrac{1 - \cos x}{2}} = \sqrt{\dfrac{1 - \left(-\dfrac{3}{5}\right)}{2}}$$

$$= \sqrt{\dfrac{\dfrac{8}{5}}{2}}$$

$$= \sqrt{\dfrac{4}{5}}$$

$$= \dfrac{2}{\sqrt{5}} \text{ or } \dfrac{2\sqrt{5}}{5}$$

(B) $\cos 2x = 2 \cos^2 x - 1 = 2\left(-\dfrac{3}{5}\right)^2 - 1 = 2\left(\dfrac{9}{25}\right) - 1 = \dfrac{18}{25} - 1 = -\dfrac{7}{25}$ *(6-3)*

95. (A) $2 \sin^2 x = 3 \cos x$ $0 \le x < 2\pi$

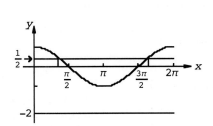

$2(1 - \cos^2 x) = 3 \cos x$

$2 - 2 \cos^2 x = 3 \cos x$

$\qquad\qquad 0 = 2 \cos^2 x + 3 \cos x - 2$

$\qquad\qquad 0 = (2 \cos x - 1)(\cos x + 2)$

$2 \cos x - 1 = 0 \qquad\qquad \cos x + 2 = 0$

$\qquad \cos x = \dfrac{1}{2} \qquad\qquad \cos x = -2$

$\qquad\qquad x = \dfrac{\pi}{3}, \dfrac{5\pi}{3} \qquad\qquad$ No solution

Solutions: $\dfrac{\pi}{3}, \dfrac{5\pi}{3}$

(B) Examining the graph of $y_1 = 2 \sin^2 x$ and $y_2 = 3 \cos x$, $0 \le x \le 2\pi$, we obtain

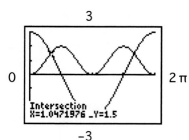

To four decimal places, the solutions are 1.0472 and 5.2360. *(6-5)*

96. (A) If $r^2 = 36 \cos 2\theta$, we can write $r = \pm\sqrt{36 \cos 2\theta}$. We then set up a table that indicates how r varies as we let 2θ vary through each set of quadrant values.

θ	2θ	$\cos 2\theta$	$36 \cos 2\theta$	$\pm\sqrt{36 \cos 2\theta}$
0 to $\frac{\pi}{4}$	0 to $\frac{\pi}{2}$	1 to 0	36 to 0	6 to 0 {-6 to 0
$\frac{\pi}{4}$ to $\frac{\pi}{2}$	$\frac{\pi}{2}$ to π	0 to -1	0 to -36	Not defined
$\frac{\pi}{2}$ to $\frac{3\pi}{4}$	π to $\frac{3\pi}{2}$	-1 to 0	-36 to 0	Not defined
$\frac{3\pi}{4}$ to π	$\frac{3\pi}{2}$ to 2π	0 to 1	0 to 36	0 to 6 {0 to -6

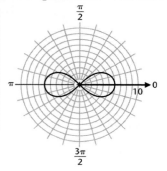

(B)

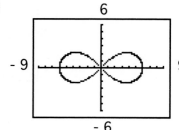

(7-4)

97. (A)

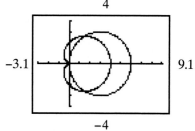

(B) 6

(C) We solve the system
$$r = 2 + 2 \cos \theta \qquad 0 \le \theta \le 2\pi$$
$$r = 6 \cos \theta$$
by equating the right sides:
$$6 \cos \theta = 2 + 2 \cos \theta$$
$$4 \cos \theta = 2$$
$$\cos \theta = \frac{1}{2}$$
$$\theta = \frac{\pi}{3}, \frac{5\pi}{3}$$

If we substitute these values of θ in either of the original equations, we get the corresponding values of r.

$$\theta = \frac{\pi}{3} \qquad r = 6 \cos \frac{\pi}{3} = 3$$

$$\theta = \frac{5\pi}{3} \qquad r = 6 \cos \frac{5\pi}{3} = 3$$

The two solutions of the system are $\left(3, \frac{\pi}{3}\right)$ and $\left(3, \frac{5\pi}{3}\right)$.

(D) The points on r_2 and r_1 arrive at the intersection points for different values of θ, except for the two found in part (C).   (7-4)

98. The linear factors of $x^3 + i$ will be $x - x_i$, where x_i are zeros of $x^3 + i$, that is, $x^3 + i = 0$ or $x^3 = -i$. So the x_i's are the three cube roots of $-i$, or $1e^{(-90°)i}$.

Using the nth-root theorem, all three cube roots of $-i$ are given by
$$1^{1/3}e^{(-90°/3 + k \cdot 360°/3)i} = 1e^{(-30° + k \cdot 120°)i} \qquad k = 0, 1, 2$$

$$w_1 = 1e^{(-30° + 0 \cdot 120°)i} = e^{-30°i} = \cos(-30°) + i \sin(-30°) = \frac{\sqrt{3}}{2} - \frac{i}{2}$$

$$w_2 = 1e^{(-30° + 1 \cdot 120°)i} = e^{90°i} = \cos 90° + i \sin 90° = i$$

$$w_3 = 1e^{(-30° + 2 \cdot 120°)i} = e^{210°i} = \cos 210° + i \sin 210° = -\frac{\sqrt{3}}{2} - \frac{i}{2}$$

$$x^3 + i = \left[x - \left(\frac{\sqrt{3}}{2} - \frac{i}{2}\right)\right](x - i)\left[x - \left(-\frac{\sqrt{3}}{2} - \frac{i}{2}\right)\right] \qquad (7-5)$$

99. In one year the line sweeps out one full revolution, or 2π radians. In 5 days the line sweeps out $\frac{5}{365}$ of a full revolution, or $\frac{5}{365} \cdot 2\pi = \frac{2\pi}{73}$ radians.   (5-1)

100. In the diagram, let E be the point from which the balloon was released. Then BDE and BCE are right triangles.

In Triangle BDE, $\cot 37° = \dfrac{DE}{h}$

In triangle BCE, $\cot 24° = \dfrac{1,000 + DE}{h}$

Then $\cot 24° - \cot 37° = \dfrac{1,000 + DE}{h} - \dfrac{DE}{h}$

$$\cot 24° - \cot 37° = \frac{1,000}{h}$$

$$h(\cot 24° - \cot 37°) = 1,000$$

$$h = \frac{1,000}{\cot 24° - \cot 37°}$$

$$h = 1,088 \text{ m} \qquad (5-3)$$

101. Sketch a figure:

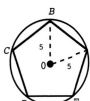

Angle $AOB = \dfrac{1}{5}(360°) = 72°$

We use the law of cosines in triangle AOB to determine side AB.

$$(AB)^2 = 5^2 + 5^2 - 2(5)(5)\cos 72°$$
$$= 34.54915\ldots$$
$$AB = 5.88 \text{ in}$$

$$(5-3, 7-2)$$

102. The three sides of triangle *ABC* are each in turn the hypotenuse of a right triangle formed with two edges of the solid. By the Pythagorean theorem,

$$AB^2 = 12^2 + 14^2 = 340 \qquad AB = 18.44 \text{ cm}$$
$$AC^2 = 12^2 + 42^2 = 1,908 \qquad AC = 43.68 \text{ cm}$$
$$BC^2 = 14^2 + 42^2 = 1,960 \qquad BC = 44.27 \text{ cm}$$

To find $\angle ABC$, we apply the law of cosines.

$$AC^2 = AB^2 + BC^2 - 2(AB)(BC)\cos ABC$$

$$\cos ABC = \frac{AB^2 + BC^2 - AC^2}{2(AB)(BC)} = \frac{340 + 1,960 - 1,908}{2(18.44)(44.27)}$$

$$\angle ABC = \cos^{-1}\left(\frac{340 + 1,960 - 1,908}{2(18.44)(44.27)}\right) = 76° \qquad (7\text{-}2)$$

103. When $t = 0$, $I = 50$. So $50 = A \cos B(0)$, that is, $A = 50$. Since the period is $\frac{1}{110}$ seconds, $\frac{2\pi}{B} = \frac{1}{110}$, $B = 220\pi$. The equation is

$$I = 50 \cos 220\pi t \qquad (5\text{-}5)$$

104. Sketch a figure.

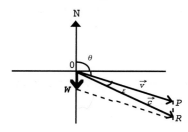

The actual course $\vec{c}$ will be the resultant of $\vec{v}$, the plane's heading, and $\vec{w}$, the wind velocity. In triangle *OPR*, we know $OP = |\vec{v}| = 260$, $PR = |\vec{w}| = 36$, and angle *OPR*, which is congruent to θ, the heading of 110°.

Given two sides and the included angle, we use the law of cosines to find

$$|\vec{c}| = OR$$
$$|\vec{c}|^2 = OP^2 + PR^2 - 2(OP)(PR)\cos OPR$$
$$= (260)^2 + (36)^2 - 2(260)(36)\cos 110°$$
$$= 75,298.617…$$
$$|\vec{c}| = 274 \text{ miles per hour.}$$

To find the actual direction of the plane, we must find angle *POR*, or ε. Then the heading will be $\theta + \varepsilon$. We use the law of sines.

$$\frac{\sin \varepsilon}{PR} = \frac{\sin OPR}{OR}$$
$$\frac{\sin \varepsilon}{36} = \frac{\sin 110°}{274}$$
$$\sin \varepsilon = \frac{36 \sin 110°}{274}$$
$$\sin \varepsilon = 0.1233$$
$$\varepsilon = \sin^{-1} 0.1233 \text{ since } \varepsilon \text{ is acute}$$
$$\varepsilon = 7°$$

The heading of the plane's actual direction = $\theta + \varepsilon$ = 110° + 7° = 117°.

$$(7\text{-}3)$$

105. First, form a force diagram with all forces in standard position at the origin.

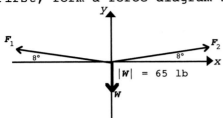

Let F_1 = the tension in left cable
F_2 = the tension in right cable

Write each force vector in terms of i and j unit vectors.
$F_1 = |F_1|(-\cos 8°)i + |F_1|(\sin 8°)j$
$F_2 = |F_2|(\cos 8°)i + |F_2|(\sin 8°)j$
$W = -65j$

For the system to be in static equilibrium, we must have
$F_1 + F_2 + W = 0$
which becomes, on addition,
$[-|F_1|\cos 8° + |F_2|\cos 8°]i + [|F_1|\sin 8° + |F_2|\sin 8° - 65]j = 0i + 0j$

Since two vectors are equal if and only if their corresponding components are equal, we are led to the following system of equations in $|F_1|$ and $|F_2|$:
$-|F_1|\cos 8° + |F_2|\cos 8° = 0$
$|F_1|\sin 8° + |F_2|\sin 8° - 65 = 0$

To solve this system, first solve the first equation for $|F_1|$, then plug the result in for $|F_1|$ in the second equation. The result is:
$|F_1| = |F_2|$ (as we expect from the symmetry of the situation)
$|F_1| = \dfrac{65}{2\sin 8°} = 234$ lb = tension in each half of the cable.   (*7-4*)

106. (A) Add the perpendicular bisector of the chord as shown in the figure. Then, $\sin \theta = \dfrac{4}{R}$ and $\theta = \dfrac{5}{R}$.
Substituting the second into the first, we obtain
$\sin \dfrac{5}{R} = \dfrac{4}{R}$.

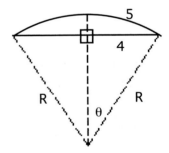

(B) R cannot be isolated on one side of the equation.

(C) Plot $y_1 = \sin \dfrac{5}{R}$ and $y_2 = \dfrac{4}{R}$ in the same viewing window and solve for R at the point of intersection using the INTERSECT command (see figure). To three decimal places, $R = 4.420$ cm.

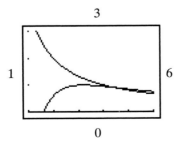

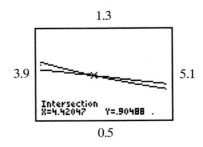

(*6-5*)

107.   (A)

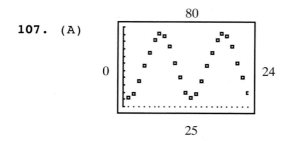

(B) $|A| = \dfrac{Y_{max} - Y_{min}}{2} = \dfrac{76 - 31}{2} = 22.5.$ So $A = 22.5$ or $-22.5.$

$k = \dfrac{Y_{max} + Y_{min}}{2} = \dfrac{76 + 31}{2} = 53.5;$ $P = 12 = \dfrac{2\pi}{B};$ then $12B = 2\pi$, and $B = \dfrac{\pi}{6}.$

The x intercept closest to the origin is estimated from the graph as 4.0. To find C, substitute $B = \dfrac{\pi}{6}$ and $x = 4.0$ into the phase shift formula $x = -\dfrac{C}{B}$ and solve for C.

$$x = -\dfrac{C}{B}$$

$$4.0 = -C \div \dfrac{\pi}{6}$$

$$C = -4.0\left(\dfrac{\pi}{6}\right)$$

$$C = -2.1$$

With this value of C, the graph is seen to be shifted up from the graph of a standard sine curve with $A = 22.5$. The equation is

$$y = 53.5 + 22.5 \sin\left(\dfrac{\pi}{6} x - 2.1\right).$$

(C)

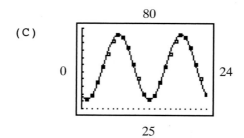

(D)

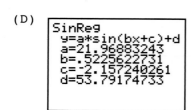

Compute the regression
equation.

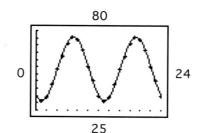

Graph the data and the
regression equation.   *(5-5)*

CHAPTER 8   Modeling with Systems of Equations and Inequalities

Section 8-1

1. Answers will vary.   3. Answers will vary.   5. Answers will vary.

7. Both lines in the given system are different, but they have the same slope $\left(\dfrac{1}{2}\right)$ and are therefore parallel. This system corresponds to (b) and has no solution.

9. In slope-intercept form, these equations are $y = 2x - 5$ and $y = -\dfrac{3}{2}x - \dfrac{3}{2}$. The first has slope 2 and y intercept -5; the other has slope $-\dfrac{3}{2}$ and y intercept $-\dfrac{3}{2}$. This system corresponds to (d) and its solution can be read from the graph as $(1, -3)$. Checking, we see that
$$2x - y = 2 \cdot 1 - (-3) = 5$$
$$3x + 2y = 3 \cdot 1 + 2(-3) = -3$$

11. $x + y = 7$
    $x - y = 3$
    If we add, we can eliminate y.
    $\begin{array}{r} x + y = 7 \\ x - y = 3 \\ \hline 2x = 10 \\ x = 5 \end{array}$
    Now substitute $x = 5$ back into the top equation and solve for y.
    $5 + y = 7$
    $y = 2$
    $(5,\ 2)$

13. $3x - 2y = 12$
    $7x + 2y = 8$
    If we add, we can eliminate y.
    $\begin{array}{r} 3x - 2y = 12 \\ 7x + 2y = 8 \\ \hline 10x = 20 \\ x = 2 \end{array}$
    Now substitute $x = 2$ back into the bottom equation and solve for y.
    $7(2) + 2y = 8$
    $2y = -6$
    $y = -3$
    $(2,\ -3)$

15. $3u + 5v = 15$
    $6u + 10v = -30$
    If we multiply the top equation by -2 and add, we eliminate both u and v.
    $\begin{array}{r} -6u - 10v = -30 \\ 6u + 10v = -30 \\ \hline 0 = -60 \end{array}$
    No solution. The equations represent parallel lines.

17. $y = 2x + 3$
    $y = 3x - 5$
    Substitute y from the first equation into the second equation to eliminate y.
    $2x + 3 = 3x - 5$
    $-x + 3 = -5$
    $-x = -8$
    $x = 8$
    Now replace x with 8 in the first equation to find y.
    $y = 2 \cdot 8 + 3$
    $y = 19$
    Solution: $x = 8,\ y = 19$

19. $x - y = 4$
    $x + 3y = 12$
    Solve the first equation for x in terms of y.
    $x = 4 + y$
    Substitute into the second equation to eliminate x.
    $(4 + y) + 3y = 12$
    $4y = 8$
    $y = 2$
    Now replace y with 2 in the first equation to find x.
    $x - 2 = 4$
    $x = 6$
    Solution: $x = 6,\ y = 2$

21. $3x - y = 7$
$2x + 3y = 1$
Solve the first equation for y in terms of x.
$-y = 7 - 3x$
$y = -7 + 3x$
Substitute into the second equation to eliminate y.
$2x + 3(-7 + 3x) = 1$
$2x - 21 + 9x = 1$
$11x = 22$
$x = 2$

Now replace x with 2 in the first equation to find y.
$3 \cdot 2 - y = 7$
$6 - y = 7$
$y = -1$
Solution: $x = 2$, $y = -1$

23. $4x + 3y = 26$
$3x - 11y = -7$
Solve the second equation for x in terms of y.
$3x = 11y - 7$
$x = \dfrac{11y - 7}{3}$
Substitute into the first equation to eliminate x.
$4\left(\dfrac{11y - 7}{3}\right) + 3y = 26$
$\dfrac{44y - 28}{3} + 3y = 26$
$44y - 28 + 9y = 78$
$53y = 106$
$y = 2$
Now replace y with 2 in the above expression for x:
$x = \dfrac{11 \cdot 2 - 7}{3}$
$x = 5$

Solution: $x = 5$, $y = 2$

25. $7m + 12n = -1$
$5m - 3n = 7$
Solve the first equation for n in terms of m.
$12n = -1 - 7m$
$n = \dfrac{-1 - 7m}{12}$
Substitute into the second equation to eliminate n.
$5m - 3\left(\dfrac{-1 - 7m}{12}\right) = 7$
$5m - \dfrac{-1 - 7m}{4} = 7$
$20m + 1 + 7m = 28$
$27m = 27$
$m = 1$
Now replace m with 1 in the above expression for n:
$n = \dfrac{-1 - 7(1)}{12}$
$n = -\dfrac{2}{3}$

Solution: $m = 1$, $n = -\dfrac{2}{3}$

27. $y = 0.08x$
$y = 100 + 0.04x$
Substitute y from the first equation into the second equation to eliminate y.
$0.08x = 100 + 0.04x$
$0.04x = 100$
$x = 2{,}500$ $y = 0.08(2{,}500)$
$y = 200$

Solution: $x = 2{,}500$, $y = 200$

29. $0.2u - 0.5v = 0.07$
$0.8u - 0.3v = 0.79$

For convenience, eliminate decimals by multiplying both sides of each equation by 100.

$20u - 50v = 7$ Don't forget to multiply right sides by 100, too!
$80u - 30v = 79$

Solve the first equation for u in terms of v and substitute into the second equation to eliminate u.

$$20u = 50v + 7$$
$$u = \frac{50v + 7}{20}$$
$$80\left(\frac{50v + 7}{20}\right) - 30v = 79$$
$$4(50v + 7) - 30v = 79$$
$$200v + 28 - 30v = 79$$
$$170v = 51$$
$$v = 0.3 \qquad u = \frac{50(0.3) + 7}{20}$$
$$u = 1.1$$

Solution: $u = 1.1$, $v = 0.3$

31. $\dfrac{2}{5}x + \dfrac{3}{2}y = 2$

$\dfrac{7}{3}x - \dfrac{5}{4}y = -5$

Eliminate fractions by multiplying both sides of the first equation by 10 and both sides of the second equation by 12.

$$10\left(\frac{2}{5}x + \frac{3}{2}y\right) = 20$$
$$4x + 15y = 20$$
$$12\left(\frac{7}{3}x - \frac{5}{4}y\right) = -60$$
$$28x - 15y = -60$$

Solve the first equation for y in terms of x and substitute into the second equation to eliminate y.

$$15y = 20 - 4x$$
$$y = \frac{20 - 4x}{15}$$
$$28x - 15\left(\frac{20 - 4x}{15}\right) = -60$$
$$28x - (20 - 4x) = -60$$
$$28x - 20 + 4x = -60$$
$$32x = -40$$
$$x = -\frac{5}{4}$$
$$y = \frac{20 - 4\left(-\frac{5}{4}\right)}{15}$$
$$y = \frac{20 + 5}{15}$$
$$y = \frac{5}{3}$$

Solution: $x = -\dfrac{5}{4}$, $y = \dfrac{5}{3}$

In Problems 33 and 35, a graphing calculator solution is presented.

33. First solve each equation for y.

$2x - 3y = -5$
$\quad -3y = -2x - 5$
$\qquad y = \dfrac{2}{3}x + \dfrac{5}{3}$

$3x + 4y = 13$
$\quad 4y = -3x + 13$
$\qquad y = -\dfrac{3}{4}x + \dfrac{13}{4}$

Next enter each equation, graph, and approximate the solution.

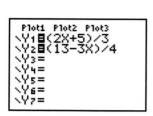

Equation definitions

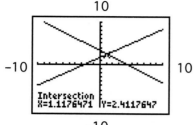

Intersection point

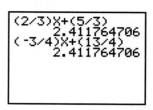

Check

To two decimal places, the solution is $x = 1.12$, $y = 2.41$ or $(1.12, 2.41)$.

35. First solve each equation for y.

$3.5x - 2.4y = 0.1$
$\quad -2.4y = -3.5x + 0.1$
$\qquad y = \dfrac{3.5}{2.4}x - \dfrac{0.1}{2.4}$

$2.6x - 1.7y = -0.2$
$\quad -1.7y = -2.6x - 0.2$
$\qquad y = \dfrac{2.6x}{1.7} + \dfrac{0.2}{1.7}$

Next enter each equation, graph, and approximate the solution.

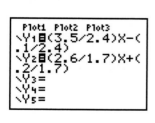

Equation definitions

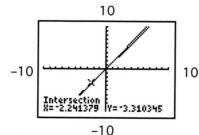

Intersection point

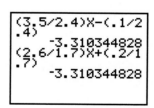

Check

To two decimal places, the solution is $x = -2.24$, $y = -3.31$ or $(-2.24, -3.31)$.

37. $x = 2 + p - 2q$
$y = 3 - p + 3q$
Solve the first equation for p in terms of q and x and substitute into the second equation to eliminate p, then solve for q in terms of x and y.

$\quad p = x - 2 + 2q$
$\quad y = 3 - (x - 2 + 2q) + 3q$
$\quad y = 3 - x + 2 - 2q + 3q$
$\quad y = 5 - x + q$
$\quad q = x + y - 5$

Now substitute this expression for q into $p = x - 2 + 2q$ to find p in terms of x and y.

$\quad p = x - 2 + 2(x + y - 5)$
$\quad p = x - 2 + 2x + 2y - 10$
$\quad p = 3x + 2y - 12$

Solution: $p = 3x + 2y - 12$, $q = x + y - 5$

To check this solution substitute into the original equations to see if true statements result:

$x = 2 + p - 2q$ $y = 3 - p + 3q$

$x \overset{?}{=} 2 + (3x + 2y - 12) - 2(x + y - 5)$ $y \overset{?}{=} 3 - (3x + 2y - 12) + 3(x + y - 5)$

$x \overset{?}{=} 2 + 3x + 2y - 12 - 2x - 2y + 10$ $y \overset{?}{=} 3 - 3x - 2y + 12 + 3x + 3y - 15$

$x \overset{\checkmark}{=} x$ $y \overset{\checkmark}{=} y$

39. $ax + by = h$

$cx + dy = k$

Solve the first equation for x in terms of y and the constants.

$ax = h - by$

$x = \dfrac{h - by}{a} \quad (a \neq 0)$

Substitute this expression into the second equation to eliminate x.

$c\left(\dfrac{h - by}{a}\right) + dy = k$ Multiply both sides by a

$ac\left(\dfrac{h - by}{a}\right) + ady = ak$

$c(h - by) + ady = ak$

$ch - bcy + ady = ak$

$ady - bcy = ak - ch$ Factor y out of the left side

$(ad - bc)y = ak - ch$

$y = \dfrac{ak - ch}{ad - bc} \quad ad - bc \neq 0$

Similarly, solve the first equation for y in terms of x and the constants.

$by = h - ax$

$y = \dfrac{h - ax}{b} \quad (b \neq 0)$

Substitute this expression into the second equation to eliminate y.

$cx + d\left(\dfrac{h - ax}{b}\right) = k$ Multiply both sides by b

$bcx + bd\left(\dfrac{h - ax}{b}\right) = bk$

$bcx + d(h - ax) = bk$

$bcx + dh - adx = bk$

$bcx - adx = bk - dh$ Factor x out of the left side

$(bc - ad)x = bk - dh$

$x = \dfrac{bk - dh}{bc - ad} \quad bc - ad \neq 0$

or, for consistency with the expression for y,

$x = \dfrac{dh - bk}{ad - bc}$

Solution: $x = \dfrac{dh - bk}{ad - bc}, \quad y = \dfrac{ak - ch}{ad - bc} \quad ad - bc \neq 0$

41. Let x = airspeed of the plane
  y = rate at which wind is blowing
Then $x - y$ = ground speed flying from Atlanta to Los Angeles (head wind)
      $x + y$ = ground speed flying from Los Angeles to Atlanta (tail wind)
Then, applying Distance = Rate × Time, we have
  $2{,}100 = 8.75(x - y)$
  $2{,}100 = 5(x + y)$
Divide both sides of the first equation by 8.75 and both sides of the second by 5:
  $x - y = 240$
  $x + y = 420$
Solve the first equation for x in terms of y and substitute into the second equation. $x = 240 + y$
  $240 + y + y = 420$
      $2y = 180$
      $y = 90$ mph = wind rate
    $x = 240 + y$
    $x = 240 + 90$
    $x = 330$ mph = airspeed

43. Let x = time rowed upstream
  y = time rowed downstream
Then $x + y = \dfrac{1}{4}$ (15 min = $\dfrac{1}{4}$ hr.)
Since   rate upstream = $20 - 2 = 18$ mph and
    rate downstream = $20 + 2 = 22$ mph,

applying Distance = Rate × Time to the equal distances upstream and downstream, we have
  $18x = 22y$
Solve the first equation for y in terms of x and substitute into the second equation.
  $y = \dfrac{1}{4} - x$
  $18x = 22\left(\dfrac{1}{4} - x\right)$
  $18x = 5.5 - 22x$
  $40x = 5.5$
  $x = 0.1375$ hr.
Then the distance rowed upstream is $18x = 18(0.1375) = 2.475$ km.

45. Let x = amount of 50% solution
  y = amount of 80% solution
100 milliliters are required, so
  $x + y = 100$
68% of the 100 milliliters must be acid, so
  $0.50x + 0.80y = 0.68(100)$
Solve the first equation for y in terms of x and substitute into the second equation.
  $y = 100 - x$
  $0.50x + 0.80(100 - x) = 0.68(100)$
  $0.5x + 80 - 0.8x = 68$
      $-0.3x = -12$
        $x = 40$ milliliters of 50% solution
    $y = 100 - x$
    $y = 100 - 40 = 60$ milliliters of 80% solution

47. "Break even" means Cost = Revenue.
Let y represent both cost and revenue (since they're equal)
Let x = number of CDs sold
$\quad y$ = Revenue = number of CDs sold × price per CD
$\quad y = x(8.00)$
$\quad y$ = Cost = Fixed Cost + Variable Cost
$\qquad\quad$ = 17,680 + number of CDs × cost per CD
$\qquad y = 17{,}680 + x(4.60)$

Substitute y from the first equation into the second equation to eliminate y.
$\quad 8.00x = 17{,}680 + 4.60x$
$\quad 3.40x = 17{,}680$
$\qquad x = \dfrac{17{,}680}{3.40}$
$\qquad x = 5{,}200$ CDs

49. Let x = amount invested at 10% $0.1x$ = yield on amount invested at 10%
$\quad\quad y$ = amount invested at 15% $0.15y$ = yield on amount invested at 15%
Then
$\qquad\qquad x + y = 12{,}000$ (amount invested)
$\quad 0.1x + 0.15y = 0.12(12{,}000)$ (total yield)

Solve the first equation for y in terms of x and substitute into the second equation.
$y = 12{,}000 - x$
$\quad 0.1x + 0.15(12{,}000 - x)\qquad\ = 0.12(12{,}000)$
$\quad 0.1x + 0.15(12{,}000) - 0.15x = 0.12(12{,}000)$
$\quad 0.1x + 1{,}800 - 0.15x\qquad\ = 1{,}440$
$\qquad\qquad -0.05x + 1{,}800 = 1{,}440$
$\qquad\qquad -0.05x\qquad\qquad = -360$
$\qquad\qquad\qquad\quad x = \$7{,}200$ invested at 10%
$\qquad y = 12{,}000 - x = \$4{,}800$ invested at 15%

51. Let x = number of hours Mexico plant is operated
$\quad\quad y$ = number of hours Taiwan plant is operated
Then (Production at Mexico plant) + (Production at Taiwan plant) = (Total Production)
$\qquad\qquad 40x \qquad\qquad + \qquad\qquad 20y \qquad\qquad = 4{,}000$ (keyboards)
$\qquad\qquad 32x \qquad\qquad + \qquad\qquad 32y \qquad\qquad = 4{,}000$ (screens)

Solve the first equation for y in terms of x and substitute into the second equation.
$\qquad\qquad 20y = 4{,}000 - 40x$
$\qquad\qquad\quad y = 200 - 2x$
$\quad 32x + 32(200 - 2x) = 4{,}000$
$\quad 32x + 6{,}400 - 64x = 4{,}000$
$\qquad\qquad\qquad -32x = -2{,}400$
$\qquad\qquad\qquad\quad x = 75$ hours Mexico plant
$\quad y = 200 - 2x = 200 - 2(75)$
$\qquad = 50$ hours Taiwan plant

53. Let x = number of grams of Mix A
$\quad\quad y$ = number of grams of Mix B
Then (Nutrition from Mix A) + (Nutrition for Mix B) = (Total Nutrition)
$\qquad\qquad 0.10x \qquad + \qquad 0.20y \qquad = \quad 20$ (Total protein)
$\qquad\qquad 0.06x \qquad + \qquad 0.02y \qquad = \quad\ 6$ (Total fat)

For convenience, eliminate decimals by multiplying both sides of the first equation by 10 and the second equation by 100.
$\qquad\qquad x + 2y = 200$
$\qquad\qquad 6x + 2y = 600$

Solve the first equation for x in terms of y and substitute into the second equation.
$$x = 200 - 2y$$
$$6(200 - 2y) + 2y = 600$$
$$1200 - 12y + 2y = 600$$
$$-10y = -600$$
$$y = 60 \text{ grams Mix } B$$
$$x = 200 - 2y = 200 - 2(60)$$
$$= 80 \text{ grams Mix } A$$

55. (A) If $p = 4$, then $4 = 0.007q + 3$, $q = \dfrac{1}{0.007} = 143$ T-shirts is the number that suppliers are willing to supply at this price.

$4 = -0.018q + 15$, $q = \dfrac{11}{0.018} = 611$ T-shirts is the number that consumers will purchase.

Demand exceeds supply and the price will rise.

(B) If $p = 8$, then $8 = 0.007q + 3$, $q = \dfrac{5}{0.007} = 714$ T-shirts is the number that suppliers are willing to supply.

$8 = -0.018q + 15$, $q = \dfrac{7}{0.018} = 389$ T-shirts is the number that consumers will purchase at this price.

Supply exceeds demand and the price will fall.

(C) Solve $p = 0.007q + 3$
$ p = -0.018q + 15$

Substitute the expression for p in terms of q from the first equation into the second equation.
$$0.007q + 3 = -0.018q + 15$$
$$0.025q + 3 = 15$$
$$0.025q = 12$$
$$q = 480 \text{ T-shirts is the}$$
$$\text{equilibrium quantity.}$$
$$p = 0.007(480) + 3 = \$6.36 \text{ is the}$$
$$\text{equilibrium price.}$$

(D)

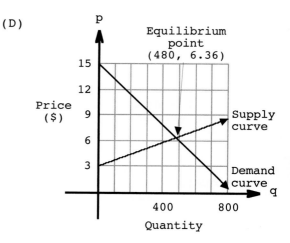

57. (A) Write $p = aq + b$.
Since $p = 0.60$ corresponds to supply $q = 450$,
$$0.60 = 450a + b$$
Since $p = 0.90$ corresponds to supply $q = 750$,
$$0.90 = 750a + b$$

Solve the first equation for b in terms of a and substitute into the second equation.
$$b = 0.60 - 450a$$
$$0.90 = 750a + 0.60 - 450a$$
$$0.30 = 300a$$
$$a = 0.001$$
$$b = 0.60 - 450a = 0.60 - 450(0.001)$$
$$= 0.15$$
The supply equation is $p = 0.001q + 0.15$.

(B)   Write $p = cq + d$.
Since $p = 0.60$ corresponds to demand $q = 645$,
$$0.60 = 645c + d$$
Since $p = 0.90$ corresponds to demand $q = 495$,
$$0.90 = 495c + d$$
Solve the first equation for d in terms of c and substitute into the second equation.
$$d = 0.60 - 645c$$
$$0.90 = 495c + 0.60 - 645c$$
$$0.30 = -150c$$
$$c = -0.002$$
$$d = 0.60 - 645c = 0.60 - 645(-0.002)$$
$$= 1.89$$
The demand equation is $p = -0.002q + 1.89$

(C)   Solve the system of equations
$$p = 0.001q + 0.15$$
$$p = -0.002q + 1.89$$
Substitute p from the first equation into the second equation to eliminate p.
$$0.001q + 0.15 = -0.002q + 1.89$$
$$0.003q = 1.74$$
$$q = 580 \text{ bushels} = \text{equilibrium quantity}$$
$$p = 0.001q + 0.15 = 0.001(580) + 0.15$$
$$= \$0.73 \text{ equilibrium price}$$

59. $s = a + bt^2$

(A)   We are given: When $t = 1$, $s = 180$
When $t = 2$, $s = 132$

Substituting these values in the given equation, we have
$$180 = a + b(1)^2$$
$$132 = a + b(2)^2 \quad \text{or} \quad 180 = a + b$$
$$132 = a + 4b$$
Solve the first equation for a in terms of b and substitute into the second equation to eliminate a.
$$a = 180 - b$$
$$132 = 180 - b + 4b$$
$$132 = 180 + 3b$$
$$-48 = 3b$$
$$b = -16 \qquad 180 = a - 16$$
$$a = 196$$

(B) The height of the building is represented by s, the distance of the object above the ground, when $t = 0$. Since we now know
$$s = 196 - 16t^2$$
from part (A), when $t = 0$, $s = 196$ feet is the height of the building.

(C) The object falls until s, its distance above the ground, is zero. Since
$$s = 196 - 16t^2$$
we substitute $s = 0$ and solve for t.
$$0 = 196 - 16t^2$$
$$16t^2 = 196$$
$$t^2 = \frac{196}{16}$$
$$t = \frac{14}{4} \quad \text{(discarding the negative solution)}$$
$$t = 3.5 \text{ seconds}$$

61. Let p = time of primary wave
   s = time for secondary wave
We know
$s - p = 16$ (time difference)
To find a second equation, we have to use Distance = Rate × Time
   $5p$ = distance for primary wave
   $3s$ = distance for secondary wave
These distances are equal, so
   $5p = 3s$

Solve the first equation for s in terms of p and substitute into the second equation to eliminate s.
   $s = p + 16$
   $5p = 3(p + 16)$
   $5p = 3p + 48$
   $2p = 48$
   $p = 24$ seconds $s = 24 + 16$
                       $s = 40$ seconds
The distance traveled is $5p = 120$ miles.

Section 8-2

1. Answers will vary.   **3.** Answers will vary.

Note: The checking steps are not shown, but should be performed by the student.

5.
$$\begin{array}{lll} -2x & = 2 & E_1 \\ x - 3y & = 2 & E_2 \\ -x + 2y + 3z & = -7 & E_3 \end{array}$$

Solve E_1 for x.
$$\begin{array}{ll} -2x = 2 & E_1 \\ x = -1 \end{array}$$

Substitute $x = -1$ in E_2 and solve for y.
$$\begin{array}{ll} x - 3y = 2 & E_2 \\ -1 - 3y = 2 \\ y = -1 \end{array}$$

Substitute $x = -1$ and $y = -1$ in E_3 and solve for z.
$$\begin{array}{ll} -x + 2y + 3z = -7 & E_3 \\ -(-1) + 2(-1) + 3z = -7 \\ z = -2 \end{array}$$

$(-1, -1, -2)$

7.
$$\begin{array}{lll} 2y - z = 2 & E_1 \\ -4y + 2z = 1 & E_2 \\ x - 2y + 3z = 0 & E_3 \end{array}$$

Multiply E_1 by 2 and add to E_2
$$\begin{array}{ll} 4y - 2z = 4 & 2E_1 \\ \underline{-4y + 2z = 1} & E_2 \\ 0 = 5 & E_4 \end{array}$$

A contradiction. No solution.

9.
$$\begin{array}{lll} x - 3y & = 2 & E_1 \\ 2y + z & = -1 & E_2 \\ x - y + z & = 1 & E_3 \end{array}$$

Multiply E_1 by -1 and add to E_3 to eliminate x.
$$\begin{array}{lll} -x + 3y & = -2 & (-1)E_1 \\ \underline{x - y + z = 1} & & E_3 \\ 2y + z & = -1 & E_4 \end{array}$$

Equivalent system:
$$\begin{array}{lll} x - 3y & = 2 & E_1 \\ 2y + z & = -1 & E_2 \\ 2y + z & = -1 & E_4 \end{array}$$

If E_2 is multiplied by -1 and added to E_4, 0 = 0 results. The system is dependent and equivalent to

$$x - 3y = 2$$
$$2y + z = -1$$

Let $y = s$. Then

$$2s + z = -1$$
$$z = -2s - 1$$
$$x - 3s = 2$$
$$x = 3s + 2$$

Solutions: $\{(3s + 2, s, -2s - 1) \mid s$ is any real number$\}$

11.
$$4y - z = -13 \quad E_1$$
$$3y + 2z = 4 \quad E_2$$
$$6x - 5y - 2z = -18 \quad E_3$$

Multiply E_1 by 2 and add to E_2 to eliminate z.

$$8y - 2z = -26 \qquad 2E_1$$
$$\underline{3y + 2z = 4} \qquad E_2$$
$$11y = -22 \qquad E_4$$
$$y = -2$$

Substitute $y = -2$ into E_2 and solve for z.

$$3y + 2z = 4 \quad E_2$$
$$3(-2) + 2z = 4$$
$$z = 5$$

Substitute $y = -2$ and $z = 5$ into E_3 and solve for x.

$$6x - 5(-2) - 2(5) = -18$$
$$x = -3$$

$(-3, -2, 5)$

13.
$$2x + y - z = 5 \qquad E_1$$
$$x - 2y - 2z = 4 \qquad E_2$$
$$3x + 4y + 3z = 3 \qquad E_3$$

Multiply E_2 by -2 and add to E_1 to eliminate x. Also multiply E_2 by -3 and add to E_3 to eliminate x.

$$2x + y - z = 5 \qquad E_1$$
$$\underline{-2x + 4y + 4z = -8} \qquad (-2)E_2$$
$$5y + 3z = -3 \qquad E_4$$

$$3x + 4y + 3z = 3 \qquad E_3$$
$$\underline{-3x + 6y + 6z = -12} \qquad (-3)E_2$$
$$10y + 9z = -9 \qquad E_5$$

Equivalent system:

$$x - 2y - 2z = 4 \qquad E_2$$
$$5y + 3z = -3 \qquad E_4$$
$$10y + 9z = -9 \qquad E_5$$

Multiply E_4 by -2 and add to E_5 to eliminate y.

$$-10y - 6z = 6 \qquad (-2)E_4$$
$$\underline{10y + 9z = -9} \qquad E_5$$
$$3z = -3$$
$$z = -1$$

Substitute $z = -1$ into E_4 and solve for y.

$$5y + 3z = -3 \qquad E_4$$
$$5y + 3(-1) = -3$$
$$y = 0$$

Substitute $y = 0$ and $z = -1$ into E_2 and solve for x.

$x - 2y - 2z = 4$ E_2

$x - 2(0) - 2(-1) = 4$

$x = 2$

$(2, 0, -1)$

15. $x - y + z = 1$ E_1

$2x + y + z = 6$ E_2

$7x - y + 5z = 15$ E_3

Multiply E_1 by -2 and add to E_2 to eliminate x. Also multiply E_1 by -7 and add to E_3 to eliminate x.

$-2x + 2y - 2z = -2$ $(-2)E_1$

$\underline{2x + y + z = 6}$ E_2

$3y - z = 4$ E_4

$-7x + 7y - 7z = -7$ $(-7)E_1$

$\underline{7x - y + 5z = 15}$ E_3

$6y - 2z = 8$ E_5

Equivalent system:

$x - y + z = 1$ E_1

$3y - z = 4$ E_4

$6y - 2z = 8$ E_5

If E_4 is multiplied by -2 and added to E_5, $0 = 0$ results. The system is dependent and equivalent to

$x - y + z = 1$

$3y - z = 4$

Let $y = s$. Then $3s - z = 4$

$z = 3s - 4$

$x - s + (3s - 4) = 1$

$x = -2s + 5$

Solutions: $\{(-2s + 5,\ s,\ 3s - 4) \mid s \text{ is any real number}\}$

17. $2a + 4b + 3c = -6$ E_1

$a - 3b + 2c = -15$ E_2

$a + 2b - c = 9$ E_3

Add E_2 to E_3 to eliminate a. Also multiply E_2 by -2 and add to E_1 to eliminate a.

$a - 3b + 2c = -15$ E_2

$\underline{-a + 2b - c = 9}$ E_3

$-b + c = -6$ E_4

$-2a + 6b - 4c = 30$ $(-2)E_2$

$\underline{2a + 4b + 3c = -6}$ E_1

$10b - c = 24$ E_5

Equivalent system:

$a - 3b + 2c = -15$ E_2

$-b + c = -6$ E_4

$10b - c = 24$ E_5

Add E_4 to E_5 to eliminate c

$-b + c = -6$ E_4

$\underline{10b - c = 24}$ E_5

$9b = 18$

$b = 2$

Substitute $b = 2$ into E_4 and solve for c.

$$-b + c = -6 \qquad E_4$$
$$-2 + c = -6$$
$$c = -4$$

Substitute $b = 2$ and $c = -4$ into E_2 and solve for a.

$$a - 3b + 2c = -15 \qquad E_2$$
$$a - 3(2) + 2(-4) = -15$$
$$a = -1$$

$(-1, 2, -4)$

19.
$$2x - 3y + 3z = -5 \qquad E_1$$
$$3x + 2y - 5z = 34 \qquad E_2$$
$$5x - 4y - 2z = 23 \qquad E_3$$

Multiply E_1 by $-\dfrac{3}{2}$ and add to E_2 to eliminate x. Also multiply E_1 by $-\dfrac{5}{2}$ and add to E_3 to eliminate x.

$$-3x + \frac{9}{2}y - \frac{9}{2}z = \frac{15}{2} \qquad \left(-\frac{3}{2}\right)E_1$$
$$\underline{3x + 2y - 5z = 34} \qquad E_2$$
$$\frac{13}{2}y - \frac{19}{2}z = \frac{83}{2} \qquad E_4$$

$$-5x + \frac{15}{2}y - \frac{15}{2}z = \frac{25}{2} \qquad \left(-\frac{5}{2}\right)E_1$$
$$\underline{5x - 4y - 2z = 23} \qquad E_3$$
$$\frac{7}{2}y - \frac{19}{2}z = \frac{71}{2} \qquad E_5$$

Equivalent system:
$$2x - 3y + 3z = -5 \qquad E_1$$
$$\frac{13}{2}y - \frac{19}{2}z = \frac{83}{2} \qquad E_4$$
$$\frac{7}{2}y - \frac{19}{2}z = \frac{71}{2} \qquad E_5$$

Multiply E_4 by -1 and add to E_5 to eliminate z.

$$-\frac{13}{2}y + \frac{19}{2}z = -\frac{83}{2} \qquad (-1)E_4$$
$$\underline{\frac{7}{2}y - \frac{19}{2}z = \frac{71}{2}} \qquad E_5$$
$$-3y = -6 \qquad E_6$$

Solve E_6 for y to obtain $y = 2$. Substitute $y = 2$ into E_4 and solve for z.

$$\frac{13}{2}y - \frac{19}{2}z = \frac{83}{2} \qquad E_4$$
$$\frac{13}{2}(2) - \frac{19}{2}z = \frac{83}{2}$$
$$-\frac{19}{2}z = \frac{57}{2}$$
$$z = -3$$

Substitute $y = 2$ and $z = -3$ into E_1 and solve for x.

$$2x - 3y + 3z = -5 \qquad E_1$$
$$2x - 3(2) + 3(-3) = -5$$
$$x = 5$$

$(5, 2, -3)$

21.
$$\begin{array}{rl} x + 2y + z = 2 & E_1 \\ -2x + 3y - 2z = -3 & E_2 \\ x - 5y + z = 2 & E_3 \end{array}$$

Multiply E_1 by 2 and add to E_2 to eliminate x. Also multiply E_1 by -1 and add to E_3 to eliminate x.

$$\begin{array}{rl} 2x + 4y + 2z = 4 & 2E_1 \\ \underline{-2x + 3y - 2z = -3} & E_2 \\ 7y \quad\quad = 1 & E_4 \end{array}$$

$$\begin{array}{rl} -x - 2y - z = -2 & (-1)E_1 \\ \underline{x - 5y + z = 2} & E_3 \\ -7y = 0 & E_5 \end{array}$$

Equivalent system

$$\begin{array}{rl} x + 2y + z = 2 & E_1 \\ 7y \quad\quad = 1 & E_4 \\ -7y = 0 & E_5 \end{array}$$

y cannot simultaneously satisfy E_4 and E_5; this gives us a contradiction (You can also obtain a contradiction by adding E_4 and E_5). There is no solution.

23.
$$\begin{array}{rl} -x + 2y - z = -4 & E_1 \\ 2x + 5y - 4z = -16 & E_2 \\ x + y - z = -4 & E_3 \end{array}$$

Multiply E_1 by 2 and add to E_2 to eliminate x. Also add E_1 to E_3 to eliminate x.

$$\begin{array}{rl} -2x + 4y - 2z = -8 & 2E_1 \\ \underline{2x + 5y - 4z = -16} & E_2 \\ 9y - 6z = -24 & E_4 \end{array}$$

$$\begin{array}{rl} -x + 2y - z = -4 & E_1 \\ \underline{x + y - z = -4} & E_3 \\ 3y - 2z = -8 & E_5 \end{array}$$

Equivalent system:

$$\begin{array}{rl} -x + 2y - z = -4 & E_1 \\ 9y - 6z = -24 & E_4 \\ 3y - 2z = -8 & E_5 \end{array}$$

If E_5 is multiplied by -3 and added to E_4, $0 = 0$ results. The system is dependent and equivalent to

$$\begin{array}{rl} -x + 2y - z = -4 \\ 3y - 2z = -8 \end{array}$$

Let $z = s$. Then

$$3y - 2s = -8$$
$$y = \frac{2s - 8}{3} \text{ or } \frac{2}{3}s - \frac{8}{3}$$

$$-x + 2\left(\frac{2}{3}s - \frac{8}{3}\right) - s = -4$$

$$-x = -\frac{4}{3}s + \frac{16}{3} + s - 4$$

$$x = \frac{1}{3}s - \frac{4}{3}$$

Solutions: $\left\{\left(\frac{1}{3}s - \frac{4}{3}, \frac{2}{3}s - \frac{8}{3}, s\right) \,\Big|\, s \text{ is any real number}\right\}$

25. $2x + 3y + 2z = 2 \qquad E_1$
$5x + 8y + 4z = 3 \qquad E_2$
$4x + 7y + 2z = 0 \qquad E_3$

Multiply E_1 by $-\dfrac{5}{2}$ and add to E_2 to eliminate x. Also multiply E_1 by -2 and add to E_3 to eliminate x.

$-5x - \dfrac{15}{2}y - 5z = -5 \qquad \left(-\dfrac{5}{2}\right)E_1$

$\underline{5x + 8y + 4z = 3} \qquad E_2$

$ \dfrac{1}{2}y - z = -2 \qquad E_4$

$-4x - 6y - 4z = -4 \qquad (-2)E_1$

$\underline{4x + 7y + 2z = 0} \qquad E_3$

$ y - 2z = -4 \qquad E_5$

Equivalent system:
$2x + 3y + 2z = 2 \qquad E_1$

$ \dfrac{1}{2}y - z = -2 \qquad E_4$

$ y - 2z = -4 \qquad E_5$

If E_4 is multiplied by -2 and added to E_5, $0 = 0$ results. The system is dependent and equivalent to
$2x + 3y + 2z = 2$
$ y - 2z = -4$
Let $z = s$. Then
$ y - 2s = -4$
$ y = 2s - 4$
$2x + 3(2s - 4) + 2s = 2$
$ 2x = -6s + 12 - 2s + 2$
$ x = -4s + 7$
Solutions: $\{(-4s + 7,\ 2s - 4,\ s) \mid s \text{ is any real number}\}$

27. If the points are on the graph, the coordinates satisfy the equation. Substituting, in turn, $(-2, 9)$, $(1, -9)$ and $(4, 9)$ into $ax^2 + bx + c = y$ yields
$4a - 2b + c = 9 \qquad E_1$
$a + b + c = -9 \qquad E_2$
$16a + 4b + c = 9 \qquad E_3$
Multiply E_2 by -1 and add to E_1 to eliminate c. Also multiply E_2 by -1 and add to E_3 to eliminate c.
$-a - b - c = 9 \qquad (-1)E_2$
$\underline{4a - 2b + c = 9} \qquad E_1$
$3a - 3b = 18 \qquad E_4$
$-a - b - c = 9 \qquad (-1)E_2$
$\underline{16a + 4b + c = 9} \qquad E_3$
$15a + 3b = 18 \qquad E_5$
Equivalent system
$a + b + c = -9 \qquad E_2$
$3a - 3b = 18 \qquad E_4$
$15a + 3b = 18 \qquad E_5$
Add E_4 to E_5 to eliminate b.
$3a - 3b = 18$
$\underline{15a + 3b = 18}$
$18a = 36$
$ a = 2$

Substitute $a = 2$ into E_5 and solve for b.

$$15a \quad + 3b = 18 \qquad E_5$$
$$15(2) + 3b = 18$$
$$b = -4$$

Substitute $a = 2$ and $b - 4$ into E_2 and solve for c.

$$a + b + c = -9 \qquad E_2$$
$$2 + (-4) + c = -9$$
$$c = -7$$

$$a = 2, \; b = -4, \; c = -7$$

29. If the points are on the graph, the coordinates satisfy the equation. Substituting, in turn, $(-1, 1)$, $(5, 1)$, and $(6, -6)$ into $x^2 + y^2 + ax + by + c = 0$ yields

$$1 + 1 - \quad a + b + c = 0$$
$$25 + 1 + 5a + b + c = 0$$
$$36 + 36 + 6a \; - 6b + c = 0$$

Simplifying yields

$$-a + \quad b + c = \quad -2 \qquad E_1$$
$$5a + \quad b + c = -26 \qquad E_2$$
$$6a - 6b + c = -72 \qquad E_3$$

Multiply E_1 by -1 and add to E_2 to eliminate c (and b).

Multiply E_1 by -1 and add to E_3 to eliminate c.

$$a - b - c = \quad 2 \qquad (-1)E_1$$
$$\underline{5a + b + c = -26} \qquad E_2$$
$$6a \qquad\qquad = -24 \qquad E_4$$

$$a -, \; b - c = \quad 2 \qquad (-1)E_1$$
$$\underline{6a - 6b + c = -72} \qquad E_3$$
$$7a - 7b \qquad = -70 \qquad E_5$$

Equivalent system:

$$-a + \quad b + c = \quad -2 \qquad E_1$$
$$6a \qquad\qquad = -24 \qquad E_4$$
$$7a - 7b \qquad = -70 \qquad E_5$$

Solve E_4 for a to obtain $a = -4$. Substitute $a = -4$ into E_5 and solve for b.

$$7a - 7b = -70 \qquad E_5$$
$$7(-4) - 7b = -70$$
$$b = \quad 6$$

Substitute $a = -4$ and $b = 6$ into E_1 and solve for c.

$$-a + b + c = \quad -2 \qquad E_1$$
$$-(-4) + 6 + c = \quad -2$$
$$c = -12$$

$$a = -4, \; b = 6, \; c = -12$$

31. Let

$$x = \text{number of lawn mowers manufactured each week}$$
$$y = \text{number of snowblowers manufactured each week}$$
$$z = \text{number of chain saws manufactured each week}$$

Then

E_1 $20x + 30y + 45z = 35{,}000$ Labor
E_2 $35x + 50y + 40z = 50{,}000$ Materials
E_3 $15x + 25y + 10z = 20{,}000$ Shipping

Multiply E_3 by -4.5 and add to E_1 to eliminate z. Also multiply E_3 by -4 and add to E_2 to eliminate z.

$$\begin{array}{rl}
20x \quad + 30y \quad + 45z = \quad 35{,}000 & E_1 \\
\underline{-67.5x - 112.5y - 45z = -90{,}000} & (-4.5)E_3 \\
-47.5x - \quad 82.5y \qquad\quad = -55{,}000 & E_4
\end{array}$$

$$\begin{array}{rl}
35x + \quad 50y \quad + 40z = \quad 50{,}000 & E_2 \\
\underline{-60x - 100y \quad - 40z = -80{,}000} & (-4)E_3 \\
-25x - \quad 50y \qquad\quad = -30{,}000 & E_5
\end{array}$$

Equivalent system:

$$\begin{array}{rl}
15x \quad + 25y + 10z = \quad 20{,}000 & E_3 \\
-47.5x \quad - 82.5y \qquad\quad = -55{,}000 & E_4 \\
-25x \quad - 50y \qquad\quad = -30{,}000 & E_5
\end{array}$$

Multiply E_5 by -1.9 and add to E_4 to eliminate x.

$$\begin{array}{rl}
47.5x + 95y \quad = \quad 57{,}000 & (-1.9)E_5 \\
\underline{47.5x - 82.5y = -55{,}000} & E_4 \\
12.5y = \quad 2{,}000 \\
y = \qquad 160
\end{array}$$

Substitute $y = 160$ into E_5 and solve for x.

$$\begin{array}{rl}
-25x - 50y \qquad = -30{,}000 & E_5 \\
-25x - 50(160) = -30{,}000 \\
x = \qquad 880
\end{array}$$

Substitute $x = 880$ and $y = 160$ into E_3 and solve for z

$$\begin{array}{rl}
15x \quad + 25y \qquad + 10z = 20{,}000 & E_3 \\
15(880) + 25(160) + 10z = 20{,}000 \\
z = 280
\end{array}$$

880 lawn mowers, 160 snowblowers, 280 chain saws.

33. Let

x = number of days operating the Michigan plant
y = number of days operating the New York plant
z = number of days operating the Ohio plant

Then

$$\begin{array}{lll}
E_1 & 10x + 70y + 60z = 2{,}150 & \text{Notebooks} \\
E_2 & 20x + 50y + 80z = 2{,}300 & \text{Desktops} \\
E_3 & 40x + 30y + 90z = 2{,}500 & \text{Servers}
\end{array}$$

Multiply E_1 by -2 and add to E_2 to eliminate x. Also multiply E_1 by -4 and add to E_3 to eliminate x.

$$\begin{array}{rl}
-20x - 140y - 120z = -4{,}300 & (-2)E_1 \\
\underline{20x + \quad 50y + \quad 80z = \quad 2{,}300} & E_2 \\
-90y - \quad 40z = -2{,}000 & E_4
\end{array}$$

$$\begin{array}{rl}
-40x - 280y - 240z = -8{,}600 & (-4)E_1 \\
\underline{40x + \quad 30y + \quad 90z = \quad 2{,}500} & E_3 \\
-250y - 150z = -6{,}100 & E_5
\end{array}$$

Equivalent system:

$$\begin{array}{rl}
10x + 70y + \quad 60z = \quad 2{,}150 & E_1 \\
-90y - \quad 40z = -2{,}000 & E_4 \\
-250y - 150z = -6{,}100 & E_5
\end{array}$$

Multiply E_5 by -0.36 and add to E_4 to eliminate y.

$$\begin{array}{rl}
-90y - 40z = -2{,}000 & E_4 \\
\underline{90y + 54z = \quad 2{,}196} & (0.36)E_5 \\
14z = \quad 196 \\
z = \qquad 14
\end{array}$$

Substitute $z = 14$ into E_4 and solve for y

$$-90y - 40z = -2{,}000 \qquad E_4$$
$$-90y - 40(14) = -2{,}000$$
$$y = 16$$

Substitute $y = 16$ and $z = 14$ into E_1 and solve for x.

$$10x + 70y + 60z = 2{,}150 \quad E_1$$
$$10x + 70(16) + 60(14) = 2{,}150$$
$$x = 19$$

19 days Michigan plant, 16 days New York plant, 14 days Ohio plant.

35. Let x = amount invested in treasury bonds
y = amount invested in municipal bonds
z = amount invested in corporate bonds

Then

$$x + y + z = 100{,}000 \quad \text{total investment}$$
$$.04x + .035y + .05z = 4{,}400 \quad \text{total income}$$
$$x + y = z \qquad\qquad \text{risk control factor}$$

Clearing of decimals and changing to standard form yields

$$x + y + z = 100{,}000 \quad E_1$$
$$40x + 35y + 50z = 4{,}400{,}000 \quad E_2$$
$$x + y - z = 0 \quad E_3$$

Multiply E_3 by -1 and add to E_1 to eliminate x and y.

$$x + y + z = 100{,}000 \qquad E_1$$
$$\underline{-x - y + z = 0} \qquad (-1)E_3$$
$$2z = 100{,}000$$
$$z = 50{,}000$$

Substituting $z = 50{,}000$ into E_1 and into E_2 yields

$$x + y + z = 100{,}000 \quad E_1$$
$$x + y + 50{,}000 = 100{,}000$$
$$40x + 35y + 50z = 4{,}400{,}000 \quad E_2$$
$$40x + 35y + 50(50{,}000) = 4{,}400{,}000$$

Simplifying yields

$$x + y = 50{,}000 \quad E_4$$
$$40x + 35y = 1{,}900{,}000 \quad E_5$$

Multiply E_4 by -35 and add to E_5 to eliminate y.

$$-35x - 35y = -1{,}750{,}000 \quad (-35)E_4$$
$$\underline{40x + 35y = 1{,}900{,}000} \quad E_5$$
$$5x = 150{,}000$$
$$x = 30{,}000$$

Substituting $x = 30{,}000$ into E_4 yields $y = 20{,}000$. $30,000 in treasury bonds, $20,000 in municipal bonds, $50,000 in corporate bonds.

37. Let x = number of pounds of peanuts
y = number of pounds of cashews
z = number of pounds of walnuts

Then

$$x + y + z = 1{,}500 \qquad \text{total weight}$$
$$1.4x + 4y + 2.9z = 3{,}500 \qquad \text{total value}$$
$$x = y + z \qquad \text{blend is half peanuts}$$

Clearing of decimals and changing to standard form yields

$$x + y + z = 1{,}500 \quad E_1$$
$$14x + 40y + 29z = 35{,}000 \quad E_2$$
$$x - y - z = 0 \quad E_3$$

Adding equations E_1 and E_3 eliminates y and z.

$$
\begin{array}{rcl}
x + y + z & = & 1{,}500 \qquad E_1 \\
\underline{x - y - z} & = & \underline{0} \qquad E_3 \\
2x & = & 1{,}500 \\
x & = & 750
\end{array}
$$

Substitute $x = 750$ into E_1 and E_2.

$$
\begin{array}{rcl}
750 + y + z & = & 1{,}500 \\
14(750) + 40y + 29z & = & 35{,}000
\end{array}
$$

Simplifying yields

$$
\begin{array}{rcll}
y + z & = & 750 & E_4 \\
40y + 29z & = & 24{,}500 & E_5
\end{array}
$$

Multiply E_4 by -40 and add to E_5 to eliminate y

$$
\begin{array}{rcll}
-40y - 40z & = & -30{,}000 & (-40)E_4 \\
\underline{40y + 29z} & = & \underline{24{,}500} & E_5 \\
-11z & = & -5{,}500 & \\
z & = & 500 &
\end{array}
$$

Substituting $z = 500$ into E_4 yields $y = 250$.

750 lbs. peanuts, 250 lbs. cashews, 500 lbs. walnuts.

39. Let $f(x) = ax^2 + bx + c$ be the quadratic model and evaluate f at each of the 3 data points.

$$
\begin{array}{lrcll}
x = 0: & c & = & 75 & E_1 \\
x = 50: & 2{,}500a + 50b + c & = & 150 & E_2 \\
x = 100: & 10{,}000a + 100b + c & = & 275 & E_3
\end{array}
$$

Substitute $c = 75$ into E_2 and E_3.

$$
\begin{array}{rcl}
2{,}500a + 50b + 75 & = & 150 \\
10{,}000a + 100b + 75 & = & 275
\end{array}
$$

Simplifying yields

$$
\begin{array}{rcll}
2{,}500a + 50b & = & 75 & E_4 \\
10{,}000a + 100b & = & 200 & E_5
\end{array}
$$

Multiply E_4 by -2 and add to E_5 to eliminate b.

$$
\begin{array}{rcll}
-5{,}000a - 100b & = & -150 & (-2)E_4 \\
\underline{10{,}000a + 100b} & = & \underline{200} & E_5 \\
5{,}000a & = & 50 & \\
a & = & 0.01 &
\end{array}
$$

Substitute $a = 0.01$ into E_5 and solve for b.

$$
\begin{array}{rcl}
10{,}000a + 100b & = & 200 \\
10{,}000(0.01) + 100b & = & 200 \\
b & = & 1
\end{array}
$$

The model is $f(x) = 0.01x^2 + x + 75$.
To find the predicted population in 2050 calculate $f(150)$.

$$f(150) = 0.01(150)^2 + 150 + 75 = 450 \text{ million}$$

41. Let $f(x) = ax^2 + bx + c$ be the quadratic model and evaluate f at each of the three data points.

$$
\begin{array}{lrcll}
x = 0: & c & = & 77.6 & E_1 \\
x = 5: & 25a + 5b + c & = & 78 & E_2 \\
x = 10: & 100a + 10b + c & = & 78.6 & E_3
\end{array}
$$

Substitute $c = 77.6$ into E_2 and E_3

$$
\begin{array}{rcl}
25a + 5b + 77.6 & = & 78 \\
100a + 10b + 77.6 & = & 78.6
\end{array}
$$

Simplifying yields

$$25a + 5b = 0.4 \qquad E_4$$
$$100a + 10b = 1 \qquad E_5$$

Multiply E_4 by -2 and add to E_5 to eliminate b.

$$
\begin{array}{r}
-50a - 10b = -0.8 \\
\underline{100a + 10b = 1} \\
50a = 0.2 \\
a = 0.004
\end{array}
$$

Substitute $a = 0.004$ into E_5 and solve for b.

$$100a + 10b = 1 \quad E_5$$
$$100(0.004) + 10b = 1$$
$$b = 0.06$$

The model is $f(x) = 0.004x^2 + 0.06x + 77.6$.

Predicted life expectancy for females born 1995–2000:

$$f(15) = 0.004(15)^2 + 0.06(15) + 77.6 = 79.4 \text{ yrs.}$$

Predicted life expectancy for females born 2000–2005:

$$f(20) = 0.004(20)^2 + 0.06(20) + 77.6 = 80.4 \text{ yrs.}$$

43. The independent variable is given by the first year in the interval minus the starting year 1980, so enter 0, 5, 10, 15, 20 as L_1. The dependent variable is life expectancy, so enter the data 77.6, 78, 78.6, 79.1, 79.7 as L_2. Then use the quadratic regression command from the STAT CALC menu and plot using the STAT PLOT screen.

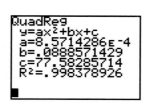 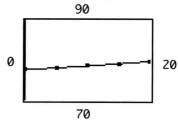

Section 8-3

1. Answers will vary.

3. Answers will vary.

5. Graph $2x - 3y = 6$ as a dashed line, since equality is not included in the original statement. The origin is a suitable test point.
$2x - 3y < 6$
$2(0) - 3(0) = 0 < 6$
So $(0, 0)$ is in the solution set. The solution region is the set of points (x, y) that are above the line $2x - 3y = 6$.

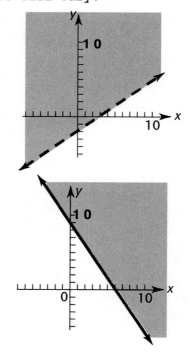

7. Graph $3x + 2y = 18$ as a solid line, since equality is included in the original statement. The origin is a suitable test point.
$3(0) + 2(0) = 0 \not\geq 18$
So $(0, 0)$ is not in the solution set. The solution region is the set of points (x, y) that are on or above the line $3x + 2y = 18$.

9. Graph $y = \dfrac{2}{3}x + 5$ as a solid line, since equality is included in the original statement. The origin is a suitable test point.

$0 \overset{?}{\le} (0) + 5$

$0 \le 5$

So $(0, 0)$ is in the solution set. The solution region is the set of points (x, y) that are on or below the line $y = \dfrac{2}{3}x + 5$.

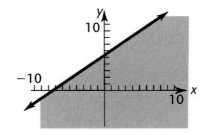

11. Graph $y = 8$ as a dashed line, since equality is not included in the original statement. The solution region consists of all points whose y-coordinates are less than 8, that is, the set of points (x, y) that are below the line $y = 8$.

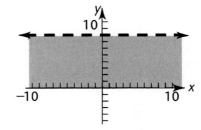

13. This system is equivalent to the system
$y \ge -3$
$y < 2$
and its graph is the intersection of the graphs of these inequalities. The solution region is the set of points (x, y) that are below the horizontal line $y = 2$, and on or above the horizontal line $y = -3$.

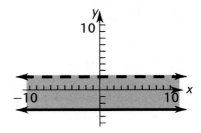

15. $x + 2y \le 8$
$3x - 2y \ge 0$
Choose a suitable test point that lies on neither line, for example, $(2, 0)$.
   $2 + 2(0) = 2 \le 8$ So $(2, 0)$ is in the solution region for the first inequality, and the solution region is *below* the graph of $x + 2y = 8$.
$3(2) - 2(0) = 6 \ge 0$ So $(2, 0)$ is also in the solution region of the second inequality, and the solution region is *below* the graph of $3x - 2y = 0$.
The region in the diagram that is below both lines is region IV.

17. $x + 2y \ge 8$
$3x - 2y \ge 0$
Choose a suitable test point that lies on neither line, for example, $(2, 0)$.
   $2 + 2(0) = 2 \ne 8$ So $(2, 0)$ is not in the solution region of the first inequality, and the solution region is *above* the graph of $x + 2y = 8$.
$3(2) - 2(0) = 6 \ge 0$ So $(2, 0)$ is in the solution region of the second inequality, and the solution region is *below* the graph of $3x - 2y = 0$.
The region in the diagram that is above $x + 2y = 8$ and below $3x - 2y = 0$ is Region I.

19.

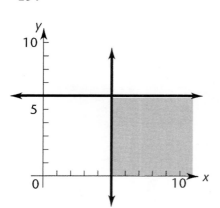

21.

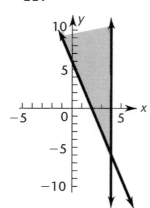

23.

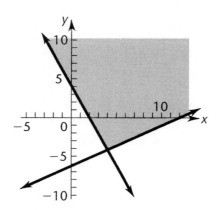

25. (A) Solution region is the double-shaded region.

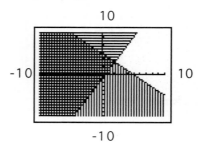

(B) Solution region is the unshaded region.

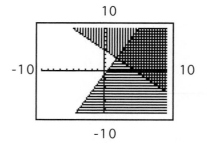

27. (A) Solution region is the double-shaded region.

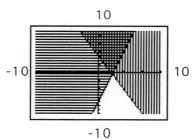

(B) Solution region is the unshaded region.

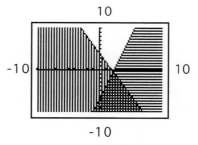

29. Choose a suitable test point that lies on none of the lines, say (5, 1).

$5 + 3(1) = 8 \leq 18$ So (5, 1) is in the solution region of the first inequality, and the solution region is *below* the graph of $x + 3y = 18$.

$2(5) + 1 = 11 \not\geq 16$ So (5, 1) is not in the solution region of the first inequality, and the solution region is *above* the graph of $2x + y = 16$.

From $x \geq 0$ and $y \geq 0$, we conclude that the solution region is in Quadrant I. So the solution region is region IV in the diagram. The corner points are the labeled points (6, 4), (8, 0), and (18, 0).

31. Choose a suitable test point that lies on none of the lines, say (5, 1).

$5 + 3(1) = 8 \not\geq 18$ So (5, 1) is not in the solution region of the first inequality, and the solution region is *above* the graph of $x + 3y = 18$.

$2(5) + 1 = 11 \not\geq 16$ So (5, 1) is not in the solution region of the first inequality, and the solution region is *above* the graph of $2x + y = 16$.

From $x \geq 0$ and $y \geq 0$, we conclude that the solution region is in Quadrant I. So the solution region is region I in the diagram. The corner points are the labeled points (0, 16), (6, 4), and (18, 0).

33. The solution region is bounded by the
2 axes and the line $2x + 3y = 6$. The
corner points are obvious from the
graph:
$(0, 0)$, $(0, 2)$, $(3, 0)$.

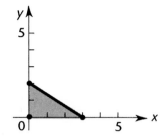

Check:

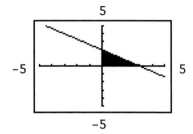

35. The solution region is unbounded.
The corner points are obvious from
the graph: $(0, 4)$ and $(5, 0)$.

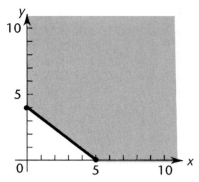

Check:

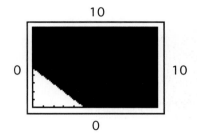

37. The solution region is bounded. Three
corner points are obvious from the
graph: $(0, 4)$, $(0, 0)$, $(4, 0)$. The
fourth corner point is obtained by
solving the system
$x + 3y = 12$

$2x + y = 8$ to obtain $\left(\dfrac{12}{5}, \dfrac{16}{5}\right)$.

(Solution left to the student.)

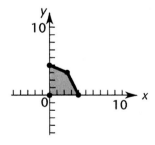

Check:

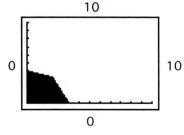

39. The solution region is unbounded.
Two corner points are obvious from
the graph: $(9, 0)$ and $(0, 8)$. The
third corner point is obtained by
solving the system
$4x + 3y = 24$
$2x + 3y = 18$ to obtain $(3, 4)$.
(Solution left to the student.)

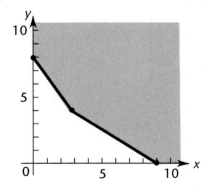

Check:

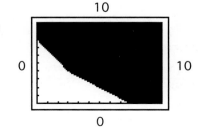

41. The solution region is bounded. Three corner points are obvious from the graph: $(6, 0)$, $(0, 0)$, and $(0, 5)$. The other corner points are obtained by solving:

$2x + y = 12$ and	$x + y = 7$
$x + y = 7$	$x + 2y = 10$
to obtain	to obtain
$(5, 2)$	$(4, 3)$

(Solutions left to the student.)

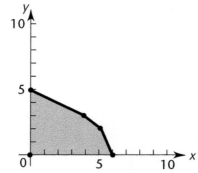

Check:

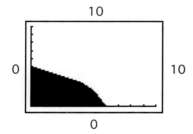

45. The solution region is bounded. The corner points are obtained by solving:

$$\begin{cases} x + y = 11 \\ 5x + y = 15 \end{cases} \text{ to obtain } (1, 10)$$

$$\begin{cases} 5x + y = 15 \\ x + 2y = 12 \end{cases} \text{ to obtain } (2, 5), \text{ and}$$

$$\begin{cases} x + y = 11 \\ x + 2y = 12 \end{cases} \text{ to obtain } (10, 1)$$

(Solutions left to the student.)

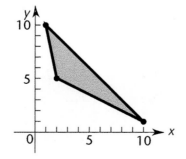

43. The solution region is unbounded. Two of the corner points are obvious from the graph: $(16, 0)$ and $(0, 14)$. The other corner points are obtained by solving:

$x + 2y = 16$ and	$x + y = 12$
$x + y = 12$	$2x + y = 14$
to obtain	to obtain
$(8, 4)$	$(2, 10)$

(Solutions left to the student.)

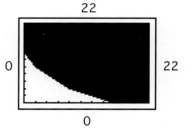

Check:

47. The lighter shaded regions are the solution regions to $3x + 2y \geq 24$ and $3x + y \leq 15$; the darker is the intersection of the two. We can see that no portion of the darker region lies to the right of $x = 4$, so there are no points that satisfy both of the first two inequalities and $x > 4$. The solution region is empty.

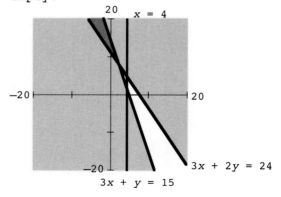

49. The solution region is bounded.
The corner points are obtained by
solving:

$$\begin{cases} x + y = 10 \\ 3x - 2y = 15 \end{cases}$$ to obtain (7, 3)

$$\begin{cases} 3x - 2y = 15 \\ 3x + 5y = 15 \end{cases}$$ to obtain (5, 0),

$$\begin{cases} 3x + 5y = 15 \\ -5x + 2y = 6 \end{cases}$$ to obtain (0, 3), and

$$\begin{cases} -5x + 2y = 6 \\ x + y = 10 \end{cases}$$ to obtain (2, 8)
(Solutions left to the student.)

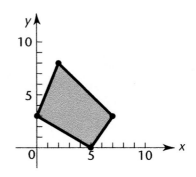

51. The solution region is bounded. The
corner points are obtained by solving:

$$\begin{cases} 16x + 13y = 119 \\ 12x + 16y = 101 \end{cases}$$ to obtain (5.91, 1.88)

$$\begin{cases} 16x + 13y = 119 \\ -4x + 3y = 11 \end{cases}$$ to obtain (2.14, 6.52)
and

$$\begin{cases} 12x + 16y = 101 \\ -4x + 3y = 11 \end{cases}$$ to obtain (1.27, 5.36)
(Solutions left to the student.)

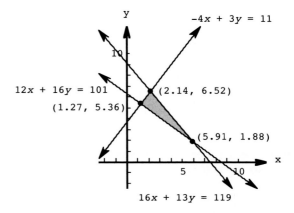

53. Let x = number of trick skis produced per day.
  y = number of slalom skis produced per day

Clearly x and y must be non-negative.
So $x \geq 0$ (1)
  $y \geq 0$ (2)
To fabricate x trick skis requires $6x$ hours.
To fabricate y slalom skis requires $4y$ hours.
108 hours are available for fabricating; so
$6x + 4y \leq 108$ (3)
To finish x trick skis requires $1x$ hours.
To finish y slalom skis requires $1y$ hours.
24 hours are available for finishing, so
$x + y \leq 24$ (4)

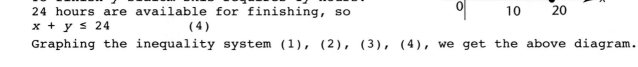

Graphing the inequality system (1), (2), (3), (4), we get the above diagram.

55. (A) All production schedules in the feasible region that are on the graph
     of $50x + 60y = 1,100$ will result in a profit of $1,100.

  (B) There are many possible choices. For example, producing 5 trick and 15
     slalom skis will produce a profit of $1,150. The graph of the line
     $50x + 60y = 1,150$ includes all the production schedules in the feasible
     region that result in a profit of $1,150.

  (C) A graphical approach would involve drawing other lines of the type
     $50x + 60y = A$. The graphs of these lines include all production schedules
     that will result in a profit of A. Increase A until the line either
     intersects the feasible region only in 1 corner point or contains an edge
     of the feasible region. This value of A will be the maximum profit
     possible. For more details, see Section 5-4 of the text.

57. Clearly x and y must be non-negative.
So $x \geq 0$ (1)
   $=y \geq 0$ (2)
x cubic yards of mix A contains $20x$ pounds of phosphoric acid.
y cubic yards of mix B contains $10y$ pounds of phosphoric acid.

At least 460 pounds of phosphoric acid are required, so
$20x + 10y \geq 460$ (3)
x cubic yards of mix A contains $30x$ pounds of nitrogen.
y cubic yards of mix B contains $30y$ pounds of nitrogen.
At least 960 pounds of nitrogen are required, so
$30x + 30y \geq 960$ (4)

x cubic yards of mix A contains $5x$ pounds of potash.
y cubic yards of mix B contains $10y$ pounds of potash.
At least 220 pounds of potash are required, so
$5x + 10y \geq 220$ (5)

Graphing the inequality system (1), (2), (3), (4), (5), we get the following diagram:

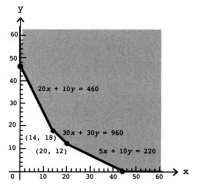

59. Clearly x and y must be non-negative.
So $x \geq 0$ (1)
   $y \geq 0$ (2)
Each sociologist will spend 10 hours collecting data: $10x$ hours.
Each research assistant will spend 30 hours collecting data: $30y$ hours.
At least 280 hours must be spent collecting data; so
$10x + 30y \geq 280$ (3)
Each sociologist will spend 30 hours analyzing data: $30x$ hours.
Each research assistant will spend 10 hours analyzing data: $10y$ hours.
At least 360 hours must be spent analyzing data; so
$30x + 10y \geq 360$ (4)
Graphing the inequality system (1), (2), (3), (4), we get the above diagram.

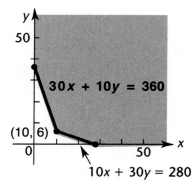

Section 8-4

1. Answers will vary.   **3.** Answers will vary.   **5.** Answers will vary.

7. Answers will vary.

9.

Corner Point (x, y)	Objective Function $z = x + y$	
(0, 12)	12	
(7, 9)	16	Maximum value
(10, 0)	10	
(0, 0)	0	

The maximum value of z on S is 16 at (7, 9).

11.

Corner Point $(x,\ y)$	Objective Function $z = 3x + 7y$		
(0, 12)	84	Minimum value	
(7, 9)	84	Minimum value	Multiple optimal solutions
(10, 0)	30		
(0, 0)	0		

The maximum value of z on S is 84 at both (0, 12) and (7, 9).

13. Plugging in zero for x, we get $z = 2y$ or $y = \dfrac{z}{2}$.

Plugging in zero for y, we get $z = x$. So the

intercepts of the line $z = x + 2y$ are $\left(0, \dfrac{z}{2}\right)$ and

$(z, 0)$. The feasible region is shown at the right with several constant z lines drawn in.

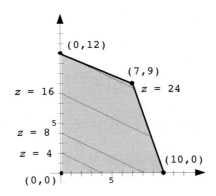

Sliding a straight edge parallel to these constant z lines in the direction of increasing z, we can see that the point in the feasible region that will intersect the constant z line for largest possible z is (7, 9). When $x = 7$ and $y = 9$, $z = 25$.

Check:

Corner Point $(x,\ y)$	Objective Function $z = x + 2y$	
(0, 12)	24	
(7, 9)	25	Maximum Value
(10, 0)	10	
(0, 0)	0	

The maximum value of z on S is 25 at (7, 9).

15. Plugging in zero for x, we get $z = 2y$ or $y = \dfrac{z}{2}$.

Plugging in zero for y, we get $z = 7x$ or $x = \dfrac{z}{7}$.

So the intercepts of the line $z = 7x + 2y$ are

$\left(0, \dfrac{z}{2}\right)$ and $\left(\dfrac{z}{7}, 0\right)$. The feasible region is shown at the right with several constant z lines drawn in.

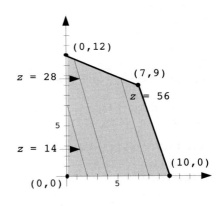

If we slide a straight edge parallel to the constant z lines as z increases, the last point in S that will intersect our lines is (10, 0). When $x = 10$ and $y = 0$, $z = 70$.

Check:

Corner Point $(x,\ y)$	Objective Function $z = 7x + 2y$	
(0, 12)	24	
(7, 9)	67	
(10, 0)	70	Maximum Value
(0, 0)	0	

The maximum value of z on S is 70 at (10, 0).

17.

Corner Point (x, y)	Objective Function $z = 7x + 4y$	
(0, 12)	48	
(12, 0)	84	
(4, 3)	40	
(0, 8)	32	Minimum value

The minimum value of z on S is 32 at (0, 8).

19.

Corner Point (x, y)	Objective Function $z = 3x + 8y$	
(0, 12)	96	
(12, 0)	36	Minimum value ⎫ Multiple optimal solutions
(4, 3)	36	Minimum value ⎬
(0, 8)	64	⎭

The minimum value of z on S is 36 at both (12, 0) and (4, 3).

21. If $x = 5$ and $y = 5$, $z = 5 + 2(5) = 15$, so the constant-value line we need is $x + 2y = 15$. The feasible region is shown at the right with the constant-value line.

If we slide a straightedge parallel to this constant-value line in the direction of decreasing z (downward), the last point in T that will intersect our line is (4, 3). When $x = 4$ and $y = 3$, $z = 10$.

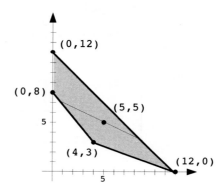

Check:

Corner Point (x, y)	Objective Function $z = x + 2y$	
(0, 12)	24	
(12, 0)	12	
(0, 8)	16	
(4, 3)	10	Minimum Value

The minimum value of z on T is 10 at (4, 3).

23. If $x = 5$ and $y = 5$, $z = 5(5) + 4(5) = 45$, so the constant-value line we need is $5x + 4y = 45$. The feasible region is shown at the right with the constant-value line.

The constant-value line appears to be parallel to the edge connecting (0, 8) and (4, 3). So the minimum could occur at either (0, 8) or (4, 3) when $x = 0$ and $y = 8$, $z = 32$. When $x = 4$ and $y = 3$, $z = 32$ also.

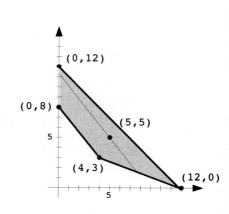

Check:

Corner Point (x, y)	Objective Function $z = 5x + 4y$	
(0, 12)	48	
(12, 0)	60	
(0, 8)	32	Minimum Value
(4, 3)	32	Minimum Value

The minimum value of 32 occurs at both (0, 8) and (4, 3).

25. The feasible region is graphed as follows:
The corner points (0, 5), (5, 0) and (0, 0) are obvious from the graph. The corner point (4, 3) is obtained by solving the system
$x + 2y = 10$
$3x + y = 15$
We now evaluate the objective function at each corner point.

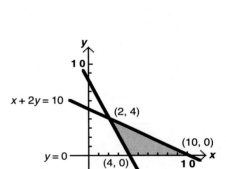

Corner Point (x, y)	Objective Function $z = 3x + 2y$	
(0, 5)	10	
(0, 0)	0	
(5, 0)	15	
(4, 3)	18	Maximum value

The maximum value of z on S is 18 at (4, 3).

27. The feasible region is graphed as follows:
The corner points (4, 0) and (10, 0) are obvious from the graph. The corner point (2, 4) is obtained by solving the system
$x + 2y = 10$
$2x + y = 8$
We now evaluate the objective function at each corner point.

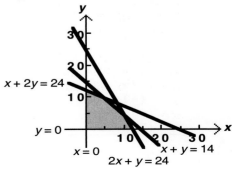

Corner Point (x, y)	Objective Function $z = 3x + 4y$	
(4, 0)	12	Minimum value
(10, 0)	30	
(2, 4)	22	

The minimum value of z on S is 12 at (4, 0).

29. The feasible region is graphed below. The corner points (0, 12), (0, 0), and (12, 0) are obvious from the graph. The other corner points are obtained by solving:
$x + 2y = 24$ and $x + y = 14$
 $x + y = 14$ to obtain (4, 10) $2x + y = 24$ to obtain (10, 4)
We now evaluate the objective function at each corner point.

Corner Point (x, y)	Objective Function $z = 3x + 4y$	
(0, 12)	48	
(0, 0)	0	
(12, 0)	36	
(10, 4)	46	
(4, 10)	52	Maximum value

The maximum value of z on S is 52 at (4, 10).

31. The feasible region is graphed as follows:
The corner points (0, 20) and (20, 0) are obvious
from the graph. The third corner point is obtained
by solving:
$x + 4y = 20$
$4x + y = 20$ to obtain (4, 4)
We now evaluate the objective function at each
corner point.

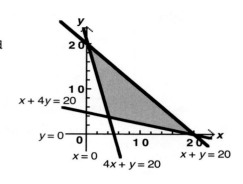

Corner Point (x, y)	Objective Function z = 5x + 6y	
(0, 20)	120	
(20, 0)	100	
(4, 4)	44	Minimum value

The minimum value of z on S is 44 at (4, 4).

33. The feasible region is graphed as follows:
The corner points (60, 0) and (120, 0) are
obvious from the graph. The other corner
points are obtained by solving:

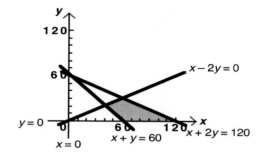

$x + y = 60$	and	$x + 2y = 120$
$x - 2y = 0$		$x - 2y = 0$
to obtain (40, 20)		to obtain (60, 30)

We now evaluate the objective function at
each corner point.

Corner Point (x, y)	Objective Function 25x + 50y	
(60, 0)	1,500	Minimum value
(40, 20)	2,000	
(60, 30)	3,000	Maximum value ⎫
(120, 0)	3,000	Maximum value ⎬ Multiple optimal solutions

The minimum value of z on S is 1,500 at (60, 0). The maximum value of z on S is
3,000 at (60, 30) and (120, 0).

35. The feasible region is graphed as follows:
The corner points (0, 45), (0, 20), (25, 0), and (60, 0) are obvious from the
graph shown below. The other corner points are obtained by solving:

$3x + 4y = 240$ and $3x + 4y = 240$
       $y = 45$ to obtain (20, 45) $x = 60$ to obtain (60, 15)
We now evaluate the objective function at each corner point.

Corner Point (x, y)	Objective Function 25x + 15y	
(0, 45)	675	
(0, 20)	300	Minimum value
(25, 0)	625	
(60, 0)	1,500	
(60, 15)	1,725	Maximum value
(20, 45)	1,175	

The minimum value of z on S is 300 at $(0, 20)$. The maximum value of z on S is 1,725 at $(60, 15)$.

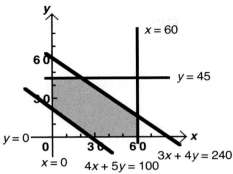

37. The feasible region is graphed as shown at the right.

    The corner point $(0, 0)$ is obvious from the graph. The other corner points are obtained by solving:

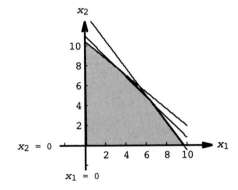

    $x_1 = 0$ $350x_1 + 340x_2 = 3,762$
    $275x_1 + 322x_2 = 3,381$ $275x_1 + 322x_2 = 3,381$
    to obtain $(0, 10.5)$ to obtain $(3.22, 7.75)$

    $350x_1 + 340x_2 = 3,762$ $425x_1 + 306x_2 = 4,114$
    $425x_1 + 306x_2 = 4,114$ $x_2 = 0$
    to obtain $(6.62, 4.25)$ to obtain $(9.68, 0)$

    We now evaluate the objective function at each corner point.

Corner Point (x_1, x_2)	Objective Function $525x_1 + 478x_2$	
$(0, 0)$	0	Minimum value
$(0, 10.5)$	5,019	
$(3.22, 7.75)$	5,395	
$(6.62, 4.25)$	5,507	Maximum value
$(9.68, 0)$	5,082	

    The maximum value of P is 5,507 at the corner point $(6.62, 4.25)$.

39. The feasible region is graphed as follows:
    Consider the objective function $x + y$. We can see that it takes on the value 2 along $x + y = 2$, the value 4 along $x + y = 4$, the value 7 along $x + y = 7$, and so on. The maximum value of the objective function, then, on S, is 7, which occurs at B. Graphically this occurs when the line $x + y = c$ coincides with the boundary of S. So to answer questions (A)-(E) we must determine values of a and b so that the appropriate line $ax + by = c$ coincides with the boundary of S only at the specified points.

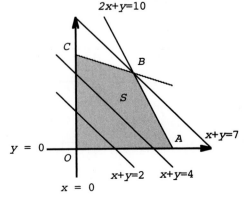

    (A) The line $ax + by = c$ must have slope negative, but greater in absolute value than that of line segment AB, $2x + y = 10$. Therefore $a > 2b$.

(B) The line $ax + by = c$ must have slope negative but between that of $x + 3y = 15$ and $2x + y = 10$. Therefore $\frac{1}{3}b < a < 2b$.

(C) The line $ax + by = c$ must have slope greater than that of line segment BC, $x + 3y = 15$. Therefore $a < \frac{1}{3}b$ or $b > 3a$

(D) The line $ax + by = c$ must be parallel to line segment AB, therefore $a = 2b$.

(E) The line $ax + by = c$ must be parallel to line segment BC, therefore $b = 3a$.

41. We let x = the number of trick skis
y = the number of slalom skis
The problem constraints were
$6x + 4y \leq 108$
$x + y \leq 24$
The non-negative constraints were
$x \geq 0$
$y \geq 0$
The feasible region was graphed there.
(A) We note now: the linear objective function
$P = 40x + 30y$ represents the profit.

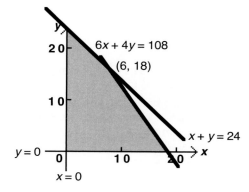

Three of the corner points are obvious from the graph: $(0, 24)$, $(0, 0)$, and $(18, 0)$. The fourth corner point is obtained by solving:
$6x + 4y = 108$
$x + y = 24$ to obtain $(6, 18)$.
Summarizing: the mathematical model for this problem is:
Maximize $P = 40x + 30y$
subject to: $6x + 4y \leq 108$
$x + y \leq 24$
$x, y \geq 0$
We now evaluate the objective function $40x + 30y$ at each corner point.

Corner Point (x, y)	Objective Function $40x + 30y$	
(0, 0)	0	
(18, 0)	720	
(6, 18)	780	Maximum value
(0, 24)	720	

The optimal value is 780 at the corner point $(6, 18)$. So 6 trick skis and 18 slalom skis should be manufactured to obtain the maximum profit of $780.

(B) The objective function now becomes $40x + 25y$. We evaluate this at each corner point.

Corner Point (x, y)	Objective Function $40x + 25y$	
(0, 0)	0	
(18, 0)	720	Maximum value
(6, 18)	690	
(0, 24)	600	

The optimal value is now 720 at the corner point $(18, 0)$. So 18 trick skis and no slalom skis should be produced to obtain a maximum profit of $720.

(C) The objective function now becomes $40x + 45y$. We evaluate this at each corner point.

Corner Point (x, y)	Objective Function $40x + 45y$	
(0, 0)	0	
(18, 0)	720	
(6, 18)	1,050	
(0, 24)	1,080	Maximum value

The optimal value is now 1,080 at the corner point (0, 24). So no trick skis and 24 slalom skis should be produced to obtain a maximum profit of $1,080.

43. Let x = number of model A trucks
     y = number of model B trucks
We form the linear objective function
$C = 15,000x + 24,000y$
We wish to minimize C, the cost of buying x trucks @ $15,000 and y trucks @ $24,000, subject to the constraints.
  $x + y \le 15$ maximum number of trucks constraint
$2x + 3y \ge 36$ capacity constraint
  $x, y \ge 0$ non-negative constraints.

Solving the system of constraint inequalities graphically, we obtain the feasible region S shown in the diagram.
Next we evaluate the objective function at each corner point.

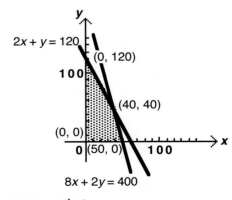

Corner Point (x, y)	Objective Function $C = 15,000x + 24,000y$	
(0, 12)	288,000	
(0, 15)	360,000	
(9, 6)	279,000	Minimum value

The optimal value is $279,000 at the corner point (9, 6). So the company should purchase 9 model A trucks and 6 model B trucks to realize the minimum cost of $279,000.

45. (A) Let x = number of tables
         y = number of chairs
We form the linear objective function
$P = 90x + 25y$
We wish to maximize P, the profit from x tables @ $90 and y chairs @ $25, subject to the constraints
$8x + 2y \le 400$ assembly department constraint
 $2x + y \le 120$ finishing department constraint
   $x, y \ge 0$ non-negative constraints

Solving the system of constraint inequalities graphically, we obtain the feasible region S shown in the diagram.
Next we evaluate the objective function at each corner point.

Corner Point (x, y)	Objective Function $P = 90x + 25y$	
(0, 0)	0	
(50, 0)	4,500	
(40, 40)	4,600	Maximum value
(0, 120)	3,000	

The optimal value is 4,600 at the corner point (40, 40). So the company should manufacture 40 tables and 40 chairs for a maximum profit of $4,600.

(B) We are faced with the further condition
     that $y \geq 4x$. We wish, then, to maximize
     $P = 90x + 25y$ under the constraints
$$8x + 2y \leq 400$$
$$2x + y \leq 120$$
$$y \geq 4x$$
$$x, y \geq 0$$
The feasible region is now S' as graphed.

Note that the new condition has the effect
of excluding (40, 40) from the feasible
region.

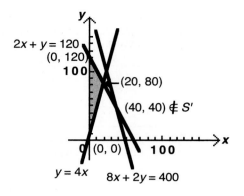

We now evaluate the objective function at
the new corner points.

Corner Point (x, y)	Objective Function $P = 90x + 25y$	
(0, 120)	3,000	
(0, 0)	0	
(20, 80)	3,800	Maximum value

The optimal value is now 3,800 at the corner point (20, 80). So the company
should manufacture 20 tables and 80 chairs for a maximum profit of $3,800.

47. Let x = number of gallons produced using the old process
        y = number of gallons produced using the new process
     We form the linear objective function
        $P = 0.6x + 0.2y$

(A) We wish to maximize P, the profit from x gallons using the old process and
     y gallons using the new process, subject to the constraints
$$20x + 5y \leq 16,000 \qquad \text{sulfur dioxide constraint}$$
$$40x + 20y \leq 30,000 \qquad \text{particulate matter constraint}$$
$$x, y \geq 0 \qquad \text{non-negative constraints}$$

Solving the system of constraint
inequalities graphically, we obtain the
feasible region S shown in the diagram.
Note that no corner points are
determined by the sulfur dioxide
constraint.

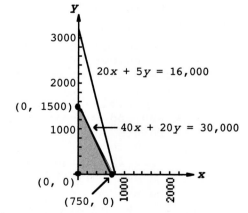

We evaluate the objective function at each corner point.

Corner Point (x, y)	Objective Function $P = 0.6x + 0.2y$	
(0, 0)	0	
(0, 1500)	300	
(750, 0)	450	Maximum value

The optimal value is 450 at the corner point (750, 0). So the company
should manufacture 750 gallons by the old process exclusively, for a
profit of $450.

(B) The sulfur dioxide constraint is now
$20x + 5y \leq 11,500$. We now wish to
maximize P subject to the constraints

$$20x + 5y \leq 11,500$$
$$40x + 20y \leq 30,000$$
$$x, y \geq 0$$

The feasible region is now S_1 as shown.

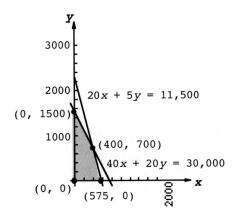

We evaluate the objective function at the new corner points.

Corner Point (x, y)	Objective Function $P = 0.6x + 0.2y$	
(0, 0)	0	
(575, 0)	345	
(400, 700)	380	Maximum value
(0, 1,500)	300	

The optimal value is now 380 at the corner point (400, 700). So the
company should manufacture 400 gallons by the old process and 700 gallons
by the new process, for a profit of $380.

(C) The sulfur dioxide constraint is now
$20x + 5y \leq 7,200$. We now wish to
maximize P subject to the constraints

$$20x + 5y \leq 7,200$$
$$40x + 20y \leq 30,000$$
$$x, y \geq 0$$

The feasible region is now S_2 as shown.

Note that now no corner points are
determined by the particulate matter
constraint.

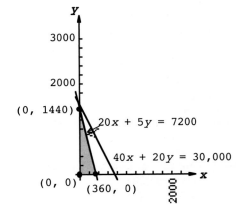

We evaluate the objective function at the new corner points.

Corner Point (x, y)	Objective Function $P = 0.6x + 0.2y$	
(0, 0)	0	
(360, 0)	216	
(0, 1,440)	288	Maximum value

The optimal value is now 288 at the corner point (0, 1440). So the
company should manufacture 1,440 gallons by the new process exclusively,
for a profit of $288.

49. (A) Let x = number of bags of Brand A
y = number of bags of Brand B
We form the objective function
$N = 6x + 7y$
N represents the amount of nitrogen in x bags @ 6 pounds per bag and y bags @ 7 pounds per bag.

We wish to optimize N subject to the constraints
$2x + 4y \geq 480$ phosphoric acid constraint
$6x + 3y \geq 540$ potash constraint
$3x + 4y \leq 620$ chlorine constraint
   $x, y \geq 0$ non-negative contraints
Solving the system of constraint inequalities graphically, we obtain the feasible region S shown in the diagram.

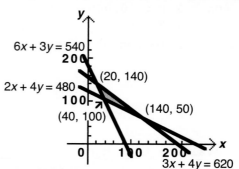

Next we evaluate the objective function at the corner points.

Corner Point (x, y)	Objective Function $N = 6x + 7y$	
$(20, 140)$	1,100	
$(40, 100)$	940	Minimum value
$(140, 50)$	1,190	Maximum value

So the nitrogen will range from a minimum of 940 pounds when 40 bags of brand A and 100 bags of Brand B are used to a maximum of 1,190 pounds when 140 bags of brand A and 50 bags of brand B are used.

Chapter 8 Review

1. $y = 4x - 9$
$y = -x + 6$
Substitute y from the first equation into the second equation to eliminate y.
$4x - 9 = -x + 6$
$5x - 9 = 6$
   $5x = 15$
    $x = 3$
     $y = -x + 6 = -3 + 6 = 3$

$x = 3, \ y = 3$ *(8-1)*

2. $3x + 2y = 5$
$4x - y = 14$
Solve the first equation for y in terms of x and substitute into the second equation
   $2y = 5 - 3x$
    $y = \dfrac{5 - 3x}{2}$
  $4x - \left(\dfrac{5 - 3x}{2}\right) = 14$
  $8x - \dfrac{2}{1}\left(\dfrac{5 - 3x}{2}\right) = 28$
  $8x - (5 - 3x) = 28$
   $8x - 5 + 3x = 28$
        $11x = 33$
          $x = 3$
       $y = \dfrac{5 - 3x}{2} = \dfrac{5 - 3 \cdot 3}{2}$
       $y = -2$
$x = 3, \ y = -2$ *(8-1)*

3. Graph both lines on the same set of axes and find the intersection point:

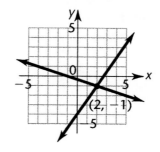

The solution is $x = 2, \ y = -1$.

                        (8-1)

4. First solve each equation for y.

$$2x + y = 4 \qquad\qquad -2x + 7y = 9$$
$$y = 4 - 2x \qquad\qquad 7y = 2x + 9$$
$$y = \frac{2}{7}x + \frac{9}{7}$$

Next, enter each equation, graph, and approximate the solution.

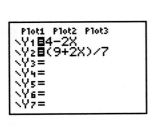

Equation definitions

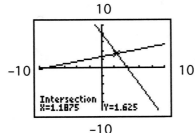

Intersection point

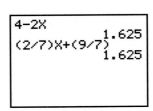

Check

The solution is $x = 1.1875$, $y = 1.625$.   (*8–1*)

5. Graph $3x - 4y = 24$ as a solid line since equality is included in the original statement. $(0, 0)$ is a suitable test point.
$3(0) - 4(0) = 0 \neq 24$
Therefore $(0, 0)$ is not in the solution set. The line $3x - 4y = 24$ and the half-plane not containing $(0, 0)$ form the graph.

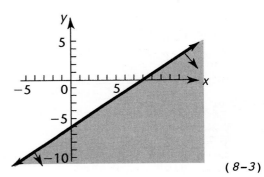

(*8–3*)

6. Graph both $2x + y = 2$ and $x + 2y = -2$ as solid lines since equality is included in each. $(0, 0)$ is a suitable test point.

$2(0) + 0 = 0 \leq 2$ so $(0, 0)$ satisfies the first inequality, and the solution region is below the line $2x + y = 2$.

$0 + 2(0) = 0 \geq -2$ so $(0, 0)$ satisfies the second inequality, and the solution region is above the line $x + 2y = -2$. The solution region is graphed at the right.

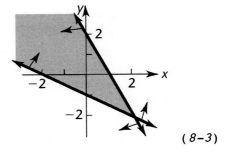

(*8–3*)

7. $2x + y = 7$
$3x - 2y = 0$
We multiply the top equation by 2 and add.

$$4x + 2y = 14$$
$$3x - 2y = 0$$
$$\overline{7x = 14}$$
$$x = 2$$

Substituting $x = 2$ in the top equation, we have
$2(2) + y = 7$
$y = 3$
Solution: $(2, 3)$ (*8–1*)

8. $3x - 6y = 5$
$-2x + 4y = 1$
We multiply the top equation by 2, the bottom by 3, and add.

$$6x - 12y = 10$$
$$-6x + 12y = 3$$
$$\overline{ 0 = 13}$$

No solution   (*8–1*)

9. $4x - 3y = -8$

$-2x + \dfrac{3}{2}y = 4$

We multiply the bottom equation by 2 and add.

$$
\begin{array}{r}
4x - 3y = -8 \\
-4x + 3y = 8 \\
\hline
0 = 0
\end{array}
$$

There are infinitely many solutions. For any real number t, $4t - 3y = -8$, and
$-3y = -4t - 8$

$y = \dfrac{4t + 8}{3}$

Then $\left(t, \dfrac{4t + 8}{3}\right)$ is a solution for any real number t.   *(8-1)*

10.
$$
\begin{array}{rll}
x - 3y + z = 4 & E_1 \\
-x + 4y - 4z = 1 & E_2 \\
2x - y + 5z = -3 & E_3
\end{array}
$$

Add E_1 to E_2 to eliminate x.
Also multiply E_1 by -2 and add to E_3 to eliminate x.

$$
\begin{array}{rll}
x - 3y + z = 4 & E_1 \\
-x + 4y - 4z = 1 & E_2 \\
\hline
y - 3z = 5 & E_4
\end{array}
$$

$$
\begin{array}{rll}
-2x + 6y - 2z = -8 & (-2)E_1 \\
2x - y + 5z = -3 & E_3 \\
\hline
5y + 3z = -11 & E_5
\end{array}
$$

Equivalent system:
$$
\begin{array}{rll}
x - 3y + z = 4 & E_1 \\
y - 3z = 5 & E_4 \\
5y + 3z = -11 & E_5
\end{array}
$$

Add E_4 to E_5 to eliminate z.
$$
\begin{array}{rll}
y - 3z = 5 & E_4 \\
5y + 3z = -11 & E_5 \\
\hline
6y = -6 & \\
y = -1 &
\end{array}
$$

Substitute $y = -1$ into E_5 and solve for z.
$$
\begin{array}{rl}
5y + 3z = -11 & E_5 \\
5(-1) + 3z = -11 & \\
z = -2 &
\end{array}
$$

Substitute $y = -1$ and $z = -2$ into E_1 and solve for x.
$$
\begin{array}{rl}
x - 3y + z = 4 & E_1 \\
x - 3(-1) + (-2) = 4 & \\
x = 3 &
\end{array}
$$

$(3, -1, -2)$ *(8-2)*

11.
$$2x + y - z = 5 \quad E_1$$
$$x - 2y - 2z = 4 \quad E_2$$
$$3x + 4y + 3z = 3 \quad E_3$$

Multiply E_2 by -2 and add to E_1 to eliminate x. Also multiply E_2 by -3 and add to E_3 to eliminate x.

$$
\begin{array}{rl}
2x + y - z = 5 & E_1 \\
\underline{-2x + 4y + 4z = -8} & (-2)E_2 \\
5y + 3z = -3 & E_4 \\
3x + 4y + 3z = 3 & E_3 \\
\underline{-3x + 6y + 6z = -12} & (-3)E_2 \\
10y + 9z = -9 & E_5
\end{array}
$$

Equivalent system:
$$
\begin{array}{rl}
x - 2y - 2z = 4 & E_2 \\
5y + 3z = -3 & E_4 \\
10y + 9z = -9 & E_5
\end{array}
$$

Multiply E_4 by -2 and add to E_5 to eliminate y.

$$
\begin{array}{rl}
-10y - 6z = 6 & (-2)E_4 \\
\underline{10y + 9z = -9} & E_5 \\
3z = -3 & \\
z = -1 &
\end{array}
$$

Substitute $z = -1$ into E_4 and solve for y.
$$
\begin{array}{rl}
5y + 3z = -3 & E_4 \\
5y + 3(-1) = -3 & \\
y = 0 &
\end{array}
$$

Substitute $y = 0$ for $z = -1$ into E_2 and solve for x.
$$
\begin{array}{rl}
x - 2y - 2z = 4 & E_2 \\
x - 2(0) - 2(-1) = 4 & \\
x = 2 &
\end{array}
$$
$(2, 0, -1)$ \hfill *(8-2)*

12. The boundary line goes through $(0, -4)$ and $(6, 0)$:
$$m = \frac{0 - (-4)}{6 - 0} = \frac{4}{6} = \frac{2}{3}$$
$$y - 0 = \frac{2}{3}(x - 6)$$
Multiply both sides by 3
$$3y = 2(x - 6)$$
$$3y = 2x - 12$$
$$2x - 3y = 12$$

Now note that $(0, 0)$ is in the shaded region, and $2(0) - 3(0) = 0 < 12$, so the inequality is $2x - 3y \le 12$ (equality included since the boundary line is solid). \hfill *(8-3)*

13. The boundary line goes through $(0, 8)$ and $(2, 0)$:
$$m = \frac{0 - 8}{2 - 0} = \frac{-8}{2} = -4$$
$$y - 0 = -4(x - 2)$$
$$y = -4x + 8$$
$$4x + y = 8$$

Now note that $(4, 0)$ is in the shaded region, and $4(4) - 3(0) = 16 > 8$, so the inequality is $4x + y \ge 8$ (equality included since the boundary line is solid). \hfill *(8-3)*

14.

Corner Point (x, y)	Objective Function $z = 5x + 3y$	
$(0, 10)$	30	
$(0, 6)$	18	Minimum value
$(4, 2)$	26	
$(6, 4)$	42	Maximum value

The maximum value of z on S is 42 at $(6, 4)$. The minimum value of z on S is 18 at $(0, 6)$. \hfill *(8-4)*

15.

$$x + 2y + 3z = 1 \qquad E_1$$
$$2x + 3y + 4z = 3 \qquad E_2$$
$$x + 2y + z = 3 \qquad E_3$$

Multiply E_1 by -1 and add to E_3 to eliminate x and y:

$$\begin{array}{ll} -x - 2y - 3z = -1 & (-1)E_1 \\ \underline{x + 2y + z = 3} & E_2 \\ -2z = 2 \\ z = -1 \end{array}$$

Substitute $z = -1$ into E_1 and E_2 to obtain a simpler system:

$$x + 2y + 3(-1) = 1$$
$$2x + 3y + 4(-1) = 3$$

or

$$x + 2y = 4 \qquad E_4$$
$$2x + 3y = 7 \qquad E_5$$

Multiply E_4 by -2 and add to E_5 to eliminate x.

$$\begin{array}{ll} -2x - 4y = -8 & (-2)E_4 \\ \underline{2x + 3y = 7} & E_5 \\ -y = -1 \\ y = 1 \end{array}$$

Substitute $y = 1$ into E_4 to find x.

$$x + 2(1) = 4$$
$$x = 2$$
$$(2, 1, -1)$$

$(8-2)$

16.

$$x + 2y - z = 2 \qquad E_1$$
$$2x + 3y + z = -3 \qquad E_2$$
$$3x + 5y = -1 \qquad E_3$$

Add E_1 and E_2 to eliminate z.

$$\begin{array}{ll} x + 2y - z = 2 & E_1 \\ \underline{2x + 3y + z = -3} & E_2 \\ 3x + 5y = -1 & E_4 \end{array}$$

Multiply E_3 by -1 and add to E_4.

$$\begin{array}{ll} -3x - 5y = 1 & (-1)E_3 \\ \underline{3x + 5y = -1} & E_4 \\ 0 = 0 \end{array}$$

There are infinitely many solutions; the system is dependent and equivalent to:

$$x + 2y - z = 2 \qquad E_1$$
$$3x + 5y = -1 \qquad E_3$$

Multiply E_1 by -3 and add to E_3.

$$\begin{array}{ll} -3x - 6y + 3z = -6 & (-3)E_1 \\ \underline{3x + 5y = -1} & E_3 \\ -y + 3z = -7 \\ y = 3z + 7 \end{array}$$

Let $z = t$. Then $y = 3t + 7$. Substitute this into E_3.

$$3x + 5(3t + 7) = -1$$
$$3x + 15t + 35 = -1$$
$$3x = -15t - 36$$
$$x = -5t - 12$$

Solutions: $\{(-5t - 12, 3t + 7, t) \mid t \text{ is any real number}\}$

$(8-2)$

17.

$$\begin{array}{rcll} x - y &=& 1 & E_1 \\ 2x - y &=& 0 & E_2 \\ x - 3y &=& -2 & E_3 \end{array}$$

Multiply E_1 by -2 and add to E_2 to eliminate x.

$$\begin{array}{rcll} -2x + 2y &=& -2 & (-2)E_1 \\ \underline{2x - y} &=& \underline{0} & E_2 \\ y &=& -2 & \end{array}$$

Substitute $y = -2$ back into E_1 and E_2.

E_1 : $x - (-2) = 1$; $x = -1$

E_2 : $2x - (-2) = 0$; $x = 1$

Since x cannot be both 1 and -1, the system is inconsistent, and there is no solution. $(8\text{-}2)$

18.

$$\begin{array}{rcll} x + 2y - z &=& 2 & E_1 \\ 3x - y + 2z &=& -3 & E_2 \\ 4x + y + z &=& 7 & E_3 \end{array}$$

Add E_2 and E_3 to eliminate y.

$$\begin{array}{rcll} 3x - y + 2z &=& -3 & E_2 \\ \underline{4x + y + z} &=& \underline{7} & E_3 \\ 7x + 3z &=& 4 & E_4 \end{array}$$

Multiply E_2 by 2 and add to E_1 to eliminate y again:

$$\begin{array}{rcll} 6x - 2y + 4z &=& -6 & 2E_2 \\ \underline{x + 2y - z} &=& \underline{2} & E_1 \\ 7x + 3z &=& -4 & E_5 \end{array}$$

Now multiply E_4 by -1 and add to E_5.

$$\begin{array}{rcll} -7x - 3z &=& -4 & (-1)E_4 \\ \underline{7x + 3z} &=& \underline{-4} & E_5 \\ 0 &=& -8 & \end{array}$$

The system is inconsistent, and there is no solution. $(8\text{-}2)$

19. The solution region is bounded. Three corner points are obvious from the graph: $(0, 4)$, $(0, 0)$, and $(4, 0)$. The fourth corner point is obtained by solving the system
$2x + y = 8$
$2x + 3y = 12$ to obtain $(3, 2)$.

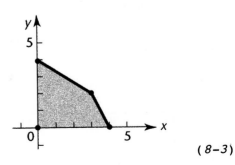

$(8\text{-}3)$

20. The solution region is unbounded. Two corner points are obvious from the graph: $(0, 8)$ and $(12, 0)$. The third corner point is obtained by solving the system
$2x + y = 8$
$x + 3y = 12$
to obtain $\left(\dfrac{12}{5}, \dfrac{16}{5}\right)$.

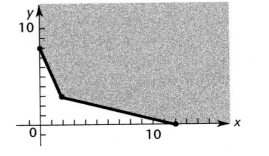

$(8\text{-}3)$

21. The solution region is bounded. The corner point (20, 0) is obvious from the graph. The other corner points are found by solving the system

$$\begin{cases} x + y = 20 \\ x - y = 0 \end{cases} \text{ to obtain (10, 10)}$$

and

$$\begin{cases} x - y = 0 \\ x + 4y = 20 \end{cases} \text{ to obtain (4, 4)}$$

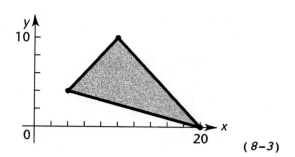

(8-3)

22. The feasible region is graphed as follows. The corner points (0, 4), (0, 0) and (5, 0) are obvious from the graph. The corner point (4, 2) is obtained by solving the system

$x + 2y = 8$
$2x + y = 10$

We now evaluate the objective function at each corner point.

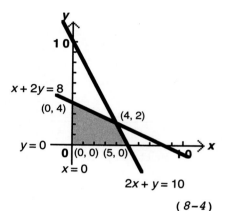

Corner Point (x, y)	Objective Function $z = 7x + 9y$	
(0, 4)	36	
(0, 0)	0	
(5, 0)	35	
(4, 2)	46	Maximum value

The maximum value of z on S is 46 at (4, 2).

(8-4)

23. The feasible region is graphed as follows. The corner points (0, 20), (0, 15), (15, 0) and (20, 0) are obvious from the graph. The corner point (3, 6) is obtained by solving the system

$3x + y = 15$
$x + 2y = 15$

We now evaluate the objective function at each corner point.

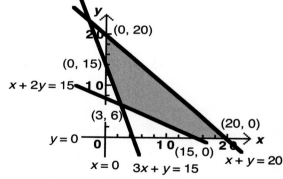

Corner Point (x, y)	Objective Function $z = 5x + 10y$	
(0, 20)	200	
(0, 15)	150	
(3, 6)	75	Minimum value ⎱
(15, 0)	75	Minimum value ⎰ Multiple optimal solutions
(20, 0)	100	

The minimum value of z on S is 75 at (3, 6) and (15, 0).

(8-4)

24. The feasible region is graphed below. The corner points $(0, 10)$ and $(0, 7)$ are obvious from the graph. The other corner points are obtained by solving the systems:

$x + 2y = 20$ and $3x + y = 15$
$3x + y = 15$ to obtain $(2, 9)$ $x + y = 7$ to obtain $(4, 3)$

We now evaluate the objective function at each corner point.

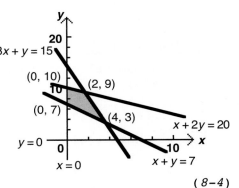

Corner Point (x, y)	Objective Function $z = 5x + 8y$	
$(0, 10)$	80	
$(0, 7)$	56	
$(4, 3)$	44	Minimum value
$(2, 9)$	82	Maximum value

The minimum value of z on S is 44 at $(4, 3)$.
The maximum value of z on S is 82 at $(2, 9)$.

$(8-4)$

25.
$$x_1 + x_2 + x_3 = 7,000 \quad E_1$$
$$0.04x_1 + 0.05x_2 + 0.06x_3 = 360 \quad E_2$$
$$0.04x_1 + 0.05x_2 - 0.06x_3 = 120 \quad E_3$$

Multiply E_2 by -1 and add to E_3 to eliminate x_1 and x_2.

$$-0.04x_1 - 0.05x_2 - 0.06x_3 = -360 \quad (-1)E_2$$
$$\underline{0.04x_1 + 0.05x_2 - 0.06x_3 = 120} \quad E_3$$
$$-0.12x_3 = -240$$
$$x_3 = 2,000$$

Substitute $x_3 = 2,000$ back into E_1 and E_2 to obtain a simpler system.

$$x_1 + x_2 + 2,000 = 7,000$$
$$0.04x_1 + 0.05x_2 + 0.06(2,000) = 360$$
or
$$x_1 + x_2 = 5,000 \quad E_4$$
$$0.04x_1 + 0.05x_2 = 240 \quad E_5$$

Multiply E_4 by -0.04 and add to E_5.

$$-0.04x_1 - 0.04x_2 = -200 \quad (-0.04)E_4$$
$$\underline{0.04x_1 + 0.05x_2 = 240} \quad E_5$$
$$0.01x_2 = 40$$
$$x_2 = 4,000$$

Substitute $x_2 = 4,000$ back into E_4.

$$x_1 + 4,000 = 5,000$$
$$x_1 = 1,000$$
$(1,000, 4,000, 2,000)$

$(8-2)$

26. For convenience we rewrite the constraint conditions:
$1.2x + 0.6y \le 960$ becomes $2x + y \le 1,600$
$0.04x + 0.03y \le 36$ becomes $4x + 3y \le 3,600$
$0.2x + 0.3y \le 270$ becomes $2x + 3y \le 2,700$
$x, y \ge 0$ are unaltered.

The feasible region is graphed in the diagram. The corner points $(0, 900)$, $(0, 0)$, and $(800, 0)$ are obvious from the graph. The other corner points are obtained by solving the system:

$2x + y = 1,600$ and $4x + 3y = 3,600$
$4x + 3y = 3,600$ to obtain $(600, 400)$ $2x + 3y = 2,700$ to obtain $(450, 600)$

We now evaluate the objective function at each corner point.

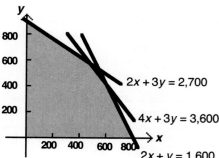

$2x + 3y = 2{,}700$

$4x + 3y = 3{,}600$

$2x + y = 1{,}600$

Corner Point (x, y)	Objective Function $z = 30x + 20y$	
(0, 900)	18,000	
(0, 0)	0	
(800, 0)	24,000	
(600, 400)	26,000	Maximum value
(450, 600)	25,500	

The maximum value of z on S is 26,000 at (600, 400).

(8-4)

27. Let x = number of $\frac{1}{2}$-pound packages

   y = number of $\frac{1}{3}$-pound packages

There are 120 packages. So

$x + y = 120$ (1)

Since x $\frac{1}{2}$-pound packages weigh $\frac{1}{2}x$ pounds and y $\frac{1}{3}$-pound packages weigh $\frac{1}{3}y$ pounds, we have

$\frac{1}{2}x + \frac{1}{3}y = 48$ (2)

We solve the system (1), (2) using elimination by addition. We multiply the second equation by -3 and add.

$x + y = 120$

$-\frac{3}{2}x - y = -144$

$-\frac{1}{2}x = -24$

$x = 48$

Substituting into equation (1), we have

$48 + y = 120$

$y = 72$

48 $\frac{1}{2}$-pound packages and 72 $\frac{1}{3}$-pound packages.

(8-1)

28. Using the formulas for perimeter and area of a rectangle, we have

$2a + 2b = 28$

$ab = 48$

Solving the first equation for b and substituting into the second equation, we have

$2b = 28 - 2a$

$b = 14 - a$

$a(14 - a) = 48$

$-a^2 + 14a = 48$

$a^2 - 14a + 48 = 0$

$(a - 6)(a - 8) = 0$

$a = 6, 8$

For $a = 6$ For $a = 8$

$b = 14 - 6$ $b = 14 - 8$

$= 8$ $= 6$

Dimensions: 6 meters by 8 meters.

(8-1)

29. Let x = the number of grams of mix A
y = the number of grams of mix B
z = the number of grams of mix C

Then using the percentages in the table, we get
$0.30x + 0.20y + 0.10z = 27$ (protein)
$0.03x + 0.05y + 0.04z = 5.4$ (fat)
$0.10x + 0.20y + 0.10z = 19$ (moisture)

We can clear decimals easily by multiplying the 1st and 3rd equation by 10, and the 2nd by 100:
$3x + 2y + z = 270$ E_1
$3x + 5y + 4z = 540$ E_2
$x + 2y + z = 190$ E_3

Multiply E_1 by -4 and add to E_2.
$-12x - 8y - 4z = -1{,}080$ $(-4)E_1$
$\underline{3x + 5y + 4z = \phantom{-1{,}0}540}$ E_2
$-9x - 3y = \phantom{-1{,}0}-540$ E_4

Multiply E_1 by -1 and add to E_3.
$-3x - 2y - z = -270$ $(-1)E_1$
$\underline{x + 2y + z = 190}$ E_3
$-2x = -80$
$x = 40$

Substitute $x = 40$ back into E_4.
$-9(40) - 3y = -540$
$-3y = -180$
$y = 60$

Substitute $x = 40$, $y = 60$ back into E_1.
$3(40) + 2(60) + z = 270$
$240 + z = 270$
$z = 30$

They should use 40 grams of mix A, 60 grams of mix B, and 30 grams of mix C.
$(8-2)$

30. We begin by writing an expression for the amount of time needed in each machine. Let x = the number of calculator boards produced, and let y = the number of toaster boards produced. Now we can write an expression for the total minutes in each machine for each type of board:

Selective machine: Calculator boards $4x$ (4 minutes per board times x boards), toaster boards $3y$.

Wave machine: Calculator boards $2x$, toaster boards y

Total minutes in selective machine: $4x + 3y$

Total minutes in wave machine: $2x + y$

(A) The total minutes in the selective machine must be no more than 300:
$4x + 3y \leq 300$. Both x and y must be non-negative: $x \geq 0$, $y \geq 0$.

The graph is shown at the right.

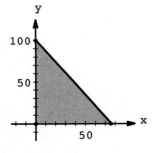

(B) The total minutes in the wave machine
must be no more than 120:
$2x + y \le 120$.   Both x and y must be
non-negative: $x \ge 0$, $y \ge 0$.

The graph is shown at the right.

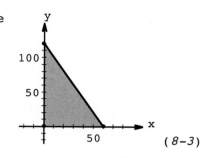

(8-3)

31. Let x = number of regular sails
y = number of competition sails

(A) We form the linear objective function
$P = 60x + 100y$
We wish to maximize P, the profit from
x regular sails @ \$60 and y competition
sails @ \$100, subject to the constraints

$x + 2y \le 140$ cutting department constraint
$3x + 4y \le 360$ sewing department constraint
$x, y \ge 0$ non-negative constraints

Solving the system of constraint
inequalities graphically, we obtain the
feasible region S shown.

Next we evaluate the objective function at
each corner point.

Corner Point (x, y)	Objective Function $P = 60x + 100y$	
(0, 0)	0	
(0, 70)	7,000	
(80, 30)	7,800	Maximum value
(120, 0)	7,200	

The optimal value is 7,800 at the corner point (80, 30). So the company should
manufacture 80 regular sails and 30 competition sails for a maximum profit
of \$7,800.

(B) The objective function now becomes $60x + 125y$. We evaluate this at each
corner point.

Corner Point (x, y)	Objective Function $P = 60x + 125y$	
(0, 0)	0	
(0, 70)	8,750	Maximum value
(80, 30)	8,550	
(120, 0)	7,200	

The optimal value is now 8,750 at the corner point (0, 70). So the
company should manufacture 70 competition sails and no regular sails
for a maximum profit of \$8,750.

(C) The objective function now becomes $60x + 75y$. We evaluate this at each
corner point.

Corner Point (x, y)	Objective Function $P = 60x + 75y$	
(0, 0)	0	
(0, 70)	5,250	
(80, 30)	7,050	
(120, 0)	7,200	Maximum value

The optimal value is now 7,200 at the corner point (120, 0). So the
company should manufacture 120 regular sails and no competition sails
for a maximum profit of \$7,200.

(8-4)

32. Let x = number of grams of mix A
y = number of grams of mix B

(A) We form the objective function
$$C = 0.07x + 0.04y$$

C represents the cost of x grams at \$0.07 per gram and y grams at \$0.04 per gram. We wish to minimize C subject to

$5x + 2y \geq 800$	vitamin constraint
$2x + 4y \geq 800$	mineral constraint
$4x + 4y \leq 1300$	calorie constraint
$x, y \geq 0$	non-negative constraints

Solving the system of constraint inequalities graphically, we obtain the feasible region S shown in the diagram.

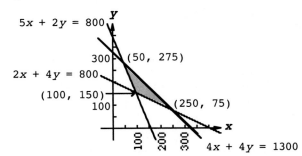

Next we evaluate the objective function at each corner point.

Corner Point (x, y)	Objective Function $C = 0.07x + 0.04y$	
(100, 150)	13	Minimum value
(250, 75)	20.5	
(50, 275)	14.5	

The optimal value is 13 at the corner point (100, 150). So 100 grams of mix A and 150 grams of mix B should be used for a cost of \$13.

(B) The objective function now becomes $0.07x + 0.02y$. We evaluate this at each corner point.

Corner Point (x, y)	Objective Function $C = 0.07x + 0.02y$	
(100, 150)	10	
(250, 75)	19	
(50, 275)	9	Minimum value

The optimal value is now 9 at the corner point (50, 275). So 50 grams of mix A and 275 grams of mix B should be used for a cost of \$9.

(C) The objective function now becomes $0.07x + 0.15y$. We evaluate this at each corner point.

Corner Point (x, y)	Objective Function $C = 0.07x + 0.15y$	
(100, 150)	29.5	
(250, 75)	28.75	Minimum value
(50, 275)	44.75	

The optimal value is now 28.75 at the corner point (250, 75). So 250 grams of mix A and 75 grams of mix B should be used for a cost of \$28.75.

(8-4)

CHAPTER 9   Matrices & Determinants

Section 9-1

1. Answers will vary.   3. Answers will vary.   **5.** Answers will vary.

7. Answers will vary.   9. Answers will vary.

11. No. Condition 2 is violated: The first nonzero entry in row 2 is not 1.

13. Yes

15. No. Condition 4 is violated: The first nonzero entry in row 3 is to the left of the first nonzero entry in row 2.

17. Yes

19. $x_1 \qquad = -2$
    $\qquad x_2 \quad = 3$
    $\qquad\qquad x_3 = 0$

    The system is already solved.

21. $x_1 \qquad - 2x_3 = 3$
    $\qquad x_2 + \quad x_3 = -5$

    Solution:
    $x_3 = t$
    $x_2 = -5 - x_3 = -5 - t$
    $x_1 = 3 + 2x_3 = 3 + 2t$
    So the solution is $x_1 = 2t + 3$,
    $x_2 = -t - 5$,
    $x_3 = t$ for any real number t.

23. $x_1 \qquad = 0$
    $\qquad x_2 \quad = 0$
    $\qquad\qquad 0 = 1$

    The system has no solution.

25. $x_1 - 2x_2 \qquad - 3x_4 = -5$
    $\qquad\qquad x_3 + 3x_4 = 2$

    Solution:
    $x_4 = t$
    $x_3 = 2 - 3x_4 = 2 - 3t$
    $x_2 = s$
    $x_1 = -5 + 2x_2 + 3x_4 = -5 + 2s + 3t$
    So the solution is $x_1 = 2s + 3t - 5$,
    $x_2 = s$, $x_3 = -3t + 2$, $x_4 = t$ for any
    real numbers s and t.

27. $\begin{bmatrix} 1 & 2 & | & -1 \\ 0 & 1 & | & 3 \end{bmatrix}$ $(-2)R_2 + R_1 \rightarrow R_1$

    Need a 0 here

    $\sim \begin{bmatrix} 1 & 0 & | & -7 \\ 0 & 1 & | & 3 \end{bmatrix}$

29. $\begin{bmatrix} 1 & 0 & -3 & | & 1 \\ 0 & 1 & 2 & | & 0 \\ 0 & 0 & 3 & | & -6 \end{bmatrix}$ $\frac{1}{3} R_3 \rightarrow R_3$

    $\uparrow$
    Need a 1 here
    Need 0's here
    $\downarrow$

    $\sim \begin{bmatrix} 1 & 0 & -3 & | & 1 \\ 0 & 1 & 2 & | & 0 \\ 0 & 0 & 1 & | & -2 \end{bmatrix}$ $\begin{matrix} 3R_3 + R_1 \rightarrow R_1 \\ (-2)R_3 + R_2 \rightarrow R_2 \end{matrix}$

    $\sim \begin{bmatrix} 1 & 0 & 0 & | & -5 \\ 0 & 1 & 0 & | & 4 \\ 0 & 0 & 1 & | & -2 \end{bmatrix}$

31. $\begin{bmatrix} 1 & 2 & -2 & | & -1 \\ 0 & 3 & -6 & | & 1 \\ 0 & -1 & 2 & | & -\frac{1}{3} \end{bmatrix} \frac{1}{3} R_2 \to R_2$

Need a 1 here

$\sim \begin{bmatrix} 1 & 2 & -2 & | & -1 \\ 0 & 1 & -2 & | & \frac{1}{3} \\ 0 & -1 & 2 & | & -\frac{1}{3} \end{bmatrix} \begin{array}{l} (-2)R_2 + R_1 \to R_1 \\ \\ R_3 + R_2 \to R_3 \end{array}$

Need 0's here

$\sim \begin{bmatrix} 1 & 0 & 2 & | & -\frac{5}{3} \\ 0 & 1 & -2 & | & \frac{1}{3} \\ 0 & 0 & 0 & | & 0 \end{bmatrix}$

33. $\begin{bmatrix} 2 & 4 & -10 & | & -2 \\ 3 & 9 & -21 & | & 0 \\ 1 & 5 & -12 & | & 1 \end{bmatrix} R_1 \leftrightarrow R_3$

Need a 1 here

$\sim \begin{bmatrix} 1 & 5 & -12 & | & 1 \\ 3 & 9 & -21 & | & 0 \\ -2 & 4 & -10 & | & -2 \end{bmatrix} \begin{array}{l} (-3)R_1 + R_2 \to R_2 \\ (-2)R_1 + R_3 \to R_3 \end{array}$

Need 0's here

$\sim \begin{bmatrix} 1 & 5 & -12 & | & 1 \\ 0 & -6 & 15 & | & -3 \\ 0 & -6 & 14 & | & -4 \end{bmatrix} -\frac{1}{6} R_2 \to R_2$

Need a 1 here

$\sim \begin{bmatrix} 1 & 5 & -12 & | & 1 \\ 0 & 1 & -\frac{5}{2} & | & \frac{1}{2} \\ 0 & -6 & 14 & | & -4 \end{bmatrix} \begin{array}{l} (-5)R_2 + R_1 \to R_1 \\ \\ 6R_2 + R_3 \to R_3 \end{array}$

Need 0's here

$\sim \begin{bmatrix} 1 & 0 & \frac{1}{2} & | & -\frac{3}{2} \\ 0 & 1 & -\frac{5}{2} & | & \frac{1}{2} \\ 0 & 0 & -1 & | & -1 \end{bmatrix} -R_3 \to R_3$

Need a 1 here

$\sim \begin{bmatrix} 1 & 0 & \frac{1}{2} & | & -\frac{3}{2} \\ 0 & 1 & \frac{5}{2} & | & \frac{1}{2} \\ 0 & 0 & 1 & | & 1 \end{bmatrix} \begin{array}{l} (-\frac{1}{2})R_3 + R_1 \to R_1 \\ \frac{5}{2} R_3 + R_2 \to R_2 \end{array}$

Need 0's here

$\sim \begin{bmatrix} 1 & 0 & 0 & | & -2 \\ 0 & 1 & 0 & | & 3 \\ 0 & 0 & 1 & | & 1 \end{bmatrix}$

Solution: $x_1 = -2$, $x_2 = 3$, and $x_3 = 1$.

35. $\begin{bmatrix} 3 & 8 & -1 & | & -18 \\ 2 & 1 & 5 & | & 8 \\ 2 & 4 & 2 & | & -4 \end{bmatrix} \begin{array}{l} \frac{1}{2} R_3 \to R_3 \\ \\ R_3 \leftrightarrow R_1 \end{array}$

Need a 1 here

$\sim \begin{bmatrix} 1 & 2 & 1 & | & -2 \\ 2 & 1 & 5 & | & 8 \\ 3 & 8 & -1 & | & -18 \end{bmatrix} \begin{array}{l} (-2)R_1 + R_2 \to R_2 \\ (-3)R_1 + R_3 \to R_3 \end{array}$

Need 0's here

$\sim \begin{bmatrix} 1 & 2 & 1 & | & -2 \\ 0 & -3 & 3 & | & 12 \\ 0 & 2 & -4 & | & -12 \end{bmatrix} -\frac{1}{3} R_2 \to R_2$

Need a 1 here

$\sim \begin{bmatrix} 1 & 2 & 1 & | & -2 \\ 0 & 1 & -1 & | & -4 \\ 0 & 2 & -4 & | & -12 \end{bmatrix} \begin{array}{l} (-2)R_2 + R_1 \to R_1 \\ \\ (-2)R_2 + R_3 \to R_3 \end{array}$

Need 0's here

$\sim \begin{bmatrix} 1 & 0 & 3 & | & 6 \\ 0 & 1 & -1 & | & -4 \\ 0 & 0 & -2 & | & -4 \end{bmatrix} -\frac{1}{2} R_3 \to R_3$

Need a 1 here

$\sim \begin{bmatrix} 1 & 0 & 3 & | & 6 \\ 0 & 1 & -1 & | & -4 \\ 0 & 0 & 1 & | & 2 \end{bmatrix} \begin{array}{l} (-3)R_3 + R_1 \to R_1 \\ R_3 + R_2 \to R_2 \end{array}$

Need 0's here

$\begin{bmatrix} 1 & 0 & 0 & | & 0 \\ 0 & 1 & 0 & | & -2 \\ 0 & 0 & 1 & | & 2 \end{bmatrix}$ Solution: $x_1 = 0$, $x_2 = -2$, and $x_3 = 2$.

37. $\begin{bmatrix} 2 & -1 & -3 & | & 8 \\ 1 & -2 & 0 & | & 7 \end{bmatrix}$ $R_1 \leftrightarrow R_2$

$\sim \begin{bmatrix} 1 & -2 & 0 & | & 7 \\ 2 & -1 & -3 & | & 8 \end{bmatrix}$ $(-2)R_1 + R_2 \rightarrow R_2$

$\sim \begin{bmatrix} 1 & -2 & 0 & | & 7 \\ 0 & 3 & -3 & | & -6 \end{bmatrix}$ $\frac{1}{3} R_2 \rightarrow R_2$

$\sim \begin{bmatrix} 1 & -2 & 0 & | & 7 \\ 0 & 1 & -1 & | & -2 \end{bmatrix}$ $2R_2 + R_1 \rightarrow R_1$

$\sim \begin{bmatrix} 1 & 0 & -2 & | & 3 \\ 0 & 1 & -1 & | & -2 \end{bmatrix}$

Let $x_3 = t$. Then

$x_2 - x_3 = -2$

$\quad x_2 = x_3 - 2 = t - 2$

$x_1 - 2x_3 = 3$

$\quad x_1 = 2x_3 + 3 = 2t + 3$

Solution: $x_1 = 2t + 3$, $x_2 = t - 2$, $x_3 = t$, t any real number.

41. $\begin{bmatrix} 3 & -4 & -1 & | & 1 \\ 2 & -3 & 1 & | & 1 \\ 1 & -2 & 3 & | & 2 \end{bmatrix}$ $R_1 \leftrightarrow R_3$

$\sim \begin{bmatrix} 1 & -2 & 3 & | & 2 \\ 2 & -3 & 1 & | & 1 \\ 3 & -4 & -1 & | & 1 \end{bmatrix}$ $\begin{matrix} (-2)R_1 + R_2 \rightarrow R_2 \\ (-3)R_1 + R_3 \rightarrow R_3 \end{matrix}$

$\sim \begin{bmatrix} 1 & -2 & 3 & | & 2 \\ 0 & 1 & -5 & | & -3 \\ 0 & 2 & -10 & | & -5 \end{bmatrix}$ $(-2)R_2 + R_3 \rightarrow R_3$

$\sim \begin{bmatrix} 1 & -2 & 3 & | & 2 \\ 0 & 1 & -5 & | & -3 \\ 0 & 0 & 0 & | & 1 \end{bmatrix}$

Since the last row corresponds to the equation $0x_1 + 0x_2 + 0x_3 = 1$, there is no solution.

45. $\begin{bmatrix} 2 & -2 & -4 & | & -2 \\ -3 & 3 & 6 & | & 3 \end{bmatrix}$ $\begin{matrix} \frac{1}{2} R_1 \rightarrow R_1 \\ \frac{1}{3} R_2 \rightarrow R_2 \end{matrix}$

$\sim \begin{bmatrix} 1 & -1 & -2 & | & -1 \\ -1 & 1 & 2 & | & 1 \end{bmatrix}$ $R_1 + R_2 \rightarrow R_2$

$\sim \begin{bmatrix} 1 & -1 & -2 & | & -1 \\ 0 & 0 & 0 & | & 0 \end{bmatrix}$

39. $\begin{bmatrix} 2 & -1 & | & 0 \\ 3 & 2 & | & 7 \\ 1 & -1 & | & -1 \end{bmatrix}$ $R_1 \leftrightarrow R_3$

$\sim \begin{bmatrix} 1 & -1 & | & -1 \\ 3 & 2 & | & 7 \\ 2 & -1 & | & 0 \end{bmatrix}$ $\begin{matrix} (-3)R_1 + R_2 \rightarrow R_2 \\ (-2)R_1 + R_3 \rightarrow R_3 \end{matrix}$

$\sim \begin{bmatrix} 1 & -1 & | & -1 \\ 0 & 5 & | & 10 \\ 0 & 1 & | & 2 \end{bmatrix}$ $R_2 \leftrightarrow R_3$

$\sim \begin{bmatrix} 1 & -1 & | & -1 \\ 0 & 1 & | & 2 \\ 0 & 5 & | & 10 \end{bmatrix}$ $\begin{matrix} R_2 + R_1 \rightarrow R_1 \\ (-5)R_2 + R_3 \rightarrow R_3 \end{matrix}$

$\sim \begin{bmatrix} 1 & 0 & | & 1 \\ 0 & 1 & | & 2 \\ 0 & 0 & | & 0 \end{bmatrix}$

Solution: $x_1 = 1$ and $x_2 = 2$.

43. $\begin{bmatrix} -2 & 1 & 3 & | & -7 \\ 1 & -4 & 2 & | & 0 \\ 1 & -3 & 1 & | & 1 \end{bmatrix}$ $R_1 \leftrightarrow R_2$

$\sim \begin{bmatrix} 1 & -4 & 2 & | & 0 \\ -2 & 1 & 3 & | & -7 \\ 1 & -3 & 1 & | & 1 \end{bmatrix}$ $\begin{matrix} 2R_1 + R_2 \rightarrow R_2 \\ (-1)R_1 + R_3 \rightarrow R_3 \end{matrix}$

$\sim \begin{bmatrix} 1 & -4 & 2 & | & 0 \\ 0 & -7 & 7 & | & -7 \\ 0 & 1 & -1 & | & 1 \end{bmatrix}$ $R_2 \leftrightarrow R_3$

$\sim \begin{bmatrix} 1 & -4 & 2 & | & 0 \\ 0 & 1 & -1 & | & 1 \\ 0 & -7 & 7 & | & -7 \end{bmatrix}$ $\begin{matrix} 4R_2 + R_1 \rightarrow R_1 \\ 7R_2 + R_3 \rightarrow R_3 \end{matrix}$

$\sim \begin{bmatrix} 1 & 0 & -2 & | & 4 \\ 0 & 1 & -1 & | & 1 \\ 0 & 0 & 0 & | & 0 \end{bmatrix}$

Let $x_3 = t$. Then

$\quad x_2 - x_3 = 1$

$\qquad x_2 = x_3 + 1 = t + 1$

$\quad x_1 - 2x_3 = 4$

$x_1 = 2x_3 + 4 = 2t + 4$

Solution: $x_1 = 2t + 4$, $x_2 = t + 1$, $x_3 = t$, t any real number.

Let $x_3 = t$, $x_2 = s$. Then

$\quad x_1 - x_2 - 2x_3 = -1$

$\quad x_1 \qquad\qquad = x_2 + 2x_3 - 1$

$\qquad\qquad\qquad = s + 2t - 1$

Solution: $x_1 = s + 2t - 1$, $x_2 = s$, $s_3 = t$, s and t any real numbers.

47. $\begin{bmatrix} 4 & -1 & 2 & | & 3 \\ -4 & 1 & -3 & | & -10 \\ 8 & -2 & 9 & | & -1 \end{bmatrix} \begin{matrix} \\ R_1 + R_2 \rightarrow R_2 \\ (-2)R_1 + R_3 \rightarrow R_3 \end{matrix}$

$\sim \begin{bmatrix} 4 & -1 & 2 & | & 3 \\ 0 & 0 & -1 & | & -7 \\ 0 & 0 & 5 & | & -7 \end{bmatrix} \begin{matrix} \\ \\ 5R_2 + R_3 \rightarrow R_3 \end{matrix}$

$\sim \begin{bmatrix} 4 & -1 & 2 & | & 3 \\ 0 & 0 & -1 & | & -7 \\ 0 & 0 & 0 & | & -42 \end{bmatrix}$

Since the last row corresponds to the equation $0x_1 + 0x_2 + 0x_3 = -42$, there is no solution.

49. $\begin{bmatrix} 2 & -5 & -3 & | & 7 \\ -4 & 10 & 2 & | & 6 \\ 6 & -15 & -1 & | & -19 \end{bmatrix} \begin{matrix} \\ 2R_1 + R_2 \rightarrow R_2 \\ (-3)R_1 + R_3 \rightarrow R_3 \end{matrix}$

$\sim \begin{bmatrix} 2 & -5 & -3 & | & 7 \\ 0 & 0 & -4 & | & 20 \\ 0 & 0 & 8 & | & -40 \end{bmatrix} \begin{matrix} \frac{1}{2} R_1 \rightarrow R_1 \\ -\frac{1}{4} R_2 \rightarrow R_2 \\ \end{matrix}$

$\sim \begin{bmatrix} 1 & -2.5 & -1.5 & | & 3.5 \\ 0 & 0 & 1 & | & -5 \\ 0 & 0 & 8 & | & -40 \end{bmatrix} \begin{matrix} 1.5R_2 + R_1 \rightarrow R_1 \\ \\ (-8)R_2 + R_3 \rightarrow R_3 \end{matrix}$

$\sim \begin{bmatrix} 1 & -2.5 & 0 & | & -4 \\ 0 & 0 & 1 & | & -5 \\ 0 & 0 & 0 & | & 0 \end{bmatrix}$

Let $x_2 = t$. Then $x_3 = -5$ and

$\quad x_1 - 2.5x_2 = -4$

$\qquad\qquad x_1 = 2.5x_2 - 4$

$\qquad\qquad x_1 = 2.5t - 4$

Solution: $x_1 = 2.5t - 4$, $x_2 = t$, $x_3 = -5$, t any real number.

51. $\begin{bmatrix} 5 & -3 & 2 & | & 13 \\ 2 & -1 & -3 & | & 1 \\ 4 & -2 & 4 & | & 12 \end{bmatrix} \begin{matrix} \\ \\ \frac{1}{4} R_3 \rightarrow R_3 \end{matrix}$

$\sim \begin{bmatrix} 5 & -3 & 2 & | & 13 \\ 2 & -1 & -3 & | & 1 \\ 1 & -\frac{1}{2} & 1 & | & 3 \end{bmatrix} \begin{matrix} \\ \\ R_1 \leftrightarrow R_3 \end{matrix}$

$\sim \begin{bmatrix} 1 & -\frac{1}{2} & 1 & | & 3 \\ 2 & -1 & -3 & | & 1 \\ 5 & -3 & 2 & | & 13 \end{bmatrix} \begin{matrix} \\ (-2)R_1 + R_2 \rightarrow R_2 \\ (-5)R_1 + R_3 \rightarrow R_3 \end{matrix}$

$\sim \begin{bmatrix} 1 & -\frac{1}{2} & 1 & | & 3 \\ 0 & 0 & -5 & | & -5 \\ 0 & -\frac{1}{2} & -3 & | & -2 \end{bmatrix} \begin{matrix} (-1)R_3 + R_1 \rightarrow R_1 \\ -\frac{1}{5} R_2 \rightarrow R_2 \\ \end{matrix}$

$\sim \begin{bmatrix} 1 & 0 & 4 & | & 5 \\ 0 & 0 & 1 & | & 1 \\ 0 & -\frac{1}{2} & -3 & | & -2 \end{bmatrix} \begin{matrix} (-4)R_2 + R_1 \rightarrow R_1 \\ 3R_2 + R_3 \rightarrow R_3 \\ \end{matrix}$

$\sim \begin{bmatrix} 1 & 0 & 0 & | & 1 \\ 0 & 0 & 1 & | & 1 \\ 0 & -\frac{1}{2} & 0 & | & 1 \end{bmatrix} \begin{matrix} \\ \\ (-2)R_3 \rightarrow R_3 \end{matrix}$

$\sim \begin{bmatrix} 1 & 0 & 0 & | & 1 \\ 0 & 0 & 1 & | & 1 \\ 0 & 1 & 0 & | & -2 \end{bmatrix} \begin{matrix} \\ R_2 \leftrightarrow R_3 \\ \end{matrix}$

$\sim \begin{bmatrix} 1 & 0 & 0 & | & 1 \\ 0 & 1 & 0 & | & -2 \\ 0 & 0 & 1 & | & 1 \end{bmatrix}$

Solution: $x_1 = 1$, $x_2 = -2$, $x_3 = 1$.

53. **(A)** The reduced form matrix will have the form
$$\begin{bmatrix} 1 & a & b & | & c \\ 0 & 0 & 0 & | & 0 \\ 0 & 0 & 0 & | & 0 \end{bmatrix}$$
So the system is equivalent to
$$x_1 + ax_2 + bx_3 = c$$
$$0 = 0$$
$$0 = 0$$

The system is dependent, and x_2 and x_3 can assume any real values.
There are two parameters in the solution.

(B) The reduced form matrix will have the form
$$\begin{bmatrix} 1 & 0 & a & | & b \\ 0 & 1 & c & | & d \\ 0 & 0 & 0 & | & 0 \end{bmatrix}$$
So the system is equivalent to
$$x_1 + ax_3 = b$$
$$x_2 + cx_3 = d$$
$$0 = 0$$

The system is dependent, with a solution for any real value of x_3.
There is one parameter in the solution.

(C) The reduced form matrix will have the form $\begin{bmatrix} 1 & 0 & 0 & | & a \\ 0 & 1 & 0 & | & b \\ 0 & 0 & 1 & | & c \end{bmatrix}$
So there is only one solution, $x_1 = a$, $x_2 = b$, $x_3 = c$, and the system is
independent.

(D) This is impossible; there are only 3 equations.

55. The augmented matrix of the system is
$$\begin{bmatrix} 1 & -1 & | & 4 \\ 3 & k & | & 7 \end{bmatrix} (-3)R_1 + R_2 \rightarrow R_2$$
$$\sim \begin{bmatrix} 1 & -1 & | & 4 \\ 0 & k+3 & | & -5 \end{bmatrix}$$
If $k = -3$, this is $\begin{bmatrix} 1 & -1 & | & 4 \\ 0 & 0 & | & -5 \end{bmatrix}$ and the system has no solutions.
If $k \neq -3$, the second matrix shown can be reduced further to
$$\sim \begin{bmatrix} 1 & 0 & | & 4 - \frac{5}{k+3} \\ 0 & 1 & | & \frac{-5}{k+3} \end{bmatrix}$$
(steps not shown). The system has one solution: $x_1 = 4 - \dfrac{5}{k+3}$, $x_2 = \dfrac{-5}{k+3}$.

57. The augmented matrix of the system is
$$\begin{bmatrix} 1 & k & | & 3 \\ 2 & 6 & | & 6 \end{bmatrix} (-2)R_1 + R_2 \rightarrow R_2$$
$$\sim \begin{bmatrix} 1 & k & | & 3 \\ 0 & 6-2k & | & 0 \end{bmatrix}$$

If $k = 3$, this is $\begin{bmatrix} 1 & 3 & | & 3 \\ 0 & 0 & | & 0 \end{bmatrix}$ and the system has infinite solutions: Let $x_2 = s$.

Then

$x_1 + 3s = 3$

$x_1 = 3 - 3s$

The solutions are $x_1 = 3 - 3s$, $x_2 = s$, s any real number.

If $k \neq 3$, the second matrix shown can be reduced further to

$\sim \begin{bmatrix} 1 & 0 & | & 3 \\ 0 & 1 & | & 0 \end{bmatrix}$

(steps not shown). The system has one solution: $x_1 = 3$, $x_2 = 0$.

59. $\begin{bmatrix} 1 & 2 & -4 & -1 & | & 7 \\ 2 & 5 & -9 & -4 & | & 16 \\ 1 & 5 & -7 & -7 & | & 13 \end{bmatrix} \begin{array}{l} \\ (-2)R_1 + R_2 \rightarrow R_2 \\ (-1)R_1 + R_3 \rightarrow R_3 \end{array}$

$\sim \begin{bmatrix} 1 & 2 & -4 & -1 & | & 7 \\ 0 & 1 & -1 & -2 & | & 2 \\ 0 & 3 & -3 & -6 & | & 6 \end{bmatrix} \begin{array}{l} (-2)R_2 + R_1 \rightarrow R_1 \\ \\ (-3)R_2 + R_3 \rightarrow R_3 \end{array}$

$\sim \begin{bmatrix} 1 & 0 & -2 & 3 & | & 3 \\ 0 & 1 & -1 & -2 & | & 2 \\ 0 & 0 & 0 & 0 & | & 0 \end{bmatrix}$

Let $x_4 = t$, $x_3 = s$. Then

$x_2 - x_3 - 2x_4 = 2$

$x_2 = s + 2t + 2 + 2$

$x_1 - 2x_3 + 3x_4 = 3$

$x_1 = 2x_3 - 3x_4 + 3$

Solution: $x_1 = 2s - 3t + 3$,

$x_2 = s + 2t + 2$, $x_3 = s$, $x_4 = t$,

s and t any real numbers.

61. $\begin{bmatrix} 1 & -1 & 3 & -2 & | & 1 \\ -2 & 4 & -3 & 1 & | & 0.5 \\ 3 & -1 & 10 & -4 & | & 2.9 \\ 4 & -3 & 8 & -2 & | & 0.6 \end{bmatrix} \begin{array}{l} \\ 2R_1 + R_2 \rightarrow R_2 \\ (-3)R_1 + R_3 \rightarrow R_3 \\ (-4)R_1 + R_4 \rightarrow R_4 \end{array}$

$\sim \begin{bmatrix} 1 & -1 & 3 & -2 & | & 1 \\ 0 & 2 & 3 & -3 & | & 2.5 \\ 0 & 2 & 1 & 2 & | & -0.1 \\ 0 & 1 & -4 & 6 & | & -3.4 \end{bmatrix} R_4 \leftrightarrow R_2$

$\sim \begin{bmatrix} 1 & -1 & 3 & -2 & | & 1 \\ 0 & 1 & -4 & 6 & | & -3.4 \\ 0 & 2 & 1 & 2 & | & -0.1 \\ 0 & 2 & 3 & -3 & | & 2.5 \end{bmatrix} \begin{array}{l} R_2 + R_1 \rightarrow R_1 \\ \\ (-2)R_2 + R_3 \rightarrow R_3 \\ (-2)R_2 + R_4 \rightarrow R_4 \end{array}$

$\sim \begin{bmatrix} 1 & 0 & -1 & 4 & | & -2.4 \\ 0 & 1 & -4 & 6 & | & -3.4 \\ 0 & 0 & 9 & -10 & | & 6.7 \\ 0 & 0 & 11 & -15 & | & 9.3 \end{bmatrix} (-1)R_4 + R_3 \rightarrow R_3$

$\sim \begin{bmatrix} 1 & 0 & -1 & 4 & | & -2.4 \\ 0 & 1 & -4 & 6 & | & -3.4 \\ 0 & 0 & -2 & 5 & | & -2.6 \\ 0 & 0 & 11 & -15 & | & 9.3 \end{bmatrix} -\frac{1}{2}R_3 \leftrightarrow R_3$

$\sim \begin{bmatrix} 1 & 0 & -1 & 4 & | & -2.4 \\ 0 & 1 & -4 & 6 & | & -3.4 \\ 0 & 0 & 1 & -2.5 & | & 1.3 \\ 0 & 0 & 11 & -15 & | & 9.3 \end{bmatrix} \begin{array}{l} R_3 + R_1 \rightarrow R_1 \\ 4R_3 + R_2 \rightarrow R_2 \\ \\ (-11)R_3 + R_4 \rightarrow R_4 \end{array}$

$$\sim \begin{bmatrix} 1 & 0 & 0 & 1.5 & | & -1.1 \\ 0 & 1 & 0 & -4 & | & 1.8 \\ 0 & 0 & 1 & -2.5 & | & 1.3 \\ 0 & 0 & 0 & 12.5 & | & -5 \end{bmatrix} \quad \frac{1}{12.5} R_4 \rightarrow R_4$$

$$\sim \begin{bmatrix} 1 & 0 & 0 & 1.5 & | & -1.1 \\ 0 & 1 & 0 & -4 & | & 1.8 \\ 0 & 0 & 1 & -2.5 & | & 1.3 \\ 0 & 0 & 0 & 1 & | & -0.4 \end{bmatrix} \quad \begin{matrix} (-1.5)R_4 + R_1 \rightarrow R_1 \\ 4R_4 + R_2 \rightarrow R_2 \\ 2.5R_4 + R_3 \rightarrow R_3 \end{matrix}$$

$$\sim \begin{bmatrix} 1 & 0 & 0 & 0 & | & -0.5 \\ 0 & 1 & 0 & 0 & | & 0.2 \\ 0 & 0 & 1 & 0 & | & 0.3 \\ 0 & 0 & 0 & 1 & | & -0.4 \end{bmatrix} \quad \text{Solution: } x_1 = -0.5, \ x_2 = 0.2, \ x_3 = 0.3, \ x_4 = -0.4$$

63.
$$\begin{bmatrix} 1 & -2 & 1 & 1 & 2 & | & 2 \\ -2 & 4 & 2 & 2 & -2 & | & 0 \\ 3 & -6 & 1 & 1 & 5 & | & 4 \\ -1 & 2 & 3 & 1 & 1 & | & 3 \end{bmatrix} \quad \begin{matrix} 2R_1 + R_2 \rightarrow R_2 \\ (-3)R_1 + R_3 \rightarrow R_3 \\ R_1 + R_4 \rightarrow R_4 \end{matrix}$$

$$\sim \begin{bmatrix} 1 & -2 & 1 & 1 & 2 & | & 2 \\ -2 & 4 & 2 & 2 & -2 & | & 0 \\ 3 & -6 & 1 & 1 & 5 & | & 4 \\ -1 & 2 & 3 & 1 & 1 & | & 3 \end{bmatrix} \quad \frac{1}{4} R_2 \rightarrow R_2$$

$$\sim \begin{bmatrix} 1 & -2 & 1 & 1 & 2 & | & 2 \\ 0 & 0 & 1 & 1 & 0.5 & | & 1 \\ 0 & 0 & -2 & -2 & -1 & | & -2 \\ 0 & 0 & 4 & 2 & 3 & | & 5 \end{bmatrix} \quad \begin{matrix} (-1)R_2 + R_1 \rightarrow R_1 \\ 2R_2 + R_3 \rightarrow R_3 \\ (-4)R_2 + R_4 \rightarrow R_4 \end{matrix}$$

$$\sim \begin{bmatrix} 1 & -2 & 0 & 0 & 1.5 & | & 1 \\ 0 & 0 & 1 & 1 & 0.5 & | & 1 \\ 0 & 0 & 0 & 0 & 0 & | & 0 \\ 0 & 0 & 0 & -2 & 1 & | & 1 \end{bmatrix} \quad R_3 \leftrightarrow R_4$$

$$\sim \begin{bmatrix} 1 & -2 & 0 & 0 & 1.5 & | & 1 \\ 0 & 0 & 1 & 1 & 0.5 & | & 1 \\ 0 & 0 & 0 & -2 & 1 & | & 1 \\ 0 & 0 & 0 & 0 & 0 & | & 0 \end{bmatrix} \quad \left(-\frac{1}{2}\right) R_3 \rightarrow R_3$$

$$\sim \begin{bmatrix} 1 & -2 & 0 & 0 & 1.5 & | & 1 \\ 0 & 0 & 1 & 1 & 0.5 & | & 1 \\ 0 & 0 & 0 & 1 & -0.5 & | & -0.5 \\ 0 & 0 & 0 & 0 & 0 & | & 0 \end{bmatrix} \quad (-1)R_3 + R_2 \rightarrow R_2$$

$$\sim \begin{bmatrix} 1 & -2 & 0 & 0 & 1.5 & | & 1 \\ 0 & 0 & 1 & 0 & 1 & | & 1.5 \\ 0 & 0 & 0 & 1 & -0.5 & | & -0.5 \\ 0 & 0 & 0 & 0 & 0 & | & 0 \end{bmatrix}$$

Let $x_5 = t$. Then

$x_4 - 0.5x_5 = -0.5$

$\qquad x_4 = 0.5x_5 - 0.5 = 0.5t - 0.5$

$\quad x_3 + x_5 = 1.5$

$\qquad x_3 = -x_5 + 1.5 = -t + 1.5$

Let $x_2 = s$. Then

$\quad x_1 - 2x_2 + 1.5x_5 = 1$

$\qquad\qquad x_1 = 2x_2 - 1.5x_5 + 1 = 2s - 1.5t + 1$

Solution: $x_1 = 2s - 1.5t + 1$, $x_2 = s$, $x_3 = -t + 1.15$, $x_4 = 0.5t - 0.5$, $x_5 = t$,
s and t any real numbers.

65. Let x_1 = number of 15-cent stamps

x_2 = number of 20-cent stamps

x_3 = number of 35-cent stamps

Then $x_1 + x_2 + x_3 = 45$ (total number of stamps)

$15x_1 + 20x_2 + 35x_3 = 1400$ (total value of stamps)

We write the augmented matrix and solve by Gauss-Jordan elimination.

$\begin{bmatrix} 1 & 1 & 1 & | & 45 \\ 15 & 20 & 35 & | & 1,400 \end{bmatrix}$ $(-15)R_1 + R_2 \rightarrow R_2$

$\sim \begin{bmatrix} 1 & 1 & 1 & | & 45 \\ 0 & 5 & 20 & | & 725 \end{bmatrix}$ $\frac{1}{5}R_2 \rightarrow R_2$

$\sim \begin{bmatrix} 1 & 1 & 1 & | & 45 \\ 0 & 1 & 4 & | & 145 \end{bmatrix}$ $(-1)R_2 + R_1 \rightarrow R_1$

$\sim \begin{bmatrix} 1 & 0 & -3 & | & -100 \\ 0 & 1 & 4 & | & 145 \end{bmatrix}$

This augmented matrix is in reduced form. It corresponds to the system:

$x_1 - 3x_3 = -100$

$x_2 + 4x_3 = 145$

Let $x_3 = t$. Then $x_2 = -4x_3 + 145$

$= -4t + 145$

$x_1 = 3x_3 - 100$

$= 3t - 100$

A solution is achieved, not for every real value of t, but for integer values of t that give rise to non-negative x_1, x_2, x_3.

$x_1 \geq 0$ means $3t - 100 \geq 0$ or $t \geq 33\frac{1}{3}$

$x_2 \geq 0$ means $-4t + 145 \geq 0$ or $t \leq 36\frac{1}{4}$

The only integer values of t that satisfy these conditions are 34, 35, 36. So we have the solutions:

$x_1 = (3t - 100)$ 15-cent stamps

$x_2 = (145 - 4t)$ 20-cent stamps

$x_3 = t$ 35-cent stamps

where $t = 34$, 35, or 36.

67. Let x_1 = number of 500-cc containers of 10% solution

x_2 = number of 500-cc containers of 20% solution

x_3 = number of 1,000-cc containers of 50% solution

Then $500x_1 + 500x_2 + 1,000x_3 = 12,000$ (total number of cc)

$0.10(500x_1) + 0.20(500x_2) + 0.50(1,000x_3) = 0.30(12,000)$

(total amount of ingredient in solution.)

After dividing the first equation by 500 and the second by 50, we get

$x_1 + x_2 + 2x_3 = 24$

$x_1 + 2x_2 + 10x_3 = 72$

We write the augmented matrix and solve by Gauss-Jordan elimination:

$\begin{bmatrix} 1 & 1 & 2 & | & 24 \\ 1 & 2 & 10 & | & 72 \end{bmatrix}$ $(-1)R_1 + R_2 \rightarrow R_2$

$\sim \begin{bmatrix} 1 & 1 & 2 & | & 24 \\ 0 & 1 & 8 & | & 48 \end{bmatrix}$ $(-1)R_2 + R_1 \rightarrow R_1$

$\sim \begin{bmatrix} 1 & 0 & -6 & | & -24 \\ 0 & 1 & 8 & | & 48 \end{bmatrix}$

This augmented matrix is in reduced form. It corresponds to the system:

$x_1 - 6x_3 = -24$

$x_2 + 8x_3 = 48$

Let $x_3 = t$. Then

$x_2 = -8x_3 + 48$

$\quad = -8t + 48$

$x_1 = 6x_3 - 24$

$\quad = 6t - 24$

A solution is achieved, not for every real value of t, but for integer values of t that give rise to non-negative x_1, x_2, x_3.

$x_1 \geq 0$ means $6t - 24 \geq 0$ or $t \geq 4$

$x_2 \geq 0$ means $-8t + 48 \geq 0$ or $t \leq 6$

So we have the solution:

$x_1 = (6t - 24)$ 500-cc containers of 10% solution

$x_2 = (48 - 8t)$ 500-cc containers of 20% solution

$x_3 = t$ 1000-cc containers of 50% solution

where $t = 4, 5,$ or 6.

69. If the curve passes through a point, the coordinates of the point satisfy the equation of the curve. So

$3 = a + b(-2) + c(-2)^2$

$2 = a + b(-1) + c(-1)^2$

$6 = a + b(1) + c(1)^2$

After simplification, we have

$a - 2b + 4c = 3$

$a - b + c = 2$

$a + b + c = 6$

We write the augmented matrix and solve by Gauss-Jordan elimination.

$$\begin{bmatrix} 1 & -2 & 4 & | & 3 \\ 1 & -1 & 1 & | & 2 \\ 1 & 1 & 1 & | & 6 \end{bmatrix} \begin{matrix} \\ (-1)R_1 + R_2 \rightarrow R_2 \\ (-1)R_1 + R_3 \rightarrow R_3 \end{matrix}$$

$$\sim \begin{bmatrix} 1 & -2 & 4 & | & 3 \\ 0 & 1 & -3 & | & -1 \\ 0 & 3 & -3 & | & 3 \end{bmatrix} \begin{matrix} 2R_2 + R_1 \rightarrow R_1 \\ \\ (-3)R_2 + R_3 \rightarrow R_3 \end{matrix}$$

$$\sim \begin{bmatrix} 1 & 0 & -2 & | & 1 \\ 0 & 1 & -3 & | & -1 \\ 0 & 0 & 6 & | & 6 \end{bmatrix} \begin{matrix} \\ \\ \frac{1}{6}R_3 \rightarrow R_3 \end{matrix}$$

$$\sim \begin{bmatrix} 1 & 0 & -2 & | & 1 \\ 0 & 1 & -3 & | & -1 \\ 0 & 0 & 1 & | & 1 \end{bmatrix} \begin{matrix} 2R_3 + R_1 \rightarrow R_1 \\ 3R_3 + R_2 \rightarrow R_2 \\ \end{matrix}$$

$$\sim \begin{bmatrix} 1 & 0 & 0 & | & 3 \\ 0 & 1 & 0 & | & 2 \\ 0 & 0 & 1 & | & 1 \end{bmatrix}$$

Solution: $a = 3$, $b = 2$, $c = 1$.

71. If the curve passes through a point, the coordinates of the point satisfy the equation of the curve. So

$$6^2 + 2^2 + a(6) + b(2) + c = 0$$
$$4^2 + 6^2 + a(4) + b(6) + c = 0$$
$$(-3)^2 + (-1)^2 + a(-3) + b(-1) + c = 0$$

After simplification, we have

$$6a + 2b + c = -40$$
$$4a + 6b + c = -52$$
$$-3a - b + c = -10$$

We write the augmented matrix and solve by Gauss-Jordan elimination.

$$\left[\begin{array}{ccc|c} 6 & 2 & 1 & -40 \\ 4 & 6 & 1 & -52 \\ -3 & -1 & 1 & -10 \end{array}\right] \frac{1}{6}R_1 \to R_1$$

$$\sim \left[\begin{array}{ccc|c} 1 & \frac{1}{3} & \frac{1}{6} & -\frac{20}{3} \\ 4 & 6 & 1 & -52 \\ -3 & -1 & 1 & -10 \end{array}\right] \begin{array}{l} (-4)R_1 + R_2 \to R_2 \\ 3R_1 + R_3 \to R_3 \end{array}$$

$$\sim \left[\begin{array}{ccc|c} 1 & \frac{1}{3} & \frac{1}{6} & -\frac{20}{3} \\ 0 & \frac{14}{3} & \frac{1}{3} & -\frac{76}{3} \\ -3 & 0 & \frac{3}{2} & -30 \end{array}\right] \frac{2}{3}R_3 \to R_3$$

$$\sim \left[\begin{array}{ccc|c} 1 & \frac{1}{3} & \frac{1}{6} & -\frac{20}{3} \\ 0 & \frac{14}{3} & \frac{1}{3} & -\frac{76}{3} \\ 0 & 0 & 1 & -20 \end{array}\right] \begin{array}{l} (-\frac{1}{6})R_3 + R_1 \to R_1 \\ (-\frac{1}{3})R_3 + R_2 \to R_2 \end{array}$$

$$\sim \left[\begin{array}{ccc|c} 1 & \frac{1}{3} & 0 & -\frac{10}{3} \\ 0 & \frac{14}{3} & 0 & -\frac{56}{3} \\ 0 & 0 & 1 & -20 \end{array}\right] \frac{3}{14}R_2 \to R_2$$

$$\sim \left[\begin{array}{ccc|c} 1 & \frac{1}{3} & 0 & -\frac{10}{3} \\ 0 & 1 & 0 & -4 \\ 0 & 0 & 1 & -20 \end{array}\right] (-\frac{1}{3})R_2 + R_1 \to R_1$$

$$\left[\begin{array}{ccc|c} 1 & 0 & 0 & -2 \\ 0 & 1 & 0 & -4 \\ 0 & 0 & 1 & -20 \end{array}\right]$$

Solution: $a = -2$, $b = -4$, and $c = -20$.

73. Let x_1 = number of one-person boats
x_2 = number of two-person boats
x_3 = number of four-person boats

(A) We have

$0.5x_1 + 1.0x_2 + 1.5x_3 = 380$ cutting department
$0.6x_1 + 0.9x_2 + 1.2x_3 = 330$ assembly department
$0.2x_1 + 0.3x_2 + 0.5x_3 = 120$ packing department

> **Common Error:**
> The facts in this problem do not justify the equation
> $0.5x_1 + 0.6x_2 + 0.2x_3 = 380$

Clearing of decimals for convenience:

$x_1 + 2x_2 + 3x_3 = 760$ (First equation multiplied by 2)
$6x_1 + 9x_2 + 12x_3 = 3,300$ (Second equation multiplied by 10)
$2x_1 + 3x_2 + 5x_3 = 1,200$ (Third equation multiplied by 10)

We write the augmented matrix and solve by Gauss-Jordan elimination:

$$\begin{bmatrix} 1 & 2 & 3 & | & 760 \\ 6 & 9 & 12 & | & 3,300 \\ 2 & 3 & 5 & | & 1,200 \end{bmatrix} \begin{matrix} \\ (-6)R_1 + R_2 \to R_2 \\ (-2)R_1 + R_3 \to R_3 \end{matrix}$$

$$\sim \begin{bmatrix} 1 & 2 & 3 & | & 760 \\ 0 & -3 & -6 & | & -1,260 \\ 0 & -1 & -1 & | & -320 \end{bmatrix} -\tfrac{1}{3}R_2 \to R_2$$

$$\sim \begin{bmatrix} 1 & 2 & 3 & | & 760 \\ 0 & 1 & 2 & | & 420 \\ 0 & -1 & -1 & | & -320 \end{bmatrix} \begin{matrix} (-2)R_2 + R_1 \to R_1 \\ \\ R_2 + R_3 \to R_3 \end{matrix}$$

$$\sim \begin{bmatrix} 1 & 0 & -1 & | & -80 \\ 0 & 1 & 2 & | & 420 \\ 0 & 0 & 1 & | & 100 \end{bmatrix} \begin{matrix} R_3 + R_1 \to R_1 \\ (-2)R_3 + R_2 \to R_2 \\ \\ \end{matrix}$$

$$\begin{bmatrix} 1 & 0 & 0 & | & 20 \\ 0 & 1 & 0 & | & 220 \\ 0 & 0 & 1 & | & 100 \end{bmatrix}$$

Therefore
$x_1 = 20$ one-person boats
$x_2 = 220$ two-person boats
$x_3 = 100$ four-person boats

(B) This assumption discards the third equation. The system, cleared of decimals, reads

$$x_1 + 2x_2 + 3x_3 = 760$$
$$6x_1 + 9x_2 + 12x_3 = 3,300$$

The augmented matrix becomes

$$\begin{bmatrix} 1 & 2 & 3 & | & 760 \\ 6 & 9 & 12 & | & 3,300 \end{bmatrix}$$

We solve by Gauss-Jordan elimination. We start by introducing a 0 into the lower left corner using $(-6)R_1 + R_2$ as in the previous problem:

$$\sim \begin{bmatrix} 1 & 2 & 3 & | & 760 \\ 0 & -3 & -6 & | & -1,260 \end{bmatrix} -\tfrac{1}{3}R_2 \to R_2$$

$$\sim \begin{bmatrix} 1 & 2 & 3 & | & 760 \\ 0 & 1 & 2 & | & 420 \end{bmatrix} (-2)R_2 + R_1 \to R_1$$

$$\sim \begin{bmatrix} 1 & 0 & -1 & | & -80 \\ 0 & 1 & 2 & | & 420 \end{bmatrix}$$

This augmented matrix is in reduced form. It corresponds to the system:

$$x_1 - x_3 = -80$$
$$x_2 + 2x_3 = 420$$

Let $x_3 = t$. Then $x_2 = -2x_3 + 420$
$$= -2t + 420$$
$$x_1 = x_3 - 80$$
$$= t - 80$$

A solution is achieved, not for every real value of t, but for integer values of t that give rise to non-negative x_1, x_2, x_3.

$x_1 \geq 0$ means $t - 80 \geq 0$ or $t \geq 80$

$x_2 \geq 0$ means $-2t + 420 \geq 0$ or $210 \geq t$

So we have the solution

$x_1 = (t - 80)$ one-person boats

$x_2 = (-2t + 420)$ two-person boats

$x_3 = t$ four-person boats

$80 \leq t \leq 210$, t an integer

(C) In this case we have $x_3 = 0$ from the beginning. The three equations of part (A), cleared of decimals, read:

$$x_1 + 2x_2 = 760$$
$$6x_1 + 9x_2 = 3,300$$
$$2x_1 + 3x_2 = 1,200$$

The augmented matrix becomes:

$$\begin{bmatrix} 1 & 2 & | & 760 \\ 6 & 9 & | & 3,300 \\ 2 & 3 & | & 1,200 \end{bmatrix}$$

Notice that the row operation

$$(-3)R_3 + R_2 \rightarrow R_2$$

transforms this into the equivalent augmented matrix: $\begin{bmatrix} 1 & 2 & | & 760 \\ 0 & 0 & | & -300 \\ 2 & 3 & | & 1,200 \end{bmatrix}$.

Therefore, since the second row corresponds to the equation

$$0x_1 + 0x_2 = -300$$

there is no solution.

No production schedule will use all the work-hours in all departments.

75. Let x_1 = number of ounces of food A.

x_2 = number of ounces of food B.

x_3 = number of ounces of food C.

> **Common Error:**
> The facts in this problem do not justify the equation
> $30x_1 + 10x_2 + 10x_3 = 340$

(A) Then

$$30x_1 + 10x_2 + 20x_3 = 340 \text{ (calcium)}$$
$$10x_1 + 10x_2 + 20x_3 = 180 \text{ (iron)}$$
$$10x_1 + 30x_2 + 20x_3 = 220 \text{ (vitamin A)}$$

or

$$3x_1 + x_2 + 2x_3 = 34$$
$$x_1 + x_2 + 2x_3 = 18$$
$$x_1 + 3x_2 + 2x_3 = 22$$

is the system to be solved. We form the augmented matrix and solve by Gauss-Jordan elimination.

$$\begin{bmatrix} 3 & 1 & 2 & | & 34 \\ 1 & 1 & 2 & | & 18 \\ 1 & 3 & 2 & | & 22 \end{bmatrix} R_1 \leftrightarrow R_2$$

$$\sim \begin{bmatrix} 1 & 1 & 2 & | & 18 \\ 3 & 1 & 2 & | & 34 \\ 1 & 3 & 2 & | & 22 \end{bmatrix} \begin{matrix} \\ (-3)R_1 + R_2 \rightarrow R_2 \\ (-1)R_1 + R_3 \rightarrow R_3 \end{matrix}$$

$$\sim \begin{bmatrix} 1 & 1 & 2 & | & 18 \\ 0 & -2 & -4 & | & -20 \\ 0 & 2 & 0 & | & 4 \end{bmatrix} -\tfrac{1}{2}R_2 \rightarrow R_2$$

$$\sim \begin{bmatrix} 1 & 1 & 2 & | & 18 \\ 0 & 1 & 2 & | & 10 \\ 0 & 2 & 0 & | & 4 \end{bmatrix} \begin{matrix} (-1)R_2 + R_1 \rightarrow R_1 \\ \\ (-2)R_2 + R_3 \rightarrow R_3 \end{matrix}$$

$$\sim \begin{bmatrix} 1 & 0 & 0 & | & 8 \\ 0 & 1 & 2 & | & 10 \\ 0 & 0 & -4 & | & -16 \end{bmatrix} -\tfrac{1}{4}R_3 \rightarrow R_3$$

$$\sim \begin{bmatrix} 1 & 0 & 0 & | & 8 \\ 0 & 1 & 2 & | & 10 \\ 0 & 0 & 1 & | & 4 \end{bmatrix} (-2)R_3 + R_2 \rightarrow R_2$$

$$\sim \begin{bmatrix} 1 & 0 & 0 & | & 8 \\ 0 & 1 & 0 & | & 2 \\ 0 & 0 & 1 & | & 4 \end{bmatrix}$$

Solution: $x_1 = 8$ ounces food A

$x_2 = 2$ ounces food B

$x_3 = 4$ ounces food C

(B) In this case we have $x_3 = 0$ from the beginning. The three equations of part (A) become $30x_1 + 10x_2 = 340$

$$10x_1 + 10x_2 = 180$$
$$10x_1 + 30x_2 = 220$$

or

$$3x_1 + x_2 = 34$$
$$x_1 + x_2 = 18$$
$$x_1 + 3x_2 = 22$$

The augmented matrix becomes $\begin{bmatrix} 3 & 1 & | & 34 \\ 1 & 1 & | & 18 \\ 1 & 3 & | & 22 \end{bmatrix}$

We solve by Gauss-Jordan elimination, starting by the row operation

$R_1 \leftrightarrow R_2$

$$\begin{bmatrix} 1 & 1 & | & 18 \\ 3 & 1 & | & 34 \\ 1 & 3 & | & 22 \end{bmatrix} \begin{matrix} \\ (-3)R_1 + R_2 \rightarrow R_2 \\ (-1)R_1 + R_3 \rightarrow R_3 \end{matrix}$$

$$\sim \begin{bmatrix} 1 & 1 & | & 18 \\ 0 & -2 & | & -20 \\ 0 & 2 & | & 4 \end{bmatrix} R_2 + R_3 \rightarrow R_3$$

$$\sim \begin{bmatrix} 1 & 1 & | & 18 \\ 0 & -2 & | & -20 \\ 0 & 0 & | & -16 \end{bmatrix}$$

Since the third row corresponds to the equation

$0x_1 + 0x_2 = -16$

there is no solution.

(C) In this case we discard the third equation. The system becomes

$$30x_1 + 10x_2 + 20x_3 = 340 \qquad \text{or} \qquad 3x_1 + x_2 + 2x_3 = 34$$
$$10x_1 + 10x_2 + 20x_3 = 180 \qquad\qquad x_1 + x_2 + 2x_3 = 18$$

The augmented matrix becomes

$$\begin{bmatrix} 3 & 1 & 2 & | & 34 \\ 1 & 1 & 2 & | & 18 \end{bmatrix}$$

We solve by Gauss-Jordan elimination, starting by the row operation $R_1 \leftrightarrow R_2$.

$$\begin{bmatrix} 1 & 1 & 2 & | & 18 \\ 3 & 1 & 2 & | & 34 \end{bmatrix} (-3)R_1 + R_2 \rightarrow R_2$$

$$\sim \begin{bmatrix} 1 & 1 & 2 & | & 18 \\ 0 & -2 & -4 & | & -20 \end{bmatrix} -\tfrac{1}{2}R_2 \rightarrow R_2$$

$$\sim \begin{bmatrix} 1 & 1 & 2 & | & 18 \\ 0 & 1 & 2 & | & 10 \end{bmatrix} (-1)R_2 + R_1 \to R_1$$

$$\sim \begin{bmatrix} 1 & 0 & 0 & | & 8 \\ 0 & 1 & 2 & | & 10 \end{bmatrix}$$

This augmented matrix is in reduced form. It corresponds to the system

$$x_1 = 8$$
$$x_2 + 2x_3 = 10$$

Let $x_3 = t$

Then $x_2 = -2x_3 + 10$
$$= -2t + 10$$

A solution is achieved, not for every real value t, but for values of t that give rise to non-negative x_2, x_3.

$x_3 \geq 0$ means $t \geq 0$

$x_2 \geq 0$ means $-2t + 10 \geq 0$, $5 \geq t$

So we have the solution

$x_1 = 8$ ounces food A

$x_2 = -2t + 10$ ounces food B

$x_3 = t$ ounces food C

$0 \leq t \leq 5$

77. Let x_1 = number of barrels of Mix A

$\quad x_2$ = number of barrels of Mix B

$\quad x_3$ = number of barrels of Mix C

$\quad x_4$ = number of barrels of Mix D

Then $30x_1 + 30x_2 + 30x_3 + 60x_4 = 900$ (phosphoric acid)

$\quad 50x_1 + 75x_2 + 25x_3 + 25x_4 = 750$ (nitrogen)

$\quad 30x_1 + 20x_2 + 20x_3 + 50x_4 = 700$ (potash)

or $\quad x_1 + x_2 + x_3 + 2x_4 = 30$

$\quad 2x_1 + 3x_2 + x_3 + x_4 = 30$

$\quad 3x_1 + 2x_2 + 2x_3 + 5x_4 = 70$

is the system to be solved. We form the augmented matrix and solve by Gauss-Jordan elimination.

$$\begin{bmatrix} 1 & 1 & 1 & 2 & | & 30 \\ 2 & 3 & 1 & 1 & | & 30 \\ 3 & 2 & 2 & 5 & | & 70 \end{bmatrix} \begin{matrix} \\ (-2)R_1 + R_2 \to R_2 \\ (-3)R_1 + R_3 \to R_3 \end{matrix}$$

$$\sim \begin{bmatrix} 1 & 1 & 1 & 2 & | & 30 \\ 0 & 1 & -1 & -3 & | & -30 \\ 0 & -1 & -1 & -1 & | & -20 \end{bmatrix} \begin{matrix} (-1)R_2 + R_1 \to R_1 \\ \\ R_2 + R_3 \to R_3 \end{matrix}$$

$$\sim \begin{bmatrix} 1 & 0 & 2 & 5 & | & 60 \\ 0 & 1 & -1 & -3 & | & -30 \\ 0 & 0 & -2 & -4 & | & -50 \end{bmatrix} -\tfrac{1}{2} R_3 \to R_3$$

$$\sim \begin{bmatrix} 1 & 0 & 2 & 5 & | & 60 \\ 0 & 1 & -1 & -3 & | & -30 \\ 0 & 0 & 1 & 2 & | & 25 \end{bmatrix} \begin{matrix} (-2)R_3 + R_1 \to R_1 \\ R_3 + R_2 \to R_2 \end{matrix}$$

$$\sim \begin{bmatrix} 1 & 0 & 0 & 1 & | & 10 \\ 0 & 1 & 0 & -1 & | & -5 \\ 0 & 0 & 1 & 2 & | & 25 \end{bmatrix}$$

This augmented matrix is in reduced form. It corresponds to the system

$$x_1 + x_4 = 10$$
$$x_2 - x_4 = -5$$
$$x_3 + 2x_4 = 25$$

Let $x_4 = t.$ Then $x_3 = -2x_4 + 25$
$$= -2t + 25$$
$$x_2 = x_4 - 5$$
$$= t - 5$$
$$x_1 = -x_4 + 10$$
$$= -t + 10$$

A solution is achieved, not for every real value t, but for values of t that give rise to non-negative x_1, x_2, x_3, x_4.

$x_4 \geq 0$ means $t \geq 0$

$x_3 \geq 0$ means $-2t + 25 \geq 0$ $t \leq 12.5$

$x_2 \geq 0$ means $t - 5 \geq 0$ $t \geq 5$

$x_1 \geq 0$ means $-t + 10 \geq 0$ $t \leq 10$

So we have the solution: $x_1 = 10 - t$ barrels of mix A,
$$x_2 = t - 5 \text{ barrels of mix } B,$$
$$x_3 = 25 - 2t \text{ barrels of mix } C,$$
$$x_4 = t \text{ barrels of mix } D$$

where $5 \leq t \leq 10$.

Section 9-2

1. Answers will vary.   3. Answers will vary.   5. Answers will vary.

7. Answers will vary.   9. Answers will vary.

11. $\begin{bmatrix} 5 & -2 \\ 3 & 0 \end{bmatrix} + \begin{bmatrix} -3 & 7 \\ 1 & -6 \end{bmatrix} = \begin{bmatrix} 5 + (-3) & (-2) + 7 \\ 3 + 1 & 0 + (-6) \end{bmatrix} = \begin{bmatrix} 2 & 5 \\ 4 & -6 \end{bmatrix}$

13. $\begin{bmatrix} 4 & 0 \\ -2 & 3 \\ 8 & 1 \end{bmatrix} + \begin{bmatrix} -1 & 2 \\ 0 & 5 \\ 4 & -6 \end{bmatrix} = \begin{bmatrix} 4 + (-1) & 0 + 2 \\ (-2) + 0 & 3 + 5 \\ 8 + 4 & 1 + (-6) \end{bmatrix} = \begin{bmatrix} 3 & 2 \\ -2 & 8 \\ 12 & -5 \end{bmatrix}$

15. These matrices have different sizes, so the sum is not defined.

17. $\begin{bmatrix} 5 & -1 & 0 \\ 4 & 6 & 3 \end{bmatrix} - \begin{bmatrix} 2 & 4 & -6 \\ 3 & 5 & -5 \end{bmatrix} = \begin{bmatrix} 5 - 2 & (-1) - 4 & 0 - (-6) \\ 4 - 3 & 6 - 5 & 3 - (-5) \end{bmatrix} = \begin{bmatrix} 3 & -5 & 6 \\ 1 & 1 & 8 \end{bmatrix}$

19. $\begin{bmatrix} 12 & -16 & 28 \\ -8 & 36 & 20 \end{bmatrix}$ 21. $[5 \quad 3]\begin{bmatrix} 4 \\ 7 \end{bmatrix} = [5 \cdot 4 + 3 \cdot 7] = [41]$

23. $\begin{bmatrix} -6 & 3 \\ 2 & -5 \end{bmatrix}\begin{bmatrix} 1 \\ 3 \end{bmatrix} = \begin{bmatrix} (-6)1 + 3 \cdot 3 \\ 2 \cdot 1 + (-5)3 \end{bmatrix} = \begin{bmatrix} 3 \\ -13 \end{bmatrix}$

25. $\begin{bmatrix} 5 & 1 \\ 4 & 6 \end{bmatrix}\begin{bmatrix} 2 & 0 \\ 3 & 8 \end{bmatrix} = \begin{bmatrix} 5 \cdot 2 + 1 \cdot 3 & 5 \cdot 0 + 1 \cdot 8 \\ 4 \cdot 2 + 6 \cdot 3 & 4 \cdot 0 + 6 \cdot 8 \end{bmatrix} = \begin{bmatrix} 13 & 8 \\ 26 & 48 \end{bmatrix}$

27. $\begin{bmatrix} 8 & -3 \\ -5 & 3 \end{bmatrix}\begin{bmatrix} 2 & 0 \\ 0 & 6 \end{bmatrix} = \begin{bmatrix} 8 \cdot 2 + (-3)0 & 8 \cdot 0 + (-3)6 \\ (-5)2 + 3 \cdot 0 & (-5)0 + 3 \cdot 6 \end{bmatrix} = \begin{bmatrix} 16 & -18 \\ -10 & 18 \end{bmatrix}$

29. $[4 \quad -2]\begin{bmatrix} -5 \\ -3 \end{bmatrix} = [4(-5) + (-2)(-3)] = [-14]$

31. $\begin{bmatrix} -5 \\ -3 \end{bmatrix}[4 \quad -2] = \begin{bmatrix} (-5)4 & (-5)(-2) \\ (-3)4 & (-3)(-2) \end{bmatrix} = \begin{bmatrix} -20 & 10 \\ -12 & 6 \end{bmatrix}$

33. $[3 \quad -2 \quad -4]\begin{bmatrix} 1 \\ 2 \\ -3 \end{bmatrix} = [3 \cdot 1 + (-2)2 + (-4)(-3)] = [11]$

35. $\begin{bmatrix} 1 \\ 2 \\ -3 \end{bmatrix}[3 \quad -2 \quad -4] = \begin{bmatrix} 1 \cdot 3 & 1(-2) & 1(-4) \\ 2 \cdot 3 & 2(-2) & 2(-4) \\ (-3)3 & (-3)(-2) & (-3)(-4) \end{bmatrix} = \begin{bmatrix} 3 & -2 & -4 \\ 6 & -4 & -8 \\ -9 & 6 & 12 \end{bmatrix}$

37. C has 3 columns. A has 2 rows. Therefore, CA is not defined.

39. $BA = \begin{bmatrix} -3 & 1 \\ 2 & 5 \end{bmatrix}\begin{bmatrix} 2 & -1 & 3 \\ 0 & 4 & -2 \end{bmatrix} = \begin{bmatrix} (-3)2 + 1 \cdot 0 & (-3)(-1) + 1 \cdot 4 & (-3)3 + 1(-2) \\ 2 \cdot 2 + 5 \cdot 0 & 2(-1) + 5 \cdot 4 & 2 \cdot 3 + 5(-2) \end{bmatrix}$

$$= \begin{bmatrix} -6 & 7 & -11 \\ 4 & 18 & -4 \end{bmatrix}$$

41. $C^2 = \begin{bmatrix} -1 & 0 & 2 \\ 4 & -3 & 1 \\ -2 & 3 & 5 \end{bmatrix}\begin{bmatrix} -1 & 0 & 2 \\ 4 & -3 & 1 \\ -2 & 3 & 5 \end{bmatrix}$

$$= \begin{bmatrix} (-1)(-1) + 0 \cdot 4 + 2(-2) & (-1)0 + 0(-3) + 2 \cdot 3 & (-1)2 + 0 \cdot 1 + 2 \cdot 5 \\ 4(-1) + (-3)4 + 1(-2) & 4 \cdot 0 + (-3)(-3) + 1 \cdot 3 & 4 \cdot 2 + (-3)1 + 1 \cdot 5 \\ (-2)(-1) + 3 \cdot 4 + 5(-2) & (-2)0 + 3(-3) + 5 \cdot 3 & (-2)2 + 3 \cdot 1 + 5 \cdot 5 \end{bmatrix}$$

$$= \begin{bmatrix} -3 & 6 & 8 \\ -18 & 12 & 10 \\ 4 & 6 & 24 \end{bmatrix}$$

43. $DA = \begin{bmatrix} 3 & -2 \\ 0 & -1 \\ 1 & 2 \end{bmatrix}\begin{bmatrix} 2 & -1 & 3 \\ 0 & 4 & -2 \end{bmatrix}$

$$= \begin{bmatrix} 3 \cdot 2 + (-2)0 & 3(-1) + (-2)4 & 3 \cdot 3 + (-2)(-2) \\ 0 \cdot 2 + (-1)0 & 0(-1) + (-1)4 & 0 \cdot 3 + (-1)(-2) \\ 1 \cdot 2 + 2 \cdot 0 & 1(-1) + 2 \cdot 4 & 1 \cdot 3 + 2(-2) \end{bmatrix} = \begin{bmatrix} 6 & -11 & 13 \\ 0 & -4 & 2 \\ 2 & 7 & -1 \end{bmatrix}$$

$C + DA = \begin{bmatrix} -1 & 0 & 2 \\ 4 & -3 & 1 \\ -2 & 3 & 5 \end{bmatrix} + \begin{bmatrix} 6 & -11 & 13 \\ 0 & -4 & 2 \\ 2 & 7 & -1 \end{bmatrix} = \begin{bmatrix} 5 & -11 & 15 \\ 4 & -7 & 3 \\ 0 & 10 & 4 \end{bmatrix}$

45. $0.2CD = 0.2\begin{bmatrix} -1 & 0 & 2 \\ 4 & -3 & 1 \\ -2 & 3 & 5 \end{bmatrix}\begin{bmatrix} 3 & -2 \\ 0 & -1 \\ 1 & 2 \end{bmatrix}$

$$= 0.2\begin{bmatrix} (-1)3 + 0 \cdot 0 + 2 \cdot 1 & (-1)(-2) + 0(-1) + 2 \cdot 2 \\ 4 \cdot 3 + (-3)0 + 1 \cdot 1 & 4(-2) + (-3)(-1) + 1 \cdot 2 \\ (-2)3 + 3 \cdot 0 + 5 \cdot 1 & (-2)(-2) + 3(-1) + 5 \cdot 2 \end{bmatrix}$$

$$= 0.2\begin{bmatrix} -1 & 6 \\ 13 & -3 \\ -1 & 11 \end{bmatrix} = \begin{bmatrix} -0.2 & 1.2 \\ 2.6 & -0.6 \\ -0.2 & 2.2 \end{bmatrix}$$

47. $DB = \begin{bmatrix} 3 & -2 \\ 0 & -1 \\ 1 & 2 \end{bmatrix}\begin{bmatrix} -3 & 1 \\ 2 & 5 \end{bmatrix} = \begin{bmatrix} 3(-3) + (-2)2 & 3 \cdot 1 + (-2)5 \\ 0(-3) + (-1)2 & 0 \cdot 1 + (-1)5 \\ 1(-3) + 2 \cdot 2 & 1 \cdot 1 + 2 \cdot 5 \end{bmatrix} = \begin{bmatrix} -13 & -7 \\ -2 & -5 \\ 1 & 11 \end{bmatrix}$

$CD = \begin{bmatrix} -1 & 6 \\ 13 & -3 \\ -1 & 11 \end{bmatrix}$ (see problem 45)

$2DB + 5CD = 2\begin{bmatrix} -13 & -7 \\ -2 & -5 \\ 1 & 11 \end{bmatrix} + 5\begin{bmatrix} -1 & 6 \\ 13 & -3 \\ -1 & 11 \end{bmatrix} = \begin{bmatrix} -26 & -14 \\ -4 & -10 \\ 2 & 22 \end{bmatrix} + \begin{bmatrix} -5 & 30 \\ 65 & -15 \\ -5 & 55 \end{bmatrix} = \begin{bmatrix} -31 & 16 \\ 61 & -25 \\ -3 & 77 \end{bmatrix}$

49. $(-1)AC$ is a matrix of size 2×3. $3DB$ is a matrix of size 3×2.
So, $(-1)AC + 3DB$ is not defined.

51. $CD = \begin{bmatrix} -1 & 6 \\ 13 & -3 \\ -1 & 11 \end{bmatrix}$ (see problem 45)

$CDA = \begin{bmatrix} -1 & 6 \\ 13 & -3 \\ -1 & 11 \end{bmatrix} \begin{bmatrix} 2 & -1 & 3 \\ 0 & 4 & -2 \end{bmatrix}$

$= \begin{bmatrix} (-1)2 + 6 \cdot 0 & (-1)(-1) + 6 \cdot 4 & (-1)3 + 6(-2) \\ 13 \cdot 2 + (-3)0 & 13(-1) + (-3)4 & 13 \cdot 3 + (-3)(-2) \\ (-1)2 + 11 \cdot 0 & (-1)(-1) + 11 \cdot 4 & (-1)3 + 11(-2) \end{bmatrix} = \begin{bmatrix} -2 & 25 & -15 \\ 26 & -25 & 45 \\ -2 & 45 & -25 \end{bmatrix}$

53. $DB = \begin{bmatrix} -13 & -7 \\ -2 & -5 \\ 1 & 11 \end{bmatrix}$ (see problem 47)

$DBA = \begin{bmatrix} -13 & -7 \\ -2 & -5 \\ 1 & 11 \end{bmatrix} \begin{bmatrix} 2 & -1 & 3 \\ 0 & 4 & -2 \end{bmatrix}$

$= \begin{bmatrix} (-13)2 + (-7)0 & (-13)(-1) + (-7)4 & (-13)3 + (-7)(-2) \\ (-2)2 + (-5)0 & (-2)(-1) + (-5)4 & (-2)3 + (-5)(-2) \\ 1 \cdot 2 + 11 \cdot 0 & 1(-1) + 11 \cdot 4 & 1 \cdot 3 + 11(-2) \end{bmatrix} = \begin{bmatrix} -26 & -15 & -25 \\ -4 & -18 & 4 \\ 2 & 43 & -19 \end{bmatrix}$

55. Entering matrix B in a graphing calculator, we obtain the results

```
[B]²
    [[.28 .72]]
     [.24 .76]]
[B]³
   [[.256 .744]
    [.248 .752]]
```
```
[B]^4
   [[.2512 .7488]
    [.2496 .7504]]
[B]^5
   [[.25024 .74976…
    [.24992 .75008…
```
```
[B]^6
   [[.250048 .7499…
    [.249984 .7500…
[B]^7
   [[.2500096 .749…
    [.2499968 .750…
```

It appears that $B^n \to \begin{bmatrix} 0.25 & 0.75 \\ 0.25 & 0.75 \end{bmatrix}$

We calculate AB, AB^2, AB^3, … and obtain the results

```
[A]*[B]
      [[.26 .74]]
[A]*[B]²
     [[.252 .748]]
[A]*[B]³
  [[.2504 .7496]]
```
```
[A]*[B]^4
   [[.25008 .74992…
[A]*[B]^5
   [[.250016 .7499…
[A]*[B]^6
   [[.2500032 .749…
```

It appears that $AB^n \to [0.25 \quad 0.75]$.

57. $\begin{bmatrix} a & b \\ c & d \end{bmatrix} + \begin{bmatrix} 2 & -3 \\ 0 & 1 \end{bmatrix} = \begin{bmatrix} a + 2 & b - 3 \\ c & d + 1 \end{bmatrix}$

This matrix will be equal to $\begin{bmatrix} 1 & -2 \\ 3 & -4 \end{bmatrix}$ if all of the corresponding elements are equal.

$a + 2 = 1 \qquad b - 3 = -2 \qquad c = 3 \qquad d + 1 = -4$
$\quad\;\; a = -1 \qquad\quad b = 1 \qquad\quad c = 3 \qquad\quad d = -5$

59. $\begin{bmatrix} 3x & 5 \\ -1 & 4x \end{bmatrix} + \begin{bmatrix} 2y & -3 \\ -6 & -y \end{bmatrix} = \begin{bmatrix} 3x + 2y & 2 \\ -7 & 4x - y \end{bmatrix}$

This matrix will be equal to $\begin{bmatrix} 7 & 2 \\ -7 & 2 \end{bmatrix}$ if all of the corresponding elements are equal.

$3x + 2y = 7$ $2 \overset{\vee}{=} 2$ Two conditions are already met.
$-7 \overset{\vee}{=} -7$ $4x - y = 2$

To find x and y, we solve the system:

$3x + 2y = 7$
$4x - y = 2$ to obtain $x = 1$, $y = 2$. (Solution left to the student.)

61. Compute the square of matrix A:

$$A^2 = \begin{bmatrix} a & b \\ c & -a \end{bmatrix}\begin{bmatrix} a & b \\ c & -a \end{bmatrix} = \begin{bmatrix} a^2 + bc & ab + (-ab) \\ ac + (-ac) & cb + a^2 \end{bmatrix} = \begin{bmatrix} a^2 + bc & 0 \\ 0 & a^2 + bc \end{bmatrix}$$

Two of the entries are already zero and the other two are both $a^2 + bc$. So if $a^2 + bc = 0$, then $A^2 = 0$.

If $a = 1$, $b = 1$, $c = -1$, then $a^2 + bc = 0$, so the matrix $A = \begin{bmatrix} 1 & 1 \\ -1 & -1 \end{bmatrix}$

will have $A^2 = 0$. If $a = 2$, $b = 4$, $c = -1$, then $a^2 + bc = 0$, so the matrix

$A = \begin{bmatrix} 2 & 4 \\ -1 & -2 \end{bmatrix}$ will have $A^2 = 0$. (There are many other possible examples.)

63. Compute the product AB:

$$AB = \begin{bmatrix} a & b \\ c & d \end{bmatrix}\begin{bmatrix} 1 & 1 \\ 1 & -1 \end{bmatrix} = \begin{bmatrix} a + b & a + b \\ c + d & c + d \end{bmatrix}$$

Two of the entries are $a + b$ and the other two are $c + d$, so if $a = -b$ and $c = -d$, then $AB = 0$.

The following are a couple of examples of matrices A that will satisfy $AB = 0$:

$$\begin{bmatrix} 2 & -2 \\ 4 & -4 \end{bmatrix} \quad \begin{bmatrix} -5 & 5 \\ 1 & -1 \end{bmatrix}$$

65. $\begin{bmatrix} 1 & 3 \\ -2 & -2 \end{bmatrix}\begin{bmatrix} x & 1 \\ 3 & 2 \end{bmatrix} = \begin{bmatrix} x + 9 & 7 \\ -2x - 6 & -6 \end{bmatrix}$

This matrix will be equal to $\begin{bmatrix} y & 7 \\ y & -6 \end{bmatrix}$ if all of the corresponding elements are equal.

$x + 9 = y$ $7 \overset{\vee}{=} 7$ Two conditions are already met.
$-2x - 6 = y$ $-6 \overset{\vee}{=} -6$

To find x and y, we solve the system:

$x + 9 = y$
$-2x - 6 = y$ to obtain $x = -5$, $y = 4$. (Solution left to the student.)

67. $\begin{bmatrix} 1 & 3 \\ 1 & 4 \end{bmatrix}\begin{bmatrix} a & b \\ c & d \end{bmatrix} = \begin{bmatrix} a + 3c & b + 3d \\ a + 4c & b + 4d \end{bmatrix}$

This matrix will be equal to $\begin{bmatrix} 6 & -5 \\ 7 & -7 \end{bmatrix}$ if all of the corresponding elements are equal.

$a + 3c = 6$ $b + 3d = -5$
$a + 4c = 7$ $b + 4d = -7$

Solving these systems we obtain $a = 3$, $b = 1$, $c = 1$, $d = -2$. (Solution left to the student.)

69. (A) Since $\begin{bmatrix} a_1 & 0 \\ 0 & d_1 \end{bmatrix} + \begin{bmatrix} a_2 & 0 \\ 0 & d_2 \end{bmatrix} = \begin{bmatrix} a_1 + a_2 & 0 \\ 0 & d_1 + d_2 \end{bmatrix}$, the statement is true.

(B) $A + B = B + A$ is true for any matrices for which $A + B$ is defined, as it is in this case.

(C) Since $\begin{bmatrix} a_1 & 0 \\ 0 & d_1 \end{bmatrix}\begin{bmatrix} a_2 & 0 \\ 0 & d_2 \end{bmatrix} = \begin{bmatrix} a_1 a_2 & 0 \\ 0 & d_1 d_2 \end{bmatrix}$, the statement is true.

(D) Since $\begin{bmatrix} a_1 & 0 \\ 0 & d_1 \end{bmatrix} \begin{bmatrix} a_2 & 0 \\ 0 & d_2 \end{bmatrix} = \begin{bmatrix} a_1 a_2 & 0 \\ 0 & d_1 d_2 \end{bmatrix} = \begin{bmatrix} a_2 a_1 & 0 \\ 0 & d_2 d_1 \end{bmatrix} = \begin{bmatrix} a_2 & 0 \\ 0 & d_2 \end{bmatrix} \begin{bmatrix} a_1 & 0 \\ 0 & d_2 \end{bmatrix}$,
the statement is true.

71. $\frac{1}{2}(A + B) = \frac{1}{2}\left(\begin{bmatrix} 30 & 25 \\ 60 & 80 \end{bmatrix} + \begin{bmatrix} 36 & 27 \\ 54 & 74 \end{bmatrix} \right) = \frac{1}{2} \begin{bmatrix} 66 & 52 \\ 114 & 154 \end{bmatrix} = \begin{matrix} & \text{Guitar Banjo} \\ & \begin{bmatrix} 33 & 26 \\ 57 & 77 \end{bmatrix} \begin{matrix} \text{Materials} \\ \text{Labor} \end{matrix} \end{matrix}$

73. If a quantity is increased by 15%, the result is a multiplication by 1.15.
If a quantity is increased by 10%, the result is a multiplication by 1.1.
So we must calculate $1.1N - 1.15M$. The mark-up matrix is:

$1.1N - 1.15M = 1.1 \begin{bmatrix} 13,900 & 783 & 263 & 215 \\ 15,000 & 838 & 395 & 236 \\ 18,300 & 967 & 573 & 248 \end{bmatrix} - 1.15 \begin{bmatrix} 10,400 & 682 & 215 & 182 \\ 12,500 & 721 & 295 & 182 \\ 16,400 & 827 & 443 & 192 \end{bmatrix}$

$= \begin{bmatrix} 15,290 & 861.3 & 289.3 & 236.5 \\ 16,500 & 921.8 & 434.5 & 259.6 \\ 20,130 & 1,063.7 & 630.3 & 272.8 \end{bmatrix} - \begin{bmatrix} 11,960 & 784.3 & 247.25 & 209.3 \\ 14,375 & 829.15 & 339.25 & 209.3 \\ 18,860 & 951.05 & 509.45 & 220.8 \end{bmatrix}$

$= \begin{matrix} & \begin{matrix} \text{Basic} & & \text{AM/FM} & \text{Cruise} \\ \text{Car} & \text{Air} & \text{Radio} & \text{Control} \end{matrix} \\ \begin{matrix} \text{Model } A \\ \text{Model } B \\ \text{Model } C \end{matrix} & \begin{bmatrix} \$3,330 & \$77 & \$42 & \$27 \\ \$2,125 & \$93 & \$95 & \$50 \\ \$1,270 & \$113 & \$121 & \$52 \end{bmatrix} \end{matrix} = \text{Mark up}$

75. (A) $[0.6 \quad 0.6 \quad 0.2] \begin{bmatrix} 8 \\ 10 \\ 5 \end{bmatrix} = (0.6)8 + (0.6)10 + (0.2)5 = 11.80$ dollars per boat

(B) $[1.5 \quad 1.2 \quad 0.4] \begin{bmatrix} 9 \\ 12 \\ 6 \end{bmatrix} = (1.5)9 + (1.2)12 + (0.4)6 = 30.30$ dollars per boat

(C) The matrix NM has no clear meaning, but the matrix MN gives the labor
costs per boat at each plant.

(D) $MN = \begin{bmatrix} 0.6 & 0.6 & 0.2 \\ 1.0 & 0.9 & 0.3 \\ 1.5 & 1.2 & 0.4 \end{bmatrix} \begin{bmatrix} 8 & 9 \\ 10 & 12 \\ 5 & 6 \end{bmatrix}$

$= \begin{bmatrix} (0.6)8 + (0.6)10 + (0.2)5 & (0.6)9 + (0.6)12 + (0.2)6 \\ (1.0)8 + (0.9)10 + (0.3)5 & (1.0)9 + (0.9)12 + (0.3)6 \\ (1.5)8 + (1.2)10 + (0.4)5 & (1.5)9 + (1.2)12 + (0.4)6 \end{bmatrix}$

$= \begin{matrix} \begin{matrix} \text{Plant I} & \text{Plant II} \end{matrix} \\ \begin{bmatrix} \$11.80 & \$13.80 \\ \$18.50 & \$21.60 \\ \$26.00 & \$30.30 \end{bmatrix} \begin{matrix} \text{One-person boat} \\ \text{Two-person boat} \\ \text{Four-person boat} \end{matrix} \end{matrix}$

This matrix gives the labor costs for each type of boat at each plant.

77. (A) $A^2 = AA = \begin{bmatrix} 0 & 1 & 0 & 1 & 0 \\ 0 & 0 & 1 & 0 & 0 \\ 1 & 0 & 0 & 0 & 1 \\ 0 & 0 & 1 & 0 & 0 \\ 0 & 0 & 0 & 1 & 0 \end{bmatrix} \begin{bmatrix} 0 & 1 & 0 & 1 & 0 \\ 0 & 0 & 1 & 0 & 0 \\ 1 & 0 & 0 & 0 & 1 \\ 0 & 0 & 1 & 0 & 0 \\ 0 & 0 & 0 & 1 & 0 \end{bmatrix} = \begin{bmatrix} 0 & 0 & 2 & 0 & 0 \\ 1 & 0 & 0 & 0 & 1 \\ 0 & 1 & 0 & 2 & 0 \\ 1 & 0 & 0 & 0 & 1 \\ 0 & 0 & 1 & 0 & 0 \end{bmatrix}$

The 1 in row 2 and column 1 of A^2 indicates that there is one way to
travel from Baltimore to Atlanta with one intermediate connection. The 2 in
row 1 and column 3 indicates that there are two ways to travel from Atlanta
to Chicago with one intermediate connection. In general, the elements in
A^2 indicate the number of different ways to travel from the ith city to
the jth city with one intermediate connection.

(B) $A^3 = A^2A = \begin{bmatrix} 0 & 0 & 2 & 0 & 0 \\ 1 & 0 & 0 & 0 & 1 \\ 0 & 1 & 0 & 2 & 0 \\ 1 & 0 & 0 & 0 & 1 \\ 0 & 0 & 1 & 0 & 0 \end{bmatrix} \begin{bmatrix} 0 & 1 & 0 & 1 & 0 \\ 0 & 0 & 1 & 0 & 0 \\ 1 & 0 & 0 & 0 & 1 \\ 0 & 0 & 1 & 0 & 0 \\ 0 & 0 & 0 & 1 & 0 \end{bmatrix} = \begin{bmatrix} 2 & 0 & 0 & 0 & 2 \\ 0 & 1 & 0 & 2 & 0 \\ 0 & 0 & 3 & 0 & 0 \\ 0 & 1 & 0 & 2 & 0 \\ 1 & 0 & 0 & 0 & 1 \end{bmatrix}$

The 1 in row 4 and column 2 of A^3 indicates that there is one way to travel from Denver to Baltimore with two intermediate connections. The 2 in row 1 and column 5 indicates that there are two ways to travel from Atlanta to El Paso with two intermediate connections. In general, the elements in A^3 indicate the number of different ways to travel from the ith city to the jth city with two intermediate connections.

(C) A is given above.

$A + A^2 = \begin{bmatrix} 0 & 1 & 0 & 1 & 0 \\ 0 & 0 & 1 & 0 & 0 \\ 1 & 0 & 0 & 0 & 1 \\ 0 & 0 & 1 & 0 & 0 \\ 0 & 0 & 0 & 1 & 0 \end{bmatrix} + \begin{bmatrix} 0 & 0 & 2 & 0 & 0 \\ 1 & 0 & 0 & 0 & 1 \\ 0 & 1 & 0 & 2 & 0 \\ 1 & 0 & 0 & 0 & 1 \\ 0 & 0 & 1 & 0 & 0 \end{bmatrix} = \begin{bmatrix} 0 & 1 & 2 & 1 & 0 \\ 1 & 0 & 1 & 0 & 1 \\ 1 & 1 & 0 & 2 & 1 \\ 1 & 0 & 1 & 0 & 1 \\ 0 & 0 & 1 & 1 & 0 \end{bmatrix}$

$A + A^2 + A^3 = \begin{bmatrix} 0 & 1 & 2 & 1 & 0 \\ 1 & 0 & 1 & 0 & 1 \\ 1 & 1 & 0 & 2 & 1 \\ 1 & 0 & 1 & 0 & 1 \\ 0 & 0 & 1 & 1 & 0 \end{bmatrix} + \begin{bmatrix} 2 & 0 & 0 & 0 & 2 \\ 0 & 1 & 0 & 2 & 0 \\ 0 & 0 & 3 & 0 & 0 \\ 0 & 1 & 0 & 2 & 0 \\ 1 & 0 & 0 & 0 & 1 \end{bmatrix} = \begin{bmatrix} 2 & 1 & 2 & 1 & 2 \\ 1 & 1 & 1 & 2 & 1 \\ 1 & 1 & 3 & 2 & 1 \\ 1 & 1 & 1 & 2 & 1 \\ 1 & 0 & 1 & 1 & 1 \end{bmatrix}$

A zero element remains, so we must compute A^4.

$A^4 = A^3A = \begin{bmatrix} 2 & 0 & 0 & 0 & 2 \\ 0 & 1 & 0 & 2 & 0 \\ 0 & 0 & 3 & 0 & 0 \\ 0 & 1 & 0 & 2 & 0 \\ 1 & 0 & 0 & 0 & 1 \end{bmatrix} \begin{bmatrix} 0 & 1 & 0 & 1 & 0 \\ 0 & 0 & 1 & 0 & 0 \\ 1 & 0 & 0 & 0 & 1 \\ 0 & 0 & 1 & 0 & 0 \\ 0 & 0 & 0 & 1 & 0 \end{bmatrix} = \begin{bmatrix} 0 & 2 & 0 & 4 & 0 \\ 0 & 0 & 3 & 0 & 0 \\ 3 & 0 & 0 & 0 & 3 \\ 0 & 0 & 3 & 0 & 0 \\ 0 & 1 & 0 & 2 & 0 \end{bmatrix}$

Then $A + A^2 + A^3 + A^4 = \begin{bmatrix} 2 & 1 & 2 & 1 & 2 \\ 1 & 1 & 1 & 2 & 1 \\ 1 & 1 & 3 & 2 & 1 \\ 1 & 1 & 1 & 2 & 1 \\ 1 & 0 & 1 & 1 & 1 \end{bmatrix} + \begin{bmatrix} 0 & 2 & 0 & 4 & 0 \\ 0 & 0 & 3 & 0 & 0 \\ 3 & 0 & 0 & 0 & 3 \\ 0 & 0 & 3 & 0 & 0 \\ 0 & 1 & 0 & 2 & 0 \end{bmatrix} = \begin{bmatrix} 2 & 3 & 2 & 5 & 2 \\ 1 & 1 & 4 & 2 & 1 \\ 4 & 1 & 3 & 2 & 4 \\ 1 & 1 & 4 & 2 & 1 \\ 1 & 1 & 1 & 3 & 1 \end{bmatrix}$

This matrix indicates that it is possible to travel from any origin to any destination with at most 3 intermediate connections.

79. (A) $[1{,}000 \quad 500 \quad 5{,}000] \begin{bmatrix} \$0.80 \\ \$1.50 \\ \$0.40 \end{bmatrix} = 1{,}000(\$0.80) + 500(\$1.50) + 5{,}000(\$0.40),$
$= \$3{,}550$

(B) $[2{,}000 \quad 800 \quad 8{,}000] \begin{bmatrix} \$0.80 \\ \$1.50 \\ \$0.40 \end{bmatrix} = 2{,}000(\$0.80) + 800(\$1.50) + 8{,}000(\$0.40)$
$= \$6{,}000$

(C) The matrix MN has no obvious interpretations, but the matrix NM represents the total cost of all contacts in each town.

(D) $NM = \begin{bmatrix} 1{,}000 & 500 & 5{,}000 \\ 2{,}000 & 800 & 8{,}000 \end{bmatrix} \begin{bmatrix} \$0.80 \\ \$1.50 \\ \$0.40 \end{bmatrix} = \begin{bmatrix} 1{,}000(0.80) + 500(1.50) + 5{,}000(0.40) \\ 2{,}000(0.80) + 800(1.50) + 8{,}000(0.40) \end{bmatrix}$

$= \begin{bmatrix} \$3{,}550 \\ \$6{,}000 \end{bmatrix} \begin{matrix} \text{Berkeley} \\ \text{Oakland} \end{matrix} = $ cost of all contacts in each town.

(E) The matrix $[1 \quad 1]N$ can be used to find the total number of each of the three types of contact:

$[1 \quad 1] \begin{bmatrix} 1{,}000 & 500 & 5{,}000 \\ 2{,}000 & 800 & 8{,}000 \end{bmatrix} = [1{,}000 + 2{,}000 \quad 500 + 800 \quad 5{,}000 + 8{,}000]$

[Telephone House Letter] = $[3{,}000 \quad 1{,}300 \quad 13{,}000]$

(F) The matrix $N\begin{bmatrix}1\\1\\1\end{bmatrix}$ can be used to find the total number of contacts in each town:

$$\begin{bmatrix}1,000 & 500 & 5,000\\2,000 & 800 & 8,000\end{bmatrix}\begin{bmatrix}1\\1\\1\end{bmatrix}$$

$$=\begin{bmatrix}1,000+500+5,000\\2,000+800+8,000\end{bmatrix}=\begin{bmatrix}6,500\\10,800\end{bmatrix}=\begin{bmatrix}\text{Berkeley contacts}\\\text{Oakland contacts}\end{bmatrix}$$

81. (A) Since player 1 did not defeat player 1, a 0 is placed in row 1, column 1.
Since player 1 did not defeat player 2, a 0 is placed in row 1, column 2.
Since player 1 defeated player 3, a 1 is placed in row 1, column 3.
Since player 1 defeated player 4, a 1 is placed in row 1, column 4.
Since player 1 defeated player 5, a 1 is placed in row 1, column 5.
Since player 1 did not defeat player 6, a 0 is placed in row 1, column 6.
Proceeding in this manner, we obtain

$$A=\begin{bmatrix}0 & 0 & 1 & 1 & 1 & 0\\1 & 0 & 0 & 1 & 1 & 0\\0 & 1 & 0 & 1 & 0 & 0\\0 & 0 & 0 & 0 & 0 & 1\\0 & 0 & 1 & 1 & 0 & 1\\1 & 1 & 1 & 0 & 0 & 0\end{bmatrix}$$

(B) $A^2=\begin{bmatrix}0 & 0 & 1 & 1 & 1 & 0\\1 & 0 & 0 & 1 & 1 & 0\\0 & 1 & 0 & 1 & 0 & 0\\0 & 0 & 0 & 0 & 0 & 1\\0 & 0 & 1 & 1 & 0 & 1\\1 & 1 & 1 & 0 & 0 & 0\end{bmatrix}+\begin{bmatrix}0 & 0 & 1 & 1 & 1 & 0\\1 & 0 & 0 & 1 & 1 & 0\\0 & 1 & 0 & 1 & 0 & 0\\0 & 0 & 0 & 0 & 0 & 1\\0 & 0 & 1 & 1 & 0 & 1\\1 & 1 & 1 & 0 & 0 & 0\end{bmatrix}=\begin{bmatrix}0 & 1 & 1 & 2 & 0 & 2\\0 & 0 & 2 & 2 & 1 & 2\\1 & 0 & 0 & 1 & 1 & 1\\1 & 1 & 1 & 0 & 0 & 0\\1 & 2 & 1 & 1 & 0 & 1\\1 & 1 & 1 & 3 & 2 & 0\end{bmatrix}$

$$A+A^2=\begin{bmatrix}0 & 1 & 2 & 3 & 1 & 2\\1 & 0 & 2 & 3 & 2 & 2\\1 & 1 & 0 & 2 & 1 & 1\\1 & 1 & 1 & 0 & 0 & 1\\1 & 2 & 2 & 2 & 0 & 2\\2 & 2 & 2 & 3 & 2 & 0\end{bmatrix}=B$$

Check (using graphing calculator):

```
[A]
[[0 0 1 1 1 0]
 [1 0 0 1 1 0]
 [0 1 0 1 0 0]
 [0 0 0 0 0 1]
 [0 0 1 1 0 1]
 [1 1 1 0 0 0]]
```

```
[A]²
[[0 1 1 2 0 2]
 [0 0 2 2 1 2]
 [1 0 0 1 1 1]
 [1 1 1 0 0 0]
 [1 2 1 1 0 1]
 [1 1 1 3 2 0]]
```

```
[A]+[A]²→[B]
[[0 1 2 3 1 2]
 [1 0 2 3 2 2]
 [1 1 0 2 1 1]
 [1 1 1 0 0 1]
 [1 2 2 2 0 2]
 [2 2 2 3 2 0]]
```

(C) To add up the rows of AB, we can multiply by a column matrix of all ones: that is, if

$$C=\begin{bmatrix}1\\1\\1\\1\\1\\1\end{bmatrix}\quad\text{Then }BC=\begin{bmatrix}0 & 1 & 2 & 3 & 1 & 2\\1 & 0 & 2 & 3 & 2 & 2\\1 & 1 & 0 & 2 & 1 & 1\\1 & 1 & 1 & 0 & 0 & 1\\1 & 2 & 2 & 2 & 0 & 2\\2 & 2 & 2 & 3 & 2 & 0\end{bmatrix}\begin{bmatrix}1\\1\\1\\1\\1\\1\end{bmatrix}=\begin{bmatrix}9\\10\\6\\4\\9\\11\end{bmatrix}$$

(D) BC measures the relative strength of the players, with the larger numbers representing greater strength. Player 6 is the strongest and player 4 the weakest; ranking:

Frank, Bart, Aaron & Elvis (tie), Charles, Dan.

Section 9-3

1. Answers will vary.   **3.** Answers will vary.   **5.** Answers will vary.

7. $\begin{bmatrix} 2 & -3 \\ 4 & 5 \end{bmatrix}$ **9.** $\begin{bmatrix} 2 & -3 \\ 4 & 5 \end{bmatrix}$ **11.** $\begin{bmatrix} -2 & 1 & 3 \\ 2 & 4 & -2 \\ 5 & 1 & 0 \end{bmatrix}$ **13.** $\begin{bmatrix} -2 & 1 & 3 \\ 2 & 4 & -2 \\ 5 & 1 & 0 \end{bmatrix}$

15. $\begin{bmatrix} 3 & -4 \\ -2 & 3 \end{bmatrix}\begin{bmatrix} 3 & 4 \\ 2 & 3 \end{bmatrix} = \begin{bmatrix} 3\cdot3+(-4)2 & 3\cdot4+(-4)3 \\ (-2)3+3\cdot2 & (-2)4+3\cdot3 \end{bmatrix} = \begin{bmatrix} 1 & 0 \\ 0 & 1 \end{bmatrix}$
These two matrices are inverses of each other.

17. $\begin{bmatrix} 2 & 2 \\ -1 & -1 \end{bmatrix}\begin{bmatrix} 1 & 1 \\ -1 & -1 \end{bmatrix} = \begin{bmatrix} 2\cdot1+2(-1) & 2\cdot1+2(-1) \\ (-1)1+(-1)(-1) & (-1)1+(-1)(-1) \end{bmatrix} = \begin{bmatrix} 0 & 0 \\ 0 & 0 \end{bmatrix}$
These two matrices are not inverses of each other.

19. $\begin{bmatrix} -5 & 2 \\ -8 & 3 \end{bmatrix}\begin{bmatrix} 3 & -2 \\ 8 & -5 \end{bmatrix} = \begin{bmatrix} (-5)3+2\cdot8 & (-5)(-2)+2(-5) \\ (-8)3+3\cdot8 & (-8)(-2)+3(-5) \end{bmatrix} = \begin{bmatrix} 1 & 0 \\ 0 & 1 \end{bmatrix}$
These two matrices are inverses of each other.

21. $\begin{bmatrix} 1 & 2 & 0 \\ 0 & 1 & 0 \\ -1 & -1 & 1 \end{bmatrix}\begin{bmatrix} 1 & -2 & 0 \\ 0 & 1 & 0 \\ 1 & -1 & 0 \end{bmatrix} =$

$\begin{bmatrix} 1\cdot1+2\cdot0+0\cdot1 & 1(-2)+2\cdot1+0(-1) & 1\cdot0+2\cdot0+0\cdot0 \\ 0\cdot1+1\cdot1+0(-1) & 0(-2)+1\cdot1+0(-1) & 0\cdot0+1\cdot0+0\cdot0 \\ (-1)1+(-1)1+1\cdot1 & (-1)(-2)+(-1)1+1(-1) & (-1)0+(-1)0+0\cdot0 \end{bmatrix}$

$= \begin{bmatrix} 1 & 0 & 0 \\ 1 & 1 & 0 \\ -1 & 0 & 0 \end{bmatrix}$

These two matrices are not inverses of each other.

23. $\begin{bmatrix} 1 & -1 & 1 \\ 0 & 2 & -1 \\ 2 & 3 & 0 \end{bmatrix}\begin{bmatrix} 3 & 3 & -1 \\ -2 & -2 & 1 \\ -4 & -5 & 2 \end{bmatrix}$

$= \begin{bmatrix} 1\cdot3+(-1)(-2)+1(-4) & 1\cdot3+(-1)(-2)+1(-5) & 1\cdot(-1)+(-1)\cdot1+1\cdot2 \\ 0\cdot3+2(-2)+(-1)(-4) & 0\cdot3+2(-2)+(-1)(-5) & 0(-1)+2\cdot1+(-1)\cdot2 \\ 2\cdot3+3(-2)+0(-4) & 2\cdot3+3(-2)+0(-5) & 2(-1)+3\cdot1+0\cdot2 \end{bmatrix}$

$= \begin{bmatrix} 1 & 0 & 0 \\ 0 & 1 & 0 \\ 0 & 0 & 1 \end{bmatrix}$

These two matrices are inverses of each other.

25. $\begin{bmatrix} 1 & 9 & | & 1 & 0 \\ 0 & 1 & | & 0 & 1 \end{bmatrix}$ $(-9)R_2 + R_1 \rightarrow R_1$

$\sim \begin{bmatrix} 1 & 0 & | & 1 & -9 \\ 0 & 1 & | & 0 & 1 \end{bmatrix}$

The inverse is $\begin{bmatrix} 1 & -9 \\ 0 & 1 \end{bmatrix}$

Check: $M^{-1}M = \begin{bmatrix} 1 & -9 \\ 0 & 1 \end{bmatrix}\begin{bmatrix} 1 & 9 \\ 0 & 1 \end{bmatrix} = \begin{bmatrix} 1\cdot1+(-9)0 & 1\cdot9+(-9)1 \\ 0\cdot1+1\cdot0 & 0\cdot9+1\cdot1 \end{bmatrix} = \begin{bmatrix} 1 & 0 \\ 0 & 1 \end{bmatrix}$

27. $\begin{bmatrix} -1 & -2 & | & 1 & 0 \\ 2 & 5 & | & 0 & 1 \end{bmatrix} 2R_1 + R_2 \rightarrow R_2$

$\sim \begin{bmatrix} -1 & -2 & | & 1 & 0 \\ 0 & 1 & | & 2 & 1 \end{bmatrix} 2R_2 + R_1 \rightarrow R_1$

$\sim \begin{bmatrix} -1 & -2 & | & 1 & 0 \\ 0 & 1 & | & 2 & 1 \end{bmatrix} 2R_2 + R_1 \rightarrow R_1$

$\sim \begin{bmatrix} -1 & 0 & | & 5 & 2 \\ 0 & 1 & | & 2 & 1 \end{bmatrix} (-1)R_1 \rightarrow R_1$

$\sim \begin{bmatrix} 1 & 0 & | & -5 & -2 \\ 0 & 1 & | & 2 & 1 \end{bmatrix}$

The inverse is $\begin{bmatrix} -5 & -2 \\ 2 & 1 \end{bmatrix}$

Check: $M^{-1}M = \begin{bmatrix} -5 & -2 \\ 2 & 1 \end{bmatrix}\begin{bmatrix} -1 & -2 \\ 2 & 5 \end{bmatrix} = \begin{bmatrix} (-5)(-1) + (-2)2 & (-5)(-2) + (-2)5 \\ 2(-1) + 1 \cdot 2 & 2(-2) + 1 \cdot 5 \end{bmatrix}$

$$= \begin{bmatrix} 1 & 0 \\ 0 & 1 \end{bmatrix}$$

29. $\begin{bmatrix} -5 & 7 & | & 1 & 0 \\ 2 & -3 & | & 0 & 1 \end{bmatrix} 3R_2 + R_1 \rightarrow R_1$

$\sim \begin{bmatrix} 1 & -2 & | & 1 & 3 \\ 2 & -3 & | & 0 & 1 \end{bmatrix} (-2)R_1 + R_2 \rightarrow R_2$

$\sim \begin{bmatrix} 1 & -2 & | & 1 & 3 \\ 0 & 1 & | & -2 & -5 \end{bmatrix} 2R_2 + R_1 \rightarrow R_1$

$\sim \begin{bmatrix} 1 & 0 & | & -3 & -7 \\ 0 & 1 & | & -2 & -5 \end{bmatrix}$

The inverse is $\begin{bmatrix} -3 & -7 \\ -2 & -5 \end{bmatrix}$

Check: $M^{-1}M = \begin{bmatrix} -3 & -7 \\ -2 & -5 \end{bmatrix}\begin{bmatrix} -5 & 7 \\ 2 & -3 \end{bmatrix} = \begin{bmatrix} (-3)(-5) + (-7)2 & (-3)7 + (-7)(-3) \\ (-2)(-5) + (-5)2 & (-2)7 + (-5)(-3) \end{bmatrix} = \begin{bmatrix} 1 & 0 \\ 0 & 1 \end{bmatrix}$

31. If the inverse existed we would find it by row operations on the following matrix:

$= \begin{bmatrix} 3 & 9 & | & 1 & 0 \\ 2 & 6 & | & 0 & 1 \end{bmatrix}$

But consider what happens if we perform $(-\frac{2}{3})R_1 + R_2 \rightarrow R_2$

$= \begin{bmatrix} 3 & 9 & | & 1 & 0 \\ 0 & 0 & | & -\frac{2}{3} & 1 \end{bmatrix}$

Since a row of zeros results to the left of the vertical line, no inverse exists.

33. $\begin{bmatrix} 2 & 3 & | & 1 & 0 \\ 3 & 5 & | & 0 & 1 \end{bmatrix} \frac{1}{2}R_1 \rightarrow R_1$

$\sim \begin{bmatrix} 1 & 1.5 & | & 0.5 & 0 \\ 3 & 5 & | & 0 & 1 \end{bmatrix} (-3)R_1 + R_2 \rightarrow R_2$

$\sim \begin{bmatrix} 1 & 1.5 & | & 0.5 & 0 \\ 0 & 0.5 & | & -1.5 & 1 \end{bmatrix} 2R_2 \rightarrow R_2$

$\sim \begin{bmatrix} 1 & 1.5 & | & 0.5 & 0 \\ 0 & 1 & | & -3 & 2 \end{bmatrix} (-1.5)R_2 + R_1 \rightarrow R_1$

$$\sim \begin{bmatrix} 1 & 0 & | & 5 & -3 \\ 0 & 1 & | & -3 & 2 \end{bmatrix}$$

The inverse is $\begin{bmatrix} 5 & -3 \\ -3 & 2 \end{bmatrix}$

The checking steps are left to the student in this and some subsequent problems.

35. $\begin{bmatrix} 1 & -1 & 0 & | & 1 & 0 & 0 \\ -1 & 1 & -1 & | & 0 & 1 & 0 \\ 0 & -1 & 1 & | & 0 & 0 & 1 \end{bmatrix} \quad R_1 + R_2 \rightarrow R_2$

$\sim \begin{bmatrix} 1 & -1 & 0 & | & 1 & 0 & 0 \\ 0 & 0 & -1 & | & 1 & 1 & 0 \\ 0 & -1 & 1 & | & 0 & 0 & 1 \end{bmatrix} \quad R_2 \leftrightarrow R_3$

$\sim \begin{bmatrix} 1 & -1 & 0 & | & 1 & 0 & 0 \\ 0 & -1 & 1 & | & 0 & 0 & 1 \\ 0 & 0 & -1 & | & 1 & 1 & 0 \end{bmatrix} \quad (-1)R_2 + R_1 \rightarrow R_1$

$\sim \begin{bmatrix} 1 & 0 & -1 & | & 1 & 0 & -1 \\ 0 & -1 & 1 & | & 0 & 0 & 1 \\ 0 & 0 & -1 & | & 1 & 1 & 0 \end{bmatrix} \quad \begin{matrix} (-1)R_3 + R_1 \rightarrow R_1 \\ R_3 + R_2 \rightarrow R_2 \end{matrix}$

$\sim \begin{bmatrix} 1 & 0 & 0 & | & 0 & -1 & -1 \\ 0 & -1 & 0 & | & 1 & 1 & 1 \\ 0 & 0 & -1 & | & 1 & 1 & 0 \end{bmatrix} \quad \begin{matrix} (-1)R_2 \rightarrow R_2 \\ (-1)R_3 \rightarrow R_3 \end{matrix}$

$\sim \begin{bmatrix} 1 & 0 & 0 & | & 0 & -1 & -1 \\ 0 & 1 & 0 & | & -1 & -1 & -1 \\ 0 & 0 & 1 & | & -1 & -1 & 0 \end{bmatrix}$

The inverse is $\begin{bmatrix} 0 & -1 & -1 \\ -1 & 1 & -1 \\ -1 & -1 & 0 \end{bmatrix}$

Check: $M^{-1}M = \begin{bmatrix} 0 & -1 & -1 \\ -1 & 1 & -1 \\ -1 & -1 & 0 \end{bmatrix}\begin{bmatrix} 1 & -1 & 0 \\ -1 & 1 & -1 \\ 0 & -1 & 1 \end{bmatrix}$

$= \begin{bmatrix} 0 \cdot 1 + (-1)(-1) + (-1)0 & 0(-1) + (-1)1 + (-1)(-1) & 0 \cdot 0 + (-1)(-1) + (-1)1 \\ (-1)1 + (-1)(-1) + (-1)0 & (-1)(-1) + (-1)1 + (-1)(-1) & (-1)0 + (-1)(-1) + (-1)1 \\ (-1)1 + (-1)(-1) + 0 \cdot 0 & (-1)(-1) + (-1)1 + 0(-1) & (-1)0 + (-1)(-1) + 0 \cdot 1 \end{bmatrix}$

$= \begin{bmatrix} 1 & 0 & 0 \\ 0 & 1 & 0 \\ 0 & 0 & 1 \end{bmatrix}$

37. $\begin{bmatrix} 1 & 2 & 5 & | & 1 & 0 & 0 \\ 3 & 5 & 9 & | & 0 & 1 & 0 \\ 1 & 1 & -2 & | & 0 & 0 & 1 \end{bmatrix} \quad \begin{matrix} (-3)R_1 + R_2 \rightarrow R_2 \\ (-1)R_1 + R_3 \rightarrow R_3 \end{matrix}$

$\sim \begin{bmatrix} 1 & 2 & 5 & | & 1 & 0 & 0 \\ 0 & -1 & -6 & | & -3 & 1 & 0 \\ 0 & -1 & -7 & | & -1 & 0 & 1 \end{bmatrix} \quad \begin{matrix} 2R_2 + R_1 \rightarrow R_1 \\ (-1)R_2 + R_3 \rightarrow R_3 \end{matrix}$

$\sim \begin{bmatrix} 1 & 0 & -7 & | & -5 & 2 & 0 \\ 0 & -1 & -6 & | & -3 & 1 & 0 \\ 0 & 0 & -1 & | & 2 & -1 & 1 \end{bmatrix} \quad \begin{matrix} (-7)R_3 + R_1 \rightarrow R_1 \\ (-6)R_3 + R_2 \rightarrow R_2 \end{matrix}$

$$\sim \begin{bmatrix} 1 & 0 & 0 & | & -19 & 9 & -7 \\ 0 & -1 & 0 & | & -15 & 7 & -6 \\ 0 & 0 & -1 & | & 2 & -1 & 1 \end{bmatrix} \begin{array}{l} (-1)R_2 \rightarrow R_2 \\ (-1)R_3 \rightarrow R_3 \end{array}$$

$$\sim \begin{bmatrix} 1 & 0 & 0 & | & -19 & 9 & -7 \\ 0 & 1 & 0 & | & 15 & -7 & 6 \\ 0 & 0 & 1 & | & -2 & 1 & -1 \end{bmatrix}$$

The inverse is $\begin{bmatrix} -19 & 9 & -7 \\ 15 & -7 & 6 \\ -2 & 1 & -1 \end{bmatrix}$

Check: $M^{-1}M = \begin{bmatrix} -19 & 9 & -7 \\ 15 & -7 & 6 \\ -2 & 1 & -1 \end{bmatrix} \begin{bmatrix} 1 & 2 & 5 \\ 3 & 5 & 9 \\ 1 & 1 & -2 \end{bmatrix}$

$$= \begin{bmatrix} (-19)1 + 9 \cdot 3 + (-7)1 & (-19)2 + 9 \cdot 5 + (-7)1 & (-19)5 + 9 \cdot 9 + (-7)(-2) \\ 15 \cdot 1 + (-7)3 + 6 \cdot 1 & 15 \cdot 2 + (-7)5 + 6 \cdot 1 & 15 \cdot 5 + (-7)9 + 6(-2) \\ (-2)1 + 1 \cdot 3 + (-1)1 & (-2)2 + 1 \cdot 5 + (-1)1 & (-2)5 + 1 \cdot 9 + (-1)(-2) \end{bmatrix}$$

$$= \begin{bmatrix} 1 & 0 & 0 \\ 0 & 1 & 0 \\ 0 & 0 & 1 \end{bmatrix}$$

39. $\begin{bmatrix} 2 & 2 & -1 & | & 1 & 0 & 0 \\ 0 & 4 & -1 & | & 0 & 1 & 0 \\ -1 & -2 & 1 & | & 0 & 0 & 1 \end{bmatrix} R_1 \leftrightarrow R_3$

$$\sim \begin{bmatrix} -1 & -2 & 1 & | & 0 & 0 & 1 \\ 0 & 4 & -1 & | & 0 & 1 & 0 \\ 2 & 2 & -1 & | & 1 & 0 & 0 \end{bmatrix} 2R_1 + R_3 \rightarrow R_3$$

$$\sim \begin{bmatrix} -1 & -2 & 1 & | & 0 & 0 & 1 \\ 0 & 4 & -1 & | & 0 & 1 & 0 \\ 0 & -2 & 1 & | & 1 & 0 & 2 \end{bmatrix} \begin{array}{l} (-1)R_3 + R_1 \rightarrow R_1 \\ 2R_3 + R_2 \rightarrow R_2 \end{array}$$

$$\sim \begin{bmatrix} -1 & 0 & 0 & | & -1 & 0 & -1 \\ 0 & 0 & 1 & | & 2 & 1 & 4 \\ 0 & -2 & 1 & | & -1 & 0 & 2 \end{bmatrix} (-1)R_2 + R_3 \rightarrow R_3$$

$$\sim \begin{bmatrix} -1 & 0 & 0 & | & -1 & 0 & -1 \\ 0 & 0 & 1 & | & 2 & 1 & 4 \\ 0 & -2 & 0 & | & -1 & -1 & -2 \end{bmatrix} R_2 \leftrightarrow R_3$$

$$\sim \begin{bmatrix} -1 & 0 & 0 & | & -1 & 0 & -1 \\ 0 & -2 & 0 & | & -1 & -1 & -2 \\ 0 & 0 & 1 & | & 2 & 1 & 4 \end{bmatrix} \begin{array}{l} (-1)R_1 \rightarrow R_1 \\ (-\frac{1}{2})R_2 \rightarrow R_2 \end{array}$$

$$\sim \begin{bmatrix} 1 & 0 & 0 & | & 1 & 0 & 1 \\ 0 & 1 & 0 & | & \frac{1}{2} & \frac{1}{2} & 1 \\ 0 & 0 & 1 & | & 2 & 1 & 4 \end{bmatrix}$$

The inverse is $\begin{bmatrix} 1 & 0 & 1 \\ \frac{1}{2} & \frac{1}{2} & 1 \\ 2 & 1 & 4 \end{bmatrix}$

41. If the inverse existed we would find it by row operations on the following matrix:

$$\begin{bmatrix} 2 & 1 & 1 & | & 1 & 0 & 0 \\ 1 & 1 & 0 & | & 0 & 1 & 0 \\ -1 & -1 & 0 & | & 0 & 0 & 1 \end{bmatrix}$$

But consider what happens if we perform $R_2 + R_3 \to R_2$

$$\begin{bmatrix} 2 & 1 & 1 & | & 1 & 0 & 0 \\ 0 & 0 & 0 & | & 0 & 1 & 1 \\ -1 & -1 & 0 & | & 0 & 0 & 1 \end{bmatrix}$$

Since a row of zeros results to the left of the vertical line, no inverse exists.

43.

$$\begin{bmatrix} 1 & 5 & 10 & | & 1 & 0 & 0 \\ 0 & 1 & 4 & | & 0 & 1 & 0 \\ 1 & 6 & 15 & | & 0 & 0 & 1 \end{bmatrix} (-1)R_1 + R_3 \to R_3$$

$$\sim \begin{bmatrix} 1 & 5 & 10 & | & 1 & 0 & 0 \\ 0 & 1 & 4 & | & 0 & 1 & 0 \\ 0 & 1 & 5 & | & -1 & 0 & 1 \end{bmatrix} \begin{array}{l}(-5)R_2 + R_1 \to R_1 \\ (-1)R_2 + R_3 \to R_3 \end{array}$$

$$\sim \begin{bmatrix} 1 & 0 & -10 & | & 1 & -5 & 0 \\ 0 & 1 & 4 & | & 0 & 1 & 0 \\ 0 & 0 & 1 & | & -1 & -1 & 1 \end{bmatrix} \begin{array}{l}10R_3 + R_1 \to R_1 \\ (-4)R_3 + R_2 \to R_2 \end{array}$$

$$\sim \begin{bmatrix} 1 & 0 & 0 & | & -9 & -15 & 10 \\ 0 & 1 & 0 & | & 4 & 5 & -4 \\ 0 & 0 & 1 & | & -1 & -1 & 1 \end{bmatrix}$$

The inverse is $\begin{bmatrix} -9 & -15 & 10 \\ 4 & 5 & -4 \\ -1 & -1 & 1 \end{bmatrix}$

45. Try to find A^{-1}:

$$\begin{bmatrix} a & 0 & | & 1 & 0 \\ 0 & d & | & 0 & 1 \end{bmatrix} \frac{1}{a}R_1 \to R_1$$

$$\sim \begin{bmatrix} 1 & 0 & | & \frac{1}{a} & 0 \\ 0 & d & | & 0 & 1 \end{bmatrix} \frac{1}{d}R_2 \to R_2$$

$$\sim \begin{bmatrix} 1 & 0 & | & \frac{1}{a} & 0 \\ 0 & 1 & | & 0 & \frac{1}{d} \end{bmatrix} \quad \text{The inverse is } A^{-1} = \begin{bmatrix} \frac{1}{a} & 0 \\ 0 & \frac{1}{d} \end{bmatrix}$$

This will exist unless either a or d is zero, which would make $\frac{1}{a}$ or $\frac{1}{d}$ undefined. So A^{-1} exists exactly when both a and d are non-zero.

47. (A) A^{-1}:

$$\begin{bmatrix} 3 & 2 & | & 1 & 0 \\ -4 & -3 & | & 0 & 1 \end{bmatrix} R_2 + R_1 \to R_1$$

$$\sim \begin{bmatrix} -1 & -1 & | & 1 & 1 \\ -4 & -3 & | & 0 & 1 \end{bmatrix} (-1)R_1 \to R_1$$

$$\sim \begin{bmatrix} 1 & 1 & | & -1 & -1 \\ -4 & -3 & | & 0 & 1 \end{bmatrix} 4R_1 + R_2 \to R_2$$

$$\sim \begin{bmatrix} 1 & 1 & | & -1 & -1 \\ 0 & 1 & | & -4 & 1-3 \end{bmatrix} -1R_2 + R_1 \to R_1$$

$$\sim \begin{bmatrix} 1 & 0 & | & 3 & 2 \\ 0 & 1 & | & -4 & -3 \end{bmatrix} \quad A^{-1} = \begin{bmatrix} 3 & 2 \\ -4 & -3 \end{bmatrix}$$

A^2:

$$\begin{bmatrix} 3 & 2 \\ -4 & -3 \end{bmatrix} \begin{bmatrix} 3 & 2 \\ -4 & -3 \end{bmatrix}$$

$$= \begin{bmatrix} 3 \cdot 3 + 2(-4) & 3 \cdot 2 + 2(-3) \\ (-4)3 + (-3)(-4) & (-4)2 + (-3)(-3) \end{bmatrix}$$

$$= \begin{bmatrix} 1 & 0 \\ 0 & 1 \end{bmatrix}$$

(B) A^{-1}:

$$\begin{bmatrix} -2 & -1 & | & 1 & 0 \\ 3 & 2 & | & 0 & 1 \end{bmatrix} R_2 + R_1 \rightarrow R_1$$

$$\sim \begin{bmatrix} 1 & 1 & | & 1 & 1 \\ 3 & 2 & | & 0 & 1 \end{bmatrix} -3R_1 + R_2 \rightarrow R_2$$

$$\sim \begin{bmatrix} 1 & 1 & | & 1 & 1 \\ 0 & -1 & | & -3 & -2 \end{bmatrix} R_2 + R_1 \rightarrow R_1$$

$$\sim \begin{bmatrix} 1 & 0 & | & -2 & -1 \\ 0 & -1 & | & -3 & -2 \end{bmatrix} -1R_2 \rightarrow R_2$$

$$\sim \begin{bmatrix} 1 & 0 & | & -2 & -1 \\ 0 & 1 & | & 3 & 2 \end{bmatrix} \quad A^{-1} = \begin{bmatrix} -2 & -1 \\ 3 & 2 \end{bmatrix}$$

A^2:

$$\begin{bmatrix} -2 & -1 \\ 3 & 2 \end{bmatrix} \begin{bmatrix} -2 & -1 \\ 3 & 2 \end{bmatrix}$$

$$= \begin{bmatrix} (-2)(-2) + (-1)3 & (-2)(-1) + (-1)2 \\ 3(-2) + 2 \cdot 3 & 3(-1) + 2 \cdot 2 \end{bmatrix}$$

$$= \begin{bmatrix} 1 & 0 \\ 0 & 1 \end{bmatrix}$$

Note that in both cases $A^{-1} = A$ and $A^2 = I$.

49. (A) We calculate A^{-1} by row operations on

$$\begin{bmatrix} 4 & 2 & | & 1 & 0 \\ 1 & 3 & | & 0 & 1 \end{bmatrix} R_1 \leftrightarrow R_2$$

$$\sim \begin{bmatrix} 1 & 3 & | & 0 & 1 \\ 4 & 2 & | & 1 & 0 \end{bmatrix} (-4)R_1 + R_2 \rightarrow R_2$$

$$\sim \begin{bmatrix} 1 & 3 & | & 0 & 1 \\ 0 & -10 & | & 1 & -4 \end{bmatrix} \left(-\frac{1}{10}\right)R_2 \rightarrow R_2$$

$$\sim \begin{bmatrix} 1 & 3 & | & 0 & 1 \\ 0 & 1 & | & -\frac{1}{10} & \frac{2}{5} \end{bmatrix} (-3)R_2 + R_1 \rightarrow R_1$$

$$\sim \begin{bmatrix} 1 & 0 & | & \frac{3}{10} & -\frac{1}{5} \\ 0 & 1 & | & -\frac{1}{10} & \frac{2}{5} \end{bmatrix}$$

$$A^{-1} = \begin{bmatrix} \frac{3}{10} & -\frac{1}{5} \\ -\frac{1}{10} & \frac{2}{5} \end{bmatrix}$$

We calculate $(A^{-1})^{-1}$ by row operations on

$$\begin{bmatrix} \frac{3}{10} & -\frac{1}{5} & | & 1 & 0 \\ -\frac{1}{10} & \frac{2}{5} & | & 0 & 1 \end{bmatrix} \frac{10}{3} R_1 \rightarrow R_1$$

$$\sim \begin{bmatrix} 1 & -\frac{2}{3} & | & \frac{10}{3} & 0 \\ -\frac{1}{10} & \frac{2}{5} & | & 0 & 1 \end{bmatrix} \frac{1}{10} R_1 + R_2 \rightarrow R_2$$

$$\sim \begin{bmatrix} 1 & -\frac{2}{3} & | & \frac{10}{3} & 0 \\ 0 & \frac{1}{3} & | & \frac{1}{3} & 1 \end{bmatrix} 2R_2 + R_1 \rightarrow R_1$$

$$\sim \begin{bmatrix} 1 & 0 & | & 4 & 2 \\ 0 & \frac{1}{3} & | & \frac{1}{3} & 1 \end{bmatrix} 3R_2 \rightarrow R_2$$

$$\sim \begin{bmatrix} 1 & 0 & | & 4 & 2 \\ 0 & 1 & | & 1 & 3 \end{bmatrix}$$

$$(A^{-1})^{-1} = \begin{bmatrix} 4 & 2 \\ 1 & 3 \end{bmatrix}$$

(B) We calculate A^{-1} by row operations on

$$\begin{bmatrix} 5 & 5 & | & 1 & 0 \\ -1 & 3 & | & 0 & 1 \end{bmatrix} R_1 \leftrightarrow R_2$$

$$\sim \begin{bmatrix} -1 & 3 & | & 0 & 1 \\ 5 & 5 & | & 1 & 0 \end{bmatrix} 5R_1 + R_2 \rightarrow R_2$$

$$\sim \begin{bmatrix} -1 & 3 & | & 0 & 1 \\ 0 & 20 & | & 1 & 5 \end{bmatrix} \frac{1}{20} R_2 \rightarrow R_2$$

$$\sim \begin{bmatrix} -1 & 3 & | & 0 & 1 \\ 0 & 1 & | & \frac{1}{20} & \frac{1}{4} \end{bmatrix} (-3)R_2 + R_1 \rightarrow R_1$$

$$\sim \begin{bmatrix} -1 & 0 & | & -\frac{3}{20} & \frac{1}{4} \\ 0 & 1 & | & \frac{1}{20} & \frac{1}{4} \end{bmatrix} (-1)R_1 \rightarrow R_1$$

$$\sim \begin{bmatrix} 1 & 0 & | & \frac{3}{20} & -\frac{1}{4} \\ 0 & 1 & | & \frac{1}{20} & \frac{1}{4} \end{bmatrix}$$

$$A^{-1} = \begin{bmatrix} \frac{3}{20} & -\frac{1}{4} \\ \frac{1}{20} & \frac{1}{4} \end{bmatrix}$$

We calculate A^{-1} by row operations on

$$\begin{bmatrix} \frac{3}{20} & -\frac{1}{4} & | & 1 & 0 \\ \frac{1}{20} & \frac{1}{4} & | & 0 & 1 \end{bmatrix} \frac{20}{3} R_1 \rightarrow R_1$$

$$\sim \begin{bmatrix} 1 & -\frac{5}{3} & | & \frac{20}{3} & 0 \\ \frac{1}{20} & \frac{1}{4} & | & 0 & 1 \end{bmatrix} -\frac{1}{20} R_1 + R_2 \rightarrow R_2$$

$$\sim \begin{bmatrix} 1 & -\frac{5}{3} & | & \frac{20}{3} & 0 \\ 0 & \frac{1}{3} & | & -\frac{1}{3} & 1 \end{bmatrix} 5R_2 + R_1 \rightarrow R_1$$

$$\sim \begin{bmatrix} 1 & 0 & | & 5 & 5 \\ 0 & \frac{1}{3} & | & -\frac{1}{3} & 1 \end{bmatrix} 3R_2 \rightarrow R_2$$

$$\sim \begin{bmatrix} 1 & 0 & | & 5 & 5 \\ 0 & 1 & | & -1 & 3 \end{bmatrix}$$

$$(A^{-1})^{-1} = \begin{bmatrix} 5 & 5 \\ -1 & 3 \end{bmatrix}$$

Note that in both cases the inverse of the inverse works out to be the original matrix.

51. (A) Using a graphing calculator, we enter
A and B and calculate A^{-1} and B^{-1}

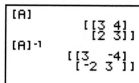

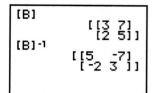

Then $(AB)^{-1}$ and
$A^{-1}B^{-1}$ are
calculated as

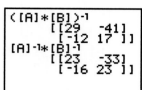

Finally $B^{-1}A^{-1}$ is calculated as

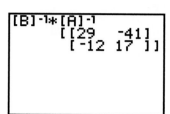

(B) Using a graphing calculator, we enter
A and B and calculate A^{-1} and B^{-1}

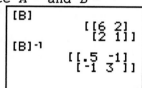

Then $(AB)^{-1}$ and
$A^{-1}B^{-1}$ are
calculated as

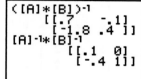

Finally $B^{-1}A^{-1}$ is calculated as

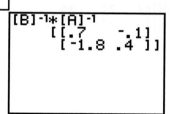

Notice that in each case $(AB)^{-1} = B^{-1}A^{-1}$ but $(AB)^{-1} \neq A^{-1}B^{-1}$.

53. Using the assignation numbers 1 to 27 with the letters of the alphabet and a blank as in the text, write

```
C   A   T       I   N       T   H   E       H   A   T
3   1   20  27  9   14  27  20  8   5   27  8   1   20
```

and calculate

$$\begin{bmatrix} 3 & 5 \\ 1 & 2 \end{bmatrix}\begin{bmatrix} 3 & 20 & 9 & 27 & 8 & 27 & 1 \\ 1 & 27 & 14 & 20 & 5 & 8 & 20 \end{bmatrix} = \begin{bmatrix} 14 & 195 & 97 & 181 & 49 & 121 & 103 \\ 5 & 74 & 37 & 67 & 18 & 43 & 41 \end{bmatrix}$$

The encoded message is

```
14  5  195  74  97  37  181  67  49  18  121  43  103  41
```

55. The inverse of matrix A is easily calculated to be

$$A^{-1} = \begin{bmatrix} 2 & -5 \\ -1 & 3 \end{bmatrix} \text{ (Details omitted)}$$

Putting the coded message into matrix form and multiplying by A^{-1} yields:

$$\begin{bmatrix} 2 & -5 \\ -1 & 3 \end{bmatrix}\begin{bmatrix} 111 & 40 & 177 & 50 & 116 & 86 & 62 & 121 & 68 \\ 43 & 15 & 68 & 19 & 45 & 29 & 22 & 43 & 27 \end{bmatrix}$$

$$= \begin{bmatrix} 7 & 5 & 14 & 5 & 7 & 27 & 14 & 27 & 1 \\ 18 & 5 & 27 & 7 & 19 & 1 & 4 & 8 & 13 \end{bmatrix}$$

This decodes to   7 18 5 5 14 27 5 7 7 19 27 1 14 4 27 8   1 13
              G   R   E E   N   E   G   G   S   A   N   D   H   A M

57. Using the assignation of numbers 1 to 27 with the letters of the alphabet and a blank as in the text, write

```
D  W  I  G  H  T     D  A  V  I  D     E  I  S  E  N  H  O  W  E  R
4 23  9  7  8 20  27  4  1 22  9  4 27  5  9 19  5 14  8 15 23  5 18 27 27
```

and calculate

$$
\begin{bmatrix}
1 & 0 & 1 & 0 & 1\\
0 & 1 & 1 & 0 & 3\\
2 & 1 & 1 & 1 & 1\\
0 & 0 & 1 & 0 & 2\\
1 & 1 & 1 & 2 & 1
\end{bmatrix}
\begin{bmatrix}
4 & 20 & 9 & 19 & 23\\
23 & 27 & 4 & 5 & 5\\
9 & 4 & 27 & 14 & 18\\
7 & 1 & 5 & 8 & 27\\
8 & 22 & 9 & 15 & 27
\end{bmatrix}
=
\begin{bmatrix}
21 & 46 & 45 & 48 & 68\\
56 & 97 & 58 & 64 & 104\\
55 & 94 & 63 & 80 & 123\\
25 & 48 & 45 & 44 & 72\\
58 & 75 & 59 & 69 & 127
\end{bmatrix}
$$

The encoded message is

```
21  56  55  25  58   46   97  94  48  75  45  58  63  45  59  48  64
80  44  69  68  104  123  72  127
```

59. The inverse of B is calculated to be

$$
B^{-1} =
\begin{bmatrix}
-2 & -1 & 2 & 2 & -1\\
3 & 2 & -2 & -4 & 1\\
6 & 2 & -4 & -5 & 2\\
-2 & -1 & 1 & 2 & 0\\
-3 & -1 & 2 & 3 & -1
\end{bmatrix}
\quad \text{(Details omitted)}
$$

Putting the coded message into matrix form and multiplying by B^{-1} yields

$$
\begin{bmatrix}
-2 & -1 & 2 & 2 & -1\\
3 & 2 & -2 & -4 & 1\\
6 & 2 & -4 & -5 & 2\\
-2 & -1 & 1 & 2 & 0\\
-3 & -1 & 2 & 3 & -1
\end{bmatrix}
\begin{bmatrix}
41 & 25 & 43 & 44 & 68\\
84 & 56 & 54 & 67 & 135\\
82 & 67 & 89 & 86 & 136\\
44 & 20 & 39 & 44 & 81\\
74 & 54 & 102 & 90 & 149
\end{bmatrix}
=
\begin{bmatrix}
12 & 14 & 14 & 15 & 14\\
25 & 27 & 5 & 8 & 27\\
14 & 2 & 19 & 14 & 27\\
4 & 1 & 27 & 19 & 27\\
15 & 9 & 10 & 15 & 27
\end{bmatrix}
$$

This decodes to

```
12 25 14  4 15 14 27  2  1  9 14  5 19 27 10 15  8 14 19 15 14 27 27 27
 L  Y  N  D  O  N     B  A  I  N  E  S     J  O  H  N  S  O  N
```

Section 9-4

1. Answers will vary.   **3.** Answers will vary.   **5.** Answers will vary.

7. Answers will vary.   **9.** Answers will vary.

11. $2x_1 - x_2 = 3$
$x_1 + 3x_2 = -2$

13. $-2x_1 + x_3 = 3$
$x_1 + 2x_2 + x_3 = -4$
$x_2 - x_3 = 2$

15. $\begin{bmatrix} 4 & -3\\ 1 & 2 \end{bmatrix}\begin{bmatrix} x_1\\ x_2 \end{bmatrix} = \begin{bmatrix} 2\\ 1 \end{bmatrix}$

17. $\begin{bmatrix} 1 & -2 & 1\\ -1 & 1 & 0\\ 2 & 3 & 1 \end{bmatrix}\begin{bmatrix} x_1\\ x_2\\ x_3 \end{bmatrix} = \begin{bmatrix} -1\\ 2\\ -3 \end{bmatrix}$

19. Since $\begin{bmatrix} 3 & -2\\ 1 & 4 \end{bmatrix}\begin{bmatrix} -2\\ 1 \end{bmatrix} = \begin{bmatrix} 3(-2) + (-2)1\\ 1(-2) + 4 \cdot 1 \end{bmatrix}$
$= \begin{bmatrix} -8\\ 2 \end{bmatrix}, \begin{bmatrix} x_1\\ x_2 \end{bmatrix} = \begin{bmatrix} -8\\ 2 \end{bmatrix}$ if and only if $x_1 = -8$ and $x_2 = 2$.

21. Since $\begin{bmatrix} -2 & 3\\ 2 & -1 \end{bmatrix}\begin{bmatrix} 3\\ 2 \end{bmatrix} = \begin{bmatrix} (-2)3 + 3 \cdot 2\\ 2 \cdot 3 + (-1)2 \end{bmatrix} = \begin{bmatrix} 0\\ 4 \end{bmatrix}$
$\begin{bmatrix} x_1\\ x_2 \end{bmatrix} = \begin{bmatrix} 0\\ 4 \end{bmatrix}$ if and only if $x_1 = 0$ and $x_2 = 4$.

23. $\begin{bmatrix} 1 & -1 \\ 1 & -2 \end{bmatrix} \begin{bmatrix} x_1 \\ x_2 \end{bmatrix} = \begin{bmatrix} 5 \\ 7 \end{bmatrix}$

      A X $=$ B

$AX = B$ has solution $X = A^{-1}B$.
To find A^{-1}, we perform row
operations on

$\begin{bmatrix} 1 & -1 & | & 1 & 0 \\ 1 & -2 & | & 0 & 1 \end{bmatrix} (-1)R_1 + R_2 \rightarrow R_2$

$\sim \begin{bmatrix} 1 & -1 & | & 1 & 0 \\ 0 & -1 & | & -1 & 1 \end{bmatrix} (-1)R_2 + R_1 \rightarrow R_1$

$\sim \begin{bmatrix} 1 & 0 & | & 2 & -1 \\ 0 & -1 & | & -1 & 1 \end{bmatrix} (-1)R_2 \rightarrow R_2$

$\sim \begin{bmatrix} 1 & 0 & | & 2 & -1 \\ 0 & 1 & | & 1 & -1 \end{bmatrix}$

$A^{-1} = \begin{bmatrix} 2 & -1 \\ 1 & -1 \end{bmatrix}$

$X = A^{-1}B = \begin{bmatrix} 2 & -1 \\ 1 & -1 \end{bmatrix} \begin{bmatrix} 5 \\ 7 \end{bmatrix} = \begin{bmatrix} 2 \cdot 5 + (-1)(7) \\ 1 \cdot 5 + (-1)(7) \end{bmatrix} = \begin{bmatrix} 3 \\ -2 \end{bmatrix}$

So $x_1 = 3$, $x_2 = -2$

25. $\begin{bmatrix} 1 & 1 \\ 2 & -3 \end{bmatrix} \begin{bmatrix} x_1 \\ x_2 \end{bmatrix} = \begin{bmatrix} 15 \\ 10 \end{bmatrix}$

      A X $=$ B

$AX = B$ has solution $X = A^{-1}B$.
To find A^{-1}, we perform row
operations on

$\begin{bmatrix} 1 & 1 & | & 1 & 0 \\ 2 & -3 & | & 0 & 1 \end{bmatrix} (-2)R_1 + R_2 \rightarrow R_2$

$\sim \begin{bmatrix} 1 & 1 & | & 1 & 0 \\ 0 & -5 & | & -2 & 1 \end{bmatrix} \left(-\frac{1}{5}\right)R_2 \rightarrow R_2$

$\sim \begin{bmatrix} 1 & 1 & | & 1 & 0 \\ 0 & 1 & | & \frac{2}{5} & -\frac{1}{5} \end{bmatrix} (-1)R_2 + R_1 \rightarrow R_1$

$\sim \begin{bmatrix} 1 & 0 & | & \frac{3}{5} & \frac{1}{5} \\ 0 & 1 & | & \frac{2}{5} & -\frac{1}{5} \end{bmatrix}$

$A^{-1} = \begin{bmatrix} \frac{3}{5} & \frac{1}{5} \\ \frac{2}{5} & -\frac{1}{5} \end{bmatrix}$

$X = A^{-1}B = \begin{bmatrix} \frac{3}{5} & \frac{1}{5} \\ \frac{2}{5} & -\frac{1}{5} \end{bmatrix} \begin{bmatrix} 15 \\ 10 \end{bmatrix} = \begin{bmatrix} \frac{3}{5} \cdot 15 + \frac{1}{5} \cdot 10 \\ \frac{2}{5} \cdot 15 + \left(-\frac{1}{5}\right)10 \end{bmatrix} = \begin{bmatrix} 11 \\ 4 \end{bmatrix}$

So $x_1 = 11$, $x_2 = 4$

27. $\begin{bmatrix} -1 & -2 \\ 2 & 5 \end{bmatrix} \begin{bmatrix} x_1 \\ x_2 \end{bmatrix} = \begin{bmatrix} k_1 \\ k_2 \end{bmatrix}$

$AX = K$ has solution $X = A^{-1}K$.

We find $A^{-1}K$ for each given K. From problem 27 in Section 9-3, $A^{-1} = \begin{bmatrix} -5 & -2 \\ 2 & 1 \end{bmatrix}$

(A) $K = \begin{bmatrix} 2 \\ 5 \end{bmatrix}$ $A^{-1}K = \begin{bmatrix} -5 & -2 \\ 2 & 1 \end{bmatrix} \begin{bmatrix} 2 \\ 5 \end{bmatrix} = \begin{bmatrix} -20 \\ 9 \end{bmatrix}$ $x_1 = -20$, $x_2 = 9$

(B) $K = \begin{bmatrix} -4 \\ 1 \end{bmatrix}$ $A^{-1}K = \begin{bmatrix} -5 & -2 \\ 2 & 1 \end{bmatrix} \begin{bmatrix} -4 \\ 1 \end{bmatrix} = \begin{bmatrix} 18 \\ -7 \end{bmatrix}$ $x_1 = 18$, $x_2 = -7$

(C) $K = \begin{bmatrix} -3 \\ -2 \end{bmatrix}$ $A^{-1}K = \begin{bmatrix} -5 & -2 \\ 2 & 1 \end{bmatrix} \begin{bmatrix} -3 \\ -2 \end{bmatrix} = \begin{bmatrix} 19 \\ -8 \end{bmatrix}$ $x_1 = 19$, $x_2 = -8$

29. $\begin{bmatrix} -5 & 7 \\ 2 & -3 \end{bmatrix} \begin{bmatrix} x_1 \\ x_2 \end{bmatrix} = \begin{bmatrix} k_1 \\ k_2 \end{bmatrix}$

$AX = K$ has solution $X = A^{-1}K$.

We find $A^{-1}K$ for each given K. From problem 29 in Section 9-3, $A^{-1} = \begin{bmatrix} -3 & -7 \\ -2 & -5 \end{bmatrix}$

(A) $K = \begin{bmatrix} -5 \\ 1 \end{bmatrix}$ $A^{-1}K = \begin{bmatrix} -3 & -7 \\ -2 & -5 \end{bmatrix} \begin{bmatrix} -5 \\ 1 \end{bmatrix} = \begin{bmatrix} 8 \\ 5 \end{bmatrix}$ $x_1 = 8$, $x_2 = 5$

(B) $K = \begin{bmatrix} 8 \\ -4 \end{bmatrix}$ $A^{-1}K = \begin{bmatrix} -3 & -7 \\ -2 & -5 \end{bmatrix} \begin{bmatrix} 8 \\ -4 \end{bmatrix} = \begin{bmatrix} 4 \\ 4 \end{bmatrix}$ $x_1 = 4$, $x_2 = 4$

(C) $K = \begin{bmatrix} 6 \\ 0 \end{bmatrix}$ $A^{-1}K = \begin{bmatrix} -3 & -7 \\ -2 & -5 \end{bmatrix} \begin{bmatrix} 6 \\ 0 \end{bmatrix} = \begin{bmatrix} -18 \\ -12 \end{bmatrix}$ $x_1 = -18$, $x_2 = -12$

31. $\begin{bmatrix} 1 & -1 & 0 \\ -1 & 1 & -1 \\ 0 & -1 & 1 \end{bmatrix} \begin{bmatrix} x_1 \\ x_2 \\ x_3 \end{bmatrix} = \begin{bmatrix} k_1 \\ k_2 \\ k_3 \end{bmatrix}$

$AX = K$ has solution $X = A^{-1}K$.
We find $A^{-1}K$ for each given K. From problem 35 in Section 9-3,

$A^{-1} = \begin{bmatrix} 0 & -1 & -1 \\ -1 & -1 & -1 \\ -1 & -1 & 0 \end{bmatrix}$

(A) $K = \begin{bmatrix} 1 \\ 1 \\ 2 \end{bmatrix}$ $A^{-1}K = \begin{bmatrix} 0 & -1 & -1 \\ -1 & -1 & -1 \\ -1 & -1 & 0 \end{bmatrix} \begin{bmatrix} 1 \\ 1 \\ 2 \end{bmatrix} = \begin{bmatrix} -3 \\ -4 \\ -2 \end{bmatrix}$ $x_1 = -3,\ x_2 = -4,\ x_3 = -2$

(B) $K = \begin{bmatrix} -1 \\ 0 \\ -4 \end{bmatrix}$ $A^{-1}K = \begin{bmatrix} 0 & -1 & -1 \\ -1 & -1 & -1 \\ -1 & -1 & 0 \end{bmatrix} \begin{bmatrix} -1 \\ 0 \\ -4 \end{bmatrix} = \begin{bmatrix} 4 \\ 5 \\ 1 \end{bmatrix}$ $x_1 = 4,\ x_2 = 5,\ x_3 = 1$

(C) $K = \begin{bmatrix} 3 \\ -2 \\ 0 \end{bmatrix}$ $A^{-1}K = \begin{bmatrix} 0 & -1 & -1 \\ -1 & -1 & -1 \\ -1 & -1 & 0 \end{bmatrix} \begin{bmatrix} 3 \\ -2 \\ 0 \end{bmatrix} = \begin{bmatrix} 2 \\ -1 \\ -1 \end{bmatrix}$ $x_1 = 2,\ x_2 = -1,\ x_3 = -1$

33. $\begin{bmatrix} 1 & 2 & 5 \\ 3 & 5 & 9 \\ 1 & 1 & -2 \end{bmatrix} \begin{bmatrix} x_1 \\ x_2 \\ x_3 \end{bmatrix} = \begin{bmatrix} k_1 \\ k_2 \\ k_3 \end{bmatrix}$

$AX = K$ has solution $X = A^{-1}K$.
We find $A^{-1}K$ for each given K. From problem 37 in Section 9-3,

$A^{-1} = \begin{bmatrix} -19 & 9 & -7 \\ 15 & -7 & 6 \\ -2 & 1 & -1 \end{bmatrix}$

(A) $K = \begin{bmatrix} 0 \\ 1 \\ 4 \end{bmatrix}$ $A^{-1}K = \begin{bmatrix} -19 & 9 & -7 \\ 15 & -7 & 6 \\ -2 & 1 & -1 \end{bmatrix} \begin{bmatrix} 0 \\ 1 \\ 4 \end{bmatrix} = \begin{bmatrix} -19 \\ 17 \\ -3 \end{bmatrix}$ $x_1 = -19,\ x_2 = 17,\ x_3 = -3$

(B) $K = \begin{bmatrix} 5 \\ -1 \\ 0 \end{bmatrix}$ $A^{-1}K = \begin{bmatrix} -19 & 9 & -7 \\ 15 & -7 & 6 \\ -2 & 1 & -1 \end{bmatrix} \begin{bmatrix} 5 \\ -1 \\ 0 \end{bmatrix} = \begin{bmatrix} -104 \\ 82 \\ -11 \end{bmatrix}$ $x_1 = -104,\ x_2 = 82,\ x_3 = -11$

(C) $K = \begin{bmatrix} -6 \\ 0 \\ 2 \end{bmatrix}$ $A^{-1}K = \begin{bmatrix} -19 & 9 & -7 \\ 15 & -7 & 6 \\ -2 & 1 & -1 \end{bmatrix} \begin{bmatrix} -6 \\ 0 \\ 2 \end{bmatrix} = \begin{bmatrix} 100 \\ -78 \\ 10 \end{bmatrix}$ $x_1 = 100,\ x_2 = -78,\ x_3 = 10$

35. The system cannot be solved by matrix inverse methods because the matrix of the system, $\begin{bmatrix} -2 & 4 \\ 6 & -12 \end{bmatrix}$, is singular. (The second row is -3 times the first, so row-reduction will result in a row of zeros.) It can be solved by Gauss-Jordan elimination, as follows.

$\begin{bmatrix} -2 & 4 & | & -5 \\ 6 & -12 & | & 15 \end{bmatrix} 3R_1 + R_2 \rightarrow R_2$

$\sim \begin{bmatrix} -2 & 4 & | & -5 \\ 0 & 0 & | & 0 \end{bmatrix} (-0.5)R_1 \rightarrow R_1$

$\sim \begin{bmatrix} 1 & -2 & | & 2.5 \\ 0 & 0 & | & 0 \end{bmatrix}$

Let $x_2 = t$.   Then $x_1 - 2x_2 = 2.5$
$$x_1 = 2.5 + 2x_2$$
$$= 2.5 + 2t$$
Solution: $x_1 = 2.5 + 2t$, $x_2 = t$, t any real number

37. The system cannot be solved by matrix inverse methods because it does not have the same number of variables as equations. So the coefficient matrix is not square and doesn't have an inverse. Applying Gauss-Jordan elimination, we obtain

$$\begin{bmatrix} 1 & -3 & -2 & | & -1 \\ -2 & 6 & 4 & | & 3 \end{bmatrix} 2R_1 + R_2 \rightarrow R_2$$

$$\sim \begin{bmatrix} 1 & -3 & -2 & | & -1 \\ 0 & 0 & 0 & | & 1 \end{bmatrix}$$

Since the last row corresponds to the equation $0x_1 + 0x_2 + 0x_3 = 1$, there is no solution.

39. The system cannot be solved by matrix inverse methods because the matrix of the system, $\begin{bmatrix} 1 & -2 & 3 \\ 2 & -3 & -2 \\ 1 & -1 & -5 \end{bmatrix}$, is singular. (Note that the reduction below results in a row of zeros.) It can be solved by Gauss-Jordan elimination, as follows

$$\begin{bmatrix} 1 & -2 & 3 & | & 1 \\ 2 & -3 & -2 & | & 3 \\ 1 & -1 & -5 & | & 2 \end{bmatrix} \begin{matrix} \\ (-2)R_1 + R_2 \rightarrow R_2 \\ (-1)R_1 + R_3 \rightarrow R_3 \end{matrix}$$

$$\sim \begin{bmatrix} 1 & -2 & 3 & | & 1 \\ 0 & 1 & -8 & | & 1 \\ 0 & 1 & -8 & | & 1 \end{bmatrix} \begin{matrix} 2R_2 + R_1 \rightarrow R_1 \\ \\ (-1)R_2 + R_3 \rightarrow R_3 \end{matrix}$$

$$\sim \begin{bmatrix} 1 & 0 & -13 & | & 3 \\ 0 & 1 & -8 & | & 1 \\ 0 & 0 & 0 & | & 0 \end{bmatrix}$$

Let $x_3 = t$. Then
$$x_2 - 8x_3 = 1$$
$$x_2 = 8x_3 + 1$$
$$= 8t + 1$$
$$x_1 - 13x_3 = 3$$
$$x_1 = 13x_3 + 3$$
$$= 13t + 3$$
Solution: $x_1 = 13t + 3$, $x_2 = 8t + 1$, $x_3 = t$, t any real number

41. $AX = BX + C$

$AX + (-BX) = (-BX) + BX + C$	Addition property of equality
$AX - BX = 0 + C$	Additive inverse property; definition of subtraction
$AX - BX = C$	
$(A - B)X = C$	Right distributive property

[To be more careful, we should write
$AX - BX = AX + -BX$ by the definition of subtraction
$\quad\quad\quad\quad = [A + (-B)]X$ by the right distributive property
$\quad\quad\quad\quad = (A - B)X$ by the definition of subtraction
but the distributive properties are generally understood as applying to subtraction also.]

$(A - B)^{-1}[(A - B)X] = (A - B)^{-1}C$	Left multiplication property of equality
$[(A - B)^{-1}(A - B)]X = (A - B)^{-1}C$	Associative property
$IX = (A - B)^{-1}C$	Multiplicative inverse property
$X = (A - B)^{-1}C$	Multiplicative identity property

43.
$$X = AX + C$$
$$X + (-AX) = (-AX) + AX + C \quad \text{Addition property of equality}$$
$$X - AX = 0 + C \quad \text{Additive inverse property;}$$
$$\text{definition of subtraction}$$
$$X - AX = C \quad \text{Additive identity property}$$
$$IX - AX = C \quad \text{Multiplicative identity property}$$
$$(I - A)X = C \quad \text{Right distributive property}$$
$$(I - A)^{-1}[(I - A)X] = (I - A)^{-1} C \quad \text{Left multiplication property of equality}$$
$$[(I - A)^{-1}(I - A)]X = (I - A)^{-1} C \quad \text{Associative property}$$
$$IX = (I - A)^{-1} C \quad \text{Multiplicative inverse property}$$
$$X = (I - A)^{-1} C \quad \text{Multiplicative identity property}$$

45.
$$AX + C = 3X$$
$$AX + (-AX) + C = 3X + (-AX) \quad \text{Addition property of equality}$$
$$0 + C = 3X - AX \quad \text{Additive inverse property;}$$
$$\text{definition of subtraction}$$
$$C = 3X - AX \quad \text{Additive identity property}$$
$$C = 3IX - AX \quad \text{Multiplicative identity property}$$
$$C = (3I - A)X \quad \text{Right distributive property}$$
$$(3I - A)^{-1} C = (3I - A)^{-1}[(3I - A)X] \quad \text{Left multiplication property of equality}$$
$$(3I - A)^{-1}C = [(3I - A)^{-1}(3I - A)]X \quad \text{Associative property}$$
$$(3I - A)^{-1}C = IX \quad \text{Multiplicative inverse property}$$
$$(3I - A)^{-1}C = X \quad \text{Multiplicative inverse property}$$

47.
$$\begin{bmatrix} 1 & 2.001 \\ 1 & 2 \end{bmatrix} \begin{bmatrix} x_1 \\ x_2 \end{bmatrix} = \begin{bmatrix} k_1 \\ k_2 \end{bmatrix}$$

$AX = K$ has solution $X = A^{-1}K$
We find A^{-1} as usual.

$$\begin{bmatrix} 1 & 2.001 & | & 1 & 0 \\ 1 & 2 & | & 0 & 1 \end{bmatrix} \quad -R_1 + R_2 \to R_2$$

$$\sim \begin{bmatrix} 1 & 2.001 & | & 1 & 0 \\ 0 & -0.001 & | & -1 & 1 \end{bmatrix} \quad -1{,}000R_2 \to R_2$$

$$\sim \begin{bmatrix} 1 & 2.001 & | & 1 & 0 \\ 0 & 1 & | & 1000 & -1000 \end{bmatrix} \quad -2.001R_2 \to R_2$$

$$\sim \begin{bmatrix} 1 & 0 & | & -2{,}000 & 2{,}001 \\ 0 & 1 & | & 1{,}000 & -1{,}000 \end{bmatrix} \quad A^{-1} = \begin{bmatrix} -2{,}000 & 2{,}001 \\ 1{,}000 & -1{,}000 \end{bmatrix}$$

We find $A^{-1}K$ for each given K.

(A) $K = \begin{bmatrix} 1 \\ 1 \end{bmatrix}$ $A^{-1}K = \begin{bmatrix} -2{,}000 & 2{,}001 \\ 1{,}000 & -1{,}000 \end{bmatrix} \begin{bmatrix} 1 \\ 1 \end{bmatrix} = \begin{bmatrix} 1 \\ 0 \end{bmatrix} = \begin{bmatrix} x_1 \\ x_2 \end{bmatrix}$ $x_1 = 1,\ x_2 = 0$

(B) $K = \begin{bmatrix} 1 \\ 0 \end{bmatrix}$ $A^{-1}K = \begin{bmatrix} -2{,}000 & 2{,}001 \\ 1{,}000 & -1{,}000 \end{bmatrix} \begin{bmatrix} 1 \\ 0 \end{bmatrix} = \begin{bmatrix} -2{,}000 \\ 1{,}000 \end{bmatrix} = \begin{bmatrix} x_1 \\ x_2 \end{bmatrix}$ $x_1 = -2{,}000,\ x_2 = 1{,}000$

(C) $K = \begin{bmatrix} 0 \\ 1 \end{bmatrix}$ $A^{-1}K = \begin{bmatrix} -2{,}000 & 2{,}001 \\ 1{,}000 & -1{,}000 \end{bmatrix} \begin{bmatrix} 0 \\ 1 \end{bmatrix} = \begin{bmatrix} 2{,}001 \\ -1{,}000 \end{bmatrix} = \begin{bmatrix} x_1 \\ x_2 \end{bmatrix}$ $x_1 = 2{,}001,\ x_2 = -1{,}000$

Because of the size of the entries of A^{-1}, a small change in K has a magnified effect on the solutions of $AX = K$. Geometrically, the lines are very close to being parallel, so a small change in the position (y intercept) of one line displaces their point of intersection a great distance.

49. The system to be solved, for an arbitrary return, is derived as follows:

Let x_1 = number of \$4 tickets sold

$\qquad x_2$ = number of \$8 tickets sold

Then $x_1 + x_2 = 10,000$ number of seats

$\qquad 4x_1 + 8x_2 = k_2$ return required

We solve the system by writing it as a matrix equation.

$$\overset{A}{\begin{bmatrix} 1 & 1 \\ 4 & 8 \end{bmatrix}} \overset{X}{\begin{bmatrix} x_1 \\ x_2 \end{bmatrix}} = \overset{B}{\begin{bmatrix} 10,000 \\ k_2 \end{bmatrix}}$$

If A^{-1} exists, then $X = A^{-1}B$. To find A^{-1}, we perform row operations on

$$\begin{bmatrix} 1 & 1 & | & 1 & 0 \\ 4 & 8 & | & 0 & 1 \end{bmatrix} (-4)R_1 + R_2 \rightarrow R_2$$

$$\sim \begin{bmatrix} 1 & 1 & | & 1 & 0 \\ 0 & 4 & | & -4 & 1 \end{bmatrix} 0.25R_2 \rightarrow R_2$$

$$\sim \begin{bmatrix} 1 & 1 & | & 1 & 0 \\ 0 & 1 & | & -1 & 0.25 \end{bmatrix} (-1)R_2 + R_1 \rightarrow R_1$$

$$\sim \begin{bmatrix} 1 & 0 & | & 2 & -0.25 \\ 0 & 1 & | & -1 & 0.25 \end{bmatrix}$$

$$A^{-1} = \begin{bmatrix} 2 & -0.25 \\ -1 & 0.25 \end{bmatrix}$$

Check: $A^{-1}A = \begin{bmatrix} 2 & -0.25 \\ -1 & 0.25 \end{bmatrix}\begin{bmatrix} 1 & 4 \\ 1 & 8 \end{bmatrix} = \begin{bmatrix} 1 & 0 \\ 0 & 1 \end{bmatrix}$

We can now solve the system as

$$\overset{X}{\begin{bmatrix} x_1 \\ x_2 \end{bmatrix}} = \overset{A^{-1}}{\begin{bmatrix} 2 & -0.25 \\ -1 & 0.25 \end{bmatrix}} \overset{B}{\begin{bmatrix} 10,000 \\ k_2 \end{bmatrix}}$$

If $k_2 = 56,000$ (Concert 1),

$$\begin{bmatrix} x_1 \\ x_2 \end{bmatrix} = \begin{bmatrix} 2 & -0.25 \\ -1 & 0.25 \end{bmatrix}\begin{bmatrix} 10,000 \\ 56,000 \end{bmatrix} = \begin{bmatrix} 6,000 \\ 4,000 \end{bmatrix}$$

Concert 1: 6,000 \$4 tickets and 4,000 \$8 tickets

If $k_2 = 60,000$ (Concert 2),

$$\begin{bmatrix} x_1 \\ x_2 \end{bmatrix} = \begin{bmatrix} 2 & -0.25 \\ -1 & 0.25 \end{bmatrix}\begin{bmatrix} 10,000 \\ 60,000 \end{bmatrix} = \begin{bmatrix} 5,000 \\ 5,000 \end{bmatrix}$$

Concert 2: 5,000 \$4 tickets and 5,000 \$8 tickets

If $k_2 = 68,000$ (Concert 3),

$$\begin{bmatrix} x_1 \\ x_2 \end{bmatrix} = \begin{bmatrix} 2 & -0.25 \\ -1 & 0.25 \end{bmatrix}\begin{bmatrix} 10,000 \\ 68,000 \end{bmatrix} = \begin{bmatrix} 3,000 \\ 7,000 \end{bmatrix}$$

Concert 3: 3,000 \$4 tickets and 7,000 \$8 tickets

51. We solve the system, for arbitrary V_1 and V_2, by writing it as a matrix equation.

$$\overset{A}{\begin{bmatrix} 1 & -1 & 1 \\ 1 & 1 & 0 \\ 0 & 1 & 2 \end{bmatrix}} \overset{J}{\begin{bmatrix} I_1 \\ I_2 \\ I_3 \end{bmatrix}} = \overset{B}{\begin{bmatrix} 0 \\ V_1 \\ V_2 \end{bmatrix}}$$

If A^{-1} exists, then $J = A^{-1}B$. To find A^{-1}, we perform row operations on

$$\begin{bmatrix} 1 & -1 & 1 & | & 1 & 0 & 0 \\ 1 & 1 & 0 & | & 0 & 1 & 0 \\ 0 & 1 & 2 & | & 0 & 0 & 1 \end{bmatrix} (-1)R_1 + R_2 \rightarrow R_2$$

$$\sim \begin{bmatrix} 1 & -1 & 1 & | & 1 & 0 & 0 \\ 0 & 2 & -1 & | & -1 & 1 & 0 \\ 0 & 1 & 2 & | & 0 & 0 & 1 \end{bmatrix} R_2 \leftrightarrow R_3$$

$$\sim \begin{bmatrix} 1 & -1 & 1 & | & 1 & 0 & 0 \\ 0 & 1 & 2 & | & 0 & 0 & 1 \\ 0 & 2 & -1 & | & -1 & 1 & 0 \end{bmatrix} \begin{matrix} R_2 + R_1 \rightarrow R_1 \\ \\ (-2)R_2 + R_3 \rightarrow R_3 \end{matrix}$$

$$\sim \begin{bmatrix} 1 & 0 & 3 & | & 1 & 0 & 1 \\ 0 & 1 & 2 & | & 0 & 0 & 1 \\ 0 & 0 & -5 & | & -1 & 1 & -2 \end{bmatrix} -\frac{1}{5} R_3 \rightarrow R_3$$

$$\sim \begin{bmatrix} 1 & 0 & 3 & | & 1 & 0 & 1 \\ 0 & 1 & 2 & | & 0 & 0 & 1 \\ 0 & 0 & 1 & | & \frac{1}{5} & -\frac{1}{5} & \frac{2}{5} \end{bmatrix} \begin{matrix} (-3)R_3 + R_1 \rightarrow R_1 \\ (-2)R_3 + R_2 \rightarrow R_2 \end{matrix}$$

$$\sim \begin{bmatrix} 1 & 0 & 0 & | & \frac{2}{5} & \frac{3}{5} & -\frac{1}{5} \\ 0 & 1 & 0 & | & -\frac{2}{5} & \frac{2}{5} & \frac{1}{5} \\ 0 & 0 & 1 & | & \frac{1}{5} & -\frac{1}{5} & \frac{2}{5} \end{bmatrix}$$

$$A^{-1} = \frac{1}{5} \begin{bmatrix} 2 & 3 & -1 \\ -2 & 2 & 1 \\ 1 & -1 & 2 \end{bmatrix}$$

Check: $A^{-1}A = \frac{1}{5} \begin{bmatrix} 2 & 3 & -1 \\ -2 & 2 & 1 \\ 1 & -1 & 2 \end{bmatrix} \begin{bmatrix} 1 & -1 & 1 \\ 1 & 1 & 0 \\ 0 & 1 & 2 \end{bmatrix} = \begin{bmatrix} 1 & 0 & 0 \\ 0 & 1 & 0 \\ 0 & 0 & 1 \end{bmatrix}$

We can now solve the system as

$$\begin{matrix} J & \quad & A^{-1} & \quad & B \end{matrix}$$
$$\begin{bmatrix} I_1 \\ I_2 \\ I_3 \end{bmatrix} = \frac{1}{5} \begin{bmatrix} 2 & 3 & -1 \\ -2 & 2 & 1 \\ 1 & -1 & 2 \end{bmatrix} \begin{bmatrix} 0 \\ V_1 \\ V_2 \end{bmatrix}$$

(A) $V_1 = 10$ $V_2 = 10$

$$\begin{bmatrix} I_1 \\ I_2 \\ I_3 \end{bmatrix} = \frac{1}{5} \begin{bmatrix} 2 & 3 & -1 \\ -2 & 2 & 1 \\ 1 & -1 & 2 \end{bmatrix} \begin{bmatrix} 0 \\ 10 \\ 10 \end{bmatrix} = \begin{bmatrix} 4 \\ 6 \\ 2 \end{bmatrix} \qquad I_1 = 4, \; I_2 = 6, \; I_3 = 2 \text{ (amperes)}$$

(B) $V_1 = 10$ $V_2 = 15$

$$\begin{bmatrix} I_1 \\ I_2 \\ I_3 \end{bmatrix} = \frac{1}{5} \begin{bmatrix} 2 & 3 & -1 \\ -2 & 2 & 1 \\ 1 & -1 & 2 \end{bmatrix} \begin{bmatrix} 0 \\ 10 \\ 15 \end{bmatrix} = \begin{bmatrix} 3 \\ 7 \\ 4 \end{bmatrix} \qquad I_1 = 3, \; I_2 = 7, \; I_3 = 4 \text{ (amperes)}$$

(C) $V_1 = 15$ $V_2 = 10$

$$\begin{bmatrix} I_1 \\ I_2 \\ I_3 \end{bmatrix} = \frac{1}{5} \begin{bmatrix} 2 & 3 & -1 \\ -2 & 2 & 1 \\ 1 & -1 & 2 \end{bmatrix} \begin{bmatrix} 0 \\ 15 \\ 10 \end{bmatrix} = \begin{bmatrix} 7 \\ 8 \\ 1 \end{bmatrix} \qquad I_1 = 7, \; I_2 = 8, \; I_3 = 1 \text{ (amperes)}$$

53. If the graph of $f(x) = ax^2 + bx + c$ passes through a point, the coordinates of the point must satisfy the equation of the graph. Then

$$k_1 = a(1)^2 + b(1) + c$$

$$k_2 = a(2)^2 + b(2) + c$$

$$k_3 = a(3)^2 + b(3) + c$$

After simplification, we obtain:

$$a + b + c = k_1$$

$$4a + 2b + c = k_2$$

$$9a + 3b + c = k_3$$

We solve this system, for arbitrary k_1, k_2, k_3, by writing it as a matrix equation.

$$\begin{array}{ccc} A & X & B \end{array}$$
$$\begin{bmatrix} 1 & 1 & 1 \\ 4 & 2 & 1 \\ 9 & 3 & 1 \end{bmatrix} \begin{bmatrix} a \\ b \\ c \end{bmatrix} = \begin{bmatrix} k_1 \\ k_2 \\ k_3 \end{bmatrix}$$

If A^{-1} exists, then $X = A^{-1}B$. To find A^{-1} we perform row operations on

$$\begin{bmatrix} 1 & 1 & 1 & | & 1 & 0 & 0 \\ 4 & 2 & 1 & | & 0 & 1 & 0 \\ 9 & 3 & 1 & | & 0 & 0 & 1 \end{bmatrix} \begin{array}{l} \\ (-4)R_1 + R_2 \rightarrow R_2 \\ (-9)R_1 + R_3 \rightarrow R_3 \end{array}$$

$$\sim \begin{bmatrix} 1 & 1 & 1 & | & 1 & 0 & 0 \\ 0 & -2 & -3 & | & -4 & 1 & 0 \\ 0 & -6 & -8 & | & -9 & 0 & 1 \end{bmatrix} -\tfrac{1}{2}R_2 \rightarrow R_2$$

$$\sim \begin{bmatrix} 1 & 1 & 1 & | & 1 & 0 & 0 \\ 0 & 1 & \tfrac{3}{2} & | & 2 & -\tfrac{1}{2} & 0 \\ 0 & -6 & -8 & | & -9 & 0 & 1 \end{bmatrix} \begin{array}{l} (-1)R_2 + R_1 \rightarrow R_1 \\ \\ 6R_2 + R_3 \rightarrow R_3 \end{array}$$

$$\sim \begin{bmatrix} 1 & 0 & -\tfrac{1}{2} & | & -1 & \tfrac{1}{2} & 0 \\ 0 & 1 & \tfrac{3}{2} & | & 2 & -\tfrac{1}{2} & 0 \\ 0 & 0 & 1 & | & 3 & -3 & 1 \end{bmatrix} \begin{array}{l} \tfrac{1}{2}R_3 + R_1 \rightarrow R_1 \\ (-\tfrac{3}{2})R_3 + R_2 \rightarrow R_2 \end{array}$$

$$\sim \begin{bmatrix} 1 & 0 & 0 & | & \tfrac{1}{2} & -1 & \tfrac{1}{2} \\ 0 & 1 & 0 & | & -\tfrac{5}{2} & 4 & -\tfrac{3}{2} \\ 0 & 0 & 1 & | & 3 & -3 & 1 \end{bmatrix}$$

$$A^{-1} = \tfrac{1}{2}\begin{bmatrix} 1 & -2 & 1 \\ -5 & 8 & -3 \\ 6 & -6 & 2 \end{bmatrix}$$

Check: $A^{-1}A = \tfrac{1}{2}\begin{bmatrix} 1 & -2 & 1 \\ -5 & 8 & -3 \\ 6 & -6 & 2 \end{bmatrix}\begin{bmatrix} 1 & 1 & 1 \\ 4 & 2 & 1 \\ 9 & 3 & 1 \end{bmatrix} = \begin{bmatrix} 1 & 0 & 0 \\ 0 & 1 & 0 \\ 0 & 0 & 1 \end{bmatrix}$

We can now solve the system as

$$\begin{array}{ccc} X & A^{-1} & B \end{array}$$
$$\begin{bmatrix} a \\ b \\ c \end{bmatrix} = \tfrac{1}{2}\begin{bmatrix} 1 & -2 & 1 \\ -5 & 8 & -3 \\ 6 & -6 & 2 \end{bmatrix}\begin{bmatrix} k_1 \\ k_2 \\ k_3 \end{bmatrix}$$

(A) $\begin{bmatrix} a \\ b \\ c \end{bmatrix} = \tfrac{1}{2}\begin{bmatrix} 1 & -2 & 1 \\ -5 & 8 & -3 \\ 6 & -6 & 2 \end{bmatrix}\begin{bmatrix} -2 \\ 1 \\ 6 \end{bmatrix} = \begin{bmatrix} 1 \\ 0 \\ -3 \end{bmatrix}$ $a = 1$, $b = 0$, $c = -3$

(B) $\begin{bmatrix} a \\ b \\ c \end{bmatrix} = \frac{1}{2}\begin{bmatrix} 1 & -2 & 1 \\ -5 & 8 & -3 \\ 6 & -6 & 2 \end{bmatrix}\begin{bmatrix} 4 \\ 3 \\ -2 \end{bmatrix} = \begin{bmatrix} -2 \\ 5 \\ 1 \end{bmatrix}$ $a = -2, \ b = 5, \ c = 1$

(C) $\begin{bmatrix} a \\ b \\ c \end{bmatrix} = \frac{1}{2}\begin{bmatrix} 1 & -2 & 1 \\ -5 & 8 & -3 \\ 6 & -6 & 2 \end{bmatrix}\begin{bmatrix} 8 \\ -5 \\ 4 \end{bmatrix} = \begin{bmatrix} 11 \\ -46 \\ 43 \end{bmatrix}$ $a = 11, \ b = -46, \ c = 43$

Using Quadratic Regression

Enter 1, 2, 3 as L_1.

(A) Enter -2, 1, 6 as L_2 and use the quadratic regression command from the STAT CALC menu.

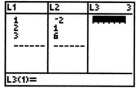 

(B) Enter 4, 3, -2 as L_2 and repeat.

(C) Enter 8, -5, 4 as L_2 and repeat.

55. The system to be solved, for an arbitrary diet, is derived as follows:
Let x_1 = amount of mix A
    x_2 = amount of mix B
Then $0.20x_1 + 0.10x_2 = k_1$ (k_1 = amount of protein)
     $0.02x_1 + 0.06x_2 = k_2$ (k_2 = amount of fat)
We solve the system by writing it as a matrix equation.

$$\overset{A}{\begin{bmatrix} 0.20 & 0.10 \\ 0.02 & 0.06 \end{bmatrix}} \overset{X}{\begin{bmatrix} x_1 \\ x_2 \end{bmatrix}} = \overset{B}{\begin{bmatrix} k_1 \\ k_2 \end{bmatrix}}$$

If A^{-1} exists, then $X = A^{-1}B$. To find A^{-1}, we perform row operations on

$\begin{bmatrix} 0.20 & 0.10 & | & 1 & 0 \\ 0.02 & 0.06 & | & 0 & 1 \end{bmatrix} \begin{matrix} 5R_1 \to R_1 \\ 50R_2 \to R_2 \end{matrix}$

$\sim \begin{bmatrix} 1 & 0.5 & | & 5 & 0 \\ 1 & 3 & | & 0 & 50 \end{bmatrix} (-1)R_1 + R_2 \to R_2$

$\sim \begin{bmatrix} 1 & 0.5 & | & 5 & 0 \\ 0 & 2.5 & | & -5 & 50 \end{bmatrix} 0.4R_2 \to R_2$

$\sim \begin{bmatrix} 1 & 0.5 & | & 5 & 0 \\ 0 & 1 & | & -2 & 20 \end{bmatrix} (-0.5)R_2 + R_1 \to R_1$

$$\sim \begin{bmatrix} 1 & 0 & | & 6 & -10 \\ 0 & 1 & | & -2 & 20 \end{bmatrix}$$

$$A^{-1} = \begin{bmatrix} 6 & -10 \\ -2 & 20 \end{bmatrix}$$

Check: $A^{-1}A = \begin{bmatrix} 6 & -10 \\ -2 & 20 \end{bmatrix} \begin{bmatrix} 0.20 & 0.10 \\ 0.02 & 0.06 \end{bmatrix} = \begin{bmatrix} 1 & 0 \\ 0 & 1 \end{bmatrix}$

We can now solve the system as

$$\begin{array}{ccc} X & A^{-1} & B \end{array}$$

$$\begin{bmatrix} x_1 \\ x_2 \end{bmatrix} = \begin{bmatrix} 6 & -10 \\ -2 & 20 \end{bmatrix} \begin{bmatrix} k_1 \\ k_2 \end{bmatrix}$$

For Diet 1, $k_1 = 20$ and $k_2 = 6$

$$\begin{bmatrix} x_1 \\ x_2 \end{bmatrix} = \begin{bmatrix} 6 & -10 \\ -2 & 20 \end{bmatrix} \begin{bmatrix} 20 \\ 6 \end{bmatrix} = \begin{bmatrix} 60 \\ 80 \end{bmatrix}$$ Diet 1: 60 ounces Mix A and 80 ounces Mix B

For Diet 2, $k_1 = 10$ and $k_2 = 4$

$$\begin{bmatrix} x_1 \\ x_2 \end{bmatrix} = \begin{bmatrix} 6 & -10 \\ -2 & 20 \end{bmatrix} \begin{bmatrix} 10 \\ 4 \end{bmatrix} = \begin{bmatrix} 20 \\ 60 \end{bmatrix}$$ Diet 2: 20 ounces Mix A and 60 ounces Mix B

For Diet 3, $k_1 = 10$ and $k_2 = 6$

$$\begin{bmatrix} x_1 \\ x_2 \end{bmatrix} = \begin{bmatrix} 6 & -10 \\ -2 & 20 \end{bmatrix} \begin{bmatrix} 10 \\ 6 \end{bmatrix} = \begin{bmatrix} 0 \\ 100 \end{bmatrix}$$ Diet 3: 0 ounces Mix A and 100 ounces Mix B

Section 9-5

1. Answers will vary.   **3.** Answers will vary.   **5.** Answers will vary.

7. $\begin{vmatrix} 5 & 4 \\ 2 & 3 \end{vmatrix} = 5 \cdot 3 - 2 \cdot 4 = 7$ **9.** $\begin{vmatrix} 3 & -7 \\ -5 & 6 \end{vmatrix} = 3 \cdot 6 - (-5)(-7) = -17$

11. $\begin{vmatrix} 4.3 & -1.2 \\ -5.1 & 3.7 \end{vmatrix} = (4.3)(3.7) - (-5.1)(-1.2) = 9.79$

13. $\begin{vmatrix} 5 & 1 & 3 \\ 3 & 4 & 6 \\ 0 & -2 & 8 \end{vmatrix} = \begin{vmatrix} 4 & 6 \\ -2 & 8 \end{vmatrix}$ **15.** $\begin{vmatrix} 5 & -1 & -3 \\ 3 & 4 & 6 \\ 0 & -2 & 8 \end{vmatrix} = \begin{vmatrix} 5 & -1 \\ 0 & -2 \end{vmatrix}$

17. $(-1)^{1+1} \begin{vmatrix} 4 & 6 \\ -2 & 8 \end{vmatrix} = (-1)^2[4 \cdot 8 - (-2)6] = 44$

19. $(-1)^{2+3} \begin{vmatrix} 5 & -1 \\ 0 & -2 \end{vmatrix} = (-1)^5[5(-2) - 0(-1)] = 10$

21. We expand by row 1

$$\begin{vmatrix} 1 & 0 & 0 \\ -2 & 4 & 3 \\ 5 & -2 & 1 \end{vmatrix} = a_{11}(\text{cofactor of } a_{11}) + a_{12}(\text{cofactor of } a_{12}) + a_{13}(\text{cofactor of } a_{13})$$

$$= 1(-1)^{1+1} \begin{vmatrix} 4 & 3 \\ -2 & 1 \end{vmatrix} + 0(\nearrow) + 0(\searrow)$$

It is unnecessary to evaluate these since they are multiplied by 0.

$$= (-1)^2[4 \cdot 1 - (-2)3]$$
$$= 10$$

23. We expand by column 1

$$\begin{vmatrix} 0 & 1 & 5 \\ 3 & -7 & 6 \\ 0 & -2 & -3 \end{vmatrix} = a_{11}(\text{cofactor of } a_{11}) + a_{21}(\text{cofactor of } a_{21}) + a_{31}(\text{cofactor of } a_{31})$$

$$= 0(\diagup) + 3(-1)^{2+1}\begin{vmatrix} 1 & 5 \\ -2 & -3 \end{vmatrix} + 0(\diagdown)$$

It is unnecessary to evaluate these since they are multiplied by 0.

$$= 3(-1)^3[1(-3) - (-2)5]$$
$$= -21$$

> **Common Error:** Neglecting the sign of the cofactor. The cofactor is often called the "signed" minor.

25. We expand by column 2

$$\begin{vmatrix} -1 & 2 & -3 \\ -2 & 0 & -6 \\ 4 & -3 & 2 \end{vmatrix} = a_{12}(\text{cofactor of } a_{12}) + a_{22}(\text{cofactor of } a_{22}) + a_{32}(\text{cofactor of } a_{32})$$

$$= 2(-1)^{1+2}\begin{vmatrix} -2 & -6 \\ 4 & 2 \end{vmatrix} + 0(\diagup) + (-3)(-1)^{3+2}\begin{vmatrix} -1 & -3 \\ -2 & -6 \end{vmatrix}$$

It is unnecessary to evaluate this since it's multiplied by zero.

$$= 2(-1)^3[(-2)2 - 4(-6)] + (-3)(-1)^5[(-1)(-6) - (-2)(-3)]$$
$$= (-2)(20) + 3(0) = -40$$

27. $(-1)^{1+1}\begin{vmatrix} a_{11} & a_{12} & a_{13} & a_{14} \\ a_{21} & a_{22} & a_{23} & a_{24} \\ a_{31} & a_{32} & a_{33} & a_{34} \\ a_{41} & a_{42} & a_{43} & a_{44} \end{vmatrix} = (-1)^{1+1}\begin{vmatrix} a_{22} & a_{23} & a_{24} \\ a_{32} & a_{33} & a_{34} \\ a_{42} & a_{43} & a_{44} \end{vmatrix}$

29. $(-1)^{4+3}\begin{vmatrix} a_{11} & a_{12} & a_{13} & a_{14} \\ a_{21} & a_{22} & a_{23} & a_{24} \\ a_{31} & a_{32} & a_{33} & a_{34} \\ a_{41} & a_{42} & a_{43} & a_{44} \end{vmatrix} = (-1)^{4+3}\begin{vmatrix} a_{11} & a_{12} & a_{14} \\ a_{21} & a_{22} & a_{24} \\ a_{31} & a_{32} & a_{34} \end{vmatrix}$

31. We expand by the second column

$$\begin{vmatrix} 3 & -2 & -8 \\ -2 & 0 & -3 \\ 1 & 0 & -4 \end{vmatrix} = a_{12}(\text{cofactor of } a_{12}) + a_{22}(\text{cofactor of } a_{22}) + a_{32}(\text{cofactor of } a_{32})$$

$$= (-2)(-1)^{1+2}\begin{vmatrix} -2 & -3 \\ 1 & -4 \end{vmatrix} + 0 + 0$$

$$= (-2)(-1)^3[(-2)(-4) - 1(-3)]$$
$$= 2(11)$$
$$= 22$$

33. We expand by the first row

$$\begin{vmatrix} 1 & 4 & 1 \\ 1 & 1 & -2 \\ 2 & 1 & -1 \end{vmatrix} = a_{11}(\text{cofactor of } a_{11}) + a_{12}(\text{cofactor of } a_{12}) + a_{13}(\text{cofactor of } a_{13})$$

$$= 1(-1)^{1+1}\begin{vmatrix} 1 & -2 \\ 1 & -1 \end{vmatrix} + 4(-1)^{1+2}\begin{vmatrix} 1 & -2 \\ 2 & -1 \end{vmatrix} + 1(-1)^{1+3}\begin{vmatrix} 1 & 1 \\ 2 & 1 \end{vmatrix}$$

$$= (-1)^2[1(-1) - 1(-2)] + 4(-1)^3[1(-1) - 2(-2)] + (-1)^4[1 \cdot 1 - 2 \cdot 1]$$
$$= 1 + (-12) + (-1)$$
$$= -12$$

35. We expand by the first row

$$\begin{vmatrix} 1 & 4 & 3 \\ 2 & 1 & 6 \\ 3 & -2 & 9 \end{vmatrix} = a_{11}(\text{cofactor of } a_{11}) + a_{12}(\text{cofactor of } a_{12}) + a_{13}(\text{cofactor of } a_{13})$$

$$= 1(-1)^{1+1}\begin{vmatrix} 1 & 6 \\ -2 & 9 \end{vmatrix} + 4(-1)^{1+2}\begin{vmatrix} 2 & 6 \\ 3 & 9 \end{vmatrix} + 3(-1)^{1+3}\begin{vmatrix} 2 & 1 \\ 3 & -2 \end{vmatrix}$$

$$= (-1)^2[1\cdot9 - (-2)6] + 4(-1)^3[2\cdot9 - 3\cdot6] + 3(-1)^4[2(-2) - 1\cdot3]$$

$$= 21 + 0 - 21$$

$$= 0$$

37. We expand by the second row. Clearly the only non-zero term will be a_{22} (cofactor of a_{22}), which is

$$3(-1)^{2+2}\begin{vmatrix} 2 & 1 & 7 \\ 3 & 2 & 5 \\ 0 & 0 & 2 \end{vmatrix}$$

The order 3 determinant is expanded by the third row. Again there is only one non-zero term, a_{33} (cofactor of a_{33}). So the original determinant is reduced to

$$3(-1)^{2+2}\ 2(-1)^{3+3}\begin{vmatrix} 2 & 1 \\ 3 & 2 \end{vmatrix} = 6(-1)^{10}(2\cdot2 - 3\cdot1) = 6$$

39. In every case, we'll be expanding by the first row.

$$\begin{vmatrix} -2 & 0 & 0 & 0 & 0 \\ 9 & -1 & 0 & 0 & 0 \\ 2 & 1 & 3 & 0 & 0 \\ -1 & 4 & 2 & 2 & 0 \\ 7 & -2 & 3 & 5 & 5 \end{vmatrix} = (-2)(-1)^{1+1}\begin{vmatrix} -1 & 0 & 0 & 0 \\ 1 & 3 & 0 & 0 \\ 4 & 2 & 2 & 0 \\ -2 & 3 & 5 & 5 \end{vmatrix} + 0 \text{ terms}$$

$$= -2\begin{vmatrix} -1 & 0 & 0 & 0 \\ 1 & 3 & 0 & 0 \\ 4 & 2 & 2 & 0 \\ -2 & 3 & 5 & 5 \end{vmatrix} = (-2)\left[(-1)(-1)^{1+1}\begin{vmatrix} 3 & 0 & 0 \\ 2 & 2 & 0 \\ 3 & 5 & 5 \end{vmatrix} + 0 \text{ terms}\right]$$

$$= (-2)(-1)\begin{vmatrix} 3 & 0 & 0 \\ 2 & 2 & 0 \\ 3 & 5 & 5 \end{vmatrix} = (-2)(-1)\left[3(-1)^{1+1}\begin{vmatrix} 2 & 0 \\ 5 & 5 \end{vmatrix} + 0 \text{ terms}\right]$$

$$= (-2)(-1)3\begin{vmatrix} 2 & 0 \\ 5 & 5 \end{vmatrix}$$

$$= (-2)(-1)(3)[2\cdot5 - 5\cdot0]$$

$$= (-2)(-1)(3)(2)(5)$$

$$= 60$$

41.

$$\begin{vmatrix} 2 & 6 & -1 & 2 & 6 \\ 5 & 3 & -7 & 5 & 3 \\ -4 & -2 & 1 & -4 & -2 \end{vmatrix}$$

$$2\cdot3\cdot1 + 6(-7)(-4) + (-1)(5)(-2) - (-4)(3)(-1) - (-2)(-7)2 - 6\cdot5\cdot1$$

$$= 6 + 168 + 10 - 12 - 28 - 30 = 114$$

43. False. $\begin{vmatrix} 10 & 10 \\ 0 & 0 \end{vmatrix}$ is a counterexample.

45. True. Expanding $\begin{vmatrix} a_{11} & a_{12} & a_{13} & a_{14} \\ 0 & a_{22} & a_{23} & a_{24} \\ 0 & 0 & a_{33} & a_{34} \\ 0 & 0 & 0 & a_{44} \end{vmatrix}$ by the first column, we obtain successively

$$a_{11} \begin{vmatrix} a_{22} & a_{23} & a_{24} \\ 0 & a_{33} & a_{34} \\ 0 & 0 & a_{44} \end{vmatrix} = a_{11}a_{22} \begin{vmatrix} a_{33} & a_{34} \\ 0 & a_{44} \end{vmatrix} = a_{11}a_{22}a_{33}a_{44}.$$

Similarly for the determinant of an $n \times n$ upper triangular matrix, we would obtain $a_{11}a_{22}a_{33} \cdot \ldots \cdot a_{nn}$ as proposed.

47. $\begin{vmatrix} a & b \\ c & d \end{vmatrix} = ad - bc \qquad \begin{vmatrix} c & d \\ a & b \end{vmatrix} = cb - ad = -(ad - bc)$

We can see that $\begin{vmatrix} a & b \\ c & d \end{vmatrix} = -\begin{vmatrix} c & d \\ a & b \end{vmatrix}$; interchanging the rows of this determinant changes its sign.

49. $\begin{vmatrix} a & b \\ c & d \end{vmatrix} = ad - bc \qquad \begin{vmatrix} ka & b \\ kc & d \end{vmatrix} = kad - kcb = k(ad - bc)$

We can see that $\begin{vmatrix} ka & b \\ kc & d \end{vmatrix} = k\begin{vmatrix} a & b \\ c & d \end{vmatrix}$; multiplying a column of this determinant by a number k multiplies the value of the determinant by k.

51. $\begin{vmatrix} a & b \\ c & d \end{vmatrix} = ad - bc \qquad \begin{vmatrix} kc + a & kd + b \\ c & d \end{vmatrix} = (kc + a)d - (kd + b)c$

$$= kcd + ad - kdc - bc$$
$$= ad - bc$$

We can see that $\begin{vmatrix} kc + a & kd + b \\ c & d \end{vmatrix} = \begin{vmatrix} a & b \\ c & d \end{vmatrix}$; adding a multiple of one row to the other row does not change the value of this determinant.

53. Expanding by the first column

$\begin{vmatrix} a_{11} & a_{12} & a_{13} \\ a_{21} & a_{22} & a_{23} \\ a_{31} & a_{32} & a_{33} \end{vmatrix}$

$= a_{11}(-1)^{1+1}\begin{vmatrix} a_{22} & a_{23} \\ a_{32} & a_{33} \end{vmatrix} + a_{21}(-1)^{2+1}\begin{vmatrix} a_{12} & a_{13} \\ a_{32} & a_{33} \end{vmatrix} + a_{31}(-1)^{3+1}\begin{vmatrix} a_{12} & a_{13} \\ a_{22} & a_{23} \end{vmatrix}$

$= a_{11}\begin{vmatrix} a_{22} & a_{23} \\ a_{32} & a_{33} \end{vmatrix} - a_{21}\begin{vmatrix} a_{12} & a_{13} \\ a_{32} & a_{33} \end{vmatrix} + a_{31}\begin{vmatrix} a_{12} & a_{13} \\ a_{22} & a_{23} \end{vmatrix}$

$= a_{11}(a_{22}a_{33} - a_{32}a_{23}) - a_{21}(a_{12}a_{33} - a_{32}a_{13}) + a_{31}(a_{12}a_{23} - a_{22}a_{13})$

$= a_{11}a_{22}a_{33} - a_{11}a_{32}a_{23} - a_{21}a_{12}a_{33} + a_{21}a_{32}a_{13} + a_{31}a_{12}a_{23} - a_{31}a_{22}a_{13}$

     ①   ②   ③   ④   ⑤   ⑥

Expanding by the third row

$\begin{vmatrix} a_{11} & a_{12} & a_{13} \\ a_{21} & a_{22} & a_{23} \\ a_{31} & a_{32} & a_{33} \end{vmatrix}$

$= a_{31}(-1)^{3+1}\begin{vmatrix} a_{12} & a_{13} \\ a_{22} & a_{23} \end{vmatrix} + a_{32}(-1)^{3+2}\begin{vmatrix} a_{11} & a_{13} \\ a_{21} & a_{23} \end{vmatrix} + a_{33}(-1)^{3+3}\begin{vmatrix} a_{11} & a_{12} \\ a_{21} & a_{22} \end{vmatrix}$

$$= a_{31} \begin{vmatrix} a_{12} & a_{13} \\ a_{22} & a_{23} \end{vmatrix} - a_{32} \begin{vmatrix} a_{11} & a_{13} \\ a_{21} & a_{23} \end{vmatrix} + a_{33} \begin{vmatrix} a_{11} & a_{12} \\ a_{21} & a_{22} \end{vmatrix}$$

$$= a_{31}(a_{12}a_{23} - a_{13}a_{22}) - a_{32}(a_{11}a_{23} - a_{13}a_{21}) + a_{33}(a_{11}a_{22} - a_{12}a_{21})$$

$$= a_{31}a_{12}a_{23} - a_{31}a_{13}a_{22} - a_{32}a_{11}a_{23} + a_{32}a_{13}a_{21} + a_{33}a_{11}a_{22} - a_{33}a_{12}a_{21}$$

⑤   ⑥   ②   ④   ①   ③

Comparing the two expressions, with the aid of the numbers under the terms, shows that the expressions are the same.

55. $A = \begin{bmatrix} 2 & 3 \\ 1 & -2 \end{bmatrix}$ $B = \begin{bmatrix} -1 & 3 \\ 2 & 1 \end{bmatrix}$

We calculate $AB = \begin{bmatrix} 2 & 3 \\ 1 & -2 \end{bmatrix} \begin{bmatrix} -1 & 3 \\ 2 & 1 \end{bmatrix} = \begin{bmatrix} 2(-1) + 3 \cdot 2 & 2 \cdot 3 + 3 \cdot 1 \\ 1(-1) + (-2)2 & 1 \cdot 3 + (-2) \cdot 1 \end{bmatrix}$

$$= \begin{bmatrix} 4 & 9 \\ -5 & 1 \end{bmatrix}$$

$$\det (AB) = \begin{vmatrix} 4 & 9 \\ -5 & 1 \end{vmatrix} = 4 \cdot 1 - (-5)9 = 49$$

$$\det A = \begin{vmatrix} 2 & 3 \\ 1 & -2 \end{vmatrix} = 2(-2) - 1 \cdot 3 = -7$$

$$\det B = \begin{vmatrix} -1 & 3 \\ 2 & 1 \end{vmatrix} = (-1)1 - 2 \cdot 3 = -7$$

57. The matrix $xI - A$ is calculated as

$$x\begin{bmatrix} 1 & 0 \\ 0 & 1 \end{bmatrix} - \begin{bmatrix} 5 & -4 \\ 2 & -1 \end{bmatrix} = \begin{bmatrix} x & 0 \\ 0 & x \end{bmatrix} - \begin{bmatrix} 5 & -4 \\ 2 & -1 \end{bmatrix} = \begin{bmatrix} x - 5 & 4 \\ -2 & x + 1 \end{bmatrix}$$

The characteristic polynomial is the determinant of this matrix:

$$\begin{bmatrix} x - 5 & 4 \\ -2 & x + 1 \end{bmatrix} = (x - 5)(x + 1) - (4)(-2) = x^2 - 4x - 5 + 8 = x^2 - 4x + 3$$

The zeros of this polynomial are the solutions of $x^2 - 4x + 3 = 0$

$$x^2 - 4x + 3 = 0$$
$$(x - 1)(x - 3) = 0$$
$$x = 1, 3$$

The eigenvalues of this matrix are 1 and 3.

59. The matrix $xI - A$ is calculated as

$$x\begin{bmatrix} 1 & 0 & 0 \\ 0 & 1 & 0 \\ 0 & 0 & 1 \end{bmatrix} - \begin{bmatrix} 4 & -4 & 0 \\ 2 & -2 & 0 \\ 4 & -8 & -4 \end{bmatrix} = \begin{bmatrix} x & 0 & 0 \\ 0 & x & 0 \\ 0 & 0 & x \end{bmatrix} - \begin{bmatrix} 4 & -4 & 0 \\ 2 & -2 & 0 \\ 4 & -8 & -4 \end{bmatrix} = \begin{bmatrix} x - 4 & 4 & 0 \\ -2 & x + 2 & 0 \\ -4 & 8 & x + 4 \end{bmatrix}$$

The characteristic polynomial is the determinant of this matrix.

$$\begin{bmatrix} x - 4 & 4 & 0 \\ -2 & x + 2 & 0 \\ -4 & 8 & x + 4 \end{bmatrix} = (x + 4)(-1)^{3+3}\begin{bmatrix} x - 4 & 4 \\ -2 & x + 2 \end{bmatrix} \text{ expanding by the third column}$$

$$= (x + 4)(1)[(x - 4)(x + 2) - 4(-2)]$$
$$= (x + 4)(x^2 - 2x - 8 + 8)$$
$$= (x + 4)(x^2 - 2x)$$
$$= x^3 + 2x^2 - 8x$$

The zeros of this polynomial are the solutions of

$$x^3 + 2x^2 - 8x = 0$$
$$x(x^2 + 2x - 8) = 0$$
$$x(x + 4)(x - 2) = 0$$
$$x = 0, -4, 2$$

The eigenvalues of this matrix are 0, -4, and 2.

Section 9-6

1. Answers will vary.

3. Theorem 1   5. Theorem 1   7. Theorem 2   9. Theorem 3   11. Theorem 5

13. $3C_1 + C_2 \rightarrow C_2$ has been used, $x = 3(-1) + 3 = 0$

15. $3C_1 + C_3 \rightarrow C_3$ has been used, $x = 3(1) + 2 = 5$

17. Interchanging two rows of a determinant changes the sign of the determinant (Theorem 3).

$$\begin{vmatrix} c & d \\ a & b \end{vmatrix} = -10$$

19. Adding a multiple (in this case a multiple by 1) of a row to another row does not change the value of the determinant (Theorem 5).

$$\begin{vmatrix} a + c & b + d \\ c & d \end{vmatrix} = 10$$

21. Adding a multiple of a column to another column does not change the value of the determinant (Theorem 5).

$$\begin{vmatrix} a & a - b \\ c & c - d \end{vmatrix} = \begin{vmatrix} a & -b \\ c & -d \end{vmatrix}$$

If every element of a column of a determinant is multiplied by -1, the value of the determinant is -1 times the original.

$$\begin{vmatrix} a & a - b \\ c & c - d \end{vmatrix} = \begin{vmatrix} a & -b \\ c & -d \end{vmatrix} = (-1)\begin{vmatrix} a & b \\ c & d \end{vmatrix} = -1(10) = -10$$

23. $\begin{vmatrix} -1 & 0 & 3 \\ 2 & 5 & 4 \\ 1 & 5 & 2 \end{vmatrix} = \begin{vmatrix} -1 & 0 & 3 \\ 1 & 0 & 2 \\ 1 & 5 & 2 \end{vmatrix}$ $-1R_3 + R_2 \rightarrow R_2$

$$= 5(-1)^{3+2}\begin{vmatrix} -1 & 3 \\ 1 & 2 \end{vmatrix} = -5[(-1)2 - 1(3)] = 25$$

25. $\begin{vmatrix} 3 & 5 & 0 \\ 1 & 1 & -2 \\ 2 & 1 & -1 \end{vmatrix} = \begin{vmatrix} 3 & 5 & 0 \\ -3 & -1 & 0 \\ 2 & 1 & -1 \end{vmatrix}$ $-2R_3 + R_2 \rightarrow R_2$

$$= (-1)(-1)^{3+3}\begin{vmatrix} 3 & 5 \\ -3 & -1 \end{vmatrix} = (-1)[3(-1) - (-3)5] = -12$$

27. Theorem 1   29. Theorem 2   31. Theorem 5

33. $2C_3 + C_1 \rightarrow C_1$ has been used, $x = 2 \cdot 1 + 3 = 5$
    $C_3 + C_2 \rightarrow C_2$ has been used, $y = (-2) + 2 = 0$

35. $(-4)R_2 + R_1 \rightarrow R_1$ has been used, $x = (-4)3 + 9 = -3$
    $2R_2 + R_3 \rightarrow R_3$ has been used, $y = 2 \cdot 3 + 4 = 10$

37. We will generate zeros in the first column by row operations.

$$\begin{vmatrix} 1 & 5 & 3 \\ 4 & 2 & 1 \\ 3 & 1 & 2 \end{vmatrix} = \begin{vmatrix} 1 & 5 & 3 \\ 0 & -18 & -11 \\ 0 & -14 & -7 \end{vmatrix} \begin{matrix} (-4)R_1 + R_2 \rightarrow R_2 \\ (-3)R_1 + R_3 \rightarrow R_3 \end{matrix}$$

$$= 1(-1)^{1+1} \begin{vmatrix} -18 & -11 \\ -14 & -7 \end{vmatrix} + 0 + 0$$

$$= (-1)^2[(-18)(-7) - (-14)(-11)] = -28$$

39. We will generate zeros in the second row by column operations.

$$\begin{vmatrix} 5 & 2 & -3 \\ -2 & 4 & 4 \\ 1 & -1 & 3 \end{vmatrix} = \begin{vmatrix} 5 & 12 & 7 \\ -2 & 0 & 0 \\ 1 & 1 & 5 \end{vmatrix} \begin{matrix} 2C_1 + C_2 \rightarrow C_2 \\ 2C_1 + C_3 \rightarrow C_3 \end{matrix}$$

$$= (-2)(-1)^{2+1} \begin{vmatrix} 12 & 7 \\ 1 & 5 \end{vmatrix} + 0 + 0$$

$$= (-2)(-1)^3[12 \cdot 5 - 1 \cdot 7] = 106$$

41. The column operation $(-3)C_3 + C_1 \rightarrow C_1$ transforms this determinant into

$$\begin{vmatrix} 0 & -4 & 1 \\ 0 & -1 & 2 \\ 0 & 2 & 3 \end{vmatrix}$$

By Theorem 2, the value of this determinant is 0.

43. We start by generating one more zero in the first row.

$$\begin{vmatrix} 0 & 1 & 0 & 1 \\ 1 & -2 & 4 & 3 \\ 2 & 1 & 5 & 4 \\ 1 & 2 & 1 & 2 \end{vmatrix} = \begin{vmatrix} 0 & 0 & 0 & 1 \\ 1 & -5 & 4 & 3 \\ 2 & -3 & 5 & 4 \\ 1 & 0 & 1 & 2 \end{vmatrix} (-1)C_4 + C_2 \rightarrow C_2$$

$$= 1(-1)^{1+4} \begin{vmatrix} 1 & -5 & 4 \\ 2 & -3 & 5 \\ 1 & 0 & 1 \end{vmatrix} + 0 + 0 + 0$$

$$= (-1) \begin{vmatrix} 1 & -5 & 4 \\ 2 & -3 & 5 \\ 1 & 0 & 1 \end{vmatrix}$$

$$= \begin{vmatrix} 1 & 5 & 4 \\ 2 & 3 & 5 \\ 1 & 0 & 1 \end{vmatrix} \text{ by Theorem 1}$$

We now generate one more zero in the third row.

$$\begin{vmatrix} 1 & 5 & 4 \\ 2 & 3 & 5 \\ 1 & 0 & 1 \end{vmatrix} = \begin{vmatrix} 1 & 5 & 3 \\ 2 & 3 & 3 \\ 1 & 0 & 0 \end{vmatrix} (-1)C_1 + C_3 \rightarrow C_3$$

$$= 1(-1)^{3+1} \begin{vmatrix} 5 & 3 \\ 3 & 3 \end{vmatrix} = (-1)^4[5 \cdot 3 - 3 \cdot 3] = 6$$

45. We start by generating zeros in the third row.

$$\begin{vmatrix} 3 & 2 & 3 & 1 \\ 3 & -2 & 8 & 5 \\ 2 & 1 & 3 & 1 \\ 4 & 5 & 4 & -3 \end{vmatrix} = \begin{vmatrix} 1 & 1 & 0 & 1 \\ -7 & -7 & -7 & 5 \\ 0 & 0 & 0 & 1 \\ 10 & 8 & 13 & -3 \end{vmatrix} \begin{matrix} (-2)C_4 + C_1 \rightarrow C_1 \\ (-1)C_4 + C_2 \rightarrow C_2 \\ (-3)C_4 + C_3 \rightarrow C_3 \end{matrix}$$

$$= 1(-1)^{3+4} \begin{vmatrix} 1 & 1 & 0 \\ -7 & -7 & -7 \\ 10 & 8 & 13 \end{vmatrix} + 0 + 0 + 0$$

$$= (-1) \begin{vmatrix} 1 & 1 & 0 \\ -7 & -7 & -7 \\ 10 & 8 & 13 \end{vmatrix}$$

$$= \begin{vmatrix} 1 & 1 & 0 \\ 7 & 7 & 7 \\ 10 & 8 & 13 \end{vmatrix} \text{ by Theorem 1}$$

We now generate one more zero in the first row.

$$\begin{vmatrix} 1 & 1 & 0 \\ 7 & 7 & 7 \\ 10 & 8 & 13 \end{vmatrix} = \begin{vmatrix} 1 & 0 & 0 \\ 7 & 0 & 7 \\ 10 & -2 & 13 \end{vmatrix} \quad (-1)C_1 + C_2 \rightarrow C_2$$

$$= (-2)(-1)^{3+2} \begin{vmatrix} 1 & 0 \\ 7 & 7 \end{vmatrix} + 0 + 0$$

$$= (-2)(-1)^5[1 \cdot 7 - 0 \cdot 7] = 14$$

47. Expand, for example, by the first column.

$$\begin{vmatrix} a & b & a \\ d & e & d \\ g & h & g \end{vmatrix} = a(-1)^{1+1} \begin{vmatrix} e & d \\ h & g \end{vmatrix} + d(-1)^{2+1} \begin{vmatrix} b & a \\ h & g \end{vmatrix} + g(-1)^{3+1} \begin{vmatrix} b & a \\ e & d \end{vmatrix}$$

$$= a \begin{vmatrix} e & d \\ h & g \end{vmatrix} - d \begin{vmatrix} b & a \\ h & g \end{vmatrix} + g \begin{vmatrix} b & a \\ e & d \end{vmatrix}$$

$$= a(eg - hd) - d(bg - ha) + g(bd - ea)$$

$$= aeg - adh - bdg + adh + bdg - aeg$$

$$= 0$$

49. We expand the left side by the first column, and the right side by the second column.

$$\begin{vmatrix} a_1 & b_1 & c_1 \\ a_2 & b_2 & c_2 \\ a_3 & b_3 & c_3 \end{vmatrix} = a_1(-1)^{1+1} \begin{vmatrix} b_2 & c_2 \\ b_3 & c_3 \end{vmatrix} + a_2(-1)^{2+1} \begin{vmatrix} b_1 & c_1 \\ b_3 & c_3 \end{vmatrix} + a_3(-1)^{3+1} \begin{vmatrix} b_1 & c_1 \\ b_2 & c_2 \end{vmatrix}$$

$$= a_1 \begin{vmatrix} b_2 & c_2 \\ b_3 & c_3 \end{vmatrix} - a_2 \begin{vmatrix} b_1 & c_1 \\ b_3 & c_3 \end{vmatrix} + a_3 \begin{vmatrix} b_1 & c_1 \\ b_2 & c_2 \end{vmatrix}$$

$$-\begin{vmatrix} b_1 & a_1 & c_1 \\ b_2 & a_2 & c_2 \\ b_3 & a_3 & c_3 \end{vmatrix} = -\left[a_1(-1)^{1+2} \begin{vmatrix} b_2 & c_2 \\ b_3 & c_3 \end{vmatrix} + a_2(-1)^{2+2} \begin{vmatrix} b_1 & c_1 \\ b_3 & c_3 \end{vmatrix} + a_3(-1)^{3+2} \begin{vmatrix} b_1 & c_1 \\ b_2 & c_3 \end{vmatrix} \right]$$

$$= -\left[-a_1 \begin{vmatrix} b_2 & c_2 \\ b_3 & c_3 \end{vmatrix} + a_2 \begin{vmatrix} b_1 & c_1 \\ b_3 & c_3 \end{vmatrix} - a_3 \begin{vmatrix} b_1 & c_1 \\ b_2 & c_2 \end{vmatrix} \right]$$

$$= a_1 \begin{vmatrix} b_2 & c_2 \\ b_3 & c_3 \end{vmatrix} - a_2 \begin{vmatrix} b_1 & c_1 \\ b_3 & c_3 \end{vmatrix} + a_3 \begin{vmatrix} b_1 & c_1 \\ b_2 & c_2 \end{vmatrix}$$

The two original expressions are equal.

51. The statements: (2, 5) satisfies the equation $\begin{vmatrix} x & y & 1 \\ 2 & 5 & 1 \\ -3 & 4 & 1 \end{vmatrix} = 0$

and (-3, 4) satisfies the equation $\begin{vmatrix} x & y & 1 \\ 2 & 5 & 1 \\ -3 & 4 & 1 \end{vmatrix} = 0$

are equivalent to the statements $\begin{vmatrix} 2 & 5 & 1 \\ 2 & 5 & 1 \\ -3 & 4 & 1 \end{vmatrix} = 0$ and $\begin{vmatrix} -3 & 4 & 1 \\ 2 & 5 & 1 \\ -3 & 4 & 1 \end{vmatrix} = 0$.

The latter statements are true by Theorem 4.

53. The statement $\begin{vmatrix} x & y & 1 \\ x_1 & y_1 & 1 \\ x_2 & y_2 & 1 \end{vmatrix} = 0$ is the equation of a line because, expanding by the

first row, we have $x(-1)^{1+1}\begin{vmatrix} y_1 & 1 \\ y_2 & 1 \end{vmatrix} + y(-1)^{1+2}\begin{vmatrix} x_1 & 1 \\ x_2 & 1 \end{vmatrix} + 1(-1)^{1+3}\begin{vmatrix} x_1 & y_1 \\ x_2 & y_2 \end{vmatrix} = 0$

This is in the standard form for the equation of a line
$Ax + By + C = 0$

To show that the line passes through (x_1, y_1) and (x_2, y_2), we note that (x_1, y_1) and (x_2, y_2) satisfy the equation because

$\begin{vmatrix} x_1 & y_1 & 1 \\ x_2 & y_2 & 1 \\ x_3 & y_3 & 1 \end{vmatrix} = 0$ and $\begin{vmatrix} x_2 & y_2 & 1 \\ x_1 & y_1 & 1 \\ x_2 & y_2 & 1 \end{vmatrix} = 0$ are true by Theorem 4.

55. Using the result stated in problem 54, we have
$\begin{vmatrix} x_1 & y_1 & 1 \\ x_2 & y_2 & 1 \\ x_3 & y_3 & 1 \end{vmatrix} = 2 \times$ (area of triangle formed by the three points).

If the determinant is 0, then the area of the triangle formed by the three points is zero. The only way this can happen is if the three points are on the same line; that is, the points are collinear.

Section 9-7

1. Answers will vary.

3. No. If the coefficient matrix is not square, it does not have a determinant, and Cramer's rule cannot be used.

5. $D = \begin{vmatrix} 1 & 2 \\ 1 & 3 \end{vmatrix} = 1; \quad x = \dfrac{\begin{vmatrix} 1 & 2 \\ -1 & 3 \end{vmatrix}}{D} = \dfrac{5}{1} = 5; \quad y = \dfrac{\begin{vmatrix} 1 & 1 \\ 1 & -1 \end{vmatrix}}{D} = \dfrac{-2}{1} = -2$

7. $D = \begin{vmatrix} 2 & 1 \\ 5 & 3 \end{vmatrix} = 1; \quad x = \dfrac{\begin{vmatrix} 1 & 1 \\ 2 & 3 \end{vmatrix}}{D} = \dfrac{1}{1} = 1; \quad y = \dfrac{\begin{vmatrix} 2 & 1 \\ 5 & 2 \end{vmatrix}}{D} = \dfrac{-1}{1} = -1$

9. $D = \begin{vmatrix} 2 & -1 \\ -1 & 3 \end{vmatrix} = 5; \quad x = \dfrac{\begin{vmatrix} -3 & -1 \\ 3 & 3 \end{vmatrix}}{D} = \dfrac{-6}{5} = -\dfrac{6}{5}; \quad y = \dfrac{\begin{vmatrix} 2 & -3 \\ -1 & 3 \end{vmatrix}}{D} = \dfrac{3}{5}$

11. $D = \begin{vmatrix} 4 & -3 \\ 3 & 2 \end{vmatrix} = 17; \quad x = \dfrac{\begin{vmatrix} 4 & -3 \\ -2 & 2 \end{vmatrix}}{D} = \dfrac{2}{17}; \quad y = \dfrac{\begin{vmatrix} 4 & 4 \\ 3 & -2 \end{vmatrix}}{D} = \dfrac{-20}{17} = -\dfrac{20}{17}$

13. $D = \begin{vmatrix} 0.9925 & -0.9659 \\ 0.1219 & 0.2588 \end{vmatrix} = 0.37460$

$x = \dfrac{\begin{vmatrix} 0 & -0.9659 \\ 2,500 & 0.2588 \end{vmatrix}}{D} = \dfrac{2,414.75}{0.37460} = 6,400$ to two significant digits

$y = \dfrac{\begin{vmatrix} 0.9925 & 0 \\ 0.1219 & 2,500 \end{vmatrix}}{D} = \dfrac{2,481.25}{0.37460} = 6,600$ to two significant digits

15. $D = \begin{vmatrix} 0.9954 & -0.9942 \\ 0.0958 & 0.1080 \end{vmatrix} = 0.20275$

$x = \dfrac{\begin{vmatrix} 0 & -0.9942 \\ 155 & 0.1080 \end{vmatrix}}{D} = \dfrac{154.10}{0.20275} = 760$ to two significant digits

$y = \dfrac{\begin{vmatrix} 0.9954 & 0 \\ 0.0958 & 155 \end{vmatrix}}{D} = \dfrac{154.29}{0.20275} = 760$ to two significant digits

17. $D = \begin{vmatrix} 1 & 1 & 0 \\ 0 & 2 & 1 \\ -1 & 0 & 1 \end{vmatrix} = 1;$ $x = \dfrac{\begin{vmatrix} 0 & 1 & 0 \\ -5 & 2 & 1 \\ -3 & 0 & 1 \end{vmatrix}}{D} = \dfrac{2}{1} = 2;$ $y = \dfrac{\begin{vmatrix} 1 & 0 & 0 \\ 0 & -5 & 1 \\ -1 & -3 & 1 \end{vmatrix}}{D} = \dfrac{-2}{1} = -2$

$z = \dfrac{\begin{vmatrix} 1 & 1 & 0 \\ 0 & 2 & -5 \\ -1 & 0 & -3 \end{vmatrix}}{D} = \dfrac{-1}{1} = -1$

19. $D = \begin{vmatrix} 1 & 1 & 0 \\ 0 & 2 & 1 \\ 0 & -1 & 1 \end{vmatrix} = 3;$ $x = \dfrac{\begin{vmatrix} 1 & 1 & 0 \\ 0 & 2 & 1 \\ 1 & -1 & 1 \end{vmatrix}}{D} = \dfrac{4}{3};$ $y = \dfrac{\begin{vmatrix} 1 & 1 & 0 \\ 0 & 0 & 1 \\ 0 & 1 & 1 \end{vmatrix}}{D} = \dfrac{-1}{3} = -\dfrac{1}{3}$

$z = \dfrac{\begin{vmatrix} 1 & 1 & 0 \\ 0 & 2 & 0 \\ 0 & -1 & 1 \end{vmatrix}}{D} = \dfrac{2}{3}$

21. $D = \begin{vmatrix} 0 & 3 & 1 \\ 1 & 0 & 2 \\ 1 & -3 & 0 \end{vmatrix} = 3;$ $x = \dfrac{\begin{vmatrix} -1 & 3 & 1 \\ 3 & 0 & 2 \\ -2 & -3 & 0 \end{vmatrix}}{D} = \dfrac{-27}{3} = -9;$ $y = \dfrac{\begin{vmatrix} 0 & -1 & 1 \\ 1 & 3 & 2 \\ 1 & -2 & 0 \end{vmatrix}}{D} = \dfrac{-7}{3} = -\dfrac{7}{3}$

$z = \dfrac{\begin{vmatrix} 0 & 3 & -1 \\ 1 & 0 & 3 \\ 1 & -3 & -2 \end{vmatrix}}{D} = \dfrac{18}{3} = 6$

23. $D = \begin{vmatrix} 0 & 2 & -1 \\ 1 & -1 & -1 \\ 1 & -1 & 2 \end{vmatrix} = -6;$ $x = \dfrac{\begin{vmatrix} -3 & 2 & -1 \\ 2 & -1 & -1 \\ 4 & -1 & 2 \end{vmatrix}}{D} = \dfrac{-9}{-6} = \dfrac{3}{2};$ $y = \dfrac{\begin{vmatrix} 0 & -3 & -1 \\ 1 & 2 & -1 \\ 1 & 4 & 2 \end{vmatrix}}{D} = \dfrac{7}{-6} = -\dfrac{7}{6}$

$z = \dfrac{\begin{vmatrix} 0 & 2 & -3 \\ 1 & -1 & 2 \\ 1 & -1 & 4 \end{vmatrix}}{D} = \dfrac{-4}{-6} = \dfrac{2}{3}$

25. Compute the coefficient determinant:

$D = \begin{vmatrix} a & 3 \\ 2 & 4 \end{vmatrix} = 4a - 3(2) = 4a - 6$

If $D \neq 0$, there is a unique solution:

$\begin{aligned} 4a - 6 &= 0 \\ 4a &= 6 \\ a &= \frac{6}{4} = \frac{3}{2} \end{aligned}$

So if $a \neq \dfrac{3}{2}$ there is one solution. If $a = \dfrac{3}{2}$ we need to use Gauss-Jordan elimination to determine the nature of the solutions after plugging in $\frac{3}{2}$ for a.

$$\begin{bmatrix} \frac{3}{2} & 3 & \Big| & b \\ 2 & 4 & \Big| & 5 \end{bmatrix} \frac{2}{3} R_1 \rightarrow R_1$$

$$\sim \begin{bmatrix} 1 & 2 & \Big| & \frac{2b}{3} \\ 2 & 4 & \Big| & 5 \end{bmatrix} - 2R_1 + R_2 \rightarrow R_2$$

$$\sim \begin{bmatrix} 1 & 2 & \Big| & \frac{2b}{3} \\ 0 & 0 & \Big| & -\frac{4b}{3} + 5 \end{bmatrix}$$

If the bottom row is all zeros, there are infinitely many solutions. If the bottom row has zero in the third position, there is no solution. We need to know when $\frac{-4b}{3} + 5 = 0$.

$$\frac{-4b}{3} + 5 = 0$$
$$-4b + 15 = 0$$
$$-4b = -15$$
$$b = \frac{15}{4}$$

So there are infinitely many solutions if $a = \frac{3}{2}$ and $b = \frac{15}{4}$, and no solutions if $a = \frac{3}{2}$ and $b \neq \frac{15}{4}$.

27. $x = \dfrac{\begin{vmatrix} -3 & -3 & 1 \\ -11 & 3 & 2 \\ 3 & -1 & -1 \end{vmatrix}}{\begin{vmatrix} 2 & -3 & 1 \\ -4 & 3 & 2 \\ 1 & -1 & -1 \end{vmatrix}} = \dfrac{20}{5} = 4$

29. $y = \dfrac{\begin{vmatrix} 12 & 5 & 11 \\ 15 & -13 & -9 \\ 5 & 0 & 2 \end{vmatrix}}{\begin{vmatrix} 12 & -14 & 11 \\ 15 & 7 & -9 \\ 5 & -3 & 2 \end{vmatrix}} = \dfrac{28}{14} = 2$

31. $z = \dfrac{\begin{vmatrix} 3 & -4 & 18 \\ -9 & 8 & -13 \\ 5 & -7 & 33 \end{vmatrix}}{\begin{vmatrix} 3 & -4 & 5 \\ -9 & 8 & 7 \\ 5 & -7 & 10 \end{vmatrix}} = \dfrac{5}{2}$

33. $D = \begin{vmatrix} 1 & -4 & 9 \\ 4 & -1 & 6 \\ 1 & -1 & 3 \end{vmatrix} = \begin{vmatrix} -2 & -1 & 0 \\ 2 & 1 & 0 \\ 1 & -1 & 3 \end{vmatrix} \begin{array}{l} (-3)R_3 + R_1 \rightarrow R_1 \\ (-2)R_3 + R_2 \rightarrow R_2 \end{array} = -\begin{vmatrix} 2 & 1 & 0 \\ 2 & 1 & 0 \\ 1 & -1 & 3 \end{vmatrix}$ by Theorem 1

$$= 0 \text{ by Theorem 4}$$

Since $D = 0$, the system either has no solution or infinitely many. Since $x = 0$, $y = 0$, $z = 0$ is a solution, the second case must hold.

35. We start with

$$a_{11}x + a_{12}\,y = k_1$$
$$a_{21}x + a_{22}\,y = k_2$$

To solve for y, we will eliminate x. We multiply the top equation by $-a_{21}$ and the bottom equation by a_{11}, then add.

$$\begin{array}{rcl} -a_{11}a_{21}x - a_{21}a_{12}y &=& -k_1a_{21} \\ \underline{a_{11}a_{21}x + a_{11}a_{22}y} &=& \underline{k_2a_{11}} \\ 0x + a_{11}a_{22}y + a_{21}a_{12}y &=& k_2a_{11} - k_1a_{21} \\ (a_{11}a_{22} - a_{21}a_{12})y &=& a_{11}k_2 - a_{21}k_1 \end{array}$$

$$y = \frac{a_{11}k_2 - a_{21}k_1}{a_{11}a_{22} - a_{21}a_{12}} = \frac{\begin{vmatrix} a_{11} & k_1 \\ a_{21} & k_2 \end{vmatrix}}{\begin{vmatrix} a_{11} & a_{12} \\ a_{21} & a_{22} \end{vmatrix}}$$

37. (A) $R = xp + yq = (200 - 6p + 4q)p + (300 + 2p - 3q)q$

$= 200p - 6p^2 + 4pq + 300q + 2pq - 3q^2 = 200p + 300q - 6p^2 + 6pq - 3q^2$

(B) Rewrite the demand equations as

$6p - 4q = 200 - x$

$-2p + 3q = 300 - y$

Apply Cramer's rule: $D = \begin{vmatrix} 6 & -4 \\ -2 & 3 \end{vmatrix} = 10$

$p = \dfrac{\begin{vmatrix} 200 - x & -4 \\ 300 - y & 3 \end{vmatrix}}{D} = \dfrac{1800 - 3x - 4y}{10} = -0.3x - 0.4y + 180$

$q = \dfrac{\begin{vmatrix} 6 & 200 - x \\ -2 & 300 - y \end{vmatrix}}{D} = \dfrac{2200 - 2x - 6y}{10} = -0.2x - 0.6y + 220$

Then

$R = xp + yq = x(-0.3x - 0.4y + 180) + y(-0.2x - 0.6y + 220)$

$= -0.3x^2 - 0.4xy + 180x - 0.2xy - 0.6y^2 + 220y$

$= 180x + 220y - 0.3x^2 - 0.6xy - 0.6y^2$

CHAPTER 9 REVIEW

1. $R_1 \leftrightarrow R_2$ means interchange Rows 1 and 2.

$\begin{bmatrix} 3 & -6 & | & 12 \\ 1 & -4 & | & 5 \end{bmatrix}$ *(9-1)*

2. $\frac{1}{3}R_2 \rightarrow R_2$ means multiply Row 2 by $\frac{1}{3}$.

$\begin{bmatrix} 1 & -4 & | & 5 \\ 1 & -2 & | & 4 \end{bmatrix}$ *(9-1)*

3. $(-3)R_1 + R_2 \rightarrow R_2$ means replace Row 2 by itself plus -3 times Row 1.

$\begin{bmatrix} 1 & -4 & | & 5 \\ 0 & 6 & | & -3 \end{bmatrix}$ *(9-1)*

4. $x_1 = 4$

$x_2 = -7$

The solution is $(4, -7)$ *(9-1)*

5. $x_1 - x_2 = 4$

$\qquad 0 = 1$

No solution   *(9-1)*

6. $x_1 - x_2 = 4$

$\qquad 0 = 0$

Solution:

$x_2 = t$

$x_1 = x_2 + 4 = t + 4$

The solution is $x_1 = t + 4$, $x_2 = t$ for any real number t.   *(9-1)*

7. $AB = \begin{bmatrix} 4 & -2 \\ 0 & 3 \end{bmatrix}\begin{bmatrix} -1 & 5 \\ -4 & 6 \end{bmatrix} = \begin{bmatrix} 4(-1) + (-2)(-4) & 4 \cdot 5 + (-2)6 \\ 0(-1) + 3(-4) & 0 \cdot 5 + 3 \cdot 6 \end{bmatrix} = \begin{bmatrix} 4 & 8 \\ -12 & 18 \end{bmatrix}$ *(9-2)*

8. $CD = \begin{bmatrix} -1 & 4 \end{bmatrix}\begin{bmatrix} 3 \\ -2 \end{bmatrix} = [(-1)3 + 4(-2)] = [-11]$ *(9-2)*

9. $CB = \begin{bmatrix} -1 & 4 \end{bmatrix}\begin{bmatrix} -1 & 5 \\ -4 & 6 \end{bmatrix} = [(-1)(-1) + 4(-4) \quad (-1)5 + 4 \cdot 6] = \begin{bmatrix} -15 & 19 \end{bmatrix}$ *(9-2)*

10. $AD = \begin{bmatrix} 4 & -2 \\ 0 & 3 \end{bmatrix}\begin{bmatrix} 3 \\ -2 \end{bmatrix} = \begin{bmatrix} 4 \cdot 3 + (-2)(-2) \\ 0 \cdot 3 + 3(-2) \end{bmatrix} = \begin{bmatrix} 16 \\ -6 \end{bmatrix}$ *(9-2)*

11. $A + B = \begin{bmatrix} 4 & -2 \\ 0 & 3 \end{bmatrix} + \begin{bmatrix} -1 & 5 \\ -4 & 6 \end{bmatrix} = \begin{bmatrix} 4 + (-1) & (-2) + 5 \\ 0 + (-4) & 3 + 6 \end{bmatrix} = \begin{bmatrix} 3 & 3 \\ -4 & 9 \end{bmatrix}$ *(9-2)*

12. $C + D$ is not defined   *(9-2)*   **13.** $A + C$ is not defined   *(9-2)*

14. $2A - 5B = 2\begin{bmatrix} 4 & -2 \\ 0 & 3 \end{bmatrix} - 5\begin{bmatrix} -1 & 5 \\ -4 & 6 \end{bmatrix} = \begin{bmatrix} 8 & -4 \\ 0 & 6 \end{bmatrix} - \begin{bmatrix} -5 & 25 \\ -20 & 30 \end{bmatrix} = \begin{bmatrix} 13 & -29 \\ 20 & -24 \end{bmatrix}$ *(9-2)*

15. $CA + C = [-1 \quad 4]\begin{bmatrix} 4 & -2 \\ 0 & 3 \end{bmatrix} + [-1 \quad 4] = [(-1)4 + 4 \cdot 0 \quad (-1)(-2) + 4 \cdot 3] + [-1 \quad 4]$

$\qquad\qquad\qquad = [-4 \quad 14] + [-1 \quad 4] = [-5 \quad 18]$ *(9-2)*

16. $\begin{bmatrix} 4 & 7 & | & 1 & 0 \\ -1 & -2 & | & 0 & 1 \end{bmatrix}$ $R_1 \leftrightarrow R_2$

$\sim \begin{bmatrix} -1 & -2 & | & 1 & 0 \\ 4 & 7 & | & 0 & 1 \end{bmatrix}$ $4R_1 + R_2 \rightarrow R_2$

$\sim \begin{bmatrix} -1 & -2 & | & 0 & 1 \\ 0 & -1 & | & 1 & 4 \end{bmatrix}$ $(-2)R_2 + R_1 \rightarrow R_1$

$\sim \begin{bmatrix} -1 & 0 & | & -2 & -7 \\ 0 & -1 & | & 1 & 4 \end{bmatrix}$ $\begin{matrix}(-1)R_1 \rightarrow R_1 \\ (-1)R_2 \rightarrow R_2\end{matrix}$

$\sim \begin{bmatrix} 1 & 0 & | & 2 & 7 \\ 0 & 1 & | & -1 & -4 \end{bmatrix}$

The inverse is $\begin{bmatrix} 2 & 7 \\ -1 & -4 \end{bmatrix}$

Check: $A^{-1}A = \begin{bmatrix} 2 & 7 \\ -1 & -4 \end{bmatrix}\begin{bmatrix} 4 & 7 \\ -1 & -2 \end{bmatrix} = \begin{bmatrix} 2 \cdot 4 + 7(-1) & 2 \cdot 7 + 7(-2) \\ (-1)4 + (-4)(-1) & (-1)7 + (-4)(-2) \end{bmatrix}$

$\qquad = \begin{bmatrix} 1 & 0 \\ 0 & 1 \end{bmatrix} = I$ *(9-3)*

17. As a matrix equation the system becomes

$\qquad\quad A \quad\ X \quad\ B$
$\begin{bmatrix} 3 & 2 \\ 4 & 3 \end{bmatrix}\begin{bmatrix} x_1 \\ x_2 \end{bmatrix} = \begin{bmatrix} k_1 \\ k_2 \end{bmatrix}$

The solution of $AX = B$ is $X = A^{-1}B$.

Applying matrix reduction methods, we obtain $A^{-1} = \begin{bmatrix} 3 & -2 \\ -4 & 3 \end{bmatrix}$ (details omitted).

Applying this inverse, we have

$X = \begin{bmatrix} 3 & -2 \\ -4 & 3 \end{bmatrix}\begin{bmatrix} k_1 \\ k_2 \end{bmatrix}$

(A) $\begin{bmatrix} x_1 \\ x_2 \end{bmatrix} = \begin{bmatrix} 3 & -2 \\ -4 & 3 \end{bmatrix}\begin{bmatrix} 3 \\ 5 \end{bmatrix} = \begin{bmatrix} 3 \cdot 3 + (-2)5 \\ (-4)3 + 3 \cdot 5 \end{bmatrix} = \begin{bmatrix} -1 \\ 3 \end{bmatrix}$ $x_1 = -1, \; x_2 = 3$

(B) $\begin{bmatrix} x_1 \\ x_2 \end{bmatrix} = \begin{bmatrix} 3 & -2 \\ -4 & 3 \end{bmatrix}\begin{bmatrix} 7 \\ 10 \end{bmatrix} = \begin{bmatrix} 3 \cdot 7 + (-2)10 \\ (-4)7 + 3 \cdot 10 \end{bmatrix} = \begin{bmatrix} 1 \\ 2 \end{bmatrix}$ $x_1 = 1, \; x_2 = 2$

(C) $\begin{bmatrix} x_1 \\ x_2 \end{bmatrix} = \begin{bmatrix} 3 & -2 \\ -4 & 3 \end{bmatrix}\begin{bmatrix} 4 \\ 2 \end{bmatrix} = \begin{bmatrix} 3 \cdot 4 + (-2)2 \\ (-4)4 + 3 \cdot 2 \end{bmatrix} = \begin{bmatrix} 8 \\ -10 \end{bmatrix}$ $x_1 = 8, \; x_2 = -10$ *(9-4)*

18. $\begin{vmatrix} 2 & -3 \\ -5 & -1 \end{vmatrix} = 2(-1) - (-5)(-3) = -17$ *(9-5)*

19. $\begin{vmatrix} 2 & 3 & -4 \\ 0 & 5 & 0 \\ 1 & -4 & -2 \end{vmatrix} = 0 + 5(-1)^{2+2}\begin{vmatrix} 2 & -4 \\ 1 & -2 \end{vmatrix} + 0 = 5(-1)^4[2(-2) - 1(-4)] = 0$ *(9-5, 9-6)*

20. $D = \begin{vmatrix} 3 & -2 \\ 1 & 3 \end{vmatrix} = 11$

$x = \dfrac{\begin{vmatrix} 8 & -2 \\ -1 & 3 \end{vmatrix}}{D} = \dfrac{22}{11} = 2 \qquad y = \dfrac{\begin{vmatrix} 3 & 8 \\ 1 & -1 \end{vmatrix}}{D} = \dfrac{-11}{11} = -1$ (9-7)

21. (A) Interchanging two rows of a determinant changes the sign of the determinant (Theorem 3, Section 9-6).

$\begin{vmatrix} g & h & i \\ d & e & f \\ a & b & c \end{vmatrix} = -2$

(B) If every element of a column of a determinant is multiplied by 3, (Theorem 1, Section 9-6), the value of the determinant is 3 times the original.

$\begin{vmatrix} a & 3b & c \\ d & 3e & f \\ g & 3h & i \end{vmatrix} = 3 \cdot 2 = 6$

(C) Adding a multiple of a column to another column does not change the value of the determinant (Theorem 5, Section 9-6).

$\begin{vmatrix} a & b & a+b+c \\ d & e & d+e+f \\ g & h & g+h+i \end{vmatrix} = \begin{vmatrix} a & b & b+c \\ d & e & e+f \\ g & h & h+i \end{vmatrix} = \begin{vmatrix} a & b & c \\ d & e & f \\ g & h & i \end{vmatrix} = 2$ (9-6)

22. We write the augmented matrix:

$\begin{bmatrix} 1 & -1 & | & 4 \\ 2 & 1 & | & 2 \\ -2 & 2 & -8 \end{bmatrix} \quad (-2)R_1 + R_2 \rightarrow R_2$

Need a 0 here

$\sim \begin{bmatrix} 1 & -1 & | & 4 \\ 0 & 3 & | & -6 \end{bmatrix} \quad \tfrac{1}{3}R_2 \rightarrow R_2 \quad$ corresponds to the linear system $\quad \begin{array}{r} x_1 - x_2 = 4 \\ 3x_2 = -6 \end{array}$

Need a 1 here
Need a 0 here

$\sim \begin{bmatrix} 1 & -1 & | & 4 \\ 0 & 1 & | & -2 \end{bmatrix} \quad R_2 + R_1 \rightarrow R_1 \;$ corresponds to the linear system $\quad \begin{array}{r} x_1 - x_2 = 4 \\ x_2 = -2 \end{array}$

$\sim \begin{bmatrix} 1 & 0 & | & 2 \\ 0 & 1 & | & -2 \end{bmatrix} \qquad$ corresponds to the linear system $\quad \begin{array}{r} x_1 = 2 \\ x_2 = -2 \end{array}$

The solution is $x_1 = 2$, $x_2 = -2$. Each pair of lines graphed below has the same intersection point, $(2, -2)$.

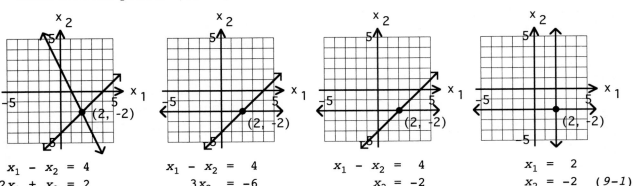

$\begin{array}{l} x_1 - x_2 = 4 \\ 2x_1 + x_2 = 2 \end{array} \qquad \begin{array}{l} x_1 - x_2 = 4 \\ 3x_2 = -6 \end{array} \qquad \begin{array}{l} x_1 - x_2 = 4 \\ x_2 = -2 \end{array} \qquad \begin{array}{l} x_1 = 2 \\ x_2 = -2 \end{array}$ (9-1)

23. $\begin{bmatrix} 3 & 2 & | & 3 \\ 1 & 3 & | & 8 \end{bmatrix} R_1 \leftrightarrow R_2$

$\sim \begin{bmatrix} 1 & 3 & | & 8 \\ 3 & 2 & | & 3 \end{bmatrix} (-3)R_1 + R_2 \rightarrow R_2$

$\quad\quad -3 \quad -9 \quad -24 \;\swarrow$

$\sim \begin{bmatrix} 1 & 3 & | & 8 \\ 0 & -7 & | & -21 \end{bmatrix} -\frac{1}{7}R_2 \rightarrow R_2$

$\sim \begin{bmatrix} 1 & 3 & | & 8 \\ 0 & 1 & | & 3 \end{bmatrix} (-3)R_2 + R_1 \rightarrow R_1$

$\quad\quad 0 \quad -3 \quad -9 \;\swarrow$

$\sim \begin{bmatrix} 1 & 0 & | & -1 \\ 0 & 1 & | & 3 \end{bmatrix}$

Solution: $x_1 = -1$, $x_2 = 3$ (*9-1*)

24. $\begin{bmatrix} 1 & 1 & 0 & | & 1 \\ 1 & 0 & -1 & | & -2 \\ 0 & 1 & 2 & | & 4 \end{bmatrix} (-1)R_1 + R_2 \rightarrow R_2$

$\sim \begin{bmatrix} 1 & 1 & 0 & | & 1 \\ 0 & -1 & -1 & | & -3 \\ 0 & 1 & 2 & | & 4 \end{bmatrix} (-1)R_2 \rightarrow R_2$

$\sim \begin{bmatrix} 1 & 1 & 0 & | & 1 \\ 0 & 1 & 1 & | & 3 \\ 0 & 1 & 2 & | & 4 \end{bmatrix} \begin{matrix}(-1)R_2 + R_1 \rightarrow R_1 \\ (-1)R_2 + R_3 \rightarrow R_3\end{matrix}$

$\sim \begin{bmatrix} 1 & 0 & -1 & | & -2 \\ 0 & 1 & 1 & | & 3 \\ 0 & 0 & 1 & | & 1 \end{bmatrix} \begin{matrix}R_3 + R_1 \rightarrow R_1 \\ (-1)R_3 + R_2 \rightarrow R_2\end{matrix}$

$\sim \begin{bmatrix} 1 & 0 & 0 & | & -1 \\ 0 & 1 & 0 & | & 2 \\ 0 & 0 & 1 & | & 1 \end{bmatrix}$

Solution: $x_1 = -1$, $x_2 = 2$, $x_3 = 1$
     (*9-1*)

25. $\begin{bmatrix} 1 & 2 & 3 & | & 1 \\ 2 & 3 & 4 & | & 3 \\ 1 & 2 & 1 & | & 3 \end{bmatrix} \begin{matrix}(-2)R_1 + R_3 \rightarrow R_3 \\ (-1)R_1 + R_3 \rightarrow R_3\end{matrix}$

$\sim \begin{bmatrix} 1 & 2 & 3 & | & 1 \\ 0 & -1 & -2 & | & 1 \\ 0 & 0 & -2 & | & 2 \end{bmatrix} \begin{matrix}(-1)R_2 \rightarrow R_2 \\ -\frac{1}{2}R_3 \rightarrow R_3\end{matrix}$

$\sim \begin{bmatrix} 1 & 2 & 3 & | & 1 \\ 0 & 1 & 2 & | & -1 \\ 0 & 0 & 1 & | & -1 \end{bmatrix} (-2)R_2 + R_1 \rightarrow R_1$

$\sim \begin{bmatrix} 1 & 0 & -1 & | & 3 \\ 0 & 1 & 2 & | & -1 \\ 0 & 0 & 1 & | & -1 \end{bmatrix} \begin{matrix}R_3 + R_1 \rightarrow R_1 \\ (-2)R_3 + R_2 \rightarrow R_2\end{matrix}$

$\sim \begin{bmatrix} 1 & 0 & 0 & | & 2 \\ 0 & 1 & 0 & | & 1 \\ 0 & 0 & 1 & | & -1 \end{bmatrix}$

Solution: $x_1 = 2$, $x_2 = 1$, $x_3 = -1$
     (*9-1*)

26. $\begin{bmatrix} 1 & 2 & -1 & | & 2 \\ 2 & 3 & 1 & | & -3 \\ 3 & 5 & 0 & | & -1 \end{bmatrix} \begin{matrix}(-2)R_1 + R_2 \rightarrow R_2 \\ (-3)R_1 + R_3 \rightarrow R_3\end{matrix}$

$\sim \begin{bmatrix} 1 & 2 & -1 & | & 2 \\ 0 & -1 & 3 & | & -7 \\ 0 & -1 & 3 & | & -7 \end{bmatrix} (-1)R_2 + R_3 \rightarrow R_3$

$\sim \begin{bmatrix} 1 & 2 & -1 & | & 2 \\ 0 & -1 & 3 & | & -7 \\ 0 & 0 & 0 & | & 0 \end{bmatrix} 2R_2 + R_1 \rightarrow R_1$

$\sim \begin{bmatrix} 1 & 0 & 5 & | & -12 \\ 0 & -1 & 3 & | & -7 \\ 0 & 0 & 0 & | & 0 \end{bmatrix} (-1)R_2 \rightarrow R_2$

$\sim \begin{bmatrix} 1 & 0 & 5 & | & -12 \\ 0 & 1 & -3 & | & 7 \\ 0 & 0 & 0 & | & 0 \end{bmatrix}$

This corresponds to the system
$x_1 + \quad\quad 5x_3 = -12$
$\quad x_2 - 3x_3 = 7$

Let $x_3 = t$
Then $x_2 = 3x_3 + 7$
$\quad\quad = 3t + 7$
$\quad x_1 = -5x_3 - 12$
$\quad\quad = -5t - 12$
Solution: $x_1 = -5t - 12$,
$x_2 = 3t + 7$, $x_3 = t$ for every real
number t. There are infinitely many
solutions.   (*9-1*)

27.
$$\begin{bmatrix} 1 & -2 & | & 1 \\ 2 & -1 & | & 0 \\ 1 & -3 & | & -2 \end{bmatrix} \begin{matrix} \\ (-2)R_1 + R_2 \to R_2 \\ (-1)R_1 + R_3 \to R_3 \end{matrix}$$

$$\sim \begin{bmatrix} 1 & -2 & | & 1 \\ 0 & 3 & | & -2 \\ 0 & -1 & | & -3 \end{bmatrix} 3R_3 + R_2 \to R_2$$

$$\sim \begin{bmatrix} 1 & -2 & | & 1 \\ 0 & 0 & | & -11 \\ 0 & -1 & | & -3 \end{bmatrix}$$

The second row corresponds to the equation

$$0x_1 + 0x_2 = -11,$$

so there is no solution. *(9-1)*

28.
$$\begin{bmatrix} 1 & 2 & -1 & | & 2 \\ 3 & -1 & 2 & | & -3 \end{bmatrix} (-3)R_1 + R_2 \to R_2$$

$$\sim \begin{bmatrix} 1 & 2 & -1 & | & 2 \\ 0 & -7 & 5 & | & -9 \end{bmatrix} -\tfrac{1}{7}R_2 \to R_2$$

$$\sim \begin{bmatrix} 1 & 2 & -1 & | & 2 \\ 0 & 1 & -\tfrac{5}{7} & | & \tfrac{9}{7} \end{bmatrix} (-2)R_2 + R_1 \to R_1$$

$$\sim \begin{bmatrix} 1 & 0 & \tfrac{3}{7} & | & -\tfrac{4}{7} \\ 0 & 1 & -\tfrac{5}{7} & | & \tfrac{9}{7} \end{bmatrix}$$

This corresponds to the system
$$x_1 \quad\quad + \tfrac{3}{7}x_3 = -\tfrac{4}{7}$$
$$x_2 - \tfrac{5}{7}x_3 = \tfrac{9}{7}$$

Let $x_3 = t$

Then $x_2 = \tfrac{5}{7}x_3 + \tfrac{9}{7}$
$$= \tfrac{5}{7}t + \tfrac{9}{7}$$
$$x_1 = -\tfrac{3}{7}x_3 - \tfrac{4}{7}$$
$$= -\tfrac{3}{7}t - \tfrac{4}{7}$$

Solution: $x_1 = -\tfrac{3}{7}t - \tfrac{4}{7}$, $x_2 = \tfrac{5}{7}t + \tfrac{9}{7}$, $x_3 = t$ for every real number t. There are infinitely many solutions. *(9-1)*

29. $AD = \begin{bmatrix} 1 & 2 \\ 4 & 5 \\ -3 & -1 \end{bmatrix} \begin{bmatrix} 7 & 0 & -5 \\ 0 & 8 & -2 \end{bmatrix} = \begin{bmatrix} 1\cdot 7 + 2\cdot 0 & 1\cdot 0 + 2\cdot 8 & 1(-5) + 2(-2) \\ 4\cdot 7 + 5\cdot 0 & 4\cdot 0 + 5\cdot 8 & 4(-5) + 5(-2) \\ (-3)7 + (-1)0 & (-3)0 + (-1)8 & (-3)(-5) + (-1)(-2) \end{bmatrix}$

$$= \begin{bmatrix} 7 & 16 & -9 \\ 28 & 40 & -30 \\ -21 & -8 & 17 \end{bmatrix} \qquad (9-2)$$

30. $DA = \begin{bmatrix} 7 & 0 & -5 \\ 0 & 8 & -2 \end{bmatrix} \begin{bmatrix} 1 & 2 \\ 4 & 5 \\ -3 & -1 \end{bmatrix} = \begin{bmatrix} 7\cdot 1 + 0\cdot 4 + (-5)(-3) & 7\cdot 2 + 0\cdot 5 + (-5)(-1) \\ 0\cdot 1 + 8\cdot 4 + (-2)(-3) & 0\cdot 2 + 8\cdot 5 + (-2)(-1) \end{bmatrix}$

$$= \begin{bmatrix} 22 & 19 \\ 38 & 42 \end{bmatrix} \qquad (9-2)$$

31. $BC = \begin{bmatrix} 6 \\ 0 \\ -4 \end{bmatrix} [2 \quad 4 \quad -1] = \begin{bmatrix} 6 \cdot 2 & 6 \cdot 4 & 6(-1) \\ 0 \cdot 2 & 0 \cdot 4 & 0(-1) \\ (-4)2 & (-4)4 & (-4)(-1) \end{bmatrix} = \begin{bmatrix} 12 & 24 & -6 \\ 0 & 0 & 0 \\ -8 & -16 & 4 \end{bmatrix}$ (9-2)

32. $CB = [2 \quad 4 \quad -1] \begin{bmatrix} 6 \\ 0 \\ -4 \end{bmatrix} = [2 \cdot 6 + 4 \cdot 0 + (-1)(-4)] = [16]$ (9-2)

33. Since D has 3 columns and E has 2 rows, DE is not defined.   (9-2)

34. $ED = \begin{bmatrix} 9 & -3 \\ -6 & 2 \end{bmatrix} \begin{bmatrix} 7 & 0 & -5 \\ 0 & 8 & -2 \end{bmatrix}$

$= \begin{bmatrix} 9 \cdot 7 + (-3)0 & 9 \cdot 0 + (-3)8 & 9(-5) + (-3)(-2) \\ (-6)7 + 2 \cdot 0 & (-6)0 + 2 \cdot 8 & (-6)(-5) + 2(-2) \end{bmatrix} = \begin{bmatrix} 63 & -24 & -39 \\ -42 & 16 & 26 \end{bmatrix}$ (9-2)

35. $\begin{bmatrix} 1 & 0 & 4 & | & 1 & 0 & 0 \\ -2 & 1 & 0 & | & 0 & 1 & 0 \\ 4 & -1 & 4 & | & 0 & 0 & 1 \end{bmatrix}$ $\begin{array}{l} 2R_1 + R_1 \rightarrow R_2 \\ (-4)R_1 + R_3 \rightarrow R_3 \end{array}$

$\sim \begin{bmatrix} 1 & 0 & 4 & | & 1 & 0 & 0 \\ 0 & 1 & 8 & | & 2 & 1 & 0 \\ 0 & -1 & -12 & | & -4 & 0 & 1 \end{bmatrix}$ $R_2 + R_3 \rightarrow R_3$

$\sim \begin{bmatrix} 1 & 0 & 4 & | & 1 & 0 & 0 \\ 0 & 1 & 8 & | & 2 & 1 & 0 \\ 0 & 0 & -4 & | & -2 & 1 & 1 \end{bmatrix}$ $\begin{array}{l} R_3 + R_1 \rightarrow R_1 \\ 2R_3 + R_2 \rightarrow R_2 \end{array}$

$\sim \begin{bmatrix} 1 & 0 & 0 & | & -1 & 1 & 1 \\ 0 & 1 & 0 & | & -2 & 3 & 2 \\ 0 & 0 & -4 & | & -2 & 1 & 1 \end{bmatrix} \left(-\frac{1}{4}\right)R_3 \rightarrow R_3$

$\sim \begin{bmatrix} 1 & 0 & 0 & | & -1 & 1 & 1 \\ 0 & 1 & 0 & | & -2 & 3 & 2 \\ 0 & 0 & 1 & | & \frac{1}{2} & -\frac{1}{4} & -\frac{1}{4} \end{bmatrix}$

The inverse is $\begin{bmatrix} -1 & 1 & 1 \\ -2 & 3 & 2 \\ \frac{1}{2} & -\frac{1}{4} & -\frac{1}{4} \end{bmatrix}$

Check:

$A^{-1}A = \begin{bmatrix} -1 & 1 & 1 \\ -2 & 3 & 2 \\ \frac{1}{2} & -\frac{1}{4} & -\frac{1}{4} \end{bmatrix} \begin{bmatrix} 1 & 0 & 4 \\ -2 & 1 & 0 \\ 4 & -1 & 4 \end{bmatrix}$

$= \begin{bmatrix} (-1)1 + 1(-2) + 1 \cdot 4 & (-1)0 + 1 \cdot 1 + 1(-1) & (-1)4 + 1 \cdot 0 + 1 \cdot 4 \\ (-2)1 + 3(-2) + 2 \cdot 4 & (-2)0 + 3 \cdot 1 + 2(-1) & (-2)4 + 3 \cdot 0 + 2 \cdot 4 \\ (\frac{1}{2})1 + (-\frac{1}{4})(-2) + (-\frac{1}{4})4 & (\frac{1}{2})0 + (-\frac{1}{4})1 + (-\frac{1}{4})(-1) & (\frac{1}{2})4 + (-\frac{1}{4})0 + (-\frac{1}{4})4 \end{bmatrix}$

$= \begin{bmatrix} 1 & 0 & 0 \\ 0 & 1 & 0 \\ 0 & 0 & 1 \end{bmatrix}$ (9-3)

36. $\overset{A}{\begin{bmatrix} 1 & 2 & 3 \\ 2 & 3 & 4 \\ 1 & 2 & 1 \end{bmatrix}} \overset{X}{\begin{bmatrix} x_1 \\ x_2 \\ x_3 \end{bmatrix}} = \overset{B}{\begin{bmatrix} k_1 \\ k_2 \\ k_3 \end{bmatrix}}$

The solution to $AX = B$ is $X = A^{-1}B$.

Applying reduction methods, we obtain $A^{-1} = \begin{bmatrix} -\frac{5}{2} & 2 & -\frac{1}{2} \\ 1 & -1 & 1 \\ \frac{1}{2} & 0 & -\frac{1}{2} \end{bmatrix}$ (details omitted).

Applying the inverse, we get

(A) $B = \begin{bmatrix} 1 \\ 3 \\ 3 \end{bmatrix}$ $X = \begin{bmatrix} x_1 \\ x_2 \\ x_3 \end{bmatrix} = \begin{bmatrix} -\frac{5}{2} & 2 & -\frac{1}{2} \\ 1 & -1 & 1 \\ \frac{1}{2} & 0 & -\frac{1}{2} \end{bmatrix} \begin{bmatrix} 1 \\ 3 \\ 3 \end{bmatrix} = \begin{bmatrix} 2 \\ 1 \\ -1 \end{bmatrix}$ $x_1 = 2,\ x_2 = 1,\ x_3 = -1$

(B) $B = \begin{bmatrix} 0 \\ 0 \\ -2 \end{bmatrix}$ $X = \begin{bmatrix} x_1 \\ x_2 \\ x_3 \end{bmatrix} = \begin{bmatrix} -\frac{5}{2} & 2 & -\frac{1}{2} \\ 1 & -1 & 1 \\ \frac{1}{2} & 0 & -\frac{1}{2} \end{bmatrix} \begin{bmatrix} 0 \\ 0 \\ -2 \end{bmatrix} = \begin{bmatrix} 1 \\ -2 \\ 1 \end{bmatrix}$ $x_1 = 1,\ x_2 = -2,\ x_3 = 1$

(C) $B = \begin{bmatrix} -3 \\ -4 \\ 1 \end{bmatrix}$ $X = \begin{bmatrix} x_1 \\ x_2 \\ x_3 \end{bmatrix} = \begin{bmatrix} -\frac{5}{2} & 2 & -\frac{1}{2} \\ 1 & -1 & 1 \\ \frac{1}{2} & 0 & -\frac{1}{2} \end{bmatrix} \begin{bmatrix} -3 \\ -4 \\ 1 \end{bmatrix} = \begin{bmatrix} -1 \\ 2 \\ -2 \end{bmatrix}$ $x_1 = -1,\ x_2 = 2,\ x_3 = -2$ (9-4)

37. $\begin{vmatrix} -\frac{1}{4} & \frac{3}{2} \\ \frac{1}{2} & \frac{2}{3} \end{vmatrix} = \left(-\frac{1}{4}\right)\left(\frac{2}{3}\right) - \left(\frac{1}{2}\right)\left(\frac{3}{2}\right) = -\frac{1}{6} - \frac{3}{4} = -\frac{11}{12}$ (9-4)

38. $\begin{vmatrix} 2 & -1 & 1 \\ -3 & 5 & 2 \\ 1 & -2 & 4 \end{vmatrix} = \begin{vmatrix} 0 & 0 & 1 \\ -7 & 7 & 2 \\ -7 & 2 & 4 \end{vmatrix}$ $\begin{array}{l} (-2)C_3 + C_1 \rightarrow C_1 \\ C_3 + C_2 \rightarrow C_2 \end{array}$

$= 0 + 0 + 1(-1)^{1+3} \begin{vmatrix} -7 & 7 \\ -7 & 2 \end{vmatrix} = (-1)^4[(-7)2 - (-7)7] = 35$ (9-5, 9-6)

39. $y = \dfrac{\begin{vmatrix} 1 & -6 & 1 \\ 0 & 4 & -1 \\ 2 & 2 & 1 \end{vmatrix}}{\begin{vmatrix} 1 & -2 & 1 \\ 0 & 1 & -1 \\ 2 & 2 & 1 \end{vmatrix}} = \dfrac{\begin{vmatrix} 1 & -6 & 1 \\ 0 & 4 & -1 \\ 0 & 14 & -1 \end{vmatrix}}{\begin{vmatrix} 1 & -2 & 1 \\ 0 & 1 & 0 \\ 2 & 2 & 3 \end{vmatrix}} = \dfrac{1(-1)^{1+1}\begin{vmatrix} 4 & -1 \\ 14 & -1 \end{vmatrix}}{1(-1)^{2+2}\begin{vmatrix} 1 & -1 \\ 2 & 3 \end{vmatrix}} = \dfrac{(-1)^2[4(-1) - 14(-1)]}{(-1)^4[1\cdot3 - 2(-1)]}$

$= \frac{10}{5} = 2$ (9-7)

40. (A) If the coefficient matrix has an inverse, then the system can be written as $AX = B$ and its solution can be written $X = A^{-1}B$. The system has one solution.

(B) If the coefficient matrix does not have an inverse, then the system can be solved by Gauss-Jordan elimination, but it will not have exactly one solution. The other possibilities are that the system has no solution or an infinite number of solutions, and either possibility may occur. (9-4)

41. If we assume that A is a non-zero matrix with an inverse A^{-1}, then if $A^2 = 0$ we can write $A^{-1}A^2 = A^{-1}0$ or $A^{-1}AA = 0$ or $IA = 0$ or $A = 0$. But A was assumed non-zero, so there is a contradiction. Therefore A^{-1} cannot exist for such a matrix. (9-4)

42. (A) The system is independent. There is one solution.

(B) The matrix is then

$$\begin{bmatrix} 1 & 0 & -3 & | & 4 \\ 0 & 1 & 2 & | & 5 \\ 0 & 0 & 0 & | & n \end{bmatrix}$$

The third row corresponds to the equation $0x_1 + 0x_2 + 0x_3 = n$. This is impossible. The system has no solution.

(C) The matrix is then

$$\begin{bmatrix} 1 & 0 & -3 & | & 4 \\ 0 & 1 & 2 & | & 5 \\ 0 & 0 & 0 & | & 0 \end{bmatrix}$$

There are an infinite number of solutions ($x_3 = t$, $x_2 = 5 - 2t$, $x_1 = 4 + 3t$, for t any real number.)   (9-4)

43.

$$AX - B = CX$$
$$AX - B + B - CX = CX - CX + B \quad \text{Addition property}$$
$$AX + 0 - CX = 0 + B \quad \quad M + (-M) = 0$$
$$AX - CX = B \quad \quad \quad M + 0 = M$$
$$(A - C)X = B \quad \quad \quad \text{Right distributive property}$$

(applied to subtraction of matrices; see problem 39, Exercise 6-4)

$$(A - C)^{-1}[(A - C)X] = (A - C)^{-1}B \quad \text{Left multiplication property}$$
$$[(A - C)^{-1}(A - C)]X = (A - C)^{-1}B \quad \text{Associative property}$$
$$IX = (A - C)^{-1}B \quad \quad A^{-1}A = I$$
$$X = (A - C)^{-1}B \quad \quad IX = X$$

Common Errors: $(A - C)^{-1}B \neq B(A - C)^{-1}$
$(A - C)^{-1} \neq A^{-1} - C^{-1}$

(9-4)

44.
$$\begin{bmatrix} 4 & 5 & 6 & | & 1 & 0 & 0 \\ 4 & 5 & -6 & | & 0 & 1 & 0 \\ 1 & 1 & 1 & | & 0 & 0 & 1 \end{bmatrix} \quad (-1)R_1 + R_2 \rightarrow R_2$$

$$\sim \begin{bmatrix} 4 & 5 & 6 & | & 1 & 0 & 0 \\ 0 & 0 & -12 & | & -1 & 1 & 0 \\ 1 & 1 & 1 & | & 0 & 0 & 1 \end{bmatrix} \quad R_1 \leftrightarrow R_3$$

$$\sim \begin{bmatrix} 1 & 1 & 1 & | & 0 & 0 & 1 \\ 0 & 0 & -12 & | & -1 & 1 & 0 \\ 4 & 5 & 6 & | & 1 & 0 & 0 \end{bmatrix} \quad (-4)R_1 + R_3 \rightarrow R_3$$

$$\sim \begin{bmatrix} 1 & 1 & 1 & | & 0 & 0 & 1 \\ 0 & 0 & -12 & | & -1 & 1 & 0 \\ 0 & 1 & 2 & | & 1 & 0 & -4 \end{bmatrix} \quad R_2 \leftrightarrow R_3$$

$$\sim \begin{bmatrix} 1 & 1 & 1 & | & 0 & 0 & 1 \\ 0 & 1 & 2 & | & 1 & 0 & -4 \\ 0 & 0 & -12 & | & -1 & 1 & 0 \end{bmatrix} \quad (-1)R_2 + R_1 \rightarrow R_1$$

$$\sim \begin{bmatrix} 1 & 0 & -1 & | & -1 & 0 & 5 \\ 0 & 1 & 2 & | & 1 & 0 & -4 \\ 0 & 0 & -12 & | & -1 & 1 & 0 \end{bmatrix} \quad -\frac{1}{12} R_3 \rightarrow R_3$$

$$\sim \begin{bmatrix} 1 & 0 & -1 & | & -1 & 0 & 5 \\ 0 & 1 & 2 & | & 1 & 0 & -4 \\ 0 & 0 & 1 & | & \frac{1}{12} & -\frac{1}{12} & 0 \end{bmatrix} \quad \begin{array}{l} R_3 + R_1 \rightarrow R_1 \\ (-2)R_3 + R_2 \rightarrow R_2 \end{array}$$

$$\sim \begin{bmatrix} 1 & 0 & 0 & \vline & -\frac{11}{12} & -\frac{1}{12} & 5 \\ 0 & 1 & 0 & \vline & \frac{10}{12} & \frac{2}{12} & -4 \\ 0 & 0 & 1 & \vline & \frac{1}{12} & -\frac{1}{12} & 0 \end{bmatrix}$$

The inverse is $\begin{bmatrix} -\frac{11}{12} & -\frac{1}{12} & 5 \\ \frac{10}{12} & \frac{2}{12} & -4 \\ \frac{1}{12} & -\frac{1}{12} & 0 \end{bmatrix}$ or $\frac{1}{12}\begin{bmatrix} -11 & -1 & 60 \\ 10 & 2 & -48 \\ 1 & -1 & 0 \end{bmatrix}$

Check: $A^{-1}A = \frac{1}{12}\begin{bmatrix} -11 & -1 & 60 \\ 10 & 2 & -48 \\ 1 & -1 & 0 \end{bmatrix}\begin{bmatrix} 4 & 5 & 6 \\ 4 & 5 & -6 \\ 1 & 1 & 1 \end{bmatrix}$

$$= \frac{1}{12}\begin{bmatrix} (-11)4 + (-1)4 + 60\cdot1 & (-11)5 + (-1)5 + 60\cdot1 & (-11)6 + (-1)(-6) + 60\cdot1 \\ 10\cdot4 + 2\cdot4 + (-48)1 & 10\cdot5 + 2\cdot5 + (-48)1 & 10\cdot6 + 2(-6) + (-48)1 \\ 1\cdot4 + (-1)4 + 0\cdot1 & 1\cdot5 + (-1)5 + 0\cdot1 & 1\cdot6 + (-1)(-6) + 0\cdot1 \end{bmatrix}$$

$$= \frac{1}{12}\begin{bmatrix} 12 & 0 & 0 \\ 0 & 12 & 0 \\ 0 & 0 & 12 \end{bmatrix} = \begin{bmatrix} 1 & 0 & 0 \\ 0 & 1 & 0 \\ 0 & 0 & 1 \end{bmatrix} = I \qquad\qquad (9\text{-}3)$$

45. Multiplying the first two equations by 100, the system becomes

$$4x_1 + 5x_2 + 6x_3 = 36,000$$
$$4x_1 + 5x_2 - 6x_3 = 12,000$$
$$x_1 + x_2 + x_3 = 7,000$$

As a matrix equation, we have

$$\begin{matrix} \quad A & X & \quad B \end{matrix}$$
$$\begin{bmatrix} 4 & 5 & 6 \\ 4 & 5 & -6 \\ 1 & 1 & 1 \end{bmatrix}\begin{bmatrix} x_1 \\ x_2 \\ x_3 \end{bmatrix} = \begin{bmatrix} 36,000 \\ 12,000 \\ 7,000 \end{bmatrix}$$

The solution to $AX = B$ is $X = A^{-1}B$. Using A^{-1} from problem 44, we have

$$X = \begin{bmatrix} x_1 \\ x_2 \\ x_3 \end{bmatrix} = \frac{1}{12}\begin{bmatrix} -11 & -1 & 60 \\ 10 & 2 & -48 \\ 1 & -1 & 0 \end{bmatrix}\begin{bmatrix} 36,000 \\ 12,000 \\ 7,000 \end{bmatrix}$$

$$= \frac{1}{12}\begin{bmatrix} (-11)(36,000) + (-1)(12,000) + (60)(7,000) \\ (10)(36,000) + (2)(12,000) + (-48)(7,000) \\ 1(36,000) + (-1)(12,000) + (0)(7,000) \end{bmatrix} = \frac{1}{12}\begin{bmatrix} 12,000 \\ 48,000 \\ 24,000 \end{bmatrix} = \begin{bmatrix} 1,000 \\ 4,000 \\ 2,000 \end{bmatrix}$$

Solution: $x_1 = 1,000$, $x_2 = 4,000$, $x_3 = 2,000$ $\qquad\qquad (9\text{-}4)$

46. $\begin{vmatrix} -1 & 4 & 1 & 1 \\ 5 & -1 & 2 & -1 \\ 2 & -1 & 0 & 3 \\ -3 & 3 & 0 & 3 \end{vmatrix} = \begin{vmatrix} -1 & 4 & 1 & 1 \\ 7 & -9 & 0 & -3 \\ 2 & -1 & 0 & 3 \\ -3 & 3 & 0 & 3 \end{vmatrix}$ $\quad (-2)R_1 + R_2 \rightarrow R_2$

$$= 1(-1)^{1+3}\begin{vmatrix} 7 & -9 & -3 \\ 2 & -1 & 3 \\ -3 & 3 & 3 \end{vmatrix} + 0 + 0 + 0$$

$$= \begin{vmatrix} 7 & -9 & -3 \\ 2 & -1 & 3 \\ -3 & 3 & 3 \end{vmatrix} \quad \begin{matrix} R_1 + R_2 \rightarrow R_2 \\ R_1 + R_3 \rightarrow R_3 \end{matrix}$$

$$= \begin{vmatrix} 7 & -9 & -3 \\ 9 & -10 & 0 \\ 4 & -6 & 0 \end{vmatrix} = (-3)(-1)^{1+3}\begin{vmatrix} 9 & -10 \\ 4 & -6 \end{vmatrix} + 0 + 0$$

$$= (-3)(-1)^4[9(-6) - 4(-10)]$$
$$= (-3)(-14) = 42 \qquad\qquad (9\text{-}6)$$

47. $\begin{vmatrix} u + kv & v \\ w + kx & x \end{vmatrix} = (u + kv)x - (w + kx)v = ux + kvx - wv - kvx = ux - wv$

$$= \begin{vmatrix} u & v \\ w & x \end{vmatrix} \tag{9-6}$$

48. The statements: $(1, 2)$ satisfies the equation $\begin{vmatrix} x & y & 1 \\ 1 & 2 & 1 \\ -1 & 5 & 1 \end{vmatrix} = 0$ and $(-1, 5)$

satisfies the equation $\begin{vmatrix} x & y & 1 \\ 1 & 2 & 1 \\ -1 & 5 & 1 \end{vmatrix} = 0$ are equivalent to the statements

$\begin{vmatrix} 1 & 2 & 1 \\ 1 & 2 & 1 \\ -1 & 5 & 1 \end{vmatrix} = 0$ and $\begin{vmatrix} -1 & 5 & 1 \\ 1 & 2 & 1 \\ -1 & 5 & 1 \end{vmatrix} = 0$. The latter statements are true by Theorem 4,

Section 9-6. All other points on the line through the given points will also
satisfy the equation. (9-6)

49. Let x_1 = number of grams of mix A

$\qquad x_2$ = number of grams of mix B

$\qquad x_3$ = number of grams of mix C

We have

$0.30x_1 + 0.20x_2 + 0.10x_3 = 27$ (protein)

$0.03x_1 + 0.05x_2 + 0.04x_3 = 5.4$ (fat)

$0.10x_1 + 0.20x_2 + 0.10x_3 = 19$ (moisture)

Multiply the first and third equations by 10, and the second equation by 100 to
clear decimals:

$3x_1 + 2x_2 + x_3 = 270$

$3x_1 + 5x_2 + 4x_3 = 540$

$x_1 + 2x_2 + x_3 = 190$

Form the augmented matrix and solve by Gauss-Jordan elimination.

$\begin{bmatrix} 3 & 2 & 1 & | & 270 \\ 3 & 5 & 4 & | & 540 \\ 1 & 2 & 1 & | & 190 \end{bmatrix} R_3 \leftrightarrow R_1$

$\sim \begin{bmatrix} 1 & 2 & 1 & | & 190 \\ 3 & 5 & 4 & | & 540 \\ 3 & 2 & 1 & | & 270 \end{bmatrix} \begin{matrix} \\ (-3)R_1 + R_2 \rightarrow R_2 \\ (-3)R_1 + R_3 \rightarrow R_3 \end{matrix}$

$\sim \begin{bmatrix} 1 & 2 & 1 & | & 190 \\ 0 & -1 & 1 & | & -30 \\ 0 & -4 & -2 & | & -300 \end{bmatrix} \begin{matrix} \\ (-1)R_2 \rightarrow R_2 \\ \\ \end{matrix}$

$\sim \begin{bmatrix} 1 & 2 & 1 & | & 190 \\ 0 & 1 & -1 & | & 30 \\ 0 & -4 & -2 & | & -300 \end{bmatrix} \begin{matrix} (-2)R_2 + R_1 \rightarrow R_1 \\ \\ 4R_2 + R_3 \rightarrow R_3 \end{matrix}$

$\sim \begin{bmatrix} 1 & 0 & 3 & | & 130 \\ 0 & 1 & -1 & | & 30 \\ 0 & 0 & -6 & | & -180 \end{bmatrix} -\frac{1}{6} R_3 \rightarrow R_3$

$\sim \begin{bmatrix} 1 & 0 & 3 & | & 130 \\ 0 & 1 & -1 & | & 30 \\ 0 & 0 & 1 & | & 30 \end{bmatrix} \begin{matrix} (-3)R_3 + R_1 \rightarrow R_1 \\ R_3 + R_2 \rightarrow R_2 \\ \\ \end{matrix}$

$$\sim \begin{bmatrix} 1 & 0 & 0 & | & 40 \\ 0 & 1 & 0 & | & 60 \\ 0 & 0 & 1 & | & 30 \end{bmatrix}$$

Therefore

x_1 = 40 grams Mix A
x_2 = 60 grams Mix B
x_3 = 30 grams Mix C (*9-1*)

50. (A) Let x_1 = number of nickels
            x_2 = number of dimes
        Then $x_1 + x_2 = 30$ (total number of coins)
            $5x_1 + 10x_2 = 190$ (total value of coins)
We form the augmented matrix and solve by Gauss-Jordan elimination.

$$\begin{bmatrix} 1 & 1 & | & 30 \\ 5 & 10 & | & 190 \end{bmatrix} (-5)R_1 + R_2 \to R_2$$

$$\sim \begin{bmatrix} 1 & 1 & | & 30 \\ 0 & 5 & | & 40 \end{bmatrix} \tfrac{1}{5}R_2 \to R_2$$

$$\sim \begin{bmatrix} 1 & 1 & | & 30 \\ 0 & 1 & | & 8 \end{bmatrix} (-1)R_2 + R_1 \to R_1$$

$$\sim \begin{bmatrix} 1 & 0 & | & 22 \\ 0 & 1 & | & 8 \end{bmatrix}$$

The augmented matrix is in reduced form. It corresponds to the system
x_1 = 22 nickels
x_2 = 8 dimes

   (B) Let x_1 = number of nickels
            x_2 = number of dimes
            x_3 = number of quarters
        Then $x_1 + x_2 + x_3 = 30$ (total number of coins)
            $5x_1 + 10x_2 + 25x_3 = 190$ (total value of coins)

We form the augmented matrix and solve by Gauss-Jordan elimination

$$\begin{bmatrix} 1 & 1 & 1 & | & 30 \\ 5 & 10 & 25 & | & 190 \end{bmatrix} (-5)R_1 + R_2 \to R_2$$

$$\sim \begin{bmatrix} 1 & 1 & 1 & | & 30 \\ 0 & 5 & 20 & | & 40 \end{bmatrix} \tfrac{1}{5}R_2 \to R_2$$

$$\sim \begin{bmatrix} 1 & 1 & 1 & | & 30 \\ 0 & 1 & 4 & | & 8 \end{bmatrix} (-1)R_2 + R_1 \to R_1$$

$$\sim \begin{bmatrix} 1 & 0 & -3 & | & 22 \\ 0 & 1 & 4 & | & 8 \end{bmatrix}$$

The augmented matrix is in reduced form. It corresponds to the system:
$x_1 - 3x_3 = 22$
$ x_2 + 4x_3 = 8$

Let $x_3 = t$. Then $x_2 = -4x_3 + 8$
$$= -4t + 8$$
$$x_1 = 3x_3 + 22$$
$$= 3t + 22$$

A solution is achieved, not for every real value of t, but for integer values of t that give rise to non-negative x_1, x_2, x_3.

$x_1 \geq 0$ means $3t + 22 \geq 0$ or $t \geq -7\frac{1}{3}$

$x_2 \geq 0$ means $-4t + 8 \geq 0$ or $t \leq 2$

$x_3 \geq 0$ means $t \geq 0$

The only integer values of t that satisfy these conditions are 0, 1, 2. So we have the solutions

$x_1 = 3t + 22$ nickels

$x_2 = 8 - 4t$ dimes

$x_3 = t$ quarters

where $t = 0$, 1, or 2

(9-1)

51. Let x_1 = number of tons at Big Bend

x_2 = number of tons at Saw Pit

Then

$0.05x_1 + 0.03x_2$ = number of tons of nickel at both mines = k_1

$0.07x_1 + 0.04x_2$ = number of tons of copper at both mines = k_2

We solve

$0.05x_1 + 0.03x_2 = k_1$

$0.07x_1 + 0.04x_4 = k_2$,

For arbitrary k_1 and k_2, by writing the system as a matrix equation.

$$\overset{A}{\begin{bmatrix} 0.05 & 0.03 \\ 0.07 & 0.04 \end{bmatrix}} \overset{X}{\begin{bmatrix} x_1 \\ x_2 \end{bmatrix}} = \overset{B}{\begin{bmatrix} k_1 \\ k_2 \end{bmatrix}}$$

If A^{-1} exists, then $X = A^{-1}B$. To find A^{-1}, we perform row operations on

$$\begin{bmatrix} 0.05 & 0.03 & | & 1 & 0 \\ 0.07 & 0.04 & | & 0 & 1 \end{bmatrix} 20R_1 \rightarrow R_1$$

$$\sim \begin{bmatrix} 1 & 0.6 & | & 20 & 0 \\ 0.07 & 0.04 & | & 0 & 1 \end{bmatrix} -0.07R_1 + R_2 \rightarrow R_2$$

$$\sim \begin{bmatrix} 1 & 0.6 & | & 20 & 0 \\ 0 & -0.002 & | & -1.4 & 1 \end{bmatrix} -500R_2 \rightarrow R_2$$

$$\sim \begin{bmatrix} 1 & 0.6 & | & 20 & 0 \\ 0 & 1 & | & 700 & -500 \end{bmatrix} -0.6R_2 + R_1 \rightarrow R_1$$

$$\sim \begin{bmatrix} 1 & 0 & | & -400 & 300 \\ 0 & 1 & | & 700 & -500 \end{bmatrix}$$

$$A^{-1} = \begin{bmatrix} -400 & 300 \\ 700 & -500 \end{bmatrix}$$

Check: $A^{-1}A = \begin{bmatrix} -400 & 300 \\ 700 & -500 \end{bmatrix} \begin{bmatrix} 0.05 & 0.03 \\ 0.07 & 0.04 \end{bmatrix} = \begin{bmatrix} 1 & 0 \\ 0 & 1 \end{bmatrix}$

We can now solve the system as:

$$\overset{X}{\begin{bmatrix} x_1 \\ x_2 \end{bmatrix}} = \overset{A^{-1}}{\begin{bmatrix} -400 & 300 \\ 700 & -500 \end{bmatrix}} \overset{B}{\begin{bmatrix} k_1 \\ k_2 \end{bmatrix}}$$

(A) If $k_1 = 3.6$, $k_2 = 5$,

$$\begin{bmatrix} x_1 \\ x_2 \end{bmatrix} = \begin{bmatrix} -400 & 300 \\ 700 & -500 \end{bmatrix} \begin{bmatrix} 3.6 \\ 5 \end{bmatrix} = \begin{bmatrix} 60 \\ 20 \end{bmatrix}$$

60 tons of ore must be produced at Big Bend, 20 tons of ore at Saw Pit.

(B) If $k_1 = 3$, $k_2 = 4.1$,

$$\begin{bmatrix} x_1 \\ x_2 \end{bmatrix} = \begin{bmatrix} -400 & 300 \\ 700 & -500 \end{bmatrix} \begin{bmatrix} 3 \\ 4.1 \end{bmatrix} = \begin{bmatrix} 30 \\ 50 \end{bmatrix}$$

30 tons of ore must be produced at Big Bend, 50 tons of ore at Saw Pit.

(C) If $k_1 = 3.2$, $k_2 = 4.4$,

$$\begin{bmatrix} x_1 \\ x_2 \end{bmatrix} = \begin{bmatrix} -400 & 300 \\ 700 & -500 \end{bmatrix} \begin{bmatrix} 3.2 \\ 4.4 \end{bmatrix} = \begin{bmatrix} 40 \\ 40 \end{bmatrix}$$

40 tons of ore must be produced at Big Bend, 40 tons of ore at Saw Pit.   (9-4)

52. (A) The labor cost of producing one printer stand at the South Carolina plant is the product of the stand row of L with South Carolina column of H.

$$[0.9 \quad 1.8 \quad 0.6] \begin{bmatrix} 10.00 \\ 8.50 \\ 4.50 \end{bmatrix} = 27 \text{ dollars}$$

(B) The matrix HL has no clear meaning, but the matrix LH represents the total labor costs for each item at each plant.

(C) $LH = \begin{bmatrix} 1.7 & 2.4 & 0.8 \\ 0.9 & 1.8 & 0.6 \end{bmatrix} \begin{bmatrix} 11.50 & 10.00 \\ 9.50 & 8.50 \\ 5.00 & 4.50 \end{bmatrix}$

$= \begin{bmatrix} 1.7 \cdot 11.50 + 2.4 \cdot 9.50 + 0.8 \cdot 5.00 & 1.7 \cdot 10.00 + 2.4 \cdot 8.50 + 0.8 \cdot 4.50 \\ 0.9 \cdot 11.50 + 1.8 \cdot 9.50 + 0.6 \cdot 5.00 & 0.9 \cdot 10.00 + 1.8 \cdot 8.50 + 0.6 \cdot 4.50 \end{bmatrix}$

 N.C.　　S.C.

$\begin{bmatrix} \$46.35 & \$41.00 \\ \$30.45 & \$27.00 \end{bmatrix} \begin{matrix} \text{Desks} \\ \text{Stands} \end{matrix}$

(9-2)

53. (A) The average monthly production for the months of January and February is represented by the matrix $\frac{1}{2}(J + F)$

$\frac{1}{2}(J + F) = \frac{1}{2}\left(\begin{bmatrix} 1,500 & 1,650 \\ 850 & 700 \end{bmatrix} + \begin{bmatrix} 1,700 & 1,810 \\ 850 & 740 \end{bmatrix}\right) = \frac{1}{2}\begin{bmatrix} 3,200 & 3,460 \\ 1,780 & 1,440 \end{bmatrix}$

 N.C.　　S.C.

$= \begin{bmatrix} 1,600 & 1,730 \\ 890 & 720 \end{bmatrix} \begin{matrix} \text{Desks} \\ \text{Stands} \end{matrix}$

(B) The increase in production from January to February is represented by the matrix $F - J$.

$F - J = \begin{bmatrix} 1,700 & 1,810 \\ 930 & 740 \end{bmatrix} - \begin{bmatrix} 1,500 & 1,650 \\ 850 & 700 \end{bmatrix}$

 N.C.　S.C.

$= \begin{bmatrix} 200 & 160 \\ 80 & 40 \end{bmatrix} \begin{matrix} \text{Desks} \\ \text{Stands} \end{matrix}$

(C) $J\begin{bmatrix} 1 \\ 1 \end{bmatrix} = \begin{bmatrix} 1,500 & 1,650 \\ 850 & 700 \end{bmatrix}\begin{bmatrix} 1 \\ 1 \end{bmatrix} = \begin{bmatrix} 3,150 \\ 1,550 \end{bmatrix} \begin{matrix} \text{Desks} \\ \text{Stands} \end{matrix}$

This matrix represents the total production of each item in January.

(9-2)

54. The inverse of matrix B is calculated to be

$$B^{-1} = \begin{bmatrix} 1 & 1 & -1 \\ 0 & -1 & 1 \\ -1 & 0 & 1 \end{bmatrix}$$

Putting the coded message into matrix form and multiplying by B^{-1} yields

$$\begin{bmatrix} 1 & 1 & -1 \\ 0 & -1 & 1 \\ -1 & 0 & 1 \end{bmatrix}\begin{bmatrix} 25 & 24 & 21 & 41 & 21 & 52 \\ 8 & 25 & 41 & 30 & 32 & 52 \\ 26 & 33 & 48 & 50 & 41 & 79 \end{bmatrix} = \begin{bmatrix} 7 & 16 & 14 & 21 & 12 & 25 \\ 18 & 8 & 7 & 20 & 9 & 27 \\ 1 & 9 & 27 & 9 & 20 & 27 \end{bmatrix}$$

This decodes to

  7 18 1 16 8 9 14 7 27 21 20 9 12 9 20 25 27 27
  G R   A P   H I N   G   U   T   I L   I T   Y   (*9-3*)

Chapters 8 & 9 Cumulative Review

1. We choose elimination by addition.
We multiply the top equation by 3,
the bottom by 5, and add.

$$9x - 15y = 33$$
$$10x + 15y = \ \ 5$$
$$\overline{\hspace{0.5cm}}$$
$$19x \hspace{1.2cm} = 38$$
$$x = \ \ 2$$

Substituting $x = 2$ in the bottom
equation, we have
$$2(2) + 3y = 1$$
$$4 + 3y = 1$$
$$3y = -3$$
$$y = -1$$
Solution: $(2, -1)$ *(8-1)*

2.

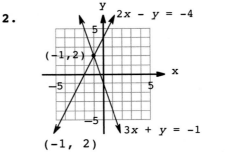

$(-1, \ 2)$ *(8-1)*

3. $-6x + 3y = 2$
$\ \ 2x - \ \ y = 1$
Solve the second equation for y
and substitute into the first.
$$2x - y = 1$$
$$2x - 1 = y$$
$$-6x + 3(2x - 1) = 2$$
$$-6x + 6x - 3 = 2$$
$$-3 = 2$$
This is a contradiction, so the
system has no solution.   *(8-1)*

4.

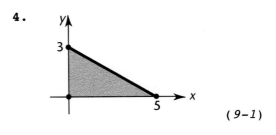

(9-1)

5.

Corner Point $(x, \ y)$	Objective Function $z = 2x + 3y$	
$(0, \ 4)$	12	
$(5, \ 0)$	10	Minimum value
$(6, \ 7)$	33	Maximum value
$(0, \ 10)$	30	

The minimum value of z on S is 10 at $(5, \ 0)$.
The maximum value of z on S is 33 at $(6, \ 7)$.   *(8-5)*

6. (A) $M - 2N = \begin{bmatrix} 2 & 1 \\ 1 & -3 \end{bmatrix} - 2\begin{bmatrix} 1 & 2 \\ -1 & 3 \end{bmatrix} = \begin{bmatrix} 2 & 1 \\ 1 & -3 \end{bmatrix} - \begin{bmatrix} 2 & 4 \\ -2 & 6 \end{bmatrix} = \begin{bmatrix} 0 & -3 \\ 3 & -9 \end{bmatrix}$

(B) $P + Q$ is not defined

(C) $PQ = \begin{bmatrix} 1 & 2 \end{bmatrix}\begin{bmatrix} -1 \\ 2 \end{bmatrix} = [1(-1) + 2 \cdot 2] = [3]$

(D) $MN = \begin{bmatrix} 2 & 1 \\ 1 & -3 \end{bmatrix}\begin{bmatrix} 1 & 2 \\ -1 & 3 \end{bmatrix} = \begin{bmatrix} 2 \cdot 1 + 1(-1) & 2 \cdot 2 + 1 \cdot 3 \\ 1 \cdot 1 + (-3)(-1) & 1 \cdot 2 + (-3)3 \end{bmatrix} = \begin{bmatrix} 1 & 7 \\ 4 & -7 \end{bmatrix}$

(E) $PN = \begin{bmatrix} 1 & 2 \end{bmatrix}\begin{bmatrix} 1 & 2 \\ -1 & 3 \end{bmatrix} = \begin{bmatrix} 1 \cdot 1 + 2(-1) & 1 \cdot 2 + 2 \cdot 3 \end{bmatrix} = \begin{bmatrix} -1 & 8 \end{bmatrix}$

(F) QM is not defined   *(9-2)*

7. $\begin{vmatrix} 0 & 2 & 0 \\ 1 & 3 & 2 \\ -1 & 4 & 3 \end{vmatrix} = 0 + 2(-1)^{1+2}\begin{vmatrix} 1 & 2 \\ -1 & 3 \end{vmatrix} + 0 = 2(-1)^3[1 \cdot 3 - (-1)2] = -10$ *(9-5)*

8. (A) $x_1 = 3$

$\quad\quad x_2 = -4$

(B) $x_1 - 2x_2 = 3$

$\quad$ Let $x_2 = t$

$\quad$ Then $x_1 = 2x_2 + 3$

$\quad\quad\quad\quad = 2t + 3$

Solution: $x_1 = 2t + 3$, $x_2 = t$ for every real number t.

(C) $\quad x_1 - 2x_2 = 3$

$\quad\quad 0x_1 + 0x_2 = 1$

No solution. $\quad\quad\quad\quad\quad\quad\quad\quad$ *(9-1)*

9. (A) $\begin{bmatrix} 1 & 1 & | & 3 \\ -1 & 1 & | & 5 \end{bmatrix}$

(B) $\begin{bmatrix} 1 & 1 & | & 3 \\ -1 & 1 & | & 5 \end{bmatrix} R_1 + R_2 \rightarrow R_2$

$\sim \begin{bmatrix} 1 & 1 & | & 3 \\ 0 & 2 & | & 8 \end{bmatrix} \frac{1}{2}R_2 \rightarrow R_2$

$\sim \begin{bmatrix} 1 & 1 & | & 3 \\ 0 & 1 & | & 4 \end{bmatrix} (-1)R_2 + R_1 \rightarrow R_1$

$\sim \begin{bmatrix} 1 & 0 & | & -1 \\ 0 & 1 & | & 4 \end{bmatrix}$

(C) Solution: $x_1 = -1$, $x_2 = 4$

$\quad\quad\quad\quad\quad\quad$ *(9-1, 9-2)*

10. (A) As a matrix equation the system becomes

$\quad\quad\quad A \quad\quad\quad X \quad\quad B$

$\begin{bmatrix} 1 & -3 \\ 2 & -5 \end{bmatrix} \begin{bmatrix} x_1 \\ x_2 \end{bmatrix} = \begin{bmatrix} k_1 \\ k_2 \end{bmatrix}$

(B) To find A^{-1}, we perform row operations on

$\begin{bmatrix} 1 & -3 & | & 1 & 0 \\ 2 & -5 & | & 0 & 1 \end{bmatrix} (-2)R_1 + R_2 \rightarrow R_2$

$\sim \begin{bmatrix} 1 & -3 & | & 1 & 0 \\ 0 & 1 & | & -2 & 1 \end{bmatrix} 3R_2 + R_1 \rightarrow R_1$

$\sim \begin{bmatrix} 1 & 0 & | & -5 & 3 \\ 0 & 1 & | & -2 & 1 \end{bmatrix}$

$A^{-1} = \begin{bmatrix} -5 & 3 \\ -2 & 1 \end{bmatrix}$

Check: $A^{-1}A = \begin{bmatrix} -5 & 3 \\ -2 & 1 \end{bmatrix} \begin{bmatrix} 1 & -3 \\ 2 & -5 \end{bmatrix} = \begin{bmatrix} (-5)1 + 3 \cdot 2 & (-5)(-3) + 3(-5) \\ (-2)1 + 1 \cdot 2 & (-2)(-3) + 1(-5) \end{bmatrix} = \begin{bmatrix} 1 & 0 \\ 0 & 1 \end{bmatrix}$

(C) The solution of $AX = B$ is $X = A^{-1}B$. Using the result of (B), we get

$X = \begin{bmatrix} -5 & 3 \\ -2 & 1 \end{bmatrix} \begin{bmatrix} k_1 \\ k_2 \end{bmatrix}$

If $k_1 = -2$, $k_2 = 1$, we get

$\begin{bmatrix} x_1 \\ x_2 \end{bmatrix} = \begin{bmatrix} -5 & 3 \\ -2 & 1 \end{bmatrix} \begin{bmatrix} -2 \\ 1 \end{bmatrix} = \begin{bmatrix} (-5)(-2) + 3 \cdot 1 \\ (-2)(-2) + 1 \cdot 1 \end{bmatrix} = \begin{bmatrix} 13 \\ 5 \end{bmatrix}$ $x_1 = 13$, $x_2 = 5$

(D) If $k_1 = 1$, $k_2 = -2$, reasoning as in (C) we get

$\begin{bmatrix} x_1 \\ x_2 \end{bmatrix} = \begin{bmatrix} -5 & 3 \\ -2 & 1 \end{bmatrix} \begin{bmatrix} 1 \\ -2 \end{bmatrix} = \begin{bmatrix} (-5)1 + 3(-2) \\ (-2)1 + 1(-2) \end{bmatrix} = \begin{bmatrix} -11 \\ -4 \end{bmatrix}$ $x_1 = -11$, $x_2 = -4$ $\quad\quad$ *(9-4)*

11. (A) $D = \begin{vmatrix} 2 & -3 \\ 4 & -5 \end{vmatrix} = 2(-5) - 4(-3) = 2$

(B) $x = \dfrac{\begin{vmatrix} 1 & -3 \\ 2 & -5 \end{vmatrix}}{D} = \dfrac{1}{2} \quad\quad y = \dfrac{\begin{vmatrix} 2 & 1 \\ 4 & 2 \end{vmatrix}}{D} = \dfrac{0}{2} = 0$ $\quad\quad\quad\quad\quad\quad$ *(9-7)*

12. We write the augmented matrix:

$$\begin{bmatrix} 1 & 3 & | & 10 \\ 2 & -1 & | & -1 \end{bmatrix} (-2)R_1 + R_2 \rightarrow R_2$$

$$-2 \quad -6 \quad -20$$

↑
Need a 0 here

$$\sim \begin{bmatrix} 1 & 3 & | & 10 \\ 0 & -7 & | & -21 \end{bmatrix} -\frac{1}{7}R_2 \rightarrow R_2 \qquad \text{corresponds to the linear system} \qquad \begin{array}{r} x_1 + 3x_2 = 10 \\ -7x_2 = -21 \end{array}$$

↑
Need a 1 here

Need a 0 here
↓

$$\begin{bmatrix} 1 & 3 & | & 10 \\ 0 & 1 & | & 3 \end{bmatrix} (-3)R_2 + R_1 \rightarrow R_1 \qquad \text{corresponds to the linear system} \qquad \begin{array}{r} x_1 + 3x_2 = 10 \\ x_2 = 3 \end{array}$$

$$0 \quad -3 \quad -9$$

$$\sim \begin{bmatrix} 1 & 0 & | & 1 \\ 0 & 1 & | & 3 \end{bmatrix} \text{corresponds to the linear system} \quad \begin{array}{r} x_1 = 1 \\ x_2 = 3 \end{array}$$

The solution is $x_1 = 1$, $x_2 = 3$. Each pair of lines graphed below has the same intersection point, (1, 3).

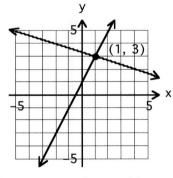

$$\begin{array}{r} x_1 + 3x_2 = 10 \\ 2x_1 - x_2 = -1 \end{array}$$

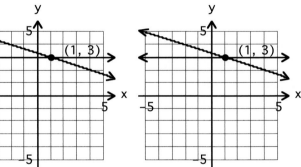

$$\begin{array}{r} x_1 + 3x_2 = 10 \\ -7x_2 = -21 \end{array}$$

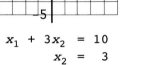

$$\begin{array}{r} x_1 + 3x_2 = 10 \\ x_2 = 3 \end{array}$$

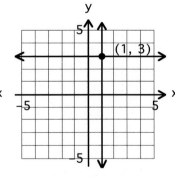

$$\begin{array}{rr} x_1 & = 1 \\ x_2 & = 3 \end{array} \quad (8\text{-}1)$$

13. Here is a computer-generated graph of the system, entered as

$$y = \frac{7 + 2x}{3}$$

$$y = \frac{18 - 3x}{4}$$

After zooming in (graph not shown) the intersection point is located at (1.53, 3.35) to two decimal places.

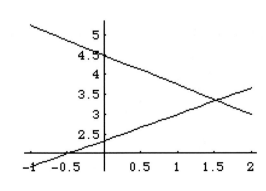

$$(8\text{-}1)$$

14. $\begin{bmatrix} 1 & 2 & -1 & | & 3 \\ 0 & 1 & 1 & | & -2 \\ 2 & 3 & 1 & | & 0 \end{bmatrix}$ $(-2)R_1 + R_3 \rightarrow R_3$

$\sim \begin{bmatrix} 1 & 2 & -1 & | & 3 \\ 0 & 1 & 1 & | & -2 \\ 0 & -1 & 3 & | & -6 \end{bmatrix}$ $\begin{array}{l}(-2)R_2 + R_1 \rightarrow R_1 \\ \\ R_2 + R_3 \rightarrow R_3 \end{array}$

$\sim \begin{bmatrix} 1 & 0 & -3 & | & 7 \\ 0 & 1 & 1 & | & -2 \\ 0 & 0 & 4 & | & -8 \end{bmatrix}$ $\frac{1}{4}R_3 \rightarrow R_3$

$\sim \begin{bmatrix} 1 & 0 & -3 & | & 7 \\ 0 & 1 & 1 & | & -2 \\ 0 & 0 & 1 & | & -2 \end{bmatrix}$ $\begin{array}{l} 3R_3 + R_1 \rightarrow R_1 \\ (-1)R_3 + R_2 \rightarrow R_2 \end{array}$

$\sim \begin{bmatrix} 1 & 0 & 0 & | & 1 \\ 0 & 1 & 0 & | & 0 \\ 0 & 0 & 1 & | & -2 \end{bmatrix}$

Solution: $x_1 = 1$, $x_2 = 0$, $x_3 = -2$

$(9-1)$

15. $\begin{bmatrix} 1 & 1 & -2 & | & 2 \\ 0 & 4 & 6 & | & -1 \\ 0 & 6 & 9 & | & 0 \end{bmatrix}$ $(-\frac{2}{3})R_2 + R_3 \rightarrow R_3$

$\sim \begin{bmatrix} 1 & 1 & -2 & | & 2 \\ 0 & 4 & 6 & | & -1 \\ 0 & 0 & 0 & | & \frac{2}{3} \end{bmatrix}$

The last row corresponds to the equation
$$0x_1 + 0x_2 + 0x_3 = \tfrac{2}{3} ,$$
so there is no solution. $(9-1)$

16. $\begin{bmatrix} 1 & -2 & 1 & | & 1 \\ 3 & -2 & -1 & | & -5 \end{bmatrix}$ $(-3)R_1 + R_2 \rightarrow R_2$

$\sim \begin{bmatrix} 1 & -2 & 1 & | & 1 \\ 0 & 4 & -4 & | & -8 \end{bmatrix}$ $\frac{1}{4}R_2 \rightarrow R_2$

$\sim \begin{bmatrix} 1 & -2 & 1 & | & 1 \\ 0 & 1 & -1 & | & -2 \end{bmatrix}$ $2R_2 + R_1 \rightarrow R_1$

$\sim \begin{bmatrix} 1 & 0 & -1 & | & -3 \\ 0 & 1 & -1 & | & -2 \end{bmatrix}$

This corresponds to the system

$x_1 \qquad - x_3 = -3$

$\qquad x_2 - x_3 = -2$

Let $x_3 = t$

Then $x_2 = x_3 - 2$

$\qquad = t - 2$

$\quad x_1 = x_3 - 3$

$\qquad = t - 3$

Solution: $x_1 = t - 3$, $x_2 = t - 2$, $x_3 = t$ for every real number t. $(9-1)$

17. (A) $MN = \begin{bmatrix} 1 & 2 & -1 \end{bmatrix} \begin{bmatrix} 1 \\ -1 \\ 2 \end{bmatrix} = \begin{bmatrix} 1 \cdot 1 + 2(-1) + (-1)2 \end{bmatrix} = \begin{bmatrix} -3 \end{bmatrix}$

(B) $NM = \begin{bmatrix} 1 \\ -1 \\ 2 \end{bmatrix} \begin{bmatrix} 1 & 2 & -1 \end{bmatrix} = \begin{bmatrix} 1 \cdot 1 & 1 \cdot 2 & 1(-1) \\ (-1)1 & (-1)2 & (-1)(-1) \\ 2 \cdot 1 & 2 \cdot 2 & 2(-1) \end{bmatrix} = \begin{bmatrix} 1 & 2 & -1 \\ -1 & -2 & 1 \\ 2 & 4 & -2 \end{bmatrix}$ $(9-2)$

18. (A) $LM = \begin{bmatrix} 2 & -1 & 0 \\ 1 & 2 & 1 \end{bmatrix} \begin{bmatrix} 1 & 2 \\ -1 & 0 \\ 1 & 1 \end{bmatrix} = \begin{bmatrix} 2 \cdot 1 + (-1)(-1) + 0 \cdot 1 & 2 \cdot 2 + (-1)0 + 0 \cdot 1 \\ 1 \cdot 1 + 2(-1) + 1 \cdot 1 & 1 \cdot 2 + 2 \cdot 0 + 1 \cdot 1 \end{bmatrix}$

$= \begin{bmatrix} 3 & 4 \\ 0 & 3 \end{bmatrix}$

$LM - 2N = \begin{bmatrix} 3 & 4 \\ 0 & 3 \end{bmatrix} - 2 \begin{bmatrix} 2 & 1 \\ -1 & 0 \end{bmatrix} = \begin{bmatrix} 3 & 4 \\ 0 & 3 \end{bmatrix} - \begin{bmatrix} 4 & 2 \\ -2 & 0 \end{bmatrix} = \begin{bmatrix} -1 & 2 \\ 2 & 3 \end{bmatrix}$

(B) Since ML is a 3×3 matrix and N is a 2×2 matrix, $ML + N$ is not defined.

$(9-2)$

19. The solution region is unbounded.
Two corner points are obvious from
the graph: (0, 6) and (8, 0). The
third corner point is obtained by
solving the system
$3x + 2y = 12$
$x + 2y = 8$ to obtain (2, 3).

(Solution left to the student.)

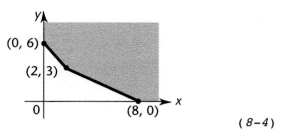

(8-4)

20. The feasible region is graphed as follows:
The corner points (0, 7), (0, 0) and (8, 0)
are obvious from the graph. The corner
point (6, 4) is obtained by solving the
system
$x + 2y = 14$
$2x + y = 16$ (Solution left to the student.)
We now evaluate the objective function at
each corner point.

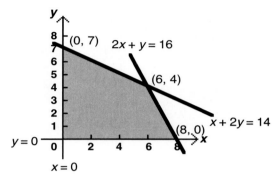

Corner Point (x, y)	Objective Function $z = 4x + 9y$	
(0, 7)	63	Maximum value
(0, 0)	0	
(8, 0)	32	
(6, 4)	60	

(8-5)

21. (A) As a matrix equation the system becomes

$$\underset{A}{\begin{bmatrix} 1 & 4 & 2 \\ 2 & 6 & 3 \\ 2 & 5 & 2 \end{bmatrix}} \underset{X}{\begin{bmatrix} x_1 \\ x_2 \\ x_3 \end{bmatrix}} = \underset{B}{\begin{bmatrix} k_1 \\ k_2 \\ k_3 \end{bmatrix}}$$

(B) To find A^{-1}, we perform row operations on

$$\begin{bmatrix} 1 & 4 & 2 & | & 1 & 0 & 0 \\ 2 & 6 & 3 & | & 0 & 1 & 0 \\ 2 & 5 & 2 & | & 0 & 0 & 1 \end{bmatrix} \begin{array}{l} (-2)R_1 + R_2 \to R_2 \\ (-2)R_1 + R_3 \to R_3 \end{array}$$

$$\sim \begin{bmatrix} 1 & 4 & 2 & | & 1 & 0 & 0 \\ 0 & -2 & -1 & | & -2 & 1 & 0 \\ 0 & -3 & -2 & | & -2 & 0 & 1 \end{bmatrix} -\frac{1}{2}R_2 \to R_2$$

$$\sim \begin{bmatrix} 1 & 4 & 2 & | & 1 & 0 & 0 \\ 0 & 1 & \frac{1}{2} & | & 1 & -\frac{1}{2} & 0 \\ 0 & -3 & -2 & | & -2 & 0 & 1 \end{bmatrix} (-4)R_2 + R_1 \to R_1$$

$$\sim \begin{bmatrix} 1 & 0 & 0 & | & -3 & 2 & 0 \\ 0 & 1 & \frac{1}{2} & | & 1 & -\frac{1}{2} & 0 \\ 0 & 0 & -\frac{1}{2} & | & 1 & -\frac{3}{2} & 1 \end{bmatrix} R_3 + R_2 \to R_2$$

$$\sim \begin{bmatrix} 1 & 0 & 0 & | & -3 & 2 & 0 \\ 0 & 1 & 0 & | & 2 & -2 & 1 \\ 0 & 0 & -\frac{1}{2} & | & 1 & -\frac{3}{2} & 1 \end{bmatrix} -2R_3 \rightarrow R_3$$

$$\sim \begin{bmatrix} 1 & 0 & 0 & | & -3 & 2 & 0 \\ 0 & 1 & 0 & | & 2 & -2 & 1 \\ 0 & 0 & 1 & | & -2 & 3 & -2 \end{bmatrix}$$

$$A^{-1} = \begin{bmatrix} -3 & 2 & 0 \\ 2 & -2 & 1 \\ -2 & 3 & -2 \end{bmatrix}$$

Check: $A^{-1}A = \begin{bmatrix} -3 & 2 & 0 \\ 2 & -2 & 1 \\ -2 & 3 & -2 \end{bmatrix} \begin{bmatrix} 1 & 4 & 2 \\ 2 & 6 & 3 \\ 2 & 5 & 2 \end{bmatrix}$

$$= \begin{bmatrix} (-3)1 + 2 \cdot 2 + 0 \cdot 2 & (-3)4 + 2 \cdot 6 + 0 \cdot 5 & -3 \cdot 2 + 2 \cdot 3 + 0 \cdot 2 \\ 2 \cdot 1 + (-2)2 + 1 \cdot 2 & 2 \cdot 4 + (-2)6 + 1 \cdot 5 & 2 \cdot 2 + (-2)3 + 1 \cdot 2 \\ (-2)1 + 3 \cdot 2 + (-2)2 & (-2)4 + 3 \cdot 6 + (-2)5 & (-2)2 + 3 \cdot 3 + (-2)2 \end{bmatrix}$$

$$= \begin{bmatrix} 1 & 0 & 0 \\ 0 & 1 & 0 \\ 0 & 0 & 1 \end{bmatrix}$$

(C) The solution of $AX = B$ is $X = A^{-1}B$. Using the result of (B), we get

$$X = \begin{bmatrix} -3 & 2 & 0 \\ 2 & -2 & 1 \\ -2 & 3 & -2 \end{bmatrix} \begin{bmatrix} k_1 \\ k_2 \\ k_3 \end{bmatrix}$$

If $k_1 = -1$, $k_2 = 2$, $k_3 = 1$, we get

$$\begin{bmatrix} x_1 \\ x_2 \\ x_3 \end{bmatrix} = \begin{bmatrix} -3 & 2 & 0 \\ 2 & -2 & 1 \\ -2 & 3 & -2 \end{bmatrix} \begin{bmatrix} -1 \\ 2 \\ 1 \end{bmatrix} = \begin{bmatrix} 7 \\ -5 \\ 6 \end{bmatrix} \quad x_1 = 7, \; x_2 = -5, \; x_3 = 6$$

(D) If $k_1 = 2$, $k_2 = 0$, $k_3 = -1$, reasoning as in (C) we get

$$\begin{bmatrix} x_1 \\ x_2 \\ x_3 \end{bmatrix} = \begin{bmatrix} -3 & 2 & 0 \\ 2 & -2 & 1 \\ -2 & 3 & -2 \end{bmatrix} \begin{bmatrix} 2 \\ 0 \\ -1 \end{bmatrix} = \begin{bmatrix} -6 \\ 3 \\ 2 \end{bmatrix} \quad x_1 = -6, \; x_2 = 3, \; x_3 = -2 \qquad \textbf{(9-4)}$$

22. (A) $D = \begin{vmatrix} 1 & 2 & -1 \\ 2 & 8 & 1 \\ -1 & 3 & 5 \end{vmatrix} = 1(-1)^{1+1}\begin{vmatrix} 8 & 1 \\ 3 & 5 \end{vmatrix} + 2(-1)^{1+2}\begin{vmatrix} 2 & 1 \\ -1 & 5 \end{vmatrix} + (-1)(-1)^{1+3}\begin{vmatrix} 2 & 8 \\ -1 & 3 \end{vmatrix}$

$= (-1)^2(8\cdot5 - 3\cdot1) + 2(-1)^3[2\cdot5 - (-1)1] + (-1)^5[2\cdot3 - (-1)8]$

$= 37 - 22 - 14 = 1$

(B) $z = \dfrac{\begin{vmatrix} 1 & 2 & 1 \\ 2 & 8 & -2 \\ -1 & 3 & 2 \end{vmatrix}}{D} = \dfrac{1(-1)^{1+1}\begin{vmatrix} 8 & 1 \\ 3 & 5 \end{vmatrix} + 2(-1)^{1+2}\begin{vmatrix} 2 & -2 \\ -1 & 2 \end{vmatrix} + 1(-1)^{1+3}\begin{vmatrix} 2 & 8 \\ -1 & 3 \end{vmatrix}}{1}$

$= 32$

$\qquad\qquad\qquad\qquad\qquad\qquad\qquad\qquad\qquad\qquad\qquad\qquad\qquad\qquad \textbf{(9-6, 9-7)}$

23. (A) If m and n are both zero, the matrix is

$$\begin{bmatrix} 1 & 0 & -5 & | & 2 \\ 0 & 1 & 3 & | & 6 \\ 0 & 0 & 0 & | & 0 \end{bmatrix}$$

Then there are an infinite number of solutions ($x_3 = t$, $x_2 = 6 - 3t$, $x_1 = 2 + 5t$, for t any real number.)

(B) If $m = 0$, the matrix is
$$\begin{bmatrix} 1 & 0 & -5 & | & 2 \\ 0 & 1 & 3 & | & 6 \\ 0 & 0 & 0 & | & n \end{bmatrix}$$
The third row corresponds to the equation $0x_1 + 0x_2 + 0x_3 = n$. This is impossible (since $n \neq 0$). The system has no solution.

(C) The system is independent. There is one solution. $(9\text{-}1)$

24. If $A^2 = A$ and A^{-1} exists, then

$A^{-1}A^2 = A^{-1}A$

$A^{-1}AA = I$ (since $A^{-1}A$ is the identity matrix for any A)

  $IA = I$ (same reason applied to the left side)

   $A = I$ (since the identity matrix times A is just A)

This shows that A must be equal to the identity matrix of the same size. $(9\text{-}4)$

25. L, M, and P are in reduced form.
N is not in reduced form; it violates condition 1 of the definition of reduced matrix: there is a row of 0's above rows having non-zero elements. $(9\text{-}1)$

26. $k\begin{vmatrix} a & b \\ c & d \end{vmatrix} = k(ad - bc) = kad - kbc = kad - kcb = \begin{vmatrix} ka & b \\ kc & d \end{vmatrix}$ $(9\text{-}6)$

27. $\begin{vmatrix} a & b \\ c & d \end{vmatrix} = ad - bc = ad + akb - akb - bc = ad + akb - (bc + bka)$

$= a(d + kb) - b(c + ka) = \begin{vmatrix} a & b \\ c + ka & d + kb \end{vmatrix}$ $(9\text{-}6)$

28. We assume that $\det M = ad - bc \neq 0$ and find M^{-1} by applying row operations to

$\begin{bmatrix} a & b & | & 1 & 0 \\ c & d & | & 0 & 1 \end{bmatrix}$ Case I: $a \neq 0$

                           $(-\frac{c}{a})R_1 + R_2 \rightarrow R_2$

$\sim \begin{bmatrix} a & b & | & 1 & 0 \\ 0 & \frac{ad-bc}{a} & | & -\frac{c}{a} & 1 \end{bmatrix} \frac{a}{ad-bc} R_2 \rightarrow R_2$

$\sim \begin{bmatrix} a & b & | & 1 & 0 \\ 0 & 1 & | & \frac{-c}{ad-bc} & \frac{a}{ad-bc} \end{bmatrix} (-b)R_2 + R_1 \rightarrow R_1$

$\sim \begin{bmatrix} a & 0 & | & 1 + \frac{bc}{ad-bc} & \frac{-ab}{ad-bc} \\ 0 & 1 & | & \frac{-c}{ad-bc} & \frac{a}{ad-bc} \end{bmatrix}$

$\sim \begin{bmatrix} a & 0 & | & \frac{ad}{ad-bc} & \frac{-ab}{ad-bc} \\ 0 & 1 & | & \frac{-c}{ad-bc} & \frac{a}{ad-bc} \end{bmatrix} \frac{1}{a}R_1 \rightarrow R_1$

$\sim \begin{bmatrix} 1 & 0 & | & \frac{d}{ad-bc} & \frac{-b}{ad-bc} \\ 0 & 1 & | & \frac{-c}{ad-bc} & \frac{a}{ad-bc} \end{bmatrix}$

Case II: $a = 0$ (b, $c \neq 0$, since $b = 0$ or $c = 0$ would imply det $M = ad - bc = 0$)

$$\begin{bmatrix} 0 & b & | & 1 & 0 \\ c & d & | & 0 & 1 \end{bmatrix} \quad R_1 \leftrightarrow R_2$$

$$\sim \begin{bmatrix} c & d & | & 0 & 1 \\ 0 & b & | & 1 & 0 \end{bmatrix} \quad \frac{1}{c} R_1 \to R_1$$

$$\sim \begin{bmatrix} 1 & \frac{d}{c} & | & 0 & \frac{1}{c} \\ 0 & b & | & 1 & 0 \end{bmatrix} \quad \left(-\frac{d}{bc}\right) R_2 + R_1 \to R_1$$

$$\sim \begin{bmatrix} 1 & 0 & | & -\frac{d}{bc} & \frac{1}{c} \\ 0 & b & | & 1 & 0 \end{bmatrix} \quad \frac{1}{b} R_2 \to R_2$$

$$\sim \begin{bmatrix} 1 & 0 & | & -\frac{d}{bc} & \frac{1}{c} \\ 0 & 1 & | & \frac{1}{b} & 0 \end{bmatrix}$$

$$\sim \begin{bmatrix} 1 & 0 & | & \frac{d}{ad-bc} & \frac{-b}{ad-bc} \\ 0 & 1 & | & \frac{-c}{ad-bc} & \frac{a}{ad-bc} \end{bmatrix} \quad \text{since } a = 0.$$

In either case,

$$M^{-1} = \begin{bmatrix} \frac{d}{ad-bc} & \frac{-b}{ad-bc} \\ \frac{-c}{ad-bc} & \frac{a}{ad-bc} \end{bmatrix} = \frac{1}{ad-bc}\begin{bmatrix} d & -b \\ -c & a \end{bmatrix} = \frac{1}{\det M}\begin{bmatrix} d & -b \\ -c & a \end{bmatrix} \tag{9-6}$$

29. True. For example,

$$\begin{bmatrix} a_{11} & a_{12} & a_{13} \\ 0 & a_{22} & a_{23} \\ 0 & 0 & a_{33} \end{bmatrix} + \begin{bmatrix} b_{11} & b_{12} & b_{13} \\ 0 & b_{22} & b_{23} \\ 0 & 0 & b_{33} \end{bmatrix} = \begin{bmatrix} a_{11} + b_{11} & a_{12} + b_{12} & a_{13} + b_{13} \\ 0 & a_{22} + b_{22} & a_{23} + b_{23} \\ 0 & 0 & a_{33} + b_{33} \end{bmatrix}$$

proves the statement for 3×3 matrices. A similar proof can be given for any order matrix. $\qquad (9-2)$

30. True. For example, $\begin{bmatrix} a_{11} & 0 \\ a_{21} & a_{22} \end{bmatrix}\begin{bmatrix} b_{11} & 0 \\ b_{21} & b_{22} \end{bmatrix} = \begin{bmatrix} a_{11}b_{11} & 0 \\ a_{21}b_{11} + a_{22}b_{21} & a_{22}b_{22} \end{bmatrix}$ proves the statement for 2×2 matrices. A similar proof can be given for any order matrix. $\qquad (9-1)$

31. False. For example, $\begin{bmatrix} 1 & 1 \\ 0 & 1 \end{bmatrix} + \begin{bmatrix} 1 & 0 \\ 1 & 1 \end{bmatrix} = \begin{bmatrix} 2 & 1 \\ 1 & 2 \end{bmatrix}$ which is not a diagonal matrix.

$\qquad (9-2)$

32. False. For example, $\begin{bmatrix} 1 & 1 \\ 0 & 1 \end{bmatrix}\begin{bmatrix} 1 & 0 \\ 1 & 1 \end{bmatrix} = \begin{bmatrix} 2 & 1 \\ 1 & 1 \end{bmatrix}$ which is not a diagonal matrix.

$\qquad (9-2)$

33. True. If all elements above and all elements below the principal diagonal are zero, then all elements not on the principal diagonal are zero. $\qquad (9-3)$

34. True. For example, the inverse of $\begin{bmatrix} a_{11} & 0 & 0 \\ 0 & a_{22} & 0 \\ 0 & 0 & a_{33} \end{bmatrix}$ is $\begin{bmatrix} \frac{1}{a_{11}} & 0 & 0 \\ 0 & \frac{1}{a_{22}} & 0 \\ 0 & 0 & \frac{1}{a_{33}} \end{bmatrix}$.

This is defined if none of elements on the principal diagonal are zero. A similar situation holds for any order matrix. $\qquad (9-3)$

35. True. For example, $\begin{vmatrix} a_{11} & 0 & 0 & 0 \\ 0 & a_{22} & 0 & 0 \\ 0 & 0 & a_{33} & 0 \\ 0 & 0 & 0 & a_{44} \end{vmatrix}$ can be expanded by the first column to

yield, successively, $a_{11} \begin{vmatrix} a_{22} & 0 & 0 \\ 0 & a_{33} & 0 \\ 0 & 0 & a_{44} \end{vmatrix} = a_{11}a_{22} \begin{vmatrix} a_{33} & 0 \\ 0 & a_{44} \end{vmatrix} = a_{11}a_{12}a_{33}a_{44}.$

A similar result holds for any order matrix.   (9-5)

36. True. For example, $\begin{vmatrix} a_{11} & 0 & 0 & 0 \\ a_{22} & a_{22} & 0 & 0 \\ a_{31} & a_{32} & a_{33} & 0 \\ a_{41} & a_{42} & a_{43} & a_{44} \end{vmatrix}$ can be expanded by the first row to

yield, successively, $a_{11} \begin{vmatrix} a_{22} & 0 & 0 \\ a_{32} & a_{33} & 0 \\ a_{42} & a_{43} & a_{44} \end{vmatrix} = a_{11}a_{22} \begin{vmatrix} a_{33} & 0 \\ a_{43} & a_{44} \end{vmatrix} = a_{11}a_{22}a_{33}a_{44}.$

A similar result holds for any order matrix.   (9-5)

37. Let x = amount invested at 8%
   y = amount invested at 14%
Then $x + y = 12,000$ (total amount invested)   (1)
$0.08x + 0.14y = 0.10(12,000)$ (total yield on investment)   (2)
We solve the system of equations (1), (2) using elimination by addition.
$-0.08x - 0.08y = -0.08(12,000)$ $- 0.08$ [equation (1)]
$\ \ 0.08x + 0.14y = \ \ 0.10(12,000)$ equation (2)
_____
$\ \ \ \ \ \ \ \ \ 0.06y = -0.08(12,000) + 0.10(12,000)$

$\ \ \ \ \ \ \ \ \ 0.06y = 240$
$\ \ \ \ \ \ \ \ \ \ \ \ \ y = 4,000$
$\ \ \ \ \ x + y = 12,000$
$\ \ \ \ \ \ \ \ \ \ \ \ \ x = 8,000$
$8,000 at 8% and $4,000 at 14%.   (8-1)

38. Let x_1 = number of ounces of mix A.
   x_2 = number of ounces of mix B.
   x_3 = number of ounces of mix C.
Then
   $0.2x_1 + \ \ 0.1x_2 + 0.15x_3 = 23$ (protein)
   $0.02x_1 + 0.06x_2 + 0.05x_3 = 6.2$ (fat)
   $0.15x_1 + \ \ 0.1x_2 + 0.05x_3 = 16$ (moisture)
or
   $4x_1 + 2x_2 + 3x_3 = 460$ (1st equation multiplied by 20)
   $2x_1 + 6x_2 + 5x_3 = 620$ (2nd equation multiplied by 100)
   $3x_1 + 2x_2 + \ \ x_3 = 320$ (3rd equation multiplied by 20)
is the system to be solved. We form the augmented matrix and solve by Gauss-Jordan elimination.

$\begin{bmatrix} 4 & 2 & 3 & | & 460 \\ 2 & 6 & 5 & | & 620 \\ 3 & 2 & 1 & | & 320 \end{bmatrix} \frac{1}{4}R_1 \to R_1$

$\sim \begin{bmatrix} 1 & \frac{1}{2} & \frac{3}{4} & | & 115 \\ 2 & 6 & 5 & | & 620 \\ 3 & 2 & 1 & | & 320 \end{bmatrix} \begin{matrix} \\ (-2)R_1 + R_2 \to R_2 \\ (-3)R_1 + R_3 \to R_3 \end{matrix}$

$$\sim \begin{bmatrix} 1 & \frac{1}{2} & \frac{3}{4} & \bigm| & 115 \\ 0 & 5 & \frac{7}{2} & \bigm| & 390 \\ 0 & \frac{1}{2} & -\frac{5}{4} & \bigm| & -25 \end{bmatrix} R_2 \leftrightarrow R_3$$

$$\sim \begin{bmatrix} 1 & \frac{1}{2} & \frac{3}{4} & \bigm| & 115 \\ 2 & \frac{1}{2} & -\frac{5}{4} & \bigm| & -25 \\ 3 & 5 & \frac{7}{2} & \bigm| & 390 \end{bmatrix} \begin{array}{l}(-1)R_2 + R_1 \rightarrow R_1 \\ \\ (-10)R_2 + R_3 \rightarrow R_3\end{array}$$

$$\sim \begin{bmatrix} 1 & 0 & 2 & \bigm| & 140 \\ 0 & \frac{1}{2} & -\frac{5}{4} & \bigm| & -25 \\ 0 & 0 & 16 & \bigm| & 640 \end{bmatrix} \begin{array}{l} 2R_2 \rightarrow R_2 \\ \\ \frac{1}{16}R_3 \rightarrow R_3 \end{array}$$

$$\sim \begin{bmatrix} 1 & 0 & 2 & \bigm| & 140 \\ 0 & 1 & -\frac{5}{2} & \bigm| & -50 \\ 0 & 0 & 1 & \bigm| & 40 \end{bmatrix} \begin{array}{l}(-2)R_3 + R_1 \rightarrow R_1 \\ \left(-\frac{5}{2}\right)R_3 + R_2 \rightarrow R_2 \end{array}$$

$$\sim \begin{bmatrix} 1 & 0 & 0 & \bigm| & 60 \\ 0 & 1 & 0 & \bigm| & 50 \\ 0 & 0 & 1 & \bigm| & 40 \end{bmatrix}$$

Solution: $x_1 = 60$g Mix A,
$x_2 = 50$g Mix B, $x_3 = 40$g Mix C (9-1)

39. Let x_1 = number of model A trucks
x_2 = number of model B trucks
x_3 = number of model C trucks
Then $x_1 + x_2 + x_3 = 12$ (total number of trucks)
$18,000x_1 + 22,000x_2 + 30,000x_3 = 300,000$ (total funds needed)

We form the augmented matrix and solve by Gauss-Jordan elimination.

$$\begin{bmatrix} 1 & 1 & 1 & \bigm| & 12 \\ 18,000 & 22,000 & 30,000 & \bigm| & 300,000 \end{bmatrix} \frac{1}{2,000}R_2 \rightarrow R_2$$

$$\sim \begin{bmatrix} 1 & 1 & 1 & \bigm| & 12 \\ 9 & 11 & 15 & \bigm| & 150 \end{bmatrix} (-9)R_1 + R_2 \rightarrow R_2$$

$$\sim \begin{bmatrix} 1 & 1 & 1 & \bigm| & 12 \\ 0 & 2 & 6 & \bigm| & 42 \end{bmatrix} -\frac{1}{2}R_2 + R_1 \rightarrow R_1$$

$$\sim \begin{bmatrix} 1 & 0 & -2 & \bigm| & -9 \\ 0 & 2 & 6 & \bigm| & 42 \end{bmatrix} \frac{1}{2}R_2 \rightarrow R_2$$

$$\sim \begin{bmatrix} 1 & 0 & -2 & \bigm| & -9 \\ 0 & 1 & 3 & \bigm| & 21 \end{bmatrix}$$

Let $x_3 = t$. Then $x_2 = -3x_3 + 21$
$= -3t + 21$
$x_1 = 2x_3 - 9$
$= 2t - 9$

A solution is achieved, not for every value of t, but for integer values of t that give rise to non-negative x_1, x_2, x_3.

$x_1 \geq 0$ means $2t - 9 \geq 0$ or $t \geq 4\frac{1}{2}$
$x_2 \geq 0$ means $-3t + 21 \geq 0$ or $t \leq 7$

The only integer values of t that satisfy these conditions are 5, 6, 7. So we get the solutions
$x_1 = 2t - 9$ model A trucks
$x_2 = -3t + 21$ model B trucks
$x_3 = t$ model C trucks

where $t = 5$, 6, or 7. The distributor can purchase
1 model A truck, 6 model B trucks, and 5 model C trucks, or
3 model A trucks, 3 model B trucks, and 6 model C trucks, or
5 model A trucks and 7 model C trucks.   (9-1)

40. Let x = number of standard day packs
   y = number of deluxe day packs

(A) We form the linear objective function
   $P = 8x + 12y$
   We wish to maximize P, the profit from x standard packs @ \$8 and y deluxe packs @ \$12, subject to the constraints
   $0.5x + 0.5y \leq 300$ fabricating constraint
   $0.3x + 0.6y \leq 240$ sewing constraint
   $x, y \geq 0$ non-negative constraints

Solving the system of constraint inequalities graphically, we obtain the feasible region S shown in the diagram.

The three corner points $(0, 400)$, $(0, 0)$, and $(600, 0)$ are obvious from the diagram. The corner point $(400, 200)$ is found by solving the system
   $0.5x + 0.5y = 300$
   $0.3x + 0.6y = 240$ (Left to the student.)

Next we evaluate the objective function at each corner point.

Corner Point (x, y)	Objective Function $P = 8x + 12y$	
$(0, 0)$	0	
$(600, 0)$	4,800	
$(400, 200)$	5,600	Maximum value
$(0, 400)$	4,800	

The optimal value is 5,600 at the corner point $(400, 200)$. So the company should manufacture 400 standard packs and 200 deluxe packs for a maximum profit of \$5,600.

(B) The objective function now becomes $5x + 15y$. We evaluate this at each corner point.

Corner Point (x, y)	Objective Function $P = 5x + 15y$	
$(0, 0)$	0	
$(600, 0)$	3,000	
$(400, 200)$	5,000	
$(0, 400)$	6,000	Maximum value

The optimal value is now 6,000 at the corner point $(0, 400)$. So the company should manufacture no standard packs and 400 deluxe packs for a maximum profit of \$6,000.

(C) The objective function now becomes $11x + 9y$. We evaluate this at each corner point.

Corner Point (x, y)	Objective Function $P = 11x + 9y$	
$(0, 0)$	0	
$(600, 0)$	6,600	Maximum value
$(400, 200)$	6,200	
$(0, 400)$	3,600	

The optimal value is now 6,600 at the corner point $(600, 0)$. So the company should manufacture 600 standard packs and no deluxe packs for a maximum profit of \$6,600.
   (8–5)

41. (A) $M\begin{bmatrix} 0.25 \\ 0.25 \\ 0.25 \\ 0.25 \end{bmatrix} = \begin{bmatrix} 78 & 84 & 81 & 86 \\ 91 & 65 & 84 & 92 \\ 95 & 90 & 92 & 91 \\ 75 & 82 & 87 & 91 \\ 83 & 88 & 81 & 76 \end{bmatrix} \begin{bmatrix} 0.25 \\ 0.25 \\ 0.25 \\ 0.25 \end{bmatrix}$

$= \begin{bmatrix} 0.25(78 + 84 + 81 + 86) \\ 0.25(91 + 65 + 84 + 92) \\ 0.25(95 + 90 + 92 + 91) \\ 0.25(75 + 82 + 87 + 91) \\ 0.25(83 + 88 + 81 + 76) \end{bmatrix} = \begin{bmatrix} 82.25 \\ 83 \\ 92 \\ 83.75 \\ 82 \end{bmatrix} \begin{matrix} \text{Ann} \\ \text{Bob} \\ \text{Carol} \\ \text{Dan} \\ \text{Eric} \end{matrix}$

(B) $M\begin{bmatrix} 0.2 \\ 0.2 \\ 0.2 \\ 0.4 \end{bmatrix} = \begin{bmatrix} 78 & 84 & 81 & 86 \\ 91 & 65 & 84 & 92 \\ 95 & 90 & 92 & 91 \\ 75 & 82 & 87 & 91 \\ 83 & 88 & 81 & 76 \end{bmatrix} \begin{bmatrix} 0.2 \\ 0.2 \\ 0.2 \\ 0.4 \end{bmatrix}$

$= \begin{bmatrix} 0.2(78 + 84 + 81) + 0.4 \cdot 86 \\ 0.2(91 + 65 + 84) + 0.4 \cdot 92 \\ 0.2(95 + 90 + 92) + 0.4 \cdot 91 \\ 0.2(75 + 82 + 87) + 0.4 \cdot 91 \\ 0.2(83 + 88 + 81) + 0.4 \cdot 76 \end{bmatrix} = \begin{bmatrix} 83 \\ 84.8 \\ 91.8 \\ 85.2 \\ 80.8 \end{bmatrix} \begin{matrix} \text{Ann} \\ \text{Bob} \\ \text{Carol} \\ \text{Dan} \\ \text{Eric} \end{matrix}$

(C) $[0.2 \quad 0.2 \quad 0.2 \quad 0.2 \quad 0.2]M = [0.2 \quad 0.2 \quad 0.2 \quad 0.2 \quad 0.2] \begin{bmatrix} 78 & 84 & 81 & 86 \\ 91 & 65 & 84 & 92 \\ 95 & 90 & 92 & 91 \\ 75 & 82 & 87 & 91 \\ 83 & 88 & 81 & 76 \end{bmatrix}$

$= [\,0.2(78 + 91 + 95 + 75 + 83) \quad 0.2(84 + 65 + 90 + 82 + 88) \quad 0.2(81 + 84 + 92 + 87 + 81) \quad 0.2(86 + 92 + 91 + 91 + 76)\,]$

Test 1	Test 2	Test 3	Test 4

$= [84.4 \qquad 81.8 \qquad 85 \qquad 87.2]$

(9-2)

CHAPTER 10   Sequences, Induction, and Probability

Section 10-1

1. Answers will vary.

3. Answers will vary.

5. $a_1 = 1 - 2 = -1;$ $a_2 = 2 - 2 = 0;$ $a_3 = 3 - 2 = 1;$ $a_4 = 4 - 2 = 2$

7. $a_1 = \dfrac{1-1}{1+1} = 0;$ $a_2 = \dfrac{2-1}{2+1} = \dfrac{1}{3};$ $a_3 = \dfrac{3-1}{3+1} = \dfrac{2}{4} = \dfrac{1}{2};$ $a_4 = \dfrac{4-1}{4+1} = \dfrac{3}{5}$

9. $a_1 = (-2)^{1+1} = (-2)^2 = 4;$ $a_2 = (-2)^{2+1} = (-2)^3 = -8$
$a_3 = (-2)^{3+1} = (-2)^4 = 16;$ $a_4 = (-2)^{4+1} = (-2)^5 = -32$

11. $a_8 = 8 - 2 = 6$

13. $a_{100} = \dfrac{100 - 1}{100 + 1} = \dfrac{99}{101}$

15. $S_5 = 1 + 2 + 3 + 4 + 5$

17. $S_3 = \dfrac{1}{10^1} + \dfrac{1}{10^2} + \dfrac{1}{10^3} = \dfrac{1}{10} + \dfrac{1}{100} + \dfrac{1}{1,000}$

19. $S_4 = (-1)^1 + (-1)^2 + (-1)^3 + (-1)^4 = (-1) + 1 + (-1) + 1 = -1 + 1 - 1 + 1$

21. $a_1 = (-1)^{1+1}\,1^2 = 1$ $a_2 = (-1)^{2+1}\,2^2 = -4$ $a_3 = (-1)^{3+1}\,3^2 = 9$
$a_4 = (-1)^{4+1}\,4^2 = -16$ $a_5 = (-1)^{5+1}\,5^2 = 25$

23.
$$a_1 = \frac{1}{3}\left[1 - \frac{1}{10^1}\right] = \frac{1}{3}\cdot\frac{9}{10} = \frac{3}{10} = 0.3$$
$$a_2 = \frac{1}{3}\left[1 - \frac{1}{10^2}\right] = \frac{1}{3}\cdot\frac{99}{100} = \frac{33}{100} = 0.33$$
$$a_3 = \frac{1}{3}\left[1 - \frac{1}{10^3}\right] = \frac{1}{3}\cdot\frac{999}{1,000} = \frac{333}{1,000} = 0.333$$
$$a_4 = \frac{1}{3}\left[1 - \frac{1}{10^4}\right] = \frac{1}{3}\cdot\frac{9,999}{10,000} = \frac{3,333}{10,000} = 0.3333$$
$$a_5 = \frac{1}{3}\left[1 - \frac{1}{10^5}\right] = \frac{1}{3}\cdot\frac{99,999}{100,000} = \frac{33,333}{100,000} = 0.33333$$

25. $a_1 = \left(-\dfrac{1}{2}\right)^{1-1} = \left(-\dfrac{1}{2}\right)^0 = 1$

$a_2 = \left(-\dfrac{1}{2}\right)^{2-1} = -\dfrac{1}{2}$

$a_3 = \left(-\dfrac{1}{2}\right)^{3-1} = \dfrac{1}{4}$

$a_4 = \left(-\dfrac{1}{2}\right)^{4-1} = -\dfrac{1}{8}$

$a_5 = \left(-\dfrac{1}{2}\right)^{5-1} = \dfrac{1}{16}$

27. $a_1 = 7$
$a_2 = a_{2-1} - 4 = a_1 - 4 = 7 - 4 = 3$
$a_3 = a_{3-1} - 4 = a_2 - 4 = 3 - 4 = -1$
$a_4 = a_{4-1} - 4 = a_3 - 4 = -1 - 4 = -5$
$a_5 = a_{5-1} - 4 = a_4 - 4 = -5 - 4 = -9$

> **Common Error:**
> $a_2 \neq a_2 - 1 - 4.$ The 1 is in
> the subscript:
> $a_{2-1};$ that is, a_1

29. $a_1 = 4$

$a_2 = \frac{1}{4} a_{2-1} = \frac{1}{4} a_1 = \frac{1}{4} \cdot 4 = 1$

$a_3 = \frac{1}{4} a_{3-1} = \frac{1}{4} a_2 = \frac{1}{4} \cdot 1 = \frac{1}{4}$

$a_4 = \frac{1}{4} a_{4-1} = \frac{1}{4} a_3 = \frac{1}{4} \cdot \frac{1}{4} = \frac{1}{16}$

$a_5 = \frac{1}{4} a_{5-1} = \frac{1}{4} a_4 = \frac{1}{4} \cdot \frac{1}{16} = \frac{1}{64}$

31. $a_1 = 1$

$a_2 = 2$

$a_3 = a_1 + 2a_2 = 1 + 2(2) = 5$

$a_4 = a_2 + 2a_3 = 2 + 2(5) = 12$

$a_5 = a_3 + 2a_4 = 5 + 2(12) = 29$

$a_6 = a_4 + 2a_5 = 12 + 2(29) = 70$

$a_7 = a_5 + 2a_6 = 29 + 2(70) = 169$

The graphic demonstrates use of the RECUR program on a TI-84.

```
PrgmRECUR
FIRST TERM? 1
COEF OF 1ST TERM
1
SECOND TERM? 2
COEF OF 2ND TERM
2
NBR OF TERMS? 7
{1 2 5 12 29 70 169}
                  Done
```

33. $a_1 = -1$

$a_2 = 2$

$a_3 = 2a_1 + a_2 = 2(-1) + 2 = 0$

$a_4 = 2a_2 + a_3 = 2(2) + 0 = 4$

$a_5 = 2a_3 + a_4 = 2(0) + 4 = 4$

$a_6 = 2a_4 + a_5 = 2(4) + 4 = 12$

$a_7 = 2a_5 + a_6 = 2(4) + 12 = 20$

The graphic demonstrates use of the RECUR program on a TI-84.

```
PrgmRECUR
FIRST TERM? -1
COEF OF 1ST TERM
2
SECOND TERM? 2
COEF OF 2ND TERM
1
NBR OF TERMS? 7
{-1 2 0 4 4 12 20}
                  Done
```

35. a_n: 4, 5, 6, 7, ···

$n = 1, 2, 3, 4, ···$

Comparing a_n with n, we see that
$a_n = n + 3$

37. a_n: 3, 6, 9, 12, ···

$n = 1, 2, 3, 4, ···$

Comparing a_n with n, we see that
$a_n = 3n$

39. a_n: $\frac{1}{2}$, $\frac{2}{3}$, $\frac{3}{4}$, $\frac{4}{5}$, ···

$n = 1, 2, 3, 4, ···$

Comparing a_n with n, we see that

$a_n = \frac{n}{n + 1}$

41. a_n: 1, -1, 1, -1, ···

$n = 1, 2, 3, 4, ···$

Comparing a_n with n, we see that a_n involves (-1) to successively even and odd powers, so to a power that depends on n. We could write

$a_n = (-1)^{n-1}$ or $a_n = (-1)^{n+1}$

or other choices. $a_n = (-1)^{n+1}$ is one of many correct answers.

43. a_n: -2, 4, -8, 16, $\cdots$

$n = 1$, 2, 3, 4, $\cdots$

Comparing a_n with n, we see that a_n involves -1 and 2 to successively higher powers, so to powers that depend on n. We write

$a_n = (-1)^n(2)^n$ or $a_n = (-2)^n$

45. a_n: x, $\dfrac{x^2}{2}$, $\dfrac{x^3}{3}$, $\dfrac{x^4}{4}$, $\cdots$ or $\dfrac{x^1}{1}$, $\dfrac{x^2}{2}$, $\dfrac{x^3}{3}$, $\dfrac{x^4}{4}$, $\cdots$

Comparing a_n with n, we see that $a_n = \dfrac{x^n}{n}$

47. The most obvious formula is $a_n = 2^n$ since $a_1 = 1$ (or 2^0), $a_2 = 2$ (or 2^1) and $a_3 = 4$ (or 2^2). We'll use quadratic regression to find another, treating n as independent variable and a_n as dependent variable. We enter 1, 2, 3 as L_1 and 1, 2, 4 as L_2 then use the quadratic regression command.

The quadratic regression is $a_n = 0.5n^2 - 0.5n + 1$. Using this formula, $a_1 = 1$, $a_2 = 2$, $a_3 = 4$, so this fits the desired criteria.

49. The first three terms are the cubes of 1, 2, and 3, so one choice is $a_n = n^3$. We'll use quadratic regression to find another, treating n as independent variable and a_n as dependent variable. We enter 1, 2, 3 as L_1 and 1, 8, 27 as L_2, then use the quadratic regression command.

The quadratic regression is $a_n = 6n^2 - 11n + 6$. Using this formula: $a_1 = 1$, $a_2 = 8$, $a_3 = 27$, so this fits the desired criteria.

51. Compare to problem 37: this sequence is the same as $\{3n\}$ with alternating signs. We accomplish this by including $(-1)^n$, so one choice is $a_n = (-1)^n \cdot 3n$. We'll use quadratic regression to find another, treating n as independent variable and a_n as dependent variable. We enter 1, 2, 3 as L_1 and 1, 8, 27 as L_2, then use the quadratic regression command.

The quadratic regression is $a_n = -12n^2 + 45n - 36$. Using this formula, $a_1 = -3$, $a_2 = 6$, $a_3 = -9$, so this fits the desired criteria.

53.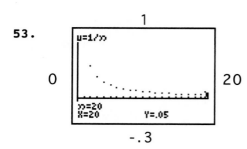

55.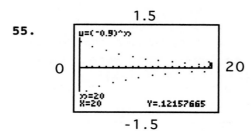

57. $S_4 = \dfrac{(-2)^{1+1}}{1} + \dfrac{(-2)^{2+1}}{2} + \dfrac{(-2)^{3+1}}{3} + \dfrac{(-2)^{4+1}}{4}$

$= \dfrac{4}{1} - \dfrac{8}{2} + \dfrac{16}{3} - \dfrac{32}{4}$

59. $S_3 = \dfrac{1}{1}x^{1+1} + \dfrac{1}{2}x^{2+1} + \dfrac{1}{3}x^{3+1}$

$= x^2 + \dfrac{x^3}{2} + \dfrac{x^4}{3}$

61. $S_5 = \dfrac{(-1)^{1+1}}{1}x^1 + \dfrac{(-1)^{2+1}}{2}x^2 + \dfrac{(-1)^{3+1}}{3}x^3 + \dfrac{(-1)^{4+1}}{4}x^4 + \dfrac{(-1)^{5+1}}{5}x^5$

$= x - \dfrac{x^2}{2} + \dfrac{x^3}{3} - \dfrac{x^4}{4} + \dfrac{x^5}{5}$

63. $S_4 = 1^2 + 2^2 + 3^2 + 4^2$

$k = 1, 2, 3, 4, \cdots$

$a_k = k^2, \quad k = 1, 2, 3, 4$

$S_4 = \displaystyle\sum_{k=1}^{4} k^2$

65. $S_5 = \dfrac{1}{2^1} + \dfrac{1}{2^2} + \dfrac{1}{2^3} + \dfrac{1}{2^4} + \dfrac{1}{2^5}$

$k = 1, 2, 3, 4, \cdots$

$a_k = \dfrac{1}{2^k}, \quad k = 1, 2, 3, 4, 5$

$S_5 = \displaystyle\sum_{k=1}^{5} \dfrac{1}{2^k}$

67. $S_n = 1 + \dfrac{1}{2^2} + \dfrac{1}{3^2} + \cdots + \dfrac{1}{n^2}$

$k = 1, 2, 3, \cdots, n$

$a_k = \dfrac{1}{k^2}, \quad k = 1, 2, 3, \cdots, n$

$S_n = \displaystyle\sum_{k=1}^{n} \dfrac{1}{k^2}$

69. $S_n = 1 - 4 + 9 + \cdots + (-1)^{n+1}n^2$

$k = 1, 2, 3, \cdots, n$

$a_k = (-1)^{k+1}k^2, \quad k = 1, 2, 3, \cdots, n$

$S_n = \displaystyle\sum_{k=1}^{n} (-1)^{k+1}k^2$

71. (A) $a_1 = 3$

$a_2 = \dfrac{a_{2-1}^2 + 2}{2a_{2-1}} = \dfrac{a_1^2 + 2}{2a_1} = \dfrac{3^2 + 2}{2 \cdot 3} \approx 1.83$

$a_3 = \dfrac{a_{3-1}^2 + 2}{2a_{3-1}} = \dfrac{a_2^2 + 2}{2a_2} = \dfrac{(1.83)^2 + 2}{2(1.83)} \approx 1.46$

$a_4 = \dfrac{a_{4-1}^2 + 2}{2a_{4-1}} = \dfrac{a_3^2 + 2}{2a_3} = \dfrac{(1.46)^2 + 2}{2(1.46)} \approx 1.415$

(B) Calculator $\sqrt{2} = 1.4142135\ldots$

(C) $a_1 = 1$

$a_2 = \dfrac{a_1^2 + 2}{2a_1} = \dfrac{1^2 + 2}{2 \cdot 1} = 1.5$

$a_3 = \dfrac{a_2^2 + 2}{2a_2} = \dfrac{(1.5)^2 + 2}{2(1.5)} \approx 1.417$

$a_4 = \dfrac{a_3^2 + 2}{2a_3} = \dfrac{(1.417)^2 + 2}{2(1.417)} \approx 1.414$

73. The first ten terms of the Fibonacci sequence a_n are

1, 1, 2, 3, 5, 8, 13, 21, 34, 55

The first ten terms of b_n are

1, 3, 4, 7, 11, 18, 29, 47, 76, 123

The first ten terms of $c_n = \dfrac{b_n}{a_n}$ are

$$\frac{1}{1}, \ \frac{3}{1}, \ \frac{4}{2}, \ \frac{7}{3}, \ \frac{11}{5}, \ \frac{18}{8}, \ \frac{29}{13}, \ \frac{47}{21}, \ \frac{76}{34}, \ \frac{123}{55}$$

In decimal notation this becomes

1, 3, 2, 2.33..., 2.2, 2.25, 2.23..., 2.238..., 2.235..., 2.236...

75. $e^{0.2} = 1 + \dfrac{0.2}{1!} + \dfrac{(0.2)^2}{2!} + \dfrac{(0.2)^3}{3!} + \dfrac{(0.2)^4}{4!}$

$\quad\quad = 1.2214000$

$e^{0.2} = 1.2214028$ (calculator—direct evaluation)

77. $\displaystyle\sum_{k=1}^{n} ca_k = ca_1 + ca_2 + ca_3 + \cdots + ca_n = c(a_1 + a_2 + a_3 + \cdots a_n) = c \sum_{k=1}^{n} a_k$

79. (A) The height of the fifth bounce is a_5:

$$a_5 = 10\left(\frac{1}{2}\right)^{5-1} = 10\left(\frac{1}{2}\right)^4 = 10\left(\frac{1}{16}\right) = \frac{10}{16} = \frac{5}{8} = 0.625 \text{ feet}$$

The height of the tenth bounce is a_{10}:

$$a_{10} = 10\left(\frac{1}{2}\right)^{10-1} = 10\left(\frac{1}{2}\right)^9 = 10\left(\frac{1}{512}\right) = \frac{10}{512} = \frac{5}{256} = 0.01953 \text{ feet}$$

(B) $\displaystyle\sum_{n=1}^{10} 10\left(\frac{1}{2}\right)^{n-1} = 10\left(\frac{1}{2}\right)^{1-1} + 10\left(\frac{1}{2}\right)^{2-1} + 10\left(\frac{1}{2}\right)^{3-1} + 10\left(\frac{1}{2}\right)^{4-1} + 10\left(\frac{1}{2}\right)^{5-1} + 10\left(\frac{1}{2}\right)^{6-1} + 10\left(\frac{1}{2}\right)^{7-1}$

$$+ \ 10\left(\frac{1}{2}\right)^{8-1} + 10\left(\frac{1}{2}\right)^{9-1} + 10\left(\frac{1}{2}\right)^{10-1}$$

$= 10 + 5 + 2.5 + 1.25 + 0.625 + 0.3125 + 0.1563 + 0.0781 + 0.0391 + 0.0195$

$= 91.98$ feet

This is the sum of the heights of all ten bounces.

81. (A) First year: $\$40,000 = a_1$

Second year: $\$40,000 + 0.04(\$40,000) = \$41,600 = a_2$

Third year: $\$41,600 + 0.04(\$41,600) = \$43,264 = a_3$

Fourth year: $\$43,264 + 0.04(\$43,264) = \$44,995 = a_4$

Fifth year: $\$44,995 + 0.04(\$44,995) = \$46,795 = a_5$

Sixth year: $\$46,795 + 0.04(\$46,795) = \$48,667 = a_6$

(B) The salary for each year is calculated by multiplying the previous year's salary by 0.04 and adding to the previous year's salary. This can easily be written as a recursive sequence: $a_n = a_{n-1} + 0.04a_{n-1}$, which can be simplified a bit as $a_n = 1.04a_{n-1}$. But we can also write as a non-recursive formula: at each stage, we multiply by one more factor of 1.04, so we can write

$a_n = 40,000\left(1.04\right)^{n-1}$.

(C) $\displaystyle\sum_{n=1}^{6} a_n = 40,000 + 41,600 + 43,264 + 44,995 + 46,795 + 48,667 = \$265,321$

This is the total amount earned in the first six years.

Section 10-2

1. Answers will vary.

3. $(3 + 5)^1 = 3^1 + 5^1$ True
   $(3 + 5)^2 = 3^2 + 5^2$ False
   Fails at $n = 2$

5. $1^2 = 3 \cdot 1 - 2$ True
   $2^2 = 3 \cdot 2 - 2$ True
   $3^2 = 3 \cdot 3 - 2$ False
   Fails at $n = 3$

7. (A) P_1: $2 = 2 \cdot 1^2$ $2 = 2$
       P_2: $2 + 6 = 2 \cdot 2^2$ $8 = 8$
       P_3: $2 + 6 + 10 = 2 \cdot 3^2$ $18 = 18$

   (B) P_k: $2 + 6 + 10 + \cdots + (4k - 2) = 2k^2$
       P_{k+1}: $2 + 6 + 10 + \cdots + (4k - 2) + [4(k + 1) - 2] = 2(k + 1)^2$
       or $2 + 6 + 10 + \cdots + (4k - 2) + (4k + 2) = 2(k + 1)^2$

   (C) Prove: $2 + 6 + 10 + \cdots + (4n - 2) = 2n^2$, $n \in N$

       Write: P_n: $2 + 6 + 10 + \cdots + (4n - 2) = 2n^2$

   Part 1: Show that P_1 is true:
   P_1: $2 = 2 \cdot 1^2$
       $= 2$ P_1 is true

   Part 2: Show that if P_k is true, then P_{k+1} is true:
   Write out P_k and P_{k+1}.
       P_k: $2 + 6 + 10 + \cdots + (4k - 2) = 2k^2$
       P_{k+1}: $2 + 6 + 10 + \cdots + (4k - 2) + (4k + 2) = 2(k + 1)^2$

   We start with P_k:
       $2 + 6 + 10 + \cdots + (4k - 2) = 2k^2$

   Adding $(4k + 2)$ to both sides, we get
       $2 + 6 + 10 + \cdots + (4k - 2) + (4k + 2) = 2k^2 + 4k + 2$
       $= 2(k^2 + 2k + 1)$
       $= 2(k + 1)^2$

   We have shown that if P_k is true, then P_{k+1} is true.

   Conclusion: P_n is true for all positive integers n.

   > **Common Errors:**
   > P_1 is not the nonsensical statement
   > $2 + 6 + 10 + \cdots + (4 \cdot 1 - 2) = 2 \cdot 1^2$
   > achieved by "substitution" of 1 for n. The left side of P_n has n terms; the left side of P_1 must
   > have 1 term. Also, $P_{k+1} \neq P_k + 1$
   > Nor is P_{k+1} simply its k + 1st term. $4k + 2 \neq 2(k + 1)^2$

9. (A) P_1: $a^5 a^1 = a^{5+}$ $a^6 = a^6$

       P_2: $a^5 a^2 = a^{5+2}$ $a^5 a^2 = a^5(a^1 a) = (a^5 a)a = a^6 a = a^7 = a^{5+2}$

       P_3: $a^5 a^3 = a^{5+3}$ $a^5 a^3 = a^5(a^2 a) = a^5(a^1 a)a = [(a^5 a)a]a = a^8 = a^{5+3}$

(B) P_k: $a^5 a^k = a^{5+k}$

   P_{k+1}: $a^5 a^{k+1} = a^{5+k+1}$

(C) Prove: $a^5 a^n = a^{5+n}$, $n \in N$
   Write: P_n: $a^5 a^n = a^{5+n}$

Part 1: Show that P_1 is true:
P_1: $a^5 a^1 = a^{5+1}$

   $a^5 a = a^{5+1}$. True by the recursive definition of a^n.

Part 2: Show that if P_k is true, then P_{k+1} is true:
Write out P_k and P_{k+1}.

   P_k: $a^5 a^k = a^{5+k}$
   P_{k+1}: $a^5 a^{k+1} = a^{5+k+1}$

We start with P_k:
   $a^5 a^k = a^{5+k}$

Multiply both sides by a:
   $a^5 a^k a = a^{5+k} a$
   $a^5 a^{k+1} = a^{5+k+1}$ by the recursive definition of a^n.

We have shown that if P_k is true, then P_{k+1} is true.

Conclusion: P_n is true for all positive integers n.

11. (A) P_1: $9^1 - 1 = 8$ is divisible by 4

   P_2: $9^2 - 1 = 80$ is divisible by 4

   P_3: $9^3 - 1 = 728$ is divisible by 4

(B) P_k: $9^k - 1 = 4r$ for some integer r

   P_{k+1}: $9^{k+1} - 1 = 4s$ for some integer s

(C) Prove: $9^n - 1$ is divisible by 4.
   Write: P_n: $9^n - 1$ is divisible by 4

Part 1: Show that P_1 is true:

P_1: $9^1 - 1 = 8 = 4 \cdot 2$ is divisible by 4. Clearly true.

Part 2: Show that if P_k is true, then P_{k+1} is true:
Write out P_k and P_{k+1}.

   P_k: $9^k - 1 = 4r$ for some integer r
   P_{k+1}: $9^{k+1} - 1 = 4s$ for some integer s

We start with P_k:
   $9^k - 1 = 4r$ for some integer r

Now, $9^{k+1} - 1 = 9^{k+1} - 9^k + 9^k - 1$
$= 9^k (9 - 1) + 9^k - 1$
$= 9^k \cdot 8 + 9^k - 1$
$= 4 \cdot 9^k \cdot 2 + 4r$
$= 4(9^k \cdot 2 + r)$

Therefore,

$9^{k+1} - 1 = 4s$ for some integer s ($= 2 \cdot 9^k + r$)
$9^{k+1} - 1$ is divisible by 4.

We have shown that if P_k is true, then P_{k+1} is true.

Conclusion: P_n is true for all positive integers n.

13. The polynomial $x^4 + 1$ is a counterexample, since it has degree 4, and no real zeros.

15. 23 is a counterexample, since there are no prime numbers p such that $23 < p < 29$.

17. P_n: $2 + 2^2 + 2^3 + \cdots + 2^n = 2^{n+1} - 2$

Part 1: Show that P_1 is true:
$$
\begin{aligned}
P_1: 2 &= 2^{1+1} - 2 \\
&= 2^2 - 2 \\
&= 4 - 2 \\
&= 2 \qquad P_1 \text{ is true}
\end{aligned}
$$

Part 2: Show that if P_k is true, then P_{k+1} is true:
Write out P_k and P_{k+1}.

P_k: $2 + 2^2 + 2^3 + \cdots + 2^k = 2^{k+1} - 2$

P_{k+1}: $2 + 2^2 + 2^3 + \cdots + 2^k + 2^{k+1} = 2^{k+2} - 2$

We start with P_k:

$2 + 2^2 + 2^3 + \cdots + 2^k = 2^{k+1} - 2$

Adding 2^{k+1} to both sides,
$$
\begin{aligned}
2 + 2^2 + 2^3 + \cdots + 2^k + 2^{k+1} &= 2^{k+1} - 2 + 2^{k+1} \\
&= 2^{k+1} + 2^{k+1} - 2 \\
&= 2 \cdot 2^{k+1} - 2 \\
&= 2^1 \cdot 2^{k+1} - 2 \\
&= 2^{k+2} - 2
\end{aligned}
$$

We have shown that if P_k is true, then P_{k+1} is true.

Conclusion: P_n is true for all positive integers n.

19. P_n: $1^2 + 3^2 + 5^2 + \cdots + (2n - 1)^2 = \dfrac{1}{3}(4n^3 - n)$

Part 1: Show that P_1 is true:
$$
\begin{aligned}
P_1: 1^2 &= \frac{1}{3}(4 \cdot 1^3 - 1) \\
&= \frac{1}{3}(4 - 1) \\
&= 1 \qquad \text{True.}
\end{aligned}
$$

Part 2: Show that if P_k is true, then P_{k+1} is true:
Write out P_k and P_{k+1}.

P_k: $1^2 + 3^2 + 5^2 + \cdots + (2k - 1)^2 = \dfrac{1}{3}(4k^3 - k)$

P_{k+1}: $1^2 + 3^2 + 5^2 + \cdots + (2k - 1)^2 + (2k + 1)^2 = \dfrac{1}{3}[4(k + 1)^3 - (k + 1)]$

We start with P_k:

$$1^2 + 3^2 + 5^2 + \cdots + (2k - 1)^2 = \frac{1}{3}(4k^3 - k)$$

Adding $(2k + 1)^2$ to both sides,

$$1^2 + 3^2 + 5^2 + \cdots + (2k - 1)^2 + (2k + 1)^2 = \frac{1}{3}(4k^3 - k) + (2k + 1)^2$$

$$= \frac{1}{3}[4k^3 - k + 3(2k + 1)^2]$$

$$= \frac{1}{3}[4k^3 - k + 3(4k^2 + 4k + 1)]$$

$$= \frac{1}{3}[4k^3 - k + 12k^2 + 12k + 3]$$

$$= \frac{1}{3}[4k^3 + 12k^2 + 12k - k + 3]$$

$$= \frac{1}{3}[4k^3 + 12k^2 + 12k + 4 - k - 1]$$

$$= \frac{1}{3}[4(k^3 + 3k^2 + 3k + 1) - (k + 1)]$$

$$= \frac{1}{3}[4(k + 1)^3 - (k + 1)]$$

We have shown that if P_k is true, then P_{k+1} is true.

Conclusion: P_n is true for all positive integers n.

21. P_n: $1^2 + 2^2 + 3^2 + \cdots + n^2 = \dfrac{n(n + 1)(2n + 1)}{6}$

Part 1: Show that P_1 is true:

$$P_1: \quad 1^2 = \frac{1(1 + 1)(2 \cdot 1 + 1)}{6}$$

$$= \frac{1 \cdot 2 \cdot 3}{6}$$

$$= 1 \qquad P_1 \text{ is true.}$$

Part 2: Show that if P_k is true, then P_{k+1} is true:

Write out P_k and P_{k+1}.

$$P_k: \quad 1^2 + 2^2 + 3^2 + \cdots + k^2 = \frac{k(k + 1)(2k + 1)}{6}$$

$$P_{k+1}: \quad 1^2 + 2^2 + 3^2 + \cdots + k^2 + (k + 1)^2 = \frac{(k + 1)(k + 2)(2k + 3)}{6}$$

We start with P_k:

$$1^2 + 2^2 + 3^2 + \cdots + k^2 = \frac{k(k + 1)(2k + 1)}{6}$$

Adding $(k + 1)^2$ to both sides,

$$1^2 + 2^2 + 3^2 + \cdots + k^2 + (k + 1)^2 = \frac{k(k + 1)(2k + 1)}{6} + (k + 1)^2$$

$$= \frac{k(k + 1)(2k + 1)}{6} + \frac{6(k + 1)^2}{6}$$

$$= \frac{k(k + 1)(2k + 1) + 6(k + 1)^2}{6}$$

$$= \frac{(k + 1)[k(2k + 1) + 6(k + 1)]}{6}$$

$$= \frac{(k + 1)[2k^2 + 7k + 6]}{6}$$

$$= \frac{(k + 1)(k + 2)(2k + 3)}{6}$$

We have shown that if P_k is true, then P_{k+1} is true.

Conclusion: P_n is true for all positive integers n.

23. P_n: $\dfrac{a^n}{a^3} = a^{n-3}$ $n > 3$

Part 1: Show that P_4 is true:

P_4: $\dfrac{a^4}{a^3} = a^{4-3}$

But $\dfrac{a^4}{a^3} = \dfrac{a^3 a}{a^3} = a = a^1 = a^{4-3}$ So, P_4 is true.

Part 2: Show that if P_k is true, then P_{k+1} is true:
Write out P_k and P_{k+1}.

$\qquad P_k$: $\dfrac{a^k}{a^3} = a^{k-3}$

$\qquad P_{k+1}$: $\dfrac{a^{k+1}}{a^3} = a^{k-2}$

We start with P_k:

$\qquad \dfrac{a^k}{a^3} = a^{k-3}$

Multiply both sides by a:

$\qquad \dfrac{a^k}{a^3} a = a^{k-3} a$

$\qquad \dfrac{a^{k+1}}{a^3} = a^{k-3+1}$ by the recursive definition of a^n

$\qquad \dfrac{a^{k+1}}{a^3} = a^{k-2}$

We have shown that if P_k is true, then P_{k+1} is true.

Conclusion: P_n is true for all integers $n > 3$.

25. Write: P_n: $a^m a^n = a^{m+n}$ m an arbitrary element of N.

Part 1: Show that P_1 is true:

P_1: $a^m a^1 = a^{m+1}$ This is true by the recursive definition of a^n.

Part 2: Show that if P_k is true, then P_{k+1} is true:
Write out P_k and P_{k+1}.

$\qquad P_k$: $a^m a^k = a^{m+k}$

$\qquad P_{k+1}$: $a^m a^{k+1} = a^{m+k+1}$

We start with P_k:

$\qquad a^m a^k = a^{m+k}$

Multiplying both sides by a:

$\qquad a^m a^k a = a^{m+k} a$

$\qquad a^m a^{k+1} = a^{m+k+1}$ by the recursive definition of a^n.

We have shown that if P_k is true, then P_{k+1} is true.

Conclusion: $a^m a^n = a^{m+n}$ is true for arbitrary m, n positive integers.

27. Write: P_n: $x^n - 1 = (x - 1)Q_n(x)$ for some polynomial $Q_n(x)$.

Part 1: Show that P_1 is true:

P_1: $x^1 - 1 = (x - 1)Q_1(x)$ for some $Q_1(x)$. This is true since we can choose $Q_1(x) = 1$.

Part 2: Show that if P_k is true, then P_{k+1} is true:

Write out P_k and P_{k+1}.

P_k: $x^k - 1 = (x - 1)Q_k(x)$ for some $Q_k(x)$

P_{k+1}: $x^{k+1} - 1 = (x - 1)Q_{k+1}(x)$ for some $Q_{k+1}(x)$

We start with P_k:

$x^k - 1 = (x - 1)Q_k(x)$ for some $Q_k(x)$

Now, $x^{k+1} - 1 = x^{k+1} - x^k + x^k - 1$

$$= x^k(x - 1) + x^k - 1$$

$$= x^k(x - 1) + (x - 1)Q_k(x)$$

$$= (x - 1)(x^k + Q_k(x))$$

$$= (x - 1)Q_{k+1}(x), \text{ where } Q_{k+1}(x) = x^k + Q_k(x)$$

Conclusion: $x^n - 1 = (x - 1)Q_n(x)$ for all positive integers n. Therefore $x^n - 1$ is divisible by $x - 1$ for all positive integers n.

29. Write: P_n: $x^{2n} - 1 = (x - 1)Q_n(x)$ for some polynomial $Q_n(x)$.

Part 1: Show that P_1 is true:

P_1: $x^2 - 1 = (x - 1)Q_1(x)$ for some polynomial $Q_1(x)$. This is true since we can choose $Q_1(x) = x + 1$.

Part 2: Show that if P_k is true, then P_{k+1} is true:

Write out P_k and P_{k+1}.

P_k: $x^{2k} - 1 = (x - 1)Q_k(x)$ for some $Q_k(x)$

P_{k+1}: $x^{2k+2} - 1 = (x - 1)Q_{k+1}(x)$ for some $Q_{k+1}(x)$

Now, $x^{2k+2} - 1 = x^{2k+2} - x^{2k} + x^{2k} - 1$

$$= x^{2k}(x^2 - 1) + x^{2k} - 1$$

$$= x^{2k}(x + 1)(x - 1) + (x - 1)Q_k(x)$$

$$= (x - 1)[x^{2k}(x + 1) + Q_k(x)]$$

$$= (x - 1)Q_{k+1}(x), \text{ where } Q_{k+1}(x) = x^{2k}(x + 1) + Q_k(x)$$

Conclusion: $x^{2n} - 1 = (x - 1)Q_n(x)$ for all positive integers n. Therefore $x^{2n} - 1$ is divisible by $x - 1$ for all positive integers n.

31. P_n: $1^3 + 2^3 + 3^3 + \cdots + n^3 = (1 + 2 + 3 + \cdots + n)^2$

Part 1: Show that P_1 is true:

P_1: $1^3 = 1^2$ True

Part 2: Show that if P_k is true, then P_{k+1} is true:

Write out P_k and P_{k+1}.

P_k: $1^3 + 2^3 + 3^3 + \cdots + k^3 = (1 + 2 + 3 + \cdots + k)^2$

P_{k+1}: $1^3 + 2^3 + 3^3 + \cdots + k^3 + (k + 1)^3 = (1 + 2 + 3 + \cdots + k + k + 1)^2$

We take it as proved in text Matched Problem 2 that

$$1 + 2 + 3 + \cdots + n = \frac{n(n + 1)}{2} \quad \text{for} \quad n \in N$$

So $1 + 2 + 3 + \cdots + k = \dfrac{k(k+1)}{2}$

$1 + 2 + 3 + \cdots + k + (k+1) = \dfrac{(k+1)(k+2)}{2}$ are known

We start with P_k:

$1^3 + 2^3 + 3^3 + \cdots + k^3 = (1 + 2 + 3 + \cdots + k)^2$

Adding $(k+1)^3$ to both sides:

$$1^3 + 2^3 + 3^3 + \cdots + k^3 + (k+1)^3 = (1 + 2 + 3 + \cdots + k)^2 + (k+1)^3$$

$$= \left[\frac{k(k+1)}{2}\right]^2 + (k+1)^3$$

$$= (k+1)^2\left[\left(\frac{k}{2}\right)^2 + (k+1)\right]$$

$$= (k+1)^2\left[\frac{k^2}{4} + \frac{4(k+1)}{4}\right]$$

$$= (k+1)^2\left[\frac{k^2 + 4k + 4}{4}\right]$$

$$= \frac{(k+1)^2(k+2)^2}{4}$$

$$= \left[\frac{(k+1)(k+2)}{2}\right]^2$$

$$= [1 + 2 + 3 + \cdots + k + (k+1)]^2$$

Conclusion: P_n is true for all positive integers n.

33. We note:

$$
\begin{array}{rcccl}
2 &=& 2 &=& 1\cdot 2 \quad n = 1 \\
2 + 4 &=& 6 &=& 2\cdot 3 \quad n = 2 \\
2 + 4 + 6 &=& 12 &=& 3\cdot 4 \quad n = 3 \\
2 + 4 + 6 + 8 &=& 20 &=& 4\cdot 5 \quad n = 4
\end{array}
$$

Hypothesis: P_n: $2 + 4 + 6 + \cdots + 2n = n(n+1)$

Proof: Part 1: Show P_1 is true

P_1: $2 = 1\cdot 2$ True

Part 2: Show that if P_k is true, then P_{k+1} is true:

Write out P_k and P_{k+1}.

P_k: $2 + 4 + 6 + \cdots + 2k = k(k+1)$

P_{k+1}: $2 + 4 + 6 + \cdots + 2k + (2k+2) = (k+1)(k+2)$

We start with P_k:

$2 + 4 + 6 + \cdots + 2k = k(k+1)$

Adding $2k + 2$ to both sides:

$$2 + 4 + 6 + \cdots + 2k + (2k+2) = k(k+1) + 2k + 2$$

$$= (k+1)k + (k+1)2$$

$$= (k+1)(k+2)$$

Conclusion: The hypothesis P_n is true for all positive integers n.

35. $n = 1$: no line is determined.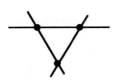
$n = 2$: one line is determined.

$n = 3$: three lines are determined.

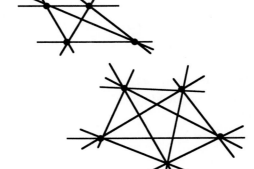

$n = 4$: six lines are determined.

$n = 5$: ten lines are determined.

Hypothesis: P_n: n points (no three collinear) determine

$$1 + 2 + 3 + \cdots + (n - 1) = \frac{n(n - 1)}{2} \text{ lines, } n \geq 2.$$

Proof: Part 1: Show that P_2 is true.

P_2: 2 points determine one line:

$$1 = \frac{2(2 - 1)}{2}$$
$$= \frac{2 \cdot 1}{2}$$
$$= 1 \text{ is true}$$

Part 2: Show that if P_k is true, then P_{k+1} is true:

Write out P_k and P_{k+1}.

P_k: k points determine $1 + 2 + 3 + \cdots + (k - 1) = \dfrac{k(k - 1)}{2}$ lines

P_{k+1}: $k + 1$ points determine $1 + 2 + 3 + \cdots + (k - 1) + k = \dfrac{(k + 1)k}{2}$ lines

We start with P_k:

k points determine $1 + 2 + 3 + \cdots + (k - 1) = \dfrac{k(k - 1)}{2}$ lines

Now, the $k + 1$st point will determine a total of k new lines, one with each of the previously existing k points. These k new lines will be added to the previously existing lines. Therefore, $k + 1$ points determine

$$1 + 2 + 3 + \cdots + (k - 1) + k = \frac{k(k - 1)}{2} + k \text{ lines}$$
$$= k\left[\frac{k - 1}{2} + 1\right] \text{ lines}$$
$$= k\left[\frac{k - 1 + 2}{2}\right] \text{ lines}$$
$$= \frac{k(k + 1)}{2} \text{ lines}$$

Conclusion: The hypothesis P_n is true for $n \geq 2$.

37. P_n: $a > 1 \Rightarrow a^n > 1$, $n \in N$.

Part 1: Show that P_1 is true:

P_1: $a > 1 \Rightarrow a^1 > 1$. This is automatically true.

Part 2: Show that if P_k is true, then P_{k+1} is true:
Write out P_k and P_{k+1}.

$\quad P_k$: $a > 1 \Rightarrow a^k > 1$

$\quad P_{k+1}$: $a > 1 \Rightarrow a^{k+1} > 1$

We start with P_k. Further, assume $a > 1$ and try to derive $a^{k+1} > 1$. If this succeeds, we have proved P_{k+1}.
Assume $a > 1$.

From P_k we know that $a^k > 1$, also $1 > 0$, so $a > 0$. We may therefore multiply both sides of the inequality $a^k > 1$ by a without changing the direction of the inequality.

$\quad a^k a > 1a$

$\quad a^{k+1} > a$

But, $a > 1$ by assumption, so $a^{k+1} > 1$. We have derived this from P_k and $a > 1$. So if P_k is true, then P_{k+1} is true.

Conclusion: P_n is true for all $n \in N$.

39. P_n: $n^2 > 2n$, $n \geq 3$.

Part 1: Show that P_3 is true:

P_3: $3^2 > 2 \cdot 3$

$\quad\quad 9 > 6$ True

Part 2: Show that if P_k is true, then P_{k+1} is true:
Write out P_k and P_{k+1}.

$\quad P_k$: $k^2 > 2k$

$\quad P_{k+1}$: $(k + 1)^2 > 2(k + 1)$

We start with P_k:

$\quad k^2 > 2k$

Adding $2k + 1$ to both sides: $k^2 + 2k + 1 > 2k + 2k + 1$

$\quad (k + 1)^2 > 2k + 2 + 2k - 1$

Now, $2k - 1 > 0$, $k \in N$, so $2k + 2 + 2k - 1 > 2k + 2$.

Therefore, $(k + 1)^2 > 2k + 2$

$\quad\quad\quad (k + 1)^2 > 2(k + 1)$

So if P_k is true, then P_{k+1} is true.

Conclusion: P_n is true for all integers $n \geq 3$.

41. $3^4 + 4^4 + 5^4 + 6^4 = 2,258$

$\quad\quad\quad\quad 7^4 = 2,401$

$3^4 + 4^4 + 5^4 + 6^4 \neq 7^4$

There is no true obvious generalization of the given facts.

43. To prove $a_n = b_n$, $n \in N$, write:

P_n: $a_n = b_n$.

Proof: Part 1: Show P_1 is true.

P_1: $a_1 = b_1$; $a_1 = 1$; $b_1 = 2 \cdot 1 - 1 = 1$
So $a_1 = b_1$

Part 2: Show that if P_k is true, then P_{k+1} is true:
Write out P_k and P_{k+1}.

P_k: $a_k = b_k$
P_{k+1}: $a_{k+1} = b_{k+1}$

We start with P_k:

$a_k = b_k$

Now, $a_{k+1} = a_{k+1-1} + 2 = a_k + 2 = b_k + 2 = 2k - 1 + 2 = 2k + 1$
$= 2(k + 1) - 1 = b_{k+1}$

Therefore, $a_{k+1} = b_{k+1}$

So if P_k is true, then P_{k+1} is true.

Conclusion: P_n is true for all $n \in N$, and $\{a_n\} = \{b_n\}$.

45. To prove: $a_n = b_n$, $n \in N$, write:

P_n: $a_n = b_n$.

Proof: Part 1: Show P_1 is true.

P_1: $a_1 = b_1$; $a_1 = 2$; $b_1 = 2^{2 \cdot 1 - 1} = 2^1 = 2$
so $a_1 = b_1$

Part 2: Show that if P_k is true, then P_{k+1} is true:
Write out P_k and P_{k+1}.

P_k: $a_k = b_k$
P_{k+1}: $a_{k+1} = b_{k+1}$

We start with P_k:

$a_k = b_k$

Now, $a_{k+1} = 2^2 a_{k+1-1} = 2^2 a_k = 2^2 b_k = 2^2 2^{2k-1} = 2^{2+2k-1} = 2^{2(k+1)-1} = b_{k+1}$

Therefore, $a_{k+1} = b_{k+1}$

So if P_k is true, then P_{k+1} is true.

Conclusion: P_n is true for all $n \in N$, and $\{a_n\} = \{b_n\}$.

Section 10-3

1. Answers will vary.   **3.** Answers will vary.

5. (A) Since $(-16) - (-11) = -5$ and $(-21) - (-16) = -5$, the given terms can start an arithmetic sequence with $d = -5$. Then next terms are then $-21 + (-5) = -26$, and $(-26) + (-5) = -31$.

(B) Since $(-4) - 2 = -6$ and $8 - (-4) = 12$, there is no common difference. Since $(-4) \div 2 = -2$ and $8 \div (-4) = -2$, the given terms can start a geometric sequence with $r = -2$. The next terms are then $8 \cdot (-2) = -16$, and $(-16) \cdot (-2) = 32$.

(C) Since $4 - 1 \neq 9 - 4$, there is no common difference, so the sequence is not an arithmetic sequence. Since $4 \div 1 \neq 9 \div 4$, there is no common ratio, so the sequence is not geometric either.

(D) Since $\frac{1}{6} - \frac{1}{2} = -\frac{1}{3}$ and $\frac{1}{18} - \frac{1}{6} = -\frac{1}{9}$, there is no common difference. Since $\frac{1}{6} \div \frac{1}{2} = \frac{1}{3}$ and $\frac{1}{18} \div \frac{1}{6} = \frac{1}{3}$, the given terms can start a geometric sequence with $r = \frac{1}{3}$. The next terms are then $\frac{1}{18} \cdot \frac{1}{3} = \frac{1}{54}$ and $\frac{1}{54} \cdot \frac{1}{3} = \frac{1}{162}$.

7. $a_2 = a_1 + d = -5 + 4 = -1$
$a_3 = a_2 + d = -1 + 4 = 3$
$a_4 = a_3 + d = 3 + 4 = 7$
$a_n = a_1 + (n - 1)d$

9. $a_{15} = a_1 + 14d = -3 + 14 \cdot 5 = 67$
$S_n = \frac{n}{2}[2a_1 + (n - 1)d]$
$S_{11} = \frac{11}{2}[2(-3) + (11 - 1)5]$
$= \frac{11}{2}(44)$
$= 242$

11. $a_2 - a_1 = 5 - 1 = d = 4$
$S_n = \frac{n}{2}[2a_1 + (n - 1)d]$
$S_{21} = \frac{21}{2}[2 \cdot 1 + (21 - 1)4]$
$= \frac{21}{2}(82)$
$= 861$

13. $a_2 - a_1 = 5 - 7 = -2 = d$
$a_n = a_1 + (n - 1)d$
$a_{15} = 7 + (15 - 1)(-2)$
$= -21$

15. $a_2 = a_1 r = (-6)\left(-\frac{1}{2}\right) = 3$
$a_3 = a_2 r = 3\left(-\frac{1}{2}\right) = -\frac{3}{2}$
$a_4 = a_3 r = \left(-\frac{3}{2}\right)\left(-\frac{1}{2}\right) = \frac{3}{4}$

17. $a_n = a_1 r^{n-1}$
$a_{10} = 81\left(\frac{1}{3}\right)^{10-1}$
$= \frac{1}{243}$

19. $S_n = \frac{a_1 - ra_n}{1 - r}$
$S_7 = \frac{3 - 3(2,187)}{1 - 3}$
$= 3,279$

21. $a_n = a_1 + (n - 1)d$
$a_{20} = a_1 + 19d$
$117 = 3 + 19d$
$114 = 19d$
So, $d = 6$. Therefore
$a_{101} = a_1 + (100)d$
$= 3 + 100(6) = 603$

23. $S_n = \frac{n}{2}(a_1 + a_n)$
$S_{40} = \frac{40}{2}(-12 + 22)$
$= 200$

25. $a_2 - a_1 = d$
$\frac{1}{2} - \frac{1}{3} = \frac{1}{6} = d$
$a_n = a_1 + (n - 1)d$
$a_{11} = \frac{1}{3} + (11 - 1)\frac{1}{6}$
$= 2$
$S_n = \frac{n}{2}(a_1 + a_n)$
$S_{11} = \frac{11}{2}\left(\frac{1}{3} + 2\right) = \frac{77}{6}$

27. $a_n = a_1 + (n - 1)d$
$a_{10} = a_1 + 9d$
$a_3 = a_1 + 2d$
Eliminating d between these two statements by addition, we have
$2a_{10} = 2a_1 + 18d$
$-9a_3 = -9a_1 - 18d$
$\overline{2a_{10} - 9a_3 = -7a_1}$
$a_1 = \frac{2a_{10} - 9a_3}{-7} = \frac{2(55) - 9(13)}{-7} = 1$

29.
$$a_n = a_1 r^{n-1}$$
$$1 = 100 r^{6-1}$$
$$\frac{1}{100} = r^5$$
$$r = \sqrt[5]{0.01} = 10^{-2/5}$$
$$r = 0.398$$

31. $S_n = \dfrac{a_1 - a_1 r^n}{1 - r}$

$$S_{10} = \frac{5 - 5(-2)^{10}}{1 - (-2)}$$
$$= \frac{-5,115}{3}$$
$$= -1,705$$

33. First find r:
$$a_n = a_1 r^{n-1}$$
$$\frac{8}{3} = 9 r^{4-1}$$
$$\frac{8}{27} = r^3$$
$$r = \frac{2}{3}$$
$$a_2 = r a_1 = \frac{2}{3}(9) = 6$$
$$a_3 = r a_2 = \frac{2}{3}(6) = 4$$

35.
$$a_n = a_1 + (n - 1)d$$
$$d = a_2 - a_1$$
$$= (3 \cdot 2 + 3) - (3 \cdot 1 + 3) = 3$$
$$a_{51} = a_1 + (51 - 1)d$$
$$= (3 \cdot 1 + 3) + 50 \cdot 3$$
$$= 156$$
$$S_n = \frac{n}{2}(a_1 + a_n)$$
$$S_{51} = \frac{51}{2}(6 + 156)$$
$$= \frac{51}{2}[162]$$
$$= 4,131$$

37. $S_n = \dfrac{a_1 - a_1 r^n}{1 - r}$

First, note $a_1 = (-3)^{1-1} = 1$
$$r = \frac{a_2}{a_1} = \frac{(-3)^{2-1}}{(-3)^{1-1}} = -3$$
$$S_7 = \frac{1 - 1(-3)^7}{1 - (-3)}$$
$$= \frac{2,188}{4}$$
$$= 547$$

39.
$$g(t) = 5 - t$$
$$g(1) = 5 - 1 = 4$$
$$g(51) = 5 - 51 = -46$$
$$g(1) + g(2) + g(3) + \cdots + g(51) = S_{51}$$
$$S_n = \frac{n}{2}(a_1 + a_n)$$
$$S_{51} = \frac{51}{2}(g(1) + g(51))$$
$$= \frac{51}{2}[4 + (-46)]$$
$$= -1,071$$

41. $g(1) + g(2) + \cdots + g(10)$ is a geometric series, with

$$g(1) = a_1 = \left(\frac{1}{2}\right)^1 = \frac{1}{2}$$
$$r = \frac{g(2)}{g(1)} = \frac{\left(\frac{1}{2}\right)^2}{\left(\frac{1}{2}\right)^1} = \frac{1}{2}$$
$$S_n = \frac{a_1 - a_1 r^n}{1 - r}$$
$$S_{10} = \frac{\frac{1}{2} - \frac{1}{2}\left(\frac{1}{2}\right)^{10}}{1 - \frac{1}{2}}$$
$$= \frac{1 - \left(\frac{1}{2}\right)^{10}}{2 - 1}$$
$$= \frac{1,023}{1,024}$$

43. First, find n:
$$a_n = a_1 + (n - 1)d$$
$$134 = 22 + (n - 1)2$$
$$n = 57$$
Now, find S_{57}
$$S_n = \frac{n}{2}(a_1 + a_n)$$
$$S_{57} = \frac{57}{2}(22 + 134)$$
$$= 4,446$$

45. To prove:

$$1 + 3 + 5 + \cdots + (2n - 1) = n^2$$

The sequence 1, 3, 5, $\cdots$ is an arithmetic sequence, with $d = 2$. We are to find S_n.

But, $S_n = \dfrac{n}{2}(a_1 + a_n)$

$\qquad = \dfrac{n}{2}[1 + (2n - 1)]$

$\qquad = \dfrac{n}{2} \cdot 2n$

So, $S_n = n^2$

47. $\dfrac{a_2}{a_1} = \dfrac{a_3}{a_2} = r$ for $a_1 + a_2 + a_3$ to be a geometric series. Then,

$\dfrac{x}{-2} = \dfrac{-6}{x}$

$x^2 = 12$

$x = 2\sqrt{3}$,

since x is specified positive.

49. Note that $a_n - a_{n-1} = 3$, so this is an arithmetic sequence, with $d = 3$. We are to find a_n.

But $a_n = a_1 + (n - 1)d$

So $a_n = -3 + (n - 1)3$ or $3n - 6$

51. Here's what an arithmetic sequence with $d = 0$ would look like:

$a_1, a_1, a_1, a_1, \cdots$

So the sum S_n is given by $S_n = a_1 + a_1 + \cdots + a_1$ where there are n terms in the sum. But this is exactly how we define multiplication! A simple form is $S_n = n \cdot a_1$.

53.

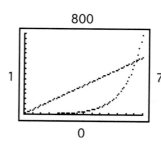

Enter the sequences in the sequence editor.

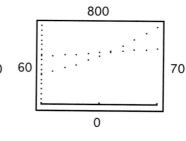

Graph the sequences.

Display the points (in the appropriate interval) in a table.

From the second graph, 66 is the least positive integer n such that $a_n < b_n$. The table confirms this.

55.

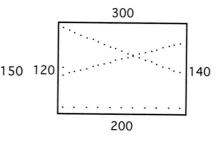

Enter the sequences in the sequence editor.

Graph the sequences.

Display the points (in the appropriate interval) in a table.

From the second graph, 133 is the least positive integer n such that $a_n < b_n$. The table confirms this.

57. $\dfrac{a_2}{a_1} = 1 \div 3 = \dfrac{1}{3} = r.$ $|r| < 1$

Therefore, this infinite geometric series has a sum.

$$S_\infty = \dfrac{a_1}{1 - r}$$

$$= \dfrac{3}{1 - \frac{1}{3}} = \dfrac{9}{2}$$

59. $\dfrac{a_2}{a_1} = \dfrac{4}{2} = 2 = r \geq 1$

Therefore, this infinite geometric series has no sum.

61. $\dfrac{a_2}{a_1} = \left(-\dfrac{1}{2}\right) \div 2 = -\dfrac{1}{4} = r.$ $|r| < 1$

Therefore, this infinite geometric series has a sum.

$$S_\infty = \dfrac{a_1}{1 - r}$$

$$= \dfrac{2}{1 - \left(-\frac{1}{4}\right)}$$

$$= \dfrac{8}{5}$$

63. $0.\overline{7} = 0.777\ldots = 0.7 + 0.07 + 0.007$
$+ 0.0007 + \ldots$

This is an infinite geometric series with
$a_1 = 0.7$ and $r = 0.1$.
So

$$S_\infty = \dfrac{a_1}{1 - r} = \dfrac{0.7}{1 - 0.1} = \dfrac{0.7}{0.9} = \dfrac{7}{9}$$

65. $0.\overline{54}\ldots = 0.54 + 0.0054 + 0.000054 + \ldots$
This is an infinite geometric series with $a_1 = 0.54$ and $r = 0.01$. So

$$S_\infty = \dfrac{a_1}{1 - r} = \dfrac{0.54}{1 - 0.01} = \dfrac{0.54}{0.99} = \dfrac{6}{11}$$

67. $3.\overline{216} = 3.216216216\ldots = 3 + 0.216 + 0.000216 + 0.000000216 + \ldots$
Therefore, we note: $0.216 + 0.000216 + 0.000000216 + \ldots$ is an infinite geometric series with $a_1 = 0.216$ and $r = 0.001$. So

$$3.\overline{216} = 3 + S_\infty = 3 + \dfrac{a_1}{1 - r} = 3 + \dfrac{0.216}{1 - 0.001} = 3 + \dfrac{0.216}{0.999} = 3\tfrac{8}{37} \text{ or } \tfrac{119}{37}$$

69. Write: P_n: $a_n = a_1 + (n - 1)d$

Proof: Part 1: Show that P_1 is true:
P_1: $a_1 = a_1 + (1 - 1)d = a_1$ P_1 is true.

Part 2: Show that if P_k is true, then P_{k+1} is true:
Write out P_k and P_{k+1}.
   P_k: $a_k = a_1 + (k - 1)d$
   P_{k+1}: $a_{k+1} = a_1 + kd$
We start with P_k:
   $a_k = a_1 + (k - 1)d$
Adding d to both sides:
   $a_k + d = a_1 + (k - 1)d + d$
      $a_{k+1} = a_1 + d[(k - 1) + 1]$
      $a_{k+1} = a_1 + kd$
So if P_k is true, then P_{k+1} is true.

Conclusion: P_n is true for all positive integers n.

71. This is a geometric sequence with ratio $\dfrac{a_n}{a_{n-1}} = -3$.

$$a_n = a_1 r^{n-1}$$
$$= (-2)(-3)^{n-1}$$

73. Assume x, y, z are consecutive terms of an arithmetic progression.
Then, $y - x = d$, $z - y = d$
To show $a = x^2 + xy + y^2$, $b = z^2 + xz + x^2$, $c = y^2 + yz + z^2$, are consecutive terms of an arithmetic progression, we need only show that
$b - a = c - b$
$$\begin{aligned}
b - a &= z^2 + xz + x^2 - (x^2 + xy + y^2) \\
&= z^2 + xz - xy - y^2 \\
&= z^2 + x(z - y) - y^2 \\
&= z^2 - y^2 + x(z - y) \\
&= (z - y)(z + y + x) \\
&= d(x + y + z) \\
c - b &= y^2 + yz + z^2 - (z^2 + xz + x^2) \\
&= y^2 + yz - xz - x^2 \\
&= y^2 - x^2 + yz - xz \\
&= y^2 - x^2 + z(y - x) \\
&= (y - x)[y + x + z] \\
&= d(x + y + z)
\end{aligned}$$
Therefore, a, b, c, that is, $x^2 + xy + y^2$, $z^2 + xz + x^2$, $y^2 + yz + z^2$ are consecutive terms of an arithmetic progression.

75. Write: P_n: $a_n = a_1 r^{n-1}$, $n \in N$

Proof: Part 1: Show that P_1 is true:
$$\begin{aligned}
P_1:\ a_1 &= a_1 r^{1-1} \\
&= a_1 r^0 \\
&= a_1 \quad \text{True.}
\end{aligned}$$

Part 2: Show that if P_k is true, then P_{k+1} is true:
Write out P_k and P_{k+1}.
$$P_k:\ a_k = a_1 r^{k-1}$$
$$P_{k+1}:\ a_{k+1} = a_1 r^k$$
We start with P_k:
$$a_k = a_1 r^{k-1}$$

Multiply both sides by r:
$$a_k\, r = a_1 r^{k-1} r$$
$$a_k r = a_1 r^k$$
But, we are given that $a_{k+1} = r a_k = a_k r$.
Then $a_{k+1} = a_1 r^k$
So if P_k is true, then P_{k+1} is true.

Conclusion: P_n is true for all $n \in N$.

77. Given a, b, c, d, e, f is an arithmetic progression, let D be the common difference. Then, assuming $D \neq 0$:

$$b = a + D, \quad c = a + 2D, \quad d = a + 3D, \quad e = a + 4D, \quad f = a + 5D$$

From the solution to Section 8-1, problem 39, we know that there will be a solution of

$$ax + by = c$$
$$dx + ey = f$$

if $ae - bd \neq 0$

In this case, $ae - bd = a(a + 4D) - (a + 3D)(a + D)$

$$= a^2 + 4aD - (a^2 + 4aD + 3D^2)$$

$$= -3D^2 \neq 0$$

Then, from the same problem, we have

$$x = \frac{ce - bf}{ae - bd} = \frac{(a + 2D)(a + 4D) - (a + D)(a + 5D)}{-3D^2}$$

$$= \frac{a^2 + 6aD + 8D^2 - (a^2 + 6aD + 5D^2)}{-3D^2}$$

$$= \frac{3D^2}{-3D^2}$$

$$= -1$$

$$y = \frac{af - dc}{ae - bd} = \frac{a(a + 5D) - (a + 3D)(a + 2D)}{-3D^2}$$

$$= \frac{a^2 + 5aD - (a^2 + 5aD + 6D^2)}{-3D^2}$$

$$= \frac{-6D^2}{-3D^2}$$

$$= 2$$

79. We can use the formula for a sum from Theorem 5.

$$S_5 = \frac{1 - 2^5}{1 - 2} = \frac{-31}{-1} = 31$$

$$S_{10} = \frac{1 - 2^{10}}{1 - 2} = \frac{-1,023}{-1} = 1,023$$

$$S_{15} = \frac{1 - 2^{15}}{1 - 2} = \frac{-32,767}{-1} = 32,767$$

$$S_{20} = \frac{1 - 2^{20}}{1 - 2} = \frac{-1,048,575}{-1} = 1,048,575$$

It looks like S_n continues to get larger and approaches ∞ as n approaches ∞.

81. With each firm, the salaries form an arithmetic sequence. We are asked for the sum of fifteen terms, or S_{15}, given a_1 and d.

$$S_n = \frac{n}{2}[2a_1 + (n - 1)d\,]$$

$$S_{15} = \frac{15}{2}[2a_1 + 14d\,] = 15(a_1 + 7d)$$

Firm A: $a_1 = 25,000 \quad d = 1,200 \quad S_{15} = 15(25,000 + 7 \cdot 1,200) = \$501,000$

Firm B: $a_1 = 28,000 \quad d = 800 \quad S_{15} = 15(28,000 + 7 \cdot 800) = \$504,000$

83. We are asked for the sum of an infinite geometric series.

$a_1 = \$800,000$

$r = 0.8 \quad |r| \leq 1,$

so the series has a sum,

$S_\infty = \dfrac{a_1}{1 - r}$

$ = \dfrac{\$800,000}{1 - 0.8}$

$ = \dfrac{\$800,000}{0.2}$

$ = \$4,000,000$

85. After one year, $P(1 + r)$ is present, so the geometric sequence has $a_1 = P(1 + r)$. The ratio is given as $(1 + r)$, so

$a_n = a_1(1 + r)^{n-1}$

$ = P(1 + r)(1 + r)^{n-1}$

$ = P(1 + r)^n$ is the amount present after n years.

The time taken for P to double is represented by n years. We set $A = 2P$, $r = 0.06$, then solve

$2P = P(1 + 0.06)^n$

$2 = (1.06)^n$

$\log 2 = n \log 1.06$

$n = \dfrac{\log 2}{\log 1.06}$

$ \approx 12$ years

87. This involves an arithmetic sequence. Let d = increase in earnings each year.

$a_1 = 7,000$

$a_{11} = 14,000$

$a_n = a_1 + (n - 1)d$

$14,000 = 7,000 + (11 - 1)d$

$d = \$700$

The amount of money received over the 11 years is

$S_{11} = a_1 + a_2 + \cdots + a_{11}$

$S_n = \dfrac{n}{2}(a_1 + a_n)$

$S_{11} = \dfrac{11}{2}(7,000 + 14,000) = \$115,500$

89. We are asked for the sum of an infinite geometric series.

a_1 = number of revolutions in the first minute = 300

$r = \dfrac{2}{3} \quad |r| < 1$, so the series has a sum,

$S_\infty = \dfrac{a_1}{1 - r} = \dfrac{300}{1 - \frac{2}{3}} = 900$ revolutions.

91. This involves a geometric sequence. Let $a_n = 2,000$ calories. There are five stages: $n = 5$. We require a_1 on the assumption that $r = 20\% = \frac{1}{5}$.

$$a_n = a_1 r^{n-1}$$
$$2000 = a_1 \left(\frac{1}{5}\right)^{5-1}$$
$$a_1 = 2,000 \cdot 5^4$$
$$= 1,250,000 \text{ calories}$$

93. We have an arithmetic sequence, 16, 48, 80, ···, with $a_1 = 16$, $d = 32$.

(A) This requires a_{11}: $a_n = a_1 + (n - 1)d$
$$a_{11} = 16 + (11 - 1)32$$
$$= 336 \text{ feet}$$

(B) This requires s_{11}: $s_n = \frac{n}{2}(a_1 + a_n)$
$$s_{11} = \frac{11}{2}(16 + 336)$$
$$= 1,936 \text{ feet}$$

(C) This requires s_t: $s_n = \frac{n}{2}[2a_1 + (n - 1)d]$
$$s_t = \frac{t}{2}[2a_1 + (t - 1)32]$$
$$= \frac{t}{2}[32 + (t - 1)32]$$
$$= \frac{t}{2} \cdot 32t$$
$$= 16t^2 \text{ feet}$$

95. This involves a geometric sequence. Let $a_1 = 2A_0$ = number present after 1 half-hour period. In t hours, $2t$ half-hours will have elapsed, so $n = 2t$; $r = 2$, since the number of bacteria doubles in each period.

$$a_n = a_1 r^{n-1}$$
$$a_{2t} = 2A_0 2^{2t-1}$$
$$= A_0 2^1 2^{2t-1}$$
$$= A_0 2^{2t}$$

97. If b_n is the brightness of an nth-magnitude star, we find r for the geometric progression b_1, b_2, b_3, ···, given $b_1 = 100b_6$.

$$b_n = b_1 r^{n-1}$$
$$b_6 = b_1 r^{6-1}$$
$$b_6 = b_1 r^5$$
$$b_6 = 100b_6 r^5$$
$$\frac{1}{100} = r^5$$
$$r = \sqrt[5]{0.01} = 10^{-0.4} = 0.398$$

99. This involves a geometric sequence. a_1 = amount of money on first square; $a_1 = \$0.01$, $r = 2$; a_n = amount of money on nth square.

$$a_n = a_1 r^{n-1}$$
$$a_{64} = 0.01 \cdot 2^{64-1}$$
$$= 9.22 \times 10^{16} \text{ dollars}$$

The amount of money on the whole board = $a_1 + a_2 + a_3 + \cdots + a_{64} = S_{64}$.

$$S_n = \frac{a_1 - ra_n}{1 - r}$$
$$S_{64} = \frac{0.01 - 2(9.223 \times 10^{16})}{1 - 2}$$
$$= 1.845 \times 10^{17} \text{ dollars}$$

101. This involves a geometric sequence. Let $a_1 = 15 =$ pressure at sea level, $r = \dfrac{1}{10} =$ factor of decrease. We require $a_5 =$ pressure after four 10-mile increases in altitude.

$$a_5 = 15\left(\frac{1}{10}\right)^{5-1}$$

$$= 0.0015 \text{ pounds per square inch.}$$

103. This involves an infinite geometric series. $a_1 =$ perimeter of first triangle $= 1$, $r = \dfrac{1}{2}$, $|r| < 1$, so the series has a sum.

$$S_\infty = \frac{a_1}{1 - r} = \frac{1}{1 - \frac{1}{2}} = 2$$

105.

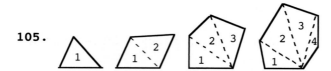

From the figure, and ordinary induction, we can see that the interior angles of an $n + 2$-sided polygon, $n = 1, 2, 3, \cdots$, are those of n triangles, that is, $180°n$. So for the sequence of interior angles $\{a_n\}$,

$a_n - a_{n-1} = 180°n - 180°(n - 1) = 180° = d$.

A detailed proof by mathematical induction is omitted.

For a 21-sided polygon, we use $a_n = a_1 + (n - 1)d$ with $a_1 = 180°$, $d = 180°$, and $n + 2 = 21$, in which case $n = 19$.

$a_{19} = 180° + (19 - 1)180°$
$\quad = 3,420°$

Section 10-4

1. Answers will vary. **3.** Answers will vary. **5.** Answers will vary.

7. $9! = 362,880$ **9.** $11! = 39,916,800$

11. $\dfrac{11!}{8!} = \dfrac{11 \cdot 10 \cdot 9 \cdot 8!}{8!} = 990$ **13.** $\dfrac{5!}{2! \, 3!} = \dfrac{5 \cdot 4 \cdot 3!}{2 \cdot 1 \cdot 3!} = 10$

15. $\dfrac{7!}{4!(7 - 4)!} = \dfrac{7!}{4! \, 3!} = \dfrac{7 \cdot 6 \cdot 5 \cdot 4!}{4! \, 3 \cdot 2 \cdot 1} = 35$ **17.** $\dfrac{7!}{7!(7 - 7)!} = \dfrac{7!}{7! \, 0!} = \dfrac{7!}{7!(1)} = 1$

19. The alleged solution for number 11 is incorrect. The function "11 n Pr8" represents the permutations of 11 objects taken 8 at a time, which is calculated by

$$\frac{11!}{(11 - 8)!} = \frac{11!}{3!}$$

This is different from problem 11, which is $11!/8!$.

The alleged solution for number 13 is incorrect also. The calculator interprets the input 5!/2!3! as $\dfrac{5!}{2!} \cdot 3!$. The correct calculation is $5!/(2!3!)$.

The solution for number 15 is correct.

21. $P_{5,3} = \dfrac{5!}{(5-3)!} = \dfrac{5!}{2!} = \dfrac{5 \cdot 4 \cdot 3 \cdot 2!}{2!} = 60$

23. $P_{52,4} = \dfrac{52!}{(52-4)!} = \dfrac{52!}{48!} = \dfrac{52 \cdot 51 \cdot 50 \cdot 49 \cdot 48!}{48!} = 6,497,400$

25. $C_{5,3} = \dfrac{5!}{3!(5-3)!} = \dfrac{5!}{3!\,2!} = 10$

(See problem 13)

27. $C_{52,4} = \dfrac{52!}{4!(52-4)!} = \dfrac{P_{52,4}}{4!} = \dfrac{6,497,400}{4 \cdot 3 \cdot 2 \cdot 1} = 270,725$

29. (A) Order is important. The selection is a permutation.

(B) Order is not important. The selection is a combination.

31. Order is important here. We use permutations, selecting, in order, three horses out of ten:

$P_{10,3} = 10 \cdot 9 \cdot 8 = 720$ different finishes

33. For the subcommittee, order is not important. We use combinations, selecting three persons out of seven:

$C_{7,3} = \dfrac{7!}{3!(7-3)!} = \dfrac{7!}{3!\,4!} = \dfrac{7 \cdot 6 \cdot 5 \cdot 4!}{3 \cdot 2 \cdot 1 \cdot 4!} = 35$ subcommittees

In choosing a president, a vice-president, and a secretary, we can use permutations, or apply the multiplication principle.

O_1: Selecting the president N_1: 7 ways

O_2: Selecting the vice-president N_2: 6 ways (the president is not considered)

O_3: Selecting the secretary N_3: 5 ways (the president and vice-president are not considered)

There are $N_1 \cdot N_2 \cdot N_3 = 7 \cdot 6 \cdot 5\ (= P_{7,3}) = 210$ ways.

35. For each game, we are selecting two teams out of ten to be opponents. Since the order of the opponents does not matter (this has nothing to do with the order in which the games might be played, which is not under discussion here), we use combinations.

$C_{10,2} = \dfrac{10!}{2!(10-2)!} = \dfrac{10!}{2!\,8!} = \dfrac{10 \cdot 9 \cdot 8!}{2 \cdot 1 \cdot 8!} = 45$ games

37.

	No letter can be repeated	Allowing letters to repeat
O_1: Selecting first letter	N_1: 6 ways	N_1: 6 ways
O_2: Selecting second letter	N_2: 5 ways	N_2: 6 ways
O_3: Selecting third letter	N_3: 4 ways	N_3: 6 ways
O_4: Selecting fourth letter	N_4: 3 ways	N_4: 6 ways
	$P_{6,4} = 6 \cdot 5 \cdot 4 \cdot 3 = 360$ possible code words	$6 \cdot 6 \cdot 6 \cdot 6 = 1,296$ possible code words

39.

	No digit can be repeated	Allowing digits to repeat
O_1: Selecting first digit	N_1: 10 ways	N_1: 10 ways
O_2: Selecting second digit	N_2: 9 ways	N_2: 10 ways
$\vdots$	$\vdots$	$\vdots$
O_5: Selecting fifth digit	N_5: 6 ways	N_5: 10 ways

$$P_{10,5} = 10 \cdot 9 \cdot 8 \cdot 7 \cdot 6 = \qquad 10 \cdot 10 \cdot 10 \cdot 10 \cdot 10 = 100,000$$

30,240 lock combinations   lock combinations

41. We are selecting five cards out of the 13 hearts in the deck. The order is not important, so we use combinations.

$$C_{13,5} = \frac{13!}{5!(13-5)!} = \frac{13!}{5!\,8!} = \frac{13 \cdot 12 \cdot 11 \cdot 10 \cdot 9 \cdot 8!}{5 \cdot 4 \cdot 3 \cdot 2 \cdot 1 \cdot 8!} = 1,287$$

43.

	Repeats allowed	No repeats allowed
O_1: Selecting first letter	N_1: 26 ways	N_1: 26 ways
O_2: Selecting second letter	N_2: 26 ways	N_2: 25 ways
O_3: Selecting third letter	N_3: 26 ways	N_3: 24 ways
O_4: Selecting first digit	N_4: 10 ways	N_4: 10 ways
O_5: Selecting second digit	N_5: 10 ways	N_5: 9 ways
O_6: Selecting third digit	N_6: 10 ways	N_6: 8 ways

$26 \cdot 26 \cdot 26 \cdot 10 \cdot 10 \cdot 10 =$ $26 \cdot 25 \cdot 24 \cdot 10 \cdot 9 \cdot 8 =$
17,576,000 license plates   11,232,000 license plates

45. O_1: Choosing 5 spades out of 13 possible (order is not important)
N_1: $C_{13,5}$
O_2: Choosing 2 hearts out of 13 possible (order is not important)
N_2: $C_{13,2}$
Using the multiplication principle, we have:

$$\text{Number of hands} = C_{13,5} \cdot C_{13,2} = \frac{13!}{5!(13-5)!} \cdot \frac{13!}{2!(13-2)!}$$

$$= 1,287 \cdot 78$$

$$= 100,386$$

47. O_1: Selecting the color N_1: 5 ways
O_2: Selecting the transmission N_2: 3 ways
O_3: Selecting the interior N_3: 4 ways
O_4: Selecting the engine N_4: 2 ways
Applying the multiplication principle, there are $5 \cdot 3 \cdot 4 \cdot 2 = 120$ variations of the car.

49. O_1: Choosing 3 appetizers out of 8 possible (order is not important)

N_1: $C_{8,3}$

O_2: Choosing 4 main courses out of 10 possible (order is not important)

N_2: $C_{10,4}$

O_3: Choosing 2 desserts out of 7 possible (order is not important)

N_3: $C_{7,2}$

Using the multiplication principle, we have:

$$\text{Number of banquets} = C_{8,3} \cdot C_{10,4} \cdot C_{7,2} = \frac{8!}{3!(8-3)!} \cdot \frac{10!}{4!(10-4)!} \cdot \frac{7!}{2!(7-2)!}$$

$$= 56 \cdot 210 \cdot 21$$

$$= 246,960$$

51. O_1: Choosing a left glove out of 12 possible

N_1: 12

O_2: Choosing a right glove out of all the right gloves that do *not* match the left glove already chosen

N_2: 11

Using the multiplication principle, we have:

Number of ways to mismatch gloves = $12 \cdot 11 = 132$.

53. The number of ways to deal a hand containing exactly 1 king is computed as follows:

O_1: Choosing 1 king out of 4 possible (order is not important)

N_1: $C_{4,1}$

O_2: Choosing 4 remaining cards out of 48 cards that are not kings (order is not important)

N_2: $C_{48,4}$

Using the multiplication principle, we have

$$N_1 \cdot N_2 = C_{4,1} \cdot C_{48,4} = \frac{4!}{1!(4-1)!} \cdot \frac{48!}{4!(48-4)!}$$

$$= \frac{4!}{1!\,3!} \cdot \frac{48!}{4!\,44!}$$

$$= 4 \cdot 194,580$$

$$= 778,320 \text{ ways}$$

The number of ways to deal a hand containing no hearts is

$$C_{39,5} = \frac{39!}{5!(39-5)!} = \frac{39!}{5!\,34!} = 575,757 \text{ ways}$$

The hand that contains exactly one king is more likely.

55. To seat two people, we can seat one person, then the second person.

O_1: Seat the first person in any chair

N_1: 5 ways

O_2: Seat the second person in any remaining chair

N_2: 4 ways

Applying the multiplication principle, we can seat two persons in

$N_1 \cdot N_2 = 5 \cdot 4 = 20$ ways.

We can continue this reasoning for a third person.

O_3: Seat the third person in any of the three remaining chairs

N_3: 3 ways

We can seat 3 persons in $5 \cdot 4 \cdot 3 = 60$ ways.

For a fourth person:

O_4: Seat the fourth person in any of the two remaining chairs

N_4: 2 ways

We can seat 4 persons in $5 \cdot 4 \cdot 3 \cdot 2 = 120$ ways.

For a fifth person:

O_5: Seat the fifth person. There will be only one chair remaining.

N_5: 1 way

We can seat 5 persons in $5 \cdot 4 \cdot 3 \cdot 2 \cdot 1 = 120$ ways

57. (A) Order is important, so we use permutations, selecting, in order, 5 persons out of 8:

$$P_{8,5} = 8 \cdot 7 \cdot 6 \cdot 5 \cdot 4 = 6{,}720 \text{ teams.}$$

(B) Order is not important, so we use combinations, selecting 5 persons out of 8

$$C_{8,5} = \frac{8!}{5!(8-5)!} = \frac{8!}{5! \, 3!} = \frac{8 \cdot 7 \cdot 6 \cdot 5!}{5! \cdot 3 \cdot 2 \cdot 1} = 56 \text{ teams}$$

(C) O_1: Selecting either Mike or Ken out of {Mike, Ken}

N_1: $C_{2,1}$

O_2: Selecting the 4 remaining players out of the 6 possibilities that do not include either Mike or Ken.

N_2: $C_{6,4}$

Using the multiplication principle, we have

$$
\begin{aligned}
N_1 \cdot N_2 = C_{2,1} \cdot C_{6,4} &= \frac{2!}{1!(2-1)!} \cdot \frac{6!}{4!(6-4)!} \\
&= \frac{2!}{1! \, 1!} \cdot \frac{6!}{4! \, 2!} \\
&= 2 \cdot 15 \\
&= 30 \text{ teams}
\end{aligned}
$$

59. (A) We are choosing 2 points out of the 8 to join by a chord. Order is not important.

$$C_{8,2} = \frac{8!}{2!(8-2)!} = 28 \text{ chords}$$

(B) No three of the points can be collinear, since no line intersects a circle at more than two points. So we can select any three of the 8 to use as vertices of the triangle. Order is not important.

$$C_{8,3} = \frac{8!}{3!(8-3)!} = 56 \text{ triangles}$$

(C) We can select any four of the eight points to use as vertices of a quadrilateral. Order is not important.

$$C_{8,4} = \frac{8!}{4!(8-4)!} = 70 \text{ quadrilaterals}$$

61. (A) $P_{10,0} = 1$ $0! = 1$

         $P_{10,1} = 10$ $1! = 1$

         $P_{10,2} = 90$ $2! = 2$

         $P_{10,3} = 720$ $3! = 6$

         $P_{10,4} = 5{,}040$ $4! = 24$

$$P_{10,5} = 30,240 \qquad\qquad 5! = 120$$
$$P_{10,6} = 151,200 \qquad\quad 6! = 720$$
$$P_{10,7} = 604,800 \qquad\quad 7! = 5,040$$
$$P_{10,8} = 1,814,400 \qquad 8! = 40,320$$
$$P_{10,9} = 3,628,800 \qquad 9! = 362,880$$
$$P_{10,10} = 3,628,800 \qquad 10! = 3,628,800$$

We can see that $P_{10,r} \geq r!$ for $r = 0, 1, \ldots, 10$

(B) If $r = 0$, $P_{10,0} = 0! = 1$. If $r = 10$, $P_{10,10} = 10! = 3,628,800$

(C) $P_{n,r}$ and $r!$ are each the product of r consecutive integers, the largest of which is n for $P_{n,r}$ and r for $r!$. So if $r \leq n$, $P_{n,r} \geq r!$

Section 10-5

1. Answers will vary.   3. Answers will vary.   5. Answers will vary.

7. Event E is certain to occur.

9. (A) Must be rejected. It violates condition 1; no probability can be negative.
   (B) Must be rejected. It violates condition 2, since $P(R) + P(G) + P(Y) + P(B) \neq 1$.
   (C) This is an acceptable probability assignment.

11. This is a compound event made up of the simple events R and Y.
    $P(E) = P(R) + P(Y) = .26 + .30 = .56$

13. $P(E) \approx \dfrac{f(E)}{n} = \dfrac{25}{250} = .1$

15. $P(E) \approx \dfrac{f(E)}{n} = \dfrac{189}{420} = .45$

17. A probability assignment in this case would be based on observed data, so it would be an empirical probability.

19. A probability assignment in this case would be based on the assumption that each of the combinations has a certain likelihood. This is an example of theoretical probability.

21. The sample space S is the set of all possible sequences of three digits with no repeats. $n(S) = P_{10,3} = 10 \cdot 9 \cdot 8 = 720$. The event is the simple event of one particular sequence. $n(E) = 1$. Then $P(E) = \dfrac{n(E)}{n(S)} = \dfrac{1}{720} \approx .0014$.

23. The sample space S is the set of all possible 5-card hands, chosen from a deck of 52 cards, $n(S) = C_{52,5}$. The event E is the set of all possible 5-card hands that are all black, so are chosen from the 26 black cards. $n(E) = C_{26,5}$.
    $$P(E) = \frac{n(E)}{n(S)} = \frac{C_{26,5}}{C_{52,5}} = \frac{26!}{5!(26-5)!} \div \frac{52!}{5!(52-5)!} \approx .025$$

25. The sample space S is the set of all possible 5-card hands, chosen from a deck of 52 cards. $n(S) = C_{52,5}$. The event E is the set of all possible 5-card hands that are all face cards, so are chosen from the 4 face cards in each of 4 suits, or 16 face cards. $n(E) = C_{16,5}$.

$$P(E) = \frac{n(E)}{n(S)} = \frac{C_{16,5}}{C_{52,5}} = \frac{16!}{5!(16-5)!} \div \frac{52!}{5!(52-5)!} \approx .0017$$

27. It is acceptable. There are four aces in a deck so a 5-card hand could have anywhere from zero to four aces. But it is not an equally likely sample space. Drawing four aces in a 5-card hand is far less likely than drawing none.

29. The sample space S is the set of all possible four digit numbers less than 5,000 formed from the digits 1, 3, 5, 7, 9.
$n(S) = N_1 \cdot N_2 \cdot N_3 \cdot N_4$
where
   $N_1 = 2$ (digits 1 or 3 only for the first digit since the number is
      less than 5,000)
   $N_2 = N_3 = N_4 = 5$ (digits 1, 3, 5, 7, 9)
$n(S) = 2 \cdot 5 \cdot 5 \cdot 5 = 250$
The event E is the set of those numbers in S which are divisible by 5. These must end in 5.
$n(E) = N_1 \cdot N_2 \cdot N_3 \cdot N_4'$
where
   $N_4' = 1$ (digit 5 only)
$n(E) = 2 \cdot 5 \cdot 5 \cdot 1 = 50$
$$P(E) = \frac{n(E)}{n(S)} = \frac{50}{250} = .2$$

31. The sample space S is the set of all arrangements of the five notes in the five envelopes. $n(S) = P_{5,5} = 5! = 120$. The event E is the one correct arrangement.

$n(E) = 1.$ $P(E) = \dfrac{n(E)}{n(S)} = \dfrac{1}{120} \approx .008$

In problems 33—47, using Figure 1 of the text, we have that S = the set of all possible ordered pairs of dots showing, n(S) = 36.

33. The sum of the dots can be 2 in exactly one way. $E = \{(1, 1)\}$
$n(E) = 1$
$P(E) = \dfrac{n(E)}{n(S)} = \dfrac{1}{36}$

35. The sum of the dots can be 6 in 5 ways:
$E = \{(5, 1), (4, 2), (3, 3), (2, 4), (1, 5)\}$
$n(E) = 5$
$P(E) = \dfrac{n(E)}{n(S)} = \dfrac{5}{36}$

37. The sum of the dots can be less than 5 in 6 ways. (upper left-hand corner of figure) $n(E) = 6$.
$P(E) = \dfrac{n(E)}{n(S)} = \dfrac{6}{36} = \dfrac{1}{6}$

39. The sum of the dots can be 7 in 6 ways and 11 in 2 ways:
$n(E) = 36 - (6 + 2) = 28$
$P(E) = \dfrac{n(E)}{n(S)} = \dfrac{28}{36} = \dfrac{7}{9}$

41. The sum of the dots cannot be 1: $P(E) = 0$.

43. The sum of the dots is divisible by 3 if the sum is 3, 6, 9, or 12. 3 can occur in 2 ways, 6 in 5 ways, 9 in 4 ways, and 12 in 1 way:
$n(E) = 2 + 5 + 4 + 1 = 12$.
$P(E) = \dfrac{n(E)}{n(S)} = \dfrac{12}{36} = \dfrac{1}{3}$

45. The sum of the dots can be 7 in 6 ways and 11 in 2 ways:
$n(E) = 6 + 2 = 8$.
$$P(E) = \frac{n(E)}{n(S)} = \frac{8}{36} = \frac{2}{9}$$

47. The sum of the dots will *not* be divisible by 2 or 3 if the sum is 5, 7, or 11. 5 can occur in 4 ways, 7 in 6 ways, and 11 in 2 ways. Otherwise the sum will be divisible by 2 or 3:
$n(E) = 36 - (4 + 6 + 2) = 36 - 12 = 24$.
$$P(E) = \frac{n(E)}{n(S)} = \frac{24}{36} = \frac{2}{3}$$

49. The sample space is the set of events consisting of choosing a person with birthday on day 1 through day 365 of the year. (February 29 is neglected.)
$S = \{1, 2, 3, \ldots, 365\}$

Since each of these days is essentially equally likely, $P(e_i) = \frac{1}{365}$.

51. (A) Use $P(E) \approx \frac{f(E)}{n}$ with $n = 500$, where $f(E)$ is the observed frequency of an event.

$$P(2) = \frac{11}{500} = .022 \qquad P(3) = \frac{35}{500} = .07 \qquad P(4) = \frac{44}{500} = .088$$

$$P(5) = \frac{50}{500} = .1 \qquad P(6) = \frac{71}{500} = .142 \qquad P(7) = \frac{89}{500} = .178$$

$$P(8) = \frac{72}{500} = .144 \qquad P(9) = \frac{52}{500} = .104 \qquad P(10) = \frac{36}{500} = .072$$

$$P(11) = \frac{26}{500} = .052 \qquad P(12) = \frac{14}{500} = .028$$

(B) Use Figure 1 of the text, with S = the set of all possible ordered pairs of dots showing, $n(S) = 36$.

$$P(2) = \frac{1}{36} \qquad P(3) = \frac{2}{36} \qquad P(4) = \frac{3}{36}$$

$$P(5) = \frac{4}{36} \qquad P(6) = \frac{5}{36} \qquad P(7) = \frac{6}{36}$$

$$P(8) = \frac{5}{36} \qquad P(9) = \frac{4}{36} \qquad P(10) = \frac{3}{36}$$

$$P(11) = \frac{2}{36} \qquad P(12) = \frac{1}{36}$$

(C) The expected frequency of the occurrence of E is $n \cdot P(E)$. So the expected frequency of the occurrence of sum 2 is $500 \cdot P(2) = 500 \cdot \frac{1}{36}$ or 13.9.

The expected frequency of the occurrence of sum 3 is $500 \cdot P(3) = 500 \cdot \frac{2}{36} = 27.8$.

The expected frequency of the occurrence of sum 4 is $500 \cdot P(4) = 500 \cdot \frac{3}{36} = 41.7$.

The expected frequency of the occurrence of sum 5 is $500 \cdot P(5) = 500 \cdot \frac{4}{36} = 55.6$.

The expected frequency of the occurrence of sum 6 is $500 \cdot P(6) = 500 \cdot \frac{5}{36} = 69.4$.

The expected frequency of the occurrence of sum 7 is $500 \cdot P(7) = 500 \cdot \frac{6}{36} = 83.3$.

The expected frequency of the occurrence of sum 8 is $500 \cdot P(8) = 500 \cdot \frac{5}{36} = 69.4$.

The expected frequency of the occurrence of sum 9 is $500 \cdot P(9) = 500 \cdot \frac{4}{36} = 55.6$.

The expected frequency of the occurrence of sum 10 is $500 \cdot P(10) = 500 \cdot \frac{3}{36} = 41.7$.

The expected frequency of the occurrence of sum 11 is $500 \cdot P(11) = 500 \cdot \dfrac{2}{36} = 27.8$.

The expected frequency of the occurrence of sum 12 is $500 \cdot P(12) = 500 \cdot \dfrac{1}{36} = 13.9$.

(D) Use the command r and Int (i, k, n). Let 1 to 6 represent the possible outcomes in tossing one die; then rand Int $(1, 6, 500)$ will give a list of 500 possible outcomes in tossing one die. rand Int $(1, 6, 500)$ + rand Int $(1, 6, 500)$ will give a list of 500 possible outcomes in tossing 2 dice. A possible result is outlined here, but the student's result will almost certainly be different.

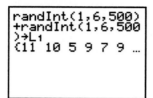

Generating the random numbers.

Setting up the histogram.

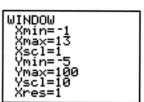

Selecting the window variables.

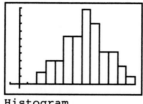

Histogram

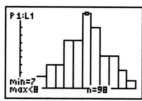

Use trace to display any result.

53. (A) It is possible, but very unlikely, to get 29 heads in 30 flips of a coin. If one thinks of the 30 flips as a sequence of words with 30 letters chosen from $\{H, T\}$, then there are 2 ways to choose each of the letters and therefore 2^{30} such words. Only 30 of these words have exactly 29 H's because only 30 have exactly one T. So the probability of getting 29 heads in 30 flips of a fair coin is

$$P(E) = \frac{n(E)}{n(S)} = \frac{30}{2^{30}} \approx 2.8 \times 10^{-8}$$

(B) There is a strong suspicion that the coin is unfair, because this event is unlikely, although not as unlikely as in part A. If the coin is unfair, the empirical probabilities are

$$P(H) \approx \frac{f(H)}{n} = \frac{42}{50} = .84 \qquad P(T) \approx \frac{f(T)}{n} = \frac{8}{50} = .16$$

55. The sample space for this experiment is $\{HTH, HHH, THH, TTH\}$, assuming that the third coin is the one with a head on both sides:
$n(S) = 4$. E is the event $\{TTH\}$: $n(E) = 1$.
$$P(E) = \frac{n(E)}{n(S)} = \frac{1}{4}$$

57. See problem 55. $n(S) = 4$. E is the event $\{HHH\}$: $n(E) = 1$
$$P(E) = \frac{n(E)}{n(S)} = \frac{1}{4}$$

59. See problem 55. $n(S) = 4$. E is the event $\{HTH, HHH, THH\}$: $n(E) = 3$
$$P(E) = \frac{n(E)}{n(S)} = \frac{3}{4}$$

61. Since it is equally likely that each die shows a 1, a 2 or a 3, we can draw the following figure to represent S, the set of all possibilities:

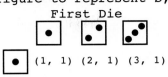

First Die

| | (1, 1) (2, 1) (3, 1) |
Second Die | (1, 2) (2, 2) (3, 2) |
| (1, 3) (2, 3) (3, 3) |

$n(S) = 9$
The sum of the dots can be 2 in exactly one way. $n(E) = 1$
$$P(E) = \frac{n(E)}{n(S)} = \frac{1}{9}$$

63. See figure, problem 61, $n(S) = 9$. The sum of the dots can be 4 in 3 ways:
$E = \{(1, 3), (2, 2), (3, 1)\}$. $n(E) = 3$
$$P(E) = \frac{n(E)}{n(S)} = \frac{3}{9} = \frac{1}{3}$$

65. See figure, problem 61. $n(S) = 9$. The sum of the dots can be 6 in exactly one way: $E = \{(3, 3)\}$. $n(E) = 1$
$$P(E) = \frac{n(E)}{n(S)} = \frac{1}{9}$$

67. See figure, problem 61. $n(S) = 9$. The sum of the dots will be odd if the sum is 3 or 5.
$E = \{(1, 2), (2, 1)(2, 3), (3, 2)\}$. $n(E) = 4$
$$P(E) = \frac{n(E)}{n(S)} = \frac{4}{9}$$

69. The sample space S is the set of all possible 5-card hands, chosen from a deck of 52 cards. $n(S) = C_{52,5} = 2{,}598{,}960$. The event E is the set of all possible 5-card hands that are all jacks through aces, so are chosen from the 16 cards (4 jacks, 4 queens, 4 kings, 4 aces) of this type.
$n(E) = C_{16,5} = 4{,}368$.
$$P(E) = \frac{n(E)}{n(S)} = \frac{C_{16,5}}{C_{52,5}} = \frac{4{,}368}{2{,}598{,}960} \approx .00168$$

71. The sample space S is the set of all possible 5-card hands chosen from a deck of 52 cards. $n(S) = C_{52,5}$. The event E is the set of all possible 5-card hands of the type $AAAAx$ where x is any of the $52 - 4 = 48$ cards that are not aces.
$n(E) = 48$
$$P(E) = \frac{n(E)}{n(S)} = \frac{48}{C_{52,5}} = \frac{48}{2{,}598{,}960} \approx .000\,0185$$

73. The sample space S is the set of all possible 5-card hands, chosen from a deck of 52 cards. $n(S) = C_{52,5}$. The event E is the set consisting of the cards ace, king, queen, jack, ten, in each of the 4 suits, abbreviated:
{club AKQJ10, diamond AKQJ10, heart AKQJ10, spade AKQJ10}. $n(E) = 4$.
$$P(E) = \frac{n(E)}{n(S)} = \frac{4}{C_{52,5}} = \frac{4}{2{,}598{,}960} \approx .000\,0015$$

75. The sample space S is the set of all possible 5-card hands, chosen from a deck of 52 cards. $n(S) = C_{52,5}$. The event E is the set of all possible 5-card hands chosen as follows:

O_1: Choose 2 aces out of the set of 4 aces

N_1: $C_{4,2}$

O_2: Choose 3 queens out of the set of 4 queens

N_2: $C_{4,3}$

Applying the multiplication principle, $n(E) = N_1 N_2 = C_{4,2} C_{4,3}$

$$P(E) = \frac{n(E)}{n(S)} = \frac{C_{4,2} \cdot C_{4,3}}{C_{52,5}} = \frac{6 \cdot 4}{2,598,960} = .000\ 009$$

77. (A) P (this event) $= \dfrac{15}{1,000} = .015$

(B) P (this event) $= \dfrac{130 + 80 + 12}{1,000} = .222$

(C) P (this event) $= \dfrac{30 + 32 + +28 + 25 + 20 + 21 + 12 + 1}{1,000} = .169$

(D) P (this event) $= 1 - P$ (owning zero television sets)

$$= 1 - \frac{2 + 10 + 30}{1,000} = .958$$

Section 10-6

1. Answers will vary.   **3.** Answers will vary.

For Problems 5—15, the first eight lines of Pascal's triangle are as follows:

```
0                        1
1                     1     1
2                  1     2     1
3               1     3     3     1
4            1     4     6     4     1
5         1     5    10    10     5     1
6      1     6    15    20    15     6    1
7   1     7    21    35    35    21    7    1
```

5. Using the row of the triangle marked "4", we get

$(a + b)^4 = a^4 + 4a^3b + 6a^2b^2 + 4ab^3 + b^4$

7. Using the row of the triangle marked "5", we get

$(x + 2)^5 = x^5 + 5x^4 \cdot 2^1 + 10x^3 \cdot 2^2 + 10x^2 \cdot 2^3 + 5x \cdot 2^4 + 2^5$

$\qquad = x^5 + 10x^4 + 40x^3 + 80x^2 + 80x + 32$

9. $\dbinom{5}{3}$ is the fourth entry in line 5 (start counting from 0) of the triangle, 10.

11. $\dbinom{4}{2}$ is the third entry in line 4 (start counting from 0) of the triangle, 6.

13. $C_{6,3}$ is the fourth entry in line 6 (start counting from 0) of the triangle, 20.

15. $C_{7,4}$ is the fifth entry in line 7 (start counting from 0) of the triangle, 35.

Problems 17 and 19 are shown on one screen of a graphing calculator.

```
9 nCr 3
            84
12 nCr 10
            66
```

Problems 21 and 23 are shown on one screen of a graphing calculator.

```
17 nCr 13
             2380
50 nCr 4
          230300
```

25. $(m + n)^3 = \sum_{k=0}^{3} \binom{3}{k} m^{3-k} n^k$

$\qquad = \binom{3}{0} m^3 + \binom{3}{1} m^2 n + \binom{3}{2} mn^2 + \binom{3}{3} n^3$

$\qquad = m^3 + 3m^2 n + 3mn^2 + n^3$

27. $(2x - 3y)^3 = [2x + (-3y)]^3$

$\qquad = \sum_{k=0}^{3} \binom{3}{k} (2x)^{3-k}(-3y)^k$

$\qquad = \binom{3}{0}(2x)^3 + \binom{3}{1}(2x)^2(-3y)^1 + \binom{3}{2}(2x)(-3y)^2 + \binom{3}{3}(-3y)^3$

$\qquad = 8x^3 + 3(4x^2)(-3y) + 3(2x)(9y^2) + (-27y^3)$

$\qquad = 8x^3 - 36x^2 y + 54xy^2 - 27y^3$

29. $(x - 2)^4 = [x + (-2)]^4 = \sum_{k=0}^{4} \binom{4}{k} x^{4-k}(-2)^k$

$\qquad = \binom{4}{0} x^4 + \binom{4}{1} x^3(-2)^1 + \binom{4}{2} x^2(-2)^2 + \binom{4}{3} x(-2)^3 + \binom{4}{4}(-2)^4$

$\qquad = x^4 - 8x^3 + 24x^2 - 32x + 16$

31. $(m + 3n)^4 = \sum_{k=0}^{4} \binom{4}{k} m^{4-k}(3n)^k$

$\qquad = \binom{4}{0} m^4 + \binom{4}{1} m^3(3n)^1 + \binom{4}{2} m^2(3n)^2 + \binom{4}{3} m(3n)^3 + \binom{4}{4}(3n)^4$

$\qquad = m^4 + 12m^3 n + 54m^2 n^2 + 108mn^3 + 81n^4$

33. $(2x - y)^5 = [2x + (-y)]^5 = \sum_{k=0}^{5} \binom{5}{k} (2x)^{5-k}(-y)^k$

$\qquad = \binom{5}{0}(2x)^5 + \binom{5}{1}(2x)^4(-y)^1 + \binom{5}{2}(2x)^3(-y)^2 + \binom{5}{3}(2x)^2(-y)^3$

$\qquad\qquad\qquad + \binom{5}{4}(2x)(-y)^4 + \binom{5}{5}(-y)^5$

$\qquad = 32x^5 - 80x^4 y + 80x^3 y^2 - 40x^2 y^3 + 10xy^4 - y^5$

35. $(m + 2n)^6 = \sum_{k=0}^{6} \binom{6}{k} m^{6-k}(2n)^k$

$$= \binom{6}{0} m^6 + \binom{6}{1} m^5(2n)^1 + \binom{6}{2} m^4(2n)^2 + \binom{6}{3} m^3(2n)^3 + \binom{6}{4} m^2(2n)^4$$

$$+ \binom{6}{5} m(2n)^5 + \binom{6}{6}(2n)^6$$

$$= m^6 + 12m^5n + 60m^4n^2 + 160m^3n^3 + 240m^2n^4 + 192mn^5 + 64n^6$$

37. In the expansion

$$(x + 1)^7 = \sum_{k=0}^{7} \binom{7}{k} x^{7-k}1^k$$

the exponent of x is 4 when $k = 3$. So the term containing x^4 is

$$\binom{7}{3} x^4 1^3 = 35x^4$$

39. In the expansion

$$(2x - 1)^{11} = [2x + (-1)]^{11} = \sum_{k=0}^{11} \binom{11}{k} (2x)^{11-k}(-1)^k$$

the exponent of x is 6 when $k = 5$. So the term containing x^6 is

$$\binom{11}{5} (2x)^6(-1)^5 = (462)(64x^6)(-1) = -29,568x^6$$

41. In the expansion

$$(2x + 3)^{18} = \sum_{k=0}^{18} \binom{18}{k} (2x)^{18-k}(3)^k$$

the exponent of x is 14 when $k = 4$. So the term containing x^{14} is

$$\binom{18}{4} (2x)^{14}(3)^4 = 3,060(2^{14}x^{14})(3^4) = 4,060,938,240x^{14}$$

43. In the expansion

$$(x^2 - 1)^6 = \sum_{k=0}^{6} \binom{6}{k} (x^2)^{6-k}(-1)^k = \sum_{k=0}^{6} \binom{6}{k} x^{12-2k}(-1)^k$$

the exponent of x, $12 - 2k$, is 8, when $12 - 2k = 8$ and $2k = 4$, $k = 2$. So the term containing x^8 is

$$\binom{6}{2} x^8(-1)^2 = 15x^8$$

45. In the expansion

$$(x^2 + 1)^9 = \sum_{k=0}^{9} \binom{9}{k} (x^2)^{9-k}(1)^k = \sum_{k=0}^{9} \binom{9}{k} x^{18-2k}(1)^k$$

the exponent of x, $18 - 2k$, is even and can never equal 11.

47. In the expansion of $(a + b)^n$, the exponent of b in the rth term is $r - 1$ and the exponent of a is $n - (r - 1)$. Here, $r = 7$, $n = 15$.

$$\text{Seventh term} = \binom{15}{6} u^9 v^6$$

$$= \frac{15!}{9!\,6!} u^9 v^6$$

$$= \frac{15 \cdot 14 \cdot 13 \cdot 12 \cdot 11 \cdot 10}{6 \cdot 5 \cdot 4 \cdot 3 \cdot 2 \cdot 1} u^9 v^6$$

$$= 5{,}005 u^9 v^6$$

49. In the expansion of $(a + b)^n$, the exponent of b in the rth term is $r - 1$ and the exponent of a is $n - (r - 1)$. Here, $r = 11$, $n = 12$.

$$\text{Eleventh term} = \binom{12}{10} (2m)^2 n^{10}$$

$$= \frac{12!}{10!\,2!} 4m^2 n^{10}$$

$$= \frac{12 \cdot 11}{2 \cdot 1} 4m^2 n^{10}$$

$$= 264 m^2 n^{10}$$

51. In the expansion of $(a + b)^n$, the exponent of b in the rth term is $r - 1$ and the exponent of a is $n - (r - 1)$. Here, $r = 7$, $n = 12$.

$$\text{Seventh term} = \binom{12}{6} \left(\frac{w}{2}\right)^6 (-2)^6$$

$$= \frac{12!}{6!\,6!} \frac{w^6}{2^6} 2^6$$

$$= \frac{12 \cdot 11 \cdot 10 \cdot 9 \cdot 8 \cdot 7 \cdot 6!}{6 \cdot 5 \cdot 4 \cdot 3 \cdot 2 \cdot 1 \cdot 6!} w^6$$

$$= 924 w^6$$

53. In the expansion of $(a + b)^n$, the exponent of b in the rth term is $r - 1$ and the exponent of a is $n - (r - 1)$. Here, $r = 6$, $n = 8$.

$$\text{Sixth term} = \binom{8}{5} (3x)^3 (-2y)^5$$

$$= \frac{8!}{5!\,3!} (3x)^3 (-2y)^5$$

$$= \frac{8 \cdot 7 \cdot 6 \cdot 5!}{5! \cdot 3 \cdot 2 \cdot 1} (27x^3)(-32y^5)$$

$$= -(56)(27)(32) x^3 y^5$$

$$= -48{,}384 x^3 y^5$$

55.

$$\frac{f(x + h) - f(x)}{h} = \frac{(x + h)^3 - x^3}{h} = \frac{x^3 + 3x^2h + 3xh^2 + h^3 - x^3}{h}$$

$$= \frac{3x^2h + 3xh^2 + h^3}{h} = \frac{h(3x^2 + 3xh + h^2)}{h} = 3x^2 + 3xh + h^2$$

As h approaches 0, this quantity approaches $3x^2$.

57.

$$\frac{f(x + h) - f(x)}{h} = \frac{(x + h)^5 - x^5}{h} = \frac{x^5 + 5x^4h + 10x^3h^2 + 10x^2h^3 + 5xh^4 + h^5 - x^5}{h}$$

$$= \frac{5x^4h + 10x^3h^2 + 10x^2h^3 + 5xh^4 + h^5}{h}$$

$$= \frac{h(5x^4 + 10x^3h + 10x^2h^2 + 5xh^3 + h^4)}{h}$$

$$= 5x^4 + 10x^3h + 10x^2h^2 + 5xh^3 + h^4$$

As h approaches 0, this approaches $5x^4$.

59. Graphing the sequence, we obtain

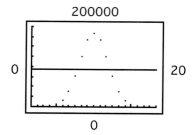

The solid line indicates $\frac{1}{2}(184{,}756) = 92{,}378$, that is, half the largest term. The graph shows five terms greater than half the largest term. The table display confirms this:

$$\binom{20}{8} = 125{,}970 \quad \binom{20}{9} = 167{,}960 \quad \binom{20}{10} = 184{,}756 \quad \binom{20}{11} = 167{,}960 \quad \binom{20}{12} = 125{,}970$$

are all greater than 92,378, but the other terms are less than 92,378.

61. (A) Graphing the sequence, we obtain the graph at the right.

From the table display we find the largest term to be $\binom{10}{4}(.6)^6(.4)^4 = 0.251$ as displayed in the graph.

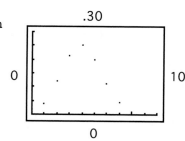

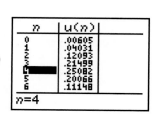

(B) According to the binomial formula,

$$\sum_{k=0}^{n} \binom{n}{k} a^{n-k}b^k = (a + b)^n$$

Then $a_0 + a_1 + a_2 + \cdots + a_{10} = \displaystyle\sum_{k=0}^{10} \binom{10}{k}(0.6)^{10-k}(0.4)^k = (0.6 + 0.4)^{10} = 1^{10} = 1$

63. $(1.01)^{10} = (1 + 0.01)^{10} = \sum_{k=0}^{10} \binom{10}{k} 1^{10-k}(0.01)^k, \ 1^{10-k} = 1$

$$= \sum_{k=0}^{10} \binom{10}{k}(0.01)^k$$

$$= \binom{10}{0} + \binom{10}{1}(0.01)^1 + \binom{10}{2}(0.01)^2 + \binom{10}{3}(0.01)^3$$

$$+ \binom{10}{4}(0.01)^4 + \binom{10}{5}(0.01)^5 + \binom{10}{6}(0.01)^6 + \binom{10}{7}(0.01)^7$$

$$+ \binom{10}{8}(0.01)^8 + \binom{10}{9}(0.01)^9 + \binom{10}{10}(0.01)^{10}$$

$$= 1 + 0.1 + 0.0045 + 0.00012 + 0.0000021 + \text{(other terms}$$
$$\text{with no effect in fourth decimal place)}$$
$$= 1.1046$$

65. $\binom{n}{r} = \dfrac{n!}{r!(n-r)!} = \dfrac{n!}{(n-r)!\,r!} = \dfrac{n!}{(n-r)![n-(n-r)]!} = \binom{n}{n-r}$

67. $\binom{k}{r-1} + \binom{k}{r} = \dfrac{k!}{(r-1)!(k-r+1)!} + \dfrac{k!}{r!(k-r)!}$

$$= \dfrac{rk! + (k-r+1)k!}{r!(k-r+1)!} = \dfrac{(r+k-r+1)k!}{r!(k-r+1)!}$$

$$= \dfrac{(k+1)k!}{r!(k-r+1)!} = \dfrac{(k+1)!}{r!(k-r+1)!}$$

$$= \binom{k+1}{r}$$

69. $\binom{k}{k} = \dfrac{k!}{k!(k-k)!} = 1 = \dfrac{(k+1)!}{(k+1)![(k+1)-(k+1)]!} = \binom{k+1}{k+1}$

71. $2^n = (1+1)^n = \sum_{k=0}^{n} \binom{n}{k} 1^{n-k}1^k, \ 1^{n-k} = 1^k = 1$

$$= \sum_{k=0}^{n} \binom{n}{k}$$

$$= \binom{n}{0} + \binom{n}{1} + \binom{n}{2} + \cdots + \binom{n}{n}$$

CHAPTER 10 REVIEW

1. (A) Since $\dfrac{-8}{16} = \dfrac{4}{-8} = -\dfrac{1}{2}$, this could start a geometric sequence.

(B) Since $7 - 5 = 9 - 7 = 2$, this could start an arithmetic sequence.

(C) Since $-5 - (-8) = -2 - (-5) = 3$, this could start an arithmetic sequence.

(D) Since $\dfrac{3}{2} \neq \dfrac{5}{3}$ and $3 - 2 \neq 5 - 3$, this could start neither an arithmetic nor a geometric sequence.

(E) Since $\dfrac{2}{-1} = \dfrac{-4}{2} = -2$, this could start a geometric sequence.   *(10-1, 10-3)*

2. $a_n = 2n + 3$

(A) $a_1 = 2 \cdot 1 + 3 = 5$
$a_2 = 2 \cdot 2 + 3 = 7$
$a_3 = 2 \cdot 3 + 3 = 9$
$a_4 = 2 \cdot 4 + 3 = 11$

(B) This is an arithmetic sequence with $d = 2$.
$$a_n = a_1 + (n - 1)d$$
$$a_{10} = 5 + (10 - 1)2$$
$$= 23$$

(C) $S_n = \dfrac{n}{2}(a_1 + a_n)$

$S_{10} = \dfrac{10}{2}(5 + 23)$
$= 140$

$(10\text{-}1,\ 10\text{-}3)$

3. $a_n = 32\left(\dfrac{1}{2}\right)^n$

(A) $a_1 = 32\left(\dfrac{1}{2}\right)^1 = 16$

$a_2 = 32\left(\dfrac{1}{2}\right)^2 = 8$

$a_3 = 32\left(\dfrac{1}{2}\right)^3 = 4$

$a_4 = 32\left(\dfrac{1}{2}\right)^4 = 2$

(B) This is a geometric sequence with $r = \dfrac{1}{2}$.
$$a_n = a_1 r^{n-1}$$
$$a_{10} = 16\left(\dfrac{1}{2}\right)^{10-1}$$
$$= \dfrac{1}{32}$$

(C) $S_n = \dfrac{a_1 - ra_n}{1 - r}$

$S_{10} = \dfrac{16 - \frac{1}{2}\left(\frac{1}{32}\right)}{\frac{1}{2}} = \dfrac{16 - \frac{1}{64}}{\frac{1}{2}} = 31\frac{31}{32}$

$(10\text{-}1,\ 10\text{-}3)$

4. $a_1 = -8,\ a_n = a_{n-1} + 3,\ n \geq 2$

(A) $a_1 = -8$
$a_2 = a_1 + 3 = -8 + 3 = -5$
$a_3 = a_2 + 3 = -5 + 3 = -2$
$a_4 = a_3 + 3 = -2 + 3 = 1$

(B) This is an arithmetic sequence with $d = 3$.
$$a_n = a_1 + (n - 1)d$$
$$a_{10} = -8 + (10 - 1)3$$
$$= 19$$

(C) $S_n = \dfrac{n}{2}(a_1 + a_n)$

$S_{10} = \dfrac{10}{2}(-8 + 19)$
$= 55$

$(10\text{-}1,\ 10\text{-}3)$

5. $a_1 = -1,\ a_n = (-2)a_{n-1},\ n \geq 2$

(A) $a_1 = -1$
$a_2 = (-2)a_1 = (-2)(-1) = 2$
$a_3 = (-2)a_2 = (-2)2 = -4$
$a_4 = (-2)a_3 = (-2)(-4) = 8$

(B) This is a geometric sequence with $r = -2$.
$$a_n = a_1 r^{\,n-1}$$
$$a_{10} = (-1)(-2)^{10-1}$$
$$= 512$$

(C) $S_n = \dfrac{a_1 - ra_n}{1 - r}$

$\qquad S_{10} = \dfrac{-1 - (-2)(512)}{1 - (-2)}$

$\qquad\quad = 341$ $\hfill (10\text{-}1,\ 10\text{-}3)$

6. This is a geometric sequence with $a_1 = 16$ and $r = \dfrac{1}{2} < 1$, so the sum exists:

$S_\infty = \dfrac{a_1}{1 - r}$

$\quad = \dfrac{16}{1 - \frac{1}{2}}$

$\quad = 32$ $\hfill (10\text{-}3)$

7. $6! = 6 \cdot 5 \cdot 4 \cdot 3 \cdot 2 \cdot 1 = 720$ $\hfill (10\text{-}4)$

8. $\dfrac{22!}{19!} = \dfrac{22 \cdot 21 \cdot 20 \cdot 19!}{19!} = 9{,}240$ $\hfill (10\text{-}4)$

9. $\dfrac{7!}{2!(7 - 2)!} = \dfrac{7!}{2!\,5!} = \dfrac{7 \cdot 6 \cdot 5!}{2 \cdot 1 \cdot 5!} = 21$ $\hfill (10\text{-}4)$

10. $C_{6,2} = \dfrac{6!}{2!(6 - 2)!} = \dfrac{6!}{2!\,4!} = \dfrac{6 \cdot 5 \cdot 4!}{2 \cdot 1 \cdot 4!} = 15 \quad P_{6,2} = 6 \cdot 5 = 30$ $\hfill (10\text{-}5)$

11. (A) The outcomes can be displayed in a tree diagram as follows:

	H	(1, H)
1	T	(1, T)
	H	(2, H)
2	T	(2, T)
	H	(3, H)
3	T	(3, T)
	H	(4, H)
4	T	(4, T)
	H	(5, H)
5	T	(5, T)
	H	(6, H)
6	T	(6, T)

(B) O_1: Rolling the die
   N_1: 6 outcomes
   O_2: Flipping the coin
   N_2: 2 outcomes

Applying the multiplication principle, there are $6 \cdot 2 = 12$ combined outcomes.
$\hfill (10\text{-}5)$

12. O_1: Seating the first person N_1: 6 ways
O_2: Seating the second person N_2: 5 ways
O_3: Seating the third person N_3: 4 ways
O_4: Seating the fourth person N_4: 3 ways
O_5: Seating the fifth person N_5: 2 ways
O_6: Seating the sixth person N_6: 1 way

Applying the multiplication principle, there are $6 \cdot 5 \cdot 4 \cdot 3 \cdot 2 \cdot 1 = 720$ arrangements.
$\hfill (10\text{-}5)$

13. Order is important here. We use permutations to determine the number of arrangements of 6 objects. $P_{6,6} = 6! = 720$ *(10-5)*

14. The sample space S is the set of all possible 5-card hands, chosen from a deck of 52 cards. $n(S) = C_{52,5}$

The event E is the set of all possible 5-card hands that are all clubs, so are chosen from the 13 clubs. $n(E) = C_{13,5}$

$$P(E) = \frac{n(E)}{n(S)} = \frac{C_{13,5}}{C_{52,5}} = \frac{13!}{5!(13-5)!} \div \frac{52!}{5!(52-5)!} \approx .0005$$ *(10-5)*

15. The sample space S is the set of all possible arrangements of two persons, chosen from a set of 15 persons. Order is important here, so we use permutations. $n(S) = P_{15,2}$. The event E is one of these arrangements. $n(E) = 1$.

$$P(E) = \frac{n(E)}{n(S)} = \frac{1}{P_{15,2}} = \frac{1}{15 \cdot 14} \approx .005$$ *(10-5)*

16. $P(E) \approx \dfrac{f(E)}{n} = \dfrac{50}{1,000} = .05$ *(10-5)*

17. P_1: $5 = 1^2 + 4 \cdot 1 = 5$
P_2: $5 + 7 = 2^2 + 4 \cdot 2$
$\qquad 12 = 12$
P_3: $5 + 7 + 9 = 3^2 + 4 \cdot 3$
$\qquad\quad 21 = 21$ *(10-2)*

18. P_1: $2 = 2^{1+1} - 2 = 4 - 2 = 2$
P_2: $2 + 4 = 2^{2+1} - 2$
$\qquad\quad 6 = 6$
P_3: $2 + 4 + 8 = 2^{3+1} - 2$
$\qquad\qquad 14 = 14$ *(10-2)*

19. P_1: $49^1 - 1$ is divisible by 6
$48 = 6 \cdot 8$ true
P_2: $49^2 - 1$ is divisible by 6
$2,400 = 6 \cdot 400$ true
P_3: $49^3 - 1$ is divisible by 6
$117,648 = 19,608 \cdot 6$ true *(10-2)*

20. P_k: $5 + 7 + 9 + \cdots + (2k + 3) = k^2 + 4k$
P_{k+1}: $5 + 7 + 9 + \cdots + (2k + 3) + (2k + 5) = (k + 1)^2 + 4(k + 1)$ *(10-2)*

21. P_k: $2 + 4 + 8 + \cdots + 2^k = 2^{k+1} - 2$
P_{k+1}: $2 + 4 + 8 + \cdots + 2^k + 2^{k+1} = 2^{k+2} - 2$ *(10-2)*

22. P_k: $49^k - 1 = 6r$ for some integer r
P_{k+1}: $49^{k+1} - 1 = 6s$ for some integer s *(10-2)*

23. Although 1 is less than 4, $1 + \dfrac{1}{2}$ is less than 4, $1 + \dfrac{1}{2} + \dfrac{1}{3}$ is less than 4, and so on, the statement is false. In fact,

$$1 + \frac{1}{2} + \frac{1}{3} + \cdots + \frac{1}{31} \approx 4.027245$$

so $n = 31$ is a counterexample. *(10-2)*

24. $S_{10} = (2 \cdot 1 - 8) + (2 \cdot 2 - 8) + (2 \cdot 3 - 8) + (2 \cdot 4 - 8) + (2 \cdot 5 - 8) + (2 \cdot 6 - 8)$

$\qquad\qquad + (2 \cdot 7 - 8) + (2 \cdot 8 - 8) + (2 \cdot 9 - 8) + (2 \cdot 10 - 8)$

$\qquad = (-6) + (-4) + (-2) + 0 + 2 + 4 + 6 + 8 + 10 + 12$

$\qquad = 30$ $\hfill (10\text{-}3)$

25. $S_7 = \dfrac{16}{2^1} + \dfrac{16}{2^2} + \dfrac{16}{2^3} + \dfrac{16}{2^4} + \dfrac{16}{2^5} + \dfrac{16}{2^6} + \dfrac{16}{2^7}$

$\qquad = 8 + 4 + 2 + 1 + \dfrac{1}{2} + \dfrac{1}{4} + \dfrac{1}{8}$

$\qquad = 15\dfrac{7}{8}$ $\hfill (10\text{-}3)$

26. This is an infinite geometric sequence with $a_1 = 27$.

$r = \dfrac{-18}{27} = -\dfrac{2}{3} \quad \left|-\dfrac{2}{3}\right| < 1$, so the sum exists

$S_\infty = \dfrac{a_1}{1 - r}$

$\qquad = \dfrac{27}{1 - \left(-\dfrac{2}{3}\right)}$

$\qquad = \dfrac{81}{5}$ $\hfill (10\text{-}3)$

27. $S_n = \displaystyle\sum_{k=1}^{n} \dfrac{(-1)^{k+1}}{3^k}$

This geometric sequence has $a_1 = \dfrac{1}{3}$.

$r = \left(-\dfrac{1}{9}\right) \div \dfrac{1}{3} = -\dfrac{1}{3} \quad \left|-\dfrac{1}{3}\right| < 1$, so the sum exists.

$S_\infty = \dfrac{a_1}{1 - r}$

$\qquad = \dfrac{\dfrac{1}{3}}{1 - \left(-\dfrac{1}{3}\right)}$

$\qquad = \dfrac{1}{4}$ $\hfill (10\text{-}3)$

28. The probability of an event cannot be negative, but $P(e_2)$ is given as negative. The sum of the probabilities of the simple events must be 1, but it is given as 2.5. The probability of an event cannot be greater than 1, but $P(e_4)$ is given as 2. $\hfill (10\text{-}5)$

29. We can select any three of the six points to use as vertices of the triangle. Order is not important. $C_{6,3} = \dfrac{6!}{3!(6-3)!} = 20$ triangles $\hfill (10\text{-}5)$

30. First, find d:

$a_n = a_1 + (n - 1)d$

$31 = 13 + (7 - 1)d$

$31 = 13 + 6d$

$\ d = 3$

So $a_5 = 13 + (5 - 1)3$

$\qquad = 25$ $\hfill (10\text{-}3)$

31.

	Case 1	Case 2	Case 3
O_1: select the first letter	N_1: 8 ways	8 ways	8 ways
O_2: select the second letter	N_2: 7 ways	8 ways	7 ways
O_3: select the third letter	N_3: 6 ways	8 ways	7 ways
	(exclude first and second letter)		(exclude second letter.)
	$8 \cdot 7 \cdot 6 = 336$ words	$8 \cdot 8 \cdot 8 = 512$ words	$8 \cdot 7 \cdot 7 = 392$ words

$\hfill (10\text{-}4)$

32. (A) $P(2 \text{ heads}) \approx \dfrac{210}{1,000} = .21$

(B) A sample space of equally likely events is:

$$S = \{HH, \ HT, \ TH, \ TT\}$$

$\quad\quad\quad P(1 \text{ head}) \approx \dfrac{480}{1,000} = .48$

Let $E_1 = 2 \text{ heads} = \{HH\}$

$\quad\quad\quad E_2 = 1 \text{ head} = \{HT, \ TH\}$

$\quad\quad\quad P(0 \text{ heads}) \approx \dfrac{310}{1,000} = .31$

$\quad\quad\quad E_3 = 0 \text{ heads} = \{TT\}$

then $P(E_1) = \dfrac{n(E_1)}{n(S)} = \dfrac{1}{4} = .25$

$\quad\quad P(E_2) = \dfrac{n(E_2)}{n(S)} = \dfrac{2}{4} = .5$

$\quad\quad P(E_3) = \dfrac{n(E_3)}{n(S)} = \dfrac{1}{4} = .25$

(C) $\quad\quad\quad\quad$ Expected frequency $= P(E) \cdot$number of trials

Expected frequency of 2 heads $= P(E_1) \cdot 1,000 = .25(1,000) = 250$

Expected frequency of 1 head $\ = P(E_2) \cdot 1,000 = .5(1,000) = 500$

Expected frequency of 0 heads $= P(E_3) \cdot 1,000 = .25(1,000) = 250$ *(10-5)*

33. The sample space S is the set of all possible 5-card hands chosen from a deck of 52 cards. $n(S) = C_{52,5}$.

(A) The event E is the set of all possible 4-card hands that are all diamonds, so are chosen from the 13 diamonds. $n(E) = C_{13,5}$

$$P(E) = \frac{n(E)}{n(S)} = \frac{C_{13,5}}{C_{52,5}}$$

(B) The event E is the set of all possible 5-card hands chosen as follows:

O_1: Choose 3 diamonds out of the 13 in the deck. N_1: $C_{13,3}$

O_2: Choose 2 spades out of the 13 in the deck. N_2: $C_{13,2}$

Applying the multiplication principle, $n(E) = N_1 \cdot N_2 = C_{13,3} \cdot C_{13,2}$

$$P(E) = \frac{n(E)}{n(S)} = \frac{C_{13,3} \cdot C_{13,2}}{C_{52,5}}$$ *(10-5)*

34. The sample space S is the set of all possible 4-person choices out of the 10 people. $n(S) = C_{10,4}$

The event E is the set of all of those choices that include 2 particular people, so

O_1: Choose the 2 married people $\quad\quad\quad\quad\quad\quad N_1$: 1

O_2: Choose 2 more people out of the 8 remaining N_2: $C_{8,2}$

Applying the multiplication principle, $n(E) = N_1 \cdot N_2 = 1 \cdot C_{8,2} = C_{8,2}$

$$P(E) = \frac{n(E)}{n(S)} = \frac{C_{8,2}}{C_{10,4}} = \frac{8!}{2!(8-2)!} \div \frac{10!}{4!(10-4)!} = 28 \div 210 = \frac{2}{15}$$ *(10-5)*

35. The sample space S is the set of all possible results of spinning the device twice:

$\{(1, \ 1), \ (1, \ 2), \ (1, \ 3), \ (2, \ 1), \ (2, \ 2), \ (2, \ 3), \ (3, \ 1), \ (3, \ 2), \ (3, \ 3)\}$.
$n(S) = 9$.

(A) The event E is the set of all possible results in which both spins are the same.

$E = \{(1, \ 1), \ (2, \ 2), \ (3, \ 3)\}$. $n(E) = 3$.

$$P(E) = \frac{n(E)}{n(S)} = \frac{3}{9} = \frac{1}{3}$$

(B) The event E is the set of those spin sequences that add up to 5:
$E = \{(2, 3), (3, 2)\}$. $n(E) = 2$.

$$P(E) = \frac{n(E)}{n(S)} = \frac{2}{9} \tag{10-5}$$

36. $0.\overline{72} = 0.72 + 0.0072 + 0.000072 + \cdots$
This is an infinite geometric sequence with $a_1 = 0.72$ and $r = 0.01$.

$$
\begin{aligned}
0.\overline{72} = S_\infty &= \frac{a_1}{1-r} \\
&= \frac{0.72}{1-0.01} \\
&= \frac{0.72}{0.99} \quad \text{or} \quad \frac{72}{99} \\
&= \frac{8}{11}
\end{aligned}
\tag{10-3}
$$

37. (A) Order is important here. We are selecting 3 digits out of 6 possible.
$P_{6,3} = 6 \cdot 5 \cdot 4 = 120$ lock combinations

(B) Order is not important here. We are selecting 2 players out of 5.

$$C_{5,2} = \frac{5!}{2!(5-2)!} = 10 \text{ games} \tag{10-4}$$

38.
$$
\begin{aligned}
\frac{20!}{18!(20-18)!} &= \frac{20!}{18!\,2!} \\
&= \frac{20 \cdot 19 \cdot 18!}{18!\,2 \cdot 1} = 190
\end{aligned}
\tag{10-6}
$$

39.
$$
\begin{aligned}
\binom{16}{12} &= \frac{16!}{12!(16-12)!} = \frac{16!}{12!\,4!} \\
&= \frac{16 \cdot 15 \cdot 14 \cdot 13 \cdot 12!}{12! \cdot 4 \cdot 3 \cdot 2 \cdot 1} = 1{,}820
\end{aligned}
\tag{10-6}
$$

40.
$$\binom{11}{11} = \frac{11!}{11!(11-11)!} = \frac{11!}{11!\,0!} = 1 \tag{10-6}$$

41.
$$
\begin{aligned}
(x-y)^5 = [x + (-y)]^5 &= \sum_{k=0}^{5} \binom{5}{k}(x)^{5-k}(-y)^k \\
&= \binom{5}{0}x^5 + \binom{5}{1}x^4(-y)^1 + \binom{5}{2}x^3(-y)^2 + \binom{5}{3}x^2(-y)^3 \\
&\qquad\qquad + \binom{5}{4}x(-y)^4 + \binom{5}{5}(-y)^5 \\
&= x^5 - 5x^4y + 10x^3y^2 - 10x^2y^3 + 5xy^4 - y^5
\end{aligned}
\tag{10-6}
$$

42. In the expansion
$$(x+2)^9 = \sum_{k=0}^{9} \binom{9}{k}x^{9-k}\,2^k$$

the exponent of x is 6 when $k = 3$, and the term containing x^6 is

$$\binom{9}{3}x^6 2^3 = 84x^6 \cdot 8 = 672x^6 \tag{10-6}$$

43. In the expansion of $(a + b)^n$, the exponent of b in the rth term is $r - 1$ and the exponent of a is $n - (r - 1)$. Here, $r = 10$, $n = 12$.

$$\text{Tenth term} = \binom{12}{9}(2x)^3(-y)^9$$

$$= \frac{12!}{9!\ 3!}(8x^3)(-y^9)$$

$$= \frac{12 \cdot 11 \cdot 10 \cdot 9!}{9! \cdot 3 \cdot 2 \cdot 1}(-8x^3y^9)$$

$$= -1{,}760x^3y^9 \qquad (10\text{-}6)$$

44. Write: P_n: $5 + 7 + 9 + \cdots + (2n + 3) = n^2 + 4n$

Proof: Part 1: Show that P_1 is true:

P_1: $5 = 1^2 + 4 \cdot 1$

$\qquad = 1 + 4$ Clearly true.

Part 2: Show that if P_k is true, then P_{k+1} is true:

Write out P_k and P_{k+1}.

$\quad P_k$: $5 + 7 + 9 + \cdots + (2k + 3) = k^2 + 4k$

$\quad P_{k+1}$: $5 + 7 + 9 + \cdots + (2k + 3) + (2k + 5) = (k + 1)^2 + 4(k + 1)$

We start with P_k:

$\quad 5 + 7 + 9 + \cdots + (2k + 3) = k^2 + 4k$

Adding $2k + 5$ to both sides:

$\quad 5 + 7 + 9 + \cdots + (2k + 3) + (2k + 5) = k^2 + 4k + 2k + 5$

$$= k^2 + 6k + 5$$
$$= k^2 + 2k + 1 + 4k + 4$$
$$= (k + 1)^2 + 4(k + 1)$$

We have shown that if P_k is true, then P_{k+1} is true.

Conclusion: P_n is true for all positive integers n. $\qquad (10\text{-}2)$

45. Write: P_n: $2 + 4 + 8 + \cdots + 2^n = 2^{n+1} - 2$

Proof: Part 1: Show that P_1 is true:

P_1: $2 = 2^{1+1} - 2$

$\qquad = 4 - 2$ Clearly true.

Part 2: Show that if P_k is true, then P_{k+1} is true:

Write out P_k and P_{k+1}.

$\quad P_k$: $2 + 4 + 8 + \cdots + 2^k = 2^{k+1} - 2$

$\quad P_{k+1}$: $2 + 4 + 8 + \cdots + 2^k + 2^{k+1} = 2^{k+2} - 2$

We start with P_k:

$\quad 2 + 4 + 8 + \cdots + 2^k = 2^{k+1} - 2$

Adding 2^{k+1} to both sides:

$\quad 2 + 4 + 8 + \cdots + 2^k + 2^{k+1} = 2^{k+1} - 2 + 2^{k+1}$

$$= 2^{k+1} + 2^{k+1} - 2$$
$$= 2 \cdot 2^{k+1} - 2$$
$$= 2^{k+2} - 2$$

We have shown that if P_k is true, then P_{k+1} is true.

Conclusion: P_n is true for all positive integers n. $\qquad (10\text{-}2)$

46. Write: P_n: $49^n - 1 = 6r$ for some r in N.

Part 1: Show that P_1 is true:

P_1: $49^1 - 1 = 6r$ is true ($r = 8$).

Part 2: Show that if P_k is true, then P_{k+1} is true:

Write out P_k and P_{k+1}.

P_k: $49^k - 1 = 6r$, for some integer r

P_{k+1}: $49^{k+1} - 1 = 6s$ for some integer s

We start with P_k:

$49^k - 1 = 6r$ for some integer r

Now, $49^{k+1} - 1 = 49^{k+1} - 49^k + 49^k - 1$

$= 49^k(49 - 1) + 49^k - 1$

$= 49^k \cdot 8 \cdot 6 + 6r$

$= 6(49^k \cdot 8 + r)$

$= 6s$ with $s = 49^k \cdot 8 + r$

We have shown that if P_k is true, then P_{k+1} is true.

Conclusion: P_n is true for all positive integers n.   (10-2)

47.

Enter the sequences in the sequence editor.

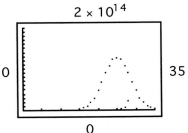

Graph the sequences.

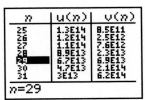

Display the points (in the appropriate interval) in a table.

From the second graph, 29 is the least positive integer n such that $a_n < b_n$.
The table confirms this.   (10-6)

48.

Enter the sequences in the sequence editor.

Graph the sequences.

Display the points in a table.

From the graph, the least positive integer n such that $a_n < b_n$ is between 25 and 30. The table confirms that $n = 26$.   (10-1)

49. In the first case, the order matters. We have five successive events, each of which can happen two ways (girl or boy). Applying the multiplication principle, there are $2 \cdot 2 \cdot 2 \cdot 2 \cdot 2 = 32$ possible families. In the second case, the possible families can be listed as {0 girls, 1 girl, 2 girls, 3 girls, 4 girls, 5 girls}, so there are 6 possibilities.   (10-4)

50. An arithmetic sequence is involved, with $a_1 = \dfrac{g}{2}$, $d = \dfrac{3g}{2} - \dfrac{g}{2} = g$.

Distance fallen during the twenty-fifth second $= a_{25}$.

$$a_n = a_1 + (n - 1)d$$

$$a_{25} = \frac{g}{2} + (25 - 1)g$$

$$= \frac{49g}{2} \text{ feet}$$

Total distance fallen after twenty-five seconds $= a_1 + a_2 + a_3 + \cdots + a_{25} = S_{25}$

$$S_n = \frac{n}{2}(a_1 + a_n)$$

$$S_{25} = \frac{25}{2}\left(\frac{g}{2} + \frac{49g}{2}\right)$$

$$= \frac{625g}{2} \text{ feet}$$

(10-3)

51. To seat two people, we can seat one person, then the second person.

O_1: Seat the first person in any chair.

N_1: 4 ways

O_2: Seat the second person in any remaining chair.

N_2: 3 ways

Applying the multiplication principle, there are $4 \cdot 3 = 12$ ways to seat two persons.

(10-4)

52. $(x + i)^6 = \displaystyle\sum_{k=0}^{6} \binom{6}{k} x^{6-k} i^k$

$$= \binom{6}{0} x^6 + \binom{6}{1} x^5 i^1 + \binom{6}{2} x^4 i^2 + \binom{6}{3} x^3 i^3 + \binom{6}{4} x^2 i^4 + \binom{6}{5} x i^5 + \binom{6}{6} i^6$$

$$= x^6 + 6ix^5 - 15x^4 - 20ix^3 + 15x^2 + 6ix - 1$$

(10-6)

53. Since there are several ways in which at least one woman can be selected, it is simplest to note: probability of selecting at least one woman = 1 - probability of selecting 0 women. To calculate the probability of selecting 0 women, that is, three men, we note:

The sample space S is the set of all 3-person subsets, chosen from the ten people. $n(S) = C_{10,3}$

The event E is the set of all 3-man subsets, chosen from the 7 men. $n(E) = C_{7,3}$

Then $P(E) = \dfrac{C_{7,3}}{C_{10,3}}$ and

$$P(\text{at least one woman}) = 1 - P(0 \text{ women}) = 1 - P(E)$$

$$= 1 - \frac{C_{7,3}}{C_{10,3}} = 1 - \frac{7}{24} = \frac{17}{24}$$

(10-5)

54. (A) $P(E) \approx \dfrac{f(E)}{n} = \dfrac{350}{1,000} = .350$

(B) A sample space of equally likely events is:
$S = \{HHH, HHT, HTH, HTT, THH, THT, TTH, TTT\}$
Let $E = 2$ heads $= \{HHT, HTH, THH\}$
then $P(E) = \dfrac{n(E)}{n(S)} = \dfrac{3}{8} = .375$

(C) Expected frequency $= P(E) \cdot$ number of trials $= .375(1,000) = 375$

(10-5)

55. Write: P_n: $\sum\limits_{k=1}^{n} k^3 = \left(\sum\limits_{k=1}^{n} k\right)^2$

Proof: Part 1: Show that P_1 is true.

P_1: $\sum\limits_{k=1}^{1} k^3 = 1^3 = 1^2 = \left(\sum\limits_{k=1}^{1} k\right)^2$

Part 2: Show that if P_j is true, then P_{j+1} is true.
Write out P_j and P_{j+1}.

P_j: $\sum\limits_{k=1}^{j} k^3 = \left(\sum\limits_{k=1}^{j} k\right)^2$

P_{j+1}: $\sum\limits_{k=1}^{j+1} k^3 = \left(\sum\limits_{k=1}^{j+1} k\right)^2$

We start with P_j:

$$\sum\limits_{k=1}^{j} k^3 = \left(\sum\limits_{k=1}^{j} k\right)^2$$

Adding $(j + 1)^3$ to both sides:

$$\sum\limits_{k=1}^{j} k^3 + (j + 1)^3 = \left(\sum\limits_{k=1}^{j} k\right)^2 + (j + 1)^3$$

$$\sum\limits_{k=1}^{j+1} k^3 = (1 + 2 + 3 + \ldots + j)^2 + (j + 1)^3$$

$$= \left[\frac{j(j + 1)}{2}\right]^2 + (j + 1)^3 \text{ using Matched Problem 2, Section 10-2}$$

$$= (j + 1)^2 \frac{j^2}{4} + (j + 1)^2(j + 1)$$

$$= (j + 1)^2 \left[\frac{j^2}{4} + j + 1\right]$$

$$= (j + 1)^2 \left[\frac{j^2 + 4j + 4}{4}\right]$$

$$= (j + 1)^2 \frac{(j + 2)^2}{2^2}$$

$$= \left[\frac{(j + 1)(j + 2)}{2}\right]^2$$

$$= [1 + 2 + 3 + \ldots + (j + 1)]^2 \text{ using Matched Problem 2, Section 10-2}$$

$$= \left(\sum\limits_{k=1}^{j+1} k\right)^2$$

We have shown that if P_j is true, then P_{j+1} is true.

Conclusion: P_n is true for all positive integers n. *(10-2)*

56. Write: P_n: $x^{2n} - y^{2n} = (x - y)Q_n(x, y)$, where $Q_n(x, y)$ denotes some polynomial in x and y.

Proof: Part 1: Show that P_1 is true.

P_1: $x^{2 \cdot 1} - y^{2 \cdot 1} = (x - y)(x + y) = (x - y)Q_1(x, y)$ P_1 is true.

Part 2: Show that if P_k is true, then P_{k+1} is true.
Write out P_k and P_{k+1}.

P_k: $x^{2k} - y^{2k} = (x - y)Q_k(x, y)$
P_{k+1}: $x^{2k+2} - y^{2k+2} = (x - y)Q_{k+1}(x, y)$

We start with P_k:

$x^{2k} - y^{2k} = (x - y)Q_k(x, y)$

Now, $x^{2k+2} - y^{2k+2} = x^{2k+2} - x^{2k}y^2 + x^{2k}y^2 - y^{2k+2}$

$\qquad\qquad\qquad\quad = x^{2k}(x^2 - y^2) + y^2(x^{2k} - y^{2k})$

$\qquad\qquad\qquad\quad = (x - y)x^{2k}(x + y) + y^2(x - y)Q_k(x, y)$ by P_k

$\qquad\qquad\qquad\quad = (x - y)[x^{2k}(x + y) + y^2 Q_k(x, y)]$

$\qquad\qquad\qquad\quad = (x - y)Q_{k+1}(x, y)$

We have shown that if P_k is true, then P_{k+1} is true.

Conclusion: P_n is true for all positive integers n.   *(10-2)*

57. Write: P_n: $\dfrac{a^n}{a^m} = a^{n-m}$, m an arbitrary positive integer, $n > m$.

Proof: Part 1: Show P_{m+1} is true.

P_{m+1}: $\dfrac{a^{m+1}}{a^m} = a^{m+1-m}$

$\qquad \dfrac{a^m a}{a^m} = a^1$ by the recursive definition of a^n

$\qquad\quad a = a$ P_{m+1} is true.

Part 2: Show that if P_k is true, then P_{k+1} is true.
Write out P_k and P_{k+1}.

P_k: $\dfrac{a^k}{a^m} = a^{k-m}$

P_{k+1}: $\dfrac{a^{k+1}}{a^m} = a^{k-m+1}$

We start with P_k:

$\dfrac{a^k}{a^m} = a^{k-m}$

Multiplying both sides by a:

$\dfrac{a^k}{a^m} a = a^{k-m}a$

$\dfrac{a^k a}{a^m} = a^{k-m+1}$

$\dfrac{a^{k+1}}{a^m} = a^{k-m+1}$

We have shown that if P_k is true, then P_{k+1} is true, for m an arbitrary positive integer. Conclusion: P_n is true for all positive integers m, n.   *(10-2)*

58. To prove $a_n = b_n$, n a positive integer, write:

P_n: $a_n = b_n$

Proof: Part 1: Show P_1 is true.

$a_1 = -3$ $b_1 = -5 + 2 \cdot 1 = -3$.

$a_1 = b_1$

Part 2: Show that if P_k is true, then P_{k+1} is true.

Write out P_k and P_{k+1}.

P_k: $a_k = b_k$

P_{k+1}: $a_{k+1} = b_{k+1}$

We start with P_k:

$a_k = b_k$

Now, $a_{k+1} = a_k + 2 = b_k + 2$

$$= -5 + 2k + 2$$
$$= -5 + 2(k + 1)$$
$$= b_{k+1}$$

Therefore, $a_{k+1} = b_{k+1}$.

So if P_k is true, then P_{k+1} is true.

Conclusion: P_n is true for all $n \in N$. That is $\{a_n\} = \{b_n\}$ (10-2)

59. Write: P_n: $(1!)1 + (2!)2 + (3!)3 + \cdots + (n!)n = (n + 1)! - 1$.

Proof: Part 1: Show that P_1 is true.

P_1: $(1!)1 = 1 = 2 - 1 = 2! - 1$ is true.

Part 2: Show that if P_k is true, then P_{k+1} is true.

Write out P_k and P_{k+1}.

P_k: $(1!)1 + (2!)2 + (3!)3 + \cdots + (k!)k = (k + 1)! - 1$

P_{k+1}: $(1!)1 + (2!)2 + (3!)3 + \cdots + (k!)k + (k + 1)!(k + 1) = (k + 2)! - 1$

We start with P_k:

$(1!)1 + (2!)2 + (3!)3 + \cdots + (k!)k = (k + 1)! - 1$

Adding $(k + 1)!(k + 1)$ to both sides:

$(1!)1 + (2!)2 + (3!)3 + \cdots + (k!)k + (k + 1)!(k + 1)$

$$= (k + 1)! - 1 + (k + 1)!(k + 1)$$
$$= (k + 1)!(1 + k + 1) - 1$$
$$= (k + 2)(k + 1)! - 1$$
$$= (k + 2)! - 1$$

So if P_k is true, then P_{k+1} is true.

Conclusion: P_n is true for all positive integers n.   (10-2)

60. The unpaid balance starts at \$7,200, and decreases each month by \$300. So the interest starts at $7,200(0.01)$ and decreases each month by $300(0.01)$. This is an arithmetic sequence with

$a_1 = 7,200(0.01)$, $a_{24} = 300(0.01)$, $n = 24$.

Then the total interest paid is

$$S_n = \frac{n}{2}(a_1 + a_n)$$

$$S_{24} = \frac{24}{2}[7,200(0.01) + 300(0.01)]$$

$$= \$900$$ (10-3)

61. This involves an infinite geometric series. $a_1 = (0.75)2,400.$ $r = 0.75.$

$|r| < 1$, so the series has a sum.

$$S_\infty = \frac{a_1}{1-r} = \frac{(0.75)2400}{1-0.75} = \$7,200 \qquad (10\text{-}3)$$

62. Since $A(n)$ is a geometric sequence with common ratio 1.06, we can write

$$A(n) = a_n = (1.06)^n\,500$$

If $n = 10$, $a_{10} = (1.06)^{10}\,500 = \$895.42$

If $n = 20$, $a_{10} = (1.06)^{20}\,500 = \$1,603.57 \qquad (10\text{-}3)$

63. A route plan can be regarded as a series of choices of stores, which is really an arrangement of the 5 stores. Since the order matters, we use permutations: $P_{5,5} = 5! = 120$ route plans. $\qquad (10\text{-}4)$

64. (A) $P(\text{this event}) = \dfrac{40}{1,000} = .04$

(B) $P(\text{this event}) = \dfrac{100+60}{1,000} = .16$

(C) $P(\text{this event}) = 1 - P\,(\text{not } 12 - 18 \text{ and buys 0 or 1 cassette annually})$

$$= 1 - \frac{60 + 70 + 70 + 110 + 100 + 50}{1,000} = 1 - \frac{460}{1,000} = .54 \qquad (10\text{-}5)$$

65. $P\,(\text{shipment returned}) = 1 - P\,(\text{no substandard part found}).$
The sample space S is the set of all possible 4-part subsets of the 12 part set.
$n(S) = C_{12,4}$
The event, no substandard part found, is the set of all possible 4-part subsets of the 10 acceptable parts. $n(E) = C_{10,4}.$

Then $P(\text{shipment returned}) = 1 - \dfrac{n(E)}{n(S)} = 1 - \dfrac{C_{10,4}}{C_{12,4}} = 1 - \dfrac{14}{33} \approx .576. \qquad (10\text{-}5)$

CHAPTER 11   Additional Topics in Analytic Geometry

Section 11-1

1. Answers will vary.   **3.** Answers will vary.   **5.** Answers will vary.

7. To graph $y^2 = 4x$, assign x values that make the right side a perfect square (x must be non-negative for y to be real) and solve for y. Since the coefficient of x is positive, a must be positive, and the parabola opens right.

x	0	1	4
y	0	±2	±4

To find the focus and directrix,
solve
$4a = 4$
$a = 1$
Focus: (1, 0) Directrix: $x = -1$

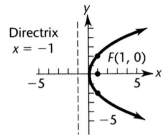

9. To graph $x^2 = 8y$, assign y values that make the right side a perfect square (y must be non-negative for x to be real) and solve for x. Since the coefficient of y is positive, a must be positive, and the parabola opens up.

x	0	±4	±2
y	0	2	$\frac{1}{2}$

To find the focus and directrix,
solve
$4a = 8$
$a = 2$
Focus: (0, 2) Directrix: $y = -2$

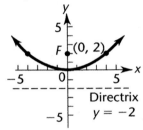

11. To graph $y^2 = -12x$, assign x values that make the right side a perfect square (x must be zero or negative for y to be real) and solve for y. Since the coefficient of x is negative, a must be negative, and the parabola opens left.

x	0	-3	$-\frac{1}{3}$
y	0	±6	±2

To find the focus and directrix,
solve
$4a = -12$
$a = -3$
Focus: (-3, 0)
Directrix: $x = -(-3) = 3$

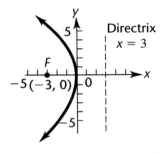

13. To graph $x^2 = -4y$, assign y values that make the right side a perfect square (y must be zero or negative for x to be real) and solve for x. Since the coefficient of y is negative, a must be negative, and the parabola opens down.

x	0	±2	±4
y	0	-1	-4

To find the focus and directrix,
solve
$4a = -4$
$a = -1$
Focus: (0, -1)
Directrix: $y = -(-1) = 1$

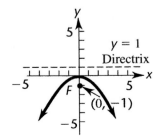

15. To graph $y^2 = -20x$, we may proceed as in problem 11. Alternatively, after noting that x must be zero or negative for y to be real, we may pick convenient values for x and solve for y using a calculator. Since the coefficient of x is negative, a must be negative, and the parabola opens left.

x	0	−1	−2
y	0	$\pm\sqrt{20} \approx \pm4.5$	$\pm\sqrt{40} \approx \pm6.3$

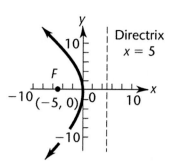

To find the focus and directrix, solve
$4a = -20$
$\ a = -5$
Focus: $(-5, 0)$
Directrix: $x = -(-5) = 5$

17. To graph $x^2 = 10y$, we may proceed as in Problem 9. Alternatively, after noting that y must be non-negative for x to be real, we may pick convenient values for y and solve for x using a calculator. Since the coefficient of y is positive, a must be positive, and the parabola opens up.

x	0	$\pm\sqrt{10} \approx \pm3.2$	$\pm\sqrt{20} \approx \pm4.5$
y	0	1	2

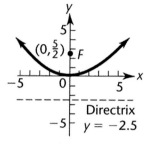

To find the focus and directrix, solve

$4a = 10$
$\ a = 2.5$
Focus: $(0, 2.5)$
Directrix: $y = -2.5$

19. Comparing $y^2 = 39x$ with $y^2 = 4ax$, the standard equation of a parabola symmetric with respect to the x axis, we have
$4a = 39$ Focus on x axis
$\ a = 9.75$
Focus: $(9.75, 0)$

21. Comparing $x^2 = -105y$ with $x^2 = 4ay$, the standard equation of a parabola symmetric with respect to the y axis, we have
$4a = -105$ Focus on y axis
$\ a = -26.25$
Focus: $(0, -26.25)$

23. Comparing $y^2 = -77x$ with $y^2 = 4ax$, the standard equation of a parabola symmetric with respect to the x axis, we have
$4a = -77$ Focus on x axis
$\ a = -19.25$
Focus: $(-19.25, 0)$

25. Comparing directrix $y = -3$ with the information in the chart of standard equations for a parabola, we see:
$a = 3$.   Axis: the y axis.   Equation: $x^2 = 4ay$.
The equation of the parabola must be $x^2 = 4 \cdot 3y$, or $x^2 = 12y$.

27. Comparing focus $(0, -7)$ with the information in the chart of standard equations for a parabola, we see:
$a = -7$ Axis: the y-axis.   Equation: $x^2 = 4ay$.
The equation of the parabola must be $x^2 = 4(-7)y$, or $x^2 = -28y$.

29. Comparing directrix $x = 6$ with the information in the chart of standard equations for a parabola, we see:

$6 = -a$. $a = -6$. Axis: the x-axis. Equation: $y^2 = 4ax$.

The equation of the parabola must be $y^2 = 4(-6)x$, or $y^2 = -24x$.

31. Comparing focus $(2, 0)$ with the information in the chart of standard equations for a parabola, we see:

$a = 2$ Axis: the x-axis. Equation: $y^2 = 4ax$.

The equation of the parabola must be $y^2 = 4(2)x$, or $y^2 = 8x$.

33. The parabola is opening up and has an equation of the form $x^2 = 4ay$. Since $(4, 2)$ is on the graph, we have:

$$x^2 = 4ay$$
$$(4)^2 = 4a(2)$$
$$16 = 8a$$
$$2 = a$$

The equation of the parabola is

$$x^2 = 4(2)y$$
$$x^2 = 8y$$

35. The parabola is opening left and has an equation of the form $y^2 = 4ax$. Since $(-3, 6)$ is on the graph, we have:

$$y^2 = 4ax$$
$$(6)^2 = 4a(-3)$$
$$-3 = a$$

The equation of the parabola is

$$y^2 = 4(-3)x$$
$$y^2 = -12x$$

37. The parabola is opening down and has an equation of the form $x^2 = 4ay$. Since $(-6, -9)$ is on the graph, we have:

$$x^2 = 4ay$$
$$(-6)^2 = 4a(-9)$$
$$36 = -36a$$
$$-1 = a$$

The equation of the parabola is $x^2 = 4(-1)y$
$$x^2 = -4y$$

39. $x^2 = 4y$
$y^2 = 4x$

Solve for y in the first equation, then substitute into the second equation.

$$y = \frac{x^2}{4}$$
$$\left(\frac{x^2}{4}\right)^2 = 4x$$
$$\frac{x^4}{16} = 4x$$
$$x^4 = 64x$$
$$x^4 - 64x = 0$$
$$x(x^3 - 64) = 0$$
$$x(x - 4)(x^2 + 4x + 8) = 0$$
$$x = 0 \quad x - 4 = 0 \quad x^2 + 4x + 8 = 0$$
$$x = 4 \quad \text{No real solutions}$$

For $x = 0$ For $x = 4$
$$y = 0 \qquad y = \frac{4^2}{4}$$
$$y = 4$$

Solutions: $(0, 0)$, $(4, 4)$

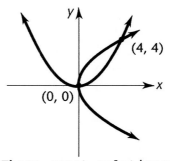

These exact solutions yield more than 3-digit accuracy.

41. $y^2 = 6x$
$x^2 = 5y$

Solve for x in the first equation, then substitute into the second equation.

$$x = \frac{y^2}{6}$$

$$\left(\frac{y^2}{6}\right)^2 = 5y$$

$$\frac{y^4}{36} = 5y$$

$$y^4 = 180y$$

$$y^4 - 180y = 0$$

$$y(y^3 - 180) = 0$$

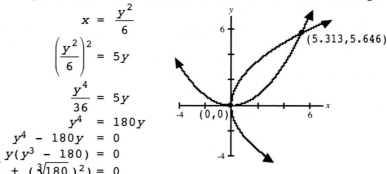

(5.313,5.646)

$y(y - \sqrt[3]{180})(y^2 + y\sqrt[3]{180} + (\sqrt[3]{180})^2) = 0$

$y = 0 \quad y - \sqrt[3]{180} = 0 \qquad y^2 + y\sqrt[3]{180} + (\sqrt[3]{180})^2 = 0$

$\qquad\qquad y = \sqrt[3]{180} \qquad$ No real solutions

$\qquad\qquad \approx 5.646$

For $y = 0$ For $y = \sqrt[3]{180}$

$\quad x = 0 \qquad\qquad x = \dfrac{(\sqrt[3]{180})^2}{6}$

$\qquad\qquad\qquad\qquad \approx 5.313$

Solutions: $(0, 0)$, $(5.313, 5.646)$

43. (A) The line $x = 0$ intersects the parabola $x^2 = 4ay$ only at $(0, 0)$.
The line $y = 0$ intersects the parabola $x^2 = 4ay$ only at $(0, 0)$.
Only these 2 lines intersect the parabola at exactly one point (see part (B)).

(B) A line through $(0, 0)$ with slope $m \neq 0$ has equation $y = mx$.
Solve the system:
$$y = mx$$
$$x^2 = 4ay$$
by substituting y from the first equation into the second equation.
$$x^2 = 4amx$$
$$x^2 - 4amx = 0$$
$$x(x - 4am) = 0$$
$$x = 0 \quad x = 4am$$
For $x = 0$ For $x = 4am$
$\quad y = 0 \qquad\qquad y = m(4am)$
$\qquad\qquad\qquad\qquad = 4am^2$

Solutions: $(0, 0)$, $(4am, 4am^2)$ are the required coordinates.

45. Since A and B lie on the curve $x^2 = 4ay$, their coordinates must satisfy the equation of the curve. Clearly, the y coordinate of A, F, and B is a. Substituting a for y, we have

$$x^2 = 4aa$$
$$x^2 = 4a^2$$
$$x = \pm 2a$$

Therefore, A has coordinates $(-2a, a)$ and B has coordinates $(2a, a)$.

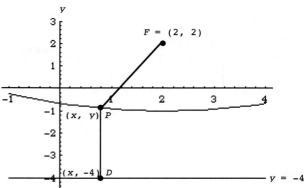

47. Let $P = (x, y)$ be a point on the parabola. Then, by the definition of the parabola, the distance from $P = (x, y)$ to the focus $F = (2, 2)$ must equal the perpendicular distance from P to the directrix at $D = (x, -4)$. Applying the distance formula, we have

$$d(P, F) = d(P, D)$$
$$\sqrt{(x - 2)^2 + (y - 2)^2} = \sqrt{(x - x)^2 + [y - (-4)]^2}$$
$$(x - 2)^2 + (y - 2)^2 = (x - x)^2 + (y + 4)^2$$
$$x^2 - 4x + 4 + y^2 - 4y + 4 = 0 + y^2 + 8y + 16$$
$$x^2 - 4x - 12y - 8 = 0$$

49. Let $P = (x, y)$ be a point on the parabola.

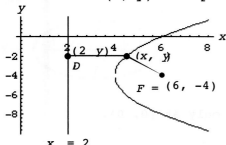

$x = 2$

Then, by the definition of the parabola, the distance from $P = (x, y)$ to the focus $F = (6, -4)$ must equal the perpendicular distance from P to the directrix at $D = (2, y)$.

Applying the distance formula, we have

$$d(P, F) = d(P, D)$$
$$\sqrt{(x - 6)^2 + [y - (-4)]^2} = \sqrt{(x - 2)^2 + (y - y)^2}$$
$$(x - 6)^2 + (y + 4)^2 = (x - 2)^2 + (y - y)^2$$
$$x^2 - 12x + 36 + y^2 + 8y + 16 = x^2 - 4x + 4 + 0$$
$$y^2 + 8y - 8x + 48 = 0$$

51. Using Figure 5 in Section 11-1, we note:
The point $P = (x, y)$ is a point on the parabola if and only if

$$d_1 = d_2$$
$$d(P, N) = d(P, F)$$
$$\sqrt{(x - x)^2 + (y + a)^2} = \sqrt{(x - 0)^2 + (y - a)^2}$$
$$(y + a)^2 = x^2 + (y - a)^2$$
$$y^2 + 2ay + a^2 = x^2 + y^2 - 2ay + a^2$$
$$x^2 = 4ay$$

53. From the figure, we see that the coordinates of P must be $(-100, -50)$. The parabola is opening down with axis the y axis, so it has an equation of the form $x^2 = 4ay$. Since $(-100, -50)$ is on the graph, we have

$$(-100)^2 = 4a(-50)$$
$$10,000 = -200a$$
$$a = -50$$

The equation of the parabola is
$$x^2 = 4(-50)y$$
$$x^2 = -200y$$

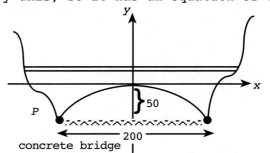

concrete bridge

55. (A) From the figure, we see that the parabola is opening up with axis the y axis, so it has an equation of the form $x^2 = 4ay$. Since the focus is at $(0, a) = (0, 100)$, $a = 100$ and the equation of the parabola is $x^2 = 400y$ or $y = 0.0025x^2$

(B) Since the depth represents the y coordinate y_1 of a point on the parabola with $x = 100$, we have $y_1 = 0.0025(100)^2$
depth = 25 feet

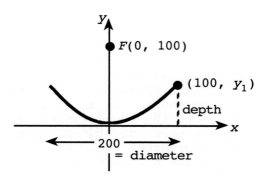

Section 11-2

1. No. Explanations will vary. **3.** No. Explanations will vary.

5. Answers will vary.

Except as otherwise noted, the graphing calculator drawings in this and the following sections of this chapter have been plotted in a squared window (zoom menu 5 on the TI-84) to avoid distorting the shapes of the conic sections.

7. When $y = 0$, $\dfrac{x^2}{25} = 1$. x intercepts: ± 5

When $x = 0$, $\dfrac{y^2}{4} = 1$. y intercepts: ± 2

So $a = 5$, $b = 2$, and the major axis is on the x axis.
Foci: $c^2 = a^2 - b^2$
$c^2 = 25 - 4$
$c^2 = 21$
$c = \sqrt{21}$
Foci: $F' = (-\sqrt{21}, 0)$, $F = (\sqrt{21}, 0)$
Major axis length = $2(5) = 10$; Minor axis length = $2(2) = 4$

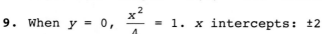

9. When $y = 0$, $\dfrac{x^2}{4} = 1$. x intercepts: ± 2

When $x = 0$, $\dfrac{y^2}{25} = 1$. y intercepts: ± 5

So $a = 5$, $b = 2$, and the major axis is on the y axis.
Foci: $c^2 = a^2 - b^2$
$c^2 = 25 - 4$
$c^2 = 21$
$c = \sqrt{21}$ Foci: $F' = (0, -\sqrt{21})$, $F = (0, \sqrt{21})$
Major axis length = $2(5) = 10$ Minor axis length = $2(2) = 4$

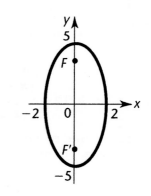

11. First, write the equation in standard form by dividing both sides by 9.
$x^2 + 9y^2 = 9$
$\dfrac{x^2}{9} + \dfrac{y^2}{1} = 1$
Locate the intercepts.

When $y = 0$, $\dfrac{x^2}{9} = 1$. x intercepts: ± 3

When $x = 0$, $\dfrac{y^2}{1} = 1$. y intercepts: ± 1

So $a = 3$, $b = 1$, and the major axis is on the x axis.
Foci: $c^2 = a^2 - b^2$
$c^2 = 9 - 1$
$c^2 = 8$
$c = \sqrt{8}$ Foci: $F' = (-\sqrt{8}, 0)$, $F = (\sqrt{8}, 0)$
Major axis length = $2(3) = 6$ Minor axis length = $2(1) = 2$

13. When $y = 0$, $9x^2 = 144$. x intercepts: ± 4
When $x = 0$, $16y^2 = 144$. y intercepts: ± 3
This corresponds to graph (b).

15. When $y = 0$, $4x^2 = 16$. x intercepts: ± 2
When $x = 0$, $y^2 = 16$. y intercepts: ± 4
This corresponds to graph (a).

17. First, write the equation in standard form by dividing both sides by 225.
$$25x^2 + 9y^2 = 225$$
$$\frac{x^2}{9} + \frac{y^2}{25} = 1$$
Locate the intercepts.

When $y = 0$, $\frac{x^2}{9} = 1$. x intercepts: ± 3

When $x = 0$, $\frac{y^2}{25} = 1$. y intercepts: ± 5

So $a = 5$, $b = 3$, and the major axis is on the y axis.
Foci: $c^2 = a^2 - b^2$
     $c^2 = 25 - 9$
     $c^2 = 16$
      $c = 4$ Foci: $F' = (0, -4)$, $F = (0, 4)$
Major axis length = $2(5) = 10$ Minor axis length = $2(3) = 6$

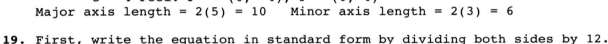

19. First, write the equation in standard form by dividing both sides by 12.
$$2x^2 + y^2 = 12$$
$$\frac{x^2}{6} + \frac{y^2}{12} = 1$$
Locate the intercepts.

When $y = 0$, $\frac{x^2}{6} = 1$. x intercepts: $\pm\sqrt{6}$

When $x = 0$, $\frac{y^2}{12} = 1$. y intercepts: $\pm\sqrt{12}$

So $a = \sqrt{12}$, $b = \sqrt{6}$, and the major axis is on the y axis.
Foci: $c^2 = a^2 - b^2$
     $c^2 = 12 - 6$
     $c^2 = 6$
      $c = \sqrt{6}$ Foci: $F' = (0, -\sqrt{6})$, $F = (0, \sqrt{6})$
Major axis length = $2\sqrt{12} \approx 6.93$ Minor axis length = $2\sqrt{6} \approx 4.90$

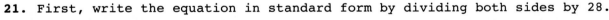

21. First, write the equation in standard form by dividing both sides by 28.
$$4x^2 + 7y^2 = 28$$
$$\frac{x^2}{7} + \frac{y^2}{4} = 1$$
Locate the intercepts.

When $y = 0$, $\frac{x^2}{7} = 1$. x intercepts: $\pm\sqrt{7}$

When $x = 0$, $\frac{y^2}{4} = 1$. y intercepts: ± 2

So $a = \sqrt{7}$, $b = 2$, and the major axis is on the x axis.

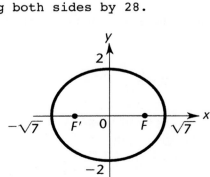

Foci: $c^2 = a^2 - b^2$

$\quad\quad c^2 = 7 - 4$

$\quad\quad c^2 = 3$

$\quad\quad\; c = \sqrt{3}$ Foci: $F' = (-\sqrt{3}, 0)$, $F = (\sqrt{3}, 0)$

Major axis length $= 2\sqrt{7} \approx 5.29$

Minor axis length $= 2(2) = 4$

23. The x intercepts are ± 5; if $y = 0$, $\dfrac{(\pm 5)^2}{M} = 1$, so $M = 25$.

The y intercepts are ± 4; if $x = 0$, $\dfrac{(\pm 4)^2}{N} = 1$, so $N = 16$.

Equation: $\dfrac{x^2}{25} + \dfrac{y^2}{16} = 1$

25. The x intercepts are ± 3; if $y = 0$, $\dfrac{(\pm 3)^2}{M} = 1$, so $M = 9$.

The y intercepts are ± 6; if $x = 0$, $\dfrac{(\pm 6)^2}{N} = 1$, so $N = 36$.

Equation: $\dfrac{x^2}{9} + \dfrac{y^2}{36} = 1$

27. Make a rough sketch of the ellipse and compute x and y intercepts.

$\dfrac{x^2}{a^2} + \dfrac{y^2}{b^2} = 1$

$a = \dfrac{10}{2} = 5$, $b = \dfrac{6}{2} = 3$

$\dfrac{x^2}{25} + \dfrac{y^2}{9} = 1$

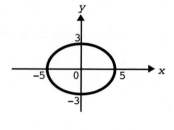

29. Make a rough sketch of the ellipse and compute x and y intercepts.

$\dfrac{x^2}{b^2} + \dfrac{y^2}{a^2} = 1$

$a = \dfrac{22}{2} = 11$, $b = \dfrac{16}{2} = 8$

$\dfrac{x^2}{64} + \dfrac{y^2}{121} = 1$

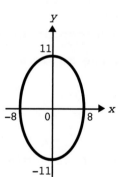

31. Make a rough sketch of the ellipse, locate focus and x intercepts, then determine y intercepts using the Pythagorean Theorem.

$\dfrac{x^2}{a^2} + \dfrac{y^2}{b^2} = 1$

$a = \dfrac{16}{2} = 8$

$b^2 = 8^2 - 6^2 = 64 - 36 = 28$

$b = \sqrt{28}$

$\dfrac{x^2}{64} + \dfrac{y^2}{28} = 1$

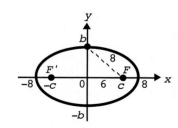

33. Make a rough sketch of the ellipse, locate focus and x intercepts, then determine y intercepts using the Pythagorean Theorem.

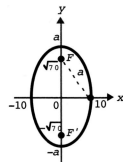

$$\frac{x^2}{b^2} + \frac{y^2}{a^2} = 1$$

$$b = \frac{20}{2} = 10$$

$$a^2 = 10^2 + (\sqrt{70})^2 = 100 + 70 = 170$$

$$a = \sqrt{170}$$

$$\frac{x^2}{100} + \frac{y^2}{170} = 1$$

35. The graph does not pass the vertical line test; most vertical lines that intersect an ellipse do so in two places. The equation does not define a function.

37. From the figure, we see that the point $P = (x, y)$ is a point on the curve if and only if

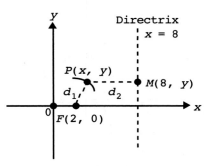

$$d_1 = \frac{1}{2}d_2$$

$$d(P, F) = \frac{1}{2}d(P, M)$$

$$\sqrt{(x - 2)^2 + (y - 0)^2} = \frac{1}{2}\sqrt{(x - 8)^2 + (y - y)^2}$$

$$(x - 2)^2 + y^2 = \frac{1}{4}(x - 8)^2$$

$$x^2 - 4x + 4 + y^2 = \frac{1}{4}(x^2 - 16x + 64)$$

$$4x^2 - 16x + 16 + 4y^2 = x^2 - 16x + 64$$

$$3x^2 + 4y^2 = 48$$

$$\frac{x^2}{16} + \frac{y^2}{12} = 1$$

The curve must be an ellipse since its equation can be written in standard form for an ellipse.

39. The only points that could satisfy this condition are those on the line segment $F'F$, and all of those would qualify.

41. $d_1 + d_2 = 2a$ Use the distance formula:

$\sqrt{(x + c)^2 + y^2} = 2a - \sqrt{(x - c)^2 + y^2}$ Square both sides:

$(x + c)^2 + y^2 = 4a^2 - 4a\sqrt{(x - c)^2 + y^2} + (x - c)^2 + y^2$ Isolate the square root:

$\sqrt{(x - c)^2 + y^2} = a - \frac{cx}{a}$ Square both sides:

$(x - c)^2 + y^2 = a^2 - 2cx + \frac{c^2x^2}{a^2}$ Simplify and collect terms:

$\left(1 - \frac{c^2}{a^2}\right)x^2 + y^2 = a^2 - c^2$ Use $a^2 - c^2 = b^2$, $1 - \frac{c^2}{a^2} = \frac{b^2}{a^2}$, and divide both sides by b^2:

$$\frac{x^2}{a^2} + \frac{y^2}{b^2} = 1$$

43. From the figure we see that the x and y intercepts of the ellipse must be 20 and 12 respectively. The equation of the ellipse must be

$$\frac{x^2}{(20)^2} + \frac{y^2}{(12)^2} = 1$$

or

$$\frac{x^2}{400} + \frac{y^2}{144} = 1$$

To find the clearance above the water 5 feet from the bank, we need the y coordinate y_1 of the point P whose x coordinate is $a - 5 = 15$. Since P is on the ellipse, we have

$$\frac{15^2}{400} + \frac{y_1^2}{144} = 1$$

$$\frac{225}{400} + \frac{y_1^2}{144} = 1$$

$$0.5625 + \frac{y_1^2}{144} = 1$$

$$\frac{y_1^2}{144} = 0.4375$$

$$y_1^2 = 144(0.4375)$$

$$y_1^2 = 63$$

$$y_1 \approx \pm 7.94$$

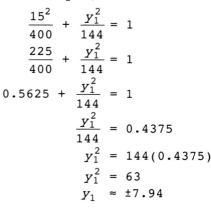
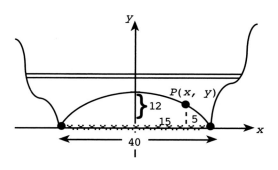

Therefore, the clearance is 7.94 feet, approximately.

45. (A) From the figure we see that the x intercept of the ellipse must be 24.0. The equation of the ellipse must have the form

$$\frac{x^2}{(24.0)^2} + \frac{y^2}{b^2} = 1$$

Since the point (23.0, 1.14) is on the ellipse, its coordinates must satisfy the equation of the ellipse. So

$$\frac{(23.0)^2}{(24.0)^2} + \frac{(1.14)^2}{b^2} = 1$$

$$\frac{(1.14)^2}{b^2} = 1 - \frac{(23.0)^2}{(24.0)^2}$$

$$\frac{(1.14)^2}{b^2} = \frac{47}{576}$$

$$b^2 = \frac{576(1.14)^2}{47}$$

$$b^2 = 15.9$$

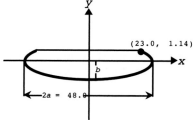

The equation of the ellipse must be $\dfrac{x^2}{576} + \dfrac{y^2}{15.9} = 1$

(B) From the figure, we can see that the width of the wing must equal
$1.14 + b = 1.14 + \sqrt{15.9} = 5.13$ feet.

Section 11-3

1. Answers will vary.   3. Answers will vary.   5. Answers will vary.

7. When $y = 0$, $x^2 = 1$. x intercepts: ± 1
When $x = 0$, $-y^2 = 1$. There are no y intercepts.
This corresponds to graph (d).

9. When $y = 0$, $-x^2 = 4$.   There are no x intercepts.
   When $x = 0$, $y^2 = 4$. y intercepts: ± 2
   This corresponds to graph (c).

11. When $y = 0$, $\dfrac{x^2}{9} = 1$. x intercepts: ± 3 $a = 3$

   When $x = 0$, $-\dfrac{y^2}{4} = 1$. There are no y intercepts, but $b = 2$.

   Sketch the asymptotes using the
   asymptote rectangle, then sketch in
   the hyperbola.
   Foci: $c^2 = 3^2 + 2^2$
   $\qquad c^2 = 13$
   $\qquad\ \ c = \sqrt{13}$
   $F' = (-\sqrt{13},\ 0)$, $F = (\sqrt{13},\ 0)$

   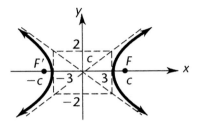

   Transverse axis length $= 2(3) = 6$
   Conjugate axis length $= 2(2) = 4$

13. When $y = 0$, $-\dfrac{x^2}{9} = 1$. There are no x intercepts, but $b = 3$.

   When $x = 0$, $\dfrac{y^2}{4} = 1$. y intercepts: ± 2, $a = 2$

   Sketch the asymptotes using the
   asymptote rectangle, then sketch in
   the hyperbola.
   Foci: $c^2 = 2^2 + 3^2$
   $\qquad c^2 = 13$
   $\qquad\ \ c = \sqrt{13}$
   $F' = (0,\ -\sqrt{13})$, $F = (0,\ \sqrt{13})$

   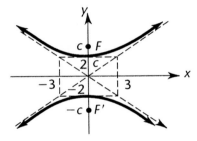

   Transverse axis length $= 2(2) = 4$
   Conjugate axis length $= 2(3) = 6$

15. First, write the equation in standard form by dividing both sides by 16.
   $\qquad 4x^2 - y^2 = 16$
   $\qquad \dfrac{x^2}{4} - \dfrac{y^2}{16} = 1$
   Locate intercepts:
   When $y = 0$, $x = \pm 2$. x intercepts: ± 2 $a = 2$

   When $x = 0$, $-\dfrac{y^2}{16} = 1$. There are no y intercepts,
   $\qquad\qquad\qquad\qquad$ but $b = 4$.
   Sketch the asymptotes using the asymptote
   rectangle, then sketch in the hyperbola.
   Foci: $c^2 = 2^2 + 4^2$
   $\qquad c^2 = 20$
   $\qquad\ \ c = \sqrt{20}$
   $F' = (-\sqrt{20},\ 0)$, $F = (\sqrt{20},\ 0)$

   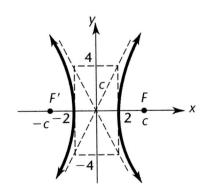

   Transverse axis length $= 2(2) = 4$
   Conjugate axis length $= 2(4) = 8$

17. First, write the equation in standard form by dividing both sides by 144.
$9y^2 - 16x^2 = 144$

$$\frac{y^2}{16} - \frac{x^2}{9} = 1$$

Locate intercepts:

When $y = 0$, $-\dfrac{x^2}{9} = 1$. There are no x intercepts, but $b = 3$.

When $x = 0$, $\dfrac{y^2}{16} = 1$, $y = \pm 4$.

y intercepts: ± 4 $a = 4$

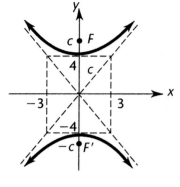

Sketch the asymptotes using the asymptote rectangle, then sketch in the hyperbola.

Foci: $c^2 = 4^2 + 3^2$
$\qquad c^2 = 25$
$\qquad\ c = 5$

$F' = (0, -5)$, $F = (0, 5)$

Transverse axis length $= 2(4) = 8$. Conjugate axis length $= 2(3) = 6$.

19. First, write the equation in standard form by dividing both sides by 12.
$3x^2 - 2y^2 = 12$

$$\frac{x^2}{4} - \frac{y^2}{6} = 1$$

Locate intercepts: When $y = 0$, $\dfrac{x^2}{4} = 1$. $x = \pm 2$.

$\qquad\qquad\qquad\qquad x$ intercepts: ± 2 $a = 2$

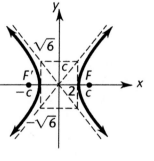

When $x = 0$, $-\dfrac{y^2}{6} = 1$. There are no y intercepts, but $b = \sqrt{6}$.

Sketch the asymptotes using the asymptote rectangle, then sketch in the hyperbola.

Foci: $c^2 = 2^2 + (\sqrt{6})^2$
$\qquad c^2 = 10$
$\qquad\ c = \sqrt{10}$

$F' = (-\sqrt{10}, 0)$, $F = (\sqrt{10}, 0)$

Transverse axis length $= 2(2) = 4$ Conjugate axis length $= 2\sqrt{6} \approx 4.90$

21. First, write the equation in standard form by dividing both sides by 28.
$7y^2 - 4x^2 = 28$

$$\frac{y^2}{4} - \frac{x^2}{7} = 1$$

Locate intercepts:

When $y = 0$, $-\dfrac{x^2}{7} = 1$. There are no x intercepts, but $b = \sqrt{7}$.

When $x = 0$, $\dfrac{y^2}{4} = 1$, $y = \pm 2$. y intercepts: ± 2 $a = 2$

Sketch the asymptotes using the asymptote rectangle, then sketch in the hyperbola.

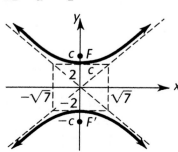

Foci: $c^2 = 2^2 + (\sqrt{7})^2$
$\qquad c^2 = 11$
$\qquad\ c = \sqrt{11}$

$F' = (0, -\sqrt{11})$, $F = (0, \sqrt{11})$

Transverse axis length $= 2(2) = 4$
Conjugate axis length $= 2\sqrt{7} \approx 5.29$

23. Since the graph has x intercepts ($x = \pm 3$) but no y intercepts, the equation must be in the form

$$\frac{x^2}{M} - \frac{y^2}{N} = 1$$

and, when $y = 0$, $x = \pm 3$, so

$$\frac{9}{M} - \frac{0}{N} = 1$$
$$M = 9$$

Since the point $(5, 4)$ is on the graph, its coordinates must satisfy the equation, which is known to be of form

$$\frac{x^2}{9} - \frac{y^2}{N} = 1$$

So we get

$$\frac{5^2}{9} - \frac{4^2}{N} = 1$$
$$\frac{25}{9} - \frac{16}{N} = 1$$
$$-\frac{16}{N} = -\frac{16}{9}$$
$$N = 9$$

The equation is therefore

$$\frac{x^2}{9} - \frac{y^2}{9} = 1$$

25. Since the graph has y intercepts ($y = \pm 4$) but no x intercepts, the equation must be in the form

$$\frac{y^2}{N} - \frac{x^2}{M} = 1$$

and, when $x = 0$, $y = \pm 4$, so

$$\frac{16}{N} - \frac{0}{M} = 1$$
$$N = 16$$

Since the point $(3, 5)$ is on the graph, its coordinates must satisfy the equation, which is known to be of form

$$\frac{y^2}{16} - \frac{x^2}{M} = 1$$

So we get

$$\frac{5^2}{16} - \frac{3^2}{M} = 1$$
$$\frac{25}{16} - \frac{9}{M} = 1$$
$$-\frac{9}{M} = -\frac{9}{16}$$
$$M = 16$$

The equation is therefore

$$\frac{y^2}{16} - \frac{x^2}{16} = 1$$

27. Since the transverse axis is on the x axis, start with

$$\frac{x^2}{a^2} - \frac{y^2}{b^2} = 1$$

and find a and b

$$a = \frac{14}{2} = 7 \text{ and } b = \frac{10}{2} = 5$$

The equation is

$$\frac{x^2}{49} - \frac{y^2}{25} = 1$$

29. Since the transverse axis is on the y axis, start with

$$\frac{y^2}{a^2} - \frac{x^2}{b^2} = 1$$

and find a and b

$$a = \frac{24}{2} = 12 \text{ and } b = \frac{18}{2} = 9$$

The equation is

$$\frac{y^2}{144} - \frac{x^2}{81} = 1$$

31. Since the transverse axis is on the x axis, start with

$$\frac{x^2}{a^2} - \frac{y^2}{b^2} = 1$$

and find a and b

$$a = \frac{18}{2} = 9$$

To find b, sketch the asymptote rectangle, label known parts, and use the Pythagorean Theorem.

$$b^2 = 11^2 - 9^2$$
$$b^2 = 40$$
$$b = \sqrt{40}$$

The equation is

$$\frac{x^2}{81} - \frac{y^2}{40} = 1$$

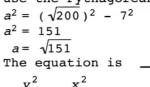

33. Since the conjugate axis is on the x axis, start with

$$\frac{y^2}{a^2} - \frac{x^2}{b^2} = 1$$

and find a and b

$$b = \frac{14}{2} = 7$$

To find a, sketch the asymptote rectangle, label known parts, and use the Pythagorean Theorem.

$$a^2 = (\sqrt{200})^2 - 7^2$$
$$a^2 = 151$$
$$a = \sqrt{151}$$

The equation is

$$\frac{y^2}{151} - \frac{x^2}{49} = 1$$

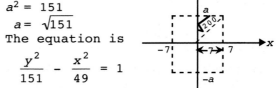

35. The equation $\dfrac{x^2}{25} - \dfrac{y^2}{4} = 1$ is in standard form with $a^2 = 25$, $b^2 = 4$. Since the intercepts are on the x axis, the asymptotes have equations $y = \pm\dfrac{b}{a}x = \pm\dfrac{2}{5}x$.

37. The equation $\dfrac{y^2}{4} - \dfrac{x^2}{16} = 1$ is in standard form with $a^2 = 4$, $b^2 = 16$. Since the intercepts are on the y axis, the asymptotes have equations $y = \pm\dfrac{a}{b}x = \pm\dfrac{2}{4}x$ or $y = \pm\dfrac{1}{2}x$.

39. $9x^2 - y^2 = 9$. In standard form, this becomes
$$\dfrac{x^2}{1} - \dfrac{y^2}{9} = 1$$
Then $a^2 = 1$, $b^2 = 9$. Since the intercepts are on the x axis, the asymptotes have equations $y = \pm\dfrac{b}{a}x = \pm\dfrac{3}{1}x$ or $y = \pm 3x$.

41. $2y^2 - 3x^2 = 1$. In standard form, this becomes
$$\dfrac{y^2}{1/2} - \dfrac{x^2}{1/3} = 1$$
Then $a^2 = \dfrac{1}{2}$, $b^2 = \dfrac{1}{3}$. Since the intercepts are on the y axis, the asymptotes have equations $y = \pm\dfrac{a}{b}x$. Write $a = \sqrt{\dfrac{1}{2}}$, $b = \sqrt{\dfrac{1}{3}}$, $\dfrac{a}{b} = \sqrt{\dfrac{1}{2}} \div \sqrt{\dfrac{1}{3}} = \sqrt{\dfrac{3}{2}}$ or $\dfrac{\sqrt{3}}{\sqrt{2}}$.
Asymptotes: $y = \pm\dfrac{\sqrt{3}}{\sqrt{2}}x$.

43. (A) If a hyperbola has center at $(0, 0)$ and a focus at $(1, 0)$, its equation must be of form
$$\dfrac{x^2}{a^2} - \dfrac{y^2}{b^2} = 1$$
with $c = 1$, so $a^2 + b^2 = 1$ or $b^2 = 1 - a^2$
Therefore there are an infinite number of such hyperbolas.
Each has an equation of form
$$\dfrac{x^2}{a^2} - \dfrac{y^2}{1 - a^2} = 1 \qquad \text{Note that since } 0 < a < c, \text{ we require } 0 < a < 1.$$

(B) If an ellipse has center at $(0, 0)$ and a focus at $(1, 0)$, its equation must be of form
$$\dfrac{x^2}{a^2} + \dfrac{y^2}{b^2} = 1$$
with $c = 1$, so $a^2 - 1 = b^2$
Therefore there are an infinite number of such ellipses. Each has an equation of form
$$\dfrac{x^2}{a^2} + \dfrac{y^2}{a^2 - 1} = 1$$
Note that since $a > c > 0$, we require $a > 1$.

(C) If a parabola has vertex at $(0, 0)$ and focus at $(1, 0)$, its equation must be of form
$$y^2 = 4ax$$
with $a = 1$.

Therefore there is one such parabola; its equation is $y^2 = 4x$.

45. (A) The points of intersection are the solutions of the system

$$x^2 - y^2 = 1$$

$$y = 0.5x = \frac{1}{2}x$$

We solve using substitution:

$$x^2 - \left(\frac{1}{2}x\right)^2 = 1$$

$$x^2 - \frac{1}{4}x^2 = 1$$

$$\frac{3}{4}x^2 = 1$$

$$x^2 = \frac{4}{3}$$

$$x = \pm\frac{2}{\sqrt{3}}$$

$y = \frac{1}{2}x$, so when $x = \frac{2}{\sqrt{3}}$, $y = \frac{1}{2} \cdot \frac{2}{\sqrt{3}} = \frac{1}{\sqrt{3}}$

when $x = \frac{-2}{\sqrt{3}}$, $y = \frac{1}{2} \cdot \frac{-2}{\sqrt{3}} = \frac{-1}{\sqrt{3}}$

The points of intersection are $\left(\frac{2}{\sqrt{3}}, \frac{1}{\sqrt{3}}\right)$ and $\left(\frac{-2}{\sqrt{3}}, \frac{-1}{\sqrt{3}}\right)$.

(B) $x^2 - y^2 = 1$

$$y = 2x$$

$$x^2 - (2x)^2 = 1$$

$$x^2 - 4x^2 = 1$$

$$-3x^2 = 1$$

$$x^2 = -\frac{1}{3}$$

No solution.
The graphs do not intersect.
Repeat the above calculations using $y = mx$ instead of $y = 2x$.

$$x^2 - y^2 = 1$$

$$y = mx$$

$$x^2 - (mx)^2 = 1$$

$$x^2 - m^2x^2 = 1$$

$$x^2(1 - m^2) = 1$$

$$x^2 = \frac{1}{1 - m^2}$$

If the right side is negative the equation will have no solution, so the graphs intersect only if $1 - m^2 > 0$. This occurs only if m^2 is between zero and 1; that is, for $-1 < m < 1$. In this case, $x = \pm\frac{1}{\sqrt{1 - m^2}}$; $y = mx = \pm\frac{m}{\sqrt{1 - m^2}}$.

So if $-1 < m < 1$, the graphs intersect at $\left(\frac{1}{\sqrt{1 - m^2}}, \frac{m}{\sqrt{1 - m^2}}\right)$ and $\left(\frac{-1}{\sqrt{1 - m^2}}, \frac{-m}{\sqrt{1 - m^2}}\right)$.

47. (A) The points of intersection are the solutions of the system

$$y^2 - 4x^2 = 1$$

$$y = x$$

$$x^2 - 4x^2 = 1$$

$$-3x^2 = 1$$

$$x^2 = -\frac{1}{3}$$

No solution. The graphs do not intersect.

(B)
$$y^2 - 4x^2 = 1$$
$$y = 3x$$
$$(3x)^2 - 4x^2 = 1$$
$$9x^2 - 4x^2 = 1$$
$$5x^2 = 1$$
$$x^2 = \frac{1}{5}$$
$$x = \pm\frac{1}{\sqrt{5}}$$

$y = 3x$, so when $x = \dfrac{1}{\sqrt{5}}$, $y = \dfrac{3}{\sqrt{5}}$ and when $x = \dfrac{-1}{\sqrt{5}}$, $y = \dfrac{-3}{\sqrt{5}}$. The points of intersection are $\left(\dfrac{1}{\sqrt{5}}, \dfrac{3}{\sqrt{5}}\right)$ and $\left(\dfrac{-1}{\sqrt{5}}, \dfrac{-3}{\sqrt{5}}\right)$.

Repeat the above calculations with $y = mx$ in place of $y = 3x$.
$$y^2 - 4x^2 = 1$$
$$y = mx$$
$$(mx)^2 - 4x^2 = 1$$
$$m^2x^2 - 4x^2 = 1$$
$$x^2(m^2 - 4) = 1$$
$$x^2 = \frac{1}{m^2 - 4}$$

This equation will have solutions only if $m^2 - 4$ is positive; this occurs when $m > 2$ or $m < -2$. In this case,
$$x = \pm\frac{1}{\sqrt{m^2 - 4}}; \quad y = mx = \pm\frac{m}{\sqrt{m^2 - 4}}.$$

So if $m > 2$ or $m < -2$ the points of intersection are
$$\left(\frac{1}{\sqrt{m^2 - 4}}, \frac{m}{\sqrt{m^2 - 4}}\right) \quad \text{and} \quad \left(\frac{-1}{\sqrt{m^2 - 4}}, \frac{-m}{\sqrt{m^2 - 4}}\right).$$

49. (A) $\dfrac{x^2}{a^2} - \dfrac{y^2}{b^2} = 1$ Solve for y.

$$-\frac{y^2}{b^2} = 1 - \frac{x^2}{a^2}$$

$$\frac{y^2}{b^2} = \frac{x^2}{a^2} - 1$$

$$y^2 = \frac{b^2x^2}{a^2} - b^2$$

$$y^2 = \frac{b^2}{a^2}x^2\left(1 - \frac{a^2}{x^2}\right)$$

$$y = \pm\frac{b}{a}x\sqrt{1 - \frac{a^2}{x^2}}$$

(B) As $|x| \to \infty$, $\dfrac{a}{|x|} \to 0$, $\dfrac{a^2}{x^2} \to 0$, $1 - \dfrac{a^2}{x^2} \to 1$, and $\sqrt{1 - \dfrac{a^2}{x^2}} \to 1$.

So as $|x| \to \infty$, $y \to \pm\dfrac{b}{a}x$. Therefore these lines form asymptotes for the hyperbola.

(C) Since $\sqrt{1 - \dfrac{a^2}{x^2}} < 1$ for $|x| > a$, $\dfrac{b}{a}x\sqrt{1 - \dfrac{a^2}{x^2}} < \dfrac{b}{a}x$, for $x > 0$. So in quadrant I, the hyperbola is below its asymptote $y = \dfrac{b}{a}x$.

However, $-\dfrac{b}{a}x < -\dfrac{b}{a}x\sqrt{1 - \dfrac{a^2}{x^2}}$, for $x > 0$, so, in quadrant IV, the hyperbola is above its asymptote, $y = -\dfrac{b}{a}x$. Similar arguments applied for $x < 0$ show that in quadrant II the hyperbola is below its asymptote $y = -\dfrac{b}{a}x$, and in quadrant III, it is above its asymptote $y = \dfrac{b}{a}x$.

51. By the triangle inequality, no points can satisfy this condition. The set is empty.

53.
$$|d_1 - d_2| = 2a$$
$$d_1 - d_2 = \pm 2a$$
$$d_1 = \pm 2a + d_2 \qquad \text{Use the distance formula}$$
$$\sqrt{(x + c)^2 + y^2} = \pm 2a + \sqrt{(x - c)^2 + y^2} \qquad \text{Square both sides}$$
$$(x + c)^2 + y^2 = 4a^2 \pm 4a\sqrt{(x - c)^2 + y^2} + (x - c)^2 + y^2 \qquad \text{Isolate the square root.}$$
$$\pm\sqrt{(x - c)^2 + y^2} = a - \dfrac{cx}{a} \qquad \text{Square both sides}$$
$$(x - c)^2 + y^2 = a^2 - 2cx + \dfrac{c^2x^2}{a^2} \qquad \text{Simplify and collect terms}$$
$$\left(1 - \dfrac{c^2}{a^2}\right)x^2 + y^2 = a^2 - c^2 \quad \text{Use } a^2 - c^2 = b^2,\ 1 - \dfrac{c^2}{a^2} = \dfrac{b^2}{a^2} \text{ and divide both sides by } b^2.$$
$$\dfrac{x^2}{a^2} - \dfrac{y^2}{b^2} = 1$$

55. From the figure, we see that the point $P(x, y)$ is a point on the curve if and only if
$$d_1 = \dfrac{3}{2}d_2$$
$$d(P, F) = \dfrac{3}{2}d(P, M)$$
$$\sqrt{(x - 3)^2 + (y - 0)^2} = \dfrac{3}{2}\sqrt{\left(x - \dfrac{4}{3}\right)^2 + (y - y)^2}$$
$$(x - 3)^2 + y^2 = \dfrac{9}{4}\left(x - \dfrac{4}{3}\right)^2$$
$$x^2 - 6x + 9 + y^2 = \dfrac{9}{4}\left(x^2 - \dfrac{8}{3}x + \dfrac{16}{9}\right)$$
$$4x^2 - 24x + 36 + 4y^2 = 9x^2 - 24x + 16$$
$$-5x^2 + 4y^2 = -20$$
$$\dfrac{x^2}{4} - \dfrac{y^2}{5} = 1$$

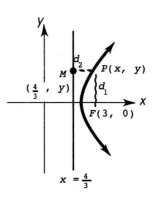

The curve must be a hyperbola since its equation can be written in standard form for a hyperbola.

57. From the figure below, we see that the transverse axis of the hyperbola must be on the y axis and $a = 4$. So the equation of the hyperbola must have the form

$$\frac{y^2}{4^2} - \frac{x^2}{b^2} = 1$$

To find b, we note that the point $(8, 12)$ is on the hyperbola, so its coordinates satisfy the equation. Substituting, we have

$$\frac{12^2}{4^2} - \frac{8^2}{b^2} = 1$$

$$9 - \frac{64}{b^2} = 1$$

$$-\frac{64}{b^2} = -8$$

$$-64 = -8b^2$$

$$b^2 = 8$$

The equation required is

$$\frac{y^2}{16} - \frac{x^2}{8} = 1$$

Using this equation, we can compute y when $x = 6$ to answer the question asked (see figure).

$$\frac{y^2}{16} - \frac{6^2}{8} = 1$$

$$\frac{y^2}{16} - \frac{36}{8} = 1$$

$$y^2 - 72 = 16$$

$$y^2 = 88$$

$$y = 9.38 \text{ to two decimal places}$$

The height above the vertex

$$= y - \text{height of vertex}$$
$$= 9.38 - 4$$
$$= 5.38 \text{ feet}$$

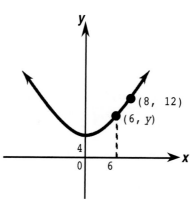

Hyperbola part of dome

59. From the figure below, we can see:

$FF' = 2c = 120 - 20 = 100$

So $c = 50$

$FV = c - a = 120 - 110 = 10$

So $a = c - 10 = 50 - 10 = 40$

Since $c^2 = a^2 + b^2$

$$50^2 = 40^2 + b^2$$

$$b = 30$$

So the equation of the hyperbola, in standard form, is

$$\frac{y^2}{40^2} - \frac{x^2}{30^2} = 1$$

Expressing y in terms of x, we have

$$\frac{y^2}{40^2} = 1 + \frac{x^2}{30^2}$$

$$\frac{y^2}{40^2} = \frac{1}{30^2}(30^2 + x^2)$$

$$y^2 = \frac{40^2}{30^2}(30^2 + x^2)$$

$$y = \frac{4}{3}\sqrt{x^2 + 30^2}$$

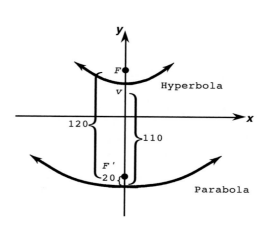

discarding the negative solution, since the reflecting hyperbola is above the x axis.

Section 11-4

1. (A) Since $(h,\ k) = (3,\ 5)$, use
    translation formulas
    $$x' = x - h = x - 3$$
    $$y' = y - k = y - 5$$

    (B) $x'^2 + y'^2 = 81$

    (C) Circle

3. (A) Since $(h,\ k) = (-7,\ 4)$, use
    translation formulas
    $$x' = x - h = x + 7$$
    $$y' = y - k = y - 4$$

    (B) $\dfrac{x'^2}{9} + \dfrac{y'^2}{16} = 1$

    (C) Ellipse

5. (A) Since $(h,\ k) = (4,\ -9)$, use
    translation formulas
    $$x' = x - h = x - 4$$
    $$y' = y - k = y + 9$$

    (B) $y'^2 = 16x'$

    (C) Parabola

7. (A) Since $(h,\ k) = (-8,\ -3)$, use
    translation formulas
    $$x' = x - h = x + 8$$
    $$y' = y - k = y + 3$$

    (B) $\dfrac{x'^2}{12} + \dfrac{y'^2}{8} = 1$

    (C) Ellipse

9. (A) Divide both sides by 144

    $$\frac{(x-3)^2}{9} - \frac{(y+2)^2}{16} = 1$$

    (B) This is the equation of a
    hyperbola.

11. (A) Divide both sides by 30.

    $$\frac{(x+5)^2}{5} + \frac{(y+7)^2}{6} = 1$$

    (B) This is the equation of an
    ellipse.

13. (A) Subtract $24(y - 4)$ from both sides.
    $$(x + 6)^2 = -24(y - 4)$$

    (B) This is the equation of a parabola.

For question 15 and 17, plug the given x and y coordinates into the formulas
$x' = x \cos\ \theta + y \sin\ \theta,\ y' = -x \sin\ \theta + y \cos\ \theta.$

15. $(1, 0)$: $x' = 1 \cos 30° + 0 \sin 30°,\quad y' = -1 \sin 30° + 0 \cos 30°$

    $$x' = \frac{\sqrt{3}}{2} \qquad\qquad\qquad y' = -\frac{1}{2}$$

    $\left(\dfrac{\sqrt{3}}{2}, -\dfrac{1}{2}\right)$

    $(0, 1)$: $x' = 0 \cos 30° + 1 \sin 30°,\quad y' = -0 \sin 30° + 1 \cos 30°$

    $$x' = \frac{1}{2} \qquad\qquad\qquad y' = \frac{\sqrt{3}}{2}$$

    $\left(\dfrac{1}{2}, \dfrac{\sqrt{3}}{2}\right)$

    $(1, -1)$: $x' = 1 \cos 30° + (-1) \sin 30°,\quad y' = -1 \sin 30° + (-1) \cos 30°$

    $$x' = \frac{\sqrt{3}}{2} - \frac{1}{2} = \frac{\sqrt{3} - 1}{2} \qquad y' = -\frac{1}{2} - \frac{\sqrt{3}}{2} = \frac{-1 - \sqrt{3}}{2}$$

    $\left(\dfrac{\sqrt{3} - 1}{2}, \dfrac{-1 - \sqrt{3}}{2}\right)$

$(-3, 4)$: $x' = -3 \cos 30° + 4 \sin 30°$, $\quad y' = -(-3) \sin 30° + 4 \cos 30°$

$$x' = -3 \cdot \frac{\sqrt{3}}{2} + 4 \cdot \frac{1}{2} \qquad\qquad y' = 3 \cdot \frac{1}{2} + 4 \cdot \frac{\sqrt{3}}{2}$$

$$x' = \frac{-3\sqrt{3} + 4}{2} \qquad\qquad y' = \frac{3 + 4\sqrt{3}}{2}$$

$$\left(\frac{-3\sqrt{3} + 4}{2}, \frac{3 - 4\sqrt{3}}{2} \right)$$

17. $(1, 0)$: $x' = 1 \cos 45° + 0 \sin 45°$, $\quad y' = -1 \sin 45° + 0 \cos 45°$

$$x' = \frac{\sqrt{2}}{2} \qquad\qquad y' = -\frac{\sqrt{2}}{2}$$

$$\left(\frac{\sqrt{2}}{2}, -\frac{\sqrt{2}}{2} \right)$$

$(0, 1)$: $x' = 0 \cos 45° + 1 \sin 45°$, $\quad y' = -0 \sin 45° + 1 \cos 45°$

$$x' = \frac{\sqrt{2}}{2} \qquad\qquad y' = \frac{\sqrt{2}}{2}$$

$$\left(\frac{\sqrt{2}}{2}, \frac{\sqrt{2}}{2} \right)$$

$(-1, -2)$: $x' = -1 \cos 45° + (-2) \sin 45°$, $\quad y' = -(-1) \sin 45° + (-2) \cos 45°$

$$x' = -\frac{\sqrt{2}}{2} - \frac{2\sqrt{2}}{2} \qquad\qquad y' = \frac{\sqrt{2}}{2} - \frac{2\sqrt{2}}{2}$$

$$x' = \frac{3\sqrt{2}}{2} \qquad\qquad y' = -\frac{\sqrt{2}}{2}$$

$$\left(\frac{-3\sqrt{2}}{2}, -\frac{\sqrt{2}}{2} \right)$$

$(1, -3)$: $x' = 1 \cos 45° + (-3) \sin 45°$, $\quad y' = -1 \sin 45° + (-3) \cos 45°$

$$x' = \frac{\sqrt{2}}{2} - \frac{3\sqrt{2}}{2} \qquad\qquad y' = -\frac{\sqrt{2}}{2} - \frac{3\sqrt{2}}{2}$$

$$x' = -\frac{2\sqrt{2}}{2} = -\sqrt{2} \qquad\qquad y' = -\frac{4\sqrt{2}}{2} = -2\sqrt{2}$$

$(-\sqrt{2}, -2\sqrt{2})$

19. We need to use the rotation identities to rewrite the equations of the x axis ($y' = 0$) and the y' axis ($x' = 0$).

First, plug in $\theta = 30°$:

$$x' = x \cos 30° + y \sin 30° \qquad\qquad y' = -x \sin 30° + y \cos 30°$$

$$x' = \frac{\sqrt{3}}{2}x + \frac{1}{2}y \qquad\qquad y' = -\frac{1}{2}x + \frac{\sqrt{3}}{2}y$$

The x' axis corresponds to $y' = 0$ so we set $-\frac{1}{2}x + \frac{\sqrt{3}}{2}y$ equal to zero and solve for y:

$$-\frac{1}{2}x + \frac{\sqrt{3}}{2}y = 0$$

$$\frac{\sqrt{3}}{2}y = \frac{1}{2}x$$

$$y = \frac{1}{\sqrt{3}}x$$

To find the equation of the y' axis set $\frac{\sqrt{3}}{2}x + \frac{1}{2}y$ equal to zero:

$$\frac{\sqrt{3}}{2}x + \frac{1}{2}y = 0$$

$$\frac{1}{2}y = -\frac{\sqrt{3}}{2}x$$

$$y = -\sqrt{3}\,x$$

21. We need to use the rotation identities to rewrite the equations of the x axis ($y' = 0$) and the y' axis ($x' = 0$).

First plug in $\theta = 45°$:

$$x' = x \cos 45° + y \sin 45° \qquad\qquad y' = -x \sin 45° + y \cos 45°$$

$$x' = \frac{\sqrt{2}}{2}x + \frac{\sqrt{2}}{2}y \qquad\qquad\qquad y' = -\frac{\sqrt{2}}{2}x + \frac{\sqrt{2}}{2}y$$

The x' axis corresponds to $y' = 0$ so we set $-\frac{\sqrt{2}}{2}x + \frac{\sqrt{2}}{2}y$ equal to zero and solve for y:

$$-\frac{\sqrt{2}}{2}x + \frac{\sqrt{2}}{2}y = 0$$

$$\frac{\sqrt{2}}{2}y = \frac{\sqrt{2}}{2}x$$

$$y = x$$

To find the equation of the y' axis set $\frac{\sqrt{2}}{2}x + \frac{\sqrt{2}}{2}y$ equal to zero:

$$\frac{\sqrt{2}}{2}x + \frac{\sqrt{2}}{2}y = 0$$

$$\frac{\sqrt{2}}{2}y = -\frac{\sqrt{2}}{2}x$$

$$y = -x$$

23.

$$4x^2 + 9y^2 - 16x - 36y + 16 = 0$$

$$4x^2 - 16x + 9y^2 - 36y = -16$$

$$4(x^2 - 4x + ?) + 9(y^2 - 4y + ?) = -16 \qquad \left(\frac{-4}{2}\right)^2 = 4$$

$$4(x^2 - 4x + 4) + 9(y^2 - 4y + 4) = -16 + 16 + 36$$

$$4(x - 2)^2 + 9(y - 2)^2 = 36$$

$$\frac{(x - 2)^2}{9} + \frac{(y - 2)^2}{4} = 1$$

This is the equation of an ellipse with center at $(2, 2)$. The equations of translation are $x' = x - 2$, $y' = y - 2$. Making these substitutions, we obtain

$$\frac{x'^2}{9} + \frac{y'^2}{4} = 1$$

We graph this in the $x'y'$ system, following the process discussed in Section 11-3.

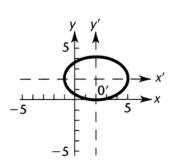

25. $x^2 + 8x + 8y = 0$

$x^2 + 8x = -8y$

$x^2 + 8x + 16 = -8y + 16$

$(x + 4)^2 = -8(y - 2)$

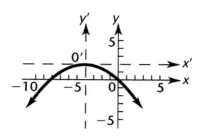

This is the equation of a parabola opening down with vertex at $(h, k) = (-4, 2)$. The equations of translation are $x' = x + 4$, $y' = y - 2$. Making these substitutions, we obtain

$x'^2 = -8y'$

We graph this in the $x'y'$ system, following the process discussed in Section 11-1.

27. $x^2 + y^2 + 12x + 10y + 45 = 0$

$x^2 + 12x + y^2 + 10y = -45$

$x^2 + 12x + 36 + y^2 + 10y + 25 = -45 + 36 + 25$

$(x + 6)^2 + (y + 5)^2 = 16$

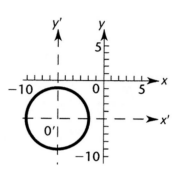

This is the equation of a circle with center at $(-6, -5)$ and radius 4. The equations of translation are $x' = x + 6$, $y' = y + 5$. Making these substitutions, we obtain

$x'^2 + y'^2 = 16$

We graph this in the $x'y'$ system.

29. $-9x^2 + 16y^2 - 72x - 96y - 144 = 0$

$-9x^2 - 72x + 16y^2 - 96y = 144$

$-9(x^2 + 8x + ?) + 16(y^2 - 6y + ?) = 144$

$-9(x^2 + 8x + 16) + 16(y^2 - 6y + 9) = 144 - 144 + 144$

$-9(x + 4)^2 + 16(y - 3)^2 = 144$

$\dfrac{(y - 3)^2}{9} - \dfrac{(x + 4)^2}{16} = 1$

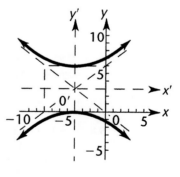

This is the equation of a hyperbola with center at $(-4, 3)$. The equations of translation are $x' = x + 4$, $y' = y - 3$. Making these substitutions, we obtain

$\dfrac{y'^2}{9} - \dfrac{x'^2}{16} = 1$

We graph this in the $x'y'$ system, following the process discussed in Section 11-3.

31. First find the coordinates of the foci in the translated system.

$c'^2 = 3^2 - 2^2 = 5$

$c' = \sqrt{5}$

$-c' = -\sqrt{5}$

The coordinates in the translated system are

$F' = (-\sqrt{5}, 0)$ and $F = (\sqrt{5}, 0)$

Now use

$x = x' + h = x' + 2$

$y = y' + k = y' + 2$

to obtain

$F' = (-\sqrt{5} + 2, 2)$ and $F = (\sqrt{5} + 2, 2)$

as the coordinates of the foci in the original system.

33. First find the coordinates of the focus in the translated system. Since $a = -2$, and the parabola opens down, the coordinates are $(0, -2)$. Now use
$$x = x' + h = x' - 4 = 0 - 4$$
$$y = y' + k = y' + 2 = -2 + 2$$
to obtain $(-4, 0)$ as the coordinates of the focus in the original system.

35. First find the coordinates of the foci in the translated system.
$$c'^2 = 3^2 + 4^2 = 25$$
$$c' = 5$$
$$-c' = -5$$
The coordinates in the translated system are
$$F' = (0, -5) \text{ and } F = (0, 5)$$
Now use
$$x = x' + h = x' - 4$$
$$y = y' + k = y' + 3$$
to obtain
$$F' = (0 - 4, -5 + 3) = F' = (-4, -2) \text{ and } F = (0 - 4, 5 + 3) = F = (-4, 8) \text{ as}$$
the coordinates of the foci in the original system.

37.
$$x^2 - 2x + y^2 + 4y + 5 = 0$$
$$x^2 - 2x + y^2 + 4y = -5$$
$$(x^2 - 2x + ?) + (y^2 + 4y + ?) = -5 \qquad \left(\frac{-2}{1}\right)^2 = 1; \ \left(\frac{4}{2}\right)^2 = 4$$
$$(x^2 - 2x + 1) + (y^2 + 4y + 4) = -5 + 1 + 4$$
$$(x - 1)^2 + (y + 2)^2 = 0$$

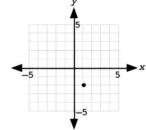

This looks like the equation of a circle with center $(1, -2)$. But the radius would be zero, so this is a degenerate circle which is really just the point $(1, -2)$.

39. $x^2 + 8x - 4y^2 + 8y + 12 = 0$
$$x^2 + 8x - 4y^2 + 8y = -12$$
$$x^2 + 8x - 4(y^2 - 2y) = -12$$
$$(x^2 + 8x + ?) - 4(y^2 - 2y + ?) = -12 \qquad \left(\frac{8}{2}\right)^2 = 16; \ \left(\frac{-2}{2}\right)^2 = 1$$
$$(x^2 + 8x + 16) - 4(y^2 - 2y + 1) = -12 + 16 - 4$$
$$(x + 4)^2 - 4(y - 1)^2 = 0$$

This looks like the equation of a hyperbola but it is degenerate because of the zero. To see what the graph is, we solve the equation for y:
$$4(y - 1)^2 = (x + 4)^2$$
$$2(y - 1) = \pm(x + 4)$$

$2y - 2 = x + 4$ or $2y - 2 = -(x + 4) = -x - 4$
$$2y = x + 6 \qquad\qquad 2y = -x - 2$$
$$y = \frac{1}{2}x + 6 \qquad\qquad y = -\frac{1}{2}x - 1$$

The graph is 2 lines that intersect at $(-4, 1)$ (which would have been the center of the hyperbola.)

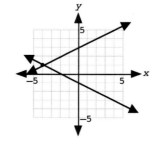

41. If $A \neq 0$, $C = 0$, and $E \neq 0$, write
$$Ax^2 + Dx + Ey + F = 0$$
Complete the square relative to x:

$$A\left(x^2 + \frac{D}{A}x + ?\right) = -Ey - F$$

$$A\left(x^2 + \frac{D}{A}x + \frac{D^2}{4A^2}\right) = -Ey - F + \frac{D^2}{4A}$$

$$\left(x + \frac{D}{2A}\right)^2 = -\frac{E}{A}y - \frac{F}{A} + \frac{D^2}{4A^2}$$

$$\left(x + \frac{D}{2A}\right)^2 = 4 \cdot \left(-\frac{E}{4A}\right)\left[y + \frac{F}{E} - \frac{D^2}{4AE}\right]$$

$$\left[x - \left(-\frac{D}{2A}\right)\right]^2 = 4\left(-\frac{E}{4A}\right)\left[y - \frac{D^2 - 4AF}{4AE}\right]$$

The equations of translation are
$$x' = x - \left(-\frac{D}{2A}\right) \qquad y' = y - \frac{D^2 - 4AF}{4AE}$$

So $h = -\dfrac{D}{2A}$, $k = \dfrac{D^2 - 4AF}{4AE}$

The equation would become $x'^2 = 4\left(-\dfrac{E}{4A}\right)y'$, that is, the equation of a parabola.

43. $x^2 + y^2 = 49$
The rotation equations are

$x = \cos 45°x' - \sin 45°y'$ $\qquad\qquad$ $y = \sin 45°x' + \cos 45°y'$

$x = \dfrac{\sqrt{2}}{2}x' - \dfrac{\sqrt{2}}{2}y'$ $\qquad\qquad$ $y = \dfrac{\sqrt{2}}{2}x' + \dfrac{\sqrt{2}}{2}y'$

Substitute into the original equation and simplify:

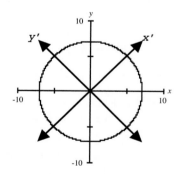

$$\left(\frac{\sqrt{2}}{2}x' - \frac{\sqrt{2}}{2}y'\right)^2 + \left(\frac{\sqrt{2}}{2}x' + \frac{\sqrt{2}}{2}y'\right)^2 = 49$$

$$\frac{1}{2}x'^2 - x'y' + \frac{1}{2}y'^2 + \frac{1}{2}x'^2 + x'y' + \frac{1}{2}y'^2 = 49$$

$$x'^2 + y'^2 = 49$$

We got the original equation back! This makes sense because the graph is a circle with center $(0, 0)$. Rotating the axes will not change the graph.

45. $2x^2 + \sqrt{3}\,xy + y^2 - 10 = 0$
The rotation equations are

$x = \cos 30°x' - \sin 30°y'$ $\qquad\qquad$ $y = \sin 30°x' + \cos 30°y'$

$x = \dfrac{\sqrt{3}}{2}x' - \dfrac{1}{2}y'$ $\qquad\qquad$ $y = \dfrac{1}{2}x' + \dfrac{\sqrt{3}}{2}y'$

Substitute into the original equation and simplify:

$$2\left(\frac{\sqrt{3}}{2}x' - \frac{1}{2}y'\right)^2 + \sqrt{3}\left(\frac{\sqrt{3}}{2}x' - \frac{1}{2}y'\right)^2\left(\frac{1}{2}x' + \frac{\sqrt{3}}{2}y'\right) + \left(\frac{1}{2}x' + \frac{\sqrt{3}}{2}y'\right)^2 - 10 = 0$$

$$2\left(\frac{3}{4}\,x'^2 - \frac{\sqrt{3}}{2}\,x'y' + \frac{1}{4}\,y'^2\right) + \sqrt{3}\left(\frac{\sqrt{3}}{4}\,x'^2 + \frac{1}{2}\,x'y' - \frac{\sqrt{3}}{4}\,y'^2\right) + \frac{1}{4}x'^2 + \frac{\sqrt{3}}{2}\,x'y' + \frac{3}{4}\,y'^2 = 10$$

$$\frac{3}{2}x'^2 - \sqrt{3}\,x'y' + \frac{1}{2}y'^2 + \frac{3}{4}x'^2 + \frac{\sqrt{3}}{2}x'y' - \frac{3}{4}y'^2 + \frac{1}{4}x'^2 + \frac{\sqrt{3}}{2}x'y' + \frac{3}{4}y'^2 = 10$$

$$\frac{5}{2}x'^2 + \frac{1}{2}y'^2 = 10 \quad \text{(Divide both sides by 10)}$$

$$\frac{x'^2}{4} + \frac{y'^2}{20} = 1$$

This is the equation of an ellipse with intercepts
(2, 0), (-2, 0), (0, $\sqrt{20}$), and (0, $-\sqrt{20}$) rotated
through 30°.

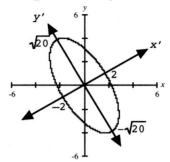

47. $x^2 - 4xy + y^2 = 12$

Use the formula $\cot 2\theta = \dfrac{A - C}{B}$ to find the appropriate angle.

$$\cot 2\theta = \frac{1 - 1}{-4} = 0$$

Since $\cot 90° = \dfrac{\cos 90°}{\sin 90°} = \dfrac{0}{1} = 0$, $2\theta = 90°$, and $\theta = 45°$.

The rotation equations are
$$x = \cos 45°x' - \sin 45°y' \qquad y = \sin 45°x' + \cos 45°y'$$
$$x = \frac{\sqrt{2}}{2}x' - \frac{\sqrt{2}}{2}y' \qquad\qquad y = \frac{\sqrt{2}}{2}x' + \frac{\sqrt{2}}{2}y'$$

Substitute into the original equation and simplify.

$$\left(\frac{\sqrt{2}}{2}\,x' - \frac{\sqrt{2}}{2}\,y'\right)^2 - 4\left(\frac{\sqrt{2}}{2}\,x' - \frac{\sqrt{2}}{2}\,y'\right)\left(\frac{\sqrt{2}}{2}\,x' - \frac{\sqrt{2}}{2}\,y'\right) + \left(\frac{\sqrt{2}}{2}\,x' + \frac{\sqrt{2}}{2}\,y'\right)^2 = 12$$

$$\frac{1}{2}x'^2 - x'y' + \frac{1}{2}y'^2 - 4\left(\frac{1}{2}x'^2 - \frac{1}{2}y'^2\right) + \frac{1}{2}x'^2 + x'y' + \frac{1}{2}y'^2 = 12$$

$$\frac{1}{2}x'^2 - x'y' + \frac{1}{2}y'^2 - 2x'^2 + 2y'^2 + \frac{1}{2}x'^2 + x'y' + \frac{1}{2}y'^2 = 12$$

$$-x'^2 + 3y'^2 = 12 \quad \text{(Divide both sides by 12)}$$

$$\frac{y'^2}{4} - \frac{x'^2}{12} = 1$$

This is the equation of a hyperbola with transverse
axis length 4 and conjugate axis length $2\sqrt{12}$
rotated through 45°.

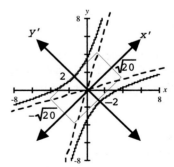

49. $8x^2 - 4xy + 5y^2 = 36$

Use the formula $\cot 2\theta = \dfrac{A - C}{B}$ to find the appropriate angle.

$$\cot 2\theta = \frac{8 - 5}{-4} = -\frac{3}{4}$$

2θ is an angle in the second quadrant and, using the

reference triangle at the right, we see that $\cos 2\theta = -\dfrac{3}{5}$.

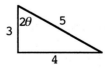

We can find the rotation equations using the half-angle identities

$$\sin \theta = \sqrt{\frac{1 - \cos 2\theta}{2}} \quad \text{and} \quad \cos \theta = \sqrt{\frac{1 + \cos 2\theta}{2}}$$

$$\sin \theta = \sqrt{\frac{1 - \left(-\frac{3}{5}\right)}{2}} \qquad \cos \theta = \sqrt{\frac{1 + \left(-\frac{3}{5}\right)}{2}}$$

$$\sin \theta = \sqrt{\frac{8}{10}} \quad \text{or} \quad \frac{2}{\sqrt{5}} \qquad \cos \theta = \sqrt{\frac{1}{5}} \quad \text{or} \quad \frac{1}{\sqrt{5}}$$

The rotation equations are

$$x = \frac{1}{\sqrt{5}} x' - \frac{2}{\sqrt{5}} y' \qquad y = \frac{2}{\sqrt{5}} x' + \frac{1}{\sqrt{5}} y'$$

Substitute into the original equation and simplify:

$$8\left(\frac{1}{\sqrt{5}} x' - \frac{2}{\sqrt{5}} y'\right)^2 - 4\left(\frac{1}{\sqrt{5}} x' - \frac{2}{\sqrt{5}} y'\right)\left(\frac{2}{\sqrt{5}} x' + \frac{1}{\sqrt{5}} y'\right) + 5\left(\frac{2}{\sqrt{5}} x' + \frac{1}{\sqrt{5}} y'\right)^2 = 36$$

$$8\left(\frac{1}{5} x'^2 - \frac{4}{5} x'y' + \frac{4}{5} y'^2\right) - 4\left(\frac{2}{5} x'^2 - \frac{3}{5} x'y' - \frac{2}{5} y'^2\right) + 5\left(\frac{4}{5} x'^2 + \frac{4}{5} x'y' + \frac{1}{5} y'^2\right) = 36$$

$$\frac{8}{5} x'^2 - \frac{32}{5} x'y' + \frac{32}{5} y'^2 - \frac{8}{5} x'^2 + \frac{12}{5} x'y' + \frac{8}{5} y'^2 + 4x'^2 + 4x'y' + y'^2 = 36$$

$$4x'^2 + 9y'^2 = 36 \quad \text{(Divide both sides by 36)}$$

$$\frac{x'^2}{9} + \frac{y'^2}{4} = 1$$

This is the equation of an ellipse with intercepts (3, 0), (-3, 0), (0, 2), and (0, -2). To find the angle of rotation, note that $\cos 2\theta = -\frac{3}{5}$, and using the inverse cosine function on a calculator, we get $2\theta = 126.87°$ so $\theta = 63.4°$.

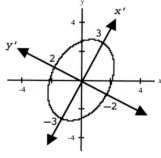

51. $x^2 - 2\sqrt{3}\,xy + 3y^2 - 16\sqrt{3}\,x - 16y = 0$

Use the formula $\cot 2\theta = \dfrac{A - C}{B}$ to find the appropriate angle.

$$\cot 2\theta = \frac{1 - 3}{-2\sqrt{3}} = \frac{-2}{-2\sqrt{3}} = \frac{1}{\sqrt{3}}$$

Note that $\cot 60° = \dfrac{\cos 60°}{\sin 60°} = \dfrac{\frac{1}{2}}{\frac{\sqrt{3}}{2}} = \dfrac{1}{\sqrt{3}}$, so $2\theta = 60°$ and $\theta = 30°$.

The rotation equations are

$$x = \cos 30°x' - \sin 30°y' \qquad y = \sin 30°x' + \cos 30°y'$$

$$x = \frac{\sqrt{3}}{2} x' - \frac{1}{2} y' \qquad y = \frac{1}{2} x' + \frac{\sqrt{3}}{2} y'$$

Substitute into the original equation and simplify:

$$\left(\frac{\sqrt{3}}{2} x' - \frac{1}{2} y'\right)^2 - 2\sqrt{3}\left(\frac{\sqrt{3}}{2} x' - \frac{1}{2} y'\right)\left(\frac{1}{2} x' + \frac{\sqrt{3}}{2} y'\right) + 3\left(\frac{1}{2} x' + \frac{\sqrt{3}}{2} y'\right)^2$$

$$- 16\sqrt{3}\left(\frac{\sqrt{3}}{2} x' - \frac{1}{2} y'\right) - 16\left(\frac{1}{2} x' + \frac{\sqrt{3}}{2} y'\right) = 0$$

$$\frac{3}{4} x'^2 - \frac{\sqrt{3}}{2} x'y' + \frac{1}{4} y'^2 - 2\sqrt{3}\left(\frac{\sqrt{3}}{4} x'^2 + \frac{1}{2} x'y' - \frac{\sqrt{3}}{4} y'^2\right) + 3\left(\frac{1}{4} x'^2 + \frac{\sqrt{3}}{2} x'y' - \frac{3}{4} y'^2\right)$$

$$- 24x' + 8\sqrt{3}\,y' - 8x' - 8\sqrt{3}\,y' = 0$$

$$\frac{3}{4} x'^2 - \frac{\sqrt{3}}{2} x'y' + \frac{1}{4} y'^2 - \frac{3}{2} x'^2 - \sqrt{3}\,x'y' + \frac{3}{2} y'^2 + \frac{3}{4} x'^2 + \frac{3\sqrt{3}}{2} x'y' + \frac{9}{4} y'^2$$

$$- 24x' + 8\sqrt{3}\,y' - 8x' - 8\sqrt{3}\,y' = 0$$

$$4y'^2 - 32x' = 0$$

$$4y'^2 = 32x'$$

$$y'^2 = 8x'$$

This is the equation of a parabola with vertex (0, 0) rotated through 30°.

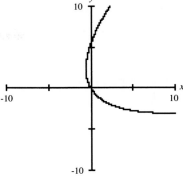

53. This is the equation of a hyperbola with center at (3, -2). The equations of translation are $x' = x - 3$, $y' = y + 2$. Making these substitutions, we obtain
$$x'^2 - y'^2 = 1$$
This equation is in standard form with $a = 1$, $b = 1$. Since the intercepts are on the x' axis, the asymptotes have equations
$$y' = \pm \frac{b}{a} x' = \pm x'$$
In the original coordinate system, these become
$$y + 2 = \pm(x - 3)$$

55. This is the equation of a hyperbola with center at (0, -1). The equations of translation are $x' = x$, $y' = y + 1$. Making these substitutions, we obtain
$$\frac{x'^2}{4} - \frac{y'^2}{25} = 1$$
This equation is in standard form with $a = 2$, $b = 5$. Since the intercepts are on the x' axis, the asymptotes have equations
$$y' = \pm \frac{b}{a} x' = \pm \frac{5}{2} x'$$
In the original coordinate system, these become
$$y + 1 = \pm \frac{5}{2} x$$

57. In standard form, this becomes
$$\frac{(y - 5)^2}{16} - \frac{(x + 2)^2}{9} = 1$$
This is the equation of a hyperbola with center at (-2, 5). The equations of translation are $x' = x + 2$, $y' = y - 5$. Making these substitutions we obtain
$$\frac{y'^2}{16} - \frac{x'^2}{9} = 1$$
This equation in standard form with $a = 4$, $b = 3$. Since the intercepts are on the y' axis, the asymptotes have equations
$$y' = \pm \frac{a}{b} x' = \pm \frac{4}{3} x'$$
In the original coordinate system, these become
$$y - 5 = \pm \frac{4}{3}(x + 2)$$

59. In standard form, this becomes
$$\frac{(y + 4)^2}{1/3} - \frac{x^2}{1} = 1$$
This is the equation of a hyperbola with center (0, -4). The equations of translation are $x' = x$, $y' = y' + 4$. Making these substitutions, we obtain
$$\frac{y'^2}{1/3} - \frac{x'^2}{1} = 1$$

This equation is in standard form with $a = \dfrac{1}{\sqrt{3}}$, $b = 1$. Since the intercepts are on the y' axis, the asymptotes have equations

$$y' = \pm\frac{a}{b}x' = \pm\frac{1}{\sqrt{3}}x'$$

In the original coordinate system, these become

$$y + 4 = \pm\frac{1}{\sqrt{3}}x$$

61. Applying the methods of this section (outline solution)
$A = C = 0$, $B = 1$, so $\cot 2\theta = 0$, $\theta = 45°$.

The rotation formulas are $x = \dfrac{x' - y'}{\sqrt{2}}$, $y = \dfrac{x' + y'}{\sqrt{2}}$

In the rotated coordinate system, the equation becomes

$$\frac{x'^2 - y'^2}{2} - 9 = 0 \quad\text{or}\quad \frac{x'^2}{18} - \frac{y'^2}{18} = 1$$

This equation is in standard form and the hyperbola has asymptotes $y' = \pm x'$ in the rotated system. In the original system, applying the reverse rotation formulas,

$$x' = \frac{x + y}{\sqrt{2}} \qquad y' = -\frac{x + y}{\sqrt{2}}$$

the asymptotes become

$$-\frac{x + y}{\sqrt{2}} = \pm\frac{x + y}{\sqrt{2}}$$

which simplifies to $x = 0$, $y = 0$.

It would be much simpler, however, to note that $xy - 9 = 0$ is the equation of the rational function $y = \dfrac{9}{x}$, which has asymptotes $x = 0$ and $y = 0$, (Section 3-5).

63. Locate the vertex and axis in the original coordinate system, then sketch the parabola and translate the origin to the vertex of the parabola. Next write the equation of the parabola in the translated system:

$$x'^2 = 4ay'$$

The origin in the translated system is at $(h, k) = (2, 5)$ and the translation formulas are

$$x' = x - h = x - 2$$
$$y' = y - k = y - 5$$

The equation of the parabola in the original system is

$$(x - 2)^2 = 4a(y - 5)$$

Since the point $(-2, 1)$ is on the parabola, its coordinates must satisfy the equation of the parabola, so

$$[(-2) - 2]^2 = 4a(1 - 5)$$
$$16 = -16a$$
$$a = -1$$

The equation of the parabola is

$$(x - 2)^2 = -4(y - 5)$$
$$x^2 - 4x + 4 = -4y + 20$$
$$x^2 - 4x + 4y - 16 = 0$$

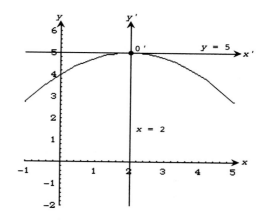

65. Locate the vertices in the original coordinate system, then sketch the ellipse and translate the origin to the center of the ellipse. Next write the equation of the ellipse in the translated system. Since $2a = 8$, and $2b = 4$, we know $a = 4$, $b = 2$, and the equation is

$$\frac{x'^2}{4^2} + \frac{y'^2}{2^2} = 1$$

The origin in the translated system is at $(h, k) = (-2, -3)$ and the translation formulas are

$$x' = x - h = x - (-2) = x + 2$$
$$y' = y - k = y - (-3) = y + 3$$

The equation of the ellipse in the original system is

$$\frac{(x + 2)^2}{16} + \frac{(y + 3)^2}{4} = 1$$
$$(x + 2)^2 + 4(y + 3)^2 = 16$$
$$x^2 + 4x + 4 + 4y^2 + 24y + 36 = 16$$
$$x^2 + 4y^2 + 4x + 24y + 24 = 0$$

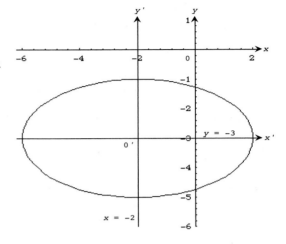

67. Locate the vertices in the original coordinate system, then sketch the ellipse and translate the origin to the center of the ellipse. Since $2a = 3 - (-7) = 10$, $a = 5$. Since $2c = 2 - (-6) = 8$, $c = 4$. Then $b = \sqrt{a^2 - c^2} = \sqrt{25 - 16} = 3$. The endpoints of the minor axis are symmetrically placed with respect to the center $(4, -2)$, that is, at $(1, -2)$ and $(7, -2)$.

Next write the equation of the ellipse in the translated system.

$$\frac{x'^2}{9} + \frac{y'^2}{25} = 1$$

The origin in the translated system is at $(h, k) = (4, -2)$ and the translation formulas are

$$x' = x - h = x - 4$$
$$y' = y - k = y - (-2) = y + 2$$

The equation of the ellipse in the original system is

$$\frac{(x - 4)^2}{9} + \frac{(y + 2)^2}{25} = 1$$
$$25(x - 4)^2 + 9(y + 2)^2 = 225$$
$$25(x^2 - 8x + 16) + 9(y^2 + 4y + 4) = 225$$
$$25x^2 - 200x + 400 + 9y^2 + 36y + 36 = 225$$
$$25x^2 + 9y^2 - 200x + 36y + 211 = 0$$

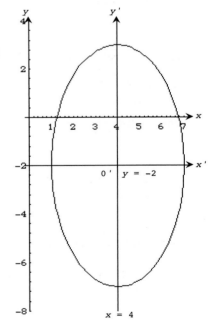

69. Locate the vertices, asymptote rectangle, and asymptotes in the original coordinate system, then sketch the hyperbola and translate the origin to the center of the hyperbola. Next write the equation of the hyperbola in the translated system.

$$\frac{y'^2}{4} - \frac{x'^2}{1} = 1$$

The origin in the translated system is at $(h, k) = (2, 3)$ and the translation formulas are

$$x' = x - h = x - 2$$
$$y' = y - k = y - 3$$

The equation of the hyperbola in the original system is

$$\frac{(y - 3)^2}{4} - \frac{(x - 2)^2}{1} = 1$$
$$(y - 3)^2 - 4(x - 2)^2 = 4$$
$$y^2 - 6y + 9 - 4(x^2 - 4x + 4) = 4$$
$$y^2 - 6y + 9 - 4x^2 + 16x - 16 = 4$$
$$-4x^2 + y^2 + 16x - 6y - 11 = 0$$
$$4x^2 - y^2 - 16x + 6y + 11 = 0$$

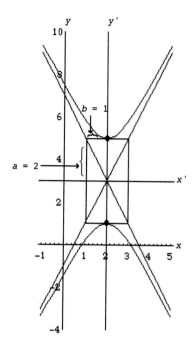

71. The center of the ellipse is at $(-2, 1) = (h, k)$. The major axis is vertical, with length $4 - (-2) = 6 = 2a$, so $a = 3$. The minor axis is horizontal, with length $(-1) - (-3) = 2 = 2b$, so $b = 1$. Substituting in

$$\frac{(x - h)^2}{b^2} + \frac{(y - k)^2}{a^2} = 1$$

we obtain

$$\frac{[x - (-2)]^2}{1^2} + \frac{(y - 1)^2}{3^2} = 1$$

The equation of the ellipse is

$$(x + 2)^2 + \frac{(y - 1)^2}{9} = 1$$
$$9(x + 2)^2 + (y - 1)^2 = 9$$
$$9(x^2 + 4x + 4) + (y^2 - 2y + 1) = 9$$
$$9x^2 + 36x + 36 + y^2 - 2y + 1 = 9$$
$$9x^2 + y^2 + 36x - 2y + 28 = 0$$

73. The center of the hyperbola is at $(1, 2) = (h, k)$. The transverse axis is horizontal, with length $2 = 2a$, so $a = 1$. Substituting in

$$\frac{(x - h)^2}{a^2} - \frac{(y - k)^2}{b^2} = 1$$

we obtain

$$\frac{(x - 1)^2}{1^2} - \frac{(y - 2)^2}{b^2} = 1$$

Since the point $(4, 4)$ is on the graph, its coordinates satisfy the equation.

Substituting, we obtain

$$\frac{(4-1)^2}{1} - \frac{(4-2)^2}{b^2} = 1$$

$$9 - \frac{4}{b^2} = 1$$

$$-\frac{4}{b^2} = -8$$

$$b^2 = \frac{1}{2} \quad \text{(It is not necessary to solve for } b.)$$

The equation of the hyperbola is

$$\frac{(x-1)^2}{1^2} - \frac{(y-2)^2}{1/2} = 1$$

$$(x-1)^2 - 2(y-2)^2 = 1$$

$$x^2 - 2x + 1 - 2(y^2 - 4y + 4) = 1$$

$$x^2 - 2x + 1 - 2y^2 + 8y - 8 = 1$$

$$x^2 - 2y^2 - 2x + 8y - 8 = 0$$

75. $13x^2 + 10xy + 13y^2 - 72 = 0$

$B^2 - 4AC = 100 - 4(13)(13) = 100 - 676 = -576$

Since the discriminant is negative, the equation is an ellipse. To graph on a graphing calculator, we need to solve for y using the quadratic formula:

$$13y^2 + 10xy + 13x^2 - 72 = 0 \quad a = 13,\ b = 10x,\ c = 13x^2 - 72$$

$$y = \frac{-10x \pm \sqrt{100x^2 - 4(13)(13x^2 - 72)}}{2(13)}$$

$$y = \frac{-10x \pm \sqrt{100x^2 - 676x^2 + 3,744}}{26}$$

$$y = \frac{-10x \pm \sqrt{3,744 - 576x^2}}{26}$$

Enter $y_1 = (-10x + \sqrt{3,744 - 576x^2}) \div 26$, $y_2 = (-10x - \sqrt{3,744 - 576x^2}) \div 26$.

 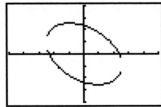 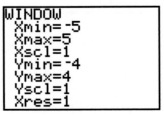

77. $x^2 - 6\sqrt{3}\,xy - 5y^2 - 8 = 0$

$B^2 - 4AC = (-6\sqrt{3})^2 - 4(1)(-5) = 108 + 20 = 128$

Since the discriminant is positive, the equation is a hyperbola. To graph on a graphing calculator we need to solve for y using the quadratic formula:

$$-5y^2 - 6\sqrt{3}\,xy + x^2 - 8 = 0 \quad a = -5,\ b = -6\sqrt{3}\,x,\ c = x^2 - 8$$

$$y = \frac{6\sqrt{3}x \pm \sqrt{\left(-6\sqrt{3}x\right)^2 - 4(-5)(x^2 - 8)}}{2(-5)}$$

$$y = \frac{6\sqrt{3}x \pm \sqrt{108x^2 + 20x^2 - 32}}{-10}$$

$$y = \frac{6\sqrt{3}x \pm \sqrt{128x^2 - 32}}{-10}$$

Enter $y_1 = (6\sqrt{3}x + \sqrt{128x^2 - 32}) \div -10$, $y_2 = (6\sqrt{3}x - \sqrt{128x^2 - 32}) \div -10$.

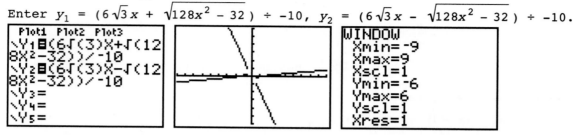

79. $16x^2 - 24xy + 9y^2 + 60x + 80y = 0$
$B^2 - 4AC = (-24)^2 - 4(16)(9) = 576 - 576 = 0$

Since the discriminant is zero, the equation is a parabola. To graph on a graphing calculator we need to solve for y using the quadratic formula:

$$9y^2 + 80y - 24xy + 16x^2 + 60x = 0$$
$$9y^2 + (80 - 24x)y + 16x^2 + 60x = 0 \qquad a = 9, \ b = 80 - 24x, \ c = 16x^2 + 60x$$

$$y = \frac{-(80 - 24x) \pm \sqrt{(80 - 24x)^2 - 4(9)(16x^2 + 60x)}}{2(9)}$$

$$y = \frac{-80 + 24x \pm \sqrt{6,400 - 3,840x + 576x^2 - 576x^2 - 2,160x}}{18}$$

$$y = \frac{-80 - 24x \pm \sqrt{6,400 - 6,000x}}{18}$$

Enter $y_1 = (-80 + 24x + \sqrt{6,400 - 6,000x}) \div 18$,
$\qquad y_2 = (-80 + 24x - \sqrt{6,400 - 6,000x}) \div 18$.

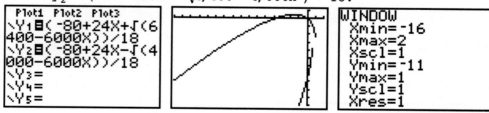

81. $x^2 + 2\sqrt{3}xy + 3y^2 - 8\sqrt{3}x - 8y - 4 = 0$

Use the formula $\cot 2\theta = \dfrac{A - C}{B}$ to find the appropriate angle.

$$\cot 2\theta = \frac{1 - 3}{2\sqrt{3}} = \frac{-2}{2\sqrt{3}} = \frac{-1}{\sqrt{3}}$$

Note that $\cot 120° = \dfrac{\cos 120°}{\sin 120°} = \dfrac{-\frac{1}{2}}{\frac{\sqrt{3}}{2}} = \dfrac{-1}{\sqrt{3}}$, so $2\theta = 120°$ and $\theta = 60°$.

The rotation equations are
$x = \cos 60°x' - \sin 60°y' \qquad y = \sin 60°x' + \cos 60°y'$

$$x = \frac{1}{2}x' - \frac{\sqrt{3}}{2}y' \qquad\qquad y = \frac{\sqrt{3}}{2}x' + \frac{1}{2}y'$$

Substitute into the original equation and simplify:

$$\left(\frac{1}{2}x' - \frac{\sqrt{3}}{2}y'\right)^2 + 2\sqrt{3}\left(\frac{1}{2}x' - \frac{\sqrt{3}}{2}y'\right)\left(\frac{\sqrt{3}}{2}x' + \frac{1}{2}y'\right) + 3\left(\frac{\sqrt{3}}{2}x' + \frac{1}{2}y'\right)^2$$

$$- 8\sqrt{3}\left(\frac{1}{2}x' - \frac{\sqrt{3}}{2}y'\right) - 8\left(\frac{\sqrt{3}}{2}x' + \frac{1}{2}y'\right) - 4 = 0$$

$$\frac{1}{4}x'^2 - \frac{\sqrt{3}}{2}x'y' + \frac{3}{4}y'^2 + 2\sqrt{3}\left(\frac{\sqrt{3}}{4}x'^2 - \frac{1}{2}x'y' - \frac{\sqrt{3}}{4}y'^2\right) + 3\left(\frac{3}{4}x'^2 + \frac{\sqrt{3}}{2}x'y' + \frac{1}{4}y'^2\right)$$

$$- 4\sqrt{3}x' + 12y' - 4\sqrt{3}x' - 4y' - 4 = 0$$

$$\frac{1}{4}x'^2 - \frac{\sqrt{3}}{2}x'y' + \frac{3}{4}y'^2 + \frac{3}{2}x'^2 - \sqrt{3}x'y' - \frac{3}{2}y'^2 + \frac{9}{4}x'^2 + \frac{3\sqrt{3}}{2}x'y' + \frac{3}{4}y'^2$$

$$- 4\sqrt{3}x' + 12y' - 4\sqrt{3}x' - 4y' - 4 = 0$$

$4x'^2 - 8\sqrt{3}x' + 8y' - 4 = 0$ (Divide both sides by 4)
$x'^2 - 2\sqrt{3}x' + 2y' - 1 = 0$

This is the equation of a parabola. Next, we complete the square relative to x'.

$$x'^2 - 2\sqrt{3}x' + 2y' - 1 = 0$$

$(x'^2 - 2\sqrt{3}x' + ?) + 2y' - 1 = 0$ $\left(\frac{-2\sqrt{3}}{2}\right)^2 = 3$

$(x'^2 - 2\sqrt{3}x' + 3) + 2y' - 1 = 3$

$(x' - \sqrt{3})^2 = -2y' + 4$

$(x - \sqrt{3})^2 = -2(y' - 2)$ (*)

This is the equation of a parabola with vertex $(\sqrt{3}, 2)$.

The equations of translation are
   $x'' = x' - \sqrt{3}$ $y'' = y' - 2$

Substituting into (*), we get
   $x''^2 = -2y''$
which is a parabola with vertex $(0, 0)$ opening down. To graph, we first translate the origin to $(\sqrt{3}, 2)$, then rotate through $60°$.

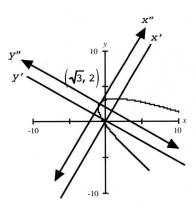

Section 11-5

1. Answers will vary.

3. $x^2 + y^2 = 169$
             $x = -12$
   $(-12)^2 + y^2 = 169$
             $y^2 = 25$
             $y = \pm 5$
   Solution: $(-12, 5)$, $(-12, -5)$

   Check: $-12 \overset{\checkmark}{=} -12$
   $(-12)^2 + (\pm 5)^2 \overset{\checkmark}{=} 169$

5. $8x^2 - y^2 = 16$
          $y = 2x$

   Substitute y from the second equation into the first equation.
   $8x^2 - (2x)^2 = 16$
     $8x^2 - 4x^2 = 16$
           $4x^2 = 16$
            $x^2 = 4$
             $x = \pm 2$
   For $x = 2$ For $x = -2$
      $y = 2(2)$ $y = 2(-2)$
      $y = 4$ $y = -4$

Solutions: $(2, 4)$, $(-2, -4)$

Check:

For $(2, 4)$ For $(-2, -4)$

$$4 \overset{\checkmark}{=} 2 \cdot 2 \qquad\qquad -4 \overset{\checkmark}{=} 2(-2)$$

$$8(2)^2 - 4^2 \overset{\checkmark}{=} 16 \qquad 8(-2)^2 - (-4)^2 \overset{\checkmark}{=} 16$$

7. $3x^2 - 2y^2 = 25$

    $x + y = 0$

Solve for y in the first degree equation

$y = -x$

Substitute into the second degree equation.

$3x^2 - 2(-x)^2 = 25$

         $x^2 = 25$

         $x = \pm 5$

For $x = 5$ For $x = -5$

   $y = -5$ $y = 5$

Solutions: $(5, -5)$, $(-5, 5)$

Check:

For $(5, -5)$ For $(-5, 5)$

$$5 + (-5) \overset{\checkmark}{=} 0 \qquad (-5) + 5 \overset{\checkmark}{=} 0$$

$$3(5)^2 - 2(-5)^2 \overset{\checkmark}{=} 25 \quad 3(-5)^2 - 2(5)^2 \overset{\checkmark}{=} 25$$

From this point on we will not show the checking steps. The student should still perform these checking steps, however.

9. $y^2 = x$

 $x - 2y = 2$

Solve for x in the first degree equation.

$x = 2y + 2$

Substitute into the second degree equation.

$$y^2 = 2y + 2$$

$y^2 - 2y - 2 = 0$

$$y = \frac{-b \pm \sqrt{b^2 - 4ac}}{2a} \qquad a = 1, \; b = -2, \; c = -2$$

$$y = \frac{-(-2) \pm \sqrt{(-2)^2 - 4(1)(-2)}}{2(1)}$$

$$y = \frac{2 \pm \sqrt{12}}{2}$$

$$y = 1 \pm \sqrt{3}$$

For $y = 1 + \sqrt{3}$ For $y = 1 - \sqrt{3}$

  $x = 2(1 + \sqrt{3}) + 2$ $x = 2(1 - \sqrt{3}) + 2$

  $x = 4 + 2\sqrt{3}$ $x = 4 - 2\sqrt{3}$

Solutions: $(4 + 2\sqrt{3}, \; 1 + \sqrt{3})$, $(4 - 2\sqrt{3}, \; 1 - \sqrt{3})$

11. $2x^2 + y^2 = 24$

   $x^2 - y^2 = -12$

Solve using elimination by addition. Adding, we obtain:

$3x^2 = 12$

 $x^2 = 4$

  $x = \pm 2$

For $x = 2$ For $x = -2$

$4 - y^2 = -12$ $4 - y^2 = -12$

   $-y^2 = -16$ $-y^2 = -16$

    $y^2 = 16$ $y = 16$

    $y = \pm 4$ $y = \pm 4$

Solutions: $(2, 4)$, $(2, -4)$, $(-2, 4)$, $(-2, -4)$

13. $x^2 + y^2 = 10$
$16x^2 + y^2 = 25$

Solve using elimination by addition. Multiply the top equation by -1 and add.

$-x^2 - y^2 = -10$
$\underline{16x^2 + y^2 =\ \ \ 25}$
$15x^2\ \ \ \ \ \ \ \ \ =\ \ \ 15$

$x^2\ \ \ \ =\ \ \ 1$
$x\ \ =\ \ \pm 1$

For $x = 1$　　For $x = -1$
$1 + y^2 = 10$　　$1 + y^2 = 10$
$y^2 = 9$　　　$y = \pm 3$
$y = \pm 3$

Solutions: $(1, 3)$, $(1, -3)$,
$(-1, 3)$, $(-1, -3)$

15. $xy - 4 = 0$
$x - y = 2$

Solve for x in the first degree equation.
$x = y + 2$
Substitute into the second degree equation
$(y + 2)y - 4 = 0$
$y^2 + 2y - 4 = 0$

$y = \dfrac{-b \pm \sqrt{b^2 - 4ac}}{2a}$　$a = 1, b = 2,$
　　　　　　　　　　　$c = -4$

$y = \dfrac{-2 \pm \sqrt{(2)^2 - 4(1)(-4)}}{2(1)}$

$y = \dfrac{-2 \pm \sqrt{20}}{2}$

$y = -1 \pm \sqrt{5}$

For $y = -1 + \sqrt{5}$　　For $y = -1 - \sqrt{5}$
$x = -1 + \sqrt{5} + 2$　　$x = -1 - \sqrt{5} + 2$
$x = 1 + \sqrt{5}$　　　$x = 1 - \sqrt{5}$

Solutions: $(1 + \sqrt{5}, -1 + \sqrt{5})$,
$(1 - \sqrt{5}, -1 - \sqrt{5})$

17. $x^2 + 2y^2 = 6$
$xy = 2$

Solve for y in the second equation

$y = \dfrac{2}{x}$

Substitute into the first equation

$$x^2 + 2\left(\dfrac{2}{x}\right)^2 = 6$$

$$x^2 + \dfrac{8}{x^2} = 6 \qquad x \neq 0$$

$$x^2 \cdot x^2 + x^2 \cdot \dfrac{8}{x^2} = 6x^2$$

$$x^4 + 8 = 6x^2$$

$$x^4 - 6x^2 + 8 = 0$$

$$(x^2 - 2)(x^2 - 4) = 0$$

$$(x - \sqrt{2})(x + \sqrt{2})(x - 2)(x + 2) = 0$$

$$x = \sqrt{2}, -\sqrt{2}, 2, -2$$

For $x = \sqrt{2}$　　For $x = -\sqrt{2}$　　For $x = 2$　　For $x = -2$
$y = \dfrac{2}{\sqrt{2}}$　　$y = -\dfrac{2}{\sqrt{2}}$　　$y = \dfrac{2}{2}$　　$y = \dfrac{2}{-2}$
$y = \sqrt{2}$　　$y = -\sqrt{2}$　　$y = 1$　　$y = -1$

Solutions: $(\sqrt{2}, \sqrt{2})$, $(-\sqrt{2}, -\sqrt{2})$, $(2, 1)$, $(-2, -1)$

19. $2x^2 + 3y^2 = -4$
$4x^2 + 2y^2 = 8$

Solve using elimination by addition. Multiply the second equation by $-\dfrac{1}{2}$ and add.

$2x^2 + 3y^2 = -4$
$\underline{-2x^2 - y^2 = -4}$
$2y^2 = -8$

$\underline{y^2 = -4}$
$y = \pm 2i$

For $y = 2i$ For $y = -2i$
$2x^2 + 3(2i)^2 = -4$ $2x^2 + 3(-2i)^2 = -4$
$2x^2 - 12 = -4$ $2x^2 - 12 = -4$
$2x^2 = 8$ $2x^2 = 8$
$x^2 = 4$ $x^2 = 4$
$x = \pm 2$ $x = \pm 2$
Solutions: $(2, 2i)$, $(-2, 2i)$, $(2, -2i)$, $(-2, -2i)$

21. $x^2 - y^2 = 2$
$y^2 = x$
Substitute y^2 from the second equation into the first equation.
$x^2 - x = 2$
$x^2 - x - 2 = 0$
$(x - 2)(x + 1) = 0$
$x = 2, -1$
For $x = 2$ For $x = -1$
$y^2 = 2$ $y^2 = -1$
$y = \pm\sqrt{2}$ $y = \pm i$
Solutions: $(2, \sqrt{2})$, $(2, -\sqrt{2})$, $(-1, i)$, $(-1, -i)$

23. $x^2 + y^2 = 9$
$x^2 = 9 - 2y$
Substitute x^2 from the second equation into the first equation.
$9 - 2y + y^2 = 9$
$y^2 - 2y = 0$
$y(y - 2) = 0$
$y = 0, 2$
For $y = 0$ For $y = 2$
$x^2 = 9 - 2(0)$ $x^2 = 9 - 2(2)$
$x^2 = 9$ $x^2 = 5$
$x = \pm 3$ $x = \pm\sqrt{5}$
Solutions: $(3, 0)$, $(-3, 0)$, $(\sqrt{5}, 2)$, $(-\sqrt{5}, 2)$

25. $x^2 - y^2 = 3$
$xy = 2$

Solve for y in the second equation.
$y = \dfrac{2}{x}$
Substitute into the first equation:
$$x^2 - \left(\dfrac{2}{x}\right)^2 = 3$$
$$x^2 - \dfrac{4}{x^2} = 3 \quad x \neq 0$$
$$x^4 - 4 = 3x^2$$
$x^4 - 3x^2 - 4 = 0$
$(x^2 - 4)(x^2 + 1) = 0$
$x^2 - 4 = 0$ $x^2 + 1 = 0$
$x^2 = 4$ $x^2 = -1$
$x = \pm 2$ $x = \pm i$
For $x = 2$ For $x = -2$ For $x = i$ For $x = -i$
$y = \dfrac{2}{2}$ $y = \dfrac{2}{-2}$ $y = \dfrac{2}{i}$ $y = \dfrac{2}{-i}$
$y = 1$ $y = -1$ $y = -2i$ $y = 2i$
Solutions: $(2, 1)$, $(-2, -1)$, $(i, -2i)$, $(-i, 2i)$

27. $y = 5 - x^2$
$y = 2 - 2x$
Substitute y from the first equation into the second equation.
$5 - x^2 = 2 - 2x$
$0 = x^2 - 2x - 3$
$0 = (x - 3)(x + 1)$
$x = 3, -1$
For $x = 3$ For $x = -1$
$y = 2 - 2(3)$ $y = 2 - 2(-1)$
$y = -4$ $y = 4$
Solutions: $(3, -4)$, $(-1, 4)$

29. $y = x^2 - x$
$y = 2x$
Substitute y from the first equation into the second equation.
$x^2 - x = 2x$
$x^2 - 3x = 0$
$x(x - 3) = 0$
$x = 0, 3$
For $x = 0$ For $x = 3$
$y = 2(0)$ $y = 2(3)$
$y = 0$ $y = 6$
Solutions: $(0, 0)$, $(3, 6)$

31. $y = x^2 - 6x + 9$
$y = 5 - x$

Substitute y from the first equation into the second equation.

$x^2 - 6x + 9 = 5 - x$
$x^2 - 5x + 4 = 0$
$(x - 1)(x - 4) = 0$
$x = 1, 4$

For $x = 1$ For $x = 4$
   $y = 5 - 1$ $y = 5 - 4$
   $y = 4$ $y = 1$

Solutions: $(1, 4)$, $(4, 1)$

33. $y = 8 + 4x - x^2$
$y = x^2 - 2x$

Substitute y from the first equation into the second equation.

$8 + 4x - x^2 = x^2 - 2x$
$0 = 2x^2 - 6x - 8$
$0 = x^2 - 3x - 4$
$0 = (x - 4)(x + 1)$
$x = 4, -1$

For $x = 4$ For $x = -1$
   $y = 4^2 - 2(4)$ $y = (-1)^2 - 2(-1)$
   $y = 8$ $y = 3$

Solutions: $(4, 8)$, $(-1, 3)$

35. (A) The lines are tangent to the circle.

(B) To find values of b such that
    $x^2 + y^2 = 5$
    $2x - y = b$
has exactly one solution, we solve the system for arbitrary b. Solve for y in the second equation.
    $y = 2x - b$

Substitute into the first equation:
     $x^2 + (2x - b)^2 = 5$
   $x^2 + 4x^2 - 4bx + b^2 = 5$
   $5x^2 - 4bx + b^2 - 5 = 0$

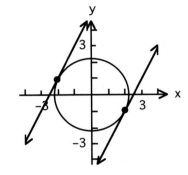

This quadratic equation will have one solution if the discriminant $B^2 - 4AC = (-4b)^2 - 4(5)(b^2 - 5)$ is equal to 0.

This will occur when
$16b^2 - 20b^2 + 100 = 0$
$-4b^2 + 100 = 0$
$b^2 = 25$
$b = \pm 5$

Consider $b = 5$
Then the solution of the system
   $x^2 + y^2 = 5$
   $2x - y = 5$
will be given by solving
   $5x^2 - 4bx + b^2 - 5 = 0$
for $b = 5$.

   $5x^2 - 4 \cdot 5x + 5^2 - 5 = 0$
     $5x^2 - 20x + 20 = 0$
      $5(x - 2)^2 = 0$
        $x - 2 = 0$
            $x = 2$

Since $2x - y = 5$
    $2 \cdot 2 - y = 5$
         $y = -1$

The intersection point is $(2, -1)$ for $b = 5$.

Consider $b = -5$
Then the solution of the system
   $x^2 + y^2 = 5$
   $2x - y = -5$
will be given by solving
   $5x^2 - 4bx + b^2 - 5 = 0$
for $b = -5$.

   $5x^2 - 4(-5)x + (-5)^2 - 5 = 0$
     $5x^2 + 20x + 20 = 0$
      $5(x + 2)^2 = 0$
        $x + 2 = 0$
            $x = -2$

Since $2x - y = -5$
    $2(-2) - y = -5$
          $y = 1$

The intersection point is $(-2, 1)$ for $b = -5$.

(C) The line $x + 2y = 0$ is perpendicular to all the lines in the family and intersects the circle at the intersection points found in part B, since this line passes through the center of the circle and therefore includes a diameter of the circle, which is perpendicular to the tangent line at their mutual point of intersection with the circle. Solving the system $x^2 + y^2 = 5$, $x + 2y = 0$ would determine the intersection points.

37. $2x + 5y + 7xy = 8$
$$xy - 3 = 0$$

Solve for y in the second equation.
$$xy = 3$$
$$y = \frac{3}{x}$$

Substitute into the first equation.
$$2x + 5\left(\frac{3}{x}\right) + 7x\left(\frac{3}{x}\right) = 8$$
$$2x + \frac{15}{x} + 21 = 8 \qquad x \neq 0$$
$$2x^2 + 15 + 21x = 8x$$
$$2x^2 + 13x + 15 = 0$$
$$(2x + 3)(x + 5) = 0$$
$$x = -\frac{3}{2}, \ -5$$

For $x = -\dfrac{3}{2}$ \qquad For $x = -5$

$$y = 3 \div \left(-\frac{3}{2}\right) \qquad y = \frac{3}{-5}$$

$$y = -2 \qquad\qquad y = -\frac{3}{5}$$

Solutions: $\left(-\dfrac{3}{2}, -2\right)$, $\left(-5, -\dfrac{3}{5}\right)$

39. $x^2 - 2xy + y^2 = 1$
$$x - 2y = 2$$

Solve for x in terms of y in the first-degree equation.
$$x = 2y + 2$$
Substitute into the second-degree equation.
$$(2y + 2)^2 - 2(2y + 2)y + y^2 = 1$$
$$4y^2 + 8y + 4 - 4y^2 - 4y + y^2 = 1$$
$$y^2 + 4y + 3 = 0$$
$$(y + 1)(y + 3) = 0$$
$$y = -1, \ -3$$

For $y = -1$ \qquad For $y = -3$
$$x = 2(-1) + 2 \qquad x = 2(-3) + 2$$
$$= 0 \qquad\qquad = -4$$

Solutions: $(0, -1)$, $(-4, -3)$

41. $2x^2 - xy + y^2 = 8$
$$x^2 - y^2 = 0$$

Factor the left side of the equation that has a zero constant term.
$$(x - y)(x + y) = 0$$
$$x = y \text{ or } x = -y$$

> **Common Error:**
> It is incorrect to replace
> $x^2 - y^2 = 0$ or $x^2 = y^2$ by $x = y$.
> This neglects the possibility $x = -y$.

The original system is equivalent to the two systems
$$2x^2 - xy + y^2 = 8 \qquad 2x^2 - xy + y^2 = 8$$
$$x = y \qquad\qquad\qquad x = -y$$

These systems are solved by substitution.

First system: \qquad\qquad Second system:
$$2x^2 - xy + y^2 = 8 \qquad\quad 2x^2 - xy + y^2 = 8$$
$$x = y \qquad\qquad\qquad\qquad x = -y$$
$$2y^2 - yy + y^2 = 8 \qquad 2(-y)^2 - (-y)y + y^2 = 8$$
$$2y^2 = 8 \qquad\qquad 2y^2 + y^2 + y^2 = 8$$
$$y^2 = 4 \qquad\qquad\qquad 4y^2 = 8$$
$$y = \pm 2 \qquad\qquad\qquad y^2 = 2$$
$$\qquad\qquad\qquad\qquad y = \pm\sqrt{2}$$

For $y = 2$ \qquad For $y = -2$ \qquad For $y = \sqrt{2}$ \qquad For $y = -\sqrt{2}$
$$x = 2 \qquad\quad x = -2 \qquad\quad x = -\sqrt{2} \qquad\quad x = \sqrt{2}$$

Solutions: $(2, 2)$, $(-2, -2)$, $(-\sqrt{2}, \sqrt{2})$, $(\sqrt{2}, -\sqrt{2})$

43. $x^2 + xy - 3y^2 = 3$
$x^2 + 4xy + 3y^2 = 0$

Factor the left side of the equation that has a zero constant term.
$(x + y)(x + 3y) = 0$
$$x = -y \text{ or } x = -3y$$
The original system is equivalent to the two systems

$x^2 + xy - 3y^2 = 3$ $\qquad\qquad$ $x^2 + xy - 3y^2 = 3$
$\qquad x = -y$ $\qquad\qquad\qquad\qquad$ $x = -3y$

These systems are solved by substitution.

First system: $\qquad\qquad\qquad$ Second system:

$$x^2 + xy - 3y^2 = 3 \qquad\qquad x^2 + xy - 3y^2 = 3$$
$$x = -y \qquad\qquad\qquad\qquad x = -3y$$
$$(-y)^2 + (-y)y - 3y^2 = 3 \qquad (-3y)^2 + (-3y)y - 3y^2 = 3$$
$$y^2 - y^2 - 3y^2 = 3 \qquad\qquad 9y^2 - 3y^2 - 3y^2 = 3$$
$$-3y^2 = 3 \qquad\qquad\qquad\qquad 3y^2 = 3$$
$$y^2 = -1 \qquad\qquad\qquad\qquad y^2 = 1$$
$$y = \pm i \qquad\qquad\qquad\qquad y = \pm 1$$

For $y = i$ $\quad$ For $y = -i$ $\qquad$ For $y = 1$ $\qquad$ For $y = -1$
$\quad x = -i$ $\qquad\quad x = i$ $\qquad\qquad\quad x = -3$ $\qquad\qquad\quad x = 3$

Solutions: $(-i, i)$, $(i, -i)$, $(-3, 1)$, $(3, -1)$

45. Before we can enter these equations in our graphing calculator, we must solve
for y:

$\quad -x^2 + 2xy + y^2 = 1$ $\qquad\qquad\qquad$ $3x^2 - 4xy + y^2 = 2$
$y^2 + 2xy - 1 - x^2 = 0$ $\qquad\qquad\qquad$ $y^2 - 4xy + 3x^2 - 2 = 0$

Applying the quadratic formula to each equation, we have

$$y = \frac{-2x \pm \sqrt{4x^2 - 4(-1 - x^2)}}{2} \qquad\qquad y = \frac{4x \pm \sqrt{16x^2 - 4(3x^2 - 2)}}{2}$$

$$y = \frac{-2x \pm \sqrt{8x^2 + 4}}{2} \qquad\qquad\qquad y = \frac{4x \pm \sqrt{4x^2 + 8}}{2}$$

$$y = -x \pm \sqrt{2x^2 + 1} \qquad\qquad\qquad y = 2x \pm \sqrt{x^2 + 2}$$

Entering each of these four equations into
a graphing calculator produces the graph
shown at the right.

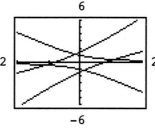

Using the INTERSECT command (details
omitted), yields $(-1.41, -0.82)$, $(-0.13,$
$1.15)$, $(0.13, -1.15)$, and $(1.41, 0.82)$ to
two decimal places.

47. Before we can enter these equations in our graphing calculator, we must solve
for y:

$\quad 3x^2 - 4xy - y^2 = 2$ $\qquad\qquad\qquad$ $2x^2 + 2xy + y^2 = 9$
$y^2 + 4xy + 2 - 3x^2 = 0$ $\qquad\qquad\qquad$ $y^2 + 2xy + 2x^2 - 9 = 0$

Applying the quadratic formula to each equation, we have

$$y = \frac{-4x \pm \sqrt{16x^2 - 4(2 - 3x^2)}}{2} \qquad\qquad y = \frac{-2x \pm \sqrt{4x^2 - 4(2x^2 - 9)}}{2}$$

$$y = \frac{-4x \pm \sqrt{28x^2 - 8}}{2} \qquad\qquad\qquad y = \frac{-2x \pm \sqrt{36 - 4x^2}}{2}$$

$$y = -2x \pm \sqrt{7x^2 - 2} \qquad\qquad\qquad y = -x \pm \sqrt{9 - x^2}$$

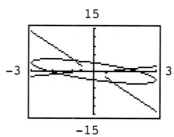

Entering each of these four equations into a graphing calculator produces the graph shown at the right.

Using the INTERSECT command (details omitted), yields $(-1.66, -0.84)$, $(-0.91, 3.77)$, $(0.91, -3.77)$, and $(1.66, 0.84)$ to two decimal places.

49. Before we can enter these equations in our graphing calculator, we must solve for y:

$$2x^2 - 2xy + y^2 = 9 \qquad\qquad 4x^2 - 4xy + y^2 + x = 3$$
$$y^2 - 2xy + 2x^2 - 9 = 0 \qquad\qquad y^2 - 4xy + 4x^2 + x - 3 = 0$$

Applying the quadratic formula to each equation, we have

$$y = \frac{2x \pm \sqrt{4x^2 - 4(2x^2 - 9)}}{2} \qquad\qquad y = \frac{4x \pm \sqrt{16x^2 - 4(4x^2 + x - 3)}}{2}$$

$$y = \frac{2x \pm \sqrt{36 - 4x^2}}{2} \qquad\qquad y = \frac{4x \pm \sqrt{12 - 4x}}{2}$$

$$y = x \pm \sqrt{9 - x^2} \qquad\qquad y = 2x \pm \sqrt{3 - x}$$

Entering each of these four equations into a graphing calculator produces the graph shown at the right.

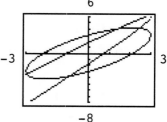

Using the INTERSECT command (details omitted), yields $(-2.96, -3.47)$, $(-0.89, -3.76)$, $(1.39, 4.05)$, and $(2.46, 4.18)$ to two decimal places.

51. Let x and y equal the two numbers. We have the system
$$x + y = 3$$
$$xy = 1$$
Solve the first equation for y in terms of x, then substitute into the second degree equation.
$$y = 3 - x$$
$$x(3 - x) = 1$$
$$3x - x^2 = 1$$
$$-x^2 + 3x - 1 = 0$$
$$x^2 - 3x + 1 = 0$$

$$x = \frac{-b \pm \sqrt{b^2 - 4ac}}{2a} \qquad a = 1,\ b = -3,\ c = 1$$

$$x = \frac{-(-3) \pm \sqrt{(-3)^2 - 4(1)(1)}}{2(1)}$$

$$x = \frac{3 \pm \sqrt{5}}{2}$$

For $x = \dfrac{3 + \sqrt{5}}{2}$ $\qquad\qquad$ For $x = \dfrac{3 - \sqrt{5}}{2}$

$\quad y = 3 - x$ $\qquad\qquad\qquad\quad y = 3 - x$

$\quad = 3 - \dfrac{3 + \sqrt{5}}{2}$ $\qquad\qquad\quad = 3 - \dfrac{3 - \sqrt{5}}{2}$

$\quad = \dfrac{6 - 3 - \sqrt{5}}{2}$ $\qquad\qquad\quad = \dfrac{6 - 3 + \sqrt{5}}{2}$

$\quad = \dfrac{3 - \sqrt{5}}{2}$ $\qquad\qquad\qquad = \dfrac{3 + \sqrt{5}}{2}$

The two numbers are $\dfrac{1}{2}(3 - \sqrt{5})$ and $\dfrac{1}{2}(3 + \sqrt{5})$.

53. Sketch a figure.
Let x and y represent the lengths of the two legs.

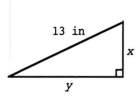

From the Pythagorean Theorem we have
$$x^2 + y^2 = 13^2$$
From the formula for the area of a triangle we have
$$\frac{1}{2}xy = 30$$
So the system of equations is
$$x^2 + y^2 = 169$$
$$\frac{1}{2}xy = 30$$

Solve the second equation for y in terms of x, then substitute into the first equation.

$$xy = 60$$
$$y = \frac{60}{x}$$
$$x^2 + \left(\frac{60}{x}\right)^2 = 169$$
$$x^2 + \frac{3600}{x^2} = 169 \qquad x \neq 0$$
$$x^4 + 3600 = 169x^2$$
$$x^4 - 169x^2 + 3600 = 0$$
$$(x^2 - 144)(x^2 - 25) = 0$$
$$(x - 12)(x + 12)(x - 5)(x + 5) = 0$$
$$x = \pm 12, \ \pm 5$$

Discarding the negative solutions, we have
$x = 12$ or $x = 5$

For $x = 12$ For $x = 5$
$$y = \frac{60}{x} \qquad\qquad y = \frac{60}{x}$$
$$y = 5 \qquad\qquad\quad y = 12$$
The lengths of the legs are 5 inches and 12 inches.

55. Let x = width of screen.
    y = height of screen.
From the Pythagorean Theorem, we have
$$x^2 + y^2 = (7.5)^2$$
From the formula for the area of a rectangle we have
$$xy = 27$$
So the system of equations is:
$$x^2 + y^2 = 56.25$$
$$xy = 27$$

Solve the second equation for y in terms of x, then substitute into the first equation.
$$y = \frac{27}{x}$$
$$x^2 + \left(\frac{27}{x}\right)^2 = 56.25 \qquad x \neq 0$$
$$x^2 + \frac{729}{x^2} = 56.25 \qquad x \neq 0$$
$$x^4 + 729 = 56.25x^2$$
$$x^4 - 56.25x^2 + 729 = 0 \qquad \text{quadratic in } x^2$$

$$x^2 = \frac{-b \pm \sqrt{b^2 - 4ac}}{2a} \qquad a = 1, \; b = -56.25, \; c = 729$$

$$x^2 = \frac{-(-56.25) \pm \sqrt{(-56.25)^2 - 4(1)(729)}}{2(1)}$$

$$x^2 = \frac{56.25 \pm 15.75}{2}$$

$$x = 6, \; 4.5 \quad \text{(discarding the negative solutions)}$$

For $x = 6$

$$y = \frac{27}{6} = 4.5$$

For $x = 4.5$

$$y = \frac{27}{4.5} = 6$$

The dimensions of the screen must be 6 inches by 4.5 inches.

57.

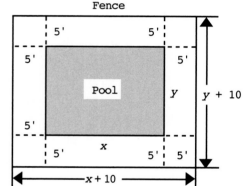

Redrawing and labeling the figure as shown, we have

Area of pool = 572

$$xy = 572$$

Area enclosed by fence = 1,152

$$(x + 10)(y + 10) = 1,152$$

We solve this system by solving for y in terms of x in the first equation, then substituting into the second equation.

$$y = \frac{572}{x}$$

$$(x + 10)\left(\frac{572}{x} + 10\right) = 1,152$$

$$572 + 10x + \frac{5,720}{x} + 100 = 1,152$$

$$10x + \frac{5,720}{x} - 480 = 0 \qquad x \neq 0$$

$$10x^2 + 5,720 - 480x = 0$$

$$x^2 - 48x + 572 = 0$$

$$(x - 26)(x - 22) = 0$$

$$x = 26, \; 22$$

For $x = 26$

$$y = \frac{572}{26}$$

$$y = 22$$

For $x = 22$

$$y = \frac{572}{22}$$

$$y = 26$$

The dimensions of the pool are 22 feet by 26 feet.

59.   Let x = average speed of Boat B
Then $x + 5$ = average speed of Boat A

Let y = time of Boat B, then $y - \dfrac{1}{2}$ = time of Boat A

Using Distance = rate × time, we have
$$75 = xy$$
$$75 = (x + 5)\left(y - \frac{1}{2}\right)$$

Note: The *faster* boat, A, has the *shorter* time. It is a common error to confuse the signs here. Another common error: if rates are expressed in miles per hour, then $y - 30$ is not the correct time for boat A. Times must be expressed in hours.

Solve the first equation for y in terms of x, then substitute into the second equation.

$$y = \frac{75}{x}$$
$$75 = (x + 5)\left(\frac{75}{x} - \frac{1}{2}\right)$$
$$75 = 75 - \frac{1}{2}x + \frac{375}{x} - \frac{5}{2}$$
$$0 = -\frac{1}{2}x + \frac{375}{x} - \frac{5}{2} \qquad x \neq 0$$
$$2x(0) = 2x\left(-\frac{1}{2}x\right) + 2x\left(\frac{375}{x}\right) - 2x\left(\frac{5}{2}\right)$$
$$0 = -x^2 + 750 - 5x$$
$$x^2 + 5x - 750 = 0$$
$$(x - 25)(x + 30) = 0$$
$$x = 25, -30$$

Discarding the negative solution, we have
$$x = 25 \text{ mph} = \text{average speed of Boat } B$$
$$x + 5 = 30 \text{ mph} = \text{average speed of Boat } A$$

CHAPTER 11 REVIEW

1. First write the equation in standard form by dividing both sides by 225.
$$9x^2 + 25y^2 = 225$$
$$\frac{x^2}{25} + \frac{y^2}{9} = 1$$
In this form the equation is identifiable as that of an ellipse. Locate the intercepts.

Common Error:
The relationship $c^2 = a^2 + b^2$ applies to a, b, c as defined for hyperbolas but not for ellipses.

When $y = 0$, $\dfrac{x^2}{25} = 1$. x intercepts: ±5

When $x = 0$, $\dfrac{x^2}{9} = 1$. y intercepts: ±3

So $a = 5$, $b = 3$, and the major axis is on the x axis.

Foci: $c^2 = a^2 - b^2$
$c^2 = 25 - 9$
$c^2 = 16$
$c = 4$
Foci: $F'(-4, 0)$, $F(4, 0)$
Major axis length = $2(5) = 10$
Minor axis length = $2(3) = 6$

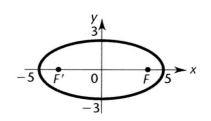

(*11-2*)

2. $x^2 = -12y$ is the equation of a parabola. To graph, assign y values that make the right side a perfect square (y must be zero or negative for x to be real) and solve for x. Since the coefficient of y is negative, a must be negative, and the parabola opens down.

x	0	±6	±2
y	0	−3	$-\frac{1}{3}$

To find the focus and directrix, solve
$4a = -12$
$\quad a = -3$
Focus: $(0, -3)$ Directrix: $y = -(-3) = 3$

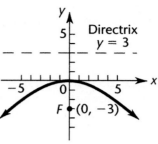

(*11–1*)

3. First, write the equation in standard form by dividing both sides by 225.
$25y^2 - 9x^2 = 225$

$$\frac{y^2}{9} - \frac{x^2}{25} = 1$$

In this form the equation is identifiable as that of a hyperbola.

When $y = 0$, $-\dfrac{x^2}{25} = 1$. There are no

x intercepts, but $b = 5$.

When $x = 0$, $\dfrac{y^2}{9} = 1$. y intercepts: ±3

Sketch the asymptotes using the asymptote rectangle, then sketch in the hyperbola.
Foci: $c^2 = 3^2 + 5^2$
$\quad c^2 = 34$
$\quad\quad c = \sqrt{34}$
Foci: $F'(0, -\sqrt{34})$, $F(0, \sqrt{34})$
Transverse axis length = $2(3) = 6$
Conjugate axis length = $2(5) = 10$

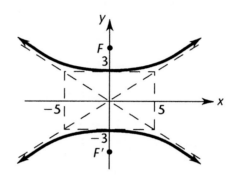

(*11–3*)

4. First, write the equation in standard form by dividing both sides by 16.
$x^2 - y^2 = 16$

$$\frac{x^2}{16} - \frac{y^2}{16} = 1$$

In this form the equation is identifiable as a hyperbola.

When $x = 0$, $-\dfrac{y^2}{16} = 1$. There are no y intercepts, but $b = 4$.

When $y = 0$, $\dfrac{x^2}{16} = 1$. x intercepts: ±4.

Sketch the asymptotes using the asymptote rectangle, then sketch in the hyperbola.
Foci: $c^2 = 4^2 + 4^2$
$\quad c^2 = 32$
$\quad\quad c = 4\sqrt{2}$
Foci: $F' = (-4\sqrt{2}, 0)$, $F = (4\sqrt{2}, 0)$
Transverse axis length = $2(4) = 8$
Conjugate axis length = $2(4) = 8$

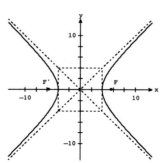

(*11–3*)

5. $y^2 = 8x$ is the equation of a parabola. To graph, assign x values that make the
    right side a perfect square (x must be zero or
    positive for y to be real) and solve for y.
    Since the coefficient of x is positive, a must be
    positive, and the parabola opens right.

x	0	2	8
y	0	± 4	± 8

    To find the focus and directrix, solve $4a = 8$
    $$a = 2.$$
    Focus: $(2, 0)$ Directrix: $x = -2$

    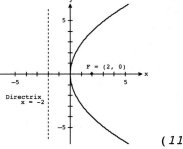

    $(11\text{-}1)$

6. First, write the equation in standard form by dividing both sides by 8.
    $$2x^2 + y^2 = 8$$
    $$\frac{x^2}{4} + \frac{y^2}{8} = 1$$
    In this form the equation is identifiable as that of an ellipse.
    Locate the intercepts.

    When $y = 0$, $\dfrac{x^2}{4} = 1$ x intercepts: ± 2

    When $x = 0$, $\dfrac{y^2}{8} = 1$ y intercepts: $\pm 2\sqrt{2}$

    So $a = 2\sqrt{2}$, $b = 2$, and the major axis
    is on the y axis.
    Foci: $c^2 = a^2 - b^2$
    $$c^2 = 8 - 4$$
    $$c^2 = 4$$
    $$c = 2$$
    Foci: $F' = (0, -2)$, $F = (0, 2)$
    Major axis length $= 2(2\sqrt{2}) = 4\sqrt{2}$
    Minor axis length $= 2(2) = 4$

    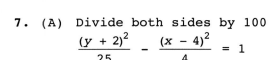

    $(11\text{-}2)$

7. (A) Divide both sides by 100
    $$\frac{(y + 2)^2}{25} - \frac{(x - 4)^2}{4} = 1$$

    (B) This is the equation of a
    hyperbola. $(11\text{-}4)$

8. (A) Subtract $12(y + 4)$ from both
    sides.
    $$(x + 5)^2 = -12(y + 4)$$

    (B) This is the equation of a
    parabola. $(11\text{-}4)$

9. (A) Divide both sides by 144
    $$\frac{(x - 6)^2}{9} + \frac{(y - 4)^2}{16} = 1$$

    (B) This is the equation of an
    ellipse. $(11\text{-}4)$

10. Use the rotation identities $x' = x \cos \theta + y \sin \theta$, $y' = -x \sin \theta + y \cos \theta$.
    (A) $x' = 3 \cos 30° + 4 \sin 30°$, $y' = -3 \sin 30° + 4 \cos 30°$

    $$x' = 3 \cdot \frac{\sqrt{3}}{2} + 4 \cdot \frac{1}{2} = \frac{3\sqrt{3} + 4}{2} \qquad y' = -3 \cdot \frac{1}{2} + 4 \cdot \frac{\sqrt{3}}{2} = \frac{-3 + 4\sqrt{3}}{2}$$

    $$\left(\frac{3\sqrt{3} + 4}{2}, \frac{-3 + 4\sqrt{3}}{2} \right)$$

(B) $x' = 3 \cos 45° + 4 \sin 45°,$ $y' = -3 \sin 45° + 4 \cos 45°$

$x' = 3 \cdot \dfrac{\sqrt{2}}{2} + 4 \cdot \dfrac{\sqrt{2}}{2} = \dfrac{7\sqrt{2}}{2}$ $y' = -3 \cdot \dfrac{\sqrt{2}}{2} + 4 \cdot \dfrac{\sqrt{2}}{2} = \dfrac{\sqrt{2}}{2}$

$\left(\dfrac{7\sqrt{2}}{2}, \dfrac{\sqrt{2}}{2}\right)$

(C) $x' = 3 \cos 60° + 4 \sin 60°,$ $y' = -3 \sin 60° + 4 \cos 60°$

$x' = 3 \cdot \dfrac{1}{2} + 4 \cdot \dfrac{\sqrt{3}}{2} = \dfrac{3 + 4\sqrt{3}}{2}$ $y' = -3 \cdot \dfrac{\sqrt{3}}{2} + 4 \cdot \dfrac{1}{2} = \dfrac{-3\sqrt{3} + 4}{2}$

$\left(\dfrac{3 + 4\sqrt{3}}{2}, \dfrac{-3\sqrt{3} + 4}{2}\right)$ *(11-4)*

11. We need to use the rotation identities to rewrite the equations of the x axis
   ($y' = 0$) and the y' axis ($x' = 0$).
      First, plug in $\theta = 75°$:
      $x' = x \cos 75° + y \sin 75°$ $y' = -x \sin 75° + y \cos 75°$
      $x' = 0.259x + 0.966y$ $y' = -0.966x + 0.259y$

   The x' axis corresponds to $y' = 0$ so we set $-0.966x + 0.259y$ equal to zero and
   solve for y:
      $-0.966x + 0.259y = 0$
      $0.259y = 0.966x$
      $y = 3.73x$

   To find the equation of the y' axis set $0.259x + 0.966y$ equal to zero:
      $0.259x + 0.966y = 0$
      $0.966y = -0.259x$
      $y = -.268x$ *(11-4)*

12. Substitute the second equation in for y in the first equation.
      $y = x^2 - 5x - 3$
      $y = -x + 2$
      $-x + 2 = x^2 - 5x - 3$
      $0 = x^2 - 4x - 5$
      $0 = (x - 5)(x + 1)$
      $x - 5 = 0$ or $x + 1 = 0$
      $x = 5$ $x = -1$
   When $x = 5$, $y = -5 + 2 = -3$. When $x = -1$, $y = -(1) + 2 = 3$. The solutions are
   (5, -3) and (-1, 3).   *(11-5)*

13. Solve the second equation for y:
      $2x - y = 3$
      $2x = 3 + y$
      $y = 2x - 3$

   Now substitute into the first equation:
      $x^2 + y^2 = 2$
      $x^2 + (2x - 3)^2 = 2$
      $x^2 + 4x^2 - 12x + 9 = 2$
      $5x^2 - 12x + 7 = 0$
      $(5x - 7)(x - 1) = 0$
      $5x - 7 = 0$ or $x - 1 = 0$
      $5x = 7$ $x = 1$
      $x = \dfrac{7}{5}$

   When $x = \dfrac{7}{5}$, $y = 2\left(\dfrac{7}{5}\right) - 3 = \dfrac{14}{5} - 3 = \dfrac{14}{5} - \dfrac{15}{5} = -\dfrac{1}{5}$.

   When $x = 1$, $y = 2(1) - 3 = -1$. The solutions are $\left(\dfrac{7}{5}, -\dfrac{1}{5}\right)$ and (1, -1).   *(11-5)*

14. Multiply the first equation by 3 and add to the second to eliminate y:

$$3x^2 - y^2 = -6$$
$$2x^2 + 3y^2 = 29$$

$$
\begin{array}{rcl}
9x^2 - 3y^2 &=& -18 \\
\underline{2x^2 + 3y^2} &=& \underline{29} \\
11x^2 &=& 11 \\
x^2 &=& 1 \\
x &=& \pm 1
\end{array}
$$

When $x = 1$, $3(1)^2 - y^2 = -6$ When $x = -1$, $3(-1)^2 - y^2 = -6$

$$
\begin{array}{rcl}
-y^2 &=& -9 \\
y^2 &=& 9 \\
y &=& 3, -3
\end{array}
\qquad
\begin{array}{rcl}
-y^2 &=& -9 \\
y^2 &=& 9 \\
y &=& 3, -3
\end{array}
$$

The solutions are $(1, 3)$, $(1, -3)$, $(-1, 3)$, and $(-1, -3)$.   *(11-5)*

15. The parabola is opening either left or right and has an equation of the form $y^2 = 4ax$. Since $(-4, -2)$ is on the graph, we have:

$$(-2)^2 = 4a(-4)$$
$$4 = -16a$$
$$-\frac{1}{4} = a$$

The equation of the parabola is

$$y^2 = 4\left(-\frac{1}{4}\right)x$$
$$y^2 = -x$$

          (11-5)

16. Make a rough sketch of the ellipse and compute x and y intercepts:

$$\frac{x^2}{a^2} + \frac{y^2}{b^2} = 1$$
$$a = \frac{12}{2} = 6 \qquad b = \frac{10}{2} = 5$$
$$\frac{x^2}{36} + \frac{y^2}{25} = 1$$

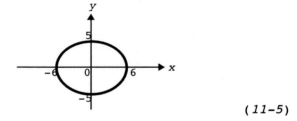

          (11-5)

17. Make a rough sketch of the ellipse, locate focus and x intercepts, then determine y intercepts using the Pythagorean theorem.

$$\frac{x^2}{b^2} + \frac{y^2}{a^2} = 1$$
$$b = \frac{12}{2} = 6$$
$$a^2 = 6^2 + 8^2 = 100$$
$$a = 10$$
$$\frac{x^2}{36} + \frac{y^2}{100} = 1$$

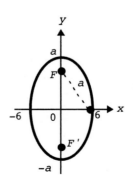

          (11-5)

18. Since the transverse axis is on the y axis, start with

$$\frac{y^2}{a^2} - \frac{x^2}{b^2} = 1$$

and find a and b.

$$b = \frac{6}{2} = 3, \quad c = \frac{8}{2} = 4$$

To find a, sketch the asymptote rectangle, label known parts, and use the Pythagorean Theorem.

$$a^2 = c^2 - b^2$$
$$a^2 = 4^2 - 3^2$$
$$a^2 = 7$$
$$a = \sqrt{7}$$

The equation is

$$\frac{y^2}{7} - \frac{x^2}{9} = 1$$

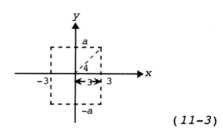

(*11-3*)

19. Since the transverse axis is on the x axis, start with

$$\frac{x^2}{a^2} - \frac{y^2}{b^2} = 1$$

and find a and b.

$$a = \frac{14}{2} = 7 \quad b = \frac{16}{2} = 8$$

The equation is

$$\frac{x^2}{49} - \frac{y^2}{64} = 1$$

(*11-3*)

20. $x^2 + 4y^2 = 32$
    $x + 2y = 0$

We solve for x in terms of y in the second equation, then substitute the expression for x into the first equation.

$$x = -2y$$
$$(-2y)^2 + 4y^2 = 32$$
$$4y^2 + 4y^2 = 32$$
$$8y^2 = 32$$
$$y^2 = 4$$
$$y = \pm 2$$

For $y = 2$ For $y = -2$
$x = -2(2)$ $x = -2(-2)$
   $= -4$ $= 4$

Solutions: $(-4, 2)$, $(4, -2)$

(*11-5*)

21. $16x^2 + 25y^2 = 400$
    $16x^2 - 45y = 0$

We eliminate x by multiplying the second equation by -1, then adding the two equations.

$$16x^2 + 25y^2 = 400$$
$$\underline{-16x^2 + 45y = 0}$$
$$25y^2 + 45y = 400 \quad \text{(Divide both sides by 5)}$$
$$5y^2 + 9y - 80 = 0$$
$$(5y - 16)(y + 5) = 0$$
$$y = \frac{16}{5} \quad \text{or} \quad y = -5$$

For $y = \dfrac{16}{5}$ For $y = -5$

$16x^2 = 45y$ $16x^2 = 45(-5)$

$16x^2 = 45\left(\dfrac{16}{5}\right)$ $x^2 = -\dfrac{225}{16}$

$x^2 = 9$ No real solution

$x = \pm 3$

Solutions: $\left(3, \dfrac{16}{5}\right)$, $\left(-3, \dfrac{16}{5}\right)$ *(11-5)*

22. $x^2 + y^2 = 10$

       $16x^2 + y^2 = 25$

We eliminate y by multiplying the first equation by -1, then adding the two equations.

$-x^2 - y^2 = -10$

$\underline{16x^2 + y^2 = 25}$

$15x^2 = 15$

        $x^2 = 1$

        $x = \pm 1$

For $x = \pm 1$

$x^2 + y^2 = 10$

$1 + y^2 = 10$

     $y^2 = 9$

      $y = \pm 3$

Solutions: $(1, 3)$, $(1, -3)$, $(-1, 3)$, $(-1, -3)$ *(11-5)*

23. $x^2 - y^2 = 2$

        $y^2 = x$

Substitute y^2 in the second equation into the first equation.

        $x^2 - x = 2$

    $x^2 - x - 2 = 0$

$(x - 2)(x + 1) = 0$

            $x = 2, -1$

For $x = 2$ For $x = -1$

 $y^2 = 2$ $y^2 = -1$

 $y = \pm\sqrt{2}$ $y = \pm i$

Solutions: $(2, \sqrt{2})$, $(2, -\sqrt{2})$, $(-1, i)$, $(-1, -i)$ *(11-5)*

24. $x^2 + 2xy + y^2 = 1$

           $xy = -2$

Solve for y in the second equation.

$y = \dfrac{-2}{x}$

Substitute into the first equation

$$x^2 + 2x\left(\dfrac{-2}{x}\right) + \left(\dfrac{-2}{x}\right)^2 = 1$$

$$x^2 - 4 + \dfrac{4}{x^2} = 1 \qquad x \neq 0$$

$$x^4 - 4x^2 + 4 = x^2$$

$$x^4 - 5x^2 + 4 = 0$$

$$(x^2 - 4)(x^2 - 1) = 0$$

$(x - 2)(x + 2)(x - 1)(x + 1) = 0$

$x = 2, -2, 1, -1$

For $x = 2$	For $x = -2$	For $x = 1$	For $x = -1$
$y = \dfrac{-2}{2}$	$y = \dfrac{-2}{-2}$	$y = \dfrac{-2}{1}$	$y = \dfrac{-2}{-1}$
$y = -1$	$y = 1$	$y = -2$	$y = 2$

Solutions: $(2, -1)$, $(-2, 1)$, $(1, -2)$, $(-1, 2)$ *(11-5)*

25. $2x^2 + xy + y^2 = 8$
$x^2 - y^2 = 0$

We factor the left side of the equation that has a zero constant term.
$(x - y)(x + y) = 0$
$x = y$ or $x = -y$
So the original system is equivalent to the two systems
$2x^2 + xy + y^2 = 8$ $2x^2 + xy + y^2 = 8$
$x = y$ $x = -y$
These systems are solved by substitution.

First System:
$2x^2 + xy + y^2 = 8$
$x = y$
$2y^2 + yy + y^2 = 8$
$4y^2 = 8$
$y^2 = 2$
$y = \pm\sqrt{2}$

Second System:
$2x^2 + xy + y^2 = 8$
$x = -y$
$2(-y)^2 + (-y)y + y^2 = 8$
$2y^2 = 8$
$y^2 = 4$
$y = \pm 2$

For $y = \sqrt{2}$	For $y = -\sqrt{2}$	For $y = 2$	For $y = -2$
$x = \sqrt{2}$	$x = -\sqrt{2}$	$x = -2$	$x = 2$

Solutions: $(\sqrt{2}, \sqrt{2})$, $(-\sqrt{2}, -\sqrt{2})$, $(2, -2)$, $(-2, 2)$ *(11-5)*

26. Comparing focus $(0, -5)$ with the information in the chart of standard equations for a parabola, we see:
$a = -5$ Axis of symmetry: the y axis.   Equation: $x^2 = 4ay$.
The equation of the parabola must be $x^2 = 4(-5)y$ or $x^2 = -20y$.   *(11-1)*

27. Since the major axis is on the x axis, the foci must be at $(\pm c, 0)$ and the equation of the ellipse is of form
$$\frac{x^2}{a^2} + \frac{y^2}{b^2} = 1$$
Since the major axis has twice the length of the minor axis, $a = 2b$. The equation then becomes
$$\frac{x^2}{4b^2} + \frac{y^2}{b^2} = 1$$
Since $(-6, 0)$ is on the graph, its coordinates satisfy the equation, so
$$\frac{(-6)^2}{4b^2} + \frac{0^2}{b^2} = 1$$
$$\frac{36}{4b^2} = 1$$
$$b^2 = 9$$
Then $a^2 = 4b^2 = 36$ and $c^2 = a^2 - b^2 = 36 - 9 = 27$; $c = \sqrt{27}$. The foci are at $(\pm\sqrt{27}, 0)$, that is, $(-3\sqrt{3}, 0)$ and $(3\sqrt{3}, 0)$.   *(11-2)*

28. Since the conjugate axis has length 4, $2b = 4$, and $b = 2$. Since $(0, -3)$ is a focus, $c = 3$. Then $a^2 = c^2 - b^2 = 9 - 4 = 5$ and $a = \sqrt{5}$. The y intercepts are $(0, \pm a)$, that is $(0, \sqrt{5})$ and $(0, -\sqrt{5})$.   *(11-3)*

29. Since the focus is at $(-4, 0)$, the axis of symmetry is the x axis, the parabola opens left, and the directrix has equation $x = -(-4)$ or $x = 4$.   *(11-1)*

30. $x^2 = 8y$

$y^2 = -x$

Solve for y in the first equation, then substitute into the second equation

$$y = \frac{x^2}{8}$$

$$\left(x^2 / 8\right)^2 = -x$$

$$\frac{x^4}{64} = -x$$

$$x^4 = -64x$$

$$x^4 + 64x = 0$$

$$x(x^3 + 64) = 0$$

$$x(x + 4)(x^2 - 4x + 16) = 0$$

$x = 0$ $x + 4 = 0$ $x^2 - 4x + 16 = 0$

$x = -4$ No real solutions.

For $x = 0$ For $x = -4$

$y = 0$ $y = \dfrac{(-4)^2}{8}$

$y = 2$

Solutions: $(0, 0)$, $(-4, 2)$

$(11\text{-}1)$

31. Since the major axis has length 14, $2a = 14$, and $a = 7$. Since $(0, -1)$ is a focus, $c = 1$. Then $b^2 = a^2 - c^2 = 49 - 1 = 48$; $b = \sqrt{48}$. The x intercepts are at $(\pm\sqrt{48}, 0)$, that is $(4\sqrt{3}, 0)$ and $(-4\sqrt{3}, 0)$. $(11\text{-}2)$

32. Since the transverse axis is on the y axis, the foci must be at $(0, \pm c)$ and the equation of the hyperbola is of the form

$$\frac{y^2}{a^2} - \frac{x^2}{b^2} = 1$$

Since the conjugate axis has twice the length of the transverse axis, $b = 2a$. The equation then becomes

$$\frac{y^2}{a^2} - \frac{x^2}{4a^2} = 1$$

Since $(0, -4)$ is on the graph,, its coordinates satisfy the equation:

$$\frac{(-4)^2}{a^2} - \frac{0^2}{4a^2} = 1$$

$$\frac{16}{a^2} = 1$$

$$a^2 = 16$$

Then $b^2 = 4a^2 = 64$ and $c^2 = a^2 + b^2 = 16 + 64 = 80$; $c = \sqrt{80}$. The foci are at $(0, \pm\sqrt{80})$, that is $(0, 4\sqrt{5})$ and $(0, -4\sqrt{5})$. $(11\text{-}3)$

33.
$$16x^2 + 4y^2 + 96x - 16y + 96 = 0$$

$$16x^2 + 96x + 4y^2 - 16y = -96$$

$$16(x^2 + 6x + ?) + 4(y^2 - 4y + ?) = -96$$

$$16(x^2 + 6x + 9) + 4(y^2 - 4y + 4) = -96 + 144 + 16$$

$$16(x + 3)^2 + 4(y - 2)^2 = 64$$

$$\frac{(x + 3)^2}{4} + \frac{(y - 2)^2}{16} = 1$$

This is the equation of an ellipse with center at $(-3, 2)$. The equations of translation are $x' = x + 3$, $y' = y - 2$. Making these substitutions, we obtain

$$\frac{x'^2}{4} + \frac{y'^2}{16} = 1$$

We graph this in the $x'y'$ system, following the process discussed in Section 11-2.

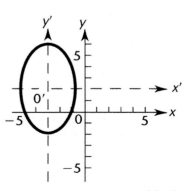

$(11\text{-}4)$

34. $x^2 - 4x - 8y - 20 = 0$

$$x^2 - 4x = 8y + 20$$
$$x^2 - 4x + 4 = 8y + 24$$
$$(x - 2)^2 = 4(2)(y + 3)$$

This is the equation of a parabola opening up with vertex at $(h, k) = (2, -3)$. The equations of translation are $x' = x - 2$, $y' = y + 3$. Making these substitutions, we obtain

$$x'^2 = 8y'$$

We graph this in the $x'y'$ system, following the process discussed in Section 11-1.

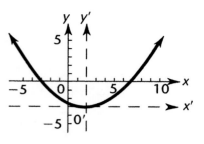

$(11-4)$

35. $4x^2 - 9y^2 + 24x - 36y - 36 = 0$

$$4x^2 + 24x - 9y^2 - 36y = 36$$
$$4(x^2 + 6x + ?) - 9(y^2 + 4y + ?) = 36$$
$$4(x^2 + 6x + 9) - 9(y^2 + 4y + 4) = 36 + 36 - 36$$
$$4(x + 3)^2 - 9(y + 2)^2 = 36$$
$$\frac{(x + 3)^2}{9} - \frac{(y + 2)^2}{4} = 1$$

This is the equation of a hyperbola with center at $(-3, -2)$. The equations of translation are $x' = x + 3$, $y' = y + 2$. Making these substitutions, we obtain

$$\frac{x'^2}{9} - \frac{y'^2}{4} = 1$$

We graph this in the $x'y'$ system, following the process discussed in Section 11-3.

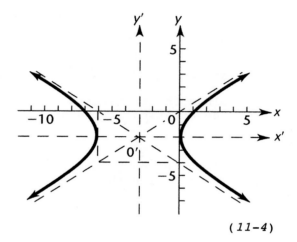

$(11-4)$

36. The rotation equations are

$x = \cos 30°x' - \sin 30°y'$ $y = \sin 30°x' + \cos 30°y'$

$x = \dfrac{\sqrt{3}}{2}x' - \dfrac{1}{2}y'$ $y = \dfrac{1}{2}x' + \dfrac{\sqrt{3}}{2}y'$

Substitute into the original equation:

$x^2 - \sqrt{3}xy + 2y^2 - 10 = 0$

$$\left(\frac{\sqrt{3}}{2}x' - \frac{1}{2}y'\right)^2 - \sqrt{3}\left(\frac{\sqrt{3}}{2}x' - \frac{1}{2}y'\right)\left(\frac{1}{2}x' + \frac{\sqrt{3}}{2}y'\right) + 2\left(\frac{1}{2}x' + \frac{\sqrt{3}}{2}y'\right)^2 - 10 = 0$$

$$\frac{3}{4}x'^2 - \frac{\sqrt{3}}{2}x'y' + \frac{1}{4}y'^2 - \sqrt{3}\left(\frac{\sqrt{3}}{4}x'^2 + \frac{3}{4}x'y' - \frac{1}{4}x'y' - \frac{\sqrt{3}}{4}y'^2\right)$$

$$+ 2\left(\frac{1}{4}x'^2 + \frac{\sqrt{3}}{2}x'y' + \frac{3}{4}y'^2\right) = 10$$

$$\frac{3}{4}x'^2 - \frac{\sqrt{3}}{2}x'y' + \frac{1}{4}y'^2 - \frac{3}{4}x'^2 - \frac{\sqrt{3}}{2}x'y' + \frac{3}{4}y'^2 + \frac{1}{2}x'^2 + \sqrt{3}x'y' + \frac{3}{2}y'^2 = 10$$

$\frac{1}{2}x'^2 + \frac{5}{2}y'^2 = 10$ (Divide both sides by 10)

$\frac{x'^2}{20} - \frac{y'^2}{4} = 1$

This is an ellipse with major axis length $2\sqrt{20}$ and minor axis length 4 rotated through 30°.

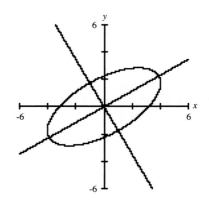

(*11-4*)

37. $5x^2 + 26xy + 5y^2 + 72 = 0$

Use the formula $\cot 2\theta = \frac{A - C}{B}$ to find the appropriate angle.

$\cot 2\theta = \frac{5 - 5}{26} = 0$

Since $\cot 90° = \frac{\cos 90°}{\sin 90°} = \frac{0}{1}$, $2\theta = 90°$ and $\theta = 45°$.

The rotation equations are
$\quad x = \cos 45°x' - \sin 45°y' \qquad y = \sin 45°x' + \cos 45°y'$

$\quad x = \frac{\sqrt{2}}{2}x' - \frac{\sqrt{2}}{2}y' \qquad\qquad y = \frac{\sqrt{2}}{2}x' + \frac{\sqrt{2}}{2}y'$

Substitute into the original equation and simplify:

$5\left(\frac{\sqrt{2}}{2}x' - \frac{\sqrt{2}}{2}y'\right)^2 + 26\left(\frac{\sqrt{2}}{2}x' - \frac{\sqrt{2}}{2}y'\right)\left(\frac{\sqrt{2}}{2}x' + \frac{\sqrt{2}}{2}y'\right) + 5\left(\frac{\sqrt{2}}{2}x' + \frac{\sqrt{2}}{2}y'\right)^2 + 72 = 0$

$5\left(\frac{1}{2}x'^2 - x'y' + \frac{1}{2}y'^2\right) + 26\left(\frac{1}{2}x'^2 - \frac{1}{2}y'^2\right) + 5\left(\frac{1}{2}x'^2 + x'y' + \frac{1}{2}y'^2\right) = -72$

$\frac{5}{2}x'^2 - 5x'y' + \frac{5}{2}y'^2 + 13x'^2 - 13y'^2 + \frac{5}{2}x'^2 + 5x'y' + \frac{5}{2}y'^2 = -72$

$18 - 8y'^2 = -72$ (Divide both sides by -72)

$\frac{y'^2}{9} - \frac{x'^2}{4} = 1$

This is a hyperbola with intercepts (0, 3) and (0, -3) rotated through 45°.

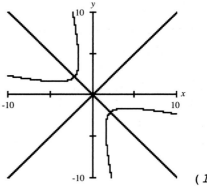

(*11-4*)

38. To identify the curve, we find the discriminant of the equation $3x^2 + 4xy + 2y^2 - 20 = 0$, which has $A = 3$, $B = 4$, and $C = 2$.
$\quad B^2 - 4AC = 16 - 4(3)(2) = -8 < 0$
The curve is an ellipse.

(*11-4*)

39. From the figure, we see that the point $P = (x, y)$ is a point on the curve if and only if

$$d_1 = d_2$$
$$d(F, P) = d(M, P)$$
$$\sqrt{(x - 2)^2 + (y - 4)^2} = \sqrt{(x - 6)^2 + (y - y)^2}$$
$$(x - 2)^2 + (y - 4)^2 = (x - 6)^2$$
$$x^2 - 4x + 4 + (y - 4)^2 = x^2 - 12x + 36$$
$$(y - 4)^2 = -8x + 32$$
$$(y - 4)^2 = -8(x - 4) \text{ or}$$
$$y^2 - 8y + 16 = -8x + 32$$
$$y^2 - 8y + 8x - 16 = 0$$

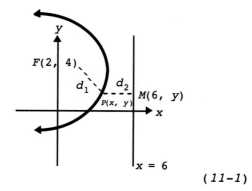

(*11-1*)

40. From the figure, we see that the point $P = (x, y)$ is a point on the curve if and only if

$$d_1 = 2d_2$$
$$d(F, P) = 2d(N, P)$$
$$\sqrt{(x - 4)^2 + (y - 0)^2} = 2\sqrt{(x - 1)^2 + (y - y)^2}$$
$$(x - 4)^2 + y^2 = 4(x - 1)^2$$
$$x^2 - 8x + 16 + y^2 = 4(x^2 - 2x + 1)$$
$$x^2 - 8x + 16 + y^2 = 4x^2 - 8x + 4$$
$$-3x^2 + y^2 = -12$$
$$\frac{x^2}{4} - \frac{y^2}{12} = 1$$

This is the equation of a hyperbola.

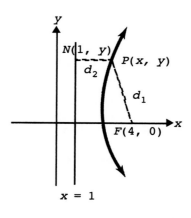

(*11-3*)

41. From the figure, we see that the point $P = (x, y)$ is a point on the curve if and only if

$$d_1 = \frac{2}{3}d_2$$
$$d(F, P) = \frac{2}{3}d(M, P)$$
$$\sqrt{(x - 4)^2 + (y - 0)^2} = \frac{2}{3}\sqrt{(x - 9)^2 + (y - y)^2}$$
$$(x - 4)^2 + y^2 = \frac{4}{9}(x - 9)^2$$
$$9[(x - 4)^2 + y^2] = 4(x - 9)^2$$
$$9(x^2 - 8x + 16 + y^2) = 4(x^2 - 18x + 81)$$
$$9x^2 - 72x + 144 + 9y^2 = 4x^2 - 72x + 324$$
$$5x^2 + 9y^2 = 180$$
$$\frac{x^2}{36} + \frac{y^2}{20} = 1$$

This is the equation of an ellipse.

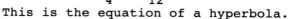

(*11-2*)

42. First find the coordinates of the foci in the translated system.
$$c'^2 = 4^2 - 2^2 = 12$$
$$c' = \sqrt{12}$$
$$-c' = -\sqrt{12}$$
The coordinates in the translated system are
$F'(0, -\sqrt{12})$ and $F(0, \sqrt{12})$

Now use
$$x = x' + h = x' - 3$$
$$y = y' + k = y' + 2$$
to obtain
$$F'(-3, -\sqrt{12} + 2) \text{ and } F(-3, \sqrt{12} + 2)$$
as the coordinates of the foci in the original system.   *(11-4)*

43. First find the coordinates of the focus in the translated system. Since $a = 2$
and the parabola opens up, they are $(0, 2)$. Now use
$$x = x' + h = x' + 2 = 0 + 2 = 2$$
$$y = y' + k = y' - 3 = 2 - 3 = -1$$
to obtain $(2, -1)$ as the coordinates of the focus in the original system.
                                                                                (11-4)

44. First find the coordinates of the foci in the original system.
$$c'^2 = 3^2 + 2^2 = 13$$
$$c' = \sqrt{13}$$
$$-c' = -\sqrt{13}$$
The coordinates in the translated system are
$$F'(-\sqrt{13}, 0) \text{ and } F(\sqrt{13}, 0)$$
Now use
$$x = x' + h = x' - 3$$
$$y = y' + k = y' - 2$$
to obtain
$$F'(-\sqrt{13} - 3, -2) \text{ and } F(\sqrt{13} - 3, -2) \text{ as the coordinates of the foci in the}$$
original system.   *(11-4)*

45. The equation $\dfrac{x^2}{49} - \dfrac{y^2}{25} = 1$ is in standard form with $a^2 = 49$, $b^2 = 25$. Since the

intercepts are on the x axis, the asymptotes have equations $y = \pm\dfrac{b}{a}x = \pm\dfrac{5}{7}x$.
                                                                                (11-3)

46. The equation $\dfrac{y^2}{64} - \dfrac{x^2}{4} = 1$ is in standard form with $a^2 = 64$, $b^2 = 4$. Since the

intercepts are on the y axis, the asymptotes have equations $y = \pm\dfrac{a}{b}x = \pm\dfrac{8}{2}x$ or

$y = \pm 4x$.   *(11-3)*

47. $4x^2 - y^2 = 1$. In standard form, this becomes
$$\frac{x^2}{1/4} - \frac{y^2}{1} = 1$$

Then $a^2 = \dfrac{1}{4}$, $b^2 = 1$. Since the intercepts are on the x axis, the asymptotes

have equations $y = \pm\dfrac{b}{a}x = \pm\dfrac{1}{1/2}x$ or $y = \pm 2x$.   *(11-3)*

48. From the figure, we see that the parabola opens up, so its equation must be of the form $x^2 = 4ay$. Since $(4, 1)$ is on the graph we have

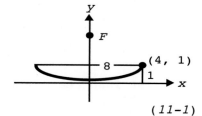

$$4^2 = 4a \cdot 1$$
$$16 = 4a$$
$$a = 4$$

So a, the distance of the focus from the vertex, is 4 feet.

$(11\text{-}1)$

49. From the figure, we see that the x intercepts must be at $(-5, 0)$ and $(5, 0)$, the foci at $(-4, 0)$ and $(4, 0)$. So $a = 5$ and $c = 4$. We can determine the y intercepts using the special triangle relationship

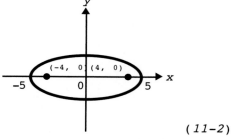

$$5^2 = 4^2 + b^2$$
$$25 = 16 + b^2$$
$$b = 3$$

The equation of the ellipse is

$$\frac{x^2}{5^2} + \frac{y^2}{3^2} = 1$$

$(11\text{-}2)$

50. From the figure, we can see:

$d + a = y_1 =$ the y coordinate of the point on the hyperbola with x coordinate 15. From the equation of the hyperbola $\dfrac{y^2}{40^2} - \dfrac{x^2}{30^2} = 1$, we have $a = 40$, and $d + 40 = y_1$, or $d = y_1 - 40$. Since the point $(15, y_1)$ is on the hyperbola, its coordinates must satisfy the equation of the hyperbola.

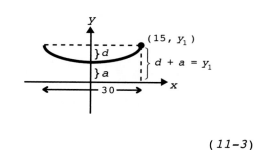

$$\frac{y_1^2}{40^2} - \frac{15^2}{30^2} = 1$$
$$\frac{y_1^2}{40^2} = 1 + \frac{15^2}{30^2}$$
$$\frac{y_1^2}{40^2} = \frac{5}{4}$$
$$y_1^2 = 2000$$
$$y_1 = 44.72$$

depth $= y_1 - 40 = 4.72$ feet.

$(11\text{-}3)$

Chapters 10 & 11 Cumulative Review

1. (A) Since 15 - 20 = 10 - 15 = -5, this could start an arithmetic sequence.

   (B) Since $\dfrac{25}{5} = \dfrac{125}{25} = 5$, this could start a geometric sequence.

   (C) Since $\dfrac{25}{5} \neq \dfrac{50}{25}$ and 25 - 5 ≠ 50 - 25, this could start neither an arithmetic nor a geometric sequence.

   (D) Since $\dfrac{-9}{27} = \dfrac{3}{-9} = -\dfrac{1}{3}$, this could start a geometric sequence.

   (E) Since (-6) - (-9) = (-3) - (-6) = 3, this could start an arithmetic sequence.   (10-3)

2. $a_n = 2 \cdot 5^n$

   (A) $a_1 = 2 \cdot 5^1 = 10$
   $a_2 = 2 \cdot 5^2 = 50$
   $a_3 = 2 \cdot 5^3 = 250$
   $a_4 = 2 \cdot 5^4 = 1{,}250$

   (B) This is a geometric sequence with $r = 5$.
   $a_n = a_1 r^{\,n-1}$
   $a_8 = 10(5)^{8-1}$
   $= 781{,}250$

   (C) $S_n = \dfrac{a_1 - r a_n}{1 - r}$
   $S_8 = \dfrac{10 - 5(781{,}250)}{1 - 5}$
   $= 976{,}560$ (10-3)

3. $a_n = 3n - 1$

   (A) $a_1 = 3 \cdot 1 - 1 = 2$
   $a_2 = 3 \cdot 2 - 1 = 5$
   $a_3 = 3 \cdot 3 - 1 = 8$
   $a_4 = 3 \cdot 4 - 1 = 11$

   (B) This is an arithmetic sequence with $d = 3$.
   $a_n = a_1 + (n - 1)d$
   $a_8 = 2 + (8 - 1)3$
   $= 23$

   (C) $S_n = \dfrac{n}{2}(a_1 + a_n)$
   $S_8 = \dfrac{8}{2}(2 + 23)$
   $= 100$ (10-3)

4. $a_1 = 100 \quad a_n = a_{n-1} - 6 \quad n \geq 2$

   (A) $a_1 = 100$
   $a_2 = a_1 - 6 = 94$
   $a_3 = a_2 - 6 = 88$
   $a_4 = a_3 - 6 = 82$

   (B) This is an arithmetic sequence with $d = -6$.
   $a_n = a_1 + (n - 1)d$
   $a_8 = 100 + (8 - 1)(-6)$
   $= 58$

   (C) $S_n = \dfrac{n}{2}(a_1 + a_n)$
   $S_8 = \dfrac{8}{2}(100 + 58)$
   $= 632$ (10-3)

5. (A) $8! = 8 \cdot 7 \cdot 6 \cdot 5 \cdot 4 \cdot 3 \cdot 2 \cdot 1 = 40{,}320$.

   (B) $\dfrac{32!}{30!} = \dfrac{32 \cdot 31 \cdot 30!}{30!} = 992$

   (C) $\dfrac{9!}{3!(9-3)!} = \dfrac{9!}{3! \, 6!}$
   $= \dfrac{9 \cdot 8 \cdot 7 \cdot 6!}{3 \cdot 2 \cdot 1 \cdot 6!} = 84$ (10-4)

6. (A) $\dbinom{7}{2} = \dfrac{7!}{2!(7-2)!} = \dfrac{7!}{2! \, 5!} = \dfrac{7 \cdot 6 \cdot 5!}{2 \cdot 1 \cdot 5!} = 21$ (B) $C_{7,2} = \dfrac{7!}{2!(7-2)!} = 21$

   (C) $P_{7,2} = \dfrac{7!}{(7-2)!} = \dfrac{7!}{5!} = \dfrac{7 \cdot 6 \cdot 5!}{5!} = 42$ (10-5)

7. First, write the equation in standard form by dividing both sides by 900.
    $25x^2 - 36y^2 = 900$
    $$\frac{x^2}{36} - \frac{y^2}{25} = 1$$
    In this form the equation is identifiable as that of a hyperbola.

    When $x = 0$, $-\dfrac{y^2}{25} = 1$. There are no y intercepts, but $b = 5$.

    When $y = 0$, $\dfrac{x^2}{36} = 1$. x intercepts: ±6.

    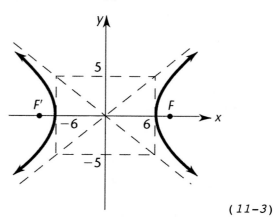

    Sketch the asymptotes using the asymptote rectangle, then sketch in the hyperbola.
    Foci: $c^2 = 5^2 + 6^2$
    $\qquad c^2 = 61$
    $\qquad\quad c = \sqrt{61}$
    Foci: $F' = (-\sqrt{61}, 0)$, $F = (\sqrt{61}, 0)$
    Transverse axis length = 2(6) = 12
    Conjugate axis length = 2(5) = 10

    (11-3)

8. First, write the equation in standard form by dividing both sides by 900.
    $25x^2 + 36y^2 = 900$
    $$\frac{x^2}{36} + \frac{y^2}{25} = 1$$
    In this form the equation is identifiable as that of an ellipse. Locate the intercepts:

    When $y = 0$, $\dfrac{x^2}{36} = 1$. x intercepts: ±6.

    When $x = 0$, $\dfrac{y^2}{25} = 1$. y intercepts: ±5

    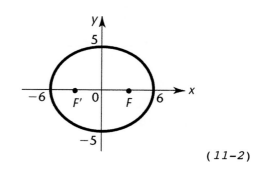

    So $a = 6$, $b = 5$, and the major axis is on the x axis.
    Foci: $c^2 = a^2 - b^2$
    $\qquad c^2 = 36 - 25$
    $\qquad c^2 = 11$
    $\qquad\quad c = \sqrt{11}$
    Foci: $F' = (-\sqrt{11}, 0)$, $F = (\sqrt{11}, 0)$
    Major axis length = 2(6) = 12
    Minor axis length = 2(5) = 10

    (11-2)

9. $25x^2 - 36y = 0$ is the equation of a parabola. For convenience, we rewrite this as $25x^2 = 36y$. To graph, assign y values that make $25x^2$ a perfect square (y must be positive or zero for x to be real) and solve for x. Since the coefficient of x is positive, a must be positive, and the parabola opens up.

    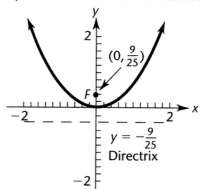

x	0	$\pm\frac{6}{5}$	$\pm\frac{12}{5}$
y	0	1	4

    To find the focus and directrix, solve
    $4a = \dfrac{36}{25}$
    $\ \ a = \dfrac{9}{25}$

    Focus: $\left(0, \dfrac{9}{25}\right)$. Directrix: $y = -\dfrac{9}{25}$

    (11-1)

10. $x^2 + y^2 = 2$
$2x - y = 1$

We choose substitution, solving the first-degree equation for y in terms of x, then substituting into the second-degree equation.

$$2x - y = 1$$
$$-y = 1 - 2x$$
$$y = 2x - 1$$
$$x^2 + (2x - 1)^2 = 2$$
$$x^2 + 4x^2 - 4x + 1 = 2$$
$$5x^2 - 4x - 1 = 0$$
$$(5x + 1)(x - 1) = 0$$
$$x = -\frac{1}{5}, \; 1$$

For $x = -\frac{1}{5}$ For $x = 1$

$\quad y = 2\left(-\frac{1}{5}\right) - 1 \qquad\qquad y = 2(1) - 1$

$\quad\;\; = -\frac{7}{5} \qquad\qquad\qquad\quad\;\; = 1$

Solutions: $\left(-\frac{1}{5}, -\frac{7}{5}\right)$, $(1, 1)$

The checking steps are omitted for lack of space.   *(11-5)*

11. Find the discriminant with $A = 3$, $B = -4$, $C = 2$:
$$B^2 - 4AC = 16 - 4(3)(2) = -8$$
Since the discriminant is negative, this is an ellipse.   *(11-4)*

12. (A) The outcomes can be displayed in a tree diagram as follows:

HHH
HHT
HTH
HTT
THH
THT
TTH
TTT

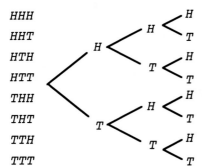

(B) O_1: First flip of the coin
N_1: 2 outcomes
O_2: Second flip of the coin
N_2: 2 outcomes
O_3: Third flip of the coin
N_3: 2 outcomes

Applying the multiplication principle, there are $2 \cdot 2 \cdot 2 = 8$ possible outcomes.

(10-5)

13. (A) O_1: Place first book N_1: 4 ways
O_2: Place second book N_2: 3 ways
O_3: Place third book N_3: 2 ways
O_4: Place fourth book N_4: 1 way

Applying the multiplication principle, there are $4 \cdot 3 \cdot 2 \cdot 1$, or 24, arrangements.

(B) Order is important here. We use permutations to determine the number of arrangements of 4 objects. $P_{4,4} = 4! = 24$.   *(10-5)*

14. The sample space S is the set of all possible 3-card hands, chosen from a deck of 52 cards. $n(S) = C_{52,3}$.

The event E is the set of all possible 3-card hands that are all diamonds, and are chosen from the 13 diamonds. $n(E) = C_{13,3}$

$$P(E) = \frac{n(E)}{n(S)} = \frac{C_{13,3}}{C_{52,3}} = \frac{13!}{3!(13-3)!} \div \frac{52!}{3!(52-3)!} \approx .0129 \qquad (10\text{-}5)$$

15. In the first case, order is important. The sample space is the set of all possible arrangements of four objects, drawn from a set of 10 objects.
$n(S) = P_{10,4}$.
The event E is one of these arrangements. $n(E) = 1$.
$$P(E) = \frac{n(E)}{n(S)} = \frac{1}{P_{10,4}} = \frac{1}{10 \cdot 9 \cdot 8 \cdot 7} \approx .0002$$

In the second case, order is not important. The sample space is the set of all possible ways of drawing four objects from a set of 10 objects $n(S) = C_{10,4}$.
The event E is one of these ways. $n(E) = 1$.
$$P(E) = \frac{n(E)}{n(S)} = \frac{1}{C_{10,4}} = 1 \div \frac{10!}{4!(10-4)!} \approx .0048 \qquad (10\text{-}5)$$

16. $P(E) \approx \dfrac{f(E)}{n} = \dfrac{100 - 38}{100} = .62$ $\qquad (10\text{-}5)$

17. P_1: $1 = 1(2 \cdot 1 - 1) = 1 \cdot 1 = 1$
$\quad P_2$: $1 + 5 = 2(2 \cdot 2 - 1)$
$\qquad\quad 6 = 6$
$\quad P_3$: $1 + 5 + 9 = 3(2 \cdot 3 - 1)$
$\qquad\qquad 15 = 15$ $\qquad (10\text{-}2)$

18. P_1: $1^2 + 1 + 2$ is divisible by 2
$\qquad 4 = 2 \cdot 2$ true
$\quad P_2$: $2^2 + 2 + 2$ is divisible by 2
$\qquad 8 = 2 \cdot 4$ true
$\quad P_3$: $3^2 + 3 + 2$ is divisible by 2
$\qquad 14 = 2 \cdot 7$ true $\qquad (10\text{-}2)$

19. P_k: $1 + 5 + 9 + \cdots + (4k - 3) = k(2k - 1)$
$\quad P_{k+1}$: $1 + 5 + 9 + \cdots + (4k - 3) + (4k + 1) = (k + 1)(2k + 1)$ $\qquad (10\text{-}2)$

20. P_k: $k^2 + k + 2 = 2r$ for some integer r
$\quad P_{k+1}$: $(k + 1)^2 + (k + 1) + 2 = 2s$ for some integer s $\qquad (10\text{-}2)$

21. The parabola is opening either up or down and has an equation of the form $x^2 = 4ay$. Since $(2, -8)$ is on the graph, we have:
$2^2 = 4a(-8)$
$\quad 4 = -32a$
$\quad a = -\dfrac{1}{8}$
The equation of the parabola is
$$x^2 = 4\left(-\frac{1}{8}\right)y$$
$$x^2 = -\frac{1}{2}y$$
$$y = -2x^2$$
$\qquad (11\text{-}1)$

22. Make a rough sketch of the ellipse, locate focus and x intercepts, then determine y intercepts using the Pythagorean theorem.
$$\frac{x^2}{a^2} + \frac{y^2}{b^2} = 1$$
$$a = \frac{10}{2} = 5$$
$$b^2 = a^2 - c^2 = 5^2 - 3^2 = 25 - 9 = 16$$
$$b = 4$$
$$\frac{x^2}{25} + \frac{y^2}{16} = 1$$
$\qquad (11\text{-}2)$

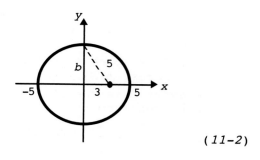

23. Start with $\dfrac{x^2}{a^2} - \dfrac{y^2}{b^2} = 1$ and find a and b.

$a = \dfrac{16}{2} = 8$

To find b, sketch the asymptote rectangle, label known parts, and use the Pythagorean Theorem.

$b^2 = (\sqrt{89})^2 - 8^2$

$\quad = 89 - 64$

$b^2 = 25$

$\quad b = 5$

The equation is $\dfrac{x^2}{64} - \dfrac{y^2}{25} = 1$

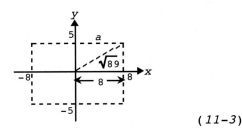

(11-3)

24. $2\sqrt{3}\,xy + 2y^2 + 3 = 0$

Use the formula $\cot 2\theta = \dfrac{A - C}{B}$ to find the appropriate angle.

$\cot 2\theta = \dfrac{0 - 2}{2\sqrt{3}} = \dfrac{-2}{2\sqrt{3}} = \dfrac{-1}{\sqrt{3}}$

Note that $\cot 120° = \dfrac{\cos(120°)}{\sin(120°)} = \dfrac{-\frac{1}{2}}{\frac{\sqrt{3}}{2}} = -\dfrac{1}{\sqrt{3}}$, so $2\theta = 120°$, and $\theta = 60°$.

The rotation formulas are:

$\quad x = \cos 60°x' - \sin 60°y' \qquad\qquad y = \sin 60°x' + \cos 60°y'$

$\quad x = \dfrac{1}{2}x' - \dfrac{\sqrt{3}}{2}y' \qquad\qquad\qquad y = \dfrac{\sqrt{3}}{2}x' + \dfrac{1}{2}y'$

Substituting these into the original equation, we get:

$2\sqrt{3}\left(\dfrac{1}{2}x' - \dfrac{\sqrt{3}}{2}y'\right)\left(\dfrac{\sqrt{3}}{2}x' + \dfrac{1}{2}y'\right) + 2\left(\dfrac{\sqrt{3}}{2}x' + \dfrac{1}{2}y'\right)^2 + 3 = 0$

$2\sqrt{3}\left(\dfrac{\sqrt{3}}{4}x'^2 + \dfrac{1}{4}x'y' - \dfrac{3}{4}x'y' - \dfrac{\sqrt{3}}{4}y'^2\right) + 2\left(\dfrac{3}{4}x'^2 + \dfrac{\sqrt{3}}{4}x'y' + \dfrac{\sqrt{3}}{4}x'y' + \dfrac{1}{4}y'^2\right) + 3 = 0$

$2\sqrt{3}\left(\dfrac{3}{4}x'^2 - \dfrac{1}{2}x'y' - \dfrac{\sqrt{3}}{4}y'^2\right) + 2\left(\dfrac{3}{4}x'^2 + \dfrac{\sqrt{3}}{2}x'y' + \dfrac{1}{4}y'^2\right) + 3 = 0$

$\dfrac{3}{2}x'^2 - \sqrt{3}x'y' - \dfrac{3}{2}y'^2 + \dfrac{3}{2}x'^2 + \sqrt{3}x'y' + \dfrac{1}{2}y'^2 + 3 = 0$

$3x'^2 - y'^2 + 3 = 0$

This can be rewritten as:

$\quad y'^2 - 3x'^2 = 3$ or

$\quad \dfrac{y'^2}{3} - x'^2 = 1$

which is the equation of a hyperbola.

To find the graph on a graphing utility, we solve the original equation for y using the quadratic formula:

$2y^2 + 2\sqrt{3}\,xy + 3 = 0 \qquad a = 2,\ b = 2\sqrt{3}\,x,\ c = 3$

$y = \dfrac{-2\sqrt{3}x \pm \sqrt{12x^2 - 4(2)(3)}}{2(2)} = \dfrac{-2\sqrt{3}x \pm \sqrt{12x^2 - 24}}{4}$

Enter $y_1 = (-2\sqrt{3}\,x + \sqrt{12x^2 - 24}\,) \div 4$ and $y_2 = (-2\sqrt{3}\,x - \sqrt{12x^2 - 24}\,) \div 4$.

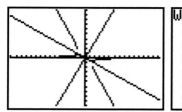 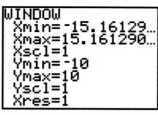 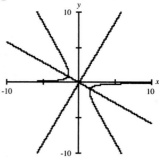

The graph is shown in a square viewing window with the rotated $x' + y'$ axes drawn in as well.　　　　　　　　　　　　　　　　　　　　　　　　*(11-4)*

25.　$x^2 + 2xy + y^2 + 4\sqrt{2}\,x - 4\sqrt{2}\,y = 0$

Use the formula $\cot 2\theta = \dfrac{A - C}{B}$ to find the appropriate angle.

$$\cot 2\theta = \frac{1 - 1}{2} = 0$$

Since $\cot 90° = \dfrac{\cos 90°}{\sin 90°} = \dfrac{0}{1} = 0$, $2\theta = 90°$, and $\theta = 45°$.

The rotation formulas are:

$$x = \cos 45°x' - \sin 45°y' \qquad\qquad y = \sin 45°x' + \cos 45°y'$$

$$x = \frac{\sqrt{2}}{2}x' - \frac{\sqrt{2}}{2}y' \qquad\qquad y = \frac{\sqrt{2}}{2}x' + \frac{\sqrt{2}}{2}y'$$

Substituting these into the original equation, we get:

$$\left(\frac{\sqrt{2}}{2}x' - \frac{\sqrt{2}}{2}y'\right)^2 + 2\left(\frac{\sqrt{2}}{2}x' + \frac{\sqrt{2}}{2}y'\right) + \left(\frac{\sqrt{2}}{2}x' + \frac{\sqrt{2}}{2}y'\right)^2$$

$$+ 4\sqrt{2}\left(\frac{\sqrt{2}}{2}x' - \frac{\sqrt{2}}{2}y'\right) - 4\sqrt{2}\left(\frac{\sqrt{2}}{2}x' + \frac{\sqrt{2}}{2}y'\right) = 0$$

$$\frac{1}{2}x'^2 - \frac{1}{2}x'y' - \frac{1}{2}x'y' + \frac{1}{2}y'^2 + 2\left(\frac{1}{2}x'^2 - \frac{1}{2}y'^2\right) + \frac{1}{2}x'^2 + \frac{1}{2}x'y' + \frac{1}{2}x'y' + \frac{1}{2}y'^2$$

$$+ 4x' - 4y' - 4x' - 4y' = 0$$

$$\frac{1}{2}x'^2 - x'y' + \frac{1}{2}y'^2 + x'^2 - y'^2 + \frac{1}{2}x'^2 + x'y' + \frac{1}{2}y'^2$$

$$+ 4x' - 4y' - 4x' - 4y' = 0$$

$$2x'^2 - 8y' = 0$$
$$8y' = 2x'^2$$
$$y = \frac{1}{4}x'^2$$

This is the equation of a parabola with vertex at the origin rotated through 45°. To graph, find the intercepts:

x intercept: $x^2 + 4\sqrt{2}\,x = 0$　　　y intercept: $y^2 - 4\sqrt{2}\,y = 0$

$\phantom{x \text{ intercept: }}x(x + 4\sqrt{2}\,) = 0$　　　　　$\phantom{y \text{ intercept: }}y(y - 4\sqrt{2}\,) = 0$

$\phantom{x \text{ intercept: }x(}x = 0,\ -4\sqrt{2}$　　　　　　$\phantom{y \text{ intercept: }y(}y = 0,\ 4\sqrt{2}$

The graph is shown below with the rotated axes.

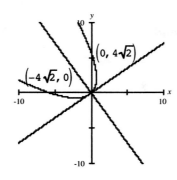

(*11-4*)

26. $x^2 - 3xy + 3y^2 = 1$
$\qquad\qquad xy = 1$

Solve for y in the second equation.

$$y = \frac{1}{x}$$

$$x^2 - 3x\left(\frac{1}{x}\right) + 3\left(\frac{1}{x}\right)^2 = 1$$

$$x^2 - 3 + \frac{3}{x^2} = 1 \quad x \neq 0$$

$$x^4 - 3x^2 + 3 = x^2$$

$$x^4 - 4x^2 + 3 = 0$$

$$(x^2 - 1)(x^2 - 3) = 0$$

$$x^2 - 1 = 0 \qquad\qquad x^2 - 3 = 0$$
$$x^2 = 1 \qquad\qquad\quad x^2 = 3$$
$$x = \pm 1 \qquad\qquad\quad x = \pm\sqrt{3}$$

For $x = 1$	For $x = -1$	For $x = \sqrt{3}$	For $x = -\sqrt{3}$
$y = \dfrac{1}{1}$	$y = \dfrac{1}{-1}$	$y = \dfrac{1}{\sqrt{3}}$	$y = \dfrac{1}{-\sqrt{3}}$
$y = 1$	$y = -1$	$y = \dfrac{\sqrt{3}}{3}$	$y = -\dfrac{\sqrt{3}}{3}$

Solutions: $(1, 1)$, $(-1, -1)$, $\left(\sqrt{3}, \dfrac{\sqrt{3}}{3}\right)$, $\left(-\sqrt{3}, -\dfrac{\sqrt{3}}{3}\right)$ (*11-5*)

27. $x^2 - 3xy + y^2 = -1$
$\qquad x^2 - xy = 0$

Factor the left side of the equation that has a zero constant term.

$x(x - y) = 0$

$\qquad x = 0 \quad$ or $\quad x = y$

The original system is equivalent to the two systems

$x^2 - 3xy + y^2 = -1$	$x^2 - 3xy + y^2 = -1$
$x = 0$	$x = y$

These systems are solved by substitution.

First system: Second system:

First system:	Second system:
$x^2 - 3xy + y^2 = -1$	$x^2 - 3xy + y^2 = -1$
$x = 0$	$x = y$
$0^2 - 3(0)y + y^2 = -1$	$y^2 - 3yy + y^2 = -1$
$y^2 = -1$	$-y^2 = -1$
$y = \pm i$	$y^2 = 1$
$x = 0$	$y = \pm 1$

For $y = 1$	For $y = -1$
$x = 1$	$x = -1$

Solutions: $(1, 1)$, $(-1, -1)$, $(0, i)$, $(0, -i)$ (*11-5*)

28. Before we can enter these equations in our graphing calculator, we must solve
for y:

$$x^2 + 2xy - y^2 = 1 \qquad\qquad 9x^2 + 4xy + y^2 = 15$$
$$y^2 - 2xy - x^2 + 1 = 0 \qquad\qquad y^2 + 4xy + 9x^2 - 15 = 0$$

Applying the quadratic formula to each equation, we have

$$y = \frac{2x \pm \sqrt{4x^2 - 4(-x^2 + 1)}}{2} \qquad\qquad y = \frac{-4x \pm \sqrt{16x^2 - 4(9x^2 - 15)}}{2}$$

$$y = \frac{2x \pm \sqrt{8x^2 - 4}}{2} \qquad\qquad y = \frac{-4x \pm \sqrt{60 - 20x^2}}{2}$$

$$y = x \pm \sqrt{2x^2 - 1} \qquad\qquad y = -2x \pm \sqrt{15 - 5x^2}$$

Entering each of these four equations into a
graphing calculator produces the graph shown at
the right.

Using the INTERSECT command (details omitted),
yields (-1.35, 0.28), (-0.87, -1.60), (0.87,
1.60), and (1.35, -0.28) to two decimal places.

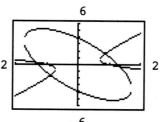

(*11-5*)

29. $\displaystyle\sum_{k=1}^{5} k^k = 1^1 + 2^2 + 3^3 + 4^4 + 5^5 = 1 + 4 + 27 + 256 + 3{,}125 = 3{,}413$ (*10-1*)

30. $S_6 = \dfrac{2}{2!} - \dfrac{2^2}{3!} + \dfrac{2^3}{4!} - \dfrac{2^4}{5!} + \dfrac{2^5}{6!} - \dfrac{2^6}{7!}$

$k = 1, 2, 3, 4, 5, 6$

Noting that the terms alternate in sign, we can rewrite as follows:

$S_6 = (-1)^2 \dfrac{2}{2!} + (-1)^3 \dfrac{2^2}{3!} + (-1)^4 \dfrac{2^3}{4!} + (-1)^5 \dfrac{2^4}{5!} + (-1)^6 \dfrac{2^5}{6!} + (-1)^7 \dfrac{2^6}{7!}$

In this form, we can see that $a_k = (-1)^{k+1} \dfrac{2^k}{(k + 1)!}$

$S_6 = \displaystyle\sum_{k=1}^{6} (-1)^{k+1} \dfrac{2^k}{(k + 1)!}$ (*10-1*)

31. $\dfrac{a_2}{a_1} = \dfrac{-36}{108} = -\dfrac{1}{3} = r. \;\; |r| < 1.$

Therefore, this infinite geometric series has a sum.

$S_\infty = \dfrac{a_1}{1 - r}$

$ = \dfrac{108}{1 - \left(-\frac{1}{3}\right)}$

$ = 81$ (*10-3*)

32.

	Case 1	Case 2	Case 3
O_1: Select the first letter	N_1: 6 ways	6 ways	6 ways
O_2: Select the second letter	N_2: 5 ways	6 ways	5 ways (exclude first letter)
O_3: Select the third letter	N_3: 4 ways	6 ways	5 ways (exclude second letter)
O_4: Select the fourth letter	N_4: 3 ways	6 ways	5 ways (exclude third letter)
	$6 \cdot 5 \cdot 4 \cdot 3 =$ 360 words	$6 \cdot 6 \cdot 6 \cdot 6 =$ 1,296 words	$6 \cdot 5 \cdot 5 \cdot 5 = 750$ words

(*10-4*)

33. The sample space is the set of all possible ways to choose 5 players out of 12.
$n(S) = C_{12,5}$
The event E is the set of all those choices that include 2 particular people, so

O_1: Choose the two centers N_1: 1

O_2: Choose 3 more people out N_2: $C_{10,3}$
    of the 10 remaining

Applying the multiplication principle, $n(E) = N_1 \cdot N_2 = 1 \cdot C_{10,3} = C_{10,3}$

$$P(E) = \frac{n(E)}{n(S)} = \frac{C_{10,3}}{C_{12,5}} = \frac{10!}{3!(10-3)!} \div \frac{12!}{5!(12-5)!} = 120 \div 792 = \frac{5}{33} = .\overline{15} \qquad (10\text{-}5)$$

34. (A) $P \text{ (this event)} = \dfrac{195 + 170}{1000} = .365$

(B) A sample space of equally likely events is:
$S = \{1, 2, 3, 4, 5, 6\}$ $n(S) = 6$
The event E is the set of those outcomes that are divisible by 3.
$E = \{3, 6\}$. $n(E) = 2$
Then $P(E) = \dfrac{n(E)}{n(S)} = \dfrac{2}{6} = \dfrac{1}{3}$ $\qquad (10\text{-}5)$

35.

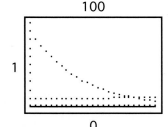

Enter the sequences   Graph the   Display the points
in the sequence editor.   sequences.   in a table.

From the graph, 22 is the least positive integer n such that $a_n < b_n$.
The table confirms this. $\qquad (10\text{-}3)$

36. (A) $P_{25,5} = 25 \cdot 24 \cdot 23 \cdot 22 \cdot 21 = 6,375,600$

(B) $C_{25,5} = \dfrac{25!}{5!(25-5)!} = \dfrac{25!}{5!\,20!} = \dfrac{25 \cdot 24 \cdot 23 \cdot 22 \cdot 21 \cdot 20!}{5 \cdot 4 \cdot 3 \cdot 2 \cdot 1 \cdot 20!} = 53,130$

(C) $\dbinom{25}{5} = \dfrac{25!}{5!(25-5)!} = 53,130$ $\qquad (10\text{-}4, \ 10\text{-}6)$

37. $\left(a + \dfrac{1}{2}b\right)^6 = \displaystyle\sum_{k=0}^{6} \binom{6}{k} a^{6-k} \left(\dfrac{1}{2}b\right)^k$

$$= \binom{6}{0}a^6 + \binom{6}{1}a^5\left(\dfrac{1}{2}b\right)^1 + \binom{6}{2}a^4\left(\dfrac{1}{2}b\right)^2 + \binom{6}{3}a^3\left(\dfrac{1}{2}b\right)^3 + \binom{6}{4}a^2\left(\dfrac{1}{2}b\right)^4$$

$$+ \binom{6}{5}a\left(\dfrac{1}{2}b\right)^5 + \binom{6}{6}\left(\dfrac{1}{2}b\right)^6$$

$$= a^6 + 3a^5b + \dfrac{15}{4}a^4b^2 + \dfrac{5}{2}a^3b^3 + \dfrac{15}{16}a^2b^4 + \dfrac{3}{16}ab^5 + \dfrac{1}{64}b^6 \qquad (10\text{-}6)$$

38. In the expansion of $(a + b)^n$, the exponent of b in the rth term is $r - 1$ and the exponent of a is $n - (r - 1)$. In the first case, $r = 5$, $n = 10$.

$$\text{Fifth term} = \binom{10}{4}(3x)^6(-y)^4$$

$$= 153,090x^6y^4$$

In the second case, $r = 8$, n is still 10.

$$\text{Eighth term} = \binom{10}{7}(3x)^3(-y)^7$$

$$= -3,240x^3y^7 \qquad (10\text{-}6)$$

39. P_n: $1 + 5 + 9 + \cdots + (4n - 3) = n(2n - 1)$
Part 1: Show that P_1 is true.
P_1: $1 = 1(2 \cdot 1 - 1) = 1$ True
Part 2: Show that if P_k is true, then P_{k+1} is true.
Write out P_k and P_{k+1}.

P_k: $1 + 5 + 9 + \cdots + (4k - 3) = k(2k - 1)$

P_{k+1}: $1 + 5 + 9 + \cdots + (4k - 3) + (4k + 1) = (k + 1)(2k + 1)$

We start with P_k:

$1 + 5 + 9 + \cdots + (4k - 3) = k(2k - 1)$

Adding $4k + 1$ to both sides:

$$1 + 5 + 9 + \cdots + (4k - 3) + (4k + 1) = k(2k - 1) + 4k + 1$$
$$= 2k^2 - k + 4k + 1$$
$$= 2k^2 + 3k + 1$$
$$= (k + 1)(2k + 1)$$

We have shown that if P_k is true, then P_{k+1} is true.
Conclusion: P_n is true for all positive integers n. $\qquad (10\text{-}2)$

40. P_n: $n^2 + n + 2 = 2p$ for some integer p.
Part 1: Show that P_1 is true.
P_1: $1^2 + 1 + 2 = 4 = 2 \cdot 2$ is true.

Part 2: Show that if P_k is true, then P_{k+1} is true.
Write out P_k and P_{k+1}.

P_k: $k^2 + k + 2 = 2r$ for some integer r

P_{k+1}: $(k + 1)^2 + (k + 1) + 2 = 2s$ for some integer s

We start with P_k:

$k^2 + k + 2 = 2r$ for some integer r

Now, $(k + 1)^2 + (k + 1) + 2 = k^2 + 2k + 1 + k + 1 + 2$
$$= k^2 + k + 2 + 2k + 2$$
$$= 2r + 2k + 2$$
$$= 2(r + k + 2)$$

Therefore,

$(k + 1)^2 + (k + 1) + 2 = 2s$ for some integer s $(= r + k + 2)$

$(k + 1)^2 + (k + 1) + 2$ is divisible by 2.

We have shown that if P_k is true, then P_{k+1} is true.
Conclusion: P_n is true for all positive integers n. $\qquad (10\text{-}2)$

41. We are to find $51 + 53 + \cdots + 499$. This is the sum of an arithmetic sequence, S_n, with $d = 2$.

First, find n:

$a_n = a_1 + (n - 1)d$

$499 = 51 + (n - 1)2$

$n = 225$

Now, find S_{225}

$S_n = \dfrac{n}{2}(a_1 + a_n)$

$S_{225} = \dfrac{225}{2}(51 + 499) = 61{,}875$

(10-3)

42. $2.\overline{45} = 2.454545\ldots$

$\qquad = 2 + 0.454545\ldots$

$0.454545\ldots = 0.45 + 0.0045 + 0.000045 + \cdots$

This is an infinite geometric series with $a_1 = 0.45$ and $r = 0.01$.

$S_\infty = \dfrac{a_1}{1 - r} = \dfrac{0.45}{1 - 0.01} = \dfrac{0.45}{0.99} = \dfrac{5}{11}$

Then $2.\overline{45} = 2 + \dfrac{5}{11} = \dfrac{27}{11}$

(10-3)

43. Graphing the sequence, we obtain the graph at the right. The solid line indicates 0.01. The largest term of the sequence is a_{27}. From the table display,

$a_{27} = \dbinom{30}{27}(0.1)^{30-27}(0.9)^{27} = 0.236088$

There are 8 terms larger than 0.01, as can be seen from the graph or the table display (details omitted).

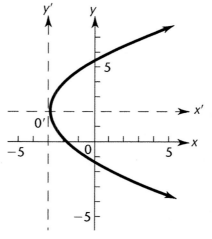

(10-6)

44. $4x + 4y - y^2 + 8 = 0$

$\qquad 4x + 8 = y^2 - 4y$

$\quad 4x + 8 + 4 = y^2 - 4y + 4$

$\qquad 4x + 12 = (y - 2)^2$

$\qquad 4(x + 3) = (y - 2)^2$

This is the equation of a parabola opening right with vertex at $(h, k) = (-3, 2)$.

The equations of translation are $x' = x + 3$, $y' = y - 2$. Making these substitutions, we obtain

$4x' = y'^2$

We graph this in the $x'y'$ system.

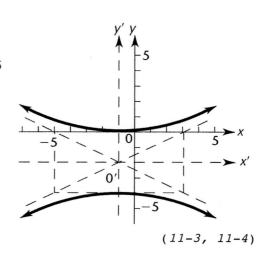

(11-1, 11-4)

45. $\qquad\qquad x^2 + 2x - 4y^2 - 16y + 1 = 0$

$\qquad\qquad x^2 + 2x - 4y^2 - 16y = -1$

$x^2 + 2x + ? - 4(y^2 + 4y + ?) = -1$

$x^2 + 2x + 1 - 4(y^2 + 4y + 4) = -1 + 1 - 16$

$\qquad\quad (x + 1)^2 - 4(y + 2)^2 = -16$

$\qquad\quad \dfrac{(y + 2)^2}{4} - \dfrac{(x + 1)^2}{16} = 1$

This is the equation of a hyperbola with center at $(-1, -2)$. The equations of translation are $x' = x + 1$, $y' = y + 2$. Making these substitutions, we obtain

$\dfrac{y'^2}{4} - \dfrac{x'^2}{16} = 1$

We graph this in the $x'y'$ system.

(11-3, 11-4)

46.
$$4x^2 - 16x + 9y^2 + 54y + 61 = 0$$
$$4x^2 - 16x + 9y^2 + 54y = -61$$
$$4(x^2 - 4x + ?) + 9(y^2 + 6y + ?) = -61$$
$$4(x^2 - 4x + 4) + 9(y^2 + 6y + 9) = -61 + 16 + 81$$
$$4(x - 2)^2 + 9(y + 3)^2 = 36$$
$$\frac{(x - 2)^2}{9} + \frac{(y + 3)^2}{4} = 1$$

This is the equation of an ellipse with center
at $(2, -3)$. The equations of translation are
$x' = x - 2$, $y' = y + 3$. Making these
substitutions, we obtain
$$\frac{x'^2}{9} + \frac{y'^2}{4} = 1$$
We graph this in the $x'y'$ system.

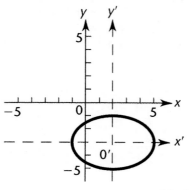

(11-2, 11-4)

47.

	Allowing digits to repeat	No digit can be repeated
O_1: Selecting first digit	N_1: 10 ways	N_1: 10 ways
O_2: Selecting second digit	N_2: 10 ways	N_2: 9 ways
$\vdots$	$\vdots$	$\vdots$
O_9: Selecting ninth digit	N_9: 10 ways	N_9: 2 ways

$$10\cdot10\cdot10\cdot10\cdot10\cdot10\cdot10\cdot10\cdot10$$
$$= 10^9 \text{ zip codes}$$

$$P_{10,9} = 10\cdot9\cdot8\cdot7\cdot6\cdot5\cdot4\cdot3\cdot2$$
$$= 3,628,800 \text{ zip codes}$$

(10-1)

48. Prove P_n: $\dfrac{1}{1 \cdot 3} + \dfrac{1}{3 \cdot 5} + \dfrac{1}{5 \cdot 7} + \cdots + \dfrac{1}{(2n - 1)(2n + 1)} = \dfrac{n}{2n + 1}$

Part 1: Show that P_1 is true.
$$P_1: \quad \frac{1}{1 \cdot 3} = \frac{1}{2 \cdot 1 + 1}$$
$$\frac{1}{3} = \frac{1}{3} \quad P_1 \text{ is true}$$

Part 2: Show that if P_k is true, then P_{k+1} is true.
Write out P_k and P_{k+1}:

$$P_k: \quad \frac{1}{1 \cdot 3} + \frac{1}{3 \cdot 5} + \frac{1}{5 \cdot 7} + \cdots + \frac{1}{(2k - 1)(2k + 1)} = \frac{k}{2k + 1}$$

$$P_{k+1}: \quad \frac{1}{1 \cdot 3} + \frac{1}{3 \cdot 5} + \frac{1}{5 \cdot 7} + \cdots + \frac{1}{(2k - 1)(2k + 1)} + \frac{1}{(2k + 1)(2k + 3)} = \frac{k + 1}{2k + 3}$$

We start with P_k:

$$\frac{1}{1 \cdot 3} + \frac{1}{3 \cdot 5} + \frac{1}{5 \cdot 7} + \cdots + \frac{1}{(2k - 1)(2k + 1)} = \frac{k}{2k + 1}$$

Adding $\dfrac{1}{(2k + 1)(2k + 3)}$ to both sides, we get

$$\dfrac{1}{1 \cdot 3} + \dfrac{1}{3 \cdot 5} + \dfrac{1}{5 \cdot 7} + \cdots + \dfrac{1}{(2k - 1)(2k + 1)} + \dfrac{1}{(2k + 1)(2k + 3)}$$

$$= \dfrac{k}{2k + 1} + \dfrac{1}{(2k + 1)(2k + 3)}$$

$$= \dfrac{k(2k + 3)}{(2k + 1)(2k + 3)} + \dfrac{1}{(2k + 1)(2k + 3)}$$

$$= \dfrac{k(2k + 3) + 1}{(2k + 1)(2k + 3)}$$

$$= \dfrac{2k^2 + 3k + 1}{(2k + 1)(2k + 3)}$$

$$= \dfrac{(2k + 1)(k + 1)}{(2k + 1)(2k + 3)}$$

$$= \dfrac{k + 1}{2k + 3}$$

We have shown that if P_k is true, then P_{k+1} is true.

Conclusion: P_n is true for positive integers n.   (*10-2*)

49. Case 1: No repetition.
The sample space S is the set of all possible sequences of three digits, chosen from {1, 2, 3, 4, 5} with no repeats.
$n(S) = P_{5,3} = 5 \cdot 4 \cdot 3 = 60$
The event E is the set of those sequences that end in 2 or 4.
$n(E) = N_1 \cdot N_2 \cdot N_3$
where
$N_1 = 2$ (digits 2 or 4 only for the last digit)
$N_2 = 4$ (digits remaining for middle digit after last digit is chosen)
$N_3 = 3$ (digits remaining for first digit after others are chosen)
$n(E) = 2 \cdot 4 \cdot 3 = 24$
$P(E) = \dfrac{n(E)}{n(S)} = \dfrac{24}{60} = \dfrac{2}{5}$
Case 2: Allowing repetition
The sample space S is the set of all possible three digit numbers chosen from {1, 2, 3, 4, 5}.
$n(S) = 5 \cdot 5 \cdot 5 = 125$
The event E is the set of those numbers that end in 2 or 4.
$n(E) = 5 \cdot 5 \cdot 2 = 50$
$P(E) = \dfrac{n(E)}{n(S)} = \dfrac{50}{125} = \dfrac{2}{5}$ (*10-5*)

50. $(x - 2i)^6 = [x + (-2i)]^6 = \displaystyle\sum_{k=0}^{6} \binom{6}{k} x^{6-k} (-2i)^k$

$$= \binom{6}{0} x^6 + \binom{6}{1} x^5 (-2i)^1 + \binom{6}{2} x^4 (-2i)^2 + \binom{6}{3} x^3 (-2i)^3$$

$$+ \binom{6}{4} x^2 (-2i)^4 + \binom{6}{5} x(-2i)^5 + \binom{6}{6} (-2i)^6$$

$$= x^6 - 12ix^5 - 60x^4 + 160ix^3 + 240x^2 - 192ix - 64$$

(*10-6*)

51. Let $P = (x, y)$ be a point on the parabola. Then by the definition of the parabola, the distance from $P = (x, y)$ to the focus $F = (6, 1)$ must equal the perpendicular distance from P to the directrix at $D = (x, 3)$. Applying the distance formula, we have

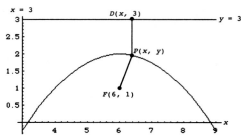

$$d(P, F) = d(P, D)$$
$$\sqrt{(x - 6)^2 + (y - 1)^2} = \sqrt{(x - x)^2 + (y - 3)^2}$$
$$(x - 6)^2 + (y - 1)^2 = (x - x)^2 + (y - 3)^2$$
$$x^2 - 12x + 36 + y^2 - 2y + 1 = 0 + y^2 - 6y + 9$$
$$x^2 - 12x + 4y + 28 = 0$$

(11-1)

52. Make a rough sketch of the ellipse, locate focus and x intercepts, then determine y intercepts using the special triangle relationship.

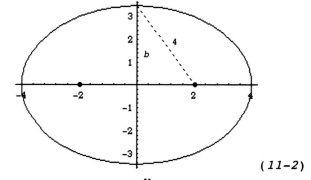

$a = 4$
$b^2 = a^2 - c^2$
$\quad = 4^2 - 2^2$
$\quad = 12$
$b = \sqrt{12} = 2\sqrt{3}$
y intercepts: $\pm 2\sqrt{3}$

(11-2)

53. Sketch the asymptote rectangle, label known parts, and use the Pythagorean Theorem.

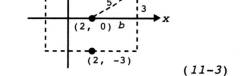

$a = 3 \qquad c = 5$
$b^2 = c^2 - a^2$
$\quad = 5^2 - 3^2$
$\quad = 16$
$b = 4$
So the length of the conjugate axis is $2b = 8$.

(11-3)

54. We can select any three of the seven points to use as vertices of the triangle. Order is not important.

$$C_{7,3} = \frac{7!}{3!(7 - 3)!} = 35 \text{ triangles}$$

(10-4)

55. P_n: $2^n < n!$ $\quad n \geq 4$ (for n an integer, this is identical with the condition $n > 3$)

Part 1: Show that P_4 is true

P_4: $2^4 < 4!$
$\quad\quad 16 < 24$ True.

Part 2: Show that if P_k is true, then P_{k+1} is true.
Write out P_k and P_{k+1}:

$\quad P_k$: $2^k < k!$

$\quad P_{k+1}$: $2^{k+1} < (k + 1)!$

We start with P_k:

$\quad 2^k < k!$

Multiplying both sides by 2: $2 \cdot 2^k < 2 \cdot k!$
Now, $k > 3$, so $1 < k$, $2 < k + 1$, so $2 \cdot k! < (k + 1)k!$

Therefore, $2^{k+1} = 2 \cdot 2^k < 2 \cdot k! < (k+1)k! = (k+1)!$
$$2^{k+1} < (k+1)!$$
So if P_k is true, then P_{k+1} is true.

Conclusion: P_n is true for all $n \geq 4$. (10-2)

56. To prove $a_n = b_n$ for all positive integers n, write:

P_n: $a_n = b_n$

Proof: Part 1: Show P_1 is true.

P_1: $a_1 = b_1$. $a_1 = 3$. $b_1 = 2^1 + 1 = 3$.
So, $a_1 = b_1$

Part 2: Show that if P_k is true, then P_{k+1} is true.
Write out P_k and P_{k+1}.

$\quad P_k$: $a_k = b_k$
$\quad\quad P_{k+1}$: $a_{k+1} = b_{k+1}$

We start with P_k:

$\quad a_k = b_k$

Now, $a_{k+1} = 2a_{k+1-1} - 1 = 2a_k - 1 = 2b_k - 1 = 2(2^k + 1) - 1$
$$\quad\quad = 2 \cdot 2^k + 2 - 1 = 2^1 \cdot 2^k + 1 = 2^{k+1} + 1 = b_{k+1}$$

Therefore, $a_{k+1} = b_{k+1}$
So if P_k is true, then P_{k+1} is true.

Conclusion: P_n is true for all positive integers n. So, $\{a_n\} = \{b_n\}$. (10-2)

57. Make a rough sketch of the situation.

We are given $d(P, A) = 3d(P, B)$. Applying the distance formula, we have

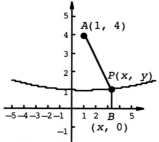

$$\sqrt{(x-1)^2 + (y-4)^2} = 3\sqrt{(x-x)^2 + (y-0)^2}$$
$$(x-1)^2 + (y-4)^2 = 9[(x-x)^2 + (y-0)^2]$$
$$x^2 - 2x + 1 + y^2 - 8y + 16 = 9y^2$$
$$x^2 - 2x - 8y^2 - 8y + 17 = 0$$

is the equation of the curve. Completing the square relative to x and y, we have

$$x^2 - 2x + 1 - 8\left(y^2 + y + \frac{1}{4}\right) - 1 + 2 + 17 = 0$$

$$(x-1)^2 - 8\left(y + \frac{1}{2}\right)^2 = -18$$

$$\frac{\left(y + \frac{1}{2}\right)^2}{\frac{18}{8}} - \frac{(x-1)^2}{18} = 1$$

This is the equation of a hyperbola. (11-3)

58. The sample space S is the set of all possible ways of selecting 3 bulbs out of 12.

$n(S) = C_{12,3}$

P(at least one defective bulb) $= 1 - P$(no defective bulbs chosen)

The event, no defective bulbs chosen, is the set of all possible 3-bulb subsets of the 8 acceptable bulbs. $n(E) = C_{8,3}$

Then P(at least one defective bulb) $= 1 - \dfrac{n(E)}{n(S)} = 1 - \dfrac{C_{8,3}}{C_{12,3}}$

$$= 1 - \dfrac{56}{220} = \dfrac{41}{55} = .7\overline{45} \qquad (10\text{-}5)$$

59. We are asked for the sum of an infinite geometric series.
$a = \$2,000,000(0.75)$
$r = 0.75 \quad |r| \le 1$, so the series has a sum,

$$S_\infty = \dfrac{a_1}{1 - r}$$

$$= \dfrac{\$2,000,000(0.75)}{1 - 0.75}$$

$$= \dfrac{\$2,000,000(0.75)}{0.25}$$

$$= \$6,000,000 \qquad (10\text{-}3)$$

60. Let x = the length and y = the width.
The perimeter is $2x + 2y$ so $2x + 2y = 24$.
The area is xy so $xy = 32$.

We solve the second equation for y and substitute into the first:

$$xy = 32 \qquad\qquad 2x + 2\left(\dfrac{32}{x}\right) = 24$$

$$y = \dfrac{32}{x} \qquad\qquad 2x + \dfrac{64}{x} = 24 \quad \text{(Multiply both sides by } x.)$$

$$2x^2 + 64 = 24x$$
$$2x^2 - 24x + 64 = 0$$
$$2(x^2 - 12x + 32) = 0$$
$$2(x - 8)(x - 4) = 0$$
$$x = 8 \quad \text{or} \quad x = 4$$

If $x = 8$, $y = \dfrac{32}{8} = 4$. If $x = 4$, $y = \dfrac{32}{4} = 8$. The dimensions are 4m × 8m. $(11\text{-}5)$

61.

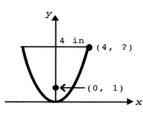

From the figure, we see that the parabola can be positioned so that it is opening up with axis the y axis, and it has an equation of the form $x^2 = 4ay$. Since the focus is at $(0, a) = (0, 1)$, $a = 1$ and the equation of the parabola is $x^2 = 4y$.

Since the depth represents the y coordinate y_1 of a point on the parabola with $x = 4$, we have
$$4^2 = 4y_1$$
$$y_1 = \text{depth} = 4 \text{ in.} \qquad (11\text{-}1)$$

62.

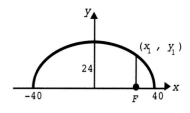

From the figure, we see that the x and y intercepts of the ellipse must be 40 and 24 respectively. The equation of the ellipse must be

$$\dfrac{x^2}{(40)^2} + \dfrac{y^2}{(24)^2} = 1$$

or $\dfrac{x^2}{1,600} + \dfrac{y^2}{576} = 1$

To find the distance c of each focus from the center of the arch, we use the special triangle relationship. $a = 40$, $b = 24$

$$c^2 = a^2 - b^2$$
$$= 40^2 - 24^2$$
$$c^2 = 1024$$
$$c = 32 \text{ ft.}$$

To find the height of the arch above each focus, we need the y coordinate y_1 of the point P whose x coordinate is 32. Since P is on the ellipse, we have

$$\frac{32^2}{1,600} + \frac{y_1^2}{576} = 1$$

$$\frac{1,024}{1,600} + \frac{y_1^2}{576} = 1$$

$$0.64 + \frac{y_1^2}{576} = 1$$

$$\frac{y_1^2}{576} = 0.36$$

$$y_1^2 = 576(0.36)$$

$$y_1^2 = 207.36$$

$$y_1 = 14.4 \text{ ft.}$$

(*11-2*)

63. (A) P (this event) $= \dfrac{130}{1,000} = .13$

(B) P (this event) $= \dfrac{80 + 90}{1,000} = .17$

(C) P (this event) $= P$ (being an independent) $+ P$ (being over 59) $- P$ (being both)

$= \dfrac{120}{1,000} + \dfrac{230}{1,000} - \dfrac{30}{1,000} = .32$ (*10-5*)

APPENDIX A   BASIC ALGEBRA REVIEW

Section A-1

1. True   3. False   5. False   7. True

9. False   11. True   13. 1

15. Division by zero is not defined. Undefined.

17. $0 \div (9 \cdot 8) = 0 \div 72 = 0$

19. $0 \cdot \left(100 + \dfrac{1}{100}\right) = 0$ since $a \cdot 0 = 0 \cdot a = 0$ for all real numbers a.

21. $\dfrac{4}{9} + \dfrac{12}{5} = \dfrac{4 \cdot 5 + 12 \cdot 9}{9 \cdot 5} = \dfrac{20 + 108}{45} = \dfrac{128}{45}$

23. $-\left(\dfrac{1}{100} + \dfrac{4}{25}\right) = -\left(\dfrac{1}{100} + \dfrac{16}{100}\right) = -\dfrac{17}{100}$ or $\dfrac{-17}{100}$

25. $\left(\dfrac{3}{8}\right)^{-1} + 2^{-1} = \dfrac{8}{3} + \dfrac{1}{2} = \dfrac{8 \cdot 2 + 1 \cdot 3}{3 \cdot 2} = \dfrac{16 + 3}{6} = \dfrac{19}{6}$

27. Commutative ($\cdot$)   29. Distributive   31. Inverse ($\cdot$)

33. Inverse (+)   35. Identity (+)   37. Negatives (Theorem 1)

39. The even integers between –3 and 5 are –2, 0, 2, and 4. The set is written {–2, 0, 2, 4}.

41. The letters in "status" are: s, t, a, u. The set is written $\{s, t, a, u\}$, or, equivalently, $\{a, s, t, u\}$.

43. Since there are no months starting with B, the set is empty. $\varnothing$

45. (A) The empty set, $\varnothing$, is a subset of every set.
$\{a\}$ and $\{b\}$ are subsets of S_2.
S_2 is a subset of itself.
There are four subsets of S_2.

(B) The empty set, $\varnothing$, is a subset of every set.
$\{a\}$, $\{b\}$, and $\{c\}$ are one-member subsets of S_3.
$\{b, c\}$, $\{a, c\}$, and $\{a, b\}$ are two-member subsets of S_3.
S_3 is a subset of itself.
There are $1 + 3 + 3 + 1 = 8$ subsets of S_3.

(C) The empty set, $\varnothing$, is a subset of every set.
$\{a\}$, $\{b\}$, $\{c\}$, and $\{d\}$ are one-member subsets of S_4.
$\{a, b\}$, $\{a, c\}$, $\{a, d\}$, $\{b, c\}$, $\{b, d\}$, and $\{c, d\}$ are two-member subsets of S_4.
$\{b, c, d\}$, $\{a, c, d\}$, $\{a, b, d\}$, and $\{b, c, d\}$ are three-member subsets of S_4.
S_4 is a subset of itself.
There are $1 + 4 + 6 + 4 + 1 = 16$ subsets of S_4.

47. Yes. This restates Zero Property (2). (Theorem 2, Part 2)

49. (A) True.

(B) False, $\dfrac{2}{3}$ is an example of a real number that is not irrational.

(C) True.

51. $\frac{3}{5}$ and -1.43 are two examples of infinitely many.

53. (A) $\{1, \sqrt{144}\}$ (B) $\{-3, 0, 1, \sqrt{144}\}$ (C) $\left\{-3, -\frac{2}{3}, 0, 1, \frac{9}{5}, \sqrt{144}\right\}$ (D) $\{\sqrt{3}\}$

55. (A) 0.888 888…; repeating; repeated digit: 8
(B) 0.272 727…; repeating; repeated digits: 27
(C) 2.236 067 977…; nonrepeating and nonterminating
(D) 1.375; terminating

57. (A) True; commutative property for addition.
(B) False; for example $3 - 5 \neq 5 - 3$.
(C) True; commutative property for multiplication.
(D) False; for example $9 \div 3 \neq 3 \div 9$.

59. F $3 - 8 = -5$ is one of many counterexamples.   **61.** T

63. F $\sqrt{2} \cdot \frac{\sqrt{2}}{2} = 1$ is one of many counterexamples.   **65.** T

67. (A) List each element of A. Follow these with each element of B that is not yet listed. $\{1,2,3,4,6\}$
(B) List each element of A that is also an element of B $\{2,4\}$.

69.   Let $c = 0.090909…$
Then $100c = 9.0909…$
$100c - c = (9.0909…) - (0.090909…)$
$99c = 9$
$c = \frac{9}{99} = \frac{1}{11}$

71.   23
$\underline{12}$
46      $23 \cdot 2$
$\underline{230}$ $23 \cdot 10$
276

$23 \cdot 12 = 23(2 + 10)$
$= 23 \cdot 2 + 23 \cdot 10$
$= 46 + 230$
$= 276$

Section A-2

1. 256   **3.** $\left(\frac{2}{3}\right)^4 = \frac{2^4}{3^4} = \frac{16}{81}$ **5.** $4^{-4} = \frac{1}{4^4} = \frac{1}{256}$ **7.** 625

9. $(-3)^{-1} = \frac{1}{-3} = \frac{-1}{3}$ **11.** $-7^{-2} = -\frac{1}{7^2} = \frac{-1}{49}$ **13.**   1

15. 10   **17.** 5   **19.** $9^{-3/2} = \frac{1}{9^{3/2}} = \frac{1}{\left(9^{1/2}\right)^3} = \frac{1}{3^3} = \frac{1}{27}$

21. Irrational   **23.** $1^{-3} + 3^{-1} = \frac{1}{1^3} + \frac{1}{3^1} = \frac{1}{1} + \frac{1}{3} = \frac{3+1}{3} = \frac{4}{3}$

25. $\frac{5}{5^4} = \frac{1}{5^3} = \frac{1}{125}$ **27.** $x^5 x^{-2} = x^{5+(-2)} = x^3$

29. $(2y)(3y^2)(5y^4) = 2 \cdot 3 \cdot 5 yy^2 y^4 = 30 y^{1+2+4} = 30 y^7$

31. $(a^2 b^3)^5 = (a^2)^5 (b^3)^5 = a^{10} b^{15}$ **33.** $u^{1/3} u^{5/3} = u^{(1/3)+(5/3)} = u^{6/3} = u^2$

35. $(x^{-3})^{1/6} = x^{-3/6} = x^{-1/2} = \frac{1}{x^{1/2}}$

37. $(49 a^4 b^{-2})^{1/2} = 49^{1/2} (a^4)^{1/2} (b^{-2})^{1/2} = 7a^2 b^{-1} = \frac{7a^2}{b}$

39. $45,320,000 = 4.5320000. \times 10^7$

7 places left

$= 4.532 \times 10^7$

41. $0.066 = 0.06.6 \times 10^{-2} = 6.6 \times 10^{-2}$

2 places right

negative exponent

43. $0.000\ 000\ 084 = 8.4 \times 10^{-8}$

8 places right

45. $9 \times 10^{-5} = 0.00009. = 0.000\ 09$

5 places left

47. $3.48 \times 10^6 = 3.480\ 000. = 3,480,000$

6 places right

49. $4.2 \times 10^{-9} = 0.000\ 000\ 004.2 = 0.000\ 000\ 004\ 2$

9 places left

51. $\left(\dfrac{x^4 y^{-1}}{x^{-2} y^3}\right)^2 = \dfrac{x^8 y^{-2}}{x^{-4} y^6} = x^{8-(-4)} y^{-2-6} = x^{12} y^{-8} = \dfrac{x^{12}}{y^8}$

53. $\left(\dfrac{2x^{-3} y^2}{4xy^{-1}}\right)^{-2} = \left(\dfrac{x^{-3} y^2}{2xy^{-1}}\right)^{-2} = \dfrac{x^6 y^{-4}}{2^{-2} x^{-2} y^2} = 2^2 x^{6-(-2)} y^{-4-2} = 4x^8 y^{-6} = \dfrac{4x^8}{y^6}$

55. $\left(\dfrac{a^{-3}}{b^4}\right)^{1/12} = \dfrac{a^{-3/12}}{b^{4/12}} = \dfrac{a^{-1/4}}{b^{1/3}} = \dfrac{1}{a^{1/4} b^{1/3}}$

57. $\left(\dfrac{4x^{-2}}{y^4}\right)^{-1/2} = \dfrac{4^{-1/2} x^1}{y^{-2}} = \dfrac{xy^2}{4^{1/2}} = \dfrac{xy^2}{2}$

Common Error: $\dfrac{4x^1}{y^{-2}}$. This is wrong; the exponent $(-\frac{1}{2})$ applies to constant as well as variable factors.

59. $\left(\dfrac{8a^{-4} b^3}{27a^2 b^{-3}}\right)^{1/3} = \dfrac{8^{1/3} a^{-4/3} b^1}{27^{1/3} a^{2/3} b^{-1}} = \dfrac{2b^{1-(-1)}}{3a^{2/3-(-4/3)}} = \dfrac{2b^2}{3a^2}$

61. $-3(x^3 + 3)^{-4}(3x^2) = -3 \cdot \dfrac{1}{(x^3 + 3)^4} \cdot 3x^2$

$\qquad\qquad = \dfrac{-9x^2}{(x^3 + 3)^4}$

Common Error: $(x^3 + 3)^{-4} \neq x^{-12} + 3^{-4}$

63. $2^{3^2} = 64$ on the assumption that 2^3^2 is entered, without parentheses.

65. The identity property for multiplication states that for any real number x, $x(1) = x$, and 1 is the only real number with that property. Therefore, if $a^m a^0 = a^m$, since $a^m(1) = a^m$, a^0 should equal 1.

67. $\dfrac{(32.7)(0.000\ 000\ 008\ 42)}{(0.0513)(80,700,000,000)} = \dfrac{(32.7)(8.42 \times 10^{-9})}{(0.0513)(8.07 \times 10^{10})} = 6.65 \times 10^{-17}$

69. $\dfrac{(5,760,000,000)}{(527)(0.000\ 007\ 09)} = \dfrac{5.76 \times 10^9}{(527)(7.09 \times 10^{-6})} = 1.54 \times 10^{12}$

71. $15^{5/4} = 15^{1.25} = 29.52$

73. $103^{-3/4} = 103^{-0.75} = 0.03093$

75. $2.876^{8/5} = 2.876^{1.6} = 5.421$

77. $(0.000\ 000\ 077\ 35)^{-2/7} = (7.735 \times 10^{-8})^{(-2 \div 7)} = 107.6$

79. There are many examples; one simple choice is $x = y = 1$. Then $(x + y)^{1/2} = (1 + 1)^{1/2} = 2^{1/2} = \sqrt{2}$, and $x^{1/2} + y^{1/2} = 1^{1/2} + 1^{1/2} = 1 + 1 = 2$; the left side is not equal to the right side.

81. There are many examples; one simple choice is $x = y = 1$. Then $(x + y)^{1/3} = (1 + 1)^{1/3} = 2^{1/3} = \sqrt[3]{2}$ and $\dfrac{1}{(x + y)^3} = \dfrac{1}{(1 + 1)^3} = \dfrac{1}{2^3} = \dfrac{1}{8}$; the left side is not equal to the right side.

83. $(a^{3/n}b^{3/m})^{1/3} = a^{(1/3)(3/n)}b^{(1/3)(3/m)} = a^{1/n}\ b^{1/m}$

85. $(x^{m/4}y^{n/3})^{-12} = x^{-12(m/4)}y^{-12(n/3)} = x^{-3m}\ y^{-4n} = \dfrac{1}{x^{3m}y^{4n}}$

87. (A) Since $(x^2)^{1/2}$ represents the positive real square root of x^2, it will not equal x if x is negative. So any negative value of x, for example $x = -2$, will make the left side $\neq$ right side. $[(-2)^2]^{1/2} = 4^{1/2} = 2 \neq -2$.

(B) Similarly any positive (or zero) value of x, for example $x = 2$, will make the left side = right side. $[2^2]^{1/2} = 4^{1/2} = 2 = 2$.

(C) Any value of x will make the left side = right side.
$[2^3]^{1/3} = 8^{1/3} = 2 \qquad [(-2)^3]^{1/3} = [-8]^{1/3} = -2 \qquad [0^3]^{1/3} = 0^{1/3} = 0$
There is no real value of x such that $(x^3)^{1/3} \neq x$.

89. No. Any negative value of b, for example -4, will make $(b^m)^{1/n}$ not a real number if n is even and m is odd. So $[(-4)^3]^{1/2} = [-64]^{1/2}$ is not a real number.

91. mass of earth in pounds
= mass of earth in grams $\times$ number of pounds/gram
= $6.1 \times 10^{27} \times 2.2 \times 10^{-3}$
= 13.42×10^{24}
= 1.342×10^{25}
= 1.3×10^{25} pounds to two significant digits

93. 1 operation in 10^{-8} seconds means $1 \div 10^{-8}$ operations/second, that is 10^8 operations in 1 second, that is, 100,000,000 or 100 million. Similarly, 1 operation in 10^{-10} seconds means $1 \div 10^{-10}$, that is 10,000,000,000 or 10 billion operations in 1 second.

Since 1 minute is 60 seconds, multiply each number by 60 to get $60 \times 10^8 = 6 \times 10^9$ or 6,000,000,000 or 6 billion operations in 1 minute and $60 \times 10^{10} = 6 \times 10^{11}$ or 600,000,000,000 or 600 billion operations in 1 minute.

95. One person's share of national debt
= amount of national debt $\div$ number of persons = 6,760,000,000,000 $\div$ 291,000,000
= $6.76 \times 10^{12} \div 2.91 \times 10^8$
= 2.32×10^4 or \$23,200 per person to three significant digits

97. We are asked to calculate
$N = 10x^{3/4}y^{1/4}$ given $x = 256$ units of labor and $y = 81$ units of capital.
$$N = 10(256)^{3/4}(81)^{1/4}$$
$$= 10(256^{1/4})^3(81)^{1/4}$$
$$= 10(4^3)(3)$$
$$= 1{,}920 \text{ units of finished product.}$$

99. We are asked to calculate
$d = 0.0212v^{7/3}$ given $v = 70$ miles/hour.
$$d = 0.0212(70)^{7/3}$$
$$= 0.0212(70)^{(7 \div 3)}$$
$$= 428 \text{ feet (to the nearest ft)}$$

Section A-3

1. $\sqrt[5]{32}$

3. $8\sqrt[3]{x^2}$

5. $4x^{-1/2} = 4\sqrt{x^{-1}}$ or $\dfrac{4}{\sqrt{x}}$

7. $\sqrt[3]{x} - \sqrt[3]{y}$

9. $361^{1/2}$

11. $4xy^{3/5}$

13. $(x^2 + y^2)^{1/3}$

15. -2

17. $-\sqrt{128} = -\sqrt{64 \cdot 2} = -\sqrt{8^2 \cdot 2} - \sqrt{8^2}\sqrt{2} = -8\sqrt{2}$

19. $\sqrt{27} - 5\sqrt{3} = \sqrt{9 \cdot 3} - 5\sqrt{3} = \sqrt{3^2 \cdot 3} - 5\sqrt{3} = \sqrt{3^2}\sqrt{3} - 5\sqrt{3} = 3\sqrt{3} - 5\sqrt{3} = -2\sqrt{3}$

21. $\sqrt[3]{5} - \sqrt[3]{25} + \sqrt[3]{125 \cdot 5} = \sqrt[3]{5} - \sqrt[3]{25} + \sqrt[3]{5^3 \cdot 5} = \sqrt[3]{5} - \sqrt[3]{25} + \sqrt[3]{5^3}\sqrt[3]{5}$
$$= \sqrt[3]{5} - \sqrt[3]{25} + 5\sqrt[3]{5} = 6\sqrt[3]{5} - \sqrt[3]{25}$$

23. $\sqrt[3]{25}\sqrt[3]{10} = \sqrt[3]{250} = \sqrt[3]{125 \cdot 2} = \sqrt[3]{5^3 \cdot 2} = \sqrt[3]{5^3}\sqrt[3]{2} = 5\sqrt[3]{2}$

25. $\sqrt{9x^8y^4} = \sqrt{9}\sqrt{x^8}\sqrt{y^4} = 3x^4y^2$

27. $\sqrt[4]{16m^4n^8} = \sqrt[4]{16}\sqrt[4]{m^4}\sqrt[4]{n^8} = 2mn^2$

29. $\sqrt[4]{m^2} = \sqrt[2 \cdot 2]{m^{2 \cdot 1}} = \sqrt[2]{m^1} = \sqrt{m}$

31. $\sqrt[5]{\sqrt[3]{xy}} = \sqrt[15]{xy}$

33. $\sqrt[3]{9x^2}\sqrt[3]{9x} = \sqrt[3]{(9x^2)(9x)} = \sqrt[3]{81x^3} = \sqrt[3]{27x^3 \cdot 3} = \sqrt[3]{27x^3}\sqrt[3]{3} = 3x\sqrt[3]{3}$

35. $\dfrac{1}{2\sqrt{5}} = \dfrac{1\sqrt{5}}{2\sqrt{5}\sqrt{5}} = \dfrac{\sqrt{5}}{2 \cdot 5} = \dfrac{\sqrt{5}}{10}$

37. $\dfrac{1}{\sqrt[3]{7}} = \dfrac{1\sqrt[3]{7^2}}{\sqrt[3]{7}\sqrt[3]{7^2}} = \dfrac{\sqrt[3]{49}}{\sqrt[3]{7^3}} = \dfrac{\sqrt[3]{49}}{7}$

39. $\dfrac{2}{\sqrt{44}} = \dfrac{2}{\sqrt{2^2 \cdot 11}} = \dfrac{2\sqrt{11}}{\sqrt{2^2}\sqrt{11}\sqrt{11}} = \dfrac{2\sqrt{11}}{2 \cdot 11} = \dfrac{\sqrt{11}}{11}$

41. $\dfrac{6x}{\sqrt{3x}} = \dfrac{6x}{\sqrt{3x}}\dfrac{\sqrt{3x}}{\sqrt{3x}} = \dfrac{6x\sqrt{3x}}{3x} = 2\sqrt{3x}$

43. $\dfrac{2}{\sqrt{2} - 1} = \dfrac{2}{(\sqrt{2} - 1)}\dfrac{(\sqrt{2} + 1)}{(\sqrt{2} + 1)} = \dfrac{2(\sqrt{2} + 1)}{2 - 1} = \dfrac{2(\sqrt{2} + 1)}{1} = 2(\sqrt{2} + 1) = 2\sqrt{2} + 2$

45. $\dfrac{\sqrt{2}}{\sqrt{6} + 2} = \dfrac{\sqrt{2}}{(\sqrt{6} + 2)}\dfrac{(\sqrt{6} - 2)}{(\sqrt{6} - 2)} = \dfrac{\sqrt{2}(\sqrt{6} - 2)}{6 - 4} = \dfrac{\sqrt{2}\sqrt{6} - 2\sqrt{2}}{2} = \dfrac{\sqrt{12} - 2\sqrt{2}}{2} = \dfrac{2\sqrt{3} - 2\sqrt{2}}{2}$
$$= \dfrac{2(\sqrt{3} - \sqrt{2})}{2} = \sqrt{3} - \sqrt{2}$$

47. $x\sqrt[5]{3^6x^7y^{11}} = x\sqrt[5]{(3^5x^5y^{10})(3x^2y)} = x\sqrt[5]{3^5x^5y^{10}}\sqrt[5]{3x^2y} = x(3xy^2)\sqrt[5]{3x^2y} = 3x^2y^2\sqrt[5]{3x^2y}$

49. $\dfrac{\sqrt[4]{32m^7n^9}}{2mn} = \dfrac{\sqrt[4]{16m^4n^8 \cdot 2m^3n}}{2mn} = \dfrac{2mn^2\sqrt[4]{2m^3n}}{2mn} = n\sqrt[4]{2m^3n}$

51. $\sqrt[3]{\sqrt[4]{a^9 b^3}} = \sqrt[3\cdot4]{a^{3\cdot3} b^{3\cdot1}} = \sqrt[4]{a^3 b}$

53. $\sqrt[3]{2x^2 y^4} \sqrt[3]{3x^5 y} = \sqrt[3]{(2x^2 y^4)(3x^5 y)} = \sqrt[3]{6x^7 y^5} = \sqrt[3]{(x^6 y^3)(6xy^2)} = \sqrt[3]{x^6 y^3} \sqrt[3]{6xy^2} = x^2 y \sqrt[3]{6xy^2}$

55. $\dfrac{\sqrt{2m}\sqrt{5}}{\sqrt{20m}} = \dfrac{\sqrt{10m}}{\sqrt{20m}} = \sqrt{\dfrac{10m}{20m}} = \sqrt{\dfrac{1}{2}} = \sqrt{\dfrac{1}{2}\cdot\dfrac{2}{2}} = \sqrt{\dfrac{2}{4}} = \dfrac{\sqrt{2}}{2}$ or $\dfrac{1}{2}\sqrt{2}$

57. $\dfrac{3\sqrt{y}}{2\sqrt{y}-3} = \dfrac{3\sqrt{y}}{(2\sqrt{y}-3)}\dfrac{(2\sqrt{y}+3)}{(2\sqrt{y}+3)} = \dfrac{3\sqrt{y}(2\sqrt{y}+3)}{(2\sqrt{y})^2-(3)^2} = \dfrac{6y+9\sqrt{y}}{4y-9}$

59. $\dfrac{2\sqrt{5}+3\sqrt{2}}{5\sqrt{5}+2\sqrt{2}} = \dfrac{(2\sqrt{5}+3\sqrt{2})}{(5\sqrt{5}+2\sqrt{2})}\dfrac{(5\sqrt{5}-2\sqrt{2})}{(5\sqrt{5}-2\sqrt{2})}$

$= \dfrac{2\sqrt{5}\cdot5\sqrt{5}-2\sqrt{5}\cdot2\sqrt{2}+3\sqrt{2}\cdot5\sqrt{5}-3\sqrt{2}\cdot2\sqrt{2}}{(5\sqrt{5})^2+(2\sqrt{2})^2}$

$= \dfrac{(10)(5)-4\sqrt{10}+15\sqrt{10}-(6)(2)}{(25)(5)-(4)(2)} = \dfrac{50+11\sqrt{10}-12}{125-8} = \dfrac{38+11\sqrt{10}}{117}$

61. Here we want to remove the radical terms from the *numerator*.

$\dfrac{\sqrt{t}-\sqrt{x}}{t-x} = \dfrac{(\sqrt{t}-\sqrt{x})}{t-x}\cdot\dfrac{(\sqrt{t}+\sqrt{x})}{\sqrt{t}+\sqrt{x}} = \dfrac{(\sqrt{t})^2-(\sqrt{x})^2}{(t-x)(\sqrt{t}+\sqrt{x})}$

$= \dfrac{\overset{1}{\cancel{t-x}}}{(\cancel{t-x})(\sqrt{t}+\sqrt{x})}_{1} = \dfrac{1}{\sqrt{t}+\sqrt{x}}$

63. Here we want to remove the radical terms from the *numerator*.

$\dfrac{\sqrt{x+h}-\sqrt{x}}{h} = \dfrac{(\sqrt{x+h}-\sqrt{x})}{h}\cdot\dfrac{(\sqrt{x+h}+\sqrt{x})}{(\sqrt{x+h}+\sqrt{x})} = \dfrac{(\sqrt{x+h})^2-(\sqrt{x})^2}{h(\sqrt{x+h}+\sqrt{x})} = \dfrac{x+h-x}{h(\sqrt{x+h}+\sqrt{x})}$

$= \dfrac{\overset{1}{\cancel{h}}}{\cancel{h}(\sqrt{x+h}+\sqrt{x})}_{1} = \dfrac{1}{\sqrt{x+h}+\sqrt{x}}$

65. $\sqrt{0.032\,965} = 0.1816$ **67.** $\sqrt[3]{45.0218} = (45.0218)^{1/3} = 45.0218^{(1\div3)} = 3.557$

69. $\sqrt[8]{5.477\times10^{-9}} = (5.477\times10^{-9})^{(1\div8)} = 0.09275$

71. $\sqrt[5]{9+\sqrt[5]{9}} = (9+9^{(1\div5)})^{(1\div5)} = 1.602$

73. $\sqrt{x^2}$ is the principal square root of x^2, that is, the positive (or zero) square root of x^2. The two square roots of x^2 are x and $-x$. $\sqrt{x^2} = -x$ if and only if $-x$ is positive or zero. So $\sqrt{x^2} = -x$ if and only if x is negative or zero. $x \le 0$.

75. $\sqrt[3]{x^3}$ is the real third root of x^3. This is always equal to x if x is real. All real numbers.

77. (A) $\sqrt{3}+\sqrt{5} \approx 3.968\,118\,785$ (B) $\sqrt{2+\sqrt{3}}+\sqrt{2-\sqrt{3}} \approx 2.449\,489\,743$

(C) $1+\sqrt{3} \approx 2.732\,050\,808$ (D) $\sqrt[3]{10+6\sqrt{3}} \approx 2.732\,050\,808$

(E) $\sqrt{8 + \sqrt{60}} \approx 3.968\ 118\ 785$ (F) $\sqrt{6} \approx 2.449\ 489\ 743$

So (A) and (E), (B) and (F), and (C) and (D) have the same value to 9 decimal places. To show that they are actually equal:

(A) and (E): $(\sqrt{3} + \sqrt{5})^2 = (\sqrt{3})^2 + 2\sqrt{3}\sqrt{5} + (\sqrt{5})^2 = 3 + \sqrt{4}\sqrt{3}\sqrt{5} + 5$
$= 8 + \sqrt{60}$ Therefore, since $\sqrt{3} + \sqrt{5}$ is positive, $\sqrt{3} + \sqrt{5}$ is the positive square root of $8 + \sqrt{60}$, and $\sqrt{8 + \sqrt{60}} = \sqrt{3} + \sqrt{5}$.

(B) and (F):

$$\left(\sqrt{2 + \sqrt{3}} + \sqrt{2 - \sqrt{3}}\right)^2 = \left(\sqrt{2 + \sqrt{3}}\right)^2 + 2\sqrt{2 + \sqrt{3}}\sqrt{2 - \sqrt{3}} + \left(\sqrt{2 - \sqrt{3}}\right)^2$$
$$= 2 + \sqrt{3} + 2\sqrt{(2 + \sqrt{3})(2 - \sqrt{3})} + 2 - \sqrt{3}$$
$$= 4 + 2\sqrt{2^2 - (\sqrt{3})^2}$$
$$= 4 + 2\sqrt{4 - 3}$$
$$= 4 + 2 \cdot 1$$
$$= 6$$

Therefore, since $\sqrt{2 + \sqrt{3}} + \sqrt{2 - \sqrt{3}}$ is positive, $\sqrt{2 + \sqrt{3}} + \sqrt{2 - \sqrt{3}}$ is the positive square root of 6, and $\sqrt{2 + \sqrt{3}} + \sqrt{2 - \sqrt{3}} = \sqrt{6}$.

(C) and (D):
$$(1 + \sqrt{3})^3 = (1 + \sqrt{3})(1 + \sqrt{3})(1 + \sqrt{3})$$
$$= (1^2 + 2\sqrt{3} + \sqrt{3}^2)(1 + \sqrt{3})$$
$$= (4 + 2\sqrt{3})(1 + \sqrt{3})$$
$$= 4 + 4\sqrt{3} + 2\sqrt{3} + 2\sqrt{3}\sqrt{3}$$
$$= 10 + 6\sqrt{3}$$

Therefore, $1 + \sqrt{3}$ is the real cube root of $10 + 6\sqrt{3}$, and $1 + \sqrt{3} = \sqrt[3]{10 + 6\sqrt{3}}$.

79. $\dfrac{1}{\sqrt[3]{a} - \sqrt[3]{b}} = \dfrac{1}{(\sqrt[3]{a} - \sqrt[3]{b})}\dfrac{\left[(\sqrt[3]{a})^2 + \sqrt[3]{a}\sqrt[3]{b} + (\sqrt[3]{b})^2\right]}{\left[(\sqrt[3]{a})^2 + \sqrt[3]{a}\sqrt[3]{b} + (\sqrt[3]{b})^2\right]}$

using $(x - y)(x^2 + xy + y^2) = x^3 - y^3$

$= \dfrac{1\left[\sqrt[3]{a^2} + \sqrt[3]{ab} + \sqrt[3]{b^2}\right]}{(\sqrt[3]{a})^3 - (\sqrt[3]{b})^3} = \dfrac{\sqrt[3]{a^2} + \sqrt[3]{ab} + \sqrt[3]{b^2}}{a - b}$

81. $\dfrac{1}{\sqrt{x} - \sqrt{y} + \sqrt{z}} = \dfrac{1}{\left[(\sqrt{x} - \sqrt{y}) + \sqrt{z}\right]}\dfrac{\left[(\sqrt{x} - \sqrt{y}) - \sqrt{z}\right]}{\left[(\sqrt{x} - \sqrt{y}) - \sqrt{z}\right]} = \dfrac{\sqrt{x} - \sqrt{y} - \sqrt{z}}{(\sqrt{x} - \sqrt{y})^2 + (\sqrt{z})^2}$

$= \dfrac{\sqrt{x} - \sqrt{y} - \sqrt{z}}{(\sqrt{x})^2 - 2\sqrt{x}\sqrt{y} + (\sqrt{y})^2 - (\sqrt{z})^2} = \dfrac{\sqrt{x} - \sqrt{y} - \sqrt{z}}{x - 2\sqrt{xy} + y - z}$

$= \dfrac{\sqrt{x} - \sqrt{y} - \sqrt{z}}{x - 2\sqrt{xy} + y - z} = \dfrac{\sqrt{x} - \sqrt{y} - \sqrt{z}}{x + y - z - 2\sqrt{xy}}$

$= \dfrac{(\sqrt{x} - \sqrt{y} - \sqrt{z})}{\left[(x + y - z) - 2\sqrt{xy}\right]}\dfrac{\left[(x + y - z) + 2\sqrt{xy}\right]}{\left[(x + y - z) + 2\sqrt{xy}\right]}$

$= \dfrac{(\sqrt{x} - \sqrt{y} - \sqrt{z})(x + y - z) + 2\sqrt{xy}}{(x + y - z)^2 - (2\sqrt{xy})^2}$

$= \dfrac{(\sqrt{x} - \sqrt{y} - \sqrt{z})(x + y - z) + 2\sqrt{xy}}{(x + y - z)^2 - 4xy}$

83. $\dfrac{\sqrt[3]{x+h}-\sqrt[3]{x}}{h} = \dfrac{\left(\sqrt[3]{x+h}-\sqrt[3]{x}\right)\left[(\sqrt[3]{x+h})^2+\sqrt[3]{x+h}\,\sqrt[3]{x}+(\sqrt[3]{x})^2\right]}{h\left[(\sqrt[3]{x+h})^2+\sqrt[3]{x+h}\,\sqrt[3]{x}+(\sqrt[3]{x})^2\right]}$

$= \dfrac{(\sqrt[3]{x+h})^3-(\sqrt[3]{x})^3}{h\left[\sqrt[3]{(x+h)^2}+\sqrt[3]{x(x+h)}+\sqrt[3]{x^2}\right]}$

Using $(a-b)(a^2+ab+b^2)=a^3-b^3$

$= \dfrac{x+h-x}{h\left[\sqrt[3]{(x+h)^2}+\sqrt[3]{x(x+h)}+\sqrt[3]{x^2}\right]}$

$= \dfrac{h}{h\left[\sqrt[3]{(x+h)^2}+\sqrt[3]{x(x+h)}+\sqrt[3]{x^2}\right]}$

$= \dfrac{1}{\sqrt[3]{(x+h)^2}+\sqrt[3]{x(x+h)}+\sqrt[3]{x^2}}$

85. $\sqrt[kn]{x^{km}}=(x^{km})^{1/kn}=x^{km/kn}=x^{m/n}=\sqrt[n]{x^m}$

87. $\dfrac{M_0}{\sqrt{1-\dfrac{v^2}{c^2}}} = M_0 \div \sqrt{1-\dfrac{v^2}{c^2}}$

$= M_0 \div \sqrt{\dfrac{c^2-v^2}{c^2}} = M_0 \div \dfrac{\sqrt{c^2-v^2}}{c}$

$= M_0 \cdot \dfrac{c}{\sqrt{c^2-v^2}}$

Now rationalize the denominator

$= M_0 \cdot \dfrac{c}{\sqrt{c^2-v^2}}\dfrac{\sqrt{c^2-v^2}}{\sqrt{c^2-v^2}}$

$= \dfrac{M_0c\sqrt{c^2-v^2}}{c^2-v^2}$

Section A-4

1. 4

3. $(x^4-2x^2+3)+(x^3-1)=x^4-2x^2+3+x^3-1=x^4+x^3-2x^2+2$
This polynomial has degree 4.

5. $(x^4-2x^2+3)(x^3-1)=x^7-x^4-2x^5+2x^2+3x^3-3$
$=x^7-2x^5-x^4+3x^3+2x^2-3$

7. $(x^4-2x^2+3)-(x^3-1)=x^4-2x^2+3-x^3+1=x^4-x^3-2x^2+4$

9. Yes. 2   **11.** No   **13.** Yes. 3

15. $2(x-1)+3(2x-3)-(4x-5)=2x-2+6x-9-4x+5=4x-6$

17. $2y-3y[4-2(y-1)]=2y-3y[4-2y+2]$
$=2y-3y[6-2y]$
$=2y-18y+6y^2$
$=6y^2-16y$

19. $m^2 - n^2$

21. $(4t - 3)(t - 2) = 4t^2 - 8t - 3t + 6 = 4t^2 - 11t + 6$

23. $(5y - 1)(3 - 2y) = 15y - 10y^2 - 3 + 2y = -10y^2 + 17y - 3$

25. $(3x + 2y)(x - 3y) = 3x^2 - 9xy + 2xy - 6y^2 = 3x^2 - 7xy - 6y^2$

27. $4m^2 - 49$

29. $(6x - 4y)(5x + 3y) = 30x^2 + 18xy - 20xy - 12y^2 = 30x^2 - 2xy - 12y^2$

31. $9x^2 - 4y^2$

33. $(4x - y)^2 = (4x)^2 - 2(4x)y + y^2 = 16x^2 - 8xy + y^2$

35. $(a + b)(a^2 - ab + b^2) = a^3 - a^2b + ab^2 + a^2b - ab^2 + b^3 = a^3 + b^3$

37.
$$\begin{aligned}
2x - 3\{x + 2[x - (x + 5)] + 1\} &= 2x - 3\{x + 2[x - x - 5] + 1\} \\
&= 2x - 3\{x + 2(-5) + 1\} \\
&= 2x - 3\{x - 10 + 1\} \\
&= 2x - 3\{x - 9\} \\
&= 2x - 3x + 27 \\
&= -x + 27
\end{aligned}$$

39.
$$\begin{aligned}
2\{3[a - 4(1 - a)] - (5 - a)\} &= 2\{3[a - 4 + 4a] - 5 + a\} \\
&= 2\{3[5a - 4] - 5 + a\} \\
&= 2\{15a - 12 - 5 + a\} \\
&= 2\{16a - 17\} \\
&= 32a - 34
\end{aligned}$$

41.
$$\begin{aligned}
(2x^2 - 3x + 1)(x^2 + x - 2) &= 2x^4 + 2x^3 - 4x^2 - 3x^3 - 3x^2 + 6x + x^2 + x - 2 \\
&= 2x^4 - x^3 - 6x^2 + 7x - 2
\end{aligned}$$

43.
$$\begin{aligned}
(x - 2y)^2(x + 2y)^2 &= [(x - 2y)(x + 2y)]^2 \\
&= [x^2 - 4y^2]^2 \\
&= x^4 - 8x^2y^2 + 16y^4
\end{aligned}$$

45.
$$\begin{aligned}
(3u - 2v)^2 - (2u - 3v)(2u + 3v) &= (9u^2 - 12uv + 4v^2) - (4u^2 - 9v^2) \\
&= 9u^2 - 12uv + 4v^2 - 4u^2 + 9v^2 \\
&= 5u^2 - 12uv + 13v^2
\end{aligned}$$

47.
$$\begin{aligned}
(z + 2)(z^2 - 2z + 3) + z - 7 &= z^3 - 2z^2 + 3z + 2z^2 - 4z + 6 + z - 7 \\
&= z^3 - 1
\end{aligned}$$

49.
$$\begin{aligned}
(2m - n^3) &= (2m - n)^2(2m - n) \\
&= (4m^2 - 4mn + n^2)(2m - n) \\
&= 8m^3 - 4m^2n - 8m^2n + 4mn^2 + 2mn^2 - n^3 \\
&= 8m^3 - 12m^2n + 6mn^2 - n^3
\end{aligned}$$

51. $3(x + h) - 7 - (3x - 7) = 3x + 3h - 7 - 3x + 7 = 3h$

53.
$$\begin{aligned}
2(x + h)^2 - 3(x + h) - (2x^2 - 3x) &= 2(x^2 + 2xh + h^2) - 3x - 3h - 2x^2 + 3x \\
&= 2x^2 + 4xh + 2h^2 - 3x - 3h - 2x^2 + 3x \\
&= 4xh + 2h^2 - 3h
\end{aligned}$$

55.
$$\begin{aligned}
2(x + h)^2 &- 4(x + h) - 9 - (2x^2 - 4x - 9) \\
&= 2(x^2 + 2xh + h^2) - 4x - 4h - 9 - 2x^2 + 4x + 9 \\
&= 2x^2 + 4xh + 2h^2 - 4x - 4h - 9 - 2x^2 + 4x + 9 \\
&= 4xh + 2h^2 - 4h
\end{aligned}$$

57. $(x + h)^3 - 2(x + h)^2 - (x^3 - 2x^2)$
$$= (x + h)^2(x + h) - 2(x + h)^2 - (x^3 - 2x^2)$$
$$= (x^2 + 2xh + h^2)(x + h) - 2(x^2 + 2xh + h^2) - x^3 + 2x^2$$
$$= x^3 + hx^2 + 2xh^2 + 2h^2x + h^2x + h^3 - 2x^2 - 4xh - 2h^2 - x^3 + 2x^2$$
$$= 3hx^2 + 3h^2x + h^3 - 4xh - 2h^2$$

59. The sum of the first two polynomials:
$(3m^2 - 2m + 5) + (4m^2 - m) = 3m^2 - 2m + 5 + 4m^2 - m = 7m^2 - 3m + 5$

The sum of the last two polynomials:
$(3m^2 - 3m - 2) + (m^3 + m^2 + 2) = 3m^2 - 3m - 2 + m^3 + m^2 + 2 = m^3 + 4m^2 - 3m$

Subtract from the sum of the last two polynomials the sum of the first two.
$(m^3 + 4m^2 - 3m) - (7m^2 - 3m + 5) = m^3 + 4m^2 - 3m - 7m^2 + 3m - 5$
$$= m^3 - 3m^2 - 5$$

61. The area of the entire region is given by $(a + b)(c + d)$. The sum of the areas of the four small rectangles is given by $ac + ad + bc + bd$. Since the area of the entire region must equal the sum of the areas of the four small rectangles, $(a + b)(c + d) = ac + ad + bc + bd$.

63. The area of the large square is given by c^2. The area of the small square is given by $(a - b)^2$. The area of each of the four triangles is given by $\frac{1}{2}ab$. Therefore,
$$(a - b)^2 + 4\left(\frac{1}{2}ab\right) = c^2$$
Simplifying, we obtain
$$a^2 - 2ab + b^2 + 2ab = c^2$$
$$a^2 + b^2 = c^2$$
This is the Pythagorean Theorem.

65. $2(x - 2)^3 - (x - 2)^2 - 3(x - 2) - 4$
$$= 2(x - 2)^2(x - 2) - (x - 2)^2 - 3(x - 2) - 4$$
$$= 2(x^2 - 4x + 4)(x - 2) - (x^2 - 4x + 4) - 3(x - 2) - 4$$
$$= 2(x^3 - 2x^2 - 4x^2 + 8x + 4x - 8) - (x^2 - 4x + 4) - 3(x - 2) - 4$$
$$= 2x^3 - 4x^2 - 8x^2 + 16x + 8x - 16 - x^2 + 4x - 4 - 3x + 6 - 4$$
$$= 2x^3 - 13x^2 + 25x - 18$$

67. $-3x\{x[x - x(2 - x)] - (x + 2)(x^2 - 3)\}$
$$= -3x\{x[x - 2x + x^2] - (x^3 - 3x + 2x^2 - 6)\}$$
$$= -3x\{x[-x + x^2] - (x^3 - 3x + 2x^2 - 6)\}$$
$$= -3x\{-x^2 + x^3 - x^3 + 3x - 2x^2 + 6\}$$
$$= -3x\{-3x^2 + 3x + 6\}$$
$$= 9x^3 - 9x^2 - 18x$$

69. One example is given by choosing $a = 1$ and $b = 1$. Then $(a + b)^2 = (1 + 1)^2 = 2^2 = 4$, but $a^2 + b^2 = 1^2 + 1^2 = 1 + 1 = 2$. So in general, $(a + b)^2 \neq a^2 + b^2$. In fact, since $(a + b)^2 = a^2 + 2ab + b^2$, this quantity can only equal $a^2 + b^2$ if $2ab = 0$. By the properties of 0, either $a = 0$ or $b = 0$. In these cases $(a + b)^2$ would equal $a^2 + b^2$, but only in these.

71. The non-zero term with the highest degree in the polynomial of degree m has degree m. This term is not changed during the addition, since there is no term in the polynomial of degree n with the degree $m(m > n)$. It will still be the highest degree term in the sum, so the sum will be a polynomial of degree m.

73. Now the non-zero term with the highest degree in each polynomial has degree m. It is possible for the coefficients of these terms to be equal in absolute value but opposite in sign, in which case they would add to 0, leaving a term of degree less than m as the term with the highest degree in the sum. Otherwise the terms of degree m will combine to give another term of degree m, which will be the highest degree term in the sum. So $(2x^4 + x^2 + 1) + (-2x^4 + x^2 + 1) = 2x^2 + 2$, but $(2x^4 + x^2 + 1) + (2x^4 + 2x^2 + 1) = 4x^4 + 3x^2 + 2$. Summarizing, the degree of the sum may be less than or equal to m.

75. There are three quantities in this problem, perimeter, length, and width. They are related by the perimeter formula $P = 2\ell + 2w$. Since x = length of the rectangle, and the width is 5 meters less than the length, $x - 5$ = width of the rectangle. So $P = 2x + 2(x - 5)$ represents the perimeter of the rectangle. Simplifying: $P = 2x + 2x - 10 = 4x - 10$ (meters)

77. There are several quantities involved in this problem. It is important to keep them distinct by using enough words. We write:

$$x = \text{number of nickels}$$
$$x - 5 = \text{number of dimes}$$
$$(x - 5) + 2 = \text{number of quarters}$$

This follows because there are five fewer dimes than nickels $(x - 5)$ and 2 more quarters than dimes (2 more than $x - 5$). Each nickel is worth 5 cents, each dime worth 10 cents, and each quarter worth 25 cents. So the value of the nickels is 5 times the number of nickels, the value of the dimes is 10 times the number of dimes, and the value of the quarters is 25 times the number of quarters.

$$\text{value of nickels} = 5x$$
$$\text{value of dimes} = 10(x - 5)$$
$$\text{value of quarters} = 25[(x - 5) + 2]$$

The value of the pile = (value of nickels) + (value of dimes)
+ (value of quarters)
$$= 5x + 10(x - 5) + 25[(x - 5) + 2]$$

Simplifying this expression, we get:

The value of the pile $= 5x + 10x - 50 + 25[x - 5 + 2]$
$$= 5x + 10x - 50 + 25[x - 3]$$
$$= 5x + 10x - 50 + 25x - 75$$
$$= 40x - 125 \text{ (cents)}$$

79. The volume of the plastic shell is equal to the volume of the larger sphere ($V = \frac{4}{3}\pi r^3$) minus the volume of the hole. Since the radius of the hole is x cm and the plastic is 0.3 cm thick, the radius of the larger sphere is $x + 0.3$ cm.

$$\begin{pmatrix} \text{Volume of} \\ \text{shell} \end{pmatrix} = \begin{pmatrix} \text{Volume of} \\ \text{larger sphere} \end{pmatrix} - \begin{pmatrix} \text{Volume of} \\ \text{hole} \end{pmatrix}$$

$$\begin{aligned}
\text{Volume} &= \tfrac{4}{3}\pi(x + 0.3)^3 - \tfrac{4}{3}\pi x^3 \\
&= \tfrac{4}{3}\pi(x + 0.3)(x + 0.3)^2 - \tfrac{4}{3}\pi x^3 \\
&= \tfrac{4}{3}\pi(x + 0.3)(x^2 + 0.6x + 0.09) - \tfrac{4}{3}\pi x^3 \\
&= \tfrac{4}{3}\pi(x^3 + 0.6x^2 + 0.09x + 0.3x^2 + 0.18x + 0.027) - \tfrac{4}{3}\pi x^3 \\
&= \tfrac{4}{3}\pi(x^3 + 0.9x^2 + 0.27x + 0.027) - \tfrac{4}{3}\pi x^3 \\
&= \tfrac{4}{3}\pi x^3 + 1.2\pi x^2 + 0.36\pi x + 0.036\pi - \tfrac{4}{3}\pi x^3 \\
&= 1.2\pi x^2 + 0.36\pi x + 0.036\pi \ (\text{cm}^3)
\end{aligned}$$

Section A-5

1. $2x^2(3x^2 - 4x - 1)$　**3.** $5xy(2x^2 + 4xy - 3y^2)$　**5.** $(x + 1)(5x - 3)$　**7.** $(y - 2z)(2w - x)$

9. $x^2 - 2x + 3x - 6 = (x^2 - 2x) + (3x - 6)$
$$= x(x - 2) + 3(x - 2)$$
$$= (x - 2)(x + 3)$$

11. $6m^2 + 10m - 3m - 5 = (6m^2 + 10m) - (3m + 5) = 2m(3m + 5) - 1(3m + 5)$
$$= (3m + 5)(2m - 1)$$

13. $2x^2 - 4xy - 3xy + 6y^2 = (2x^2 - 4xy) - (3xy - 6y^2) = 2x(x - 2y) - 3y(x - 2y)$
$$= (x - 2y)(2x - 3y)$$

15. $8ac + 3bd - 6bc - 4ad = 8ac - 4ad - 6bc + 3bd$
$$= (8ac - 4ad) - (6bc - 3bd)$$
$$= 4a(2c - d) - 3b(2c - d)$$
$$= (2c - d)(4a - 3b)$$

17. $(2x + 3)(x - 1)$　　**19.** Prime

21. $(x + 7y)(x - 2y)$

23. $(2a + 3b)(2a - 3b)$

25. $4x^2 - 20x + 25 = (2x)^2 - 2(2x)(5) + 5^2 = (2x - 5)^2$　　**27.** Prime

29. Prime

31. $6x^2 + 48x + 72 = 6(x^2 + 8x + 12) = 6(x + 2)(x + 6)$

33. $2y^3 - 22y^2 + 48y = 2y(y^2 - 11y + 24) = 2y(y - 3)(y - 8)$

35. $16x^2y - 8xy + y = y(16x^2 - 8x + 1) = y[(4x)^2 - 2(4x) + 1] = y(4x - 1)^2$

37. $(3m + 4n)(2m - 3n)$

39. $x^3y - 9xy^3 = xy(x^2 - 9y^2) = xy(x - 3y)(x + 3y)$

41. $3m(m^2 - 2m + 5)$　　　**43.** $(m + n)(m^2 - mn + n^2)$

45. $8x^3 - 125 = (2x)^3 - 5^3 = (2x - 5)[(2x)^2 + (2x)5 + 5^2]$
$$= (2x - 5)(4x^2 + 10x + 25)$$

47. $2x(x + 1)^4 + 4x^2(x + 1)^3 = 2x(x + 1)^3[(x + 1) + 2x] = 2x(x + 1)^3(3x + 1)$

49. $6(3x - 5)(2x - 3)^2 + 4(3x - 5)^2(2x - 3) = 2(3x - 5)(2x - 3)[3(2x - 3) + 2(3x - 5)]$
$$= 2(3x - 5)(2x - 3)[6x - 9 + 6x - 10]$$
$$= 2(3x - 5)(2x - 3)(12x - 19)$$

51. $5x^4(9 - x)^4 - 4x^5(9 - x)^3 = x^4(9 - x)^3[5(9 - x) - 4x]$
$$= x^4(9 - x)^3[45 - 5x - 4x]$$
$$= x^4(9 - x)^3(45 - 9x) \quad \text{or} \quad 9x^4(9 - x)^3(5 - x)$$

53. $2(x + 1)(x^2 - 5)^2 + 4x(x + 1)^2(x^2 - 5) = 2(x + 1)(x^2 - 5)[x^2 - 5 + 2x(x + 1)]$
$$= 2(x + 1)(x^2 - 5)[x^2 - 5 + 2x^2 + 2x]$$
$$= 2(x + 1)(x^2 - 5)(3x^2 + 2x - 5)$$
$$= 2(x + 1)(x^2 - 5)(3x + 5)(x - 1)$$

55. $(a - b)^2 - 4(c - d)^2 = (a - b)^2 - [2(c - d)]^2$
$= [(a - b) - 2(c - d)][(a - b) + 2(c - d)]$

57. $2am - 3an + 2bm - 3bn = (2am - 3an) + (2bm - 3bn) = a(2m - 3n) + b(2m - 3n)$
$= (2m - 3n)(a + b)$

59. Prime

61. $x^3 - 3x^2 - 9x + 27 = (x^3 - 3x^2) - (9x - 27) = x^2(x - 3) - 9(x - 3) = (x - 3)(x^2 - 9)$
$= (x - 3)(x - 3)(x + 3) = (x - 3)^2(x + 3)$

63. $a^3 - 2a^2 - a + 2 = (a^3 - 2a^2) - (a - 2) = a^2(a - 2) - 1(a - 2) = (a - 2)(a^2 - 1)$
$= (a - 2)(a + 1)(a - 1)$

65. Prime

67. $(x^2 + 2)(x^2 + 4)$

69. $m^4 - n^4 = (m^2)^2 - (n^2)^2 = (m^2 - n^2)(m^2 + n^2) = (m - n)(m + n)(m^2 + n^2)$

71. $y^4 - 11y^2 + 18 = (y^2 - 2)(y^2 - 9) = (y^2 - 2)(y - 3)(y + 3)$

73. $m^2 + 2mn + n^2 - m - n = (m^2 + 2mn + n^2) - (m + n) = (m + n)^2 - 1(m + n)$
$= (m + n)(m + n - 1)$

75. $18a^3 - 8a(x^2 + 8x + 16) = 2a[9a^2 - 4(x^2 + 8x + 16)] = 2a\{(3a)^2 - [2(x + 4)]^2\}$
$= 2a[3a - 2(x + 4)][3a + 2(x + 4)]$

77. $x^4 + 2x^2 + 1 - x^2 = (x^4 + 2x^2 + 1) - x^2 = (x^2 + 1)^2 - x^2$
$= (x^2 + 1 - x)(x^2 + 1 + x) = (x^2 - x + 1)(x^2 + x + 1)$

79. $u^7 - v^7 = (u - v)(u^6 + u^5v + u^4v^2 + u^3v^3 + u^2v^4 + uv^5 + v^6)$
Verifying: The right side =
$(u - v)u^6 + (u - v)u^5v + (u - v)u^4v^2 + (u - v)u^3v^3 + (u - v)u^2v^4$
$+ (u - v)uv^5 + (u - v)v^6$

$= u^7 - u^6v + u^6v - u^5v^2 + u^5v^2 - u^4v^3 + u^4v^3 - u^3v^4 + u^3v^4 - u^2v^5 + u^2v^5 - uv^6$
$+ uv^6 - v^7$

$= u^7 - v^7$

81. 2, 3, 5, 7, 11, 13, 17, 19, 23, 29, 31, 37, 41, 43, 47, 53, 59, 61, 67, 71, 73, 79, 83, 89, 97, 101, 103, 107, 109, 113, 127, 131, 137, 139, 149, 151, 157, 163, 167, 173, 179, 181, 191, 193, 197, 199.

83. (A) If $\sqrt{2} = \dfrac{a}{b}$, than $2 = \dfrac{a^2}{b^2}$ and $a^2 = 2b^2$.

 (B) Any factor of a must appear in the factorization of a^2 twice for each time it appears in the factorization of a. This means that it appears an even number of times (possibly 0 times).

 (C) Any factor of b^2 appears an even number of times by the reasoning of (B). Then the number of times 2 appears must be 1 more than this even number, that is, an odd number of times.

 (D) To have two different numbers of appearances of a factor contradicts the uniqueness of a prime factorization.

85. **(A)** The area of the cardboard can be written as (Original area) - (Removed area), where the original area = 20^2 = 400 and the removed area consists of 4 squares of area x^2 each;

(Original area) - (Removed area) = 400 - $4x^2$ in expanded form.

In factored form, $400 - 4x^2 = 4(100 - x^2) = 4(10 - x)(10 + x)$.

(B) See figure.

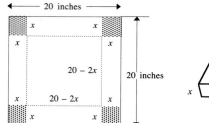 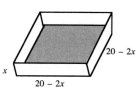

The volume of the box = length × width × height = $x(20 - 2x)(20 - 2x)$
$$= 4x(10 - x)(10 - x) \text{ in factored form}$$

In expanded form, $4x(10 - x)(10 - x) = 4x[10^2 - 2(10)x + x^2]$
$$= 4x[100 - 20x + x^2]$$
$$= 400x - 80x^2 + 4x^3$$

Section A-6

1. $\dfrac{42}{105} = \dfrac{21 \cdot 2}{21 \cdot 5} = \dfrac{2}{5}$ **3.** $\dfrac{x + 1}{x^2 + 3x + 2} = \dfrac{\overset{1}{\cancel{x + 1}}}{\cancel{(x + 1)}\,(x + 2)} = \dfrac{1}{x + 2}$
$\underset{1}{}$

5. $\dfrac{x^2 - 9}{x^2 + 3x - 18} = \dfrac{\overset{1}{\cancel{(x - 3)}}\,(x + 3)}{\underset{1}{\cancel{(x - 3)}}\,(x + 6)} = \dfrac{x + 3}{x + 6}$

7. $\dfrac{3x^2 y^3}{x^4 y} = \dfrac{x^2 y \cdot 3y^2}{x^2 y \cdot x^2} = \dfrac{3y^2}{x^2}$

9. $\dfrac{5}{6} - \dfrac{11}{15} = \dfrac{25}{30} - \dfrac{22}{30} = \dfrac{3}{30} = \dfrac{1}{10}$

11. $\left(\dfrac{b^2}{2a} \div \dfrac{b}{a^2}\right) \cdot \dfrac{a}{3b} = \left(\dfrac{b^2}{2a} \cdot \dfrac{a^2}{b}\right) \cdot \dfrac{a}{3b} = \left(\dfrac{\overset{b}{\cancel{b^2}}}{\underset{1}{2\cancel{a}}} \cdot \dfrac{\overset{a}{\cancel{a^2}}}{\underset{1}{\cancel{b}}}\right) \cdot \dfrac{a}{3b}$

$ = \dfrac{ba}{2} \cdot \dfrac{a}{3b}$

$ = \dfrac{a^2}{6}$

13. $\dfrac{1}{n} - \dfrac{1}{m} = \dfrac{m}{mn} - \dfrac{n}{mn} = \dfrac{m - n}{mn}$

15. $\dfrac{x^2 - 1}{x + 2} \div \dfrac{x + 1}{x^2 - 4} = \dfrac{x^2 - 1}{x + 2} \cdot \dfrac{x^2 - 4}{x + 1}$

$\qquad\qquad\qquad = \dfrac{(x - 1)(x + 1)}{x + 2} \cdot \dfrac{(x - 2)(x + 2)}{x + 1}$

$\qquad\qquad\qquad = \dfrac{(x - 1)\overset{1}{\cancel{(x + 1)}}}{\underset{1}{\cancel{x + 2}}} \cdot \dfrac{(x - 2)\overset{1}{\cancel{(x + 2)}}}{\underset{1}{\cancel{x + 1}}}$

$\qquad\qquad\qquad = (x - 1)(x - 2)$

17. $\dfrac{1}{c} + \dfrac{1}{b} + \dfrac{1}{a} = \dfrac{ab}{abc} + \dfrac{ac}{abc} + \dfrac{bc}{abc} = \dfrac{ab + ac + bc}{abc}$

19. $\dfrac{2a - b}{a^2 - b^2} - \dfrac{2a + 3b}{a^2 + 2ab + b^2} = \dfrac{2a - b}{(a + b)(a - b)} - \dfrac{2a + 3b}{(a + b)(a + b)}$

$\qquad\qquad\qquad\qquad = \dfrac{(2a - b)(a + b)}{(a + b)(a + b)(a - b)} - \dfrac{(2a + 3b)(a - b)}{(a + b)(a + b)(a - b)}$

$\qquad\qquad\qquad\qquad = \dfrac{(2a - b)(a + b) - (2a + 3b)(a - b)}{(a + b)^2(a - b)}$

$\qquad\qquad\qquad\qquad = \dfrac{2a^2 + ab - b^2 - (2a^2 + ab - 3b^2)}{(a + b)^2(a - b)}$

$\qquad\qquad\qquad\qquad = \dfrac{2a^2 + ab - b^2 - 2a^2 - ab + 3b^2}{(a + b)^2(a - b)}$

$\qquad\qquad\qquad\qquad = \dfrac{2b^2}{(a + b)^2(a - b)}$

21. $m + 2 - \dfrac{m - 2}{m - 1} = \dfrac{m + 2}{1} - \dfrac{m - 2}{m - 1} = \dfrac{(m + 2)(m - 1)}{m - 1} - \dfrac{(m - 2)}{m - 1}$

$\qquad\qquad\qquad = \dfrac{(m^2 + m - 2) - (m - 2)}{m - 1} = \dfrac{m^2 + m - 2 - m + 2}{m - 1}$

$\qquad\qquad\qquad = \dfrac{m^2}{m - 1}$

23. $\dfrac{3}{x - 2} - \dfrac{2}{2 - x} = \dfrac{3}{x - 2} - \dfrac{-2}{x - 2} = \dfrac{3 + 2}{x - 2} = \dfrac{5}{x - 2}$

25. $\dfrac{3}{y + 2} + \dfrac{2}{y - 2} - \dfrac{4y}{y^2 - 4} = \dfrac{3(y - 2)}{(y + 2)(y - 2)} + \dfrac{2(y + 2)}{(y + 2)(y - 2)} - \dfrac{4y}{(y + 2)(y - 2)}$

$\qquad\qquad\qquad\qquad = \dfrac{3(y - 2) + 2(y + 2) - 4y}{(y + 2)(y - 2)} = \dfrac{3y - 6 + 2y + 4 - 4y}{(y + 2)(y - 2)}$

$\qquad\qquad\qquad\qquad = \dfrac{y - 2}{(y + 2)(y - 2)} = \dfrac{\overset{1}{\cancel{(y - 2)}}}{(y + 2)\underset{1}{\cancel{(y - 2)}}} = \dfrac{1}{y + 2}$

27. $\dfrac{\dfrac{x^2}{y^2} - 1}{\dfrac{x}{y} + 1} = \dfrac{y^2\left(\dfrac{x^2}{y^2} - 1\right)}{y^2\left(\dfrac{x}{y} + 1\right)} = \dfrac{x^2 - y^2}{xy + y^2} = \dfrac{(x - y)\overset{1}{\cancel{(x + y)}}}{y\underset{1}{\cancel{(x + y)}}} = \dfrac{x - y}{y}$

29. $\dfrac{6x^3(x^2+2)^2 - 2x(x^2+2)^3}{x^4} = \dfrac{2x(x^2+2)^2[3x^2-(x^2+2)]}{x^4}$

$$= \dfrac{2x(x^2+2)^2[3x^2-x^2-2]}{x^4}$$

$$= \dfrac{2\overset{1}{\cancel{x}}(x^2+2)^2(2x^2-2)}{\underset{x^3}{\cancel{x^4}}}$$

$$= \dfrac{2(x^2+2)^2(2x^2-2)}{x^3}$$

$$= \dfrac{2(x^2+2)^2\,2(x^2-1)}{x^3}$$

$$= \dfrac{4(x^2+2)^2(x+1)(x-1)}{x^3}$$

31. $\dfrac{2x(1-3x)^3 + 9x^2(1-3x)^2}{(1-3x)^6} = \dfrac{x(1-3x)^2[2(1-3x)+9x]}{(1-3x)^6}$

$$= \dfrac{x(1-3x)^2[2-6x+9x]}{(1-3x)^6}$$

$$= \dfrac{x\cancel{(1-3x)}^2(2+3x)}{\underset{(1-3x)^4}{\cancel{(1-3x)^6}}}$$

$$= \dfrac{x(2+3x)}{(1-3x)^4}$$

33. $\dfrac{-2x(x+4)^3 - 3(3-x^2)(x+4)^2}{(x+4)^6} = \dfrac{(x+4)^2[-2x(x+4)-3(3-x^2)]}{(x+4)^6}$

$$= \dfrac{(x+4)^2[-2x^2-8x-9+3x^2]}{(x+4)^6}$$

$$= \dfrac{\cancel{(x+4)}^2(x^2-8x-9)}{\underset{(x+4)^4}{\cancel{(x+4)^6}}}$$

$$= \dfrac{x^2-8x-9}{(x+4)^4}$$

$$= \dfrac{(x+1)(x-9)}{(x+4)^4}$$

35. $\dfrac{y}{y^2-2y-8} - \dfrac{2}{y^2-5y+4} + \dfrac{1}{y^2+y-2}$

$$= \dfrac{y}{(y-4)(y+2)} - \dfrac{2}{(y-4)(y-1)} + \dfrac{1}{(y+2)(y-1)}$$

$$= \dfrac{y(y-1)}{(y-1)(y-4)(y+2)} - \dfrac{2(y+2)}{(y-4)(y-1)(y+2)} + \dfrac{1(y-4)}{(y-4)(y-1)(y+2)}$$

$$= \frac{y^2 - y - 2y - 4 + y - 4}{(y - 4)(y - 1)(y + 2)} \quad = \frac{y^2 - 2y - 8}{(y - 4)(y - 1)(y + 2)} \quad = \frac{\overset{1}{(y - 4)}\,\overset{1}{(y + 2)}}{\underset{1}{(y - 4)}(y - 1)\underset{1}{(y + 2)}}$$

$$= \frac{1}{y - 1}$$

37. $\dfrac{16 - m^2}{m^2 + 3m - 4} \cdot \dfrac{m - 1}{m - 4} = \dfrac{\overset{-1}{(4 - m)}\,\overset{1}{(4 + m)}}{\underset{1}{(m + 4)}\,\underset{1}{(m - 1)}} \cdot \dfrac{\overset{1}{m - 1}}{\underset{1}{m - 4}} = \dfrac{-1}{1} = -1$

39. $\dfrac{x + 7}{ax - bx} + \dfrac{y + 9}{by - ay} = \dfrac{x + 7}{x(a - b)} + \dfrac{y + 9}{y(b - a)} = \dfrac{y(x + 7)}{xy(a - b)} + \dfrac{-x(y + 9)}{xy(a - b)}$

$$= \frac{xy + 7y - xy - 9x}{xy(a - b)} = \frac{7y - 9x}{xy(a - b)}$$

41. $\dfrac{x^2 - 16}{2x^2 + 10x + 8} \div \dfrac{x^2 - 13x + 36}{x^3 + 1} = \dfrac{x^2 - 16}{2x^2 + 10x + 8} \cdot \dfrac{x^3 + 1}{x^2 - 13x + 36}$

$$= \frac{\overset{1}{(x - 4)}\,\overset{1}{(x + 4)}}{2\underset{1}{(x + 1)}\,\underset{1}{(x + 4)}} \cdot \frac{\overset{1}{(x + 1)}(x^2 - x + 1)}{\underset{1}{(x - 4)}(x - 9)} = \frac{x^2 - x + 1}{2(x - 9)}$$

43. $\dfrac{x^2 - xy}{xy + y^2} \div \left(\dfrac{x^2 - y^2}{x^2 + 2xy + y^2} \div \dfrac{x^2 - 2xy + y^2}{x^2y + xy^2} \right)$

$$= \frac{x^2 - xy}{xy + y^2} \div \left(\frac{x^2 - y^2}{x^2 + 2xy + y^2} \cdot \frac{x^2y + xy^2}{x^2 - 2xy + y^2} \right)$$

$$= \frac{x^2 - xy}{xy + y^2} \div \left(\frac{\overset{1}{(x - y)}\,\overset{1}{(x + y)}}{\underset{1}{(x + y)}\,\underset{1}{(x + y)}} \cdot \frac{\overset{1}{xy(x + y)}}{(x - y)(x - y)} \right)$$

$$= \frac{x^2 - xy}{xy + y^2} \div \frac{xy}{x - y} = \frac{x^2 - xy}{xy + y^2} \cdot \frac{x - y}{xy}$$

$$= \frac{\overset{1}{x}(x - y)}{y(x + y)} \cdot \frac{x - y}{\underset{1}{xy}} = \frac{(x - y)^2}{y^2(x + y)}$$

45. $\left(\dfrac{x}{x^2 - 16} - \dfrac{1}{x + 4} \right) \div \dfrac{4}{x + 4} = \left(\dfrac{x}{(x - 4)(x + 4)} - \dfrac{1}{x + 4} \right) \div \dfrac{4}{x + 4}$

$$= \left(\frac{x}{(x - 4)(x + 4)} - \frac{(x - 4)}{(x - 4)(x + 4)} \right) \div \frac{4}{x + 4}$$

$$= \frac{x - x + 4}{(x - 4)(x + 4)} \div \frac{4}{x + 4} = \frac{4}{(x - 4)(x + 4)} \div \frac{4}{x + 4}$$

$$= \frac{\overset{1}{4}}{(x - 4)\underset{1}{(x + 4)}} \cdot \frac{\overset{1}{x + 4}}{\underset{1}{4}} = \frac{1}{x - 4}$$

47. $\dfrac{1 + \frac{2}{x} - \frac{15}{x^2}}{1 + \frac{4}{x} - \frac{5}{x^2}} = \dfrac{x^2\left(1 + \frac{2}{x} - \frac{15}{x^2}\right)}{x^2\left(1 + \frac{4}{x} - \frac{5}{x^2}\right)} = \dfrac{x^2 + 2x - 15}{x^2 + 4x - 5} = \dfrac{\overset{1}{\cancel{(x + 5)}}\,(x - 3)}{\underset{1}{\cancel{(x + 5)}}\,(x - 1)} = \dfrac{x - 3}{x - 1}$

49. $\dfrac{\frac{1}{x + h} - \frac{1}{x}}{h} = \dfrac{\frac{x}{x(x + h)} - \frac{x + h}{x(x + h)}}{h} = \dfrac{\frac{x - x - h}{x(x + h)}}{h} = \dfrac{\frac{-h}{x(x + h)}}{h} = \dfrac{-h}{x(x + h)} \div h$

$= \dfrac{-h}{x(x + h)} \div \dfrac{h}{1} = \dfrac{\overset{-1}{\cancel{-h}}}{x(x + h)} \cdot \dfrac{1}{\underset{1}{\cancel{h}}} = \dfrac{-1}{x(x + h)}$

51. $\dfrac{\frac{(x + h)^2}{x + h + 2} - \frac{x^2}{x + 2}}{h} = \dfrac{\frac{(x + h)^2(x + 2)}{(x + h + 2)(x + 2)} - \frac{x^2(x + h + 2)}{(x + h + 2)(x + 2)}}{h} = \dfrac{\frac{(x + h)^2(x + 2) - x^2(x + h + 2)}{(x + h + 2)(x + 2)}}{h}$

$= \dfrac{\frac{(x^2 + 2xh + h^2)(x + 2) - x^3 - x^2h - 2x^2}{(x + h + 2)(x + 2)}}{h}$

$= \dfrac{\frac{x^3 + 2x^2h + h^2x + 2x^2 + 4xh + 2h^2 - x^3 - x^2h - 2x^2}{(x + h + 2)(x + 2)}}{h}$

$= \dfrac{\frac{x^2h + h^2x + 4xh + 2h^2}{(x + h + 2)(x + 2)}}{h} = \dfrac{x^2h + h^2x + 4xh + 2h^2}{(x + h + 2)(x + 2)} \div h$

$= \dfrac{x^2h + h^2x + 4xh + 2h^2}{(x + h + 2)(x + 2)} \div \dfrac{h}{1} = \dfrac{\overset{1}{\cancel{h}}(x^2 + hx + 4x + 2h)}{(x + h + 2)(x + 2)} \cdot \dfrac{1}{\underset{1}{\cancel{h}}}$

$= \dfrac{x^2 + hx + 4x + 2h}{(x + h + 2)(x + 2)}$

53. (A) The solution is incorrect, because in the first step the quantity 4 has been removed from numerator and denominator. But 4 is a term, and only factors can be cancelled. A correct solution would factor numerator and denominator, then cancel common factors, if any.

(B) $\dfrac{x^2 + 5x + 4}{x + 4} = \dfrac{(x + 1)\overset{1}{\cancel{(x + 4)}}}{\underset{1}{\cancel{x + 4}}} = x + 1$

55. (A) The solution is incorrect, because in the first step the quantity h has been removed from numerator and denominator. Although h is a factor of the denominator it is not a factor of the numerator as shown and only common factors of numerator and denominator can be cancelled. A correct solution would factor numerator and denominator, revealing that h is actually a factor of the numerator in factored form and allowing a correct cancellation of h.

(B) $\dfrac{(x + h)^2 - x^2}{h} = \dfrac{[(x + h) - x][(x + h) + x]}{h}$ using the difference of two squares

$= \dfrac{[x + h - x][x + h + x]}{h}$

$= \dfrac{\overset{1}{\cancel{h}}[2x + h]}{\underset{1}{\cancel{h}}} = 2x + h$

57. (A) The solution is incorrect. In adding a quantity to a fractional expression, the quantity cannot simply be placed in the numerator. A correct solution would rewrite $x - 2$ as a fractional expression, first with denominator 1, and then with denominator equal to the LCD of all denominators in the addition.

(B)
$$\frac{x^2 - 2x}{x^2 - x - 2} + x - 2 = \frac{\overset{1}{x(x - 2)}}{(x + 1)\underset{1}{(x - 2)}} + \frac{x - 2}{1}$$

$$= \frac{x}{x + 1} + \frac{x - 2}{1}$$

$$= \frac{x}{x + 1} + \frac{(x - 2)(x + 1)}{x + 1}$$

$$= \frac{x + x^2 - x - 2}{x + 1}$$

$$= \frac{x^2 - 2}{x - 1}$$

59. (A)(B) The solution is correct.

61.
$$\frac{y - \dfrac{y^2}{y - x}}{1 + \dfrac{x^2}{y^2 - x^2}} = \frac{(y^2 - x^2)[y - \dfrac{y^2}{y - x}]}{(y^2 - x^2)[1 + \dfrac{x^2}{y^2 - x^2}]} = \frac{(y^2 - x^2)y - \overset{(y + x)}{(y^2 - x^2)}\,\dfrac{y^2}{\underset{1}{y - x}}}{y^2 - x^2 + \underset{1}{(y^2 - x^2)}\,\dfrac{x^2}{\underset{1}{y^2 - x^2}}}$$

$$= \frac{y^3 - x^2y - y^3 - xy^2}{y^2 - x^2 + x^2} = \frac{-x^2y - xy^2}{y^2} = \frac{-x\overset{1}{y}(x + y)}{\underset{y}{y^2}} = \frac{-x(x + y)}{y}$$

63.
$$2 - \frac{1}{1 - \dfrac{2}{a + 2}} = 2 - \frac{(a + 2)\cdot 1}{(a + 2)[1 - \dfrac{2}{a + 2}]} = 2 - \frac{a + 2}{a + 2 - \underset{1}{(a + 2)}\,\dfrac{2}{\underset{1}{a+2}}} = 2 - \frac{a + 2}{a + 2 - 2}$$

$$= 2 - \frac{a + 2}{a} = \frac{2}{1} - \frac{a + 2}{a} = \frac{2a}{a} - \frac{(a + 2)}{a} = \frac{2a - a - 2}{a} = \frac{a - 2}{a}$$

65. (A) The multiplicative inverse of x is the unique real number $\frac{1}{x}$ such that $x(\frac{1}{x}) = 1$. Since $\frac{c}{d} \cdot \frac{d}{c} = \frac{cd}{dc} = \frac{cd}{cd} = 1$, $\frac{d}{c}$ is the multiplicative inverse of $\frac{c}{d}$.
(B) By Definition 1 of Section A-1 of the appendix
$a \div b = a(\frac{1}{b}) = a \cdot$ (multiplicative inverse of b).

$$\frac{a}{b} \div \frac{c}{d} = \frac{a}{b} \cdot \text{(multiplicative inverse of } \frac{c}{d})$$

$$= \frac{a}{b} \cdot \frac{d}{c} \text{ by part A.}$$

APPENDIX A REVIEW

1. (A) Since 3 is an element of {1, 2, 3, 4, 5}, the statement is true. T.
(B) Since 5 is not an element of {4, 1, 2}, the statement is true. T.
(C) Since B is not an element of {1, 2, 3, 4, 5}, the statement is false. F.

(D) Since each element of the set {1, 2, 4} is also an element of the set {1, 2, 3, 4, 5}, the statement is true. T.

(D) Since sets B and C have exactly the same elements, the sets are equal. The statement is false. F.

(E) The statement is false, since not every element of the set {1, 2, 3, 4, 5} is an element of the set {1, 2, 4}. For example, 3 is an element of the first, but not the second. F.

$(A-1)$

2. (A) The commutative property ($\cdot$) states that, in general, $xy = yx$.
Comparing this with
$x(y + z) = ?$
we see that $x(y + z) = (y + z)x$
$(y + z)x$

(B) The associative property (+) states that, in general,
$(x + y) + z = x + (y + z)$
Comparing this with
$? = 2 + (x + y)$
we see that
$(2 + x) + y = 2 + (x + y)$
$(2 + x) + y$

(C) The distributive property states that, in general,
$(x + y)z = xz + yz$
Comparing this with
$(2 + 3)x = ?$
we see that
$(2 + 3)x = 2x + 3x$
$2x + 3x$

$(A-1)$

3. $\left(\dfrac{2}{3} \cdot \dfrac{15}{14}\right)^{-1} = \left(\dfrac{5}{7}\right)^{-1} = \dfrac{7}{5}$

$(A-1)$

4. $\dfrac{9}{10} + \dfrac{5}{12} = \dfrac{54}{60} + \dfrac{25}{60} = \dfrac{79}{60}$

$(A-1)$

5. $-\left(\dfrac{8}{7} + 2^{-1}\right) = -\left(\dfrac{8}{7} + \dfrac{1}{2}\right) = -\left(\dfrac{16}{14} + \dfrac{7}{14}\right) = -\left(\dfrac{23}{14}\right) = -\dfrac{23}{14}$

$(A-1)$

6. $(4^{-1}9^{-1})^{-1} = \left(\dfrac{1}{4} \cdot \dfrac{1}{9}\right)^{-1} = \left(\dfrac{1}{36}\right)^{-1} = 36$

$(A-1)$

7. 3 $(A-4)$

8. 5 $(A-4)$

9. $(x^3 + 2) + (x^5 - x^3 + 1) = x^3 + 2 + x^5 - x^3 + 1 = x^5 + 3$. This polynomial has degree 5.

$(A-4)$

10. $(x^3 + 2)(x^5 - x^3 + 1) = x^8 + 2x^5 - x^6 - 2x^3 + x^3 + 2 = x^8 - x^6 + 2x^5 - x^3 + 2$. This polynomial has degree 8.

$(A-4)$

11. See problem 10. $x^8 - x^6 + 2x^5 - x^3 + 2$.

$(A-4)$

12. See problem 9. $x^5 + 3$

$(A-4)$

13. 17 $(A-2)$ 14. 6 $(A-2)$ 15. $8^{-2/3} = \dfrac{1}{8^{2/3}} = \dfrac{1}{(8^{1/3})^2} = \dfrac{1}{2^2} = \dfrac{1}{4}$

$(A-2)$

16. $(-64)^{5/3} = [(-64)^{1/3}]^5 = (-4)^5 = -1,024$

$(A-2)$

17. $\left(\dfrac{9}{16}\right)^{-1/2} = 1 \div \left(\dfrac{9}{16}\right)^{1/2} = 1 \div \dfrac{3}{4} = 1 \cdot \dfrac{4}{3} = \dfrac{4}{3}$

$(A-2)$

18. $(121^{1/2} + 25^{1/2})^{-3/4} = (11 + 5)^{-3/4} = 16^{-3/4} = \dfrac{1}{16^{3/4}} = \dfrac{1}{(16^{1/4})^3} = \dfrac{1}{2^3} = \dfrac{1}{8}$ $\qquad$ $(A$-$2)$

19. $5x^2 - 3x[4 - 3(x - 2)] = 5x^2 - 3x[4 - 3x + 6]$
$\qquad\qquad\qquad\qquad\quad = 5x^2 - 3x[10 - 3x]$
$\qquad\qquad\qquad\qquad\quad = 5x^2 - 30x + 9x^2$
$\qquad\qquad\qquad\qquad\quad = 14x^2 - 30x$ $\qquad$ $(A$-$4)$

20. $9m^2 - 25n^2$ $\qquad$ $(A$-$4)$

21. $(2x + y)(3x - 4y) = 6x^2 - 8xy + 3xy - 4y^2 = 6x^2 - 5xy - 4y^2$ $\qquad$ $(A$-$4)$

22. $(2a - 3b)^2 = (2a)^2 - 2(2a)(3b) + (3b)^2 = 4a^2 - 12ab + 9b^2$ $\qquad$ $(A$-$4)$

23. $(3x - 2)^2$ $\qquad$ $(A$-$5)$

24. Prime $\qquad$ $(A$-$5)$

25. $6n^3 - 9n^2 - 15n = 3n(2n^2 - 3n - 5) = 3n(2n - 5)(n + 1)$ $\qquad$ $(A$-$5)$

26. $\dfrac{2}{5b} - \dfrac{4}{3a^3} - \dfrac{1}{6a^2b^2} = \dfrac{12a^3b}{30a^3b^2} - \dfrac{40b^2}{30a^3b^2} - \dfrac{5a}{30a^3b^2} = \dfrac{12a^3b - 40b^2 - 5a}{30a^3b^2}$ $\qquad$ $(A$-$6)$

27. $\dfrac{3x}{3x^2 - 12x} + \dfrac{1}{6x} = \dfrac{3x}{3x(x - 4)} + \dfrac{1}{6x} = \dfrac{2(3x)}{6x(x - 4)} + \dfrac{1(x - 4)}{6x(x - 4)} = \dfrac{6x + x - 4}{6x(x - 4)} = \dfrac{7x - 4}{6x(x - 4)}$ $(A$-$6)$

28. $\dfrac{y - 2}{y^2 - 4y + 4} \div \dfrac{y^2 + 27}{y^2 + 4y + 4} = \dfrac{y - 2}{y^2 - 4y + 4} \cdot \dfrac{y^2 + 4y + 4}{y^2 + 2y}$

$\qquad\qquad\qquad\qquad\qquad = \dfrac{\overset{1}{\cancel{(y - 2)}}}{\cancel{(y - 2)}(y - 2)} \cdot \dfrac{\overset{1}{\cancel{(y + 2)}}(y + 2)}{y\cancel{(y + 2)}}$

$\qquad\qquad\qquad\qquad\qquad = \dfrac{y + 2}{y(y - 2)}$ $\qquad$ $(A$-$6)$

29. $\dfrac{u - \frac{1}{u}}{1 - \frac{1}{u^2}} = \dfrac{u^2\left(u - \frac{1}{u}\right)}{u^2\left(1 - \frac{1}{u^2}\right)} = \dfrac{u^3 - u}{u^2 - 1} = \dfrac{u\overset{1}{\cancel{(u^2 - 1)}}}{\underset{1}{\cancel{u^2 - 1}}} = u$ $\qquad$ $(A$-$6)$

30. $6(xy^3)^5 = 6x^5(y^3)^5 = 6x^5y^{15}$ $\qquad$ $(A$-$2)$

31. $\dfrac{9u^8v^6}{3u^4v^8} = \dfrac{3u^4}{v^2}$ $\qquad$ $(A$-$2)$

32. $(2 \times 10^5)(3 \times 10^{-3}) = (2 \cdot 3) \times 10^{5+(-3)} = 6 \times 10^2$ $\qquad$ $(A$-$2)$

33. $(x^{-3}y^2)^{-2} = (x^{-3})^{-2}(y^2)^{-2} = x^6y^{-4} = \dfrac{x^6}{y^4}$ $\qquad$ $(A$-$2)$

34. $u^{5/3}u^{2/3} = u^{5/3+2/3} = u^{7/3}$ $\qquad$ $(A$-$2)$

35. $(9a^4b^{-2})^{1/2} = 9^{1/2}(a^4)^{1/2}(b^{-2})^{1/2} = 3a^2b^{-1} = \dfrac{3a^2}{b}$ $\qquad$ $(A$-$3)$

36. $3\sqrt[5]{x^2}$ $\qquad$ $(A$-$3)$

37. $-3(xy)^{2/3}$ $\qquad$ $(A$-$3)$

38. $3x\sqrt[3]{x^5y^4} = 3x\sqrt[3]{x^3y^3 \cdot x^2y} = 3x\sqrt[3]{x^3y^3}\sqrt[3]{x^2y} = 3x(xy)\sqrt[3]{x^2y} = 3x^2y\sqrt[3]{x^2y}$ $\qquad$ $(A$-$3)$

39. $\sqrt{2x^2y^5}\,\sqrt{18x^3y^2} = \sqrt{(2x^2y^5)(18x^3y^2)} = \sqrt{36x^2y^7} = \sqrt{36x^4y^6 \cdot xy} = \sqrt{36x^4y^6}\,\sqrt{xy}$

$= 6x^2y^3\,\sqrt{xy}$ *(A-3)*

40. $\dfrac{6ab}{\sqrt{3a}} = \dfrac{6ab}{\sqrt{3a}}\,\dfrac{\sqrt{3a}}{\sqrt{3a}} = \dfrac{\overset{2}{\cancel{6ab}}\sqrt{3a}}{\underset{1}{\cancel{3a}}} = 2b\sqrt{3a}$ *(A-3)*

41. $\dfrac{\sqrt{5}}{3 - \sqrt{5}} = \dfrac{\sqrt{5}}{(3 - \sqrt{5})}\dfrac{(3 + \sqrt{5})}{(3 + \sqrt{5})} = \dfrac{\sqrt{5}(3 + \sqrt{5})}{3^2 - (\sqrt{5})^2} = \dfrac{\sqrt{5}(3 + \sqrt{5})}{9 - 5} = \dfrac{\sqrt{5}(3 + \sqrt{5})}{4} = \dfrac{3\sqrt{5} + 5}{4}$ *(A-3)*

42. $\sqrt[8]{y^6} = \sqrt[2\cdot4]{y^{2\cdot3}} = \sqrt[4]{y^3}$ *(A-3)*

43. The odd integers between -4 and 2 are -3, -1, and 1. The set is written $\{-3, -1, 1\}$ *(A-1)*

44. Subtraction *(A-1)* **45.** Commutative (+) *(A-1)* **46.** Distributive *(A-1)*

47. Associative ($\cdot$) *(A-1)* **48.** Negatives *(A-1)* **49.** Identity (+) *(A-1)*

50. (A) T (B) F *(A-1)* **51.** 0 and -3 are two examples of infinitely many. *(A-1)*

52. (A) a and d (B) None *(A-4)*

53. $(2x - y)(2x + y) - (2x - y)^2 = 4x^2 - y^2 - (4x^2 - 4xy + y^2)$

$= 4x^2 - y^2 - 4x^2 + 4xy - y^2$

$= 4xy - 2y^2$ *(A-4)*

54. This can be simplified directly using the distributive property or:
$(m^2 + 2mn - n^2)(m^2 - 2mn - n^2) = [(m^2 - n^2) + 2mn][(m^2 - n^2) - 2mn]$

$= (m^2 - n^2)^2 - (2mn)^2$

$= m^4 - 2m^2n^2 + n^4 - 4m^2n^2$

$= m^4 - 6m^2n^2 + n^4$ *(A-4)*

55. $5(x + h)^2 - 7(x + h) - (5x^2 - 7x) = 5(x^2 + 2hx + h^2) - 7(x + h) - (5x^2 - 7x)$

$= 5x^2 + 10hx + 5h^2 - 7x - 7h - 5x^2 + 7x$

$= 10hx + 5h^2 - 7h$ *(A-4)*

56. $-2x\{(x^2 + 2)(x - 3) - x[x - x(3 - x)]\}$

$= -2x\{(x^2 + 2)(x - 3) - x[x - 3x + x^2]\}$

$= -2x\{(x^2 + 2)(x - 3) - x[-2x + x^2]\}$

$= -2x\{x^3 - 3x^2 + 2x - 6 + 2x^2 - x^3\}$

$= -2x\{-x^2 + 2x - 6\}$

$= 2x^3 - 4x^2 + 12x$ *(A-4)*

57. $(x - 2y)^3 = (x - 2y)^2(x - 2y)$

$= (x^2 - 4xy + 4y^2)(x - 2y)$

$= x^3 - 2x^2y - 4x^2y + 8xy^2 + 4xy^2 - 8y^3$

$= x^3 - 6x^2y + 12xy^2 - 8y^3$ *(A-4)*

58. $(4x - y)^2 - 9x^2 = (4x - y)^2 - (3x)^2$
$= [(4x - y) - 3x][(4x - y) + 3x]$
$= (x - y)(7x - y)$ $(A-5)$

59. Prime $(A-5)$ **60.** $3xy(2x^2 + 4xy - 5y^2)$ $(A-5)$

61. $(y - b)^2 - y + b = (y - b)(y - b) - 1(y - b) = (y - b)(y - b - 1)$ $(A-5)$

62. $3x^3 + 24y^3 = 3(x^3 + 8y^3) = 3[x^3 + (2y)^3] = 3(x + 2y)[x^2 - x(2y) + (2y)^2]$
$= 3(x + 2y)(x^2 - 2xy + 4y^2)$ $(A-5)$

63. $y^3 + 2y^2 - 4y - 8 = y^2(y + 2) - 4(y + 2)$
$= (y + 2)(y^2 - 4) = (y + 2)(y + 2)(y - 2)$
$= (y - 2)(y + 2)^2$ $(A-5)$

64. $2x(x - 4)^3 + 3x^2(x - 4)^2 = x(x - 4)^2[2(x - 4) + 3x]$
$= x(x - 4)^2[2x - 8 + 3x]$
$= x(x - 4)^2(5x - 8)$ $(A-5)$

65. $\dfrac{3x^2(x + 2)^2 - 2x(x + 2)^3}{x^4} = \dfrac{x(x + 2)^2[3x - 2(x + 2)]}{x^4}$

$= \dfrac{\overset{1}{\cancel{x}}(x + 2)^2[3x - 2x - 4]}{\underset{x^3}{\cancel{x^4}}}$

$= \dfrac{(x + 2)^2(x - 4)}{x^3}$ $(A-6)$

66. $\dfrac{m - 1}{m^2 - 4m + 4} + \dfrac{m + 3}{m^2 - 4} + \dfrac{2}{2 - m}$

$= \dfrac{(m - 1)}{(m - 2)(m - 2)} + \dfrac{m + 3}{(m - 2)(m + 2)} + \dfrac{-2}{m - 2}$

$= \dfrac{(m - 1)(m + 2)}{(m - 2)(m - 2)(m + 2)} + \dfrac{(m + 3)(m - 2)}{(m - 2)(m - 2)(m + 2)}$
$\qquad\qquad + \dfrac{-2(m - 2)(m + 2)}{(m - 2)(m - 2)(m + 2)}$

$= \dfrac{(m - 1)(m + 2) + (m + 3)(m - 2) - 2(m - 2)(m + 2)}{(m - 2)^2(m + 2)}$

$= \dfrac{m^2 + m - 2 + m^2 + m - 6 - 2(m^2 - 4)}{(m - 2)^2(m + 2)}$

$= \dfrac{2m^2 + 2m - 8 - 2m^2 + 8}{(m - 2)^2(m + 2)} = \dfrac{2m}{(m - 2)^2(m + 2)}$ $(A-6)$

67. $\dfrac{y}{x^2} \div \left(\dfrac{x^2 + 3x}{2x^2 + 5x - 3} \div \dfrac{x^3y - x^2y}{2x^2 - 3x + 1} \right) = \dfrac{y}{x^2} \div \left(\dfrac{x^2 + 3x}{2x^2 + 5x - 3} \cdot \dfrac{2x^2 - 3x + 1}{x^3y - x^2y} \right)$

$$= \dfrac{y}{x^2} \div \left(\dfrac{\cancel{x}(x + 3)}{(2x - 1)(x + 3)} \cdot \dfrac{(2x - 1)(x - 1)}{\cancel{x}^2 y(x - 1)} \right)$$

$$= \dfrac{y}{x^2} \div \dfrac{1}{xy}$$

$$= \dfrac{y}{\cancel{x}^2} \cdot \dfrac{\cancel{x}y}{1}$$

$$= \dfrac{y^2}{x} \qquad\qquad (A\text{-}6)$$

68. $\dfrac{1 - \dfrac{1}{1 + \frac{x}{y}}}{1 - \dfrac{1}{1 - \frac{x}{y}}} = \dfrac{1 - \dfrac{y(1)}{y\left(1 + \frac{x}{y}\right)}}{1 - \dfrac{y(1)}{y\left(1 - \frac{x}{y}\right)}} = \dfrac{1 - \dfrac{y}{y + x}}{1 - \dfrac{y}{y - x}} = \dfrac{\dfrac{y + x}{y + x} - \dfrac{y}{y + x}}{\dfrac{y - x}{y - x} - \dfrac{y}{y - x}}$

$$= \dfrac{\dfrac{x}{y + x}}{\dfrac{-x}{y - x}} = \dfrac{x}{y + x} \div \dfrac{-x}{y - x} = \dfrac{\cancel{x}}{y + x} \cdot \dfrac{y - x}{-\cancel{x}} = \dfrac{-1(y - x)}{y + x} = \dfrac{x - y}{x + y} \qquad (A\text{-}6)$$

69. $\dfrac{a^{-1} - b^{-1}}{ab^{-2} - ba^{-2}} = \dfrac{\dfrac{1}{a} - \dfrac{1}{b}}{\dfrac{a}{b^2} - \dfrac{b}{a^2}} = \dfrac{a^2b^2\left(\dfrac{1}{a} - \dfrac{1}{b}\right)}{a^2b^2\left(\dfrac{a}{b^2} - \dfrac{b}{a^2}\right)}$

$$= \dfrac{ab^2 - a^2b}{a^3 - b^3} = \dfrac{ab\cancel{(b - a)}}{\cancel{(a - b)}(a^2 + ab + b^2)} = \dfrac{-ab}{a^2 + ab + b^2} \qquad (A\text{-}6,\ A\text{-}2)$$

70. The solution is incorrect. In adding a quantity to a fractional expression, the quantity cannot simply be placed in the numerator. A correct solution would rewrite $x + 2$ as a fractional expression, first with denominator 1, and then with denominator equal to the LCD of all denominators in the expression.

$$\dfrac{x^2 + 2x}{x^2 + x - 2} + x + 2 = \dfrac{x(x + 2)}{(x - 1)(x + 2)} + \dfrac{x + 2}{1}$$

$$= \dfrac{x}{x - 1} + \dfrac{x + 2}{1}$$

$$= \dfrac{x}{x - 1} + \dfrac{(x + 2)(x - 1)}{x - 1}$$

$$= \dfrac{x + x^2 + x - 2}{x - 1}$$

$$= \dfrac{x^2 + 2x - 2}{x - 1} \qquad\qquad (A\text{-}6)$$

71. $\left(\dfrac{8u^{-1}}{2^2 u^2 v^0} \right)^{-2} \left(\dfrac{u^{-5}}{u^{-3}} \right)^3 = \left(\dfrac{2^3 u^{-1}}{2^2 u^2} \right)^{-2} \left(\dfrac{u^{-5}}{u^{-3}} \right)^3 = (2u^{-3})^{-2} \left(\dfrac{u^{-5}}{u^{-3}} \right)^3 = 2^{-2} u^6 \dfrac{u^{-15}}{u^{-9}}$

$$= \dfrac{1}{2^2} u^6 u^{-15 - (-9)} = \dfrac{1}{2^2} u^6 u^{-6} = \dfrac{1}{4} \qquad\qquad (A\text{-}2)$$

72. $\dfrac{5^0}{3^2} + \dfrac{3^{-2}}{2^{-2}} = \dfrac{1}{3^2} + \dfrac{2^2}{3^2} = \dfrac{1}{9} + \dfrac{4}{9} = \dfrac{5}{9}$ *(A-2)*

73. $\left(\dfrac{27x^2y^{-3}}{8x^{-4}y^3}\right)^{1/3} = \left(\dfrac{3^3x^{2-(-4)}y^{-3-3}}{2^3}\right)^{1/3} = \left(\dfrac{3^3x^6y^{-6}}{2^3}\right)^{1/3} = \dfrac{3x^2y^{-2}}{2} = \dfrac{3x^2}{2y^2}$ *(A-2)*

74. $(a^{-1/3}b^{1/4})(9a^{1/3}b^{-1/2})^{3/2} = a^{-1/3}b^{1/4}9^{3/2}a^{1/2}b^{-3/4}$

$\qquad = a^{-1/3+1/2}b^{1/4-3/4}9^{3/2} = 27a^{1/6}b^{-1/2}$

$\qquad = \dfrac{27a^{1/6}}{b^{1/2}}$ *(A-2)*

75. $(x^{1/2} + y^{1/2})^2 = (x^{1/2})^2 + 2(x^{1/2})(y^{1/2}) + (y^{1/2})^2$

$\qquad = x + 2x^{1/2}y^{1/2} + y$

> **Common Error:**
> $(x^{1/2} + y^{1/2})^2 \neq x + y.$
> (Exponents do not distribute over addition.)

(A-2)

76. $(3x^{1/2} - y^{1/2})(2x^{1/2} + 3y^{1/2}) = (3x^{1/2})(2x^{1/2}) + (3x^{1/2})(3y^{1/2})$

$\qquad\qquad - (2x^{1/2})(y^{1/2}) - (y^{1/2})(3y^{1/2})$

$\qquad = 6x + 9x^{1/2}y^{1/2} - 2x^{1/2}y^{1/2} - 3y$

$\qquad = 6x + 7x^{1/2}y^{1/2} - 3y$ *(A-2)*

77. $\dfrac{0.000\,000\,000\,52}{(1,300)(0.000\,002)} = \dfrac{5.2 \times 10^{-10}}{(1.3 \times 10^3)(2 \times 10^{-6})}$

$\qquad = \dfrac{\overset{2}{\cancel{5.2}} \times 10^{-10}}{\underset{1}{\cancel{2.6}} \times 10^{-3}} = 2 \times 10^{-10-(3)}$

$\qquad = 2 \times 10^{-7}$ *(A-2)*

78. $\dfrac{(20,410)(0.000\,003\,477)}{0.000\,000\,022\,09} = \dfrac{(2.041 \times 10^4)(3.477 \times 10^{-6})}{(2.209 \times 10^{-8})} = 3.213 \times 10^6$

Note that the input is given to 4 significant digits, so we round the calculator answer to 4 significant digits. *(A-2)*

79. 4.434×10^{-5} *(A-2)*

80. -4.541×10^{-6} *(A-2)*

81. $82.45^{8/3} = (82.45)^{(8\div3)} = 128,800$ *(A-2)*

82. $(0.000\,000\,419\,9)^{2/7} = (4.199 \times 10^{-7})^{(2\div7)} = 0.01507$ *(A-2)*

83. $\sqrt[5]{0.006\,604} = (6.604 \times 10^{-3})^{1/5} = (6.604 \times 10^{-3})^{0.2} = 0.3664$ *(A-3)*

84. $\sqrt[3]{3 + \sqrt{2}} = (3 + \sqrt{2})^{(1\div3)} = 1.640$ *(A-3)*

85. $\dfrac{2^{-1/2} - 3^{-1/2}}{2^{-1/3} + 3^{-1/3}} = [2^{(-0.5)} - 3^{(-0.5)}] \div [2^{(-1\div3)} + 3^{(-1\div3)}] = 0.08726$ *(A-2)*

86. $-2x\sqrt[5]{3^6x^7y^{11}} = -2x\sqrt[5]{3^5x^5y^{10} \cdot 3x^2y} = -2x\sqrt[5]{3^5x^5y^{10}}\sqrt[5]{3x^2y}$

$\qquad = -2x \cdot 3xy^2 \sqrt[5]{3x^2y} = -6x^2y^2 \sqrt[5]{3x^2y}$ *(A-3)*

87. $\dfrac{2x^2}{\sqrt[3]{4x}} = \dfrac{2x^2}{\sqrt[3]{4x}}\dfrac{\sqrt[3]{2x^2}}{\sqrt[3]{2x^2}} = \dfrac{2x^2\sqrt[3]{2x^2}}{\sqrt[3]{8x^3}} = \dfrac{\overset{x}{\cancel{2x^2}}\sqrt[3]{2x^2}}{\underset{1}{\cancel{2x}}} = x\sqrt[3]{2x^2}$ *(A-3)*

88. $\sqrt[5]{\dfrac{3y^2}{8x^2}} = \sqrt[5]{\dfrac{3y^2}{8x^2}\dfrac{4x^3}{4x^3}} = \sqrt[5]{\dfrac{12x^3y^2}{32x^5}} = \dfrac{\sqrt[5]{12x^3y^2}}{\sqrt[5]{32x^5}} = \dfrac{\sqrt[5]{12x^3y^2}}{2x}$ ⁣(A-3)

89. $\sqrt[9]{8x^6y^{12}} = \sqrt[9]{2^3x^6y^{12}} = \sqrt[3\cdot3]{2^{3\cdot1}x^{3\cdot2}y^{3\cdot4}} = \sqrt[3]{2x^2y^4} = \sqrt[3]{y^3\cdot2x^2y} = \sqrt[3]{y^3}\sqrt[3]{2x^2y} = y\sqrt[3]{2x^2y}$ ⁣(A-3)

90. $\sqrt{\sqrt[3]{4x^4}} = \sqrt[2\cdot3]{2^{2\cdot1}x^{2\cdot2}} = \sqrt[3]{2x^2}$ ⁣(A-3)

91. $(2\sqrt{x} - 5\sqrt{y})(\sqrt{x} + \sqrt{y}) = 2\sqrt{x}\sqrt{x} + 2\sqrt{x}\sqrt{y} - 5\sqrt{x}\sqrt{y} - 5\sqrt{y}\sqrt{y}$
$$= 2x - 3\sqrt{x}\sqrt{y} - 5y = 2x - 3\sqrt{xy} - 5y$$ ⁣(A-3)

92. $\dfrac{3\sqrt{x}}{2\sqrt{x} - \sqrt{y}} = \dfrac{3\sqrt{x}}{(2\sqrt{x} - \sqrt{y})}\dfrac{(2\sqrt{x} + \sqrt{y})}{(2\sqrt{x} + \sqrt{y})} = \dfrac{3\sqrt{x}(2\sqrt{x} + \sqrt{y})}{(2\sqrt{x})^2 - (\sqrt{y})^2}$
$$= \dfrac{3\sqrt{x}(2\sqrt{x} + \sqrt{y})}{4x - y} = \dfrac{6x + 3\sqrt{xy}}{4x - y}$$ ⁣(A-3)

93. $\dfrac{2\sqrt{u} - 3\sqrt{v}}{2\sqrt{u} + 3\sqrt{v}} = \dfrac{(2\sqrt{u} - 3\sqrt{v})}{(2\sqrt{u} + 3\sqrt{v})}\dfrac{(2\sqrt{u} - 3\sqrt{v})}{(2\sqrt{u} - 3\sqrt{v})} = \dfrac{(2\sqrt{u})^2 - 2(2\sqrt{u})(3\sqrt{v}) + (3\sqrt{v})^2}{(2\sqrt{u})^2 - (3\sqrt{v})^2} = \dfrac{4u - 12\sqrt{uv} + 9v}{4u - 9v}$ ⁣(A-3)

94. $\dfrac{y^2}{\sqrt{y^2 + 4} - 2} = \dfrac{y^2}{(\sqrt{y^2 + 4} - 2)}\dfrac{(\sqrt{y^2 + 4} + 2)}{(\sqrt{y^2 + 4} + 2)} = \dfrac{y^2(\sqrt{y^2 + 4} + 2)}{(\sqrt{y^2 + 4})^2 - (2)^2} = \dfrac{y^2(\sqrt{y^2 + 4} + 2)}{y^2 + 4 - 4}$
$$= \dfrac{y^2(\sqrt{y^2 + 4} + 2)}{y^2} = \sqrt{y^2 + 4} + 2$$ ⁣(A-3)

95. $\dfrac{\sqrt{t} - \sqrt{5}}{t - 5} = \dfrac{(\sqrt{t} - \sqrt{5})}{(t - 5)}\dfrac{(\sqrt{t} + \sqrt{5})}{\sqrt{t} + \sqrt{5}} = \dfrac{(\sqrt{t})^2 - (\sqrt{5})^2}{(t - 5)(\sqrt{t} + \sqrt{5})} = \dfrac{\overset{1}{\cancel{t - 5}}}{\underset{1}{\cancel{(t - 5)}}(\sqrt{t} + \sqrt{5})} = \dfrac{1}{\sqrt{t} + \sqrt{5}}$ ⁣(A-3)

96. $\dfrac{4\sqrt{x} - 3}{2\sqrt{x}} = \dfrac{4\sqrt{x}}{2\sqrt{x}} - \dfrac{3}{2\sqrt{x}} = 2 - \dfrac{3}{2x^{1/2}} = 2 - \dfrac{3}{2}x^{-1/2} = 2x^0 - \dfrac{3}{2}x^{-1/2}$ ⁣(A-3)

97. Let $c = 0.54545454\ldots$ Then
$$100c = 54.545454\ldots$$
So $(100c - c) = (54.545454\ldots) - (0.54545454\ldots)$
$$= 54$$
$$99c = 54$$
$$c = \dfrac{54}{99} = \dfrac{6}{11}$$

The number can be written as the quotient of two integers, so it is rational. ⁣(A-1)

98. (A) List each element of M. Follow these with each element of N that is not yet listed. $\{-4,-3,2,0\}$, or, in increasing order, $\{-4,-3,0,2\}$.
(B) List each element of M that is also an element of N. $\{-3,2\}$. ⁣(A-1)

99. $x^2 - 4x + 1 = (2 - \sqrt{3})^2 - 4(2 - \sqrt{3}) + 1 = (2)^2 - 2(2)\sqrt{3} + (\sqrt{3})^2 - 8 + 4\sqrt{3} + 1$
$$= 4 - 4\sqrt{3} + 3 - 8 + 4\sqrt{3} + 1 = 0$$ ⁣(A-3)

100. $x(2x - 1)(x + 3) = x(2x^2 + 6x - x - 3)$
$$= x(2x^2 + 5x - 3)$$
$$= 2x^3 + 5x^2 - 3x$$

$(x - 1)^3 = (x - 1)^2(x - 1)$
$$= (x^2 - 2x + 1)(x - 1)$$
$$= x^3 - x^2 - 2x^2 + 2x + x - 1$$
$$= x^3 - 3x^2 + 3x - 1$$

Therefore,
$x(2x - 1)(x + 3) - (x - 1)^3 = (2x^3 + 5x^2 - 3x) - (x^3 - 3x^2 + 3x - 1)$
$$= 2x^3 + 5x^2 - 3x - x^3 + 3x^2 - 3x + 1$$
$$= x^3 + 8x^2 - 6x + 1 \qquad (A-4)$$

101. $4x(a^2 - 4a + 4) - 9x^3 = x[4(a^2 - 4a + 4) - 9x^2]$
$$= x[2^2(a - 2)^2 - (3x)^2]$$
$$= x[2(a - 2) + 3x][2(a - 2) - 3x]$$
$$= x(2a - 4 + 3x)(2a - 4 - 3x)$$
$$= x(2a + 3x - 4)(2a - 3x - 4) \qquad (A-5)$$

102. (A) $\sqrt{3 + \sqrt{5}} + \sqrt{3 - \sqrt{5}} \approx 3.162\ 277\ 660$

(B) $\sqrt{4 + \sqrt{15}} + \sqrt{4 - \sqrt{15}} \approx 3.162\ 277\ 660$

(C) $\sqrt{10} \approx 3.162\ 277\ 660$

(A), (B), and (C) all have the same value to 9 decimal places. To show that they are actually equal:

$\left(\sqrt{3 + \sqrt{5}} + \sqrt{3 - \sqrt{5}} \right)^2 = \left(\sqrt{3 + \sqrt{5}} \right)^2 + 2\sqrt{3 + \sqrt{5}} \sqrt{3 - \sqrt{5}} + \left(\sqrt{3 - \sqrt{5}} \right)^2$
$$= 3 + \sqrt{5} + 2\sqrt{(3 + \sqrt{5})(3 - \sqrt{5})} + 3 - \sqrt{5}$$
$$= 6 + 2\sqrt{3^2 - (\sqrt{5})^2}$$
$$= 6 + 2\sqrt{9 - 5}$$
$$= 6 + 4$$
$$= 10$$

Similarly,

$\left(\sqrt{4 + \sqrt{15}} \sqrt{4 - \sqrt{15}} \right)^2 = \left(\sqrt{4 + \sqrt{15}} \right)^2 + 2\sqrt{4 + \sqrt{15}} \sqrt{4 - \sqrt{15}} + \left(\sqrt{4 - \sqrt{15}} \right)^2$
$$= 4 + \sqrt{15} + 2\sqrt{(4 + \sqrt{15})(4 - \sqrt{15})} + 4 - \sqrt{15}$$
$$= 8 + 2\sqrt{4^2 - (\sqrt{15})^2}$$
$$= 8 + 2\sqrt{16 - 15}$$
$$= 8 + 2$$
$$= 10$$

Therefore, since both $\sqrt{3 + \sqrt{5}} + \sqrt{3 - \sqrt{5}}$ and $\sqrt{4 + \sqrt{15}} + \sqrt{4 - \sqrt{15}}$ are positive, they are both equal to the positive square root of 10, and
$$\sqrt{3 + \sqrt{5}} + \sqrt{3 - \sqrt{5}} = \sqrt{4 + \sqrt{15}} + \sqrt{4 - \sqrt{15}} = \sqrt{10} \qquad (A-3)$$

103. $\dfrac{8(x - 2)^{-3}(x + 3)^2}{12(x - 2)^{-4}(x + 3)^{-2}} = \dfrac{2(x - 2)^{-3-(-4)}(x + 3)^{2-(-2)}}{3} = \dfrac{2(x - 2)^1(x + 3)^4}{3} = \dfrac{2(x - 2)(x + 3)^4}{3}$
$$= \frac{2}{3}(x - 2)(x + 3)^4 \qquad (A-2)$$

104. $\left(\dfrac{a^{-2}}{b^{-1}} + \dfrac{b^{-2}}{a^{-1}}\right)^{-1} = \left(\dfrac{b}{a^2} + \dfrac{a}{b^2}\right)^{-1} = \left(\dfrac{b^3}{a^2b^2} + \dfrac{a^3}{a^2b^2}\right)^{-1} = \left(\dfrac{b^3 + a^3}{a^2b^2}\right)^{-1} = \dfrac{a^2b^2}{b^3 + a^3}$ or $\dfrac{a^2b^2}{a^3 + b^3}$

$(A-2)$

105. $(x^{1/3} - y^{1/3})(x^{2/3} + x^{1/3}y^{1/3} + y^{2/3}) = (x^{1/3})^3 - (y^{1/3})^3 = x - y$ $(A-2)$

106. $\left(\dfrac{x^{m^2}}{x^{2m-1}}\right)^{1/(m-1)} = (x^{m^2-(2m-1)})^{1/(m-1)} = (x^{m^2-2m+1})^{1/(m-1)} = x^{(m^2-2m+1)/(m-1)} = x^{(m-1)^2/(m-1)}$

$$= x^{\cancel{(m-1)}^2 / \underset{1}{\cancel{m-1}}} = x^{m-1}$$

$(A-2)$

107. $\dfrac{1}{1 - \sqrt[3]{x}} = \dfrac{1}{(1 - \sqrt[3]{x})} \dfrac{[1 + \sqrt[3]{x} + (\sqrt[3]{x})^2]}{[1 + \sqrt[3]{x} + (\sqrt[3]{x})^2]} = \dfrac{1 + \sqrt[3]{x} + \sqrt[3]{x^2}}{(1)^3 - (\sqrt[3]{x})^3}$

Common Error:
$(1 - \sqrt[3]{x})(1 + \sqrt[3]{x}) \neq 1 - x$ This would not be a correct application of $(a - b)(a + b) = a^2 - b^2$

$$= \dfrac{1 + \sqrt[3]{x} + \sqrt[3]{x^2}}{1 - x}$$

$(A-3)$

108. $\dfrac{\sqrt[3]{t} - \sqrt[3]{5}}{t - 5} = \dfrac{(\sqrt[3]{t} - \sqrt[3]{5})}{(t - 5)} \dfrac{[(\sqrt[3]{t})^2 + \sqrt[3]{t}\sqrt[3]{5} + (\sqrt[3]{5})^2]}{[(\sqrt[3]{t})^2 + \sqrt[3]{t}\sqrt[3]{5} + (\sqrt[3]{5})^2]} = \dfrac{(\sqrt[3]{t})^3 - (\sqrt[3]{5})^3}{(t - 5)[\sqrt[3]{t^2} + \sqrt[3]{5t} + \sqrt[3]{5^2}]}$

$$= \dfrac{\underset{1}{\cancel{t-5}}}{\cancel{(t-5)}[\sqrt[3]{t^2} + \sqrt[3]{5t} + \sqrt[3]{25}]} = \dfrac{1}{\sqrt[3]{t^2} + \sqrt[3]{5t} + \sqrt[3]{25}}$$

$(A-3)$

109. $\sqrt[(n+1)]{x^{n^2}x^{2n+1}} = \sqrt[(n+1)]{x^{n^2}x^{2n+1}}$

$$= (x^{n^2+2n+1})^{1/(n+1)} = x^{(n^2+n+1)/(n+1)} = x^{(n+1)^2/(n+1)} = x^{\cancel{(n+1)}^2 / \cancel{(n+1)}} = x^{n+1}$$

$(A-3)$

110. The volume of the concrete wall is equal to the volume of the outer cylinder ($V = \pi r^2 h$) minus the volume of the basin. Since the radius of the basin is x ft and the concrete is 2 ft thick, the radius of the outer cylinder is $x + 2$ ft.

$$\begin{pmatrix}\text{Volume of}\\ \text{concrete}\\ \text{wall}\end{pmatrix} = \begin{pmatrix}\text{Volume of}\\ \text{outer}\\ \text{cylinder}\end{pmatrix} - \begin{pmatrix}\text{Volume}\\ \text{of}\\ \text{basin}\end{pmatrix}$$

$$\begin{aligned}\text{Volume} &= \pi(x + 2)^2\,3 - \pi x^2 3\\ &= \pi(x^2 + 4x + 4)3 - \pi x^2 3\\ &= 3\pi x^2 + 12\pi x + 12\pi - 3\pi x^2\\ &= 12\pi x + 12\pi \ (\text{ft}^3)\end{aligned}$$

$(A-4)$

111. average personal income = total personal income ÷ number of persons

$$\begin{aligned}&= 9{,}208{,}000{,}000{,}000 \div 291{,}000{,}000\\ &= 9.208 \times 10^{12} \div 2.91 \times 10^8\\ &= 3.16 \times 10^4 \ \ \text{or} \ \ \$31{,}600 \text{ to three significant digits}\end{aligned}$$

$(A-2)$

112. (A) We are asked to estimate $N = 20x^{1/2}y^{1/2}$ given $x = 1,600$ units of capital and $y = 900$ units of labor.
$N = 20(1,600)^{1/2}(900)^{1/2} = 20(40)(30) = 24,000$ units produced.

(B) Given $x = 3,200$ units of capital and $y = 1,800$ units of labor, then
$$N = 20(3,200)^{1/2}(1,800)^{1/2} = 20(2 \cdot 1,600)^{1/2}(2 \cdot 900)^{1/2}$$
$$= 20 \cdot 2^{1/2}(1,600)^{1/2}\, 2^{1/2}(900)^{1/2}$$
$$= 20 \cdot 2^{1/2}(40) \cdot 2^{1/2}(30)$$
$$= 20(40)(30) \cdot 2^{1/2} \cdot 2^{1/2}$$
$$= 24,000 \cdot 2^1$$
$$= 48,000 \text{ units produced}$$

(C) The effect of changing x to $2x$ and y to $2y$ is to replace N by
$$20(2x)^{1/2}(2y)^{1/2} = 20 \cdot 2^{1/2}x^{1/2}2^{1/2}y^{1/2}$$
$$= 20 \cdot 2^{1/2} \cdot 2^{1/2}x^{1/2}y^{1/2}$$
$$= 2^1 \cdot 20x^{1/2}y^{1/2}$$
$$= 2N$$
The production is doubled at any production level. *(A-6)*

113. $\dfrac{1}{\dfrac{1}{R_1} + \dfrac{1}{R_2} + \dfrac{1}{R_3}} = \dfrac{R_1 R_2 R_3 \cdot 1}{R_1 R_2 R_3 \left(\dfrac{1}{R_1} + \dfrac{1}{R_2} + \dfrac{1}{R_3}\right)} = \dfrac{R_1 R_2 R_3}{R_2 R_3 + R_1 R_3 + R_1 R_2}$ *(A-4)*

114. (A) The area of the cardboard can be written as (Original area) − (Removed area), where the original area = 16 × 30 = 480 and the removed area consists of 6 squares of area x^2 each;

(Original area) − (Removed area) = $480 - 6x^2$ in expanded form.
In factored form, $480 - 6x^2 = 6(80 - x^2)$.

(B) See figure.

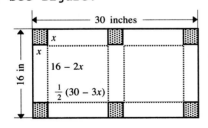 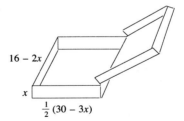

The volume of the box = length × width × height = $x(16 - 2x)\dfrac{1}{2}(30 - 3x)$
$$= x(16 - 2x)(15 - 1.5x)$$
or
$$3x(8 - x)(10 - x) \quad \text{in factored form.}$$

In expanded form, $3x(8 - x)(10 - x) = 3x(80 - 18x + x^2)$
$$= 3x^3 - 54x^2 + 240x \qquad (A-5)$$

APPENDIX B   Review of Equations and Graphing

Section B-1

1. $x + 5 = 12$
  $x + 5 - 5 = 12 - 5$
      $x = 7$

3. $2s - 7 = -2$
  $2s - 7 + 7 = -2 + 7$
        $2s = 5$
        $\dfrac{2s}{2} = \dfrac{5}{2}$
        $s = 2.5$

5. $2m + 8 = 5m - 7$
  $2m + 8 - 5m = 5m - 7 - 5m$
      $8 - 3m = -7$
  $8 - 3m - 8 = -7 - 8$
        $-3m = -15$
        $\dfrac{-3m}{-3} = \dfrac{-15}{-3}$
        $m = 5$

7. $-8 \le x \le 7$

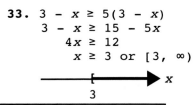

9. $-6 \le x < 6$

11. $x \ge -6$

13. $(-2, 6]$

15. $(-7, 8)$

17. $(-\infty, -2]$

19. $[-7, 2)$; $-7 \le x < 2$

21. $(-\infty, 0]$; $x \le 0$

23. $12 > 6$ and $12 + 5 > 6 + 5$

25. $-6 > -8$ and $-6 - 3 > -8 - 3$

27. $2 > -1$ and $-2(2) > -2(-1)$

29. $2 < 6$ and $\dfrac{2}{2} < \dfrac{6}{2}$

31. $7x - 8 < 4x + 7$
    $3x < 15$
    $x < 5$ or $(-\infty, 5)$

33. $3 - x \ge 5(3 - x)$
    $3 - x \ge 15 - 5x$
    $4x \ge 12$
    $x \ge 3$ or $[3, \infty)$

35. $\dfrac{N}{-2} > 4$
    $N < -8$ or $(-\infty, -8)$

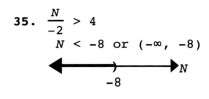

37. $3 - m < 4(m - 3)$
    $3 - m < 4m - 12$
    $-5m < -15$
    $m > 3$ or $(3, \infty)$

> Common Error :
> Neglecting to reverse the
> order after division by -5.

39. $-2 - \dfrac{B}{4} \le \dfrac{1 + B}{3}$

  $12\left(-2 - \dfrac{B}{4}\right) \le 12\dfrac{(1 + B)}{3}$
    $-24 - 3B \le 4(1 + B)$
    $-24 - 3B \le 4 + 4B$
        $-7B \le 28$
        $B \ge -4$ or $[-4, \infty)$

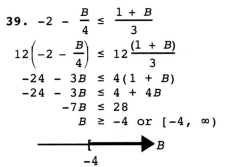

41. $3 - \dfrac{2x - 3}{3} = \dfrac{5 - x}{2}$ LCD = 6

  $6 \cdot 3 - 6\dfrac{(2x - 3)}{3} = 6\dfrac{(5 - x)}{2}$
  $18 - 2(2x - 3) = 3(5 - x)$
    $18 - 4x + 6 = 15 - 3x$
    $-4x + 24 = 15 - 3x$
        $-x = -9$
        $x = 9$

> Common Error :
> After line 2, students often write
> $18 - 2\dfrac{2x - 3}{3} = \ ...$
> $18 - 4x - 3 = \ ...$
> forgetting to distribute the -2.
> Put compound numerators in parentheses to avoid this.

43.
$$0.1(x - 7) + 0.05x = 0.8$$
$$0.1x - 0.7 + 0.05x = 0.8$$
$$0.15x - 0.7 = 0.8$$
$$0.15x = 1.5$$
$$x = \frac{1.5}{0.15}$$
$$x = 10$$

Problems of this type can also be solved by elimination of decimals: we could multiply every term on both sides by 100 to get $10(x - 7) + 5x = 80$ and proceed from here.

45.
$$0.3x - 0.04(x + 1) = 2.04$$
$$0.3x - 0.04x - 0.04 = 2.04$$
$$0.26x - 0.04 = 2.04$$
$$0.26x = 2.08$$
$$x = \frac{2.08}{0.26}$$
$$x = 8$$

47.
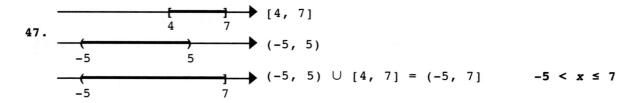

[4, 7]

(−5, 5)

(−5, 5) ∪ [4, 7] = (−5, 7]   **−5 < x ≤ 7**

49.
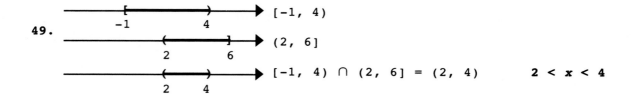

[−1, 4)

(2, 6]

[−1, 4) ∩ (2, 6] = (2, 4)   **2 < x < 4**

51.
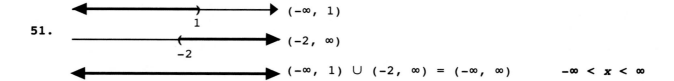

(−∞, 1)

(−2, ∞)

(−∞, 1) ∪ (−2, ∞) = (−∞, ∞)   **−∞ < x < ∞**

53.
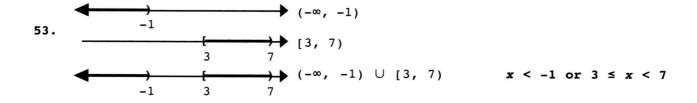

(−∞, −1)

[3, 7)

(−∞, −1) ∪ [3, 7)   **x < −1 or 3 ≤ x < 7**

55.

[2, 3]

(1, 5)

[2, 3] ∪ (1, 5) = (1, 5)   **1 < x < 5**

57.

$(-\infty,\ 4)$

$(-1,\ 6]$

$(-\infty,\ 4)\ \cup\ (-1,\ 6] = (-\infty,\ 6]$ $\quad$ **$x \leq 6$**

59. $\dfrac{q}{7}\ -\ 3\ >\ \dfrac{q\ -\ 4}{3}\ +\ 1$

$21\left(\dfrac{q}{7}\ -\ 3\right)\ >\ 21\left(\dfrac{(q\ -\ 4)}{3}\right)$

$3q\ -\ 63\ >\ 7(q\ -\ 4)\ +\ 21$

$3q\ -\ 63\ >\ 7q\ -\ 28\ +\ 21$

$3q\ -\ 63\ >\ 7q\ -\ 7$

$\qquad -4q\ >\ 56$

$\qquad\quad q\ <\ -14\ \text{or}\ (-\infty,\ -14)$ q

61. $\dfrac{2x}{5}\ -\ \dfrac{1}{2}(x\ -\ 3)\ \leq\ \dfrac{2x}{3}\ -\ \dfrac{3}{10}(x\ +\ 2)$

LCD = 30

$12x\ -\ 15(x\ -\ 3)\ \leq\ 20x\ -\ 9(x\ +\ 2)$

$12x\ -\ 15x\ +\ 45\ \leq\ 20x\ -\ 9x\ -\ 18$

$\qquad -3x\ +\ 45\ \leq\ 11x\ -\ 18$

$\qquad\qquad -14x\ \leq\ -63$

$\qquad\qquad\quad x\ \geq\ 4.5\ \text{or}\ [4.5,\ \infty)$

x

63. $-4\qquad\quad \leq\ \dfrac{9}{5}x\ +\ 32\ \leq\ 68$

$-36\ \leq\ \dfrac{9}{5}x\ \leq\ 36$

$\dfrac{5}{9}(-36)\ \leq\ x\ \leq\ \dfrac{5}{9}(36)$

$-20\ \leq\ x\ \leq\ 20\ \text{or}\ [-20,\ 20]$

x

65. $16\ <\ 7\ -\ 3x\ \leq\ 31$

$\quad 9\ <\ -3x\ \leq\ 24$

$-3\ >\ x\ \geq\ -8$

$-8\ \leq\ x\ <\ -3\ \text{or}\ [-8,\ -3)$

x

67. $-6\ <\ -\dfrac{2}{5}(1\ -\ x)\ \leq\ 4$

$-30\ <\ -2(1\ -\ x)\ \leq\ 20$

$-30\ <\ -2\ +\ 2x\ \leq\ 20$

$-28\ <\ 2x\ \leq\ 22$

$-14\ <\ x\ \leq\ 11\ \text{or}\ (-14,\ 11]$

x

69. (A) F (B) T (C) T

71. If $a < b$, then by definition of $<$, there exists a positive number p such that $a + p = b$. Then, adding c to both sides, we obtain $(a + c) + p = b + c$, where p is positive. So by definition of $<$, $a + c < b + c$

73. (A) If $a < b$, then by definition of $<$, there exists a positive number p such that $a + p = b$. If we multiply both sides of this by the positive number c, we obtain $(a + p)c = bc$, or $ac + pc = bc$, where pc is positive. So by definition of $<$, $ac < bc$.

(B) If $a < b$, then by definition of $<$, there exists a positive number p such that $a + p = b$. If we multiply both sides of this by the negative number c, we obtain $(a + p)c = bc$, or $ac + pc = bc$, where pc is negative. So by definition of $<$, we have $ac > bc$.

Section B-2

1.

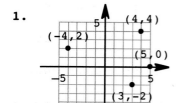

3.

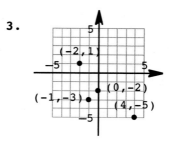

5. $A = (2, 4)$, $B = (3, -1)$, $C = (-4, 0)$, $D = (-5, 2)$

7. $A = (-3, -3)$, $B = (0, 4)$, $C = (-3, 2)$, $D = (5, -1)$

9.

x	$y = x + 1$
-3	-2
-2	-1
-1	0
0	1
1	2
2	3
3	4

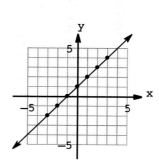

11.

x	$y = x^2 - 5$
-3	4
-2	-1
-1	-4
0	-5
1	-4
2	-1
3	4

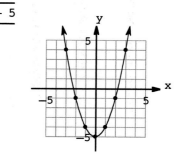

13.

x	$y = 3 + x - 0.5x^2$
-3	-4.5
-2	-1
-1	1.5
0	3
1	3.5
2	3
3	1.5

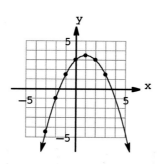

15. (A) When $x = 8$, the corresponding y value on the graph is 6, to the nearest integer.

(B) When $x = -5$, the corresponding y value on the graph is -5, to the nearest integer.

(C) When $x = 0$, the corresponding y value on the graph is -1, to the nearest integer.

(D) When $y = 6$, the corresponding x value on the graph is 8, to the nearest integer.

(E) When $y = -5$, the corresponding x value on the graph is -5, to the nearest integer.

(F) When $y = 0$, the corresponding x value on the graph is 5, to the nearest integer.

17. (A) When $x = 1$, the corresponding y value on the graph is 6, to the nearest integer.

(B) When $x = -8$, the corresponding y value on the graph is 4, to the nearest integer.

(C) When $x = 0$, the corresponding y value on the graph is 4, to the nearest integer.

(D) When $y = -6$, the corresponding x value on the graph is 8, to the nearest integer.

(E) Three values of x correspond to $y = 4$ on the graph. To the nearest integer they are -8, 0, and 6.

(F) Three values of x correspond to $y = 0$ on the graph. To the nearest integer they are -7, -2, and 7.

19. (A) Plot the points (-2, -2), (0, 0), and (2, 2). There are many possible graphs connecting these points. The simplest is:

(B) Plot the points (-1, 2), (0, 0), and (1, -2). Again, the simplest graph connecting the points is a line:

(C)

x	-2	-1	0	1	2
y	-2	2	0	-2	2

Plot the points (-2, -2), (-1, 2), (0, 0), (1, -2), and (2, 2).
The graph of $y = x^3 - 3x$ is shown below.

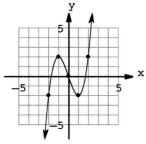

(D) All of the points on the two graphs in (A) and (B) are on the graph of $y = x^3 - 3x$. If you only plot certain points, it's possible, even likely, to draw an incorrect conclusion about the shape of the graph.

21.

x	$y = x^{1/3}$
-5	$\sqrt[3]{5} \approx -1.7$
-3	$\sqrt[3]{-3} \approx -1.4$
-1	-1
0	0
1	1
3	$\sqrt[3]{3} \approx 1.4$
5	$\sqrt[3]{5} \approx 1.7$

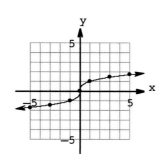

23.

x	$y = x^3$
-2	-8
-1.5	-3.375
-1	-1
0	0
1	1
1.5	3.375
2	8

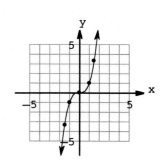

25.

x	$y = \sqrt{x-1}$
1	0
2	1
3	$\sqrt{2} \approx 1.4$
4	$\sqrt{3} \approx 1.7$
5	2

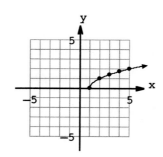

27.

x	$y = \sqrt{1+x^2}$
-5	$\sqrt{26} \approx 5.1$
-3	$\sqrt{10} \approx 3.2$
-1	$\sqrt{2} \approx 1.4$
0	1
1	$\sqrt{2} \approx 1.4$
3	$\sqrt{30} \approx 3.2$
5	$\sqrt{26} \approx 5.1$

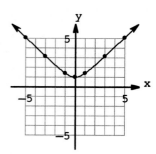

29. (A) and (B)

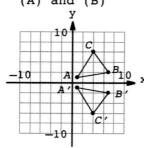

(C) Triangle $A'B'C'$ is the reflection of triangle ABC in the x axis, and triangle ABC is the reflection of triangle $A'B'C'$ in the x axis. In general, changing the sign of the y coordinate of all the points on a graph has the effect of reflecting the graph in the x axis.

31. (A) and (B)

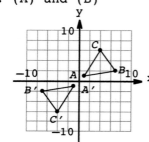

(C) Triangle $A'B'C'$ is the reflection of triangle ABC through the origin, and triangle ABC is the reflection of triangle $A'B'C'$ through the origin. In general, changing the signs of the x and y coordinates of all the points on a graph has the effect of reflecting the graph through the origin.

33.

P	$R = (10-p)p$
5	25
6	24
7	21
8	16
9	9
10	0

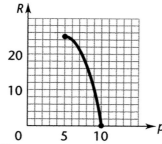

35. (A)

x	$v = 0.5\sqrt{2-x}$
0	0.7
0.5	0.6
1	0.5
1.5	0.35
2	0

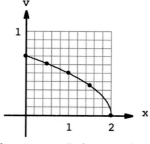

(B) When the ball is at the top of its swing (maximum displacement) $x = 2$, the velocity $v = 0$. As the ball moves to the bottom of its swing $x = 0$, the velocity increases to approximately 0.7 meters per second. As the ball swings past the bottom up to its maximum displacement, the velocity decreases to 0 again.

37. (A) 3,000 cases

(B) If the price is increased from \$6 to \$6.30 a case, the demand decreases from 3,000 cases to 2,600 cases, that is, by 400 cases.

(C) If the price is decreased from \$6 to \$5.70 a case, the demand increases from 3,000 cases to 3,600 cases, that is, by 600 cases.

(D) As the price varies from \$5.60 to \$6.90 a case, the demand varies from 4,000 to 2,000 cases, increasing with decreasing price and decreasing with increasing price.

39. (A) The temperature at 9 AM is approximately 53°.

(B) The highest temperature occurs at approximately 3 PM. This temperature is approximately 68°.

(C) The temperature is 49° at approximately 1 AM, 7 AM and 11 PM.

Section B-3

1. $d = \sqrt{(4 - 1)^2 + (4 - 0)^2} = \sqrt{9 + 16} = \sqrt{25} = 5$

$M = \left(\dfrac{1 + 4}{2}, \dfrac{0 + 4}{2}\right) = \left(\dfrac{5}{2}, 2\right)$

3. $d = \sqrt{(5 - 0)^2 + (10 - (-2))^2} = \sqrt{5^2 + 12^2} = \sqrt{25 + 144} = \sqrt{169} = 13$

$M = \left(\dfrac{0 + 5}{2}, \dfrac{-2 + 10}{2}\right) = \left(\dfrac{5}{2}, 4\right)$

5. $d = \sqrt{(3 - (-6))^2 + (4 - (-4))^2} = \sqrt{9^2 + 8^2} = \sqrt{81 + 64} = \sqrt{145}$

$M = \left(\dfrac{-6 + 3}{2}, \dfrac{-4 + 4}{2}\right) = \left(\dfrac{-3}{2}, 0\right)$

7. $d = \sqrt{(4 - 6)^2 + (-2 - 6)^2} = \sqrt{(-2)^2 + (-8)^2} = \sqrt{4 + 64} = \sqrt{68}$

$M = \left(\dfrac{6 + 4}{2}, \dfrac{6 + (-2)}{2}\right) = (5, 2)$

9. $(x - 0)^2 + (y - 0)^2 = 7^2$

$\phantom{(x - 0)^2 + {}} x^2 + y^2 = 49$

11. $(x - 2)^2 + (y - 3)^2 = 6^2$

$\phantom{(x - 2)^2 + {}} (x - 2)^2 + (y - 3)^2 = 36$

13. $[x - (-4)]^2 + (y - 1)^2 = (\sqrt{7})^2$

$ (x + 4)^2 + (y - 1)^2 = 7$

> **Common Error:** *not* $(x - 4)^2$

15. $[x - (-3)]^2 + [y - (-4)]^2 = (\sqrt{2})^2$

$ (x + 3)^2 + (y + 4)^2 = 2$

17. $\dfrac{b}{2} = \dfrac{8}{2} = 4; \ (x + 4)^2 = x^2 + 8x + 16$

19. $\dfrac{b}{2} = \dfrac{-3}{2}; \ \left(x - \dfrac{3}{2}\right)^2 = x^2 - 3x + \dfrac{9}{4}$

21. This is a circle with center $(0, 0)$ and radius 2.

$$x^2 + y^2 = 4$$

23. This is a circle with center $(1, 0)$ and radius 1.

$$(x - 1)^2 + y^2 = 1$$

25. This is a circle with center $(-2, 1)$ and radius 3.

$$(x - (-2))^2 + (y - 1)^2 = 9$$
$$\text{or } (x + 2)^2 + (y - 1)^2 = 9$$

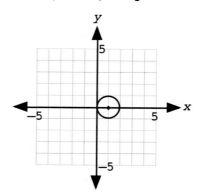

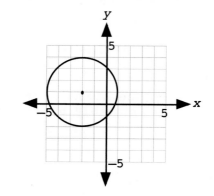

27. (A) $-2 = \dfrac{a_1 + 1}{2}$ (B) $6 = \dfrac{a_2 + 3}{2}$

   $-4 = a_1 + 1$ $12 = a_2 + 3$

   $-5 = a_1$ $9 = a_2$

(C) From parts (A) and (B), $A = (-5, 9)$

$$d(A, M) = \sqrt{(-2(-5)^2 + (6 - 9)^2} = \sqrt{3^2 + (-3)^2} = \sqrt{9 + 9} = \sqrt{18}$$

$$d(M, B) = \sqrt{(1 - (-2)^2 + (3 - 6)^2} = \sqrt{3^2 + (-3)^2} = \sqrt{9 + 9} = \sqrt{18}$$

As expected, the distances are the same.

29. This is a circle with center $(0, 2)$ and radius 2. That is, the set of all points that are 2 units away from $(0, 2)$.

$$x^2 + (y - 2)^2 = 4$$

31. This is a circle with center $(1, 1)$ and radius 4. That is, the set of all points that are 4 units away from $(1, 1)$.

$$(x - 1)^2 + (y - 1)^2 = 16$$

33. $M = \left(\dfrac{-4.3 + 9.6}{2}, \dfrac{5.2 + (-1.7)}{2}\right) = \left(\dfrac{5.3}{2}, \dfrac{3.5}{2}\right) = (2.65, 1.75)$

$$\begin{aligned} d(A, M) &= \sqrt{(2.65 - (-4.3))^2 + (1.75 - 5.2)^2} &= \sqrt{6.95^2 + (-3.45)^2} \\ & &= \sqrt{48.3025 + 11.9025} \\ & &= \sqrt{60.205} = 7.76 \end{aligned}$$

$$\begin{aligned} d(M, B) &= \sqrt{(4.6 - 2.65)^2 + (-1.7 - 1.75)^2} &= \sqrt{6.95^2 + (-3.45)^2} \\ & &= \sqrt{48.3025 + 11.9025} \\ & &= \sqrt{60.205} = 7.76 \end{aligned}$$

$$\begin{aligned} d(A, B) &= \sqrt{(9.6 - (-4.3))^2 + (-1.7 - 5.2)^2} &= \sqrt{13.9^2 + (-6.9)^2} \\ & &= \sqrt{193.21 + 47.61} \\ & &= \sqrt{240.82} = 15.52 \end{aligned}$$

$$\frac{1}{2}d(A, B) = \frac{1}{2}(15.52) = 7.76$$

35. Write $B = (b_1, b_2)$. -5 is the average of 25 and b_1, so $-5 = \dfrac{25 + b_1}{2}$

$$-10 = 25 + b_1$$
$$-35 = b_1$$

-2 is the average of 10 and b_2, so $-2 = \dfrac{10 + b_2}{2}$

$$-4 = 10 + b_2$$
$$-14 = b_2$$

So $B = (-35, -14)$.

$d(A, M) = \sqrt{(-5 - 25)^2 + (-2 - 10)^2} = \sqrt{(-30)^2 - (12)^2} = \sqrt{900 + 144} = \sqrt{1,044} = 32.3$

$d(M, B) = \sqrt{(-35 - (-5))^2 + (-14 - (-2))^2} = \sqrt{(-30)^2 - (12)^2} = \sqrt{900 + 144} = \sqrt{1,044} = 32.3$

$d(A, B) = \sqrt{(-35 - 25)^2 + (-14 - 10)^2} = \sqrt{(-60)^2 + (-24)^2} = \sqrt{3,600 + 576} = \sqrt{4,176} = 64.6$

$\dfrac{1}{2} d(A, B) = \dfrac{1}{2}(64.6) = 32.3$

37. Write $A = (a_1, a_2)$. -8 is the average of a_1 and 2, so $-8 = \dfrac{a_1 + 2}{2}$

$$-16 = a_1 + 2$$
$$-18 = a_1$$

-6 is the average of a_2 and 4, so $-6 = \dfrac{a_2 + 4}{2}$

$$-12 = a_2 + 4$$
$$-16 = a_2$$

So $A = (-18, -16)$.

$d(A, M) = \sqrt{(-8 - (-18))^2 + (-6 - (-16))^2} = \sqrt{10^2 + 10^2} = \sqrt{100 + 100} = \sqrt{200} = 14.14$

$d(M, B) = \sqrt{(2 - (-8))^2 + (4 - (-6))^2} = \sqrt{10^2 + 10^2} = \sqrt{100 + 100} = \sqrt{200} = 14.14$

$d(A, B) = \sqrt{(2 - (-18))^2 + (4 - (-16))^2} = \sqrt{20^2 + 20^2} = \sqrt{400 + 400} = \sqrt{800} = 28.28$

$\dfrac{1}{2} d(A, B) = \dfrac{1}{2}(28.28) = 14.14$

39. We need to complete the square with respect to y.

$x^2 + y^2 - 2y + ? = 0 \qquad \left(\dfrac{-2}{2}\right)^2 = (-1)^2 = 1$

$x^2 + y^2 - 2y + 1 = 1$
$x^2 + (y - 1)^2 = 1$
center = $(0, 1)$, $r = 1$

41. We need to complete the square with respect to x and y.

$x^2 - 2x + ? + y^2 + 6y + ? = 6 \qquad \left(\dfrac{-2}{2}\right)^2 = 1; \ \left(\dfrac{6}{2}\right)^2 = 9$

$x^2 - 2x + 1 + y^2 + 6y + 9 = 6 + 1 + 9$
$(x - 1)^2 + (y + 3)^2 = 16$
center = $(1, -3)$, $r = 4$

43. We need to complete the square with respect to x and y.

$$x^2 + x + ? + y^2 + 3y + ? = -2 \qquad \left(\frac{1}{2}\right)^2 = \frac{1}{4}; \quad \left(\frac{3}{2}\right)^2 = \frac{9}{4}$$

$$x^2 + x + \frac{1}{4} + y^2 + 3y + \frac{9}{4} = -2 + \frac{1}{4} + \frac{9}{4}$$

$$\left(x + \frac{1}{2}\right)^2 + \left(y + \frac{3}{2}\right)^2 = -\frac{8}{4} + \frac{1}{4} + \frac{9}{4} = \frac{2}{4} = \frac{1}{2}$$

$$\left(x + \frac{1}{2}\right)^2 + \left(y + \frac{3}{2}\right)^2 = \frac{1}{2}$$

center $= \left(-\frac{1}{2}, -\frac{3}{2}\right)$, $r = \sqrt{\frac{1}{2}}$

45. Let $A = (-3, 2)$, $B = (1, -2)$, $C = (8, 5)$

$d(A, B) = \sqrt{(1 - (-3))^2 + (-2 - 2)^2} = \sqrt{4^2 + (-4)^2} = \sqrt{16 + 16} = \sqrt{32}$

$d(B, C) = \sqrt{(8 - 1)^2 + (5 - (-2))^2} = \sqrt{7^2 + 7^2} = \sqrt{49 + 49} = \sqrt{98}$

$d(A, C) = \sqrt{(8 - (-3))^2 + (5 - 2)^2} = \sqrt{11^2 + 3^2} = \sqrt{121 + 9} = \sqrt{130}$

Notice that $(d(A, B))^2 + (d(B, C))^2 = 32 + 98 = 130 = (d(A, C))^2$

Since these distances satisfy the Pythagorean Theorem, the three points are vertices of a right triangle. The segment connecting A and C is the hypotenuse (it's the longest side) so we need to find its midpoint.

$$M = \left(\frac{-3 + 8}{2}, \frac{2 + 5}{2}\right) = \left(\frac{5}{2}, \frac{7}{2}\right)$$

The vertex opposite the hypotenuse is B.

$d(M, B) = \sqrt{\left(1 - \frac{5}{2}\right)^2 + \left(-2 - \frac{7}{2}\right)^2} = \sqrt{\left(\frac{-3}{2}\right)^2 + \left(-\frac{11}{2}\right)^2} = \sqrt{\frac{9}{4} + \frac{121}{4}} = \sqrt{\frac{130}{4}} = \sqrt{32.5}$

47. Perimeter = sum of lengths of all three sides

$= \sqrt{[1 - (-3)]^2 + [(-2) - 1]^2} + \sqrt{(4 - 1)^2 + [3 - (-2)]^2} + \sqrt{[4 - (-3)]^2 + (3 - 1)^2}$

$= \sqrt{16 + 9} + \sqrt{9 + 25} + \sqrt{49 + 4}$

$= \sqrt{25} + \sqrt{34} + \sqrt{53}$

$= 18.11$ to two decimal places

49. $d(P_1, M) = \sqrt{\left(\frac{x_1 + x_2}{2} - x_1\right)^2 + \left(\frac{y_1 + y_2}{2} - y_1\right)^2}$

$= \sqrt{\left(\frac{x_1 + x_2 - 2x_1}{2}\right)^2 + \left(\frac{y_1 + y_2 - 2y_1}{2}\right)^2}$

$= \sqrt{\left(\frac{x_2 - x_1}{2}\right)^2 + \left(\frac{y_2 - y_1}{2}\right)^2}$

$d(M, P_2) = \sqrt{\left(x_2 - \frac{x_1 + x_2}{2}\right)^2 + \left(y_1 - \frac{y_1 + y_2}{2}\right)^2}$

$= \sqrt{\left(\frac{2x_2 - x_1 - x_2}{2}\right)^2 + \left(\frac{2y_2 - y_1 - y_2}{2}\right)^2}$

$= \sqrt{\left(\frac{x_2 - x_1}{2}\right)^2 + \left(\frac{y_2 - y_1}{2}\right)^2}$

$$\frac{1}{2} d(P_1, \; P_2) \; = \; \frac{1}{2}\sqrt{(x_2 - x_1)^2 + (y_2 - y_1)^2}$$

$$= \; \sqrt{\frac{1}{4}\left[(x_2 - x_1)^2 + (y_2 - y_1)^2\right]} \quad \text{Note: The } \frac{1}{2} \text{ becomes } \frac{1}{4} \text{ when moved inside the root.}$$

$$= \; \sqrt{\frac{(x_2 - x_1)^2}{4} + \frac{(y_2 - y_1)^2}{4}}$$

$$= \; \sqrt{\left(\frac{x_2 - x_1}{2}\right)^2 + \left(\frac{y_2 - y_1}{2}\right)^2}$$

All three of these distances are equal.

51. The center of a circle is the midpoint of its diameter. The radius of a circle is half the length of its diameter. Using the midpoint formula, the center of this circle has coordinates given by

$$\left(\frac{x_1 + x_2}{2}, \frac{y_1 + y_2}{2}\right) \; = \; \left(\frac{7 + 1}{2}, \frac{-3 + 7}{2}\right) \; = \; (4, \; 2)$$

The radius of the circle is given by

$$r \; = \; \frac{1}{2}\sqrt{(x_2 - x_1)^2 + (y_2 - y_1)^2}$$

$$= \; \frac{1}{2}\sqrt{(1 - 7)^2 + [7 - (-3)]^2}$$

$$= \; \frac{1}{2}\sqrt{36 + 100} \; = \; \frac{1}{2}\sqrt{136} \; = \; \frac{1}{2}\sqrt{4 \cdot 34} \; = \; \sqrt{34}$$

So the equation of the circle is given by

$$(x - 4)^2 + (y - 2)^2 = (\sqrt{34})^2$$
$$(x - 4)^2 + (y - 2)^2 = 34$$

53. The radius of a circle is the distance from the center to any point on the circle. Since the center of this circle is (2, 2) and (3, -5) is a point on the circle, the radius is given by

$$r \; = \; \sqrt{(3 - 2)^2 + [(-5) - 2]^2} \; = \; \sqrt{1 + 49} \; = \; \sqrt{50}$$

So the equation of the circle is given by

$$(x - 2)^2 + (y - 2)^2 = (\sqrt{50})^2$$
$$(x - 2)^2 + (y - 2)^2 = 50$$

55.

$$d(B, \; D) \; = \; \sqrt{(60 - 0)^2 + (27 - 13.5)^2} \; = \; \sqrt{60^2 + 13.5^2} \; = \; \sqrt{3600 + 182.25}$$

$$= \; \sqrt{3,782.25} \; = \; 66 \text{ ft.}$$

$$d(F, \; C) \; = \; \sqrt{(78 - 0)^2 + (13.5 - 27)^2} \; = \; \sqrt{78^2 + (-13.5)^2} \; = \; \sqrt{6,084 + 182.25}$$

$$= \; \sqrt{6,266.25} \; = \; 79 \text{ ft.}$$

57. Using the hint, we note that $(2, r - 1)$ must satisfy $x^2 + y^2 = r^2$, that is

$$2^2 + (r - 1)^2 = r^2$$
$$4 + r^2 - 2r + 1 = r^2$$
$$-2r + 5 = 0$$
$$r = \frac{5}{2} \text{ or } 2.5 \text{ ft.}$$

59. (A) From the drawing, we can write:

$$\binom{\text{Distance from tower}}{\text{to town } B} = 2 \times \binom{\text{Distance from tower}}{\text{to town } A}$$

$$\binom{\text{Distance from } (x, y)}{\text{to } (36, 15)} = 2 \times \binom{\text{Distance from } (x, y)}{\text{to } (0, 0)}$$

$$\sqrt{(36 - x)^2 + (15 - y)^2} = 2\sqrt{(0 - x)^2 + (0 - y)^2}$$
$$\sqrt{(36 - x)^2 + (15 - y)^2} = 2\sqrt{x^2 + y^2}$$
$$(36 - x)^2 + (15 - y)^2 = 4(x^2 + y^2)$$
$$1{,}296 - 72x + x^2 + 225 - 30y + y^2 = 4x^2 + 4y^2$$
$$1{,}521 = 3x^2 + 3y^2 + 72x + 30y$$
$$507 = x^2 + y^2 + 24x + 10y$$
$$144 + 25 + 507 = x^2 + 24x + 144 + y^2 + 10y + 25$$
$$676 = (x + 12)^2 + (y + 5)^2$$

The circle has center $(-12, -5)$ and radius 26.

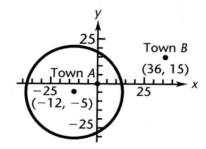

(B) All points due east of Town A have y coordinate 0 in this coordinate system. The points on the circle for which $y = 0$ are found by substituting $y = 0$ into the equation of the circle and solving for x.

$$(x + 12)^2 + (y + 5)^2 = 676$$
$$(x + 12)^2 + 25 = 676$$
$$(x + 12)^2 = 651$$
$$x + 12 = \pm\sqrt{651}$$
$$x = -12 \pm \sqrt{651}$$

Choosing the positive square root so that x is greater than -12 (east rather than west) we have $x = -12 + \sqrt{651} \approx 13.5$ miles.

APPENDIX C   Special Topics

Section C-1

1. 123,005   3. 20,040   5. 6.0   7. 80.000   9. 0.012   11. 0.000 960

13. 3.08   15. 924,000   17. 23.7   19. 2.82×10^3 21. 6.78×10^{-4}

23.

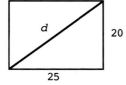

Using the Pythagorean Theorem,
$$25^2 + 20^2 = d^2$$
$$625 + 400 = d^2$$
$$1{,}025 = d^2$$
$$32.0156 = d$$

Rounded to 1 significant digit, which is the smallest number of significant digits in the original measurements, the diagonal is 30 ft.

Section C-2

1. $$\frac{7x - 14}{(x - 4)(x + 3)} = \frac{A}{x - 4} + \frac{B}{x + 3} = \frac{A(x + 3) + B(x - 4)}{(x - 4)(x + 3)}$$

The numerators must be equal, so
$$7x - 14 = A(x + 3) + B(x - 4)$$

Graphical Solution:

$$7x - 14 = A(x + 3) + B(x - 4) \Rightarrow 7x - 14 = (A + B)x + 3A - 4B$$

For this equation to be true, the coefficients of x on each side must be equal; that is, $7 = A + B$. Also, the constant terms must be equal; that is, $-14 = 3A - 4B$. Now solve each of these equations for B and graph, treating A as independent variable and B as dependent variable.

$$\begin{matrix} A + B = 7 \\ 3A - 4B = -14 \end{matrix} \quad \Rightarrow \quad \begin{matrix} B = 7 - A \\ B = \dfrac{14 + 3A}{4} \end{matrix}$$

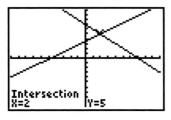

$A = 2$, $B = 5$

Algebraic Solution:

If $x = -3$, then
$$-35 = -7B$$
$$B = 5$$
If $x = 4$, then
$$14 = 7A$$
$$A = 2$$
$$A = 2, \quad B = 5$$

3. $\dfrac{17x - 1}{(2x - 3)(3x - 1)} = \dfrac{A}{2x - 3} + \dfrac{B}{3x - 1} = \dfrac{A(3x - 1) + B(2x - 3)}{(2x - 3)(3x - 1)}$

The numerators must be equal, so

$17x - 1 = A(3x - 1) + B(2x - 3)$

<u>Graphical Solution:</u>

(See Problem 1 for details.)

$17x - 1 = A(3x - 1) + B(2x - 3) \Rightarrow 17x - 1 = (3A + 2B)x - A - 3B$

$\begin{array}{l} 3A + 2B = 17 \\ -A - 3B = -1 \end{array} \Rightarrow \begin{array}{l} B = \dfrac{17 - 3A}{2} \\ B = \dfrac{1 - A}{3} \end{array}$

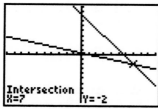

$A = 7,\ B = -2$

<u>Algebraic Solution:</u> If $x = \dfrac{1}{3}$, then

$17\left(\dfrac{1}{3}\right) - 1 = B\left[2\left(\dfrac{1}{3}\right) - 3\right]$

$\dfrac{14}{3} = -\dfrac{7}{3}B$

$B = -2$

If $x = \dfrac{3}{2}$, then

$17\left(\dfrac{3}{2}\right) - 1 = A\left[3\left(\dfrac{3}{2}\right) - 1\right]$

$\dfrac{49}{2} = \dfrac{7}{2}A$

$A = 7$

$A = 7,\ B = -2$

5. $\dfrac{3x^2 + 7x + 1}{x(x + 1)^2} = \dfrac{A}{x} + \dfrac{B}{x + 1} + \dfrac{C}{(x + 1)^2} = \dfrac{A(x + 1)^2 + Bx(x + 1) + Cx}{x(x + 1)^2}$

$3x^2 + 7x + 1 = A(x + 1)^2 + Bx(x + 1) + Cx$

$3x^2 + 7x + 1 = A(x^2 + 2x + 1) + Bx^2 + Bx + Cx$

$3x^2 + 7x + 1 = Ax^2 + 2Ax + A + Bx^2 + Bx + Cx$

$3x^2 + 7x + 1 = (A + B)x^2 + (2A + B + C)x + A$

Equating coefficients, we know $A = 1$. Also, $A + B = 3$, so $1 + B = 3$ and $B = 2$. Finally, $2A + B + C = 7$, so $2 + 2 + C = 7$ and $C = 3$.

Check by addition:

$\dfrac{1}{x} + \dfrac{2}{x + 1} + \dfrac{3}{(x + 1)^2} = \dfrac{(x + 1)^2}{x(x + 1)^2} + \dfrac{2x(x + 1)}{x(x + 1)^2} + \dfrac{3x}{x(x + 1)^2}$

$$= \frac{x^2 + x + 1 + 2x^2 + 2x + 3x}{x(x + 1)^2}$$

$$= \frac{3x^2 + 7x + 1}{x(x + 1)^2}$$

Check with graphing calculator:

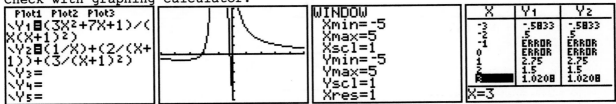

The graphs appear to be identical and the tables are the same.

7. $\frac{3x^2 + x}{(x - 2)(x^2 + 3)} = \frac{A}{x - 2} + \frac{Bx + C}{x^2 + 3} = \frac{A(x^2 + 3) + (Bx + C)(x - 2)}{(x - 2)(x^2 + 3)}$

$3x^2 + x = A(x^2 + 3) + (Bx + C)(x - 2)$

First, plug in $x = 2$:

$3(2)^2 + 2 = A(4 + 3) + 0$

$\qquad 14 = 7A$

$\qquad A = 2$

$3x^2 + x = 2(x^2 + 3) + (Bx + C)(x - 2)$

$3x^2 + x = 2x^2 + 6 + Bx^2 - 2Bx + Cx - 2C$

$3x^2 + x = (2 + B)x^2 + (C - 2B)x + 6 - 2C$

Equating coefficients,

$2 + B = 3$, so $B = 1$

$6 - 2C = 0$, so $6 = 2C$ and $C = 3$.

Check by addition:

$\frac{x}{x - 2} + \frac{x + 3}{x^2 + 3} = \frac{2(x^2 + 3) + (x + 3)(x - 2)}{(x - 2)(x^2 + 3)}$

$$= \frac{2x^2 + 6 + x^2 + x - 6}{(x - 2)(x^2 + 3)}$$

$$= \frac{3x^2 + x}{(x - 2)(x^2 + 3)}$$

Check with graphing calculator:

The graphs appear to be identical and the tables are the same.

9. $\frac{2x^2 + 4x - 1}{(x^2 + x + 1)^2} = \frac{Ax + B}{x^2 + x + 1} + \frac{Cx + D}{(x^2 + x + 1)^2} = \frac{(Ax + B)(x^2 + x + 1) + (Cx + D)}{(x^2 + x + 1)^2}$

$2x^2 + 4x - 1 = (Ax + B)(x^2 + x + 1) + Cx + D$

$2x^2 + 4x - 1 = Ax^3 + Ax^2 + Ax + Bx^2 + Bx + B + Cx + D$

$2x^2 + 4x - 1 = Ax^3 + (A + B)x^2 + (A + B + C)x + B + D$

Equating coefficients,
$A = 0$
$A + B = 2$, so $B = 2$
$A + B + C = 4$, so $0 + 2 + C = 4$, and $C = 2$
$B + D = -1$, so $2 + D = -1$, and $D = -3$

Check by addition:

$$\frac{2}{x^2 + x + 1} + \frac{2x - 3}{(x^2 + x + 1)^2} = \frac{2(x^2 + x + 1)}{(x^2 + x + 1)^2} + \frac{2x - 3}{(x^2 + x + 1)^2}$$

$$= \frac{2x^2 + 2x + 2 + 2x - 3}{(x^2 + x + 1)^2}$$

$$= \frac{2x^2 + 4x - 1}{(x^2 + x + 1)^2}$$

Check with graphing calculator:

The graphs appear to be identical and the tables are the same.

11. Since $x^2 - 2x - 8 = (x + 2)(x - 4)$, we write

$$\frac{-x + 22}{x^2 - 2x - 8} = \frac{A}{x + 2} + \frac{B}{x - 4} = \frac{A(x - 4) + B(x + 2)}{(x + 2)(x - 4)}$$

The numerators must be equal, so

$-x + 22 = A(x - 4) + B(x + 2)$
If $x = 4$
$18 = 6B$
$B = 3$

If $x = -2$
$24 = -6A$
$A = -4$

So $\dfrac{-x + 22}{x^2 - 2x - 8} = \dfrac{-4}{x + 2} + \dfrac{3}{x - 4}$.

13. Since $6x^2 - x - 12 = (3x + 4)(2x - 3)$, we write

$$\frac{3x - 13}{6x^2 - x - 12} = \frac{A}{3x + 4} + \frac{B}{2x - 3} = \frac{A(2x - 3) + B(3x + 4)}{(3x + 4)(2x - 3)}$$

The numerators must be equal, so
$3x - 13 = A(2x - 3) + B(3x + 4)$
If $x = \dfrac{3}{2}$

$$3\left(\frac{3}{2}\right) - 13 = B\left[3\left(\frac{3}{2}\right) + 4\right]$$

$$-\frac{17}{2} = \frac{17}{2}B$$

$$B = -1$$

If $x = -\dfrac{4}{3}$

$$3\left(-\frac{4}{3}\right) - 13 = A\left[2\left(-\frac{4}{3}\right) - 3\right]$$

$$-17 = -\frac{17}{3}A$$

$$A = 3$$

So $\dfrac{3x - 13}{6x^2 - x - 12} = \dfrac{3}{3x + 4} - \dfrac{1}{2x - 3}$

15. Since $x^3 - 6x^2 + 9x = x(x^2 - 6x + 9) = x(x - 3)^2$, we write

$$\frac{x^2 - 12x + 18}{x^3 - 6x^2 + 9x} = \frac{A}{x} + \frac{B}{x - 3} + \frac{C}{(x - 3)^2} = \frac{A(x - 3)^2 + Bx(x - 3) + Cx}{x(x - 3)^2}$$

The numerators must be equal, so

$x^2 - 12x + 18 = A(x - 3)^2 + Bx(x - 3) + Cx$

If $x = 3$
$-9 = 3C$
$C = -3$

If $x = 0$
$18 = 9A$
$A = 2$

If $x = 2$
$-2 = A - 2B + 2C$
$-2 = 2 - 2B - 6$ using $A = 2$ and $C = -3$
$2 = -2B$
$B = -1$

So $\dfrac{x^2 - 12x + 18}{x^3 - 6x^2 + 9x} = \dfrac{2}{x} - \dfrac{1}{x - 3} - \dfrac{3}{(x - 3)^2}$

17. Since $x^3 + 2x^2 + 3x = x(x^2 + 2x + 3)$, we write

$$\frac{5x^2 + 3x + 6}{x^3 + 2x^2 + 3x} = \frac{A}{x} + \frac{Bx + C}{x^2 + 2x + 3} = \frac{A(x^2 + 2x + 3) + (Bx + C)x}{x(x^2 + 2x + 3)}$$

> **Common Error:**
> Writing $\dfrac{B}{x^2 + 2x + 3}$. Since $x^2 + 2x + 3$ is quadratic, the numerator must be first degree.

The numerators must be equal, so
$5x^2 + 3x + 6 = A(x^2 + 2x + 3) + (Bx + C)x = (A + B)x^2 + (2A + C)x + 3A$
Equating coefficients of like terms, we have
$5 = A + B$
$3 = 2A + C$
$3A = 6$
We can now see that $A = 2$; use the 2nd and 3rd equations to find B and C: $B = 3$ and $C = -1$.

So $\dfrac{5x^2 + 3x + 6}{x^3 + 2x^2 + 3x} = \dfrac{2}{x} + \dfrac{3x - 1}{x^2 + 2x + 3}$

19. Since $x^4 + 4x^2 + 4 = (x^2 + 2)^2$, we write

$$\frac{2x^3 + 7x + 5}{x^4 + 4x^2 + 4} = \frac{Ax + B}{x^2 + 2} + \frac{Cx + D}{(x^2 + 2)^2} = \frac{(Ax + B)(x^2 + 2) + Cx + D}{(x^2 + 2)^2}$$

The numerators must be equal, so

$$2x^3 + 7x + 5 = (Ax + B)(x^2 + 2) + Cx + D = Ax^3 + Bx^2 + (2A + C)x + 2B + D$$

Equating coefficients of like terms, we have

$2 = A$
$0 = B$
$7 = 2A + C$
$5 = 2B + D$

Now we know that $A = 2$ and $B = 0$; use the 3rd and 4th equations to find C and D:
$C = 3$ and $D = 5$.

$$\frac{2x^3 + 7x + 5}{x^4 + 4x^2 + 4} = \frac{2x}{x^2 + 2} + \frac{3x + 5}{(x^2 + 2)^2}$$

21. First we divide to obtain a polynomial plus a proper fraction.

$$
\begin{array}{r}
x - 2 \\
x^2 - 5x + 6 \overline{\smash{\big)}\ x^3 - 7x^2 + 17x - 17} \\
\underline{x^3 - 5x^2 + 6x} \\
-2x^2 + 11x - 17 \\
\underline{-2x^2 + 10x - 12} \\
x - 5
\end{array}
$$

So, $\dfrac{x^3 - 7x^2 + 17x - 17}{x^2 - 5x + 6} = x - 2 + \dfrac{x - 5}{x^2 - 5x + 6}$

To decompose the proper fraction, we note $x^2 - 5x + 6 = (x - 2)(x - 3)$ and we write:

$$\frac{x - 5}{x^2 - 5x + 6} = \frac{A}{x - 2} + \frac{B}{x - 3} = \frac{A(x - 3) + B(x - 2)}{(x - 2)(x - 3)}$$

The numerators must be equal, so
$x - 5 = A(x - 3) + B(x - 2)$
If $x = 3$
$-2 = B$
If $x = 2$
$-3 = -A$
$A = 3$

So $\dfrac{x^3 - 7x^2 + 17x - 17}{x^2 - 5x + 6} = x - 2 + \dfrac{3}{x - 2} - \dfrac{2}{x - 3}$

23. First, we must factor $x^3 - 6x - 9$. Possible rational zeros of this polynomial are ± 1, ± 3, ± 9. Forming a synthetic division table, we see

	1	0	-6	-9
1	1	1	-5	-14
3	1	3	3	0

So $x^3 - 6x - 9 = (x - 3)(x^2 + 3x + 3)$
$x^2 + 3x + 3$ cannot be factored further in the real numbers.

So $\dfrac{4x^2 + 5x - 9}{x^3 - 6x - 9} = \dfrac{A}{x - 3} + \dfrac{Bx + C}{x^2 + 3x + 3} = \dfrac{A(x^2 + 3x + 3) + (Bx + C)(x - 3)}{(x - 3)(x^2 + 3x + 3)}$

The numerators must be equal, so
$$4x^2 + 5x - 9 = A(x^2 + 3x + 3) + (Bx + C)(x - 3)$$
$$= Ax^2 + 3Ax + 3A + Bx^2 - 3Bx + Cx - 3C$$
$$= (A + B)x^2 + (3A - 3B + C)x + 3A - 3C$$

Before equating coefficients of like terms, we note that if $x = 3$
$$4(3)^2 + 5(3) - 9 = A(3^2 + 3 \cdot 3 + 3)$$
$$\text{So} \quad 42 = 21A$$
$$A = 2$$
$$\text{Since} \quad 4 = A + B, \ B = 2$$
$$5 = 3A - 3B + C = 6 - 6 + C, \text{ so } C = 5$$

We have $A = 2$, $B = 2$, $C = 5$

So $\dfrac{4x^2 + 5x - 9}{x^3 - 6x - 9} = \dfrac{2}{x - 3} + \dfrac{2x + 5}{x^2 + 3x + 3}$

25. First, we must factor $x^3 + 2x^2 - 15x - 36$. The possible rational zeros are ±1, ±2, ±3, ±4, ±6, ±9, ±12, ±18, ±36. Forming a synthetic division table, we see

	1	2	-15	-36
1	1	3	-12	-48
2	1	4	-7	-50
3	1	5	0	-36
4	1	6	9	0

So $x^3 + 2x^2 - 15x - 36 = (x - 4)(x^2 + 6x + 9) = (x - 4)(x + 3)^2$.

So $\dfrac{x^2 + 16x + 18}{x^3 + 2x^2 - 15x - 36} = \dfrac{A}{x - 4} + \dfrac{B}{x + 3} + \dfrac{C}{(x + 3)^2}$

$$= \dfrac{A(x + 3)^2 + B(x - 4)(x + 3) + C(x - 4)}{(x - 4)(x + 3)^2}$$

The numerators must be equal, so
$$x^2 + 16x + 18 = A(x + 3)^2 + B(x - 4)(x + 3) + C(x - 4)$$

If $x = -3$
$$-21 = -7C$$
$$C = 3$$

If $x = 4$
$$98 = 49A$$
$$A = 2$$

If $x = 5$
$$123 = 64A + 8B + C$$
$$= 128 + 8B + 3 \text{ using } A = 2 \text{ and } C = 3$$
$$-8 = 8B$$
$$B = -1$$

So $\dfrac{x^2 + 16x + 18}{x^3 + 2x^2 - 15x - 36} = \dfrac{2}{x - 4} - \dfrac{1}{x + 3} + \dfrac{3}{(x + 3)^2}$

27. First, we must factor $x^4 - 5x^3 + 9x^2 - 8x + 4$. The possible rational zeros are ±1, ±2, ±4. We form a synthetic division table:

	1	-5	9	-8	4
1	1	-4	5	-3	1
2	1	-3	3	-2	0

We examine $x^3 - 3x^2 + 3x - 2$ to see if 2 is a double zero.

	1	-3	3	-2
2	1	-1	1	0

So $x^4 - 5x^3 + 9x^2 - 8x + 4 = (x - 2)^2(x^2 - x + 1)$. $x^2 - x + 1$ cannot be factored further in the real numbers, so

$$\frac{-x^2 + x - 7}{x^4 - 5x^3 + 9x^2 - 8x + 4} = \frac{A}{x - 2} + \frac{B}{(x - 2)^2} + \frac{Cx + D}{x^2 - x + 1}$$

$$= \frac{A(x - 2)(x^2 - x + 1) + B(x^2 - x + 1) + (Cx + D)(x - 2)^2}{(x - 2)^2(x^2 - x + 1)}$$

The numerators must be equal, so

$-x^2 + x - 7 = A(x - 2)(x^2 - x + 1) + B(x^2 - x + 1) + (Cx + D)(x - 2)^2$

If $x = 2$

$-9 = 3B$

$B = -3$

$-x^2 + x - 7 = A(x^3 - 3x^2 + 3x - 2) + B(x^2 - x + 1) + (Cx + D)(x^2 - 4x + 4)$

$\qquad = (A + C)x^3 + (-3A + B - 4C + D)x^2 + (3A - B + 4C - 4D)x - 2A + B + 4D$

We have already $B = -3$, so equating coefficients of like terms,

$\quad 0 = A + C$

$-1 = -3A - 3 - 4C + D$

$\quad 1 = 3A + 3 + 4C - 4D$

$-7 = -2A - 3 + 4D$

Since $C = -A$, we can write

$-1 = -3A - 3 + 4A + D$

$\quad 1 = 3A + 3 - 4A - 4D$

$-7 = -2A - 3 + 4D$

So $D = 0$ (adding the first equations), $A = 2$, $C = -2$.

$$\frac{-x^2 + x - 7}{x^4 - 5x^3 + 9x^2 - 8x + 4} = \frac{2}{x - 2} - \frac{3}{(x - 2)^2} - \frac{2x}{x^2 - x + 1}$$

29. First we divide to obtain a polynomial plus a proper fraction.

$$
\begin{array}{r}
x + 2 \\
4x^4 + 4x^3 - 5x^2 + 5x - 2 \overline{\smash{\big)}\, 4x^5 + 12x^4 - x^3 + 7x^2 - 4x + 2} \\
\underline{4x^5 + 4x^4 - 5x^3 + 5x^2 - 2x} \\
8x^4 + 4x^3 + 2x^2 - 2x + 2 \\
\underline{8x^4 + 8x^3 - 10x^2 + 10x - 4} \\
-4x^3 + 12x^2 - 12x + 6
\end{array}
$$

Now we must decompose $\dfrac{-4x^3 + 12x^2 - 12x + 6}{4x^4 + 4x^3 - 5x^2 + 5x - 2}$, starting by factoring $4x^4 + 4x^3 - 5x^2 + 5x - 2$.

Possible rational zeros are ± 1, ± 2, $\pm\dfrac{1}{2}$, $\pm\dfrac{1}{4}$. We form a synthetic division table:

	4	4	-5	5	-2	
1	4	8	3	8	6	1 is an upper bound, eliminating 2
-1	4	0	-5	10	-12	
-2	4	-4	3	-1	0	-2 is a zero

We investigate $4x^3 - 4x^2 + 3x - 1$; the only remaining rational zeros are $\pm\frac{1}{2}$, $\pm\frac{1}{4}$.

We form a synthetic division table:

	4	-4	3	-1
$\frac{1}{2}$	4	-2	2	0

So $4x^4 + 4x^3 - 5x^2 + 5x - 2 = (x + 2)\left(x - \frac{1}{2}\right)(4x^2 - 2x + 2)$

$$= (x + 2)\left(x - \frac{1}{2}\right)2(2x^2 - x + 1)$$

$$= (x + 2)(2x - 1)(2x^2 - x + 1)$$

$2x^2 - x + 1$ cannot be factored further in the real numbers, so we write:

$$\frac{-4x^3 + 12x^2 - 12x + 6}{4x^4 + 4x^3 - 5x^2 + 5x - 2}$$

$$= \frac{A}{x + 2} + \frac{B}{2x - 1} + \frac{Cx + D}{2x^2 - x + 1}$$

$$= \frac{A(2x - 1)(2x^2 - x + 1) + B(x + 2)(2x^2 - x + 1) + (Cx + D)(x + 2)(2x - 1)}{(x + 2)(2x - 1)(2x^2 - x + 1)}$$

The numerators must be equal, so
$-4x^3 + 12x^2 - 12x + 6$
$\quad = A(2x - 1)(2x^2 - x + 1) + B(x + 2)(2x^2 - x + 1) + (Cx + D)(x + 2)(2x - 1)$

If $x = -2$
$-4(-2)^3 + 12(-2)^2 - 12(-2) + 6 = A(-5)[2(-2)^2 - (-2) + 1]$
$$32 + 48 + 24 + 6 = -55A$$
$$110 = -55A$$
$$A = -2$$

If $x = \frac{1}{2}$

$$-4\left(\frac{1}{2}\right)^3 + 12\left(\frac{1}{2}\right)^2 - 12\left(\frac{1}{2}\right) + 6 = B\left(\frac{5}{2}\right)\left[2\left(\frac{1}{2}\right)^2 - \frac{1}{2} + 1\right]$$

$$-\frac{1}{2} + 3 - 6 + 6 = B\left(\frac{5}{2}\right)(1)$$

$$B = 1$$

If $x = 0$
$6 = A(-1)(1) + B(2)(1) + D(2)(-1)$
$6 = -A + 2B - 2D$
$6 = 2 + 2 - 2D$
$D = -1$
If $x = 1$
$-4 + 12 - 12 + 6 = A(1)(2) + B(3)(2) + (C + D)(3)(1)$
$$2 = 2A + 6B + 3C + 3D$$
$$= -4 + 6 + 3C - 3$$
$$C = 1$$

So $\dfrac{4x^5 + 12x^4 - x^3 + 7x^2 - 4x + 2}{4x^4 + 4x^3 - 5x^2 + 5x - 2} = x + 2 - \dfrac{2}{x + 2} + \dfrac{1}{2x - 1} + \dfrac{x - 1}{2x^2 - x + 1}$.

Section C-3

1. $P(x) = x^3 + 2x + 7$; signs of coefficients: $+$ $+$ $+$
   variations in sign: none

   $P(-x) = (-x)^3 + 2(-x) + 7 = -x^3 - 2x + 7$; signs of coefficients: $-$ $-$ $+$
   variations in sign: one

3. $P(x) = x^3 - 3x^2 - 9$; signs of coefficients: + - -
variations in sign: one

$P(-x) = (-x)^3 - 3(-x)^2 - 9 = -x^3 - 3x^2 - 9$; signs of coefficients: - - -
variations in sign: none

5. $P(x) = x^4 + 2x^3 + 3x - 5$; signs of coefficients: + + + -
variations in sign: one

$P(-x) = (-x)^4 + 2(-x)^3 + 3(-x) - 5 = x^4 - 2x^3 - 3x - 5$;
signs of coefficients: + - - -
variations in sign: one

7. $P(x) = x^3 + 2x - 4$ One variation in sign
$P(-x) = (-x)^3 + 2(-x) - 4 = -x^3 - 2x - 4$ No variations in sign

$P(x)$ has either 1 positive zero or the number of positive zeros is less than 1 by an even number. Since there obviously can't be less than zero positive zeros, we know for sure there is one. Since $P(-x)$ has no variations in sign, $P(x)$ has no negative zeros. As a third degree polynomial, $P(x)$ has a total of three zeros: one is real and two are imaginary.

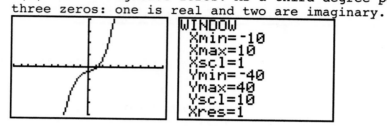

9. $f(x) = x^3 + 2x^2 + 1$ No variations in sign.
$f(-x) = (-x)^3 + 2(-x)^2 + 1 = -x^3 + 2x^2 + 1$ One variation in sign.

Since $f(x)$ has no variations in sign, $f(x)$ has no positive zeros. Since $f(-x)$ has one variation in sign, $f(x)$ has exactly one negative zero. (The number of negative zeros can't be less than 1 by an even number because then it would be negative.) As a third degree polynomial, $f(x)$ has a total of three zeros: one is real and two are imaginary.

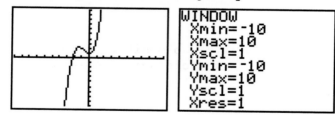

11. $S(x) = x^4 - 2x^3 - 7x - 8$ One variation in sign
$S(-x) = (-x)^4 - 2(-x)^3 - 7(-x) - 8 = x^4 + 2x^3 + 7x - 8$ One variation in sign

Since both $S(x)$ and $S(-x)$ have one variation in sign, $S(x)$ has exactly one positive and one negative zero. (The number of either can't be less than 1 by an even number because then it would be negative.) As a fourth degree polynomial, $S(x)$ has a total of four zeros: two are real and two are imaginary.

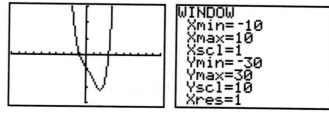

13. $P(x) = x^3 - 3x^2 - 2x + 4$ Two variations in sign; two or zero positive zeros
$P(-x) = (-x)^3 - 3(-x)^2 - 2(-x) + 4$
$\quad\quad\quad = -x^3 - 3x^2 + 2x + 4$ One variation in sign; one negative zero

Zeros	Pos.	Neg.	Imag.
	2	1	0
	0	1	2

15. $t(x) = x^4 - 2x^3 - 4x^2 - 2x + 3$ Two variations in sign; two or zero positive zeros

$t(-x) = (-x)^4 - 2(-x)^3 - 4(-x)^2 - 2(-x) + 3$
$\quad\quad\quad = x^4 + 2x^3 - 4x^2 + 2x + 3$ Two variations in sign; two or zero negative zeros

Zeros	Pos.	Neg.	Imag.
	2	2	0
	2	0	2
	0	2	2
	0	0	4

17. $f(x) = x^5 - x^4 + 3x^3 + 9x^2 - x + 5$
Four variations in sign; four, two, or zero positive zeros
$f(-x) = (-x)^5 - (-x)^4 + 3(-x)^3 + 9(-x)^2 - (-x) + 5$
$\quad\quad\quad = -x^5 - x^4 - 3x^3 + 9x^2 + x + 5$
One variation in sign; one negative zero

Zeros	Pos.	Neg.	Imag.
	4	1	0
	2	1	2
	0	1	4

19. $P(x) = x^6 - 12$ One variation in sign; one positive zero
$P(-x) = (-x)^6 - 12 = x^6 - 12$ One variation in sign; one negative zero

Zeros	Pos.	Neg.	Imag.
	1	1	4

21. $r(x) = x^7 + 32$ No variations in sign; no positive zeros
$r(-x) = (-x)^7 + 32 = -x^7 + 32$ One variation in sign, one negative zero

Zeros	Pos.	Neg.	Imag.
	0	1	6

23. $P(x) = x^2 + ax + b$
If $a > 0$ and $b > 0$, then $P(x)$ has no variations in sign, so there are no positive zeros.
$P(-x) = (-x)^2 + a(-x) + b = x^2 - ax + b$
In this case, the coefficients look like + - +, so there are two variations in sign and either two or zero negative zeros. If there are no negative zeros, there are two imaginary zeros.

25. $P(x) = x^2 + ax + b$
If $a > 0$ and $b < 0$, the signs of the coefficients are + + -, so there is one variation in sign and one positive zero.
$P(-x) = (-x)^2 + a(-x) + b = x^2 - ax + b$
In this case, the signs of the coefficients are + - -, and there is one variation in sign, so one negative zero. In summary, there is one positive zero and one negative zero.

27. $f(x) = x^6 - 3x^5 + 4x^4 + 3x^3 - 2x + 5$ Four variations in sign; four, two, or zero positive zeros.

$f(-x) = (-x)^6 - 3(-x)^5 + 4(-x)^4 + 3(-x)^3 - 2(-x) + 5$

$= x^6 + 3x^5 + 4x^4 - 3x^3 + 2x + 5$ Two variations in sign; two or zero negative zeros.

Zeros	Pos.	Neg.	Imag.
	4	2	0
	4	0	2
	2	2	2
	2	0	4
	0	2	4
	0	0	6

29. $S(x) = x^7 + 3x^5 + 4x^2 - 3x + 5$ Two variations in sign; two or zero positive zeros

$S(-x) = (-x)^7 + 3(-x)^5 + 4(-x)^2 - 3(-x) + 5$

$= -x^7 - 3x^5 + 4x^2 + 3x + 5$ One variation in sign; one negative zero

Zeros	Pos.	Neg.	Imag.
	2	1	4
	0	1	6

31. $P(x) = x^8 - x - 1$ One variation in sign; one positive zero

$P(-x) = (-x)^8 - (-x) - 1$

$= x^8 + x - 1$ One variation in sign; one negative zero

Zeros	Pos.	Neg.	Imag.
	1	1	6

33. $P(x) = x^3 + ax + b$

If a and b are both positive, there are no variations in sign so there are no positive zeros.

$P(-x) = (-x)^3 + a(-x) + b = -x^3 - ax + b$

In this case, there is one variation in sign so there is one negative zero. The remaining two zeros are imaginary.

35. $P(x) = x^3 + ax + b$

If $a < 0$ and $b > 0$, the signs of the coefficients are + - +, so there are two variations in sign and either two or zero positive zeros.

$P(-x) = (-x)^3 + a(-x) + b = -x^3 - ax + b$ with a negative, the signs of the coefficients are - + +, so there is one variation in sign and one negative zero. The remaining two zeros are either positive or imaginary.

Section C-4

1. Note that if you eliminate the parameter from the parametric equations by substituting x for t^2 in $y = t^2 - 2$, the result is the equation $y = x - 2$. Since $x = t^2$, x must be non-negative, and the graph of the parametric equations will consist only of a ray, the portion of the line $y = x - 2$ for $x \geq 0$.

3. Using trace and the forward and backward arrows, we can move the cursor along the graph to locate each of the points identified in the table.

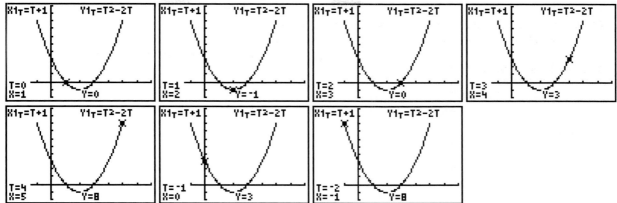

5. After applying ZDecimal, the "window" screen looks like this:

The values of x and y were affected but the values of t are unchanged.

7. After re-entering the original window variables, choosing Zoom In, and zooming on the portion of the graph near the y-intercept, the graph and "window" screens look like this:

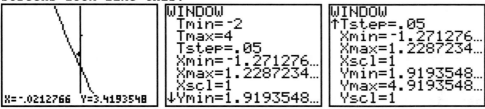

The values of x and y were affected but the values of t are unchanged.

9. Construct a table and graph.

t	0	1	2	3	−1	−2	−3
x	0	−1	−2	−3	1	2	3
y	−2	0	2	4	−4	−6	−8

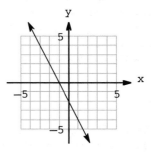

To eliminate the parameter t we solve
$x = -t$ for t to obtain
$$t = -x$$
Then we substitute the expression for t into
$y = 2t - 2$ to obtain
$$y = 2(-x) - 2$$
$$y = -2x - 2$$
This is the equation of a straight line.

11. Construct a table and graph.

t	0	1	2	3	−1	−2	−3	
x	0	−1	−4	−9	−1	−4	−9	Note: $x \le 0$
y	−2	0	6	16	0	6	16	

To eliminate the parameter t we solve
$x = -t^2$ for t^2 (we do not need t) to obtain
$$t^2 = -x$$
Then we substitute the expression for t^2 into
$y = 2t^2 - 2$ to obtain
$$y = 2(-x) - 2$$
$$y = -2x - 2$$

This is the equation of a straight line. However,
since $x = -t^2$, x is restricted so that $x \le 0$.
Therefore the equation of the curve is actually
$$y = -2x - 2, \quad x \le 0$$
This is the equation of a ray (part of a straight line).

13. Construct a table and graph.

t	0	1	2	3	−1	−2	−3
x	0	3	6	9	−3	−6	−9
y	0	−2	−4	−6	2	4	6

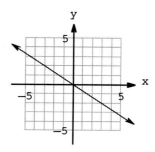

To eliminate the parameter t we solve
$x = 3t$ for t to obtain
$$t = \frac{x}{3}$$
Then we substitute the expression for t into
$y = -2t$ to obtain
$$y = -2\left(\frac{x}{3}\right)$$
$$y = -\frac{2}{3}x$$
This is the equation of a straight line.

15. Construct a table of values and graph.

t	0	1	2	3	−1	−2	−3
x	0	$\frac{1}{4}$	1	$\frac{9}{4}$	$\frac{1}{4}$	1	$\frac{9}{4}$
y	0	1	2	3	−1	−2	−3

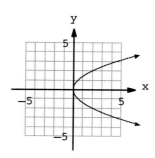

To eliminate the parameter t we substitute $t = y$
(from $y = t$) into the expression for x to obtain
$$x = \frac{1}{4}y^2$$
$$y^2 = 4x$$
This is the equation of a parabola.

17. Construct a table of values and graph.

t	0	1	2	3	-1	-2	-3
x	0	$\frac{1}{4}$	4	$\frac{81}{4}$	$\frac{1}{4}$	4	$\frac{81}{4}$
y	0	1	4	9	1	4	9

To eliminate the parameter we substitute $t^2 = y$
(from $y = t^2$) into the expression for x to obtain

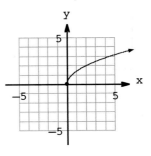

$$x = \frac{1}{4}t^4$$
$$x = \frac{1}{4}(t^2)^2$$
$$x = \frac{1}{4}y^2$$
$$y^2 = 4x$$

This is the equation of a parabola. However, since $y = t^2$, y is restricted to
nonnegative values only, and the equation of the curve is actually
 $y^2 = 4x$, $y \geq 0$
This is the equation of the upper half of a parabola.

19. We eliminate the parameter t as follows:
  $x = t - 2$
$x + 2 = t$

Substituting this expression for t into the expression for y, we obtain
$y = 4 - 2t$
$y = 4 - 2(x + 2)$
$y = 4 - 2x - 4$
$y = -2x$
This is the equation of a line.

21. We eliminate the parameter t as follows:
Solve $x = t - 1$ for t to obtain $t = x + 1$. Substitute the expression for t into
$y = \sqrt{t}$ to obtain
$y = \sqrt{x + 1}$ or
$y^2 = x + 1$, $y \geq 0$.

Since t is restricted so that $t \geq 0$, we also have
$x + 1 \geq 0$ or $x \geq -1$. This graph is a parabola with vertex at $(-1, 0)$, opening
right. Since $y \geq 0$ only the upper half of the parabola is traced out.

23. We eliminate the parameter t as follows:
Solve $x = \sqrt{t}$ for t to obtain $t = x^2$.
Substitute the expression for t into $y = 2\sqrt{16 - t}$ to obtain

$$y = 2\sqrt{16 - x^2} \text{ or}$$
$$y^2 = 4(16 - x^2)$$
$$y^2 = 64 - 4x^2$$
$$4x^2 + y^2 = 64$$
$$\frac{x^2}{16} + \frac{y^2}{64} = 1 \quad \text{(ellipse in standard form)}$$

Since $0 \leq t \leq 16$ and $x = \sqrt{t}$, it follows that $0 \leq x \leq 4$.
Since $0 \leq t \leq 16$ and $t = x^2$, it follows that $0 \leq x^2 \leq 16$, $0 \leq 16 - x^2 \leq 16$,
$0 \leq \sqrt{16 - x^2} \leq 4$, and $0 \leq 2\sqrt{16 - x^2} \leq 8$. So $0 \leq y \leq 8$.

The graph of $\frac{x^2}{16} + \frac{y^2}{64} = 1$, $0 \leq x \leq 4$ and $0 \leq y \leq 8$, is the portion of an
ellipse lying in the first quadrant.

25. We eliminate the parameter t as follows:
Solve $x = -\sqrt{t + 1}$ for t to obtain
$$x^2 = t + 1$$
$$t = x^2 - 1$$

Substitute the expression for t into $y = -\sqrt{t - 1}$ to obtain
$$y = -\sqrt{x^2 - 1 - 1}$$
$$y = -\sqrt{x^2 - 2}$$
$$y^2 = x^2 - 2$$
$$x^2 - y^2 = 2 \quad \text{(hyperbola)}$$

Since $t \geq 1$ it follows that $t + 1 \geq 2$, $\sqrt{t + 1} \geq \sqrt{2}$ and $x = -\sqrt{t + 1} \leq -\sqrt{2}$.
Also it follows that $t - 1 \geq 0$, $\sqrt{t - 1} \geq 0$, and $y = -\sqrt{t - 1} \leq 0$.
The graph of $x^2 - y^2 = 2$, $x \leq -\sqrt{2}$ and $y \leq 0$, is the portion of a hyperbola lying in the third quadrant.

27. If $A \neq 0$, $C = 0$, and $E \neq 0$, write
$$Ax^2 + Dx + Ey + F = 0$$
There are many possible parametric equations for this curve. A simple approach is to set $x = t$ and solve for y in terms of t.
$$At^2 + Dt + Ey + F = 0$$
$$At^2 + Dt + F = -Ey$$
$$\frac{At^2 + Dt + F}{-E} = y$$
Then $x = t$, $y = \dfrac{At^2 + Dt + F}{-E}$, $-\infty < t < \infty$, are parametric equations for this curve. The curve is a parabola.

29. We eliminate the parameter t as follows:
Square the expressions for x and y.
$$x^2 = t^2 + 1 \qquad y^2 = t^2 + 9$$

Solve the first equation for t^2 and substitute the result into the second equation.
$$t^2 = x^2 - 1$$
$$y^2 = x^2 - 1 + 9$$
$$y^2 = x^2 + 8$$
$$y^2 - x^2 = 8$$

This is the equation of a hyperbola.

Since $t^2 \geq 0$, $t^2 + 1 \geq 1$, $\sqrt{t^2 + 1} \geq 1$. It follows that $x \geq 1$.
Since $t^2 \geq 0$, $t^2 + 9 \geq 9$, $\sqrt{t^2 + 9} \geq 3$. It follows that $y \geq 3$.
The curve is the part of the hyperbola $y^2 - x^2 = 8$ for $x \geq 1$ and $y \geq 3$.

31. We eliminate the parameter t as follows:
Square the expressions for x and y.

$$x^2 = \frac{4}{t^2 + 1} \qquad y^2 = \frac{4t^2}{t^2 + 1}$$

$$x^2 + y^2 = \frac{4}{t^2 + 1} + \frac{4t^2}{t^2 + 1}$$

$$x^2 + y^2 = \frac{4 + 4t^2}{t^2 + 1}$$

$$x^2 + y^2 = \frac{4(1 + t^2)}{t^2 + 1}$$

$$x^2 + y^2 = 4$$

This is the equation of a circle.

Since $t^2 \geq 0$, $t^2 + 1 \geq 1$, $\sqrt{t^2 + 1} \geq 1$, $0 < \dfrac{1}{\sqrt{t^2 + 1}} \leq 1$, $0 < \dfrac{2}{\sqrt{t^2 + 1}} \leq 2$.

It follows that $0 < x \leq 2$. Therefore $0 < x^2 \leq 4$, $0 > -x^2 \geq -4$, $4 > 4 - x^2 \geq 0$, $2 > \sqrt{4 - x^2} \geq 0$, $-2 < -\sqrt{4 - x^2} \leq 0$. Since $y = \pm\sqrt{4 - x^2}$ on the circle $x^2 + y^2 = 4$, it follows that $-2 < y < 2$ (since $x \neq 0$). So the curve is the portion of the circle $x^2 + y^2 = 4$ for which $0 < x \leq 2$, $-2 < y < 2$. This is the right-hand semicircle, excluding the points $(0, \pm 2)$.

33. We eliminate the parameter t as follows: First note that since $x = \dfrac{8}{t^2 + 4}$, $(t^2 + 4)x = 8$ or $t^2 + 4 = \dfrac{8}{x}$. Therefore x cannot be 0. Now square the expressions for x and y.

$$x^2 = \frac{64}{(t^2 + 4)^2}$$

$$y^2 = \frac{16t^2}{(t^2 + 4)^2}$$

$$x^2 + y^2 = \frac{64}{(t^2 + 4)^2} + \frac{16t^2}{(t^2 + 4)^2}$$

$$x^2 + y^2 = \frac{64 + 16t^2}{(t^2 + 4)^2}$$

$$x^2 + y^2 = \frac{16(4 + t^2)}{(t^2 + 4)^2}$$

$$x^2 + y^2 = \frac{16}{t^2 + 4}$$

Since $x = \dfrac{8}{t^2 + 4}$, $2x = \dfrac{16}{t^2 + 4}$, so

$$x^2 + y^2 = 2x \quad x \neq 0$$

This is the equation of a circle. Completing the square we see

$$x^2 - 2x + y^2 = 0$$
$$x^2 - 2x + 1 + y^2 = 1$$
$$(x - 1)^2 + y^2 = 1, \; x \neq 0$$

So the circle has center $(1, 0)$ and radius 1. The origin is not part of the curve since x cannot equal zero, so there is a hole in the graph there.

35. Here are the graphs in a squared viewing window (shown together with the line $y = x$).

(A) The graphs are reflections of each other in the line $y = x$.

(B) Since $t = x_1$, $y_1 = e^{x_1}$. Since $t = y_2$, $x_2 = e^{y_2}$, $y_2 = \ln x_2$. The functions are inverses of each other.

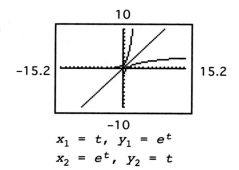

$x_1 = t,\ y_1 = e^t$

$x_2 = e^t,\ y_2 = t$

37. At the instant the supplies are dropped, the vertical speed is 0, the horizontal speed is 125 meters per second, and the altitude is 1,000 meters. Substituting these values in
$$x = h_0 t$$
$$y = a_0 + v_0 t - 4.9 t^2 \qquad 0 \le t \le b$$
the parametric equations for the path of the supplies are
$$x = 125t$$
$$y = 1000 - 4.9 t^2$$

The supplies strike the ground when $y = 0$. Using the parametric equations for y, we obtain
$$y = 1{,}000 - 4.9 t^2 = 0$$
$$-4.9 t^2 = -1{,}000$$
$$t = \sqrt{\frac{-1{,}000}{-4.9}} \approx 14.2857 \text{ sec.}$$

The distance inland required is the value of x for this value of t. Substituting $t = 14.2857$ in $x = 125t$, the distance is
$$x = 125(14.2857) \approx 1{,}786 \text{ meters}$$

Notes

Notes

Notes

Notes